T0340082

Metal Cutting Theory and Practice

Third Edition

Metal Cutting Theory and Practice

Third Edition

David A. Stephenson
John S. Agapiou

CRC Press
Taylor & Francis Group
Boca Raton   London   New York

CRC Press is an imprint of the
Taylor & Francis Group, an **informa** business

CRC Press
Taylor & Francis Group
6000 Broken Sound Parkway NW, Suite 300
Boca Raton, FL 33487-2742

First issued in paperback 2019

© 2016 by Taylor & Francis Group, LLC
CRC Press is an imprint of Taylor & Francis Group, an Informa business

No claim to original U.S. Government works

ISBN-13: 978-1-4665-8753-3 (hbk)
ISBN-13: 978-0-367-86819-2 (pbk)

Visit the Taylor & Francis Web site at
http://www.taylorandfrancis.com

and the CRC Press Web site at
http://www.crcpress.com

Contents

Preface to the Third Edition

In this new edition, we revised and expanded Chapters 1, 3, and 10 from the previous edition, and updated material and references in the remaining chapters, especially in Chapters 6, 9, 11, and 14. Chapters 15 and 16 from the second edition, on high-throughput and agile machining and design for machining, respectively, have been replaced with new chapters on minimum quantity lubrication machining and accuracy and error compensation of CNC machining systems. Finally, we are pleased to announce that the chapter on gear machining deferred from the second edition enters the current lineup as Chapter 17.

We are grateful to Paul Bojanowski, Emenike Chukwuma, David N. Dilley, William Dowling, Trevor Hill, Tim Hull, Jack Knapke, Dr. Heinrich Schwenke, Dr. Herman Stadtfeld, James Stead, Alexander Stoll, Kalvis Terauds, David Wall, Tim Walker, Douglas Watts, and Michael Williams for providing technical input and figures for the current edition. We are also grateful to Cindy Carelli and her team at CRC Press for their usual expert editing work. Finally, we thank our families for their forbearance with our sporadic but longstanding preoccupation with this work.

When we began writing the first edition of this book in 1992, we could not have anticipated the project's longevity or that we would still be working on it more than 20 years later. We thank our readers for their patience, salute them for their fortitude, and wish them success in their metal-cutting endeavors.

David A. Stephenson
John S. Agapiou

MATLAB® is a registered trademark of The MathWorks, Inc. For product information, please contact:

The MathWorks, Inc.
3 Apple Hill Drive
Natick, MA 01760-2098 USA
Tel: 508-647-7000
Fax: 508-647-7001
E-mail: info@mathworks.com
Web: www.mathworks.com

Preface to the Second Edition

We were pleased with the reception of the first edition of this book, and we are very glad to have an opportunity to correct some of its deficiencies in this new version.

In addition to updating material throughout, we made several structural changes. Chapter 2 in the first edition, covering both machining operations and machine tools, has been split into two separate chapters in the current edition. Additionally, Chapter 5, on chip formation, was eliminated, with the material being distributed between the current Chapters 6 and 11. We added three chapters, on cutting fluids, agile and high-throughput machining, and design for machining. These are areas of significant recent development, and also reflect areas of emphasis in our recent professional practice. Finally, since the first edition was unexpectedly used as a university textbook, we have added examples and problems at the end of chapters to make the new edition more suitable for this purpose.

Due to lack of space, we reluctantly decided not to include a planned chapter on gear machining. We apologize to any readers disappointed by this omission.

We are grateful to John Rutz, David Yen, and Albert Shih for useful feedback on the first edition, and to David N. Dilley and Mikhail Lundblad for valuable technical input for the new edition. We are also grateful to Rita Lazazzaro and Barbara Mathieu of Marcel Dekker Inc. and Cindy Carelli and Preethi Cholmondeley of CRC Press for their patient editing. Finally, and most importantly, we thank our families for their forbearance during our second attempt at this project.

David A. Stephenson
John S. Agapiou

Preface to the First Edition

Metal cutting is a subject as old as the Industrial Revolution, but one that evolved continuously as technology advanced. The first metal-cutting machine tools, built some 450 years ago, were powered by water and employed iron and carbon steel tools. Over time, these gave way to machines powered by steam and leather belts employing high-speed steel tools, to electrically powered machines using sintered carbide tools, and, most recently, to computer-controlled machines using ceramic and diamond tools. The pace of change seems to have increased over the last 20 years, with progressively rapid advances in materials science and computer technology. For example, since the beginning of our careers, typical production rates have doubled in many operations and numerous new tool materials, work materials, and machine architectures have been introduced—and we are still many years from retirement.

Our purpose in writing this book is twofold. First, many of the books from which we learned much of our trade were written in the 1970s or earlier, and despite recent updated editions, they are showing inevitable signs of age. We hoped to write a reference book that would provide a fuller treatment of recent developments than is currently available. Second, the literature in this field is somewhat dichotomous, consisting, on the one hand, of scientific books and articles read largely by academics and researchers, and, on the other hand, of trade journals, handbooks, and sales brochures read by practicing engineers. We also hoped to write a book that would appeal to both audiences by covering both research results and the current industrial practice. To make the project manageable, we had to limit the technical topics to be covered. We have chosen to consider only metallic work materials, to concentrate on the traditional chip-forming cutting processes with limited material on abrasive processes, and to largely ignore subjects such as machine tool control, which could fill entire books in their own right. Even with these limitations, we recognize that we have taken on an ambitious subject. Readers, of course, will have to decide how well we have covered it.

We have been fortunate over the years to have worked with, and learned from, many fine engineers in academia and in the automotive, machine tool, and cutting tool industries. We received valuable feedback on drafts of portions of this book from Robin Stevenson, Pulak Bandyopadhyay, Yhu-Tin Lin, David W. Yen, I. S. Jawahir, Jochen S. Zenker, and Ellen D. Kock. We are grateful to these colleagues for correcting many errors and inconsistencies in the manuscript; we are responsible for all those that still remain. We also thank Simon Yates, Dawn Wechsler, Walter Brownfield, and Vivian Jao of Marcel Dekker Inc. for their courteous and helpful editing.

Finally, and most importantly, we thank our wives, Maria Clelia Milletti and Christina Agapiou, and our children, Stylianos Ioannis Agapiou, Alexandra Maria Agapiou, Adonis Ioannis Agapiou, and Daphne Elizabeth Agapiou, and Francesca Laura Stephenson and Luke Andrew Stephenson, for their patience during the many times our preoccupation with this project inconvenienced them.

David A. Stephenson
John S. Agapiou

Authors

David A. Stephenson is a technical specialist at Ford Powertrain Advanced Manufacturing Engineering in Livonia, Michigan. Dr. Stephenson worked for several years at General Motors Research and General Motors Powertrain; he has also worked at Third Wave Systems Inc., D3 Vibrations Inc., the University of Michigan, and Fusion Coolant Systems. He is a member of the American Society of Mechanical Engineers (ASME) and a fellow of the Society of Manufacturing Engineers (SME). He has served as a journal technical editor for both societies and on the ASME Manufacturing Science and Engineering Division Executive Committee from 2002 to 2007. Dr. Stephenson received the ASME Blackall Machine Tool and Gage Award in 1994, the SME Outstanding Young Manufacturing Engineer Award in 1994, and the M. Eugene Merchant Manufacturing Medal of ASME/SME in 2004. He earned his bachelor and master degrees in mechanical engineering at the Massachusetts Institute of Technology in 1981 and 1983, respectively, and his PhD from the University of Wisconsin in 1985.

John S. Agapiou is a technical fellow at the Manufacturing Systems Research Lab at General Motors R&D Center, Warren, Michigan. He is also a part-time professor in the Department of Mechanical Engineering at Wayne State University, Detroit, Michigan. He received the Society of Manufacturing Engineers (SME) Outstanding Young Manufacturing Engineer Award in 1992 and the SME S. M. Wu Research Implementation Award in 2015. His research focus is on developing and implementing world-class manufacturing, quality, and process validation strategies in the production and development of automotive powertrains. Dr. Agapiou's research and teaching interests include modeling and optimization of metal-cutting operations, including cutting tools and machining systems, and modeling manufacturing part quality for machining lines to improve part quality, process, and productivity. He earned his bachelor and master degrees in mechanical engineering from the University of Louisville in 1980 and 1981, respectively, and his PhD from the University of Wisconsin in 1985.

1 Introduction

1.1 SCOPE OF THE SUBJECT

Metal cutting processes are industrial processes in which metal parts are shaped by the removal of unwanted material. In this book, we will primarily consider traditional chip-forming processes such as turning, boring, drilling, and milling. In these operations, metal is removed as a plastically deformed chip of appreciable dimensions, and a fairly unified physical analysis can be carried out using basic orthogonal and oblique cutting models (Figure 1.1).

Related metal removal processes include abrasive processes, such as grinding and honing, and nontraditional machining processes, such as electrodischarge, ultrasonic, electrochemical, and laser machining. In abrasive processes, metal is removed in the form of small chips produced by a combination of cutting, plowing, and friction mechanisms; in nontraditional processes, metal is removed on a much smaller scale by mechanical, thermal, electrical, or chemical means. In all cases, the physical mechanisms of removal differ considerably from those of chip formation, so different physical analyses are required. Basic information on abrasive processes, tools, and surface finish capabilities is included in Chapters 2, 4, and 10. Physical analyses of abrasive and nontraditional machining processes are not considered in this book but are available in the literature [1–6].

Metal cutting processes can also be applied to nonmetallic work materials such as polymers, wood, and ceramics. When these applications are considered, the subject is more commonly called machining. Because of differences in thermomechanical properties, the analyses of metal cutting discussed in this book provide only limited insight into the machining of nonmetals. More relevant information can be found in the literature on the machining of specific classes of materials [7,8].

The objective in this book is to provide a physical understanding of conventional and high-speed cutting processes applied to metallic workpieces. The mechanics of chip formation, temperature generation, tribology, dynamics, and material interactions are emphasized. We also include significant descriptive information on modern machinery, tooling, and coolant systems. Hopefully this information, summarized with reference to the large research and trade literature on this subject, will provide the reader with sufficient physical insight and understanding to design, operate, troubleshoot, and improve high-quality, cost-effective metal cutting operations.

1.2 HISTORICAL DEVELOPMENT

1.2.1 ANCIENT AND MEDIEVAL PREDECESSORS

Metalworking is an old human activity. Native metals, especially copper, and meteoric iron were worked wherever they were encountered by Neolithic peoples, and the mining and smelting of metal ores dates to preliterate times [9–12]. The techniques used to process metals during this period would today be classified as casting and forming processes. Casting has been practiced since at least 5000 BCE [11], and many ancient peoples, notably the Egyptians and the Chinese, were expert metal founders. Forging by hammer (an open die operation) was the chief activity of the smiths and goldbeaters of antiquity and was the basis for deformation processes such as coining, wire drawing, and rolling, all of which were developed in ancient or early medieval times [13–16].

Metal-cutting operations have a more modest pedigree. Ancient peoples certainly ground and filed metals but did not machine them. (Vitruvius and Philo of Byzantium both report that Ktesibios of Alexandria, who lived in about c. 300 BCE, bored the brass or bronze cylinders of pump-like

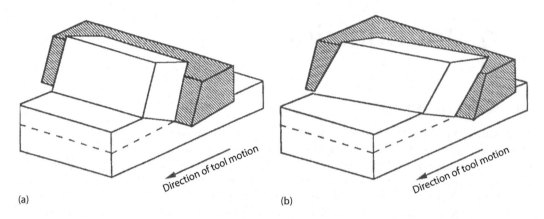

FIGURE 1.1 Orthogonal (a) and oblique (b) cutting of a flat workpiece by a wedge-shaped tool.

machines [17,18], but the engineering details of their descriptions are not credible.) As will be shown, with the isolated exception of canon boring, metal cutting as contemporary engineers would understand it was not practiced until the Industrial Revolution of the late eighteenth century, and no machine resembling a contemporary machine tool was developed until around the year 1800.

Most contemporary machining operations, however, were based on analogous earlier methods for shaping wood and stone. Grinding and drilling both have prehistoric roots. Prehistoric peoples shaped stone objects such as querns and axe heads by abrasion with harder stones (especially sandstones), and sharpened axes and other tools by similar methods [19,20]. Drilling was carried out in Neolithic times by rotating as stick in an abrasive, generally sand, to bore holes in stone objects [19]; trepanning was also practiced by substituting a hollow bone tube for a solid stick. In later practice, both Egyptian and Roman artisans sharpened tools with whetstones [21,22] and drilled holes in wood using bow drills [23,24]. Holes in stone were trepanned as before by rotating a bronze tube in abrasives [25]. Copper or iron saws with loose abrasives were also used to cut stone [21,22]. The first known representation of a cord lathe, similar to those used in contemporary nonindustrial societies, is from a Ptolemaic Egyptian tomb [26], although earlier stone and wooden objects that were clearly turned have been recovered [27]. Medieval European craftsmen developed a variety of pole and crank-driven lathes as described by Theophilus and in other contemporary manuscripts [15,27,28]. The cylindrical grinding stone and the brace and bit for drilling were medieval innovations [24,29]. Flat surface were machined in ancient times by grinding, filing, and planing using hand planes [24,30]. Early metal planers and shapers are similar in concept to hand planes. Face milling of flat surfaces, which has no ancient precedent, is a nineteenth-century invention, and surface broaching, which is broadly similar to filing, was not introduced until the 1930s [31].

Before proceeding to more modern times, it is interesting to note four ancient and medieval manufacturing methods, which bear similarities to later practice:

1. *Ancient Egyptian multispindle drilling*: The ancient Egyptians used large quantities of beads to decorate mummies and made them in high volumes in specialized workshops [32,33]. The final step in bead production was the drilling of mounting holes using a copper-tipped bow drill. Not surprisingly, this holemaking operation was the production bottleneck, and at least some craftsmen developed methods of driving multiple drills from a single bow to increase throughput (Figure 1.2) [32]. Stocks [33] describes several depictions of this practice from Theban tombs, including three-, four-, and five-spindle applications. Similar multispindle drilling methods found widespread later use in the railroad and automotive industries.

2. *Philistine tool-sharpening service*: The Hebrew Bible reports that the Philistines, having at one point conquered the Israelites, forbade them from sharpening their own tools,

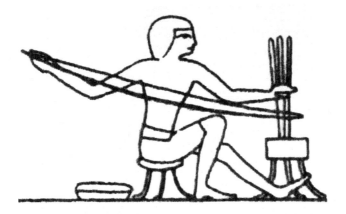

FIGURE 1.2 An early multi-spindle drilling operation in the manufacture of stone beads in ancient Egypt: simultaneous bow-drilling of three beads using a long bow to drive three drills. (Detail of a painting from the tomb of Vizier Rekh-Mi-Re, Thebes, 18th Dynasty, c. 1430 BCE.)

recognizing that this skill would also be useful in making weapons [34]. Instead, the Israelites were required to send dull tools to Philistine artisans, who charged by the piece for sharpening (a pim for a plowshare, a third of a shekel for axe, etc.). Following this biblical prototype, contemporary tool-sharpening businesses still typically use a per-piece fee structure. The Philistines did not charge for programming or setup.

3. *Da Vinci's machine tools*: Leonardo Da Vinci drew designs for a number of machine tools in his notebooks, including lathes, thread-cutting machines, boring mills, and grinding machines [27,29,31,35–38]. None of these machines was apparently ever built, but many contain design details independently developed and used in later practice. Two representative instances are shown in Figures 1.3 and 1.4. The internal grinder

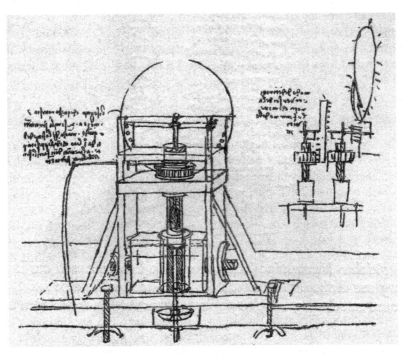

FIGURE 1.3 Leonardo's internal grinder for wooden cylinders, *Codex Atlanticus* f291r.

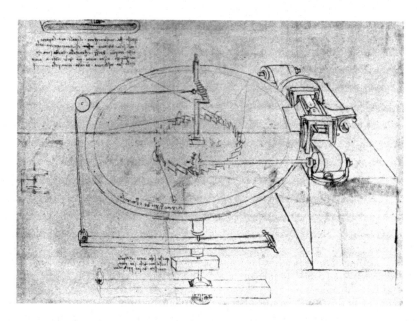

FIGURE 1.4 Leonardo's needle-grinding production system, *Codex Atlanticus* f31v. The mechanism shown at right is one of five planned stations arranged around the central rotary table.

in Figure 1.3 [29,31,36,38] looks strikingly like a contemporary honing machine, and the needle-making machine (actually a five-station production system) in Figure 1.4 [29,35,38] is similar to later rotary index and dial machines. In the notes accompanying the drawing of the needle-making system [38], Leonardo made a factor of 10 error in computing his projected profits, an early example of a flawed manufacturing business plan.

4. *Medieval clockmaker's tools*: Medieval clock and watchmakers developed a number of hand-operated tools for precisely cutting shafts, screws, gears, and complex parts for escapements [27,31,39–41]. Although the mechanisms themselves bear little resemblance to later machine tools, the kinematic methods used to generate screw threads and space gear teeth would have later impact on the development of screw-cutting lathes and gear-cutting machines.

1.2.2 CANON BORING

The first large, externally powered metal-cutting machine tools were developed in Europe for boring canon. Early canon fired stone balls shaped by masons, and since these varied significantly in shape and diameter there was little point in precisely machining the cast bore of the gun. This changed, however, with the introduction of cast iron shot of relatively uniform dimension in the fifteenth century. Gun boring to improve the accuracy and range of ordnance was soon taken up in Germany and Italy. Canon boring is mentioned in medieval German city records as early as 1373 [28], and there is a crude sketch of a vertical, horse-driven cannon boring mill in fifteenth century German manuscript [28,41–44]. The best early depiction of a canon boring mill is from a book published in 1540 by Vannoccio Biringuccio (1480–c. 1539), a master craftsman from Sienna who had traveled widely in Italy and Germany [45]. He also reported the results of experiments with a variety of solid and inserted blade iron cutters carried out in Florence, which are the earliest tooling trials known to history. Figure 1.5 shows two of Biringuccio's horizontal boring mills, together with typical iron-boring tools. This machine, and others like it employing both horizontal and (more commonly) vertical layouts, proliferated throughout France, Holland, England, Scotland, and Russia over the next 200 years [31,39,41,46–51]. In all these machines, the tool rotated while the

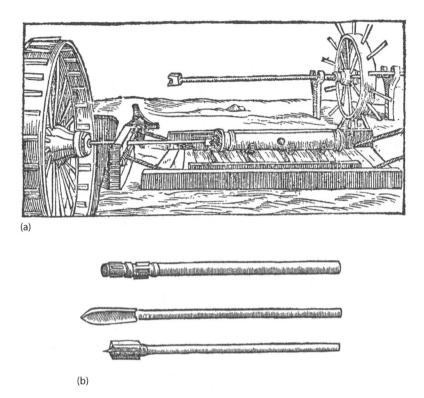

(a)

(b)

FIGURE 1.5 Canon boring mills (a) and boring tools (b). (From Biringuccio, V., *De La Pirotechnia*, Senato Veneto, Venice, Italy, 1540; English Trans. by Smith, C.S. and Gnudi, M.T., The American Institute of Mining and Metallurgical Engineers, New York, 1942, pp. 308–312.)

part was held stationary. Feed was imparted using a windlass for a horizontal setup or by lowering the gun onto the tool using pulleys in the vertical case. They lacked rigidity and were suitable only for making clean-up cuts on cored castings. They were used to cut both bronze and iron castings.

The great breakthrough in canon boring was made by the Swiss craftsman Jean Maritz (1680–1743), who invented a machine capable of boring bronze canon accurately from solid castings sometime around 1714. There is a great deal of confusion concerning Maritz's career in the literature, since he had a son also named Jean Maritz (1711–1790, normally designated Jean II), and assorted later family members named Jean were also active in the ordinance business well into the nineteenth century [52,53]. Many earlier writers treat Jean and Jean II as the same person. The original Jean Maritz developed his invention in his native Burgdorf and later moved to a larger shop in Geneva. He was approached by both the French and Dutch governments to supply guns [48,53], and ultimately entered French service in 1734 as the Gun Founder at Lyon, assisted by Jean II. Together they set up a boring machine and produced specimen parts for military acceptance tests, in what appears to be the earliest recorded instance of a machine run-off [53]. This machine was transferred to the armory at Strasbourg in 1740, where Jean II had been named Master Founder. Jean II was an influential figure after 1750, extending his responsibilities to the founding and boring of marine guns (cast from iron rather than bronze) in France and setting up foundries and boring shops in Spain [52]. Jean II was also acknowledged as a central figure in the development of the *System Gribeauval* [54,55], the eighteenth century reform of the French artillery service, which was instrumental in the development of interchangeable manufacture.

No depiction of the original Maritz borer has survived, but based on descriptions of the machine at Strasbourg (Figure 1.6) [49] and later machines based on it [48,54], it was a massive, water-powered machine with complex gearing, which incorporated two significant innovations. First, it had a

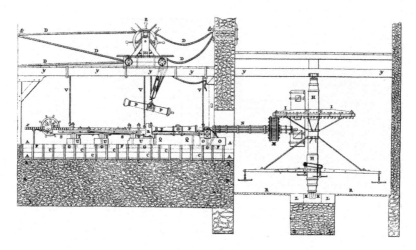

FIGURE 1.6 The Maritz borer at Strasbourg, from a book by Charles Dartien published in 1810. The machine depicted was powered by a horse mill, but the earliest Maritz borers were reportedly water powered. (After de Beer, C., Ed., *The Art of Gunfounding*, Jean Boudriot, Rotherford, East Sussex, U.K., 1991, pp. 12–15, 85–89, 93–94.)

horizontal layout that permitted use of a massive stone structure for added stiffness, and second, in contrast to previous machines, it featured a stationary tool and rotating gun. The advantages of the heavy, stiff structure are obvious, and in addition, the use of a horizontal layout simplified monitoring and adjusting the machine during boring. The advantage of rotating the gun rather than the tool is that a much straighter pilot hole could be drilled prior to boring, since any drill wander would generate radial forces, which would drive the drill back toward the axis of rotation. It is also easier to adjust the workpiece to run true (i.e., to align the axis of the part with the axis of rotation) than to minimize the runout of the drill, especially as drill length increases. Since the initial pilot hole form was followed in subsequent boring steps, this greatly improved final bore straightness and provided a gun with more uniform wall thickness.

Maritz's work had significant influence in France, Holland, and England. A book published by the republican government of France in 1794 [56] and apparently intended to stimulate backyard canon production contains plates of two- and four-spindle machines similar in principle to the Maritz borer. In Holland, Jan Verbruggen (1712–1781), Master Founder at The Hague, constructed similar machine for boring canon from the solid, which was in operation by 1758; he was assisted in this work by Johan Jacob Siegler, who had learned his trade at French foundries [47–49]. Jan and his son Pieter Verbruggen (1735–1786) moved to the Royal Brass Foundry at Woolwich in England in 1770 and immediately built two similar borers, which were in production from 1770 until 1842 [43,47–49,57]. In 1774, the English ironmaster John Wilkinson (1728–1808) patented a similar-looking machine with a rack-and-pinion feed mechanism for boring iron canon from the solid [46,58–60]. The patent was challenged by other gun founders and vacated in 1779, but the machine formed the basis for his later steam-engine boring mill discussed in the next section.

The first scientific study involving metal cutting also resulted from canon boring. Count Rumford (1753–1814) reported his experiences in boring canon in Bavaria in a paper presented in 1798 to the Royal Society in London [61]. He observed that boring produced a great deal of heat, and that water poured on boring tools as a coolant frequently boiled away. At the time heat was widely thought to be carried by a fluid called caloric, and it was assumed that the heat in boring resulted from caloric being liberated from the material being cut. Rumford modified a canon boring mill and conducted experiments in which dulled boring bars were forced against short cylinders immersed in water (Figure 1.7). He noted that the water boiled away even when the tools were so dull that they burnished rather than cut the work material, and moreover that the supply of heat was inexhaustible,

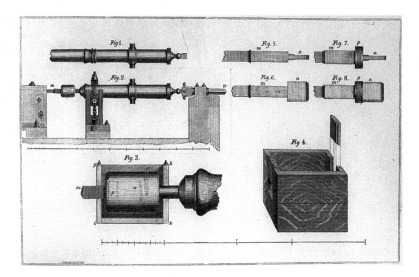

FIGURE 1.7 Illustration from Count Rumford's paper on the nature of heat, showing the boring mill and measurement apparatus used in his cannon boring experiments.

since the experiment could be continued indefinitely and still produce boiling. This contradicted the caloric theory of heat and led him to theorize that heat, like mechanical work, was a form of energy. His experimental results were not properly appreciated initially, but were ultimately important in the later development of Thermodynamics. In 1805, Rumford married the widow of Antoine Lavoisier (1743–1794), a leading proponent of the caloric theory of heat who was guillotined during the French Revolution [62].

1.2.3   THE INDUSTRIAL REVOLUTION AND THE STEAM ENGINE

The Industrial Revolution of the late eighteenth century resulted in (and to an extent resulted from) advances in steam engine design and the invention of automatic machinery for textile production. It also stimulated the development of many basic machine tools, since these were required to produce the precise cylinders, surfaces of revolution, screw threads, and flat surfaces integral to the function of the new machinery. The steam engine had additional significance because it freed machine tools from dependence on water power, so that machine shops could be built at any convenient location. The bulk of the significant early machine tool development took place in England, although it built on earlier work from other parts of Europe, especially France, and stimulated further development over a wider geographic area. The achievements of the early English machine builders are also better documented than those of engineers in most other areas, since the prolific Victorian writer Samuel Smiles wrote biographies of several leading Scottish and English engineers [63–66]. These biographies, based on extensive interviews and primary written sources, influenced later writers such as Roe [67] and Rolt [31].

The bores of atmospheric steam engines invented in the early eighteenth century, which were used especially to pump water from mines, were machined on modified canon boring mills [39,59,60,67]. These mills were not suitable, however, for the manufacture of James Watt's condensing engine, which operated at higher pressures and consequently required more precise bores. In fact, Watt, an instrument maker by trade, made small model engines but could not produce a full-sized specimen for 10 years because existing boring mills could not machine large cylinders to the required accuracy [64,67]. This problem was solved when John Wilkinson invented a more accurate horizontal boring mill in 1775 (Figure 1.8) [31,59,60,64]. Wilkinson used a much heavier boring bar than was used in cannon boring mills, and supported the bar at both ends, greatly increasing

(a)   (b)

FIGURE 1.8   Detail of a model of Wilkinson's boring mill in the Science Museum at Kensington. (a) General layout and (b) detail showing work table and cutter. (From Science and Society Picture Library, London.)

rigidity and accuracy. He also used cutters with replaceable inserts driven by a feed screw inside the bar [59]. This machine was the first recognizably modern machine tool, since it could perform heavy cuts with reasonable accuracy and employed a basic design that was replicated into the late twentieth century. Machine tools with the same basic layout are still widely used for line boring and large diameter work.

Heavy metal cutting lathes came into use in the eighteenth century for machining rolling mill rolls and cylinders for automatic looms [27]. A number of novel machines were devised in France in the latter half of the century, notably by Jacques de Vaucanson (1709–1782), the celebrated inventor of automata [27,39,64,68]. The design ancestral to later screw-cutting engine lathes, however, was developed by Henry Maudslay (1771–1831) in England [64,67]. Maudslay was born in Woolwich and apprenticed to the arsenal there at age 12. He worked with leading engineers on woodworking and lock-making machinery prior to his lathe work. His screw-cutting lathe (Figure 1.9) [27,31,39,41,44,67], built around 1797, employed Vee ways and a slide rest driven from a lead screw to cut threads. As in the case of Wilkinson's boring machine, this basic design was replicated into the twentieth century, albeit with significant improvements in gearing. Maudslay set new standards of accuracy for the profession by inventing a bench micrometer and pioneering the use of surface plates. He also trained many of the leading machine builders of the day, most notably Joseph Whitworth, who standardized screw threads, developed many feed and drive mechanisms, and was among the first to appreciate the primary importance of structural rigidity in machine tool design, and James Naysmith, inventor of the steam hammer [31,64,66,67].

FIGURE 1.9   Henry Maudslay's first screw cutting lathe, made c. 1797. (From Science Museum/Science & Society Picture Library, London.)

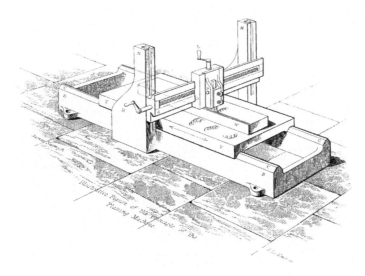

FIGURE 1.10 Typical early English planer. (From Naysmith, J., Remarks on the Introduction of the Slide Principle in Tools and Machines Employed in the Production of Machinery, in R. Buchanan, Ed., *Practical Essays on Millwork and Other Machinery*, 3rd edn., John Weale, London, U.K., 1841, Appendix B, pp. 393–418.)

Textile machinery, steam locomotives, and commercially produced machine tools required accurately machined flat surfaces, which led to the development of planers and shapers in the early nineteenth century. Both planers and shapers produced flat surfaces by moving a tool over a workpiece in a series of parallel and overlapping strokes. In a planer, the tool remained stationary and the workpiece, generally fixed to a table, moved relative to it; in a shaper, the work was stationary and the tool moved, generally on some sort of bar or arm. The nomenclature of these machines describes their traditional uses. Planers were used almost exclusively to produce flat surfaces, since the moving table was heavy and difficult to maneuver along a contour. Shapers, on the other hand, could be readily adapted to produce shaped features such as slots, keyways, and straight gear teeth. A number of planers were built in England before 1825, notably by Richard Roberts, James Fox, and Joseph Clement [31,39,67]. The general layout of these early machines is shown in Figure 1.10, from an article by Naysmith published in 1841 [69]. (Note that the portal frame in this somewhat later machine is more massive than in earlier designs.) Clement's planer was perhaps the most advanced of the early machines since it had a complex linkage, described in detail by Steeds [39], which permitted cutting on both the forward and back strokes. Shapers are of somewhat later origin. Among the first to be developed was Naysmith's steam arm shaper, built about 1836 [31,39,66]. In this machine (Figure 1.11 [66]), the tool was mounted to a ram driven by a connecting rod. This was a common design during the nineteenth century, although traveling head machines of the type designed by Whitworth between 1840 and 1850 [39], in which the tool was mounted on a head driven by reversing screws, were also prevalent.

Special drilling machines were rare during this period [39,44]. Drilling was typically performed on boring mills or lathes. The few exceptions consisted of vertical spindles mounted to structural beams of the mill building. Workpieces had to be transported to and positioned under these spindles, which would have been laborious and inaccurate for larger parts. Feed was manual or by dead weight prior to 1800 [44]. Upright and radial drill presses were introduced later as discussed in the next section.

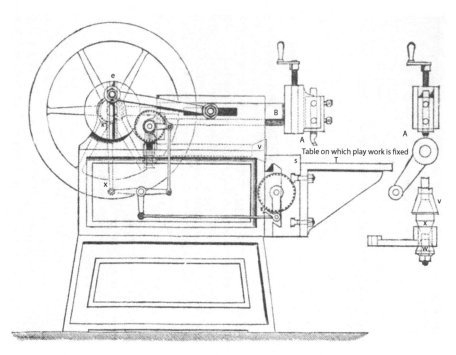

Table on which play work is fixed

FIGURE 1.11   Naysmith's steam-arm shaper, 1836.

1.2.4   Nineteenth-Century Quantity Production Industries

Machine tool development was further stimulated in the middle to late nineteenth century by the rise of quantity production methods. These methods were pioneered in the small arms industry and spread over time to the manufacture of consumer products such as sewing machines, watches, typewriters, and bicycles, as well as of capital goods such as railway locomotives. These methods differed significantly from contemporary mass production in that they relied on limit gages and physical masters and did not employ measurement datums or tolerances. Collectively, they were referred to in the older literature as interchangeable manufacture [31,70] and are now generally called the American system of manufacture [71]. Neither of these terms is strictly accurate. Parts were not truly interchangeable since part fitting and selective fit assembly of key components was often required, especially in the sewing machine and watch industries [71,72]. And although the methods were first perfected as a system by nineteenth-century American firms, key elements were derived from eighteenth-century French arms production methods (such as the *System Gribeauval* discussed in Section 1.2.2) and other European practices [55,70,71,73–75]. Quantity production methods required the rapid generation of accurate flat surfaces, holes, and gears, and led to the development of milling, drilling, grinding, and gear-cutting machines, as well as early automated and special purpose machines.

Historically, milling machines came first, since simple milling methods had already been developed by Vaucanson in France and Naysmith in England [76]. Small arms makers in Connecticut also reportedly constructed simple milling machines by 1820 [74,76]. The oldest extant machine appears to be a simple one with self-acting feed, which Roe dated to 1818 and attributed to Eli Whitney (1765–1825) [67], although this attribution has since been questioned [74,77]. The first commercially produced milling machine was the Lincoln Miller (Figure 1.12) [31,39,70,76,78], designed by F. A. Pratt and E. K. Root, which was first marketed by George S. Lincoln and Co. in 1855. Shortly thereafter Pratt's new company, Pratt and Whitney, began manufacturing them in large quantities, and they became very common machines in nineteenth-century shops. Of more significance to later machines, however, is Brown and Sharpe's universal miller, designed by J. R. Brown and introduced in 1862 (Figure 1.13) [31,39,76]. This machine featured a knee and

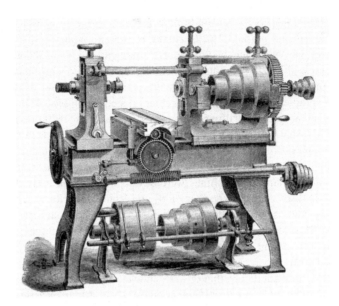

FIGURE 1.12 A Lincoln miller, first manufactured in 1855. (From Rose, J., *Modern Machine Shop Practice*, Vol. 2, Charles Scribner's Sons, New York, 1888, pp. 3, 38–54.)

FIGURE 1.13 The original Browne and Sharpe universal milling machine, 1862. (Courtesy of Brown and Sharpe, Providence, RI.)

column structure, which greatly simplified the adjustment of the work height and is the ancestor of the familiar knee-mill of twentieth-century practice.

Drill presses were introduced in the 1840s. Among the earliest were the upright or pillar drill presses introduced by Bodner and Whitworth in England in 1839 and 1847, respectively [39,79]. A similar machine was patented in Germany in 1849 by August Harmann, who had trained in

English shops [44]. Radial drill presses were built in England in the early 1840s [39], and a radial drill press was developed and marketed in Germany by Johann Mannhardt in 1848 [44]. These early machines, as well as American specimens of similar vintage, were identical in basic structure to later drill presses, although subsequent machines incorporated significant improvements in power transmission and feed mechanisms. American quantity production firms soon recognized that multiple drilling spindles could be ganged together on a common base to produce multispindle machines for rapidly and accurately drilling hole patterns. These gang drill presses were used especially in the railroad industry for drilling hole patterns in boilers and wheels [39,70]. A final significant drilling development of this period was the twist drill, patented by Stephen A. Morse in 1863 [44,80]. This now-ubiquitous tool rapidly displaced the crude spear-point drill used until that date.

Cylindrical and surface grinding machines were developed from an early date, although their evolution is complex as discussed by Woodbury [29] and Rolt [31]. Prior to 1850, they were used especially for grinding rifle barrels, hardened slides for steam engines, and pulleys and rollers for automatic machinery. Most leading tool builders in England, the United States, and Germany built machines for internal use. Many early cylindrical and surface grinders were basically modified lathes or planers fitted with abrasive wheels. Early grinders generally used natural stone wheels or natural emery abrasives embedded in leather or wooden wheels [78], and progress was limited by an inability to produce an effective bonded artificial abrasive wheel. Many early machines also lacked stiffness and required considerable skill and training to operate. The development of precision grinders began after 1860 [29,31,39] and was driven especially by the growth of the sewing machine industry, which required large volumes of precision shafts [31]. In 1876, Brown and Sharpe, an early contract manufacturer of sewing machines, brought out a universal grinder, which had the basic form of later cylindrical grinders [29,31,39], and followed with surface grinders of similarly modern form by 1883 (Figure 1.14) [39,78]. These influential machines were sold worldwide beginning in

FIGURE 1.14 A Brown and Sharpe surface grinder, introduced about 1883. (From Rose, J., *Modern Machine Shop Practice*, Vol. 2, Charles Scribner's Sons, New York, 1888, Fig. 2032.)

the 1880s. Progress in grinding development accelerated as bonded aluminum oxide and silicon carbide wheels became available between 1885 and 1910 [29,31].

Gear-cutting machines have a similarly complex history as also described by Woodbury [40] and Rolt [31]. Textile machinery and early machine tools created a demand for gears, and between 1825 and 1840 several English machine tool builders built form-cutter-type gear machines based on the milling principle. Whitworth introduced involute cutters about 1844. Templet machines, in which the teeth were not milled but cut with a shaper or slotter tool guided by a template, were also developed at an early date and after 1860 were especially popular for large gears. The golden age of gear machine development occurred after 1885 and was enabled by the development of precision grinding, which permitted manufacture of accurate complex tooling from hardened steel. The Fellows gear shaper and recognizably modern hobbing machines, such as those developed by the German firms Juengst and Reinecker, are products of the 1890s.

The requirement of large volumes of small uniform parts, coupled with a shortage of skilled labor, led to interest in labor-saving and ultimately automated machinery during this period. Turret lathes, in which a series of pre-set tools could be brought to bear on a part by a lever-activated indexing head, were pioneered in the small arms industry. The first turret lathe was built by Stephen Fitch in 1845 for the production of screws for a percussion lock (Figure 1.15) [31,39,44,67,70,81]. E. K. Root designed a similar machine at about the same time. Fitch used a cylindrical turret aligned with the work axis; vertical turrets on a rotating axis of the type designed by F. H. Howe were introduced in about 1850 at the Colt and other armories and quickly became the standard design. The first fully automatic lathe was invented by Christopher Spencer in 1873 [31,44,67,70,81,82]. Spencer, who also invented a famous repeating rifle, had a background in small arms manufacture, but his automatic lathe was first used for the production of sewing machine screws. It featured a *brain wheel*, a rotating drum with strip steel cams, which guided tool motions (Figure 1.16). Spencer disastrously did not patent the brain wheel concept [31], and it was widely copied, leading to an explosion in screw machine development as described by Ruby [81]. A parallel path of development of automatic lathes occurred in the clock and watch industries, where automatics for the production of small watch screws and shafts were developed independently by Jacob Schweizer in Switzerland and by C. V. Woerd of the Waltham Watch Company in Massachusetts [39,72,81,83]. The Waltham Watch

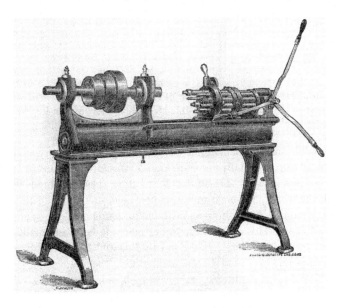

FIGURE 1.15   Stephen Fitch's turret lathe (1845) for producing screws for a percussion lock. (From Fitch, C.H., Report on the Manufactures of Interchangeable Mechanism, Report on the Manufactures of the United States at the Tenth Census, U.S. Government Printing Office, Washington, DC, 1883, p. 648.)

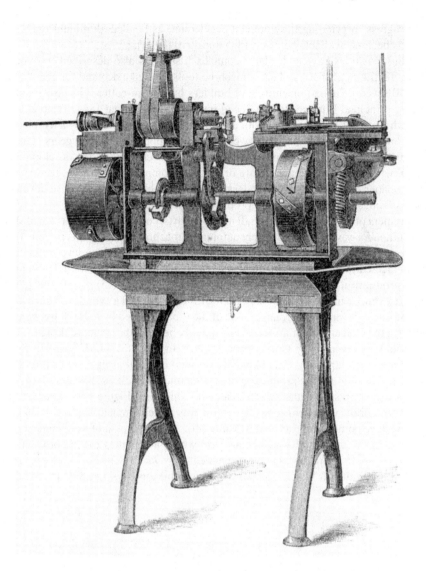

FIGURE 1.16 A *brain-wheel* style screw machine of the type invented by Christopher Spencer in 1873. (From Fitch, C.H., Report on the Manufactures of Interchangeable Mechanism, Report on the Manufactures of the United States at the Tenth Census, U.S. Government Printing Office, Washington, DC, 1883, p. 658.)

Company developed additional automated machines for the production of watch plates and cases in the 1880s, including what were essentially miniature transfer machines (Figure 1.17) [72]. Interestingly, since there was no commercial electrical service at this time, these machines were powered pneumatically. One final noteworthy advance in automated machinery during this period was made by William Sellers of Philadelphia, who demonstrated an automated gear-cutting machine at the Paris exhibition in 1867 [40]. This was a form-cutter machine with automatic feed and indexing controlled by screws and stops. Sellers was the most influential American machine builder of his time, noted for stripping machines of ornament and painting them a drab *machine gray* and more importantly for standardizing screw threads and investing heavily in scientific research as discussed in the next section [31,67].

In the 1860s, under the Tokuda government, Japanese armories began importing European machine tools. This process accelerated after the Meiji Restoration in 1868, when a western-style army was formed and military production became a priority. As described by Chokki [84], machine

FIGURE 1.17 An automatic plate facing and recessing machine from the Waltham Watch Company, c. 1880. (From Marsh, E.A., *The Evolution of Automatic Machinery as Applied to the Manufacture of Watches*, G. K. Hazzlit and Co., Chicago, IL, 1896, p. 27.)

tool imports and native machine tool construction intensified at major armories throughout Japan in the 1870s, laying the foundation for the contemporary Japanese machine tool industry.

1.2.5 EARLY SCIENTIFIC STUDIES

The first scientific studies of metal cutting were conducted between 1850 and 1885 [85–88]. Many of the early investigations concerned the mechanism of chip formation. Time [89] studied chip formation in metals and wood and apparently coined the term "chip." Tresca [90] and Mallock [91] also studied chip formation and identified it as a shearing process. Mallock used a microscope to observe chips being formed in various metals, as well as paraffin, soap, and clay, made interesting drawings of his observations (Figure 1.18) and also recognized the importance of tool–chip friction in the process. Other researchers quantified power requirements in machining. In the earliest investigation, Cocquilhat [92] measured the work required to drill a given volume of material. The most

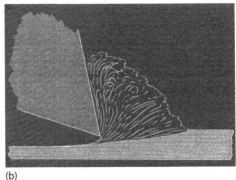

(a)   (b)

FIGURE 1.18 Drawings of steady-state chip formation when cutting wrought iron (a) and incipient chip formation when cutting copper (b). (From Mallock, A., *Proc. R. Soc. Lond.*, 33, 127, 1881.)

thorough work of this type, however, was carried out by Ernst Hartig and his students at the Royal Saxon Polytechnic Institute (now the Technical University of Dresden). Hartig had earlier designed and constructed a leaf-spring dynamometer to measure power requirements for textile machinery, and beginning in 1870 he used it to make systematic measurements of power requirements for a variety of machine tools under different speed and feed conditions, primarily for ferrous work materials, wood, and bronze. Machines at two local factories were studied, including saws, lathes, planers, milling machines, drill presses, grinders, and woodworking machines, as well as six fans and two cranes. Hartig measured both tare and cutting power and applied corrections for transmission power losses and internal friction in his dynamometer. The results were summarized in a book published in 1873 [93], which consisted largely of brief experimental descriptions accompanied by tables and formulas, and which remained the standard work on the subject for many years.

Other research was directed toward developing improved tool steels. In early machining practice, plain carbon steel tools were used, but as steelmaking technology advanced, alloy tool steels were investigated. The most successful work in this area was carried out by the English metallurgist Robert Mushet, who in the course of compulsive experimentation developed an air-hardening tungsten steel alloy in 1868 [31,94,95]. Mushet was the son of a well-known ironmaster and had already made significant practical contributions to the Bessemer process. He was secretive about his discovery, and never published or patented any details of its composition or preparation. He worked with a few picked associates in a small, remote workshop, had materials shipped through multiple intermediaries, referred to its ingredients by cyphers, and kept all records in code [94]. The new steel was marketed as "R. Mushet Special Steel," or RMS steel. (Over time various premium grades, including double Mushet, triple Mushet, and Extra Triple Mushet, were also introduced [94].) RMS steel produced a revolution in machining practice, permitting doubling or tripling turning and planing speeds in shops equipped with machines powerful enough to take advantage of its properties [95]. It was widely used in both Europe and North America in the last decades of the nineteenth century.

The great historical figure in the field of metal cutting, Frederick W. Taylor, was active at the end of the nineteenth century [31,67,95–99]. Taylor is best known as the founder of scientific management [96]. In 1880, he became foreman of the machine shop at the Midvale Iron Works near Philadelphia, where William Sellers served as president. At that time the tools and cutting speeds used to perform machining operations were selected by individual machine operators, so that practices and results varied considerably. He felt that shop productivity could be greatly increased if standard best practices were dictated by a central planning department. Putting this idea into practice, however, required the development of a quantitative understanding of the relation between speeds, feeds, tool geometries, and machining performance. Taylor embarked on a series of methodical experiments to gather the data (mainly tool life values) necessary to develop this understanding. The experiments continued over a number of years at Midvale and the nearby Bethlehem Steel Works, where he worked jointly with metallurgist Maunsel White. As a result of these experiments, Taylor was able to increase machine shop productivity at Midvale by as much as a factor of 5.

Taylor's most important practical contribution was his invention, with White, of high-speed steel-cutting tools. Starting with RMS steel, which they analyzed chemically, Taylor and White began varying the alloy composition and heat treatment. Eventually they developed alloys containing tungsten, chromium, and silicon, which were self-hardening and stable at much higher temperatures than RMS steel. Tools made of these high-speed steels were vastly superior to anything available at the time and permitted a great increase in cutting speeds for machines rigid and powerful enough to accommodate the change [95]. Taylor also established that the power required to feed the tool could equal the power required to drive the spindle at the relatively low cutting speeds then in use, especially when worn tools were used (Figure 1.19) [97]. Machine tools of the day were underpowered in the feed direction, and he had to modify all the machines at the Midvale plant to eliminate this flaw. He also demonstrated the value of coolants in metal cutting and fitted his machines with recirculating fluid systems fed from a central sump (or "suds tank"). Finally, he developed a special slide rule for determining feeds and speeds for various materials.

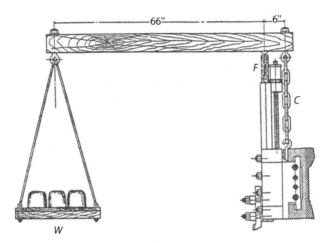

FIGURE 1.19   Balance beam dynamometer for measuring the feed force on a boring mill. *W* is the counter-weight, *F* is the fulcrum for the 11:1 lever arm, and *C* is the chain to the tooling assembly. (From Taylor, F.W., *ASME Trans.*, 28, 31, 1906.)

Taylor summarized his research results in the landmark paper "On the art of cutting metals" [97], which was delivered to a convention of the American Society of Mechanical Engineers and published in the Society Transactions in 1906. The results were based on an estimated 30,000–50,000 cutting tests conducted over a period of 26 years, consuming 800,000 lb of metal and costing 10 industrial sponsors an estimated U.S.$200,000. In conducting his tests, Taylor was fortunate to have at his disposal an essentially unlimited supply of consistent work material, as well as the backing of William Sellers to maintain funding over such a protracted period. Taylor's most important research contribution was his recognition of the importance of tool temperatures in tool life, which led to the development of his famous tool life equation (Equation 9.8).

1.2.6   TWENTIETH-CENTURY MASS PRODUCTION

Machine tools developed after 1900, the year high-speed steel tools were first exhibited in Paris [31,44], differed significantly from their predecessors. They were designed to run at much higher speeds to take advantage of the new tools. They thus required more power, stiffer structures, improved gearing, and improved lubrication and cooling for both the gears and workpiece. To supply these requirements, machine tool builders were able to borrow technologies from the simultaneously emerging electrical, chemical, and automotive industries.

Power requirements for milling machines increased from 0.4–0.8 to 3–4 kW [44], and similar increases were required for lathes and drilling machines. Coupled with the availability of commercial electric power, this led to a gradual shift in machine power from leather belts and countershafts to individual electric motors. Electric motors had first been used to power machines in Europe in the 1870s, and Taylor had fitted an experimental machine with an electric motor in 1894 to eliminate belt slip [31]. Machines driven by electric motors through V-belts became commercially available in about 1901, and as motor sizes decreased they gradually became standard throughout the industry. Machine structures also became stiffer to maintain accuracy under increased loading [100]. The cast gears used in earlier machines wore out rapidly at increased power levels and were replaced by hardened steel gears. Also, the increased speed ranges required by the new tools led to the adoption of quick change gear drives, such as W. P. Norton's 1892 tumbler drive, in place of the earlier clumsy cone pulleys and change gears [31]. Gear drives themselves were lubricated directly with oil [44], and in addition neat oil and water-based coolants began to be used routinely in standard practice. Water was used to cool grinding wheels from the early nineteenth century [29], and as noted before Taylor demonstrated the advantage of recirculating coolant systems in his research. Neat oils were used as cutting lubricants starting in the 1870s, shortly after the discovery of petroleum in the United States, and soda water mixed with fatty lubricants ("suds") was also used as a cutting fluid in the late nineteenth century. Soluble (emulsifiable) oils became available in about 1915 and were used extensively in the industry by 1925 [101].

The substantial capital invested in older tools prevented the immediate introduction of new tools in established industries. The new tools found a large market, however, in emerging mass production industries. Mass production is generally believed to have originated in the automotive industry [71], but it also has roots in World War I munitions production [100,102]. It evolved from quantity production when required production volumes increased and when parts from multiple plants had to be assembled into a single mechanism and was characterized by a shift from physical masters and limit gages to dimensioned and toleranced drawings, by a general increase in the precision of machined components, and by increased attention to factory-level organization and work standardization. The origin of tolerances is obscure, but according to Buckingham they first came into use in the United States shortly after 1900 [102].

There were two paths to achieve the required increase in precision and uniformity at increased volumes: the use of special jigs and fixture on standard machines, and the development of special purpose machines for high volume production [100]. The development of special purpose machines is of more interest in the general development of machining technology, and much of this work was centered in the automotive industry. It is best documented for the case of Ford's Model T plant in Highland Park, MI [71,103–109], but considerable similar work was carried out in other automotive plants in the United States and Europe over the decades between the two World Wars. The Model T plant retained some nineteenth-century features—for example, machines were powered by line shafting and leather belts, although the shafts were turned by electric motors [71], and part transfer between machines for heavy-machined components was often manual—but machine placement was carefully considered, and both single-purpose machines and special fixtures were designed in-house. Representative examples include special piston-turning machines [107], multi-way multispindle drilling machines [71,103,104,109], multipart and autoclamping fixtures [71,103,110], and large capacity gang multispindle milling machines (Figure 1.20). At Ford and elsewhere (including earlier railroad

FIGURE 1.20 Special three-spindle gantry machine and multipart fixtures for milling engine block oil pan faces for the Ford Model T, c. 1914. (From the Collections of the Henry Ford (P.O. 3927/THF115651).)

engine factories), specialized grinding machines also saw significant development and application during this period. Both the Norton and Landis companies in the United States produced special crankshaft and camshaft grinding machines between 1903 and 1912 [31]. The Heald Company introduced both a planetary grinder for cylinders and a piston-ring grinding machine in 1904 [31]. Landis introduced an automatic crankpin grinder in 1923 and an automatic camshaft grinder in 1928 [29].

The most striking examples of special purpose machinery, however, were transfer machines, collections of individual special machine tools (or stations) connected by an automated part transfer mechanism. The literature on the development of transfer machines is sparse [111–115]. Although the Waltham Watch Company had built small pneumatic-powered transfer machines in the 1880s (see Section 1.2.4), Lloyd [111] traced the development of automotive transfer machines to the Greenlee Company of Rockford, IL, which introduced a straight line transfer machine for boring and adzing railroad ties in 1908, and to three-way boring and similar multi-way machines used for drilling in early automotive factories. He also noted that early transfer machine development took place primarily in the United States and England. Lloyd provided a number of photographs of early in-line and rotary transfer machines; a rotary example is shown in Figure 1.21. In the in-line case, he described 39-station palletized system for transmission cases built by Archdale in 1924 (for which he provided both a photograph and an operations sequence), but this large system seems to have been an exception in this period; most early systems had three or four stations and would thus be considered transfer system segments in contemporary practice. Charles Kettering of General Motors, writing in 1927, discussed the increasing use of *multiple machines*, which from his description are clearly transfer machine segments [116]. Photographs from the 1930s also generally show short transfer machine segments with considerable part-handling between operations (Figure 1.22). This approach

FIGURE 1.21 Five-station rotary transfer machine for transmission housings, 1933. (After Lloyd, E.D., *Transfer and Unit Machines*, Industrial Press, New York, 1969, pp. 9–22.)

FIGURE 1.22   Two-station, eight-spindle transfer machine for machining valves and lifter bores in flat head V8 engine blocks, Ford Rouge River plant, 1935. (From the Collections of the Henry Ford (P.833.62054/ THF115648).)

would permit banking of parts between operations to eliminate full production shutdowns due to station downtime, an issue Kettering discussed and one which is still a basic concern for large serial systems. Hine [113], who also traced the origin of the transfer machine to the Greenlee Company, reported that only short transfer segments were used before 1940, with larger systems first being used for aircraft engine production during World War II. Early examples include a 200 ft long, 80 station system built in 1941 to machine cylinder heads for B17 and B29 bomber engines [113], and a large Greenlee engine block system delivered to the Wright Aero Corporation in the same year [111]. Large fully automated systems were not widely used in automotive and other high production industries in the United States, Europe, and Japan until the late 1940s and 1950s, when many older machine tools were replaced with high-production systems to meet increased consumer demand [111–114]. The Cross Company, a prominent post-war U.S. supplier, began building transfer systems in 1946 [115], and Toyota installed its first transfer machine, the TR1, at its Koromo plant in 1956 [117].

Tungsten carbide cutting tools were introduced by the Krupps Company of Essen, Germany in 1926 [118,119]. Krupps called the material Widia, shortened from the German *wie Diamant* (like diamond), and marketed it as Widmet in England and Carboloy in the United States [31]. As with high-speed steel previously, it enabled a significant increase in cutting speeds and was rapidly adopted both inside and outside the automotive industry [120,121].

1.2.7   NUMERICAL CONTROL

After World War II, the aircraft industry became an important market for machine tools. Since aircraft manufacture differs from automotive manufacture in that smaller batches of complex parts are required, it led to the development of more advanced general purpose machine tools. The most important innovation in this area was the introduction of numerical control (NC). Numerically controlled machines can be viewed as the descendants of cam and template-copying machines in which tools were guided along complex paths by tracing a physical master, but in the NC case the physical master was replaced by a stored computer program.

There was considerable development of servomechanism technology during World War II, and after the war, as discussed by Noble [122], various attempts were made to apply this technology to automatic machine tool control. Two basic approaches were explored: the record-playback approach, in which tool motions of a machine operated by a skilled machinist were recorded (often on magnetic tape) for later replication, and the punched tape approach, in which tool motions were controlled by a program generated by a computer and generally stored on a punched paper tape. General Electric marketed an early record-playback system, and this method was also explored by companies such as Giddings and Lewis and Grisholt. However, it had the drawback that a physical prototype had to be machined for each geometry change, which is obviously cumbersome in practice, and ultimately punched tape control proved to be more practical.

Current NC technology developed largely from work initiated by John Parsons of the Parsons Corporation. Parsons developed early numerical control concepts for producing airfoil sections, and received an Air Force contract to develop a card-controlled machine called the Cardmatic [123]. The Parsons Corporation subcontracted the servo system design to the Servomechanism Laboratory at MIT [44,122,124,125]. After considerable refinement of the initial concepts, the M.I.T. group developed and demonstrated a tape-controlled milling machine in 1952 (Figure 1.23). (The complex funding and contractual arrangements generated by this project are described by Reintjes [125].) A key contribution of the M.I.T. group was Douglas Ross's development of the APT (Automatically Programmed Tools) language for tool path code generation. This was one of the first higher-level programming languages and served as the basis for many subsequent NC Standards, including some currently used to control waterjet and laser cutting machines.

Deployment of NC machines began in the aircraft industry, which had about 100 NC machines in service by 1956 [44]. Deployment to other industries followed, with the first automotive installations occurring around 1960 [122]. American tape-controlled machines were demonstrated at a convention in Paris in 1959, and German companies marketed an NC machine in 1960 [44]. NC retrofitted machine tools became common in Europe in the mid-1960s. Deployment was initially hindered by the unreliability of tape systems, and accelerated after the introduction of Computer Numerical Control (CNC), in which the machine was equipped with an on-board computer and the

FIGURE 1.23 One of the first NC machine tools: the tape-controlled milling machine developed at MIT under an Air Force contract in 1952. The cabinets behind the machine house the controller, which used 292 vacuum tubes and had a clock speed of 512 Hz. (From MIT Museum, Cambridge, MA.)

part program was stored as a computer file rather than on a tape. CNC machines had been built as early as the late 1950s [44], but the technology did not become common until the 1980s.

With the introduction of CNC, we leave the historical era and enter the realm of current practice, to be described in the remaining chapters of this book.

REFERENCES

1. S. Malkin and C. Guo, *Grinding Technology*, 2nd edn., Industrial Press, New York, 2008.
2. G. F. Benedict, *Nontraditional Manufacturing Processes*, Marcel Dekker, New York, 1987.
3. J. E. Weller, *Nontraditional Machining Processes*, 2nd edn., Society of Manufacturing Engineers, Dearborn, MI, 1984.
4. G. Boothroyd and W. A. Knight, *Fundamentals of Machining and Machine Tools*, 3rd edn., CRC Press, Boca Raton, FL, 2005, Ch. 14.
5. J. F. Wilson, *Practice and Theory of Electrochemical Machining*, Wiley-Interscience, New York, 1971.
6. E. Kannatey-Asibu Jr., *Principles of Laser Materials Processing*, Wiley-Interscience, New York, 2009.
7. A. Kobayashi, *Machining of Plastics*, McGraw-Hill, New York, 1967.
8. S. Chandrasekar et al., Eds., *Machining of Advanced Ceramic Materials and Components*, ASME, New York, 1988.
9. R. J. Forbes, *Metallurgy in Antiquity*, E. J. Brill, Leiden, the Netherlands, 1950, pp. 9–10.
10. R. F. Tylecote, *A History of Metallurgy*, The Metals Society, London, U.K., 1976, Ch. 1.
11. R. Raymond, *Out of the Fiery Furnace*, Pennsylvania State University Press, University Park, PA, 1986, pp. 8–11, 16–22.
12. P. T. Craddock, *Early Metal Mining and Production*, Smithsonian Institution Press, Washington, DC, 1995, Ch. 3.
13. T. K. Derry and T. I. Williams, *A Short History of Technology*, Dover, New York, 1993, p. 12.
14. D. R. Cooper, The development of coinage dies from bronze to steel, *Newcomen Soc. Trans.* **67** (1995–1996) 91–108.
15. *Theophilus: On Divers Arts*, Trans. by J. G. Hawthorn and C. S. Smith, Dover, New York, 1979, pp. 87, 169, 180.
16. W. L. Roberts, *Cold Rolling of Steel*, Marcel Dekker, New York, 1978, Ch. 1.
17. *Vitruvius: Ten Books on Architecture*, Trans. by I. D. Rowland, Cambridge University Press, Cambridge, U.K., 1999, p. 10.7.
18. G. L. Irby-Massie and P. T. Keyser, *Greek Science of the Hellenistic Era*, Routledge, London, U.K., 2002, pp. 215–216.
19. J. Bordaz, *Tools of the Old and New Stone Age*, Natural History Press, Washington, DC, 1970, pp. 72, 93.
20. K. P. Oakley, *Man the Tool-Maker*, Phoenix Books, University of Chicago Press, Chicago, IL, 1959, pp. 52–54.
21. J. E. M. White, *Ancient Egypt, Its Culture and History*, Thomas Y. Crowell Co., New York, 1952, p. 112.
22. Pliny the Elder, *Natural History*, Loeb Classical Library, Harvard University Press, Cambridge, MA, 1938–1963, pp. 36.9, 36.37, 36.47.
23. P. T. Nicholson and I. Shaw, Eds., *Ancient Egyptian Materials and Technology*, Cambridge University Press, Cambridge, U.K., 2000, p. 356.
24. W. L. Goodman, *The History of Woodworking Tools*, G. Bell and Sons, London, U.K., 1964, pp. 39–44, 163.
25. D. A. Stocks, Making stone vessels in ancient Mesopotamia and Egypt, *Antiquity* **67** (1993) 596–603.
26. H. Hodges, *Technology in the Ancient World*, Barnes and Noble Books, New York, 1992, p. 188.
27. R. S. Woodbury, *History of the Lathe*, Society for the History of Technology, Cleveland, OH, 1961, pp. 20–22, 39–44, 53–54, 64–67, 82–89, 101–103.
28. F. M. Feldhaus, *Die Technik der Vorzeit, der Geschichtlichen Zeit und der Naturvölker*, Wilhelm Engelmann, Berlin, Germany, 1914, pp. 210, 391–393.
29. R. S. Woodbury, *History of the Grinding Machine*, Technology Press, Cambridge, MA, 1959, pp. 13, 17–23, 31–97, 140–142.
30. H. C. Mercer, *Ancient Carpenter's Tools*, 5th edn., Dover, New York, 2000, pp. 114–116, 293–296.
31. L. T. C. Rolt, *Tools for the Job*, Batsford, London, U.K., 1965; Reissued by HSMO Press, London, U.K., 1986, pp. 32–38, 41–47, 60–61, 93–94, 107, 111–112, 115–129, 169–170, 174–180, 184–185, 188–194, 204–116, 219–223, 229–234, 237–238, 247.
32. N. G. Davies, *Paintings from the Tomb of Rekh-Mi-Re at Thebes*, Metropolitan Museum of Art, New York, 1935, Plate XXIII.

33. D. A. Stocks, Ancient factory mass-production techniques: Indications of large-scale stone bead manufacture during the Egyptian New Kingdom period, *Antiquity* **63** (1989) 526–531.
34. 1 Samuel 13:19–22; *The New Oxford Annotated Bible with the Apocrypha*, Revised Standard Version, Oxford University Press, New York, 1977, p. 347.
35. T. Beck, *Beiträge zur Geschichte des Maschinenbaues*, Julius Springer, Berlin, Germany, 1899, pp. 104, 344–345, 427–444, 458–461.
36. F. M. Feldhaus, *Leonardo der Techniker und Erfinder*, J. Eugen Diederichs, Jena, Germany, 1913, pp. 48–54, 72–75.
37. W. B. Parsons, *Engineers and Engineering in the Renaissance*, MIT Press, Cambridge, MA, 1932, pp. 115–116, 132, 145–147.
38. B. Dibner, Leonardo: Prophet of automation, in: L. Reti and B. Dibner, Eds., *Leonardo da Vinci Technologist*, Burndy Library, Norwalk, CT, 1969, pp. 37–59.
39. W. Steeds, *A History of Machine Tools 1700–1910*, Clarendon Press, Oxford, U.K., 1969, pp. 1, 6–8, 17–18, 20–23, 38–42, 55–56, 60–73, 109–113, Plate 125.
40. R. S. Woodbury, *History of the Gear-Cutting Machine*, Technology Press, Cambridge, MA, 1958, pp. 45–59, 117–118.
41. F. N. Zagorskii, *An Outline of the History of Metal Cutting Machines to the Middle of the 19th Century*, Trans. by S. P. Pedneker, Amerind Publishing Co., New Delhi, India, 1982, pp. 68–69, 129–131, 179–192.
42. A. O. von Essenwein, *Quellen zur Geschichte der Feuerwaffen*, 2 Vols., F. A. Brockhaus, Leipzig, Germany, 1877, Vol. 1, 33–34, Vol. 2, Plate XXXVIII.
43. A. N. Kennard, *Gunfounding and Gunfounders*, Arms and Armour Press, London, U.K., 1986, pp. 17, 148–149.
44. K. Allwang, *Werkzeugmaschinen*, Deutsches Museum, Munich, Germany, 2002, pp. 21–23, 29–33, 36–38, 46–49, 54–59, 69–83.
45. V. Biringuccio, *De La Pirotechnia*, Senato Veneto, Venice, Italy, 1540; English Trans. by C. S. Smith and M. T. Gnudi, The American Institute of Mining and Metallurgical Engineers, New York, 1942, pp. 308–312.
46. D. H. P. Baird, Gun boring from the solid, *Newcomen Soc. Trans.* **58** (1986–1987) 45–58.
47. C. Ffoulkes, *The Gun-Founders of England*, 2nd edn., Arms and Armour Press, London, U.K., 1969, pp. 16–7, 65.
48. M. Jackson and C. de Beer, *Eighteenth Century Gunfounding*, David and Charles, Newton Abbot, U.K., 1973, pp. 17, 19, 35–37, 43–47, 72–74.
49. C. de Beer, Ed., *The Art of Gunfounding*, Jean Boudriot, Rotherford, East Sussex, U.K., 1991, pp. 12–15, 85–89, 93–94.
50. I. Gamel, *Description of the Tula Weapon Factory in Regard to Historical and Technical Aspects*, English translation, Amerind Publishing Co., New Delhi, India, 1988, pp. 20–26.
51. D. Diderot and J. R. d'Alembert, *Fabrication des Canons*, Bibliotheque de l'Image, Tours, France, 2002.
52. M. F. Schafroth, Die Geschützgiesser Maritz, Geschichte einer Erfindung und einer Familie, *Burgdorfer Jahrbuch* **20** (1953) 9–31; **21** (1954) 111–139; **22** (1955) 93–103.
53. F. Naulet, Les Maritz une Famille de Fondeurs a Service de la France, *Revue Internationale d'Histoire Militaire* **81** (2001) 91–100.
54. H. Rosen, The system Gribeauval: A study of technological development and institutional change in eighteenth century France, PhD thesis, University of Chicago, Chicago, IL, 1981, pp. 98–99.
55. K. Adler, *Engineering the Revolution: Arms and Enlightenment in France, 1763–1815*, Princeton University Press, Princeton, NJ, 1997, pp. 40–43, 328–338.
56. G. Monge, *Description de l'Art de Fabriquer les Canons*, Comité de Salut Public, Paris, France, 1794.
57. O. F. G. Hogg, The development of engineering at the royal arsenal, *Newcomen Soc. Trans.* **32** (1959–1960) 29–42.
58. H. W. Dickinson, John Wilkinson, *Beiträge zur Geschichte der Technik und Industrie* **3** (1911) 215–238.
59. N. C. Soldon, *John Wilkinson, 1728–1808: English Ironmaster and Inventor*, Edwin Mellon Press, Lewiston, NY, 1998, pp. 99–109.
60. E. A. Forward, The early history of the cylinder boring machine, *Newcomen Soc. Trans.* **5** (1924–1925) 24–38.
61. B. Thomson (Count Rumford), An experimental inquiry concerning the source of the heat which is excited by friction, *Philos. Trans.* **88** (1798) 80–102; see also S. C. Brown, Ed., *Collected Works of Count Rumford*, Vol. 1, Harvard University Press, Cambridge, MA, 1968, pp. 1–26.
62. S. C. Brown, *Benjamin Thompson, Count Rumford*, MIT Press, Cambridge, MA, 1979, pp. 248, 275.
63. S. Smiles, *Lives of the Engineers*, Vol. II: *Smeaton, Rennie, and Telford*, John Murray, London, U.K., 1862.

64. S. Smiles, *Industrial Biography: Iron Workers and Tool Builders*, John Murray, London, U.K., 1863, pp. 198–235, 249, 258–298.

65. S. Smiles, *Lives of Boulton and Watt*, John Murray, London, U.K., 1865, pp. 136, 212–213.

66. S. Smiles, Ed., *James Naysmith, an Autobiography*, Harper and Brothers, New York, 1884, pp. 127–156, 419.

67. J. W. Roe, *English and American Tool Builders*, Yale University Press, New Haven, CT, 1916; Reprinted by Lindsay Publications, Bradley, IL, 1987, pp. 2–3, 33–49, 53–62, 81–108, 142, 169–170, 176–177, 197, 247–251, 277.

68. S. A. Bedini, The role of automata in the history of technology, *Technol. Cult.* **5** (1964) 24–42.

69. J. Naysmith, Remarks on the introduction of the slide principle in tools and machines employed in the production of machinery, in: R. Buchanan, Ed., *Practical Essays on Millwork and Other Machinery*, 3rd edn., John Weale, London, U.K., 1841, Appendix B, pp. 393–418.

70. C. H. Fitch, Report on the manufactures of interchangeable mechanism, Report on the manufactures of the United States at the tenth census, U.S. Government Printing Office, Washington, DC, 1883, pp. 611–704.

71. D. A. Hounshell, *From the American System to Mass Production, 1800–1932*, Johns Hopkins University Press, Baltimore, MA, 1984, pp. 25–26, 217–261, 337–344.

72. E. A. Marsh, *The Evolution of Automatic Machinery as Applied to the Manufacture of Watches*, G. K. Hazzlit and Co., Chicago, IL, 1896, pp. 26–28, 100–103, 139–146.

73. K. Adler, Making things the same, *Soc. Studies Sci.* **28**:4 (August 1998) 499–545.

74. R. S. Woodbury, The legend of Eli Whitney and interchangeable parts, *Technol. Cul.* **1** (1960) 235–253.

75. M. R. Smith, *Harpers Ferry Armory and the New Technology*, Cornell University Press, Ithaca, NY, 1977, pp. 106–107.

76. R. S. Woodbury, *History of the Milling Machine*, Technology Press, Cambridge, MA, 1960, pp. 16–26, 34–35, 45–49.

77. E. A. Battison, A new look at the Whitney milling machine, *Technol. Cult.* **14** (1973) 592–598.

78. J. Rose, *Modern Machine Shop Practice*, Vol. 2, Charles Scribner's Sons, New York, 1888, pp. 3, 38–54.

79. I. Bradley, *A History of Machine Tools*, Model and Allied Publications, Hemel Hempstead, U.K., 1972, pp. 111–113.

80. S. A. Morse, Improvement in drill-bits, U.S. Patent 38119, April 7, 1863.

81. J. Ruby, *Maschinen für die Massenfertigen*, Verlag für die Geschichte der Naturwissenschaften und der Technik, Stuttgart, Germany, 1995, pp. 51–58, 82–133.

82. C. M. Spencer, Machines for making metal screws, U.S. Patent 143,306, September 10, 1873.

83. H. C. Hovey, The American watch works, *Sci. Am.* **41**:7 (August 16, 1884) 1, 102–104.

84. T. Chokki, A history of the machine tool industry in Japan, in: M. Fransman, Ed., *Machinery and Economic Development*, Macmillan Press, New York, 1986, pp. 124–152.

85. I. Finnie, Review of metal-cutting analyses of the past hundred years, *Mech. Eng.* **78** (1956) 715–721.

86. N. N. Zorev, *Metal Cutting Mechanics*, English Trans. by H. S. M. Massey, Pergamon Press, Oxford, U.K., 1966, pp. 1–2.

87. O. W. Boston, *Bibliography on the Cutting of Metals 1864–1943*, ASME, New York, 1954.

88. G. T. Smith, *Cutting Tool Technology*, Springer, London, U.K., 2008, p. 51.

89. I. A. Time, *Resistance of Metals and Wood to Cutting*, St. Petersburg, Russia, 1870 (in Russian).

90. H. Tresca, Memoire sur le Rabotage des Metaux, *Bull. Soc. d'Encouragement pour l'Industrie Nationale*, 1873, 585–685.

91. A. Mallock, The action of cutting tools, *Proc. R. Soc. Lond.* **33** (1881) 127–139.

92. M. Cocquilhat, Experiences Sur La Resistance Utile Produite Dans Le Forage, *Ann. Trav. Publ.en Belgique* **10** (1851) 199.

93. E. Hartig, *Versuche Über Leistung und Arbeits-Verbrauch der Werkzeugmaschinen*, B. G. Teubner, Leipzig, Germany, 1873.

94. R. Anstis, *Man of Iron—Man of Steel: The Lives of David and Robert Mushet*, Albion House, Gloucestershire, U.K., 1997, pp. 157–172.

95. O. M. Becker, *High-Speed Steel*, McGraw-Hill, New York, 1910, pp. 13–20.

96. F. W. Taylor, *Principles of Scientific Management*, Harper and Brothers, New York, 1913.

97. F. W. Taylor, On the art of cutting metals, *ASME Trans.* **28** (1906) 31–350.

98. R. Kanigel, *The One Best Way: Frederick Winslow Taylor and the Enigma of Efficiency*, Penguin Books, New York, 1997, pp. 387–389.

99. E. G. Thomsen and F. W. Taylor—A historical perspective, in: L. Kops and S. Ramalingam, Eds., *On The Art of Cutting Metals—75 Years Later*, ASME PED Vol. 7, ASME, New York, 1982, pp. 1–12.

100. E. Buckingham, *Principles of Interchangeable Manufacturing*, Industrial Press, New York, 1921, pp. v–vi, 121–124.

101. J. S. McCoy, Tracing the historical development of metalworking fluids, in: J. P. Beyers, Ed., *Metalworking Fluids*, 2nd edn., CRC Press, Boca Raton, FL, 2006, pp. 1–18.

102. E. Buckingham, *Dimensions and Tolerances for Mass Production*, Industrial Press, New York, 1954, pp. 1–2.

103. F. H. Colvin, Machining the Ford cylinders—I, *Am. Mach.* **38** (May 28, 1913) 841–846; Reprinted in *Automobiles 1913–1915*, Lindsay Publications, Bradley, IL, 2003, pp. 34–39.

104. F. H. Colvin, Machining the Ford cylinders—II, *Am. Mach.* **38** (1913) 971–976; reprinted in *Automobiles 1913–1915*, Lindsay Publications, Bradley, IL, 2003, pp. 39–45.

105. F. H. Colvin, Ford crank cases and transmission covers, *Am. Mach.* **39** (1913) 49–53; Reprinted in *Automobiles 1913–1915*, Lindsay Publications, Bradley, IL, 2003, pp. 53–57.

106. F. H. Colvin, Making rear axles for the ford auto, *Am. Mach.* **39** (1913) 143–149; reprinted in *Automobiles 1913–1915*, Lindsay Publications, Bradley, IL, 2003, pp. 61–67.

107. F. H. Colvin, Special machines for making pistons, *Am. Mach.* **39** (1913) 349–353; Reprinted in *Automobiles 1913–1915*, Lindsay Publications, Bradley, IL, 2003, pp. 80–84.

108. F. H. Colvin, Special machines for auto small parts, *Am. Mach.* **39** (1913) 439–443; Reprinted in *Automobiles 1913–1915*, Lindsay Publications, Bradley, IL, 2003, pp. 88–91.

109. H. L. Arnold and F. L. Faurote, *Ford Methods and the Ford Shops*, The Engineering Magazine Company, New York, 1915, pp. 5, 72–75, 159–216, 307–326.

110. Anon., *Cincinnati Milacron 1884–1984: Finding Better Ways*, Cincinnati Milacron Inc., Cincinnati, OH, 1984, p. 31.

111. E. D. Lloyd, *Transfer and Unit Machines*, Industrial Press, New York, 1969, pp. 9–22.

112. B. W. Charman, *Special Purpose Production Machines*, Crosby Lockwood and Son Ltd., London, U.K., 1968, pp. ix–x.

113. C. R. Hine, *Machine Tools for Engineers*, 2nd edn., McGraw-Hill, New York, 1959, pp. 402–411.

114. H. Holbeche and F. Griffiths, The design of transfer machines, presented to the Coventry Section of the Institution of Production Engineers, Nov. 16, 1954; excerpted in J. Baker, Austin's Built Transfer Machines, c. 2006.

115. Anon., *Of Mechanics and Machines: Commemorating the 75th Anniversary of the Cross Company*, The Cross Company, Fraser, MI, 1973.

116. C. F. Kettering, The automobile industry and machine tools, *Am. Mach.* **66** (1927) 800–801.

117. Anon., *75 Years of Toyota*, Toyota Motor Corporation, Nagoya, Japan, 2012, Section 7.

118. R. D. Prosser, Widia, its development and shop applications, *ASME Trans.* **51** (1929) 71–76.

119. H. M. Ortner, P. Ettmayer, and H. Kolaska, The history of the technological progress of hardmetals, *Int. J. Refract. Met. Hard Mater.* **44** (2014) 148–159.

120. Anon., Tungsten carbide in general motors shops, *Machinery* **37** (1930) 217–218.

121. C. Sellers 3rd, Tungsten carbide and other hard cutting materials, *ASME Trans.* **54** (1932) 225–233.

122. D. F. Noble, *Forces of Production*, Oxford University Press, Oxford, U.K., 1984, pp. 79–192, 212–264.

123. Anon., John Parsons: The man behind numerical control, *Manuf. Eng.* **88**:1 (January 1982) 127.

124. W. Pease, An automatic machine tool, *Sci. Am.* **187**:3 (March 1952) 101–115.

125. J. F. Reintjes, *Numerical Control: Making a New Technology*, Oxford University Press, Oxford, U.K., 1991.

2 Metal-Cutting Operations

2.1 INTRODUCTION

This chapter describes the common machining operations used to produce specific shapes or surface characteristics. The primary focus is on process kinematics and equations for basic cutting parameters. Brief descriptions of the general-purpose machine tools traditionally associated with each process are also included. More detailed information on machine tools is covered in Chapter 3. Additional descriptive information on basic cutting operations is available in the literature [1–10].

2.2 TURNING

In turning (Figure 2.1), a cutting tool is fed into a rotating workpiece to generate an external or internal surface concentric with the axis of rotation. Turning is carried out using a lathe, one of the most versatile conventional machine tools. The principal components and movements of a lathe are shown in Figure 2.2. The workpiece is mounted on a rotating spindle using a chuck, collet, face plate, or mandrel, or between pointed conical centers [2,4]. Lathes may have a horizontal or vertical spindle, with vertical spindle machines being used especially for large workpieces. The cutting tool is held on a translating carriage or turret or in the tailstock. The carriage or turret travels along the bedways parallel to the part axis (Z-axis). Motion perpendicular to the part axis is provided by the X-axis or a cross slide mounted on the carriage. Contours, tapers, arcs, or other shapes can be generated by motion of the X- and Z-axes.

In addition to the tool geometry, the major operating parameters to be specified in turning are the cutting speed, V, feed rate, f_r, and depth of cut (doc), d. The cutting speed is determined by the rotational speed of the spindle, N, and the initial and final workpiece diameters, D_1 and D_2:

$$V = \pi N \frac{D_1 + D_2}{2} = \pi N D_{avg}, \quad \text{for small } d : V \cong \pi N D_1 \tag{2.1}$$

The feed f is the tool advancement per revolution along its cutting path in mm/rev. Feed rate (f_r) is the speed at which the tool advances along the part longitudinally in mm/min, and is related to f through the spindle rpm N:

$$f_r = fN \tag{2.2}$$

The feed influences chip thickness and how the chip breaks. The uncut (nominal) chip thickness a is related to f through the lead angle of the tool, κ (Figure 2.1):

$$a = f \cos \kappa \tag{2.3}$$

The depth of cut, d, is the thickness of the material removed from the workpiece surface:

$$d = \frac{D_1 - D_2}{2} \tag{2.4}$$

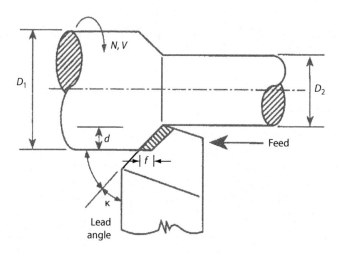

FIGURE 2.1   Turning.

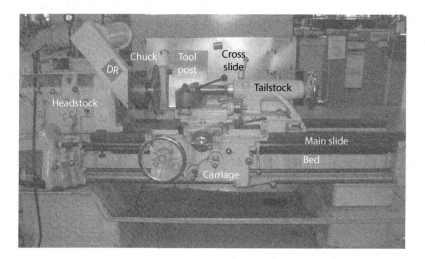

FIGURE 2.2   Principal components of an engine lathe. Rotation is provided by the spindle in the headstock. The carriage moves on the main slide on the bed to provide axial motion. Radial motion is provided by the cross slide. Speeds and feeds are adjusted through levers activating different gear sets in the headstock.

The time t_m required to cut a length L in the feed direction is

$$t_m = \frac{L + L_e}{f_r}, \quad \text{where } L_e = (\text{approach allowance}) + (\text{overtravel allowance}) \tag{2.5}$$

The material removal rate per unit time, Q, is given by the product of the cutting speed, feed, and depth of cut:

$$Q = Vfd = \frac{\pi \left(D_1^2 - D_2^2 \right)}{4} f_r \tag{2.6}$$

The spindle power P_s is

$$P_s \cong Q u_s \tag{2.7}$$

TABLE 2.1

Specific Cutting Energy u_s for Various Materials

Material	Unit Power (kW/cm³/min)	Unit Power (HP/in.³/min)
Cast irons	0.044–0.08	0.97–1.76
Steels		
Soft	0.05–0.066	1.10–1.45
$0 < R_c < 45$	0.065–0.09	1.43–1.98
$50 < R_c < 60$	0.09–0.2	1.98–4.40
Stainless steels	0.055–0.09	1.21–1.98
Magnesium alloys	0.007–0.009	0.15–0.20
Titanium	0.053–0.066	1.16–1.45
Aluminum alloys	0.012–0.022	0.26–0.48
High temperature alloys (Ni and Co based)	0.09–0.15	1.98–3.30
Free-machining brass	0.056–0.07	1.23–1.54
Copper alloys		
$R_B < 80$	0.027–0.04	0.59–0.88
$80 < R_B < 100$	0.04–0.057	0.88–1.25

Notes: For zero effective rake angle tools, 0.25 mm undeformed chip thickness, and continuous chips without BUE.

W = N m/s, 1 kW = 1.341 HP.

where u_s is the power required to cut a unit volume of work material (see Section 6.5). Typical values for u_s for common work materials are summarized in Table 2.1. These values may be multiplied by correction factors to account for tool wear or changes in tool geometry.

If the efficiency of the drive system is η, the motor power P_m required is

$$P_m = \frac{P_s}{\eta} + P_t \tag{2.8}$$

where P_t is the tare or idling power. Finally, the efficiency of the drive system η is

$$\eta = \eta_m \, \eta_b \, \eta_{br1} \, \eta_{br2} \tag{2.9}$$

where η_m, η_b, η_{br1}, and η_{br2} are the efficiencies of the motor, belt, front bearing, and rear bearing, respectively.

2.2.1 Hard Turning

A special case of turning is hard turning, in which hard metals (45–65 HR$_c$) are finish machined using ceramic or polycrystalline tools [11–18]. This process is sometimes used in place of rough turning, hardening, and finish grinding for parts made of tool steels, alloy steels, case-hardened steels, and various hard irons. Very fine finishes and tolerances can be produced by this process, and in some cases part quality is better than that can be obtained with grinding because intermediate chucking operations and associated setup errors are eliminated. Hard turning produces a different surface topography compared to grinding and better surface integrity due to reduced thermal damage in many applications. Hard turning is a more efficient process than grinding, but depending on the aspect ratio of the depth and width of cut, plunge- or creep-feed grinding with a wheel of the groove width may be better alternatives in some applications. Grinding is also better suited to interrupted cutting.

Hard turning became possible with the advent of hot-pressed ceramic and especially polycrystalline cubic boron nitride (PCBN) tools. Hard turning requires high machine and toolholder rigidity and strong insert shapes (negative rakes, large wedge angles, and special edge preparations such as chamfers, discussed in Chapter 4). Modern CNC lathes usually have adequate rigidity for hard turning. Hard turning is generally performed dry since ceramics and CBN tools are used. Typical DOCs range from 0.08 to 0.5 mm, with speeds between 50 and 150 m/min. Size tolerances of ± 0.005 mm or better are achievable with surface finish better than 0.3 μm R_a. Typical applications are bearings, gears, and axles shafts.

2.3 BORING

The boring operation (Figure 2.3) is equivalent to turning but is performed on internal surfaces. Boring is applied for roughing, semi-finishing, or finishing cast or drilled holes. Finish boring is usually a precision process characterized by small form, dimensional, and surface finish tolerances.

Boring can be performed on a number of machine tools, including lathes, drilling machines, horizontal, or vertical milling machines, and machining centers. Conventionally, however, it is performed on special horizontal or vertical boring machines. A horizontal boring machine, which can also be used for drilling and milling, is shown in Figure 2.4. Two- or three-axis machines are used depending on process requirements. The machine structure is similar to that of a milling machine with a precision spindle. High machine structural and spindle stiffness are required in order to generate quality bores. Very accurate work is often done using jig-boring machines, which are equipped with precision tables and spindles, stiff machine structures, and measurement devices built into the table.

Since boring is equivalent to turning, boring performance also depends on the cutting speed, depth of cut, and feed rate. The equations relating these parameters to part dimensions and machine variables reviewed before for turning are also applicable to boring. As will be discussed in Chapter 4, however, boring tools differ significantly from those used for turning due to their unique structural and dynamic requirements. Traditionally, moderate cutting speeds and

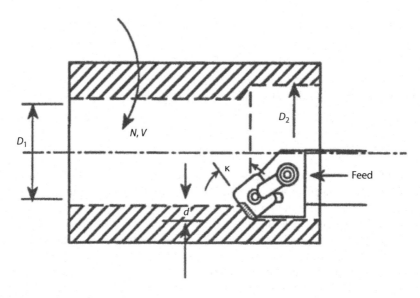

FIGURE 2.3 Boring.

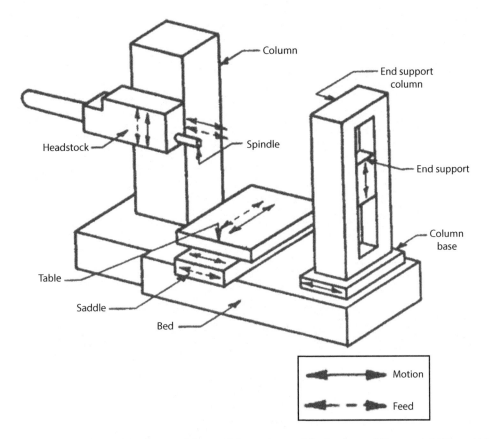

FIGURE 2.4 A horizontal boring machine, which can be used for boring, milling, and drilling. (After DeGarmo, E.P., *Materials and Processes in Manufacturing*, 5th edn., Macmillan, New York, 1979, Fig. 21-6.)

small depths of cut and feed rates are used in boring to ensure accuracy, but in more recent practice higher cutting speeds have been used to reduce errors due to mechanical and thermal distortion. Heavier depths of cut are generally performed using multipoint boring tools. The hole depth (length) to which a boring bar can cut accurately is limited by the amount of bar deflection as will be discussed in Chapter 4.

2.4 DRILLING

Drilling (Figure 2.5), the standard process for producing holes, is one of the most common metal-cutting processes.

Types of conventional drilling machines include upright or pedestal machines, radial machines, and various specialized machines such as gang drill presses [2,5]. The basic components of an upright drilling machine, shown in Figure 2.6, include the base, column, spindle, and worktable. Radial drilling machines are designed for large workpieces and consist of a large horizontal arm extending from the column; both the height and angular orientation of the arm can be adjusted. A gang drill press is made up of two or more upright drilling machines placed next to each other on a common base. Often these machines operate sequentially with each spindle carrying a different tool and the workpiece moving between them to complete all operations. Drilling can also be performed on lathes, boring mills, and milling machines.

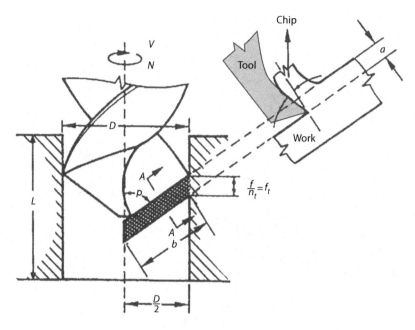

FIGURE 2.5   Drilling.

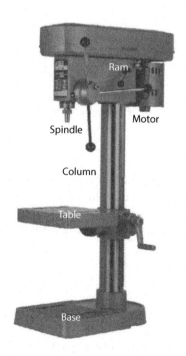

FIGURE 2.6   Principal components of an upright drill press.

Geometrically, drilling is a complex process. The difficulty of producing drills with consistent geometries has traditionally limited accuracy, although drill consistency and repeatability have greatly increased with the advent of CNC drill grinders. Also, the complexity of the tool has inhibited the introduction of new tool materials, so that productivity gains in drilling have lagged those made in turning and milling over the past 30 years. A number of manufacturers have developed diamond or CBN tipped or coated drills to attempt to address this limitation.

Drilling performance depends on the materials involved, the drill geometry, the spindle speed, and the feed rate. The cutting speed V at the periphery of the drill is given by

$$V = \pi D N \tag{2.10}$$

where
  D is the drill diameter
  N is the spindle rpm

As in turning, the feed rate f_r and feed per revolution f are related through Equation 2.2. The feed per tooth f_t depends on f and the number of flutes n_t

$$f_t = \frac{f}{n_t} \tag{2.11}$$

The effective depth of cut per flute d is

$$d = \frac{D}{2} \tag{2.12}$$

The nominal width of cut b is given by

$$b = \frac{D}{2 \sin \rho} \tag{2.13}$$

where ρ is the half point angle of the drill. The uncut (nominal) chip thickness a is

$$a = f_t \sin \rho \tag{2.14}$$

The metal removal rate Q is

$$Q = \left(\frac{\pi D^2}{4} \right) f_r \tag{2.15}$$

The time t_m required for drilling a hole of depth L is

$$t_m = \frac{L + L_e}{f_r} \tag{2.16}$$

where the approach and overtravel allowance is

$$L_e = \frac{D}{2 \tan \rho} + \Delta L \tag{2.17}$$

ΔL is the approach distance between the drill chisel edge and the entrance surface of the workpiece plus the overtravel of the drill in through holes.

In core drilling, in which a hole of initial diameter D_i is enlarged to diameter D, the nominal width of cut is

$$b = \frac{D - D_i}{2 \sin \rho} \tag{2.18}$$

And the metal removal rate is

$$Q = \left(\frac{\pi \left(D^2 - D_i^2 \right)}{4} \right) f_r \tag{2.19}$$

2.4.1 DEEP-HOLE DRILLING

A deep hole is one with a depth-to-diameter ratio of more than 5:1. Special machines are often required to drill deep holes with adequate straightness and to ensure efficient chip ejection and lubrication of the drill. Because the chips and heat generated by the process are confined, the deeper the hole, the more difficult it is to control heat buildup and remove the chips.

The traditional drilling operation, using standard or parabolic-flute twist drills, has been occasionally used for deep-hole drilling but often requires "pecking" (drilling to intermediate depths and periodically withdrawing the tool to clear chips) unless a high pressure coolant is used. Bushings with aspect ratios of 2:1 or 4:1 are also often used to support the drill at the entrance to improve location accuracy and to stabilize the drill.

Three distinct specialized deep-hole drilling methods are in use: solid drilling, trepanning, and counterboring [2,5,19,20] (Figure 2.7). Solid drilling is the most common and can be further classified

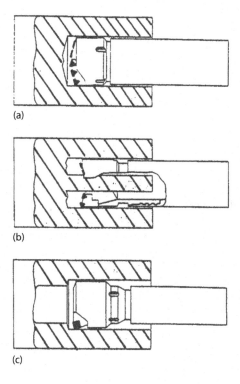

(a)

(b)

(c)

FIGURE 2.7 Deep-hole drilling operations: (a) solid drilling, (b) trepanning, and (c) counterboring. (Courtesy of Sandvik Coromant Corporation, Fair Lawn, NJ.)

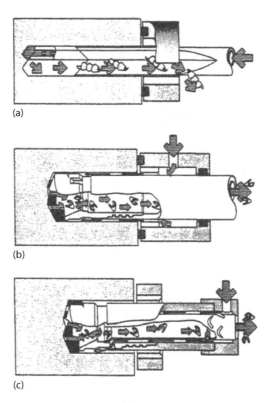

FIGURE 2.8 Solid deep-hole drilling operations: (a) gun drilling, (b) STS drilling, and (c) ejector drilling. (Courtesy of Sandvik Coromant Corporation, Fair Lawn, NJ.)

into four approaches: conventional twist drilling (discussed earlier), gun drilling, ejector drilling, and BTA (STS) drilling (Figure 2.8). In gun drilling, coolant is supplied through the center of the tool shank under high pressure, forcing the chips through the flutes. The BTA (Boring and Trepanning Association) method uses single tube system (STS) tools where the coolant is supplied under high pressure between the tool and hole surface. The coolant is then removed along with the chips through an opening above the cutting edge of the tool. The ejector or "two-tube" system feeds the coolant through a connector. Two-thirds of the coolant flows between the inner and outer tubes; the remaining coolant is drawn off through the nozzle, creating a vacuum, which causes a backward ejection of chips through the inner tube. Tools for gun drilling, BTA drilling, and ejector drilling are discussed in Chapter 4.

Ejector drilling can be used for holes larger than 18 mm in diameter and is often the best choice for cost-efficient drilling of non-precision holes. It does not require a pressure head for the coolant supply. The hole tolerance for ejector drilling is +0.075/−0.000 mm on the diameter; the hole straightness error is typically 0.05 mm/m. The BTA process can be used for holes over 10 mm in diameter. Holes with diameters between of 3 and 20 mm can be drilled with a straightness error of 0.08 mm/m using the solid and gun drilling methods. The surface finish generated in deep-hole drilling is typically between 1.0 and 1.5 μm R_a and is comparable to that obtained by reaming because the hole wall is burnished by the tool-bearing surfaces. Deep-hole drilling performance can sometimes be improved by supplying ultrasonic energy to the cutting zone [21].

The trepanning operation (Figure 2.7) removes material only at the periphery of the hole and leaves a solid core. Through holes with an aspect ratio of 2–4 can be trepanned with a hollow multi-tooth cutter and flood coolant. The depth of cut is equivalent to the width of the cutting edge and is smaller than for solid drilling. This operation, therefore, requires lower power and is well suited for through holes. Large blind holes can be trepanned using a tool with a pivoting cutoff blade, which separates the core from the base of the hole. Trepanning is less accurate than solid drilling due to misalignment caused by limited tool rigidity. It provides faster cycle times because there is no dead center area as with solid drilling.

Counterboring (Figure 2.7) enlarges an existing hole that is drilled or cast, normally for the purpose of improving its size, straightness, or surface finish. Often, a combination tool will perform a solid drilling operation while simultaneously counterboring the hole drilled ahead.

Deep-hole drilling machines typically resemble horizontal turning or milling machines (Figure 2.9), although vertical spindle machines are also used. All three drill types (gun drill, ejector, and BTA) can be used, although special attachments for the bushing and the head are required for BTA setups. Either the tool, workpiece, or both may rotate, depending on the size of the workpiece. The best hole quality is obtained when both the tool and workpiece rotate. Deep-hole drilling machines are always equipped with high pressure, high volume coolant systems.

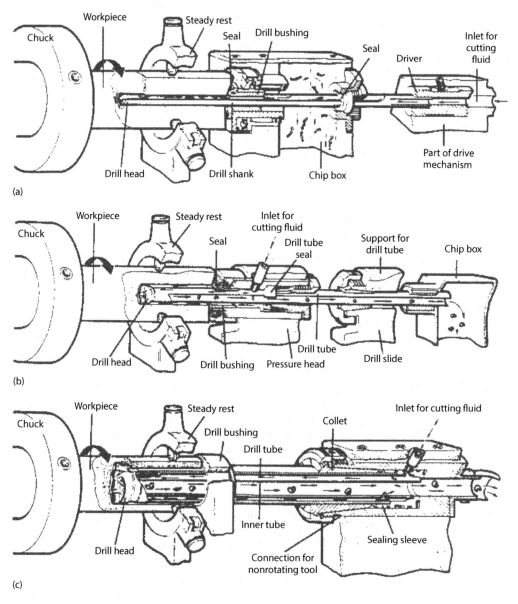

FIGURE 2.9 Deep-hole drilling machines: (a) gun drill system, (b) single tube system (STS), and (c) ejector system. (Courtesy of Sandvik Coromant Corporation, Fair Lawn, NJ.)

2.4.2 MICRODRILLING

Microdrilling is the drilling of small diameter (usually less than 0.5 mm) holes with a depth-to-diameter ratio greater than 10 [22]. Holes as small as 0.0025 mm have been successfully drilled. Microdrilling is similar to deep-hole drilling but generally presents greater problems, since coolant fed drills cannot be used. High spindle speeds are required to generate sufficient cutting speeds, especially when carbide drills are used. Low feed rates must also be used to avoid exceeding the buckling load of the drill. Feeds in the range of 0.00005–0.0005 mm/rev are common in microdrilling.

Microdrilling performance can sometimes be improved by supplying ultrasonic energy to the cutting zone [14]. High-frequency forced vibrations at frequencies between 15 and 30 kHz reduce friction and thus allow increased material removal rates. Ultrasonically assisted drilling is particularly effective for deep-hole drilling and microdrilling in cases in which chip clogging limits the allowable penetration rate. The vibration tends to break chips into smaller sections while lowering forces. In these operations, vibrations can increase throughput by a factor of two while improving tool life and hole quality. The vibration frequency must be carefully controlled because vibration at frequencies outside the useful range reduces tool life.

When chips are not easily ejected, peck drilling (frequent withdrawals of the drill) is used to clear chips from the hole and to permit intermittent cooling of the drill. Peck drilling increases cycle time and drill wear (because the drill dwells at the bottom of the hole prior to retraction). Precise feed control is necessary to avoid excessive dwelling. Peck drilling may not be necessary if high spindle speeds are available, but may be preferred in drilling composites or layers of different materials.

Very small holes can also be produced by laser cutting and by electroplating a part in which a larger hole has been drilled. In some applications, small openings can also be produced through layering, that is by drilling small holes in two thin sheets and then arranging the sheets so that the holes only partially intersect. These methods are usually slower and less accurate than drilling.

2.5 REAMING

Reaming (Figure 2.10) is used to enlarge a hole and improve its size accuracy, roundness, and surface finish. Reaming is similar to boring, and in fact both processes can be used for some operations. The differences between reaming and boring result largely from tool design as discussed in Chapter 4. The choice of performing a boring or reaming operation depends on the hole diameter and length, interruptions within the hole due to internal cavities, and the required straightness, size tolerance, surface finish, and tool life. Reaming is similar to core drilling geometrically, and the equations and cutting parameters reviewed earlier for core drilling are also applicable to reaming. However, the stock removal (doc) in reaming is generally between 0.25 and 0.7 mm, which is smaller than the stock removal in core drilling, counterboring, and rough and semi-finish boring.

Reaming machines are similar to drilling machines but have a less powerful motor and often a more precise spindle. Radial floating attachments (holders) are often used to ensure that the tool is aligned with the pre-drilled or cored hole. Special spindle heads with an attached, rotating bushing are used for squirt reaming. This improves machine capability and hole quality because the bushing travels with the spindle and supports the tool as it enters the hole.

2.6 MILLING

In milling processes, material is removed from the workpiece by a rotating cutter. The two basic milling operations are peripheral (or plain) milling and face milling (Figure 2.11). Peripheral milling generates a surface parallel to the axis of rotation, while face milling generates a surface normal to the axis of rotation. Face milling is used for relatively wide flat surfaces (usually wider than 75 mm). End milling, a type of peripheral milling operation, is generally used for profiling and slotting operations but may also be used for face milling in pocketing applications.

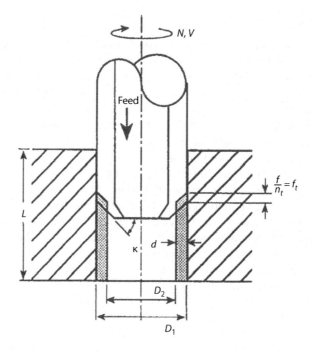

FIGURE 2.10   Reaming.

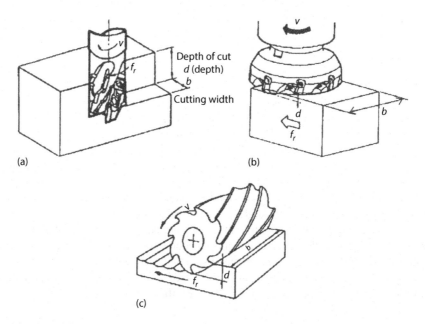

FIGURE 2.11   Types of milling operations: (a) end milling, (b) face milling, and (c) peripheral milling. (Courtesy of Toshiba Tungaloy America, Inc., Arlington Heights, IL.)

Milling processes can be further divided into up (or conventional) and down (or climb) milling operations (Figure 2.12). If the axis of the cutter does not intersect the workpiece, the motion of cutter due to rotation opposes the feed motion in up (or conventional) milling but is in the same direction as the feed motion in down-milling. When the axis of the cutter intersects the workpiece, both up- and down-milling occur at different stages of the rotation. Both up- and down-milling

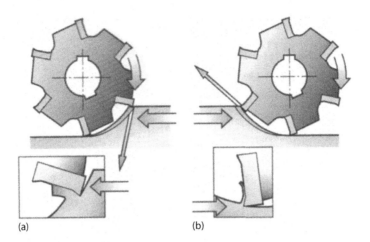

FIGURE 2.12 (a) Down (climb)- and (b) up-milling. (From Sandvik Coromant Corporation, *Technical Guide*, Fair Lawn, NJ, 2010.)

have advantages in particular applications [2,5]. Up-milling is usually preferable to down-milling when the spindle and feed drive exhibit backlash and when the part has large variations in height or a hardened outer layer due to sand casting or flame cutting. In down-milling, there is a tendency for the chip to become wedged between the insert and cutter, causing tool breakage. However, if the spindle and drive are rigid, cutting forces in peripheral down-milling tend to hold the part on the machine and reduce cutting vibrations. Down-milling is also preferred when machining nickel alloys and other materials subject to surface damage, since in down-milling the tool flank and margin do not contact and potentially burnish the machined surface.

The most common general-purpose milling machine is the knee and column milling machine or knee mill (Figure 2.13). The major components of the knee mill are the column, spindle, knee, saddle, and table. Both vertical and horizontal spindle milling machines are available. Universal mills are knee mills with a spindle head that rotates at right angles to the table's longitudinal axis, so that the spindle can be either horizontal or vertical.

The uncut chip thickness varies continuously in milling. In up-milling, the chip thickness is small at the beginning of the cut and increases as cutting progresses, while in down-milling the chip thickness is largest at the beginning of the cut. Cutting is also not continuous in milling, but rather is periodically interrupted as cutting edges enter and leave the part (Figures 2.14 and 2.15). This leads to cyclic thermal and mechanical loads on the tool, which leads to a number of fatigue failure mechanisms not encountered in continuous cutting (Chapter 9).

The cutting action of each cutting edge on a milling cutter is similar to that of a single-point tool. The cutting speed is given by Equation 2.1, and the feed rate f_r, feed per revolution f, and feed per tooth f_t are related by an equation similar to Equation 2.2:

$$f_r = Nf = n_t N f_t \qquad (2.20)$$

The variation of the uncut chip thickness in milling is complicated (Figure 2.14). Exact analyses [23–26] have shown that the uncut chip thickness varies trochoidally as the cutter rotates. For small feeds, however, a sinusoidal approximation is adequate. The uncut chip thickness a_i at an engagement angle of v_i, maximum uncut chip thickness, a_{max}, and average uncut chip thickness, a_{avg}, are given by

$$a_i = f_t \cos \kappa \sin v_i \qquad (2.21)$$

FIGURE 2.13 Components of a knee and column milling machine. The knee moves on ways on the column to adjust height; the worktable moves laterally to provide feed.

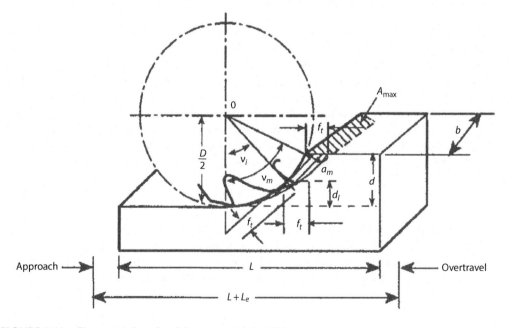

FIGURE 2.14 Characteristics of peripheral (or plain) milling.

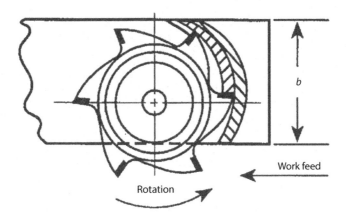

FIGURE 2.15   Face milling.

$$a_{max} = f_t \cos \kappa \sin \nu_m \tag{2.22}$$

$$a_{avg} = f_t \cos \kappa \sin \left(\frac{\nu_m}{2} \right) \tag{2.23}$$

where

$$\cos \nu_m = 1 - \frac{2d}{D} \tag{2.24}$$

and κ is the lead angle equivalent to the lead angle in turning (Figure 2.1). In peripheral milling, $\kappa = 0$; face milling, ν_m is usually 90°.

If n_i teeth are engaged at a given instant, the maximum total undeformed chip area, A_{max}, is

$$A_{max} = f_t b \cos \kappa \sum_{i=1}^{n_i} \sin \nu_i \tag{2.25}$$

The length of the arc of metal being cut, λ_c, which determines the chip length, in peripheral milling is

$$\lambda_c = (dD)^{1/2} \tag{2.26}$$

In face milling, if the cutter is wider than the workpiece, λ_c is given by

$$\lambda = (bD)^{1/2} \tag{2.27}$$

For end milling a slot (full engagement of the cutter),

$$\lambda_c = \frac{\pi D}{2} \tag{2.28}$$

The metal removal rate Q is given by

$$Q = f_r bd \tag{2.29}$$

where d and b are defined in Figures 2.11, 2.14, and 2.15. For example, d and b are the axial and radial depth of cuts for end milling, while radial and axial depth of cuts for peripheral milling. The time t_m required to cut mill a workpiece of length L is

$$t_m = \frac{L + L_e}{f_r} \tag{2.30}$$

where the approach distance L_e is peripheral milling given by

$$L_e = \sqrt{d(D-d)} + (\text{approach allowance}) + (\text{overtravel allowance}) \tag{2.31}$$

while in face milling,

$$L_e = \frac{D}{2} \tag{2.32}$$

The total travel length of the cutter is larger than the length of the workpiece due to the cutter approach and overtravel distances. The overtravel distance is normally very small, enough for the cutter axis to clear the end of the part.

When performing end milling or peripheral milling with a width of cut less than the cutter radius, chip thinning can occur, and it is very important to increase the feed per tooth (using Equations 2.21 through 2.24) to keep the uncut chip thickness at the expected value.

When performing end milling, thread milling, and related operations on a machining center, internal or external circular interpolation is often required when generating a tool path as shown in Figure 2.16. (When contour milling, the principles of straight end milling are applied, but additional checks on the feed rate, depth of cut, resultant chip thickness, and cutter edge density are required to avoid gouging when the cutter moves in a circular arc.) In these cases the feed rate for the cutter center differs from the effective peripheral feed rate of the cutting edges, and somewhat different relations between cutting parameters are required.

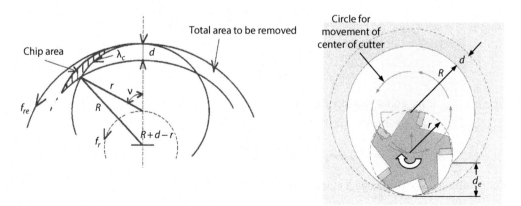

FIGURE 2.16   Tool path for circular interpolation in milling. r, cutter radius; R, workpiece radius; f_r, feed at tool center; f_{re}, peripheral feed at contour; v, angle of cutter engagement.

The depth of cut engagement angle v_i is given by

$$v_i = \cos^{-1}\left[\pm\frac{R^2 - r^2 - \left[R\pm(d-r)\right]^2}{2r\left[R\pm(d-r)\right]}\right] \tag{2.33}$$

where the plus and minus signs apply to internal and external milling, respectively. The plunge cut engagement is $2v_i$. The apparent length of contact of the cutting edges, λ_c, is given by

$$\lambda_c = (dD_e)^{1/2} \tag{2.34}$$

where D_e is the equivalent tool diameter given by

$$D_e = \frac{2r}{1 - \dfrac{r}{d\pm R}} \tag{2.35}$$

where the plus and minus signs apply to internal and external milling respectively. λ_c can also be calculated from the relation

$$\lambda_c = r v_i \tag{2.36}$$

where v_i is given in radians.

In a circular cut, the actual depth to which the edge penetrates is considerably greater than the radial depth of cut d. Therefore, it is important in circular cuts to determine the proper feed, depth of cut, and cutter diameter so that the cutter engagement is acceptable, given the available horse-power, workpiece and fixture rigidity, and chatter limits. The feed per tooth, f_t, at the center of the tool is given by Equation 2.22 or 2.23 based on a known maximum or average chip thickness, respectively. The feed rate, f_r, at the center of the tool (used for programming the contour in most CNC machines) is given by Equation 2.20. The feed can be also obtained from Equation 2.24 using an effective depth of cut, d_e, that is not the same as the depth of cut value d. The effective depth of cut is

$$d_e = \frac{d(R+d/2)}{[R+d\pm(-r)]} \tag{2.37}$$

The effective feed rate at the cutting edge is

$$f_{re} = f_r\frac{R\pm d}{R\pm(d-r)} \tag{2.38}$$

The plus and minus signs correspond to internal and external circular interpolation, respectively.

The machining time can be calculated by dividing the total volume of material to be removed by the metal removal rate:

$$t_m = \pm\frac{\pi\left[(R\pm d)^2 - R^2\right]}{n_t N\lambda_c a_{avg}} + t_{feedin} \quad \text{or} \quad t_m = \frac{2\pi\left[R\pm(d-r)\right]}{f_r} + t_{feedin} \tag{2.39}$$

where t_{feedin} is the feed-in and feed-out time required while ramping into and out of the cut. The cutter can enter and exit the cut either tangentially or radially. The tangential approach is preferable because the tool gradually ramps into the cut, improving tool life, because no marks are left on the workpiece, and because there is less tool vibration.

2.7   PLANING AND SHAPING

Planing and shaping are machining processes in which the tool moves linearly and reciprocally with respect to the workpiece. In planing, the workpiece reciprocates while the tool is fixed and indexes across the workpiece to provide the feed motion. In shaping, the tool reciprocates across the work-piece. As implied by their names, planing is used largely to produce flat surfaces, while shaping can be used to produce a variety of contoured surfaces. Workpieces of any size can be machined with the planing operation while only small or medium-sized parts can be machined by shaping. Neither process is widely used in mass production, since flat surfaces can be produced more rapidly by broaching or face milling. Planing is commonly used in machinery manufacture to produce flat surfaces on large castings or forgings. Shaping is used for keyways, gear teeth, and similar features, which in many applications may also be broached.

Single-point tools are used in both operations, and the basic elements involved, shown in Figure 2.17, are almost identical to those of turning. Therefore, the basic equations for planning and shaping are identical to those given for turning with the frequency of reciprocation or cutting strokes substituted for the rotational speed (rpm). Planing and shaping machines are generally simple in design; in planing the tool is normally mounted on a portal frame or gantry, which spans the worktable.

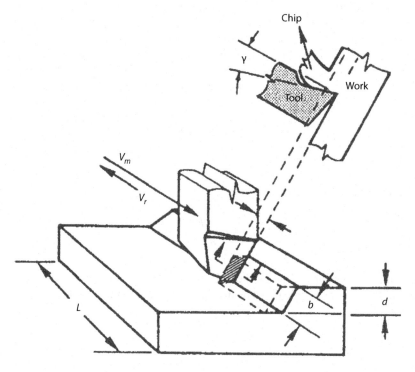

FIGURE 2.17   Cutting action in planing and shaping.

2.8 BROACHING

Broaching resembles shaping in that the tool translates linearly across the workpiece [2,5,27–29]. However, broaching employs a series of cutting edges ground or otherwise mounted on a tool body (Figure 2.18), rather than a single-point tool. By suitably varying the relative heights of successive cutting edges, a broach can be designed to perform roughing, semi-finishing, and finishing cuts in a single stroke. Broaching is consequently a highly productive process. It is also precise and can produce very fine surface finishes. Accuracies of 0.025 mm with surface finishes of 0.8–2 μm are achievable.

There are two basic types of broaching operations: external (surface) and internal broaching. External broaching is used to generate flat surfaces, contours, and various forms such as spur gears, splines, and slots. More than one surface on a workpiece can be produced in one pass. Internal broaching produces circular and noncircular holes, keyways, splines, serrations, and gear teeth. Internal broaching requires a starting hole for insertion of the tool. The broaching of gear teeth is discussed in more detail in Section 17.3.1.

Broaching machines may be of the horizontal or vertical variety. Vertical machines are very popular because they require less space. Broaching machines may also be characterized by their driving mechanism as single or twin-ram, pull-down, pull-up, push-down, push-up, continuous, or rotary types [2,5]. Broaching machines are designed with very high stiffness to support high-cutting loads. The tooling is often large and heavy, so that special designs are required to ensure easy access for inspection and changing of broaches.

An individual cutting tooth on a broach resembles a turning or shaping tool, and the basic equations for turning reviewed earlier are applicable to broaching as well, except that the linear speed of the broach, V, is substituted for the rotational speed (m/min). The feed per tooth, f_t, is determined by the broach design (as shown in Figure 2.18) and cannot be altered by changing machine settings. The front teeth are used for roughing, while the later rows are used for finishing as shown in Figure 2.18. The pitch of the broach is determined by the workpiece material and is coarser for roughing than for finishing. Each tooth removes a depth of cut that is equivalent to the feed per tooth and the total depth is removed by the summation of the individual depths of cuts (or feed per tooth). The machining time is estimated from the travel distance of the broach (which includes the roughing, semi finishing, and finishing teeth) over its cutting speed.

A process similar to broaching is turnbroaching, in which the broach is fed into a rotating workpiece. This process is used to machine crankshafts and camshafts in engine manufacture.

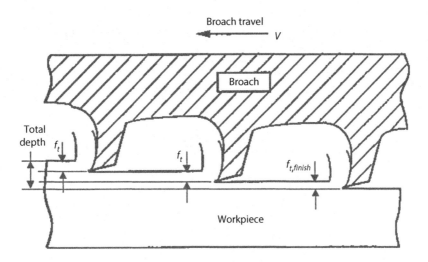

FIGURE 2.18 Broaching.

Relations for the uncut chip thickness and other parameters are similar to those for milling. Turnbroaching is sometimes referred to as skiving, a term which is also used for a gear-finishing process as discussed in Section 17.6.2.

2.9 TAPPING AND THREADING

Many workpieces require internal or external screw threads, which can be produced by a variety of cutting operations. The common cutting processes for producing internal threads are tapping, chasing, and thread milling. Cutting processes used to produce external threads include thread turning and die threading. Threads may also be produced by grinding and by various forming operations.

In *tapping*, a specially formed threading tool is fed into a hole drilled in a previous operation. The tap may either cut or plastically deform the hole wall material to form the thread. Roll form taps, which produce threads by deformation, can be used in ductile materials such as free-machining steels, soft carbon and alloy steels, austenitic stainless steels, and ductile aluminum, copper, zinc, and magnesium alloys. Roll and cut tapping produce different thread profiles as shown in Figure 2.19. In cut tapping, the minor diameter of the thread is determined by the diameter of the existing hole. The metal removal rate is governed by the tap's effective chamfer length, number of flutes, and rpm in addition to the minor diameter. The shear strength of the thread increases with the percent thread for rolled threads, but not significantly for cut threads. Roll taps produce no chips, but require consistent lubrication and control of the pre-tapped hole diameter to prevent excessive tapping torque and tap breakage. Roll-form tapping can produce stronger threads with work-hardening materials (steels and stainless steels). The most common machines used for tapping are drilling machines, milling machines, and lathes. Rigid tapping, a CNC process carried out without floating toolholders or self-reversing tapping units, requires proper synchronization of Z-axis feed to spindle revolution especially at higher speeds (>3000 rpm for a 6 mm tap).

Tapping is used for through holes and blind holes. Drilled through holes can be tapped along the entire length. Blind holes are tapped to a specified length and are sometimes called bottom holes. Blind hole tapping requires accurate depth control to avoid ramming the tap into the bottom of the hole, causing tool failure, or an insufficient number of threads. Bottoming taps must pull chips up and out of the hole, as compared to through-hole taps, which usually push chips out of the hole in front of the cutting edge.

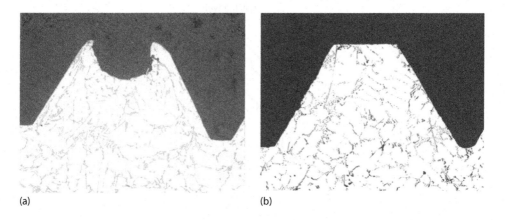

(a)   (b)

FIGURE 2.19   (a) Formed and (b) cut tapped threads.

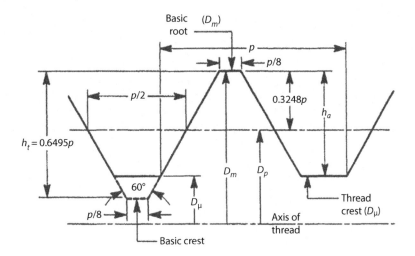

FIGURE 2.20 Thread profile parameters.

The thread height h_a or percentage thread η are important parameters in tapping. η is related to h_a, the theoretical thread height h_t, and the major and minor diameters of the thread form, D_m and D_μ (Figure 2.20) [30]:

$$\eta = \frac{\% \text{ thread}}{100} = \frac{h_a}{h_t} = \frac{D_m - D_\mu}{1.299p} \tag{2.40}$$

where

p is the thread pitch

h_t is the AN basic thread form height used in the standard method for describing the thread height

The torque required for tapping can increase by 200%–300% when the percent thread increases from 50% to 75%. The manufacturing strategy should target a minimum percentage of thread if the thread strength is essentially insensitive to this parameter [30]. The unified (UN) form and ISO standard methods are equivalent to 83% of the AN thread depth. For cut tapping, η can also be calculated from the pitch diameter, D_p, and pre-drilled hole diameter, D_d:

$$\eta = \frac{0.7698(D_p + 0.6496p - D_d)}{p} \quad \text{(cut tapping)} \tag{2.41}$$

The pre-drilled hole diameter for a target value of η is

$$D_d = D_m - 1.299\eta p \quad \left(\text{cut tapping}\right) \tag{2.42}$$

For roll form tapping, the equations corresponding to Equations 2.41 and 2.42 are

$$\eta \cong \frac{1.5396(D_d - D_\mu)}{p} \quad \text{(roll form tapping)} \tag{2.43}$$

$$D_d = D_m + 0.68\,p\eta \quad \left(\text{roll form tapping}\right) \tag{2.44}$$

Generally, tap-drill charts have been generated for $\eta = 75\%$, although the actual thread may deviate from the charted value [31,32]. Form tapping requires a larger hole size than cut tapping because the minor diameter of the thread is produced by the extrusion inward from the drilled hole.

Figure 2.21 shows a plot of the ideal linear and rotational velocity vs. time to generate a desired thread in a rigid tapping process. The cutting speed, feed rate, and depth of cut are given by

$$V = \pi D_m N \tag{2.45}$$

$$f_r = pN \tag{2.46}$$

$$d = \frac{D_m - D_d}{2} \tag{2.47}$$

The depth of cut per edge, d_i, is

$$d_i = \frac{1.299\,p\eta\cos\delta}{2n_t n_f} \tag{2.48}$$

where
  n_f is the number of threads along the chamfer
  n_t is the number of flutes
  δ is the chamfer angle of the tap measured from the tap axis

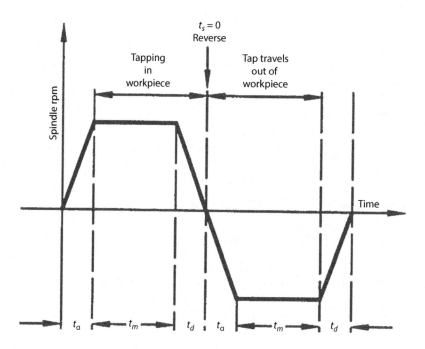

FIGURE 2.21 Variation of spindle rpm during a tapping cycle.

The metal removal rate is

$$Q = \left(\frac{p}{4} + \frac{D_m - D_d}{\tan \frac{\pi}{3}} \right) \left(\frac{D_m - D_d}{4} \right) \frac{P}{\sin \lambda} N \qquad (2.49)$$

The time to cut a thread of length L, t_m, and thread helix angle, λ, are given by

$$t_m = \frac{L}{f_r} \qquad (2.50)$$

$$\tan \lambda = \frac{p}{\pi D_p} \qquad (2.51)$$

Thread turning, the oldest and most widely used threading operation, is a process for producing external or internal threads, usually using a single point tool [11,33,34]. This process has traditionally been used on soft materials, but with the availability of PCBN tools, now also used when turning hardened steels. The tool may be fed into the workpiece either radially or axially. Radial feed cutting generates higher cutting forces and leads to greater difficulty in chip disposal and is used mainly on materials that produce short chips or with multi-toothed inserts. In flank-infeed cutting (in which the tool is fed axially) the cutting action is more like conventional turning. There are many different flank-infeed sequences, which distribute the thread form between passes (Figure 2.22) [33]. The infeed sequence can be optimized to reduce the number of passes while keeping the chip load constant between passes as illustrated in Figure 2.23. The optimum number of passes depends on the tool geometry and edge strength. In some cases, the center portion of the thread is removed using radial infeed while the remaining stock is removed using flank infeed. In other cases a significant amount of material is removed with a grooving tool, leaving only a small amount to be cut with a threading tool. The lead of the thread is determined by the longitudinal motion of the tool in relation to the rotation of the workpiece. The feed coincides with the pitch of the thread.

Multi-toothed full-profile indexable inserts are also used to turn threads. Such inserts generate the full thread profile including the crest in a single pass, eliminating the multiple passes required to produce threads with a single point tool.

Thread milling is used to generate internal or external threads using a milling cutter [11,35,45]. The cutter is fed along the axis of the workpiece as in thread turning to generate the threads in a single pass as shown in Figure 2.24. With a stationary workpiece, a rotating tool moves simultaneously along three axes to generate the helical thread (as compared to the two-axis motion used in circular interpolation). When cutting an external thread, the tool moves along the part's outside diameter; when cutting an internal thread, the tool moves inside a previously drilled hole. As in cut tapping, the feed rate is determined by the workpiece speed in a turning machine or by the helical path speed in NC machining centers or special machines. The accuracy of the thread is controlled by the accuracy of the axial and circular feed mechanisms of the machine, not by the cutting tool. It is preferable to start the thread-milling operation at the bottom of blind hole so that the tool moves outward to avoid chip recutting at the bottom of the hole. In thread milling, the tool rotates at higher speeds and lower feeds than in tapping or thread turning; the feed can be adjusted to generate the desired surface finish and is not constrained by the desired thread pitch as in other threading operations.

Compared to conventional tapping, thread milling allows more room for chip evacuation; chips are also much smaller. The power required for threading can be reduced considerably using thread milling. Percent threads approaching 100% can be generated, and tapered threads can be generated

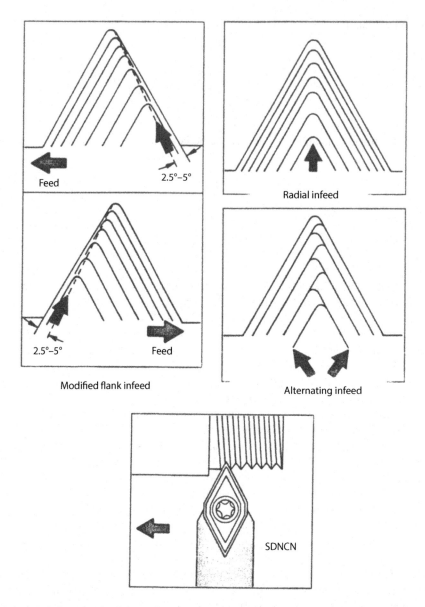

FIGURE 2.22 Schematic view of the radial, flank, and alternating infeed methods for threading. (Courtesy of Carboloy Corporation, Carboloy/Seco, Warren, MI.)

easily and accurately. Thread milling is used primarily for large holes (diameter >30 mm), while tapping is used for smaller holes (diameter <40 mm) due to tool cost.

Threads in smaller holes can be milled using a combined short-hole drilling and thread milling operation called thrilling or drill/threadmilling (Figure 2.25 [34]). Thrilling uses a combined drill-threading tool rotating continuously at a high spindle speed to drill a blind or through hole and generates the thread through a helical retraction motion. Thread milling accuracy is dependent on the machine control system generating the helical interpolation including the machine motion accuracy. Thread milling tends to generate smoother and more accurate threads than tapping and is more efficient than thread turning. Thread milling also eliminates the spindle reversal at the bottom of the hole required in tapping. However, milled threads must typically be gauged much more

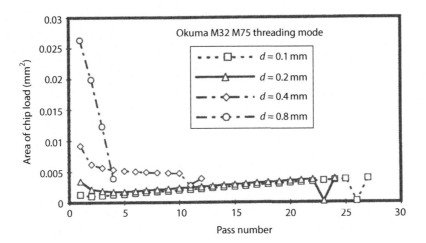

FIGURE 2.23   The effect of depth of cut on number of passes and on chip load area in thread turning.

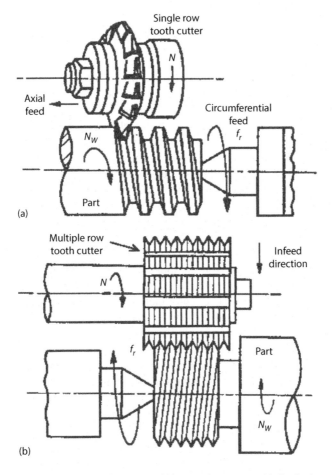

FIGURE 2.24   Thread milling operation performed in a turning center: (a) single thread milling cutter and (b) multiple thread milling cutter.

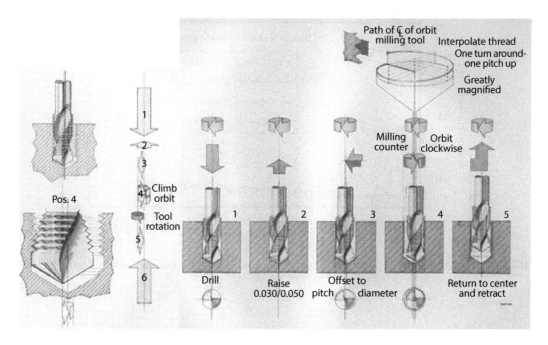

FIGURE 2.25 Drill/threadmilling using a combination drill and thread milling tool.

carefully than tapped threads. Coolant requirements in thread milling are not as critical as those in cut tapping. Tooling costs are generally higher than for tapping.

Thread-milling machines usually employ a climb-milling motion, especially when producing internal right-handed threads, because the tool travels out of the hole during the cut, which reduces chip interference. However, a conventional up-milling mode is used when the part or fixture stiffness is low to reduce tool deflections and chatter. Up-milling is also used with materials that are difficult to machine (e.g., stainless steels) to improve tool life. A thread-milling tool can cut from either the entrance or exit/bottom of the hole, compared to a tap that must start at the entry. If the force in thread milling is too high for the tool L/D ratio, multiple passes can be used to avoid tool breakage. Tapping generates the full thread form and machines to final size in one pass.

Thread milling can produce high tool pressures when milling at full thread length and depth, which can result in excessive tool deflection and tool breakage. Machine requirements also limit the applicability of thread milling; the proper speeds and feed rates must be available, and the machine must be capable of producing an accurate circular motion at high speeds and feeds, especially with nonferrous parts. Thread milling can also only be applied when the ratio of the thread length to the major diameter of the tool falls within relatively narrow limits.

There are two approaches for entering the cut in an internal hole: along a tangential arc or along a radial arc. The radial approach is simple but the cutter should enter to the full doc at 30%–50% the circular feed rate to avoid vibration. The tangential approach allows for the tool to ramp gradually into and out of the cut. It also eliminates the dwell vertical mark at the entry or exit point. The tangential approach requires more complex programming than the radial approach.

When thread milling a stationary workpiece, the tool feed rate at the cutting edge (true feed rate occurring at the perimeter) f_{rt}, and rotational speed N of the cutter are related through

$$f_{rt} = n_t f_t N \qquad (2.52)$$

When thread milling a workpiece rotating at a spindle speed of N_w on a lathe, the feed rate is given by

$$f_{rt} = \pi D_m N_w \tag{2.53}$$

and the feed per tooth is

$$f_t = \frac{\pi D_m N_w}{n_t N} \tag{2.54}$$

In an NC machine, the feed rate is programmed so that the X, Y, and Z axes are synchronized to generate the helical motion of the cutter. The chip load, which is proportional to cutter advance per tooth as in milling, can be changed by varying the tool speed, workpiece speed, or the number of teeth in the tool. The apparent contact length (chip length) for each cutting edge per tool revolution is given by either Equation 2.26 or 2.28. The cross-sectional area of the chip is trapezoidal. The uncut chip thickness does not correspond exactly to the feed per tooth because the programmed cutter feed is about its center and not at the tool periphery where the cut occurs (as explained in Equation 2.38 for circular interpolation milling applications). The centerline feed rate (on the orbiting diameter) for internal and external threads is, respectively,

$$f_r = f_{rt} \frac{D_m - D}{D_m} \quad \text{(internal threads)}$$

$$f_r = f_{rt} \frac{D_\mu + D}{D_\mu} \quad \text{(external threads)} \tag{2.55}$$

where D is the threading cutter major diameter. The metal removal rate for a tool with n_{tr} rows of teeth in contact with the workpiece is

$$Q = \left[\frac{p}{4} + \frac{D_m - D_d}{\tan 60} \right] \left[\frac{D_m - D_d}{4} \right] n_{tr} f_{rt} \tag{2.56}$$

The time required to cut a thread of axial length L in one pass is

$$t_m = 1.1 \frac{\pi D_p}{f_{rt} \cos \lambda} \tag{2.57}$$

When a single row tooth cutter is used the machining time is

$$t_m = \frac{\pi D_p L}{p f_{rt} \cos \lambda} \tag{2.58}$$

Die threading is used to generate external threads using solid or self-opening dies (chasers). Materials with hardness lower than 36 HR$_c$ can be threaded with a die. It is a slower operation than thread rolling but faster than thread turning. Die threading machines are similar to tapping machines.

Thread whirling [34] is a process used for internal and external threads. Whirling removes material in a manner similar to thread milling. This process uses a special head supporting the threading tool as illustrated in Figures 2.26 and 2.27. The whirling head is mounted in a machine spindle that rotates eccentrically at high speed around the slowly rotating workpiece, or performs

(a)   (b)

FIGURE 2.26   (a) External and (b) internal thread whirling operations. (Courtesy of Leistritz Corporation, Allendale, NJ.)

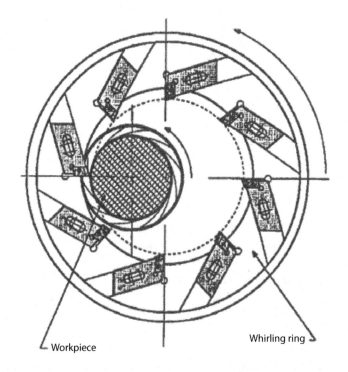

Workpiece   Whirling ring

FIGURE 2.27   An external whirling ring assembly using eight indexable inserts.

a planetary rotation along a stationary workpiece. The cutter uses inserts arranged along the inside or outside circumference of a ring for external and internal threads, respectively. The cutting occurs when the whirling unit and whirling ring are off-center with respect to the workpiece (using a helically interpolated cutting path as in thread milling). This process is faster than thread milling and may generate better surface finish, lower lead error, and shorter chips. It has been successfully used on hardened steels in place of turning or grinding to reduce the cycle time. The most common applications are the machining of screws or worms. It is a preferred process when cutting very long threads or threads with high helix angles.

Thread rolling is an external cold forming process for producing threads on a cylindrical or conical blank. It is similar to internal roll form tapping. The functional principle of thread rolling is shown in Figure 2.28. The tool or die displaces or extrudes the metal from the part surface to form the threads. Generally, thread rolling is carried out using specially designed vertical or horizontal machines. Thread rolling machines are designed with two flat die rollers in a side-by-side position. The threads on the blank, which is inserted between the rollers, are usually completed after eight revolutions. Other machines use two or three rollers, which form a radial-infeed cylindrical die. Thread rolling can be also carried out using lathes and automatic bar machines using special single- or double-roll attachments.

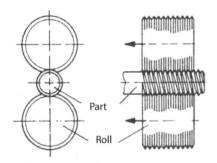

Feed in axial direction (direction of arrow)
Head stationary, component part rotating

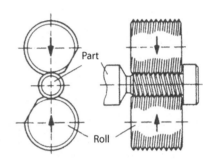

Feed is radial by means of appropriate roll geometry
Rolling head stationary, component part rotating

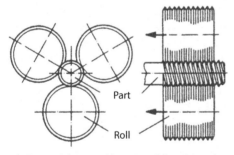

Feed in axial direction (direction of arrow)
1. Rolling head rotating, component part stationary
2. Rolling head stationary, component part rotating

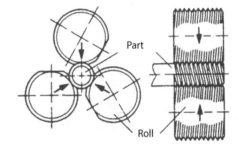

Feed is radial by means of appropriate roll geometry
1. Rolling head rotating, component part stationary
2. Rolling head stationary, component part rotating

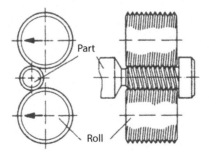

Feed is tangential (direction of arrow)
Rolling head stationary, component part rotating

FIGURE 2.28   The functional principle of thread rolling operations. (Courtesy of Fette Tool Systems, Inc., Cuyahoga Falls, OH.)

2.10   GRINDING AND RELATED ABRASIVE PROCESSES

Grinding is a general name for many similar processes in which a hard, abrasive surface is pressed against a work surface, resulting in the removal of material from both the workpiece and the abrasive. Grinding is a very common process traditionally used to produce parts with close tolerances, fine surface finishes, and small burrs and is used especially as a precision finishing process. It is also used in the general machining of hard or brittle materials and to produce complex contours. The major difference between traditional chip forming operations and grinding is the characteristics of the cutting tool. The individual grains on a grinding wheel are spaced randomly and have an irregular shape. The cutting speed of the wheel is very high while the feed is low compared to the traditional processes. Material removal in grinding takes place on a smaller scale than in conventional chip-forming processes. The process is characterized by three physical mechanisms: cutting, ploughing, and sliding at the interface between the abrasive grains and the work surface. As a result, grinding requires different physical models for detailed analysis [2,5,8,9,36–48]. Grinding process control depends on the interaction of the three mechanisms; force sensors are used to properly control some grinding parameters such as wheel and dresser sharpness. The chip thickness in grinding is very small (0.002–0.05 mm) compared to most other machining operations. Tool forces are high due to the random orientations and high negative rake angles of the individual grains in the wheel. The energy required to grind away a cubic centimeter of a given metal is typically 10 times that required to machine the same volume of material with a cutting tool as given in Table 2.1. Grinding is a primary process for machining hard materials such as super-alloys and ceramics.

Grinding processes employ a variety of kinematic motions of the wheel and workpiece, some of which are shown in Figure 2.29. Processes can be broadly categorized as traverse or plunge grinding depending on the direction of feed of the wheel. In traverse grinding the wheel moves parallel to the ground face, and only a portion of the workpiece surface is machined at a given time. Material is removed in several passes until the final depth of cut is achieved. In plunge grinding the wheel moves normal to the surface, removing material over the entire surface of the wheel. Plunge grinding generates complex profiles very easily and is faster than traverse grinding because the full wheel can be engaged. Grinding processes may be further classified as flat, cylindrical, or contour, depending on the shape of the ground surface, or as centerless, reciprocating surface, creep feed, microsizing, or honing depending on the kinematic motions of the workpiece and tool. The tool used may be a wheel, pad, disk, or belt made from a variety of bonded abrasives.

Surface grinding is carried out either by reciprocating the workpiece or by rotating it about an axis perpendicular to the grinding surface while it is fed laterally in front of the rotating wheel. This operation produces flat, angular, and irregular shapes [43]. It uses either the periphery or face of the grinding wheel or a belt. This method competes with planing and milling operations, which generally require higher cutting forces.

Surface grinders may have either a horizontal or a vertical spindle with a table, which reciprocates or rotates [2,4,5].

Cylindrical grinding removes material from external cylindrical surfaces by rotating the workpiece and the wheel in opposite directions. The workpiece is typically supported between centers. The resulting surfaces can be straight, tapered, or contoured. It is used for hard or brittle workpieces or parts requiring fine surface finishes. The wheel or belt rotates at a much higher speed than the workpiece. In addition to its more general meaning given earlier, *plunge grinding* is sometimes used to refer to a form of cylindrical grinding in which the wheel moves continuously into the workpiece rather than traversing.

Cylindrical grinding machines resemble lathes since they are equipped with a headstock, tailstock, table, and wheel head. The workpiece is held either between centers or securely in a fixture mounted on the workhead spindle.

Centerless grinding is similar to cylindrical grinding, but the workpiece is supported by a blade between the grinding wheel and a small regulating (or feed) wheel [2,4,20,49]. The regulating wheel

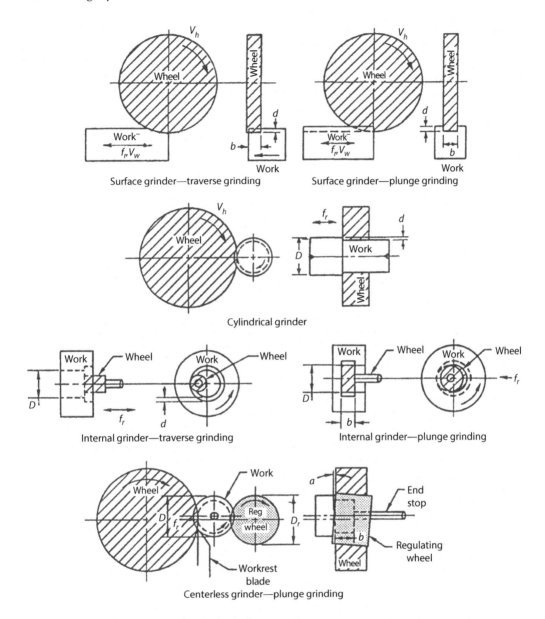

FIGURE 2.29 Schematic view of typical grinding operations.

holds the part against the grinding wheel and controls the cutting pressure and the rotation. The workpiece finds its own center as it rotates between the two wheels. Out-of-round material is pushed into the grinding wheel and ground away. The regulating wheel is usually made of a rubber bonded abrasive and serves as both a frictional driving and braking element, rotating the workpiece at a constant and uniform surface speed. Centerless grinding is used primarily for external grinding.

Centerless grinding is used instead of cylindrical grinding when it is not possible to place centering holes in the part. Centerless grinding requires less grinding stock, can take heavier cuts, and is preferred for long, slender shafts. The centerless operation produces low cylindricity errors and consistent surface finish. It provides a better size control than cylindrical grinding. This process is necessary for parts which must rotate at high accuracy in a bearing or wear situation such as valves, camshafts, spindle shafts, etc. It is important to note that in centerless grinding the ground surface, rather than the rotational axis, is the reference datum.

Centerless grinders are similar to cylindrical grinders but without centers. The work is supported on shoes or a fixed blade under pressure applied by the regulating wheel.

Internal grinding generates internal cylindrical surfaces using very small wheels. It can be used to produce straight, tapered, blind, or through holes, holes with multiple diameters, contours, and flat sections. Stock removal rates are greater than for other abrasive hole-making processes such as honing or microsizing, which are used primarily to produce special surface finishes or close tolerances.

Creep-feed grinding [50,51] is a surface or external cylindrical grinding operation that removes a full depth of cut in a single pass at a very slow feed rate. Depths of cut are typically 2–6 mm, but may be as low as 0.5 mm for hard to machine materials. The speed of conventional wheels is generally 2000 m/min, while the feed rate is often only 25–400 mm/min. When properly applied, creep-feed grinding can reduce the overall machining time by up to 50% with no loss of dimensional or geometric precision or surface quality. However, these results can be achieved only when the grinding machine is designed for creep-feed applications to provide sufficient static and dynamic stability (two to three times that of conventional grinding machines), proper dressing capabilities, and adequate coolant control. It has been applied successfully to brittle materials such as ceramics and superalloys.

In addition to the wheel parameters, input variables for grinding operations include the feed rate (down feed), wheel speed, workpiece speed, sparkout time and frequency, and depth of wheel dressing. The wheel wear rate and the stock removal rate are important output parameters. The objective in grinding process design is generally to reach a steady state with a uniform wheel wear rate, steady power consumption, constant pressure at the wheel–workpiece interface, and no apparent grinding burn or chatter. The grinding cycle starts with a heavy feed rate, which is reduced for the finish grind stage if the same wheel is used. However, the finish wheel grade often differs from the rough wheel grade. The workpiece surface is generated during the final contact between the wheel and workpiece; the maximum usable stock removal rate usually remains approximately constant over a wide range of work speed and depth of cut combinations, and depends primarily on grinding wheel characteristics, the work material, and the coolant type (oil or water). Oil coolants are often preferred because they reduce frictional heat generation and residual stresses.

The equivalent diameter of the wheel-workpiece system is used to compare different grinding operations; this parameter provides the equivalent wheel diameter corresponding in a surface grinding operation and is defined as

$$D_e = \frac{D_h}{1 \pm (D_h/D_w)} \tag{2.59}$$

where
D_h and D_w are the wheel and workpiece diameters
the + or − in the denominator is used for external (OD) or internal (ID) grinding, respectively

In the case of surface grinding, $D_e = D_h$ (or $D_w = \infty$). A large D_e usually results in a higher contact area and higher threshold forces and power.

The apparent area of contact for the abrasives is calculated the same way as in milling processes. Therefore, the contact length for each abrasive grain per wheel revolution is given by Equations 2.26 and 2.27 with $D = D_e$. The chip thickness is dependent on the grinding method and can be calculated using relations similar to those for milling processes (Equations 2.21 through 2.25). However, the cross-sectional area of the chip is nonuniform and varies between abrasive grains. An extensive analysis of this subject is given in [44].

The volumetric rate of wheel wear Q_h is

$$Q_h = D_h f_h b \tag{2.60}$$

where f_h is the rate at which the wheel diameter is decreasing. The stock removal rate from the workpiece Q_w for surface grinding is

$$Q_w = dV_w b \qquad (2.61)$$

V_w is the workpiece speed (equivalent to f_r in Figure 2.29). For cylindrical plunge grinding, Q_w is given by

$$Q_w = D_w f_r b \qquad (2.62)$$

where
  D_w is the diameter of the workpiece
  f_r is the radial feedrate of the wheel into the workpiece

For cylindrical transverse grinding, Q_w is given by

$$Q_w = D_w d f_r \qquad (2.63)$$

where f_r is the federate of the wheel along the workpiece. The grinding power is proportional to the specific grinding energy (similar to the unit power discussed for turning and in Section 6.5):

$$P_m = u_s Q_w + P_t = F_t V_h \qquad (2.64)$$

where
  F_t is the tangential force at the tool–workpiece interface
  V_h is the wheel speed

Generally, the specific grinding energy, u_s, is higher than the unit energy for other processes. A performance index used to characterize wheel wear resistance in economic analyses is the *grinding ratio G* defined as

$$G = \frac{Q_w}{Q_h} \qquad (2.65)$$

G varies in practice from 0.018 to 60,000 [43,52,53]. Higher values of G indicate that the wheel is more productive. G-ratio generally increases with wheel speeds up to a characteristic speed, after which it decreases due to a transition in wear regimes. Excessive wheel speeds may increase the G-ratio but lead to surface burning.

The processes described thus far use a hard abrasive wheel and are intended to produce surfaces with fine finishes and close dimensional tolerances. Process conditions are chosen to produce the required surface and dimensional properties as efficiently as possible without producing dynamic instability or surface damage or burn. Surface finish and integrity in grinding are discussed in Chapter 10. In troubleshooting grinding problems in practice, the grinding wheel is generally looked at first since it is relatively easy to change. Wheel characteristics of particular interest are grit size, hardness, and structure, which are described in Chapter 4. The conditioning process of the wheel, consisting of dressing and cleaning/truing, has a significant impact on material removal rates, forces, surface characteristics, and subsurface properties and is also considered in troubleshooting. Once wheel issues have been resolved, the grinding fluid is examined for proper type, concentration, and delivery. Finally the grinding infeed cycle is examined, especially when

FIGURE 2.30 A microsizing tool used in engine bore manufacture. (Courtesy of Engis Co., Wheeling, IL.)

investigating dimensional or dynamic instability problems. Contemporary CNC control and force sensing technologies allow monitoring of critical grinding parameters such as the sharpness of the dresser and wheel, wheel deflection, and the threshold force for internal grinding. Wheel dressing frequency is often controlled by force sensing.

Microsizing (superabrasive bore finishing or diamond reaming) is used to improve the accuracy of internal cylindrical surfaces. A fixed diameter tool fitted with bonded diamond or CBN abrasives on the periphery is generally used. The tool, workpiece, or both may rotate (Figure 2.30) [54]. The choice of coolant is very important, and honing oils or higher concentration water-based fluids are generally used. Since the material is gradually removed by the thousands of superabrasive particles around the periphery of the tool, it generates relatively little heat or stress, enabling excellent control of bore geometry. Stresses increase rapidly, however, with increasing stock removal. Microsized holes have low roundness errors, limited taper, and no bell-mouth at the hole entrance.

Microsizing may be a single- or multipass operation. Multiple passes may be used when more than 0.05 mm of stock must be removed, when the part is weak structurally and would distort under heavy loading, or when a fine surface finish is required. The achievable surface finish and allowable stock removal are functions of the abrasive grit size and coolant. To achieve precision bore cylindricity in a single pass, three conditions must be met: (1) the tooling design should take into account variables such as the work material, hardness, bore size and length, bore shape (solid or interrupted), bore wall thickness, finish requirements, and incoming stock; (2) the bore must be aligned with respect to the tooling to minimize side forces; and (3) proper part fixturing must be used to prevent bore distortion. A floating holder can be used to improve tool alignment at conventional speeds but usually fails at higher speeds due to centrifugal forces. Therefore, the best results are usually obtained when a floating tool holder is used in conjunction with a radially and angularly floating type part fixture to allow for perfect alignment with low forces. Two- to four-pass microsizing processes are not uncommon in high volume production due to deviations from optimum setup conditions. Tool speeds range from 8 to 90 m/min and the feed rate can vary from 0.100 to 2.5 m/min, depending on the workpiece material. The maximum stock removal is a function of the workpiece material, and the grit size of the abrasives may be as high as 0.05–0.1 mm; in steel, it is typically 0.09 mm. Tolerances of better than 0.005 mm can be obtained with straightness errors less than 20 s and roundness errors less than 0.001 mm.

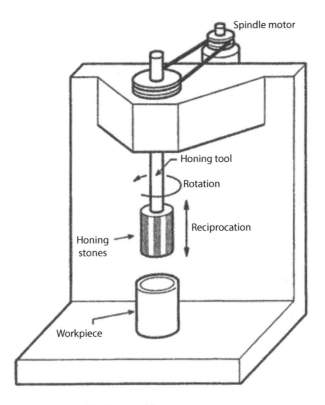

FIGURE 2.31   Schematic diagram of honing machine.

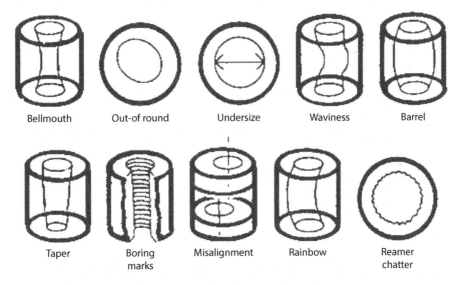

FIGURE 2.32   Bore form and size errors corrected by microsizing or honing.

Honing is a finishing operation similar to microsizing in which bonded abrasive sticks are pressed against the work surface using an expandable, reciprocating tool (Figure 2.31) [5,55–61]. Honing corrects axial and radial distortions produced by previous operations (Figure 2.32) and imparts a controlled surface texture to the bore.

The expansion pressure is a critical parameter in honing since the wall temperature and bore thermal distortion increase linearly with pressure. The lowest expansion pressure and feed rate possible should be used to increase accuracy. Temperature is also proportional to honing speed, which must be as high as possible to maintain stone sharpness. The axial speed usually ranges from 12 to 25 m/min, while the circumferential speed varies between 15 and 100 m/min. Generally, a rough honing operation is required to create a geometrically correct bore, and semi-finish and finish passes are used to improve accuracy and generate the proper surface topography. Controlled temperature flood coolants are normally applied to the bore and honing tool to control thermal errors. The metal removal rate, Q_w, and depth of cut, d, in honing are related by

$$d \cong \frac{Q_w}{n_t \lambda_h V_h} \tag{2.66}$$

where
   n_t is the number of honing stones
   λ_h is the stone length
   V_h is the peripheral speed
   Q_w depends on the expansion pressure and the abrasive grain size and sharpness

The combined rotary and reciprocating motion of the honing tool produces a cross-hatched surface pattern on the bore as shown in Figure 2.33; the cross-hatch angle κ is given by

$$\tan \frac{\kappa}{2} = 0.6366 \frac{\lambda_s N_s}{DN} \tag{2.67}$$

where
   λ_s is the stroke length
   N_s is the stroke rate in strokes/min

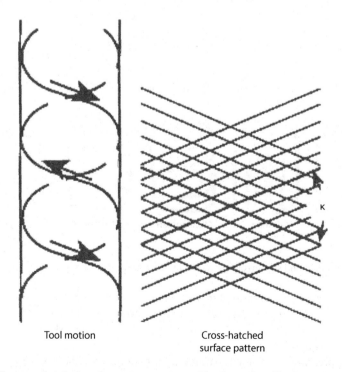

Tool motion   Cross-hatched
                                     surface pattern

FIGURE 2.33   The cross-hatched surface pattern generated by the tool motion in engine bore honing.

Abrasive *brush honing* or flexible honing is a more recently developed process using the same principle as the internal honing operation. Brush honing is carried out using flexible abrasive brush tools, which are equipped with compliant filaments which bend and press against the bore surface [62]. This process removes very little material from the surface using high production cycle times (less than 1 min); it removes the surface peaks and some of the smeared material left from previous bore finishing processes. Unlike microsizing or honing operations, it cannot generally remove the tool feed marks left by prior turning or boring operations, and cannot produce a full plateau finish and cross-hatched pattern. It also differs from conventional honing in that the resilience of the abrasive brush results in a flexible cutting action, which does not alter the bore geometry with respect to concentricity, ovality, cylindricity, and squareness. Brush honing can be carried out with or without a coolant.

External Honing, also referred to as microfinishing, superfinishing, or superhoning, is a high precision, fine-grit abrasive process for removing imperfections on external surfaces, which may result from conventional grinding. The tool is an abrasive stone that oscillates at a high rate (up to 2800 Hz) and is pressed against a rotating workpiece. A tool arm is used to position the stone and apply pressure. The stone oscillates in the direction parallel to the workpiece axis of rotation. There are four common microfinishing methods: centerless through-feed, plunge feed, plunge feed pivoting, and rotating tool/rotating workpiece. Parameters that affect part quality in microfinishing include the machine structure, work- and stone-holding methods, tool material, cutting fluid, and environmental control; requirements for these parameters are generally more stringent than for other abrasive processes [63]. External honing can be used in combination with turning on lathes by attaching a superfinishing arm behind the turning tool holder.

Stock removal is a function of stone cutting capability, pressure, oscillation amplitude and frequency, and rotational speed of the workpiece. Use of a low peripheral work speed, moderate stone pressures, and wide workpiece contact lengths limit heat generation and produce fine surface finishes; the roughness may be as low as 0.025 μm R_a, compared to 0.1–0.15 μm for conventional grinding. Smaller tolerances and surfaces with a 100% bearing ratio can also be produced since this process removes the amorphous surface layer produced by turning or grinding operations. This process has achieved roundness errors of 0.12 μm. It has been applied to a variety of work materials, including high-strength steels, carbides, and ceramics.

In *belt grinding* (or abrasive belt machining), coated abrasive belts are used to remove material at a high rate. Due to its flexibility and the availability of stronger belts, abrasive belt machining has become increasingly popular, even replacing conventional turning, milling, grinding, and polishing operations [64,65]. Belt grinding can be used in place of centerless grinding, surface grinding, and cylindrical grinding in roughing to fine finishing operations by replacing the wheel with a coated abrasive belt. Grinding can be performed using a slack belt, or with the belt riding over a contact wheel or platens, which support the belt and apply a thrust force. The contact wheel can be hard or soft, and either smooth or serrated. Smooth hard contact wheels are used for precision abrasive machining and for maximum stock removal.

Belt grinding is often more efficient than conventional grinding. The belt speed is usually between 800 and 2000 m/min, but speeds of up to 4000 m/min have been used to grind ceramics. The optimum speed is affected by the contact wheel or pressure, which supports the belt at the pressure point, regulates the cutting rate, and controls the grain breakdown as well as the grit size and work thickness. The feed is controlled either indirectly by adjusting the pressure or directly by actuating the contact wheel. Belt grinding is generally safer than standard grinding with brittle wheels. Adequate coolant flow and pressure must be used to reduce grinding temperatures and sparking.

Cycle times for belt grinding are comparatively low. It is particularly well-suited for grinding complex profiles such as the oval-shaped lobes on cam shafts [64]. Belt grinders generally produce flatness values of 0.025–0.05 mm and surface finishes from 0.25 to 2.5 μm R_a with burr-free edges.

Disk grinding is a face grinding method in which all or most of the flat face of the wheel contacts the workpiece. Disk grinding offers advantages over other grinding methods in many applications because the wide face contact area generates surfaces with low flatness errors and fine finishes. Both single- and double-disk grinders (using single or double spindle types) are available. Face grinding is attractive on bimetallic surfaces where tool life in other operations may be limited.

Double-disk grinding is carried out using two opposing disks (wheelheads) between which a part is passed. Two opposite sides of the workpiece are ground simultaneously by opposing abrasive disks, producing flat, parallel surfaces. The process can be used for both thin and thick workpieces, and only lateral support of the work is required. The stock removal per side is usually between 0.25 and 0.70 mm for roughing, and between 0.1 and 0.25 mm for finishing. In single pass operations, a maximum of 0.5 mm stock per side can be removed. Double-disk grinding can produce surfaces with flatness errors of 0.0015 mm, parallelism errors of opposing surfaces of 0.0025/25 mm, and squareness errors on the order of 0.0075/25 mm.

2.11   ROLLER BURNISHING

Roller burnishing is a surface finishing operation in which hard, smooth rollers, or balls are pressed against the work surface to generate the finished surface through plastic deformation [66–68]. Roller burnishing is used in combination with or as a replacement for lapping, honing, microsizing, grinding, or in some cases turning, milling, reaming, and boring. Burnishing is used to improve surface finish, control tolerance, increase surface hardness, and improve fatigue life.

As shown in Figure 2.34, the burnishing tool applies a pressure greater than the yield stress of the workpiece material. The rotating rollers or balls displace the surface asperities on the part and generate a plateau-like surface. Plastic deformation is induced in the surface layer of the workpiece, which work hardens from 5% to 20%. Burnishing produces a compressive residual surface stress, which often improves the wear resistance and surface fatigue life of the part. Surface finishes between 0.05 and 0.5 µm are generated in single or multipass operations; the surface bearing area can approach 100% when the preburnished surface has uniform asperities. Part dimensional accuracies can often be controlled within ±0.005 mm with proper part preparation. However, geometric errors produced in previous operations generally cannot be corrected by burnishing. Due to its dimensional consistency, burnishing is especially well suited for parts designed for press-fit assembly and face sealing.

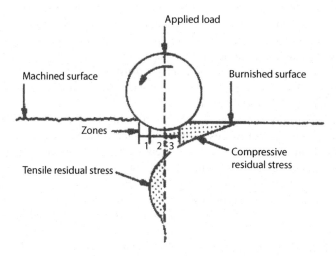

FIGURE 2.34   Schematic representation of the roller burnishing process. Zone 1, elastic compression zone; zone 2, plastic deformation zone; and zone 3, elastic recovery zone. (Courtesy of Cogsdill Tool Products, Inc., Camden, SC.)

Roller burnishing performance depends on the characteristics of the incoming surface. Turned, bored, and milled surfaces with roughnesses between 0.002 and 0.005 mm are suitable for burnishing because they have uniform asperities. Rough ground surfaces are also excellent for burnishing. Reamed surfaces are often not suitable due to surface and size inconsistencies. Special reaming tools with stationary carbide pads behind the cutting edge have been designed to ream and burnish a hole in a single operation. The surfaces generated by such reaming tools are not equivalent to those generated by roller burnishing process because the carbide pads apply a much lower surface pressure.

Roller or ball burnishing can be performed on flat surfaces, cylindrical ID and OD surfaces, tapered internal and external surfaces with taper angles up to 15°, spherical, radius, or fillet surfaces, or a combination of these surfaces. The size reduction (of external dimensions) or increase (of internal dimensions) depends on the pre- and post-burnished surface finish. The depth d to which the tool penetrates a surface is approximately

$$d = C_1(R_{a1} - R_{a2})$$ (2.68)

where

R_{a1} and R_{a2} are the surface roughnesses of the preburnished and burnished surfaces, respectively
C_1 is a constant equal to 2 theoretically; in practice, however, C_1 is between 2 and 4

The burnished hole size tolerance is half of that for the preburnished hole. Many parameters, including the burnishing force, feed rate, roller or ball material and diameter, and lubrication, affect the final finish [69,70]. The most important parameters are usually the burnishing force and the feed rate.

A knurling operation is occasionally required prior to roller burnishing to achieve a close tolerance on oversized bores or undersized shafts [71]. In these applications knurling raises the initial surface profile, providing enough material for the burnishing tool to roll down to the desired dimension.

2.12 DEBURRING

Burrs are undesired projections of material beyond the edge(s) of the workpiece due to plastic deformation during machining [72–79]. As discussed in Section 11.4, burr formation is a complex process, which depends on the workpiece material, tool geometry, part design, degree of tool wear, process parameters, and manufacturing process sequence. Burrs of excessive size often must be removed for the part to be handled safely, assembled, or function effectively. Deburring is generally a time consuming and costly operation, and in many cases is difficult to automate.

Several deburring methods are widely used. These include barrel tumbling, centrifugal barrel tumbling, vibratory deburring, water-jet or abrasive-jet deburring, sanding, brushing, abrasive-flow deburring, ice blasting, liquid-honing, chemical deburring, ultrasonic deburring, electropolishing, electrochemical deburring, thermal-energy deburring, mechanical deburring, and manual deburring. Detailed descriptions of these operations are available in the literature [72–81]. The most widely used methods are manual deburring, brushing, robotic deburing using brushes or cutting tools with or without force control, high pressure water deburing, and ultrasonic deburring. High pressure water deburing is well suited to features that are inaccessible for meachanical deburing (i.e., intersecting internal holes), and to parts that cannot be subjected to heat or corrosive chemicals.

Hole entrance and exit deburring is a very common process. Special deburring and chamfering tools have been developed to remove unwanted material created by drills or reamers. These tools generate a predetermined chamfer size at the entrance and exit of hole that remove burrs, so that a subsequent deburring operation (e.g., brushing) is not necessary. Robotic deburring strategies have also been successfully applied in holemaking [82–84].

Burr formation and control is discussed in greater detail in Section 11.4.

2.13   EXAMPLES

Example 2.1 A 50 mm diameter gray cast iron workpiece is rough turned with an uncoated carbide tool. The feed for the tool is 0.4 mm per revolution (mm/rev), the depth of cut is 4 mm, and the recommended cutting speed is 70 m/min. Calculate: (a) the material removal rate, (b) power and torque required by the spindle, and (c) the main cutting force.

Solution:

(a) The material removal rate (MRR or volumetric rate of machining) is obtained from Equation 2.6.

$$MRR = Q = Vfd = (70 \text{ m/min})(0.4 \text{ mm/rev})(4 \text{ mm})(1000 \text{ mm/m})$$

$$= 112,000 \text{ mm}^3/\text{min} = 112 \text{ cm}^3/\text{min}$$

The cutting speed occurs at the average depth of cut diameter which is $D_{avg} = 50 - 4 = 46$ mm.

(b) The specific cutting energy (unit power) is obtained from Table 2.1 as 0.065 kW/cm³/min (the average value). Hence the machining power required in this turning operation can be estimated from Equation 2.7:

$$P = Qu_s = \left(112 \text{ cm}^3/\text{min}\right)\left(0.065 \text{ kW/cm}^3/\text{min}\right) = 7.28 \text{ kW}$$

The spindle torque can then be determined from the cutting power:

$$P = T\omega = T(2\pi N) \Rightarrow T = P/(2\pi N), \quad \left(\text{W} = \text{J/s} = \text{N m/s}\right)$$

The rpm (N) is calculated from Equation 2.1. This equation yields $N = \dfrac{2 \times V}{\pi\left(D_1 + D_2\right)}$, where $D_1 = 50$ mm, $d = doc = 4$ mm, and $D_2 = D_1 - 2d = 50 - 2$ (4) = 42 mm. Hence,

$$N = \frac{2(70 \text{ m/min})(1000 \text{ mm/m})}{\pi\left(50 + 42 \text{ mm}\right)} = 485 \text{ rpm}$$

$$T = \left(7280 \text{ N m/s}\right)\left(60 \text{ s/min}\right)/\left(2\pi\right)\left(485 \text{ rpm}\right) = 143 \text{ N m}$$

Example 2.2 Calculate the machining time in Example 2.1 if the axial length of the O.D. cut is 150 mm.

Solution: The distance that the tool travels along the workpiece is the length of cut plus the approach allowance to avoid the tool edge impact with the workpiece during the rapid travel. Assuming the approach distance L_a is 3 mm, the machining time is equal to

$$t_m = \left(L + L_a\right)/f_r = \left(L + L_a\right)/\left(fN\right) = \left(150 + 3\right)/\left(0.4 \times 485\right) = 0.768 \text{ min} = 47.5 \text{ s}$$

Example 2.3 For the turning operation in Example 2.1, evaluate the uncut chip thickness and the cutting edge engagement if the lead angle (SCEA) of the tool is 20°.

Solution:
The uncut (nominal) chip thickness is given by Equation 2.3:

$$a = f \cos \kappa = (0.4 \text{ mm})(\cos 20) = 0.376 \text{ mm}$$

The cutting edge engagement (nominal width of cut) is a function of the lead angle and depth of cut (given by Equation 4.1) which is equal to

$$L_m = (\text{doc}) s\,(\kappa) = ds(\kappa) = (4 \text{ mm}) s(20) = 4.257 \text{ mm}$$

Example 2.4 In a milling application a four-fluted 75 mm diameter indexable end mill is used to machine a 200 mm wide sidewall on a 1035 steel workpiece as illustrated in Figure 2.35. Carbide coated inserts are used. The axial depth of cut is 80 mm and the radial depth of cut is 5 mm. The cutting velocity is 100 m/min with the suggested feed per tooth (or uncut chip thickness) is 0.2 mm/rev for the particular insert style and material. The cutting tool length extending out of the toolholder is 200 mm. Determine the taper on the sidewall of the workpiece generated due to tool deflection if the radial component of the force is estimated to be 30% of the cutting (tangential) force.

Solution: This is a peripheral milling cut in which the finished surface is perpendicular to the direction of feed. The end mill can be considered as an elastic cylindrical beam, cantilevered to the spindle and neglecting the compliance of the toolholder and spindle (see Chapter 5 for complete analysis of a toolholder–spindle interface). The static deflection due to the radial force (F_r) at the free end of the end mill is given by

$$\delta = \frac{F_r l^3}{3EI} \quad \text{where } I = \frac{\pi D_e^4}{64}$$

The effective diameter of the cutter is somewhat smaller than the outer diameter because the flute space is taken into consideration. The effective diameter has been found to be about 0.75–0.85 of the outer diameter. The radial cutting force is not constant but is assumed to be 30% of the cutting (tangential) force on average. In addition, there is only one cutting edge in contact with the workpiece at any time. The cutting force is estimated from the average cutting power.

From the information given, we can calculate the spindle rotational speed or the rpm of the milling cutter from Equation 2.10. Thus,

$$N = V/(\pi D) = (100 \text{ m/min})(1000 \text{ mm/m})/(3.14)(75 \text{ mm}) = 424 \text{ rpm}$$

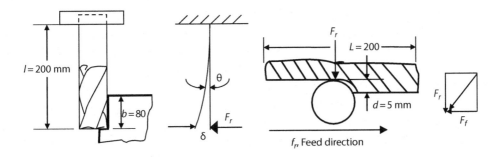

FIGURE 2.35 Schematic representation of Example 2.4.

The maximum uncut chip thickness (provided by the manufacturer of the cutter for roughing operations) is 0.2 mm. In peripheral and end milling application with the cutter diameter significantly larger than the radial DOC (d), the feed rate should be adjusted to avoid the chip thinning effect. Equation 2.22 gives the maximum uncut chip thickness with $\kappa = 0$ in this case (as is generally true in end milling) and v_m is given by Equation 2.24.

$$\cos v_m = 1 - \frac{(2)(5)}{75} \Rightarrow v_m = 30°$$

$$a_{max} = f_t \cos \kappa \, \sin v_m, \quad \text{hence } f_t = a_{max}/\sin v_m = 0.2/\sin(30°) = 0.4 \text{ mm}.$$

The feed rate for the cutter is given from Equation 2.20:

$$f_r = n_t f_t N = (4)(0.4 \text{ mm/rev/tooth})(424 \text{ rpm}) = 679 \text{ mm/min}$$

The material removal rate (MRR) is determined from Equation 2.29:

$$\text{MRR} = (\text{cross-sectional area of uncut chip})(\text{feed} - \text{rate}) = dbf_r$$
$$= (5 \text{ mm})(80 \text{ mm})(679 \text{ mm/min}) = 271,600 \text{ mm}^3/\text{min} = 272 \text{ cm}^3/\text{min}$$

The unit specific energy for the work material is estimated from Table 2.1 to be 0.08 kW/cm³/min. Hence the power required for this operation given by Equation 2.7:

$$P = \text{MRR} \, u_s = (272 \text{ cm}^3/\text{min})(0.08 \text{ W/cm}^3/\text{min}) = 21.8 \text{ kW}$$

The cutting force can be estimated from the power

$$P = FV \Rightarrow F = P/V = (21,700 \text{ N} \cdot \text{m/s})(60 \text{ s/min})/(100 \text{ m/min}) \Rightarrow F = 13,020 \text{ N}$$

Hence the $F_r = 0.3 \cdot F = 0.3 \cdot 13,020 = 3906$ N
   The deflection is then

$$\delta = \frac{F_r l^3}{3EI} = \frac{(3,906 \text{ N})(200^3 \text{ mm}^3)}{3(400,000 \text{ N/mm}^2)(1,241,895 \text{ mm}^4)} = 0.021 \text{ mm}$$

$$I = \frac{\pi D_o^4}{64} = \frac{\pi(0.8 \times 75)^4}{64} = 1,241,894.5 \text{ mm}^4$$

The slope of deflection is 0.000105, which is the taper of the sidewall. The perpendicularity error of the wall due to cutter deflection only is 0.01/100 mm.

Example 2.5 An end-mill is used to clean up the surface of a 4140-forged steel by performing face milling operation. An indexable 50 mm diameter cutter (shown in Figure 2.36) with three round inserts is selected. The IC size of the round inserts is 17 mm. The depth of cut is 3 mm. The suggested cutting conditions for this cutter are 150 m/min cutting speed and 0.2 mm/tooth feed. The specific energy for the workpiece material is 0.08 W·min/mm³. Determine the metal removal rate.

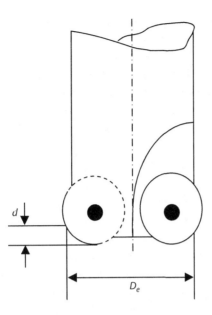

FIGURE 2.36   Schematic representation of Example 2.5.

Solution: The effective tool diameter (D_e) is not the actual cutter diameter because the axial doc is smaller than the radius of the inserts.

$$D_e = D - IC + \sqrt{IC^2 - (IC - 2d)^2} = 45.96\,\text{mm} = 46\,\text{mm}$$

Since the doc is significantly smaller than the radius of the inserts, chip thinning takes place. Therefore, from Equations 2.24 and 2.22

$$\cos v_m = 1 - \frac{(2)(3)}{17} \Rightarrow v_m = 50°$$

$$f_t = a_{max}/\sin v_m = 0.2/\sin(50°) = 0.26 \text{ mm.}$$

The feed rate for the cutter is given from Equation 2.20. At smaller axial doc, the spindle speed should be calculated based on the effective diameter in cut.

$$f_r = n_t f_t N = (3)(0.26 \text{ mm/rev/tooth})(1038 \text{ rpm})$$
$$f_r = 747 \text{ mm/min}$$

The material removal rate (MRR) can be calculated from Equation 2.29:

$$\text{MRR} = dbf_r = (3 \text{ mm})(46 \text{ mm})(747 \text{ mm/min}) = 103,136 \text{ mm}^3/\text{min}$$

Example 2.6 An indexable 25 mm two-flute ballnose end-mill is used to clean up the surface of a 4140-forged steel part. The DOC is 6 mm. The suggested cutting conditions for this cutter are 150 m/min cutting speed and 0.2 mm/tooth feed. The specific energy for the workpiece material is 0.08 W·min/mm^3. Determine the metal removal rate.

Solution: The effective tool diameter (D_e) is not the actual cutter diameter because the axial doc is smaller than the radius of the ballnose.

$$D_e = 2\sqrt{d(D-d)} = 2\sqrt{6(25-6)} = 21.4\,\text{mm}$$

Since the doc is significantly smaller than the radius of the nose, chip thinning occurs. The thickness of the chip varies depending on the doc. Using Equations 2.21 through 2.24,

$$\cos v_m = 1 - \frac{(2)(6)}{25} \Rightarrow v_m = 58.7°$$

$$f_t = a_{max}/\sin v_m = 0.2 / \sin(59°) = 0.23 \text{ mm.}$$

The feed rate for the cutter is given by Equation 2.20 and considering the effective diameter to calculate the spindle speed (rpm):

$$f_r = n_t\, f_t\, N = (2)(0.23 \text{ mm/rev/tooth})(2232 \text{ rpm})$$
$$f_r = 1027 \text{ mm/min}$$

The material removal rate (MRR) is obtained from Equation 2.29 with a radial doc of 11 mm, about half the effective diameter of the cut.

$$\text{MRR} = dbf_r = (6 \text{ mm})(11 \text{ mm})(2232 \text{ mm/min}) = 147,312 \text{ mm}^3/\text{min}$$

Example 2.7 A face milling operation is being carried out on a 150 mm wide by 700 mm long rectangular part made of aluminum as shown in Figure 2.37. A 200 mm diameter face milling cutter with 18 inserts is used to cut a 3 mm depth from the surface of the part. The lead angle of the cutter is 45°. The cutting speed is 300 m/min and the allowable uncut chip thickness is 0.3 mm. Calculate the machining time.

Solution: The cutter diameter is selected to be larger by 20%–40% then the width of the part and runs off center from the centerline of the part, following rules discussed in Chapter 4. From the information given, we can calculate the spindle rotational speed or the rpm of the milling cutter from Equation 2.10. Thus,

$$N = V/(\pi D) = (300 \text{ m/min})(1000 \text{ mm/m})/(3.14)(200 \text{ mm}) = 478 \text{ rpm}$$

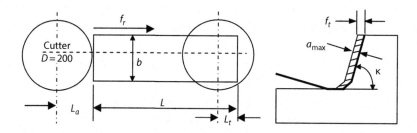

FIGURE 2.37 Schematic representation of Example 2.7.

The feed per tooth is affected by the lead angle of the cutter as illustrated in the earlier figure and explained in Equation 2.21, where $\kappa = 45°$.

$$f_t = a_i/(\cos(\kappa)) = 0.3/\cos(45°) = 0.42 \text{ mm}.$$

The feed rate for the cutter is given from Equation 2.20

$$f_r = n_t\, f_t\, N = (18)(0.42 \text{ mm/rev/tooth})(478 \text{ rpm}) = 3614 \text{ mm/min}$$

The cutting time is given by Equation 2.30. The approach and overtravel distances can be calculated from simple geometric relationship. The approach is a little larger than the radius of the cutter (see Equation 2.32) to avoid contact with the workpiece during the rapid travel motion. The overtravel distance varies depending on the surface requirements; there is a minimum negative distance (occurring when the cutter stops within the part) and a maximum distance if the surface texture is important. The selection depends on whether the trailing marks of the cutter (marts caused by a portion of the cutter's inserts that are not in the cut are rubbing on the surface already milled) should show up across the full length of the surface or not. These marks are avoided in some specially designed machines for milling with their spindle tilted slightly so that the cutter is slightly off parallel from the part surface.

$$L_e\big|_{\min} = L_a - L_t = \frac{D}{2} - \frac{1}{2}\left(\sqrt{D^2 - b^2}\right) = \frac{200}{2} - \frac{1}{2}\left(\sqrt{200^2 - 150^2}\right) = 34\,\text{mm}$$

$$L_e\big|_{\max} = L_a + L_t = \frac{D}{2} + \frac{D}{2} = \frac{200}{2} + \frac{200}{2} = 200\,\text{mm}$$

The cutting time is calculated from equation

$$t_m = (L + L_e)/f_r = (700 + 34 \text{ mm})/(3614 \text{ mm/min}) = 0.20 \text{ min} = 12 \text{ s}$$

Example 2.8 A 10 mm wide groove in a cast iron part is machined using a slot milling operation as shown in Figure 2.38. The groove is made across the full 400 mm length of the part. The depth of the slot is 15 mm. A 200 mm diameter carbide brazed side milling cutter with 16 straight teeth is

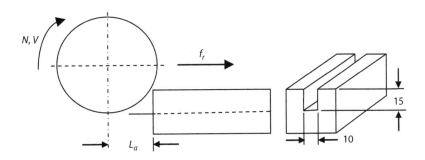

FIGURE 2.38 Schematic representation of Example 2.8.

selected. Each insert cuts the full width of the slot. The cutting speed is 80 m/min and the maximum uncut chip thickness is 0.2 mm for carbide tool material. Calculate

(a) The maximum feed rate (mm/min) possible if the machine spindle maximum available cutting power is 10 kW.
(b) The machining time to groove the part.
(c) Changes which would result if an indexable staggered tooth slotting cutter with 16 inserts is used.

Solution:

(a) The maximum feed rate can be obtained from the maximum allowable material removal rate calculated based on the available power. Estimating the specific energy for the workpiece material from Table 2.1 to be 0.07 W·min/cm³,

$$P = (\text{MRR})u_s \Rightarrow \text{MRR} = P/u_s = (10 \text{ kW})/(0.07 \text{ W}\cdot\text{min/cm}^3) = 142.9 \text{ cm}^3/\text{min}$$

The maximum MRR is given by Equation 2.29. In slotting, the radial depth of cut (d) is measured along the cutting diameter (the cutter radial portion engaged in the workpiece), whereas the axial width of cut (b) is the same as the width of the slot. Thus,

$$\text{MRR} = dbf_r \Rightarrow= (\text{MRR})/(db) = (142,900 \text{ mm}^3/min)/(15 \text{ mm})(10 \text{ mm})$$

$$f_r = 952 \text{ mm/min}$$

$$N = V/(\pi D) = (80 \text{ m/min})(1000 \text{ mm/m})/(3.14)(200 \text{ mm}) = 128 \text{ rpm}$$

The feed rate for the cutter is given from Equation 2.20:

$$f_t = f_r/(n_t N) = 952/(16)(128 \text{ rpm}) = 0.465 \text{ mm/rev/tooth}$$

However, the maximum uncut chip thickness is 0.2 mm and based on Equations 2.22 and 2.24, the maximum feed per tooth allowed for this particular cutter can be determined from

$$\cos v_{max} = 1 - [(2d)/D] = (R-d)/R = (100-15)/100 \Rightarrow n_{max} = 32°$$

$$a_{max} = f_t \sin v_m, \quad \text{hence } f_t = a_{max}/\sin v_m = 0.2/\sin(32°) = 0.377 \text{ mm/rev/tooth}.$$

Therefore, the feed per tooth of 0.38 mm/rev/tooth is selected. This feed requires 8.2 kW or 82% of the available power.

(b) The cutting time is given by Equation 2.30. The approach distance L_a is given by Equation 2.31, while the overtravel distance is zero (or very small, 1–2 mm).

$$t_m = (L + L_e)/f_r = (400 + 52.7 + 10 \text{ mm})/[(16)(0.38)(128 \text{ mm/min})] = 0.595 \text{ min} = 35.7 \text{ s}$$

A total of 10 mm for the approach and overtravel distance was added to avoid the tool hitting the workpiece during the rapid travel.

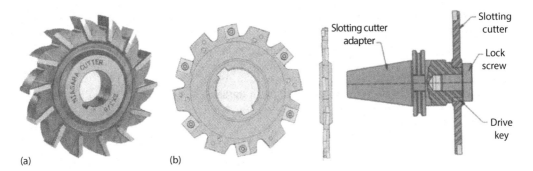

(a)   (b)

FIGURE 2.39 Comparison of slot milling cutters: (a) using an edge across the full width of the slot and (b) a staggered tooth cutter. (Courtesy of Kennametal and Carboloy, Latrobe, PA.)

(c) Figure 2.39 shows a comparison of the straight toothed carbide brazed cutter (a) with the staggered tooth cutter (b). If the cutter has 16 inserts in total, the effective number of teeth is 8 because 8 inserts on one side of the cutter and 8 overlaping inserts on the other side are used to remove the full width of 10 mm. In this case, the feed rate is half of that for the carbide brazed cutter with 16 effective teeth. Hence, $t_m = (462.7\ \text{mm})/[(8)\ (0.38)\ (128\ \text{mm/min})] = 1.19\ \text{min} = 71.3\ \text{s}$.

Example 2.9 The end face of a tube is face turned (a tube flange operation) on a lathe, using a depth of cut of 8 mm, and the inside and outside diameters of the flange are 100 and 200 mm, respectively. The tube is held with a hydraulic chuck at the other end as illustrated in Figure 2.40. The tube extends 500 mm from the lathe chuck and is made of aluminum. The maximum rotational speed for this application is 400 rpm, and the maximum feed for the particular tool geometry is 0.25 mm/rev. The side cutting-edge angle (lead angle) of the tool is 30°. Determine (a) the maximum power, (b) the cutting time, and (c) the perpendicularity of the flange (face) to the part axis.

Solution:

(a) The power required to perform a face turning operation is given by Equation 2.7. The MRR is calculated from Equation 2.6 with $f_t = 0.25$ mm/rev and doc = 8 mm. The cutting speed varies as a function of diameter and the maximum occurs at the larger workpiece diameter or at the beginning of the cut:

$$MRR = Vfd = \pi DNf_t d$$
$$= \pi(200\ \text{mm})(400\ \text{rpm})(0.25\ \text{mm/rev})(8\ \text{mm})$$
$$MRR = 502,400\ \text{mm}^3/\text{min} = 502\ \text{cm}^3/\text{min}$$

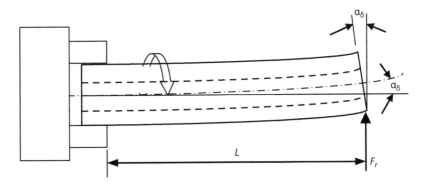

FIGURE 2.40 Schematic representation of Example 2.9.

The maximum power (P) required is

$$P = (\text{MRR})u_s = (502 \text{ cm}^3/\text{min})(0.025 \text{ W}\cdot\text{min}/\text{cm}^3) = 12.6 \text{ kW}$$

(b) The cutting time is given by Equation 2.5 with the length of cut equal to the thickness of the tube. Thus,

$$t_m = \frac{L}{f_r} = \frac{(D_o - D_i)}{2\cdot f\cdot N} = \frac{(200 - 100\,\text{mm})}{2(0.25\,\text{mm})(400\,\text{rpm})} = 0.5 \text{ min} = 30 \text{ s}$$

(c) The perpendicularity error is equivalent to the deflection of the tube under the cutting load (or the radial force). Generally, the radial force is less than 50% of the cutting force in roughing turning operations, although it can be equal to the cutting force in finishing operations. Therefore, let's assume the worse case scenario where the radial force is equal to the cutting (tangential) force. The cutting force is estimated from the power.

$$P = T_\omega = T(2\pi)N \Rightarrow T = P/(2\pi N) \quad \text{and} \quad F_c = 2T/D \quad \text{or} \quad F_c = P/(\pi N D)$$
$$F_c = (12,600 \text{ N m/s})/[\pi\times(400/60 \text{ rev/s})(200/1000 \text{ m})]$$
$$F_c = 3000 \text{ N}.$$

Thus,

$$F_r \leq F_c \quad \text{or} \quad F_r \leq 3000 \text{ N}.$$

The deflection at the tube at the end point (rigidly mounted to the machine spindle at the other end) is given by Equations 4.1 and 4.2:

$$\delta = \frac{F_r L^3}{3\cdot E\cdot I} = \frac{(3,000\,\text{N})(500^3\,\text{mm}^3)}{3(50,000\,\text{N/mm}^2)(73.594\times10^6\,\text{mm}^4)} = 0.034\,\text{mm}$$

$$I = \frac{\pi(D_o^4 - D_I^4)}{64} = \frac{\pi(200^4 - 100^4)}{64} = 73.594\times106\,\text{mm}^4$$

The face of the tube will exhibit taper (perpendicularity error with reference to the axis of the tube) if it deflects during cutting. The taper of the face angle is calculated from the deflection of the tube at the end of the face:

$$\tan a_\delta = \frac{x}{D_o} \Rightarrow x = D_o \tan a_\delta = D_o\left(\frac{\delta}{L}\right) = 200\left(\frac{0.034}{1000}\right) = 0.007\,\text{mm}$$

The taper or perpendicularity error is 0.007 mm.

Example 2.10 A slot 12 mm wide by 3 mm deep by 80 mm long is rough broached in a cast iron part. The broach has a step per tooth of 0.1 mm with 12.5 mm pitch. The cutting speed is 25 m/min. Determine (a) The number of teeth on the broach, (b) the length of the broach, (c) the machining (cutting) time, and (d) the cutting force.

Solution:

(a) Figure 2.18 shows the characteristics of a broach tool. The feed per tooth is equivalent to the step-per-tooth in a broaching operation because it determines the chip thickness. If $f_t = 0.1$ mm and $d = 3$ mm, the number of teeth is

$$n_t = d/f_t = 3/0.1 = 30 \text{ teeth.}$$

(b) The length of the broach is $L_B = pn_t = (12.5)(30) = 375$ mm
(c) The machining time is $t_m = (L_B + L_e)/V = (375 + 80)/(25 \times 1000) = 0.018$ min
(d) The workpiece is 80 mm long and the cutter has a pitch of 12.5 mm. Hence, the maximum number of teeth engage with the workpiece is

$$n_{te} = \text{int}\left[L/p\right] = \text{int}\left[80/12.5\right] = 7 \text{ teeth.}$$

The maximum depth engage of the broaching tool is $d_e = n_{te} f_t = (7)(0.1) = 0.7$ mm
The MRR is given by Equation 2.29,

$$\text{MRR} = d_e w\, V = (0.7)(12)(25)(1000 \text{ mm/m}) = 210,000 \text{ mm}^3/\text{min} = 210 \text{ cm}^3/\text{min.}$$

The force can be obtained from the power, $P = F_c V$. Hence,

$$F_c = P/V = \left(\text{MRR}\, u_s\right)/V$$
$$= \left(210 \text{ cm}^3/\text{min}\right)\left(0.07 \text{ kW min/cm}^3\right)\left(60 \text{ s/min}\right)/\left(25 \text{ m/min}\right) = 35.3 \text{ kN}$$

Example 2.11 A 100 mm diameter hole is threaded for the first 30 mm depth using a 25 mm diameter solid carbide four-flute thread-milling tool with three rows of teeth. The thread pitch is 3 mm. The allowable feed per tooth (chip load) in this case is 0.1 mm/rev, and the cutting speed is 100 m/min. Estimate the machining time.

Solution: The thread milling operation requires helical interpolation, which combines circular movement (interpolation shown in Figure 2.16) in one plane with a simultaneous linear motion perpendicular to the plane as shown in the Figure 2.41. Every tool orbit (on diameter D_x) represents a vertical movement of one pitch length.

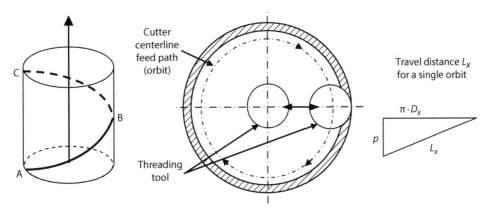

FIGURE 2.41   Schematic representation of Example 2.10.

The feed rate at the cutting edge (true feed rate) is given by Equation 2.52

$$f_{rt} = n_t\,f_t\,N = (4)(0.1\ \text{mm/rev/tooth})\big[(100)(1000)/(\pi\,25)\big](\text{rpm}) = 510\ \text{mm/min}$$

The centerline feed rate is given by Equation 2.55. The major diameter of the thread is estimated from Equation 2.40

$$\eta = (D_m - D_\mu)/(1.299\,p) \Rightarrow$$

$$D_m = \eta(1.299\,p) + D_\mu = 0.7(1.299 \times 3) + 100 = 102.7\ \text{mm}.$$

$$f_r = f_{rt}(D_m - D)/D_m = 510(102.7 - 25)/102.7 = 386\ \text{mm/min}$$

The total travel length for the tool is estimated by calculating the time to complete a single orbit (as illustrated in the above figure) during which three threads are cut and the number of orbits required to finish the full threaded depth in the hole. The length of a single orbit is

$$L_x = \sqrt{(p^2 + \pi^2 D_x^2)}, \quad D_x = D_m - D = 102.7 - 25 = 77.7\ \text{mm}, \quad \text{and} \quad L_x = 244\ \text{mm}$$

The number of orbits is

$$n_{orbits} = L/(3 \cdot p) = 30/[(3)\cdot(3)] = 3.33,$$

Hence, the number of orbits will be either three or four. The machining time is

$$t_m = L/f_r = n_{orbits}(L_x/f_r) = (3)(244/386) = 2.1\,\text{min}$$

This computed time is for cutting and does not include the approach and exit distances for each orbit.

Example 2.12 A 30 mm deep hole is being drilled in a block of magnesium alloy with a 10 mm carbide drill at a feed of 0.3 mm/rev and at 200 m/min cutting speed. The drill point angle is 120°. Calculate (a) material removal rate, (b) cutting time, and (c) torque on the drill (N m).

Solution:

(a) MRR = (cross-section-area)(feed-rate)

$$\text{MRR} = \big[p(10)^2/4\big](0.3\ \text{mm/rev})\big[(200\ \text{m/min})(1000\ \text{mm/m})/(\pi \cdot 10)\big] = 150\ \text{cm}^3/\text{min}$$

(b) The cutting time, is given by Equation 2.16, and let assume that the approach and over-travel distance for the drill is $\Delta L = 3$ mm

$$t_m = (L + L_e)/f_r = (L + D/\tan_p + \Delta L)/f_r$$

$$t_m = (30 + 10/\tan 60° + 3)/\big[(0.3\ \text{mm/rev})(6370\ \text{rpm})\big] = 0.02\ \text{min} = 1.2\ \text{s}$$

(c) The torque can be estimated from the power required to drill the hole.

$$P = (MRR)u_s = (150 \text{ cm}^3/\text{min}) (0.008 \text{ W} \cdot \text{min/cm}^3) = 1.2 \text{ kW}$$

$$P = T \omega = T (2 \pi N) \Rightarrow$$

$$T = P/(2 \pi N) = (1200 \text{ N m/s})(60 \text{ s/min})/(2\pi \cdot 6370 \text{ rpm}) = 1.8 \text{ N m}$$

Example 2.13 A vertical surface grinding operation is being carried out with a 150 mm diameter wheel at 2000 m/min cutting speed. The depth to be ground is 0.2 mm with a grinding depth of 0.01 mm per pass. The table traverse speed (feed rate of table or wheel) is 2200 mm/min. The length and width of the workpiece are 350 and 75 mm, respectively. Calculate the material removal rate and machining time.

Solution:

(a) The MRR for grinding is calculated using the same equations as for milling operations. In this surface grinding operation the width of the wheel covers the entire part width. The doc per pass is very small as is generally true in conventional grinding operations.

$$\text{MRR} = bdf_r = bdV_{work} = (75 \text{ mm})(0.01 \text{ mm/pass})(2200 \text{ mm/min})$$

$$= 1650 \text{ mm}^3/\text{min}$$

where d is the depth of the layer of material removed during one cutting pass (stroke).

(b) The time to travel across the part length (or perform one pass) is

$$t_{mp} = (L_p + D + L_a)/f_r = (350 + 150 + 40)/1650 = 0.33 \text{ min/pass}$$

where
L_p is the part length
D is the wheel diameter
L_a is the overtravel distance from both ends of the workpiece

The total time to grind the full depth is

$$t_m = \frac{(doc)_{total}}{d} t_{mp} + t_s = \frac{0.2 \text{ mm}}{0.01 \text{ mm}} 0.33 \text{ min/pass} + 0.33 = 6.93 \text{ min}$$

where t_s is the sparking-out time. In any grinding operation the doc during one cutting pass will initially be less than the nominal feed (d), in a direction normal to the work surface, setting on the machine. This infeed differential results from the deflection of the machine tool spindle, column, and workpiece under the forces generated during grinding.

Example 2.14 A 75 mm diameter cast iron bar is cylindrically ground using a 150 mm diameter wheel at 2000 m/min cutting speed. The wheel thickness is 25 mm. The depth to be ground is 0.2 mm with a grinding depth of 0.015 mm per pass. The wheel (table) traverse speed (feed rate) is 1500 mm/min. The length of the bar to be ground is 200 mm. Calculate the power required to grind and the machining time.

Solution:

(a) $\text{MRR} = \pi \cdot D_w df_r = \pi(75 \text{ mm})(0.015 \text{ mm/pass})(1500 \text{ mm/min})$

$$\text{MRR} = 5.3 \text{ cm}^3/\text{min}$$

The specific cutting energy for grinding should be much larger than that given in Table 2.1 based on turning tests. Therefore, the value of 0.08 kW·min/cm³ for cast iron from Table 2.1 should be multiplied by five times for grinding operations for a rough estimate. Hence, the estimated power required is

$$P = (\text{MRR})u_s = (5.3 \text{ cm}^3/\text{min})(5 \times 0.08 \text{ kW} \cdot \text{min/cm}^3) = 2.12 \text{ kW}$$

(b) The machining time is

$$t_m = \frac{(doc)_{total}}{d} t_{mp} + t_s = \frac{0.2 \text{ mm}}{0.015 \text{ mm}} 0.155 \text{ min/pass} + 2(0.155) = 2.38 \text{ min}$$

The time to travel across the part length (or perform one pass) is

$$t_{mp} = (L_p + W_w + L_a)/f_r = (200 + 25 + 8)/1500 = 0.155 \text{ min/pass}$$

2.14 PROBLEMS

Problem 2.1 A 1020 steel bar 150 mm in diameter is turned on an 12 kW lathe at 100 rpm and a 0.5 mm/rev feed. The lathe has a 90% efficiency. What is the maximum depth of cut that can be used with this operation?

Problem 2.2 Estimate the machining time required for rough turning an 800 mm long, 100 mm diameter annealed 4240 steel bar using a carbide tool. Estimate the time for a ceramic tool.

Problem 2.3 Estimate the machining time required for rough turning the O.D. and facing the end of a 0.6 m long, 100 mm diameter 1040 steel bar using a carbide tool. The depth of cut for both operations is 1 mm. The maximum cutting speed allowed is 70 m/min, with a feed of 0.25 mm/rev.

Problem 2.4 The OD of a 250 mm long steel bar is being reduced from 95 to 90 mm by turning on a lathe. The spindle rotates at 450 rpm, and the feed rate (axial speed) of the tool is 250 mm/min. Calculate the cutting speed, material removal rate, time of cut, power required, and the cutting force.

Problem 2.5 A shaft of 1040 steel is grooved to a final diameter of 80 mm from and initial diameter of 100 mm. The width of the groove is 5 mm. The maximum cutting speed for the cutting-tool material is 120 m/min with a feed of 0.25 mm/rev. Determine the chip area, material removal rate, power, and machining time.

Problem 2.6 A 200 mm long tapered shaft is generated from 1040 steel 80 mm round bar stock in a lathe. The small diameter is 40 mm. The taper is 10° included angle. The maximum depth-of-cut for a roughing operation is 4 mm, while for finish cut it is 0.5–1 mm. The rapid travel feedrate is 12,000 mm/min, the rough feed is 0.35 mm/rev, and the finish feed is 0.1 mm/rev. The suggested cutting speed is 100 m/min. Determine the following:

(a) The number of passes from rough to finish part
(b) The total machining time.

Problem 2.7 One thousand gray cast iron bars 100 mm diameter and 300 mm long must be turned down to 65 mm diameter for 200 mm of their length. The surface finish and accuracy requirements are such that a heavy roughing cut (removing most of the material) followed by a light finishing cut is needed. The available maximum power of the lathe used for the roughing cut is 2.5 kW with an efficiency of 85%. The finish cut has a feed of 0.13 mm/rev, a cutting speed of 90 m/min, and maximum power. Calculate the total production time in hours for the batch of work. Assume that the time taken to return the tool to the beginning of the cut is 3 s, the tool index time is 1 s, and the time taken to load and unload a workpiece is 2 min.

Problem 2.8 A 1 m diameter 6061 aluminum disc with a 300 mm diameter hole in the center is fixtured on the table in a vertical boring machine for a facing operation. The cutting tool starts to cut at the outside diameter and is fed toward the center (along a radius). A constant spindle speed of 70 rpm is used, while the tool is fed at 0.25 mm/rev with a depth of cut is 6 mm. Calculate the following: (1) the machining time and (2) the power consumption at both the beginning and just before the end of the operation.

Problem 2.9 A hole is being drilled in a block of soft steel with a 12 mm drill at 100 m/min cutting speed. The feed is 0.3 mm/rev and the hole depth (not including the drill point) is 25 mm. A standard solid carbide drill is used with a 120° point angle. Calculate the power and torque required for drilling and the machining time.

Problem 2.10 Calculate the time to drill a cored hole through a cast iron workpiece with a solid carbide three-flute drill that has a point angle of 120°. The thickness of the part at the drilling location is 50 mm. A 30 mm diameter drill is used to enlarge a 15 mm hole. The recommended cutting speed for a carbide drill is 80 m/min at a feed of 0.12 mm/rev/flute. Also calculate the material removal rate (MRR) and the torque required at the drill point.

Problem 2.11 The hole generated in Example 2.10 is reamed with a 30.5 mm eight-flute reamer. The feed for the reamer is 0.1 mm/tooth. The recommended cutting speed is 70 m/min. Calculate the machining time and power required for this operation.

Problem 2.12 A peripheral (slab-) milling operation is being carried out on a 600 mm long, 50 mm wide steel block at a feed of 0.15 mm/tooth and depth of cut of 10 mm. The hardness of the steel is approximately 30 R_c. A 200 mm diameter staggered tooth side cutter with 20 inserts is used at 250 rpm. The number of effective teeth is only one fourth the total number of teeth in the cutter (see Example 2.8). Calculate the material removal rate and cutting time, and estimate the power and torque required by the machine tool spindle.

Problem 2.13 A face milling operation is being carried out on a 500 mm long by 50 mm wide rectangular block of stainless steel. A 200 mm diameter face milling cutter with 10 inserts is used to machine the top 2 mm on the surface of the workpiece. The cutting parameters are 300 mm/min feed rate and 200 rpm on the spindle. Calculate the material removal rate, cutting time, and feed per tooth, and estimate the power required.

Problem 2.14 A face milling operation is being performed on an 80 mm wide aluminum part with a 300 mm diameter cutter. The feed per tooth is 0.2 mm and the depth of cut 8 mm. The cutter center is offset 20 mm from the workpiece centerline. The cutter has 12 inserts.

(a) Determine the uncut chip thickness
(b) Indicate in a graph the variation of uncut chip thickness on one tooth over two revolutions of the cutter.

Problem 2.15 Estimate the machining time required for tapping a hole at 20 mm deep with an M12 × 1 mm tap. A HSS cutting tap is used at 12 m/min cutting speed. What is the material removal rate?

Problem 2.16 A 100 mm diameter hole is threaded for the first 30 mm depth using a 25 mm diameter solid carbide four-flute thread milling tool with a single row of teeth. The thread pitch is 3 mm. The allowable feed per tooth (chip load) in this case is 0.1 mm/rev, while the cutting speed is 100 m/min. Estimate the machining time.

REFERENCES

1. M. Weck, *Handbook of Machine Tools*, Vols. 4, Wiley, New York, 1984.
2. T. Drozda, Ed., *Tool and Manufacturing Engineers Handbook*, Vol. 1: *Machining*, SME, Dearborn, MI, 1983.
3. A. Slocum, *Precision Machine Design*, Prentice Hall, Englewood Cliffs, NJ, 1992.
4. S. S. Heineman and G. W. Genevro, *Machine Tools—Processes and Applications*, Canfield Press, San Francisco, CA, 1979.
5. J. R. Davis, *Metals Handbook*, Vol. 16: *Machining*, 9th edn., ASM International, Metal Park, OH, 1989.
6. H. Tschätsch, *Werkzeugmaschinen der Spanlosen und Spanenden Formgebung*, 8th edn., Karl Hanser Verlag, Munich, Germany, 2003.
7. E. P. DeGarmo, *Materials and Processes in Manufacturing*, 5th edn., Macmillan, New York, 1979, Chs. 16–28.
8. E. J. A. Armarego and R. H. Brown, *The Machining of Metals*, Prentice Hall, New York, 1969.
9. G. Boothroyd and W. A. Knight, *Fundamentals of Machining and Machine Tools*, 3rd edn., CRC Press, Boca Raton, FL, 2005.
10. Anon., Sandvik Coromant Corporation, *Technical Guide*, Fair Lawn, NJ, 2010.
11. G. T. Smith, *Advanced Machining—The Handbook of Cutting Technology*, Springer Verlag, New York, 1989.
12. W. Konig et al., Machining of hard materials, *CIRP Ann.* **32** (1984) 417–427.
13. D. Stovicek, Hard part turning, *Tooling Prod.* **57**:2 (February 1992) 25–26.
14. C. Wick, Machining with PCBN tools, *Manuf. Eng.* **101**:1 (July 1988) 73–78.
15. Y. Matsumoto, Review of current hard turning technology, *Abrasives*, October/November 1996, 16–34.
16. R. Sood, C. Guo, and S. Malkin, Turning of hardened steels, *SME J. Manuf. Process.* **2** (2000) 187–193.
17. H. K. Tonshoff, H. G. Wobker, and D. Brandt, Tool wear and surface integrity in hard turning, *Prod. Eng.* **3**:1 (1996) 19–24.
18. T. G. Dawson and T. R. Kurfess, Hard turning, tool life, and surface quality, *Manuf. Eng.* **126**:4 (2001) 88–98.
19. H. J. Swinehart, *Gundrilling, Trepanning, and Deep Hole Machining*, ASTME, Dearborn, MI, 1967.
20. Machinability Data Center Technical Staff, Ed., *Machining Data Handbook*, 3rd edn., Machinability Data Center, Cincinnati, OH, 1980.
21. Sonobond Corp., Ultrasonically assisted gun drilling, *Cutting Tool Eng.* (September/October 1981) 26–27.
22. A. Feifer, Drilling micro-size deep holes, *Tooling Prod.* (October 1989) 58–60.
23. F. Koenigsberger and A. J. P. Sabberwal, Chip section and cutting force during the milling operation, *CIRP Ann.* **9** (1960) 197–203.
24. F. Koenigsberger and A. J. P. Sabberwal, An investigation into the cutting force pulsations during milling operations, *Int. J. MTDR* **1** (1961) 15.
25. O. W. Boston and C. E. Kraus, Elements of milling, *ASME Trans.* **54** (1932) 71, **56** (1934) 358.
26. M. Martellotti, Analysis of the milling process, *ASME Trans.* **63** (1941) 677, **67** (1945) 233.
27. G. Augsten and K. Schmid, Turning/turn-broaching—A new process for crankshaft and camshaft production, *Ind. Prod. Eng.* **15**:2 (1991) 36–43.
28. R. E. Roseliep, Advantages and applications in broaching, *Machining Source Book*, ASM International, Materials Park, OH, 1988, pp. 117–120.
29. L. J. Smith, Broaching: The fastest way to machine all kinds of external surfaces, *Machining Source Book*, ASM International, Materials Park, OH, 1988, pp. 121–122.
30. J. S. Agapiou, Evaluation of the effect of high speed machining on tapping, *ASME J. Eng. Ind.* **116** (1994) 457–462.
31. R. Price, Don't gamble on thread depths, *Cutting Tool Eng.* (April 1991) 40–44.
32. B. Norton, Updating tap selection, *Cutting Tool Eng.* (April 1992) 35–38.

33. D. W. Yen, Turning of precision threads on heat treated alloy steel, *Trans. NAMRI/SME* **19** (1991) 90–95.
34. F. Mason, New turns in thread cutting, *Am. Mach.* (November 1988) 137–144.
35. G. English, Thread milling: A look at the basics, *Machining Source Book*, ASM International, Materials Park, OH, 1988, pp. 123–127.
36. K. B. Lewis and W. F. Schleicher, *The Grinding Wheel*, 3rd edn., The Grinding Wheel Institute, Cleveland, OH, 1976.
37. F. H. Colvin and F. A. Stanley, *Grinding Practice*, 3rd edn., McGraw-Hill, New York, 1950.
38. C. Andrew, T. D. Howes, and T. R. A. Pearce, *Creep Feed Grinding*, Holt Rinehart and Winston, London, U.K., 1985.
39. C. P. Bhateja and R. P. Lindsay, *Grinding: Theory, Techniques, and Troubleshooting*, SME, Dearborn, MI, 1982.
40. L. J. Coes, *Abrasives*, Springer-Verlag, New York, 1971.
41. R. L. McKee, *Machining With Abrasives*, Van Nostrand Reinhold, New York, 1982.
42. F. T. Farago, *Abrasive Methods Engineering*, Vols. 1 and 2, Industrial Press, New York, 1976 (Vol. 1) and 1980 (Vol. 2).
43. R. I. King, and R. S. Hahn, *Handbook of Modern Grinding Technology*, Chapman and Hall, New York, 1986.
44. S. Malkin, *Grinding Theory and Applications*, Ellis Horwood, Chichester, U.K., 1989.
45. G. C. Sen and A. Bhattacharyya, *Principles of Metal Cutting*, 2nd edn., New Center Book Agency, Calcutta, India, 1969.
46. M. C. Shaw, *Metal Cutting Principles*, Oxford University Press, Oxford, U.K., 1984.
47. N. H. Cook, *Manufacturing Analysis*, Addison-Wesley, Reading, MA, 1966.
48. J. Kaczmarek, *Principles of Machining by Cutting, Abrasion, and Erosion*, Peter Peregrinus, Stevenage, U.K., 1976.
49. W. F. Jessup, Centerless grinding, *Handbook of Modern Grinding Technology*, Chapman and Hall, New York, 1986, Ch. 8.
50. S. C. Salmon, Creep-feed grinding, *Handbook of Modern Grinding Technology*, Chapman and Hall, New York, 1986, Ch. 12.
51. R. J. Fisher, Superabrasives and creep-feed grinding, *Machining Source Book*, ASM International, Materials Park, OH, 1988, 178.
52. L. P. Tarasov, Grindability of tool steels, *ASME Trans.* **43** (1951) 1144.
53. H. K. Tonshoff and T. Grabner, Cylindrical and profile grinding with boron nitride wheels, *Proceedings of Fifth International Conference on Production Engineering*, JSPE, Tokyo, Japan, 1984, p. 326.
54. D. W. Bouchard, Single pass superabrasive bore finishing, *Superabrasives '85 Proceedings*, Chicago, IL, 1985, pp. 5-37–5-41.
55. H. Fischer, Honing, *Handbook of Modern Grinding Technology*, Chapman and Hall, New York, 1986, Ch. 13.
56. T. Ueda and A. Yamamoto, An analytical investigation of honing mechanism, *ASME J. Eng. Ind.* **106** (1984) 237–241.
57. A. Yamamoto and T. Ueda, Honing conditions for effective use of diamond and CBN sticks, *ASME J. Eng. Ind.* **109** (1987) 179–184.
58. E. Salje, H. Mohlem, and M. von See, Comparison of grinding and honing process, *SME Manuf. Technol. Rev.* (1986) 649–653.
59. E. Salje and M. von See, Process-optimization in honing, *CIRP Ann.* **36** (1987) 235–239.
60. L. Jongchan, Fundamental study of honing, PhD dissertation, Department of Mechanical Engineering, University of Massachusetts, Cambridge, MA, 1991.
61. J. Gutowski, Honing of carbides and steel with diamond and CBN, *Superabrasives '85 Proceedings*, Chicago, IL, 1985, pp. 5-26–5-36.
62. F. J. Hettes, Which brush should you choose?, *Manuf. Eng.* (September 1991).
63. C. R. Nichols, External honing, microfinish, superfinishing, superhoning, *International Honning Clinic*, SME, Dearborn, MI, April 1992.
64. J. Lee, E. E. Wasserbaech, and C. H. Shen, Camshaft grinding using coated abrasive belts, *Trans. NAMRI/SME* **21** (1993) 215–222.
65. W. Konig, H. Tonshoff, J. Fromlowitz, and P. Dennis, Belt grinding, *CIRP Ann.* **35** (1986) 487–494.
66. W. J. Westerman, An overview of roller burnishing as a surface conditioning, SME Technical Paper MR810401, 1981.

67. J. A. Balcom, Bore to burnisher—Robot to robot, SME Technical paper MS830438, 1983.
68. N. H. Loh, S. C. Tam, and S. Miyazawa, Ball burnishing of tool steel, *Precis. Eng.* **15**:2 (1993) 100–105.
69. N. H. Loh and S. C. Tam, Effect of ball burnishing parameters on surface finish—A literature survey and discussion, *Precis. Eng.* **10** (1988) 215–220.
70. N. H. Loh, S. C. Tam, and S. Miyazawa, A study of the effects of ball burnishing parameters on surface roughness using factorial design, *J. Mech. Working Technol.* **18** (1989) 53–61.
71. W. J. Westerman, Salvage by knurling and burnishing, *Tooling Prod.* (January 1987) 60–61.
72. L. K. Gillespie, *Advances in Deburring*, SME, Dearborn, MI, 1978.
73. L. K. Gillespie, *Deburring Technology for Improved Manufacturing*, SME, Dearborn, MI, 1981.
74. A. F. Scheider and J. P. Gaser, Advanced flexible abrasive finishing tool technology, SME Technical Paper MR91-121, 1991.
75. R. Frazier, Deburring update utilizing flexible radial wheel type processes, SME Technical Paper MR91-122, 1991.
76. R. A. Tollerud, Deburring and surface conditions using nonwoven abrasives, SME Technical Paper MR91-124, 1991.
77. C. Van Sickle ansd G. Flores, Mechanical deburring and honing in the automated environment, SME Technical Paper MR91-126, 1991.
78. J. H. Indge, Edge round with flexible bonded abrasives, SME Technical Paper MR91-128, 1991.
79. J. C. Aurich, D. Dornfeld, P. J. Arrazola, V. Franke, L. Leitz, and S. Min, Burrs—Analysis, control and removal, *CIRP Ann.* **58** (2009) 519–542.
80. S. Kawamura and J. Yamakawa, Formation and growing up process of grinding burrs, *Bull. Jpn. Soc. Precis. Eng.* **23**:3 (1989) 194–199.
81. J. C. Aurich and D. Dornfeld, Burrs—Analysis, control and removal, *Proceedings of CIRP International Conference on Burrs*, University of Kaiserslautern, Kaiserslautern, Germany, April 2–3, 2009.
82. L. K. Gillespie, *Robotic Deburring Handbook*, SME, Dearborn, MI, 1987.
83. E. G. Erickson, Automated robotic deburring produces quality components, *Automation*, March 1991, 50–51.
84. K. C. Lee, H. P. Huang, and S. S. Lu, Adaptive hybrid impedance force control of robotic deburring processes, *Proceedings of the 32nd International Symposium on Robotics*, TC5-3, Seoul, Korea, 2001, pp. 1–6.

3 Machine Tools

3.1 INTRODUCTION

In addition to the workpiece and its fixture, there are three major components of a machining system: the machine tool, the cutting tool, and the toolholder. This chapter describes machine tools used to perform the various operations discussed in Chapter 2. Cutting tools are described in Chapter 4, and toolholders and fixtures are discussed in Chapter 5.

There are three principal types of machine tools. *Conventional machine tools* are designed to perform one or several operations on a variety of parts. These tools were developed early in the industrial revolution (as discussed in Chapter 1) and are still found in every machine shop, where they are used for general purpose machining of small lots of parts and for repair work. Their capabilities have been greatly enhanced by the advent of numerical control, which became available on most machine tools during the 1970s. Conventional machine tools were described in connection with basic machining operations in Chapter 2 and will not be discussed further in this chapter.

Production machine tools also called *special purpose machines* are used in high-volume manufacturing systems to perform one or a sequence of operations repetitively. They can be adapted for more than one part of the same family, but the changes required to switch from one part to another are usually time-consuming and uneconomical. Larger machines of this type are generally composed of a series of simpler machines or mechanisms, which resemble conventional machine tools, integrated with an automated materials handling system. Because of their lack of flexibility and large capital costs, they are only used when thousands of identical parts are required. Production machine tools are described further in Section 3.2.

CNC machine tools (machining and turning centers) are numerically controlled machine tools, which move cutting tools along complicated paths, often involving simultaneous motions of multiple axes, according to a stored program. CNC machine tools are flexible and can produce very complex parts in quantity with consistent quality and repeatability. They have traditionally been used for low or medium production volumes but have found increasing use in automotive and other high-volume applications in the last 15 years. Scalable manufacturing systems are easily designed using CNC machines. They are described further in Section 3.3.

Following descriptions of the principle types of machine tools, common designs of basic machine elements, including machine tool structures, slides and guideways, axis drives, spindles, coolant systems, tool changers, and pallets, are described in Sections 3.4 through 3.10. Only brief descriptions intended to acquaint end users of machine tools with the requirements of each element and the designs currently available are given. More detailed descriptions of each component are available in books on machine tool design [1–6]. Energy use in CNC machines is discussed in Section 3.11.

3.2 PRODUCTION MACHINE TOOLS

Most parts have a number of machined features and therefore must be produced on a group of machine tools arranged in a system. For parts required in high volumes, a number of specially designed machine tools may be grouped on a common base or connected by automated materials handling, coolant delivery, and swarf handling equipment to form a dedicated machining system. It is common to call these systems production machine tools, although in reality they are composed of a grouping of simpler tools or stations.

The most common types of production machine tools feature automated part transfer between stations and are called transfer machines. There are two basic classes of such machines: *rotary transfer machines* and *in-line* or *conventional transfer machines* [7–10].

A common type of rotary transfer machine is the *rotary indexing system*, in which parts are mounted on a horizontal table or dial and transferred by rotation through various machining stations arranged around the table (Figure 3.1a). Stations are configured to perform specific operations repeatability and are often numerically controlled. A trunnion or vertical drum can be used instead of a table if the parts require machining on opposite sides or at compound angles. In a *dial system*, rotations are of a standard length. Dial systems provide two access planes and a limited number of stations and cannot be expanded by adding additional stations. The *European* rotary type provides three access planes. A variation of the dial system is the *prism system* composed of fixtures that advance in a horizontal plane between workstations. Prism systems allow the rotation of individual pallets so that parts can be machined on multiple surfaces.

Center column systems (Figure 3.1b) are similar to rotary indexing systems but have machining stations mounted on a central column. In this base configuration, they have a small footprint relative to the volume of material they remove. They can accommodate more operations than conventional rotary indexing systems if additional spindles or slides are mounted around the periphery; in this configuration, they provide three access planes. They are more difficult to maintain than rotary indexing systems since mechanisms in the center column are often difficult to access, and in addition, they often have top-mounted motors. Generally, rotary indexing systems are used for small or light parts, while center column machines are favored for heavy parts or parts with greater machined content.

Conventional or in-line transfer machines (Figure 3.2) are dedicated systems designed to produce a single part in large volumes (e.g., >25,000/year). Since their output is high, the investment cost per part is relatively low, especially if the part is produced for several years. They consist of stations connected by an automated part transfer system. There are three conventional types of part transfer: sliding transfer, palletized transfer, and walking-beam or lift-and-carry transfer. In sliding transfer (Figure 3.3a), the part is moved on rails or rollers by an indexing bar and is located and clamped at each station in turn. Sliding transfer is used especially for heavy iron or steel parts, such as engine blocks and motor housings. In palletized transfer (Figure 3.3b), the part is located and clamped on a traveling pallet, which is moved between stations by an indexing bar or moving chain. This method requires more investment and (frequently) floor space since some provision for pallet return must be made but is preferred for relatively compliant parts such as aluminum transmission cases. In lift-and-carry transport, the part is moved by a gantry or linkage; this method is particularly suited to small or irregularly shaped parts such as connecting rods and crankshafts. Other parts of a transfer machine include the center bed, which contains the transfer and swarf handling mechanisms, wing bases, tool driving heads, and load/unload equipment (Figure 3.4). Types of tool driving heads include milling heads, drilling and boring heads, and multispindle drilling heads as described in older books [8–12]. The dimensions and utility connections of these components have been standardized in both the United States and Europe so that machines can, in principle, be assembled modularly from basic components (Figure 3.5).

Conventional transfer machines are well suited for manufacturing products with long market cycle lives (greater than 5 years). They use well-established machine technologies but generally have little provision for adjustment. To operate economically, transfer machines require accurate preplanning of the product design and accurate forecasts of market demand. In addition to their lack of flexibility, which can lead to excessive new product lead times, a major disadvantage of conventional transfer machines is that they are serial systems, and thus do not run when any of the stations need repair or a tool change. Due to this limitation, large systems are often operating roughly 60% of the time. One strategy to improve machine utilization is to break large systems up into smaller subsystems with intermediate parts buffers, but this increases

(a)

(b)

FIGURE 3.1 Rotary transfer machines. (a) Rotary indexing system. (From Moss Group Automation, Annesley, Nottigham, U.K.) (b) Center column system. (After Charman, B.W., *Special Purpose Production Machines*, Crosby Lockwood & Son Ltd., London, U.K., 1968.)

FIGURE 3.2   Conventional in-line transfer line. (Courtesy of MAG Automotive, Sterling Heights, MI.)

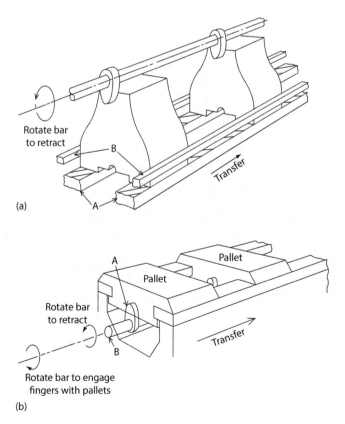

FIGURE 3.3   Automated part transfer system used in transfer lines: (a) sliding transfer and (b) palletized transfer. (After Lloyd, E.D., *Transfer and Unit Machines*, Industrial Press, New York, 1969.)

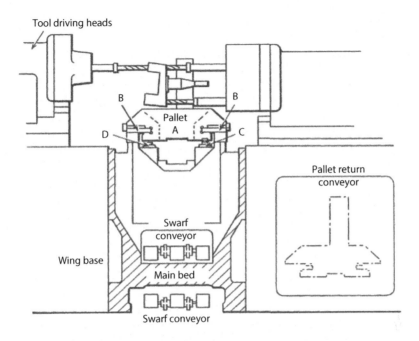

FIGURE 3.4 Cross section of a palletized transfer line showing the pallet, bed, wing bases, tool driving heads, swarf conveyor, and pallet return method. (After Lloyd, E.D., *Transfer and Unit Machines*, Industrial Press, New York, 1969.)

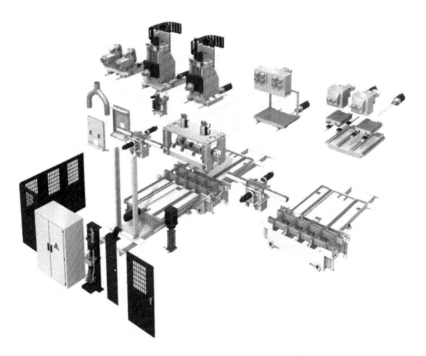

FIGURE 3.5 Basic modular elements for transfer systems, with decentralized electrical control, decentralized fluid power, and perimeter guarding. (Courtesy of MAG Automotive, Sterling Heights, MI.)

in-process inventory and may reduce quality and increase repair costs if out-of-tolerance conditions are not noticed promptly in any subsection.

In recent years, a number of technologies have been applied to in-line automated production systems to improve flexibility and dimensional capability. A *convertible transfer line* operates like a dedicated line for one part but is designed to accept a defined range or family of parts (e.g., iron and aluminum versions of an engine block, or six and eight cylinder versions of an engine head). A convertible system allows flexibility in producing parts within a family based on changing demand with minimal loss of production time. Convertible systems can typically be switched over from one part to another in a few days, compared to several months for a conventional system (assuming the conversion is even feasible). Convertible transfer machines are designed for part volumes similar to those for conventional transfer machines but have lower part life cycles times. They may also run a mix of part types simultaneously, with volumes of each part adjusted based on market demand. A convertible system requires additional initial investment and accurate preplanning. It is especially desirable to use a common locating scheme for all parts to minimize fixture rework and to standardize hole diameters so that common tooling can be used.

A *flexible transfer line* (FTL) is a production system designed for high-volume production, which is capable of producing a family of similar parts with unplanned changes or additional machined content. The life cycle of individual parts can vary from a few to several years as long as there is sufficient flexibility to fully utilize the system for 10–15 years. Such systems allow new products of the same family to be introduced quickly without major retooling. The changeover time between different products is usually a few hours, depending on the number of workstations involved and the available flexibility. Flexible transfer machines are well suited to applications in which a few similar parts are required in high volumes (e.g., >50,000/year), as is often the case in automotive powertrain and component production. Flexible transfer lines require a significant initial investment premium compared to conventional transfer machines and still require accurate part planning and market forecasts to operate economically. Currently, flexibility is commonly accomplished by using machining stations with indexable heads (turrets) or shuttle heads, each fitted with a number of different tools. CNC machines have also been used in FTLs, making them similar to the agile production systems described in the next section, although machine layout (serial vs. parallel) is generally different from that in an agile cell. In addition to requiring a significant premium in initial investment, flexible systems increase machine structural and fixture complexity and often required tooling and gaging inventories.

Product design is much more critical for FTLs than for conventional lines. Part features should be grouped and commonized so that a large number of features can be machined with single spindles. This means, for example, that the holes in the part should all have the same diameter so that they can be drilled using the same tool to reduce the number of tool changes [13]. Other methods for increasing the production rate of an FTL include the use of multiple independent spindles for machining to minimize machining time or the use of multiple spindles with multiple part loading (e.g., twin spindles with dual part loading).

Automatic lathes, such as screw machines, bar chuckers, drum- and Swiss-style automatics, and vertical turret lathes comprise another class of production machine tools still encountered in older operations. These machines, which are described in detail in the literature [2,9,10], have been replaced in many recent applications by the CNC turning centers described in the next section. CNC Swiss machines are still widely used in the manufacturing of small parts.

3.3 CNC MACHINE TOOLS AND CNC-BASED MANUFACTURING SYSTEMS

3.3.1 GENERAL

Most machine tools built since 1980 are numerically controlled machines. The basic components of an NC control system are the program of instructions (also called the part program or G-code), the controller itself, servo drives for each axis of motion, and feedback devices for each axis of motion [14,15]. The controller commands the servo drives to move the machine axes to drive the tool along the tool path specified in the part program. Both linear and rotary motions can be precisely controlled simultaneously.

NC systems can be classified by the method used to control machine slides, the number of axes, positional information, the feedback mode, the interpolation method, or the data format. Machining centers with five, seven, or more axes, which can generate very complex surfaces that cannot be produced with conventional machines, are available. Naming conventions for axes and common structural configurations are described as follows. The methods available for controlling the relative motion of axes are the following: (1) point-to-point (usually for two-axis machines such as drilling machines with single- or dual-axis control); (2) straight cut (control along a path parallel to a linear or circular machine way); and (3) continuous path or contouring (continuous control along a path in two or more axes). These three types require an increasing level of control sophistication. The positional information is either absolute (predetermined or fixed datum) or incremental (referenced from the current position).

Absolute control systems are closed-loop systems, which rely on angular encoders or linear displacement encoders to determine absolute axis positions; feedback from the encoders is continuously compared to a reference value by a microprocessor, which adjusts the slide speed to eliminate deviation between the absolute position and the reference value. Incremental control systems can be either closed- or open-loop systems. A closed-loop incremental control system uses an incremental displacement encoder, which produces pulses corresponding to the smallest measurable unit of displacement, which are counted by the microprocessor and compared to the reference distance. The slide is stopped when the counter reaches the reference distance. An open-loop incremental control system operates without feedback; a stepping motor is used as an actuator that receives the number of pulses corresponding to a specified displacement directly from the controller. If power is lost, the operation may be resumed without re-zeroing with an absolute control system; re-zeroing is required with an incremental system. An open-loop control system has better dynamic characteristics than a closed-loop system but does not provide positional verification. Closed-loop systems with dynamic error compensation are required for high-speed contour milling. The communication rate for each individual axis processor for the servo interface can be the determining factor for the maximum cutting rate.

Historically, all controllers were proprietary or closed-architecture computers. Although proprietary controllers still exist, there has been a recent trend toward open-architecture controls in which control logic software can be run on a number of platforms, including standard Windows-based PCs. Often, open control means using PC front ends and interfaces to proprietary machine tool controls that have full connectivity via standard networking and communications protocols. Open controls are enabling machine tools to take advantage of the latest software, networking, and operating system technologies, and to communicate with open-architecture factory automation software, resulting in more flexibility. More detailed descriptions of these components and NC systems are available in the literature [1,2,5,16–22]. Regardless of whether the controller has an open or closed architecture, in most contemporary machines it communicates with ancillary equipment controls or an on-board programmable logic controller (PLC).

3.3.2 Types of CNC Machines

There are two general classes of CNC machine tools: *CNC lathes* or *turning centers*, which perform turning, boring, facing, threading, profiling, and cutoff operations, and *CNC machining centers*, which are used primarily for milling, boring, drilling, and tapping.

Figure 3.6 shows a typical CNC turning center. Figure 3.7 shows the typical interior structures. In most CNC lathes, tools are held in a turret, which rotates to bring a specific tool to bear. The turret is commonly mounted on a slant bed slide, which sheds chips into the bottom of the machine. All CNC lathes have a headstock with spindle, and most have a tailstock. Turning centers usually used quick-change modular tooling. In addition to standard CNC turning centers, there are a number of advanced types with additional capabilities. *CNC automatics* are similar to turning centers

FIGURE 3.6 A CNC turning center with a main spindle and a subspindle for second-side capability, a rotating tooling turret, and a contouring *C*-axis. (Courtesy of Cincinnati Milacron, Cincinnati, OH.)

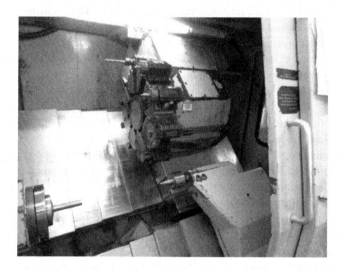

FIGURE 3.7 Interior view of slant bed CNC lathe showing turret, headstock, and tailstock.

but include more axes, rotating tooling (live tooling), and multiple slides and spindles. Figure 3.8 shows the interior of a CNC automatic with two spindles and two turrets. These machines are also called multifunctional machines or mill-turn machines when they are equipped with live spindles in the turret. On CNC automatics, a job can be divided into segments so that many tools can work on different areas of a workpiece simultaneously. Cycle times can thus be shorter than on CNC lathes, and idle time for part setup and handling is often reduced. They are especially useful for finishing smaller parts with limited machined content in a single setup. However, they often cannot be used to produce highly precise parts because they generally do not have fine controller resolution. As noted earlier, CNC lathes and automatics are modern equivalents of cam-driven screw and bar machines, which they have largely displaced.

Machining centers are usually classified by spindle orientation (vertical or horizontal) and the number of axes controlled. On a vertical spindle machine, the workpiece is mounted on a horizontal bed; on a horizontal spindle machine, the workpiece is usually mounted on a vertical fixture or table, which is more compliant. Vertical machines are preferred for large workpieces, flat parts, and especially for contoured surfaces in dies so that the thrust force is absorbed directly by the bed of the machine. Vertical gantry or bridge-type milling machines are used for very large workpieces because their

FIGURE 3.8   Interior view of CNC turning center with two turrets and two spindles.

two-column design gives greater stability to the cutting spindle(s). Horizontally configured machines are more versatile because four sides of the workpiece can be machined without re-fixturing if a rotary indexing worktable is available. Horizontals are finding increasing use in surface machining, since they provide increased access on larger complex parts and have less restriction on vertical height of the workpiece. Horizontal machines are preferred for untended use since they allow for easy chip and coolant evacuation. In North America, horizontal machines are often preferred in high-volume applications for increased ease and safety of maintenance. Horizontal machines are also often preferred for dry or minimum quantity lubrication (MQL) applications since they can be adapted to eliminate chip accumulations in the work zone more easily. Universal machines have heads that rotate to act as a horizontal or a vertical machine. The combination of tilts and swivels available in the spindles and tables allows the workpieces to be addressed at various compound angles.

Figure 3.9 shows the common nomenclature for axes and rotations. The primary axis directions on the machine are designated by the Cartesian coordinates X, Y, and Z. The corresponding rotary

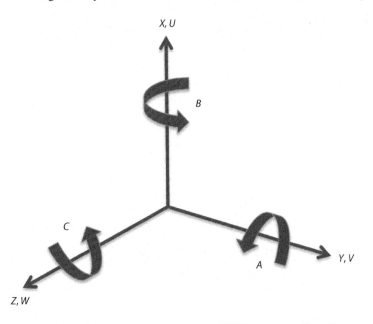

FIGURE 3.9   Axis nomenclature for CNC machines. The XYZ axes generally refer to machine column motions, while the UWV (or $X'Y'Z'$) denote motions of the table or other components.

axes are *A*, *B*, and *C*. Secondary linear axes aligned to *X*, *Y*, and *Z*, often on a table, are called *U*, *V*, and *W* or *X'*, *Y'*, and *Z'*. The *Z*-axis is commonly aligned with the spindle, whether the spindle is horizontal or vertical. On a horizontal machine, the *Z*-axis may correspond to motion of a spindle carried in a column, and the *W*-axis may refer to motion of a table toward a stationary spindle or the extension of a quill from a stationary spindle.

Conventional three-axis machines most commonly have a vertical spindle and three linear axes (*X*, *Y*, *Z*) but may have two linear and one rotational axis. Typical configurations are shown in Figures 3.10 and 3.11. Horizontal spindle three-axis machines are sometimes used for drilling, milling, and tapping large workpieces. Four-axis machines typically have three linear axes and a rotational axis on the work table. Horizontal spindle machines often have a *B*-axis table as shown in Figures 3.12 through 3.14. Horizontal spindle four-axis machines with *A*-axis tables or trunnions are used in dry and MQL machining applications since they permit the part to be machined upside down to clear chips by gravity (Figure 3.15). Vertical spindle four-axis machines commonly have an *A*-axis trunnion.

Five-axis machining centers are used for contour surface machining on components such as molds, dies, and airfoils and for positioning on workpieces requiring machining on multiple sides or at compound angles. Five-axis machining is essential for the first class of applications and often provides reduced cycle times and increased accuracy for the second. Five-axis machines have three linear and two rotational axes, with rotations being performed by a table, the spindle, or both. They are often built up by adding rotary axes to three-axes horizontal or vertical machines. Common configurations include a *B*-axis table mounted on an *A*-axis trunnion (*B* over *A*), shown in Figures 3.16 and 3.17, and a *C*-axis table mounted on a *B*-axis table (*C* over *B*), shown in Figure 3.18. These configurations are well suited to machining smaller parts, and the choice of a particular configuration depends on the workpiece dimensions and orientation, the required axis motions, fixturing, and the available base machine tools. Axes may also be added to the spindle, using a fork and swivel mechanism or a nutating head (Figures 3.19 and 3.20). This approach is common on large machines, since it is often not practical to precisely rotate large workpieces [23].

FIGURE 3.10   Solid base, fixed column vertical spindle three-axis machining center. (Courtesy of Cincinnati Milacron, Cincinnati, OH.)

FIGURE 3.11 Solid base, fixed column vertical spindle three-axis machining center with Z-axis travel (vertical movement on spindle head), X-axis travel (longitudinal movement of table), and Y-axis travel (cross movement of saddle). (Courtesy of Mori Seiki Co., Ltd., Nagoya, Japan.)

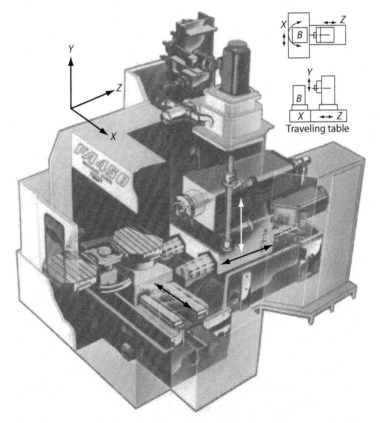

FIGURE 3.12 A horizontal spindle traveling column four-axis machining center with rotary table; Y- and Z-axis motions are performed by the column while the X-axis motion is provided by the table. (Courtesy of Toyoda Machinery USA, Inc., Arlington Heights, IL.)

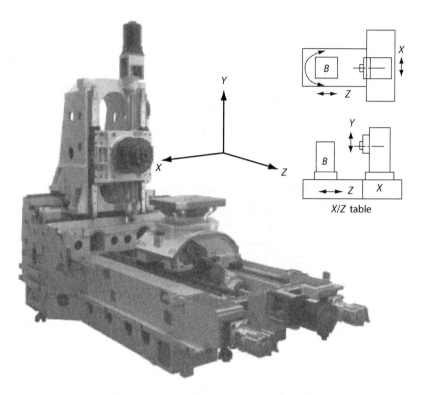

FIGURE 3.13 A horizontal spindle traveling column machining center with *X*-axis travel (longitudinal movement of the column), *Y*-axis travel (vertical movement of the spindle head), and *Z*-axis travel (cross movement of the table). (Courtesy of Mori Seiki Co., Ltd., Nagoya, Japan.)

FIGURE 3.14 A moving column high-speed four-axis machining center with the *Z*-axis on a horizontal ram and a *B*-axis table. (Courtesy of Ingersoll Milling Machine Co., Rockford, IL.)

FIGURE 3.15 Horizontal spindle four-axis machining center with *A*-axis table that can be inverted for MQL machining. (From GROB, Mindelheim, Germany.)

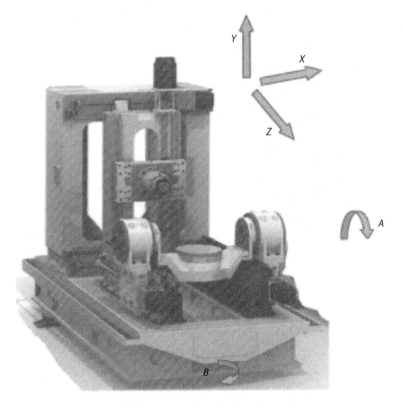

FIGURE 3.16 Horizontal spindle five-axis machining center with a *B*-axis table mounted on an *A*-axis trunnion. (Courtesy of MAG Automotive, Sterling Heights, MI.)

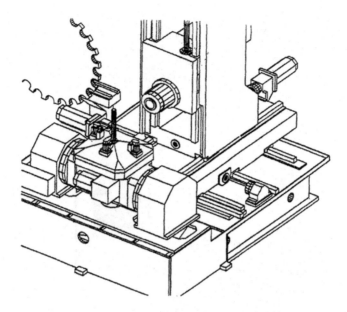

FIGURE 3.17 A five-axis machine using an integral tilt trunnion rotary table with 150° of motion on the *A*-axis (+30° to −120°) combined with a *B*-axis.

FIGURE 3.18 Five-axis bar machining center capable of five-sided machining with one clamping. The *Y*- and *Z*-axis motions are provided by the table while the *X*-axis motion is performed by the column. The workpiece carrier has a *B*-axis with a swiveling range of 270° and a *C*-axis with a swiveling range of 360°. (Courtesy of Hermle Machine Company LLC, Franklin, WI.)

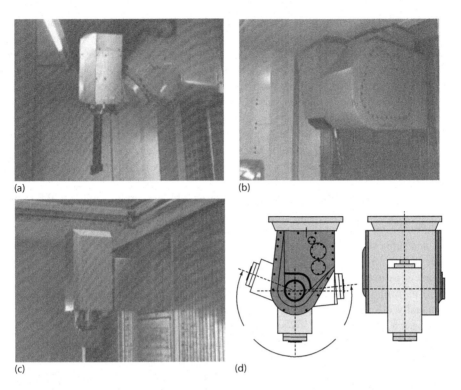

FIGURE 3.19 Spindle head styles for a five-axis system using a vertical spindle—fixed column machining center. (a) *B*-axis, (b) *B*-axis, (c) *A*-axis, and (d) Fork-head: *C*-axis and *A*-axis. (Courtesy of DMG Chicago, Inc., Hoffman Estates, IL, and Fidia Co., San Mauro Torinese, Italy.)

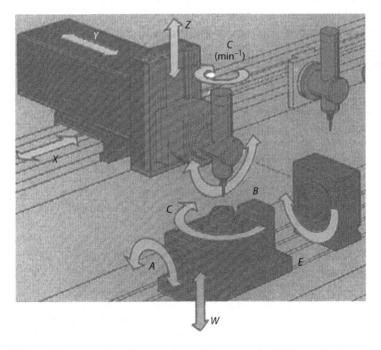

FIGURE 3.20 Schematic diagram of modular servo axes. The basic unit (*X,Y,Z*) carries the tool, the *B* and *C* axes are on the spindle, and *A/C* axes are on an indexing head with a vertical rotational axis, a horizontal swivel head, and a horizontal *E*. (Courtesy of Klinx Hochgeschwindigkeitsbearbeitung GmbH, HSC, Chemnitz, Germany.)

Hybrid machining centers combine the functions of turning centers and conventional machining centers, providing the ability to complete all machining operations for many classes of parts in a single setup. In one example (Figure 3.18), a tilting milling spindle is used to enable both horizontal and vertical machining operations as well as boring and milling on multiple faces. This type of machine can perform turning, milling, drilling, contouring with the C-axis, off-center machining with the Y-axis, milling of angled surfaces with the B-axis, grinding, and other operations. Such machines may be called multitasking turning centers because in addition to the traditional X- and Z-axes, they incorporate the Y-axis and rotary C- and B-axis for tilting the turret. As with the mill-turn machines discussed earlier, such machines can reduce cycle/lead times and work-in-process inventory, save up setup and queue time, and potentially improve part quality by eliminating refixturings.

The capabilities of machining centers are characterized by maximum spindle RPM, power, and torque versus speed curves, spindle size, and toolholder adaption (Chapter 5), axis drive motor power, rapid feed rate, fastest cutting federate, structural properties (stiffness, damping, etc.), workspace size, and support for networking.

High-speed machining centers (HSMC) operate at spindle speeds of 20,000–40,000 rpm, have high-acceleration/-deceleration (acc/dec) spindles (i.e., 1.5 s from 0 to 20,000 rpm), high-speed (>200 m/min) and high-acc/-dec (>2 g) slides, and high-speed control systems. High-speed machines usually offer lower torque at high speeds than conventional machines at lower speeds (i.e., in one case, a 30,000 rpm/20 kW specification yields 6.4 N m torque available at 30,000 rpm and 29 N m at 2,500 rpm), and therefore, the allowable depth of cut is often reduced at higher speeds. Some manufacturers offer higher torque, lower-speed spindles for hard metal machining as an option on base high-speed machines. HSMCs have tighter requirements with respect to machine tool characteristics, spindle, toolholder, and cutting tool interfaces and balance, vibration characteristics, and control systems. Toolholder and tool characteristics for HSMCs are further discussed in Chapters 4 and 5. Spindle, toolholder, and cutting tools have special requirements for coolant application to ensure the coolant is effective at the cutting zone. Dry machining is preferable when feasible in HSMCs to avoid high-pressure and high-volume coolant supply systems possibly generating unbalance at higher speeds. High-speed and high-precision machines require highly rigid feed drive systems and high-speed interpolation control.

In HSMCs, the machining time for one part feature may be significantly lower compared to a conventional machine. The machining time is typically one-third of the total time, with the remainder being used for machine travel, tool changes, spindle acceleration/deceleration, and pallet changes and rotations. As a result, high production rates are best achieved by reducing the noncutting time, and HSMCs permit reduction of noncutting time components due to their fast acceleration capabilities. Continuously raising the cutting speed proves cost-effective in just a few applications, such as aerospace applications in which parts are machined from billets or rough forgings.

As an example, a time study comparison of a standard CNC machine, a conventional (STD) HSMC, and an advanced HSMC in machining an aluminum automotive part is summarized in Table 3.1. The cutting time for the advanced HSMC is improved significantly compared to the STD HSMC because the process is changed somewhat so that different diameter holes are machined with a single end mill and multifunctional tools. The number of tools was reduced from 17 to 12. The positioning time was found to be the largest contributor to productivity improvement, with cutting time the second contributor.

Drive dynamics and the dynamic characteristics of the machine structure are also important in HSMCs. High-speed machining or five-axis machining requires control with look-ahead capability, acceleration and deceleration control, and collision detection. Look-ahead capability allows the CNC to read ahead a certain number of blocks in the program, in order to slow down the feed rate at anticipated sudden tool path direction changes. Nurbs interpolation has been used to interpolate

TABLE 3.1

Comparison of Very-High-Speed Machining for Manufacturing an Aluminum Part

Functions	STD CNC	STD HSM	Advanced HSM
Spindle (rpm)	12,000	15,000	20,000
Acceleration X-, Y-, Z-axes (g)	0.7	1.5	2
Rapid feed rate (m/min)	50	90	120
Reaches top slide speed in traveling (mm)	70	150	200
Table indexing time at 90° (s)	3.0	1.9	0.8
Table indexing time at 180° (s)	4.5	2.5	1.0
Table indexing time at 270° (s)	5.0	3.1	1.3
Tool changing time (ATC), tool-to-tool (s)	1.5	1.5	1.2
Tool changing time, chip-to-chip (s)	4.5	2.8	2.4
Number of tool used to process all features	17	17	12
Results			
Cutting time (s)	131	113	97
Positioning time (s)	130	74	55
ATC time (s)	25	25	17
Table indexing time (s)	23	14	3.6
Total time (s)	**309**	**226**	**173**
Productivity	**100%**	**137%**	**179%**

the tool path so that the control system can change direction along the curve more gradually using a high average feed rate.

CNC machines allow for operations to be combined using combination tooling since variable speeds and feeds are available. High-speed interpolation capability also increases tool flexibility and can lead to reduced tool counts.

Machine tools for large parts have special design requirements that differ substantially from those for traditional machine tools [23]. Such machines range from 2 mm to 10 m long or 1.3 to 6 m diameter. Major structural configurations for these machines are discussed in Section 3.4.

3.3.3 CNC-Based Manufacturing Systems

CNC machine tools can be grouped into serial or parallel production systems. In earlier practice, these systems were used for low to medium batch production, but they have recently become more common in mass production because they permit increased flexibility, allow for a diverse product mix with frequent design changes, and can be operated profitably at reduced part volumes.

Common types of CNC-based manufacturing systems include Cellular Manufacturing Systems and Flexible Manufacturing Systems. A *cellular manufacturing system* (CMS) is designed to produce a family of parts of similar shapes. Machine tools are arranged in cells, which may consist of one or several stations linked by a common control system. A cell can continue to function regardless of the state of the other cells and systems, as long as it has the necessary parts and tools available. A CMS is well suited for the multiproduct, small-lot-sized production requirements of a traditional machine shop, but they are also used for dedicated production of a family of parts. Various group technology concepts and schemes have been applied to manufacturing cells [24,25]. Considerable research on optimizing CMS operations to minimize intercell part flow requirements has also been carried out [26–28]. In designing a CMS, it is often useful to start with a single machine (see Figure 3.21) and add capability as required. Additional machines can be added later as long as they have standard interfaces, which allow additional software modules and material

FIGURE 3.21   A single station palletized CMS. (Courtesy of Cincinnati Milacron, Cincinnati, OH.)

handling devices to be considered. Open architecture in cell controllers has been the major contributor to the implementation of cellular systems.

A *flexible manufacturing system* (FMS) [29–39] consists of a number of flexible machining cells or CNC machines. In principle, an FMS can handle a variety of similar or dissimilar part designs and can enable new product designs to be introduced quickly without disturbing the production of other parts. An FMS should thus make possible improved machine utilization, part scheduling efficiency, and part quality, as well as reduced scrap, in-process inventory, and part setup and handling time.

FMSs require an investment premium over conventional dedicated systems, which limit their range of application. Many FMSs currently in operation have been specially designed for unique customer requirements and are used to produce a family of similar parts. They are finding increasing application in automotive powertrain machining since they can produce multiple variants simultaneously and can easily adapt to product design changes. In such high-volume applications, an FMS generally requires dedicated fixturing and materials handling equipment at significant expense. There are many technical challenges to be considered in developing increasingly useful FMSs, such as tool condition monitoring, chip control, machine diagnostics, adaptive control, automated tool and pallet handling, and flexible fixturing (Chapter 5).

An advanced type of FMS subject to recent research is a *reconfigurable manufacturing system* (RMS) [40–43]. In a conventional FMS, machine tools have fixed capabilities, with parts being routed to the necessary machines by the system automation software. It is difficult to adapt such systems to major part design changes, which require a different mix of machine capabilities. An RMS uses modular system components including reconfigurable machines and reconfigurable controllers, as well as methodologies for their systematic design and rapid ramp-up, which lead to reduced costs and times for retrofitting and conversion.

Another type of FMS is an *agile manufacturing system*, which conceptually can reallocate production line capacity to products that are in higher than expected demand, rapidly launch new products, and yet retain production ability for other products with lower expected demand [44–53]. The distinction between an agile and a flexible system is not rigorously defined in the literature, but it is generally understood that an agile system is a flexible system in which the machines can be rearranged and reused for a different product. In the broadest sense, agile manufacturing strategies are enterprise-wide, covering design, marketing, etc., in addition to production [54–57]. A number of CNC-based machining systems referred to as agile systems have been installed, primarily in automotive production [58–62] and also in munitions [63] and other sectors.

Typically cited advantages of agile systems include ease and speed of installation, scalability, reusability, and reconfigurability.

Scalability means that additional increments of production capacity (e.g., cells) can be added without disrupting ongoing production. Speed of installation (on the order of weeks, rather than months) is required to respond to market changes rapidly. Reconfigurability in this context implies that machines can be rearranged to make a different product without excessive changeover time but not necessarily reconfigured with different modular components. Reusability is required to minimize investment over time. Speed of installation, reconfigurability, and reusability are all facilitated through the use of standard machining centers as system building blocks.

Many agile systems in automotive production use four-axis machining centers with horizontal spindles and B-axis table rotation capability that allows multi-side machining as shown in Figure 3.22. (Other configurations, notably horizontal four- and five-axis A-axis machines, are also widely used.) This architecture is a good compromise between flexibility and cost. Horizontal spindles are preferred because they are comparatively easy to repair and maintain and because they simplify chip management, especially in dry or MQL machining. Four-axis capability is required for complex parts; for example, cylinder head machining requires B-axis capability to process angular intake

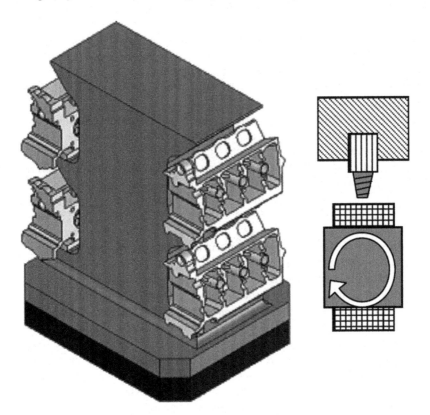

FIGURE 3.22 A tombstone fixture offers multi-side machining capability when used in a horizontal CNC machine with B-axis (table rotation).

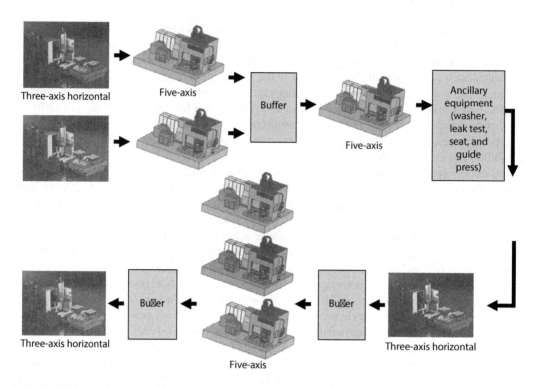

Three-axis horizontal   **Five-axis**   Buffer   **Five-axis**   Ancillary equipment (washer, leak test, seat, and guide press)

Three-axis horizontal   Buffer   Buffer   **Three-axis horizontal**

Five-axis

FIGURE 3.23   Agile cylinder head machining layout using three-axis and five-axis CNC machines.

and exhaust features without excessive refixturing as shown in 3.22. In this case, cylinder heads are located horizontally for multiple part fixturing to reduce tool change times and improve productivity. A second example is shown in Figure 3.23, in which three- and five-axis machines are used to provide the required agility at minimum cost. The same process using parallel processing with five-axis machines would require more machines but would be linearly scalable.

Flexibility, reconfigurability, and reuse of equipment have been demonstrated in existing systems. Systems that make more than one product with minimal changeover time have been in operation for many years, and machinery from some older systems has been redeployed to create new systems for new products. Production experience has, however, brought to light several difficulties not envisioned by early proponents of agile manufacturing, mainly related to high-volume production. Materials handling in these applications is often accomplished using fixed automation, such as gantries or roller conveyors. The cost and complexity of extending these systems, as well as floor space limitations, has hindered scalability in some cases. Many agile systems also necessarily incorporate relatively inflexible auxiliary equipment (such as parts washers) or dedicated stations to carry out specialized operations (e.g., cylinder hones).

Material handling between machines is an important component of FMS design [64,65]. Parts may be directly loaded into a fixture on the machine or may be transferred on a pallet. In direct load methods, parts are transferred between machines using sliding or free transfer (on a roller conveyor system), overhead crane systems, gantry transfer, or robots (Figures 3.24 through 3.26). They require relocation and reclamping of the part at each station. In a palletized system, parts are mounted on a pallet that is transferred between machines; parts are fixtured and clamped only once. Palletized systems are common for small-lot production (Figure 3.21) since they are flexible and permit multiple part loading. In large systems, a number of pallets are used, and their dimensional variations must be tracked and compensated for to avoid impacting part quality. Pallet systems used to shuttle parts in and out of the machining stations are better suited for horizontal than for vertical machines.

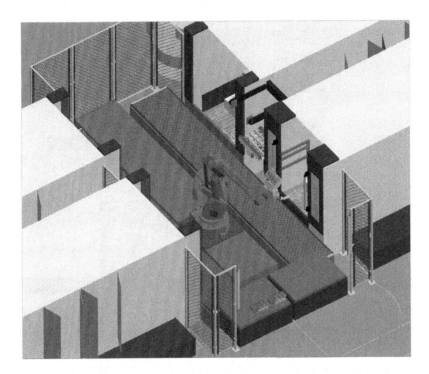

FIGURE 3.24   An agile machining cell with four machining centers and robotic part loading. (From General Motors Corp., Detroit, MI.)

Three approaches to materials handling are commonly used in large agile systems: conveyor loops, overhead gantries (with or without robot arms), and vehicular or cart transport.

When a *conveyor loop* is used, parts are loaded at a central loading or staging area and conveyed to machines on rollers or tracks (Figures 3.23 and 3.27). Part loading at the staging area may be done manually or by a robot, and parts may be loaded onto a pallet or directly onto the conveyor. Similarly, a pallet storage carousel with room for several pallets and an integrated workpiece transportation mechanism can be used with one or more CNC machines. Upon arrival at a machine, palletized parts are loaded directly, while loose parts are loaded either manually or by robot into the machine or onto an unused fixture in dual fixture setups. Conveyor loop systems are comparatively flexible, especially if manual loading is used, but conveyors must still be rebuilt if the machine configuration is changed. For systems that operate when some machines are down, a control system is required to route parts to operative but unused machines. Depending on the distances between machines, this can introduce uncertainty or excessive waiting in the system. The part routing control system should also store information on the sequence of machines visited by a particular part to facilitate troubleshooting when out-of-tolerance parts are detected. In palletized systems, this is facilitated through encoding chips on the pallets; in free transfer systems, a part marking and tracking system is required.

In a *gantry system* (Figures 3.25, 3.28, and 3.29), parts are conveyed between machines using overhead gantries. Parts are loaded into each machine through a safely door that separates the gantry from the machine tool area. A robot arm may be mounted on the gantry to manipulate parts for increased flexibility. This system allows for parts to be removed (e.g., for gaging) and reintroduced to the system. Part buffering stages can also be included in gantry systems. Gantry systems with associated control systems are well suited to simultaneous production of multiple variants. Gantry systems are fixed automation systems, which limit flexibility in part routing and machine layout changes. Complex gantry systems are expensive, often costing more than the machining centers themselves, and may also be subject to significant downtime. When operating properly, however,

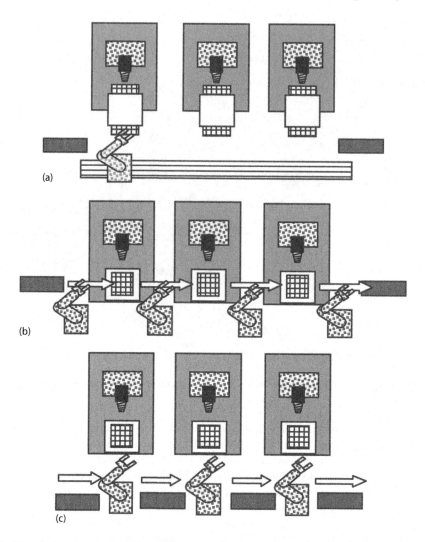

FIGURE 3.25 Various materials handling systems for agile/flexible machining. (a) A single rail-guided robot system for supporting one to three processes in a parallel machining system. (b) Serial machining system interlinked by either power roller conveyors or overhead gantries to create a cell. (c) Serial machining system interlinked with individual systems (robot, gantry, guided vehicle AGV, rail-guided transporter).

they limit misloading due to improper orientation and similar errors. Many automotive agile systems used gantry transport.

When *vehicular transport* is used, parts are conveyed in batches from a receiving or storage areas by vehicle to machines, where they are loaded manually or by robot. Vehicles used for transport include carts, fork trucks, and wire- or track-guided vehicles. Manual carts are flexible but require proper workforce training to limit overcyling and uncertainty. Fork trucks are also flexible but are often perceived as safety risks and avoided in North American plants. Guided vehicles may also be perceived as safety risks and traditionally have had lower reliability. Whatever transport method is chosen, the number of parts transported in each batch must be carefully chosen, often

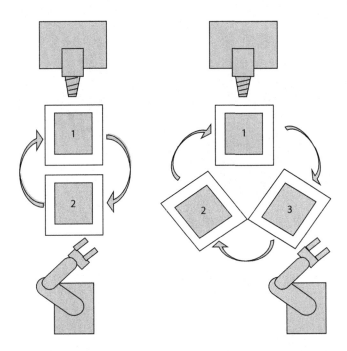

FIGURE 3.26 Agile/flexible system using a single machining center with index table for parts requiring multiple fixturing setups or machining various parts setup on different pallets.

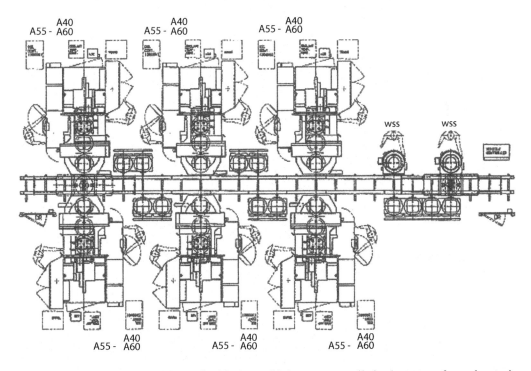

FIGURE 3.27 An agile machining cell with six machining centers, palletized part transfer, and a staging station for pallet loading. (From Makino, Mason, OH.)

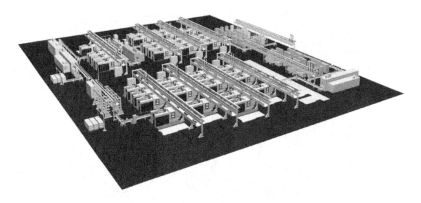

FIGURE 3.28   Agile system consisting of serial cells with overhead gantry loading. (From Halle GmbH, Edemissen, Germany.)

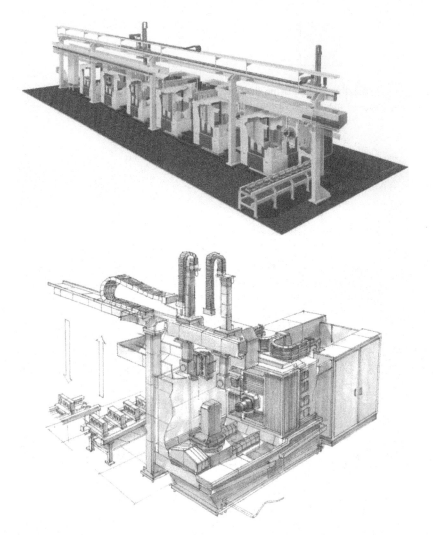

FIGURE 3.29   Agile cylinder head system based on six parallel machining centers, interlinked by an overhead gantry and power roller conveyors to form a cell. (Courtesy of Cross Huller, Eislingen, Germany.)

using discrete event simulation programs, to minimize machine starving and in-process inventory. Vehicular transport is the most flexible materials handling option but to date has seen relatively limited use due to reliability concerns. Recent developed automated vehicles that navigate via GPS systems rather than tracks or wires in the floor increase flexibility and show promise in increasing system reliability [66].

Lean machining systems represent a different philosophy in applying CNC and conventional machine tools to high-volume production [67,68]. A lean system is similar to a dedicated transfer line in that they employ one-piece flow. Part transfer between machines is manual; in a sense, the operator provides the flexibility lacking in fixed automation. Lean systems work best when they operate to a fixed takt time and have carefully defined work standards. They also typically employ pull systems with minimal in-process buffers. Conscientious machine maintenance and operator training are required for optimum results.

The design of a CMS or FMS is complex because several alternate configurations may be capable of producing a given part. For example, some configurations replacing long serial lines are multiple serial lines in parallel, parallel lines-with-crossover, hybrid, or agile configurations (Figure 3.30) [69,70]. A cell controller can direct an automated work handling system to deliver the part to any

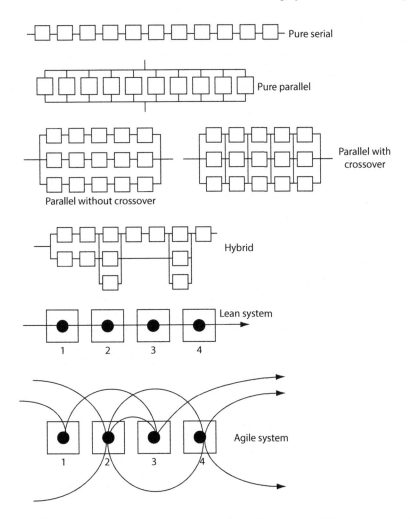

FIGURE 3.30 Differences between manufacturing system configurations and between lean and agile processes. The arrows illustrate the key distinction among the lean and agile systems (the part can take alternate routes in the agile process). Automotive agile systems often employ a parallel with crossover layout.

one of a number of interchangeable machines, with each machine performing all of the necessary machining for whatever workpiece it happens to see. Computer simulations are sometimes used to compare alternative designs. The input parameters for simulation include the part geometry, size, material, and tolerances, descriptions of each machining station, tool data, stand-alone availability, and mean times to failure and repair. In addition, constraints between operations such as precedence relations and contiguous allocations must be described. The process sequence can be specified or optimized. The available machine motions and tool data are used to determine the individual machine motions and cutting times. The simulation system should be capable of incorporating various machining centers, automatic pallet changers, material handling systems, buffers, and load/unload stations. A number of heuristic optimization algorithms have been proposed [71–80].

An important technical problem in developing easily integrated manufacturing systems is the lack of standards, which govern the type and form of information which manufacturing systems must exchange. In 1981, an Automated Manufacturing Research Facility (AMRF) was established to verify new concepts for automated manufacturing [81]. Later, the manufacturing systems integration (MSI) project at the National Institute for Standards and Technology (NIST) was created based on the AMRF technology for developing a system architecture that concentrated on the integration of manufacturing systems [82–84]. In 2007, the Association for Manufacturing Technology (AMT) initiated the development of the MTConnect standard, an open access standard based on XML (Extensible Markup Language) [85]. The first version was released in 2008. This flexible standard is used in a number of current applications [86,87]. The MTConnect Institute in McLean, Virginia releases and fosters further development of the standard. Machines often communicate with automation and auxiliary systems (such as MQL units) using PROFIBUS or PROFINET protocols.

3.4 MACHINE TOOL STRUCTURES

The machine tool structure supports the various parts of the machine tool, as well as the part and fixture, and provides rigidity to ensure accuracy. In general, a machine tool structure consists of a bed or base, column, ram, or saddle (carriage). Fixed components such as the base are most commonly made of cast iron, nodular iron, steel weldments, or composites with polymer, metal, or ceramic matrices. Castings must be aged and stress relieved. Reinforced or polymer concrete and epoxy granite are sometimes used for machines subjected to high levels of vibration; their application has been limited mainly to grinding machines due to warping, thermal gradients, and the absorption of coolant. Moving components are made of cast iron, steel, aluminum, and sheet metal. The design of moving columns is becoming increasingly critical as machine travel speeds increase, since the weight must be reduced to reduce inertia while maintaining high stiffness. Steel weldments can reduce the structural weight significantly but require careful design to resist vibration and deformation. Welded bases can be designed with high stiffness and good control of damping because welds in the bases block vibration transmission. Machines using weldment structures are often classified as light-duty machine tools unsuitable for rough cutting or precision applications. Metal-matrix composites, ceramics, and reinforced concrete materials are most often applied in precision or high-speed applications [88–94]. The machine damping obtained from the structural material itself can be improved significantly by internal means, including (1) filling structural cavities with oil, lead, sand, or concrete [1,2,6]; (2) circulating coolant through the machine structure; (3) allowing for microslip at the joints; and (4) attaching a viscous material layer between joints. Damping can also be improved by using shear plates, tuned mass dampers, viscous shear dampers, and active dampers [6,95–98]. Polymer or concrete-filled bases are preferred for grinding machines because they increase system damping and stability. The ideal machine configuration is application dependent as discussed in standards [99].

In designing the machine tool structure, system rigidity and inertia are the primary considerations. The machine structure must be rigid in order to resist deflections and vibrations due to cutting loads. The structural components should also be as light as possible to minimize the force required for acceleration or deceleration, to increase the maximum acceleration rate, to reduce jerk in machine motions, and to reduce stopping distances for increased machine accuracy. The damping characteristics of the machine structure are also important, as discussed earlier, because vibration energy is absorbed into the structure; as discussed in Chapter 12, in chatter theory the chatter limit depends on the product of modal stiffness and damping. The static and dynamic loads including forces of acceleration, deceleration, and cutting must be analyzed. The hardness and elasticity of the material must be balanced in order to withstand impact and allow elastic deformations while preventing cracking or permanent deformations. Thermal expansions and distortions of the machine frame due to external or internal heat sources must also be considered. A common current design approach is to start with an initial design based on experience, and refine this design through finite element analysis prior to machine building. Once a prototype is built, the design is further refined using experimental modal analysis and thermal mapping.

Machine tool structures can be broadly classified as open or closed as shown in Figure 3.31 [6]. In an open structure, the cutting force is supported by a structural loop consisting of the column, spindle housing, and table. This support arrangement puts a large bending moment on the column. In a closed or bridge-type machine, the load is supported by a balanced structural loop through two columns, which distributes the load and reduces the bending moments on both columns. Most general purpose CNC machine tools have open structures. Closed structures are common on larger machines, such as bridge or skin mills used in aerospace applications.

A conventional engine lathe (Figure 2.2) has a main slide mounted on a horizontal bed and a cross slide mounted on the main slide. Horizontal CNC turning centers generally have a similar structure, with the tool turret(s) mounted on a main and cross slides for axial and radial motion. The most common turret lathes are horizontal lathes using one or more turrets with several sides. The bed of a horizontal lathe can be horizontal, vertical, or slanted. The vertical and slant bed designs (Figures 3.7 and 3.8) reduce thermal distortions and floor-space requirements and simplify chip handling. The gravitational force on the workpiece is a concern in a horizontal lathe since it causes workpiece sag and uneven loading of the spindle bearings. Hence, vertical turret lathes are often used for heavy parts. Vertical lathes are becoming more common for smaller parts as well because they provide easier and more precise workpiece loading. Horizontals allow better chip removal for blind bores and more

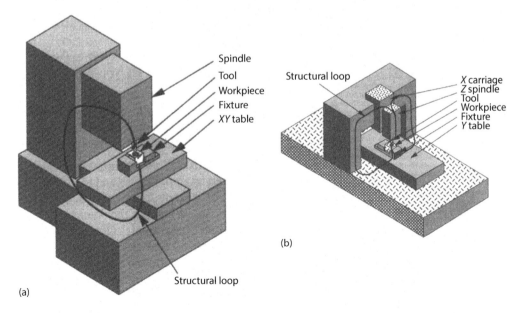

(b)

(a)

FIGURE 3.31 (a) Open machine tool structure. (b) Closed or bridge machine tool structure. (After Slocum, A., *Precision Machine Design*, Prentice Hall, Englewood Cliffs, NJ, 1992, chapters 4, 6, 7, 8, 9, and 10.)

access for long shaft-type workpieces. Hybrid turret *turn-mill* NC lathes have milling or machining center capabilities. For milling operations, the workpiece is held fixed or slowly rotated while rotating tools are brought to it. In these machines, rotary tools may be located on a separate slide or saddle that moves on the main ways or an auxiliary set, or rotary tools may be used in live spindles mounted in the turret itself. The powered turret design using both radially and axially mounted tools has been widely used. Multiple-spindle turning centers are also designed to perform operations on both ends of a part using either side-by-side or face-to-face machine configurations.

Cylindrical grinding machines are usually similar to a lathe with a grinding wheel feed mechanism substituted for the turret. For heavy (i.e., creep feed) or ultra-precision applications, special designs are required to achieve adequate rigidity and dynamic stiffness.

Typical conventional machining center structures are shown in Figures 3.11 through 3.13, 3.16, and 3.32 through 3.35; these figures illustrate traditional design variations in the base, column, slides, and spindle support. Most structures have a serial kinematic axis architecture, with each axis of movement supporting the following axis and providing its motion. Traveling column designs (Figures 3.14, 3.17, and 3.32) have all three axes located on the column, with low rigidity on the cross slide, resulting in significant spindle droop during boring. X/Z table horizontal-type or X/Y table vertical-type machines (Figure 3.11) eliminate spindle droop; in this configuration, the table is the weakest component. The traveling column design with a fixed table generally provides higher rigidity, while the traveling table design increases flexibility and reduces cost. The traveling column design also simplifies chip control somewhat since a centrally mounted chip conveyor can be used. However, traveling column designs restrict access to the workzone when the column is moved forward toward the pallet. A horizontal spindle machine is generally more accurate than a vertical spindle machine because the spindle is not cantilevered off a large C-shaped support structure, which is subject to greater deformation. The loads on a spindle create a bending moment on the column of a vertical machine in contrast to the point force on the column of a horizontal machine. The spindle can be mounted on a ram (quill), as shown in Figures 3.14 and 3.35, to reduce the weight carried by the Z slide located on the ram; this improves the traveling speed and acc/dec of the slide. The ram can also be mounted on rails, which increases stiffness.

FIGURE 3.32 The structural components of the machining center. (Courtesy of Toyoda Machinery USA, Inc., Arlington Heights, IL.)

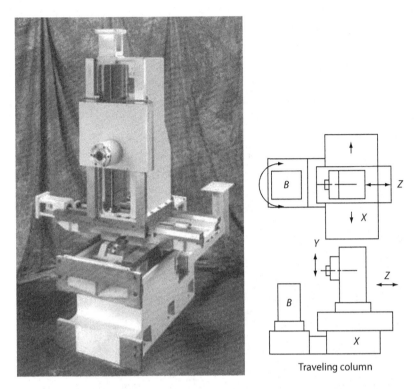

Traveling column

FIGURE 3.33 Structural components of a three- or four-axis horizontal spindle machining center with traveling column and either fixed or rotary table. All X-, Y-, and Z-axis motions are provided in the column. (Courtesy of Cellular Concepts, Detroit, MI.)

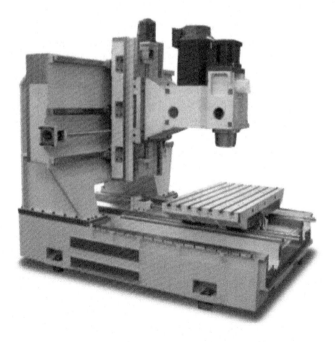

FIGURE 3.34 Typical vertical CNC machine tool structure, with heavy stationary base and column supporting lighter moving components. (Courtesy of MAG Automotive, Sterling Heights, MI.)

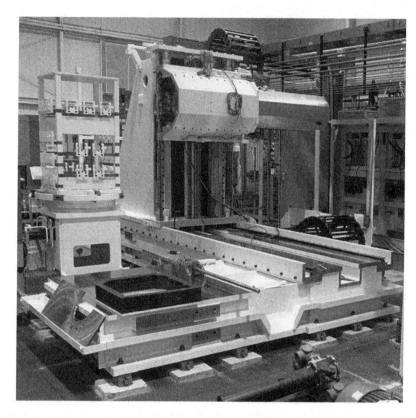

FIGURE 3.35 The structural components of the machining center in Figure 3.14, using magnetic linear motors for all axes. (Courtesy of Ingersoll Milling Machine Company, Rockford, IL.)

A fundamental drawback of serial architecture (one axis located on top of another) is that the structure of all axes must be heavy enough to provide the stiffness necessary to limit distortions that lower machine accuracy. This restricts dynamic performance and reduces operating flexibility.

Some new machine structures use a parallel kinematic-link mechanism (PKM) in place of rectangular-coordinate serial link mechanisms, eliminating the machine slides [100,101]. There are two classes of PKM structures: those having joints fixed on base and platform and extensible struts and those having movable joints with fixed length struts. Fully parallel machines are called hexapod or Stewart Platform (octahedral frame) machines [102]. The simple structure of the octahedral hexapod machine (shown in Figures 3.36 through 3.38) consists of a moveable spindle platform connected to a rigid base through six variable length links (telescoping legs/struts providing six degree-of-freedom [DOF] workspace) that control the position and orientation of the platform. The struts are connected to the platform and to the base using universal, spherical, or wobbling joints. Expanding and contracting the corresponding number of ball screws control the attitude and position of the spindle. Additional newer configurations not pictured include the open frame (G type), closed frame (portal type), tetrahedral (Lindsey's Tetraform), and spherical (NIST's M3) [6]. The interest of PKMs is largely due to their great flexibility in production using more than 3DOF (resulting in more than three-axis machining and in some cases full five-sided machining) and the force-to-weight ratio characteristics of these machines (potential for high dynamic capabilities). Some of the conceptual advantages for PKM relative to conventional machine tools are higher stiffness-to-mass ratio (due to closed kinematic loops and because struts are inherently stiff and light in weight), higher speeds, higher accuracy, modular design (reduced installation requirements), and mechanical simplicity (e.g., linear drives used for rotary movements) [103,104].

FIGURE 3.36 The structural configuration of the octahedral hexapod machining system (a 6DOF PKM). (Courtesy of Ingersoll Milling Machine Company, Rockford, IL.)

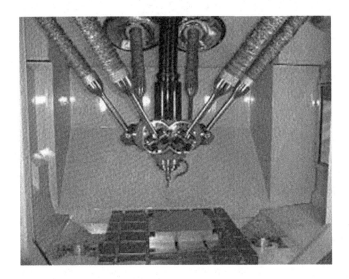

FIGURE 3.37 The structural configuration of the hexapod machining system. (A six-axis control "Cosmo Center PM-600," courtesy of Okuma America Corporation.)

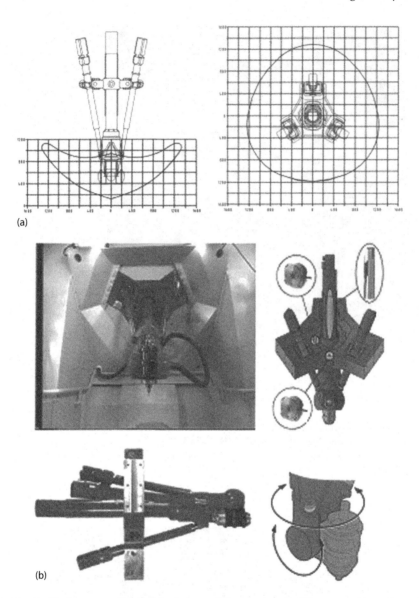

FIGURE 3.38 (a) Workspace sections of the Tricept PKM system. (b) The structural configuration of the Tricept machining system (5DOF hybrid using a 3DOF tripod and two rotary axes). (Courtesy of Tricept Inc., Charlotte, NC.)

These structures provide accelerations ranging from 1 to 3 g and feed rates up to 100 m/min. The limiting factors for hexapods are [105] (1) poor ratio of system size to workspace and (2) requirement of 6DOF passive joints having high stiffness and minimal backlash.

Structures with fixed length struts are less common for applications in machine tools due to their load characteristics [105–107]. Such non-hexapod fully parallel machines include the HexaM (Figure 3.39) and the Sprint-Z3 spindle (Figure 3.40) [103,105,108]. The HexaM has six fixed length struts and the Sprint-Z3 has three fixed length struts (three-axis spindle). The actuators (either ball screws or linear motors) are fixed to the frame to reduce the maximum inertia effects. Single-DOF hinges have been used at the base end of fixed length struts to prevent the rotation of the platform.

Unlike a serial machine, the working envelope of a PKM is not cubic and depends on the geometry of the structural components (as illustrated in Figure 3.38a). However, five-sided machining

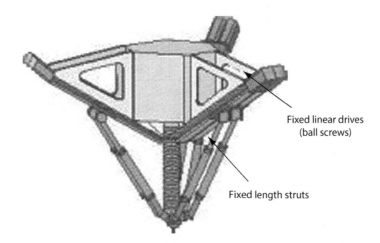

FIGURE 3.39 The structural configuration of a 6DOF PKM with six fixed length struts. This five-axis machine has non-coplanar linear joints fixed to the frame. (From www.toyoda-kouki.co.jp.)

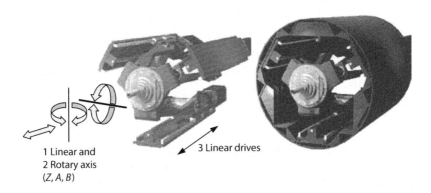

FIGURE 3.40 The structural configuration of the Sprint-Z3 spindle on parallel mechanism (3DOF PKM with three fixed length struts). This 3DOF head is used in a five-axis machine with X- and Y-axis with conventional drives on the head or workpiece carrier. (From DST Machine GmbH, Mönchengladbach, Germany.)

(almost full five axis) has been found possible using a machine with five struts [109]. There are still many problems remaining concerning rigidity and accuracy. Rigidity requires highly rigid joints and accuracy requires proper calibration technology and a suitable control system. PKMs are primarily used for niche applications due to workspace size limitations. Parallel-serial hybrid machines provide a better blend of the advantages of the PKMs and serial machines [104,110–113]. Although several three- to six-axis commercial PKM machines have been available since 1994, especially for milling applications [103], the trend is toward hybrid solutions [101,103,105]. Two such systems are the PKM spindle system (that provides the Z-axis and two rotary A- and C-axis as shown in Figure 3.40) on top of a serial X- and Y-axis system to provide 5DOF machine [103], and a 5DOF configuration using a tripod (3DOF parallel) and two rotary axes (in series) under the spindle head as shown in Figure 3.38b [112,113]. The tripod mechanism consists of four kinematic chains, three variable length legs, and one passive leg, connecting the moving platform to the fixed base that contains the bearings of the ball screws. The platform is joined rigidly with the passive middle strut, which is connected with the base via a linear and universal joint. The first rotary axis aligns to the middle strut; the second rotary axis is vertical to the first axis and carries the spindle.

The two major design parameters affecting the performance characteristics are the selection of the appropriate geometric dimensions and kinematic topology (affecting the stiffness and accuracy). The design of a PKM is very complex compared to serial kinematics and several approaches

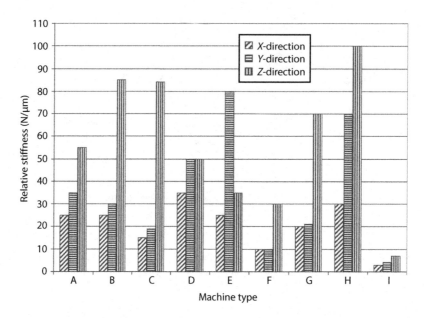

FIGURE 3.41 Relative static stiffness of different parallel and serial (PKM and SKM) machine tools. All machines are measured between a table and a dummy tool. A, midsize three-axis VMC; B, midsize three-axis HMC 12,000 rpm spindle; C, midsize three-axis HMC 24,000 rpm 24 kW spindle; D, large-size four-axis HMC; E, midsize five-axis VMC; F, large five-axis HMC; G, five-axis large hybrid PKM horizontal spindle on a PKM structure carried by a serial X- and Y-axis; H, six-axis large PKM vertical spindle; I, five-axis hybrid PKM vertical spindle.

have been proposed [103]. The tool point stiffness and accuracy for hexapods are often found to be lower than conventional machines when using the same drive components because the strut stiffness is significantly reduced by the flexibilities of the joints at each end [114–119]. Each extensible strut has a spherical joint at each end, which consists of a combination of pretensioned roller bearings ($k = 200$ N/μm), ball bearings, or gimbal ($k = 25$ N/μm). The stiffness for a hexapod varies significantly (two to six times) across the workspace, while orthogonal three-axis machine tools have almost uniform stiffness over the workspace. Generally, the structural stiffness of hexapod machines (at least in one direction is less than the stiffness of one of the links) is significantly less than that of a three-axis conventional machine but equal or better than the five-axis conventional machines as illustrated in Figure 3.41. In general, the stiffness in the Z-direction is greater than that of each strut itself as long as the tool is located within the space of the base joints. PKMs with fixed length struts can provide higher stiffness than machines with extensible struts.

Another very important parameter for machine tools is the chatter resistance criterion K (the product of the stiffness and damping), which should be greater than the spindle/tool stiffness [114]. The K term for a conventional machining center is generally greater than 6 N/μm but for PKMs it varies between 0.5 and 4 N/μm. Generally, high-speed machine tools should have structural and drive stiffness $K > 3$ N/μm.

Building accuracy into a machining system is essential and requires building accuracy into both the machine and the process. The machine accuracy is usually expressed in terms of linear positioning accuracies of the individual axes. The linear positional accuracies of many CNC machines vary from 2 to 15 μm depending on the measurement standard. Even though the positioning accuracies of many CNC machines are similar, their volumetric positioning accuracies vary significantly (5–10 μm) due to compensation methods, geometric and kinematic behavior, static and dynamic behavior, thermal distortion, and workpiece fixture characteristics as illustrated in Figure 3.42. It is often difficult for a user to determine which CNC machine has the volumetric positioning accuracy

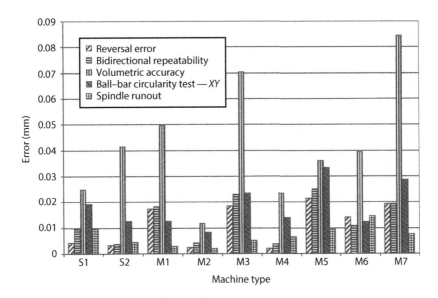

FIGURE 3.42 Comparison of accuracy for various small and medium horizontal CNC machines (the table sizes for the S and M machines are 500 and 630 mm diameter, respectively). All machines were measured with the same instrumentation as defined in ASME B5.54.

necessary for a particular precision part because the straightness and squareness or the volumetric accuracy are not specified in the standard specifications of machine tools. The 3D volumetric positioning accuracy is defined as the root mean square of the sum of all errors in all axes of motion including the straightness errors, squareness errors, and angular errors. The machine's accuracy improves by calibrating and compensating volumetrically. The ASME B5.54 [99] body diagonal displacement tests or the telescoping ball bar performance test can be used to evaluate the volumetric positional accuracy. If the machine is not accurate, however, additional tests are required to define the straightness, squareness, and angular errors requiring compensation. Error measurement and compensation methods are discussed in more detail in Chapter 16.

The volumetric accuracy of parallel kinematic machines depends on how accurately the controller model describes the real kinematic behavior of the machine [120] because the position of the spindle nose is indirectly estimated from the rotational positions of the servo motors for the struts. In addition, the thermal effects are more difficult to model than for orthogonal structures. Due to the highly nonlinear kinematics and dynamics inherent to PKMs, the dynamic response errors are difficult to determine and cannot be assumed constant over the entire work volume. The interpretation of the data obtained when testing a PKM is often quite different than when the same equipment is applied to conventional machines [121]. The major concerns with PKM accuracy are the accurate identification of geometrical (kinematic) parameters, thermally insensitive strut length measurement, and deformation due to gravity [121]. The volumetric accuracy could be as high as ±0.05 mm with reversal error of ±0.01 mm. Three-axis PKMs have achieved an accuracy of 0.01–0.02 mm [103].

Passive means for temperature control should be considered in designing a machine structure. Minimizing and isolating heat sources, minimizing the coefficient of thermal expansion or maximizing the thermal diffusivity of the structural material, and isolating critical components can accomplish passive temperature control. For example, if each steel sheet of the bed or column frames has an equal heat capacity at any location, the structure will be free from any strain caused by changes in the room temperature or machine temperatures. The heat generated by the motors (especially the spindle head), by friction between moving components (spindle, ball screws, etc.), at the cutting zone, and in the machined chips must be contained and dissipated properly to minimize distortions in critical directions. The thermal expansion of the machine structure components should be considered especially when dissimilar materials, such as composites and metals, are joined together. Active temperature control by circulating a temperature-controlled fluid, air showers, or proportional control can be also used. For example, circulating a temperature-controlled oil within ±0.4°C of room temperature, throughout the main structure of the machine including the bed, column, ball screw or its bracket, and spindle head, can be used to stabilize the machine thermal state and geometry during operation. Wet machining tends to minimize heat buildup in the tool and chips, especially when through-spindle coolant is used. Temperature sensors mounted on the saddle, base, column, and spindle provide feedback to assure machine accuracy and stability in spite of changes in machine temperature. The thermal characteristics of the machine should be evaluated after the machine is anchored in the ground if anchors are used for the installation. The thermal stability of the machine structure is a significant engineering challenge in MQL machining as discussed in Chapter 15.

Machines should generally be fully enclosed by guards and safety devices. This ensures a clean machine environment and adequate protection of the operator from chip buildup or tool breakage. Windows on high-speed machines are made of polycarbonate and require periodic replacement due to etching from coolant [122,123]. Optimum designs provide large gradient angles; the way telescoping covers and bed surfaces should be slanted as sharply as possible to allow free fall of chips and coolant to an oversized chip conveyor as shown in Figure 3.43. Single or dual swarf disposal conveyor systems should be designed to cover the tool's entire working range and should ascend steeply to the ejection level to minimize coolant loss. In addition, noise from belts and motors can be reduced by isolating their guarding with foam.

Active or passive isolation of the machine from environmental vibrations through proper design of the machine foundation is also important, especially for high-precision machines [1]. Small machines with sufficient stiffness do not require special foundations. Larger machine tools usually require a separate concrete foundation to provide additional stiffness and to minimize the transmission of vibrations from neighboring equipment [124–126]. Separate foundations should be thick enough to provide adequate stiffness, support all machine elements, and be stable on the local soil. There are various types of mounting elements, such as machinery mounts and anchoring/alignment

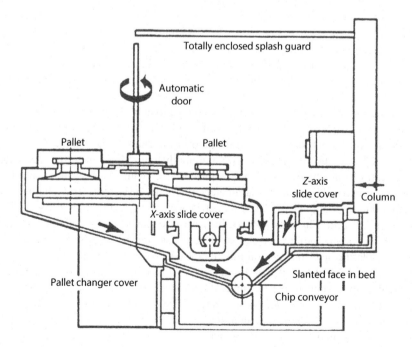

FIGURE 3.43 Chip evacuation approach using a single swarf disposal system. (Courtesy of Toyoda Machinery USA, Inc., Arlington Heights, IL.)

systems, designed for leveling, alignment, and damping of machines [1,127]. Machines are commonly (1) anchored on a concrete or steel plate, (2) anchored on isolation pads, which are located on concrete or steel plates, or (3) positioned on isolation pads without anchoring. Mounts that provide vibration isolation include pads of special materials, leveling mounts, press mounts, and inertia blocks [126]. Rubber mounts with enhanced electrorheological fluids [127] maybe used to reduce the settling time of high-precision, high-speed machines in which accelerations produce large inertial forces. Such mounts become stiffer upon the activation of an electric current during accelerations and decelerations.

3.5   SLIDES AND GUIDEWAYS

Slides are machine components, which move a workpiece or tool on guideways (or ways) to a specified position and hold it in position under machining loads. Their traveling speed and positional accuracy determine machine tool productivity and part quality in many applications. The dynamic and chatter behavior of the machine is affected significantly by the way system because it transfers the cutting forces into the mass of the machine structure. In addition to stiffness and damping, parameters considered in the selection and design of guideways are machine speed and acceleration/deceleration requirements, load capacity, friction, thermal performance, material compatibility, environmental sensitivity, accuracy, repeatability, resolution, preload, and maintenance.

Common guideway (rail) designs include rectangular, cylindrical, vee, flat, vee and flat, dovetail, and circular-groove monorail (Figure 3.44). Guideways are commonly used in pairs or double

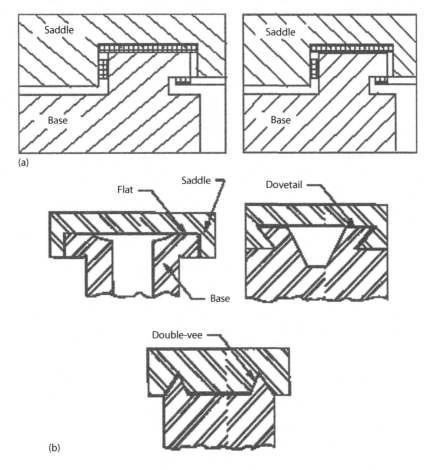

FIGURE 3.44   Common guideway designs. (a) Rectangular ways; (b) vee-and-flat and vee-ways.

tracks to ensure straightness. Dovetail and vee and flat guideways are accurate and have high stiffness. Features that are easily machined in place are normally integrally cast in the structure, while more complex features, such as hardened steel cylindrical guideways, are bolted to the structure. Integrally cast flame-hardened ways are more difficult to manufacture and are usually hand scraped compared to the *bolt-on* type ways, which are easily manufactured and reconditioned. Cast-in-place ways are less expensive but cannot be replaced and are generally used only in machines with short life expectancies. Traditionally, guideways have been made of steel or cast iron, but aluminum, ceramic, and composite guideways have been developed. Various low-friction coatings have been applied on ways to reduce friction and wear [128,129].

The surface finish and flatness of the ways are controlled by the final manufacturing operation, which may be scraping, super finish milling, or grinding plus etching or peening. Scraping (hand lapping) a surface generates microgrooves for oil retention. The straightness and parallelism of the ways are very critical since they determine the pitch, roll, and yaw errors of each machine axis. The straightness error of the guideways is dependent on the final characteristics of the material used, the bearing type and its preload, and the roughness of contacting surfaces. For example, a light preload can lead to lower stiffness, error motion, and tool chatter, while high preload could increase bearing wear. In addition, the preload changes (decreases) with time due to the wear of the bearing elements. Box guideways made of through-hardened tool steel (about 60 HRC) and precision ground and scraped can be precisely bolted onto scraped bed or column surfaces, resulting in superior performance because they eliminate micro vibrations (which are typical in roller guideways) while providing good damping characteristics. The ways should typically be parallel to each other and perpendicular to the bed within 0.01 mm, while the tops and bottoms of the ways should be flat and parallel within 0.005 mm [130]. Typical machine straightness accuracies are 0.03–0.1 mm/m for a single axis. The straightness accuracy of precise machines may be as low as 0.002 mm/m.

The selection of the form for the guideways depends strongly on the type of bearing used to support and move the slide on the way [131]. Bearing systems can be classified as friction or antifriction types as illustrated in Figure 3.44a. Friction systems use a fluid film as a bearing media (e.g., hydrodynamic, hydrostatic, and aerostatic bearings), while antifriction systems use rolling element (linear) bearings.

Sliding contact (or plain sliding) bearings, sometimes called hydrodynamic or box-type bearings, are the oldest and most common type. They consist of two mating surfaces that have been machined, ground, or scraped to obtain the desired coefficient of friction; they provide the best combination of speed and load-carrying capacity for general applications. They are used on flat (T way), vee-flat, double-vee, and dovetail guideways (Figure 3.44b). Flat guideways have the best load-carrying capacity. They add to the rigidity of the base or column and provide very stable support and damping for large loads spread over a large area. They are insensitive to crashes. The vee shape is very accurate but the loads must be vertical. They use hydrodynamic lubrication supplied by oil under pressure. Sliding contact bearings usually exhibit significant static friction depending on the materials, surface finish, and lubricant; the static friction changes to dynamic friction as motion begins. Typical static friction coefficients range from 0.03 to 0.3, while dynamic coefficients are usually between 0.01 and 0.1. Stick-slip or stiction action due to higher static than dynamic friction causes positioning errors.

Sliding contact and stick-slip friction can be reduced by coating the ways with a thin layer of low-friction material made of Teflon-impregnated sheets (polytetrafluorethylene-PTEE, Turcite, Rulon) or molybdenum disulfide and graphite poured into place (Diamond DWh310, Moglite, SKC) [6]. For example, Turcite has a friction coefficient of 0.04 compared to 0.2 for cast iron. Pulse-metered permanent lubrication is often used to reduce friction as well. The speed and acceleration of sliding contact bearings are limited to about 25 m/min and 0.1 g, respectively. Their accuracy and repeatability range from 6 to 10 μm and from 2 to 10 μm, respectively, in the axial direction depending on the drive system. Their straightness accuracy and repeatability fall in the range of 0.1–10 μm and are affected by the quality and alignment of the guideways. A preload of about 10% of the allowable load must be used to obtain good stiffness for bidirectional loading. Details of the design and analysis of sliding contact bearings and of material considerations for slide and guideway design are discussed in [1,6,128,132–134].

Currently, slide speeds are generally restricted to below 20 m/min, although machines with slide speeds between 40 and 50 m/min are not uncommon and speeds between 70 and 100 m/min have been achieved on some high-speed machines. The maximum feed rate is typically between 40% and 50% of the maximum rapid travel rate.

Rolling element bearings (or linear guides shown in Figure 3.44a) have also been widely used in guideways. The three major types of linear rolling element bearings are nonrecirculating balls or rollers, recirculating balls, and recirculating rollers. Nonrecirculating systems are used where short travels and compact designs are needed. Cylindrical roller bearings provide higher stiffness and load-carrying capacity than ball bearings, due to increased contact area, and are normally used for accurate machines. Recirculating roller bearings provide good load capacity and stiffness, but their alignment is critical.

Generally, rolling linear guides require lower power for motion and eliminate the stick-slip action characteristic of sliding contact bearings, but exhibit less stiffness and damping and lower load-carrying capacity. The static and dynamic friction coefficients for rolling element bearings are roughly equal and are typically between 0.001 and 0.01, leading to improved resolution and controllability. Linear guides reduce axis reversal error. The upper limits for speed and acceleration are about 120 m/min and 1–2 g for adequate bearing life. The upper speed limit for cylindrical roller bearings is generally 20%–40% lower than for ball bearings. Their axial accuracy can be better than for sliding contact bearings and is the range of 1–5 μm, while their straightness is similar and depends on the tolerance class of the rolling elements and rails. Each machine axis typically has four bearing blocks, two on each rail, although many more may be used in high load applications. Rolling element bearings are sensitive to crashes. Generally, sliding contact bearings are used for heavy-duty applications with integrally cast ways (box-type ways), while linear bearing ways are used in lighter-duty or higher-speed applications. Correct linear guide selection allows a linear guide machine to approach a slideway machine's rigidity. Details of the design, analysis, and applications of rolling elements for guides are given in [1,6,35–140]. Hybrid guideways are a combination of box guideways and linear bearing packs.

Hydrostatic and aerostatic bearings are used for guideways in precision machines such as grinding and hard turning machines. These bearings have no mechanical contact between elements; the load is supported by a thin film of high-pressure oil or air that flows continuously out of the bearings. Static friction is eliminated and dynamic friction is insignificant at most speeds. Hydrostatic bearings are used at moderate speeds where high load capacity and stiffness are required, while aerostatic bearings provide moderate load capacity and stiffness and are preferable at higher speeds. Hydrostatic bearings provide better vibration and shock resistance with superior damping characteristics. The dynamic stiffness is very high due to squeeze film damping. The damping is very good normal to the bearing surface but low along the direction of motion. A great deal of research and development work on analysis and development of hydrostatic and aerostatic bearings has been carried out [1,6,141–149]. Self-compensating or gap-compensating bearings, which can be used with water as a bearing fluid, have recently been developed [6]. Using water rather than oil as the sliding medium results in more stable temperatures, higher permissible speeds, and fewer fluid contamination problems. High-performance hydrostatic slides have been made entirely from alumina ceramics. The Hydroguide and HydroRail are designs using a profile similar to linear rolling element bearings (Figure 3.45) and have been used in grinding and hard turning lathes [150].

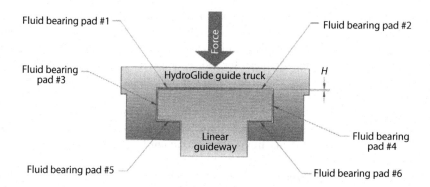

FIGURE 3.45 Example of the self-compensating HydroGlide™ system with fluid gap *H*. (Courtesy of Aesop, Inc., Concord, NH, and MIT, Cambridge, MA.)

3.6 AXIS DRIVES

Relative motion between the tool and part in a machine tool is achieved by moving the machine table, spindle, or column separately or in combination. In any case, an axis drive (or servo) system is required. Desirable features in an axis drive system include accuracy, repeatability, and high dynamic stiffness. Other significant performance characteristics include the bandwidth or response time, peak drive current, servo gain, motion control, and torque-to-inertia ratio (acc/dec capability), as well as smoothness and accuracy.

An axis drive system (Figures 3.46 and 3.47) consists of a drive motor or actuator, mechanical transmission elements, sensors, and a controller. The principle types of linear motion actuators are hydraulic or pneumatic pistons, conventional electric motors, and linear motors.

Hydraulic systems exhibit high dynamic stiffness and damping, can sustain high machine and cutting loads, and have good acceleration and power-to-weight ratios due to their low mass. They are most commonly used on single axis machines, especially on transfer lines. They are not widely used in machining centers because they are relatively inaccurate, generate excessive heat, and are often difficult to control.

Electric motors [151–157] have proven much more versatile and are the most common drive actuating systems. Both AC and DC drive motors are used. AC induction motors are more common because they require less maintenance and provide a greater bandwidth, higher gains, and better repeatability.

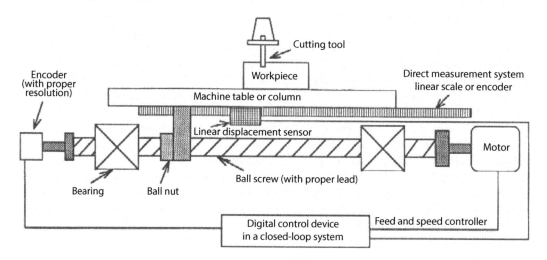

FIGURE 3.46 Conventional closed-loop machine control system.

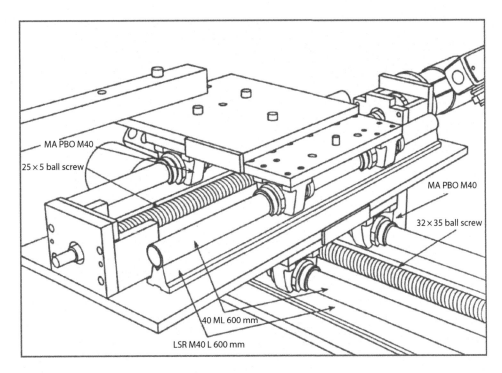

FIGURE 3.47 *X–Y* machine table system using linear motion systems for both rail bearings and the ball screw. (Courtesy of Thomson Industries, Inc., Buffalo, NY.)

AC motors, however, are less efficient and provide lower torque-to-inertia ratios. (Brushless, frequency-controlled AC drives, however, deliver considerably more power than comparable three-phase or DC motors.) The principal advantage of DC motors is that they generally produce higher torques at high speeds. A hybrid type of motor, the brushless DC motor, combines the advantages of both AC and DC brush motors. Brushless DC motors, commonly known as AC servos, are more durable than AC induction or synchronous motors, generate less heat and vibration, and are generally smaller in size and more efficient. Electric motors can be further divided into stepping and nonstepping motors. Stepping motors generate a precise motion increment in response to an electric pulse. They provide limited torque but are very accurate and more easily controlled than other types of motors. The best motor for a given application depends on the torques necessary for maximum acceleration.

Linear motors or *direct drives* (Figure 3.48) eliminate the mechanical transmission system required with rotary motors [6,158–160] since they combine the functions of the ball screw and motor. They are made with permanent or induction type magnets; induction magnets generally provide better performance and reliability. The force is transmitted through a magnetic field instead of a mechanical linkage. Because they are noncontact devices (except in the way system), they exhibit little wear and require less maintenance. Also, they can run at high speeds (up to 120 m/min) with high acceleration/deceleration characteristics (1–1.5 g) and eliminate deflections, backlash, and windup since bearings and ball nuts are not used. They can provide higher tool feed rates (almost twice that of ball screws) with equivalent or increased accuracy when compared with traditional ball screw systems since they can be controlled using many times the gain of their rotary counterparts [161]. The length of travel does not affect the performance of a linear motor as it does a ball screw. However, they have a comparatively low dynamic stiffness and limited load capabilities. Linear drives have traditionally been limited to applications with relatively small cutting forces (<5000 N) [161]. Higher forces can be achieved by placing separate drives in parallel. They require special machine designs and heat dissipation approaches due to their low mechanical efficiency. Linear drives used on horizontal beds should be carefully sealed to prevent swarf contamination in

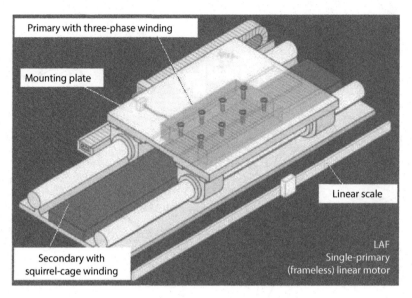

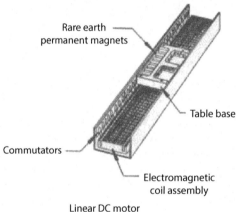

FIGURE 3.48 Linear axis drive motors. (Courtesy of The Rexroth Corporation, Indramat Division, Rochester Hills, MI.)

the track, which can degrade performance. Linear motors are not superior to ball screws in all cases [162] but are used for high-speed machining applications not possible with other systems or for large machines made for very large dies, molds, and aerospace parts.

When rotary electric motors are used to drive axes, a mechanical transmission is required to convert the rotary motion into a linear slide motion. The most common type of transmission is the lead screw (Figures 3.46 and 3.49), which is either directly coupled to the motor or connected to it by a toothed belt [1,6,130,163–165]. There are various types of lead screw mechanisms including sliding contact thread screws (power screws), recirculating or nonrecirculating ball screws, planetary roller screws, recirculating roller screws, traction drive screws, and hydrostatic nuts and screws. In each device, a nut is essentially sliding along a screw; the nut advances linearly with each rotation of the screw by an increment dependent on the screw pitch. Traditionally, the lead screw is driven and the lead screw nut is fixed rigidly to the table; the screw is supported with two bearings near the nose and two near the rear, with the rear bearings mounted on floating mounts to compensate for shaft expansion. It is essential that preload be established to eliminate backlash and increase stiffness. Preload can be provided by forcing the nut in one direction, using a doubled-nut design, or using oversized balls or shifting pitch. The preload is typically one-third of the rated maximum load.

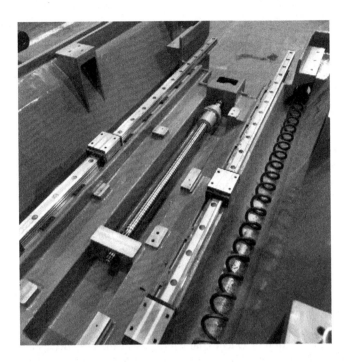

FIGURE 3.49 Typical arrangement of guideways, rolling bearing packs, and ball screw system in a machine bed. (Courtesy of Cincinnati Milacron, Cincinnati, OH.)

Special manufacturing processes are required for ball screws used in high-speed machines to prevent errors and wear due to friction [166]. Ball screws are generally better than lead screws because they are ground and can offer greater maximum speed, accuracy, and load capacity. The accuracy also depends on alignment, preload, and system stiffness. The location of the ball screw is very critical to its performance because it affects both accuracy and durability. Generally, a ball screw should be mounted as closely as possible to the guideway of the primary motion when only one of the ways is responsible for location and the other primarily supports the load. The ball screw should be precisely centered when both guideways control the location and guide functions. In a horizontal machining center, two ball screws may be used to drive the Z-axis so that the center of the base is left open for chip evacuation. Similarly, two ball screws may be used for the Y-axis to eliminate the need to counterbalance the spindle carrier. In other cases, twin ball screws are used on each linear axis to achieve high rigidity and high-speed positioning without reducing positioning accuracy. In all these cases, one of the two AC servomotors is used as the master and controls the axis movements while the other motor follows. The ball screw should also be closer to the table surface in this case. The ball screw shaft can be hollow to reduce inertia. In addition, passing coolant through the core of ball screw helps to maintain a constant temperature and prevents thermal expansion, especially during rapid feed, rapid acceleration, and prolonged operation. For increased rigidity, the end brackets of balls crews should be integrally cast with the base and column. With the present ball screw materials, the upper limit of the rapid traverse speed is 50–80 m/min with an acceleration/deceleration capability of 0.8–1.0 g. Speeds of 100 m/min at 2 g are available with optimum ball screw design and coated balls [167]. In recent development work [168], ball screw systems with rigidly mounted screws and servo-driven nuts have reportedly achieved traverse speeds up to 120 m/min and accelerations up to 4 g.

Less common types of transmissions include rack and pinion, ram and piston, and friction drive designs. *Rack and pinion* transmissions are used in large machines where long drives are needed (>4 m). *Ram and piston* drives are typically used for short displacements since they employ hydraulic cylinders. *Friction* or *traction* drives are used mainly in high-precision machines. They provide minimum backlash and low friction but have a low load capacity and moderate stiffness and damping.

The servo systems described earlier may exhibit backlash. Backlash can be eliminated when a precision gear-type speed reducer is used. Precision machines use harmonic drive gearing with elastic body mounts, rather than conventional gearing and rigid mounts. Direct drive motors for rotary axes can eliminate mechanical backlash. Lost motion due to backlash, shaft deflections, bearing looseness, and other sources is corrected for through software compensation. A feedback device at the opposite end of the ball screw (see Figure 3.46) confirms that the CNC command of a certain number of revolutions (corresponding to a specified linear distance) has been executed.

The ratio of the load inertia to rotor inertia is an important factor in servo system performance. Inertial loads that are too large for the motor often result in sluggish response, causing overshoot, slow settling time, and stalling when the rotor position falls too far behind the drive signal. This problem is avoided by limiting the load-to-rotor inertia to a maximum of 10:1. For a large speed range, the motor and drive can be designed so that the torque is a function of the motor current at low speeds and a function of the motor voltage at high speeds; this approach optimizes motor performance. The damping characteristics of the servo are also important since they affect system stability. Ensuring proper damping, however, is difficult because it depends on several factors. A more detail discussion and analysis of axis drive systems with respect to loads and static and dynamic stiffness is given in Reference 167.

The acceleration/deceleration (acc/dec) characteristics of the axis drives are especially important for high-speed machines used for rigid tapping, end milling, and similar processes. There are two basic strategies in accelerating and decelerating an axis: the "acceleration rate" and "time to acceleration" methods. The *acceleration rate* method can improve precision at higher speeds while maintaining the desired path speed but can result in machine shock at a corner in the workpiece. The *time to acceleration* method can reduce this shock but can cause path errors at corners. Therefore, the best approach is to combine both methods in high-speed, high-precision applications. The positioning distances and the value of the position controller's proportional velocity gain are the key parameters because they determine the following error and thus the time response of the system as it accererates/decelerates [161]. High acc/dec coupled with high-resolution control systems that prevent overshoot during slide reversal should be used for threading at higher cutting speeds. Acc/dec over 1 g and high proportional velocity gain are very important when machining parts with short positioning distances ($\leq$50 μm) [161].

The natural frequencies of the system are affected by the gears and ball screw as spring elements and the masses of the worktable and load. The natural frequencies of a linear drive system are significantly higher than a ball screw system due to the lack of mechanical transmission elements.

Hybrid feed drives combining a ball screw drive unit (for long travels) with a linear direct drive on top of the bell-screw table (for short movements) are under development [167]. This concept is designed for machines with long travel to avoid using very long linear motors (requiring costly permanent magnets) for the full travel. The motions of both drives are integrated kinematically so that the machining process is enhanced by the dynamic performance of the linear drive.

Sensors and controls are very important components of the axis drive because they determine the positional accuracy of the drive system. Contemporary drives use position, velocity, acceleration, and load sensors to improve positional accuracy and response bandwidth. The difference between a servo system and a simple motor is the fact that the position of the driven device for the servo is constantly being monitored. The positional accuracy depends on the feedback or closed-loop communication system, which compares the actual and programmed locations to determine the servo lag or following error. Control of the table slide's position is accomplished by attaching a transducer to the motor or the table, slide, or other driven device. The most common approaches are to attach an encoder to the servo motor shaft or a resolver on the motor and to attach linear sensors or linear encoders (scale feedback) to the moving machine elements. The latter method is more accurate because it measures the real motion produced by the motor and not just the number of shaft rotations. High-precision linear encoders use a graduated glass scale and an interferometric measuring principle to measure distances [169] (see Figure 3.48). For high-precision machines, both methods can be used to improve the axis feedback response. The speed is measured using rotary tachogenerators, eddy current sensors, or by digitally differentiating the position signal. The acceleration can be measured either directly (using piezo-based sensors) or from the second digital derivative of the position. A load-sensing system and guideway surface pressure control (compensation) system can be used to control accurately the position of heavy components at high speeds.

Control of feed drives is critical to machining accuracy because the structural flexibility of the machine tool is reflected by the drive. A proper design controller will damp the major sources of excitation of machine tool vibration [167]. Closed-loop control techniques have been developed to damp vibrations in the feed drive motor by using position, velocity, acceleration, current, and force feedback. The active damping of feed drive vibrations can be used to further improve the dynamic characteristics of the feed drives.

3.7 SPINDLES

The spindle forms the interface between the machine tool structure and the cutting tool, and its properties determine how efficiently and accurately the motion capabilities of the machine tool are transferred to the cutting tool. The spindle shaft also incorporates the tooling system, including the tool taper, drawbar mechanism, and tool release system. The spindle is therefore one of the most important machine components and must be designed to ensure accuracy and durability when subjected to the expected thrust and radial loads. Typical spindle requirements are running accuracy, speed range capability, high rigidity, minimal level of vibration, low and stable operating temperatures, long life, and minimal need for maintenance. Some of the critical factors to consider in designing a spindle are drive choices (belt driven, gear driven, motorized), bearing arrangement and mounting, bearing types (frictional characteristics and size), permissible operating conditions, external cooling conditions, weight, thermal growth, operating environment and bearing seals, method of lubrication and type of lubricant, resonant frequencies, allowable static overload, tool retention requirements, and serviceability [170,171]. Compromises must generally be made in order to provide the best combination of speed, power, stiffness, and load capacity.

Spindle performance is most often characterized by its torque and power versus speed characteristics. Spindle power-speed and torque-speed curves (Figure 3.50) are very important to machine performance and should be evaluated over the complete speed range for continuous and intermittent duty. Torque and power are related through the equation

$$P = \tau \omega \qquad (3.1)$$

where
  P is the power
  τ is the torque
  ω is the spindle rotational speed

Machines typically have full torque at 0 rpm but do not deliver full power until some minimum speed is reached, since from Equation 3.1 even a high torque yields relatively low power at a low rotational speed. Proposed speeds and feeds should be checked against the spindle characteristics. Large-diameter tools are typically run at low rpms to reduce the cutting speed, but on some machines this is not practical since the available power drops sharply at speeds less than 20% of the maximum rating.

Spindles can be broadly categorized as box spindles and motorized spindles. *Box spindles* (Figure 3.51) are driven externally by an electric motor; common drive mechanisms include belt

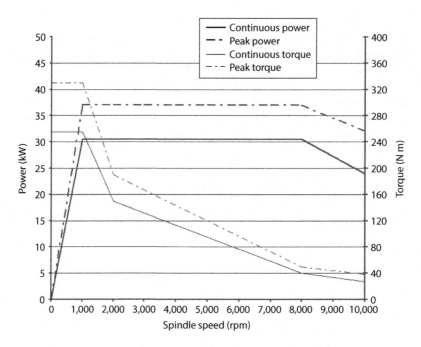

FIGURE 3.50 Typical spindle power and torque characteristics as a function of speed.

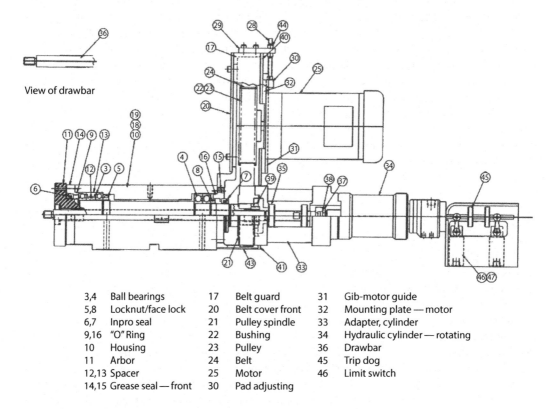

3,4	Ball bearings	17	Belt guard	31	Gib-motor guide
5,8	Locknut/face lock	20	Belt cover front	32	Mounting plate — motor
6,7	Inpro seal	21	Pulley spindle	33	Adapter, cylinder
9,16	"O" Ring	22	Bushing	34	Hydraulic cylinder — rotating
10	Housing	23	Pulley	36	Drawbar
11	Arbor	24	Belt	45	Trip dog
12,13	Spacer	25	Motor	46	Limit switch
14,15	Grease seal — front	30	Pad adjusting		

FIGURE 3.51 A box spindle assembly. (Courtesy of Setco, Cincinnati, OH.)

drives, toothed belt drives, shaft couplings, and gear case speed reducers. Belt drives are most common on general purpose machines. Adjustable belts should be used on belt-driven spindles to reduce bearing stresses and improve spindle accuracy. Gear drives are normally used for high-power spindles and machining centers. Although gear trains may exhibit backlash and lost motion due to gear profile errors, poor gear tooth surface finish, and vibration, properly designed gear drives can increase drive rigidity and reduce susceptibility to malfunction when compared with toothed belts. Planetary gear systems are also used to provide a wide torque output.

The motor is an important spindle component, which determines many performance characteristics, including the weight, inertia, size, acc/dec, torque, and power. The most common spindle drive motors are induction motors, variable speed AC motors, and DC motors [172–175]. Generally, AC motors provide quiet, vibration-free rotation at controllable speeds but have significant rotor and stator losses and comparatively low efficiencies and torque-to-inertia ratios. AC induction motors are commonly used in transfer machines and other constant speed applications because of their low cost, availability, and maintainability. Variable speed AC synchronous motors are common on NC machines. Their speed is controlled by input frequency changes; an input voltage change results in a torque change. Variable speed DC motors are used when high power or a very wide speed range are required. DC motors produce less rotor heat than AC motors. Permanent magnet (PM) brushless DC motors, switched reluctance (SR) brushless motors, and synchronous reluctance motors are the basic types of DC motors. PM brushless and SR motors are use in high end motor applications in which weight, size, and efficiency are critical. PM motors are generally smaller and more efficient than equivalent SR or AC induction motors. Brushless DC motors develop higher torques at lower speeds compared to AC motors (except vector-controlling AC induction motors) and are less sensitive to harmonics, which can create audible noise. Permanent magnet rotors result in substantially decreased rotor temperature, which improves bearing life.

A wide variety of spindle characteristics can be obtained since the spindle power, torque, and speed are largely dependent upon the driving motor so that the final specifications can be modified for a particular application by using a different motor or belt ratio. High power and torque are possible because the spindle motor is mounted externally to the spindle shaft, and therefore, it is often possible to use a very large motor without concern for space constraints. Belt-driven spindles are limited in maximum rotational speeds due to the limitations of belts and gears at higher speeds. Typically, belt-driven spindles are used to a maximum rotational speed of 12,000–15,000 rpm and power ratings up to 22.5 kW (30 HP).

In a *motorized spindle* (Figure 3.52), the electric drive motor is directly integrated into the spindle housing or body, eliminating the mechanical transmission. They are better suited to higher speed applications because they eliminate vibrations associated with gear trains and drive belts. The rotor is usually placed directly between and supported by the spindle shaft bearings. This design provides comparatively short and compact spindles with adequate stiffness; the spindle natural frequency can be very high, allowing high subcritical (below the first natural frequency) operating speeds. Occasionally, the motor is placed behind the two bearing systems at the end of the shaft (Figure 3.53). This *flywheel motor* design can provide higher stiffness since the shaft diameter between the bearing systems is not limited by the rotor bore diameter; it also decreases the spindle clamping diameter and thermal expansion of the shaft. On the other hand, the rotor mass extending in the rear limits the natural frequency of the spindle. This frequency can be increased by adding a third bearing to support the very end of the shaft, but this requires extreme alignment accuracy and control of the angular thermal expansion between the three bearing positions. New motorized spindle designs nearly match the torque/speed characteristics of gear-driven box spindles; consequently, they are finding increasing use in high-speed machining centers and machines designed for high-precision and *C*-axis capability.

The spindle's torque and power characteristics are adjusted by a frequency converter. Typically, spindles are used in a constant torque range, where power and voltage increase linearly with frequency. However, greater power at lower speeds can be obtained by programming the constant torque range to reach maximum voltage before reaching maximum frequency; from this point up to the maximum frequency, the voltage remains constant and the torque decreases (see Figure 3.50). This design

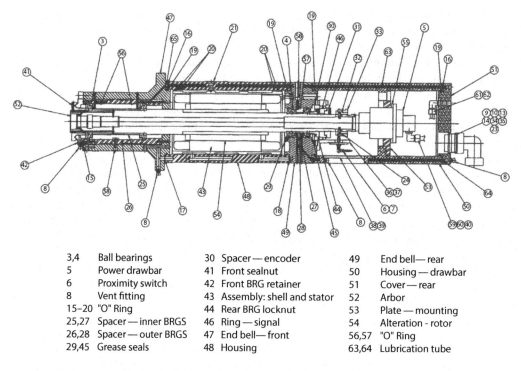

3,4	Ball bearings	30	Spacer — encoder	49	End bell— rear
5	Power drawbar	41	Front sealnut	50	Housing — drawbar
6	Proximity switch	42	Front BRG retainer	51	Cover — rear
8	Vent fitting	43	Assembly: shell and stator	52	Arbor
15–20	"O" Ring	44	Rear BRG locknut	53	Plate — mounting
25,27	Spacer — inner BRGS	46	Ring — signal	54	Alteration - rotor
26,28	Spacer — outer BRGS	47	End bell— front	56,57	"O" Ring
29,45	Grease seals	48	Housing	63,64	Lubrication tube

FIGURE 3.52   A motorized spindle. (Courtesy of Setco, Cincinnati, OH.)

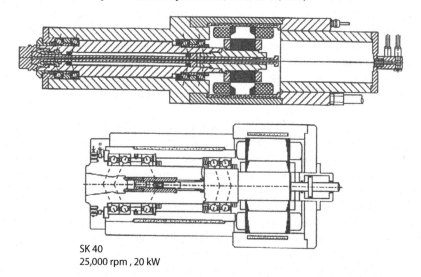

SK 40
25,000 rpm , 20 kW

FIGURE 3.53   Motorized spindles with "flywheel" motor arrangement. (Courtesy of the Precise Corporation, Racine, WI.)

approach results in lower spindle power at top speed. The spindle control loop may be open or closed (vector control). Closed-loop spindles incorporate an encoder for speed feedback. Closed-loop spindles are necessary for rigid tapping, supporting orientation, and high-torque applications.

The spindle shaft is held in position by a set of high-precision bearings. In many cases and especially for high-speed spindles, *bearings* are the critical components of the spindle. The *speed index DN*, which is determined by multiplying the bearing bore diameter (mm) with spindle speed (rpm), is used as a measure of a bearing's relative reaction to speed. The bearing

TABLE 3.2
Comparison of Various Types of Spindle Bearings

Characteristic	Rolling Element	Hydrodynamic	Hydrostatic	Aerostatic	Aerodynamic	Magnetic
Cost	Low	Medium	High	High/med	Low/med	Very high
Maintenance	Low	Medium	Med/high	Med/low	Low/med	Low
Damping	Low/med	High	Very high	Low	Medium	Med/high
Stiffness	Med/high	High	Very high	Med/low	Low/med	Med/high
Load capacity	Med/high	Med/high	High	Low	Low	Med/high
Running accuracy	Med/high	High	High	High	High	High
Speed range	Medium	Low	Med/high	High	High	Very high
Wear resistance	Medium	Medium	High	High	High/med	Very high
Power loss	Low	High	High	Medium	Medium	Very low
Cooling capacity	Medium	Medium	High	Med/low	Low/med	Med/low
Environmental factors	Low/high	Low/high	Low/high	Low	Low	Low
Reliability	Med/high	High	High	Med/high	Med/high	Med/high

speed rating is also expressed in terms of D_mN (using the bearing pitch diameter (mm) multiplied by the rpm); the DN value is usually about 50%–88% of D_mN. This factor is determined based on fatigue tests and is thus a function of the mechanical properties of the bearing materials. Actual limiting bearing speeds, however, are dependent on the bearing application (mounting arrangement and preload method) and lubrication as well as the speed factor.

As for slides and guideways, several types of spindle bearings are available including rotary bearings (angular contact ball and rolling element), hydrostatic bearings, hydrodynamic bearings, aerostatic bearings, aerodynamic bearings, and electromagnetic bearings. Table 3.2 summarizes the characteristics of the various bearing types. *Rotary bearings* for spindles are similar to linear bearings for slides (Section 3.5) and are subject to the same strengths and limitations. Similarly, *hydrostatic, hydrodynamic, aerostatic*, and *aerodynamic* spindle bearings are similar to corresponding slide bearing types. They can be installed parallel, perpendicular, or at an angle to the axis of rotation to resist radial, axial, or combined loads as shown in Figure 3.54. In *hydrodynamic* bearings, the shaft turns in a sleeve containing pressurized oil or water that generates a hydrodynamic film. True hydrodynamic bearings do not contain a pre-pressurized film; the film separating the sleeve and the shaft is established entirely by relative motion between the shaft and the sleeve, which draws the fluid into the bearing gap. The motion of the fluid film creates pressure and lifts the shaft to create a gap. Most fluid-film bearings, especially high-speed bearings, are initially hydrostatic at lower speeds so that frequent high-speed starts and stops do not damage the bearing surface but develop a significant hydrodynamic component at higher speeds. These bearings are typically referred to as hybrid-hydrostatic or hydro-stato-dynamic bearings. They have very good damping properties and are well suited for high-performance and high-precision machining and may yield useful lives exceeding 20,000 h [170]. Newer hybrid-hydrostatic bearing designs have reduced power loss and can be used at speeds up to 100 m/s, making them an attractive alternative for high-speed applications. Several hydrostatic bearing spindle systems have been used as illustrated in Figures 3.54 and 3.55. The major concern with hydrostatic bearings is frictional loss, which is typically between 35% and 50%. The damping characteristics of these spindles can increase cutting tool life and overall part accuracy and improve part surface finish. *Magnetic* bearings have been used recently for rotary applications [170,176–178], especially at very high speeds relative to their load capacity. Radial and axial magnetic bearings are used in the spindle shown in Figure 3.56. There is no shaft wear because they are frictionless bearings. They can achieve the fastest shaft surface speeds (e.g., the highest DN) currently available. Catch bearings are used to hold the shaft during a power failure and prevent catastrophic contact. The electric fields of the bearings can be controlled to

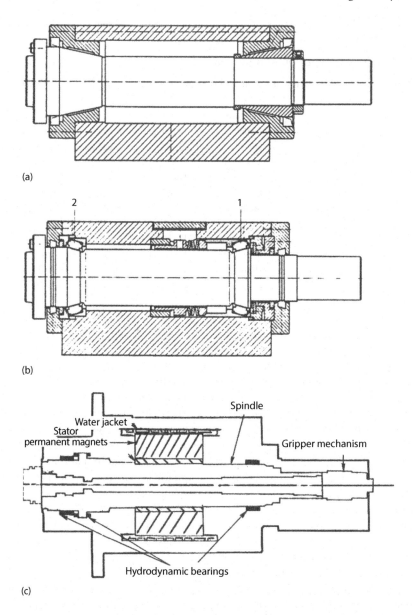

(a)

(b)

(c)

FIGURE 3.54 Spindles with hydrostatic/hydrodynamic bearings. (a) Spindle with hydrostatic sliding bearings in "O" arrangement, (b) spindle with hydrodynamic sliding pad bearings in "O" arrangement, and (c) spindle with hydrodynamic sliding axial and radial bearing. ((a) and (b) Courtesy of Pope Corporation/FAG Kugelfischer Georg Schafer KGaA; (c) courtesy of Ingersoll Milling Machine Company.)

optimize stiffness and damping characteristics. With proper control, they can also be made self-balancing. Their benefits include no mechanical contact (no wear or heat), very high accuracy, very-high-speed capability (twice that of ball bearings), and provision for process monitoring. The limitations are high cost compared to mechanical systems, limited load capability, moderate stiffness, and sensitivity to dynamic conditions. *Air bearings* are used at higher speeds, most commonly for grinding and also for milling, boring, and microdrilling. Their load capacity is low but they have excellent runout and long life and are suitable for ultra-precision machining. These bearings are extremely sensitive to contamination and moisture in the air. If the load capacity of the air film is exceeded, the film collapses and the bearing makes contact, causing catastrophic failure.

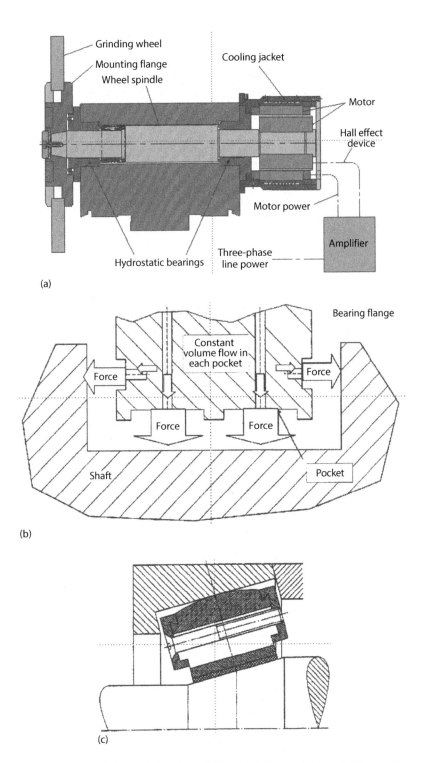

FIGURE 3.55   Spindles and hydrostatic bearings. (a) Hydrostatic spindle for grinding machines, (b) gyroscopic hydrostatic bearing construction with axial and radial capacity, and (c) hydrostatic bearing construction with axial and radial capacity. ((a) Courtesy of Landis Gardner, a UNOVA Company, (b) courtesy of FAG Kugelfischer Georg Schafer KGaA; (c) courtesy of IBAG North America.)

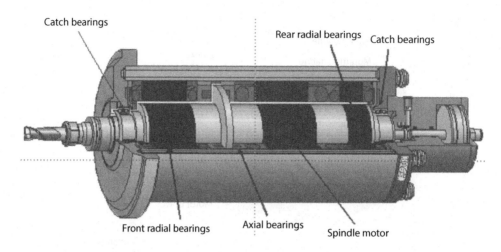

Catch bearings

Rear radial bearings   Catch bearings

Front radial bearings   Axial bearings   Spindle motor

FIGURE 3.56   An active magnetic bearing spindle. (Courtesy of IBAG, North Haven, CT.)

There are two basic types of anti-friction or rotary element bearings [179,180]: ball or point contact bearings and roller or line contact bearings. *Ball bearings* typically provide lower stiffness compared with roller bearings but generate less heat; they are widely used in all types of spindles but are especially common in high-speed, low load applications. Angular contact ball bearings are most commonly used in machine tool spindles because they help to retain the preload at higher speeds. The contact angle determines the ratio of axial to radial loading possible, with radial load capability being the primary benefit. Typically, contact angles of 12°, 15°, and 25° are available. The higher the contact angle, the greater the axial load-carrying capacity. Therefore, it may be desirable to use a bearing with a contact angle of 25° for a spindle that will be used primarily for drilling, and a contact angle of 15° for a spindle that will primarily be used for milling. The ball diameter becomes significant at high speeds due to centrifugal loads, which can damage the race and reduce bearing life. This problem is most acute at speed indices greater than 800,000 DN [181]. One response to this problem is to use smaller diameter balls to reduce ball mass and forces. Hybrid ceramic ball bearings comprise steel inner and outer races, ceramic (silicon nitride) balls, and a retainer cage. Ceramic ball bearings, being 60% lighter than steel with a higher modulus of elasticity and thus providing a smaller contact area and greater stiffness, are preferred at high speeds; ceramic balls allow up to 30% higher speeds for a given ball bearing size before the centrifugal force begins to deform the balls resulting in reduced bearing life. Ceramic ball bearings generate less heat and run about 10° cooler than all-steel bearings with grease lubrication in comparable applications [182]. The thermal expansion of ceramic bearings is negligible, resulting in longer life and better accuracy due to lower friction and linear wear. Tests have shown that spindles using hybrid ceramic bearings exhibit higher rigidity and have higher natural frequencies, making them less sensitive to vibration. Standard dimensions of hybrid and steel bearings are the same, so a switch to hybrid bearings requires no design changes. *Roller bearings* (line contact) may be of the cylindrical, spherical, tapered, or needle type [183]. Line contact bearings are stiffer than point contact bearings by a factor of 6–8 and have higher load-carrying capacities. They resist shock and impact loads better than ball bearings, but generally must be used at lower speeds and higher loads and are more subject to misalignment. Straight and tapered roller bearings are commonly used in machine tool spindle applications. Double-row cylindrical roller bearings are typically recommended to carry high loads. Single-row bearings are used at higher speeds. Special bearing types such as hydra-rib [183] and hollow roller [184] bearings have also been investigated and appear to provide advantages over conventional bearings in some applications.

Generally, the selection of a bearing type depends on many factors such as diameter, thermal stability, stiffness requirements, load-carrying capacity, speed, and lubrication as explained

TABLE 3.3

Selection Criteria on Selecting Bearings for Spindles

Spindle Requirements	Bearing Selection	Spindle Design Impact
High torque	Larger bearings	Large shaft, lower speed
High speed	Low contact angle	Small shaft, low power
High stiffness	Roller bearings, larger size	Large shaft, lower speed
Axial loading	High contact angle	Lower speed
Radial loading	Low contact angle	Higher speed
High accuracy	ABEC 9, high preload	Lower speed

in Table 3.3. This table shows that many factors that determine the final spindle design and that compromises must generally be made to meet all requirements. Bearing sizes determine the spindle shaft diameter and therefore the stiffness of the shaft. Increasing the bearing diameter increases stiffness and load-carrying capacity but reduces speed capability. Typical speed ratings for oil jet, oil mist, or oil-air (ratio 1:2) bearings are 10^6–2.5×10^6 DN for angular contact bearings and 600,000–900,000 DN for roller element bearings; speeds below these levels can be achieved comfortably with grease-packed bearings. Recent limits in standard products are about 1.5×10^6 DN for grease-packed ceramic ball bearings (average below 800,000) [170,171]. The highest DN values are obtained with radial and angular contact single-row bearings with light preloads, oil lubrication, and external cooling. When the DN value exceeds 800,000, the application is considered a high-speed application and cooling of the spindle becomes necessary.

The accuracy of a bearing is defined by ABEC number; many machine tools using ABEC-5 bearings, although high-quality and high-precision spindles use ABEC-7 or 9 bearings. The bearings are typically combined and customized to meet a machine's specific operating requirements; as noted earlier, this requires a compromise between several performance characteristics such as load-carrying capacity, rigidity, speed, preload, and tolerance.

The bearing service life at conventional speeds, at which the effects of load are dominant, is usually considered equivalent to the fatigue life, although the actual service life may be limited by wear due to environmental factors (contamination, corrosion, heat, or poor lubrication) and mounting problems. The major causes of bearing failures are contamination and lubrication breakdown due to inadequate sealing [170,171]. High-speed spindles are typically designed to yield fatigue lives of 6,000–10,000 h under specified axial and radial loads, provided the bearings do not become contaminated and vibration remains within acceptable limits.

The bearing arrangement in a spindle has a strong influence on spindle performance. Typical angular contact bearing arrangements, shown in Figure 3.57, include back-to-back "DB" or tandem "DT" arrangements. Tandem mounting does not allow forces in both directions, unless another pair of bearings is used on the spindle shaft, facing in the opposite direction. Generally, two or three bearings are used at the front in a tandem setup and another tandem set is added at the rear of the spindle. Together, the bearing sets form an overall "DB" setup (as shown in Figure 3.53). Bearings should be positioned in an arrangement that does not allow the loss of preload when the cutting forces are in the direction opposite the preload. The load capacity in radial and both axial directions is determined by the combination of bearings. The "O" arrangement (tandem "DB") is loaded against springs that limit its application to milling operations using either standard or high helix cutters. The "X" arrangement (tandem face-to-face "DF") is not subject to these limitations since the axial loads are absorbed directly in the bearings. Spindle growth with the "O" arrangement is directed out of the spindle, causing the shaft to grow longer and the tool to cut deeper. This could also lead to preload loss during thermal expansion. In contrast, bearings with an "X" arrangement yield inside, leaving additional stock on the part, which can be removed in a finishing pass, although finish pass cutting forces would be increased. One to three sets of bearings are used to support the

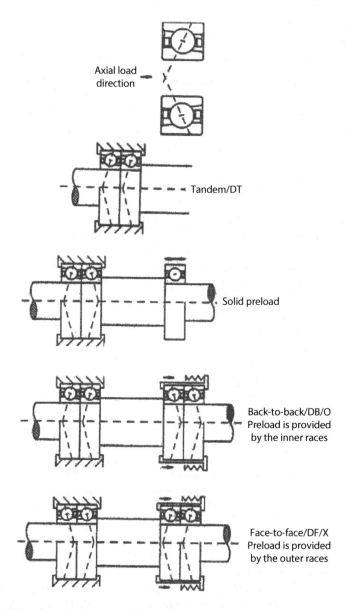

FIGURE 3.57   Ball bearing arrangements. (Courtesy of the Precise Corporation, Racine, WI.)

spindle shaft at each end. In the case of a tandem set, most of the external radial load (70%–80%) is carried by the first bearing. Since tandem sets are more difficult to lubricate, single bearing arrangements are preferred especially when the axial load is not of concern.

Most commonly, angular contact bearings are used at both ends of conventional or high-speed spindles. However, under high speed, moderate load conditions such as those typical of drilling, a tandem "O" bearing arrangement with three bearings as locating bearings on the work end and a single row of cylindrical roller bearings as a floating bearing on the drive end may be used. Under high load, average speed conditions such as those typical of milling, a bearing arrangement with two double-row cylindrical roller bearings such as radial bearings and a double direction angular contact thrust ball bearing to support the axial force may also used (Figure 3.58). Since tapered roller bearings can carry much higher loads than ball bearings, they save space because one tapered roller bearing is used at each end of the spindle (Figure 3.59), instead of two or three ball bearings.

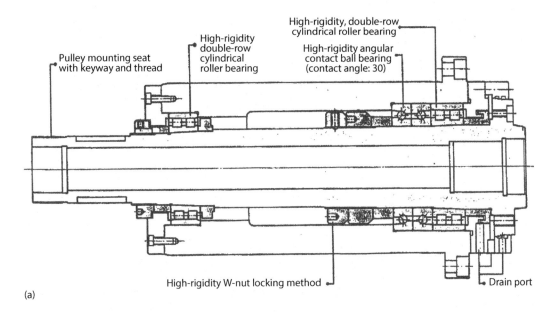

(a)

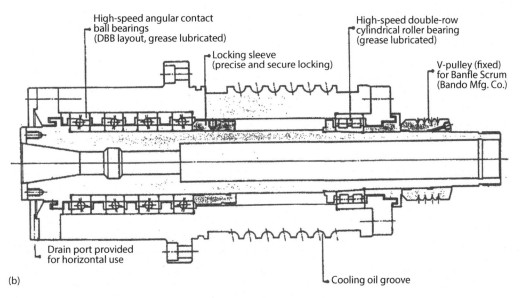

(b)

FIGURE 3.58 Machining center spindles with a combination of double-row roller bearings and angular contact ball bearings. (a) High-rigidity grease lubricated type for NC lathes spindle with rolling bearings at front and rear and (b) high-speed grease lubricated type for machining centers spindle with rolling bearings at rear. (Courtesy of NSK Corporation, Ann Arbor, MI.)

Standard tapered rolled bearings are normally used for high loads, lower speed applications, while moderate to high speeds (about 50 m/s) are achieved either with modified cage designs to direct the oil to the roller-rib contact area or by providing a secondary lubricant source to the rib.

Bearing lubrication is a critical component of the spindle system. Depending on bearing size, type, and speed, bearing lubrication may be permanent grease pack or some type of oil system. The majority of conventional spindles have permanent grease pack bearings. These are preferred where practical because they are simple, relatively inexpensive, and are maintenance free for extended periods assuming the sealing is good. They result in very low temperatures if the correct type of grease is properly applied. The grease life usually determines the bearing life in high-speed applications.

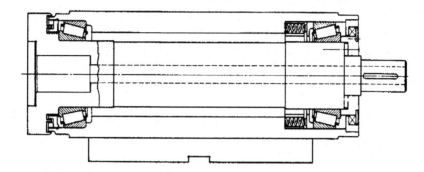

FIGURE 3.59 Spindle with tapered roller bearings (grease lubrication). (Courtesy of Timken Company, Canton, OH.)

To preserve the grease at bearing speeds above 700,000–850,000 DN, temperatures of 35°C–40°C should not be exceeded. The run-in period for temperatures and torque has an important effect on the successful operation especially for bearings used at speeds exceeding 13 m/s. Oil mist, pulsed oil–air (ratio 1:2), and occasionally oil jet or oil circulation systems are also used to improve spindle stability but are costly, pose environmental concerns, and require scheduled maintenance. Such systems require that clean, dry, and continuous air be supplied. Also, the use of the correct type, quantity, and cleanliness of lubricating oil is critical [170]. Insufficient lubricant volume results in bearing burnout, but as the volume of lubricant increases, the temperature and friction increase. Higher speeds up to about 2.5×10^6 DN can be accomplished with oil-jet lubrication even though bearing power consumption due to friction is very high. Oil circulation is regulated between 2 and 10 L/min. Water jackets containing temperature-controlled recirculating water and glycol mixtures are also used to cool bearings. These systems are effective but must be carefully sealed to prevent the contamination of the spindle grease. They may also consume considerable energy through an integral chiller.

The preload greatly affects bearing performance. Angular contact bearings are loaded in the axial direction. The preload maintains the contact points of balls at their original positions and raises the speed at which the effects of sliding become significant. Therefore, preload is increased as speed increases with external preloading arrangements. Static stiffness increases with preload, while the capacity to bear additional external loads and speed capabilities decrease. Ultimately, a load limit is reached, which gives an unacceptable fatigue life even though the mechanism of failure in very-high-speed applications is frequently wear rather than fatigue. Accuracy and repeatability increase with preload until the bearing gap is closed; after this point they decrease with increasing preload. The optimum preload setting gives the minimum compliance or the maximum dynamic stiffness. The static coefficient of friction increases with preload from 0.001 to about 0.01. Increasing the preload generally reduces the DN factor. The preload is determined by bearing specification, speed, and lubrication [95,170,171]. The preload is generally designed based on the highest speed and temperature conditions, which determine the minimum preload and material DN. A solid or fixed preload (obtained with a "DB" arrangement) allows for loads in both axial directions. This design does not perform well if the shaft changes length due to thermal expansion. A variable preload bearing mechanism enhances the characteristics of high-performance bearing arrangements, enabling them to run under a range of conditions. Variable preload, provided by springs or hydraulic pressure, allows thermal growth. A heavier preload is used at lower speeds, which increases stiffness and allows higher cutting forces. In some cases, the preload is reduced to limit the heat generated at higher speeds. It is often impossible to regulate a variable preload in an NC machine due to continuous changes in cutting conditions and loads. A self-adjustable preload for the full spindle speed range helps achieve both heavy cutting at lower speeds and accurate rotation at higher speeds. A controlled or variable preload is obtained with springs positioned at the rear of the bearing that allows the bearing set to move axially within the spindle housing. The preload must be taken into account when calculating load-life requirements because there is a trade-off between the preload and the maximum load capacity. The preload can also affect dynamic runout and hence part quality.

Thermal stability is also an important concern in spindle selection, especially for finishing operations. ANSI standard B89.6.2 [185] addresses thermal stability and thermal errors due to axial drift, radial drift, and tilting. At present, spindles with a growth of 0.025 mm or more over a 3 h time span are very common. The use of warm-up cycles can reduce but not eliminate thermal errors. Further control of errors depends on lubrication and cooling of the spindle. The best approach for improving the spindle's thermal stability is to use more thermally stable materials for the spindle components and to cool the outer bearing races and motor with a water jacket or controlled temperature oil bath [186]. Circulating a heat-exchanging fluid through the spindle controls the temperature of the spindle head to reduce thermal distortion. Steel has traditionally been used for most spindle components but can be replaced in shafts and housings by fiber-reinforced metal or plastic composites [187]. Heavy metals and refractory materials with high modulus, such as molybdenum, tungsten, ceramics, and invar, can also be used but are expensive and difficult to fabricate. Sensors in the tailstock, saddle, and column have been used to measure the temperature of these units and automatically compensate for distortion. Thermal distortion is minimized in lathes by circulating temperature-controlled oil or coolant through the headstock, saddle, and spindle.

The spindle rigidity (the static stiffness or spindle deflection measured at the spindle nose) is determined by the number, arrangement, and stiffness (type) of the bearings, the stiffness of the shaft, the bearing spacing, the housing stiffness, the preload, and the overhang distance between the front bearing and the cutting tool. Thermal expansion also has an important influence on the static and dynamic stiffness of the spindle. Optimum bearing spacing is designed for individual maximum speeds. The bearings selected should have as large a bore as practical because a 19% increase in bore size results in 100% increase in the stiffness contributed by the bearings and shaft. Material selection for the shaft and housing can improve the spindle stiffness significantly since the bearings typically account for 30%–50% spindle deflection, while the shaft and spindle housing deflections are responsible for 50%–70%. Elastic analyses of the spindle system have been effectively applied to determine the shaft stiffness as compared to the bearing stiffness and to optimize the housing design and heat transfer. Often a small stiffness improvement in a relatively stiff spindle is ineffective because the weakest, most elastic element in the system is the tool. Both the static and dynamic stiffness of the spindle are important; they differ significantly especially at higher speeds. The dynamic stiffness is influenced to a large degree by the damping characteristics and the static stiffness of the spindle.

The rotary seals at the front and rear of the spindle prevent dirt, coolant, and chips from entering the bearings and spindle housing. Seals are even more critical when through-the-spindle pressurized coolant is used because the coolant splashes against the front of the spindle with significant energy. Inadequate sealing has been found to be one of the major failure modes of spindles [171]. Spindles are usually easy to seal using lip seals, labyrinths, packing, face seals, and stuffing boxes. Some seals are designed dynamically so that they do not contact the shaft at running speeds, reducing heat generation. Other types of seals include noncontact types with air purging to prevent leakage. The mean time to failure for most seals is 500–1000 days [154]. The design of bearing seals is discussed in detail in [171,188]. Rotary unions in machines with high pressure through-spindle coolant often require a minimum coolant pressure to seal; these unions must be changed if the machine is converted to minimum quantity lubrication (MQL) as discussed in Chapter 15.

The tool retention system is also a critical spindle component for small or high-speed spindles. Figure 3.60 shows a power drawbar commonly used on NC machines. The drawbar passes through the center of the spindle shaft. The ID of the spindle nose is usually tapered to accept a tool holder; the drawbar pulls the holder securely into the spindle nose with an axial force of between 5 and 35 kN, depending on the application. (As discussed in Chapter 5, HSK tooling couplings require higher retention forces than CAT-V tapers.) An improperly designed or adjusted drawbar can apply an excessive force to the front spindle bearings, leading to stresses and bulging, which could reduce

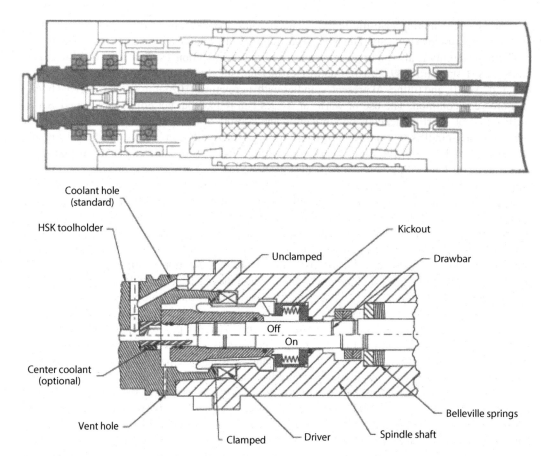

FIGURE 3.60   A power draw bar tool retention system. (Courtesy of Toyoda Machinery USA, Inc., Arlington Heights, IL.)

bearing life and performance. This is especially true for small spindles and spindles used in high-power applications requiring high drawbar forces.

Spindle performance is also affected by balance characteristics, which determine limiting rotational speeds. Balance is affected by the characteristics of the rotor, the characteristics of the machine structure and foundation, and the proximity of resonant speeds to the service speed. Precision and high-speed spindles must be dynamically balanced. The balance grade G1, representing a maximum unbalance vibration velocity of 1 mm/s, is normally sufficient for machine tool spindles, while a G0.4 may be needed for precision spindles. Details on the determination of the permissible residual unbalance over speed are given in ISO Standard 1940 [189]. Higher precision machines may require active balancing devices integrated at the front of the spindle nose in addition to conventional two-plane balancing. These devices measure the vibration (magnitude and phase) during rotation of the cutting tool (before cutting takes place) and calculate the adjustments of the pole plate rings necessary to reduce the unbalance within a specified level. More details are provided in Section 5.5.

Spindle vibration has a detrimental effect on spindle life, tool life, and part quality. The vibration can be detected in the displacement (amplitude), velocity, and acceleration in the axial and radial directions on the housing close to the bearings. Vibration of contactless bearings should be measured directly on the shaft or tool using noncontact methods. Vibration displacement and acceleration limits are spindle speed dependent; velocity limits are not and thus can be used to define one acceptance criteria for the entire range of spindle speeds. The velocity amplitude acceptance limit for gearless type spindles (box, cartridge, and multi-spindle/cluster type spindles) for up to 10,000 rpm is 0.01 mm/s; it increases at frequencies greater than four times the spindle speed frequency, to 0.12 mm/s for angular contact bearings and 0.18 mm/s for roller bearings. The acceleration limits are 0.5 g for angular contact bearings and 1 and 1.5 g for roller bearings for speeds below 1000 rpm and above 1000 rpm, respectively. The displacement limit (peak-to-peak) for speeds below 700 rpm is 25 μm. The vibration velocity acceptance limit for gear-driven spindle assemblies is 0.2 mm/s. The ASA S2/WG88 standard is being prepared to define acceptable vibration limits for high-speed spindles. Some general limits for high-speed spindles are 1 mm/s RMS for the spindle, 2 mm/s RMS for the spindle with tool, and 4–6 mm/s RMS during cutting.

Most spindles exhibit axial and radial motion errors of between 0.25 and 2 μm at the spindle nose, although precision spindles with total motion errors of less than 50 nm are available. Spindles with radial motions of up to 10 μm are common. The dynamic spindle runout is often larger than the static runout by a factor of between 10 and 20 at high spindle speeds due to increased vibration. The radial runout measured at a specific gauging distance from the spindle nose using a precision arbor is a good indication of spindle runout quality because most tools extend 100–150 mm from the nose. Systems have been designed which incorporate counterweighed rotor assemblies fixed to the machine spindle nose for active real-time balancing of the entire spindle assembly after each tool change [190–193].

The spindle/toolholder interface has an important effect on spindle performance with respect to stiffness, adequate damping, balancing characteristics, and the ability to transmit the expected torque and forces at the cutting tool. Toolholder/spindle interfaces are discussed in detail in Chapter 5.

Sensors can be integrated within the spindle for process condition monitoring. Force measurement systems based on piezo-electric force measurement sensors are used at the front of the spindle to measure the forces either in the z-axis or all three axes and estimate the torque. Single or tri-axial piezo-electric accelerometers are either integrated within or attached to the spindle housing (by the bearings) to sense the acceleration to monitor bearing wear and life and avoid early failure and to detect vibration, unbalance, and chatter in the cutting process. It is possible to detect chatter and regulate the speed to eliminate it. Special rings with piezo-electric force sensors have been developed for measuring the axial cutting force to avoid collisions and tool breakage, spindle nose failure during collisions, etc. [167]. The ring (as flange or bearing type) could include several uniaxial piezo-electric force sensors. The electrical spindle current is used to estimate the power and torque for the cutting tool. Acoustic emission (AE) sensors are used to estimate the tool performance, wear, and avoid failure. AE sensors are also used to identify chatter during machining or breakage of the tool. They have been successfully applied in grinding and honing processes. The sensors are mounted either on the housing or as a ring at the spindle nose. The measurement of temperature near or on the bearing is used to monitor the thermal expansion of the spindle, which is equivalent to that of the cutting tool. In addition, displacement sensors can be used to measure the axial displacement of the shaft.

3.8   COOLANT SYSTEMS

In a machining center, coolant is supplied to the spindle nose and to the tool either externally or internally through the spindle. Coolant is supplied externally using swiveling jets mounted on a plate at the top of the spindle, or more effectively around the spindle housing on the machine unit or on the machine top cover (Figure 3.61). These nozzles are freely adjustable to different distances between spindle and workpiece so that they are effective over the entire workspace. External coolant can also be supplied through permanent nozzles located on the face of the spindle housing; this is especially effective for a ram/quill type spindle since the nozzles travel with the spindle in the Z-direction. In this case, the coolant is supplied through the spindle housing.

Through-tool coolant can be supplied either through the center or the perimeter of the spindle shaft (Figure 3.61). When coolant is supplied through the shaft perimeter so that the coolant does not interfere with the drawbar system, it is routed through the flange of the tool holder. In the case of tools with internal coolant supply holes or when manufacturing precision bores, the coolant should be filtered to less than 40 μm, and down to 5–10 μm at high speeds to prevent wear and damage to the rotary union, tool holder/spindle interface, drawbar fingers, and tool, and to reduce stresses

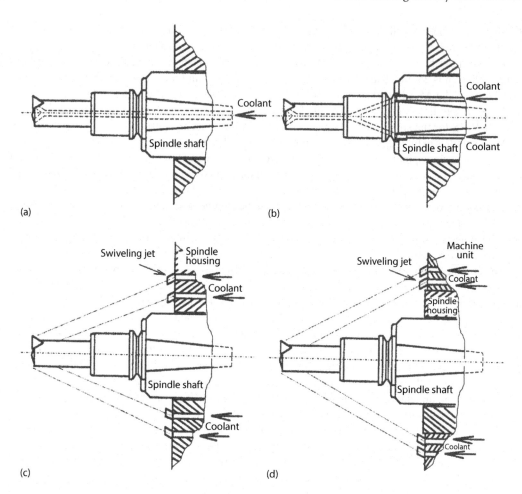

FIGURE 3.61 Coolant supply approaches. (a) Coolant through the tool and center of spindle shaft, (b) coolant through the tool with holes on the perimeter of the shaft, (c) coolant through the spindle housing, and (d) coolant through swiveling jets on the machine unit located around the spindle housing.

on seals. Filtering not only prevents the blockage of the tool oil holes but also ensures a constant surface finish in the bore and a longer tool and coolant life. Paper filters or a hydrostatic filter system with high filter capacity should be used as discussed in Chapter 14.

Remote control coolant nozzles may be used at the spindle head; these nozzles around the spindle are automatically controlled by the machine software according to the length and diameter of the tool to provide coolant at the cutting edge or workpiece.

Optimum cooling and rinsing are ensured by a combination of external cooling with annular spray nozzles and internal coolant through the tool. When high-pressure internal coolant is used, a mist collector is necessary to meet mist exposure standards. Common coolant formulations, maintenance issues, and health and safety concerns are described in detail in Chapter 14.

3.9 TOOL CHANGING SYSTEMS

CNC machine tools are usually equipped with an automated tool change (ATC) system. One or more *turrets* are always used in turning centers as shown in Figure 3.62. The number of slots in the turrets determines the number of tools that can be used in a single program and usually limits the class of parts that can be produced. Turrets are loaded by hand.

FIGURE 3.62   The turret on a turret lathe.

Machining centers have tool magazines and generally an exchange arm or similar mechanism to transfer the tool from the magazine to the spindle. Two common types of tool magazines are the chain magazine and disk magazine, shown in Figures 3.63 and 3.64. Chain magazines can have a higher capacity and can usually be loaded at the floor level. They require an exchange arm and generally cost more than disk magazines, but their added capacity may reduce machine counts in high-volume applications. Disk magazines may load directly to the spindle and may be mounted on the side for ease of loading or overhead to save space; in the overhead configuration they may require loading through the spindle. They generally cost less and have more limited capacity than chain magazines. Chain and disk magazines have similar chip-to-chip times on contemporary machines. In some applications, a tool runner system is employed, in which a monorail is used to convey tools from one or more magazines to a tool change station to increase tool storage capacity. Straight rack-type magazines are sometimes used in high-volume applications, since they permit rapid loading of a cartridge of tools at prescribed intervals to minimize production disruption.

Exchange arms are illustrated in Figure 3.65, which shows the basic designs used on general purpose machining and turning centers, including the swing arm, rack, multiple arm, wheel, and turret types shown. On machining centers, the *double arm swing system* is most common. *Single or multiple arm systems* use either a tool magazine or a tool runner for tool storage. Most ATC systems can select tools at random (rather than in a predetermined sequence), and some can store information such as the accumulated tool life for a particular tool. The setup and organization of tools in the tool magazine is often important and is part of the design process especially when large batches of identical parts are to be machined [78].

FIGURE 3.63 Chain tool magazine on a horizontal machining center. (Courtesy of MAG Automotive, Sterling Heights, MI.)

FIGURE 3.64 Horizontal *B*-axis machines with overhead disk tool magazines. (Courtesy of MAG Automotive, Sterling Heights, MI.)

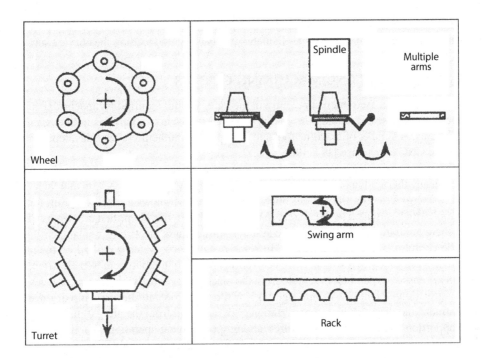

FIGURE 3.65   Types of automatic tool changers with corresponding tool change times.

The ATC cycle time varies depending on the class of machining center. The tool-to-tool change time is typically 1 s for a multiple arm system, 2–5 s for a swing arm system, and 5–10 s for a rack system. The tool change time is typically 1.5–2.5 s for a wheel and turret system, which carries a limited number of tools. Tool-to-tool change times are continuously decreasing, recently to below 0.6 s in some cases, which is leading to reduced noncutting times. The ultimate goal is to change tools without stopping the spindle, especially when operating at higher spindle speeds.

3.10   PALLETS

The pallet structure is another component of a machining system contributing to machine performance. In many cases, the workpiece is clamped to a pallet that is usually located on an indexing table, which must be clamped in the desired orientation prior to cutting. The table clamping mechanism determines accuracy and repeatability; it should provide sufficient clamp pressure with minimum pallet deflection. A good clamping design is necessary for the effective use of the full vertical axis travel in a horizontal B-axis machine. A pallet changing system improves productivity because a part is loaded while another is being machined. Pallets are used in flexible and FMS systems for high production.

A standard pallet may utilize four conic couplings. The clamping unit engages at the center of the conic coupling. This arrangement of four conical couplings has been proven accurate and rigid for up to 1.5×1.5 m pallets with load capacity up to 12 tons. It has been used successfully in multipallet systems, FMS, and other automatic systems. Acceptable results can often be obtained by hydraulically clamping the table against a curvic coupling mechanism. A curvic coupling is a pallet chucking system, which uses a given number of teeth with an engagement angle on each tooth so that it automatically locates the center of the pallet without any backlash. Curvic couplings up to 400 mm in diameter have been used successfully to secure large pallets. They provide an accuracy of 0.002 mm and repeatability better than 0.002 mm. This system is very well suited for automatic pallet changer.

The aforementioned types of pallet chucking systems can be used on top of a rotary or indexing table. Rotation can be provided by a high-precision NC system, such as double-lead worm drive system, using rotary encoder for position. Indexing accuracy of ± 2 s is achievable with NC rotary tables. $1°$ indexing tables are possible with large-diameter pallets using 360 teeth in the curvic coupling.

3.11   ENERGY USE IN CNC-MACHINING CENTERS

There has been an increased interest recently in the energy input required for machining and other manufacturing processes. Reducing the energy required to machine a part has both environmental and cost benefits. A number of machine tool and controller manufacturers have developed technologies to reduce CNC energy use [194–198].

It is not a trivial task to accurately determine how much energy a typical CNC machining center uses. As with any thermodynamic problem, it is first necessary to define a system and then consider all forms of energy crossing the system boundary. Figure 3.66 shows a block diagram for a CNC machine tool, which is treated as the system in this case. Even in this simplified description, it can be seen that there are four energy inputs to the system—three-phase electricity, coolant, chilled water (for spindle cooling), and compressed air. (In some installations, water is chilled on the machine, rather than at a central chiller, so that there is no chilled water input, but increased electricity input.) Not shown are energy expenditures remote from the machining center, which are nonetheless necessary for its function—plant HVAC and lighting, air filtering, coolant filtering, chip drying, coolant waste treatment and disposal, and transportation of parts to and from the area. It may be difficult to properly apportion these energy expenditures among specific plant operations, but they are, in many cases, significant compared to other more easily assignable energy components.

Energy use in machine tools has been studied since the 1980s [199]. The electrical energy component is the most straightforward to measure, so many studies include only this component [200–204]. Also, some studies report computed energy costs rather than direct energy usage measurements. More recent studies have included energy usage for chilled water and compressed air, generally as a percentage of the total energy expenditure [205–208]. It is often difficult to do direct component comparison due to differences in machine configurations. In some machines, for example, the same cooling system is used for spindle cooling and air conditioning of the controls cabinet, and in other cases, only the power input to a power supply feeding multiple subsystems can be easily measured.

A survey of available data indicates that the three largest sources of energy consumption in CNC machining are the machining process, the coolant system, and compressed air. The portion of total energy consumed by the cutting process varies between roughing and finishing cuts but is typically

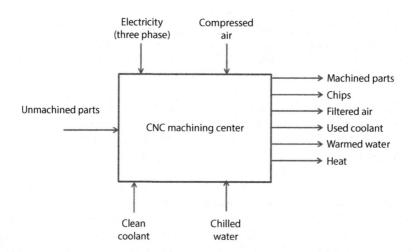

FIGURE 3.66   Block diagram showing energy flow into and out of a machining center.

roughly 25%. High-pressure coolant systems consume 30%–40% of the total energy, with low-pressure systems consuming half as much. Compressed air typically accounts for roughly 15% of total energy use. The balance of the energy is consumed by the hydraulic system, feed drives, chip transport, machine controls, and other functions. The literature on energy use does not cover a broad range of cutting conditions, so these ratios should be taken only as general guidelines subject to validation in specific applications.

3.12 EXAMPLES

Example 3.1 Estimate the stiffness of a strut used in a PKM machine tool if the stiffness of the ball screw itself is 140 N/μm, the stiffness of a Gimbal joint is 25 N/μm, and the stiffness of a preloaded roller ball/socket is 200 N/μm. Spherical joints at each end that support the ball screw should be included in the estimation of the overall strut stiffness.

Solution: The strut and the joints at its ends are in series and their stiffness is

$$\frac{1}{k_s}+2\frac{1}{k_g}=\frac{1}{k_T}; \quad \text{hence,} \quad \frac{1}{140}+2\frac{1}{25}=\frac{1}{11.5} \quad \text{or} \quad k_T=11.5\,\text{N/μm}$$

For Gimbal joints $k_T = 11.5$ N/μm and for preloaded roller ball joints $k_T = 58$ N/μm. This is the concern with PKMs compared to serial machine tools assuming that the strut is constructed using the same ball screw/nut assembly used for conventional machines.

Example 3.2 Consider a 2D PKM system (2D hexapod with variable strut length) with two struts (planar scissors pair) 600 mm apart at the base as shown in Figure 3.67. Each strut is a ball screw attached with a ball joint at the base and at the end where the spindle is mounted. Struts (*A*) and (*B*) swivel around their joints at the base and can stretch from a minimum length of 600 mm to a maximum of 1200 mm. The motions cover a large workspace. Estimate the stiffness of the PKM system at four different locations as a function of the ball screw stiffness defined in Example 3.1.

Solution: The strut stiffness is estimated in the previous Example 3.1 assuming the ball screw is supported by spherical joints at each end. Let $p = \{px;py\}$ be the location of the tip. Let a and b be the location of the ground pivots *A* and *B*, respectively. These are all 2 × 1 vectors.

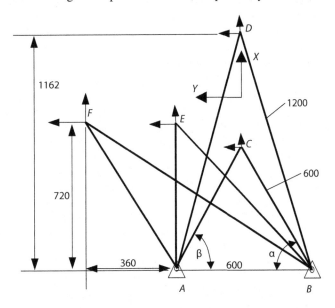

FIGURE 3.67 Illustration of several locations for a planar hexapod machine (PKM using two struts).

For shorthand, let vectors from each point in the workspace to the ground pivots be represented by the vectors

$$pa = p - a \quad \text{and} \quad pb = p - b$$

Using k_s as the stiffness of the strut, the formula for the stiffness matrix is as follows:

$$K = \begin{bmatrix} k_{xx} & k_{xy} \\ k_{xy} & k_{yy} \end{bmatrix} = k_s \left[\frac{pa \cdot pa^T}{pa^T \cdot pa} + \frac{pb \cdot pb^T}{pb^T \cdot pb} \right], \quad pa = \begin{Bmatrix} pa_x \\ pa_y \end{Bmatrix} \quad \text{and} \quad pa^T = \left\{ pa_x \quad pa_y \right\}$$

Let us now calculate the parameters for the C point in the workspace,

$$pa_x = l_s \cos(\alpha), \quad pa_y = l_s \sin(\alpha), \quad pb_x = l_s \cos(\beta), \quad \text{and} \quad pb_y = l_s \sin(\beta)$$

where
    α and β are the included angles of the struts at point C with the ground
    l_s is the length of the strut

The stiffness matrix is equal to

$$K = k_s \begin{bmatrix} \cos^2 \alpha + \cos^2 \beta & \sin \alpha \cos \alpha + \sin \beta \cos \beta \\ \sin \alpha \cos \alpha + \sin \beta \cos \beta & \sin^2 \alpha + \sin^2 \beta \end{bmatrix}$$

The angles α and β for the location C are equal to $60°$ in which case the stiffness matrix is

$$K_C = k_s \begin{bmatrix} 0.5 & 0 \\ 0 & 1.5 \end{bmatrix}$$

Therefore, $k_{xx} = 0.5k_s$ and $k_{yy} = 0.5k_s$. Note that the off-diagonal entries are zero because point C has principal directions along X- and Y-directions (its stiffness matrix is already diagonal). This is also true for point D. The angles α and β for the location F are equal to $36.9°$ and $63.4°$, respectively. The stiffness matrix for point F is

$$K_F = k_s \begin{bmatrix} 0.84 & 0.88 \\ 0.88 & 1.16 \end{bmatrix}$$

The principal stiffnesses for locations F and E are not in the X- and Y-directions and are computed using the eigenvalues, which give the principal stiffnesses. They both have weak directions due to the small angle between the struts. The values of directional stiffness k_{xx}, k_{yy}, and those along the principal directions K_{p1}, K_{p2} are given in Table 3.4.

TABLE 3.4

Stiffnesses in the Workspace of a 2D Hexapod with Variable Strut Length

Location	k_{xx}/k_s	k_{yy}/k_s	K_{p1}/k_s	K_{p2}/k_s
C	0.5	1.5	0.5	1.5
D	0.125	1.875	0.125	1.875
E	0.41	1.59	0.232	1.768
F	0.84	1.16	0.106	1.894

In summary, it is shown that the stiffness at the tip joint of the two struts of the hexapod machine varies throughout the workplace. In most places, it is much lower than the stiffness of a single strut. If it is assumed that the strut stiffness was calculated based on the stiffness of a classical machining center structure (where the ball screw/nut/thrust bearing system effectively determines the stiffness in each direction), the stiffness of the 2D hexapod is significantly lower in the weak directions. Generally, the closer the struts are to perpendicular, the better the stiffness.

Example 3.3 A conventional machine tool with 10,000 rpm/10 kW rated belt-driven spindle is being replaced with a high-speed machining (HSM) center with a 40,000 rpm/20 kW rated spindle. The new machine will be used for machining aluminum workpieces, for which it seems suited based on its rated characteristics. The conventional CNC was used to rough and finish a part. The rough process included a face milling operation using a 100 mm diameter face milling cutter at 2000 rpm. However, the new machine stalls during the face milling operation. What is the reason for such a problem with the new machine?

Solution: The power and torque required by each of the operations in the new machine should be estimated and checked against the power and torque curves. The power and torque curves were obtained upon request from the manufacturer of the HSM center as shown in Figure 3.68. This graph will help us decide if the power required by the cutting tool is available in the machine. This information is estimated using the material in Chapter 2. The power and torque required for the cut (doc = 3 mm, width of part 70 mm, feed per tooth = 0.13 mm, cutter with 5 inserts) are estimated to be 4.9 kW and 23 N m, respectively. It can be seen that the belt-driven spindle can deliver the required power and torque at 2000 rpm while the motorized high-speed spindle does not have either the power or torque for this operation. The belt-driven spindle has three times the torque of the high-speed spindle because a larger motor drives it.

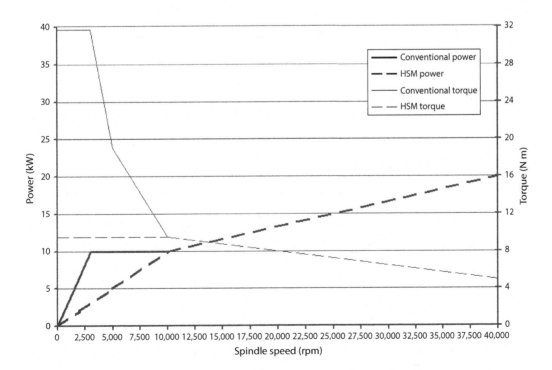

FIGURE 3.68 Machine tool spindle power and torque characteristics versus speed used in Example 3.3.

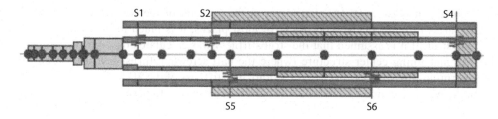

FIGURE 3.69 Two-dimensional axisymmetric finite element model of a spindle used to analyze static and dynamic characteristics in Example 3.4.

This illustrates the following:

1. The benefit of the high-speed spindle is obtained at higher speeds using lower doc to avoid exceeding the torque and power available, in which case the machining time per pass is significantly lower for the high-speed spindle.
2. The specification of the spindle by a single number (i.e., 40,000 rpm/20 kW) is not sufficient to describe the spindle characteristics; graphs of power and torque versus speed are required to properly select the cutting conditions and optimize each operation.

Example 3.4 Evaluate and compare two integral-motorized spindles of the same design but different length. Spindle (*A*) has a 264 mm long stator with 26 N m output torque, while spindle (*B*) has a 100 mm shorter stator with 15 N m output at 20,000 rpm. The shorter spindle allows reducing the distance between the front and the rear bearing by 100 mm, which affects the characteristics of the spindle. Analyze the characteristics for both spindles.

Solution: A finite element program was used to create a 2D axisymmetric model that matched the geometry of the spindle shaft and nose and rotor. The FEA program is capable of simulating both the dynamic and static behavior of a spindle and tooling system. The spindle shaft and the housing were modeled as shown in Figure 3.69. Springs S1, S2, and S4 represent the spindle bearings and springs S5 and S6 represent the connections between the spindle housing and the machine tool structure. The parameters of spring S1, S2, and S4 were determined based on the stiffness provided by the bearing manufacturer and by matching the measured dynamic response and static deflections to those generated by the model.

The model provides estimates of the difference in natural frequencies. The first natural frequency for motor *A* is 672 Hz, while for motor *B* it is 836 Hz. The modal stiffnesses for motors A and B are 2.3×10^9 and 3.0×10^8 N/m, respectively. These frequencies correspond to 40,320 and 50,160 rpm, respectively, for motors *A* and *B*. Since the natural frequency for both spindles is more than twice higher than the maximum operating speed, both spindles are safe.

A 50 mm increase in the distance between bearings in a 40,000 rpm spindle resulted in reduction of the critical spindle speed from 51,198 to 44,445 rpm [172]. This brings the 40,000 rpm operating speed close to the first natural frequency, which is not generally acceptable due to higher vibration levels resulting in lower bearing and cutting tooling life.

REFERENCES

1. M. Weck, *Handbook of Machine Tools*, Vols. 4, Wiley, New York, 1984.
2. T. J. Drozda and C. Wick, Eds., *Tool and Manufacturing Engineers Handbook*, Vol. 1: *Machining*, SME, Dearborn, MI, 1983.
3. M. P. Groover, *Automated Production Systems and Computer-Integrated Manufacturing*, Prentice-Hall, New York, 1987.
4. D. J. Williams, *Manufacturing Systems*, Halsted Press, New York, 1988.

5. J. R. Davis, *Metals Handbook*, Vol. 16: *Machining,* 9th edn., ASM International, Materials Park, OH, 1989.
6. A. Slocum, *Precision Machine Design*, Prentice Hall, Englewood Cliffs, NJ, 1992, Chs. 4, 6, 7, 8, 9, and 10.
7. G. Boothroyd and W. A. Knight, *Fundamentals of Machining and Machine Tools*, 3rd edn., CRC Press, Boca Raton, FL, 2005.
8. E. D. Lloyd, *Transfer and Unit Machines*, Industrial Press, New York, 1969.
9. B. W. Charman, *Special Purpose Production Machines*, Crosby Lockwood & Son Ltd., London, U.K., 1968.
10. O. W. Boston, *Metal Processing*, 2nd edn., Wiley, New York, 1951.
11. C. R. Hine, *Machine Tools for Engineers*, 2nd edn., McGraw-Hill, New York, 1959.
12. E. P. DeGarmo, *Materials and Processes in Manufacturing*, 5th edn., Macmillan, New York, 1979.
13. D. A. Stephenson, Design for machining, *ASM Handbook*, Vol. 20: *Materials Selection and Design*, ASM, Materials Park, OH, 1997, pp. 754–761.
14. C. T. Jones and L. A. Bryan, Programmable control, *Manufacturing High Technology Handbook*, Marcel Dekker, New York, 1987, pp. 205–254.
15. Y. Altintas, *Manufacturing Automation*, 2nd edn., Cambridge University Press, Cambridge, U.K., 2012, pp. 191–201.
16. F. J. Hilbing, Data communications, *Manufacturing High Technology Handbook*, Marcel Dekker, New York, 1987, pp. 255–298.
17. J. G. Bollinger and N. A. Duffie, *Computer Control of Machines and Processes*, Addison-Wesley, Reading, MA, 1988.
18. D. Gibbs and T. M. Crandell, *An Introduction to CNC Machining and Programming*, Industrial Press, New York, 1991.
19. J. Pusztaai and M. Sava, *Computer Numerical Control*, Reston Publishing, Reston, VA, 1983.
20. T. O. Boucher, *Computer Automation in Manufacturing: An Introduction*, Chapman & Hall, London, U.K., 1996.
21. W. W. Luggen and W. M. Luggen, *Fundamentals of Computer Numerical Control*, 3rd edn., Delmar Publishers, Albany, NY, 1994.
22. J. V. Valentino and J. Goldenberg, *Introduction to Computer Numerical Control*, 5th edn., Prentice-Hall, Englewood Cliffs, NJ, 2012.
23. L. Uriarte, M. Zatarain, D. Axinte, J. Yague-Fabra, S. Ihlenfeldt, J. Eguia, and A. Olarra, Machine tools for large parts, *CIRP Ann.* **62** (2013) 731–750.
24. I. Y. Alqattan, Systematic approach to cellular manufacturing systems design, *J. Mech. Working Tech.* **20** (1989) 415–424.
25. R. L. Diesslin, Group technology, *Manufacturing High Technology Handbook*, Marcel Dekker, New York, 1987, pp. 55–82.
26. A. Mungwattana, Design of cellular manufacturing systems for dynamic and uncertain production requirements with presence of routing flexibility, PhD thesis, Virginia Polytechnic Institute and State University, Blacksburg, VA, September 2000.
27. I. A. Shahrukh, *Handbook of Cellular Manufacturing Systems*, Wiley, New York, 1999.
28. P. A. Lewis and D. M. Love, The design of cellular manufacturing systems and whole business simulation, *Proceedings of the 30th International MATADOR Conference*, Manchester, U.K., April 1993, pp. 435–442.
29. G. Spur, Computer integrated manufacturing in Europe, *Proceedings of European Conference on Flexible Manufacturing*, Dublin, Ireland, 1988, pp. 1–21.
30. N. Alberti, U. La Commare, and S. N. la Diega, Cost efficiency: An index of operational performance of flexible automated production environments, *Proceedings of Third ORSA/TIMS Conference on Flexible Manufacturing Systems: Operations Research Models and Applications*, Amsterdam, the Netherlands, 1989, pp. 67–73.
31. W. N. Nordquist, Flexible manufacturing, *Manufacturing High Technology Handbook*, Marcel Dekker, New York, 1987, pp. 105–144.
32. Y. P. Gupta and S. Goyal, Flexibility of manufacturing systems: Concepts and measurements, *Eur. J. Oper. Res.* **43** (1989) 119–135.
33. H. T. Goranson, Agile manufacturing, in: A. Molina, J. M. Sanchez, and A. Kusiak, Eds., *Handbook of Life Cycle Engineering Concepts, Models and Technologies*, Kluwer Academic Pub., Dordrecht, the Netherlands, 1998, pp. 31–58.
34. R. Askin and C. R. Standridge, *Modeling and Analysis of Manufacturing Systems*, Wiley, New York, 1993.

35. J. K. Shim and J. G. Siegel, *Operations Management*, Barron's Educational Series, 1999. www. barronseduc.com.
36. A. W. Hallmann, Flexible manufacturing for auto parts, *Tooling Prod.* **69**:1 (January 2003) 32–33.
37. S. K. Das, The measurement of flexibility in manufacturing systems, *Int. J. Flex. Manuf. Syst.* **8** (1996) 67–93.
38. J. M. Fraser and F. F. Leimkuhler, Justification of flexible manufacturing systems, *Manufacturing High Technology Handbook*, Marcel Dekker, New York, 1987, pp. 145–174.
39. H. J. Warnecke and R. Steinhilper, *Flexible Manufacturing Systems*, Springer-Verlag, New York, 1985.
40. H. F. Lee, Optimal design for flexible manufacturing systems: Generalized analytical methods, *IIE Trans.* **31** (1999) 965–976.
41. U. Heisel and M. Mitzner, Progress in reconfigurable manufacturing systems, *CIRP Second International Conference on Reconfigurable Manufacturing*, Ann Arbor, MI, 2003.
42. A. Urbani, S. P. Negri and C. R. Boer, Example of measure of the degree of reconfigurability of a modular parallel kinematic machine, *CIRP Second International Conference on Reconfigurable Manufacturing*, Ann Arbor, MI, 2003.
43. Y. Koren et al., Reconfigurable manufacturing systems, *CIRP Ann.* **48**:2 (1999) 527–540.
44. A. Gunasekaran, Agile manufacturing: Enablers and an implementation framework, *Int. J. Prod. Res.* **36** (1998) 1223–1247.
45. A. Gunasekaran, Editorial: Design and implementation of agile manufacturing systems, *Int. J. Prod. Econ.* **62** (1999) 1–6.
46. R. N. Nagel, R. Dove, S. Goldman, and K. Preiss, *21st Century Manufacturing Enterprise Strategy: An Industry-Led View*, Iacocca Institute, Lehigh University, Bethlehem, PA, 1991.
47. Y. Y. Yusuf, M. Sarhadi, and A. Gunasekaran, Agile manufacturing: the drivers, concepts, and attributes, *Int. J. Prod. Econ.* **62** (1999) 33–43.
48. A. Gunasekaran, Agile manufacturing: A framework for research and development, *Int. J. Prod. Econ.* **62** (1999) 87–105.
49. R. DeVor, R. Graves, and J. J. Mills, Agile manufacturing research: Accomplishments and opportunities, *IIE Trans.* **29** (1997) 813–823.
50. A. Gunasekaran, *Agile Manufacturing: The 21st Century Competitive Strategy*, Elsevier, Amsterdam, the Netherlands, 2001.
51. L. M. Sanchez and R. Nagi, A review of agile manufacturing systems, *Int. J. Prod. Res.* **39** (2001) 3561–3600.
52. D. B. Sieger, A. B. Badiru, and M. Milatovic, A metric for agility measurement in product development, *IIE Trans.* **32** (2000) 637–645.
53. R. E. Giachetti, L. D. Martinez, O. A. Saenz, and C.-S. Chen, Analysis of the structural measures of flexibility and agility using a measurement theoretical framework, *Int. J. Prod. Econ.* **86** (2003) 47–62.
54. P. T. Kidd, *Agile Manufacturing: Forging New Frontiers*, Addison Wesley, Reading, MA, 1994.
55. S. Goldman, R. Nagel, and K. Preiss, *Agile Competitors and Virtual Organizations*, Van Nostrand Reinhold, New York, 1995.
56. D. Whitney et al., Agile *Pathfinders in the Aircraft and Automobile Industries—A Progress Report*, Massachusetts Institute of Technology and Lehigh University, Cambridge, MA, 1995, pp. 1–12.
57. M. P. Groover, *Automation, Production Systems, and Computer-Integrated Manufacturing*, 2nd edn., Prentice Hall, Englewood Cliffs, NJ, 2000, Ch. 27.
58. R. Dove, On cells at kelsey-hayes, *Production* (February 1995).
59. T. Beard, High speed machining: General production. Fast and flexible, *Mod. Mach. Shop* (August 1999).
60. M. Shpitalni and V. Remennik, Practical number of paths in reconfigurable manufacturing systems with crossovers, *Journal for Manufacturing Science and Production* **6** (2004) 9–20.
61. Anon., Ford furthers flexible manufacturing effort, *Manuf. Eng.* **133**:1 (July 2004).
62. R. R. Burleson and S. J. Rosenberg, *Totally Integrated Munitions Production: Affordable Munitions for the 21st Century*, Lawrence Livermore National Laboratory Report UCRL-ID-139738 rev2, September 13, 2000.
63. R. B. Aronson, High volume production, *Manuf. Eng.* **131**:1 (January 2003).
64. S. Krar and A. Gill, *Exploring Advanced Manufacturing Technology*, Industrial Press, New York, 2003.
65. R. A. Maleki, *Flexible Manufacturing Systems—The Technology and Management*, Prentice-Hall, Englewood Cliffs, NJ, 1991.
66. G. S. Vasilash, Chrysler group's approaches to advanced manufacturing engineering, *Automot. Des. Prod.* **116**:2 (February 2004).

67. J. T. Black and S. L. Hunter, *Lean Manufacturing Systems and Cell Design*, SME, Dearborn, MI, 2003.
68. C. Standard and D. Davis, *Running Today's Factory—A Proven Strategy for Lean Manufacturing*, SME and Hanser Gardner, Dearborn, MI, 1999.
69. M. Shpitalni and V. Remennik, Practical number of paths in reconfigurable manufacturing systems with crossovers, *CIRP Second International Conference on Reconfigurable Manufacturing*, Ann Arbor, MI, 2003.
70. P. Spicer, Y. Koren, M. Shpitalni, and D. Yip-Hoi, Design principles for machining system configurations, *CIRP Ann.* **51** (2002) 275–280.
71. L. O. Morgan and R. L. Daniels, Integrating product mix and technology adoption decisions: A portfolio approach for evaluating advanced technologies in the automobile industry, *J. Oper. Manag.* **19** (2001) 219–238.
72. Z. Xiaobo, Y. Jiancai, and L. Zhenbi, A stochastic model of a reconfigurable manufacturing system part 1: A framework, *Int. J. Prod. Res.* **38** (2000) 2273–2285.
73. Z. Xiaobo, Y. Jiancai, and L. Zhenbi, A stochastic model of a reconfigurable manufacturing system part 2: Optimal configurations, *Int. J. Prod. Res.* **38** (2000) 2829–2842.
74. H. F. Lee, Optimal design for flexible manufacturing systems: Generalized analytical methods, *IIE Trans.* **31** (1999) 965–976.
75. S. M. Ordoobadi and N. J. Mulvaney, Development of a justification tool for advanced manufacturing technologies: System-wide benefits value analysis, *J. Eng. Technol. Manag.* **18** (2001) 157–184.
76. D. Gupta and J. A. Buzacott, Models for first-pass FMS investment analysis, *Int. J. Flex. Manuf. Syst.* **5** (1993) 263–286.
77. B. P. Pang and K. Khodabandehloo, Modeling the reliability of flexible manufacturing cells, *Proceedings of 29th International MATADOR Conference*, 1992, pp. 183–190.
78. J. Agapiou, Sequence of operations optimization in single-stage multifunctional systems, *J. Manuf. Syst.* **10** (1991) 194–207.
79. K. Hitomi, Analysis of optimal machining conditions for flow-type automated manufacturing systems: Maximum efficiency for multi-product production, *Int. J. Prod. Res.* **28** (1990) 1153–1162.
80. J. Agapiou, Optimization of multistage machining systems, part I: Mathematical solution, and part II: The algorithm and applications, *ASME J. Eng. Ind.* **114** (1992) 524–538.
81. J. Simpson, R. Hocken, and J. Albus, The automated manufacturing research facility, *J. Manuf. Syst.* **1** (1982) 17–32.
82. M. K. Senehi, S. Wallace, and M. E. Luce, An architecture for manufacturing systems integration, *Proceedings of ASME Manufacturing International Conference*, Dallas, TX, 1992.
83. M. K. Senehi et al., Manufacturing systems integration initial architecture document, NISTIR 91-4682, September, 1991.
84. M. K. Senehi et al., Manufacturing systems integration control entity interface document, NISTIR 91-4626, June 1991.
85. S. Atluru, A. Deschpande, S. Huang, and R. Pieper, PNEUVIZ: MTConnect compliant compressed air monitoring application, *Proceedings of the ASME MSEC Conference*, 2012, Paper MSEC2012-7389.
86. MTConnect is for real, *Mod. Mach. Shop* (November 16, 2009). http://www.mmsonline.com.
87. Understanding MTConnect agents and adapters, *Mod. Mach. Shop* (December 21, 2011). http://www.mmsonline.com.
88. Z. Li and R. Katz, Theoretical study on a reconfigurable parallel kinematic high speed drilling machine, *CIRP Second International Conference on Reconfigurable Manufacturing*, Ann Arbor, MI, 2003.
89. Y. Furukawa et al., Application of a ceramic guideway in a ultra precision machine tool, *Bull. Jpn. Soc. Prec. Eng.* **20** (1986) 197–198.
90. Y. Furukawa et al., Development of ultra precision machine tool made of ceramics, *CIRP Ann.* **35** (1986) 279–282.
91. H. Sugishita et al., Development of concrete machining center and identification of dynamic and thermal structural behavior, *CIRP Ann.* **37** (1988) 377–380.
92. D. G. Lee et al., Manufacturing of a graphite epoxy composite spindle for a machine tool, *CIRP Ann.* **34** (1985) 365–369.
93. J. D. Suh and D. G. Lee, Composite machine tool structures for high speed milling machines, *CIRP Ann.* **51** (1985) 285–288.
94. K. Takada and I. Tanabe, Basic study on thermal deformation of machine tool structure composed of epoxy resin concrete and cast iron, *Bull. Jpn. Soc. Precision Eng.* **21** (1987) 173–178.
95. J. Lazan, *Damping of Materials and Members in Structural Mechanics*, Pergamon Press, London, U.K., 1968.

96. M. Tsutsumi and Y. Ito, Damping mechanism of a bolted joint in machine tools, *Proceedings of 20th International MTDR Conference*, 1979, pp. 443–448.

97. S. Haranath, K. Ganesan, and B. Rao, Dynamic analysis of machine tool structures with applied damping treatment, *Int. J. Mach. Tools Manuf.* **27** (1987) 43–55.

98. E. Rivin and H. Kang, Improvement of machining conditions for slender parts by tuned dynamic stiffness of tool, *Int. J. Mach. Tools Manuf.* **29** (1989) 361–376.

99. ASME B5.54, Methods for performance evaluation of computer numerically controlled machining centers, An American National Standard, 1992.

100. G. V. Fedewa et al., Parallel structures and their applications in reconfigurable machining systems, *Parallel Kinematic Machines International Conference*, Ann Arbor, MI, 2000, pp. 87–97.

101. M. Mandeli, T. Nagao, Y. Hatamura, M. Mitsuishi, and M. Nakao, New machine tools and systems, in: A. Dashchenko, Ed., *Manufacturing Technologies for Machines of the Future—21st Century Technologies*, Springer, Berlin, Germany, 2003, pp. 611–619.

102. D. Stewart, A platform with six degrees of freedom, *Proc. Inst. Mech. Eng.* **180** (1965) 371–386.

103. M. Weck and D. Staimer, Parallel kinematic machine tools—Current state and future potentials, *CIRP Ann.* **50**:2 (2002) 671–684.

104. B. S. El-Khasawneh and P. M. Ferreira, On using parallel link manipulators as machine tools, *Trans. NAMRI/SME* **25** (1997) 305–310.

105. F. Pierrot and O. Company, Towards non-hexapod mechanisms for high performance parallel machines, *IECON 2000, IEEE Industrial Electronics Conference*, Nagoya, Japan, October 22–28, 2000.

106. L. Molinari et al., An integrated methodology for the design of parallel kinematic machines (PKM), *CIRP Ann.* **47** (1998) 341–344.

107. H. J. Warnacke et al., Development of hexapod based machine tool, *CIRP Ann.* **47** (1998) 337–340.

108. F. Majou, P. Wenger, and D. Chablat, Design of a 3 axis parallel machine tool for high speed machining: The orthoglide, *IDMME 2002*, Clermont-Ferrand, France, May 14–16, 2002.

109. M. Schwaar, T. Jaehnert, and S. Ihlenfeld, Mechatronic design—Experimental property analysis and machining strategies for a 5-Strut-PKM, *Third Chemnitz Parallel Kinematics Seminar*, May 25, 2002, Verlag Wissenschaftliche Scriptien, Zwickau, pp. 671–682.

110. T. H. Chang et al., The development of a parallel mechanism of 5-DOF hybrid machine tool, *Parallel Kinematic Machines International Conference*, Ann Arbor, MI, 2000, pp. 79–86.

111. V. Gopalakrishnan et al., Parallel structures and their applications in reconfigurable machining systems, *Parallel Kinematic Machines International Conference*, Ann Arbor, MI, 2000, pp. 87–97.

112. H. K. Toenshoff et al., Influence of manufacturing and assembly errors on the pose accuracy of hybrid kinematics, *Parallel Kinematic Machines International Conference*, Ann Arbor, MI, 2000, pp. 255–263.

113. D. Zhang and C. M. Gosselin, Kinetostatic analysis and optimization of the tricept machine tool family, *Parallel Kinematic Machines International Conference*, Ann Arbor, MI, 2000, pp. 174–187.

114. J. Trusly, J. Ziegert, and S. Ridgeway, Fundamental comparison of the use of serial and parallel kinematics for machines tools, *CIRP Ann.* **48** (1999) 351–356.

115. S. A. Shamblin and G. J. Wiens, Characterization of dynamics in PKMs, *Parallel Kinematic Machines International Conference*, Ann Arbor, MI, 2000, pp. 24–33.

116. M. Weck and D. Stainer, Accuracy issues of parallel kinematic machine tools: Compensation and calibration. *Parallel Kinematic Machines International Conference*, Ann Arbor, MI, 2000, pp. 36–41.

117. C. C. Vong et al, On the inertial coupling effect and control of a parallel manipulator, *Parallel Kinematic Machines International Conference*, Ann Arbor, MI, 2000, pp. 136–142.

118. Y. Takeda et al., Stiffness analysis of a spatial six-degree-of-freedom in-parallel actuated mechanism with rolling spherical bearings, *Parallel Kinematic Machines International Conference*, Ann Arbor, MI, 2000, pp. 264–273.

119. C. M. Clinton et al., Stiffness modeling of a stewart-platform-based milling machine, *Trans. NAMRI/SME* **25** (1997) 335–340.

120. G. Wiens and D. Hardage, Dynamics and controls of hexapod machine tools, *Proceedings of the First European-American Forum on Parallel Kinematic Machines: Theoretical Aspects and Industrial Requirements*, Milan, Italy, 1998.

121. A. J. Wavering, Parallel kinematic machine research at NIST: Past, present, and future, *Proceedings of the First European-American Forum on Parallel Kinematic Machines: Theoretical Aspects and Industrial Requirements*, Milan, Italy, 1998.

122. D. Mewes, O. Mewes, and P. Herbst, Festigkeit von Wekrstoffen bei Aufprallbeanspuchungen, *Materialprüfung* **51** (2009) 227–233.

123. D. Mewes and R.-P. Trapp, Impact resistance of materials for guards on cutting machine tools—Requirements for future European safety standards, *JOSE* **6** (2000) 507–520.

124. T. H. N. Brogden and F. M. Stansfield, The design of machine tool foundations, *Proceedings of 11th International MTDR Conference*, Birmingham, U.K., 1970, p. 333.

125. B. S. Baghshahi and P. F. McGoldrick, Machine tool foundations—A dynamic design method, *Proceedings of 20th International MTDR Conference*, Birmingham, U.K., 1979.

126. Unisorb, Machinery installation systems, Master catalog and engineering guide, Michigan Center, MI, 1989.

127. T. C. Duclos, D. N. Acker, and J. D. Carlson, Fluids that thicken electrically, *Mach. Des.* **60**:2 (January 21, 1988) 42–46.

128. A. Devitt, Replication materials in new machine architecture, SME Technical Paper MS90-408, 1990.

129. M. W. Browne, New diamond coatings find broad application, *New York Times*, January, 1989.

130. T. Subramavian and S. P. Rangosuami, Factors influencing the positioning accuracies of CNC machine tool slides, *12th AIMTOR Conference*, IIT, Delhi, India, 1986.

131. M. Arnone, *High Performance Machining*, Hanser Garner Publications, Cincinnati, OH, 1998.

132. K. Tanaka, Y. Uchiyam, and S. Toyooka, The mechanism of wear of polytetra-fluoroethylene, *Wear* **23** (1973) 153–172.

133. B. Mortiemr and J. Lancaster, Extending the life of aerospace dry bearings by the use of hard smooth surfaces, *Wear* **121** (1988) 289–305.

134. K. Lindsey and S. Smith, Precision motion slideways, U.K. Patent 8,709,290, April 1988.

135. B. J. Hamrock and D. Dowson, *Ball Bearing Lubrication*, Wiley, New York, 1981.

136. Anon., *Hydra-Rib Bearing*, The Timken Company, Canton, OH, 1980.

137. Z. M. Levina, Research on the static stiffness of joints in machine tools, *Proceedings of Eight International MTDR Conference*, Manchester, U.K., 1967, pp. 737–758.

138. M. Dolbey and R. Bell, The contact stiffness of joints at low apparent interface pressures, *CIRP Ann.* **19** (1979) 67–79.

139. J. C. Bandrowski, Antifriction bearings for linear motion systems, *Mech. Eng.* **109** (April 1987).

140. Z. M. Levina, Main operating characteristics of antifriction slideways, *Mach. Tooling* **58** (1986).

141. W. Rowe, *Hydrostatic and Hybrid Bearing Design,* Butterworth, London, U.K., 1983.

142. F. Stansfield, *Hydrostatic Bearings for Machine Tools*, Machinery Publishing Co., London, U.K., 1970.

143. D. Fuller, *Theory and Practice of Lubrication for Engineers*, 2nd edn., Wiley, New York, 1984.

144. R. J. Welsh, *Plain Bearing Design Handbook,* Butterworths, London, U.K., 1983.

145. J. Zeleny, Servostatic guideways—A new kind of hydraulically operating guideways for machine tools, *Proceedings of 10th International MTDR Conference*, (September 1969), Pergamon Press, Oxford, U.K., (1970), 193–201.

146. J. Powell, *Design of Aerostatic Bearings*, Machinery Publishing Co., London, U.K., (1970).

147. K. J. Stout and E. G. Pink, Orifice compensated EP gas bearings: The significance of errors of manufacture, *Tribol. Int.* **13** (1980) 105.

148. H. Yabe and N. Watanabe, A study on running accuracy of an externally pressurized gas thrust bearing (load capacity fluctuation gas to machining errors of the bearing), *JSME Int. J., Ser. III* **31** (1988) 114.

149. N. K. Arakere, H. D. Nelson, and R. L. Rankin, Hydrodynamic lubrication of finite length rough gas journal bearings, *STLE Tribol. Trans.* **33** (1990) 201.

150. A. H. Slocum, Self Compensating Hydrostatic Linear Bearing, U. S. Patent No. 5,104,237, April 14, 1992.

151. H. C. Town, Control of hydraulic transmission elements, *Power Int.* (March 1986).

152. R. L. Rickert, AC servos increase machine tool productivity, *Mach. Des.* **57**:4 (February 21, 1985) 135–139.

153. B. H. Carlisle, AC drives move into DC terrtory, *Mach. Des.* **57**:4 (May 9, 1985) 61–64.

154. W. Leonhard, Adjustable-speed AC drives, *Proc. IEEE* **76**:4 (1988).

155. S. J. Bailey, Step motion control 1985: Direct digital incrementing with servo-like performance, *Control Eng.* (August 1985).

156. C. Mirra and R. Quickel, Drive for success, *Cutting Tool Eng.* (August 1989) 75–84.

157. C. Mirra, What digital AC drives bring to machining, *Tooling Prod.* **57**:4 (July 1991) 45–46.

158. D. Horn, Linear motors provide precise positioning, *Mech. Eng.* (November 1988) 70–74.

159. B. L. Triplett, Linear motors combine muscle with a fine touch, *Mach. Des.* (May 7, 1987) 94–97.

160. W. E. Barkman, Linear motor slide drive for diamond turning machine, *Prec. Eng.* **3**:1 (1981) 44–47.

161. G. Pritschow, A comparison of linear and conventional electromechanical drives, *CIRP Ann.* **47**:2 (1998) 541–548.

162. J. S. Agapiou and C. H. Shen, High speed tapping of 319 aluminum alloy, *Trans. NAMRI/SME* **20** (1992) 197–203.
163. V. G. Belyaev et al., How ballscrew diameter affects the speed of response of a machine feed drive, *Soviet Eng. Res.* **66**:5 (1986) 7–8.
164. SKF, Transrol high efficiency ball and roller screws, Publication #3369/Z U.S.
165. Anon., Basics of design engineering: Mechanical systems, *Mach. Des.*, **64**:12 (June 1992) 293–300.
166. Thyssen Corporation, Hueller Hille, Specht CNC High Speed Machining Center, Ludwigsburg, Germany, 1993.
167. Y. Antintas, A. Verl, C. Brecher, L. Uriarte, and G. Pritschow, Machine tool feed drives, *CIRP Ann.* **60** (2011) 779–796.
168. H. Weule and T. Frank, Advantages and characteristics of a dynamic feeds axis with ball screw drive and driven nut, *CIRP Ann.* **48** (1999) 303–306.
169. Anon., *Linear Encoders for Numerically Controlled Machine Tools*, Dr. Johannes Heidenhain GmbH, Traunreut, Germany, June 2007.
170. B. Popoli, Selecting the proper bearing system for your high speed spindle application, *First Machine Tool Conference*, May 28–29, 2003.
171. B. Hodge, Spindle designs for high-speed machining, *First Machine Tool Conference*, May 28–29, 2003.
172. P. Frederickson and D. Grimes, Optimizing high-speed spindle torque characteristics, *SME High Speed Machining Technical Conference*, Northbrook, IL, 2001.
173. E. G. Korolev, Motor/spindles for NC machine tools, *Soviet Eng. Res.* **6**:12 (1986) 60–61.
174. J. L. Reif, How to select cost-effective electric drive systems, *Proceedings of 21st Annual Meeting Technical Conference*, Numerical Control Society, Chicago, IL, (1984), pp. 380–395.
175. J. Hendershot, Selection of motor types for high performance machine tool spindles, *First Machine Tool Conference*, May 28–29, 2003.
176. D. Weise, Present industrial applications of active magnetic bearings, *22nd Intersociety Energy Conversion Engineering Conference*, Philadelphia, PA, August 1987.
177. W. S. Chung et al., Ultra stable magnetic suspensions for rotors in gravity experiments, *Precis. Eng.* **2**:4 (1980) 183–186.
178. E. H. M. Weck and U. Wahner, Linear magnetic bearing and levitation system for machine tools, *CIRP Ann.* **47** (1998) 311–314.
179. Y. C. Shin, K. W. Wang, and C. H. Chen, Dynamic analysis and modeling of a high speed spindle system, *Trans. NAMRI/SME* **18** (1990) 298–304.
180. T. A. Harris, *Rolling Bearing Analysis*, Wiley, New York, 1966.
181. C. Moratz, Contact angle and bearing selection for high speed spindle bearings, *SME High Speed Machining Conference*, Chicago, IL, 2003.
182. C. Moratz, *Barden/FAG Ceramic Hybrid Spindle Bearings: Optimum Bearing for Machine Tools*, The Barden Corporation, CT, 2003.
183. Timken Company, *Hydra-Rib Bearing: Machine Tool Applications*, Canton, OH, 1980.
184. C. P. Bhateja and R. D. Pine, The rotational accuracy characteristics of the preloaded hollow roller bearing, *ASME J. Lub. Technol.* **103** (1981) 6–12.
185. ANSI/ASME B89.6.2, Temperature and humidity environment for dimensional measurement, ASME, 1973.
186. W. Blewett, J. B. Bryan, R. R. Clouser, and R. R. Donaldson, Reduction of machine tool spindle growth, Report UCRL 74672, Lawrence Livermore Laboratory, Livermore, CA, 1973.
187. D. G. S. Lee, H. C. Suh, and P. Nam, Manufacturing of a graphite epoxy composite spindle for a machine tool, *CIRP Ann.* **34** (1985) 365–369.
188. Anon., Basics of design engineering: mechanical systems, *Mach. Des.* **64**:12 (June 1992) 208–237.
189. ISO 1940/1, Mechanical vibration—Balance quality requirements of rigid rotors, Part 1: Determination of permissible residual unbalance, 1986.
190. Anon., "BalaDyne" System by Balance Dynamics Corporation, Ann Arbor, MI.
191. P. Zelinski, Should you balance your tools? *Mod. Mach. Shop* (August 1997).
192. S. Zhou and J. Shi, Optimal one-plane active balancing of a rigid rotor during acceleration, *J. Sound Vibrat.* **249** (2002) 196–205.
193. V. Wowk, *Machining Vibration: Balancing*, McGraw-Hill, New York, 1995.
194. D. Watts, *Green Manufacturing Techniques and Machine Designs by MAG*, IMTS, Chicago, IL, September 2010.
195. Anon., *GROB Hydraulic Free Machine*, GROB Systems, Inc., Mindelheim, Germany, May 2012.

196. Anon., *Siemens Introduces Sinumerik CTRL-Energy*, Siemens Industry, Inc., Elk Grove, IL, August 2011.
197. M. Mori, M. Fujishima, Y. Inamasu, and Y. Oda (Mori Seiki), A study on energy efficiency improvement for machine tools, *61st CIRP General Assembly,* Budapest, Hungary, August 2011.
198. J. Laird, Lean and green: Energy savings in manufacturing, *Manuf. Eng.* (February 2012) 71–77.
199. A. De Filippi and R. Ippolito, NC machine tools as electric energy users, *CIRP Ann.* **30**:1 (1981) 323–326.
200. R. Neugebauer, M. Wabner, T. Koch, and W.-G. Drossel, *Discussion of Key Parameters and Methods for Comparison of Energy Needs of Machine Tools*, CIRP General Assembly, Boston, MA, August 2009.
201. F. Pusavec, D. Kramar, P. Krajnik, and J. Kopac, Transitioning to sustainable production—Part II: Evaluation of sustainable machining technologies, *J. Cleaner Prod.* **18** (2010) 1211–1221.
202. A. Deshpande, J. Snyder, and D. Scherrer, Feature level energy assessments for discrete part manufacturing, *Proc. NAMRI/SME* **39** (2011).
203. E. Uhlmann, B. Duchstein, and L. Arnold, Höhcre Energieproduktivität von Werkzeugmaschinen, *Werkstatt und Betrieb* **11** (2012) 26–30.
204. Y. Oda, M. Mori, K. Ogawab, S. Nishidac, M. Fujishima, and T. Kawamuraa, Study of optimal cutting condition for energy efficiency improvement in ball-end milling with tool-workpiece inclination, *CIRP Ann.* **61**:1 (2012) 119–122.
205. S. Rothenbücher and B. Kuhrke, Energiekosten bei spanenden Werkzeugmaschinen (Energy Costs for Machine Tools), *Werkstatt und Betrieb* **9** (2010) 130–137.
206. Anon., *Aspects of Energy Efficiency in Machine Tools*, Heidenhain GmbH, Traunreut, Germany, November 2010.
207. E. Abele, T. Sielaff, A. Schiffler, and S. Rothenbucher, Analyzing energy consumption of machine tool spindle units and identification of potential for improvements of efficiency, *Proceedings 18th CIRP International Conference on Life Cycle Engineering*, Brauwnsweig, Germany, May 2011.
208. Anon., *Machining Machine Tool Innovation by iMQL System*, Horkos Corporation, Hiroshima, Japan, 2009.

4 Cutting Tools

4.1 INTRODUCTION

Cutting-tool design has a strong impact on machining performance. Properly designed tools produce parts of consistent quality and have long and predictable useful lives. An improperly designed tool may wear or chip rapidly or unpredictably, reducing productivity, increasing costs, and producing parts of deteriorating quality. Tooling thus has a major influence on the productivity and economics of a process. It is important to consider all tooling geometries and material options for a given application, and especially the range of speeds and feeds for which each can be applied and their typical failure modes. In high-volume applications, perishable tooling costs are typically 3% of the component cost. Modifying tooling to increase tool life 50% reduces the total cost per component by 1%–2%. Using higher cost tool capable of running at higher material removal rates (MRRs) may be a better option, since a 20% increase in the MRR could reduce the total cost per component by 15%. In some aerospace applications, in contrast, a primary concern in tooling design is ensuring that the tool does not chip or break to avoid damaging an expensive workpiece.

Cutting tools may be broadly classified as single point tools, which have one active cutting edge, and multipoint, multifunctional, or multitasking tools, which have multiple active cutting edges. Single point tools are commonly used for turning and boring, while multipoint tools are used for drilling, milling, and in special purpose tooling. Multifunctional or multitasking tools are used to machine multistep holes or several features with one tool. Tools may be further classified based on the cutting-edge material, geometry, and clamping method. The best choice of tool material and geometry in a given operation depends on the volume of parts to be machined, the workpiece material, the required accuracy, and the capabilities of the available machine tools.

The objective of this chapter is to discuss conventional and advanced cutting-tool technologies and to explain the properties and characteristics of tools, which influence tool design or selection. The general properties of available tool materials and tool coatings are discussed in Sections 4.2 and 4.3. Specific information on the design and selection of tools for turning, boring, milling, drilling, reaming, threading, grinding, honing, microsizing, and burnishing are discussed in Sections 4.4 through 4.13.

4.2 CUTTING-TOOL MATERIALS

4.2.1 INTRODUCTION

This section discusses the basic properties of tool steels and high-speed steels, carbides, cermets, ceramics, superabrasives, and the corresponding coatings for these materials. Critical factors for performance and comparison are explained. The proper selection of a specific tool material and/or coating for a particular application is also discussed.

4.2.2 MATERIAL PROPERTIES

Cutting tools must be made of materials capable of withstanding the high stresses and temperatures generated during chip formation. Ideally, tool materials should have the following properties:

1. High penetration hardness at elevated temperatures to resist abrasive wear
2. High deformation resistance to prevent the edge from deforming or collapsing under the stresses produced by chip formation

3. High fracture toughness to resist edge chipping and breakage, especially in interrupted cutting
4. Chemical inertness (low chemical affinity) with respect to the work material to resist diffusion and chemical wear
5. High thermal conductivity to reduce cutting temperatures near the tool edge
6. High fatigue resistance, especially for tools used in interrupted cutting
7. High thermal shock resistance to prevent tool breakage in interrupted cutting
8. High stiffness to maintain accuracy
9. Adequate lubricity (low friction) with respect to the work material to prevent built-up edge, especially when cutting soft, ductile materials.

The first three properties are required to prevent sudden, catastrophic failure of the tool. Properties 1, 4, and 5 are required for the tool to resist the high temperatures generated during chip formation; as discussed in Chapter 9, elevated tool temperatures can cause rapid tool wear to do thermal softening or diffusion and chemical wear [1–4]. Properties 3, 6, and 7 are required to prevent chipping of the tool, especially in interrupted cutting. Fracture toughness is usually characterized by the transverse rupture strength; fatigue resistance is characterized by the Weibull modulus. As shown in Figure 4.1, properties 1, 4, and 5 generally determine the maximum cutting speed at which a tool can be used, while properties 3 and 6 determine allowable feed rates and depths of cut. It should be noted that fracture and fatigue strength requirements vary with the cutting speed, since cutting forces usually decrease with increasing cutting speed.

The common tool substrate materials currently in use include high-speed steels (HSS), cobalt enriched high speed steels (HSS-Co), sintered tungsten carbide (WC), cermets, ceramics, polycrystalline cubic boron nitride (PCBN), polycrystalline diamond (PCD), and single-crystal natural diamond. (Note that many production tools are coated, so that the properties of the coating as well as the substrate are important.) These materials provide a wide range of combinations of properties [5]. A relative comparison of their mechanical and physical properties, hardness, toughness, Young's modulus, thermal conductivity, specific heat, softening temperature, and Weibull moduli is given in Table 4.1 [1–9]; the magnitude of the property increases between the different materials moving from the bottom row to the top row in the table. Figure 4.2 shows a more specific

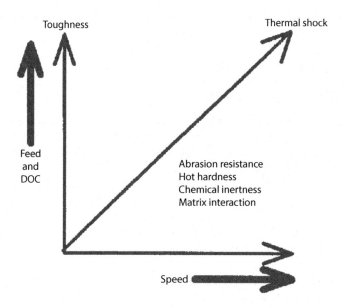

FIGURE 4.1   The influence of tool material properties on optimization of cutting conditions.

TABLE 4.1

Relative Comparison of Various Properties of Cutting Tool Materials

Micro-Hardness	Fracture Toughness	Young's Modulus	Modulus of Rupture	Thermal Conductivity	Specific Heat	Softening Temperature	Chemical Inertness	Weibull Modulus
PCD	C2 carbide	PCD	C2 carbide	PCD	Al_2O_3	PCD	Al_2O_3	C2 carbide
PCBN	(SiCw)-Al_2O_3	PCBN	PCD	PCBN	Si_3N_4, HIP	PCBN	(SiCw)-Al_2O_3	(SiCw)-Al_2O_3
Al_2O_3 + TiC	ZrO_2	C2 carbide	Sialon	C2 carbide	Si_3N_4, RB	Al_2O_3	ZrO_2	Si_3N_4, HIP
C2 carbide	Sialon	Al_2O_3	Al_2O_3 + TiC	Si_3N_4, HIP	Al_2O_3 + TiC	C2 carbide	Sialon	Si_3N_4, RB
Si_3N_4, HIP	Si_3N_4, HIP	Al_2O_3 + TiC	Si_3N_4, HIP	Sialon	Sialon		Si_3N_4	Sialon
Sialon	Al_2O_3 + TiC	(SiCw)-Al_2O_3	ZrO_2	Al_2O_3 + TiC	ZrO_2		C2 carbide	Al_2O_3 + TiC
Al_2O_3	Al_2O_3	Si_3N_4, HIP	Al_2O_3	Si_3N_4, RB				Al_2O_3
ZrO_2	Si_3N_4, RB	Sialon	PCBN	Al_2O_3				
Si_3N_4, RB		Si_3N_4, RB	Si_3N_4, RB	ZrO_2				
		ZrO_2						

Note: Materials are listed in descending order for the given property by column.

C System	Materials to be Machined	Operation	Direction of Increase in Characteristic		ISO System
			Of Cut	Of Carbide	
C1 C2 C3 C4	Cast iron (all types) Hard steel Nonferrous Nonmetallics	Roughing ⟶ Finishing	Increasing Speed ↑ ↓ Increasing Feed	Wear Resistance ↑ ↓ Toughness	K-40 K-35 K-30 K-10 K-01
C5 C6 C7 C8	Ferrous Carbon steel Alloy steel Stainless steel anneal Steel casting	Roughing ⟶ Finishing	Increasing Speed ↑ ↓ Increasing Feed	Wear Resistance ↑ ↓ Toughness	P-50 P-40 P-30 P-20 P-10 P-01
	Low-strength steels Ductile cast iron High-temper. alloys Nonferrous	Roughing ⟶ Finishing	Increasing Speed ↑ ↓ Increasing Feed	Wear Resistance ↑ ↓ Toughness	M-50 M-40 M-30 M-20 M-10 M-01

FIGURE 4.2 Comparison of hardness and toughness for various cutting tool materials. DCC, diamond-coated carbide; C-PM, coated powder metallurgy steel; SiN, silicon nitride ceramic; AlO, aluminium oxide ceramic.

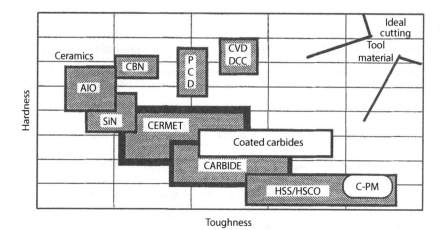

FIGURE 4.3   Relative comparison of abrasive wear resistance of tool materials.

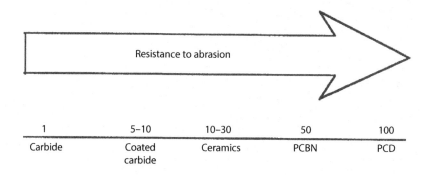

FIGURE 4.4   The effect of tool material on the allowable cutting speed.

comparison of hardness and toughness for various tool materials. Figure 4.3 shows a similar comparison of relative abrasion resistance, and Figure 4.4 shows typical allowable speed ranges for given combinations of tool and work materials, based largely on hot hardness and chemical inertness considerations. For specific grades, additional information on material properties, as well as application guidelines (e.g., allowable speed, feed, and doc ranges) can be obtained from tool suppliers. The cutting speed is the most important parameter to be selected among the cutting conditions as explained in Chapters 9 and 13. The usable cutting speed is affected by the expected tool life as discussed in Chapter 9.

As is evident from Table 4.1 and Figures 4.2 and 4.3, no single material exhibits all of the desirable properties for a tool material. Some of the desired properties, in fact, are mutually exclusive. Very hard materials, for example, tend to be brittle and thus have poor fracture toughness. The best tool material for a given application depends on several factors; as shown in Figure 4.5, these include part requirements, constraints imposed by available machine tools and tool holders, and economic considerations. For new processes, the cutting speed regime is often selected first, as this has a strong influence on the metal removal rate and tool wear characteristics as discussed in Chapter 9. The tool material selection is then made from the restricted list of materials suitable for use within that range for the given work material (Figure 4.4). In trouble-shooting existing operations to improve tool life or part quality, tool materials are often substituted based on the speed range and expected type of tool failure as shown in Figure 4.6. In either case, an understanding of the properties and application ranges of specific tool materials is required to make sound choices.

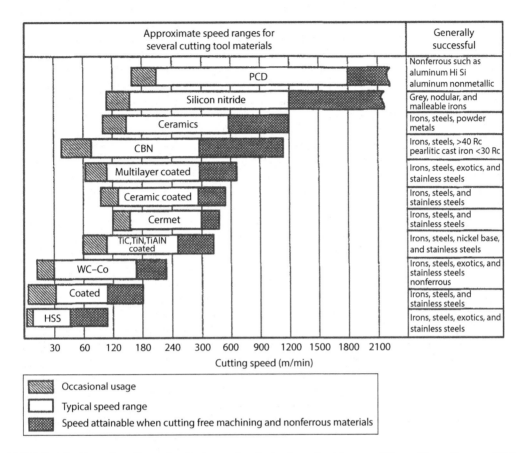

FIGURE 4.5 Procedure for tool selection and optimization of cutting conditions. (From Kramer, B.M., *ASME J. Eng. Ind.*, 109, 87, 1987.)

4.2.2.1 High-Speed Steel (HSS) and Related Materials

High-speed steels (HSS) are self-hardening steels alloyed with W, Mo, Co, V, and Cr. They exhibit *red hardness*, which permits tools to cut at a dull red heat without loss of hardness or rapid blunting of the cutting edge. HSS is inexpensive compared to the other tool materials, is easily shaped, and has excellent fracture toughness and fatigue and shock resistance. However, the hardness of HSS decreases rapidly at temperatures above 540°C–600°C, and HSSs have less wear resistance, less chemical stability, and a greater tendency to form a built-up edge than other tool materials. Their limited wear resistance and chemical stability makes HSS tools suitable for use only at limited cutting speeds. Plain HSSs are generally used at cutting speeds below 35 m/min in steel, although special alloys such as the HSS-Co alloys discussed in the following can be used at speeds up to 50 m/min. HSS can be accurately ground and resharpened using conventional abrasive wheels. HSS is very commonly used for geometrically complex rotary tools such as drills, reamers, taps, and end-mills as well as for broaches and gear hobs and form cutters. HSS tools are also widely used in multi-spindle machines (e.g., gang drill presses, screw machines, and older transfer machines) with limited rigidity and speed capabilities.

Plain HSSs are broadly classified as T-type steels, which have tungsten as the major alloying element, and M-type steels, in which the major alloying element is molybdenum. The T-types are less tough than M-type but are heat-treated more easily. M-types are more widely used for rotary tooling, especially drills, end-mills, and taps. Of the standard plain HSSs, M2 is most widely used for drills and taps, T42 has the best abrasion resistance, and M42 has the greatest hot strength.

Sintered or powder metal (P/M) HSS shows minimal distortion and improved wear resistance, hot hardness, and toughness compared with plain HSS. P/M HSS materials are especially beneficial

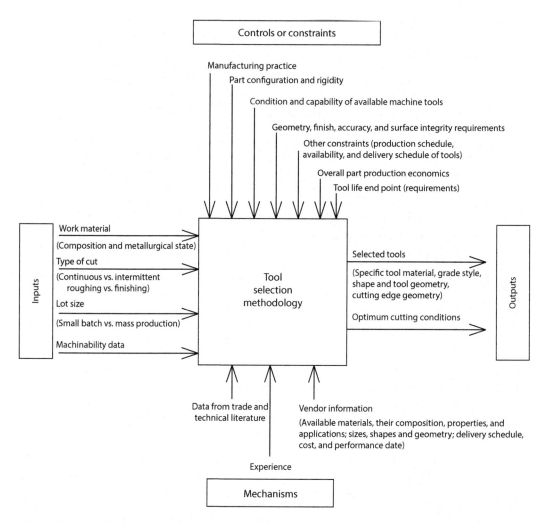

FIGURE 4.6 Substitution trends in cutting tool materials. (From Clark, I.E. and Hoffmann, J., PCD Tooling in the Automotive Industry, *Diamond & CBN Ultrahard Materials Symposium*, IDA, Windsor, Ontario, Canada, September 29–30, 1993, pp. 115–130.)

in tapping and broaching but are also used for drills and end mills because they have better toughness than carbide tools, which may chip and crack in interrupted cuts or when encountering hard spots. In proper applications, P/M tools can double tool life and remove material at twice the rate of conventional tooling [10].

In some applications, HSS is alloyed with cobalt (HSS-Co) or vanadium to produce grades with increased toughness, hot hardness, and wear resistance. HSS-Co grades are used especially for drills and taps. They are suitable for use at higher speeds, feeds, and depths of cut than plain HSSs and have better fracture toughness than sintered carbides. Similarly, stellite, a cobalt-based alloy containing chromium, tungsten, and carbon, is sometimes used for tools. Most commonly, solid stellite tool bits are used to turn difficult-to-machine materials (e.g., weldments), which cause chipping of carbide tools.

4.2.2.2   Sintered Tungsten Carbide (WC)

Sintered tungsten carbide-based hardmetals are the most common tool materials for turning, milling, threading, and boring using indexable inserts and are also common for solid round tooling.

Cemented tungsten carbide inserts and blanks are manufactured by mixing, compacting, and sintering tungsten carbide (WC) and cobalt (Co) powders. The Co acts as a binder for the hard WC grains; as discussed in the following, the grain size and binder content largely determines the insert's physical properties. Characteristics of tungsten carbides include high transverse rupture strength, high fatigue and compressive strength, and good hot hardness. The modulus of elasticity and torsional strength are twice those of HSS. Carbides conduct heat away from the tool–chip interface well and can be tailored to meet specific thermal shock requirements. By varying the cobalt content, the relative balance of hardness and toughness can be changed. Their main drawback is that they have only average chemical and thermal stability at high temperatures, which makes them unsuitable for machining steels at high cutting speeds. For nonferrous work materials, WC tools will exhibit 2–3 times the productivity and 10 times the life of HSS tools; in steels 2 times the productivity and 5 times the life.

In the United States, WC grades are often classified into eight categories denoted C1 through C8 [11]. The grades are broadly divided into two classes (C-1 through C-4 and C-5 through C-8) according to the workpiece material (see Figure 4.7). As the number increases within each class, shock resistance (toughness) decreases, hardness increases, high-temperature deformation resistance and wear-resistance increase, and carbide grain size decreases. Therefore, an increase in cutting speed or the feed load should be followed by an increase or decrease, respectively, of the classification number for the carbide grade. Hardness measurements primarily indicate the Co volume and WC grain size, and not the actual hardness of the carbide constituent. Similarly, a European classification has been adopted in ISO Standard 513 [12]. This system consists of six categories designated as P-(heavily alloyed multicarbides), M-(low-alloyed multicarbides), and K-, N-, S-, and H-grades. A summary and comparison of the characteristics of C-, K-, P-, and M-grades is given in Figure 4.7. This table can be used to determine appropriate application ranges for specific grades; for example, of K-grades suitable for use on cast iron, K01 is a wear-resistant, finishing grade suitable for finish boring with no shock, while K40 is a tough grade suitable for rough milling. The N-grades are for nonferrous metals (i.e., aluminum), the S-grades are for superalloys and titanium, and the H-grades are for hard materials. By convention, K25 represents general purpose milling. It should be noted that the same carbide grade from different producers could vary significantly since the classification systems do not rigorously describe the criteria for classifying carbide grades.

Two basic classes of carbide materials are used for cutting tools: two-phase WC-Co (straight cemented tungsten carbide; grades K01 through K40, or C1 through C4), and alloyed WC-Co grades,

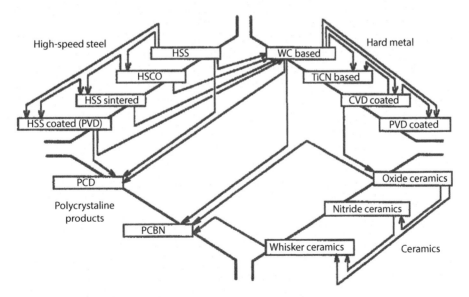

FIGURE 4.7   Classification of carbide tool materials according to use.

in which part of the WC is replaced by a solid solution of cubic carbides, such as titanium carbide (TiC), tantalum carbide (TaC), niobium carbide (NbC), or a combination of these materials (grades P01 through P50, M01 through M50, and C5 through C8). Straight WC-Co grades are used for workpiece materials that cause primarily abrasive tool wear and generate a short, discontinuous chip; typical materials include cast iron, high temperature alloys, high-silicon aluminum, other nonferrous alloys, and nonmetals. When used with steels, which generally yield longer continuous chips, straight WC-Co grades often fail due to crater wear on the rake face of the tool. As discussed in Chapter 9, crater wear is caused by diffusion of constituents of the tool into the chip at high temperatures. Alloying also improves cratering resistance; alloyed grades containing TiC or TaC were developed for improved crater wear resistance when machining steels. Compared to straight grades, alloyed grades have increased heat resistance, compressive strength, and chemical stability. TaC alloyed grades have a higher hot hardness and better thermal shock resistance than TiC alloyed grades.

The hardness, fracture toughness, and heat resistance of carbide grades depend on the Co, TiC, and TaC contents and on the carbide grain size. Increasing the Co content decreases hot hardness and edge wear, crater wear, and thermal deformation resistance, but increases fracture toughness. The compressive strength is affected significantly by the Co content, and increases with Co percentage to a maximum of about 4%. The abrasive wear resistance of cemented carbides increases with increasing TiC content and decreases with increasing TaC content. Substrates with Co content less than 6% are attractive for ferrous machining at higher speeds with smaller depths-of-cut and limited interruptions.

The WC grain size affects the tool's hardness, toughness, and edge strength. Fine WC grains generate thin sections of Co binder that are constrained by the WC grains, resulting in harder materials with lower fracture strength. Typical WC grades have 1.5–5 μm grain size and a hardness between 89 and 93 Rockwell "A." Fine grain WC is used for inserts and for solid drills, reamers, end mills, and other rotary tools. Coarse grain WC (2.5–6 μm) has higher fracture toughness but less wear resistance than finer grades and is used primarily for roughing. Since hardness increases with decreasing grain size, micrograin carbides exhibit superior resistance to crater wear, notching, and chipping in semi-roughing through finishing applications and perform well in applications where normal carbide grades tend to chip or break. Micrograin carbide grades are classified as fine grain (<1.3 μm), finest or submicron grain (<0.8 μm), ultra fine grain (<0.5 μm), and nano grain (<0.2 μm). Ultra fine grain carbide is used in special cases such as rotary tooling for high speed and/or high throughput machining applications in ferrous materials and is becoming common for high performance rotary tooling. It improves cutting edge strength and stability, preventing premature chipping and material loading and allows sharper edges because ultra-fine grain carbide is 15%–20% tougher than common carbide grades (grain size 2–3 μm with 8% Co). Nano-phase carbides are being developed for very small tools and for round high performance tools for exotic materials. An improvement of carbide performance is obtained by providing a cobalt-enriched (binder phase) layer (about 13–25 μm thick) on the surface of the tool or the tool corner. This enriched layer contains two to three times the cobalt concentrations of the bulk material. This improves the toughness of the cutting edge and resistance to chipping.

Basic guidelines for selecting carbide grades are (1) use the lowest Co content and finest grain size, provided edge chipping and tool breakage do not occur; (2) use straight WC grades when abrasive edge wear is of concern; (3) use TiC grades to prevent crater wear and/or both crater and abrasive wear; (4) use TaC grades for heavy cuts in steels.

4.2.2.3 Cermets

Cermets are TiC-, TiN-, or TiCN-based hardmetals often described as ceramic or carbide composites. The physical properties and application range of cermets generally fall between those of WC and plain ceramics. Cermets are less susceptible to diffusion wear than WC and have more favorable frictional characteristics. However, they have a lower resistance to fracture (lower strength and toughness) and a higher thermal expansion coefficient than WC, and they are more feed sensitive.

TABLE 4.2
Cermet Machining Recommendations

Turning and Boring	Grade	Speed (m/min)	Feed (mm/rev)
Material			
Steels	Tough	60–300	0.15–0.35
Steels	Hard	60–340	0.10–0.35
Stainless steels	Tough and hard	45–270	0.10–0.30
Cast and nodular iron	Tough	60–250	0.15–0.45
Cast and nodular iron	Hard	60–360	0.15–0.45
Nickel	Hard	60–200	0.10–0.25
Milling			
Steels	Tough	60–250	0.05–0.15
Steels	Hard	60–340	0.05–0.15
Stainless steels	Tough and hard	60–230	0.05–0.15
Cast iron	Hard	100–360	0.05–0.15
Nodular iron	Hard	45–170	0.05–0.15

Cermets have a higher bending strength and fracture toughness than ceramics, higher thermal shock resistance than oxide-based ceramics, and lower hardness than all ceramics [13]. In general, cermets have excellent deformation resistance and high chemical stability, but relatively poor edge strength. They can be used with sharp cutting edges in many finishing applications, which enable achievement of smooth surface finishes.

Early cermets consisted of TiC particles sintered in a Ni binder [14]. Contemporary cermets consist of TiC and TiN particles sintered with a refractory metallic binder, usually composed of nickel (Ni), cobalt (Co), tungsten (W), tantalum (Ta), or molybdenum (Mo) [15]. Ni and Ni-Mo are the most commonly used binders, and the binder volume is typically between 5% and 15%. Ni free grades with Co binders have also been developed. TiC provides hot hardness for wear resistance, oxidation resistance, chemical stability, and improved notch resistance, and reduces the tendency of the work material to adhere to the tool; TiN provides fracture toughness and thermal shock resistance. Cermets are generally available in three grades—hard, tough, and (relatively) tough but hard. Hard cermets are used in applications requiring high resistance to wear and plastic deformation, such as semi-finish and finish cutting of steels, stainless steels, free machining aluminum, and other nonferrous alloys (brass, zinc, and copper) and some cast irons as summarized in Table 4.2. Tough cermets are also used in semi-finish and finish applications (especially milling) and in some rough continuous cuts in low alloy steels, stainless steels, ductile irons, and hard steels. The tough but hard grade is used for turning and boring and for finishing milling operations. In appropriate applications, cermets provide 20%–100% longer life than coated carbides. The principal advantage of cermets over carbide is its ability to operate at much higher surface cutting speeds, with longer cutting edge life. Cermets are generally used in semi-finish to finish applications and especially high speed finishing applications. The allowable cutting speed is usually lower than that attainable with ceramics. Recommended cutting conditions for various workpiece materials are given in Table 4.2. Cermets tend to be more shock resistant than ceramics, and coolant is often recommended for finish turning, threading, and grooving with coated cermets. Micrograin cermets have much better thermal shock resistance, allowing coolant to be used in all operations.

4.2.2.4 Ceramics

Ceramic tools are hard and chemically stable and have replaced carbide tools in many high-speed machining (HSM) applications. Ceramics can withstand higher temperatures than carbides, allowing a three- to ten-fold increase in cutting speed and a two- to five-fold increase in the metal removal rates.

The mechanical properties of ceramics are superior to those of carbides only at higher temperatures (e.g., above 800°C). They retain excellent hardness and stiffness at temperatures from 1000°C to 1500°C (carbides soften appreciably at temperatures above 850°C), and do not react chemically with most workpiece materials at these temperatures. They provide better size control due to lower tool wear rates, resulting in improved quality. They do, however, have several weaknesses: relatively low strength, poor resistance to thermal and mechanical shock, and a tendency to fail by chipping. Ceramic monolith materials may not have predictable failure times, and may fail catastrophically in ways that can damage workpieces. To address this limitation, ceramic composite materials have been developed. Ceramics are usually used without coolant to avoid thermal shock. Mechanical shock should be minimized by using stiffer machine tools and low frequency interrupted cuts. This makes selecting the proper speed and edge preparation essential when using ceramic tooling.

Ceramic cutting tools, mostly made of Al_2O_3 and Si_3N_4 based materials, can be divided into four categories:

1. Aluminum oxide, Al_2O_3, sometimes mixed with zirconium oxide, Al_2O_3–ZrO_2. These tools are yellow to gray/white in color. Tools made of Al_2O_3 and phase-transformation toughened Al_2O_3–ZrO_2 exhibit high chemical inertness and resistance to wear and thermal deformation. They are used for continuous shallow cuts (semi-finishing and finishing operations) at relatively low feed rates. Typical applications include turning and hard turning carbon steels, alloy steels, tool steels (<38R_c), and gray, nodular, or malleable cast irons (<300 BHN) at speeds up 1000 m/min [7,16].

2. Alumina-titanium carbide composites, Al_2O_3–TiC, containing 30%–40% TiC. This material, which is black in color, has higher transverse rupture strength, thermal shock resistance, and hardness than conventional Al_2O_3, but still has a relatively low resistance to fracture. It is effective for continuous cuts on alloy steels, chilled and malleable cast irons, hardened ferrous materials (35–65R_c), and exotic alloys [7,16–18]. Coated Al_2O_3–TiC tools are used for finish turning of hardened steels and irons. All alumina-based (Al_2O_3, Al_2O_3–ZrO_2, and Al_2O_3–TiC) materials tend to crack and exhibit notch wear when machining steel. Also, chemically induced wear can occur, depending on the cutting temperature and the surrounding environment (air, humidity, coolant, etc.). Al_2O_3-based tools are unsuitable for machining aluminum alloys and titanium alloys because of their strong chemical affinity to these materials. They substitute for P01 to P05 or C8 carbide inserts.

3. Silicon nitride-based materials, such as reaction-bonded silicon nitride (Si_3N_4, RB), hot pressed silicon nitride (Si_3N_4, HIP), sintered reaction bonded Si_3N_4, sintered Si_3N_4, and sialon (Si_3N_4–Al_2O_3), all of which are gray in color [19–23]. Si_3N_4 RB and HIP are combinations of Si_3N_4 with yttrium, Al_2O_3, and TiC. Sialon is a ceramic alloy containing silicon, aluminum, oxygen, and nitrogen, originally developed to be simpler to fabricate into tools than monolithic silicon nitride. Compared to the materials in the first two categories, these tough ceramics exhibit superior wear and notch resistance, high red hardness, and resistance to thermal shock; tools made from them are consequently more reliable. Si_3N_4 HIP-based tools are extremely wear resistant when used to machine cast and malleable irons but are subject to excessive temperature-activated wear when machining steels and other ductile materials at high speeds. Sialon, conversely, has been applied successfully on both gray cast iron and steel at high speeds. Sialon is more chemically stable than Si_3N_4 but not quite as tough or resistant to thermal shock. Si_3N_4 is commonly used for machining cast iron at speeds up to 1200 m/min; for this work material, the speed limit for plain WC tooling is 100 m/min. More chemically stable grades, especially sintered grades, are also commonly used to machine nickel-based superalloys for aerospace and corrosion components, as well as hard steels for a variety of uses. They are not generally used for aluminum alloys due to the high solubility of silicon in aluminum. They substitute for K01 to K05 or C4 carbide inserts.

The wear rate of TiC and Ti(C,N)-coated Si_3N_4-(30%)TiC tools has been found to be significantly lower than Al_2O_3–TiC tools [25]. When a Si_3N_4–(15%)Al_2O_3–(30%)TiC tool is coated with TiC–Al_2O_3, or TiN–Al_2O_3 it outperforms Si_3N_4 and Al_2O_3 based ceramics for HSM of steels [24].

4. Silicon carbide whisker reinforced alumina (SiC_w)–Al_2O_3. This material was designed to combine reliability and superior resistance to fracture, thermal shock, and wear [16,25–27]. The tool life achievable with this material is not necessarily greater than that achievable with other ceramic tools. It is most effective when used to machine high-temperature alloys at higher cutting speeds, especially nickel-based alloys (e.g., inconel) [25]. Unlike other ceramic tools materials, it can be run with coolant. Sialon often outperforms (SiC_w)–Al_2O_3 when machining gray cast iron and 1045 and 4340 steels. (SiC_w)–Al_2O_3 substitutes for K01 to K05 or C3 and C4 carbide inserts.

Figure 4.8 shows the hot hardness and Palmquist fracture toughness of ceramic materials. Most ceramic cutting tools are presently used in the form of indexable inserts, which have been well proven for higher speed machining. They have replaced carbides in many turning and milling operations. Ceramic inserts with grooved chipbreakers are available. Operations such as drilling and end milling of ferrous materials and superalloys could be performed at higher speeds if the appropriate ceramic tools were available. Solid and indexable ceramic rotary tools are available in limited sizes and geometries, but because they are very sensitive to torsion and bending, they require cutting-edge preparations, which are difficult to produce by grinding.

When using ceramic tools, a chamfer at the entrance and exit of interrupted surfaces is recommended.

4.2.2.5 Polycrystalline Tools

Polycrystalline tools provide maximum tool life at high cutting speeds. Two materials have been developed: *polycrystalline cubic boron nitride (PCBN)* and *polycrystalline diamond (PCD)*. Both materials are manufactured using a high temperature, high pressure process in which individual diamond or CBN particles are consolidated in the presence of iron, nickel, and/or cobalt catalysts

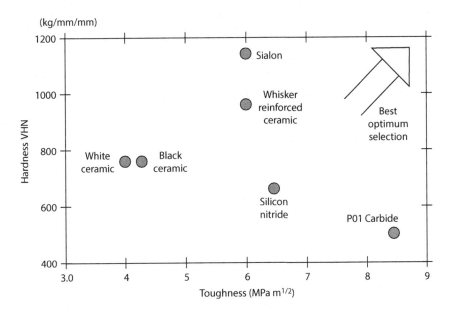

FIGURE 4.8 Comparison of microhardness (VHN-1 kg) at 1000°C and toughness of ceramic tool materials using P01 carbide as a baseline. (After Mehrotra, P.K., Productivity Improvement in Machining by Applying Ceramic Cutting Tool Materials, *A Systems Approach to Machining*, ASM, Materials Park, OH, 1993, pp. 15–20.)

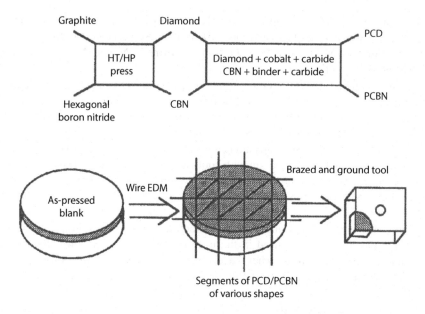

FIGURE 4.9   Manufacturing process for PCD/PCBN inserts.

that promote grain consolidation into a solid mass (solid polycrystalline) or on a WC substrate (backed polycrystalline). The grain size is small. Once manufactured, the polycrystalline compact is cut by electrodischarge machining (EDM) or laser etching into smaller pieces, which are often brazed onto WC inserts as shown in Figure 4.9 [28]. A slightly different approach presses the PCD or CBN layer on the carbide insert or round tool and bonds the two materials together during sintering. This increases edge integrity and allows for more complex geometries [29].

4.2.2.6   Polycrystalline Cubic Boron Nitride (PCBN)

At elevated speeds, PCBN, the second hardest material known, remains inert and retains high hardness and fracture toughness. Although it has superior hardness, the fracture toughness of PCBN falls between that of WC and ceramics. PCBN has a high thermal conductivity and low thermal expansion coefficient, which makes it less sensitive to thermal shock than ceramics [30]. It is thermally stable at temperatures upto 1400°C [30,31]. PCBN can wear by diffusion when cutting ferrous alloys at high speeds, but outperforms WC in this respect. The use of PCBN tooling can produce self-induced hot-cutting in hard turning because it can be used at cutting speeds sufficient to cause workpiece heating and softening.

There are several grades of PCBN insert materials and tools. The major types are listed in Table 4.3. The fracture toughness and thermal conductivity increase with increasing CBN content. However, inserts with a low CBN content have greater compressive strength.

PCBN is well suited for high-speed machining of ferrous and other hard materials with hardness between 45 and $65R_c$; such materials include sintered (P/M) irons, hard and soft pearlitic cast

TABLE 4.3
Major Characteristics of PCBN Materials

CBN Content (%) (Approx.)	Catalyst/Matrix/Binder	Format
90	Metal (Al)	Solid or carbide substrate
80	Metal/ceramic (Ti and Al)	Carbide substrate
<70	Ceramic (TiN or TiC)	Carbide substrate

irons, hardened steels, high-speed steels, stainless steels, and nickel-based alloys [32–39]. PCBN can be used to machine hard steels and superalloys at speeds roughly equal to those attainable using conventional tools to machine soft materials. The high fracture toughness of PCBN tools makes them suitable for interrupted cutting operations such as milling. Ceramic tools typically fail in these applications at medium to high speeds. The straight PCBN grade can be used for most of the aforementioned applications for roughing and finishing operations. Composite PCBN grades can be used for high speed finish machining of hardened steels and for interrupted and continuous finishing operations on hard and soft cast irons. PCBN/ceramic grades are best for turning hard steels since they have less tendency to interact chemically with the chips. The operative speed range for PCBN when machining gray cast iron is 600–1400 m/min; speed ranges for other materials are as follows: hard cast iron (>400 BHN), 80–300 m/min; superalloys (>35R_c), 180–400 m/min or higher for light finishing cuts; hardened steels (>45R_c), 70–300 m/min; sintered iron, 100–300 m/min. Case histories for cutting a variety of materials have been reported [40]. Materials that should not be cut with PCBN include soft steels, ferritic gray iron (due to a reaction with common binder materials), and ductile and malleable iron <45R_c.

In addition to cutting speed, the most important factor affecting PCBN insert performance is the cutting edge geometry or edge preparation [41,42]. It is best to use PCBN tools with a honed or chamfered edge preparation, especially for interrupted cuts. As with ceramics, PCBN tools suitable for cutting metals are available only in the form of indexable inserts.

4.2.2.7 Polycrystalline Diamond (PCD)

PCD, the hardest of all tool materials, exhibits excellent wear resistance, holds an extremely sharp edge, generates little friction in the cut, provides high fracture strength, and has good thermal conductivity. These properties contribute to PCD tooling's long life in conventional and high-speed machining of soft, nonferrous materials (aluminum, magnesium, copper, and brass alloys), advanced composites and metal-matrix composites, superalloys, and nonmetallic materials. PCD is particularly well suited for abrasive materials (i.e., drilling and reaming metal-matrix composites) where it can provide significantly better tool life than carbide. PCD is not usually recommended for ferrous materials due to the high solubility of diamond (carbon) in iron. However, they can be used to machine some of these materials under special conditions; for example, light milling cuts can be made in gray cast iron at speeds below 200 m/min.

PCD tooling requires a rigid machining system because PCD tools are very sensitive to vibration. In mass production operations, the attainable tool life may be over 1 million parts (e.g., for diamond-tipped drills or PCD milling cutters machining soft aluminum alloys). Tool lives of this length may never be achieved, however, because tooling breaks due to vibration or rough handling before wear becomes significant.

Various grades of PCD have been developed for specific applications [43–46]. The major difference between grades is the size of the individual diamond grains, which varies between 1 and 100 μm. Tool performance is affected by the microstructure of the PCD. Grades are grouped in several categories with average grain sizes of 1–4, 5–10, and 20–50 μm [43,45]. The abrasive wear resistance, thermal conductivity, and impact resistance increase with increasing grain size, but finer grained tools produce smoother machined surface finishes. For example, a coarse-grained PCD tool may provide 50% better abrasive wear resistance than a fine-grained tool, but produce a surface with 50% higher roughness. New laser-honing methods can reduce edge radii for coarse grained PCD and produce finer finishes with these grades [47]. Because of their increased impact and abrasive wear resistance, coarse grades are preferred for milling and for machining high-silicon aluminum alloys and metal-matrix composites. Multimodal PCD grades (made with bimodal, trimodal, or quadimodal distributions of PCD particles) provide the high abrasion resistance of coarse-grained unimodal grade with the high toughness and superior edge sharpness of medium-size grain tools [45]. The PCD density increases with multiple particles sizes. Multimodal grades are less prone to chipping than unimodal grades.

Laser structuring has recently been applied to flat-topped PCD inserts to produce 3-D chipbreaking grooves and similar features [48], which have proven effective in ductile material applications where chip control has traditionally been an issue.

PCD-tipped HSS or carbide rotary tools (e.g., reamers, end mills, drills, etc.) are available in a limited range of geometries due to difficulties in grinding complex geometries, particularly on small diameter tools. More complex geometries can be used on carbide rotary tools by sintering the diamond into slots (veins) located at the point and/or along the flutes [29].

The development of improved PCD drills is of interest especially for high throughput applications. Issues to be resolved include identifying the optimal cutting edge geometry for the diamond tip and the best method of pocketing the polycrystalline blank for strength and manufacturability. The point geometry, flute geometry, and web thickness have not been refined sufficiently to allow use of polycrystalline brazed drills at penetration rates comparable to the feed rates attainable in turning and milling. Although methods of brazing the polycrystalline/carbide substrate tip to the main tool body have been improving steadily (with each manufacturer using its own proprietary procedures for surface cleaning and brazing), one of the major failure modes is still the detachment of the polycrystalline tip or the wear and erosion of the braze joints intersecting the cutting edge. Wear and erosion of brazed joints is avoided when the diamond is sintered into veins within the carbide tool.

4.3 TOOL COATINGS

HSS, HSS-CO, WC, and ceramic tools are often coated to increase tool life and allowable cutting speeds. The majority of inserts in production use are coated. Coatings act as a chemical and thermal barrier between the tool and workpiece; they increase the wear resistance of the tool, prevent chemical reactions between the tool and workpiece material, reduce built-up edge formation, decrease friction between the tool and chip or the tool and workpiece, and prevent deformation of the cutting edge due to excessive heating. Coated tools therefore can be used at higher cutting speeds, provide longer tool lives than uncoated tools, and broaden the application range of a given grade. Workpiece surface finish can also be improved with coated tools. A comparison of applicable cutting speed ranges for representative coated and uncoated tools is shown in Figure 4.10.

A number of factors affect coating performance, including the coating thickness, hardness, chemical compatibility and interfacial adhesion with the substrate, crystal structure, chemical and thermal stability, elastic modulus, fracture toughness, wear resistance, thermal conductivity, diffusion stability, frictional properties, the tool geometry, and the intended application [49,50]. Post-coating treatments affect the interfacial adhesion with the substrate and the smoothness of the coated surface and are intended to improve coating adhesion and reduce coating stresses.

4.3.1 COATING METHODS

The two most common coating processes are chemical vapor deposition (CVD) and physical vapor deposition (PVD). Both are used for both single and multilayer coatings.

In the CVD process the substrate is heated and exposed to a gas stream of appropriate chemistry, which reacts with the surface to form the coating layer. For example, TiN coatings may be produced using a gas stream composed of $TiCl_4$, H_2, and N_2 [51]. CVD coatings provide optimum adhesion because the bond between a CVD coating and substrate is metallurgical and stronger than the mechanical bond produced by PVD. As a result, CVD coatings are harder than PVD coatings and provide longer tool lives when properly applied. However, the temperature requirements of traditional CVD techniques reduce the range of substrate materials to which these coatings can be applied. CVD coatings on carbide substrates are in residual tension at room temperature because the coating materials have a higher thermal expansion coefficient than carbide; this can lead to transverse cracks resulting in tool failure in interrupted cutting. The high temperatures used in the CVD process can also

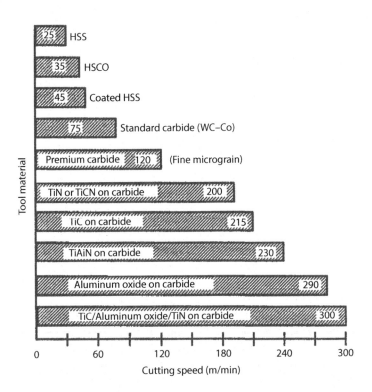

FIGURE 4.10 Comparison of cutting speeds of coated and uncoated tools.

cause degradation of WC substrates due to the formation of an eta-phase (a thin, brittle glassy layer) at the interface between the coating and substrate [52]. This process reduces the transverse rupture strength of the tool by as much as 30% and in particular embrittle, sharp edges by breaking down the substrate's cobalt binder. Hence, CVD-coated tools require a honed edge. A cobalt-enriched zone is sometimes used near the surface of a WC insert to prevent the propagation of cracks from the coating into the core. The other main disadvantage of CVD coating is that coating materials must be fed in gaseous form, which restricts the range of possible coating chemistries. The temperature limitation has been overcome by reducing the process temperature to 700°C–900°C in the medium-temperature CVD (MT-CVD) process, and to lower levels in plasma-assisted CVD (PA-CVD) [50]. This increases toughness, minimizes chipping, and improves the surface finish of the coating. MT-CVD coatings are particularly well suited for interrupted and roughing cuts. CVD coatings are typically between 5 and 15 μm thick, although thicker (>20 μm) multilayer tools are becoming more common [53].

In PVD coating the coating material is vaporized and deposited by sputtering or arc evaporation. PVD coatings are applied at lower substrate temperatures (around 500°C) and thus can be applied to a wider range of substrates. Generally, PVD coatings are better suited for precision HSS, HSS-CO, brazed WC, or solid WC tools. In fact, PVD is the only viable method for coating brazed tools because CVD methods use temperatures that melt the brazed joint and soften steel shanks. PVD coatings are essentially free of thermal cracks, and are finer-grained (effectively conforming to the sharp edges of finishing tooling) and generally smoother and more lubricious than CVD coatings, which build up on sharp corners. PVD coatings are preferred for positive rake and grooved inserts because they produce compressive stresses at the surface that reduce crack initiation and propagation. PVD coating thickness typically varies between 2 and 5 μm.

Coating/substrate compatibility is improved by applying one or more intermediate layers between the surface coating and the substrate to balance chemical bonding and thermal expansion coefficients, resulting in a multilayer coating system, which optimizes tool performance by providing resistance to several kinds of wear [50,51]. Multilayer coatings may be produced by combined

TABLE 4.4
Indication of How Various Coating Methods Withstand Different Types of Wear

Coating	Abrasion Wear	Adhesion Wear	Fatigue Wear	Plastic Deformation	Chipping/ Fracture	Try for a Sharp Edge
PVD	−	+	+	−	+	+
CVD	+	0	0	+	0	−
MT-CVD	+	0	0	+	0	−
TiAlN	+	+	0	+	−	0
TiCN	+	0	0	0	0	0
TiN	0	0	0	0	0	0
Al_2O_3	++	+	−	+	−	−

Source: Courtesy of Sandvik-Coromat, Fairlawn, NJ.
Note: +, positive impact; −, negative impact; 0, neutral impact.

CVD, MT-CVD, and PVD methods; in such cases the CVD or MT-CVD process improves adhesion between the substrate and the first coating layer, while the subsequent PVD coating layers provide a fine grained microstructure with better wear resistance and toughness or lower friction. Multilayer coatings are very common for turning and boring inserts because they provide the best combination of properties. Because machining processes result in many types of wear, multiplayer coatings can add to a tool's multipurpose capability. CVD multiplayer coatings have been also used on solid CBN inserts to improve resistance to chemical and crater wear. The effect of the coating method and type in terms of withstanding different types of wear mechanisms is summarized in Table 4.4.

There has been considerable development in nanolayer and nanocomposite or adaptive nanocrystalline coatings [53]. Nanocomposite coatings may provide high hardness (4000–5000 HV) and high heat resistance (up to 1100°C), as well as toughness and hardness comparable to nanolayers [54,55].

4.3.2 CONVENTIONAL COATING MATERIALS

Materials used for single-layer coatings include titanium nitride (TiN), titanium carbide (TiC), titanium carbo-nitride (TiCN), titanium aluminum nitride (TiAlN), aluminum oxide (Al_2O_3), chromium nitride (CrN), hafnium nitride (HfN), titanium diboride (TiB_2), boron carbide (BC), and WC/C (amorphous diamond-like carbon) hard lubricant [50,56–58]. Material combinations used in multilayer coatings include Al_2O_3 on TiC or TiCN, TiN on TiC, TiN/TiC/TiN, TiN/TiCN/TiN, TiN/TiC/TiCN, Ti(C)N/Al_2O_3/TiN, Al_2O_3/TiC, TiN/TiC/Al_2O_3/TiN, and TiAlN + WC/C [49,56,58–60]. Coatings containing Al_2O_3 are CVD grades. A concise comparison of coating material properties is given in Table 4.5, and a comparison of coating hardness levels is given in Figure 4.11.

TiN, TiCN, and TiAlN, and AlTiN (high Al content TiAlN) coatings [49,50,61–63] are very commonly used for rotary tooling since they are applied by PVD processes. The performance of these coatings depends on the work material, the type of process, the cutter geometry, and the cutting conditions as discussed in Section 9.5. TiN reduces friction and resists adhesive wear and built-up edge formation as well as increasing oxidation resistance (Figure 4.12). It also acts as a chemical barrier to diffusion wear when cutting ferrous materials with carbide tooling. TiN coatings perform best in cast, medium-alloy, and high-alloy steels. TiCN in harder than TiN and provides low friction and good abrasive wear resistance. It is used for steels, stainless steels, and nonferrous materials. Al_2O_3 provides excellent thermal heating and oxidation resistance as well as abrasive wear and adhesion (built-up edge) resistance. Al_2O_3 coatings are used especially for work materials with hard or abrasive phases. TiAlN has superior ductility and is stable at higher temperatures than TiN and TiCN. TiAlN is not as hard as TiN and TiCN but can be applied in thicker layers to compensate and provide equivalent tool life. TiAlN

TABLE 4.5
Properties of Materials and Coating Materials

Material and PVD Coatings	Hardness HVN (kg/mm²)	Friction Coefficient vs. Steel	Maximum Operating Temperature (°C)	Coating Thickness (μm)	Thermal Expansion (m/m K)
HSS	900	0.3–0.4	500		10–14
Oxides and nitrides	1500–3000	0.1–0.2	800		2.8–9.4
Carbides and borides	2000–3600	0.1–0.2	500		4.0–8.0
TiN	2200	0.4–0.5	600	1–4	
TiCN	3000	0.25–0.4	430	1–7	
TiAlN layered	3300	0.3–0.65	800	2–5	
TiAlN monolayered	4500	0.3–0.65	800	1–5	
AlTiN	3800	0.5–0.65	900	1–5	
WC/C	1000	0.1–0.2	300	1–4	
MoS₂-based	20–50	0.05–0.15	800	<1	
TiAlN + WC/C	3000	0.1–0.25	800	2–6	
CBN	4700	NA	1400		4.7
Diamond	7k–10k	0.05–0.1			1.5–4.8

Sources: McCabe, M.J., How PVD Coatings Can Improve High Speed Machining, *High Speed Machining Conference*, SME, Northbrook, IL, September 20–21, 2001; Bhat, D.G. and Woerner, P.F., *J. Metals*, 68, February 1986.
Note: Oxides and nitrides materials include Al, Si, Ti, Zr, Cr, Mo, Hf, TiAl.

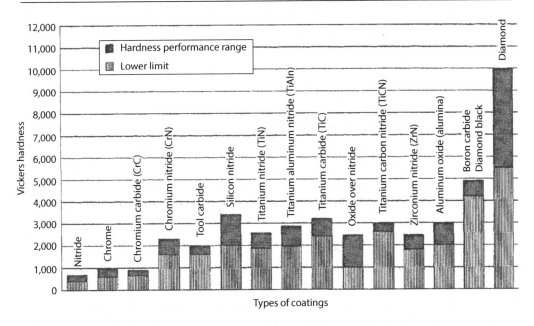

FIGURE 4.11 Comparison of microhardness for several types of coatings.

provides high resistance to oxidation, high thermal conductivity, greater hot hardness, and enhanced chemical resistance. Oxygen reacts with the aluminum in the coating and forms an amorphous Al oxide layer at the interface with the chip (where temperature can reach more than 1000°C) that is thermodynamically stable and very protective and lubricious (as shown in Figure 4.13). TiAlN coatings are available in several forms such as low stress, hard, etc. Their performance is altered by changing the aluminum content to suit specific applications. For example, high content Al or AlTiN or

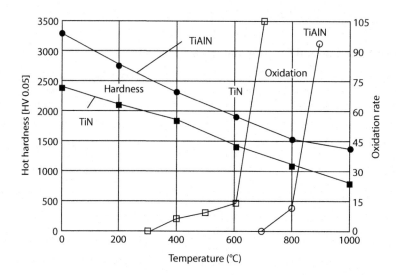

FIGURE 4.12 Hot hardness and oxidation rate of TiN and TiAlN coatings. (Courtesy of Kennametal, Inc., Latrobe, PA.)

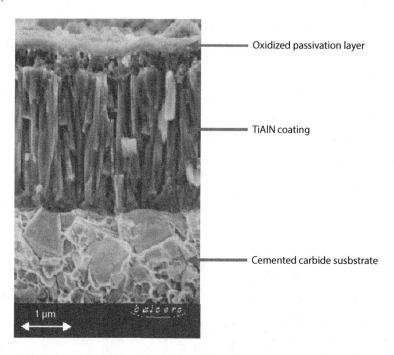

FIGURE 4.13 Photomicrograph (SEM) of a TiAlN coating on a carbide tool. (Courtesy of Oerlikon Balzers Coating AG, Principality of Liechtenstein.)

hard TiAlN coatings are intended for high-speed machining of hard and/or abrasive materials (i.e., ferrous, titanium, and exotic alloys). TiAlN can be used with all common material types. As a general rule, TiCN, and TiAlN coated tools are used at speed up to 50% and 100% higher, respectively, than uncoated tools. TiN and TiCN are very common coatings for taps.

Titanium diboride coatings [64,65] are suitable for free-machining aluminum, low silicon (hypoeutectic) aluminum alloys, and titanium. They are hard, have low affinity to aluminum, resist built-up edge formation, have high melting points, and are chemically stable at temperatures up to 1400°C. They are harder than TiN and TiAlN coatings and compete well with diamond coated carbide.

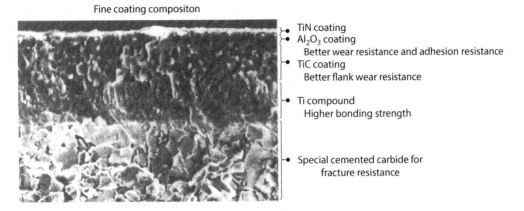

Fine coating compositon

TiN coating
Al_2O_3 coating
Better wear resistance and adhesion resistance
TiC coating
Better flank wear resistance

Ti compound
Higher bonding strength

Special cemented carbide for
fracture resistance

FIGURE 4.14   Schematic of a multilayer coating architecture. (Courtesy of Mitsubishi Materials Corporation, Tokyo, Japan.)

Lubricating coating materials, such as molybdenum disulfide (MoS_2) [66], tungsten disulfide [67], and WC/C (amorphous diamond-like carbon), are sometimes used in dry machining applications and also for drills and taps. Thin layers of these coatings are deposited on top of hard coatings, for example, MoS_2 over TiN or TiAlN and WC/C over TiAlN. These coatings improve chip flow and minimize built-up edge especially when drilling and tapping of aluminum, cast irons, and steels. WC/C is also used as a single-layer coating to reduce built-up edge at lower speeds. The friction coefficient for these softer coatings is less than half that for TiN.

As noted earlier, multilayer coatings combine the best properties of single coatings, with each coating layer contributing a specific function and application. Al_2O_3 is usually deposited over TiC or TiCN since the TiC or TiCN layer provides a good base for the top layer to adhere. For example, the first layer of TiCN can be deposited at moderate temperature of 850°C to reduce the eta phase formation at substrate/coating interface. A typical microstructure is shown in Figure 4.14. TiN, TiAlN, and TiCN multilayer coatings are especially effective for HSS and HSS-PM tools.

4.3.3   DIAMOND AND CBN COATINGS

There is long been interest in diamond and CBN coatings for cutting tools due to the potential of these very hard materials to improve wear resistance [49,68]. Diamond-coated tools are commonly used in micromachining and in the machining of composite materials and soft metals. CBN-coated tools have shown promise in experiments [69,70] but are not widely used in production [71]. This section discusses diamond coatings; since CBN coatings present similar technical challenges, the general concerns discussed for diamond coatings are applicable to both coating materials.

Diamond is deposited onto a substrate when carbon-based gases and hydrogen are disassociated at high temperature. Diamond coatings result in a fully dense layer of diamond at the tool surface, as compared to the porous layer containing cobalt phases at the surface of a PCD blank. Due to this higher density, coated diamond tools have a higher microhardness than PCD blanks. A CVD thin-film diamond coating (a layer of diamond crystals less than 50 nm thick) can be deposited on either a ceramic or tungsten carbide substrate. Coating silicon nitride substrates [72] has been successful because diamond adheres well to silicon nitride and because the thermal expansion coefficients of the coating and substrate are comparable. However, the range of application of these tools is limited by the comparatively low toughness and impact strength of the substrate, the cost of the base material, and the difficulty in manufacturing ceramic tools with complex geometries. CVD diamond thin films have not been yet fully developed for conventional aluminum PCD machining applications but have been successfully applied in the machining of phenolic resins, graphite, carbon graphite, fiberglass, and carbon graphite epoxy.

Thick CVD diamond films (0.1–1 mm thick) have also been explored to overcome the adhesion concerns with thin film coatings. Like PCD blanks, thick films are brazed onto a tool's cutting edge or corner. Brazed CVD diamond films cannot be applied easily to complex tool geometries and are also not suitable for interrupted cutting.

Much development has also been directed toward developing coated rotary WC tools, since such tools cannot yet be made of solid ceramic due to their complex geometries. Rotary WC tools, usually containing between 5% and 6% cobalt binder, are difficult to coat using a CVD process because the cobalt reacts with the diamond. In addition, the thermal expansion coefficient of WC-Co is much higher than that of diamond. A number of methods have been explored to improve diamond/WC adhesion and to improve the reliability of diamond coatings; limited success has been obtained using WC inserts with low binder contents (<4%) [73].

Resharpening of coated tools is a very important consideration in manufacturing. Resharpening usually removes the coating from the flank area of the tool while leaving the coating on the rake face. Tool life typically drops by 20%–40% after resharpening, which is still better than for uncoated tools. In some cases the coating is stripped prior to resharpening, with the sharpened tool being recoated.

4.4 BASIC TYPES OF CUTTING TOOLS

The six basic types of cutting tools are solid tools, welded or brazed tip tools, brazed head tools, sintered tools, inserted blade tools, and indexable tools (Figure 4.15).

Solid HSS, HSS-CO, WC, and ceramic tools are made from a single piece of homogeneous material. P/M, carbide, and ceramic round tools are produced using a metal injection molding process. They can be resharpened after use, but this is a precision process requiring accurate grinding equipment.

WC, PCD, and PCBN tipped tools consist of inserts or tips of hard tool materials brazed or welded to a steel or WC tool body. Tipped tools combine tool bodies that are tough, inexpensive, and easy to manufacture with tool points that are hard, wear resistant and typically brittle, and difficult

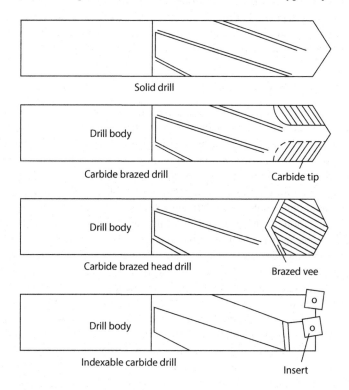

FIGURE 4.15 Types of rotary tooling.

to manufacture. They can be reground, but as with solid tooling resharpening is a precision process. The method of attaching the tip to the body affects the durability of the cutting edge(s). Welding produces the strongest bond; brazed bonds are weaker, but the brazing material (brass, bronze, or silver) acts as a cushion between the insert and tool body.

WC brazed head tools consist of a complete tool point made of WC brazed onto a steel tool body or shank. This method is used mainly for small tool sizes and in applications in which the joints of brazed inserts are not stable.

In *PCD and PCBN sintered tools*, also called veined tools, PCD or PCBN powder is sintered into slots in WC tool bodies. This method is most commonly applied to drills, taps, and end mills.

Inserted blade tools consist of blades made of HSS, HSS-CO, WC, PCD, or PCBN, which are inserted into slots around the periphery of a tool body and locked in place mechanically. These cutters are normally used in high-production applications and in gearmaking. They are ideal for close-tolerance finishing operations. Blades can sometimes be resharpened, but resharpening alters the tool size and (in some cases) geometry.

Indexable tools use standard-sized inserts of hard tool material, which are clamped into tool holders or cutter bodies. Inserts generally have more than one cutting edge; when a cutting edge is worn out, a new edge can often be brought into use by rotating or "indexing" the insert. Inserts made of almost all tool materials are available. Many are made by direct pressing processes, which allows production of complex shapes and eliminates the need for grinding. Indexable tooling offers advantages in both high and low production applications. Tool holders or cutter bodies may be made of steel, WC/heavy metal, or a combination of these materials to optimize performance. Inserts usually have uniform properties and are usually tougher and more wear resistant than other types of tools, so that they can be used at higher metal removal rates. They permit a reduction in tool inventory because a few standard inserts can replace a large number of solid tools. They eliminate regrinding, are interchangeable, and are easily changed. They are less accurate, however, than special brazed or inserted-blade cutters typically used to machine complex contours or holes with multiple diameters.

4.5 TURNING TOOLS

Turning is carried out primarily using single-point cutting tools, or tools with a single active cutting surface. Together with planing and shaping, turning is the most basic cutting process; an understanding of the cutting action and geometry of single-point turning tools therefore provides insight into the basic cutting action and geometry of more complex tools.

In the past, solid single-point tools ground from a HSS blank [74] were most commonly used for turning (Figure 4.16). At present, however, turning is most commonly carried out using indexable inserts. This section discusses the design and selection of turning tools with emphasis on the proper selection of inserts and tool holders. The discussion of tool angles and insert selection (e.g., the strengths of various shapes of inserts, relative merits of different insert holding methods, etc.) is generally applicable to other cutting operations such as milling and boring.

4.5.1 INDEXABLE INSERTS

Figure 4.17 shows a representative sample of indexable inserts used for turning. The orientation of the insert with respect to the workpiece is largely determined by the geometry of the toolholder. When selecting an insert and toolholder for a given operation, the following factors must be taken into account [4,75,76]:

1. Type of operation (roughing, finishing, etc.)
2. Continuous versus interrupted cut
3. Workpiece material and primary manufacturing. process (casting, forging, etc.)
4. Condition of the machine tool

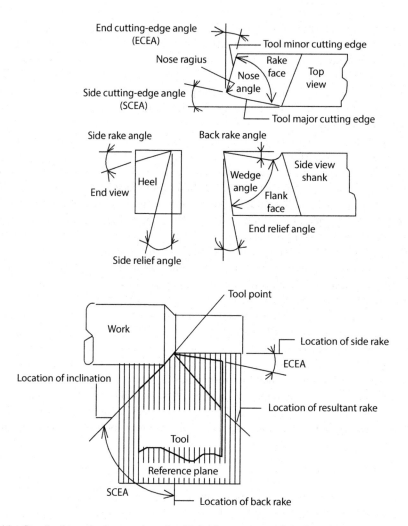

FIGURE 4.16 Standard terminology to describe the basic geometry of single-point tools.

   5. Required tolerance
   6. Feeds and speeds (production rate)

Based on these factors, the following parameters must be selected:

   7. Insert material and grade
   8. Insert shape
   9. Insert size
  10. Insert thickness
  11. Corner geometry (nose radius or flat)
  12. Groove (chipbreaker) geometry
  13. Edge preparation
  14. Edge clamping/holding method
  15. Lead, rake, relief, and inclination angles

As discussed in Sections 4.2 and 4.3, the selection of the insert material, grade, and coating depends primarily on the work material, the required production rate, and the condition of the machine tool.

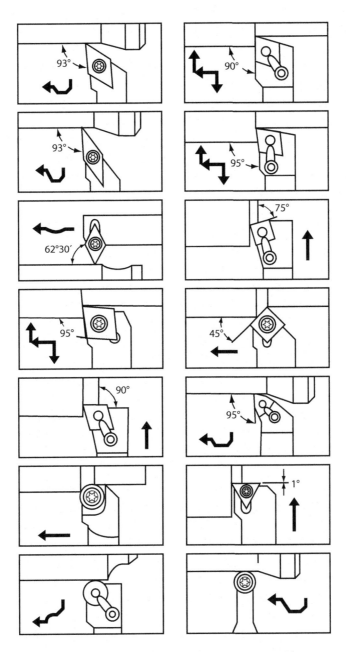

FIGURE 4.17 Representative inserts in several turning operations. (Courtesy of Mitsubishi Materials Corporation, Tokyo, Japan.)

Most of the geometric parameters are specified in the ANSI or ISO standard cutting insert nomenclature systems [77,78]. The shape of an insert is specified by the first letter of the insert designation; a TPGT-322 insert, for example, is triangular, while an SNGA-532 insert is square. Available insert shapes include diamond (C), triangle (T), square (S), octagon, round (R), and trigon (W). The shape of an insert largely determines its strength, its number of cutting edges, and its cost. As a general rule, an insert becomes stronger and dissipates heat more rapidly as its included angle is increased (Figure 4.18). The selection of the included angle is limited by the part configuration, the required tolerances, the workpiece material, and the amount of material to be removed. Therefore, insert shape selection requires a trade-off between strength and versatility.

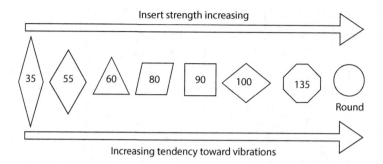

FIGURE 4.18 Relative insert strength.

For example, round inserts provide maximum edge strength and are therefore a good choice for roughing operations. They also provide a maximum number of effective cutting edges since they can be rotated (or indexed) through small angles when a given edge wears out. Round inserts thin the chip, however, and generate high radial forces; as a result, they should not be used when chatter or instability are expected, or when tight tolerances are required. Square inserts are common in general purpose applications because they provide good edge strength and a large number of cutting edges (8 for a negative rake tool, 4 for a positive rake). A 80° diamond insert is very versatile because it performs turning with 90° shoulder and facing operations. Generally, the largest included angle suitable for the workpiece geometry should be used.

The insert size is determined by the largest circle, which can be inscribed within the perimeter of the insert. The size of the inscribed circle (IC) determines the volume of tool material in the insert, and thus the insert cost. The IC size should be selected based on the depth of cut to be taken. Large IC inserts are used in heavy interrupted cuts or similar roughing operations. Smaller IC inserts perform more effectively in semi-finish and finish operations. As a rule, the cutting edge length should be 2–4 times the maximum cutting edge engagement (b or L_m), (Figure 4.19).

As with the IC, the thickness of an insert should also be selected based on edge strength requirements. In general, larger IC inserts are thicker than a smaller IC inserts. Insert manufacturers

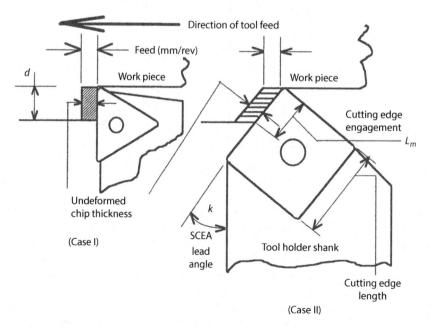

FIGURE 4.19 Effect of lead angle on undeformed chip thickness and cutting edge engagement.

provide recommendations for thickness for specific grades based on the intended application (the work material type and hardness and ranges of DOC and feed).

A nose radius is usually provided on an insert to increase edge strength; a corner with a nose radius is much stronger than a sharp corner. The nose radius also serves to improve surface finish; at a constant feed, a larger nose radius generates a smoother finish (as discussed in Chapter 10). Hence, the largest nose radius allowed by the workpiece configuration and operating conditions (feed) should be used. Deflection and chatter are the primary concerns limiting the size of the nose radius, since an increase in the nose radius increases radial cutting forces. As a rule of thumb, the feed should not exceed 2/3 the nose radius. Some workpiece materials (such as nickel-based alloys) require a nose radius larger than the depth of cut by as much as a factor of two to four for semi-finish and finish operations, since the use of large radii allows the use of higher feed rates.

The insert IC, thickness, and nose radius are specified in the first three numbers in the insert designation in the ANSI system [77]. In the examples used before, a TPGT-322 insert has an IC size of 3/8 (0.375) in., a thickness of 2/16 (0.125) in., and a nose radius of 2/64 (0.03125) in.; and SNGA-532 insert has an IC of 5/8 (0.625) in., a thickness of 3/16 (0.187) in., and a nose radius of 2/64 (0.03125) in. The ISO system [78] has a similar metric dimensioning system; insert catalogs normally include summaries of both systems.

Standard inserts are available with varying tolerances (roughing inserts, semi-finishing, and finishing inserts, etc.). Inserts can be used as sintered, ground on top and bottom faces only, or ground on all faces for precision. Ground inserts often eliminate the need for manual adjustments of the tool position and hence maintain the required tolerance with greater ease on close tolerance work (less than 0.013 mm).

4.5.2 Groove Geometry (Chip Breaker)

The chipbreaking ability of an insert is important especially with ductile work materials. The shape of the chip, cutting forces, and tool performance all are affected to a large extent by the groove geometry on the insert. Several types of molded chipbreakers are available, including single-, two- and three-stage chipbreakers, vee-type chip grooves or land-angle designs, wavy and bumpy designs, and scalloped types as shown in Figures 11.6 through 11.8. The insert groove controls the chip flow direction and can reduce cutting forces and edge wear. The groove geometry is constrained by the desired rake angle (positive, negative, or neutral) for a specific application. The geometry of a particular groove or chipbreaker has a specific feed range that will produce acceptable chips; charts of effective feed and depth of cut ranges for a given insert design are available from the insert manufacturer (see Figure 11.10). Negative/positive inserts (with a negative side rake and a positive back rake) provide better chip control than either negative/negative or positive/positive inserts (see Figure 4.20). Chip breaking is discussed in more detail in Section 11.3.

Chipbreaking grooves can be cut in PCD inserts by laser etching [48] to improve chip control when machining ductile metals.

4.5.3 Edge Preparations

Proper edge preparation is very important to the performance of a tool because it strengthens and protects the cutting edge and delays or eliminates chipping or breakage of sharp edges [42,75,79–83]. It has a significant effect on cutting forces, surface finish, residual stresses in the machined surface, burr formation, and wear rates. It also removes edge imperfections (cracked crystal layers) and prepares the edge for coating. Edge preparations are especially important for tools made of brittle materials because they affect the thermo-mechanical loads on the cutting edge. There are three basic preparations (Figure 4.21): hones (honed radius or waterfall), chamfers, and negative lands, as well as combinations of these three. The hone is either a circular (radius) hone or an oval/elliptical (waterfall) hone. In a waterfall hone, the edge surface is elliptical with the long axis parallel to the

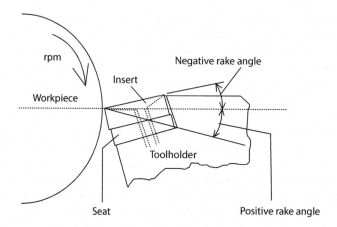

FIGURE 4.20 Grooved tool inserts that could provide a positive rake even when used in a negative rake tool holder.

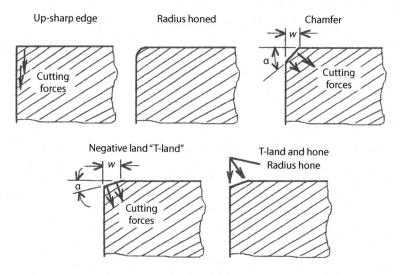

FIGURE 4.21 Cutting edge preparation methods.

rake face; in a reverse waterfall hone, the long axis is parallel to the relief face. Properly designed waterfall hones move initial wear away from the cutting edge toward the bulk of the insert. Most standard inserts have a light hone as part of their manufacturing process. An up-sharp or light hone is sufficient for machining soft, nonferrous materials such as aluminum. Nickel alloys and exotic aerospace alloys also usually require a light hone because they are cut at low feeds and cutting speeds. Steels and nodular or gray irons often require tools with medium-honed edges. Lightly honed edges are not recommended for cermet, ceramic, or CBN inserts, which are brittle and are normally provided with chamfered edges. A hone or small chamfer is generally required for the high throughput drills. The size of the hone varies with insert size, which may range from 7–12 μm (light hone) to 160 μm (heavy hone). Variation of the hone geometry is common because the hone is often applied by manual stoning, brushing, tumbling, slurry honing, or abrasive blasting [84]. The uncut chip thickness and feed should be greater than the edge hone radius. Effective hones are typically approximately 1/3–1/2 the feed.

The bevel angle of chamfered edges ranges from 20° to 45°. The negative land (T-land) is similar to the chamfer but its bevel angle is smaller (usually between 5° and 20°). Chamfered and negative land edge preparations provide an effective negative rake angle to strengthen the edge and reduce chipping with brittle tool materials. However, chamfers or T-lands increase cutting forces and can have a negative effect on both tool life and the machined surface finish. The T-land width can be smaller, equal, or larger than the feed per tooth. Generally, the chamfer or T-land width should be at maximum 70% of the feed; it should usually be kept below 30% to prevent chip jamming, high tool pressures, and heat build-up, all of which promote tool failure. T-lands are normally used on negative rather than positive rake inserts. However, a short negative land $25° × 0.05$ mm can be used with positive rake tools to redirect the forces through the tool and provide good support for the cutting edge; in this case the T-land length should be about 60% of the feed.

Edge preparations affect the edge strength, tool life, and chip control. Even though tool life reliability improves by avoiding micro-chipping or breakage, the average life could be reduced because it acts as a pre-wear feature at the cutting edge. Edge preparation is a trade-off between edge strength versus edge wear and feed versus size of edge preparation. For example, the recommended edge preparation for turning gray cast iron with PCBN is a $20° × 0.2$ mm chamfer for roughing and $20° × 0.1$ mm chamfer for finishing. In more severe applications, an additional 0.025 mm radius hone is used to achieve additional strength. In the lightest cuts, a 0.025 mm to 0.05 radius hone is often acceptable.

4.5.4 WIPER GEOMETRY

Wiper tools have additional radii behind the tool nose at the leading edge that are kept in contact with the surface to remove the peaks of the grooves made by the leading edge as discussed in Sections 10.3 and 10.4. Even though wiper inserts are designed for better finishes, they are not used much for light finishing operations because they require a semi-finishing DOC. They improve surface finishes significantly, but their main benefit is to reduce the cycle time by increasing the feed (50%–100% higher feeds of conventional inserts) while maintaining surface finish.

4.5.5 INSERT CLAMPING METHODS

The insert is usually set on a square or rectangular shank holder with adjustable screws or buttons to control the insert position. Qualified toolholders provide precise control using indexable inserts since constant adjustment of the cutting tool or machine tool compensating devices is often not necessary. The insert may be set in a pocket, which is either machined directly into a holder or into a cartridge mounted on the holder. In the former case, the rake and lead angles on the insert are designed into the pocket of the holder as shown in Figure 4.22. The cartridge style is more versatile, since various insert styles can be accommodated using different cartridges. Cartridges are often used for internal turning (or boring) operations, since they permit small tool size adjustments. The manufacturing tolerances of the insert pocket are very critical in providing proper support and interchangeability between inserts.

There are several basic insert clamping methods including top clamps for plain inserts, cam pins or screw clamps for inserts with a center hole, and combinations of cam pins with top clamps (Figure 4.22). Top clamp methods vary from manufacturer to manufacturer, but as the name implies generally hold the insert with a top clamp. One limitation of top-clamping methods is that they often require molded features on the insert surface, rendering inserts one-sided and reducing the number of available edges. The cam pin method is a very common insert locking approach in which the insert is clamped by an eccentric pin through the center hole; when the pin is rotated into the locked position, it pulls the insert down and back into the toolholder pocket.

In general, a combined top clamp and cam pin locking method provides the most rigid clamping condition, while a cam-pin clamping method alone provides the least rigid condition. A superior system is the wedge top-clamp and pin because the rear edge of the clamp locates against the rear-angled surface of the insert pocket, which ensures alignment. This method ensures accurate positioning both vertically and horizontally.

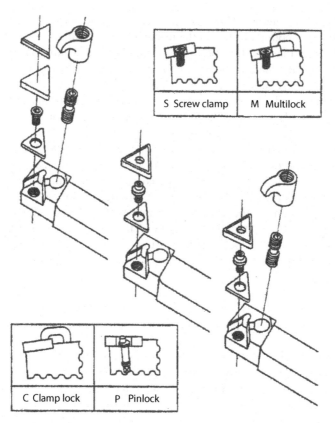

FIGURE 4.22 Basic insert locking (holding) methods. (Courtesy of Carboloy/Seco, Warren, MI, and Sumitomo Electric Corp., Osaka, Japan.)

4.5.6 Tool Angles

The lead angle, κ, is equivalent to the side cutting edge angle (SCEA) for turning and to the end cutting edge angle (ECEA) for boring and facing in the English system. The lead angle is sometimes called the bevel reciprocal angle and the cut-entering angle. The bevel angle, $\kappa_r = 90° − \kappa$, is used as the lead angle in the metric system. An increase of SCEA reduces the chip thickness and increases the chip width (see Figure 2.3). The largest lead angle allowed by the workpiece geometry should be used because (1) it thins the chip and reduces the unit pressure on the insert, allowing the feed per revolution to be increased; (2) it protects the nose radius as the insert enters a workpiece, reducing nose wear and chipping; and (3) it reduces depth of cut notching because the scale or work hardened surface of a workpiece is spread over a larger area of the cutting edge. A zero lead angle results in sudden loading and unloading of the cutting edge as it enters or exits the workpiece. Geometric considerations aside, the maximum lead angle, which can be used in a given operation, depends on the rigidity of the machining system. An increase in the lead angle results in higher radial forces and lower axial (feed) forces, and can lead to increased tool or workpiece deflections and the occurrence of chatter. 15°–30° lead angles are normally used in general purpose machining, while 0° lead angles may be used for long, slender workpieces. The lead angle often determines the size of the insert for a given depth of cut. Typical insert setups with the corresponding lead angle are shown in Figure 4.19. The ECEA must be positive to reduce the tool contact area with the metal being cut and to prevent vibration and chatter, which result in poor surface finish and tool life. The effect of SCEA and ECEA on surface finish is discussed in Chapter 10.

Other major features of the tool geometry are the relief angles and rake angles shown in Figure 4.16. The end and side relief angles protect the tool from rubbing on the cut surface.

Both these angles are a function of the workpiece material hardness and the tensile strength of the tool material; and they are usually kept to a minimum to provide sufficient support on the cutting edge especially for positive rake tools. The relief angles for tools used in hard materials are low and approach 5° when small feeds are also used. However, higher angles are required with soft materials and higher feeds. Angles of 12°–15° are not uncommon with soft materials such aluminum and occasionally cast iron. In general purpose machining, the relief angle is typically 6°–8°.

Both the back and side rake angles, which provide inclination between the face of the tool and the workpiece, have a strong influence on tool performance. The rake angles can be positive, neutral, or negative as shown in Figure 4.20. The rake angles affect the magnitude and direction of cutting forces, edge strength, chip formation, chip flow direction, chip breaking characteristics, and surface finish. A positive angle moves the chip away from the machined surface of the workpiece while a negative angle directs the chip toward the workpiece. The combination of the rake angle and relief angle determines the wedge angle. A positive rake angle reduces the wedge angle, which results in smaller shear cross sectional area of the cutting edge, reducing edge strength. However, positive rakes increase the shear angle of the chip, reducing the chip thickness and resulting in lower cutting forces. Positive rake inserts therefore require less cutting power and exert less stress on the workpiece, tool, and the machine spindle and ways. They also generate less vibration and help control chatter. Negative rake inserts alter the direction of forces in a manner that places the cutting edge under compression, which is desirable since the compressive strength of most cutting tool materials is more than twice as large as the transverse rupture strength. Traditionally, negative rake tooling is used whenever it is allowed by the workpiece and the machine tool. Typically, a 1° increase in the rake angle results in 2% decrease in power consumption. Therefore, a tool with +5° rake requires about 20% less power than a tool with −5° rake angle. A positive geometry should be considered when it can reduce production cost while improving the quality of the work. Negative side rake angles are preferred for interrupted cuts, heavy feeds, hard metals, and some nonferrous metals. Positive side rake angles are often necessary for work-hardening alloys, slow surface speeds, soft metals, and nonmetals, and when the rigidity or strength of workpiece and/or fixture is limited.

The rake angle of a flat-top surface insert is determined solely by the rake angle of the toolholder. The rake angle can also be pressed directly into the insert by using chip groove geometries, which provides a positive rake cutting action when used in a negative rake toolholder as shown in Figure 4.20. Even though the range of rake angles is generally −5° to +15°, higher positive angles are slowly being used to improve productivity. Positive-geometry inserts are most commonly made of WC; ceramic materials are very brittle and commonly available only in negative rake varieties. The rake angle of and insert is specified in the second letter of an insert designation. An SPE-422 insert can be used in a neutral or positive (P) rake holder, while and TNG-333 insert must be used in a negative (N) rake holder.

4.5.7 THREAD TURNING TOOLS

Thread turning can be performed in several ways. The three popular ways to machine thread-vees (using several passes) are discussed in Chapter 2 and illustrated in Figure 2.22. The cutting speeds for the tool material recommended in ordinary turning are usually reduced by 25% in thread turning to maintain an acceptable cutting temperature.

Typical threading type inserts are shown in Figure 4.23. Multi-toothed inserts cut the vee in a single pass, using the radial infeed method, because the succeeding teeth cut deeper than the teeth preceding them. Full-profile inserts cut a complete thread including the crest. V-profile inserts are used for various pitches with same thread angle, but the crest is generated by a pre-turning operation. The insert is inclined an amount equal to the thread's helix angle (λ) to best match the lead angle of the thread, which varies along the thread height; this ensures uniform edge wear of both flanks and therefore longer tool life as well as better surface finish. The inclination angle of an insert is between 0° and 4°; a value of 1° generally produces a satisfactory thread. The relief angles on the

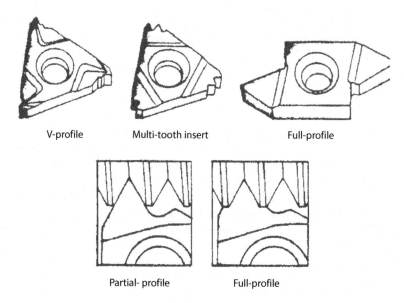

V-profile   Multi-tooth insert   Full-profile

Partial- profile   Full-profile

FIGURE 4.23   Typical threading insert forms. (Courtesy of Carboloy/Seco, Warren, MI.)

flank of the insert depend on the inclination angle; the side relief angle in the direction of the feed must be greater than the thread helix angle to prevent the insert from rubbing. The effective rake angle varies along the cutting edge; it is positive at the leading edge and negative at the trailing edge. Standard triangular turning inserts (60° included angle) are often used for roughing; they generate a thread form of slightly more than 60° when the tool is set with a negative rake angle. Cermet cutting tools produce better surface finish and size control than WC tools. Ceramic inserts are generally not as effective in conventional thread turning as in hard thread turning.

4.5.8   GROOVING AND CUTOFF TOOLS

Grooving and cutoff tools are similar to threading tools in that the tool is surrounded on three sides by the workpiece, so that the cutting forces and heat are concentrated on the weakest part of the insert. Grooving inserts and tool holders are narrow and must be relieved on all the three sides (as shown in Figure 4.24), and therefore do not provide as much support as other types of turning inserts. The lead angle can be neutral, right or left (of a few degrees). The ratio of radial depth to insert width should be less than 10 to maintain stability. The lead angle provides more stable cutting. Inserts with lead are used to reduce the small diameter protrusion left at the center in cutoff operations. Special inserts with and without chipbreaking grooves are available for these operations.

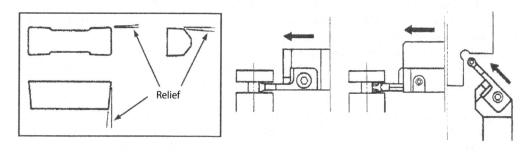

Relief

FIGURE 4.24   Multiple relief weakens grooving or cutoff inserts. (Courtesy of Valenite Corporation, Troy, MI.)

4.5.9 Form Tools

Form tools are used to cut a specific profile on a rotating workpiece in a single plunge cut (Figure 4.25). Grooving and cutoff tools are types of form tools. A form tool is fed perpendicular to the surface of the workpiece and contacts the workpiece at multiple points. Form tools produce shapes with less machining time than single-point turning tools can because multiple passes must be used with single-point tools to maintain accuracy. The surface finish obtained with form tools is usually about four times rougher than the finish on the form tool itself.

Form tools are designed in various sizes and can range in width from 4 to 200 mm. The maximum effective width of a form tool depends on its geometry, the workpiece material, infeed rate,

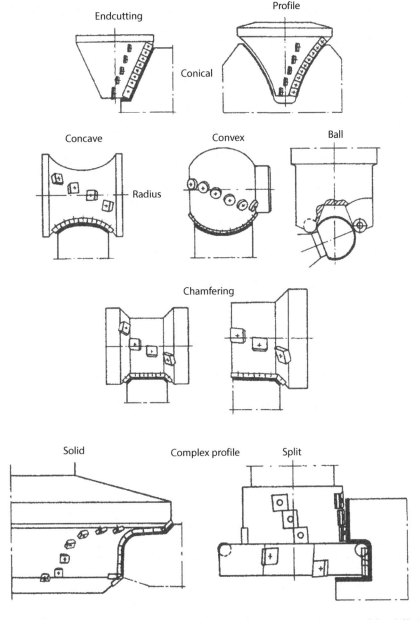

FIGURE 4.25 Form tools for complex profiles on a rotating workpiece. (Courtesy of Sandvik Coromant, Fairlawn, NJ.)

and the available machine power. Combined grooving and form tools are used for wide grooves or for generating wide shapes. One form tool can replace several single-point turning tools and provide increased dimensional accuracy and repeatability for complex profiles. Form tools are produced by electodischarge machining (EDM) or by grinding in the case of thin inserts. Tool materials used for form tools include brazed or clamped WC, HSS, and PM tool steels. Form tools can be manufactured with 5 µm dimensional repeatability using EDM.

4.6 BORING TOOLS

4.6.1 SINGLE POINT BORING TOOLS

Boring tools have the same insert or point geometry as turning tools, and most of the information reviewed in the previous section is applicable to boring as well as turning. One exception concerns the nomenclature of cutting angles; the end (rather than the side) cutting edge is the major cutting edge in boring, so that the ECEA in boring corresponds to the SCEA in turning and is generally negative. Edge clamping methods for boring are similar to those discussed in the previous section for turning.

In single point boring, however, the boring bar is long and compliant compared to a turning tool holder. In fact, due to their length to diameter ratios, boring bars and end mills are the least rigid of all cutting tools. The rigidity or compliance of the boring bar is often the primary factor determining boring process performance in single-point boring; deflections of the bar constrain its reach, affect the bored hole diameter and accuracy, and may lead to vibration and chatter. Displacement of the boring bar due to cutting forces, especially in interrupted cuts, may result from (1) bending deflection of the bar structure; (2) deformation in any joint(s) between the parts of a composite bar; (3) deformation in the joint between the spindle and the bar flange; and (4) deformation of the spindle and its bearings. Factors affecting boring system rigidity therefore include the basic machine design; the drive mechanism design; the spindle stiffness; the boring bar diameter, length, overhang, and material; and the workpiece material, structure, and fixturing.

In many general purpose applications, the deflection of the bar itself is larger than the other components. The deflection of a simple single-point bar, rigidly mounted (cantilevered) to the machine structure, with an overhang L and radial force F_r, can be calculated using the equation

$$\delta = \frac{F_r \cdot L^3}{3 \cdot E \cdot I} \tag{4.1}$$

where
  E is the modulus of elasticity of the material
  I is the cross sectional moment of inertia

For a solid round bar,

$$I = \frac{\pi \cdot D^4}{64} \tag{4.2}$$

The deflection under 100 N load for a cylindrical cross-section single or step diameter bar is illustrated in Figure 4.26; this figure illustrates the importance of minimizing tool overhang and/or L/D ratio. As a rule of thumb, deflections should be less than 0.025 mm to ensure adequate system performance.

Two major problems related to bar rigidity are often encountered with boring: springback and chatter. Bar springback is caused by insufficient stiffness and the resulting deflection or deformation of the bar due to cutting forces. Excessive springback (or elastic recovery) of the boring bar

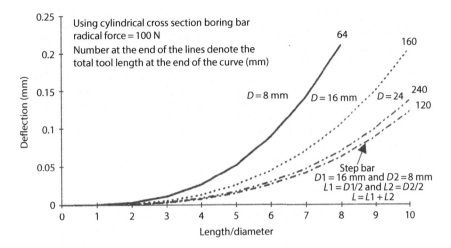

FIGURE 4.26 Deflection of a tool/toolholder system vs. length/diameter ratio.

will result in a smaller bore or scratch marks during tool retraction. Spingback and chatter during tool retraction can be eliminated by designing the tool to permit a minor (0.02–0.04 mm) shift in the cutting-edge position during the tool retraction; the shift should not change the rotational axis of the tool. Due to the small size of the tool, the shifting mechanism is difficult to package and may cause balancing problems at high rpms. In CNC applications, the shift can be accomplished by a radial offset.

Chatter can occur either during the feed or retract portions of the cycle. The resonant frequencies and chatter resistance of a bar depend strongly on the bar length to diameter (or overhang) ratio; overhang ratios greater than 4 or 5 (4 for steel bodies) are especially susceptible to chatter. On a more basic level, chatter resistance is strongly affected by the value of the quantity $k \cdot \zeta$ where k is the effective dynamic stiffness and ζ is the damping ratio of the system. The stiffness of the boring bar can be increased by using the largest diameter and shortest shank length bar permitted by the bore geometry, and by using bar materials with a high elastic modulus. Deflections can also be minimized by reducing cutting forces through the use of tip geometries designed to ensure free cutting. The damping characteristics of a boring bar system depend on the structural damping of the bar (a material property) and the interfacial frictional damping between the bar, the tool holder, and their connection or mounting technique. In a conventional setup the boring bar is mounted in a holder, which is subsequently mounted to the spindle. An improved but less flexible approach is to bolt the tool directly to the spindle via a flange. In designing a boring bar system, a trade-off between stiffness and damping characteristics may be required. Solid bars made of rigid materials enhance stiffness, but often have relatively poor damping properties, which reduce chatter resistance. Composite extended bars provide increased damping but often reduced stiffness. The stiffness and damping (chatter-resistance) characteristics of a boring tool can also be enhanced using anisotropic mandrels with directionally enhanced stiffness axes [85], and by using passive [86] or active [87] vibration control methods.

The use of optimal materials and geometries is often the preferred approach to enhancing the stability of boring bars, and cantilever tools is general [88–90]. Several composite boring bar structures using tool steels, heavy metals, hard carbides, and light alloy materials can be used effectively as shown in Figure 4.27. Composite bars generally consist of a heavy metal root segment (which is machinable) with a relatively high Young modulus. The overhang segment at the end of the tool, which holds the tool bit or insert, is designed to be light, and may be made of aluminum, titanium, or similar materials to increase the natural frequency of the cutting point without significantly affecting bar stiffness. The middle section is often made of carbide. These designs are compromises between economical solid steel and costly solid carbide bars (see Examples 4.4 through 4.6).

FIGURE 4.27   Combination structure boring bar.

Generally, solid steel boring bars are used for $L/D < 4$, composite bars are used with $L/D < 6$ or 7, and composite bars with dynamic vibration absorption are used for $L/D > 6$ [87].

Boring tools may be fixed or adjustable as shown in Figure 4.28. Fixed boring tools generate a particular diameter predetermined by the radial distance of the cutting tip from the tool axis. Such tools may be made of solid HSS, WC, PCD, or PCBN tips brazed to a solid bar, solid WC, or indexable inserts of various materials set in pocketed bars. They may have one or more cutting edges depending on the bore size and bore quality and surface finish requirements. As discussed in the following, adjustable boring tools are equipped with heads containing precision mechanisms to offset the cutting edge to compensate for tool wear or deflections. Insert size tolerances and boring bar manufacturing tolerances often make it difficult to control the size of resulting bored holes. Therefore, a size adjustment capability is often necessary. The conventional approach for adjusting the size of boring bars is to use manually or automatically adjustable boring heads mounted at the end of the boring bar.

A manual boring head uses a precision micrometer adjustment or a cartridge. The former uses a dial (micrometer spindle) that offsets the insert or the boring bar in 0.0025–0.01 mm increments per division on the diameter. The latter design does not provide any indication for size adjustment and uses a cam underneath the cartridge, which permits a small range of adjustments (Figure 4.29). The dial system has a graduated head (Figure 4.30) attached on a set screw or on a ring around the head. To set the boring diameter, the dial is turned to the desired setting. In cartridge-type systems, adjustments are made using a preloaded screw without any indication of tool offset; a gage is therefore required for any subsequent adjustments. Manual adjustable tooling is set up by trial and error and is generally unsuitable for mass production.

The automated approach for transfer of machine applications requires a complex tooling system to automatically perform small adjustments in the boring bar size, as well as precise in-line gaging capability. The boring bar requires a built-in mechanical or hydraulic tool compensating system for offsetting the tool to adjust size as shown in Figure 4.31. Mechanical systems are generally actuated by a drawbar in a spindle. The tool positioner is complex and can be fit only in boring bars with diameters larger than 50 mm. Many of the available designs have a weak link between the boring

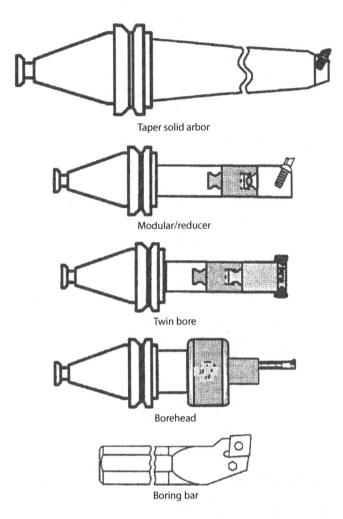

Taper solid arbor

Modular/reducer

Twin bore

Borehead

Boring bar

FIGURE 4.28 Various types of boring bars with and without diameter adjustments.

bar and the tool adjustment mechanism and therefore have significantly lower static and dynamic stiffness (20%–50%) compared to nonadjustable solid boring bars. As an alternative, an adjustable boring bar assembly in which the cutting radius can be altered by predetermined increments simply by releasing and turning the bar within a sleeve can be used. The holder sleeve is located slightly eccentric relative to the spindle axis, so that turning of the bar effectively changes the working radius of the cutting point as shown in Figure 4.32.

In CNC systems, size compensation for precision bores is accomplished using a boring head with an adjustment mechanism actuated by a screw to control radial cartridge position. In-line gaging is used to monitor bore size, and when compensation is required, the bar is moved to a set of pins or similar feature on the machine table or fixture, which engages holes or slots in the adjustment mechanism of the boring head. The machine positions the head over the feature and rotates the spindle through precise increments to index the micrometer head. This approach requires spindle clocking and rotation control capability.

Multiple-step tool bars using several cutting edges, as shown in Figure 4.33, are designed to generate concentric bores with varying diameters and/or to bore and counterbore in one pass. Similarly, a boring bar with a roughing or semifinishing tool ahead of a finishing tool (axially staggered inserts and/or radial spacing) can be used to rough and finish a bore in one pass.

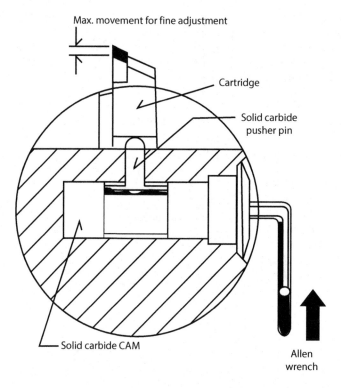

FIGURE 4.29 Manual adjustable cartridge. (Courtesy of Master Tool Corporation, Grand River, OH.)

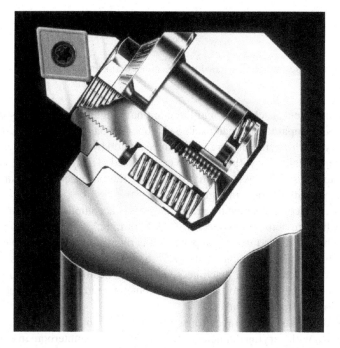

FIGURE 4.30 Single point boring bar with a diameter adjustment dial cartridge. (Courtesy of Valenite Corporation, Troy, MI.)

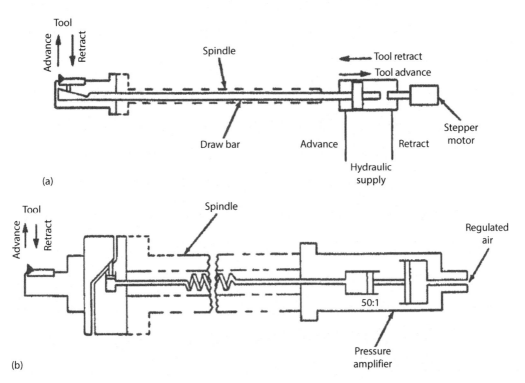

FIGURE 4.31 Schematic diagram for two automatic adjustable boring systems. (a) Mechanical system and (b) hydraulic system.

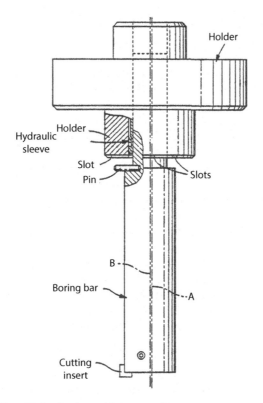

FIGURE 4.32 Manual adjustable boring bar with improved accuracy.

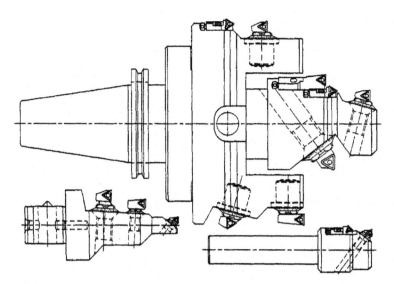

FIGURE 4.33 Multiple operation boring/counterboring/chamfering tooling. (Courtesy of Rigibore Tooling Systems, Mukwonago, WI.)

4.6.2 MULTIPOINT BORING TOOLS

Multipoint boring bars are often used to improve stability and to increase productivity in high-volume applications. Adjustable twin bars are used in general purpose machining. These bars have a caliper head on the end with a pair of opposing inserts mounted on each side. Adjusting a dial changes the span between the inserts and the cutting diameter. These bars have low radial stiffness compared to single-point bars, but the opposing inserts balance radial loading, leading to increased stability and accuracy in many applications.

Multipoint bars are used extensively in automotive cylinder boring and other high volume applications. Multipoint bars can be run at higher feed rates than single point bars, and spread tool wear over multiple tool edges, increasing tool life. In transfer line applications, semifinish and finish boring are commonly performed in one operation using a bar with multiple inserts for the down (semifinish) pass, and a single insert actuated by a drawbar for the return (finish) pass. In this system, size compensation is accomplished using methods similar to those described for single point bars earlier. In CNC applications, multipoint semifinish bars are used, generally followed by single point finishing bars. Size compensation for the multipoint finish bars is accomplished by mounting an adjustment head on the bar and actuating it through a controlled spindle rotation using pins mounted on the fixture as described earlier for single point boring.

Draglines (withdrawal or witness marks) are generated during the reversal of fixed bars from the bottom of the bore. Draglines reduce tool life and leave surface damage, which usually must be removed by later operations. In transfer line applications, a cam or pivot system activated by a drawbar may be used to withdraw inserts to a smaller radius prior to tool retraction to eliminate draglines. In coolant activated bars, the inserts are pushed clear of the machined surface by springs when the coolant pressure is removed after they exit the bore but prior to retraction. In some transfer line and many CNC applications, draglines are tolerated in rough and semifinish boring, although extra machined stock may be left for subsequent finish boring and honing operations to ensure cleanup.

4.7 MILLING TOOLS

Milling is carried out using a rotary tool with multiple cutting edges. Both solid cutters with ground cutting edges and pocketed cutter bodies fitted with inserts are used as shown in Figure 4.34. Design criteria for milling tools are similar to those for turning tools in many ways, but milling involves additional

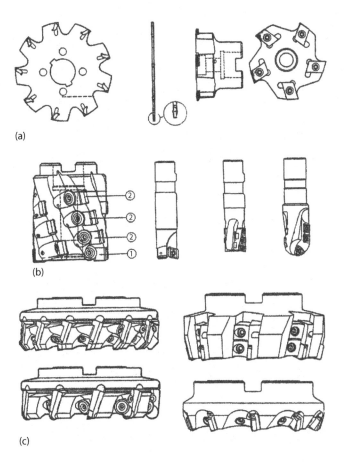

FIGURE 4.34 Indexable milling cutters. (a) Slot milling, (b) slot- and contour-milling, and (c) face milling. (Courtesy of Carboloy/Seco, Warren, MI.)

concerns because it is an interrupted cutting process. The cutting edges on a milling cutter enter and leave the cut each rotation and are actually cutting for less than half of the total machining time.

This section discusses common cutter designs and geometric factors that affect cutter and insert selection [91–96]. The nomenclature used follows the American National Standard for face and end mills [97].

4.7.1 TYPES OF MILLING CUTTERS

Several types of milling cutters are used for different operations. *Face milling cutters* are used to produce flat surfaces. They are usually fitted with indexable inserts or inserted blades. Indexable face mills are the most common and versatile type because they can be used with a variety of insert materials. Conventionally, indexable inserts are fixed in pockets in the cutter body, although they may also be mounted in cartridges. The geometry and nomenclature of a face milling cutter is shown in Figure 4.35.

Slot milling cutters are used for grooving, slotting, and side and face milling. Slot cutter designs and manufacturing methods vary depending on their size and application. Solid HSS, solid WC, and WC tipped designs are often used for small diameter cutters (<100 mm), while indexable cutters are preferred for larger diameters. Several insert geometries are available with or without chip breakers for indexable cutter bodies. The cutting edges of such cutters are similar to those used on circular saws or hole saws. Both full- and single-side slot mills are available. Full-side cutters may have

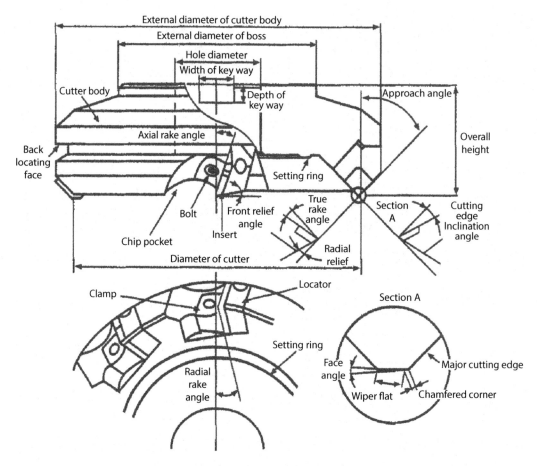

FIGURE 4.35 Milling cutter nomenclature. (Courtesy of Sumitomo Electric Corporation, Osaka, Japan.)

continuous edges across the width of the cutter, or located alternatively around the cutter periphery. The cutting edges of full side cutters can be square, trapezoidal (chamfered on both sides), pointed, or radiused. The triple-chip ground-tooth form (in which the first tooth cuts a narrow chip from the center of the cut while the second tooth, which follows the first, cuts two equal-size chips from the sides of the cut) has become popular because it removes the material much more easily. The edges are generally designed with positive or neutral rake angles.

End milling cutters generate two workpiece surfaces at the same time; cutting edges are located on both the end face and the periphery of the cutter body. They are usually used in facing, 2D or 3D profiling, slotting, shoulder cutting, slabbing, chamfering, and plunging/pocketing operations and are the most versatile milling tools. When they are used for plunging, the teeth must cut all the way to the center of the tool. They are produced in solid HSS, HSS-CO, WC, ceramic, PCD/PCBN brazed or veined construction, inserted blade, and indexable insert designs. Inserted blade end mills produce better finishes than indexable WC end mills. Much of the previous discussion about stiffness and deflection of boring bars is also applicable to end mills. Shank-type end mills are available in a wide variety of sizes, flute configurations, and lengths. The geometry of the end mill is shown in Figure 4.36 [96]; its diameter, flute length, and corner radius must match the dimensions of the pocket to be machined. Generally, two- and three-flute end mills are center cutting, while end mills with a larger number of flutes are non-center cutting. The end mill's core diameter, number of flutes, and flute space determine its rigidity in the manner discussed in the following for drills. The tooth edge geometry, radial,

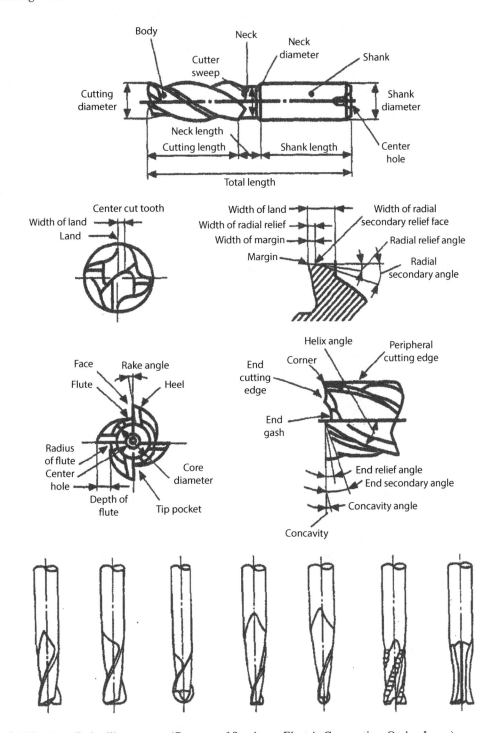

FIGURE 4.36 End-mill geometry. (Courtesy of Sumitomo Electric Corporation, Osaka, Japan.)

and axial rakes, relief angles, and flute forms are important, and for solid end mills should be designed according to the same guidelines used for single-point turning tools, indexable milling cutters, and drills. End mills with helical rather than straight flutes have a reduced tendency to chatter because one or more points of the helical flutes remain in contact with the workpiece at all times.

Rotary milling cutters can be used in some applications to improve productivity. Rotary milling cutters are equipped with round inserts, which are secured in interchangeable cartridges and which are free to rotate as they cut. The insert rotates on a bearing attached on the cartridge. The rotation of the insert alters chip formation on the rake face by increasing the effective inclination angle. The insert rotation is caused by the action of the cutting forces and chip flow over the insert face; a separate drive mechanism is not required. The contact edge length between the chip and the tool changes during cutting, leading to reduced wear distributed evenly around the periphery of the insert, and ultimately to increased tool life. Their chief attraction is that they can be run at much higher feed rates than conventional cutters, reducing cycle times and potentially machine counts in automotive applications. They can also produce smoother finishes due to the large radius on the inserts.

Research on rotary cutting tools began in the 1950s. The initial research was done for turning with a powered rotating tool [98,99], an idea which has recently been revived by Kennametal and DMG-Mori Seiki with their "Spinning Tool" product [100]. Free spinning rotary tools were developed later, mainly for milling. GE-Carboloy investigated them as part of the Defense Advanced Research Program Administration (DARPA) high speed machining project in the early 1980s [101]. There is a large patent literature on free-spinning cartridges, mainly for milling [102,103]. Rotary tooling for boring was developed in the mid-1990s by Rotary Technologies of Rancho Dominguez, CA [104]. Rotary technologies also manufactures rotary boring and turning tools.

4.7.2 Cutter Design

Factors to be selected in designing a milling cutter for a specific operation include the cutting edge geometry (rake and lead angles), cutter density, cutter diameter, entry and exit angles, and cutter construction (indexable, inserted blade, etc.) [76,105–107]. For slot and end mills, the helix angle also strongly affects cutter performance.

The standard angles used to describe the cutting geometry of rotary tools such as milling cutters are the radial (side) and axial (back) rake angles (Figure 4.35). The axial rake angle affects the chip flow direction and the thrust force, while the radial rake angle has a strong effect on the cutting power and tool life. The radial (α_f) and axial (α_p) rakes determine two additional angles, the inclination (λ_s) and orthogonal (top) rake (α_o) angles (Figure 4.37). The orthogonal rake is the true top rake, which influences cutting forces and power requirements. The true rake angle is measured from, and perpendicular to, the lead angle. The inclination angle is similar to the helix angle and is measured from the cutting edge face to the reference plane on a line parallel to the lead angle

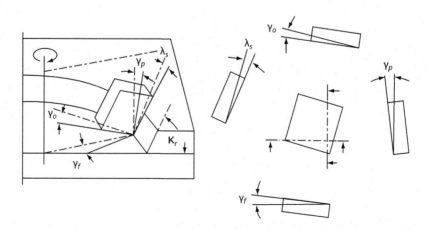

FIGURE 4.37 Radial and axial rake angle in a rotary tool. (Courtesy of Sandvik Coromant, Fairlawn, NJ.)

cutting edge. This angle determines the chip flow direction and significantly affects cutting forces and tool life. These angles are related through equations:

$$\tan \alpha_o = \tan \alpha_f \cdot \cos \kappa_r + \tan \lambda_p \cdot \sin \kappa_r \qquad (4.3)$$

$$\tan \lambda_s = \tan \alpha_p \cdot \cos \kappa_r - \tan \lambda_f \cdot \sin \kappa_r \qquad (4.4)$$

where κ_r is the bevel angle, which is related to the lead angle as discussed as follows. For example, a combination of $\alpha_p = 7°$, $\alpha_f = 0°$, and $\kappa_r = 15°$ results in $\alpha_o = 2°$ and $\lambda_s = 7°$. Three common combinations of these two angles are used on standard milling cutters as shown in Figure 4.38. Chip formation for each of the configurations is shown in Figure 4.39:

1. Cutters with positive axial and radial rake angles are called *double-positive* cutters. The positive axial rake lifts the chip and curls it away from the finished workpiece surface and toward the inside of the cutter body. The positive radial rake provides a sharper cutting edge that tends to pull the tool into the work (free cutting). Double positive cutters reduce the cutting pressure, consume less power, create less heat, reduce deflection, and result in less strain on the machine bearings, ways, and spindle. This geometry creates a true shearing action, which is suitable for finish milling of free-cutting steel, aluminum, and brass; for materials which form a continuous chip; and for work-hardening materials and many soft stainless steels.

2. Cutters with negative axial and radial rake angles are called *double-negative* cutters and are the most common type of cutters. The double-negative geometry offers several advantages: first, negative rake inserts with twice as many cutting corners as positive inserts are used; second, the strongest part of the insert, away from the cutting edge, enters the work first, whereas with positive rake cutters the edge enters first. Also, the high edge strength permits use of harder, more wear resistant insert materials. Double-negative cutters tend to push the work away from the tool, and therefore require greater power and higher system rigidity than positive cutters. Moreover, with a double negative cutter, the chip is bent forward and downward under pressure, which can cause chip evacuation problems for soft steels. Double negative cutters are effective in rough and finish milling of steel and cast iron, including hard and high-strength grades.

3. Cutters with positive axial and negative radial rake angles are called *shear-angle* cutters. These cutters combine some of the advantages of both negative and positive rake cutters. The negative radial rake provides a strong cutting edge, while the positive axial rake angle, combined with the bevel on the cutter body, lifts chips up and directs them away from the surfaces being machined. They are most often used for rough and finish milling of the tougher grades of aluminum and other nonferrous materials, and for free machining of cast iron, steels, stainless steels, and most high-temperature alloys, which are milled with difficulty using double-negative cutters.

The lead angle, shown in Figure 4.35, is equivalent to the SCEA or approach angle in turning, and is equal to 90° minus bevel angle (κ_r) shown in Figure 4.37. The best lead angle for a specific operation depends on several factors including the part configuration, the distribution of cutting forces, and the chip thickness. An increase in the lead angle results in an increase in the axial force and a reduction of the radial force. A proper lead angle also allows a cutter to enter and exit the cut more smoothly, reducing the shock load on the cutting edge. Proper selection of the lead angle is especially important for shear-angle cutters. A 15° lead is often used for face milling. Larger lead angles are used in shear-angle cutters because they provide a large inclination angle. A 45° lead provides

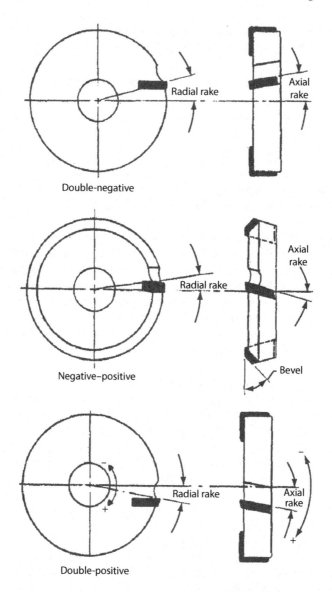

FIGURE 4.38 Three milling cutter geometries with differing radial and axial rake angles. (Courtesy of Valenite Corporation, Troy, MI.)

the strongest cutting edge for heavy duty milling; in this case the radial force will be approximately equal to the axial force. Large lead angles should also be used when the spindle is weak radially (e.g., due to a long spindle overhang), and when the radial stiffness of the part or fixture is limited.

The cutter density or tooth pitch is determined by the number of teeth in the cutter body. Cutters are classified as having either fine (3–5 inserts per 25 mm of diameter), medium, or coarse (1–1.5 inserts per 25 mm of diameter) pitch. A fine-pitch cutter has about half the chip clearance of a coarse pitch cutter. The milling cutter with the highest number of teeth allowed by the workpiece and machine tool is most efficient. However, coarse pitch cutters allow higher chip thickness or feed per tooth, resulting in fewer teeth in the cut to prevent machine overloading. Therefore, a coarse pitch is usually used for large diameter cutters and/or for heavy duty applications, while fine pitch cutters are often used in high production rate applications, when machining thin wall sections, and when the depth of cut does not exceed 7 mm. The amount of chip space required for a specific

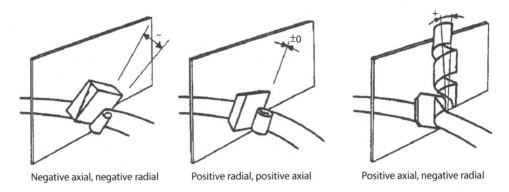

Negative axial, negative radial   Positive radial, positive axial   Positive axial, negative radial

FIGURE 4.39 Chip formation for the milling cutter geometries shown in Figure 4.38. (Courtesy of Carboloy/Seco, Warren, MI.)

operation is determined by the workpiece material and the width of cut. Two essential requirements must be considered in selecting the cutter density: (1) at least one (and preferably two) cutting edge (insert) must be in the cut at all times to prevent instability of the tool as it enters and exits the workpiece; and (2) adequate chip space and cutting edge support must be provided. Cast iron and titanium require little chip space, but steel and aluminum require relatively large chip slots. A larger number of flutes on an end mill reduces chip loads and improves the workpiece finish, but reduces radial rigidity or chip space, depending on the depth of the flutes.

Staggered or white-noise cutters, with unequal spacing of cutting edges around the cutter, can reduce or eliminate chatter by varying the chip load on each insert and reducing the effective tooth passing frequency. They effectively eliminate excitation of the tooling and part structures at the tooth passing frequency and its harmonics. Differential helix angles on successive flutes can be used for long edge end mills to reduce vibration during cutting. Roughing shell end mills may have a spirally staggered arrangement of indexable cutting edges to provide smooth cutting action and quiet operation.

The required diameter of a face, end, side, or slot mill depends on the radial depth or width of cut (or part), axial depth of cut, and feed, which also determine the required number of cutting edges per mm of diameter (e.g., deep slots require low-density cutters). The cutter diameter should allow for optimum feed per tooth and depth of cut based on the available power. The cutter diameter for a face mill should generally be at least 1.3–1.6 times the width of the part. This allows the cutter to be offset from the centerline of the workpiece and to overhang it on the tooth entry side. Symmetrical positioning of the cutter should be avoided especially when the engagement width is considerably smaller than the cutter diameter in order to prevent vibration due to pulse loading of the machine tool especially for light duty machines and positive tools. Pulse loading could lead to excitation of various structural vibration modes that may cause transient vibrations during the entry and exit of the cutter. Coarse pitch cutters are normally the first choice for diameters larger than 120 mm. The bending moment on the cutter is affected by its diameter and overhang and should be below the maximum allowed by the spindle interface to avoid cutter vibration and deflection. When a long tool overhang must be used, the smallest diameter cutter possible should be selected. Therefore, it may be preferable to use a smaller diameter cutter and to cover the width of the workpiece in two or more passes, unless this arrangement produces an uneven surface finish.

The diameter of the cutter also affects the chip thickness. As discussed in Chapter 2, the chip thinning effect is very important in milling. The maximum radial chip thickness is affected by the cutter geometry, the lead angle, and (most significantly) by the position of the cutter on the workpiece (see Section 2.6). The average chip thickness should generally be greater than 0.07 mm, and if it is less than 0.025 mm, a mixture of rubbing and cutting may occur resulting in greater heat generation and excessive flank wear. More details on the chip-thinning effect are discussed in Examples 2.5 through 2.8.

The selection of the cutter density and diameter in contour or pocket milling is more complex than in straight-line milling because the cutter engagement during convex and concave arcs must be considered. This is true, for example, when traveling around a 90° corner in pocket milling.

Cutter designs and feed rates that produce narrow chips with a heavy cross section can be used to improve metal removal rates and tool life when machining relatively deep cuts using machines with limited power and rigidity. Either step cutters or chipbreaker cutters can be used for this purpose. Inserts are stepped radially and axially in step cutters, as shown in Figure 4.40. This arrangement increases the chip thickness at reduced feedrates by breaking up each chip among several inserts in a sequence. The use of a step cutter reduces the effective number of cutting edges [108]. Chipbreaker cutters are designed with grooves around the body that are either semicircular or sinusoidal in shape. They are also called hogging cutters. Such designs are used in solid and ball nose end mills (Figure 4.41) and in indexable face, slot, and helical end mills equipped with scalloped inserts (Figure 4.42). The design of the chipbreaker grooves, specifically coarse or fine pitches, controls the chip size. The grooves are

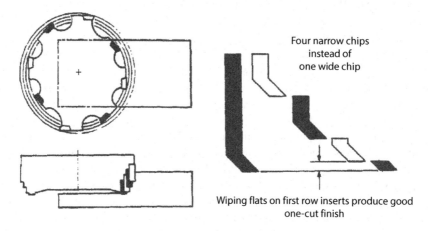

Four narrow chips
instead of
one wide chip

Wiping flats on first row inserts produce good
one-cut finish

FIGURE 4.40 Milling cutter with four radial and axial steps and the corresponding chip cross sections. (Courtesy of Ingersoll Cutting Tool Division, Rockford, IL.)

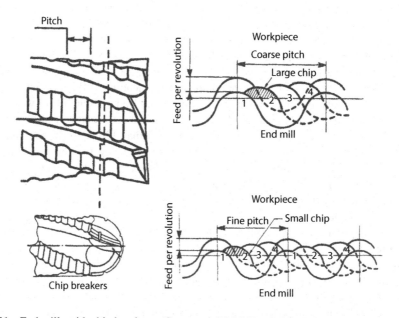

FIGURE 4.41 End mills with chip breakers. (Courtesy of OSG Tap & Die, Inc., Toyokawa, Japan.)

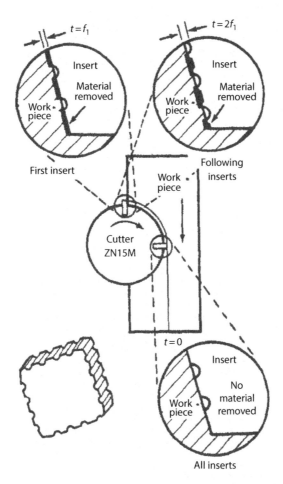

FIGURE 4.42 Sequential cutting action of a milling cutter with scalloped inserts. (Courtesy of Carboloy/Seco, Warren, MI.)

arranged so that every other edge removes the ridges left on the part by the preceding insert, as shown in Figures 4.41 and 4.42. The use of chipbreaker cutters lowers cutting forces and reduces tool deflection caused by long cutter overhangs, and therefore permits use of higher feed rates in roughing operations.

The insert contact angles during entrance (especially in down milling) and exit have a strong influence on tool life and burr formation. These angles are a function of the relative dimensions of the cutter and workpiece, the cutter overhang on the entry/exit side, and the radial rake angle of the cutting edge. The insert contact angle is

$$\varepsilon_c = 90° + \alpha - \varepsilon \tag{4.5}$$

where the insert entry angle, ε, is defined by a line through the cutter axis and perpendicular to the entrance surface and a line through the cutter axis passing through the cutting point:

$$\varepsilon = \tan^{-1}\left[\frac{\sqrt{(2 \cdot R - w_s) \cdot w_s}}{R - w_s} \right] \tag{4.6}$$

where
R is the cutter radius
w_s its overhang on the entry side (Figure 4.43)

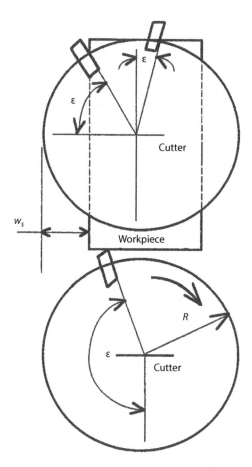

FIGURE 4.43 Schematic view of initial cutting edge contact at the entrance of a workpiece. (Courtesy of Carboloy/Seco, Warren, MI.)

The sign of the insert contact angle determines the location of the initial impact between the tool and the workpiece in down milling. The initial contact occurs on the cutting edge (the weakest section of tool) when ε_c is positive and on the rake face when ε_c is negative as shown in Figure 4.44. A positive ε_c should be avoided unless the work material is soft and ductile. The cutter position relative to the workpiece should be controlled so that the insert contact angle is negative. However, ε_c is often positive at the beginning of the workpiece cut while negative at exit and re-entry (see Figure 4.44). Tool entry angle problems are often encountered when the cutter reenters the workpiece after crossing cavities or holes in the workpiece, as is often the case in face milling automotive powertrain components. Generally, having the centerline of the cutter well inside the workpiece width is considered good practice for face milling. Reentry in peripheral shoulder and through-slot end-milling operations often results in severe entry angle conditions, which can lead to tool failure. Various tool routing approaches can be used to reduce tool entry or re-entry problems. Improper exit angles can cause burr formation as discussed in Section 11.4. Burr formation is a major concern in milling and has been the limiting factor preventing milling operations from providing part quality equivalent to grinding in many applications.

Indexable and inserted blade milling cutters can be designed for roughing and finishing in one cut. This can be done by using either a combined rough and finish cutter, or by using roughing and finishing inserts on the periphery of the tool as shown in Figure 4.45. Combination cutters use roughing inserts around the tool periphery and finishing inserts mounted on the tool face. In some cases finishing inserts with wiper flats are substituted for some of the roughing inserts.

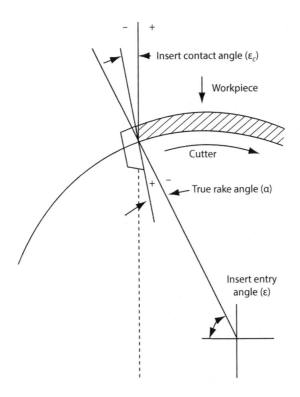

FIGURE 4.44   Schematic view of cutting edge contact during milling at the entrance and exit of a workpiece.

Finishes equivalent to grinding can be produced when milling with wipers, especially when wipers are located precisely with respect to the cutter's axial plane. These cutters perform best at light depths of cut and medium-to-high feed rates. The flat wiping surface of the insert is often crowned with a large radius to prevent the formation of a sawtooth workpiece surface profile, which may occur when the flat land is not exactly parallel to the direction of feed.

In end and slot milling, the helix angle of the cutting edges or flutes determines how rapidly cutting forces increase and decrease as the cutting edges enter and exit the cut. Higher helix angles increase the length of the cutting edge engaged in the cut and allow for coarser tooth spacings because they insure continuous cutter-workpiece contact; they often improve the surface finish and permit use of increased speeds and feed rates (see Examples in Chapter 8). Helix angles up to 60° may be used in order to reduce cutting forces for materials such as stainless steel, titanium, and inconel. However, lower helix angles provide higher edge strength in corner areas and reduce edge chipping or flaking. Left-hand helix angles are used to change the direction of the cutting forces of the workpiece in some applications. Tapered end mills (0.5°–25° per side) are used for complex profiles and mold making.

Both square and ballnose end mills are used. Ballnose end mills require consideration of the effective diameters affected by the DOC as explained in Problem 2.6. The radial depth of cut affects the surface finish because the tool leaves behind scallops on the surface that vary in height according to the stepover distance. The stepover is given by

$$\text{Stepover} = 2\left((D/2)^2 - (D/2 - h)^2\right)^{0.5} \tag{4.7}$$

where h is the height of peaks.

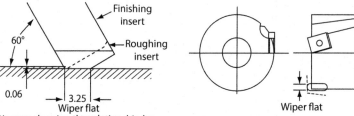

Diagram showing the relationship between roughing and finishing inserts
around periphery of MAX-1-SHEAR
cutter

Diagram of MAX-1-FLAT cutter, Inserts
are set into cutter at right angles to the
face, but they cut a bevel that is about
B9° , rather than a full 90°

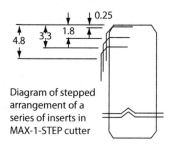

Diagram of stepped
arrangement of a
series of inserts in
MAX-1-STEP cutter

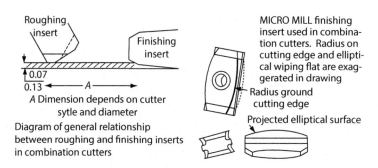

A Dimension depends on cutter
sytle and diameter

Diagram of general relationship
between roughing and finishing inserts
in combination cutters

MICRO MILL finishing
insert used in combina-
tion cutters. Radius on
cutting edge and ellipti-
cal wiping flat are exag-
gerated in drawing

Radius ground
cutting edge

Projected elliptical surface

FIGURE 4.45   Diagram showing the relationship between roughing and finishing inserts in a combination
milling cutter. (Courtesy of Ingersoll Cutting Tool Division, Rockford, IL.)

4.7.3   Milling Inserts and Edge Clamping Methods

The criteria for selecting inserts in milling are very similar to those used in turning, although
additional considerations also apply. There are several insert shapes available for milling that are
not often used for turning due to the nature of the operation. The cutting insert nomenclature is
described in Figure 4.46. The insert corner configuration can be a radius, a chamfer, a double chamfer, or a flat. The proper selection of the edge preparation (up-sharp, hone, chamfer, or combined
hone plus chamfer) has a strong influence on tool life as discussed in Section 4.5.

The methods of holding the insert in the cutter body is very critical to cutter performance
especially at higher speeds due to centrifugal force of the clamping components. The most common methods are (1) nonadjustable insert pockets, (2) axially or radially adjustable designs using
either a pocket, a precision rail and wedge lock, or a cartridge, and (3) brazed tips. Some of these
systems are illustrated in Figure 4.34. An on-edge insert held in fixed pockets by a single screw
provides secure locking against movement, permits simple, accurate indexing, and allows for the

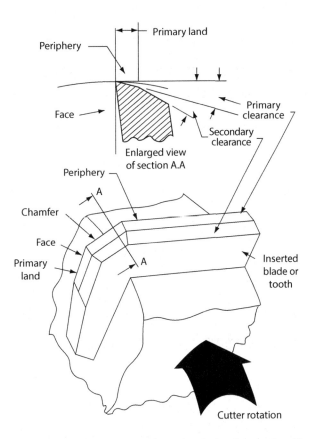

FIGURE 4.46 Insert geometry nomenclature. (Courtesy of Valenite Corporation, Troy, MI.)

strongest section of the carbide to support the cutting forces. The adjustable designs generally reduce insert strength and stiffness but permit significant reduction in insert axial and radial runout. Brazed teeth are usually used on smaller cutters and provide the lowest edge height and roundness error.

4.8 DRILLING TOOLS

A drill is an end-cutting tool with one or more straight or helical flutes, which may have a hollow body for the passage of cutting fluid and/or chips during the generation of a hole in a solid or cored part. Drills vary widely in form, dimension, and tolerance. Drills are classified according to the materials from which they are made, their lengths, shapes, number of flutes, point characteristics, shank style, and size series [109]. The best type of drill for a given application depends on the material to be drilled, its structural characteristics, the hole dimensions, whether the material to be drilled is cored or solid, whether a through or blind hole is required, the entrance and exit characteristics of the workpiece (Figure 4.47), the expected hole quality, the characteristics of the machine tool and fixture, and the cutting conditions. Selecting the proper style of drill for a given application requires consideration of all these factors.

Three types of conventional drills are widely used: regrindable drills, spade drills, and indexable insert and tipped drills. As shown in Figure 4.48, there are several types of regrindable drills including twist (or regular) drills, gun drills, counter drills, and pilot drills. Twist drills (Figure 4.49) differ widely in the number of flutes they contain, and in geometric characteristics such as the helix angle, relief (clearance) angle, point style, flute shape, web taper, web

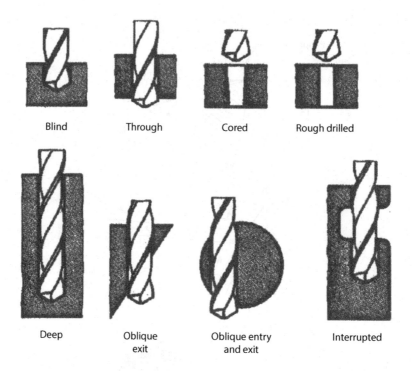

FIGURE 4.47   Influence of workpiece and hole form characteristics. (Courtesy of Guhring Inc., Brookfield, WI.)

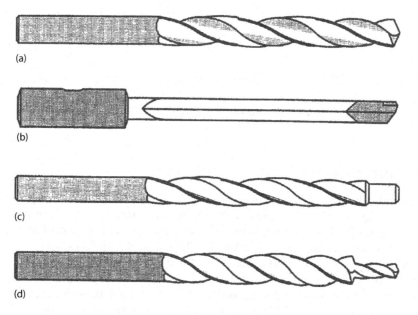

FIGURE 4.48   Regrindable drill types. (a) Regular drill, (b) gun drill, (c) pilot drill, and (d) counter drill.

thickness, and margin width. A standardized system of identifying or classifying twist drills made by different manufactures has not yet been developed.

This section describes the basic types of drills and factors affecting drill design and selection. Much of the material pertains specifically to twist drills, which are the most widely used drills in practice, but other types of drills are discussed as well.

4.8.1 TWIST DRILL STRUCTURAL PROPERTIES

The structural properties of the drill have a direct bearing on drilling performance. From a structural viewpoint, several features of a twist drill are significant as illustrated in Figure 4.49:

1. *Shank or drive type.* The butt end of the drill is generally held in a holder or the spindle and driven. Driver design is important because it determines the roundness accuracy and stiffness of the drill-holder system, as well as the limiting speeds and coolant pressures, which can be used in some applications. Maximizing the driver length improves rigidity and concentricity.

   Straight shank drills with or without ground flats are mounted in end mill holders, collets, chucks, or special hydraulically or mechanically clamped holders described in Chapter 5. *Tapered shank* drills are mounted directly into the spindle with or without intermediate sleeves or adapters.

2. *Helix angle.* The helix angle is the angle between the leading edge of the land and the drill axis.

   Standard helix drills have a helix angle of approximately 30° and are used for drilling malleable and cast irons, carbon steels, stainless steels, hard aluminum alloys, brass, and bronze. *Low (slow) helix* drills have helix angles of approximately 12°. They have increased cutting edge strength and are used for drilling high temperature alloys and other hard-to-machine materials. They are also used for brass, magnesium, aluminum alloys and similar materials, since they provide for quick ejection of chips at high penetration rates, especially for shallow holes. *High (quick) helix* drills have helix angles of approximately 40°, as well as wide, polished flutes and narrow lands. They are used for drilling low-strength nonferrous materials such as aluminum, magnesium, copper, and zinc, and for low-carbon steels. *Zero helix (straight flute)* drills have a 0° helix angle. They are used for materials that produce short chips such as brass, other nonferrous materials, and cast iron. They are especially common in horizontal drilling applications. Low or zero helix drills

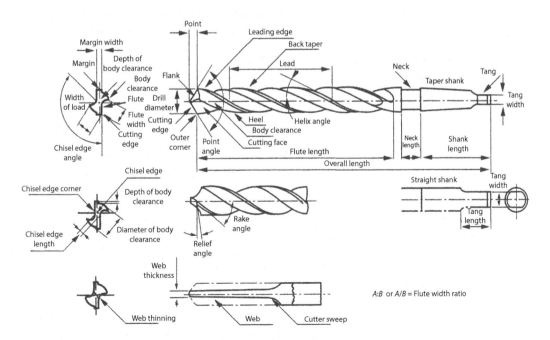

FIGURE 4.49 Drill nomenclature. (Courtesy of Sumitomo Electric Corporation, Osaka, Japan.)

can be used for holes with length-to-diameter ratios exceeding four provided pressure-fed through the tool coolants are used to evacuate chips.

The helix angle affects not only the chip ejection capability of a drill, but also its cross-sectional strength, area moments of inertia, rigidity, and rake angle. A left-hand spiral on a right-hand cutting drill resists the winding up of the shank at higher feed rates, especially for long drills. Left-hand helix is also used for taper drills to prevent chatter occurring with right-hand drills under right-hand rotation. The torsional stiffness of a drill varies parabolically with the helix angle and reaches a maximum at a helix angle of approximately 28° [110]. The radial stiffness of a drill decreases with increasing helix angle and reaches a minimum at a helix angle of approximately 35° [111]. The axial stiffness also varies parabolically with the helix angle, with a minimum occurring at approximately 20° [111]. The allowable thrust and critical cutting speed are also affected by the helix angle, especially for small diameter drills. An increase in helix angle results in increased rake angles and lower torque and thrust. The helix angle influences drill life in a complex fashion depending on other parameters and the workpiece material [112,113].

3. *Hand or direction of rotation.* The vast majority of drills are right-handed and rotate in a clockwise direction when viewed from the shank end.

4. *Number of flutes.* The number of flutes may vary from one to four, with two being the most common choice. The optimum number of flutes on a drill depends on the drill diameter, the work material, required hole quality, and hole exit conditions. Generally, one-flute drills are used for deep hole drilling, two-flute drills are used for most general purpose applications, and (as discussed below) three- and four- flute drills are used for close tolerance work and for drilling interrupted holes or through holes in workpieces with inclined exit surfaces.

5. *Coolant hole(s).* Solid drills without coolant holes are used for shallow holes (up to two diameters deep) and for conventional tool penetration rates (using feeds in the range of 0.008–0.011 mm per cutting edge per mm of drill diameter). Solid drills with coolant feeding hole(s), called coolant-fed or oil-hole drills, have passages that run through the drill body. Cutting fluid is fed through these passages to improve chip ejection and cool the cutting edges, which permits use of higher cutting speeds and penetration rates.

6. *Web and flute geometry.* The strength of a drill is largely determined by its web and flute sizes. The two main conflicting parameters in drill body design are adequate flute area for efficient chip disposal and high drill rigidity to reduce deflection and increase dynamic stability. The ratio of the web thickness to the drill diameter directly affects the drill's torsional and bending strength. For conventional two-flute drills, this ratio is usually about 0.21:1. The flute-to-land ratio also significantly affects the drill's strength. Conventional two-flute drills have a flute to land ratio of about 1.1:1, which provides a flute space area between 45% and 55% of the total cross-sectional area. This amount of flute space is sufficient for general purpose applications with most work materials. The diameter of the inscribed circle (on each side of the thin web section) tangent to the drill radius ending at the two intersections the flute forms with the drill periphery and the web of the drill has been found to be related to the measured torsional stiffness of the drill section [114]. These ratios can be optimized for specific work materials and hole depths as shown in Figure 4.50. On the other hand, the inscribed circle diameter of the flute cross section is also a critical design parameter since the drill chip has a conical shape [113]. The area moments of inertia of the drill cross section are very important because they affect drill deflection; the moments of inertia for two-flute drills are 50%–65% of the value for a solid shaft in one principal direction, and 10%–15% of the solid shaft value in the other principal direction. Two-flute drills therefore tend to deflect significantly in the weaker principal direction. A comparison of the two-, three-, and four-flute drills with respect to the area moments of inertia in both principal directions is given in Table 4.6 [115,116]; in this table,

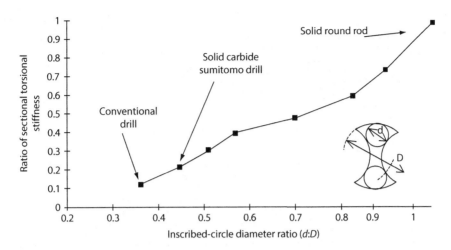

FIGURE 4.50   Effect of web thickness on drill's torsional stiffness. (From Rigidity of Twist Drills, *Metal Cuttings*, 10(3), National Twist Drill and Tool Company, Rochester, MI, July 1962; Skoglund, J., *Cutting Tool Eng.*, 35, February 1990.)

TABLE 4.6

Comparison of Area Moment of Inertia between Two-, Three-, and Four-Flute 21 mm Diameter Drills

Number of Flutes	Web Diameter (mm)	Land Width (mm)	Inscribed Circle Diameter (mm)	Ratio A_f	Moments of Inertia Principal Directions I	II
2	7.00	10.4	6.96	0.45	5180	1224
3(I)	8.73	11.2	6.10	0.40	4100	4100
3(II)	12.73	5.0	4.10	0.36	4005	4005
4	11.5	5.2	5.00	0.40	4329	4239

the ratio Af represents the ratio of the flute area over the drill diametrical area. The stiffness of the three- and four-flute drills in both principal directions is the same because the principal area moments of inertia are equal.

Parabolic (rolled-heel) flute forms increase the chip space and enhance chip ejection and are therefore widely used for dry and deep hole applications. The use of the parabolic flutes with high helix angles (greater than 30°) further improves chip ejection. Parabolic drills have a heavier core, approximately 40% of the diameter compared to 20% of standard twist drills; the heavier core adds rigidity and increases stability when drilling deep holes and/or harder materials [117]. Drills of this type are used not only for soft materials, such as aluminum, copper, and low-carbon steel, but also for stainless steel, cast iron, and nodular iron.

7. *Material.* Twist drills are most commonly made of HSS, HSS-CO, solid WC, or with WC tips or heads brazed on a steel body. Twist drills made of solid ceramics, PCD edges or tips brazed on a steel body, PCD heads brazed on a WC body, PCBN and ceramic tips, and PCD veined on a WC body are also used in specialized applications.

Significant improvements in productivity have resulted from the acceptance of solid carbide drills [118,119]. Compared to HSS drills, carbide drills permit an increase in productivity by a factor of 2–10, and/or increase in hole quality. Solid carbide drills are especially well suited for high throughput precision hole manufacturing.

Solid ceramic and CBN- and PCD-tipped drills are used at higher speeds than carbide drills. Ceramic drills can be used for fiber-reinforced composites, but their application in ferrous materials (e.g., cast iron) has been limited by a lack of machine tools with sufficient speed capability and acc/dec rates for the spindle slide reversal in blind holes. PCD-veined drills eliminate concerns about the integrity of the braze interface with the carbide blank [29]. PCD drills have been used extensively to drill aluminum alloys, other nonferrous alloys, and fiber reinforced composites at conventional and high speeds.

Generally, improved drill rigidity has a positive effect on most aspects of the drilling operation. A stiffer drill exhibits less of vibration and deflection, which allows the use of higher speeds and feed rates, and produces better hole quality and longer tool life. This property of solid carbide drills usually eliminates the need for a guide bushing, which is often required with HSS drills to maintain hole location accuracy. The use of bushings with long solid carbide drills is usually avoided because any misalignment between the drill and the bushing can result in drill breakage. Carbide head drills or brazed carbide drills should be used for deep hole applications which require a bushing.

Three-flute drills produce holes with precision hole size, finish, and roundness tolerances. The three-flute design is a compromise between the conventional two-flute and four-flute drills [116]. A larger number of flutes provide additional guidance for the drill. The flute cross section is the critical design parameter because there are three flutes compared to two flutes for standard drills. Only conical and planer (three-facet chisel) points can be ground on three-flute drills because of the odd number of cutting edges. A planar point (Figure 4.51) has three V-shaped chisel edges. Notch web thinning is used because the web (core diameter) at the point is a large percentage (usually 40%) of the drill's diameter. The residual length of the three chisel edges after web thinning is usually 5% of the diameter, which nonetheless results in much higher thrust than that obtained with a conventional two-flute drill. The residual chisel can be thinned further when drilling soft, nonferrous materials, assuming that chips do not pack in the notched areas. In some cases, a thick chisel at the center is avoided by continuing only one or two of the three edges all the way in to center [120]. However, a three-flute drill can be ground with inverted center point to which eliminate the chisel edges as shown in Figure 4.51 [116]. The inverted point reduces the drill thrust force by up to a factor of two compared to the planer notch web-thinned point. In an alternative design, all three cutting edges are brought together into a pyramid-shaped *spur* at the center [121]. This design reduces thrust but has two drawbacks: the center spur is too weak for many hard materials, and the axial rake angle of the cutting edge shifts from positive to negative at the chisel edge.

Four-flute drills generate even better hole quality than the three-flute drill because they have four margins guiding the tool. The additional margins act as a bushing and increase drill stability during interrupted cuts. The best four-flute design has two flutes that cut to center and two flutes separated from the chisel edge by an undercut as shown in Figure 4.51 [115,122]. The chisel edge is generated by the intersection of the major flanks from only two cutting edges 180° apart. Also, a wider range of point geometries can be ground on a four-flute drill than on a three-flute drill; this range includes two facet, four-facet, and helical points. The flute space is more restricted than on a three-flute drill, however, making four-flute drills most attractive for drilling materials such as cast iron, which form short chips. Either two or four coolant holes can be used in a four-flute drill to improve chip evacuation.

4.8.2 Twist Drill Point Geometries

One very important feature of a twist drill is its point geometry. The geometry of the point determines the characteristics of the drill's three cutting edges: the main cutting edges or cutting lips, the chisel edge, and the marginal cutting edges. The geometry of the main cutting edges affects the drilling torque, thrust force, radial forces, power consumption, drilling temperature, and entry

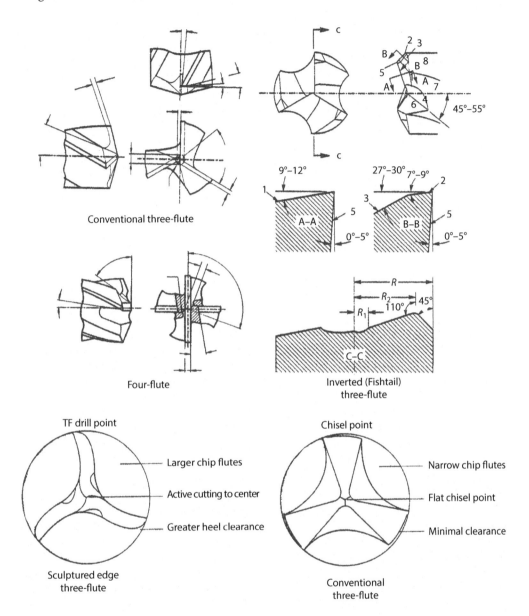

FIGURE 4.51 Configuration of three-flute and four-flute drills. (Courtesy of Guhring Inc., Brookfield, WI.)

and exit burr formation [123,124]. The chisel edge positions the drill before the main cutting edges begin to cut, and stabilizes the drill throughout the cutting process; it also affects the drill's centering characteristics (skidding and wandering at entry) and the thrust force. The margins guide and locate the drill and affect the hole straightness and roundness errors and the drilled surface finish.

The best point geometry for a given application depends most strongly on the drill and workpiece materials, hole depth and size, required hole quality, and expected chip form. Other factors to be considered include the entrance and exit surface orientation with respect to hole axis, hole interruption, burr formation and tool life concerns, and the presence or absence of a bushing (Figure 4.47). The principle geometric features of the point are the point angle, the chisel edge angle, and the relief angle.

The cutting edge length is inversely related to the point angle. An optimum point angle that yields maximum drill life and hole quality exists for every work material. A standard 118° point

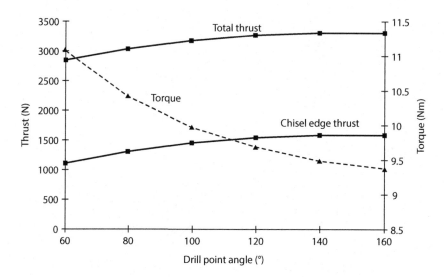

FIGURE 4.52 Effect of point angle on thrust force and torque (considering the main cutting lips only).

is used for general purpose drilling of readily machined materials. Point angles smaller than 118° are preferred for many cast irons, copper, fiber aluminum alloys, die castings, and abrasive materials. Point angles greater than 118° are used for hard steels and other difficult materials. Generally, a lower point angle reduces the thrust force while increasing the torque (Figure 4.52) and radial forces; the torque increases significantly at point angles below 118°, the value used on standard drills. The cutting edge is formed by the intersection of the flute face and the flank face; the shape of the lip is determined by the point angle, helix angle, and flute contour. The flute contour and helix angle are usually designed to provide a straight lip with a 118° point angle. A straight lip is desirable because it generally provides maximum tool life. Specialized drill designs with concave lips (e.g., racon point drills) are used for some steels, since a concave lip induces more strain in the chip and improves chip breaking.

Photographs of the point geometries of several solid carbide drills used for machining aluminum are shown in Figure 4.53. The corresponding relief and rake angles for the drills in Figure 4.53 are shown in Figure 4.54. Similarly, photographs of the point geometries of several solid carbide drills used for machining cast iron are shown in Figure 4.55. The corresponding relief and rake angles for the drills in Figure 4.55 are shown in Figure 4.56. The rake angle distribution across the main cutting edges depends on the flute helix angle. The flute helix reaches its maximum at the margin and decreases to zero at the center. Similarly, the rake angle decreases near the web; it is typically negative at the center of a drill and roughly equal to the helix angle at the outer corner as shown in Figures 4.54 and 4.56. Lip correction can be used to reduce the rake angle and increase edge strength along the main cutting edge; it generates a constant rake (helix) angle along the entire length of the cutting edge. Lip correction is used especially for inhomogeneous materials such as cast iron, and when small, discontinuous chips are desired. A 0°–5° positive rake angle produced by lip correction or the use of straight flute drills provides a strong edge for general purpose drilling of hard and brittle materials such as cast iron, aluminum metal matrix composites, stainless steel, steel alloys, nickel-chrome steel, titanium alloys, and high temperature alloys. A small or neutral rake angle will not help chip evacuation and may cause build-up at the point in softer materials. The strength of a positive rake lip can be increased by grinding a 0.025–0.1 mm hone on the edge [123,124].

The corresponding cross-sectional profiles and the characteristics of the chisel edge for the solid carbide drills in Figure 4.53 are shown in Figure 4.57. The ratio of the web or core diameter to the drill diameter is usually large. The optimal web thickness depends primarily on the

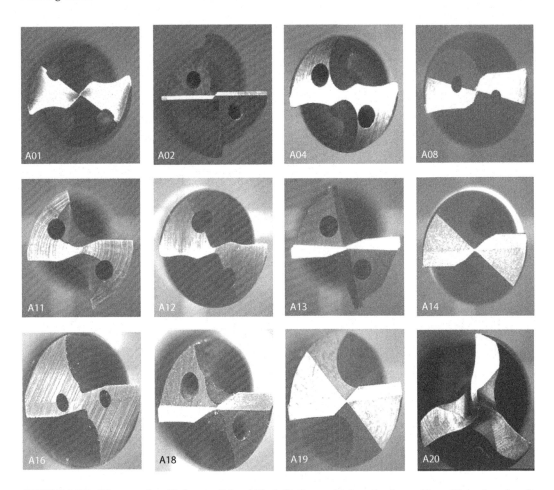

FIGURE 4.53 Photographs of 8.5 mm solid carbide drills for machining aluminum alloys. (From Agapiou, J., *High Speed Drilling of Aluminum Workpiece Material, High Speed Machining 2003 Technical Conference*, Chicago, IL, April 7–9, 2003.)

work material. The web thickness is usually about 20% of the drill diameter for large drills but may reach 50% of the diameter for small drills, which require a proportionately heavier web to maintain stiffness.

Because it cuts slowly and has a large negative rake angle, the chisel edge produces a chip by an extrusion or smearing action, rather than by cutting (Figure 4.58). Because chisel edge chips have a less direct path to the flutes, they are more likely to pack and build up in the hole. The chisel edge contributes substantially to the thrust force; the size of the contribution depends on the relative lengths of the chisel and main cutting edges. The chisel edge contributes roughly 50% of the thrust for a drill with a typical drill with a web thickness equal to 20% of the diameter (Figure 4.57). If the web thickness to diameter ratio is increased to 30%, the chisel edge thrust doubles; if this ratio is further increased to a 40%, it will increase by an additional factor of two (or to a factor of four total as compared to a 20% web drill). The three common approaches to reducing problems associated with the chisel edge are (1) reducing the chisel edge length by thinning or splitting the drill point; (2) changing the shape of the chisel edge to improve its cutting action; and (3) eliminating the chisel edge altogether.

Edge preparation follows guidelines similar to those used for turning and milling as shown in Figure 4.59 for the drills in Figure 4.55. The edge treatment is very critical to drill performance. The cutting forces will be excessive if the edge treatment is excessive (even if the rake

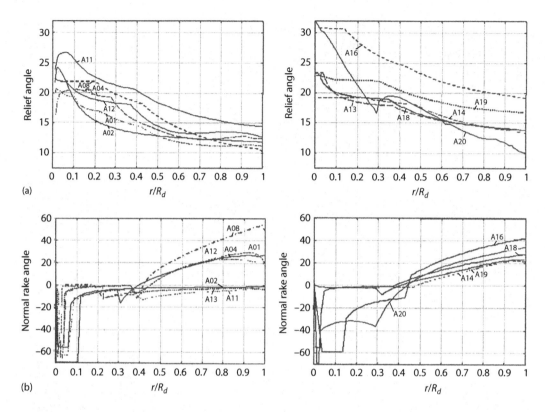

FIGURE 4.54 Measured values of (a) nominal relief angle and (b) nominal rake angle along the cutting edge of the aluminum drills in Figure 4.52. (From Agapiou, J., High Speed Drilling of Aluminum Workpiece Material, *High Speed Machining 2003 Technical Conference*, Chicago, IL, April 7–9, 2003.)

angle is large). For example, drill C01 (in Figure 4.55) with a 29° helix angle and chamfered and honing edge (shown in Figure 4.59) produces a 40% higher thrust force than drill C18 with a 14° helix angle and honed edge (also shown in Figure 4.59). In general, cutting forces depend mainly on three geometric parameters: the rake angle distribution (Figures 4.54 and 4.56), edge treatment, and chisel edge length.

A wide variety of drill point and body configurations have been developed (as shown in Figures 4.53, 4.55, and 4.57) to improve aspects of drill performance such as the drill's centering ability, thrust force, and rigidity. The effect of drill point geometry on drill performance and cutting forces is discussed in detail in the references [97,123,124] and in Chapter 8. Some common drill point geometries and their range of application are shown in Figure 4.60. Briefly, the common types of points include:

Conventional point: The conventional or conical point is the most common type of drill point ground on standard, 118° point drills. The chisel edge is usually either conventional (with conical relief) or two faceted (which results in a flat or blunt chisel) [113], and has a high negative rake angle (−50° to −60°). Conventional drills tend to *walk* or drift during entry and thus often require a centering hole. Conventional point drills are most often used in operations, which do not require high precision or production rates. The conical point can be ground to provide a small crown (0.07–0.2 mm depending on the drill diameter) along the chisel edge, which results in significant improvement in the chisel edge cutting action and centering characteristics.

Radial/Racon point: This type of drill has an arch-shaped radiused point, resulting in a more positive rake angle at the center. The radial lip provides a self-centering effect; it can therefore

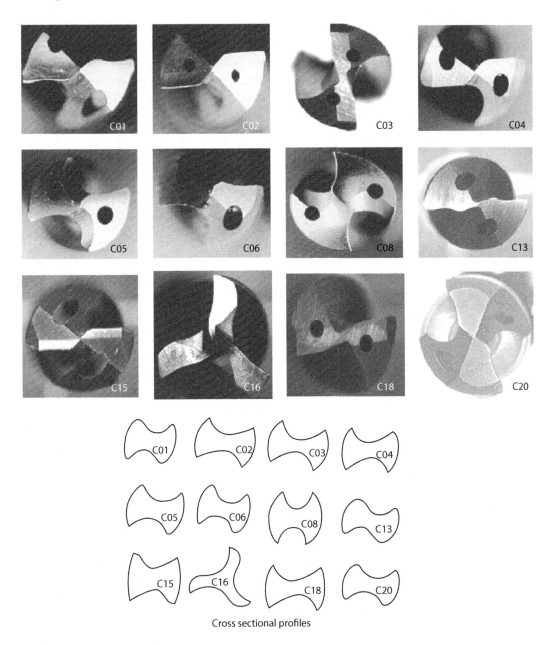

FIGURE 4.55 Photographs and cross-sectional profiles of 8 mm solid carbide drills for machining cast irons. (From Takama, H., and Hayase, T., Rotary Cutting Tool, US Patent 5,505,568, April 9, 1996.)

drill more accurate holes than a conventional drill. The cutting edge is longer than on a conventional point, resulting in slightly higher torque and/or thrust, but also lower edge temperatures and stresses since the heat and forces generated during drilling are spread out over a larger area. Radial points thin the chip at the outer corner, protecting the corner and margin from wear, reducing burr formation, and improving drilled surface finish and tool life.

Web thinned point: In this point the chisel edge is thinned by grinding a notch at the chisel edge corner with a radiused wheel. There are several variations of web thinning such as those described in DIN-Standard 1412 [125]. Web thinning reduces the chisel edge length, reducing thrust and

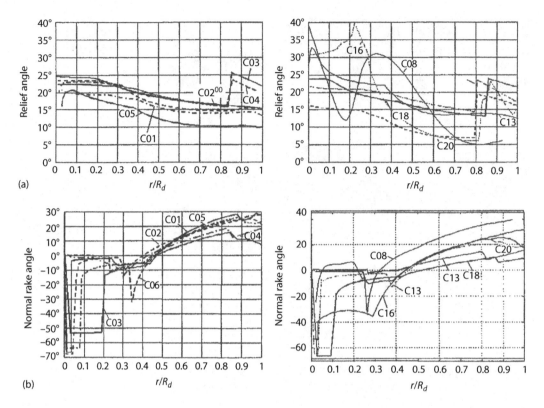

FIGURE 4.56 Measured values of (a) nominal relief angle and (b) nominal rake angle along the cutting edge for the cast iron drills in Figure 4.54. (From Agapiou, J., High Speed Drilling of Gray Cast Iron Workpiece Material, *High Speed Machining 2003 Technical Conference*, Chicago, IL, April 7–9, 2003.)

improving chip evacuation from the center of the drill. Reducing the chisel edge length also improves drill centering properties. The web is typically thinned to a diameter between 8% and 12% of the drill diameter. Lip correction can be used to thin the chisel edge as described in DIN-standard 1412 Form B.

Split point: The split point is often referred to as a crankshaft drill. It is produced by notch type web thinning as described in DIN-standard 1412 Form C. There are two or three similar variations of this point style. The most common split point type is a special case of the web-thinned point with a much smaller residual chisel edge length (typically 2%–3% of the diameter). There are however a number of drill point designs, produced by the Hosoi grinding technique, which use a modified split with an S-shaped secondary cutting edges; these include the Dijet-Hosoi, Sandvik Delta, OSG Ex-Gold, Mitsubishi New Point, Sumitomo Multi Point, Tungaloy Spiral Jet DSC, and Guhring RT 80, RT100, and Precision Twist Drill points [126]. The S-form split reduces secondary edge wear and drill failures when drilling hard materials. The notches in split point drills are prone to build-up when drilling soft materials; the split point is also prone to edge chipping when drilling tough alloys.

The notches on the crankshaft or true split point often do not reduce the chisel edge length, but generate two small cutting edges, one on each side of the chisel edge passing ahead of center. This point is self-centering and also reduces the thrust force, especially when drilling work hardening materials. The split point is especially common on long drills, such as crankshaft oil hole drills, and on small diameter (<12 mm) drills. It is also a preferred point configuration for drilling titanium, stainless steel, and high temperature alloys.

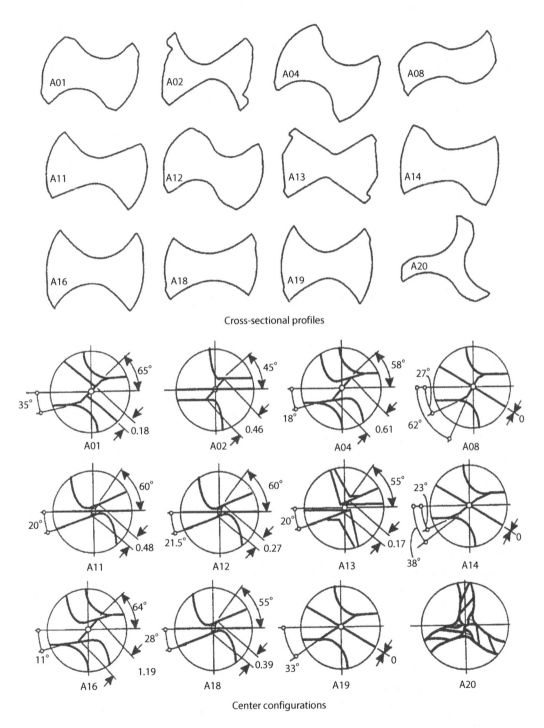

Cross-sectional profiles

Center configurations

FIGURE 4.57 Cross-sectional profiles and center configurations for the aluminum drills in Figure 4.52. (From Agapiou, J., High Speed Drilling of Aluminum Workpiece Material, *High Speed Machining 2003 Technical Conference*, Chicago, IL, April 7–9, 2003.)

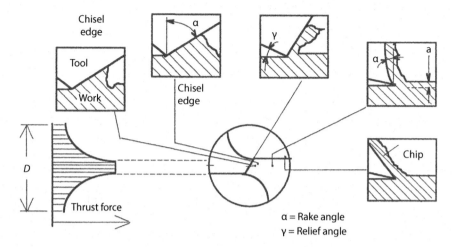

FIGURE 4.58 Chip formation at different locations along the cutting edge of a conventional drill.

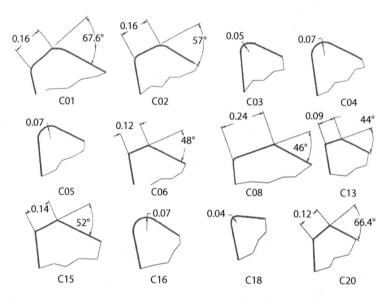

FIGURE 4.59 Edge treatment measurements for the cast iron drills in Figure 4.54. (From Agapiou, J., High Speed Drilling of Gray Cast Iron Workpiece Material, *High Speed Machining 2003 Technical Conference*, Chicago, IL, April 7–9, 2003.)

Helical/Spiral point: This point has an "S" contour with a radiused crown chisel that reduces the thrust force and makes the drills self-centering [127]. It eliminates the need for web thinning. In general, a spiral point drill has a thicker web than a conventional drill because a thin S-shape chisel limits its effectiveness when drilling soft materials. The main advantage of this point is that it reduces burr formation at drill exit. Helical points are weaker than split points and require a special drill grinder. The Hertel SE point is a special type of the helical point with concave lips.

Bickford point: The Bickford point geometry is a combination of the helical and racon point geometries. The helical point is ground on the center of the drill, while the racon point is used for the outer portion of the drill. This point combines the benefits of both the helical and racon points.

Four-Facet Chisel point: Also called the bevel ground point, this point has a chisel edge formed by the intersection of primary and secondary relief planes ground on the flank, producing a less negative rake angle as compared to the conventional point. The more favorable chisel edge geometry reduces the thrust force, improves centering accuracy, and increases drill life. This is the most common point on microdrills and can be used successfully with most workpiece materials.

Double angle points: These points are ground with a corner break (chamfer) to reduce the included point angle at the drill periphery, resulting in four-facet lips. This is the DIN-standard 1412 Form D point [125]. The reduced peripheral point angle reduces corner wear and burr formation at breakthrough and improves size accuracy; the abrupt change in the point angle also serves to split or break the chip. This point is particularly effective when drilling brittle ferrous or hard materials with severe breakthrough conditions.

Multifacet point: Several types of multifacet points are available [128]; they are most easily classified by the number of facets, or the number of primary and secondary relief surfaces ground on the flank. The six-facet Avyac point resembles a four-facet chisel point with an additional web thinning notch at the chisel edge [129]. A number of 8 and 10-facet drills were developed by S. M. Wu and his colleagues for different workpiece materials [130]. As the number of facets is increased, the point becomes increasingly difficult to grind consistently.

Brad point: The brad point (DIN-standard 1412 Form E [125]) has a web-thinned center point ground at an acute point angle (less than 120° and usually 90°) and slightly concave main lips. The outer corner functions as a trepanning tool. The center point length is usually equal to 20%–30% of the drill diameter. The brad point is designed for drilling accurate, round holes in sheet metal with minimal burr formation. A disk or slug of material is produced at the exit.

Non-Chisel edge drills: Drills can be ground without a chisel edge to reduce the thrust force. This type of point is sometimes called an inverted point because the point angle near the drill's center is inverted (greater than 180°). Grinding the inverted center section splits the cutting lips into two segments and serves to split the chip. Similar points include: (1) the *New Point* (Figure 4.60), which produces a small diameter core at the drill center that breaks off as drilling proceeds; (2) *three-flute inverted or fishtail point* (Figure 4.51, inverted three-flute), which produces less than half the thrust force of a standard three-flute drill.

Multi-margin drills: These drills have two or more margins per flute, depending on their diameter (Figure 4.61), in contrast to the conventional single margin drills. The most common double margin drill, at least for diameters less than 25 mm, resembles a multiple flute version of the pressure fed gun drill used for deep hole drilling (described as follows under multi-tip drills). They provide twice the number of contact points on the hole surface, which serves to guide the drill and results in higher drill stability and stiffness in the hole; the drill deflection therefore is significantly reduced, especially for interrupted holes. The secondary margins also burnish the hole and produce a smoother finish. Very accurate holes can be produced with diameters close to the drill size. The drill point geometry is generally independent of the number of margins. The *G-Drill* (Figure 4.60) is a double margin drill with straight flutes. This design has been successfully applied to several materials, especially aluminum and cast iron. The back taper of the margins should be two to four times that of single margin drill to maximize drill life and to prevent early margin build-up and cracking.

4.8.3 SPADE AND INDEXABLE DRILLS

Spade and indexable drills [131] have fixed lengths, which can reduce setup time. They are similar to brazed drills in that they employ a steel body. They are available in diameters from 9 to 75 mm for *L/D* ratios from 2 to 10. Many point geometries that can be ground on a twist drill are not available on spade and indexable drills.

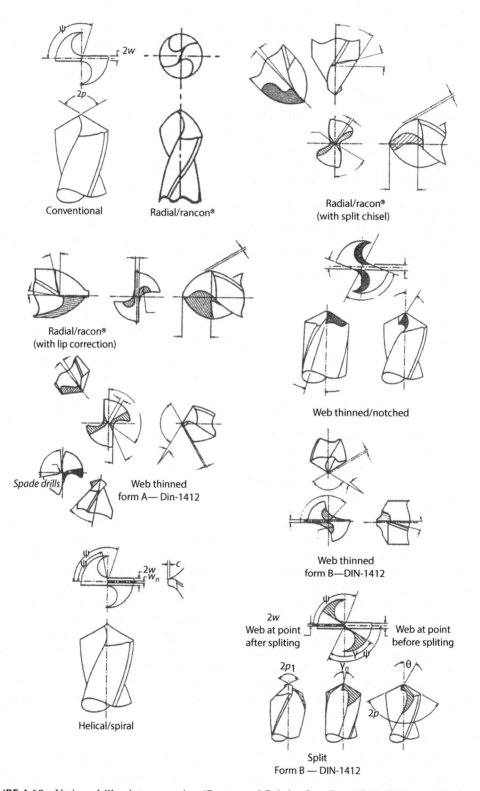

FIGURE 4.60 Various drill point geometries. (Courtesy of Guhring Inc., Brookfield, WI.)   *(Continued)*

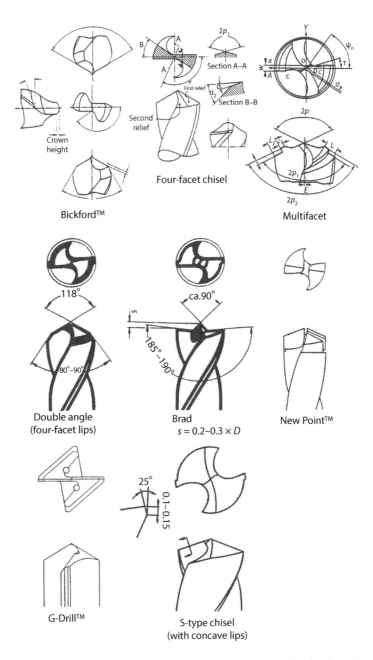

FIGURE 4.60 (*Continued*)   Various drill point geometries. (Courtesy of Guhring Inc., Brookfield, WI.)

"Spade drills" consist of a body and a removable (throw-away) cutting blade or bit, which is precisely located and clamped in a special slot at the end of the drill body (Figure 4.62). The blade may be clamped with either one or two screws, with the two-screw system usually being more stable. Spade drills are available in diameters greater than 9 mm diameter. Spade drills can be used at high penetration rates and are comparatively rigid. In general, spade drills are not used for finishing operations requiring tolerances better than 0.08 mm on the diameter unless special care is used to set the blade in the drill body. Spade drills increase the range of possible cutting edge materials as compared to conventional twist drills; the blade may be made of solid HSS, HSS-Co, WC, cermets,

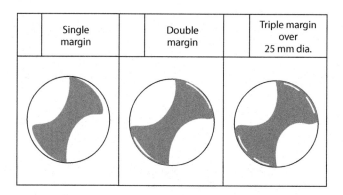

FIGURE 4.61   Cross sections of single and multiple margin drills. (Courtesy of Metcut Tools, Cincinnati, OH.)

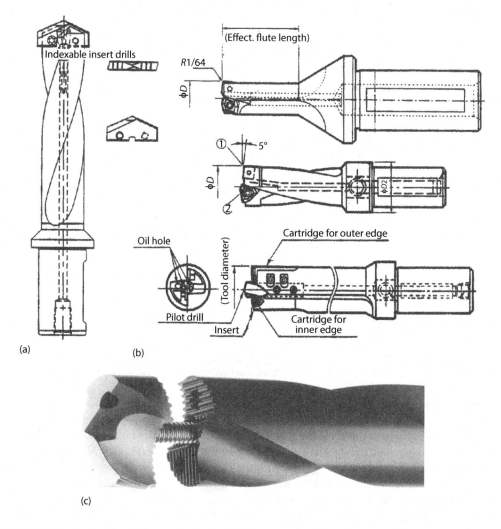

FIGURE 4.62   Indexable drills—spade, indexable insert, and replaceable head drill designs. (a) Spade. (Courtesy of Allied Machine & Engineering Corporation, Dover, OH.) (b) Indexable. (Courtesy of Toshiba Tungaloy America, Inc., Arlington Heights, IL.) (c) Replaceable. (Courtesy of Carboloy/Seco, Warren, MI.)

or ceramic, or may be PCD or PCBN tipped. This drill type offers the economic benefit of throw-away inserts, and eliminates geometric variations due to regrinding.

"Indexable insert drills" consist of a steel body and shank with indexable inserts replacing the cutting edges at the end of the body (Figure 4.62). They also eliminate the need for a pilot or center hole, which may be required with conventional twist drills. These drills have a point angle between 170° and 200° and may generate a round slug during breakthrough for through holes. Indexable drills can be used for solid drilling, core drilling, counterboring, and back boring (i.e., drilling a pilot hole and boring either a straight or tapered hole on the backstroke).

Indexable drills can be used with most workpiece materials because several insert geometries can be applied. As in turning and milling, inserts can be coated and provided with formed-in chip-breakers. Indexable drills are used mainly for holes less than two or three diameters deep. The feed per revolution used is typically half that for multiflute twist drills because the inserts on the two opposite edges do not overlap across the radius. However, the inserts can withstand higher speeds than solid twist drills because a larger selection of tool materials and coatings is available. One concern when using indexable drills is the reliability of the center insert; the cutting speed drops to zero at the center, which may result in insert breakage due to increased cutting forces and poor chip removal. To prevent such failures, the center insert should have a stronger edge geometry and be made of a tougher material than the outer corner insert.

Indexable head (replaceable tip or crown) drills consist of a steel body and shank with a replaceable carbide head (Figure 4.62). They are very similar to the carbide brazed head drills (see Figure 4.15). They provide a head with a complex point geometry in an indexable design. The head is replaced by rotating the holding clamp; the alignment repeatability is typically within 0.05 mm.

Replaceable head drills offer greater accuracy (as shown in Table 4.7) than indexable drills because the head is ground to a precise diameter; they are also more stable in large *L/D* applications and run at higher feeds than the indexable insert drills. They provide a faster tool changes and better coolant flow compared to carbide head brazed drills.

4.8.4 SUBLAND AND STEP DRILLS

Subland drills are special tools for drilling multi-diameter holes [132]. Each diameter has its own flute and land as shown in Figure 4.63; this results in a complex flute geometry, which is neces-sary for what is effectively two or more tools sharing a common axis and core. In a conventional *multistep drill* (Figure 4.64), the smaller diameter ends at the larger diameter's cutting lips, and both share common flutes, lands, and margins; it is thus a modified standard drill. The advantage of the subland drill over the step drill is they maintain a consistent geometry for all diameters after regrinding, which increases the number of possible regrinds.

TABLE 4.7

Comparison of Hole Tolerance from Nominal Diameter for Different Drill Types Based on Drill Size Tolerances

Drill Diameter (mm)	Solid Carbide X3	Brazed Carbide Tipped *L/D* < 3	Brazed Carbide Tipped *L/D* < 5	Indexable Insert[a]	Replaceable Head X1	Replaceable Head X2	Replaceable Head
<10		+0.035	+0.055				
<17	+0.030	+0.042	+0.070	+0.10	+0.042	+0.070	+0.0
<25	+0.040	+0.050	+0.080	+0.15	+0.050	+0.080	0.40

Note: X1 replaceable heads are higher quality than the X2 heads.

[a] These have the largest tolerance depending on *L/D* ratio.

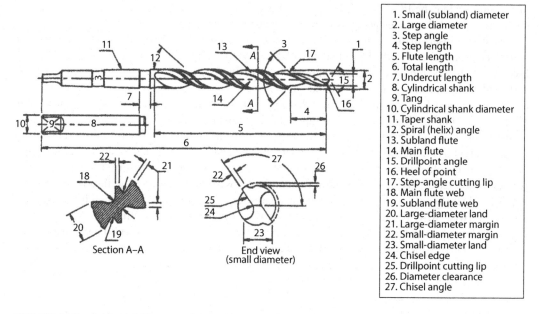

FIGURE 4.63 Subland drill nomenclature. (Courtesy of Metcut Tools, Cincinnati, OH.)

Subland and step drills are usually used for manufacturing blind multistep holes in one pass as shown in Figure 4.65, but they can also be used as a combined drill and reamer (shown in Figure 4.66) to generate close tolerance through holes in a single pass. In this case a straight or helical two-flute drill is followed by a reamer section with several straight or helical flutes; when the length of the drill section is a little longer than the part thickness to be drilled, the reamer can generate its own hole location (i.e., does not follow the drilled hole). Drill/reamers generally are used to produce relatively shallow through holes (with depths less than 2–2.5 times the reamer diameter). Though the tool coolant is usually necessary for drill/reamers to improve chip evacuation and prevent the damage of the hole surface by chips trapped between the lands and the wall.

4.8.5 Multi-Tip (Deep Hole) Drills

Multi-tip drills have segmented cutting edges as illustrated in Figure 4.67. Types of multi-tip drills include indexable insert drills (described earlier), gun drills, and BTA and ejector drills. Most are essentially two-fluted drills with one or more tips arranged on each flute; the number of tips is a function of the drill diameter. For sizes less than 35 mm, two tips, one cutting at the center and one at the periphery, are usually used. For larger diameters, a third tip cutting in the gap between the first two may be added. Multi-tip drills generally use either brazed or indexable WC tips. Multiple-lip heads provide better chip control than single tip heads and reduce power requirements due to a reduction in friction on the bearing pads. A large hole is present above the cutting edge(s) to accommodate chip disposal. The pads burnish and work-harden the hole wall surface, which acts as a pilot bushing. These tools are ideal when hole straightness is critical or when fine surface finishes are required. Multi-tip designs are classified as either balanced or unbalanced. The tip of a balanced drill is symmetric, resulting in minimal radial loads and unbalanced torque. Unbalanced drills generate

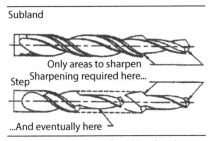

Only the cutting edges of a subland tool ever need to be resharpened. On a step tool, the small diameter's entire geometry eventually must be reground.

(a)

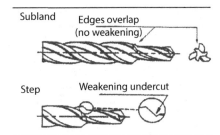

On a subland tool, the intersection between diameters produces a sharp edge. On a step tool, this intersection is a weak point.

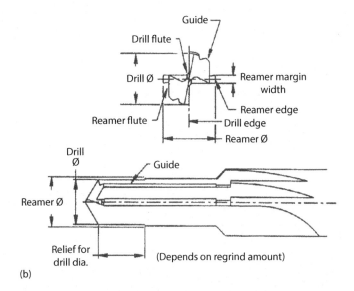

(b)

FIGURE 4.64 Comparison between subland and step drills. (a) Differences between subland and step drills. (Courtesy of Metcut Tools) and (b) subland drill-reamer geometry. (Courtesy of Kennametal Inc., Latrobe, PA.)

larger radial forces and require guide pads to support the resultant force. Some indexable drills are nominally balanced like spade or twist drills, but even in this case radial forces may result due to manufacturing tolerances in the inserts and insert pockets. Most gun drills, BTA, and ejector drills are unbalanced. The unbalanced forces generated by twist drills, gun drills, and BTA drills are shown in Figure 4.67.

BTA and Ejector drill heads [133,134] are used in single and double tube deep hole drilling systems (respectively) with high pressure coolant. The difference between these two drilling heads is illustrated in Figure 4.68. BTA heads are preferable for materials with poor chip formation characteristics such as low-carbon steels and stainless steels. They can be used at much higher penetration rates than twist drills for holes more than five diameters deep. Ejector drills are only available in sizes larger than 18 mm.

Gun drills [135] traditionally have a single lip with carbide wear (bearing) pads to support and guide the tool as shown in Figure 4.69. Two-fluted double margin or multiple-lip high pressure coolant tools, which are becoming increasingly common, are also often called gun drills. The two-fluted drill is guided by margins approximately 90° from each cutting edge as shown in Figures 4.68 and 4.69.

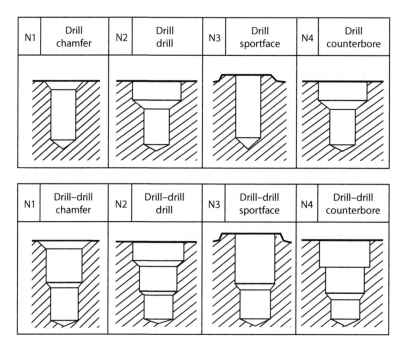

FIGURE 4.65 Hole styles generated with subland or step drills. (Courtesy of Metcut Tools, Cincinnati, OH.)

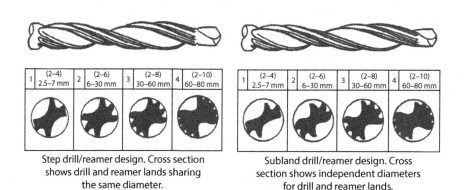

FIGURE 4.66 Comparison between step and subland drill/reamer designs. (Courtesy of Metcut Tools, Cincinnati, OH.)

The maximum practical hole depth depends on the torsional stiffness of the drill body, which is made of either single- or double-crimped 4130 aircraft quality steel tubing bodies (heat treated to 35–40 R_c) or solid single- or double-milled style bodies (Figure 4.70). The crimped tubing design provides for larger flute space and coolant holes than the milled flute type. Single- or double-crimped tube designs are used in severe applications where high coolant volume and large flute cross section are required to improve chip-removal efficiency. However, the milled style body provides higher torsional stiffness. The drill tip is supported by the drilled hole wall since it cannot be fully supported by its body. Tool life is affected by the body design because it controls the volume of the coolant, the flute space, and the drill stiffness. Although gun drills usually have straight flutes, a spiral flute may be used to reduce the drills tendency to whip.

Two types of gun drill constructions are common. In the first, a molded solid WC head is brazed onto a tubular steel body; in the other, WC tips are brazed onto a steel head attached to a

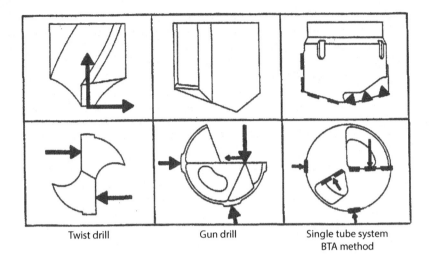

Twist drill   Gun drill   Single tube system
BTA method

FIGURE 4.67 Comparison of the radial cutting force for balanced and unbalanced drills. (Courtesy of Sandvik Coromant, Fairlawn, NJ.)

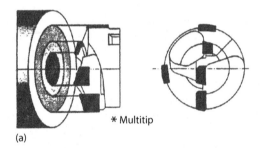

* Multitip

(a)

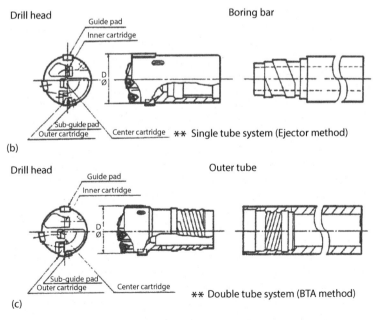

(b)

** Single tube system (Ejector method)

(c)

** Double tube system (BTA method)

FIGURE 4.68 Diagram of deep hole drill heads. (a) Multistep. (Courtesy of Sandvik Coromant, Fairlawn, NJ.) (b) Single tube (ejector). (Courtesy of Toshiba Tungaloy America, Inc., Arlington Heights, IL.) (c) Double tube (BTA).

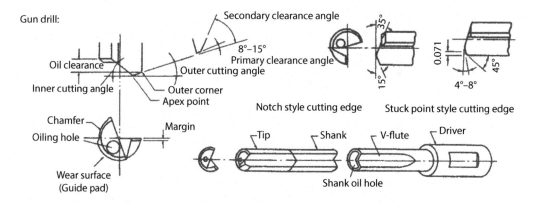

FIGURE 4.69 Characteristics of a single-flute gun drill. (Courtesy of Toshiba Tungaloy America, Inc., Arlington Heights, IL.)

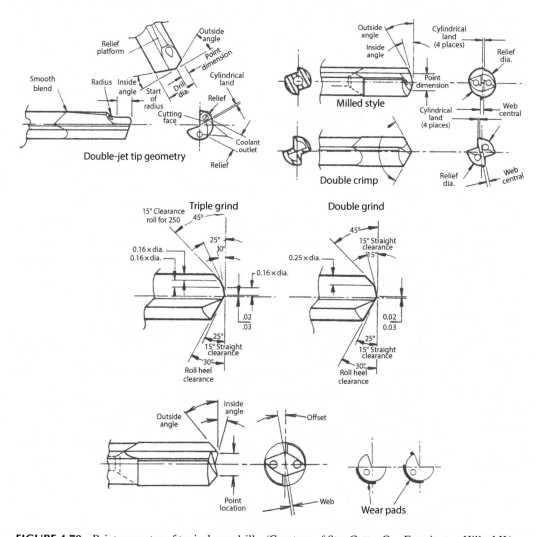

FIGURE 4.70 Point geometry of typical gun drills. (Courtesy of Star Cutter Co., Farmington Hills, MI.)

tubular body. Solid carbide heads are normally used for holes less than 37 mm in diameter; for larger holes the insert-carbide tips are often used to reduce tool cost.

Due to the nature of the deep-hole drilling process, gun drills often have distinctive point geometries. A multifacet point is often ground on single-lip gun drills in order to break the chips in two or more segments. While notch style (double ground) cutting edges are very common, a triple ground edge (shown in Figure 4.70) is often used to distribute cutting forces more evenly at the cutting edge and to permit use of higher penetration rates. The triple-ground design is used especially for tough work materials such as stainless steel and high temperature alloys. Single-flute drills are designed for materials with long and stringy chips such as tool steels. Double flute drills are preferred for materials with small or soft chips such as cast iron and aluminum.

Double-jet gun drills (Figure 4.70) have two holes for coolant through the carbide head instead of the single hole in conventional gun drills. This design improves chip ejection by increasing coolant volume and properly directing the coolant to the cutting edges.

The practical cutting speeds of gun drills are similar to that of conventional drills made of comparable tool materials. However, gun drills are used at lower penetration rates than the conventional two-flute drills since they are often a single flute drills with tubular bodies that may twist at higher feeds. Common feeds for gun drills are 0.05–0.125 mm/rev for single-flute designs and 0.125–0.25 mm/rev for two-flute designs.

4.8.6 Other Types of Drills

Half-round drills are constructed from a round rod with roughly half the diameter ground away to provide the flute area; they are normally tipped with a conical point. The ground flat is of approximate the same length as the flutes of a conventional drill, and results in a zero rake angle at the cutting edge. The conical point is ground on the centerline, resulting in no chisel edge. This drill works on the same principle as the gun drill and requires a bushing for guidance at entry unless the drill is very short or has a very small diameter (i.e., for microdrilling). This drill produces reamed hole surface quality due to its gun drilling-like support. However, its range of applicability is limited due to the difficulty of ejecting chips through the straight flat flute with only flood coolant. The practical depth-to-diameter ratio is usually below 10:1 in horizontal drilling and 3:1 in vertical drilling.

Trepanning drills cut only an annular groove at the hole periphery and leave a solid core or slug at the hole center. Trepanning drills cut more efficiently than conventional and gun drills because they cut less material overall and no material at low velocities near the center of the hole. Trepanning drills are most often used for holes larger than 40 mm even though they are available in drills down to 12 mm. Trepanning drills produce less chips and therefore require less power and thrust capability. Types of trepanning drills include single-edge with wear pads (with a solid WC head, brazed carbide tips, or indexable tip and wear pads; Figure 4.71) and multiple-edge. A *Rotabroach* drill is a trepanning drill (Figure 4.71) with several teeth depending on the tool diameter; the cut is distributed over a greater number of cutting edges resulting in longer tool life.

Microdrills are small diameter drills (<4 mm) [136]. The design of microdrills is especially critical in applications such as printed circuit board manufacturing in which accurate, high-quality holes must be drilled in composite materials. Hole damage such as delamination, fiber pull out, strain, and phase change must be avoided in these applications.

Both the drill point and body geometry are critical to microdrill performance. Complex point designs such as the spiral point, split point, notched point, etc. cannot be ground on small drills. The most common microdrill designs include conventional twist drill designs with a four-facet point [137], half-round designs, and pivot (spade) drill designs. Proper flute space is required for adequate chip ejection so that such drills can be used at comparatively large penetration rates. The microdrill should have a point angle smaller than or equal to that of the pilot drill.

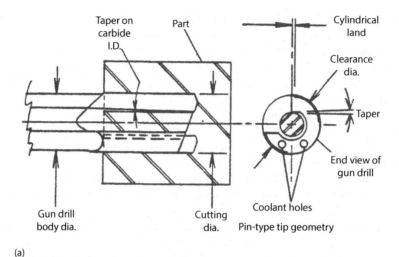

(a)

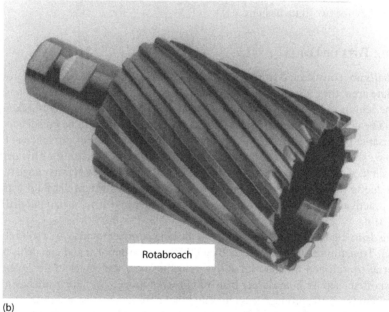

(b)

FIGURE 4.71 Point and body geometry of trepanning type drills. ((a) Courtesy of Star Cutter Co., Farmington Hills, MI; (b) Courtesy of Hougen Manufacturing, Inc., Swartz Creek, MI.)

4.8.7 Chip Removal

Chip control in drilling is critical because often determines hole quality and drill wear and reliability. The importance of chip control increases with hole depth. Chips should be rapidly ejected out of the hole to prevent heat buildup and flute packing, which can cause accelerated tool wear, plastic deformation of the cutting edge, or drill breakage.

Several techniques are used to expedite chip ejection. Proper design of the drill point geometry and flute shape can facilitate chip breakage and efficient chip flow up the flutes. For example, a parabolic flute drill, made by modifying the heel shape on the land of a high-helix twist drill, has larger flute space than standard high-helix drills, providing significantly improved chip flow and coolant access. Chip breakers or chip splitters (Section 11.3) or convex cutting edges can be used to break chips in smaller segments (if this cannot be accomplished by adjusting the feed and speed) to improve chip ejection, especially for mild and alloy steels which often generate long chips.

Chips can also be flushed from holes by properly applied coolant. Either flood coolant or internal (through the tool) coolant can be used. Internal coolant is more effective; this method cools and lubricates the cutting edge and reduces thermal shock in addition to removing chips, and therefore often results in improved tool life and productivity. Flood coolant is usually sufficient for drilling holes less than two diameters deep at moderate feeds and speeds. However, three- and four-flute drills may require through the tool coolant for holes deeper than 1–1.5 diameters depending on spindle orientation (vertical or horizontal) and penetration rate. A safe drilling length for the indexable drills with flood coolant is less than 0.75 diameters. Spade drills usually require internal coolant. High pressure coolant (300–1500 psi) is often required with soft and alloy steels or for high speed drilling of aluminum in order to clear chips. Through the tool coolant is generally required for high penetration rate and high speed drilling.

Internal coolant can be introduced either through the spindle or through a rotary coolant inducer attached between the tool holder and the spindle. The latter approach is simpler to implement but reduces the coolant pressure, the system stiffness, and (generally) the allowable rotational speed. The coolant holes on the drill may extend to the flank faces on the drill point, or may exit on the flutes. Some drills have a central coolant hole through the web and which branches near the flank, while others have one spiral coolant hole per flute. Coolant holes on the flank should be located as close to the outer diameter as possible so that the outer portion of the drill is cooled during exit for through holes, and to minimize coolant spray into the air during exit. Drill margin damage during breakthrough is also a concern when using coolant fed drills. Additional flood coolant or coolant through holes exiting in the flutes during breakthrough can help reduce or prevent outer corner wear and welding of chips on the margins.

Flood coolant is applied from nozzles located above the spindle, around the spindle housing, on the tool holder, or around the tool shank through grooves in the holder collet. The nozzles should be directed so that some apply a coolant stream to the drill point while other direct a stream behind the point; this results in more effective cooling as the drill travels into the hole.

The pressure and especially the volume of the coolant have a strong impact on drill performance. Increasing volume is required as the drill diameter increases. Larger diameter drills require increased coolant volumes or flow rates because the size of the coolant holes increases with diameter; for deep holes or vertical spindle setups, coolant pressure should also be increased. The required pressure and volume depends on the drill geometry, the size and shape of the chips produced for the particular workpiece material, and the number and size of the coolant holes through the drill [138]. As a general rule 0.5 gal/min of coolant per cutting horsepower (0.746 W) is acceptable [139]. The force of the liquid exerted on the chips (through the impulse-momentum equation) is proportional to the coolant mass flow rate and velocity in the cutting area. The velocity is proportional to the square root of pressure from the Bernoulli equation. This means that a 100% increase in pressure will result in 40% increase in the force pushing the chips. On the other hand, a 100% increase in the flow rate will increases the force by 100%. Therefore, it is preferable to increase the flow rate rather than the pressure. Figure 4.72 can be used to estimate the required coolant pressure and volume as a function of diameter for coolant fed two-flute solid carbide, brazed tipped, and replaceable head drills. The low limit is used for $L/D < 2$ while the high limit for $L/D > 3$. Indexable insert and spade drills require flow rates near the high limit and pressures closer to the lower limit.

The coolant flow rate at 7 and 4.2 MPa pressures for the various drills measured in a test is given in Table 4.8. The test was performed in a machine tool with the capability to vary the coolant pressure at the above two levels. A 40% coolant pressure drop from 7 to 4.2 MPa resulted in a 25% drop in the flow rate as expected. A 20% drop in the coolant hole diameter size resulted in 50% reduction in the flow rate, while the theoretical drop should be 36%; the higher reduction in flow rate may result from the surface roughness of the coolant holes. Straight-flute drills have larger coolant

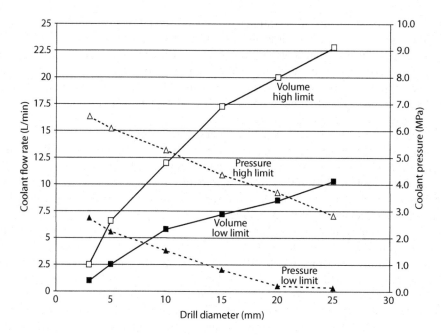

FIGURE 4.72 Coolant volume and pressure required as a function of drill diameter for coolant fed drills.

TABLE 4.8
Flow Rate for Drills with Various Coolant Hole Sizes

Drill Type	Coolant Dia. (mm)	OAL (mm)	Flow Rate (L/min) 7 MPa (1000 psi)	Flow Rate (L/min) 4.2 MPa (600 psi)
5.5 mm diameter				
G1, SF	1	145	5.1	3.9
G1b, SF	0.8	145	2.5	
G2, HF	0.88	145	3.4	2.6
M1, HF	0.60	185	1.2	
O1, HF	0.74	150	2.1	1.6
T1, HF	0.60	170	2.1	1.6
3.2 mm diameter				
G3, HF	0.60	145	0.5	0.4
G4, SF	0.88	145	3.4	2.5
M2, HF	0.60	71	0.5	0.41

Note: OAL, overall length; SF, straight flute; HF, helical flute.

holes than the corresponding helical-flute drills; this means that the straight-flute drills can deliver higher coolant flows rate than helical-flute drills as shown in the Table 4.8.

The required coolant flow rate and pressure for BTA and Ejector type drills can be estimated from Figure 4.73. The maximum conveyable flow rate as a function of pressure for different diameters and type drills can be estimated from Figure 4.74. The top Figure 4.74a shows that the two-flute solid carbide helical-flute drills with two large coolant holes can deliver higher flow rates than the three smaller coolant holes in the three-flute drills. However, straight-flute carbide drills have larger coolant holes than the corresponding helical-flute drills; this means that the

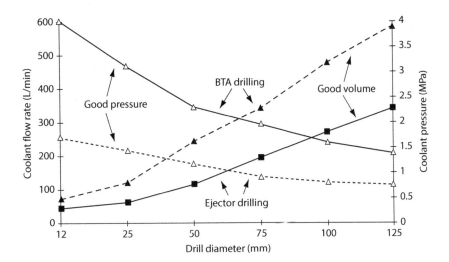

FIGURE 4.73 Coolant volume and pressure required as a function of drill diameter for gun drills.

straight-flute drills can deliver higher coolant flow rate than helical-flute drills as shown in the bottom figure. Indexable drills can deliver higher flow rates at lower pressures than solid carbide drills as shown with the replaceable head drills at the bottom Figure 4.74b. Indexable insert drills having similar coolant supply characteristics as replaceable head drills. However, a spade drill can provide a higher flow rate at lower pressures than the replaceable head or indexable insert drills, especially at the larger sizes (>20 mm). The required flow rate as a function of the drill diameter for different hole depths (applicable especially to gun drilling) can be estimated from Figure 4.75.

Spray mist systems that provide small amount of lubricant at low pressure can be used successfully on machines lacking a coolant recovery system. Compressed air can also be used to clear chips on such machines even though the noise level increases due to air flow.

An interrupted feed cycle can be used if the above approaches do not evacuate chips effectively. The most common method is peck drilling, or periodically retracting the drill to break and clear chips from the hole. Peck drilling is often necessary when drilling deep holes with flood coolant. Clearing all chips before the drill reenters is critical in peck drilling to ensure that chips do not become wedged beneath the cutting edges. An alternative method is to allow the drill to dwell for an extremely short period every two or three revolutions to break chips into short segments; this is not an optimum approach, however, because it increases drill wear.

4.8.8 Drill Life and Accuracy

From a drill life viewpoint, through holes are easier to drill than blind holes because they eliminate drill dwell at the bottom of hole when the drill reverses direction. However, drill chatter or breakage can occur during breakthrough when drilling through holes, due to feed surging (i.e., the occurrence of higher effective feed rates due to the tendency of the drilling machine to spring back because of a sudden reduction in the thrust force). In such cases, slowing the feed during exit, coolant fed drills, or a drill point geometry change may be required. A conventional practice with certain work materials that generate long, continuous chips (e.g., high strength and stainless steels) is to retract the drill before breakthrough to clear chips and relax the strain on the system; this prevents feed surging at breakthrough. Whirling vibrations may also occur during initial penetration, especially for high spindle speeds or long, slender drills [140,141].

Precision holes can be produced in a single pass (i.e., without subsequent boring and reaming) when a proper drill design is used. Although the typical diametrical tolerance for drilled holes is

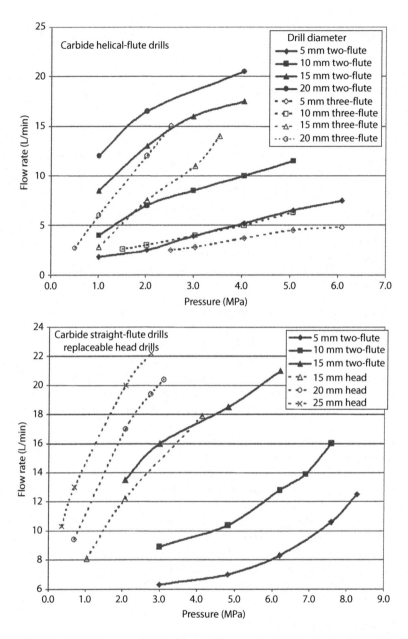

FIGURE 4.74 Relationship between coolant volume and pressure required as a function of depth of hole.

between 0.05 and 0.13 mm, tolerance of 0.015 mm can be achieved if the drill geometry, tool holder, and machine are properly designed. Some of the important parameters controlling the accuracy of the drill are the roundness accuracy or T.I.R., which should be within 0.005 mm, the lip height error, which should be less than 20% of the feed per flute, and the symmetry of the cutting edges (flute spacing, web centrality, and chisel centrality). Many drilling operations are carried out with HSS or HSS-Co drills. These are generally not produced to tight tolerances. The concentricity/runout tolerance of HSS and HSCO drills is a function of the drill diameter, size and length. Typically, a 6 mm drill with 20 mm length has a 0.06 mm runout error specification, while a 200 mm long drill of the same diameter has 0.3 mm runout specification.

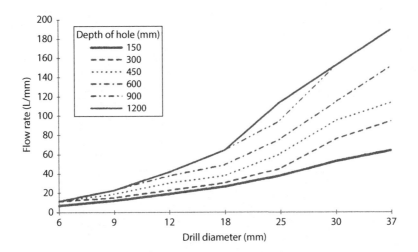

FIGURE 4.75 Relationship between coolant volume and pressure required as a function of depth of hole.

Hole location accuracy is a function of drill point geometry (especially the chisel edge), the drill stiffness, and the rigidity of the tool holder, fixture, spindle, and machine structure. Of all the drill types, solid carbide drills have the tightest tolerances and produce the best holes as summarized in Table 4.7. This makes them the best choice for extremely close tolerance holes. Brazed tipped drills and replaceable head drills produce hole tolerances about twice those of solid carbide drills. Indexable drills have the largest tolerances and hole variations, especially at high L/D ratios.

As discussed above, the web, flute geometry, chisel edge geometry, drill length, and entry condition play the most significant role in hole quality and drill life. The influences of web, flute geometry, and length have been discussed here. The entry condition (surface orientation with respect to drill axis) and chisel edge geometry are very to drill deflection and the maximum allowable thrust force. Often a starting bushing or pilot hole is used to support the drill point in order to reduce the deflection and increase the critical buckling force by four to eight times. The critical buckling Euler force is

$$F_{max} = \frac{k\pi^2 EI}{L^2} \tag{4.8}$$

where k defines the drill entry condition; k is 0.2–0.25 for drill free-point and k is 1–2 using a starting bushing or pilot hole. For this reason, long drill require either a starting bushing or pilot hole. Furthermore, carbide drills ($E = 650,000$ MPa) are three times stiffer than HSS drills ($E = 220,000$ MPa) of identical design. The optimization of the flute geometry together with the drill length has the strongest influence on drill performance. In many cases the flute geometry incorporates tapered web in order to provide a gradual increase of the moment of inertia. In other cases, when drilling hard or bridle materials, wide-web drills are used to increase the moment of inertia.

4.8.9 HOLE DEBURRING TOOLS

There are several types of general deburring tools as discussed briefly in Chapter 2. For deburring holes, special chamfering tools are used either independently or in combination on drill or reamer bodies as illustrated in Figures 4.76 and 4.77. These tools chamfer the entrance and/or exit of hole. There are several variations of such tools.

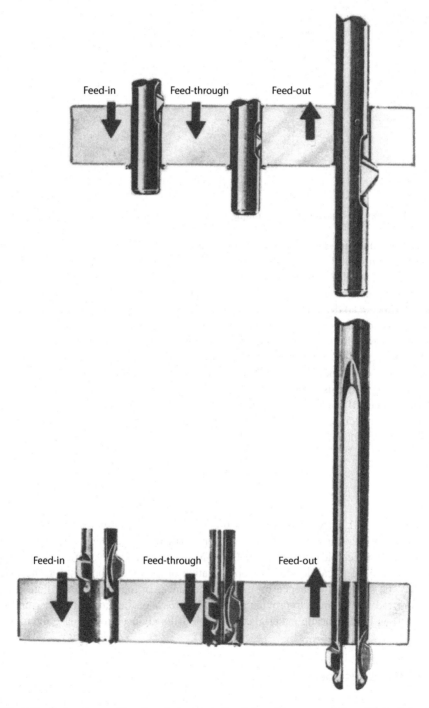

FIGURE 4.76 Deburring and/or chamfering tools for holes. (Courtesy of Cogsdill Tool Products, Inc., Camden, SC.)

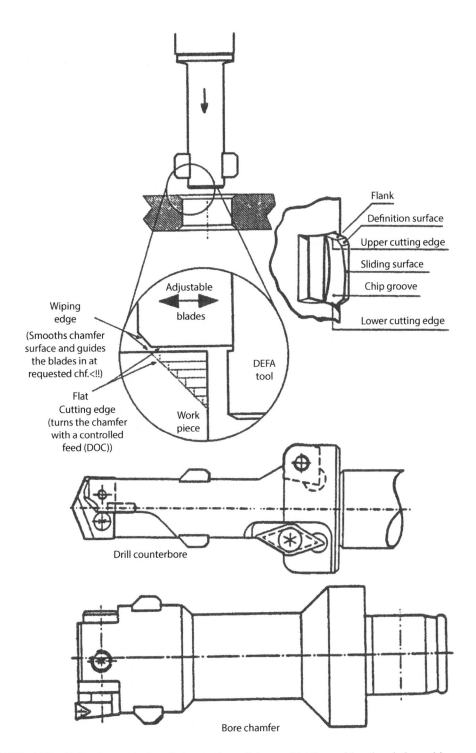

FIGURE 4.77 Deburring and chamfering tool used in combination with other hole making tooling. (Courtesy of Heley, Cincinnati, OH.)

4.9 REAMERS

Reamers are rotary single- or multi-edged cutting tools used to enlarge holes and improve hole quality. The geometry of a typical reamer is shown in Figure 4.78 [142]. Reamers are similar to drills without a chisel edge, and specifically to core drills; they are also quite similar to boring tools. Reaming tools generally do not affect the location and straightness of the starting hole since they usually have a chamfer angle and follow the centerline of the existing hole. In addition, the reamer is supported primarily by the workpiece during cutting through circular margins, which are not present on boring tools. The cutting edge geometry of reamers is specified in the same way as for boring, milling, and drilling tools. The major grinding parameters are the edge preparation at the outer corner (i.e., the dimensions of the chamfer), the axial and radial rake angles, the primary and secondary relief angles, the helix angle and direction, and the margin design.

The major difference between a reamer and a boring tool has traditionally been the way in which they are applied. The reamer traditionally is held in a radial floating holder that allows it to locate in the existing bore; it follows the centerline of the hole being reamed and does not correct the centerline of the hole in relation to the centerline of the machine spindle. However, rigid reamers are also common. Boring tools are held rigidly in order to correct the centerline of the hole being enlarged. A reamer usually operates at lower speeds and higher feedrates than a boring tool because the cutting speed must be kept low for a reamer used in a floating holder to improve hole roundness, and because reamers are usually designed with more cutting edges than boring tools.

The quality of reamed holes depends primarily on the condition of the pre-reamed hole, the geometry of the reamer, the stiffness of the reamer, machine and fixture, and on the cutting conditions, coolant application, and compatibility of the coolant with the workpiece. Reamers can produce holes with a size and roundness variation of less than 0.025 mm; a gun reamer can generate holes with size tolerances less than 0.012 mm.

Most reaming tools are of solid or brazed construction. Indexable inserts and heads have also been used in reaming tools to reduce tooling cost and eliminate regrinding. Inserted-blade reamers are also used when size control is critical, since they allow for size adjustments. As with indexable boring tools, the blade is mounted directly into the tool body for small diameter tools (<15 mm) and in a cartridge for larger tools.

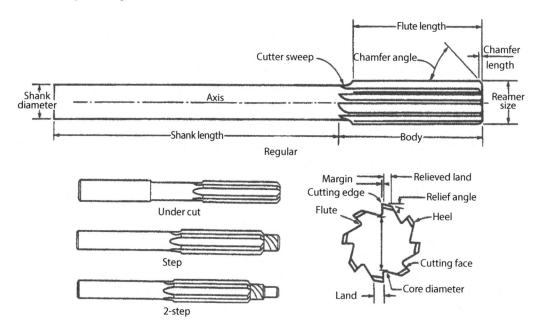

FIGURE 4.78 Reamer nomenclature. (Courtesy of Precision Twist Drill Co., Crystal Lake, IL.)

4.9.1 TYPES OF REAMERS

Different types of reamers include single- and multi-flute reamers, straight and tapered reamers, single- and multistep (multi-diameter) reamers, and expansion or adjustable reamers (Figure 4.79) [143]. This section describes mainly single-flute, multi-flute, and multistep straight reamers; design concerns for tapered reamers are similar.

A single-flute reamer is self-guiding and follows its own centerline. It therefore corrects straightness, angularity, and location errors in the initial hole within narrow limits. The WC wear pad, guide pad, and peripheral tip determine the diameter of the reamed hole and may also burnish the hole surface to improve roundness. A two-flute reamer with secondary support margins (Figure 4.80) provides similar cutting performance.

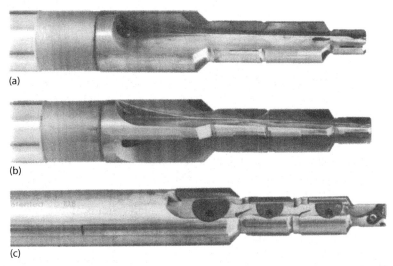

FIGURE 4.79 Configurations of reamers for short and deep holes. (a) Multistep PCD tip gun reamer, (b) multistep, multiflute PCD tip reamer, and (c) multistep inserted blade gun reamer.

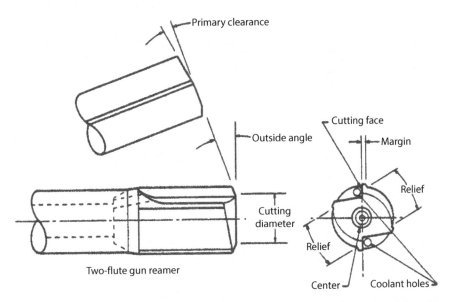

FIGURE 4.80 Configuration of two-flute double margin gun reamers. (Courtesy of Star Cutter Co., Farmington Hills, MI.)

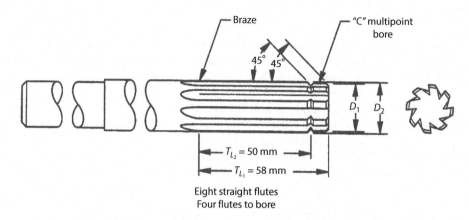

FIGURE 4.81   Configuration of a bore-reaming tool.

Single-flute reamers, often called gun reamers or gun bores, generally produce bores with lower roundness errors than multi-flute reamers. Gun reamers can often be used to produce tight tolerance holes without microsizing. A gun reamer acts as a reamer when mounted in a floating holder, as a fine boring tool when it is held in a rigid holder, and as a gun reamer when a bushing is used to control its motion at hole entry.

Multi-flute reamers remove metal faster and have a longer effective tool life than single-flute reamers, but usually produce bores with higher roundness errors. A multi-flute reamer generally produces a multicornered "lobed" hole profile due to radial vibrations of the tool axis, which result in deviations in the initial hole diameter, misalignment of the cutting portion of the reamer with respect to the hole axis, and differences in the widths of tool margins [144,145]. A reamer is often guided through a bushing or pilot surfaces to follow the desired path; when this approach is not feasible, a radially and/or axially floating holder is used to align the reamer in the hole and allow the reamer to follow the pre-reamed hole. The use of a short entering taper on the front of the reamer, or a short pilot section, guides the reamer into the hole and reduces chatter and vibration. Increasing the number of flutes on a reamer produces a hole profile with a larger number of corners or lobes; this in turn reduces the out-of-roundness errors of the reamed hole. Multi-flute reamers usually run at higher feed per revolution than a drill due to larger number of flutes. Occasionally, reamers with staggered (irregularly spaced) teeth are used to prevent chatter.

Reamers are also used to produce holes with stepped diameters concentric to each other. Step or subland designs can be used for multi-diameter reamers; the design selection depends on the required number of flutes for each step. Subland reamers provide fewer flutes per diameter than step reamers because step reamers use a single set of flutes for all diameters (Figures 4.60 and 4.61). The design of multi-diameter reamers is also influenced by the length of each step, depending on the configuration of the pre-reamed hole [146].

Multistep reamers can be also used to bore and ream a hole or to double ream a through hole in order to improve hole quality, especially when the stock remaining for reaming is excessive due to misalignment concerns between machining stations (see Figure 4.81). A drill-reamer (dreamer) is used mainly for through holes to drill and ream holes in a single pass as explained in Section 4.8.

4.9.2   REAMER GEOMETRY

Generally, a reamer has a chamfer at the outer corner of the cutting edges to guide it into the hole. The chamfer angle, shown in Figure 4.78, is complementary to the lead angle used in turning and milling tools. A standard chamfer of 45° is used for most applications. However, a chamfers between

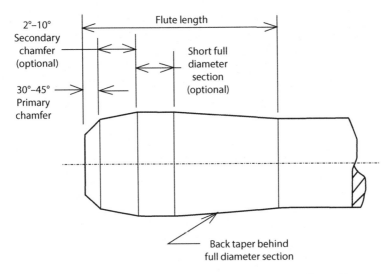

FIGURE 4.82 Several features of the point of a reamer to improve hole quality and/or surface finish.

30° and 45° can be used for steel, while chamfers up to 20° are used for cast iron and aluminum. When the reamer is being used as a boring tool (with a fixed holder), a large initial chamfer angle followed by a secondary angle smaller than 20° is used so that the tool acts as an end cutting tool (as shown in Figure 4.82), which further improves hole size and surface finish by a scraping action. A reamer with a 45° or smaller primary chamfer will not completely correct hole location and straightness errors. Higher chamfer angles result in reduced tool life when cutting ferrous materials, especially for interrupted cuts, due to rapid tool corner wear and a higher susceptibility to corner chipping. The chamfer angle also affects the rake angle. Left-hand reamers are often designed with a 30° chamfer, which results in a positive rake angle. The length of the secondary chamfer is very critical to the tool performance; it can generate chatter when it is too long depending on its relief angle, the workpiece material, and part-fixture stiffness. The chamfer width along the cutting edge should be larger than the depth of cut. Smaller secondary and primary chamfer angles and longer secondary lengths result in better surface finish.

The chamfer relief angle has a strong impact on tool life and can vary from 6° to 12° and 10° to 20°, respectively, for the primary and secondary relief angles. Reamers usually have a narrow circular land with a width that varies with the reamer diameter and the workpiece material. The width increases with diameter (i.e., 0.17–0.6 mm, respectively, for a tool diameter ranging from 6 to 50 mm).

Reamers have axial and radial rake angles defined in the same way as for end mills (Figures 4.35, 4.36, and 4.38). The axial rake is determined by the flute helix. A left-hand helix (LHH) on a right-hand cut (RHC) results in a negative axial rake; conversely, a right-hand helix (RHH) on a RHC provides a positive axial rake. A straight flute reamer has a zero axial rake angle. Reamers with a negative axial rake angle require higher feed forces and produce rougher hole surfaces. Reamers with a positive axial rake angle are more susceptible to chatter, especially when the machine tool is in poor condition or a floating holder is used. However, a freer cut is generated with a positive axial rake reamer. In a RHC, a RHH pushes the chips to the top of the hole through the flutes, while a LHH pushes the chips ahead of the tool. A LHH is therefore often used for through holes. RHH/RHC reamers are necessary for blind holes or deep holes and also produce a better finish. They are preferred for cutting nonferrous materials such as aluminum. LHH/LHC reamers are well suited for cutting steels and cast irons, and are more prone to chatter. Tapered reamers should be designed with a helix direction opposite the rotation to prevent the tool from pushing material ahead of the edge.

A back taper is a small taper (or longitudinal relief) on the diameter along the flutes. Generally, a back taper of 0.005 mm/mm of flute length is sufficient, but larger values can be used when binding is a problem.

Reamers can produce precision holes only if they run true on the spindle and are squared to the workpiece, which is most easily accomplished through the use of axially and radially adjustable holders. Radially floating holders are often used (especially on transfer lines) to ensure alignment with the pre-reamed hole. The holder for a fixed reamer may have radial and/or axial adjustments to zero out any misalignment or roundness errors between the spindle axis and the reamer as discussed in Chapter 5.

4.10 THREADING TOOLS

Types of threading tools include cut taps, roll form taps, thread mills, thread turning inserts, thread chasers, and dies. This section describes taps, chasers, and thread mills; thread turning inserts have been discussed in Section 4.5.

4.10.1 TAPS

A tap is a rotary tool with a geometry similar to that of a screw. The size and geometry style are the two major considerations in selecting the correct tap for a particular material and machine/part setup. The tap style is defined by the number of flutes, rake or hook angle, chamfer length, land, and helix angle [147,148] as shown in Figure 4.83. *Cut and roll form taps* (Figure 4.84) are the major tap styles used for internal threading. Cut taps cut and remove metal to produce threads; roll form taps displace or deform metal to form the thread profile.

A cut tap has a series of single point cutting edges arranged linearly and radially on the tool periphery as shown in Figure 4.85. Only the chamfered teeth on the front end portion the tap contribute to the cutting action as shown in Figure 4.85; the threads behind the tapered portion guide the tap by bearing on the threads already generated. The equivalent of one revolution of full thread form is produced with each revolution of the tap after all of the chamfered teeth are engaged in the workpiece. However, the actual work is spread out along the full length of the chamfer. The chamfered teeth and the tap's first full thread cut or deform the material. Each succeeding chamfered

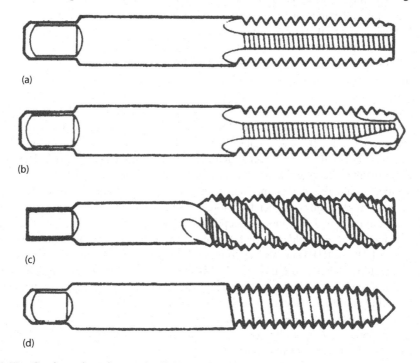

FIGURE 4.83 Configuration of cut and roll form taps. (a) Straight flute bottoming style, (b) spiral pointed plug style, (c) spiral flute bottoming style, and (d) roll form plug style.

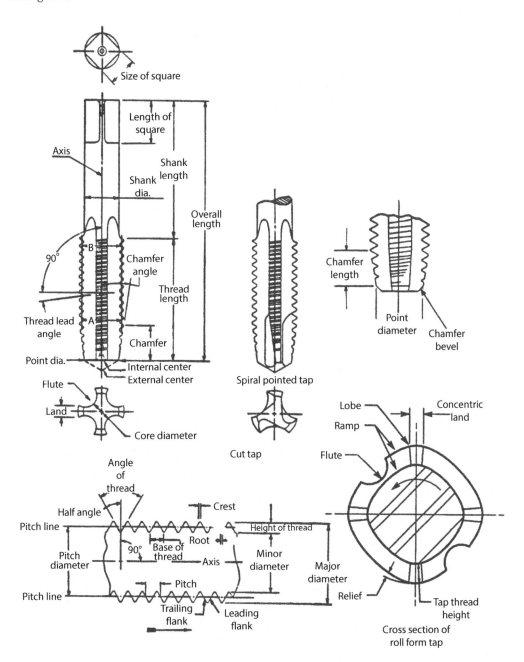

FIGURE 4.84 Nomenclature of cut and roll for taps. *Notes*: "A", pitch diameter at first full thread. This is the correct point for measuring pitch dia. Back taper, the amount pitch diameter at "A" is greater than pitch diameter at "B". (Courtesy of Besly Product Corporation, South Beloit, IL.)

tooth makes a deeper cut until the full thread is generated. The chamfer has a direct relationship to the chip load as shown in Figure 4.86. Therefore, it is important to use the proper chamfer depending on the thread pitch, workpiece material, percentage of thread and type of hole. Commonly used chamfers include taper, plug and bottoming, which are illustrated in Figure 4.87. The taper chamfer is designed for difficult-to-machine materials because it provides a better distribution of chip load per tooth, resulting in better thread quality. The chamfer should not exceed 3–5 threads for taps used in work hardening materials such as superalloys and stainless steels, so as to produce chips

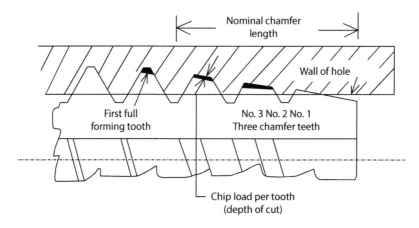

FIGURE 4.85 Chip load per individual tooth for one flute of a tap.

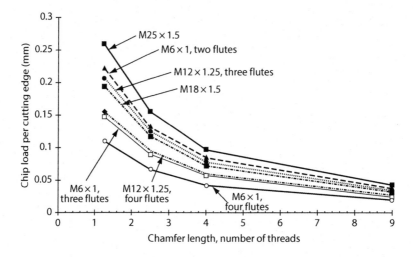

FIGURE 4.86 The effect of chamfer length on the chip load per cutting edge for several sizes of taps.

that are thick enough to allow the cutting edges to undercut any previously work hardened surface. The plug chamfer, with an average chip load per tooth, is used for through holes in most materials at conventional or higher speeds. The bottoming chamfer results in a large chip load over its few cutting teeth, which requires less torque, and is used mainly for blind holes. The longest chamfer possible should be used to improve tap's efficiency, dimensional accuracy, and life, even though a long chamfer increases the cycle time and/or requires a deeper drilled hole. Titanium alloys often require longer chamfers to prevent galling on the chamfer relief surfaces. The geometry of the transition from the chamfer to the tap's full diameter also affects tap performance.

The number of flutes generally varies from two to four and affects the chip load and the available chip and coolant space. As with drills and reamers, several factors affect the optimum number of flutes. The chip load per tooth is reduced either by increasing the flute number or increasing the chamfer length as shown in Figure 4.86; a four-flute tap removes about 20% less chips per flute than a three-flute tap. However, larger number of flutes produces a larger core diameter, reduces the land width, and reduces the chip space, which increases the likelihood of tap breakage due to chip clogging. Nevertheless, when chips are manageable (easily broken or powdery), an increase in the number of flutes should be considered; too wide a land will produce excessive friction, so that it is better to use a larger number of narrow lands on larger taps. Deeper holes require more chip space

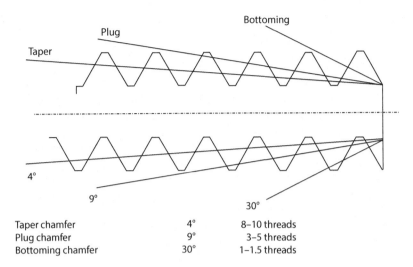

Taper chamfer	4°	8–10 threads
Plug chamfer	9°	3–5 threads
Bottoming chamfer	30°	1–1.5 threads

FIGURE 4.87 Influence of chamfer on tap's end geometry.

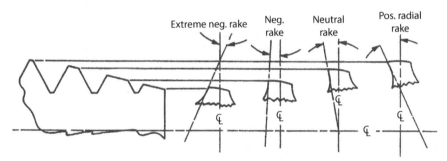

FIGURE 4.88 The effect of chamfer on the rake angle at the cutting teeth of a tap. (Courtesy of T.M. Smith Tool International Corporation, Mt Clemens MI.)

on the tool, especially when tapping blind holes or when a vertical setup is used; in these cases a three-flute tap is often the best choice. As in drilling, the nature of the chips produced often dictates the number of flutes. High speed tapping requires wider flutes, narrower lands, and a smaller core diameter than conventional tapping.

The radial rake or hook angle, defined on the face of the teeth in the flutes of the tap (Figure 4.88), is equivalent to the side rake angle on a single-point tool; the rake angle may be positive at the outer diameter, but decreases to negative values near the center. It should be selected based on the characteristics of the workpiece material. A positive rake angle provides the best cutting action in soft and ductile materials, but results in a weaker cutting edge. A positive rake reduces torque and forms chips, which are more easily disposed of. Zero or negative rake taps should be used for hard and brittle materials to prevent chipping of the thread crests, which usually occurs during tap reversal. The flute shape and rake at thread crest of conventional taps are often modified to provide increased support to the thread crest edges to reduce chipping. A spiral point at the end of the tap along each flute face, as shown in Figures 4.83 and 4.84, is used to provide a positive rake along the secondary face that covers the whole taper length. Spiral-pointed taps form the chip into a tight curl and eject chip ahead of the tap; this keeps the flutes as clean as possible, allows adequate coolant access to the cutting edge, and eliminates chip interference during tap retraction or backing out. Special shear grinds have been developed to provide a progressively changing rake angle (variable rake) over the entire thread and to improve tap performance as compared to conventional spiral points. Spiral point taps are designed primarily for tapping through holes.

The flute helix is important especially when tapping blind holes. The flute helix angle varies with the workpiece material. Slow helix angles of about 15° with low rake angles between 3° and 5° are used for titanium alloys and monel; the same helix with higher rake angles between 14° and 18° are used for tapping short holes (<1.5 × D) in structural, carbon and alloyed steels, nodular and malleable cast irons, brass, bronze, and copper. Faster helix angles between 35° and 45° with 7°–13° rake angles are used for deeper holes (<2.5 × D) in structural, carbon and alloyed steels and for all stainless steels. A medium helix angle of 25° with a small rake angle between 3° and 5° are used in hardened and tempered alloy steels and nickel-based alloys. Finally, fast spirals of about 40° with high rake angles between 15° and 25° are used for low silicon aluminum, long chip brass and thermoplastics. Spiral fluted taps are designed to draw chips smoothly out of blind holes, resulting in faster chip removal and less clogging. Spiral flutes reduce torque requirements and breakage. Spiral flute taps with multiple chamfered teeth may result in poor chip control because they produce multiple chips with different radii of curl, which can become entangled in the flutes or (eventually) the tool, resulting in damage to the threads during tap reversal. Spiral flute taps with as few flutes as possible and a short chamfer length to increase the cutting volume per tooth may alleviate some tapping problems, especially when tapping brittle materials such as cast iron, work-hardening materials such as stainless steel, or soft materials such as aluminum.

The thread relief corresponds to side clearance angle on single point cutting tools. Several types of thread relief have been developed, as shown in Figure 4.89, to provide radial clearance in the thread form at the pitch diameter of the flanks in order to reduce the contact between the tap's land and workpiece, thereby lessening the forces and heat generated. The actual thread size and quality are strongly affected by the amount of thread relief. Concentric relief provides zero relief; it supports the tap well during

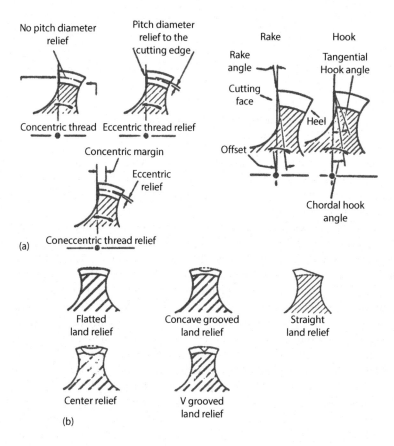

FIGURE 4.89   Types of (a) thread relief and (b) land relief for taps. (Courtesy of Besly Product Corporation, South Beloit, IL.)

cutting and produces threads that are not oversized. It is used for soft materials. Eccentric relief provides radial clearance in the threads from the cutting edge to the heel of the land; this type tends to cut oversized threads and is used mainly on tougher materials. Con-eccentric type thread relief is a combination of the above two types; the first third of the land has concentric relief, while the remaining two-thirds of the land has the eccentric relief. This design provides support to prevent oversize and results in a freer cutting tap. Double eccentric thread relief is a combination of a slight radial relief starting at the cutting edge and continuing for a portion of the land width, and a greater radial relief for the balance of the land.

There are also several types of land relief (Figure 4.89) that provide clearance on the thread crest. Flat land relief truncates the crest between the cutting edge and heel so that there is contact of the crest only at the cutting edge and the heel. Concave grooved land relief is similar to flat relief but provides a groove in the center of the land, which does not quite reach to the pitch diameter. The center relief design has a wider and deeper groove (going down to the root of the thread form) than grooved land relief; it supplies more coolant to the cutting area. V grooved land relief is similar to the other grooved types but uses a longitudinal V-notch at the center of the land. Chamfer or straight land relief provides a gradual decrease in land height from some distance of the cutting edge to heel.

The combination of the rake angle, thread relief, and land relief largely determine the cutting action and life of the tap. As with other rotary tools, back taper is also used on tap threads to minimize drag between the material being threaded and the tap. Increased back taper is required for materials with small elastic moduli.

In some cases, negative relief is substituted for the more common radial (positive) relief to effectively form the thread's final dimension. This design has a lobe or concentric land at the center of the conventional land, which results in a combined cutting and roll forming action in a single operation that is completed during the first full thread. The roll form component tends to eliminate any tap unbalance, which is often present due to radial forces imposed by grinding errors between the teeth of different flutes. In addition, it reduces the tendency of chips to re-enter the hole and damage the cut threads or the tap. A cut tap with roll forming action requires higher torque than a plain cut tap.

An interrupted thread design, produced by removing every other tooth from a standard tap, can be used to break chips in smaller sections and assist in chip ejection. In addition, it improves coolant supply to the cutting edges as compared to standard designs.

Coolant-fed cut taps have coolant exit holes in each of the tap's flutes or a center hole through the body that exits at the front end face and are used for through and deep blind holes to improve chip ejection. Coolant-fed taps are also used to improve lubrication in soft materials such as low-carbon steels.

External or internal threads can also be cut using collapsible or solid adjustable taps with blades or chasers as shown in Figure 4.90. Collapsible taps with blade or circular chasers are made in several designs and are used for cutting threads in stationary or rotating workpieces. The chasers are mounted on a die head. A single collapsible tap die head can be designed to accommodate both multiple blades or circular chasers. The circular shape of the chaser permits only enough rubbing action immediately behind the cutting edge to ensure proper lead control. Blades and chasers are designed using principles similar to those used for cut taps. They are used for both conventional threading applications and for deep hole tapping. A single chaser threading tool cuts threading time by half compared to single point threading, while its multiple cutting edges increase the tool life proportionately. More than five chasers are generally used when the workpiece has either an unusually wide flat or more than one keyway or groove on the diameter to be threaded, to provide a steady threading action and produce round, accurate threads. Figure 2.28 shows two die heads with two and three circular chasers. A range of thread diameters and pitches may be cut with each die head simply by changing chasers.

Roll form taps have neither flutes nor cutting edges. Some of the important geometric features are the number and geometry of lobes located around the periphery of the threads and vent groove (Figures 4.83 and 4.84). The torque required for roll taps is generally up to a factor of five higher than that required for cut taps. The design of roll taps significantly affects their torque requirements [149–151]. The frictional heat generated with roll taps is higher than that for cut taps, so that the lubricity of the cutting fluid becomes a critical concern, particularly at higher tapping speeds.

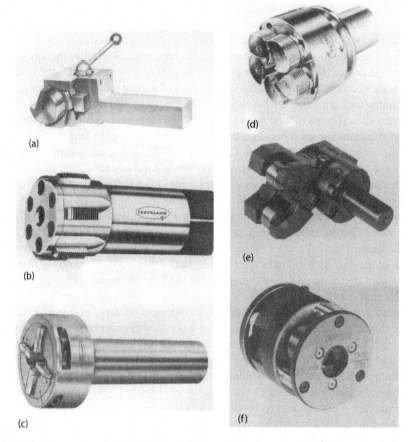

FIGURE 4.90 Configuration of collapsible and solid adjustable taps with blades or chasers. (a) Single head chaser, (b) solid adjustable tap, (c) tap with blade chasers, (d) circular chaser, (e) axial-rolling head, and (f) radial-rolling head. (Courtesy of Cleaveland Twist Drill Corporation, Cleveland, OH.)

Some medium to high carbon steels including alloy steels produce severe tool galling and high torques when threaded using roll taps. Roll form taps are best suited for use on ductile materials such as aluminum, brass, tellurium, malleable iron, low carbon steels, and die-cast zinc.

A roll tap requires a tighter size control of the pre-tapped hole than cut tap. If the pre-drilled hole is too large the tap will not generate the intended thread height; if it is too small, it will increase tapping torque. Porosity in cast materials affects the required tap drill diameter for a given roll form tap; diecast materials, for example, tend to have more porosity and thus require smaller tap drills. Tap manufacturers provide guidance on tap drill diameter requirements for specific designs.

The chamfer geometry for form taps is the same as for cut taps. The tap diameters are radially relieved to form a lobed structure as shown in Figure 4.84. The tap diameter determines the number of lobes; there are always at least two. Taps can have either a cylindrical or polygonal geometry. The lobe shape can be cylindrical or spiral. A concentric land width is critical because it contacts the formed threads and may lead to galling. One or more vent grooves are generally used to provide a path for lubrication and to act as a vent for blind hole tapping. Extra vents may be required in deep hole tapping to increase lubricant flow to the chamfer portion of the tap. There is approximately a 2%–5% variation on the generated % thread by any roll tap as compared to only 2% variation by most of cut taps. Generally, taps with either higher thread relief or lower thread length require lower torque.

Excessive torque is the most common cause of tap failure. The torque depends on the tool geometry and the tapping speed. In general, the torque is influenced more significantly by the tap design in roll form tapping than in cut tapping.

HSS is the most common material for taps, followed by HSS-Co, brazed WC, and solid WC. Brazed WC taps are used for holes larger than 10 mm in diameter. Solid micrograin WC taps were initially developed for smaller holes, but are now used in a large range of sizes. Solid micrograin WC taps perform well in abrasive workpiece materials such as sintered metals, laminated materials, cast iron, wrought and cast aluminum, brass, and titanium, but should not be used in cast materials with hard spots. Solid WC taps also perform best in high speed applications. Solid WC taps have traditionally cut cleaner, lasted longer, and achieved higher metal removal rates than taps made of other materials, although recently coating developments have allowed HSS taps to match carbide performance in some applications. Diamond taps have been manufactured either by brazing the diamond along the flute for larger size taps or sintering the diamond into a slot cut in a WC body; in the latter method, the thread is formed on the diamond using EDM.

Tap class specifications are based on the thread fit and the thread tolerance. A tap's thread tolerance depends on its basic pitch diameter. The letters H and L or D and DU are used, respectively, for unified inch or metric screw threads for high and low limits. The number next to letters H or D represent an even multiple of 0.0005 in. or 0.0127 mm, respectively, above basic diameter and likewise for L and DU below basic pitch diameter. The class fit determines the total tolerance zone of the thread for the pitch diameter and the minor diameter (drilled hole size for cut taps) in the case of an internal thread [152], which controls the percent thread. For each class of thread there are a maximum of 12 different "D" tolerance limits (D0 to D12). If no tolerances are specified, the D5 or D6 are used to meet the common 6 H class. Improved control of tap tolerances can significantly reduce tapping difficulty. The use of the full thread tolerance allows (1) an increase in tap life by specifying a higher H or D number, and (2) the selection of the proper drill size for generating the correct percent thread (thread height). For example, use of a lower "D" number in soft materials without chip ejection limitations can relax the class specification for the minor diameter. A wide tolerance range within specifications [152] can reduce the % thread and therefore the tapping torque; the percent change in torque is normally approximately equal to the percent change in % thread. This in turn can increase tool life and eliminate tap breakage and galling problems, which are generally associated with chip ejection limitations and high thread percentages.

4.10.2 Thread Mills

Thread mills for internal or external threading may be solid or indexable. Indexable thread mills (Figure 4.91) have one or more inserts peripherally depending on the tool diameter; each insert may have one or multiple teeth. More than one insert can be used, one above the other and offset symmetrically, to mill longer threads. Solid thread mills are similar to end mills with either a single row or multiple rows of teeth [153]. The geometry of solid multi-thread milling tools is similar to that of cut taps from the point of view of the helix angle, thread relief, land relief, and rake or hook angle. However, there is no backtaper or chamfer at the end point as in cut taps. Indexable internal thread mills are used for holes larger than 12 mm, and provide the same flexibility as indexable boring or drilling tools. However, solid thread mills reduce machining cycle times because they have more flutes and because their teeth generally cover the full threaded length, so that the threads are completed in one revolution or orbiting cycle. The number of orbiting cycles for an indexable tool is equal to the ratio of the threaded workpiece length to the length of the teeth on the insert. The smallest possible cutter diameter should normally be used to minimize the cutting time and the chip length; shorter chips reduce vibration and thus may improve tool life. On the other hand, a larger tool diameter is sometimes used to minimize tool deflection and produce an equal thread height from all teeth. The optimum tool diameter depends on a trade-off between these factors and can be estimated in the same way as for an end mill. Tool deflection can also be reduced by using an interrupted (staggered) thread design on a two-fluted geometry as in cut taps.

Combined drilling and thread milling tools (Figure 4.92) can be used to drill and thread a hole in a single stroke. In this case, the drill point is designed based on the principles discussed in drilling

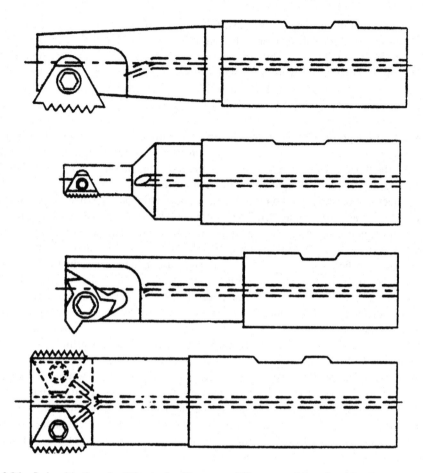

FIGURE 4.91 Indexable thread milling tools. (Courtesy of Kennametal Inc., Latrobe, PA.)

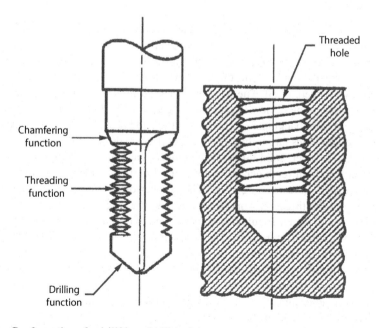

FIGURE 4.92 Configuration of a drill/threadmill tool.

section, while the thread mill is added behind the drill point. The diameter of the thread mill is slightly smaller than the drill diameter, and an undercut is generated at the bottom of the hole during thread milling. A drill/threadmill is weaker than a plain thread mill because it has deeper flutes for proper chip ejection during the drilling phase; the tool diameter in relation to the hole size is more critical especially for small holes. The tool body stiffness can be optimized by varying the diameter, web back taper along the flute length, number of flutes, and chip load. When used at high speeds in nonferrous materials, either coolant through the tool or an effective flood coolant is required.

4.11 GRINDING WHEELS

Grinding wheels are composed of abrasive grains, which provide cutting edges, and a bond material, which holds the abrasive grains and thus acts as a toolholder. Factors to consider in selecting grinding wheels include the abrasive (type, properties, particle size, distribution, and content/concentration), bond (hardness/grade, stiffness, porosity, and thermal conductivity) and wheel design (shape/size and core material). The wheel matrix exhibits porosity, which is important for the effectiveness of the coolant and chip disposal. Production speed and cutting efficiency are influenced by the abrasive type, grain size, hardness and brittleness, and dressing or sharpening properties; and by the bonding material type and its grain retaining and renewal properties. Grinding wheels must be hard and tough to withstand cutting forces, but also be designed to break down gradually, shedding worn cutting edges and exposing new ones to maintain consistent cutting efficiency. Grinding wheels range in size from 6 mm to 1 m in diameter and from 1 to 525 mm in thickness. Several wheel shapes, some of which are shown in Figure 4.93, are available.

4.11.1 ABRASIVES

Practically important properties of the abrasive grains include their hardness, abrasion resistance, crystal structure, and size/shape, which affect the relative friability and durability of the abrasive. Friability describes the ability of an abrasive grain to fracture under certain grinding conditions. The friability index increases with increasing hardness. A grain with good toughness can sustain wear and high cutting pressures but may result in excessive heat generation and (eventually) surface or wheel damage. On the other hand, easily friable grains tends to wear away very rapidly. Therefore, the objective should be to use grains with sufficient toughness that are not too friable.

In order of increasing hardness, the available abrasive grain materials include aluminum oxide (Al_2O_3), silicon carbide (SiC), zirconia alumina, cubic boron nitride (CBN), and diamond; the last two are often referred to as superabrasives, while the first three are called conventional abrasives [154,155].

Aluminum oxide is the most widely used abrasive and is used to grind carbon steel, alloy steel, malleable iron, and superalloys. There are various types of aluminum oxide grains using several impurities (such as titanium, sodium, and chromium), which affect the friability of the grains. Recently developed types include "sol-gel" and "seeded-gel" abrasives (ceramic aluminum oxide— a high purity grain manufactured in a gel sintering process), which consist of aluminum oxide grains with a randomly oriented microcrystalline structure [156]. They are used for precision grinding for steels, particularly difficult alloys. Zirconia alumina (a mixture of aluminum oxide and zirconium oxide) grains are used for rough grinding, particularly of ferrous metals. Silicon carbide is used to grind low tensile strength materials such as aluminum, copper, bronze, gray and chilled iron, cemented tungsten carbide, and nonmetallic materials such as ceramics. Silicon carbide grains are generally black, but green in purer forms; green SiC is used for heat-sensitive nonferrous materials. Silicon carbide is not effective for grinding steel due to the high chemical solubility of carbon in iron, which leads to rapid wear.

The superior hardness, abrasion resistance, thermal conductivity, and compressive strength of the superabrasives make them the best candidates for grinding hard metals, carbides, ceramics and many other hard, tough materials, especially at higher cutting speeds. Superabrasive wheels can

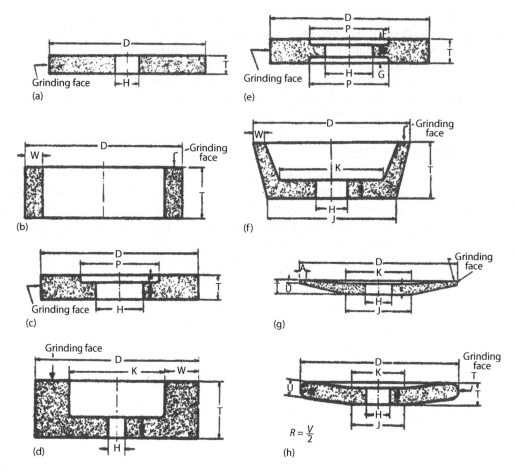

FIGURE 4.93 Shape types of grinding wheels. (a) Type 1 straight wheels, (b) type 2 cylinder wheels, (c) type 5 recessed one side wheels, (d) type 6 straight cup wheels, (e) type 7 double recessed wheels, (f) type 11 flaring cup wheels, (g) type 12 dish wheels, and (h) type 13 saucer wheels. (Courtesy of Cincinnati Milacron, Cincinnati, OH.)

be used at peripheral speeds over 150 m/s. They can be used at MRRs one to two orders of magnitude greater than those achievable with conventional abrasives without overheating the workpiece. Diamond and CBN superabrasives are much harder than the other abrasives and can be used to cut a variety of materials faster than aluminum oxide, silicon carbide, and zirconia alumina with slower wheel wear rates [157,158].

Aluminum oxide is often replaced by CBN for hardened steels (>45 HR_c), superalloys (nickel, cobalt, or iron base with hardness greater than 35 HR_c), high speed steels, and cast iron. CBN has four times the abrasion resistance of aluminum oxide. The high thermal conductivity of CBN prevents heat buildup and associated problems such as wheel glazing and workpiece metallurgical damage. CBN abrasives are available in several types to match the bond system and grinding application. Medium and high toughness uncoated CBN abrasives are used in vitrified, metal, and electroplated bond grinding wheels. Nickel coated CBN abrasives are used for resin bonds and are available in two types: medium and high toughness single-crystal (monocrystalline) and tough microcrystalline. Monocrystals (consist of single CBN crystals) tend to fracture macroscopically after dulling under high grinding forces, which provides a continuous supply of sharp cutting edges. Microcrystalline CBN particles fracture on a microscopic scale after dulling, and thus produce lower wheel wear rates.

Diamond has three times the abrasion resistance of silicon carbide and is used effectively in grinding cemented carbide tools, ceramics, and glass. Like silicon carbide, diamond is not effective for grinding steel due to the high chemical solubility of carbon in iron. Three types of diamond crystals are available: RVG, MBG, and MBS. RVG, medium friability diamond, is used in resin and vitreous wheels primarily for grinding tungsten carbide. MBG is used in metal bonded wheels, which are tougher and less friable than RVG wheels; MBG wheels are used for glass and ceramic materials. MBS is used in metal bonded saws for cutting stone and concrete. RVG diamond abrasives may be used in an uncoated state, with a nickel coating to improve grain retention in resin bonded wheels, or with a copper coating to reduce resin temperatures in dry grinding. Copper coated wheels generally contain 50% copper by weight; Nickel coated grains contain between 30% and 56% nickel by weight; the lower concentration is used for steel-carbide composites, while the 56% concentration is used for wet grinding tungsten carbide.

4.11.2 BONDS

Bonds hold the abrasive grains together in the wheel. Bonds should have sufficient rigidity and the ability to retain sharp abrasive grains during cutting, yet release dulled grains. The bond must withstand grinding forces and temperatures and resist chemical attack by the cutting fluid. The bond type also determines the wheel's maximum safe rotational speed.

There are three major types of bonds: vitrified, organic, and metal [4,155,159]. Vitrified bonds are made of inorganic materials, generally glass or silicates. Organic bond materials include resin, shellac, rubber, and oxychloride. Organic bonds are flexible on a microscopic scale, which allows grains to translate and rotate during grinding. Metal bonds are more heat resistant than organic bonds and more impact resistant than vitrified bonds. Bronze, nickel, and iron are used as metal bonds; sintered bronze is the most common material.

Vitrified and organic bonds are used with conventional abrasive grades wheels. Vitrified wheels are used mainly for precision grinding, while organic (resinoid) wheels are used in high stock removal operations and all dry applications because they have a much higher resistance to thermal shock and can sustain high pressures. High porosity bond structures can be used for aggressive removal of material without burning because the pores absorb coolant and disperse it readily in the cutting zone, improving heat removal and reducing clogging of the cutting edges with swarf [160]. Vitrified wheels traditionally can be used safely at wheel speeds up to 2000 m/min. The maximum safe speed for resinoid wheels is typically approximately 3000 m/min. Speeds between 4000 and 5000 m/min have become possible with recent advances in vitrified bonds and reinforced bonds that combine glass and ceramic materials.

Resinoid, vitrified, and metal bonds are used with superabrasive wheels. Electroplated metal bonds, usually composed of a matrix of nickel or nickel alloy, produce single-layer superabrasive wheels by attaching individual grains of diamond or CBN to a steel preform; the layer of abrasive particles is only partly submerged in the metal matrix, rather than being fully encapsulated, so that they cannot be dressed or trued on a grinding machine. This eliminates friction between the bond material and workpiece and therefore reduces grinding temperatures. Electroplating technology produces complex-shaped wheels more easily and less expensively than other methods. These wheels perform well in high speed grinding operations (with peripheral wheel speeds between 5,000 and 10,000 m/min) since grinding speeds can be increased by 20%–50% compared to other bonded wheels. Detailed discussions of the selection of abrasives and bond types is available in the literature [4,161,162].

4.11.3 WHEEL GRADES AND GRIT SIZES

The strength of the bond holding the abrasive grains in the wheel is defined as the wheel grade. The grade depends on the percent of grain and bond in the wheel. The wheel is characterized as hard or soft depending on the strength of the bond and its ability to withstand cutting forces. Hard wheels retain grains more strongly, while soft-grade wheels lose their grains easily. Several factors, such as wheel speed, work speed, workpiece hardness and ductility, grinder structural condition, and the contact area between the wheel and the workpiece, determine whether the wheel acts as a hard or soft wheel. A decrease or an increase of the infeed rate, traverse speed, work speed, and/or thrufeed rate can make a wheel act harder or softer, respectively. An increase in the workpiece or wheel diameter causes the wheel to act harder because the larger contact area distributes the stock removal over a larger number of grains, which reduces the specific force on individual grains. Soft grades are preferred for rough grinding, vertical-spindle surface grinding, and for machines that are relatively free for vibration. Hard grades are used in internal grinding, peripheral surface grinding, and peripheral cylindrical grinding.

The strength of a bond is controlled by the number and size of the microscopic flaws that are inherent in the bond material. The strength of vitrified or organic bond wheels sometimes decreases over time as the wheel is placed under stress due to stress-corrosion mechanisms that act on the bond when water in the coolant meets the bond material at the exposed edges of the flaws. The factors to be considered in selecting the proper wheel grade are the workpiece material, diameter, and size; the size of grinding wheel; the grinding contact area; the stock removal; the required surface finish; the coolant; the wheel speed; power requirements; and wheel safety. Detailed discussions of the impact of these factors on wheel selection are available in the literature [4,159,162,163].

Wheels with a finer grit and a harder bond are preferred for profile grinding of intricate shapes and/or finely detailed workpieces. Harder bond wheels are required to obtain tight workpiece size tolerances. Resinoid or rubber bonded wheels, rather than more commonly used vitrified bond wheels, may be needed when excessive variations in the size and condition of the workpiece are present in centerless grinding.

Conventional grinding wheels are characterized by their grit size; for example, 80-grit size means that the average size of the abrasive grains is approximately 80 particles per inch. Grit sizes below 50 are considered coarse, those between 50 and 90 are considered medium, and grit sizes above 90 are classified as fine. Generally, the percentage of grains coarser than the specified average grain size is smaller than the percentage of smaller grains.

Coarse grains generally remove more stock per unit time than fine grains. However, this is not always true when grinding very hard work materials; in this case a medium or fine grain may be preferred because they provide more cutting edges on the wheel. In addition, coarse grains produce scratches on the surface that it may be impossible to remove with finer grade wheels in subsequent processing.

The grain compaction, or grain spacing, density, or concentration, is a measure of the number of grains per unit volume of the grinding wheel. The compaction is not as critical for conventional wheels as for diamond and CBN wheels. Single-layer wheels offer a higher density of abrasive grains and more grain exposure than other bonded superabrasives.

A chart of the standard marking system for conventional wheels is shown in Figure 4.94. The maximum stock removal rate depends primarily on the wheel characteristics, work material, and the coolant type (oil or water); variables such as the wheel speed, truing method, and coolant application method also affect wheel capability to a lesser degree. Wheel selection becomes less critical when grinding using a high G-ratio. Some general rules are wet grinding is preferable to dry grinding, especially in heavy stock applications; a coarser grained wheel with a more open structure and a less friable abrasive can be used to increase the MRR; a harder grade with a coarser grain wheel is preferred for soft metals, while a softer grade and a finer grained wheel should be used for hard metals; a finer grain wheel with a denser structure and a less friable abrasive generates smoother finish; and finer wheel dressing results in better surface finish on the workpiece.

Small workpiece diameters require finer and harder wheels than larger diameters. In internal grinding, wheel diameters close to that of the hole that allow the proper coolant application should be used for increased wheel life, particularly for small holes (<60 mm diameter); a larger wheel diameter produces better part size and form tolerance. Organic bonded (Al_2O_3 and SiC) wheels are

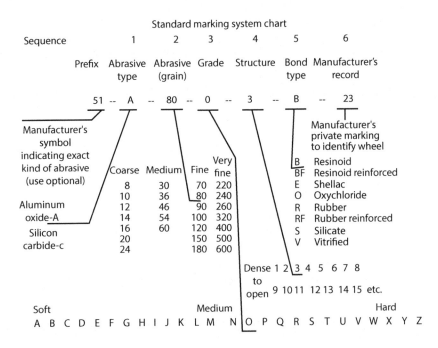

FIGURE 4.94 Standard making system for aluminum oxide and silicon carbide grinding wheels. (After Malkin, S., *Grinding Technology—Theory and Applications of Machining with Abrasives*, Ellis Horwood Limited, Chichester, U.K., 1989.)

very effective in internal grinding because they effectively withstand the high temperatures often generated in such operations; organic bonded wheels are also used when fine finishes are required. Vitrified bonded wheels are used primarily in peripheral surface grinding. Resinoid bonds are most frequently used in disc grinding because they can withstand higher speeds and greater shocks than vitrified bonds. Vitrified bonded Al_2O_3 and SiC wheels are used in conventional creep-feed grinding.

4.11.4 OPERATIONAL FACTORS

Operational factors such as wheel balancing, truing/dressing, grinding cycle design, and coolant application also affect and are affected by grinding wheel selection. Wheel truing and dressing in particular have a strong impact on wheel performance and life. The truing operation removes material from the cutting face of the wheel to generate a particular profile (or to maintain a flat profile), to minimize wheel runout, and remove lobes or irregularities. Truing is necessary if the wheel runout exceeds 0.013 mm after mounting. The dressing operation conditions the wheel surface to maintain particular cutting characteristics; it removes loaded material and glaze from a dull wheel face [4,164] to control the surface finish of the workpiece. A smooth wheel face containing dulled grains is called glazed. In some cases, the same wheel is used for rough and finish grinding a part; in these cases the course wheel is dressed to generate dull grains so that it can produce a finish normally produced by a fine grain wheel.

Truing can be carried out either when the wheel is mounted on the machine spindle or on a fixed wheel-holding device. The wheel should be trued every time it is removed from its holder or, when used without a holder, removed from the spindle. For conventional wheels, truing and dressing are commonly carried out in a single process using the same tool; in this case the combined process is simply referred to as dressing. Superabrasive wheels usually require separate truing and dressing operations.

Conventional wheels may be trued using several methods, including those employing an abrasive stick, brake controlled truing device, soft steel block, or crush rolls manufactured with conventional abrasive grains. Diamond truing tools such as single-point or multipoint diamond tools, diamond blocks, and diamond rolls are also common. In general, dressing tools are similar to truing tools. Dressing tools should be softer and finer than the wheel being dressed. Crush dressing can increase the free-cutting properties of the grinding wheel and therefore allows the use of a harder bond, finer grain, or denser structure than recommended for similar applications when dressing by conventional diamond methods.

Truing and dressing of diamond and CBN wheels are critical operations that strongly impact grinding effectiveness and wheel cost. The most common devices for truing and dressing superabrasive wheels are brake-controlled silicon carbide, aluminum oxide, or diamond-impregnated nibs. The truing frequency is controlled by the grinding G-ratio. Continuous-dress is used in some aggressive applications, such as creep-feed grinding using a diamond-roll dressing wheel, so that the wheel is dressed continuously as it grinds to maintain consistent edge sharpness. The traverse rate for a diamond truing tool is 250–500 mm/min for rough grinding and 100–180 mm/min for finish grinding; the rate of travel for metallic truing tools is 1000–1500 mm/min for rough grinding and 250–500 mm/min for finish grinding. Wet truing is often necessary to remove heat and abrasive grits. The depth per pass should be approximately 0.012 mm on average and less than 0.025 mm across the face of the wheel. In side truing, the traverse travel distance per wheel revolution should be approximately 0.005 mm. Further information on truing and dressing is available in the literature [159,161,165,166].

4.12 MICROSIZING AND HONING TOOLS

Microsizing tools for bore finishing (Figure 4.95) hold a fixed diameter during a single stroke operation with or without size compensation. The tool is usually allowed to float on a floating holder, although two universal joints in the tool body or attached to the spindle can also be used. Types of microsizing tools include one-piece arbor, adjustable arbor and auto-expansion

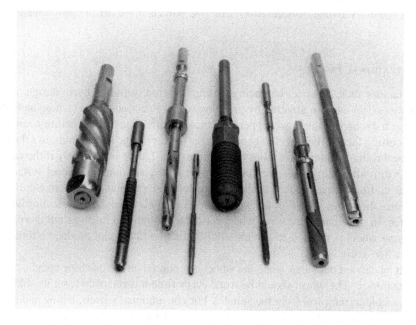

FIGURE 4.95 Various styles of superabrasive bore finishing tools. (Courtesy of Engis Corporation, Wheeling, IL.)

arbor tools [167]. They generally consist of two major components: a cast iron or steel sleeve, and a corresponding tapered arbor. The sleeve, which is the cutting portion, has a barrel shape with electro-chemically plated abrasives. The sleeve also has a helical slot for expansion of its diameter. Once the sleeve is mounted on the corresponding tapered arbor, the tool can be brought to size and adjusted for wear by moving the sleeve up the arbor. An auto-expansion arbor with a mechanism similar to those found in automatic adjustable boring bars (Figure 4.31) can also be used. One-piece arbor designs are recommended when other types are not available, such as for very small diameter bores. Various styles of coolant flutes, tool lengths, and size adjusting features can be incorporated to improve tool performance in a particular application as illustrated in Figure 4.95. The sleeve is replaced by a multiple-stone tool (carrying a number of abrasive stones around the periphery of the arbor) for larger bores between 25 and 300 mm in diameter.

Unlike microsizing tools, honing tools are generally multi-stroke tools, which are passed through the bore repeatedly. They consist of abrasive stones held in a tool body; the stones are not held at a constant diameter, but expand to exert pressure on the bore wall. Common honing tool designs [168] include single-stone, multi-stone, and Krossgrinding tool designs. Single-stone tools have a single stone, as illustrated in Figure 4.96, and are commonly used for small bores (diameter <25 mm). Multi-stone tools are used for larger bores (50–300 mm diameter) and have multiple contact points in order to provide better force distribution, better geometry, longer tool life, and faster stock removal; they are similar to large microsizing tools. Multi-stone tools are comprised of a number of guide shoes and stones. The arrangement of the shoes and stones on the tool periphery is very important to tool performance (Figure 4.97). Cone-fed expansion mechanisms (Figure 4.98) are commonly used. The angular spacing of the contact lines for the guide shoes and stones is also important. A three-point tool (two fixed guide shoes and one abrasive stone) is usually better than

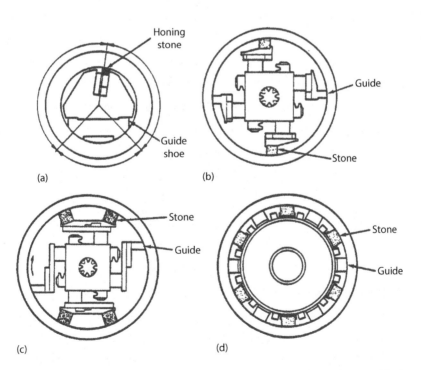

FIGURE 4.96 End view of honing tool designs. (a) Single row of stones tool with two guide shoes, (b) four-point contact tool with two rows of stones and two guide shoes, (c) six-point contact tool with four rows of stones and two guide shoes, and (d) multipoint contact tool. (After Fischer, H., Tutorial, *International Honing Conference*, SME, Dearborn, MI, April 26–27, 1994.)

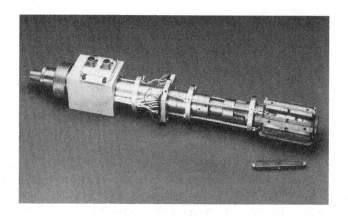

FIGURE 4.97   Typical multiple-point honing tool.

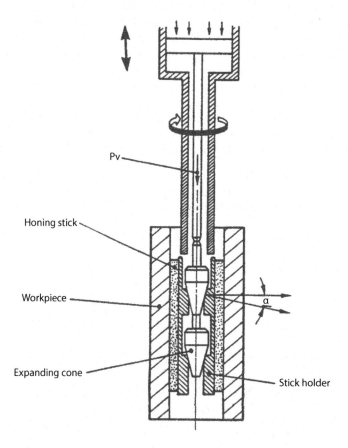

FIGURE 4.98   Honing tool design principle.

the four-point contact design (two fixed guide shoes and two sticks) [168,169]. Honing tools for automotive engine bores generally have 6–12 stones.

The Krossgrinding tool [168] is a multi-stroke superabrasive design for small bores (10–32 mm diameter), which combines the best features of multi-stroke honing tools and single-stroke micro-sizing tools; it is equipped with a plated superabrasive sleeve.

Successful honing depends not only on the tool design but also on the use of the proper honing stone. The selection of the stone is based on the same principles as the selection of a grinding wheel: grit size,

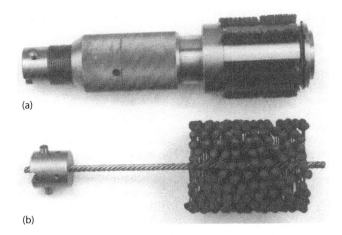

(a)

(b)

FIGURE 4.99 Brush honing tools: (a) ball, (b) bristle. (After Lin, Y.T., Honing with Abrasive Brushes, *International Honing Clinic*, SME, Dearborn, MI, April 7–9, 1992.)

hardness, bond type, and abrasive material [4,159,169]. In honing, unlike grinding, the abrasive stones cannot be dressed, and so must be self-dressing; this requires a close control of both the reciprocation and rotational speed. Abrasives are generally vitrified conventional abrasives used with oil, or bonded diamond superabrasives used with oil or water-based cutting fluids.

Finally, two types of abrasive brushes are used for brush honing [170]. The first is the ball type designed with clusters of abrasives attached to the ends of flexible filaments as shown in Figure 4.99. This tool is readily adapted to any machine tool with a rotating spindle. The second is the bristle type, made of nylon bristles impregnated with abrasives (Figure 4.99). The bristles are mounted on the same kind of honing shoes and tools as abrasive sticks; they are designed to be adjustable either manually or automatically for interference setting and tool compensation.

4.13 BURNISHING TOOLS

Roller and ball burnishing tools consist of a series of hardened, highly polished rollers positioned in slots in a hardened mandrel or cage; they resemble roller or ball bearings in construction. They are made either with fixed diameter, which remains constant during the operation (interference tools) or with a diameter that can be changed by exerting a predetermined burnishing pressure (expander tooling) [171]. Fixed diameter tools provide tolerance control, but produce surface finishes, which depend on the pre-burnished surface dimension and quality. Typical designs for various applications are shown in Figure 4.100.

Burnishing induces plastic flow of the surface asperities when the yield point of the surface material is exceeded. Therefore, asperities are flattened and compressive stresses are also induced in the surface layer, giving several improvements to mechanical properties. Burnishing will improve both the surface strength and roughness [172].

4.14 EXAMPLES

Example 4.1 A 20 mm deep hole is drilled and tapped with a M10 × 1 mm cutting tap in a cast iron workpiece. A solid carbide drill and a HSS tap are used. Estimate the machining time required to drill and tap the hole. What is the MRR for both the drill and the tap?

Solution: The cutting speed for both tools can be selected from Figure 4.4. A cutting speed of 100 m/min for an uncoated carbide drill is typical for cast iron. The cutting speed for the HSS tap

Flat surface
roller burnishing
tools

Internal expanding and external contracting
tools for roller burnishing surfaces of parts
with wide tolerance variations ... under
controlled condition

Taper roller burnishing
tool for internal application

Pressure controlled internal
and through-hole roller
burnishing tools

Pressure controlled external
roller burnishing tool
... self-feeding

FIGURE 4.100 Internal and external roller burnishing tools. (Courtesy of Cogsdill Tool Products, Inc., Camden, SC.)

is assumed to be 30 m/min. The depth of the drilled hole must be deeper than the tapping length to provide clearance for the chips and the front taper of the tap (the first few taper threads) at the bottom of the blind hole, and is assumed to be 27 mm. The spindle speeds for drilling and tapping are

$$N_d = V_d/(\pi D) = (100 \text{ m/min})(1000 \text{ mm/m})/(3.14)(9 \text{ mm}) = 3839 \text{ rpm}$$

(Assuming the drill diameter for a 10 mm tap is about 9 mm), and

$$N_t = V_t/(\pi D) = (30 \text{ m/min})(1000 \text{ mm/m})/(3.14)(10 \text{ mm}) = 955 \text{ rpm}$$

The feed for drilling cast iron is assumed to be 0.15 mm/flute. The feed rate is then

$$f_r = n_t \cdot f \cdot N = (2)(0.15 \text{ mm/rev/tooth})(3839 \text{ rpm}) = 1152 \text{ mm/min}$$

and for tapping the feed rate is determined by the pitch of the tap (given from Equation 2.46). The pitch for the M10 × 1 mm tap is 1 mm.

$$f_r = p \cdot N = (1 \text{ mm})(955 \text{ rpm}) = 955 \text{ mm/min}$$

The cutting time for drilling with 120° point drill is

$$t_{md} = \left(L_d + L_e\right)/f_r = (L + D/\tan\rho + DL)/f_r$$

$$= (27 + 9/\tan 60° + 2) / \left(1152 \text{ mm/min}\right) = 0.030 \text{ min} = 1.78 \text{ s}$$

and for tapping is

$$t_{mt} = \left(L_t + L_e\right)/f_r = (L + D/\tan\rho + \Delta L)/f_r$$

$$= (23 + 2)/(955 \text{ mm/min}) = 0.026 \text{ min} = 1.57 \text{ s}$$

L_e is assumed to be 2 mm since a hole is made. The total machining time to drill and tap the hole is the sum of the drilling and tapping times plus several other tool travel times (i.e., retract the tool from the bottom of hole, approach time, etc.) as discussed in Chapter 13. Let us assume that the total time is $t_T = 1.78 + 1.57 + \Delta t = (3.35 + \Delta t)$ s.

The MRR for the drill can be calculated from Equation 2.15:

$$\text{MRR} = \left(\text{cross-sectional area}\right)\left(\text{feed-rate}\right) = \left(\pi D^2/4\right) f_r$$

$$= [\pi(9)^2/4](1,152 \text{ mm/min}) = 73,287 \text{ mm}^3/\text{min}$$

The MRR for tapping is given by Equation 2.57

$$\text{MRR} = \left(\frac{p}{4} + \frac{D_m - D_d}{\tan 60°}\right)\left(\frac{D_m - D_d}{4}\right)\left(\frac{p \cdot N}{\sin \lambda}\right) \quad \text{where } \tan \lambda = \frac{p}{\pi \cdot D} = \frac{1}{\pi \cdot 10}$$

The helix of the thread $\lambda = 1.82°$

$$\text{MRR} = \left(\frac{1}{4} + \frac{10 - 9}{\tan 60°}\right)\left(\frac{10 - 9}{4}\right)\left(\frac{1 \cdot 955}{\sin 1.82°}\right) = 3732 \text{ mm}^3/\text{min}$$

Example 4.2 Estimate the change in machining time for drilling and tapping the hole in Example 4.1 if both the drill and tap are made from HSS.

Solution: The cutting speed for the tap is the same. The cutting time for the drill will be different because the maximum cutting speed allowed for HSS drills is much lower than carbide drills as shown in Figure 4.4. A cutting speed of 30 m/min for an uncoated HSS drill is common for cast iron.

The spindle rpm for drilling is

$$N_d = V_d/(\pi \cdot D) = (30 \text{ m/min})(1000 \text{ mm/m})/(3.14)(9 \text{ mm}) = 1152 \text{ rpm}$$

The feed for drilling cast iron with HSS drill can be the same as that with carbide drill, 0.15 mm/flute. The feed rate is then

$$f_r = n_t \cdot f \cdot N = (2)(0.15 \text{ mm/rev/tooth})(1152 \text{ rpm}) = 346 \text{ mm/min}$$

The cutting time for drilling with 120 degrees point drill is

$$t_{md} = (L + L_e)/f_r = (L + D/\tan\rho + \Delta L)/f_r$$

$$= (27 + 9/\tan 60° + 2)/(346 \text{ mm/min}) = 0.10 \text{ min} = 6 \text{ s},$$

The total machining time to drill and tap the hole is then

$$t_{T2} = 6 + 1.57 + \Delta t = (7.57 + \Delta t) \text{ s}.$$

The change in machining time between HSS and carbide drill is

$$t_{TX} = t_{T2} - t_{T1} = (7.57 + \Delta t) - (3.35 + \Delta t) = 4.22 \text{ s}.$$

Therefore, the total machining time is reduced by 53% when a carbide drill is used instead of HSS.

Example 4.3 Evaluate the effect of using CBN inserts instead of aluminum oxide CVD coated carbide inserts on a boring bar used to finish bore a 25 mm diameter hole 50 mm deep in cast iron.

Solution: A good cutting speed for Al_2O_3 multi-coating carbide inserts is 220 m/min for cast iron, while for CBN inserts it is 1000 m/min. Therefore, the carbide inserts will be run at 2,800 rpm and the CBN inserts at 12,750 rpm. The cutting time will be 5.46 times lower for CBN inserts since the feed rate ($f_r = f \cdot N$) for the CBN bar ($f_r = [f(12,750 \text{ rpm})]$ mm/min) is 5.46 higher than that of carbide bar ($f_r = [f \cdot (2800 \text{ rpm})]$ mm/min).

However, the MRR with the CBN boring bar will be 5.46 higher than that for the boring bar using carbide coated inserts. This means that the power required for boring with CBN inserts will be about 5.46 times higher than the boring bar with carbide coated inserts even though the torque and force for both boring bars will be about the same. (Generally, the torque required to remove the same material at higher speeds is slightly lower than that required at lower speeds because the chip removal is more efficient at higher speeds.) This illustrates the importance of checking the spindle power and torque availability as a function of speed as explained in Chapter 3.

Example 4.4 A 55 mm diameter hole is being bored to a 59 mm final diameter using a single point boring bar. The workpiece material is gray cast iron. A carbide insert is used in the indexable boring bar made of steel. The cutting speed for the bar is 100 m/min and the feed is 0.2 mm/rev. Two different boring bar designs (A) and (B) are used as shown in Figure 4.101.

(a) Estimate the deflection at the tool point for bar (A) due to radial force in order to determine the effective bore diameter.
(b) Check the effect of cutting conditions on improving hole size control.
(c) Calculate the hole size improvement by changing the boring bar (A) material from steel to heavy metal.
(d) Calculate the improvement on bore size by optimizing the boring bar design to (b).

The material properties are $E_{steel} = 206,700$ MPa and $E_{heavymetal} = 330,000$ MPa

Solution: In order to understand the effect of the bar design and cutting conditions on deflection, a rigidly clamped beam with a force acting at the free end is considered.

(a) In order to find the tool point deflection to determine the hole size quality, the radial cutting force acting on the boring bar should be estimated. It is known from previous work that in roughing and semi-finishing operations the feed and radial forces are smaller than half

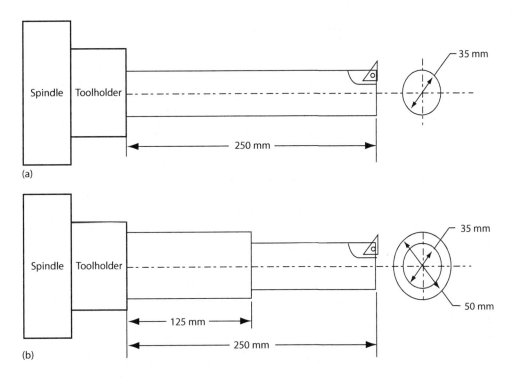

FIGURE 4.101   Boring bar configurations (a) and (b) used for machining the bore in Example 4.4.

of the cutting force along the speed vector, $(F_r \leq 0.5 \cdot F_c)$. The radial force in this problem can be equated to half the cutting force since the DOC is 2 mm (the worst case scenario is $F_r = F_c$). The cutting force is estimated from the torque and power required for the cut as explained in Chapter 2. Thus,

$$T = F_c\, R = 2\, F_r\, R \text{ and}$$

$$P = \tau\omega = 2\, F_r\, R\, (2\,\pi\, N) = (\text{MRR})\, u_s \text{ which yields } F_r = (\text{MRR})\, u_s/(4\,\pi\, R\, N)$$

$$N = V/(\pi\, D_{avg}) = (100 \text{ m/min})(1000 \text{ mm/m})/(3.14)[(59 + 54)/2] = 559 \text{ rpm}$$

$$\text{MRR} = \frac{\pi(D_o^2 - D_I^2)}{4}\, f_r = \left[\frac{\pi(59^2 - 55^2)}{4}\right](559 \text{ rpm})(0.2 \text{ mm/rev})$$

$$= 40{,}040 \text{ mm}^3/\text{min} = 40 \text{ cm}^3/\text{min}$$

$$F_r = \frac{\text{MRR}\,u_s}{4\pi RN} = \frac{\left(40{,}040\,\text{mm}^3/\text{min}\right)\left(0.06\,\text{W min/mm}^3\right)\left(\text{N m/W s}\right)}{4\pi\left(\dfrac{59+55}{4}\,\text{mm}\right)(559\,\text{rpm})\left(\dfrac{\text{min}}{60\ \text{s}}\right)\left(\dfrac{\text{m}}{1000\,\text{mm}}\right)} = 720\,\text{N}$$

The deflection at the cutting tool point for bar A is calculated using Equation 4.2

$$\delta_A = \frac{F_r L^3}{3EI} = \frac{\left(720\,\text{N}\right)(250^3\,\text{mm}^3)}{3\left(206{,}700\,\text{N/mm}^2\right)\left(73{,}624\,\text{mm}^4\right)} = 0.246\,\text{mm}$$

$$I = \frac{\pi(D_o^4 - D_I^4)}{64} = \frac{\pi(35^4)}{64} = 73{,}624 \text{ mm}^4$$

The calculated deflection is an approximation because the clamping is never absolutely rigid and it is impossible to predict the cutting and radial forces exactly. The contribution of the flexibility from the toolholder–spindle interface connection(s) is discussed in Chapter 5. The radial deflection can be compensated by offsetting the turning machine at a cutting depth equal to $D_o + 2\delta$. However, the retraction marks could be a concern especially if the spring back of the bar is significant as explained in Section 4.6.

(b) The deflection of 0.246 mm is very large and the hole diameter will be smaller by 0.492 mm and will generate the scratch marks during tool retraction. Therefore, the deflection should be reduced by reducing the radial force F_r, which is proportional to the area of cut. Hence, the doc or feed must be changed in order to reduce the force. Since a large reduction on the deflection is required, both doc and feed will be reduced by 50%. However, the reduction of the DOC by 50% will require two passes to remove the 2 mm full depth from the bore. The reduction of the feed from 0.2 to 0.1 mm/rev is acceptable but the machining time will be doubled. The reduction of the cutting conditions results in lower productivity.

$$d_2 = \frac{1}{2} d_1 \quad \text{and} \quad f_2 = \frac{1}{2} f_1, \quad \text{Hence, } \text{MRR}_2 = \frac{1}{4} d_1 \cdot f_1 \cdot V = \frac{1}{4} \text{MRR}_1$$

Hence, the power required at the above lighter cut is 25% of the power at the original conditions. Likewise, the radial force is reduced by 75%. Thus,

$$F_{r2} = 0.5\, F_{c2} = F_{r1}/4 = (720/4) = 180 \text{ N and } \delta_2 = \delta_1/4 = (0.246/4) = 0.062 \text{ mm}.$$

(c) The effect of heavy metal boring bar on deflection is

$$\delta_{hm1} = \delta_1 \,(E_{steel}/E_{heavymetal}) = 0.246 \,(206{,}700/330{,}000) = 0.155 \text{ mm}$$

or using the 50% reduced feed and DOC, the deflection is reduced to

$$\delta_{hm2} = \delta_{hm1}/4 = 0.155/4 = 0.039 \text{ mm}$$

Therefore, the substitution results in a significant reduction on deflection and in better hole size control and quality.

(d) The boring bar geometry can be changed from (A) to (B) to reduce its deflection by increasing its stiffness. The bar (B) has two different diameter cross sections along its length. The front diameter is the same as bar (A) to have an effective chip clearance between the bar and the hole since it is a horizontal boring operation. The deflection of bar (B) is:

$$\delta_B = F_r \left[\frac{L_1^3}{3 \cdot E_1 \cdot I_1} + \frac{L_2^3}{3 \cdot E_2 \cdot I_2} + \frac{L_1 L_2}{E_2 \cdot I_2} \left(L_1 + L_2\right) \right]$$

where
$L_1 = L_2 = 125$ mm
$E_1 = E_2 = 206{,}700$ MPa
$F_r = 720$ N

and the moment of inertia for the two cross sections are

$$I_1 = \frac{\pi D_1^4}{64} = \frac{\pi(35^4)}{64} = 73{,}624 \text{ mm}^4 \quad \text{and} \quad I_2 = \frac{\pi D_2^4}{64} = \frac{\pi(50^4)}{64} = 306{,}796 \text{ mm}^4$$

hence, the deflection for boring bar B is

$$\delta_B = 720 \, [4.28 + 1.03 + 6.16] \, 10^{-5} = 0.083 \text{ mm}$$

which is much lower than the deflection for bar A. If the bar is made from heavy metal the deflection is calculated to be 0.052 mm.

Example 4.5 Consider the boring bar (A) in Example 4.4 and discuss all the factors influencing the deflection due to forces acting on the free end.

Discussion: The previous Example 4.4 was a simple case of a real problem because the parameters considered were the bar material, diameter and overhang, and the radial force. However, the radial force (and therefore the tool/bar deflection) is also dependent on the geometry of the insert (lead angle, rake angle, corner radius, and edge preparation).

The cutting (tangential) force, F_c, will push the tool downward and away from the centerline by δ_T (vertical deflection) as shown in Figure 4.102. The radial force, F_r, will push the tool away from the workpiece δ_R in a radial direction. The tool point deflection in the direction of the cutting force will reduce the clearance angle Y (side relief angle in Figure 4.17) on the flank face of the insert as shown in Figure 4.103. Therefore, the clearance angle of the tool should be large enough to avoid contact between the flank face of the tool and the wall of the hole; this becomes very important with small diameter holes due to the small radius of curvature of the internal diameter. As explained in Section 4.5, the selection of the rake angle α, wedge angle w, and the relief angle Y have a decisive influence on cutting efficiency and tool strength. The deflection δ_T can be compensated for by positioning the cutting edge above the centerline of the workpiece in a turning machine. However, the change of the bore radius due to the δ_T is small and it is estimated for the boring bar (A) in Example 4.4(a) to be:

$$\delta_{Tr} = R - \left(R^2 - \delta_T^2\right)^{0.5} = \frac{59}{2} - \left[\left(\frac{59}{2}\right)^2 - (0.493)^2\right]^{0.5} = 0.004 \text{ mm}$$

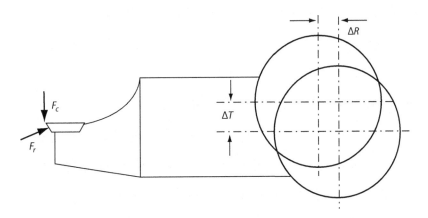

FIGURE 4.102 Free body diagram of the forces and corresponding deflections at the cutting edge of a boring bar.

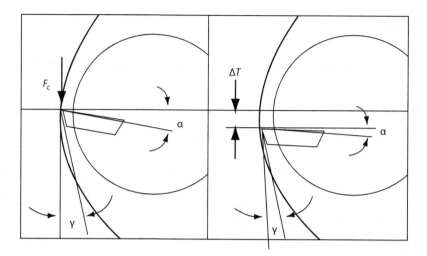

FIGURE 4.103 Free body diagram of the cutting (tangential) force and the corresponding deflection at the cutting edge of a boring bar. (From *Modern Metal Cutting—A Practical Handbook*, Sandvik Coromant, Fair Lawn, NJ, 1996.)

The δ_T of 0.004 mm is very small compared to the radial deflection of 0.246 mm due to the radial force. Therefore, the effect of the δ_T in the bore diameter is generally neglected. In addition, this error cannot be compensated for in a machining center with a rotating tool. However, the clearance angle is reduced by an amount equal to $\gamma = \sin^{-1}(\delta_T/R)$ that is equivalent to 1° for the above Example 4.4(a). This means that the 7° clearance angle of the tool is reduced to 6° during cutting.

The radial force effects the radial (horizontal) deflection, which reduces the depth of cut since the tool moves and affects the diametrical accuracy of the hole as shown in Figure 4.104 and analyzed in Example 4.4. In addition, the chip thickness will change with the varying size of the cutting forces, which causes vibration and may lead to chatter. Adjusting the DOC to be δ_r greater than the designed DOC, compensates for the deflection. Likewise, the programmed boring diameter should be adjusted to be $2\delta_r$ larger than the desired diameter. The programmed boring diameter in Example 4.4(a) should be adjusted to 59.493 mm in order to generate the desired 59 mm bore.

The lead angle affects the feed and radial components of the cutting forces as explained in Section 4.5. If the lead angle is zero, the radial force is a function of the corner radius of the insert and the DOC when it is smaller than the corner radius. As the lead angle increases, the radial force component increases while the feed force decreases. Therefore, a small lead angle is better in boring

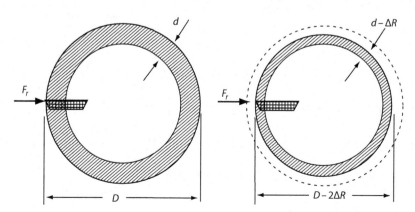

FIGURE 4.104 Free body diagram of the radial force and the corresponding radial deflection at the cutting edge of a boring bar.

even though it affects the chip thickness and the direction of the chip flow, so that a compromise often has to be made. It is suggested that a lead angle of 15° or less is used [105]; a 15° lead angle will generate twice the radial force that a 0° lead angle produces.

The noise radius of the insert also affects the radial force; the greater the nose radius, the greater the radial and cutting forces. The deflection of the tool in the radial direction is affected by the relationship between the nose radius and the depth of cut. The radial force increases most strongly with increased depth of cut as long as the nose radius is smaller than the depth of cut. The effect of the lead angle is present on the radial deflection only when the nose radius is smaller than the depth of cut. Therefore, it is better to select a nose radius somewhat smaller than the depth of cut.

Edge rounding on the primary cutting edge of the insert has a significant influence on the size of the all forces (cutting, radial, and feed forces).

Example 4.6 Optimize the static stiffness and the dynamic characteristics of the boring bar in Example 4.4(a).

Solution: The static stiffness of a solid boring bar can be increased by maximizing its diameter and/or reducing its length. The static stiffness is also dependent on the modulus of elasticity and moment of inertia as shown in Equation 4.1. The dynamic characteristics of the bar include its first (bending) natural frequency and dynamic stiffness and damping. Considering forced vibrations, the amplitude of the vibration force depends upon the static stiffness and the natural frequency of the bar. If the tooth passing frequency is near the natural frequency of the bar, resonance will occur and the amplitude of vibration tends to be very high. The natural frequency of vibration of a cylindrical cantilevered bar with a center hole is approximated by equation

$$\omega = \left(\frac{3EI}{0.23 \, m \, L^3} \right)^{0.5} = C_1 \left(\frac{D_o^2 + D_i^2}{L^3} \right)^{0.5}$$

The natural frequency of vibration for the boring bar is affected by the same parameters as the static stiffness and the mass of the bar. Therefore, a hollow bar could improve the dynamic characteristics since it reduces the mass of the bar. The stiffness at the end of the bar is defined by Equation 4.1 as

$$K = \frac{F}{\delta x} = \frac{3EI}{L^3} = C_2 \frac{D_o^4 - D_i^4}{L^3}$$

The difference in static stiffness between the solid and hollow bars is estimated as

$$\Delta K = \frac{K_s - K_h}{K_s} = \left(\frac{D_i}{D_o} \right)^4$$

The difference in natural frequency between the solid bar and the hollow bar is

$$\Delta \omega = \frac{\left(D_o^2 + D_i^2 \right)^{0.5} - D_o}{D_o}$$

There is trade-off between the static stiffness and the natural frequency when selecting the diameter of the center hole in the boring bar. Therefore, the solution is obtained by equating the above two equations ($\Delta K = \Delta \omega$). If the equations are plotted, the solution indicates that $\Delta K = \Delta \omega$ when the hole diameter (inner diameter) is approximately 67% of the bar diameter. Therefore, the 35 mm

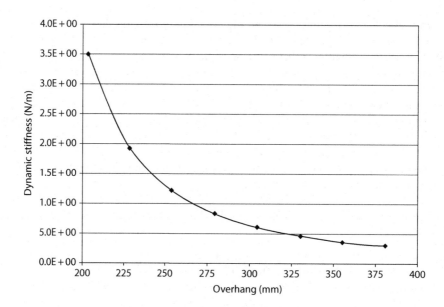

FIGURE 4.105 Dynamic stiffness of a 38.1 mm diameter boring bar. (From Alev, A., and Eversole, W., Design and Devices for Chatter Free Boring Bars, ASTME Technical Paper MR69-266, 1969.)

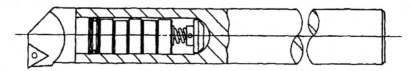

FIGURE 4.106 Illustration of a boring bar with a active vibration control. (Courtesy of Kennametal Inc., Latrobe, PA.)

diameter bar in Example 4.4(a) will require a 23.5 mm hole in which case the static stiffness will be reduced by 20%:

$$\Delta K = \left(\frac{D_i}{D_o}\right)^4 = \left(\frac{23.5}{35}\right)^4 = 0.20$$

and the corresponding natural frequency of the bar will be increased by 20%. However, it is important to note that the dynamic stiffness is also of interest in designing a boring bar. For example, the dynamic stiffness of a boring bar with an internal chatter suppression system [173] is shown in Figure 4.105 for different overhangs of a 38.1 mm diameter boring bar. This boring bar has a center hole with an internal device that consists of several spring loaded high inertia discs as illustrated in Figure 4.106. The internal device suppresses chatter by increasing the damping of the bar. The dynamic stiffness is a function of static stiffness, natural frequency, and damping characteristics. In addition, the moment of inertia of the boring bar at the end can be increased by placing the cutting edge of the insert above the neutral axis of the bored hole without affecting the cutting tool geometry [174].

Example 4.7 A 25 mm hole is made in solid aluminum with a required size tolerance of ± 0.025 mm, location tolerance of ± 0.025 mm, and a surface finish of $R_a = 0.002$ mm. The hole depth is 50 mm. Define the process and the corresponding cutting tool(s) required to manufacture this hole.

Solution: Generally, the hole is roughed with a drill and if the required tolerances are tight, a finishing operation is used such as boring, reaming, or circular interpolating with an end mill. The *L/D* is 2:1 and the tolerance requirement allows any one of the above operations to be used. Note that in general, reaming is applied to any *L/D* hole, boring for *L/D* < 10, and circular-interpolation for *L/D* < 4. In addition, boring and reaming can hold tighter tolerances than circular interpolation.

Generally, using the largest drill with the proper stock allowance for the finish pass is the fastest approach for the first operation. The second operation could be either a semi-finish or a finish operation depending on the hole quality generated by the drill. The semi-finish operation is generally not needed if a solid carbide drill is used with the proper chisel edge (web thinned) for aluminum. The semi-finish operation, if needed, could be either end milling (with 0.75–1 mm doc) or reaming (with 0.3–0.5 mm doc). The finish pass could be any one of the above three suggested operations depending on the finished hole requirements.

End milling cannot produce surface finishes equivalent to boring and reaming operations. The surface finish marks for milling are around the perimeter of the bore, while for boring/reaming they are along the bore length as explained in Chapter 10. A 15–20 mm three flute end mill will generate an ideal surface finish of about 0.0012 mm. However, the contribution of the milling tool runout in the spindle is very significant because it directly affects the surface finish. If the tool runout is 0.003–0.010 mm, the surface finish will be at best equivalent to the runout value.

The surface finish requirement of 2 µm can be obtained by boring but the reaming and end milling operations are questionable. Only multi-flute reamers with double margins or gun reamers can obtain consistently 2 µm finish (see Equation 10.8).

Finally, the location tolerance of ±0.025 mm can be obtained by all three operations, although meeting tolerance with a reamer will require the proper toolholder and cutting edge geometry unless the hole is semi-finished with an end mill because conventional reamers tend to follow the pre-drilled hole location.

Example 4.8 A thin-walled aluminum part is being machined in a 50 mm thick plate using a 10 mm solid carbide end mill. The wall height and thickness are 30 and 0.4 mm, respectively. Select the process and the corresponding cutting conditions.

Solution: The wall of the part is machined from both sides using an alternating approach as shown in Figure 4.107 to reduce the deflection of the wall. The axial doc of the first cut on one side is 1 mm.

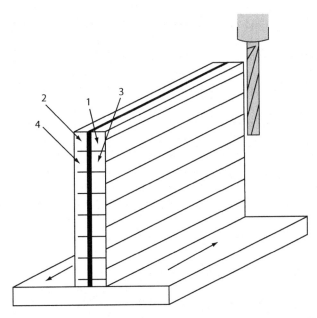

FIGURE 4.107 Illustration of the end milling tool path (process) for machining a thin wall in a part. (1–4) showing the location of the first four passes. (From *Productive Metal Cutting*, Sandvik Coromant, Fair Lawn, NJ, 1998.)

The second is done on the other side with an axial doc of 2 mm. Cutting proceeds with 2 mm axial doc until the whole 30 mm is machined. Climb (down) milling is preferable for this application with a square-shoulder end mill. Higher speeds are also preferable to reduce the contact time of the cutting edge with the wall. Therefore, the maximum spindle speed (rpm) is selected that is stable for the particular mill geometry, overhang, and cutting conditions as is explained in Chapter 12. Let us assume that a 20,000 rpm spindle is used and 18,000 rpm is acceptable for the rough end mill that corresponds to a 565 m/min cutting speed, which is acceptable for aluminum with carbide tooling. A feed of 0.12 mm/tooth is selected for the two-flute mill in the roughing passes and 0.1 mm/tooth for the four-flute mill in the finishing cuts.

The axial doc of 5 mm with a radial doc of 6–10 mm are selected for the roughing cut based on the spindle capability and tool stability. The analysis of Example 2.4 is used here to evaluate the spindle power and tool deflection. The 10 mm radial doc is used to initiate a groove through the solid material during the initial contact with the part. During the full cut the tool deflection is about 0.12 mm while for the remaining cuts with a 6 mm radial doc the deflection is 0.07 mm.

A 2 mm axial doc and 1 mm radial doc are selected for the finish cut. The tool deflection in this case using a four-flute end mill is about 0.008 mm. The axial doc can be increased as long as the deflection of the end mill and/or the wall is acceptable and the tool does not chatter. The cutter diameter for the finish cut should be also reduced to 8 mm to reduce the chip thinning effect (estimated by Equation 2.21).

Example 4.9 Machine a groove that is closed at both ends. The dimensions of the finished groove are 25 mm wide by 200 mm long by 6 mm deep. The workpiece material is low-alloy steel with hardness 180 HB. Select the process and the corresponding cutting conditions.

Solution: This operation requires an end mill rather than a side and face mill because the groove is closed at least at one end. A center-cutting end mill (having overlapping cutting edges on the axial end face) allows the end mill to be fed axially (as a drill) into the workpiece for a small depth of 6 mm in this case. A solid carbide or indexable end mill can be used for this operation. There are three alternatives with respect to the cutter pitch or number of cutting edges: two-flute (coarse pitch), three-flute, and four-flute (dense pitch). The two-flute cutter has only one tooth in cut, while the three- and four-flute have two cutting edges in cut. The general rule is to use less flutes for deeper cuts, with four or higher flutes for light cuts (depth < 0.2 D). Estimation of the cutting forces, power and stability will help in cutter selection (more detailed discussion on this is given in Examples 8.1 through 8.3 of Chapter 8). In this case, a three-flute end mill should be used in order to increase the cutting edge engagement and to avoid chip thinning that may cause vibration especially when extended tool overhang is used to reach in the slot. In addition, the three-flute cutter ensures good chip evacuation out of the grooves.

The cutting edge geometry is either a zero or 30° helix indexable end mill with a corner radius of 0.75 mm on the inserts, or an equivalent solid carbide end mill. A down milling process is used to achieve the most favorable cutting action. A feed of 0.12 mm/tooth and the cutting speed of 200 m/min are selected based on tooling manufacturer recommendations for coated inserts or a coated solid carbide end-mill. The cutter diameter is 25 mm for a conventional slot milling operation. If the slot is generated through circular interpolation, a 13 mm end mill with three flutes should be used instead.

4.15 PROBLEMS

Problem 4.1 A 40 mm diameter by 100 mm long hole (shown in Figure 4.108) is being bored to 42 mm diameter with a single point boring bar. The bar should enter from the 90 mm cavity due to the complexity of the part. The workpiece material is medium-carbon steel. A carbide insert is used in the indexable boring bar made of steel or heavy metal. The boring bar is integral to a CAT-50 toolholder and extends 220 mm. The cross section of the solid boring bar is cylindrical with 30 mm

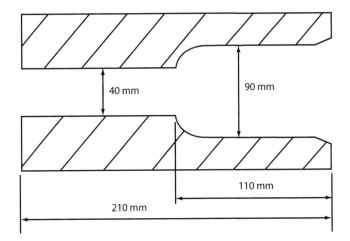

FIGURE 4.108   Illustration of the cross section of a step hole for boring the 40 mm hole in Problem 4.1.

diameter. The cutting speed for the bar is 80 m/min while the feed is 0.2 mm/rev. Design a boring bar and minimize the deflection as much as possible. What should be the size of the boring bar in order to obtain a bore with tolerance of ±0.05 mm?

Problem 4.2 A 100 mm × 100 mm × 35 mm deep cavity is machined in steel with a hardness of 46 HRC. It is suggested to use rough, semi-finish and finish passes to finish the cavity. A 25 mm diameter two-flute ballnose end mill is used for the rough cut with axial doc of 3 mm and radial doc of 10 mm. A 20 mm diameter two-flute ballnose end mill is used for the semi-finish cut with axial doc of 0.8 mm and radial doc of 0.05 mm. A 12 mm diameter four-flute ballnose end mill is used for the finish cut with axial doc of 0.3 mm and radial doc of 0.002 mm. Select the cutting conditions and estimate the machining time for each tool.

REFERENCES

1. B. M. Kramer, On tool material for high speed machining, *ASME J. Eng. Ind.* **109** (1987) 87–91.
2. S. J. Burden, J. Hong. J. W. Rue, and C. L. Stromsborg, Comparison of hot-isostatically-pressed and uniaxially hot-pressed alumina-titanium-carbide cutting tools, *Ceramic Bull.* **67** (1988) 1003–1005.
3. R. Komanduri and J. D. Desai, *Tool Materials for Machining*, Technical Information Series, Report No. 82CRD220, General Electric, Schenectady, NY, August 1982.
4. T. J. Drozda and C. Wick, *Tool and Manufacturing Engineers Handbook: Vol. I—Machining*, SME, Dearborn, MI, 1988, Ch. 1.
5. R. Edwards, *Cutting Tools*, Institute of Materials, London, U.K.; Ashgate Publishing Co., Brookfield, VT, 1993.
6. K. H. Smith, Whisker-reinforced ceramic composite cutting tools, *Carbide Tool J.* **8**:5 (1986) 8–11.
7. W. W. Gruss, Ceramic tools improve cutting performance, *Ceramic Bull.* **67** (1988) 993–996.
8. P. N. Tomlinson and R. J. Wedlake, The current status of diamond and cubic boron nitrite composites, *Proceedings of Efficient Metal Forming and Machining Symposium*, Pretoria, South Africa, November 1982.
9. I. E. Clark and J. Hoffmann, PCD tooling in the automotive industry, *Diamond & CBN Ultrahard Materials Symposium*, IDA, Windsor, Ontario, Canada, September 29–30, 1993, pp. 115–130.
10. M. Deren, P/M edge—Powder-metal end mills: The tougher roughers, *Cutting Tool Eng.* **55**:1 (January 2003).
11. J. L. Johnson, G. Runyon, and C. Morton, Powder power, *Cutting Tool Eng.* **60**:2 (February 2008). http://www.ctemag.com.
12. ISO 513:2012 Classification and application of hard cutting materials for metal removal with defined cutting edges—Designation of the main groups and groups of application.
13. W. W. Gruss, Turning and milling of steel with cermet cutting tools, *Proceedings of ASM Tool Materials for High-Speed Machining Symposium*, Scottsdale, AZ, 1987, pp. 49–62.

14. S. Novak and M. Komac, Wear of cermet cutting tools coated with physically vapour deposited TiN, *Wear* **205** (1997) 160–168.
15. S. Cardinal, A. Malchère, V. Garnier, and G. Fantozzi, Review: Microstructure and mechanical properties of TiC–TiN based cermets for tools application, *Int. J. Refract. Met. Hard Mater.* **27** (2009) 521–527.
16. D. Bordui, Hard-part machining with ceramic inserts, *Ceramic Bull.* **67** (1988) 998–1001.
17. A. Senthil Kumar, A. Raja Durai, and T. Sornakumar, Machinability of hardened steel using alumina based ceramic cutting tools, *Int. J. Refract. Met. Hard Mater.* **21** (2003) 109–117.
18. K. Broniszewski, J. Wozniak, K. Czechowski, L. Jaworska, and A. Olszyna, Al_2O_3–Mo cutting tools for machining hardened stainless steel, *Wear* **303** (2013) 87–91.
19. J. G. Baldodi and S. T. Buljan, Silicon nitride based ceramic cutting tools, SME Technical Paper MR86-912, 1986.
20. E. Besenyi, Performance of silicon nitride cutting tool inserts prepared by gas pressure sintering, *Int. J. Refract. Met. Hard Mater.* **11** (1992) 121–126.
21. B. Bitterlich, S. Bitsch, and K. Friederich, SiAlON based ceramic cutting tools, *J. Eur. Ceram. Soc.* **28** (2008) 989–994.
22. Anon., *Silicon Nitride Ceramics*, International Syalons (Newcastle) Limited, Wallsend, Tyne & Wear, U.K., nd.
23. Anon., *What Are SiAlON Ceramics?* International Syalons (Newcastle) Limited, Wallsend, Tyne & Wear, U.K., nd.
24. V. K. Sarin and S. T. Buljan, Coated ceramic cutting tools, *Proceedings of ASM High Productivity Machining Materials Symposium*, New Orleans, LA, 1985, pp. 105–111.
25. E. D. Whitney and P. N. Vaidyanathan, Engineered ceramics for high speed machining, *Proceedings of ASM Tool Materials for High-Speed Machining Symposium*, Scottsdale, AZ, 1987, pp. 77–82.
26. P. K. Mehrotra, Productivity improvement in machining by applying ceramic cutting tool materials, *A Systems Approach to Machining*, ASM, Materials Park, OH, 1993, pp. 15–20.
27. C. Koepfer, Successful application of ceramic inserts, *Mod. Mach. Shop* (April 15, 1999). http://www.mmsonline.com.
28. D. Novak, A better braze, a comparison of two processes for brazing PCBN onto carbide, *Cutting Tool Eng.* **56**:6 (June 2004).
29. K. R. Pontius, New veins of application, *Cutting Tool Eng.* **54**:8 (August 2002).
30. P. K. Bossom and J. Hoffmann, Turning and milling with PCBN in the automotive industry, *Diamond & CBN Ultrahard Materials Symposium*, IDA, Windsor, Ontario, Canada, September 29–30, 1993, pp. 99–114.
31. P. T. Heath, Structure, properties and applications of polycrystalline cubic boron nitride, *Proceedings of Superabrasives '85 Symposium*, Chicago, IL, 1985, pp. 10–47.
32. T. A. Notter, P. J. Heath, and K. Steinmetz, Amborite for machining hard ferrous materials, *Proceedings of Advances in Ultrahard Materials Application Technology,* Vol. 2, 1983, pp. 51–70.
33. Y. Kohno, T. Uchida, and A. Hara, New applications of polycrystalline CBN, SME Technical Paper MR85-283, 1985.
34. T. J. Broskea, High speed machining of cast iron with polycrystalline cubic boron nitride, *Proceedings of Tool Materials for High-Speed Machining Symposium*, Scottsdale, AZ, 1987, pp. 39–47.
35. Y. Huang and S. Y. Liang, Effect of cutting conditions on tool performance in CBN hard turning, *SME J. Manuf. Proc.* **7** (2005) 10–16.
36. D. Huddle, Extending the range of PCBN, *Mod. Mach. Shop* (January 15, 1999). http://www.mmsonline.com.
37. R. M. Arunachalam, M. A. Mannan, and A. C. Spowage, Residual stress and surface roughness when facing age hardened Inconel 718 with CBN and ceramic cutting tools, *Int. J. Machine Tools Manuf.* **44** (2004) 879–887.
38. J. P. Costes, Y. Guillet, G. Poulachon, and M. Dessoly, Tool-life and wear mechanisms of CBN tools in machining of Inconel 718, *Int. J. Machine Tools Manuf.* **47** (2007) 1081–1087.
39. M. W. Cook and P. K. Bossom, Trends and recent developments in the material manufacture and cutting tool application of polycrystalline diamond and polycrystalline cubic boron nitride, *Int. J. Refract. Met. Hard Mater.* **18** (2000) 147–152.
40. M. S. Deming, B. A. Young, and D. A. Ratliff, Turning gray cast iron with PCBN cutting tools, *Diamond & CBN Ultrahard Materials Symposium*, IDA, Windsor, Ontario, Canada, September 29–30, 1993, pp. 131–142.
41. T. Özel, Modeling of hard part machining: Effect of insert edge preparation in CBN cutting tools, *J. Mater. Process. Technol.* **141** (2003) 284–293.

42. B. Denkena, J. Köhler, and C. E. H. Ventura, Influence of grinding parameters on the quality of high-content PCBN cutting inserts, *J. Mater. Process. Technol.* **214** (2014) 276–284.

43. T. J. Broskea, New applications of polycrystalline diamond in the automotive industry, SME Technical Paper MR91-175, 1991.

44. T. J. Broskea, Superabrasives use in engine block production, *Diamond & CBN Ultrahard Materials Symposium*, IDA, Windsor, Ontario, Canada, September 29–30, 1993, pp. 143–152.

45. A. Richter, A mixed grade—The latest grades of PCD mix coarse particles to increase packing density and yield a more continuous cutting edge, *Cutting Tool Eng.* **56**:6, June 2004.

46. R. B. da Silva, A. R. Machadoa, E. O. Ezugwub, J. Bonneyc, and W. F. Salesc, Tool life and wear mechanisms in high speed machining of Ti–6Al–4V alloy with PCD tools under various coolant pressures, *J. Mater. Process. Technol.* **213** (2013) 1459–1464.

47. C. Dolda, M. Henerichsb, P. Gilgenb, and K. Wegener, Laser processing of coarse grain polycrystalline diamond (PCD) cutting tool inserts using picosecond laser pulses, *Phys. Proc.* **41** (2013) 610–616.

48. D. Korn, Laser edge preparation improves diamond tool performance, *Mod. Mach. Shop* (January 15, 2009). http://www.mmsonline.com.

49. C. Subramanian and K. N. Strafford, Review of multicomponent and multilayer coatings for tribological applications, *Wear* **165** (1993) 85–95.

50. K.-D. Bouzakis, N. Michailidis, G. Skordaris, E. Bouzakis, D. Biermann, and R. M'Saoubi, Cutting with coated tools: Coating technologies, characterization methods and performance optimization, *CIRP Ann.* **61** (2012) 703–723.

51. D. Drape, Cutting tool coating production, *Product. Mach.* (August 19, 2011). http://www.productionmachining.com.

52. J. Skgosmo and H. Norden, The formation of η phase in cemented carbides during chemical vapour deposition, *Int. J. Refract. Met. Hard Mater.* **11** (1992) 49–61.

53. F. J. Teeter, Thin wins, *Cutting Tool Eng.* **58**:10 (October 2006).

54. M. Olbrantz, Composition matters—New equipment allows deposition of hard and tough nanocomposite tool coatings, *Cutting Tool Eng.* **55**:12 (December 2003).

55. J. Martana and P. Benes, Thermal properties of cutting tool coatings at high temperatures, *Thermochim. Acta* **539** (2012) 51–55.

56. K. J. A. Brooks, *Word Directory and Handbook of Hard Metals and Hard Materials*, 5th edn., International Carbide Data, Hertsfordshire, U.K., 1992/1993.

57. J. Destefani, Cutting tools coatings, *Manuf. Eng.* **129**:4 (October 2002) 47–56.

58. S. PalDey and S. C. Deevi, Single layer and multilayer wear resistant coatings of (Ti,Al)N: A review, *Mater. Sci. Eng.* **A342** (2003) 58–79.

59. M. J. McCabe, How PVD coatings can improve high speed machining, *High Speed Machining Conference*, SME, Northbrook, IL, September 20–21, 2001.

60. D. G. Bhat and P. F. Woerner, Coatings for cutting tools, *J. Metals* (February 1986) 68–69.

61. P. C. Jindal, A. T. Santhanam, U. Schleinkofer, and A. F. Shuster, Performance of PVD TiN, TiCN, and TiAlN coated cemented carbide tools in turning, *Int. J. Refract. Met. Hard Mater.* **17** (1999) 163–170.

62. Y. J. Lin and A. Agrawal, Wear improvement with AlCrN coated tool inserts under high speed dry and wet machining of steels, ASME Paper IMECE 2006-14679, Chicago, IL, November 2006.

63. P. C. Siowa, J. A. Ghania, M. J. Ghazalia, T. R. Jaafarb, M. A. Selamatb, and C. H. Che Harona, Characterization of TiCN and TiCN/ZrN coatings for cutting tool application, *Ceram. Int.* **39** (2013) 1293–1298.

64. C. Pfohl, A. Bulak, and K.-T. Rie, Development of titanium diboride coatings deposited by PACVD, *Surf. Coat. Technol.* **131** (2000) 141–146.

65. J. Rao, R. Cruz, K. J. Lawson, and J. R. Nicholls, Carbon and titanium diboride multilayer coatings, *Diam. Relat. Mater.* **13** (2004) 2221–2225.

66. A. Richter, Top coat—Molybdenum disulfide tool coatings improve the machining of aluminum, titanium and nickel-base alloys—Even when dry, *Cutting Tool Eng.* **55**:12 (December 2003).

67. R. Dzierwa, Slippery when blue—A coating system that combines a patented surface impingement process and tungsten disulfide prolongs cutting tool life, but does it promote faster speeds? *Cutting Tool Eng.* **56**:1 (January 2003).

68. K. Kanda, S. Takehana, S. Yoshida, R. Watanabe, S. Takano, H. Ando, and F. Shimakura, Application of diamond-coated cutting tools, *Surf. Coat. Technol.* **73** (1995) 115–120.

69. E. Uhlmann, G. Bräuer, E. Wiemann, and M. Keunecke, CBN coatings on cutting tools, *Ann. German Acad. Soc. Product. Eng.* **11** (2004) 45–48.

70. E. Uhlmann, J. A. Oyanedel Fuentes, and M. Keunecke, Machining of high performance workpiece materials with CBN coated cutting tools, *Thin Solid Films* **518** (2009) 1451–1454.

71. A. Richter. Coating's Holy Grail, *Cutting Tool Eng.* **60**:10 (October 2008). http://www.ctemag.com.

72. N. Chen, B. Shen, G. Yang, and F. Sun, Tribological and cutting behavior of silicon nitride tools coated with monolayer and multilayer-microcrystalline HFCVD diamond films, *Appl. Surf. Sci.* **265** (2013) 850–859.

73. C. H. Shen, Machining performance of thin diamond coated inserts on 390 Al, *Trans. NAMRI/SME* **22** (1994) 201–208.

74. ANSI/ASME B94.50-1975 (R2003), Basic nomenclature and definitions for single-point cutting tools.

75. D. O. Wood and R. E. King, *Modern Metal Cutting: Turning Theory, Chapter 1: Turning and Boring Angles and Applications*, SME, Dearborn, MI, 1985, pp. 3–26.

76. J. Destefani, Cutting tools geometries, *Manuf. Eng.* **133**:5 (November 2002) 41–49.

77. ANSI B212.4-2002, Cutting tools—Indexable inserts—Identification system, 2002.

78. ISO 1832:2012, Indexable inserts for cutting tools—Designation, 2012.

79. G. T. Smith, *Advanced Machining: The Handbook of Cutting Technology*, IFS Publications Ltd., Kempston, U.K. and Springer-Verlag, New York, 1989.

80. Anon., Honed to perfection, *Cutting Tool Eng.* (September 2009) 59–60.

81. B. Kennedy, A better edge—Proper edge preparation improves tool performance, *Cutting Tool Eng.* **56**:2 (February 2004).

82. T. Özel, Y. Karpat, and A. Srivastava, Hard turning with variable micro-geometry PcBN tools, *CIRP Ann.* **57** (2008) 73–76.

83. D. Biermann and I. Terwey, Cutting edge preparation to improve drilling tools for HPC processes, *CIRP J. Manuf. Sci. Technol.* **1** (2008) 76–80.

84. R. J. Schimmel, J. Manjunathaiah, and W. J. Endres, Edge radius variability and force measurement considerations, *ASME J. Manuf. Sci. Eng.* **122** (2000) 590–593.

85. M. D. Thomas, W. A. Knight, and M. M. Sadek, Comparative dynamic performance of boring bars, *Proceedings of 11th International MTDR Conference*, Pergamon Press, Oxford, U.K., 1970, p. 59.

86. A. Ruud, R. Karlsen, K. Sorby, and C. Richt, Minimizing vibration tendencies in machining, *Mod. Mach. Shop* (March 19, 2003).

87. M. Kemmerling, Die aktive Dämpfung überlanger Bohrstangen, *Industrie-Anzeiger* **107**:63 (1985) 42–43.

88. E. I. Rivin, Chatter-resistant cantilever boring bar, *Proc. NAMRC* **11** (1983) 403–407.

89. E. I. Rivin and H. Kang, Improving cutting performance by using boring bar with torsionally compliant head, *Proc. NAMRC* **18** (1990) 230–236.

90. E. I. Rivin, Structural optimization of cantilever mechanical elements, *ASME J. Vibrat. Acoust. Stress Reliab Des.* **108** (1986) 427–433.

91. E. L. Sorice, Selecting milling cutters, in: B. K. Lambert, Ed., *Milling Methods and Machines*, SME, Dearborn, MI, 1982, pp. 62–63.

92. V. J. Tipnis, The working face mill, in: B. K. Lambert, Ed., *Milling Methods and Machines*, SME, Dearborn, MI, 1982, pp. 64–67.

93. V. J. Tipnis, The versatile end mill, in: B. K. Lambert, Ed., *Milling Methods and Machines*, SME, Dearborn, MI, 1982, pp. 68–73.

94. E. M. Sautel, How to select and use tools for milling slots: Part I & Part II, in: B. K. Lambert, Ed., *Milling Methods and Machines*, SME, Dearborn, MI, 1982, pp. 90–95.

95. D. Smith, Tips on how to do a better job of finish milling: Part I & Part II, in: B. K. Lambert, Ed., *Milling Methods and Machines*, SME Dearborn, MI, 1982, pp. 96–103.

96. P. C. Miller, Insert update: The right tool boosts profits, *Tool. Prod.* (February 1989) 50–53.

97. ANSI/ASME B94.19-1997, Milling cutters and end mills, 1997.

98. M. C. Shaw, P. A. Smith, and N. H. Cook, The rotary cutting tool, *ASME Trans.* **74** (1952) 1065–1073.

99. M. C. Shaw, *Metal Cutting Principles*, 2nd edn., Oxford University Press, Oxford, U.K., 2005, pp. 423–428.

100. Anon., *Spinning Tool*, DMG/Mori Seiki USA, Hoffman Estates, IL, 2012.

101. R. Komanduri, D. G. Flom, and M. Lee, Highlights of the DARPA advanced machining research program, *ASME J. Eng. Ind.* **107** (1985) 325–335.

102. R. Komanduri, R. H. Ettinger, M. P. Casey, and W. Reed, Dynamically stiffened rotary tool system, US Patent 4,515,047, May 7, 1985.

103. H. Takama and T. Hayase, Rotary cutting tool, US Patent 5,505,568, April 9, 1996.

104. H. M. Weiss, P. S. Szuba, P. M. Beecherl, and G. J. Kinsler, Metal boring with self-propelled rotary cutters, US Patent 6,073,524, June 13, 2000.

105. Anon., *Modern Metal Cutting—A Practical Handbook*, Sandvik Coromant, Fair Lawn, NJ, 1996.

106. Anon., *Productive Metal Cutting*, Sandvik Coromant, Fair Lawn, NJ, 1998.

107. Anon., Geometries at work—Optimize edge and insert placement in facemill and endmill cutters, *Cutting Tool Eng.* **55**:5 (May 2003).

108. Anon., More chips, less chatter, in: B. K. Lambert, Ed., *Milling Methods and Machines*, SME, Dearborn, MI, 1982, pp. 104–105.

109. ANSI/ASME B94.11M-1993, "Twist Drills", 1993.

110. K. Narasimha, M. O. M. Osman, S. Chandrashekhar, and J. Frazao, An investigation into the influence of helix angle on the torque-thrust coupling effect in twist drills, *Int. J. Adv. Manuf. Technol.* **2** (1987) 91–105.

111. T. R. Chandrupatla and W. Webster, Effect of drill geometry on the deformation of a twist drill, *Proceedings of the Twenty-Fifth International MTDR Conference*, Birmingham, U.K., 1985, pp. 231–235.

112. M. C. Shaw and C. J. Oxford Jr., On the drilling of metals 2—The torque and thrust in drilling, *ASME Trans.* **79** (1957) 139–148.

113. D. F. Galloway, Some experiments on the influence of various factors on drill performance, *ASME Trans.* **79** (1957) 191–231.

114. Anon., Rigidity of twist drills, *Metal Cuttings* **10**:3, National Twist Drill and Tool Company, Rochester, MI, July 1962.

115. J. S. Agapiou, Design characteristics of new types of drill and evaluation of their performance drilling cast iron—I. Drills with four major cutting edges, *Int. J. Mach. Tools Manuf.* **33** (1993) 321–341.

116. J. S. Agapiou, Design characteristics of new types of drill and evaluation of their performance drilling cast iron—II. Drills with three major cutting edges, *Int. J. Mach. Tools Manuf.* **33** (1993) 343–365.

117. M. Plankey, When the chips are down, *Cutting Tool Eng.* **55**:6 (June 2003).

118. J. S. Agapiou, An evaluation of advanced drill body and point geometries in drilling cast iron, *Trans. NAMRI/SME* **19** (1991) 79–89.

119. J. Skoglund, Carbide drills: The answer to high-productivity steel drilling, *Cutting Tool Eng.* (February 1990) 35–37.

120. A. Maier, Multigroove drill bit with angled frontal ridges, US Patent 4,594,034, June 10, 1986.

121. A. Maier, Multiple-tooth drill bit, US Patent, 4,645,389, February 24, 1987.

122. J. S. Agapiou, Four flute center cutting drill, US Patent 5,173,014, December 22, 1992.

123. J. Agapiou, High speed drilling of aluminum workpiece material, *High Speed Machining 2003 Technical Conference*, Chicago, IL, April 7–9, 2003.

124. J. Agapiou, High speed drilling of gray cast iron workpiece material, *High Speed Machining 2003 Technical Conference*, Chicago, IL, April 7–9, 2003.

125. DIN 1412:2001-03, Spiralbohrer aus Schnellarbeitsstahl—Anschliffformen (Twist drills made of high-speed steel—Shapes of points).

126. F. Mason, Whatever happened to the chisel edge, *Am. Mach. Automat. Manuf.* (February 1988) 49–52.

127. H. Ernst and W. A. Haggerty, The spiral point drill—A new concept in drill point geometry, *ASME Trans.* **80** (1958) 1059–1072.

128. H. T. Huang, C. I. Weng, and C. K. Chen, Analysis of clearance and rake angles along cutting edge for multifacet drills (MFD), *ASME, J. Eng. Ind.* **116** (1994) 8–16.

129. F. Fiesselmann, High-performance drilling, *Cutting Tool Eng.* (February 1988) 61–64.

130. S. M. Wu, Multifacet drills, in: R. I. King, Ed., *Handbook of High-Speed Machining Technology*, Chapman and Hall, New York, 1985, Chapter 13.

131. M. Deren, Check the index—Indexable drills allow higher speed and feed rates, *Cutting Tool Eng.* **54**:9, September 2002.

132. M. O'Donoghue, The simple subland solution, *Cutting Tool Eng.* (February 1994) 35–39.

133. A. Alongi, BTA systems—Finish what they start, *Cutting Tool Eng.* (October 1993) 32–36.

134. T. Yakamavich, Deep hole drilling—Now it is faster, easier, *Automation* (January 1991) 30–31.

135. H. J. Swinehart, *Gundrilling, Trepanning, and Deep Hole Machining*, ASTME, Dearborn, MI, 1967.

136. B. Kennedy, The skinny on microdrills—The correct application of microdrills requires attention to numerous operational parameters, *Cutting Tool Eng.* **55**:1 (September 2003).

137. C. Lin, S. K. Kang, and K. F. Ehmann, Planar micro-drill point design and grinding methods, *Trans. NAMRI/SME* **20** (1992) 173–179.

138. L. Jung and J. Ni, Prediction of coolant pressure and volume flow rate in the gundrilling process, *ASME J. Manuf. Sci. Eng.* **125** (2004) 696–702.

139. G. S. Antoun, The science of high-pressure coolant, *Product. Mach.* (July/August 2004) 36–39. http://www.productionmachining.com.

140. S. J. Lee, K. F. Ehmann, and S. M. Wu, An analysis of the drill wandering motion, *ASME J. Eng. Ind.* **109** (1987) 297–305.

141. H. Fuji, E. Marui, and S. Ema, Whirling vibration in drilling. Parts 1, 2, and 3, *ASME J. Eng. Ind.* **108** (1986) 157.

142. ANSI/ASME B94.2-1995, "Reamers", 1995.

143. M. Rubemeyer, Reaming right—A guide to reaming holes cost-effectively, *Cutting Tool Eng.* **56**:1 (January 2004).

144. K. Sakuma and H. Kiyota, Hole accuracy with carbide-tipped reamers (1st report)—Behavior of tool and its effect on multicornered profile of holes. *Bull. Jpn. Soc. Precis. Eng.* **19**:2 (June 1985) 89–95.

145. K. Sakuma and H. Kiyota, Hole accuracy with carbide-tipped reamers (2nd report)—Effect of alignment error of pre-bored hole on reaming action. *Bull. Jpn. Soc. Precision Eng.* **20**:2 (June 1986) 103–108.

146. J. S. Agapiou, Cutting tool strategies for multi-functional part configurations: Part II—Discussion of experimental and analytical results, *Int. J. Adv. Manuf. Technol.* **7** (1992) 70–78.

147. W. Verfurth, Taps and tapping, *Mod. Mach. Shop* (October 1980).

148. ANSI/ASME B94.9-2008, Taps: Ground thread with cut thread appendix (inch and metric sizes), 2008.

149. J. S. Agapiou and C. H. Shen, High speed tapping of 319 aluminum alloy, *Trans. NAMRI/SME* **20** (1992).

150. G. Lorenz, On tapping torque and tap geometry, *CIRP Ann.* **29** (1980) 1–14.

151. G. Lorenz, A study on the effect of tap geometry, *Mech. Eng. Trans. Inst. Eng. Aust.* **29** (1978) 1–4.

152. ISO 965-3:1998, General purpose metric screw threads—Tolerances—Part 3: Deviations for constructional screw threads, 1998.

153. A. Richter, Down to size—There is an alternative to threading small-diameter holes with taps using a thread mill, *Cutting Tool Eng.* **55**:2 (February 2003).

154. J. Sullivan, Choosing the right grinding wheel, *Mod. Mach. Shop* (December 2000).

155. E. Galen, Superabrasive grinding: Why bond selection matters, *Manuf. Eng.* **126**:2 (February 2001) 80–88.

156. M. A. Leitheiser and H. G. Sowman, Non-fused aluminum oxide-based abrasive mineral, US Patent 4,314,827, February 9, 1982.

157. S. C. Salmon, What is abrasive machining?, *Manuf. Eng.* **135**:2 (February 2010).

158. Anon., Grinding with CBN, *Canadian Industrial Machinery*, November 2009, http://www.cimindustry.com.

159. Machinability Data Center Staff, *Machining Data Handbook*, 3rd edn., Vol. 2, Machinability Data Center, Cincinnati, OH, 1980.

160. Anon., High porosity grinding up output, avoids burning, Saint-Gobain abrasives, *Manufacturingtalk*, December 8, 2004.

161. G. G. Rooney, Choosing wheels for today's jig grinding, *Machining Source Book*, ASM International, Materials Park, OH, 1988, pp. 184–188.

162. R. L. Mckee, *Machining With Abrasives*, Van Nostrand Reinhold Company, New York, 1982.

163. J. Schwarz, Precision grinding: Tips on selecting the right wheel, *Machining Source Book*, ASM International, Materials Park, OH, 1988, pp. 168–170.

164. ANSI B74.13–1990, Markings for identifying grinding wheels and other bonded abrasives, 1990.

165. S. F. Krar and E. Ratterman, *Superabrasives: Grinding and Machining With CBN and Diamond*, Glencoe/McGraw-Hill, Westerville, OH, 1990.

166. S. Malkin, *Grinding Technology—Theory and Applications of Machining with Abrasives*, Ellis Horwood Limited, Chichester, U.K., 1989.

167. R. Marvin, Achieving maximum effectiveness with single pass superabrasive bore finishing, *International Honing Clinic*, SME, Dearborn, MI, April 1992.

168. H. Fischer, Tutorial, *International Honing Conference*, SME, Dearborn, MI, April 26–27, 1994.

169. P. P. Bose, Honing with CBN, *Am. Mach. Automat. Manuf.* (January 1988) 66–69.

170. Y. T. Lin, Honing with abrasive brushes, *International Honing Clinic*, SME, Dearborn, MI, April 7–9, 1992.

171. C. Wick and R. F. Veilleux, Roller and ball finishing/burnishing, *Tool and Manufacturing Engineers Handbook*, Vol. III: *Materials, Finishing and Coating*, Ch. 16, SME, Dearborn, MI, 1980.

172. L. Luca, S. Neagu-Ventzel, and I. Marinescu, Effects of working parameters on surface finish in ball-burnishing of hardened steels, *Precision Eng.* **29** (2005) 253–256.

173. A. Alev and W. Eversole, Design and devices for chatter free boring bars, ASTME Technical Paper MR69-266, 1969.

174. B. Berdichevsky, Method of increasing rigidity of boring bars for boring small holes, SME Technical Paper TE81-187, 1981.

5 Toolholders and Workholders

5.1 INTRODUCTION

The design and structural properties of toolholders and workholders have a strong influence on machining cost, accuracy, and stability.

In considering accuracy, the entire tooling structure must be taken into account. The tooling structure consists of the tooling itself and the tool/machine interface. The tooling may be a solid structure or a composite structure composed of jointed elements; cutting tools may either be mounted directly to the spindle/turret, or may be connected through an adapter (arbor or toolholder). The spindle connection for a toolholder (or integral cutting tool) is the foundation that supports the cutting edge in any rotating or stationary tool machining system, and may take a variety of forms. The toolholder is often the weakest link in the machining system, which has limited full utilization of the potential of advanced cutting tool materials in some applications [1,2].

There are several factors influencing the design of a connection as shown in Figure 5.1. No single style of toolholder and/or cutting tool–toolholder interface is superior for all applications. Each type of interface can perform well in particular applications depending on performance requirements. Great care must be taken to ensure that the right toolholder is chosen for a particular job. All three toolholder-device components (the tool, toolholder, and tool-machine connection) must be given equal consideration.

Similarly, the fixture supports, locates, and constrains the part during machining and has a strong influence on the rigidity and dynamic characteristics of the part/fixture/machine tool structure.

This chapter discusses toolholding and fixturing methods. Section 5.2 describes toolholding systems. Sections 5.3 and 5.4 describe toolholder/spindle connections and tool clamping systems. Section 5.5 discusses balancing toolholders. Fixtures are described in Section 5.6.

5.2 TOOLHOLDING SYSTEMS

5.2.1 GENERAL

Properly engineered tool–toolholder and toolholder–spindle interfaces are critical to achieving high performance and high throughput. Toolholder quality, dimensional tolerances, and axial alignment vary broadly from manufacturer to manufacturer. This is unacceptable in many cases; if these interfaces are manufactured improperly or worn (e.g., out of tolerance or improperly fitting, resulting in toolholder tilting and out-of-roundness), the performance of the cutting operation degrades, resulting in poor accuracy, repeatability, rigidity, and tool life.

The important structural and dynamic characteristics of a tooling structure interface are the manufacturing tolerances, static and dynamic runout, radial and axial positioning accuracy and repeatability, connection rigidity (static and dynamic stiffness), force transmission capability, momentum and torque characteristics, clamping forces, balance requirements, fatigue life and durability, retention force requirements, safety, locking/unlocking forces, coolant capability, ease of connection and disconnection, chemical and thermal stability, maintenance requirements, sensitivity to contamination, and cost. Other aspects to be addressed include the efficiency of the connection over a long period, tool presetting requirements, and provisions for data storage modules.

The cutting tool body and toolholder are made either in one solid piece (a monolithic tool) or as a mechanically connected modular system (Figure 5.2). The tooling structure is composed of attachment devices for the cutting inserts or the tooling itself (Figure 5.3a and b). Integral toolholders are used (1) in dedicated machines and transfer lines that produce components that

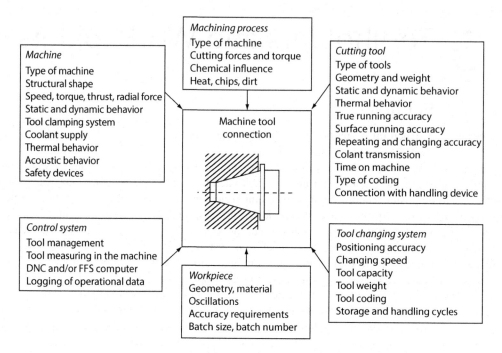

FIGURE 5.1 Factors influencing the design of the toolholder and spindle connection.

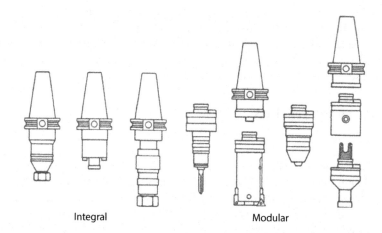

FIGURE 5.2 Comparison between monolithic (solid) and modular toolholding systems. (Courtesy of Sandvik Coromant, Fairlawn, NJ.)

will not change; (2) for tools that recur in several tooling setups, such as face mill arbors and end mill holders of fixed gage length; (3) when runout is very critical; and (4) for tooling packages for which cost is the primary consideration. Integral tooling is not versatile because it can be used only for specific applications; it is preferred when the part features in a family of components are commonized so that such tools can be applied effectively. In addition, in the event of a crash, the whole tool and holder may need to be replaced. Composite or modular systems are preferred in small lot production and other applications requiring flexibility without excessive inventory. In the event of a crash, composite/modular systems require replacement of only the damaged component or adapter. However, modular tooling provides a weaker mounting system than an integral tool, since each additional joint or interface reduces stiffness. It is also extremely difficult to achieve good

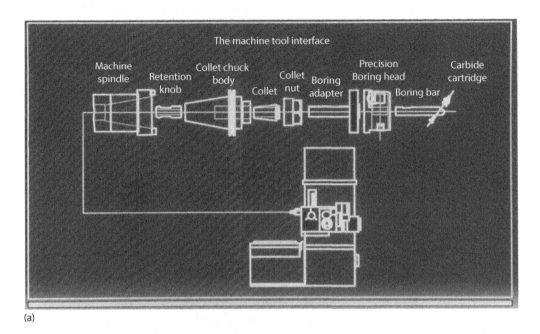

(a)

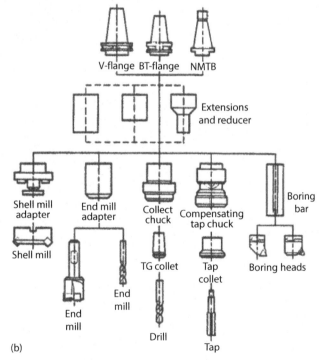

(b)

FIGURE 5.3 (a) Modular/flexible tooling system. (Courtesy of TSD.) (b) Modular Machining Center tooling system. (Courtesy of Kennametal Inc., Latrobe, PA.)

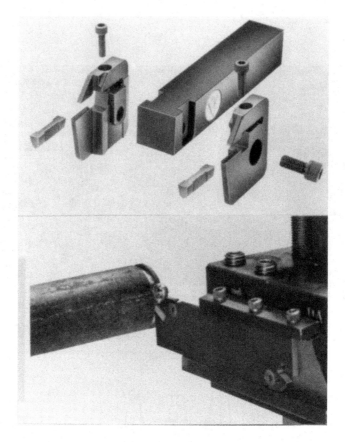

FIGURE 5.4   Manual tool system used in lathes. (Courtesy of Valenite Inc., Troy, MI.)

alignment with multiple interfaces, especially when the tolerance requirement is high. Generally, the more joints or interfaces in a system, the weaker and less accurate it becomes.

Toolholders for turning and machining centers have undergone great changes in the past few decades. The range available includes conventional square-shanked toolholders (Figure 5.4), round-shanked boring bars, and Morse taper round tools loaded manually into a turning center, as well as quick-change toolholders that are loaded either manually or automatically. Changes in toolholders for machining centers have been even more pronounced because additional design requirements have to be met for rotary toolholders. Today, the end user is confronted with a significant challenge in choosing an optimum tooling system, since there are currently more than 30 different modular and/or quick disconnect systems on the market.

5.2.2   Modular and Quick-Change Toolholding Systems

Modular toolholding systems, shown in Figures 5.2 through 5.6, consist of stationary or rotating adapters in a variety of configurations to fit various machines with a common coupling. There are several major types of connections with respect to centering and locating characteristics as shown in Figure 5.7. These include different types of cylindrical shafts including single and multiple cylinders, face and nonface contacts, as well as different types of tapers and taper/face contact systems. There are also several designs of connections with respect to torque transmission, such as polygon, straight, and spiral gear designs, the more conventional key and pin drive methods (Figure 5.8), and mounting bolt patterns or draw bars. Cylindrical shaft forms and both straight and spiral gear designs or pin configurations for torque transmission are primarily used as interfaces in

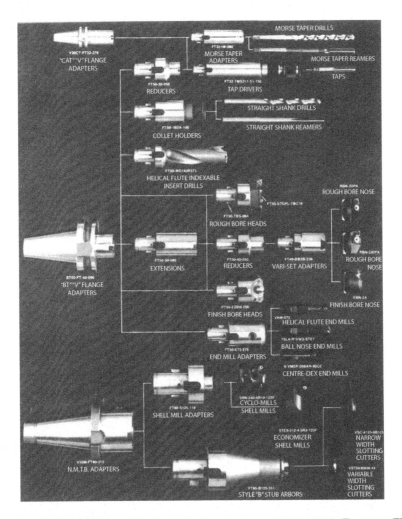

FIGURE 5.5 Automatic modular quick-change tooling system using the VTS (European FTS) adapter. (Courtesy of Valenite Inc., Troy, MI.)

manual or semiautomatic holding systems because the male and female elements of these connections are difficult to assemble automatically. Some toolholders are designed with a flange ahead of the toolholder shank to provide high positioning accuracy.

Modular components can be assembled in different configurations to make a variety of tooling systems that can be shared by multiple machines. Several assembled modular toolholder systems for boring, drilling, reaming, and tapping tooling are shown in Figure 5.9. Rotating toolholders, including collet chucks, tap chucks, various milling adapters, and rotating boring tools with a corresponding coupling, can be used on any machining center. Extension and reduction adapters facilitate assembling tools to the required gage length. Modularity reduces lead time due to the speed with which new tooling assemblies can be built from standard components, and improves tool utilization, management, and standardization. The interfaces should be designed/selected so that the decrement in stiffness due to an additional connection is not so significant as to counterbalance the benefits of such designs.

Generally, modular toolholding connections can provide precision equivalent to the H8/H9 ISO-tolerance class [3]. The standard industry tolerances for male and female adapters are +0.0 to −0.013 and +0.013 to −0.0 mm, respectively. However, some manufacturers guarantee +0.0 to −0.004 and +0.004 to −0.0 mm tolerances, respectively, for the male and female adapters, which results in a worst case eccentricity of 0.008 mm at the front of the adapter with TIR repeatability of ±0.008 mm

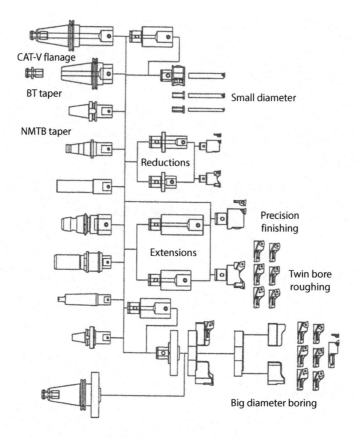

FIGURE 5.6 Advanced modular/flexible boring solutions for all applications. (Courtesy of Parlec, Inc., Fairport, NY.)

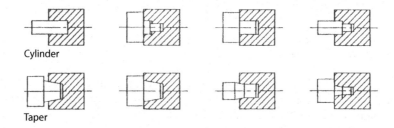

FIGURE 5.7 Configuration of centering and axial locating shaft forms available in the market under various trade names. (Courtesy of Valenite, Inc., Troy, MI.)

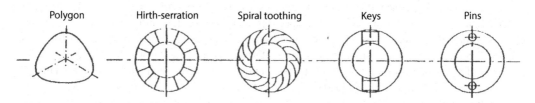

FIGURE 5.8 Configuration of tooling interface forms for torque transmission available in the market under various trade names. (Courtesy of Valenite, Inc., Troy, MI.)

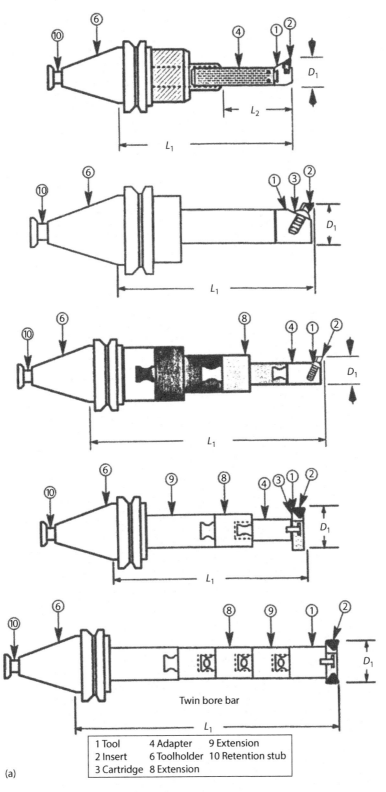

1 Tool   4 Adapter   9 Extension
2 Insert   6 Toolholder 10 Retention stub
3 Cartridge 8 Extension

(a)

FIGURE 5.9 Assembled modular tooling systems using a CAT-V toolholder. (a) Boring tooling assemblies.

(*Continued*)

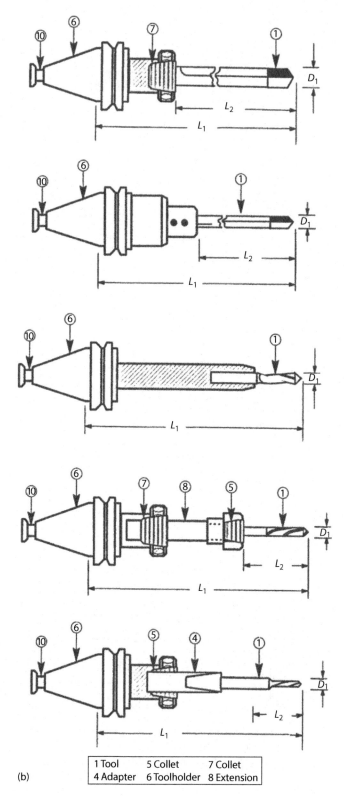

| 1 Tool | 5 Collet | 7 Collet |
| 4 Adapter | 6 Toolholder | 8 Extension |

(b)

FIGURE 5.9 (*Continued*)   Assembled modular tooling systems using a CAT-V toolholder. (b) Drilling tooling assemblies.

(Continued)

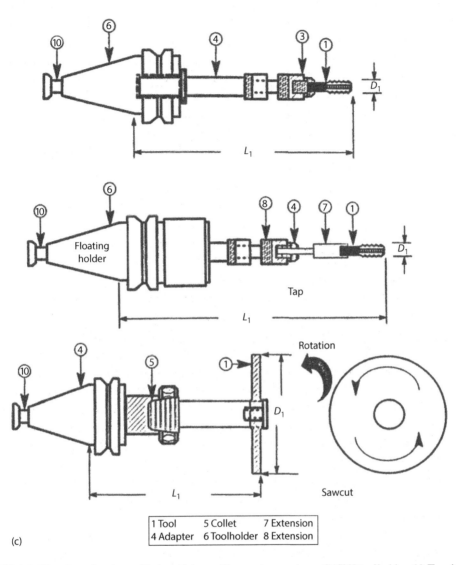

1 Tool	5 Collet	7 Extension
4 Adapter	6 Toolholder	8 Extension

(c)

FIGURE 5.9 (*Continued*)   Assembled modular tooling systems using a CAT-V toolholder. (c) Tapping and slot milling tooling assemblies.

considering a single connection. The runout is a function of the tool axis misalignment with reference to the spindle axis of rotation and is dependent on the spindle accuracy, and on the squareness of the connecting faces (for face contact systems) with respect to their centerline of rotation and/or mating surfaces for the male pilot. The runout is also a function of the total tool length, which may be comprised of one or more spacer adapters. Tilting errors are much more critical than eccentricity errors for extended tools (made of several spacer adapters). Modular tooling systems tend to lose accuracy with extended use due to dirt and chip contamination and wear, scratching, or other damage to their interfaces produced during tool change, cleaning, and storage.

Radial and angular errors, caused by the machine spindle itself, the toolholder–spindle nose interface, or toolholder–tool interface, can be corrected by an adapter located between the cutting tool and the toolholder as shown in Figure 5.10. Four bolts 90° apart are used for each of the radial and angular adjustments on individual adapters or a single, integral adapter. The importance of accurate angular adjustment increases with increasing L/D ratio (and especially when $L/D > 4$ as explained in Example 5.1).

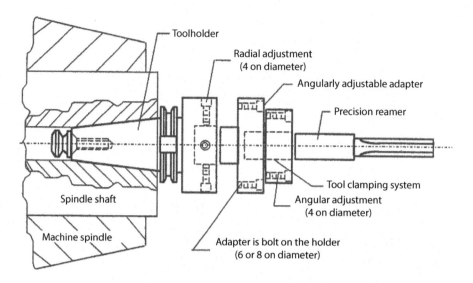

FIGURE 5.10 Adjustable adapter for radial and/or angular adjustments of a cutting tool in the spindle.

An important outgrowth of modular tooling has been *quick-change tooling*, which has led to significant improvements in productivity. A quick-change tooling system (Figures 5.4 through 5.6) consists of a tool or holder that can be changed for another as quickly as possible. It allows the correct length tool to be built off line to maintain maximum performance. The presetting capability of quick-change tooling, in conjunction with the repeatability of the coupling between the cutting unit and the toolholder, ensures that the cutting edge is properly positioned in relation to the workpiece.

5.3 TOOLHOLDER/SPINDLE CONNECTIONS

This section reviews the most common designs for the connection between the toolholder and the spindle [4].

5.3.1 GENERAL

A cutting tool is most commonly mounted in the spindle using a toolholder. In some cases, a cutting tool is mounted directly into the spindle nose to provide accurate location and high stiffness; in such cases, the spindle nose designs may include straight or tapered pilot holes with drive keys, pins (buttons), or gear teeth, and mounting bolt patterns or drawbars (Figures 5.7 and 5.8). Manually changed tooling systems are generally bolted onto the spindle face with one or more screws, supported with side screws, held hydraulically, or shrink fit as illustrated in Figure 5.11. Toolholders or cutting tools with *cylindrical shanks* (end mills, drills, reamers, etc.) are mounted directly into the spindle shaft using a Weldon connection (Types A). In other cases, the toolholder shank is flattened or has a whistle notch and one or two mounting setscrews used to lock the cutter (Types B). Toolholders with a centering (taper) shank style adaption (Types F, G, and H) are very common; they are usually driven by a radial key on the spindle nose or the drawbar and are secured to the spindle with or without face contact by a drawbar in the spindle head. They can also be bolted to the spindle through their flange, in which case they have a short tapered shank; the bolts may transmit the torque in this case. The type H interface provides much higher bending stiffness than Type G due to face contact. The flat back Type E or F interfaces have higher bending stiffness than the taper shank Type G.

The *American National Acme threaded (automotive) shank* with a whistle notch and a key drive is used for drilling, reaming, and tapping holders in transfer line applications. The whistle notch on automotive shanks or cutting tools does not provide preloaded face contact. However, a preloaded

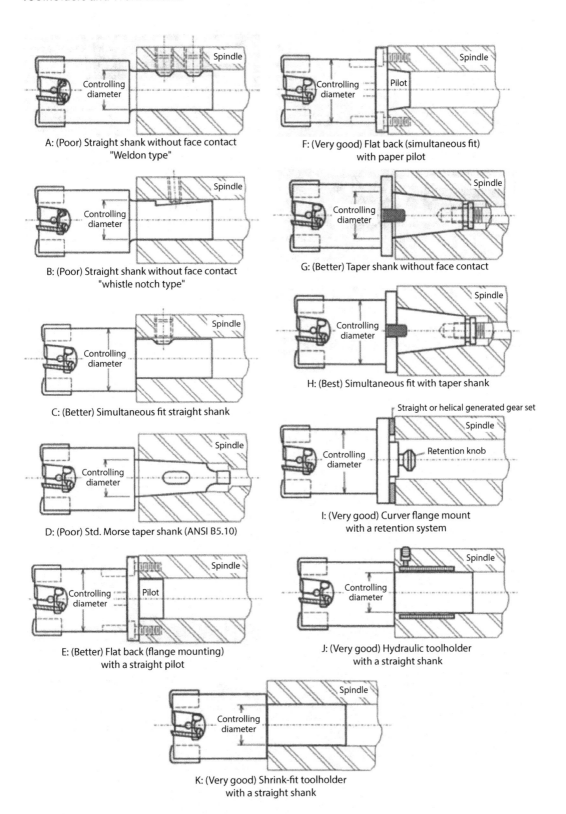

A: (Poor) Straight shank without face contact
"Weldon type"

B: (Poor) Straight shank without face contact
"whistle notch type"

C: (Better) Simultaneous fit straight shank

D: (Poor) Std. Morse taper shank (ANSI B5.10)

E: (Better) Flat back (flange mounting)
with a straight pilot

F: (Very good) Flat back (simultaneous fit)
with paper pilot

G: (Better) Taper shank without face contact

H: (Best) Simultaneous fit with taper shank

I: (Very good) Curver flange mount
with a retention system

J: (Very good) Hydraulic toolholder
with a straight shank

K: (Very good) Shrink-fit toolholder
with a straight shank

FIGURE 5.11   Spindle–toolholder connections.

face contact can be obtained using adaption Type C, in which the screw in the spindle shaft is slightly behind the center line of the seat on the shank of the tool or holder; the radial force generated by the fine pitch screw results in axial force and pushes the faces of the tool or holder and the spindle together under high pressure. Variations of this type of adapter have often been used in modular quick-change tooling systems.

The standard *Morse taper* shank with self-locking characteristics (Type D) is one of the oldest connections without face contact, a drawbar, or mounting screws. It has been replaced in many applications by the *CAT-V taper* discussed in the following. Simultaneous fit of the shank and face can be obtained by using a tapered or cylindrical pilot for centering the connection and a flat flange to provide the desired stiffness. This is illustrated in the designs E, F, and H.

Standard face milling and boring cutters use the flat back or *flange mount* design with a straight or tapered pilot center (Types E and F). The centering plug is often an integral part of the spindle shaft in which case the milling cutter or adapter is secured to the spindle with lockscrews or mounting bolts. The stiffness of the toolholder interface with flange connection is a function of the contact pressure applied by tightening of the flange bolts; the axial force (F) and contact pressure (P) acting on the contact surfaces can be calculated approximately using the formulas

$$F = \frac{5 \cdot n \cdot T}{D} \tag{5.1}$$

$$P = \frac{F}{A} \tag{5.2}$$

where
   T is the tightening torque
   D is the bolt diameter
   A is the contact area (flange surface area minus the bolt holes)
   n is the number of bolts

The flange thickness and the waviness of the joint surfaces affect the form of the interface pressure distribution [5]. A thick flange requires a larger tightening force to provide the same interface pressure as a thin flange.

Figure 5.12 shows a milling cutter with a *curvic coupling* interface (QCS), consisting of a set of mirror-image flat precision helical gears permanently attached to the spindle and to the toolholder, mounted directly to the spindle shaft through a bolt-on tool connection; their engagement assures a unique position of the toolholder relative to the spindle [6]. The curved gear design makes tool

(a)   (b)

FIGURE 5.12 Spiral toothed curvic coupling interface. (a) Manual system and (b) semiautomatic system. (Courtesy of Valenite, Inc., Troy, MI.)

mounting self-centering, highly rigid, and capable of transmitting as much torque as a steel tube of equivalent diameter; such toolholder interfaces can be mounted in a semiautomatic mode using a knob at the back of the holder and a retention system in the spindle that clamps the two halves together hydraulically (Type I). A curvic flange mount provides good radial and axial accuracy (better than 2 μm) with repeatability of 1 μm. A comparison of the curvic coupling with the CAT-V and flat joint is shown in Figure 5.13 (in which the curvic couplings A and B have, respectively, 20 and 24 teeth [6]). A flat back joint provides higher bending stiffness than an equivalent curvic coupling connection (Figure 5.11). A static axial preload markedly increases the joint stiffness of the curvic coupling, which in turn increases the natural frequency. A curvic coupling has a natural frequency lower than that of a flat joint but higher than that of a tapered connection. On the other hand, a curvic coupling has higher damping capacity than either a flat joint or a tapered connection.

A straight shank holder can be mounted in a *hydraulic expansion sleeve* in the spindle nose that is activated manually or semiautomatically (Type J). This provides 360° uniform pressure, which clamps the toolholder concentrically, and uniform contact is achieved over the full length of engagement; the clearance between the toolholder shank and the hydraulic sleeve should be smaller than

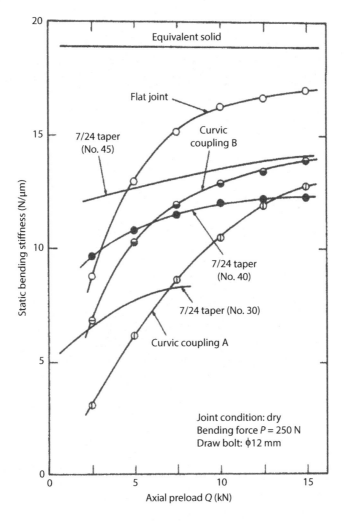

FIGURE 5.13 Effect of axial preload on static stiffness of various interface systems. (From Hazem, S. et al., A New Modular Tooling System of Curvic Coupling Type, *Proceedings of the 26th International Machine Tool Design and Research Conference*, MacMillan, New York 1987, pp. 261–267.)

0.015 mm for proper gripping. The normal amount of contraction for a 25 mm diameter arbor is 0.01–0.03 mm. Contraction is proportionately higher or lower for larger or smaller diameters. It is characterized by very low runout (TIR of 3–5 μm, 100 mm from the front face of the spindle) and high repeatability (0.0013 mm or better); it is balanced by design and is used for high precision operations or higher rotational speeds. It also provides better damping than most of the other interfaces. The clamping force for a 70–85 MPa expansion pressure can transmit a torque of 60 and 500 N m, respectively, for 10 and 32 mm tool shanks. The torque capability can be calculated from the available pressure using Equation 5.3 below. The design of the hydraulic compression sleeve could significantly affect the tool stiffness based on its geometry.

Another system comparable in performance to monolithic toolholders uses a straight shank that is secured in the spindle nose using a *shrink fit* (Type K). The advantage of this method is that the shrinking process is reversible for carbide shank tools. Standard straight shank carbide tools are shrunk into the spindle nose shaft with a high interference fit. In this case the spindle nose is induction heated; with the proper match of materials, the shank expands less than the spindle shaft, eliminating the interference and allowing for removal of the toolholder. A shrink fit provides the best possible TIR, inherent balance, and gripping force. However, the shrink fit requires the spindle shaft to be extended out of the spindle housing by an amount equal to the tool shank length shrunk into the spindle shaft pilot.

The amount of interference can be optimized based on application requirements, but generally should be less than 0.01 mm to avoid excessive compressive stress on the tool shank and to increase the holding ability and safe operating stress in the spindle nose section. The magnitude of the shrink-fit or the unit pressure and tangential stress generated by the fit can be calculated from Lame's equations [7]

$$\delta = \frac{b \cdot P}{E_h} \cdot \left[\frac{b^2 + c^2}{c^2 - b^2} + \nu_h \right] - \frac{b \cdot P \cdot \nu_t}{E_t} \tag{5.3}$$

The tangential stress at the inner surface of the ID of the holder due to the shrink is

$$\sigma_t = \frac{P \cdot \left[b^2 + c^2 \right]}{c^2 - b^2} \tag{5.4}$$

where
   δ is the diametrical interference
   P is the compressive stress (pressure) between the toolholder and the tool shank
   E_h and E_t are the moduli of elasticity for the holder and tool shank materials
   ν_h and ν_t are the Poisson's ratios for the toolholder and tool shank materials
   b and c are the inside and outside diameters of the toolholder nose where the tool shank inserted

The torque T that can be transmitted without relative displacement between the tool and toolholder is

$$T = \frac{\pi \cdot P \cdot L \cdot b^2 \cdot \mu}{2} \tag{5.5}$$

where
   L is the contact length of the shrink-fit joint
   μ is the coefficient of friction

The static coefficient of friction between steel and carbide is 0.25–0.30 with a surface finish of 0.3–0.4 and 0.2 µm, respectively, for the toolholder and tool shank materials. The use of kinetic coefficient of friction, which varies between 0.12 and 0.20, in this equation results in a high safety factor. The torsional holding power is proportional to the applied pressure at the joint, which is proportional to the diametrical interference. Torque capability between 50 and 400 N m, respectively, for 8 and 25 mm tool shank diameters can be achieved.

The amount of interference has a significant influence on the joint tangential displacement [5]. The displacement decreases with increasing interference. The relationship between the tangential displacement and the applied load is linear for a large interference but deviates from linearity for small interference values [8]. The joint tangential displacement and micro-slip increases with increasing surface roughness at the same interference value. The tangential stiffness of shrink-fit joints increases with increasing interference and decreasing surface roughness. The stiffness is approximately 75%–93% of the stiffness of and equivalent solid system for large interference values [8]. Under dynamic loading the observed tangential displacement is considerably larger than that observed under static loading.

Toolholder dimensions have a strong influence on toolholder performance. In cases without good face contact, the diameter of the toolholder at the front of the flange or the tool itself should not be any larger than the gage diameter of the toolholder shank from a bending stiffness viewpoint. The diameter of the tool or toolholder body is the controlling dimension when proper face contact at the interface is used (so that the radial and tangential loads do not separate the unit), and can be as large as the flange diameter of the toolholder.

A flowchart of the requirements for selecting machining toolholder interfaces is given in Figure 5.14. The procedure provides only guidelines of the different steps that should be used to review system components critical to the successful application of a particular toolholder interface with the machine tool. If the machine tool is new, the toolholder size and machine tool characteristics should be selected based on the application. For an existing machine tool, the cutting conditions should be selected based on the machine tool and spindle characteristics including the size of the toolholder–spindle interface.

5.3.2 Conventional Tapered "CAT-V" Connection

The Caterpillar V-flange "CAT-V" steep (7/24) taper design (Figure 5.11 Type G) is the most common toolholder interface system for CNC machines. CAT-V tapers are designated by size as 30, 40, 45, 50, and 60 tapers; the shank dimensions for each size are given in Figure 5.15. The earlier ISO 1947 standard [9] for conical surface accuracy provides for classification of the taper with respect to rated accuracy (based on the "AT" number), surface finish, roughness, and roundness of any section on a conical surface. The CAT V-flange design is included in ANSI Standard B5.50 [10], ISO Standard 7388 [11], DIN Standards 69871 [12] and 2080 [13], and Japanese standard JIS B6339 BT [14] (Figure 5.15). Some of these standards are very similar and interchangeable, while others appear similar but are not interchangeable due to differences in retention knobs and other details.

Conventional 7/24 taper interfaces have two major shortcomings: (1) radial clearance in the back part of the tapered connection, due to taper tolerancing, reduces stiffness and increases runout, and potentially generates micromotions, which cause fretting corrosion and (2) the mandated axial clearance between the flange of the toolholder and the face of the spindle creates 25–50 µm uncertainty in the axial positioning of the toolholder. However, there has been product consistency with and worldwide acceptance of the steep taper. It is well-suited for most applications including high-speed operations up to 15,000 rpm.

The main advantages are: (1) it is not self-locking and is secured by tightening the holder taper in the tapered hole of the spindle and (2) the design of the 7/24 steep tapers is relatively simple and

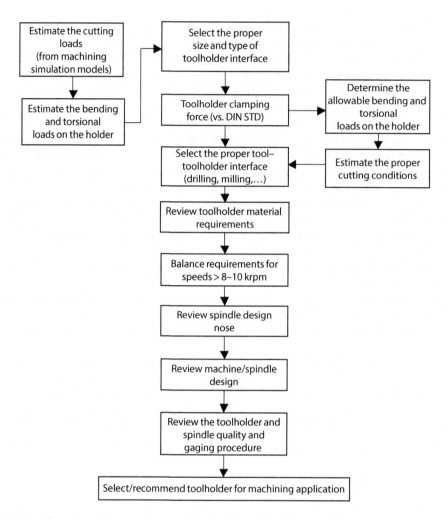

FIGURE 5.14 Flowchart of the requirements for selecting machining toolholder interfaces.

requires precision machining of only one dimension—the taper angle. The taper plays two important roles simultaneously: precise location of the toolholder relative to the spindle, and clamping to provide adequate rigidity.

The allowable tolerance for the interface taper is split between the shank (toolholder taper) and the socket (machine spindle taper); The standard tolerances specify a minus deviation of the hole angle and a plus deviation of the toolholder angle, resulting in clearance between the back of the taper and the spindle hole. The ANSI/ASME tolerance has been revised to AT4 from the middle of AT4 and AT5 (ISO 1947 STD see Figure 5.16), which is equivalent to BT holders. However, a large number of standard commercial toolholders were measured with an air gage and several were found to be between AT5 and AT6. This leads to mobility of the taper under heavy cutting forces and to runout if the draw bar force is not perfectly symmetric. The radial repetitive accuracy of #40 and #50 tapers is within 0.02 mm for new tools, but may be two or three times higher with worn tools. The repeatability of the #30 taper is generally worse than that of #40 and #50 tapers. The maximum possible runout effect from the "AT3" accuracy as compared to the ANSI specification is shown in Figure 5.17. However, if the toolholder and spindle taper tolerance are AT2 to AT3, the radial repetitive accuracy will be better than 0.006 mm.

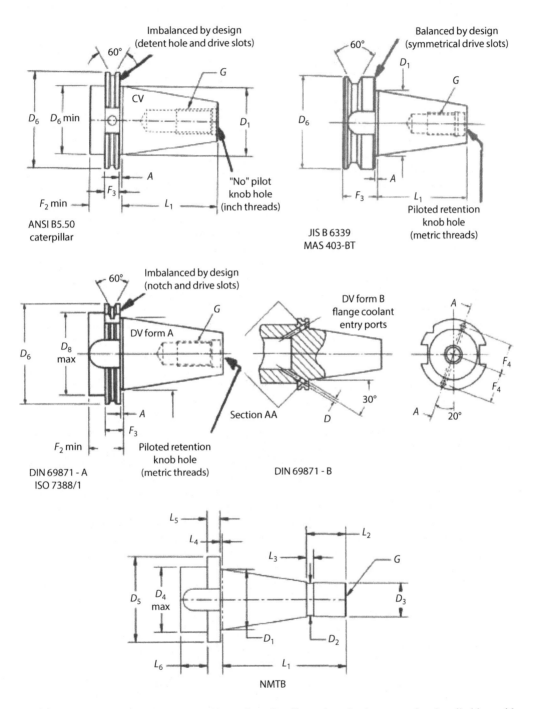

FIGURE 5.15 Characteristic features and manufacturing dimensions for the conventional toolholders with a 7/24 taper. (Courtesy of Kennametal Inc., Latrobe, PA.)

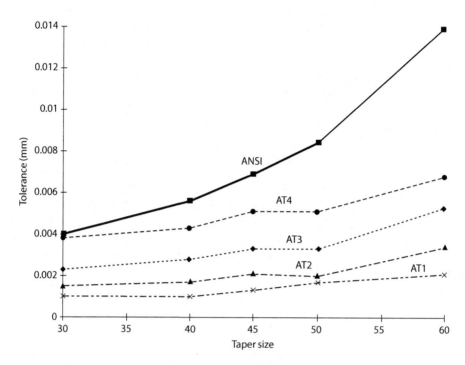

FIGURE 5.16 Tolerance for toolholder tapers based on the ISO-1947 standard.

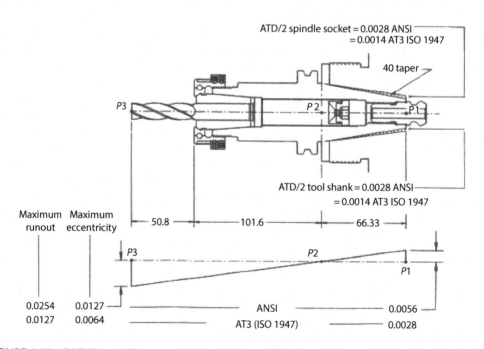

FIGURE 5.17 CAT V taper fit vs. eccentricity and runout. (Courtesy of Kennametal Inc., Latrobe, PA.)

A steep taper connection is often not suitable for high rotational speed applications (30–40 m/s peripheral speed) because the taper may expand (at speeds >10,000 rpm) due to centrifugal forces, substantially reducing the contact area between the toolholder and the spindle [8,15]. The front end of high-speed spindles expands as much as 4–5 μm at 30,000 rpm for taper #30; German researchers have demonstrated that the front end of a high-speed spindle expanded by as much as

0.2 μm at 10,000 rpm and 2.8 μm at 40,000 rpm [1]. As a consequence, the dynamic runout and dynamic stiffness of a 7/24 steep taper system progressively worsens as the spindle speed increases. As the female taper expands, the male taper is pulled further into the shaft (by as much as 0.05 mm depending on the drawbar force), changing the axial position of the tool and in some cases making removal of the toolholder difficult. The border between high speed and low speed applications, by consensus of several spindle manufacturers, is 8,000–15,000 rpm for a #50 taper, 12,000–20,000 rpm for a #40 taper, and 16,000–25,000 rpm for a #30 taper.

Many tool and machine manufacturers realize the importance of improving the accuracy of the toolholder/spindle combination from AT4/AT5 to AT3/AT3 for low to medium speed, high precision spindle systems (Figure 5.16) and AT2/AT2 for high speed, high-performance spindle systems.

Another concern in heavy machining (as in high-speed machining) is fretting corrosion (tarnishing) in the taper portion of both the toolholder and the spindle. Fretting corrosion is a frictional oxidation phenomenon caused by relative movement of the tapers under vibration or heavy cutting. For this reason, black oxide or other treatments are often not suitable for high speed and/or heavy load applications [16].

The axial, bending, and torsional stiffness of the toolholder–spindle interface are important for both static and dynamic performance characteristics. The axial movement of the holder taper in the spindle is about 0.02 mm [17,18] due to elastic deformation at the taper interface under clamping forces. Saturation of the axial displacement and better repeatability is achieved at the suggested clamping forces (drawbar preload) given in Table 5.1. For example, the CAT-40 results in 6 μm movement in the spindle nose when the drawbar force is increased from 5 to 15 kN. The maximum torque that the CAT-V interface will carry is provided in Table 5.2.

TABLE 5.1
Drawbar Forces for CAT-V and HSK Toolholders

Size	CAT-V Flange Force (kN)			Size	HSK-A Force (kN)	HSK-B Force (kN)
	Conventional	High Speed	Max Allowable			
30-T	5–6	3–6		40	6.8	
40-T	8–15	5–10	35–42	50	11	6.8
45-T	14–18	—	56–62	63	18	11
50-T	18–40	12–20	78–88	80	28	18
60-T	60–80	—		100	45	28
				120		45

TABLE 5.2
Maximum Torque (N m) Carried by the Various Toolholder Couplings

Size	CAT-V (N m)	Size	HSK-A (N m)	HSK-E (N m)	KM (N m)	Size	HSK-B (N m)	HSK-F (N m)	Size	Capto (N m)
		32	45	10–19	155	40			C3	
30		40	81	15–35	325	50		20–40	C4	
		50	185	30–75	780	63	800	45–85	C5	1600
40	800	63	360	70–150	1530	80	1200	100–175	C6	3600
45	1800	80	710		2800	100	1800		C8	7000
50	7000	100	1420		4200	125	7000		C10	13,000
55		125	2850			160				
60		160	5700							

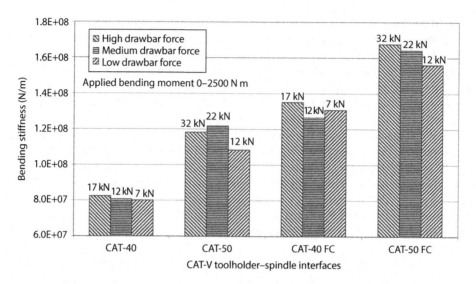

FIGURE 5.18   Effect of drawbar preload on static bending stiffness for various interfaces.

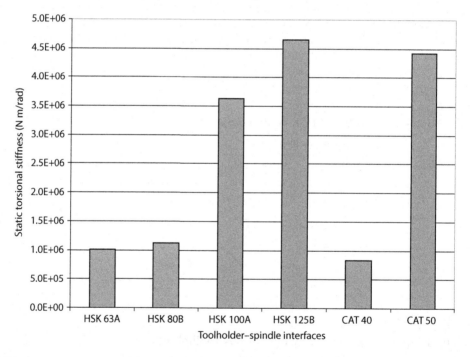

FIGURE 5.19   Static torsional stiffness for various toolholder interfaces.

The static bending and torsional stiffness (based on bench tests) for the conventional interface at three drawbar forces is given in Figures 5.18 and 5.19, respectively [19]. There was not a clear trend between the clamping force or the class of fit on the test tool taper and the static bending stiffness or torsional stiffness in the range of the applied cutting loads tested. This indicates that the static stiffness saturates at the drawbar preloads (7 and 12 kN, respectively, for the CAT-40 and CAT-50 interfaces) [18]. Other researchers found that stiffness is very sensitive to minute differences in the taper angle and to axial preload [20]. In other research, the drawbar's pull force had a strong impact on stiffness as illustrated in Figure 5.13. It shows that a 400%–800% increase in clamping force results in 20%–50% higher bending stiffness [5,20,21].

The normalized bending stiffness as a function of bending moment can be obtained from the static bending stiffness measured at very high radial loads. The rotation (tipping) angle of the toolholder flange is measured using a special setup [19]. The measurement of the rotation at the interface under an applied bending moment (as shown in Figure 5.20) provides a normalized measurement of static bending stiffness of the interfaces [21,22] (see Figure 5.21):

$$k = \frac{M_b}{\theta} = \frac{FL}{\theta} \tag{5.6}$$

The normalized stiffness is given in Figures 5.22 and 5.23 [18,21]. This normalized stiffness can be used in the selection of the interface based on the expected requirements of the cutting operations (i.e., light or heavy operations, short or long tooling, etc.). A CAT-50 interface provides about four to five times greater bending stiffness than a CAT-40 interface. The values for the CAT-V interfaces are small but still sufficient to support the bending moments for many conventional cuts since the bending moment is generally less than 200 N m. The normalized stiffness was found to be the same for all the clamping forces (7–17 kN for CAT-40 and 12–32 kN for CAT-50) for bending moments

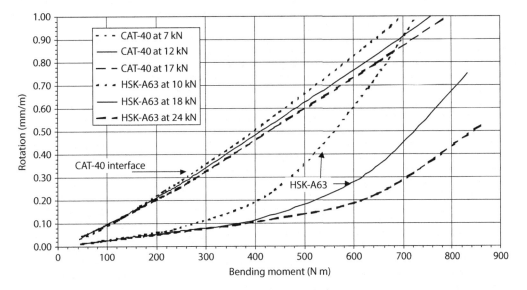

FIGURE 5.20 Comparison of toolholder tilt for CAT-40 vs. HSK-A63 interfaces at three different clamping forces (Bench tests).

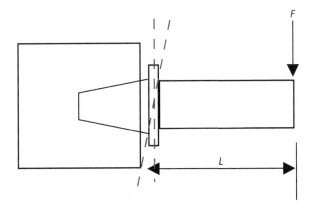

FIGURE 5.21 Normalized bending stiffness of an interface.

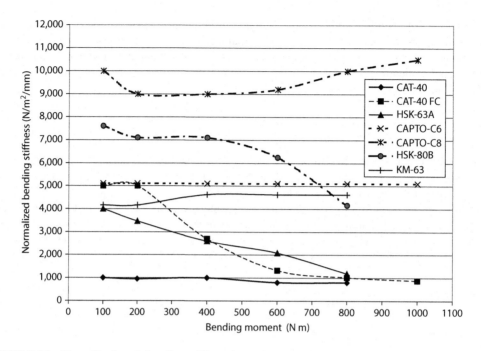

FIGURE 5.22 Normalized static bending stiffness for various small interfaces.

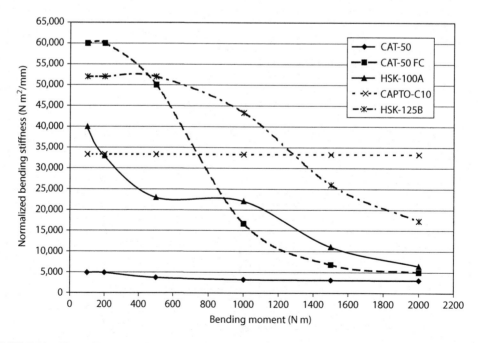

FIGURE 5.23 Normalized static bending stiffness for various large interfaces.

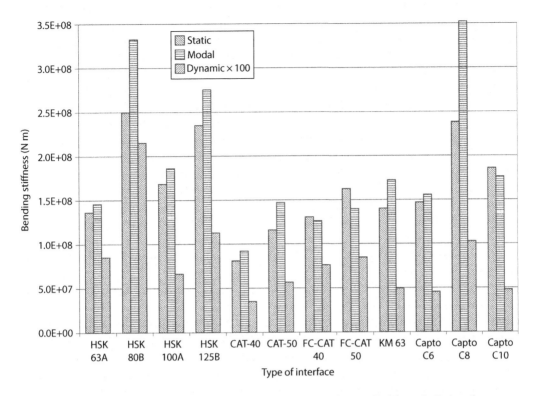

FIGURE 5.24 Static, modal, and dynamic bending stiffness of various toolholder–spindle interfaces.

less than 200 N m. The normalized stiffness increased by 10%–15% between the high- and low-clamping forces when higher than 200 N m bending moment was applied at the interface.

The results from the dynamic bending stiffness and torsional stiffness indicate that toolholder–spindle interfaces are dynamically more flexible as the drawbar preload increases [19]. In addition, the modal stiffness increases and the damping ratio decreases with the application of higher clamping forces (drawbar loads). With increasing clamping forces, the damping ratio decreases at a rate that is greater than the rate at which the modal stiffness increases, thus resulting in a dynamically more flexible system. The average static, modal, and dynamic bending, and torsional stiffness are given in Figures 5.24 and 5.25, respectively. CAT-50 provides significantly better dynamic characteristics than a CAT-40 interface. The torsional stiffness for the CAT-50 is superior to CAT-40.

Power drawbars for CAT-V style holders consist of several components, which may vary depending on the retention knob used. A clamping device extends through the spindle to grip the holder at the knob as shown in Figure 3.52 in Chapter 3. For each of these different standards, it is necessary to have a gripper, which is designed to conform to the specific retention knob [19].

5.3.3 Face-Contact CAT-V Interfaces

As noted before, the conventional 7/24 taper connection has limitations in high-speed or large bending moments applications. The advantages of the face contact "CAT-V FC" interface versus the CAT-V are: (1) interchangeability with existing toolholders, (2) applicability to higher speed machining, (3) reduction or prevention of fretting corrosion (in heavy cuts), (4) elimination of Z-axis movement at higher speeds, (5) Z-axis repeatability within one or a few microns, (6) good repeatability among a family of spindles, and (7) minimal deflection for maximum machining accuracy

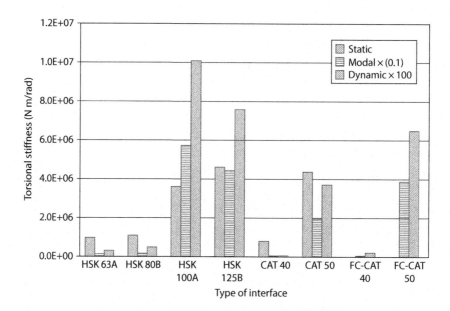

FIGURE 5.25 Static, modal, and dynamic torsional stiffness of various toolholder–spindle interfaces.

and better finish. There are several approaches for providing both taper and face contact in the CAT-V interface to improve the conventional 7/24 taper design without major changes in spindles and with full compatibility with conventional toolholders. Two main approaches are: (1) metal-to-metal contact (rigid) systems, which hold very close tolerances on both halves of the coupling, and (2) using flexible elements either on the taper or flange face of the toolholder or spindle.

The *Big-Plus* interface system offers simultaneous contact on the taper and flange face using the rigid system approach [23,24]. This system is based on the standards for CAT-V, JIS-BT, and DIN69871. Dual contact is achieved with the Big-Plus system by eliminating the gap between the machine spindle face and the toolholder flange face (as illustrated in Figure 5.26). By having tight tolerances on the gage diameter (for both toolholder and spindle) and flange face, about 20 μm axial clearance exists when the toolholder is inserted in the spindle nose, which is closed by the clamping forces given in Table 5.1. One of the important features of the Big-Plus system is interchangeability with existing toolholders as shown in Figure 5.26. A spindle with Big-Plus interface (combination A in Figure 5.26) is required to benefit from the technical advantages the Big-Plus interface offers.

The static bending stiffness of the Big-Plus CAT-40 and CAT-50 interfaces at three clamping forces is shown in Figure 5.17 (defined as Face Contact "FC") [18]; the stiffness for the face contact (Big Plus) interface is 40%–60% greater than that of the conventional CAT-V interface. The stiffness does not change significantly with increasing clamping force. The normalized static bending stiffness is superior for the face contact interface for bending moments less than 200–300 N m and 500–600 N m, respectively, for the CAT-40 FC and CAT-50 FC as shown in Figures 5.22 and 5.23. The static and dynamic bending stiffness for the CAT-V FC interface is as good or better than that of the CAT-V without FC interface (Figure 5.24). The dynamic torsional stiffness for the face contact interface is better than that without face contact as shown in Figure 5.25.

Similar characteristics of simultaneous contact have been obtained using spacers located between the toolholder flange and the spindle face. The spacers have to be precisely ground to the needed thickness. This approach will provide the advantages of a rigid face contact (discussed earlier) but will result in somewhat lower static stiffness than the Big-Plus due to the additional contact with the spacer. However, the dynamic stiffness could increase. The toolholder must be fit individually to a particular spindle and is not interchangeable with other spindles unless the gage diameter of the spindle is controlled to within a few microns.

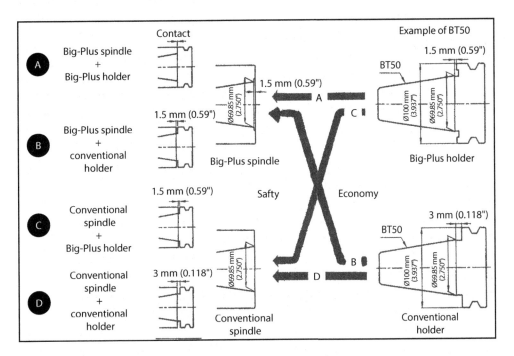

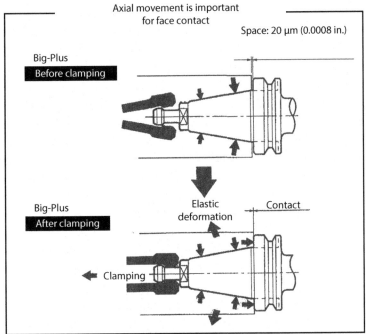

FIGURE 5.26 Big-Plus interface illustrating the clamping-unclamping position of the toolholder and interchangeability with existing CAT-V interface. (Courtesy of Daishowa Seiki Co., Osaka, Japan, and Stanley Sheppard Company, Berlin, CT.)

The rigid approach for simultaneous contact allows for spindle nose expansion (typically of about 6 μm on the diameter). This will affect the front spindle bearings unless the spindle nose female taper is extended ahead of the front bearings; this will result in lower stiffness and natural frequency.

There are several interface systems that use flexible elements to obtain simultaneous fit. The flexible elements increase damping and reduce cutting vibration, thus extending cutting tool life and the applicable cutting regimes. The *3-lock* taper interface provides simultaneous fit (see Figure 5.27) [25] using a spring loaded taper cone. The term 3-lock refers to the three locking points, which consist of the two internal *wedging or locking points* and the face contact between the machine tool spindle face and the toolholder flange. When the toolholder is inserted into the spindle, taper contact is first achieved. As the machine tool continues to pull the tool holder up into the spindle the taper is pushed down on top of the Bellville springs located below the base of the taper cone to pre-load the taper as illustrated in Figure 5.27. It counteracts the centrifugal forces generated at higher speeds. The toolholder taper is preloaded in the spindle nose [25]. The 3-lock contact ratio at taper and flange is reported to be 90%:10% compared to a solid double contact with a taper contact of 10%–50% and flange contact of 90%–50%.

Some other flexible element interfaces are the *WSU-1* and *Mono-Flex* shown in Figure 5.28 [26]. The results of an evaluation of these toolholders are given in Figure 5.29 for two clamping forces. Both systems require somewhat higher clamping force to outperform the conventional interface. The stiffness values in Figure 5.29 cannot be compared to those in Figure 5.17 because the test

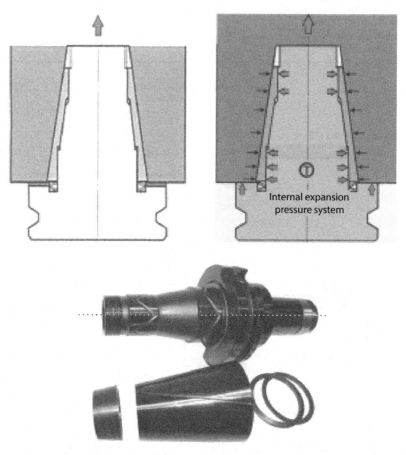

FIGURE 5.27   CAT-V FC 3-lock interfaces containing flexible elements on their taper. (Courtesy of Nikken Kosakusho Works Co., Osaka, Japan.)

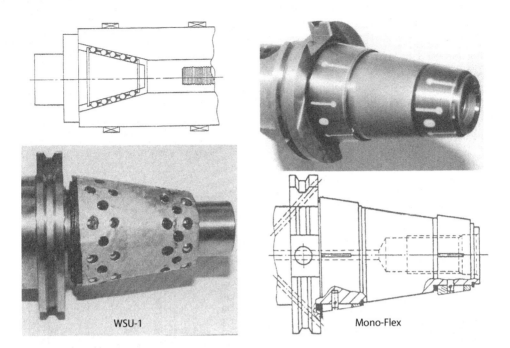

FIGURE 5.28 CAT-V FC WSU-1 and mono-flex interfaces containing flexible elements on its taper. (From Rivin, E.I., Tooling Structures—A Weak Link in Machining Centers and FMS: Some Ways for Improvement, *International Conference on Manufacturing Systems and Environment*, vol. 21, JSME, Tokyo, Japan, May 28, 1990). (Mono-Flex—Courtesy of Rigibore Tooling Systems, Mukwonago, WI.)

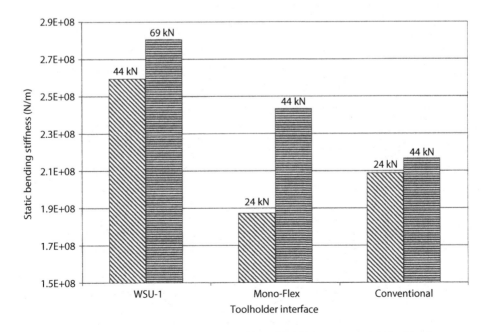

FIGURE 5.29 Static bending stiffness comparison of two CAT-50 interfaces with flexible elements.

parameters are different. The clamping force was saturated at about 44 kN for both systems. The Mono-Flex toolholder provides improvements over the conventional toolholder only at higher clamping forces. The WSU-1 system provided higher stiffness than the Mono-Flex and conventional CAT-V system.

Another interface with flexible elements on the flange face is illustrated in Figure 5.30. The flange face of the toolholder is made in two pieces with bevel washers in between; this provides the preload on the flange face during clamping. The dynamic bending stiffness of this face contact interface is much better than the nonface contact CAT-V interface as shown in Figure 5.31. However, the

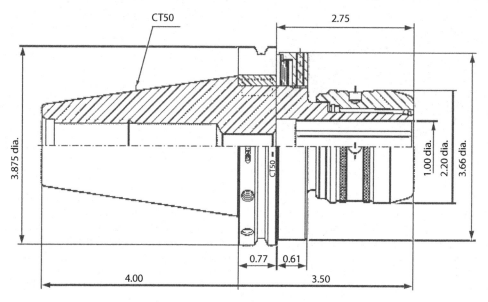

FIGURE 5.30 CAT-V FC toolholder/spindle interface with flexible elements on the flange face. (Courtesy of Richmill Manufacture Co., Osaka, Japan.)

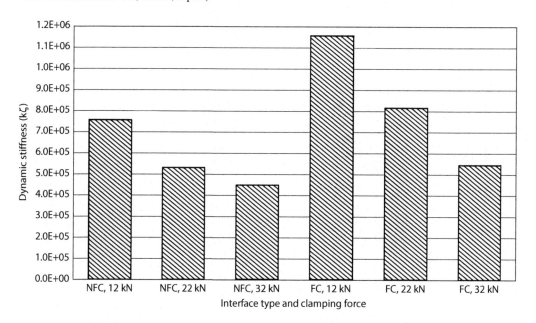

FIGURE 5.31 Comparison of dynamic bending stiffness for CAT-V with (FC) and without (NFC) face contact interface shown in Figure 5.30.

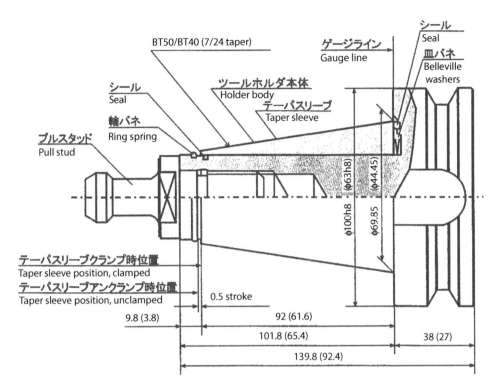

FIGURE 5.32 CAT-V FC "DFC" interface with flexible elements. (Courtesy of Showa Tool Co., Tottori, Japan.)

static bending stiffness is not as high as that of a rigid simultaneous interface but somewhat better than the conventional CAT-V interface at bending moments less than 50–100 N m. In general, the simultaneous interfaces using flexible elements will provide better dynamic bending stiffness than the conventional CAT-V interface; however, their static stiffness is improved compared to that of conventional CAT-V only at higher clamping forces.

The *CAT-V DFC* shank is another double face contact 7/24 taper interface using flexible elements to provide the simultaneous fit [27]. The toolholder consists of two pieces, the main toolholder body and a cylindrical sleeve with the 7/24 taper on the outside. Drawing the pull stud of DFC tooling results in face contact (see Figure 5.32). The clamping force requirement is similar to conventional 7/24 taper (given in Table 5.1) at the higher range. The static and dynamic bending stiffness and concentricity of DFC interface is reported to be better than the conventional 7/24 interface.

5.3.4 HSK INTERFACE

The HSK (Hohl Schaft Kegel or hollow taper shank) system has become a common toolholder/spindle connection in industry in recent years; it was developed to overcome the limitations of the steep 7/24 tapers. The shallow 1/10 taper angle and resilient thin-walled design (Figure 5.33) allows for expansion of the taper shank during clamping and results in simultaneous taper and face contact [21,22,24,28–31]. The HSK interface is intended to outperform CAT-V and other toolholder connections at high rotational speeds (>10,000 rpm) with respect to stiffness, radial and axial positioning accuracy and repeatability, and short tool exchange time.

The hollow taper HSK shank must be manufactured at very tight tolerances, and particular care must be exercised in the selection of materials, heat treatment, and gauging [32–36]. The tooling shanks and spindle receivers are described by DIN 69893 [32] and DIN 69063 [33], respectively. Other critical parameters for the HSK system are the cleanliness of the interface, the drawbar

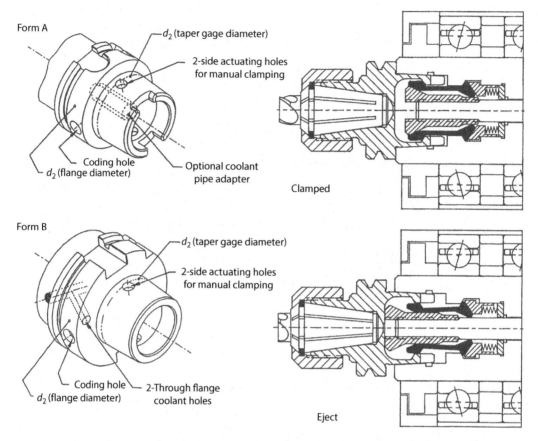

FIGURE 5.33 HSK toolholder DIN 69893 taper and face fitting and toolholder/spindle interface illustrating the clamping-unclamping position of the drawbar fingers and the knock out of the toolholder during ejection. (Courtesy of Precise Corporation, Racine, Wisconsin.)

clamping device and its draw force magnitude, the design and manufacture of the spindle nose taper, dynamic balance at higher speeds, and allowable bending and torsional loads [34,36].

The HSK design uses a 1/10 taper and is short compared to the standard 7/24 taper, resulting in faster tool exchange time. Tolerances on critical dimensions of both the toolholder and spindle are between 2 and 6 μm. The slow taper reduces the force required to produce a given amount of elastic deformation in the hollow shank to bring it into contact with the spindle hole. However, a slow taper requires a higher ejection force to separate the male and female tapers. Due to such stringent tolerances, the shallow taper angle, and resilient thin-walled design, application of an axial force results in simultaneous contact between the taper and face surfaces (simultaneous fit with face contact as illustrated in Figure 5.33). The design includes a mechanical preload on the flange face to ensure full diametrical contact. The friction resulting from the preload also helps to transmit torque. Tests have shown that positioning accuracies of 1 μm can be repeatably attained when the same toolholder is clamped and unclamped in a specific clamping unit. In applications in which more than one toolholder is used, the accuracy of each toolholder must be considered. The total indicated runout (TIR) at 100 mm from the spindle face is typically less than 5 μm.

HSK holders come in six versions (Forms A, B, C, D, E, and F) [32–35]. Forms A and B are designed for automatic tool changing; Forms C and D are for manual tool changing, and Forms E and F with symmetric drive slots are for very high-speed applications. The ratio of the V-flange diameter to the gage diameter for Forms A, C, and E is approximately 1.35:1 as compared to 1.7:1 for Forms B, D, and F, which are intended for heavier cutting load applications. Form B provides about 25% larger load capacity than Form A with the same gage line diameter. A comparison of the corresponding sizes of CAT-V and HSK interfaces is given in Table 5.2 together with their torque capability.

The drawbar in the HSK system is more complicated and expensive than that in a CAT-V system. The toolholder clamping force recommended by DIN [33,36] is given in Table 5.1. During rotation the centrifugal force acting on the fingers adds to the clamping force. The positive locking action and the centrifugal clamping enhancement greatly increases the inherent safety of the HSK interface for high-speed applications. The HSK interface provides full-face contact and partial contact along the taper as shown using Finite Element Analysis (FEA) in [34].

The static bending at the tip of a tool and torsional stiffness (based on bench tests) for the HSK interface at three drawbar forces are given in Figures 5.34 and 5.19, respectively. As with the CAT-V interface, the HSK interface shows no clear trend between the clamping forces and the static bending stiffness or torsional stiffness in the range of the applied cutting loads tested. This indicates that the static stiffness saturates at the lower clamping forces. The HSK-B80 is about 80% stiffer (in bending) than the HSK-A63 automatic change interface that is 50% stiffer than the HSK-C63 manual change interface. The HSK-B125 is about 40% stiffer (in bending) than the HSK-A100 interface. The large and small sizes interfaces cannot be compared in Figure 5.34 because of the difference in the tooling sizes and the distance of the applied load from the spindle face. The static bending stiffness for the HSK-A63 and HSK-A100 interfaces is about 60%–70% and 40%–45%, respectively, greater than that of the CAT-40 and CAT-50 interfaces (see Figure 5.24). The static torsional stiffness for the HSK-B80 interface is not significantly different (about 10%–15% greater) than that of the HSK-A63 interface (see Figure 5.19). However, the static torsional stiffness for the HSK-B125 interface is about 25%–30% greater than that of the HSK-A100 interface. On the other hand, the HSK-A63 provides equal or about 20% higher static torsional stiffness than the CAT-40 interface, while the CAT-50 shows better torsional stiffness than the HSK-A100 interface.

The normalized bending stiffness as a function of bending moment at the interface for the HSK interfaces is given in Figures 5.22 and 5.23 [30]. The HSK-B80 provides about 100% higher stiffness than the HSK-A63 that is 200%–300% better than the CAT-40 interface at bending moments

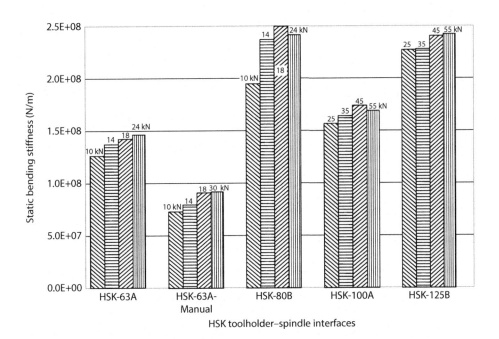

FIGURE 5.34 Effect of clamping force on static bending stiffness for HSK interfaces.

lower than 200 N m. Cat-50 provides about four to five times greater bending stiffness than the CAT-40 interface. The HSK-B125 provides about 30%–50% higher stiffness than the HSK-A100, that is, 400%–700% better than the CAT-50 interface at bending moments lower than 500 N m. The normalized bending stiffness decreases with increasing bending moment for the HSK interfaces. The normalized stiffness was found to be the same for all the clamping forces for bending moments less than 200 N m for the smaller sizes and 500 N m for the larger sized interfaces.

The maximum bending moment a toolholder can support is a function of the clamping force [19,30,31,34,36] and should always exceed the bending moment generated during cutting as explained in Example 5.2. The maximum allowable bending moment before the face joint is separated at the toolholder flange is given in Figure 5.35 [19,36] for a range of clamping forces for different sizes of HSK toolholders. For example, 320 N m is the maximum bending moment that the HSK-63-A interface can support before complete separation of the face contact occurs when the clamping force is 18 kN (see Figure 5.35). The boundary bending moment for an HSK-A63 is about 30%–35% greater than that of HSK-B63. However, the style B can be loaded at 30% higher bending moment than the same gage diameter style A interface.

The maximum allowable torque that the HSK coupling can carry is relatively low, as shown in Table 5.2, since it is carried by a single key (located at the foot of the thin cross section shank) and the friction generated between the taper and face contact areas. The HSK standard specifies two drive keys, but with the tolerances involved, it is difficult to get both drive keys in contact at the same time unless severe deformation is present [30,34,36].

The dynamic bending stiffness and torsional stiffness are better for the style-B HSK interfaces than for the style-A as illustrated in Figure 5.24. The dynamic bending stiffness for HSK-A63 interface is also much better than the CAT-40, but the HSK-A100 was marginally better than the CAT-50 interface (see Figure 5.24). The static and dynamic torsional stiffness of the larger sizes of HSK interfaces is superior to the smaller HSK sizes. In addition, the dynamic torsional stiffness for the HSK interface is better than the CAT-V interface. In addition generally, the dynamic stiffness increases with a decrease in the clamping force.

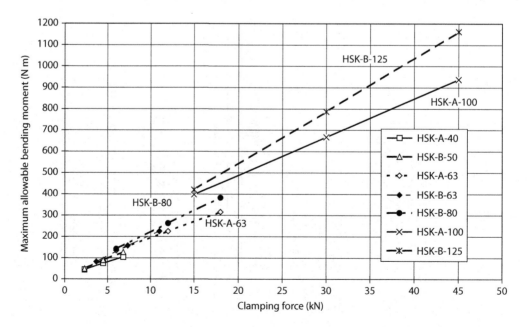

FIGURE 5.35 Maximum allowable bending moment for HSK toolholders. (From Schubert, I. et al., *Interface Machine/Tool: Testing & Optimization, Final Report on the Research Project*, WZL Laboratory for Machine Tools and Applied Economics, Aachen University, Aachen, Germany, March 31, 1994.)

The dynamic runout of HSK systems (due to the expansion of the taper spindle hole caused by the increased centrifugal forces) does not significantly increase with spindle speed as it does with 7/24 steep taper toolholders (Figure 5.36); the expansion ratio of the female to hollow tool shank for HSK is much lower than that of the 7/24 taper due to the high compliance of the hollow taper shank. The centrifugal force, being exerted on the hollow shaft of the HSK toolholder and on the fingers of the clamping system, prevents total separation between the hollow shank and the spindle socket, and therefore the holder maintains its radial (centerline) accuracy. The stress on the toolholder increases by 30%–40% while the speed increases from 1,000 to 16,000 rpm [34]. Finally, several features on the toolholder are nonsymmetrical, complicating balancing in high-speed applications. (Note that forms E and F, intended for high-speed applications, have symmetrical drive slots.) Balancing the HSK holder can be challenging because its shank is deformed to the geometry of the spindle cone. In addition, balancing the drawbar is difficult because it contains moving parts.

Through numerous tests, collected plant data, and theoretical modeling and analysis [34,36], the following technical and operational issues on the HSK toolholder system were identified:

1. The material of the toolholder is critical when used on machines with frequent automatic tool changes or used under heavy loads. A minimum requirement of 1380 MPa yield strength, 1500 MPa tensile strength, 50 HR_C shank hardness should be strictly followed. Through-hardened holders are better than case-hardened holders in these applications since the case depth can act as a stress riser, leading to fatigue failures under repeated clamping-unclampling loading.
2. The clamping force for the toolholder should at least be equal to the DIN standard recommendation but should not be larger than 130% of this value. In aggressive operations, maintaining a high clamping force in the drawbar mechanism is critical in maintaining face contact between the toolholder and the spindle. When face contact is lost, there is a dramatic decrease in interface stiffness, and a corresponding rise in toolholder stress, possibly leading to shear failure in the hollow toolholder shank.

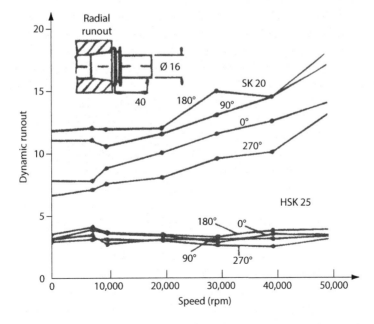

FIGURE 5.36 Dynamic runout of HSK interface versus steep taper (7/24) toolholders. (Courtesy of Precise Corporation, Racine, Wisconsin.)

3. The toolholder tapered shank and spindle tapered bore should be gauged to conform to the DIN standard using the proper gauging equipment. Typically, there is one ring of contact near the spindle nose, and another ring of contact toward the back of the taper separated by a middle region with taper-to-taper clearance. However, slightly out of tolerance taper combinations could lead to a relatively unstable situation where there is only one ring of taper-to-taper contact near the spindle nose. This will have a significant effect on the interface stiffness and stress.

4. The bending and torsional cutting loads must be checked for a specific tool geometry to avoid overloading the HSK interface. Depending on the particular loading condition, the region of maximum stress in the toolholder is typically either in the root of the long key-way or in the thin-walled segment of the toolholder shank near the edge of the 30° land gripped by the drawbar fingers. Toolholder failures in bending or in torsion will generally originate in one of these two areas.

5. Toolholder imbalance should be determined at higher speeds (>10,000 rpm) before any machining is performed. For a very well-balanced system, there is minimal stress rise in the toolholder as RPM alone is increased, even up to 15,000 rpm for the HSK-A63. Excessive imbalance, however, could cause much higher stresses.

6. The toolholder shank and the spindle nose should be maintained chip-free and clean.

5.3.5 PROPRIETARY INTERFACES

Other proprietary advanced toolholder systems include the Kennametal KM, NC5, and Sandvik Capto systems.

The *Kennametal KM (Krupp Widia—Widiaflex/UTS)* system [37–39] uses a 1/10 taper connection (Figure 5.37) very similar to HSK system. It is a common system for transfer equipment and stationary tooling (lathes) and is also used in machining centers. The gage diameter of KM toolholders is almost equal to HSK holders, but the shaft length is 20% longer. The KM interface is designed to allow for larger taper interference (about two to four times) than that of the HSK interface. The KM system

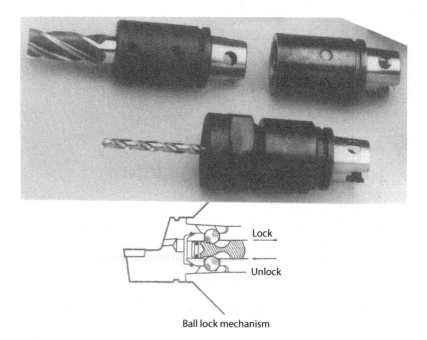

FIGURE 5.37   Hollow shank taper and face fitting KM toolholder system. (Courtesy of Kennametal Corporation, Latrobe, PA.)

generally uses large flange contact, equivalent to HSK Form-B, and a ball-track clamping mechanism with locking balls that is significantly different from that used in the HSK system (Figure 5.33). The clamping force for the KM interface is higher than that of the HSK (about 35–45 kN for the KM-63). Kennametal specifies the toolholder material in order to sustain the high stresses and deformation under high clamping forces.

A comparison of KM sizes with sizes of other style interfaces and their corresponding allowable torques is given in Table 5.2. The radial and axial accuracy and the repeatability of the KM system are similar to those obtained with the HSK system. The static and dynamic bending stiffness of KM interface falls between that of style-A and style-B of the corresponding HSK size as illustrated in Figure 5.24. The limit of bending moment is generally equivalent to the HSK interface. However, KM is designed to produce a higher clamping force than HSK, which results in a higher bending load capability. For similar clamping forces, the properties are similar. The dynamic torsional stiffness for the KM-63 is also equivalent to HSK interface as shown in Figure 5.25. The torque transmission stiffness of the KM system is larger than that of the HSK system since KM holders carry torque both with the locking balls and a key. The KM connection requires cleanliness practices equivalent to those of the HSK connection.

Kennametal has recently introduced the KM4X connection [40,41], an extension of the KM system, which uses high clamping forces and rigid interference to increase bending stiffness and bending moment capacity. It is intended to improve performance in titanium machining and other high force applications.

The *NC5 connection* is a 1/10 short taper interface with double contact obtained by flexible elements [42]. The NC5 taper toolholder is designed similar to the 3-lock 7/24 system discussed before (Figure 5.27) using a drawbar similar to 7/24 taper, but its taper angle and length are the same as the HSK. The big advantage NC5 has over HSK is that the Bellville springs act as a shock absorber. Any vibrations created due to hard milling, unbalance, or extended length are absorbed by the Bellville springs. The clamping force requirement is similar to conventional 7/24 taper (given in Table 1) at the higher range. The static bending stiffness of NC5 is somewhat lower (about 10%–15%) than that of the corresponding sizes of KM and HSK interfaces. The normalized bending stiffness for the NC5 is about 10%–20% lower than that of HSK interface (which is given in Figures 5.22 and 5.23) and the limit of bending moment for NC5 is 30% lower than the HSK system. However, the NC5 has better dynamic stiffness than the HSK and conventional 7/24 taper as shown in Table 5.3.

The *Sandvik Coromant Capto* system (Figure 5.38) uses a 1/20 tapered tri-polygon connection; it provides simultaneous taper and face spindle contact by allowing a relatively small elastic deformation of the shallow hollow taper shank in a manner similar to the HSK and KM systems. It provides self-centering and self-aligning properties and high torque stiffness due to the harmonic tri-polygon shape. The gage diameter of the Capto system is 8% smaller, while its shaft length is 20% longer than the HSK system. The polygon transmits torque through the taper shank and eliminates the need for drive keys or balls and wedges (as required with all other taper systems), whose presence may create balancing problems.

TABLE 5.3
Evaluation of NC5 Interface against the Conventional 7/24 Taper and HSK

Tool Shank	BT-40	HSK-A63	NC5–63
Static stiffness (K)	1	1.04	1.03
Damping ratio (ζ)	1	0.77	1.13
Chatter criterion ($K\zeta$)	1	0.80	1.16

Source: Courtesy of Nikken Works, LTD, Osaka, Japan.

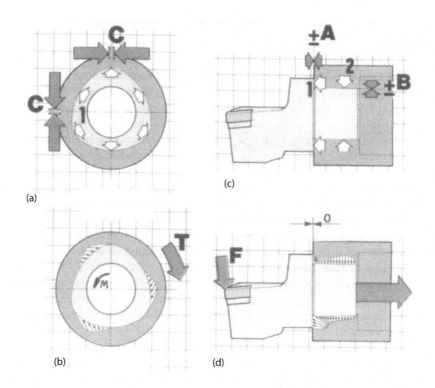

FIGURE 5.38 Hollow shank tapered polygon and face fitting Capto toolholder system. (a) Self-centering coupling, (b) symmetrical distribution of forces, (c) pretensioned coupling, and (d) large contact surfaces insure low surface pressure. (Courtesy of Sandvik Coromant, Fairlawn, NJ.)

The clamping force for the Capto system is about 75–100 kN for the three sizes C6, C8, and C10, which is higher than that of all other interfaces (see Table 5.1). The triangular form ensures that the forces applied to the coupling, particularly when rotating tools are used, will be evenly distributed in a symmetrical manner. The static and modal bending stiffness of Capto interface is equivalent (or 10%–20% higher) to that of HSK interface for the corresponding sizes as shown in Figure 5.24. The dynamic bending stiffness for the Capto system is significantly lower (30%–50%) than that of HSK interface also shown in Figure 5.24. The Capto interface also transmits higher torque than any other automatic change coupling as long as a female adapter is not used (Table 5.2). When an adapter is used, as in manual systems, the bolts holding the adapter on the spindle or turret transmit the torque. The Capto system can potentially carry larger torques than KM, HSK, and CAT systems, although torque-carrying capability is not a significant problem with these connections in many applications. The dynamic torsional stiffness of the Capto system is somewhat less than that of HSK. The radial and axial accuracy and repeatability of the Capto system are equivalent or better than those obtainable with the HSK and KM systems.

All the aforementioned advanced designs employed a *virtual taper* surface on the toolholder and provide face contact between the toolholder and the spindle and thus assure enhanced bending stiffness and high axial accuracy. A comparison between several toolholder coupling interface systems is given in Figure 5.39.

5.3.6 Quick-Change Interfaces (Toolholders/Adapters)

A variety of quick-change manual toolholding systems are also available. The most significant differences between systems are generally in the design of the connection between the adapter head (male) and the base receiver (female) unit, which is either an integral part of the spindle shaft or

Toolholder style	Bending stiffness	Torque stiffness	Accuracy	Tool change properties
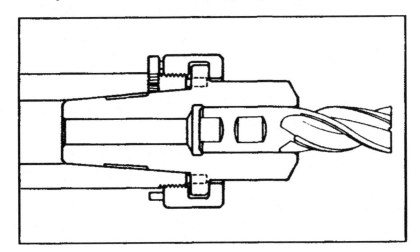	Poor	Very good	Good	Good
	Very good	Poor	Good	Poor
	Good	Poor	Poor	Good
	Very good	Poor	Good	Good
	Very good	Very good	Very good	Good
	Very good	Poor	Good	Very good

FIGURE 5.39 Comparison between toolholder coupling principles.

FIGURE 5.40 Quick-change toolholder taper connection. (Courtesy of Scully Jones Corporation, Seibert Corporation, Chicago, IL.)

an adapter that is usually removed only in case of catastrophic tool accidents. The configuration of male/female connections varies between manufacturers. In the *Pawl-Lock* design, shallow angle self-locking tapers are used in some connections for manual tool changing machines, which can hold an average positional accuracy within 0.015 mm with maximum runout of 0.03 mm (Figure 5.40, e.g., Kwick-Switch, X-Press, Tru-Taper-Smith Lock, Smith Super Taper); to change tools, the operator merely gives a quarter turn on the locknut, which secures the secondary holder in the spindle during cutting. The total indicated runout (TIR) for two such systems A and B using a precision single angle collet is listed in Table 5.4.

TABLE 5.4

Total Indicator Runout for Quick-Change Toolholders

Holder Type	Test Arbor Diameter (mm)	Test Arbor Projection in Diameters	Average TIR (mm)	Total Range TIR (mm)
A	9.5	5.3	0.014	0.030
A	9.5	9.3	0.021	0.046
B	7.9	5.3	0.015	0.025
B	7.9	8.6	0.021	0.041

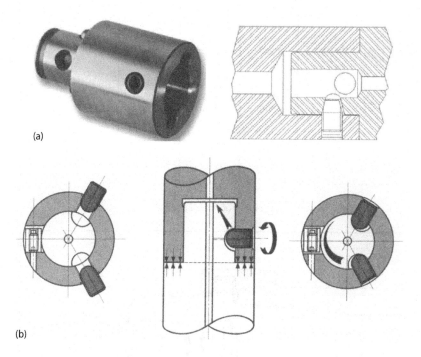

(a)

(b)

FIGURE 5.41 Quick-change interface with (a) one and (b) two side screws. ((a) Courtesy of EPB Inc. (b) Courtesy of Parlec, Inc., Fairport, NY.)

In other systems, a cylindrical male pilot in the back of the toolholder is inserted into a female bore, and one, two, or three set screws, placed symmetrically around the female bore wall, hold the mating cylindrical parts in place (Figure 5.41); the set screws often have a conical end or ballnose that seats in the rear of sockets or a notch (keyway) in the tool/holder pilot shank. When the socket is machined slightly below the setscrew centerline, the end of the setscrew pushes against the socket. This provides face contact and applies pressure to the flange face and on the periphery of the adapter and receiver interface; this creates a rigid connection. The preload of the connection at the face contact is a function of the geometry and tolerances of the setscrew end and the socket in the male pilot, as well as the tightening torque of the setscrew. Multiple setscrews are used in order to increase the preload at the face contact of the connection and possibly the torque capability.

A similar design uses a floating pin (ABS) that is axially offset from the thrust and receiving screws as shown in Figure 5.42; when the floating pin is compressed by tightening the thrust screw, forces F_1 and F_2 are compounded to $2F_A$ in the axial direction acting at the face of the connection. The axial force at the face of the interface and allowable transmission torque for the ABS system are given in Table 5.5 for a

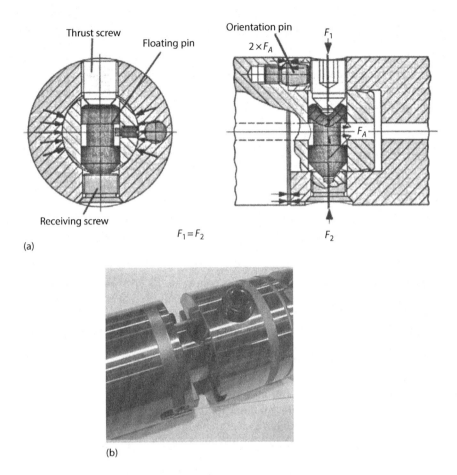

FIGURE 5.42 Quick-change interface with one side screw—(a) ABS system and (b) Beta system. ((a) Courtesy of Komet Inc.; (b) Courtesy of Command Tooling Systems, Ramsey, MN.)

TABLE 5.5
Specifications for the ABS Modular System Connection

ABS Size	Torque on Thrust Screw (N m)	Axial Force (kN)	Transmission Torque (N m)
20	2.8	9.8	51
25	4.5	19.6	73
32	18	29.4	192
40	28	36.0	294
50	36	42.7	700
63	56	48.9	1,491
80	68	55.6	2,994
100	90	75.6	5,999
125	102	91.1	10,959
160	113	104.5	14,687
200	124	113.4	17,512

specified torque on the thrust screw. A drive pin or key in such connections is used for locating and torque transmission. Other designs similar to this are the Beta interface (see Figure 5.41); their differences are in tolerancing and actuation pressures.

There are several other similar systems with side lock screws on the market. The important design and manufacturing characteristics of systems with side lock screws are the clearance tolerance between the straight pilot (adapter) and bore (receiver) and the perpendicularity of the flange face and pilot and mounting face and bore. The female bore is slightly larger than the male cylinder and should be eccentric to the spindle rotational axis, so that the male cylinder is concentric to the spindle axis after the set-screws push the tool shank against the bore; however, this is not true with most systems. The clearance should be as small as possible and usually varies between 0.004 and 0.025 mm; increasing this clearance increases the TIR of the cutting edge. Generally, modular tool-holding connections can provide precision equivalent to the H8/H9 ISO-tolerance.

The performance of all front side-locking systems suffers due to the extra length needed to incorporate the locking mechanism, which brings the cutting edge further from the gage line. Typically, a penalty of 30 mm can be expected. However, indexable tip tools with integral adapters reduce the total tooling length.

Some of these interfaces were compared and the results are given in Figures 5.43 through 5.45. Some of these interfaces can provide equal or better static bending stiffness than the HSK interface of similar size as shown in Figure 5.43. The static torsional stiffness for some of these modular connections is as good as for the HSK interface as illustrated in Figure 5.44. The dynamic stiffness for some of the modular connections was better than HSK interface as shown in Figure 5.45. Some of the simultaneous fit straight shank (side lock or side screw) interfaces perform as well or better than the simultaneous fit expandable taper shank interfaces.

Modular designs provide high vibration damping due to interface surfaces as discussed in Section 4.6, for boring bars. Therefore, modular tools, such as boring bars or shell mills, may provide better performance (much higher cutting conditions, better surface finish, etc.) than equivalent monolithic tooling. Special designs with passive dynamic damping systems (such as

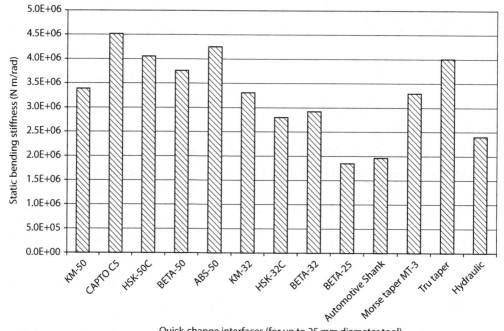

Quick-change interfaces (for up to 25 mm diameter tool)

FIGURE 5.43 Comparison of static bending stiffness for quick-change interfaces using 25 mm tool.

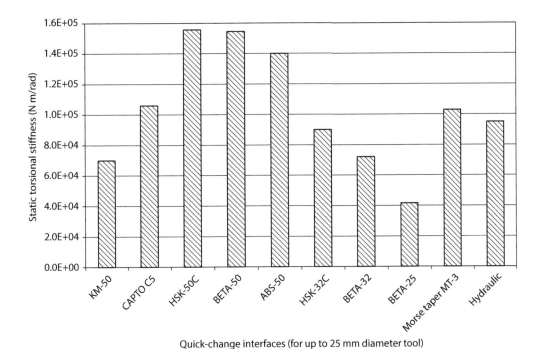

FIGURE 5.44   Comparison of static torsional stiffness for quick-change interfaces using 25 mm tool.

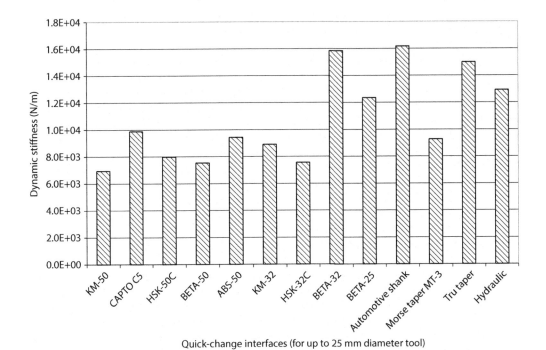

FIGURE 5.45   Comparison of dynamic bending stiffness for quick-change interfaces using 25 mm tool.

Steadyline and *Smart Damper*) significantly improve the dynamic rigidity of long milling and boring assemblies. The build-in damping system does not require tuning. Integrated dynamic damping systems eliminate unwanted vibrations inherent due to excessive toolholder length and cutting conditions. This allows much higher cutting conditions than with equivalent solid holders.

5.3.7 TOOLHOLDERS FOR TURNING MACHINES

Tool blocks are often used in turning centers, although conventional square-shanked toolholders, round-shanked boring tools (Figure 5.4), and quick-change tooling systems are also common.

Tool block systems have taken several forms, which are combined in different machine-adaptable clamping units such as the VDI system, the standard MTB adapter plate, standard or custom bolt-on blocks, and custom one piece turret discs as shown in Figure 5.46. CAT-V style toolholders have been used widely as a tool adapter system (MATS) in the United States but have limitations at heavy bending loads, since they do not provide face contact. The VDI toolholder system with a round shank and rack machined on the shank (see Figure 5.47), described in DIN/ISO 10889 [43,44], is widely used. A mating rack in the turret clamps the adapter in place. The KM and Capto modular toolholding systems were initially designed for lathes and provide good performance (Figures 5.22 through 5.24). HSK toolholders are also used for turning but have less bending load capacity than the Capto and KM systems.

The design features of quick-change tooling systems vary, but the systems available can be divided into two categories: cutting unit systems and tool adapter systems (tool blocks) as shown in Figure 5.47. The cutting unit system uses a replaceable clubhead designed for square shank holders for turning operations, and less effectively for boring and especially rotating applications. In this system, turning, boring, and rotating applications are not interchangeable. The precision coupling delivers high stiffness and excellent repeatability.

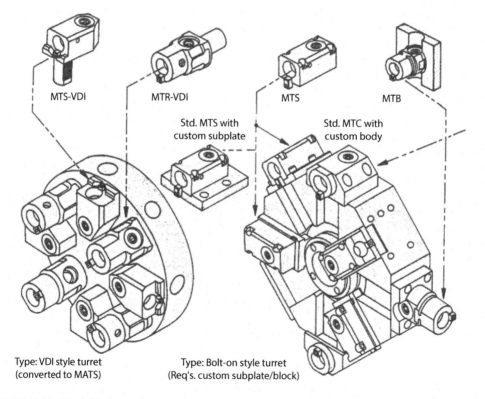

FIGURE 5.46 Tool blocks and turret designs (MATS) for quick-change tooling in turning machines. (Courtesy of Carboloy Corporation, Warren, MI.)

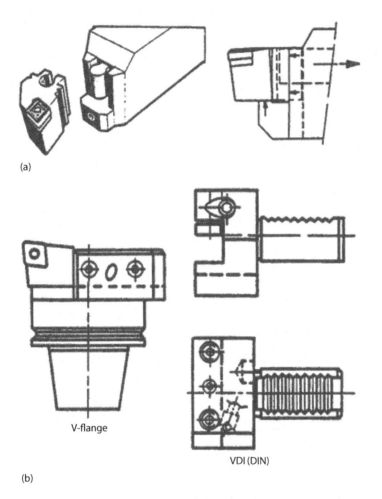

(a)

V-flange

VDI (DIN)

(b)

FIGURE 5.47 Quick-change tooling systems for turning machines. (a) Cutting unit system. (Courtesy of Kennametal Inc., Latrobe, PA.) (b) Tool adapter systems. (Courtesy of Valenite Inc., Troy, MI.)

5.3.8 EVALUATION AND COMPARISON OF TOOLHOLDER/SPINDLE INTERFACE

A comparison between several toolholder–spindle interfaces is given in Table 5.6. The static and dynamic bending stiffness and the torsional stiffness were measured from a bench test [19]. The reliability is based on an evaluation matrix using factors such as fracture fatigue, dirt/nick sensitivity, wear/corrosion sensitivity, sticking extra force, torque slippage, temperature sensitivity, maintenance, lubrication, coolant feed, and gauging. The ease of tool change is based on an evaluation matrix with the factors of holder weight, length engagement, surface cleanliness, drawbar force, critical location, and coolant through ease.

Toolholder–spindle interfaces can be evaluated and compared based on experimental and analytical/finite element results. The bench test results of the static and dynamic stiffness, shown earlier among the different interfaces (Figures 5.18 through 5.45), indicate that some toolholders are superior to others (e.g. HSK-A63 interface is superior to CAT-40 interface). However, bench test results (an approach always used in the past for comparing different sets of interfaces) do not duplicate the interfaces in a machine tool spindle. Bench tests provide a relative comparison between toolholders, but the results will be different when the same toolholders are evaluated on a spindle because the static and dynamic stiffness seen at the tip of the cutting tool also depend on the stiffness of the tool, the spindle geometry and bearings, the housing, and the overall machine structure.

TABLE 5.6

Comparison of Several Interfaces with Respect to Reliability, Ease of Change, Relative Cost, Balance, and Stiffness

Toolholder Type	Reliability[a]	Ease of Change[b]	Relative Cost	Balance[c]	Static Bending Stiffness[d] ($\times 10^8$ N/m)	Dynamic Bending Stiffness[e] ($\times 10^5$ N/m)	Static Torsional Stiffness[f] ($\times 10^6$ N m/rad)
HSK 63A	1	2	5	8	1.44	5.12	1.01
HSK 80B	1	2	2	8	1.94	2.51	1.14
HSK 100A	1	1	3	5	1.65	9.92	3.25
HSK 125B	1	1	1	5	1.99	8.05	3.62
CAT 40	10	10	10	10	0.73	5.35	0.67
CAT 50	10	10	9	7	1.15	4.26	2.35
FC-Cat 40	9	6	5	10	1.26	6.36	NA
FC-Cat 50	8	4	6	7	1.64	8.52	NA
KM 63	6	5	6	8	1.42	4.56	NA
Capto C6	6	6	6	NA	1.47	4.51	NA
Capto C8	5	5	5	NA	2.47	10.20	NA

Note: 10 = best, 1 = worst.

[a] Reliability based on evaluation matrix.

[b] Ease of tool change based on evaluation matrix.

[c] Balance of holder as made.

[d] Static bending stiffness from measurements (ratio of the applied force and the resulting deflection).

[e] Dynamic bending stiffness from measurements (product of modal stiffness and the damping ratio).

[f] Static torsion stiffness from measurements (ratio of the applied torque and the resulting rotation).

For example, a comparison of the static bending stiffness between the HSK-A63 and CAT-40 interfaces in identical machine tool spindles is shown in Figure 5.48. The tool stiffness is about equal for both interfaces especially below 600 N radial force at the tool end. In contrast, the results from the bench tests are also shown in Figure 5.48, which indicate that the HSK interface is much stiffer than the CAT system (as explained in Figures 5.22 and 5.24). The natural frequencies for the two interfaces from the machine tool spindle tests were in good agreement, with a difference of less than 5%. The modal stiffness for the CAT-40 was in fact somewhat larger than that of the HSK-A63. Bench tests results that reflect only the stiffness of the toolholder–spindle interface may thus be misleading in applications in which the properties of the spindle and bearings largely determine the structural response.

A number of researchers [45–51] proposed the characterization of the toolholder/spindle interface using the joint stiffness parameters. A methodology for estimating the joint stiffness parameters of a toolholder/spindle interface using one linear spring and one rotational spring (see Figure 5.49) has been found to be very effective [52]. A flow chart of the methodology is shown in Figure 5.50. It is based on the FRF of the interface system. It involves FEA and experimental measurements of a bench fixture. The most important assumption is that the behavior of the test system is linear. After the joint stiffness parameters are estimated, they can be used in any machine tool spindle FEM (as illustrated in Figure 5.51) to estimate the static and dynamic characteristic at the tool tip.

The Spindle Analysis program (SPA) from Manufacturing Laborites, Inc. was used to create a two-dimensional axisymmetric model of a spindle for evaluating the CAT-40 and HSK-A63 interfaces. The spindle shaft, the housing, and the toolholder structure were modeled as shown

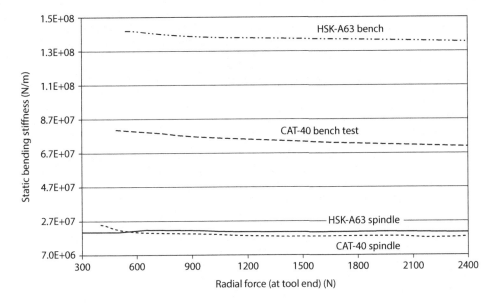

FIGURE 5.48 Comparison of static bending stiffness between CAT-40 and HSK-A63 toolholders on bench fixture and in machine tool spindle.

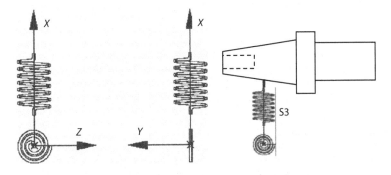

FIGURE 5.49 2-DOF (linear and rotational) spring orientations used in SPA FEA model.

in Figure 5.51. The SPA program can model a 2-DOF spring to include a linear direction and rotational direction to represent the joint between the spindle and the toolholder as illustrated by spring S3 in Figure 5.49. The SPA FEA program simulates both the dynamic and static behavior of a spindle and tooling system. The taper joint stiffness parameters for the 2-DOF spring S3 are given in Table 5.7 and obtained using the methodology in Figure 5.50 [48]. Springs S1, S2, and S4 represent the spindle bearings and springs S5 and S6 represent the connection of the spindle housing to the machine tool structure. The static and modal analysis results are given in Tables 5.8 and 5.9, respectively. The toolholder deflection and natural frequencies or mode shapes were similar for both interface styles. The first two natural frequencies from the SPA FEA (in Table 5.9) were within 5% those from the machine tool spindles. In addition, the natural frequencies are much lower when the toolholder is in the spindle than in the bench fixture [48]. These results emphasize the importance of evaluating the toolholder system on a production spindle instead of a bench test whenever possible.

The static bending deflection of the tool can also be defined analytically using the joint stiffness parameters; the bending deflection of the tool includes the elastic deflection of the bar itself,

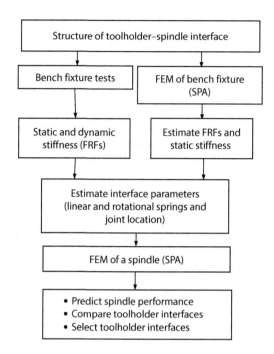

FIGURE 5.50 Flow chart of methodology for evaluating toolholder–spindle interfaces.

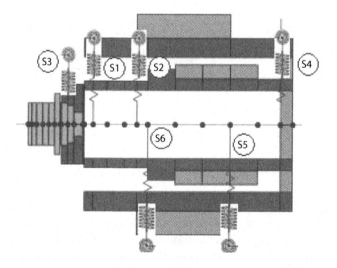

FIGURE 5.51 SPA model for CAT-40 spindle nose and test tool.

TABLE 5.7

Average Joint Characteristics for the CAT-40 and HSK-A63 Interfaces

		Spring Constants	
Interface	Joint Location (m)	Linear (N/m)	Rotational (N m/rad)
CAT-40	0.02	1.375E+09	8.108E+06
HSK-A63	0.02	2.335E+09	1.587E+07

TABLE 5.8

Comparison of Estimated Static Stiffness (SPA) in the Machine Spindle

Applied Load (N)	CAT-40 (N/m)	HSK-A63 (N/m)
1000	2.622E+07	2.706E+07

TABLE 5.9

Comparison of Estimated Modal (SPA) Parameters in the Machine Spindle

Mode	CAT-40		HSK-A63	
	ω_n (Hz)	K (N/m)	ω_n (Hz)	K (N/m)
1st	517	1.39E+08	517	1.45E+08
2nd	671	4.10E+07	675	4.22E+07
3rd	1521	3.12E+08	1532	3.33E+08

the deflection of the toolholder–spindle joint, and the spindle deflection at the front bearing(s). The stiffness of the spindle–toolholder interface is decomposed into a rotational spring and a translation (linear) spring:

$$\delta_{total} = \delta_b + \delta_h + \delta_r + \delta_s \tag{5.7}$$

where

δ_b is the deflection for a cantilevered beam
δ_h is the translation of the taper interface
δ_r is the deflection due to rotational spring of the taper interface
δ_s is the translation of the spindle shaft due to bearing stiffness

If the tool is not monolithic and there is an additional connection between the tool and the toolholder, the toolholder interface should be also considered as a sum of the deflections due to linear and rotational springs. This is further discussed in Example 5.8.

The receptance coupling substructure analysis (RCSA) method [53–57] can also be used to predict the tool point dynamic response for machine tools applications. This method considers the toolholder and tool as an assembly of two substructures. The tool interface receptance is determined analytically, while the toolholder–spindle interface receptances are determined experimentally using a standard test holder and finite difference calculations. The tool point dynamics is predicted from the RCSA by coupling rigidly the two substructures. This method includes the following features: (1) experimental identification of the toolholder–spindle interface translational and rotational receptances using a finite difference approach, (2) analytical determination of the toolholder–tool interface receptances, and (3) rigid coupling of the toolholder–spindle interface and toolholder–tool interface to determine the tool point response. This method eliminates the need to measure the tool–toolholder–spindle dynamics for each tool and holder combination in a particular machine and, therefore, significantly reduces the number of impact tests required for stability lobe diagram generation. The spindle-machine receptances can be identified experimentally in order to be coupled with the toolholder receptances using RCSA to estimate the FRF at the tool point [58].

The material removal rate that a machining center can achieve is strongly dependent upon the static and dynamic characteristics of the machine–tool–workpiece system as seen at the tip of the tool. The selection of the best spindle–toolholder interface is not simple because it depends on the intended use of the machine. The stiffness is one of the important parameters to consider in the selection. The static stiffness measures the deflection at the end of the tool in response to a static force. It provides some indication of the ability to create a surface or hole with the tool in the intended location. In milling, drilling, reaming, boring with a relatively low spindle speed, the error in location of the machined surface or hole is related to the static stiffness. The higher the static stiffness, the more accurately the surface or hole will be located with smaller form error. If static stiffness is the performance criterion, then the data show that the face contact connections perform better than the nonface contact connections in a machine tool spindle, as opposed to the results from a bench fixture showing that the HSK interface is superior to the CAT interface. Therefore, the static stiffness of the tool should be estimated by consideration of the spindle stiffness in order to properly select an interface style. The static stiffness can be estimated using the FEA or analytically assuming the joint stiffness parameters are available.

In other cases, the requirements are high metal removal rate (in milling, boring, etc.), or accurate surface location at higher spindle speeds. In both of these cases, the better selection criterion is dynamic stiffness (or chatter criterion as discussed in Chapter 12) as seen at the tip of the tool. The dynamic stiffness is a combination of stiffness and damping in a particular mode of vibration (at a particular frequency). These are the *modal* stiffness and modal damping. The damping is estimated from the FRFs while the natural frequencies and modal stiffness can be estimated using several programs, including commercial FEA solvers [24,50] and advanced codes, which consider bearing preload and other nonlinear effects [62]. For dynamic cutting performance at the tip of the tool, FEA can be also used assuming the joint parameters are available.

There are cases where the bore quality or surface flatness are very critical, in which case the quality of the interface taper and face including the quality of the tool point (runout) with respect to the taper interface are very important. In addition, the static bending stiffness becomes important assuming it is not a high-speed application. Rigid face contact interfaces provide higher static stiffness or normalized bending stiffness at higher moments assuming: (1) the taper quality for the nonface contact interfaces is worse than AT3 tolerance, and (2) the clamping force is high enough to sustain larger bending moments without losing contact between the toolholder and the spindle face.

5.4 CUTTING TOOL CLAMPING SYSTEMS

Cutting tools are generally held in the machine spindle by toolholders. Depending on the tool or cutter type, toolholders such as those shown in Figure 5.11, modular toolholders, standard collet chucks, or end milling holders, may be used. There are also several other approaches to hold round shank tools that deviate from the traditional collet chuck and eliminate the threaded locknut. These include hydraulic chucks, milling chucks, and shrink fit mountings. Toolholder dimensions outside the flange area and toolholder concentricity and tolerancing are not covered by ANSI standards and are left entirely to the manufacturer. Therefore, all toolholders are not alike.

Coolant-fed rotary toolholders are usually supplied through a center hole in the toolholder and less frequently from the flange of the toolholder (Figure 5.52). However, external coolant can be supplied through the tool using a rotary gland mounted either on the toolholder or the tool itself as shown in Figure 5.52.

5.4.1 MILLING CUTTER DRIVES

Milling cutters can be connected directly into the spindle if quick-change adapters (e.g., CAT-V, HSK, KM, Capto, or curvic coupling shanks) are attached as an integral part of the cutter body (Figure 5.53). Milling cutters with diameters up to 160 mm are generally mounted on a *stub*

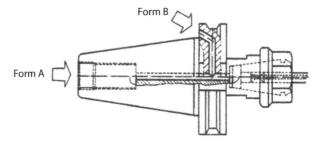

V-flange straight shank holder
Coolant: end entry through the spindle

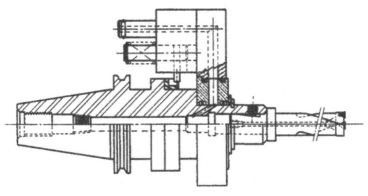

V-flange automatic tool change coolant gland
Coolant entry options:

1. End entry through the spindle
2. Coolant gland mounting for automatic tool
   change applications

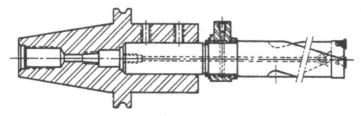

V-flange straight shank holder with rotary gland
Coolant: side entry

FIGURE 5.52 Typical coolant toolholder adapters for coolant-fed rotary cutting tools. (Courtesy of The George Whalley Company, Cleveland, OH.)

arbor toolholder and held in place by a single arbor screw (Figure 5.53). Arbors are available to fit all different spindle sockets. This method is very popular, especially for aluminum cutter bodies that must be mounted on steel toolholders. Face milling cutters with diameter over 160 mm generally mount directly to the face of the machine spindle. The two most popular methods of mounting face mills are the centering plug (flat back drive) method and the National Standard (mounting ring) method. In the centering plug method, a centering plug is used to locate the cutter on the spindle face. In the National Standard method, there is either a recessed locating diameter on the back of the cutter or a mounting ring fitted to a groove on the rear face of the cutter, which ensures proper location on the spindle face (Figure 5.53). The cutter is bolted directly to the spindle in both methods.

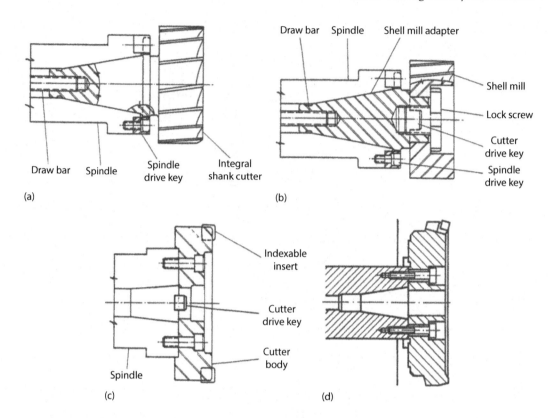

FIGURE 5.53 Types of milling cutter holding devices. (a) Integral shank shell mill drive, (b) stub arbor, (c) national standard face mill drive, (d) mounting ring. ((a–d) Courtesy of Valenite Inc., Troy, MI.)

5.4.2 Side-Lock-Type Chucks

Weldon chucks (Figure 5.11a) are used with round shank cutters with one or two Weldon flats and beveled corners. Weldon chucks are simpler and stronger but less versatile than collet chucks (described in the following). They were specifically designed to drive end-milling cutters but have also been used with indexable drills and boring tools. The standard Weldon chuck has one side-mounted clamping screw; the modified Weldon chuck has two clamping screws. The resulting close fit between the screw and the flat prevents end mill slip or pullout while providing maximum driving power. The cutter is held only by a portion of its periphery against a thin holding surface (the cylinder inside a cylinder design means that there is only one line of contact) as shown in Figure 5.54. Runout is often unavoidable and is equivalent to the sum of the manufacturing tolerances of the tool shank and the pilot bore; it may exceed 0.01 mm. Runout can be minimized by machining the pilot bore about 0.008 mm eccentrically with respect to the axis of rotation of the toolholder in the direction of the side screw. Using two clamping screws instead of one increases the static and dynamic stiffness of the system by as much as 50%.

Another type of side lock toolholding arrangement is the standard whistle notch (Figure 5.11a), a 5° inclined flat ground on the tool shank; the side set screws on the holder are not perpendicular to the tool axis, but inclined by 5°. This allows the tool to be pushed against a back stop. Back stop contact is not always achieved, which can result in loosening of the tool at high force levels. The whistle-notch method is sometimes used with drills.

Side-lock-type holders are unsuitable for high-speed operations because the pilot expands under centrifugal forces, resulting in a loosening of the tool–toolholder joint. The dynamic stiffness is low, particularly in the direction 90° offset from the set screw. Therefore, Weldon and whistle-notch

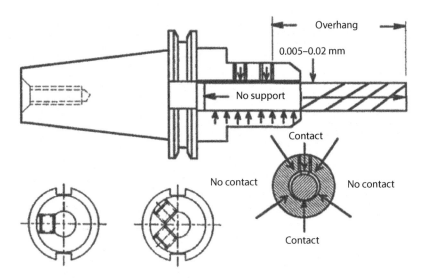

FIGURE 5.54 Clamping contact area for Weldon type holders.

system toolholders may be subject to vibration, resulting in poor surface finish and reduced bearing life. The modified Weldon system using two set screws on the same diametrical plane provides increased dynamic rigidity, and it is generally suitable for higher speed milling, but is not suitable for maximum speed and power applications.

5.4.3 COLLET CHUCKS

Collet chuck toolholders are commonly used for drills and reamers and may also be used with boring bars and end mills (Figure 5.55). These toolholders consist of a front socket, a collect, and a locknut as shown in Figures 5.56. Collets are cylindrical fixtures with one or more slots to generate flexible fingers designed to grip smooth cylindrical elements (Figure 5.57). The collet is pushed into a tapered socket, which causes the fingers to deflect and grip. *Non-pull* style end mill collets have a plug that contacts a flat on the end mill shank (i.e., a Weldon connection) to provide both positive drive and positive axial tool retention (Figures 5.57). Extended length positive pull-collets have

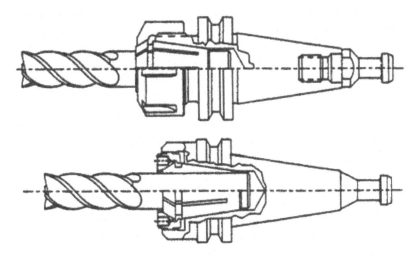

FIGURE 5.55 Single-angle collet chuck toolholders. (Courtesy of The Precise Corporation, Racine, Wisconsin.)

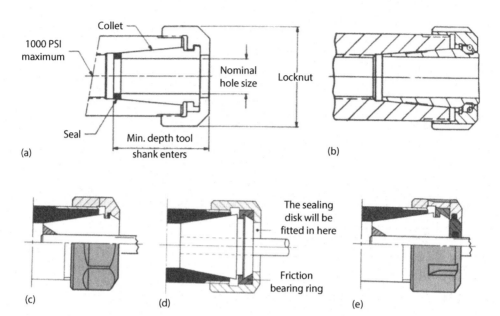

FIGURE 5.56 Single-angle collet systems: (a) STD nut, (b) ball bearing, (c) DIN 6499-D and E, and (d and e) friction bearing system. ((a,b) Courtesy of TSD DeVied-Bullard, Inc., Rockford, IL; (c,e) Courtesy of Rego-Fix Tool Corporation, Indianapolis, IN.)

been used in extended nose toolholders; in this arrangement, a draw nut or draw bolt is used to open and close the collet as shown in Figure 5.58.

Collets come is a variety of IDs to accept different tool shank sizes. The cutting tool shank is inserted into the collet fingers, which are compressed by tightening a threaded cap over the collet to secure the tool. Holding power around the shank is provided by compressive stress in concentric wedge-shaped collet sections. Holding power is increased by imposing axial motion on the tool shank as the collet is tightened. Gripping power is maximized when there is full length engagement between the tool shank and the collet bore. The grip power drops significantly when engagement becomes less than 2/3 of the full length of the collet bore.

The collet envelope size is a function of the maximum bore size and the outside shape designation. The envelopes for Erickson collets of increasing gage diameter are 50TG, 75TG, 100TG, and 150TG, which have maximum bore sizes of 12.5, 19.05, 25.4, and 38.1 mm, respectively. The minimum collet envelope relative to bore size should be used for high-speed applications because it balances the collet grip force and mass of the chuck. However, the maximum bore size in the collet envelope should be selected in general purpose applications to obtain maximum grip torque (e.g. a 12.5 mm collet provides only 60% the grip torque efficiency than that of the 19 mm shank in an Erickson 75TG envelope [60]). Using the largest envelope size collet allowed by the part and fixturing maximizes gripping power and stiffness.

Types of collets include single and double angle collets (Figure 5.57). Single angle collets are generally superior. There are several single angle collets with different cone angles; the most popular are the Rego-Fix ER style 16° (DIN 6499) included angle and the Erickson TG style 8° included angle. A lower included angle results in higher grip power (rigid tool grip) for the collet (e.g., an 8° collet provides twice the holding power of a 16° collet). The number and width of slots around the periphery determines the collapsibility or size range for the collet. A larger number of slots around the collet results in a larger collapsible range, better accuracy, and higher gripping power. When possible, a shank size equal to the nominal size of the collet should be used [60]. Collets with fewer slots designed to clamp only nominal diameter tools provide the highest concentricity. A light film of oil on the collet chuck socket cavity can increase grip power by as much as 40% and 50% using

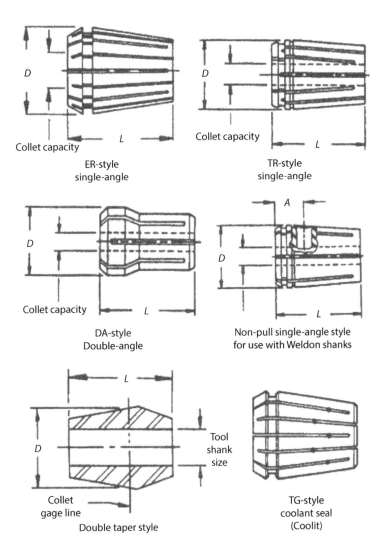

FIGURE 5.57   Collet types.

a steel and carbide tool shank, respectively. Carbide shanks generally grip slightly tighter than steel shanks. A double taper collet with one slot that does not extend through the wall thickness has low collapsibility (roughly 0.07 mm as compared to 0.3–1.0 mm of standard collets) but very high gripping power.

Double angle (DA) collet systems are used when maximum rigidity and close tolerance are not required (Figure 5.57), and when cost is a primary concern. Their runout ranges from 0.025 mm for larger collets (>3 mm) to 0.1–0.2 mm or higher for smaller collet sizes. A DA collet produces roughly half the grip force of a TG collet.

Collet accuracy is dependent on the taper fit and the roundness and surface finish of the collet and toolholder cone. The collet accuracy, concentricity tolerance, or TIR allowance is measured as an assembly as shown in Figure 5.59. The concentricity tolerance requirements for DIN 6499 [61], now replaced by ISO 15488 [62], are given in Table 5.10 measured in a chuck collet extension located on a "V" block or a precision spindle as shown in Figure 5.59. Classes I and II generally represent high precision and standard collets, respectively. The runout of standard single angle collets is normally between 0.01 and 0.02 mm but can be easily doubled or tripled when the holder is taken into account. The accuracy of precision single angle collets is about 0.008–0.010 mm.

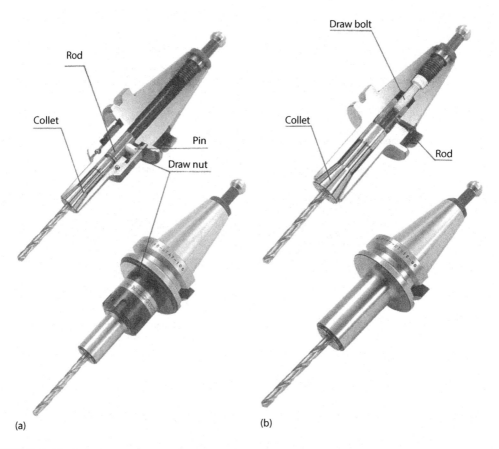

(a)   (b)

FIGURE 5.58 Extended positive pull-collet designs for extended toolholders with a 2 mm chucking range. (a) Draw nut design and (b) draw bolt design. (Courtesy of Tecnara Tooling Systems, Inc., Santa Fe Springs, CA.)

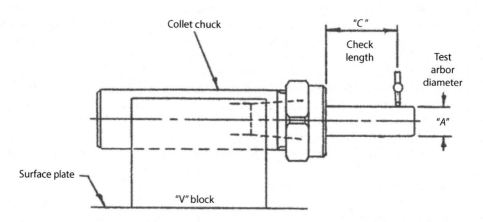

FIGURE 5.59 Collet chuck assembly total indicator runout measurement as properly noted in the DIN 6499 document. (Courtesy of TSD DeVied-Bullard, Inc., Rockford, IL.)

TABLE 5.10

Single Angle "ER" Assembled Collet Chuck Accuracy

"A" Rated Diameter (mm)		"C" Tool Overhand (mm)	Concentricity Tolerance (mm)			
Over	To		STD ER[a]	ER[a] Ultra Precision	Class I DIN 6499	Class II DIN 6499
1	1.6	6	0.010	0.005	0.008	0.015
1.6	3	10				
3	6	16				
6	10	25				
10	18	40	0.010	0.005	0.010	0.020
18	26	50				
26	34	60	0.015	0.010	0.015	0.025

Note: Using the procedure from DIN 6499 [58,59].

[a] Specified by Rego-Fix.

Gripping accuracy decreases as the shank size deviates further from the nominal collet bore; the accuracy of the extended collet in Figure 5.58 at an L/D ratio of 4 is between 0.01 and 0.02 mm, increasing as the tool shank deviates further from the nominal collet shank size. Growth during rotation affects slotted collets. For this reason, precision collets (accurate in terms of diameter and concentricity) are required for high-speed toolholders. The appropriate recommended nut tightening torque range should be used (100–135 N m for 75TG, 150–200 N m for 100TG and 200–270 N m for 150TG). As the collet is tightened on the tool by the nut, the collet must be forced into the tapered socket, which can cause the tool to move from 0.2 to as much as 0.7 mm. The grip force for the 100TG collet chuck is between 300 and 600 N m. A 40% loss of grip can occur with a metal backup screw; much lower losses occur with plastic backup screws.

Several uniquely designed coolant-fed collet chucks, which seal the adapter, and cutting tool are available. These include coolant-control collets, such as (1) collets with slots filled with silicon rubber, (2) coolit collets using rubber composite plugs in the slots (Figure 5.57), and (3) collets with slots that are open at the small end and do not intercept the groove at the front of collet. Special washers can also be used to seal the collet; typical arrangements include (4) a nylon back-up screw at the cutter end, (5) a nylon seal at the back of the collet (Figure 5.56a), and (6) a metal-nylon washer seal at the front of collet in the nut (Figure 5.56e). Types 1, 2, and 3 work well, with little leakage if any, when the collet size matches the tool diameter and minimum collet collapse is required. Type 1 is often avoided because the tool concentricity can be substandard if the silicon in the slots is not perfectly deposited. Type 4 seals the back end of cutting tool. The cutting tool should be designed to match the seal; a common center oil-coolant hole or multiple oil-coolant holes without grooves should be used. Similarly, a O-ring assembled at the tip of the backup screw can be used to seal the end of tool shank. Type 5, which compresses the nylon against the collet (Figure 5.56a), may reduce the grip power of the tool by 20%–30%.

Coolant can be supplied through the collet to the cutting area in four different ways (Figure 5.60): (1) through an oil hole in the cutting edge, (2) through adjustable nozzles mounted on the coolant cap, (3) through the space between tool shank and spacer, and (4) through the slits of collet. Methods 2, 3 and 4 are used for tools without oil holes.

The clamping locknut design also has an important impact on collet chuck performance. The clamping locknut provides the preload for the collet, and thus determines the chucking pressure. A standard nut is attached at the front of the collet by engaging the extractor spiral in the collet groove (Figure 5.56a). Another standard design described in DIN 6499-D and E [61,62] is shown in Figure 5.56c. A ball bearing clamping locknut (Figure 5.56b) is often used to provide increased chucking pressure [63,64], which can result in a 40% increase in gripping capacity and better

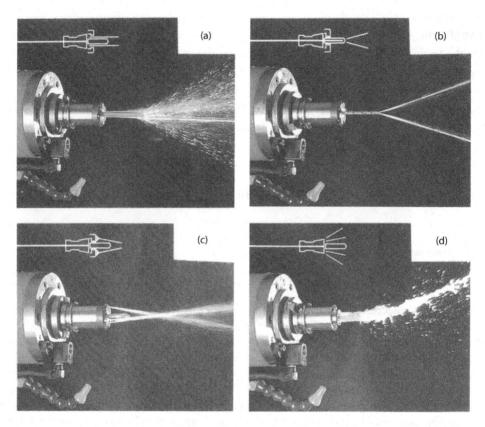

FIGURE 5.60 Coolant through systems for collet type toolholders. (a) Coolant around tool, (b) coolant through cutter, (c) coolant through nozzle, and (d) coolant through collet. (Courtesy of Tecnara Tooling Systems, Inc., Santa Fe Springs, CA.)

concentricity compared to standard locknuts with the same tightened torque. Roller bearing clamping nuts are also used to increase gripping power. A friction bearing system (Figures 5.56d and e) has also been used to improve clamping power. A comparison of the transfer torque for these nuts is shown in Figure 5.61. In high-speed applications, high-performance locknuts without an extractor spiral are used to eliminate unbalance. Although collet chucks generally provide good accuracy and dynamic stiffness, their utility at very high speeds is limited. The external collet nut expands under centrifugal force, so that the collet loses some gripping torque and may vibrate and loosen when the spindle is decelerated quickly. Better designs employ an internal nut (Figure 5.55b) [16] or no nut (Figure 5.58); at high speeds, the internal nut expands into the thread, tightening rather than loosening the connection. These designs are suitable for comparatively high power and speed applications. The clamping torque for a 13 mm tool shank in the extended positive pull-collet without nut, shown in Figure 5.58, is 95 N m.

The chucking pressure of the collet at the nose depends on the locknut type. A standard nut generally provides low pressure near to the nose, as compared to zero chucking pressure at the nose for the ball bearing locknut. The maximum chucking pressure for the ball bearing locknut occurs at about one-quarter of the collet length from the nose [63]. The collet shown in Figure 5.58 should provide significant pressure at the nose.

5.4.4 HYDRAULIC CHUCKS

In hydraulic chuck toolholders, the wall of the inside diameter of the toolholder (an internal membrane) expands when a hydraulic pressure is applied using grease or oil, then returns to its original size when the pressure is released as explained in the toolholder/spindle connection section

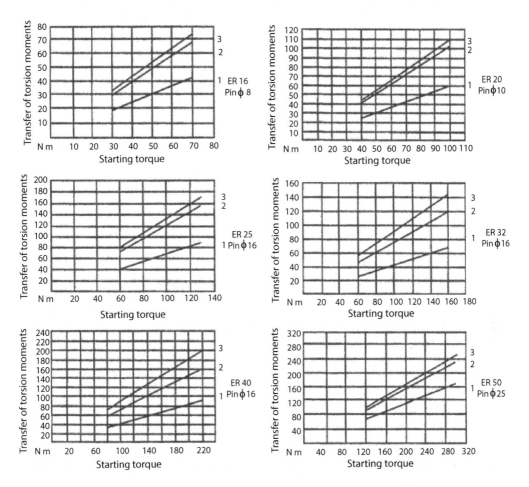

FIGURE 5.61 Torque transmitted by three different collet locknut designs: 1, standard nut; 2, ball bearing nut; and 3, friction bearing nut. (Courtesy of Rego-Fix Tool Corporation, Indianapolis, IN.)

(Figures 5.11 Type J and 5.62) [65]. A screw is turned a few revolutions to actuate a piston and force hydraulic fluid from the piston chamber into the expansion sleeve at 70–140 MPa, which clamps the tool shank with uniform pressure on its geometric centerline over the full length of engagement.

Hydraulic chucks provide very good concentricity (0.005 mm or better TIR at $2 \times D$ to $3 \times D$ from the nose of the chuck) and high repeatability (0.0013 mm or better). The runout can increase by 50%–100% when precision sleeve collets are used to step down to smaller diameters with the same chuck. The normal amount of contraction for a 25 mm diameter toolholder is 0.01–0.03 mm. Contraction is proportionately higher or lower for larger or smaller diameters. The hydraulic bore is made 0.005 mm larger than the nominal diameter, while the toolholder shank has a recommended tolerance of H6 and at maximum $D/1000$. A closer fit between the cutter shank and the sleeve increases the gripping power. This arrangement provides a balanced system for medium to higher speeds.

Hydraulic toolholders provide sufficient pressure to drive the tool shank without a flat or a tang; the clamping force for a 70–85 MPa expansion pressure can transmit a torque of 60 and 500 N m, respectively, for 10 and 32 mm toolholder shanks. The torque capability is calculated from the available pressure using Equation 5.5. In several designs, the tool shank is clamped at the top and

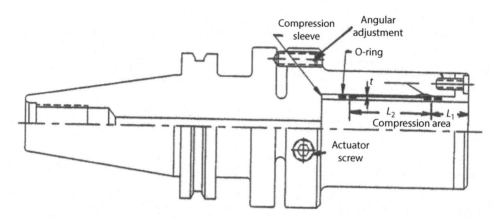

FIGURE 5.62 Hydraulic expansion toolholder with angular adjustment. (Courtesy of Hydra-Lock Corporation, Mt. Clemens, MI.)

bottom of the sleeve while the middle part provides stiffness. The hydraulic fluid in the sleeve acts as a damping agent and impact cushion and, therefore, hydraulic holding systems are relatively free of vibration and chatter. However, hydraulic chucks provide relatively low radial stiffness in the hydraulic membrane since the tool practically *floats* in the compression sleeve; they are thus not good for heavy side-milling applications. The compression sleeve (length L_2 in Figure 5.62) is some distance L_1 behind the nose of the chuck, which also reduces the stiffness of the cutter. Tools with Weldon shanks and/or whistle-notch flats are not acceptable because they would destroy the bore sleeve, unless they are used in a sleeved collet toolholder. There is no axial drawback of the tool shank as the chuck is tightened.

5.4.5 MILLING CHUCKS

Milling chucks use a broken or unbroken clamping surface principle, which permits uniform expansion and contraction of the clamping surface. Typical designs feature a Roll-Lock mechanism using a single or double set of needle rollers and an outer guide ring (Figure 5.63). The needle rollers are located between the bore, which acts like a master collet, and the movable guide ring. When the chuck is tightened, an inward uniform force is created, squeezing the chuck body inward and wrapping it around the tool shank.

The gripping power is dependent on the design characteristics and the ID size of master sleeve as shown in Figure 5.64. The chucking force is also affected by the design of the master sleeve, which is often slotted and/or grooved annularly to provide increased shrinkage of the bore and to increase the frictional resistance in the chuck by eliminating the effect of any oil film on the tool shank. The gripping force increases linearly with the guide ring travel distance. The addition of an extra row of roller bearings, five instead of four, increases the gripping power by as much as 30%. The gripping power for a double roller bearing chuck is in principle 1.5–2 times larger than that of a single roller bearing chuck, although some evidence indicates the actual gripping force may be lower [64]. It is preferable to use the smallest clearance between the tool shank and the bore of the master sleeve (<.030 mm) to obtain sufficient clamping force at the nose of chuck. The clamping range is 0.2%–0.3% of the relevant clamping diameter. The maximum shrinkage can be extended closer to the nose by using special nose slots as shown in Figure 5.63, which is especially important when using small diameter end mills. Various kinds of milling chuck toolholders are available, which differ in the collet chuck design as shown in Figure 5.64 [64]; Type B provides the highest pressure, while type A provides the widest distribution as shown in Figure 5.64.

The allowable tool runout is given in Table 5.11. Milling chucks have less than 2 μm tool length movement in the chuck during tightening.

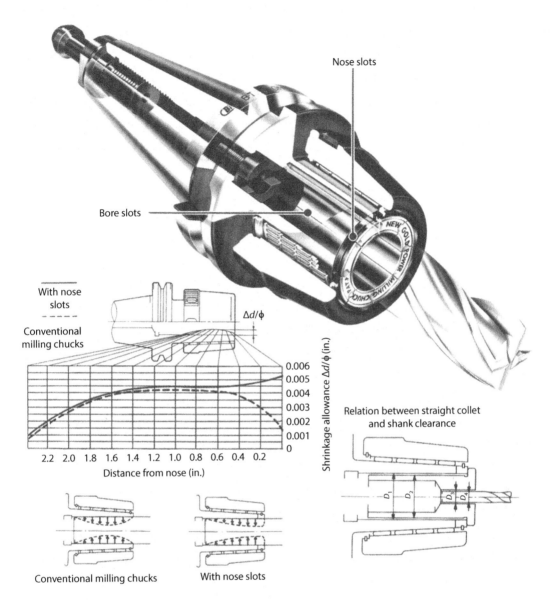

FIGURE 5.63 Milling chuck with double fine slots for better accuracy and high shrinkage allowance at nose. (Courtesy of Stanley Sheppard Company Inc., Berlin, CT.)

5.4.6 SHRINK-FIT CHUCKS

A highly reliable type of toolholder, comparable in performance to a monolithic toolholder (Figure 5.65b), is the shrink-fit chuck (Figure 5.65c). In shrink-fit chucks, standard straight shank tools are shrunk into the toolholder body with a high interference fit as explained in the toolholder/spindle connection section (Figure 5.11 Type K). To insert the tool into the holder, the holder is rapidly heated (using induction or an electromagnetic field) until the tool shank will slide into the holder. As the toolholder cools, the resulting thermal contraction exerts uniform pressure around the entire surface of the tool shank. The shrinking process is reversible for carbide shanked tools, but not for steel tools. Shrink-fit toolholders are typically designed and tested to last 20,000–50,000 tool change (shrink-unshrink) cycles. The amount of interference can be optimized based on application requirements, and the magnitude of the shrink-fit, the unit pressure, tangential stress generated by the fit, and the

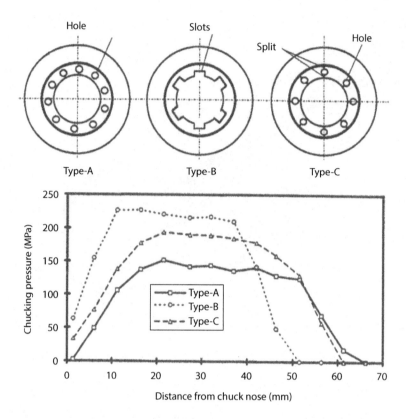

FIGURE 5.64 Various nose slot styles for milling chucks. (From Tsutsumi, M., *J. Mech. Working Technol.*, 20, 491, 1989.)

TABLE 5.11
Milling Chuck Runout Allowance and Gripping Torque Values

| Chuck Size | Maximum Allowable Runout (μm) | | Gripping Torque (N m) | |
	At Nose End	At 100–200 mm Out	Maximum	Hand Tightened
19			1000–3000	350–400
32	5	15	2500–4500	500–800
50			3000–5000	600–1100

torque transmitted can be calculated from Equations 5.3 through 5.5. Generally, the interference should be less than 0.01 mm to avoid excessive compressive stress on the tool shank and to increase the holding ability and safe operating stress in the toolholder nose section. In continuous dry cutting (end milling long travels or drilling several holes consecutively) of ferrous metals, which produces a build-up of temperatures in the tool and holder, the difference in the coefficient of thermal expansion for the tool and holder materials for carbide-shank tools will cause a decrease in the effective amount of interference; this must be considered in the design of the joint. A change in shrink pressure due to high centrifugal forces should also be considered for high-speed toolholders because the transmitted torque could drop by 70%–80% as the speed increases from zero to 40,000 rpm with an H5/H6 tool shank tolerance. The tool shank diameter must be controlled to a relatively tight tolerance to ensure effective gripping power and torque transmission. An H6 tolerance is adequate for typical applications, with H5 tolerance being recommended for higher speed operations to provide a safer minimum clamping torque.

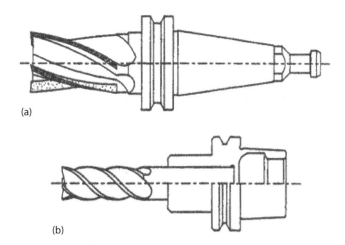

(a)

(b)

FIGURE 5.65 Advanced toolholder designs: (a) monolithic toolholder and (b) shrink-fit toolholder. (Courtesy of Precise Corporation, Racine, Wisconsin.)

The shrink-fit interface effectively provides an integral shank tool, virtually eliminating tool deflection, vibration, and slippage at the interface [7,8]. Because the thermal expansion and contraction in the toolholding system is symmetric, it provides the best possible runout, balance, and gripping torque combination characteristics. It provides inherent balance due to its symmetric design without nuts, collets, etc. Grooves around the toolholder ID can be provided for coolant access around the tool. Shrink-fit toolholders are especially advantageous in extended length or reach applications.

5.4.7 PROPRIETARY CHUCKS

There are also proprietary solutions to precision chucking, such as the Tribos, powRgrip, and CoroGrip systems, as well as various other interfaces using ultra precision collets with special nut designs (e.g., HP collet chuck shown in Figure 5.66, XT precision collet system, etc.). Several of these chucks were developed to overcome the concerns with the shrink-fit systems because they do not require high temperatures to work correctly and no cool-down period is necessary as with shrink-fit chucks. They usually employ unique mechanical activating designs that develop significantly more clamping pressure than conventional collet chucks while providing the high accuracy of shrink-fit chucks.

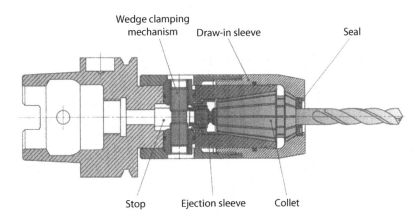

FIGURE 5.66 High performance *HP Series* collet chuck. (Courtesy of Parlec, Fairport, NY.)

1. Unclamped position

2. Application of force

3. Tool shank insertion

4. Tool clamping

FIGURE 5.67 Precision polygonal mechanical actuated *Tribos* toolholder chuck. (Courtesy of Schunk Inc., Lauffen/Neckar, Germany.)

The *Tribos* system [66,67] is a tri-polygon clamping system (shown in Figure 5.67). Its ground polygon (three-lobed) shaped bore rigidly clamps the tool shank in three places. A clamping device mechanically deforms this bore (generating precise force at three points on the toolholder) to make it round, allowing for the tool shank to be inserted; unclamping the device will return the clamping diameter to its initial three-lobed shape and clamp the tool shank at three locations under very high forces. The Tribos system provides about the same accuracy as a shrink-fit chuck (TIR of 0.003 mm at $2 \times D$ to $3 \times D$ from the nose of the chuck) because the tri-polygon bore is ground as a round bore under proper loads with the same accuracy as the shrink-fit bore. It requires tool shank tolerances similar to those used for shrink-fit chucks.

The *powRgrip* system [68] consists of a toolholder, precision collet (with 1:100 taper angle), and a hydraulic setup device to insert the collet and tool into the holder. A press is used to expand the holder nose slightly and push the collet with the tool into the toolholder nose. The maximum TIR is expected to be 0.003 mm at $4 \times D$ out from the nose [68]. The collet has a built-in set screw for presetting the tool length. The tool shank should have an H6 tolerance. This system is available for clamping diameters between 3 and 20 mm.

The *CoroGrip* system [69] consists of a thin-wall nose with a movable sleeve on a 2° taper angle. A high hydraulic pressure (10 kpsi) is used to push the outer sleeve up on the taper when the tool is clamped, and down on the taper when it is unclamped as shown in Figure 5.68. This system is similar to milling chuck or shrink-fit chuck but uses sliding surfaces to collapse the inner body of the nose onto the tool shank. The TIR is expected to be 0.002–0.005 mm at $4 \times D$ from the nose [69].

The *SINO-T* system [66,70] is a universal toolholder design more slender than the standard Weldon and collet chuck holders. Its concept is similar to the hydraulic toolholder because it consists of an expansion sleeve and an expansion chamber with elastic medium (made from a solid body for stiffness). It clamps mechanically by tightening a clamping sleeve axially, which squeezes the elastic medium that transfers a clamping force uniformly over the clamped tool shank. Torques up to 95 and 290 N m can be transferred to tool shanks with quality H6, respectively, for 12 and 20 mm chucks. Its runout accuracy at the clamping bore is 0.005 mm. Its advantage over standard

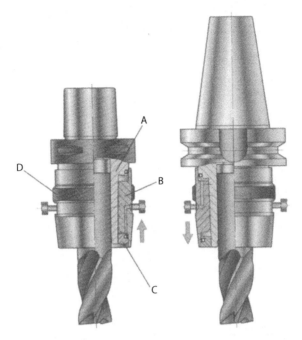

FIGURE 5.68 Precision mechanical (hydraulic actuated) *CoroGrip* chuck. (Courtesy of Sandvik Coromant, Fairlawn, NJ.)

Weldon and collet chucks results from the vibration damping characteristics of the elastic medium; it provides good stiffness for end milling. In addition, the front cup is tightened to a dead stop and cannot be overtightened, as the nuts in collet chucks can be. The axial drawback of the tool shank is a few microns as the chuck is tightened.

As discussed before, the Weldon shank has been the best gripping power over all the aforementioned toolholder–tool interfaces discussed, and tools do not pull out of side-lock toolholders with Weldon shanks. In spite of the high clamping power generated by several of the aforementioned toolholders, pullout might occur due to high loads in extreme cases or due to out of tolerance tool shanks. However, in order to overcome the disadvantages of the Weldon chuck with respect to runout (that limits tool life and part quality), some manufacturers developed special antipullout toolholders with special feature in the ID bore while either using the Weldon tool shank or special tool shank that securely holds the tool. Furthermore, the aforementioned high power clamping systems have still concerns in certain cases with high-helix end mills pulling out the toolholder; the high-helix end mills have the tendency to pull out of the toolholder when pocketing around the corners at high feeds even with shrink-fit or hydraulic toolholders [71,72]. These are

1. The *secuRgrip*, which uses threaded inserts in the Weldon shank end mill flats that match the threads in the collet toolholder similar to the powRgrip system [72],
2. The H*ydro-sure*, which uses a hydraulic toolholder that accepts the Weldon shank [72]. It incorporates the Weldon style side-lock screw in the toolholder; this provides the security of Weldon shank together with the accuracy of a hydraulic toolholder.
3. The Safe-Lock system, which uses carbide drive pins at the bottom of the toolholder. The specially shaped grooves engage with the special grooves in the tool shank preventing pullout. This design can be used in hydraulic, shrink-fit, or precision collet styles toolholders.

There are also a number of adjustable or steerable toolholders available, which can be used to minimize rotary tool runout in critical applications. In the Kyocera-Unimerco system (Figure 5.69), the toolholder is fitted with two rows of adjustment screws, which contact ground angled flats on the tool shank.

FIGURE 5.69 Steerable toolholder for transmission spool bore reamers and other rotating tools with critical runout requirements. Tightening the two rows of bolts in sequence while measuring runout with an indicator can be used to set up such tools with less than 5 μm TIR runout. (Courtesy of Kyocera Unimerco, Saline, MI.)

The screws are tightened in a prescribed procedure while checking runout with an indicator. Optimally the setup is done with the tool in the spindle rather than a setup gage. These holders can be used to reduce runout on even long tools, for example, spool bore reamers for automatic transmission valve bodies, to less than 5 μm TIR. Steerable toolholders often require nonstandard ground features on the tool shank.

5.4.8 TAPPING ATTACHMENTS

Solid toolholders with no axial or radial movement are used in synchronous tapping on CNC machining centers; the tap is typically clamped using collet chucks, shrink-fit chucks, or quick-change adapters.

On older machines without synchronous tapping capability, tapping attachments are used to assist synchronization of the machine feed (lead screw) with the spindle rpm to match precisely the pitch of the thread. The tapping device has the greatest influence on tapping performance. Tapping attachments can be held directly in the spindle shaft or toolholder or can be integrated into the toolholder body. Tapping heads are designed with features such as axial compression, axial tension, radial float, torque control, and provision for quick replacement of dull taps (Figure 5.70) [73]. Axial compression and tension cushions the tap at the entrance and bottom of a hole, while preventing double threading, and compensates for excessive feed rates when entering the hole.

Radial float can be designed into a chuck to account for location misalignment between the tap and the hole. Torque control assists in the bottom tapping of holes and prevents breakage of dull taps. The threaded depth is controlled to within one tenth of a revolution.

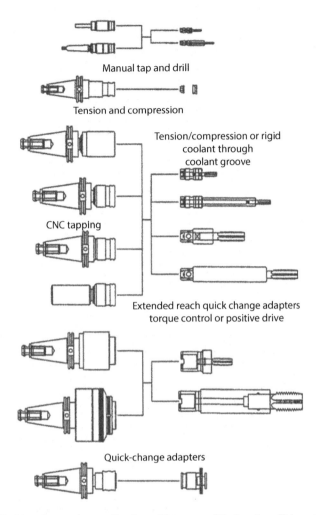

FIGURE 5.70   Toolholders for tapping applications. (Courtesy of Parlec, Inc., Fairport, NY.)

Self-reversing toolholders eliminate spindle reversals; they reverse the tap without reversing the machine spindle. The advantages of this chuck are pronounced with smaller tap sizes when the spindle rpm is high, and, therefore, the acc/dec time for the spindle is large. Self-reversing systems typically operate at under 5000 rpm.

5.4.9   REAMING ATTACHMENTS

Reaming attachments are used to assist the alignment of a reamer in a hole. Floating reamer holders are used to compensate for radial, parallel, and angular movements (centerline deviations between the machine spindle and the workpiece hole) necessary for the reamer to be positioned parallel and perpendicular to the existing hole axis; otherwise, the reamed hole squareness and angularity can be changed by the reamer.

5.4.10   COMPARISON OF CUTTING TOOL CLAMPING SYSTEMS

One of the main criteria when selecting the toolholder is the runout error at the cutting edges. The radial runout affects the equality of the radial cutting forces among the cutting edges especially in a peripheral milling cutter. It also affects tool life directly as illustrated in Figures 5.71 and 5.72. The graph in Figure 5.71 shows that precision chucks with runout at the tool point below 0.01 mm are

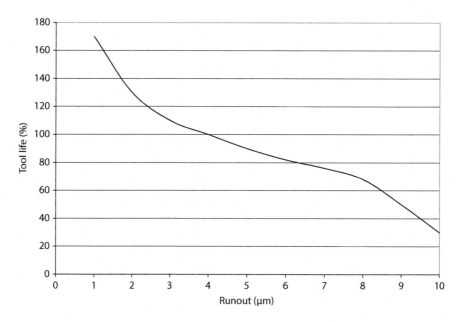

FIGURE 5.71   Influence of tool runout (TIR) on tool life. (Courtesy of Rego-Fix, Indianapolis, IN.)

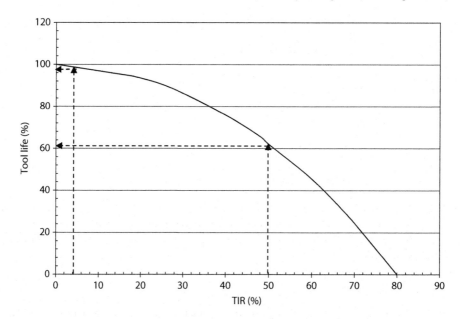

FIGURE 5.72   Influence of tool runout (TIR) as a percentage of uncut chip thickness on tool life. (Courtesy of Rego-Fix, Indianapolis, IN.)

desirable for high accuracy and/or high-speed applications. Figure 5.72 shows tool life reduction as a function of runout given as a percentage of the uncut chip thickness.

It is important to note that the runout values provided in catalogs for different toolholders represent the runout of the chuck with a perfect tool and toolholder shank or spindle. It is usually measured in a precision setup (e.g., air bearing spindle, etc.). For example, a measurement of eleven identical HSK-A63 precision toolholders (using a straight collet on a 10 mm carbide end mill with $3 \times D$ overhang) with a runout specification of 0.003–0.005 mm at $4 \times D$ was performed by the manufacturer using the proper setup, and the tool runout range was found to be 0.002–0.006 mm.

The same toolholders were measured in a presetting machine (using a low-clamping force for the HSK shank) and the runout range was 0.002–0.024 mm. The nominal runout of a complete tool–toolholder–spindle system is estimated using the best/worst case scenario approach as explain in Example 5.1 and illustrated in Figure 5.17. There are several different types of toolholder–tool interfaces within each style of interface. For example, every manufacturer has toolholder with standard collet quality and several also have precision collet types (as shown in Table 5.10). The accuracy of precision single angle collets is generally specified as 0.005 mm at $4 \times D$ from nose but on the spindle with CAT-V toolholders (with AT3 taper) the error can be as high as 0.013–0.020 mm.

The roundness of a 15.5 mm diameter carbide rod mounted in four different toolholding chuck systems, listed in Table 5.12, was compared against a gage mandrel using the HSK-80 spindle interface with all toolholders. Two toolholders were evaluated for 100 repetitions with each chuck style; the spindle was started and stopped between each trial. The minimum, maximum, range and standard deviation are given in Table 5.12 for each toolholder. The best roundness was obtained with the hydraulic chuck.

The runout for several toolholders was evaluated with a solid tool at $3 \times D$ from the nose and the results are shown in Table 5.13. These toolholders were measured by four different manufacturers,

---

TABLE 5.12

Roundness of Several Tool Chucks Using an HSK-80 Toolholders and a 15.5 mm Carbide Diameter Rod

	Roundness (µm)			
Toolholder Style	Min	Max	Range	StDev
Gage mandrel (1)	1.1	2.8	1.7	0.25
Gage mandrel (2)	0.9	2.6	1.7	0.25
Gage mandrel (3)	0.5	2.2	1.7	0.29
Standard collet, ER 40/16	8.8	31.2	22.4	3.47
Standard collet, ER 40/16	22.2	82.2	60.0	8.85
Precision collet, ER 40/15.5	1.5	10.3	8.8	1.62
Precision collet, ER 40/15.5	4.1	28.1	24.0	4.36
Hydraulic chuck	1.5	3.3	1.8	0.35
Hydraulic chuck	5.2	8.7	3.5	0.55
Whislow-notch in	6.7	9.4	2.7	0.54
ABS-FWD adapter	0.9	6.2	5.3	0.93

Source: Courtesy of Alfing Corporation, Aalen, Germany.

TABLE 5.13

Evaluation of Runout at $3 \times D$ from the Nose for Several Toolholder–Tool Interfaces

	Runout (mm)			Toolholder Accuracy (mm)	
Toolholder–Tool Interface	Minimum	Maximum	Average	Concentricity	Roundness
Milling chuck	0.0025	0.023	0.0010		
Hydraulic	0.0005	0.0076	0.0029		
Shrink fit	0.0013	0.0076	0.0040		
Collet DA-180	0.0023	0.1524	0.0386		
Collet ER-25	0.0005	0.0330	0.0087	0.00384	0.00106
Collet TG-100	0.0005	0.0127	0.0054	0.00480	0.0016
Weldon	0.0005	0.0064	0.0046	0.00450	0.00115
Collet precision	0.0005	0.0152	0.0052		

three times each, and the minimum, maximum, and average values for each toolholder–tool interface in given in Table 5.14. The runout of some of these toolholders is higher than expected based on the manufacturer specifications.

Toolholder quality was evaluated based on the concentricity and roundness of the bore I.D. to toolholder taper measurements for several end mill and collet toolholders. The results are shown in Table 5.14. The quality of the CAT-40 toolholders was 50% better than that of CAT-50 toolholder. However, the quality for the HSK-A63 was about the same as that of an HSK-A100 toolholder. In addition, the quality of the CAT-40 toolholders was better than the HSK toolholders.

The static and dynamic bending and torsional stiffness measured using bench fixture tests with a CAT-50 as the basic toolholder, and using 6.35 mm HSS and carbide tooling with an overhand length of 38.9 mm ($L/D = 6$), are shown in Figures 5.73 and 5.74, respectively. The static and torsional bending for the collet type holders (using either 6.35 or 7 mm tool size collets) was better than the other type holders (i.e. Weldon, hydraulic, milling chuck, etc.). In addition, the TG collet type had better stiffness than the ER type collet. The dynamic (bending and torsional stiffness defined as the product of modal stiffness with damping ratio) performance of the hydraulic type holder is not as good as expected compared to the other type holders, but its static bending stiffness is better than expected, even though the tool floats in the expansion sleeve. The milling chuck type holder performed unexpectedly poorly compared to the collet and Weldon type holders due to collet size reduction characteristics. The stiffness of the carbide tooling is at least twice that of HSS as expected due to carbide's higher modulus of elasticity.

Performance for 12.7 mm toolholders with 108 mm overhang ($L/D = 8$) is shown in Figures 5.75 and 5.76, respectively, for HSS and carbide tooling. Toolholders for a 12.7 mm tool performed somewhat similar to the 6.35 mm toolholders. The torsional characteristics of the shrink-fit and hydraulic holders are better than the other holder types. The wedge type end-milling holder provides better dynamic characteristics than the Weldon type holder. The Morse taper holder performed better than expected. The larger size milling chuck tends to perform better than the smaller size as also observed with the 6.35 mm tooling. The performance of the tooling in hydraulic holders with collets is worse than without collets. In addition, the modal stiffness decreases while the dynamic stiffness increases with a decrease in the clamping force for both styles of interface.

The results for 25.4 mm tooling interfaces are given in Figures 5.77 and 5.78 for the HSS and carbide tooling with $L/D = 5.6$. Again the TG collet type holders performed as well as most of the other type holders. The modular interfaces with both cylindrical and taper face contacts (i.e., HSK, CAPTO, KM, ABS, BETA, Tru-Taper) of size 50 perform somewhat better than the other interfaces but in some cases not significantly so. In addition, the expected superior performance for the milling chuck was not observed and the hydraulic holder exhibited the highest dynamic and lowest static stiffness as shown in Figure 5.78.

The transmitted torque for four different chucks as a function of tool shank diameter is shown in Figure 5.79. The clamping transmitted torque increases with shank diameter, but there is a large scatter among different manufacturers due to differing design characteristics among similar interfaces and the tolerances of the bore and the tool shank. Generally, all high-performance interfaces (shrink-fit, hydraulic, etc.) require an H6 shank tolerance to achieve optimum results. The standard collet chucks and some of the high-performance collet chucks accept H6-H11 tool shank tolerances. The collet chucks provide higher clamping torque than some of the high-performance interfaces for tool shank diameters smaller than 12–14 mm. In addition, the clamping transmitted torque for many high-performance chucks drops by about 5% for every 0.002 mm drop in tool shank diameter.

A comparison between toolholder–tool interfaces is given in Table 5.15 using several criteria. The accuracy results are based on a CAT-40 (AT3 taper) while the transmitted torque capability is based on 20 mm tool shank. The runout accuracy of shrink-fit, hydraulic, milling chucks, and high-performance collet systems (using high precision collets) is comparable and is 2–10 times better than that of standard collet and side-lock chucks. The clamping pressure and torque for shrink-fit

TABLE 5.14

Evaluation of the Concentricity and Roundness of the Bore I.D. to Toolholder Taper from Several Toolholders Measured in a Roundness Instrument

	CAT 40		CAT 50		HSK-A100		HSK-A63	
	Concentricity (mm)	Roundness (mm)	Concentricity (mm)	Roundness (mm)	Concentricity (mm)	Roundness (mm)	Concentricity (mm)	Roundness (mm)
1 EM	0.00119888	0.0016256	0.00856361	0.0012014	0.00191897	0.0006896	0.00306324	0.000428
1 EM	0.00324993	0.0010541	0.0024892	0.0006922	0.00691769	0.0013576	0.00828548	0.0024346
1 EM	0.00058928	0.0004699	0.00531114	0.0008344	0.00218567	0.0019164	0.00313563	0.000635
1 EM	0.00174879	0.0006528	0.0094996	0.0012484	0.00300609	0.0012408	0.00522986	0.0008141
1 EM	0.0033274	0.0006096					0.00180213	0.0011697
.5 EM	0.00154051	0.0012027	0.00784225	0.0007976	0.00747903	0.0005258	0.00209677	0.0016688
.5 EM	0.0022098	0.0018783	0.00217805	0.000616	0.00630555	0.006731	0.01399921	0.0013729
.5 EM	0.00221996	0.0004458	0.000077851	0.0005525	0.00246507	0.0014999	0.0053086	0.0008103
.5 EM	0.00524929	0.000997	0.01011682	0.0005893	0.00378206	0.0014719	0.00276225	0.0009258
.5 EM	0.0042926	0.0004724					0.002159	0.0008865
TG 100	0.00395605	0.0008992	0.00083185	0.000823	0.00254	0.0011925	0.00159385	0.0012916
TG 100	0.0015494	0.0012281	0.00155321	0.001632	0.00427863	0.0016764	0.0065405	0.0011341
TG 100	0.00568833	0.0036386	0.00704342	0.0006058			0.00294894	0.0007836
TG 100	0.00279908	0.0013411	0.004953	0.0006033				
TG 100	0.0013716	0.0006858						
TG 100								
ER 16	0.00536829	0.0029477	0.00477279	0.0012014	0.00296672	0.0011811	0.00401447	0.0004458
ER 16	0.00309753	0.0009068	0.0021971	0.0009182	0.00247523	0.0011887	0.00797941	0.0006693
ER 16	0.00056515	0.0003988	0.00202565	0.0015685	0.00507873	0.0008446	0.00299339	0.0010325
ER 16	0.00463169	0.0009982	0.00523875	0.0010579	0.00238633	0.0012573	0.00379857	0.0007404
ER 16	0.004064	0.0009525	0.00672592	0.0010998			0.00257302	0.0007722
1 SF					0.00838581	0.0013475		
.5 SF					0.00824865	0.0008331		
Average	0.00011559	4.607E-05	0.00031725	8.526E-05	0.00020966	6.944E-05	0.00027505	6.73E-05

Note: 1 EM = 25.4 mm end mill, 5 EM = 12.7 mm end mill, TG 100 = TG collet, ER 16 = ER collet, and 1 SF = 25.4 mm shrink fit, 5 SF = 12.7 mm shrink fit.

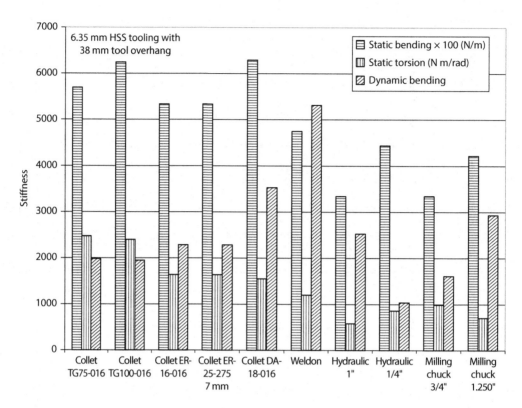

FIGURE 5.73 Comparison of static and dynamic stiffness of toolholder interfaces for 6.35 mm HSS tool.

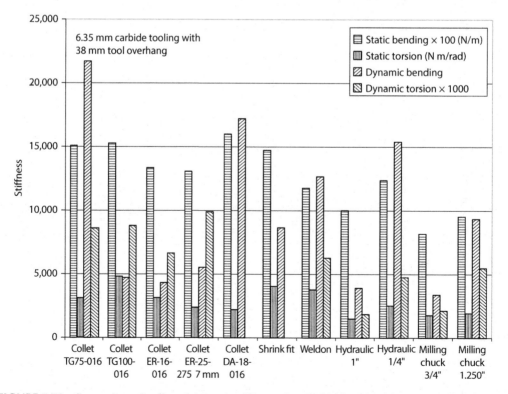

FIGURE 5.74 Comparison of static and dynamic stiffness of toolholder interfaces for 6.35 mm carbide tool.

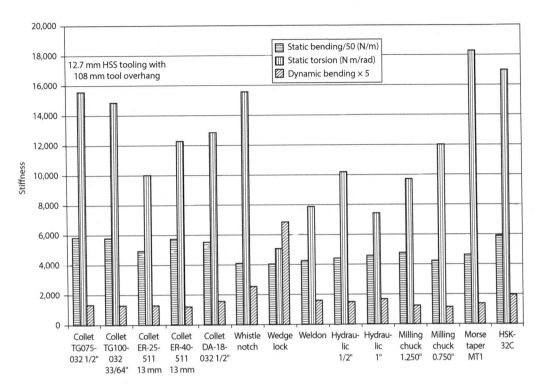

FIGURE 5.75 Comparison of static and dynamic stiffness of toolholder interfaces for 12.7 mm HSS tool.

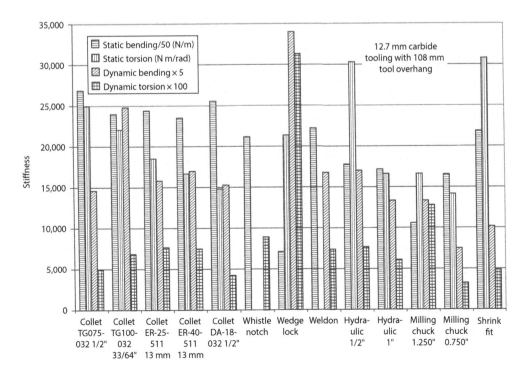

FIGURE 5.76 Comparison of static and dynamic stiffness of toolholder interfaces for 12.7 mm carbide tool.

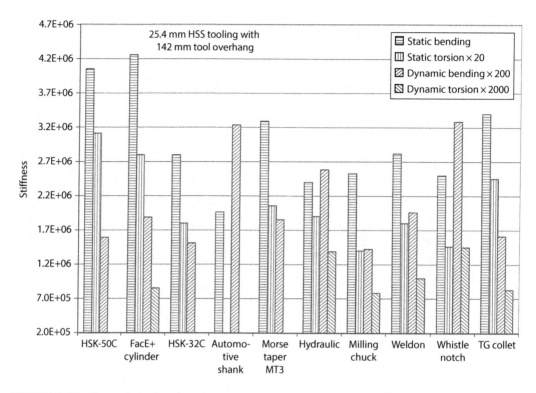

FIGURE 5.77 Comparison of static and dynamic stiffness of toolholder interfaces for 24.5 mm HSS tool.

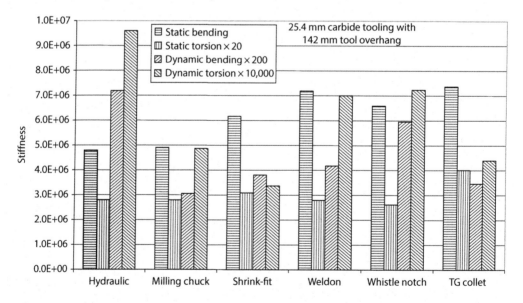

FIGURE 5.78 Comparison of static and dynamic stiffness of toolholder interfaces for 25.4 mm carbide tool.

and milling chucks are comparable, up to twice that of hydraulic expansion chucks, and up to 2.5–3 times that of collet chucks for large tool shank diameters.

A comparison of four tool–toolholder clamping systems tested on a CAT-V 50 spindle [63] showed that the parameters investigated do differ considerably between different kinds of locking systems as summarized in Table 5.16. Collet chuck locking system performance is directly proportional to the tightening torque of the locknut. A collet chuck with ball bearing locknut system

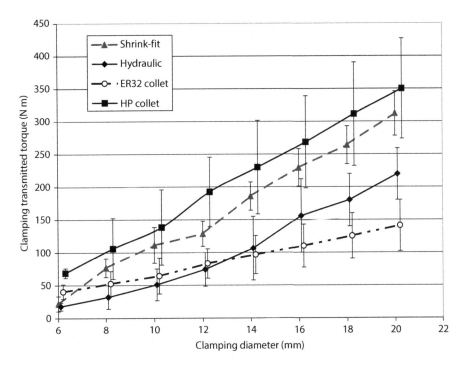

FIGURE 5.79 Transmitted torque by hydraulic (h6 tol.), shrink-fit (h6 tol.), standard collet (h6-h11 tol.), and precision collet (h6-h11 tol.) chucks. (This graph incorporates data from several manufacturers such as Parlec, Rego-Fix, Sandvik, and Schunk.)

provides higher gripping capacity compared to a standard locknut collet chucks. The static stiffness response for different systems for a 2000 N radial force is given in Table 5.16; damping characteristics, which affect chatter and forced vibrations, are also given in Table 5.16.

A comparison of a hydraulic with a precision collet chuck tool–toolholder interface is given in Table 5.17, based on dynamic tests using an HSK-63 spindle system and an *L/D* ratio of 5.9 [74]. The response of these two tool–toolholder connections is similar. The results of a second comparison of five tool–toolholder connections with an *L/D* ratio of 7.2 are given in Table 5.18. The static stiffness of the hydraulic toolholder was similar to that of the Weldon one-screw toolholder and lower than the other toolholders. The deflection response for the shrink-fit, Weldon two-screw, and collet chuck toolholders was linear with respect to overhang and had the same slope; the other two toolholders had a larger slope. The particular tested collet chuck performed exceptionally well statically and dynamically. The addition of a second screw on the Weldon type toolholder significantly improved its static and dynamic response.

Another comparison between standard collet chucks and milling chucks [64] found that the chucking pressure (and therefore the torque capability) of milling chucks is 3–20 times that of collet chucks. The maximum chucking pressure is generally only effective over the middle 40%–60% of the chuck's clamping length. Still another comparison of a 31 mm milling chuck (using an unbroken clamping surface) with a hydraulic chuck of the same size and geometry indicated that the hydraulic chuck is stiffer by 20%–30%.

The transfer functions for a 19 mm diameter tool with 50 mm overhang clamped in four different CAT-V 40 toolholders (a collet, Weldon with one screw, hydraulic chuck, and shrink-fit chuck) are summarized in Table 5.19 [75]. This comparison clearly indicates the importance of using a short holder with stiff clamping. A shrink-fit chuck provided the stiffest connection between the tool and spindle, while the hydraulic chuck provided the weakest connection. The Weldon type was weak due to its long overhang from the spindle face and its single clamping screw. The hydraulic

TABLE 5.15

Comparison between Tool–Toolholder Interfaces in a CAT-40 (AT3 Taper) for a 20 mm Bore

Application/Type of Operation	Side Lock	Collet Type DA	Collet Type ERTG	"HP" Collet Type	Hydraulic Chuck	Milling Power Chuck	Shrink Fit	Pow-Rgip Chuck	Tribos Chuck	CoroGrip Chuck
	Heavy Rough Semi-finish	Rough Semi-finish	Rough Semi-finish Finish	Rough Finish	Semi Finish and Finish	Heavy Rough and Finish	Heavy Rough Finish	Heavy Rough Finish	Heavy Rough Finish	Heavy Rough Finish
Accuracy TIR Chuck only	0.005–0.018				0.005	0.005–0.012	0.005		0.005	0.005
$4 \times D$ (mm) Chuck and collet		0.025–0.050	0.010–0.040	0.005–0.020	0.010	0.010–0.025		0.005		0.006
Rigidity	+	−	−	+	+	+++	+++			
Transmitted torque[a] (N m)	Lock screw	65–100[b]	ER32–170[c]	295	150–280	1000	200–500	360	150–240	400–580
Max krpm	12–15	30–40	30–40	30–40	30–40	20	40	40	40	40
Range of collapse (mm)	None	0.3	0.2	0.05–0.2[d]	0.008	0.012	None	None	None	0.012
Clamping range (mm)	5–50	0.5–50	1–50	3–25	5–31	6–50	6–38	3–20	6–32	6–32
Length adjustment accuracy (mm)	None or Some	0.2–0.7	0.2–0.7	0.02–0.2	±0.005	>0.005	±0.005	>.005	±0.005	±0.005
Relative cost[e]	1	1	2	4–5	10	6–10	5–7			
Balance quality as standard	−	−	−	−	+	−	+++			
Tool clamping	Allen key	C-spanner[h]	C-spanner[h]	C-spanner	Allen key	C-spanner	Device[i]	Device[j]	Device[j]	Device[j]
Maintenance	None[f]	Yes[g]	Yes[g]	Yes[g]	None[f]	Yes[g]	None[f]	Yes[g]	None[f]	Yes[g]

Notes: —, Poor; −, fair; +, good; +++, very good.

a For tolerance h6 (values indicate minimum values measured at minimum size h6 tolerance).

b 1:1 ratio of tightening torque vs. grip.

c 1:2 ratio of tightening torque vs. grip.

d High accuracy is obtained with at shank size collet.

e 1 = lowest price and 10 = highest price.

f None required (wipe bore for residual dirty coolant).

g Requires cleaning and changing collets, nuts, etc.

h C-spanner may need device to control nut tightening torque.

i Heat shrinking device.

j Special clamping device.

TABLE 5.16

Comparison of Tool–Toolholder Connections Using a 20 mm Diameter Round Shaft Extended 80 mm from the Toolholder Nose

Toolholder System (CAT-V #50 Spindle Nose)	Static Stiffness (N/mm)	Damping Ratio
Spindle nose alone	200,000	
Direct mounting of the tool with conical shank	9,090	
Weldon type chuck	6,800	0.045
Collet chuck with ball bearing locknut	6,060	0.123
Clarkson type chuck	5,850	0.091
Standard collet chuck	5,320	0.081

Source: ISO 26622-1:2008, Modular taper interface with ball track system—Part 1: Dimensions and designation of shanks.

TABLE 5.17

Comparison of Tool–Toolholder Connections Using a 10 mm Diameter Carbide Rod Extended 59 mm from the HSK-A 63 Toolholder Nose Based on a Impulse Test

Toolholder System	Hydraulic Chuck	Precision Collet Chuck
Static stiffness (N/mm)	945	867
Min. dynamic stiffness (N/mm)	41.2	38
Resonance frequency (Hz)	1530	1490

Source: Kim, T.R. et al., *ASME J. Eng. Ind.*, 111, 282, 1989.

TABLE 5.18

Comparison of Tool–Toolholder Connections Using a 12.7 mm Diameter Carbide Rod with 90 mm Overhang Based on Static Loading and Impulse Test

Toolholder System (Clamped on Rigid Block Using a Flange Mount)	Static Stiffness (N/mm)	First Natural Frequency (Hz)	Dynamic Stiffness (N/mm)
Shrink fit chuck (0.1 mm interference fit)	2500	1184	40
Hydraulic chuck	1430	810	140
Weldon chuck—one screw	1540	845	94
Weldon chuck—two screws	2170	1029	160
Standard collet chuck	2440	1161	151

Source: Kim, T.R. et al., *ASME J. Eng. Ind.*, 111, 282, 1989.

toolholder had a high modal frequency. The static stiffness of the CAT-V 40 taper is much lower than that of the CAT-V 50 taper for all chucks, as can be seen by comparing Tables 5.16 and 5.19.

Two sets of cutting tool clamping systems were evaluated with a CAT-V 50 toolholder using 19 and 25 mm diameter steel bars as cutting tools. The first 19 mm diameter set includes a hydraulic toolholder and three different collet chucks. Two 100TG collet chucks were used with 10 slits, one with a ball bearing nut (CNBB) and the other with a standard spring nut (CNSS); the third 150TG collet had three slits with a standard spring nut (CNSSL). The collet size was equal to the tool shank diameter. The second 25 mm diameter set included one end mill Weldon style holder with one set

TABLE 5.19

Comparison of Tool–Toolholder Connections Using a 19 mm Diameter Steel Rod with 50 mm Overhang on Standard 40 Taper Toolholder Based on Dynamic Tests

Toolholder System	Modes	Frequency (Hz)	Stiffness (N/mm)
Standard collet chuck (TNL = 66 mm)	1	799	1000
	2	5328	1111
Weldon chuck—one screw (TNL = 92 mm)	1	772	690
Hydraulic chuck (TNL = 18.5 mm)	1	3700	625
Shrink fit chuck (TNL= 33 mm)	Rather stiff modes	—	3333

Source: Shamime, D.M. and Shin, Y.C., *Trans. NAMRI/SME*, 28, 111, 1999.
Note: TNL = 18.5 mm toolholder nose from spindle face.

TABLE 5.20

Comparison of Tool–Toolholder Connections Using a CAT-V #50 Toolholders Based on Static (at 500 N Load at End of Tool Bar) and Dynamic Tests

Toolholder System	Symbol	Diameter (mm)	L/D	TOL (mm)	TNL (mm)	Stiffness (N/mm)	Frequency (Hz)	Damping Ratio
Hydraulic chuck	HYD19	19	4.2	171	91	1980	800	0.024
100TG collet with ball bearing nut	CNBB19	19	4.6	170	79	3400	955	0.021
100TG collet with standard spring nut	CNSS19	19	4.6	170	87	3500	908	0.014
150TG collet with standard spring nut	CNSSL19	19	4.8	170	87	3000	875	0.015
End-Mill Weldon style (set-screw 90° to the load direction)	EMW25/HS	25	3.9	175	102	7800	858	0.035
End-Mill Weldon style (set-screw along the load direction in the same side)	EMW25/VS	25	3.9	175	102	8130	830	0.0285
Hydraulic chuck	HYD25	25	3.1	178	100	8140	818	0.032

Notes: Static stiffness of spindle nose along = 350 kN/mm. TOL, total overhang (from spindle face to the end of the tool where the load is applied); TNL, toolholder nose overhang from spindle face.

screw and a hydraulic chuck. The overhang of the tool point from the spindle face, the tool L/D from the toolholder face, and the overhang of the toolholder were not exactly the same for all the tooling systems as noted in Table 5.20. The end mill toolholder was evaluated for two orientations of the set screw: (1) the set screw was displaced 90° from the applied load, and (2) the set screw was in line and on the same side as the applied load. A comparison of the static stiffnesses of all the aforementioned tooling systems is given in Table 5.20. The stiffness for all collet chucks was similar and about 50%–60% higher than for the 19 mm hydraulic chuck. The stiffnesses of the 25 mm hydraulic chuck and the end mill holder were similar. The natural frequencies and damping ratios were also measured and are given in Table 5.20. The damping ratio of the hydraulic chuck was larger than that of the collet chucks. The ball bearing nut collet had a higher damping ratio than the standard spring nut collets. The hydraulic chuck had higher damping than the end mill chuck with the set screw along the applied load direction. The damping increased for the end mill chuck when the set screw was oriented 90° from the applied load. The static stiffness results correlate well with those summarized in Table 5.16. In contrast, the damping ratios in Table 5.16 are much higher than those in Table 5.20; the difference was attributed to higher drawbar forces since the 36 kN used on the toolholders evaluated in Table 5.20 is much higher than that used in most current machine tools.

The measured deflection at the end of the tool bar and the toolholder, and the predicted deflection at the end of the tool bar based on the measured toolholder deflection and based on a monolithic tool–toolholder using 500 N load at the end of the cutting tool, are shown in Figure 5.80. The spindle deflection was 0.0025 mm. The deflection at the end of the toolholder for the earlier seven CAT-V 50 toolholders was between 0.005 and 0.013 mm. There was generally some permanent deflection of the CAT-V toolholder due to the toolholder–spindle taper fit (a maximum of 0.007 mm) that was not recovered after the load was released. The difference between the actual deflection at the tool end and the predicted deflection based on the measured toolholder deflection reflects the contribution of the tool–toolholder connection to the tool deflection response. The ratio of the actual to predicted deflections is 2.5–2.8 for the hydraulic chuck, 1.2–1.7 for the collet chucks, and 1.1 for the end mill chuck. The difference between the hydraulic tooling system and a monolithic tooling system is very significant. In contrast, the difference between a collet or end mill chuck and a monolithic tooling system is not as pronounced.

Two face-mount toolholders, one a hydraulic chuck (HYD13/F) and the other a radially adjustable end mill toolholder (SL14/F), were also evaluated; the end-mill style had a toothed clamping slit sleeve that is compressed onto the cutting tool with a set screw. The contribution of the tool–toolholder connection of the face mount toolholders to the total deflection is similar to that of the CAT-V toolholders as shown in Figure 5.80. The toolholder chuck style and the cutting tool diameter and L/D ratio are all important parameters when considering the tool stiffness. The decrease of the tooling system stiffness as a function of the cutting tool L/D is shown in Figure 5.81 for both face mount toolholders. The selection of the proper toolholder chuck style is much more important for short cutting tools.

The static bending deflection of the tool can be also defined analytically using joint stiffness parameters; the bending deflection of the tool includes the elastic deflection of the tool itself, the

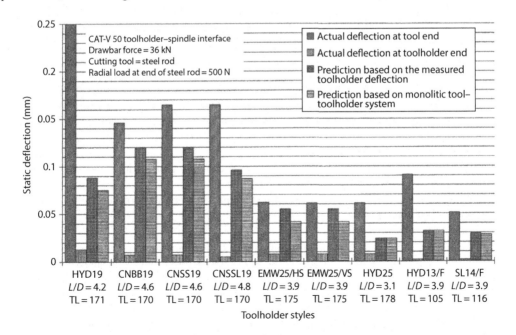

FIGURE 5.80   Comparison of several tool–toolholder connections.

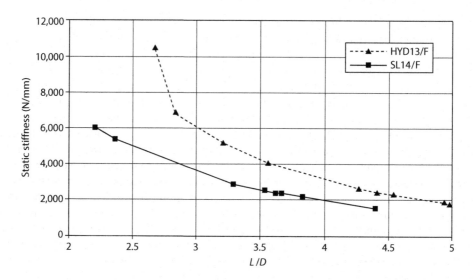

FIGURE 5.81   The effect of tool L/D ratio on the static stiffness of a tooling system.

deflection of the toolholder–tool joint, the deflection of the toolholder, deflection of the toolholder–spindle joint, and the spindle deflection at the front bearing(s) as explained in Example 5.8.

5.5   BALANCING REQUIREMENTS FOR TOOLHOLDERS

High-speed machining requires improvements in the quality and design of toolholders, specifically improved precision and rigidity and better balance. Balance is particularly important in high-speed applications since oscillating radial cutting forces and centrifugal forces (which depend on the magnitude of the unbalance) can result in a safety hazard, as well as premature spindle bearing wear or failure, holder–spindle interface fretting, poor tool life, and poor part quality or surface finish due to chatter. Toolholders must be balanced when used at spindle speeds greater than 8,000–10,000 rpm.

Unbalance is defined as the condition that exists when the principle mass axis (axis of inertia) of a tool–toolholder assembly does not coincide with the rotational axis. Unbalance is caused by several factors [76]:

1. Expansion of the spindle nose at high rpm, resulting in diminished contact between the spindle and toolholder taper. This reduces rigidity and may lead to tilting or radial movement of the toolholder within the spindle nose (due to absence of face contact), which generates unbalance
2. Uneven mass distribution in the toolholder
3. Poor tolerance in the toolholder which results in eccentric features with respect to the rotational axis
4. The use of asymmetrical components, for example, spacers, drivers, set-screws, cutting tools, or retention knobs

An unbalanced tool–toolholder assembly results in oscillating forces and movement of the cutting tool. Balancing the rotating assembly, including the toolholder and tool, distributes the mass correctly so that the rotary vibration amplitudes stay within acceptable limits at operational speeds. The ANSI S2.19 and ISO 1940 standards [77,78] define the permissible residual unbalance relative to the maximum service speed.

There are three principle types of unbalance: static, coupled, and dynamic [42,79]. *Static "single plane" unbalance* occurs when the mass axis does not coincide with the rotational axis but is parallel to the rotational axis. *Coupled unbalance* occurs when the mass axis does not coincide with the rotational axis, but does intersect the rotational axis at the center of gravity of the toolholder; the force vectors created are equal in magnitude but 180° apart. *Dynamic "two plane" unbalance* occurs when the mass axis is not coincident or parallel to the axis of rotation and does not intersect this axis; it is a combination of static and coupled unbalances.

Generally, single plane balancing is applied to toolholders that operate at speed below 15,000 rpm, which are pre-balanced by the manufacturer, and which have a length less than two times the gage line diameter (e.g., CAT V-flange and HSK side-lock holders). Two plane balancing is considered for toolholders operating at speeds greater than 20,000 rpm with lengths more than two times the gage line diameter (e.g., boring bars) [80,81].

The general equation for the allowable unbalance for a specific toolholder operating at a known rpm is

$$U = \frac{9549 \cdot G \cdot W}{\text{rpm}} \tag{5.8}$$

where
  U is the unbalance in g mm
  W is the total tool weight in kg
  rpm is the rotational speed of tool
  the "G" number represents the quality grade as defined in ANSI/S2.19 [77]

G can be calculated from

$$G = \frac{U \times \omega}{W} \tag{5.9}$$

where ω is the rotational speed in rad/s. The units for "G" are mm/s. Toolholders with unbalance of 250 g mm or more are not uncommon, although many manufacturers are reducing the unbalance to below 50 g mm. Balance requirements are determined based on the maximum radial force F_r tolerated for a given spindle or operation, given by

$$F_r = U \cdot \left[\frac{2 \cdot \pi \cdot \text{rpm}}{60} \right]^2 \tag{5.10}$$

For example, 50 g mm of unbalance produces a continuous radial force of 123 N at 15,000 rpm.

The balance quality grade "G" is very important. Once it is known and its physical significance is understood, the allowable unbalance can be calculated. The "G" number varies from case to case depending on the mass of the tool–toolholder assembly, the spindle speed to be used, and the quality of unbalance present. A tool can meet any balance specification as long as it is heavy enough or turned slowly enough. G6.3 is appropriate for machine tools and general machinery parts, G2.5 is used for machine tool drives, and G1.0 is specified for precision spindles, especially for grinding machines (Figure 5.82). G0.4 is also used for spindles. In general, toolholder balance requirements vary between G1.0 and G6.3.

Fine balance to a specific "G" tolerance is performed on balancing machines by rotating the toolholder at lower speed (about 1000 rpm). Unbalance is detected and material is drilled or milled opposite to the unbalance to reduce it. This could mean re-drilling the toolholder every time a tool is changed, which could result in the total destruction of the toolholder after several tool changes. To avoid this, a variety of toolholder balancing systems are available including pre-balanced

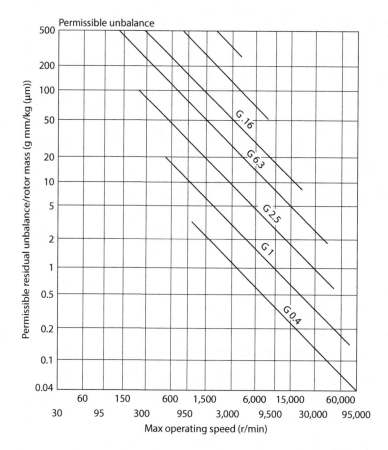

FIGURE 5.82 Maximum permissible residual specific unbalance value corresponding to various quality grades G. This graph shows permissible shaft unbalances as a function of operating speed and balancing grade. The grade number (G) represents the unbalance vibration velocity in mm/s. G1, a grade often required for precision spindles, allows a maximum rest vibration of 1 mm/s. Typically, using components of grades G1 to G6.3 leads to acceptable system balance. (Courtesy of Precise Corporation, Racine, Wisconsin.)

systems, axial, radial, or angled screws equally spaced, and weighted balls and balancing rings integrated into the toolholder body [82]. For example, set screws can be inserted into the threaded holes, which can be adjusted radially to correct for unbalance. The set screws can be held in place by a noncementing type of Loctite or by using a Nylok type of screw. Balancing rings on the toolholder have the ability to compensate for unbalance up to 60 g mm and to provide both static and dynamic balancing. All these innovative methods permit adjusting to minimum unbalance using either a balancing machine or balancing charts. Active balancing and vibration control of the spindle can be done in real-time by mounting a balancer on the spindle [83]. In one approach, two rings held in place by a permanent magnetic force are mounted on the spindle and rotate with the spindle and can also be controlled to rotate with respect to the spindle; this method is used to balance the spindle and toolholder as a system [84–86].

The tool–toolholder assembly must be balanced to a degree of accuracy that matches the machine spindle requirements. The requirements are determined based on general machinery vibration data and are often defined in terms of the velocity of the out-of-balance condition or the vibration displacement. The permissible amount of unbalance decreases with increasing speed in rpm (Figure 5.82). Generally, a balancing operation (or at least checking the balance condition) is required each time a new tool setup occurs as spindle speeds increase [57,80,87]. The cutting tool itself can cause unbalance even if the toolholder is balanced. For example, a Roller Lock milling chuck requires rebalancing after each tool change due to the repositioning of the small needle bearings in the collet actuator [87]; while these parts are small in mass, they are distant from the rotational axis and thus have a strong effect on balance. The balance consistency for a side lock end mill holder was found to be excellent while several end mills of same size were re-assembled and remounted [87]. The balance consistency of a collet chuck was not as good as that of an end mill holder but the re-assembly and remounting did not cause large balance variations. A G2.5 tolerance at 20,000 rpm is possible with precision collet chucks. However, as much as 30 g mm of unbalance can be introduced by simply loosing and retightening the nut of standard collets. The balance variation of the Roller Lock milling chuck was much higher than the side-lock end milling and collet chuck holder. If higher speeds are required (above 25,000 rpm) the chuck should be precision balanced with the cutting tool.

Mass symmetry by design is an effective approach to toolholder balance. There are controllable (fixed) and uncontrollable (variable) sources of unbalance [80]. Controllable sources, such as different depths of drive slots, set screws on end mill holders, retention knobs, unground bases of V-flanges, and other geometric characteristics, can be eliminated either through proper design or through balancing of the toolholder by the manufacturer. From a balance viewpoint, drive slots should be positioned symmetrically, although slots with different depths are sometimes used for errorproofing (e.g., to prevent insertion of a tool into the spindle backwards, as in HSK form A toolholders). Compensation must be made by adding heavy metal screws to balance geometrical unevenness. The balance corrections are made by matching the displaced and unequal masses on opposite sides of the toolholder's main body. Balancing holders for high speeds require grinding all possible exposed surfaces after heat treatment using the ground shank taper as the base reference. Repeatable balance with collet holders is attainable using a clamping nut with a fine thread with or without a coaxial extractor ring. Uncontrollable sources of unbalance generally occur after the completion of the initial balance, every time the cutting tool is changed or the collet is loosened and re-clamped. Figure 5.83 shows toolholder design principles used for a high-speed collet chucks. The possible source of unbalance is the eccentricity between the toolholder mass axis and the rotational axis, which can be minimized by using tight taper angle tolerances (AT3 per

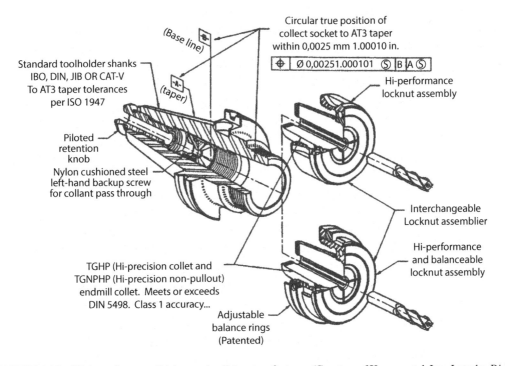

FIGURE 5.83 High-performance/high-speed collet system features. (Courtesy of Kennametal, Inc., Latrobe, PA.)

ISO 1947) for the spindle cone and for the toolholders. The toolholder should have a symmetrical design, and its ID features should be concentric to the taper within 0.005 mm TIR. Set screws in side-lock end mill holders require built-in balance compensation. Selecting the correct toolholder having excellent taper contact, accuracy (very low runout), and symmetry by design is the main goal. Toolholder bodies (usually without tool, retention knob, collet or nut) balanced by design in the neighborhood of G6.3 at 15,000 rpm are available and should be selected because they often represent the heaviest part of the toolholder assembly. Pre-balanced toolholders should be selected for speeds above 10,000 rpm.

5.6 FIXTURES

5.6.1 GENERAL

Fixtures are workholding devices used to locate, clamp, and support parts accurately and securely. Fixtures that include a build-in tool guidance device (such as bushing for drilling or reaming) are sometimes called *jigs*. A fixture or jig must accurately locate and position (clamp) the workpiece with respect to the tool in order to maintain the specified tolerances under the prevailing cutting and clamping forces. It must also not interfere with machine motions during cutting or loading/unloading. Therefore, fixtures should be given the same attention as the rest of the machining and tooling system in designing the process.

The various elements and parameters involving in the fixture design process are shown in Figure 5.84. It is important for fixture planning to be integrated with process planning, providing the link between design and manufacturing [88,89]. Generally, manufacturing practice requires an iterative process between fixture planning and process planning.

The major elements of a fixture are the *supporting structure*, *clamps*, and *locators*. Generally, as a basic *supporting structure*, subplates or tooling plates with a standard grind pattern are mounted on machine tools. In dedicated fixtures for mass production, the structure may be a complex casting connecting mounting, locating, and clamping points. Other general purpose structural elements include angle-plates and sine plates. The details of the structure of a fixture are typically part dependent;

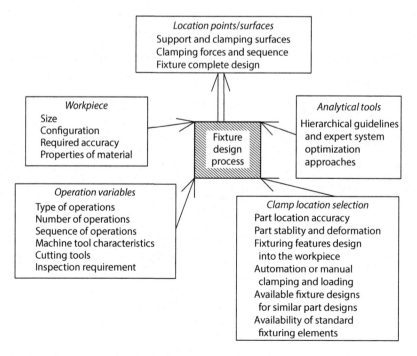

FIGURE 5.84 Factors influencing the design of a fixture.

moreover, as discussed in the following under types of fixtures, different approaches are used to construct the structure for fixtures used in mass and batch production.

Clamps are used to hold parts against locators and to withstand secondary forces. The clamp type and clamping points, forces, and sequence have a strong influence on part quality [90–94]. A variety of manual and power clamps are available. Power clamping provides better performance than manual clamping because the clamping force is more controlled and consistent and the clamping sequence is automated and thus also consistent. Hydraulic workholders (clamps and vises) are efficient and provide consistent pressures/forces; positive-locking cylinders (hydraulically activated and mechanically locked) can be used in palletized fixtures so that the hydraulic connections can be removed while the clamps are positively locked. Pneumatic and electomechanical clamps are also common. The clamping force should not exceed the level above which the part deformation is significant compared to the part tolerance. Therefore, clamping forces for finishing operations should be much lower than the forces used for rough machining operations.

It is also important to consider the response of the workpiece to clamping forces. There are two basic approaches to clamping as shown in Figure 5.85. In *bedplate clamping*, also called *toe clamping*, parts are clamped through specially designed tabs, generally against a solid subplate. In this arrangement the clamping forces are generally not supported through the main structure of the part and thus will not cause distortion of critical features. However, surfaces or features far from the clamping tabs on large parts may have relatively little support and be subject to deflection or vibration. An alternative strategy is *through clamping*, in which clamping forces are applied close to the features to be machined and supported through the main structure of the workpiece. This method leads to greater stability on critical surfaces but may result in form errors due to clamping distortion if excessive clamping forces are used.

Locators are ground pads, buttons, pins, or other dimensionally controlled feature that contacts the part at its locating points. Locators must be strong and rigid enough to resist the cutting forces exerted on the workpiece. For best locating repeatability and stability, locators should contact the workpiece on a machined surface; such surfaces can be a plane (internal machined surface), a concentric internal diameter, a precision hole (using a pin), or a combination of a surface and a pin. The most repeatable and reliable approach is to locate from an internal bore or a hole. For example, a plain surface supports/restrains the part in one direction while a surface with a round pin supports and locates the part because it restrains in three directions.

The locating points are determined by the degrees of freedom of a part. Each part has 12 possible motions, 6 linear motions (+X, +Y, +Z, −X, −Y, −Z), and 6 rotational motions (clockwise and counter-clockwise around each of the three axes). The fixture must position the workpiece in each of the three axial planes in order to fulfill a positive location criterion, the 3–2–1 (three–two–one or six-point locating principle) being the most common method of location in practice. In the 3–2–1 fixture principle (shown in Figure 5.86), the primary locating plane is provided three supporting points restricting five possible movements, a second plane is provided two supporting points restricting three possible movements, and the third plane is assigned one point to restrict one final possible movement [90–99]. The three planes must be mutually perpendicular. The 3–2–1 method, using six locators, constraints 9 out of 12 motions; the remaining three possible movements should

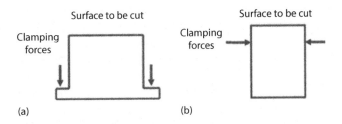

(a)   (b)

FIGURE 5.85   (a) Bedplate or toe clamping. (b) Through clamping.

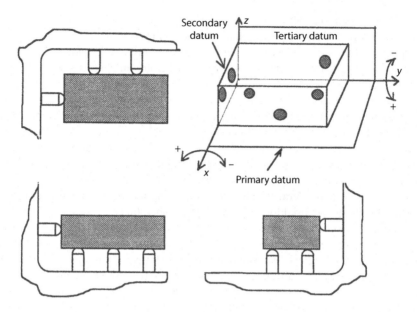

FIGURE 5.86 Six points restricting workpiece (3–2–1 locating method).

be restricted by clamps. The locators are usually selected to be as far apart as possible in order to ensure maximum stability. In practice, additional redundant supports may be necessary to avoid vibration and deflections due to cutting forces. For example, large parts may have four or more locators on the primary plane for support and stability, although this can lead to part overconstraint and deformation and can also reduce locating repeatability. The locators also establish measurement datums from which critical features are toleranced. In a typical setup, the first three points establish the primary (*A*) datum plane, the second two points define the second (*B*) plane, and the final point defines the *C* plane. For automatic clamping, it is customary to clamp against the *A*, *B*, and *C* plane locators in sequence to provide positive and repeatable location.

It is also important to consider a fixture's *operational efficiency*. Operational efficiency is a basic fixture requirement, which describes the convenience and cleanliness of the fixturing operation. The fixture should be designed for easy loading and unloading of workpieces, as well as easy chip disposal since the accumulation of chips around the fixture and especially its locators and clamps can affect part quality.

5.6.2 Types of Fixtures

Fixtures can be classified as general purpose, dedicated (permanent), modular, or flexible or reconfigurable workholders. General purpose fixtures, which have been described extensively in the literature [98–100], include plates, angle-plates, sine plates, vise-jaws, universal indexing heads, and standard clamps. They are inexpensive and reusable and thus well suited to small lot production. Common general purpose workholding devices for rotational parts include various chucks (which grip rotational parts on the outside diameter) and expanding arbors (which grip rotational parts on the inside diameter). The performance of the chuck is affected by the eccentricity between spindle and chuck and random variation caused by conditions built into the chuck (internal friction, sliding fits, etc.).

Dedicated fixtures (made from tooling plates or castings) are developed for a specific workpiece geometry and/or manufacturing operation, especially in high volume production [90–92,97]. Dedicated workholders are typically expensive because they are complex, one-of-a-kind devices with automated clamping; they also have a long construction lead time. Two typical dedicated fixtures are shown in Figures 5.87 and 5.88. Figure 5.87 shows a fixture for an engine head in a

FIGURE 5.87 Typical fixture for machining an engine head on a transfer machine. The part is mounted on a pallet with locators and clamped from the top. The transfer mechanism is at the bottom of the photo. Horizontal loads are supported by a heavy machine structure. (From General Motors Corporation, Detroit, MI.)

transfer line application. In this, case-machining loads come primarily from only one direction, and the fixture is designed to provide considerable support in that direction. Access to other faces is not a concern in this case. The fixture in Figure 5.88 is from a CNC (agile) machining system. In this case, cutting forces will generally act on multiple planes, and the structure is more compliant to accommodate tool access. The fixture shown in Figure 5.88 is a tombstone fixture, shown schematically in Figure 5.89. This type of fixture is very common for horizontal spindle machining centers. Tombstones permit multiple loading small or medium-sized parts and multi-side machining capability when the machine has B-axis table capability. Tombstones are often used with a two-position pallet changer or multiple pallets on rotary or rail-guided vehicle material handling systems. One drawback of a tombstone is that the horizontal stiffness decreases with increasing distance from the base. As a result, lower feed rates are often used on features located away from the fixture base to avoid chatter, especially in roughing operations. A tombstone fixture with thickness equal to its face width, suitable for indexing between different setups on a B-axis machine, is called a *tooling cube*.

A less common type of fixture used in CNC production systems is the window frame fixture (Figure 5.90). In this fixture the part is located on pins in the base and clamped from the top. It provides maximum access, including access on opposite faces perpendicular to the spindle, if the machine has B-axis capability. As a trade-off, part distortion may be increased because clamping forces are supported through the part. Also, for aluminum parts, the locating holes may deform if the part is clamped and unclamped three or four times, leading to a serious loss in accuracy. Tombstones with holes in the middle of the plate to accommodate large or oddly shaped parts are also sometimes called windowframes, although in this case through clamping is generally not used. In addition, large aerospace parts machined from billets often do not have a separate fixture; rather, the outer boundary of the billet is left intact to support the part, which is connected to it by tabs that are cut off when the part is finished; this approach is also sometimes called a windowframe approach.

FIGURE 5.88   Typical tombstone-type CNC fixture for an engine head. (From General Motors Corporation, Detroit, MI.)

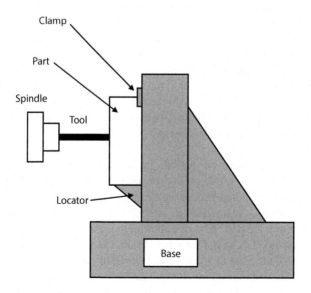

FIGURE 5.89   Schematic illustration of a tombstone fixture.

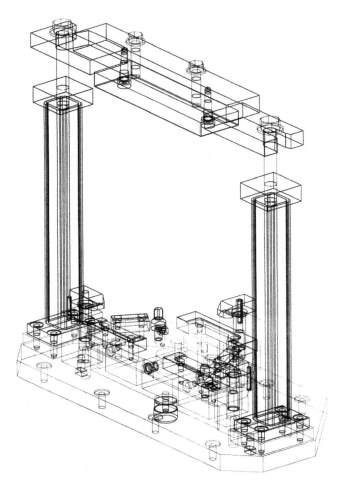

FIGURE 5.90 Exploded view of a windowframe fixture. In this type of fixture the part is located by pins on the base and clamped from the top.

Modular fixtures fill a significant gap between general and special purpose workholders. They can be used for one-time jobs, infrequent production runs, and prototype and development work. Generally, they are built up from kits of standard components as shown in Figure 5.91. Modular fixtures feature a gridwork of either accurate locating holes or precision T-slots. Grid hole designs use some combination of precisely positioned dowel holes along with tapped holes to accurately align, locate, and secure fixturing elements; some systems alternate bushing and threaded holes, while other systems have an alignment bushing plus a threaded insert in every hole. Grid hole designs provide better spatial resolution and higher clamping stiffness than T-slots. Modular fixtures built up from kits typically require more space than dedicated fixtures and lack automated clamping capability, making them unsuitable for high volume production.

The expense and lead time associated with dedicated fixtures has led to interest in the development of more flexible fixturing approaches. One approach used in high volume CNC machining systems is the use of adapter plates (Figure 5.92). An adapter plate is a precision ground pallet that locates accurately in the machine fixture, and carries additional locators on which the part is mounted. Different part types within a family can be mounted on the adapter plate; this is simplest when all parts have common locating points, but different locating schemes can be accommodated by using more than one plate type with different part locators but common machine locators. The part type and plate-specific dimensional information can be stored on a data chip on the plate. Adapter plates can be used for flexible production of two or more part types, for example, two or

FIGURE 5.91 Modular fixturing systems. (Courtesy of Carr Lane Manufacturing Co., Saint Louis, MO.)

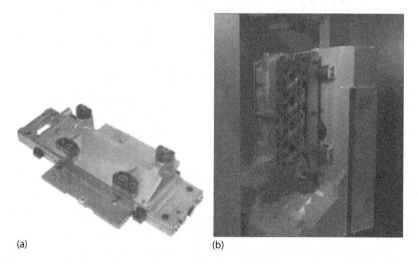

(a)   (b)

FIGURE 5.92 Adapter plate for flexible CNC production. (a) Plate showing part locators. (b) Plate and part mounted in machine fixture. (Courtesy of MAG Automotive, Sterling Heights, MI.)

more engine cylinder head designs, on the same system. Adapter plates require periodic remeasurement and maintenance to maintain accuracy.

Dedicated fixtures with modular versatility can be constructed by mounting the clamps and locators directly to a subplate or tombstone using standard bolts and T-nuts; manufacturer's specifications indicate they can hold workpieces securely with a positional accuracy between 0.01 and 0.035 mm. These types of fixtures are modular fixtures which use standard components and assemblies arranged differently for different parts but are sometimes called *flexible fixtures* if the assembly of required components is automated [101,102].

There has been significant research on automated flexible fixtures, although to date such systems have not been widely applied. Much research has focused on bed-of-nails (BON) methods, in which the part is located and clamped using protruding pins. Figure 5.93 shows one such fixture, Mazak's Form-Lok chuck, clamping an irregular workpiece. A similar concept was explored by Lamb Technicon [103,104]. This system, shown in Figure 5.94, is intended to provide automated

FIGURE 5.93   Flexible fixture using the bed-of-nails conformable surface method: Form-Lok Chuck holding an irregularly shaped workpiece. (From Mazak, Oguchi, Japan.)

FIGURE 5.94   *Bed of Nails* fixture shown holding an engine head. (From Lamb Technicon, Warren, MI.)

FIGURE 5.95 Two different engine heads clamped on a magnetic agile fixture. (From General Motors Corporation, Detroit, MI.)

reconfigurability for high volume automotive powertrain applications. In the figure, it is shown set up to support and clamp a family of similar engine cylinder heads on a pallet-like structure. An alternate approach was investigated at the General Motors Research & Development Center [105–108]. This system consists of a powerful electro-permanent magnetic chuck and several specially designed autonomous modular hydraulic elements, which are manipulated by the machine tool directly. It is capable of rearranging the locating, supporting, and clamping elements in a 3-D space quickly and automatically using appropriate modular fixture elements. Figure 5.95 shows this system holding two different aluminum engine cylinder heads with significant size and shape differences.

Research on phase change flexible fixtures has also been reported [109–111]. In these fixtures, the part is partly immersed is a special fluid, which can be made to change rapidly to a solid. Before removal of the part from the fixture, it is changed to liquid again. The phase change can be triggered by cooling/heating (using low-melting point alloys as an encapsulating medium), an electric current (using electrically active polymeric materials), or removal of air pressure (in a particulate fluid bed).

Several magnetic types of fixtures are used when grinding flat parts that would be distorted if held in other types of fixtures [112]. These fixtures are not usually very accurate and are rarely used in high-precision grinding to tolerances less than 0.003 mm.

5.6.3 FIXTURE ANALYSIS

In general purpose machining parts are normally fixtured based on experience and the available clamps and tooling plates. In high volume applications using dedicated fixtures, and in critical operations on large, expensive workpieces, it is often worthwhile to use engineering analysis methods to design fixtures. Typical methods include layout kinematic and force analysis, layout analysis and optimization, and FEA [113].

A *kinematic analysis* of the fixture-workpiece configuration is performed to ensure that the workpiece is fully constrained and easily accessible and detachable but does not account for its deformation or friction with the fixture elements. The linear motion of the workpiece is restricted by the selection of the reference surfaces, while its rotational movement due to the cutting forces is restricted by the positions of the locators. A kinematic analysis may be coupled with a force analysis, or assumptions about the magnitude of cutting forces, to refine the optimized configuration to account at least approximately for deformations due to clamping. In early work, kinematic analysis was sometimes performed using screw theory and linear or nonlinear programming [114,115], while force analysis was performed using equilibrium equations with frictional assumptions [116–119]. More recently, dynamic cutting force models have also been considered rather than assumed static values [120,121]. Kinematic and force analyses have been used to determine optimal locating and clamping points/ surfaces and the positions of support points to minimize workpiece deflection [122–125].

Kinematic and force analyses may be performed as part of a *fixture layout optimization*, which is carried out using software typically integrated with a CAD system for automatic analysis. Recent review articles [126,127] discuss the numerical methods used and CAD integration issues.

Finite element analysis is also widely used for fixture design; these analyses can take into account the deformation of the workpiece and the fixture structure. This method requires finite element models of the workpiece and fixture and proper boundary constraints to represent support, locating, and clamping elements [128–130]. It can be used to evaluate the clamping and cutting distortion including the contact deformation, fixture element stiffness/compliance, and dynamic response of fixture-workpiece system. Specific examples are given in Chapter 8.

Expert systems have also been developed for assisting the fixture design and selection. Expert systems are built on rule-based systems, which consist of strategic rules, synthesizing rules, constraining rules and heuristics (rules of thumb) for fixtures for a given family of parts [116,131–133].

As an example of fixture analysis, the fixture workpiece shown in Figure 5.96 is considered. The matrix form of the equations for this system is

$$
\begin{Bmatrix} F_x \\ F_y \\ F_z \\ M_x \\ M_y \\ M_z \end{Bmatrix} =
\begin{bmatrix}
0 & 0 & 0 & 1 & 1 & 0 \\
0 & 0 & 0 & 0 & 0 & 1 \\
1 & 1 & 1 & 0 & 0 & 0 \\
a_{41} & a_{42} & a_{43} & a_{44} & a_{45} & a_{46} \\
a_{51} & a_{52} & a_{53} & a_{54} & a_{55} & a_{56} \\
a_{61} & a_{62} & a_{63} & a_{64} & a_{65} & a_{66}
\end{bmatrix}
\cdot
\begin{Bmatrix} R_{11} \\ R_{12} \\ R_{13} \\ R_{14} \\ R_{15} \\ R_{16} \end{Bmatrix}
\tag{5.11}
$$

where

F_i and M_i are the Cartesian components of the resultant force and moment vectors
a_{ij} are the moment arms of R_{ij} contributing to moment M_i

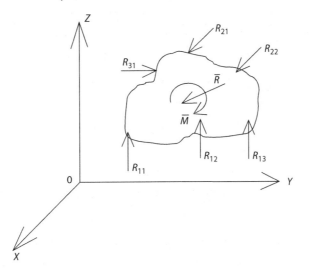

FIGURE 5.96 Free-body diagram for a fixtured workpiece.

The unknown parameters in the matrix system are the non-negative reactions R_{ij}, which are determined by solving the matrix system. These reactions are required for the design of the fixture because they form the constraints for the determination of the support points based on the 3–2–1 rule. The reaction forces also determine the deformation and resistance to machining forces as functions of applied clamping forces. The results of such an analysis may dictate the redesign of the fixture to include redundant supports and reduced clamping forces so that the proper location restraints are used and clamping loads do not deform the part significantly.

The number of clamps required is determined based on the applied resultant forces by the clamp type's rated holding capacity, which is the maximum force the clamp will sustain in the locked position. The force resisting motion in a direction perpendicular to the applied clamping force is determined using the Coulomb friction model

$$F_i \leq \mu_s \cdot P_i \; \forall i \tag{5.12}$$

where

μ is the static coefficient of friction between the workpiece and the surface of the fixture in contact

P_i and F_i are, respectively, the exerted normal clamping force and the resultant motion resisting force

P_i and F_i are related by

$$F_i = n_c \cdot P_i \tag{5.13}$$

where n_c is the number of clamps for a given part.

The part deformation due to the clamping force can be calculated using the slab method. The average normal stress p_{avg} is given approximately by

$$p_{avg} \cong \frac{2}{\sqrt{3}} \cdot Y \cdot \left[1 + \frac{\mu_s \cdot \lambda_h}{h} \right] \tag{5.14}$$

where

Y is the yield stress of a perfectly plastic material or the flow stress of a strain-hardening material

λ_h is the clamp half-width

h is the part thickness

The load to cause part deformation is

$$P_i = 2 \cdot \lambda_h \cdot p_{avg} \cdot w \tag{5.15}$$

where w is the width of the clamp.

For a solid cylindrical clamp of radius r, the average normal stress is

$$p_{avg} \cong Y \cdot \left[1 + \frac{\mu_s \cdot r}{3 \cdot h} \right] \tag{5.16}$$

and the maximum force before deformation is

$$P_i = \pi \cdot r^2 \cdot p_{avg} \tag{5.17}$$

Part quality is impacted by the flow of variation through a manufacturing system governed by fixture locaters, sequencing, orientation to the machine, and other geometric relationships. Variation simulation techniques provide a means of predicting process capability by emulating the possible errors. Variation simulation analysis assesses the impact of locator variation, workpiece datum feature variation, orientation errors, and locator and clamping sequencing [134].

5.7 EXAMPLES

Example 5.1 Estimate the nominal runout of a complete tool–toolholder–spindle system for various cases. A 10 mm diameter end mill is used with an overhang of 40 mm. Consider a precision chuck in an HSK-A63 toolholder with and without a collet, with and without an extension as shown in Figure 5.97, and a precision collet chuck in a CAT-40 toolholder.

Solution: The nominal runout for the cutting point of a tooling system can be estimated using the best/worst case scenario approach illustrated in Table 5.21. The runout of a precision collet itself is about 0.002–0.003 mm compared to 0.005–0.010 mm when the collet and the female taper in the toolholder are considered with a 30 mm tool overhang as shown in Table 5.21. The female taper

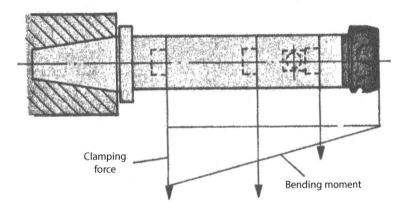

FIGURE 5.97 Standard and extended toolholder runout evaluation using the HSK-A63 and CAT-40 interfaces.

TABLE 5.21
Nominal Runout Values in Microns (μm) for Different Assemblies

System Components	HSK-A63 Toolholder			CAT-40
	Solid Holder w/t Collet	Solid Holder with Collet	Holder with Extension	Solid Holder with Collet
Spindle	3–5	3–5	6–10	6–25
Basic holder	2–6	2–6	2–6	2–6
Extension			2–4	
Clamping adapter			2–6	
Collet		2–3	2–3	2–3
Tool (end mill)	3–10	3–10	3–10	3–10
Total TIR (μm)	8–21	10–24	17–37	12–40

is considered in the basic holder in Table 5.21. The contribution of the CAT-40 toolholder–spindle interface is evaluated using the approach illustrated in Figure 5.17. The spindle contribution with the extension (in the fourth column) is twice that without extension because the runout at the end of the tool is a function of the total length of the toolholder as illustrated in Figure 5.17. In addition, the upper value of the total runout in Table 5.21 is higher than expected because it represents the worst case scenario in which the maximum tolerances of each component are summed.

However, in practice, some of the tolerance of the different components will cancel each other so that the system runout is expected to be lower than the worst case.

Note that the worst case stack-up tolerance analysis (linear stack-up of errors) is popular because of its simplicity and ease of computation. It also assumes that practically under no situation will there be a defect condition, as long as each of the errors of each component in the system has been considered within its specified error tolerance band. However, worst case tolerancing makes no assumptions on how different errors are distributed within their respective tolerance bands and may lead to conservative and unnecessarily expensive designs.

Example 5.2 Evaluate the runout of two modular tooling assemblies with long overhangs as illustrated in Figure 5.98. The difference between the two modular systems is the clamping method. One has a center threaded joint (bolt clamping) and the second system has a front clamping (rack and pinion joint). The extension spacer is 120.7 mm long by 80 mm diameter and the basic toolholder is 79.4 mm long. Note: front clamping uses a differential screw and opposite sets of serrated clamping jaws to grasp and pull the adapter/tool back into the coupling. Front clamping provides for a fast tool change. Center bolt clamping provides a stiffer joint than front clamping and is considered the optimum solution for heavy machining with long overhangs.

Solution: The runout as a function of the number of extension spacers connected together (as illustrated in Figures 5.9 and 5.98) is shown in Figure 5.99 for two modular tooling systems

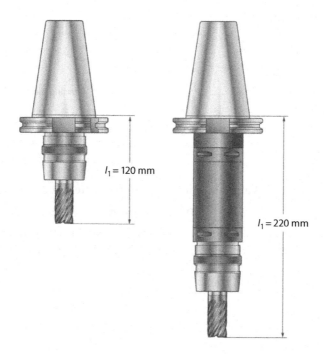

FIGURE 5.98 Bending moment and clamping force for a combination of classic and front clamping systems, which performs better for extension tooling than either system by itself. (Courtesy of Sandvik Coromant, Fairlawn, NJ.)

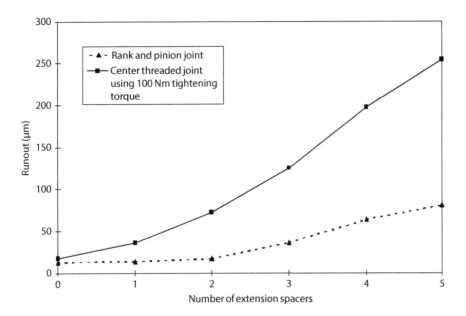

FIGURE 5.99 The effect of connection system and the number of extension adapters on the cutting tool runout.

using a CAT V-flange #50 toolholder system. The runout of the center threaded joint was higher than that of the rack and pinion joint. The difference between the two joints increases with an increasing number of spacers (or length). Large runouts may result with asymmetric key slots (slots located on one side of the contact surface only). Due to interrupted grinding conditions caused by the presence of the key slot, segments of the contact ring-shaped surface near the key slot protrude roughly 2.5–5 μm above the surface. Key slot asymmetry also results in asymmetry in contact deformation between the mating surfaces. The runout of the center threaded joint increased linearly with increasing tightening torque. The minimum runouts occur at tightening torques between 90 and 110 N m. For heavy milling applications, however, higher torques should be used if runout is not a critical concern.

The static and dynamic stiffness response of the tool was measured in both the longitudinal (x) and transverse (y) directions while the tool was mounted in the spindle of a machining center. The static stiffness as a function of the number of extensions is shown in Figure 5.100. An exponential drop in stiffness occurred as the number of joint extensions increased. There was no significant difference between the two tooling systems in either direction. The resonant frequency response (which determines the cutting speed range which should be avoided) was the same for both tooling systems and similar to that corresponding to the static stiffness as shown in Figure 5.101. In addition, it was found that the stiffness (and natural frequency) was stable at tightening torques over 120 N m.

Example 5.3 Calculate the total balancing grade of an assembled spindle–toolholder–tool system. The specified grade for each component is: spindle G1, toolholder G2.5, and tool G6.3. The mass of the spindle is 20 kg, of the toolholder is 1.27 kg, and of the tool is 0.35 kg.

Solution: The total allowable unbalance for the system is equal to the sum of the allowable unbalances for each subsystem. Hence,

$$U_{tot} = U_{spindle} + U_{holder} + U_{tool}$$

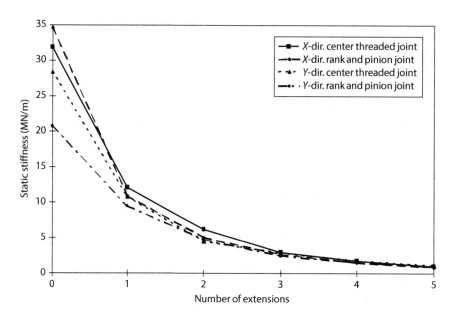

FIGURE 5.100 The effect of number of extension adapters on the static stiffness of the cutting tool at *X*- and *Y*-directions for both the classic and front clamping systems.

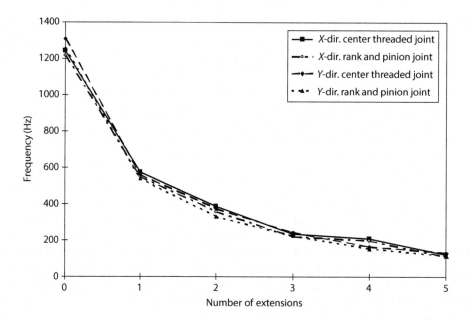

FIGURE 5.101 The effect of number of extension adapters on the frequency of the cutting tool at *X*- and *Y*-directions for both the classic and front clamping systems.

Equation 5.8 is used to calculate unbalance for each component.

$$U_{tot} = 7.639 + 1.213 + 0.842 = 9.694 \text{ g mm}$$

The *G* number is calculated from Equation 5.9 based on the total mass of the system *W* = 21.62 kg and the G_t. Hence, the G_t is calculated to be 1.17.

Example 5.4 Estimate the unbalance force reduction by selecting a CAT-40 pre-balanced tool-holder for a 15,000 rpm application.

Solution: A standard CAT-40 toolholder without pre-balance has about 30–75 g mm unbalance. The unbalance produces a centrifugal force given by Equation 5.10.

$$F_r = U \cdot \left[\frac{2 \cdot \pi \cdot \text{rpm}}{60} \right]^2 = 30 \left[\frac{2 \cdot \pi \cdot 15,000}{60} \right]^2 10^{-6} = 74\,\text{N}$$

The radial force for 30 g mm unbalance at 15,000 rpm is 74 N and for 75 g mm unbalance the force becomes 185 N. Hence, the force reduction is significant if a pre-balanced toolholder is used.

Example 5.5 A 267 mm long, 31 mm diameter solid extended integral boring bar will be operated between 10,000 and 14,000 rpm. Estimate the allowable unbalance for a balance quality grade of G2.5. In addition, determine the radial force and boring bar deflection as a function of toolholder unbalance from 0 to 55 g mm.

Solution: The allowable unbalance of the boring bar is estimated from Equation 5.8, with the weight of the bar estimated to be 3.63 kg.

$$U = \frac{9549 \cdot G \cdot W}{\text{rpm}} = \frac{9549 \cdot 2.5 \cdot (3.63\,\text{kg})}{10,000\,\text{rpm}} = 8.66\,\text{g mm}$$

So, the allowable unbalance is calculated to be 8.7 and 6.2 g mm, respectively, for 10,000 and 14,000 rpm for balance quality grade of G2.5. The radial force at the point of the boring bar is calculated using several unbalance values in the range of 0–55 g mm in Equation 5.10 as in Example 5.4. The force as a function of unbalance for three different speeds (5,000, 10,000 and 14,000 rpm) is shown in Figure 5.102. The force increases significantly with unbalance and

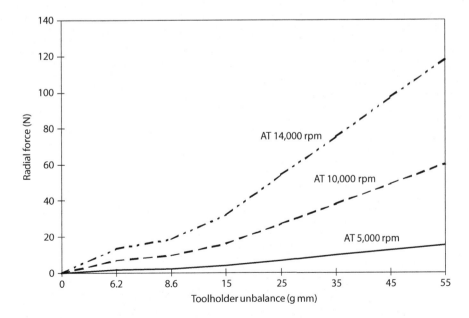

FIGURE 5.102 The effect of toolholder unbalance on the centrifugal force for different spindle speeds.

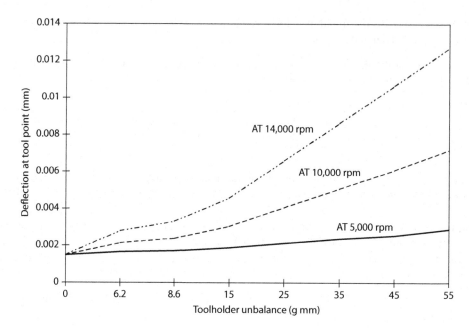

FIGURE 5.103   The effect of toolholder unbalance on the tool point deflection for different spindle speeds.

cutting speed. The tool deflection is calculated based on the deflection of the boring bar due to its weight plus the unbalance force:

$$\delta = \frac{qL^4}{8EI} + \frac{FL^3}{3EI} = \frac{(7800 \text{ kg/m}^3)(267 \text{ mm})^4}{8EI} + \frac{F(267 \text{ mm})^3}{3EI}$$

The deflection is calculated for several values of radial forces estimated above for the corresponding unbalances for three different speeds; the results are shown in Figure 5.103. The deflection increases significantly at higher speeds.

Example 5.6 Determine the effect of the toolholder clamping length on the deflection of a boring bar. Two different boring bar designs (A) and (B) can be used to bore the same hole as shown in Figure 5.104. The cutting force on the boring bar is 80 N. The clamping length of the tool shank in the holder is $2 \times D$ for bar (A) and $4 \times D$ for bar (B). The extended length of the boring bar out of the toolholder nose L_1 for both bars is equal to $7 \times D$.

Solution: The cutting force F_1 will create a reaction force at both ends of the holder in contact with the bar as shown above. The forces are calculated by the lever rule.

The lever rule for bar (A) is

$$F_1 \times L_1 = F_3 \times L_3,$$

Hence, the force acting on the back end of the boring bar is

$$F_3 = (F_1 \times L_1)/L_3 = (80 \times 7)/2 = 280 \text{ N}$$

For equilibrium: $F_1 + F_3 = F_2$

Therefore, the force acting at the front of the chuck is: $F_2 = 80 + 280 = 360$ N.

This force tends to deform the nose of the chuck at the free end. The overhang of the boring bar increases if a deformation is present.

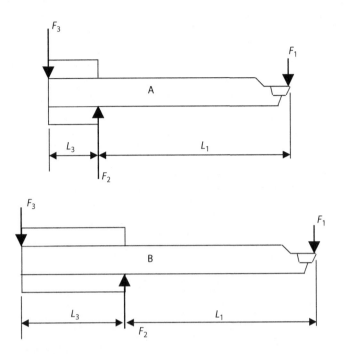

FIGURE 5.104 The effect of tool shank engagement with toolholder bore (tool clamped length) on the tool performance. (From Korn, D., *Mod. Mach. Shop*, October 12, 2006.)

The lever rule for bar (B) is

$$F_3 = (F_1 \times L_1)/L_3 = (80 \times 7)/4 = 140 \text{ N}, \ F_1 + F_3 = F_2 \text{ and } F_3 = 80 + 140 = 220 \text{ N}.$$

Hence, the reaction force acting at the free end of the chuck is smaller than that in bar (A). This will reduce the deformation of the chuck at the free end.

This example illustrates how the clamping length of the boring bar shank in the holder, L_3, affects the force, F_2, which acts on the outer end of the clamping unit as a result of the cutting force, F_1, which bends the bar. Therefore, increasing the clamping length in the shank reduces the force which acts on the rear end of the boring bar. This, in turn, means that the stress at the point of clamping is reduced, which results in higher stability. The holder–tool clamping interface is exposed to a load that can deform the surface asperities or the bore in the holder under heavy loads; therefore, the internal surfaces of the tool clamping unit should have a high level of surface finish and hardness. It is possible that the overhang of the bar (L_1) may increase due to the deformation at the outer end of the holder.

Example 5.7 An OD turning operation is performed at 3000 rpm. The diameter of the cast iron part (the chucking diameter) is 100 mm as shown in Figure 5.105. The front section of the part is turned down to 95 mm using 0.3 mm/rev feed. A 20 mm diameter hole is also drilled at the center of the part using 0.3 mm/rev feed at 2000 rpm.

Calculate the required gripping force of a power chuck.

Solution: The gripping force required depends on the type of cutting that is performed. Assuming the two operations are performed separately, each operation is examined independently.

1. The cutting forces in turning are the tangential (cutting), radial, and feed forces. The cutting forces are absorbed by the jaw faces in contact with the OD of the workpiece. The main cutting force F_c produces a moment ($F_c \cdot d_z/2$) that must be absorbed by the chuck and

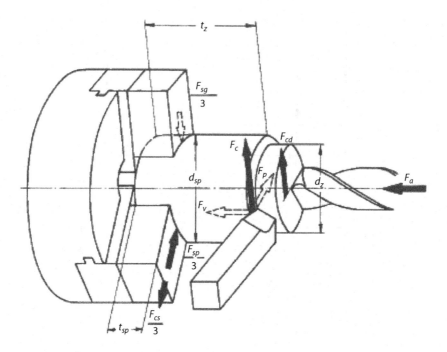

FIGURE 5.105 Clamping force consideration on a lathe chuck. (Courtesy of Rohm GmbH, Sontheim, Germany.)

transmitted by friction on the jaws through the clamping contact surfaces. The clamp force or total jaw force is the algebraic sum of the individual radial forces applied by the top jaws on the workpiece. In this case, there are three jaws and the force at each jaw is $F_{sg}/3$. The gripping force required by the jaw should be larger than the moment produced by the main cutting force during OD turning:

$$\left. \begin{array}{c} T = F_f \dfrac{d_{sp}}{2} \\[2ex] \mu = \dfrac{F_f}{F_{sg}} \\[2ex] T = F_c \dfrac{d_z}{2} \end{array} \right\} \Rightarrow F_{sg} = \dfrac{F_c \times S_z}{\mu} \times \dfrac{d_z}{d_{sp}}$$

where
    F_{sg} is the required total gripping force by the jaws without considering the effects of angular speed
    d_{sp} is the chucking diameter
    d_z is the machining diameter
    S_z is the safety factor
    l_z is the distance between machining and clamping points
    l_{sp} is the chucking length

The calculation of the torque or cutting force during a turning operation was discussed in Example 2.1 (Chapter 2).

$$P = T \cdot \omega = T\,(2\pi N) = (\text{MRR}) \cdot u_s \Rightarrow F_c = P/(2\pi \cdot R \cdot N)$$

$$T = (\text{MRR}) \cdot u_s / 2\pi N = [\pi \cdot (100^2 - 95^2)/4] \cdot (0.3 \text{ mm/rev}) \cdot N \cdot (0.065 \text{ W/mm}^3/\text{min})/2\pi \cdot N$$
$$= 142.6 \text{ N m}$$

Hence, $F_c = 2 \cdot T/d_z = 2 \cdot 142.6 \cdot 1000/100 = 2852 \text{ N}$

$$F_{sg} = \frac{F_c \cdot S_z}{\mu} \frac{d_z}{d_{sp}} = \frac{2852 \cdot 1.5}{0.20} \cdot \frac{95}{100} = 20,320 \text{ N}$$

2. Drilling the 20 mm diameter hole ($d_z = 20$ mm) at the center of the part requires
MRR = 141,000 mm³/min
$T = 58.5$ N m

$$F_{cd} = 2 \cdot T/d_z = 2\ 58.5 \cdot 1000/100 = 1170 \text{ N}$$

$$F_{sg} = \frac{F_{cd} \cdot S_z}{\mu} \frac{d_z}{d_{sp}} = \frac{1170 \cdot 1.5}{0.20} \cdot \frac{20}{100} = 1755 \text{ N}$$

This gripping force will determine the operating power for the chuck.

The gripping force of the chuck is largely influenced by the centrifugal forces (F_{cs}) of the jaws at higher speeds because all rotating parts tend to move away from the axis of rotation. The centrifugal forces is given by Equation 5.10 and the unbalance in this case is determined by how much the jaws distributed masses deviate radially from the rotating axis. This requires the knowledge of the mass (m) and the distance of the center of gravity for the jaws to the rotating axis (R). In this case, let us assume the $m = 4$ kg and $R = 0.075$ m. The centrifugal force by each jaw is:

$$F_{cs} = U \cdot \left[\frac{2 \cdot \pi \cdot \text{rpm}}{60} \right]^2 = 40.075 \left[\frac{2 \cdot \pi \cdot 2,000}{60} \right]^2 = 13,146 \text{ N}$$

This force must be taken into account when determining the initial gripping force F_{sg}. The initial gripping force with the chuck stationary is the sum of the gripping force without considering the effects of speeds and the centrifugal force of the jaws (loss of gripping force) for external gripping:

$$F_{sp} = F_{sg} + F_{cs} = 20,320 + 13,146 = 33,466 \text{ N}$$

The centrifugal force is a large percentage (65% of the gripping force required without considering the effects of speeds) of the total force required on the jaws.

Example 5.8 Consider the boring bar in Figure 5.106 for finishing a 30 mm diameter hole to 31 mm diameter with a single point. The workpiece material is gray cast iron. A carbide insert is used in the indexable steel boring bar. The cutting speed for the bar is 100 m/min and the feed is 0.15 mm/rev. The boring bar is integral to a CAT-40 toolholder. The spindle stiffness at the front (by the toolholder interface) is 2×10^8 N/m. Estimate the deflection at the tool point.

Solution: The two tapers at the toolholder–spindle interface (female taper in the spindle nose and the male taper on the holder) do not have a perfect fit because they are made with certain tolerances. Therefore, a relative motion occurs between the two tapers under cutting loads at the tool point.

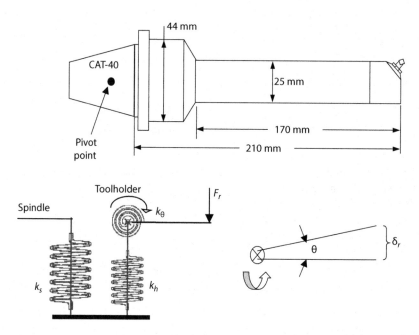

FIGURE 5.106 Analytical model of the toolholder–spindle interface with 2-DOF (linear and rotational) spring.

The static bending deflection is the first check to avoid large deflection at the tool point. The bending deflection of the tool includes the elastic deflection of the bar itself, the deflection of the toolholder–spindle joint, and the spindle deflection at the front bearing(s). The stiffness of the spindle–toolholder interface is decomposed into a rotational spring and a translational (linear) spring.

$$\delta_{total} = \delta_b + \delta_h + \delta_r + \delta_s$$

where

    δ_b is the deflection for a cantilever beam as discussed in Chapter 4
    δ_h is the translation of the taper interface
    δ_r is the deflection due to rotation of the taper interface
    δ_s is the translation of the spindle shaft due to bearing stiffness

If the tool is not monolithic and there is an additional connection between the tool and the tool-holder, the toolholder interface should be also considered as a summation of the deflection due to linear and rotational springs. The joint location between the tapers is 20 mm from the spindle face inside. The spindle stiffness is $k_s = 3 \times 10^8$ N/m. The toolholder–spindle interface has a linear spring $k_h = 1.375 \times 10^9$ n/m and a rotational spring $k_\theta = 8.108 \times 10^6$ N m/rad (from Table 5.7). The radial cutting force at the tool point was calculated based on the cutting conditions as explained in Example 4.4 (Chapter 4) to be $F_r = 185$ N.

The total linear stiffness in the direction of the radial force is equivalent to

$$\frac{1}{k} = \frac{1}{k_s} + \frac{1}{k_h} = \frac{1}{1.375 \cdot 10^9} + \frac{1}{3 \cdot 10^8} \Rightarrow k = 2.463 \cdot 10^8 \text{ N/m}$$

The deflection of the bar structure with two cross sections is

$$\delta_b = F_r \left[\frac{L_1^3}{3 \cdot E_1 \cdot I_1} + \frac{L_2^3}{3 \cdot E_2 \cdot I_2} + \frac{L_1 L_2}{E_2 \cdot I_2} (L_1 + L_2) \right]$$

where

$L_1 = 170$ mm
$L_2 = 40$ mm
$E_1 = E_2 = 206,700$ MPa, and the moment of inertia for the two cross sections are:

$$I_1 = \frac{\pi D_1^4}{64} = \frac{\pi (25^4)}{64} = 19,165 \, \text{mm}^4 \quad \text{and} \quad I_2 = \frac{\pi D_2^4}{64} = \frac{\pi (44^4)}{64} = 183,890 \, \text{mm}^4$$

Hence, $\delta_b = 0.078$ mm.

The spindle and the holder linear deflection is:

$$\delta_1 = \frac{F_r}{k} = \frac{185 \, \text{N}}{2.463 \cdot 10^8 \, \text{N/m}} = 0.00075 \, \text{mm}$$

The toolholder deflection at the tool point is affected by its shank rotation in the spindle nose taper. The toolholder rotation at the spindle nose is determined by the rotational spring at the interface located 20 mm behind the spindle face.

$$\theta = \frac{M_h}{k_\theta} = \frac{F_r \cdot L}{k_\theta} = \frac{185 \cdot (210 + 20)}{8.108 \times 10^6 \cdot 1000} = 5.3 \cdot 10^{-6} \, \text{rad}$$

$$\delta_r = L \cdot \tan \theta = (210 + 20) \cdot \tan \left(5.3 \cdot 10^{-6} \right) = 0.0012 \, \text{mm}$$

The total deflection is the sum of the above three deflections:

$$\delta_{total} = \delta_b + \delta_1 + \delta_r = 0.078 + 0.00075 + 0.0012 = 0.080 \, \text{mm}$$

The contribution of the toolholder–spindle interface compliance in the boring bar deflection is only 2.5% in this case. However, for shorter boring bar (e.g., $L = 75$ mm) the contribution of the toolholder–spindle interface compliance is increased to 29%.

Example 5.9 Evaluate the effect of the tool geometry on the performance of the toolholder–spindle interface considered in Figure 5.21.

Solution: The maximum bending moment that a toolholder can support is a function of the clamping force. The level of the clamping force is very important, especially in milling and boring operations, because it determines the limit of the torsional/bending forces at the cutting edge(s) and defines the stability and rigidity of the interface. The maximum allowable bending moment before the face joint is separated at the toolholder flange is given in Figure 5.35 for a range of clamping forces for different sizes of HSK toolholders. For example, 300 N m is the maximum bending moment that the HSK-63-A interface can support before complete separation of the face contact occurs when the clamping force is 18 kN. The torsion/bending forces at the flange must be estimated as explained in Chapters 2 and 4 (or simulated with software) based on the cutting conditions for semi and rough milling and boring applications to prevent face contact separation. The maximum value of tangential forces is a function of the two critical dimensions R and L that must be considered when designing turning tools and especially boring bars and milling cutters (Figure 5.107). The bending moment generated during cutting can be calculated from the equation:

$$M = \frac{9.55 \cdot P}{R \cdot N} \left(L^2 + c_1^2 \cdot L^2 + c_2^2 \cdot R^2 \pm 2 \cdot c_1 \cdot c_2 \cdot L \cdot R \cdot \cos\theta \pm 2 \cdot c_2 \cdot L \cdot R \cdot \sin\theta \right)^{1/2} \quad (5.18)$$

where
  M is the bending moment (N m)
  P is the cutting power (W)
  R is the radius of the milling cutter or the distance of the cutting edge to the center axis of the
     bar for boring tools (mm)
  L is the length of the cutting edge from the spindle nose (mm)
  N is the spindle rpm

The + or − in the parenthesis is used for the axial force being away and into the spindle direction, respectively

In addition, c_1 and c_2 are the ratios of the tangential force to the radial force and axial force, respectively:

$$c_1 = \frac{F_X}{F_Y} \quad \text{and} \quad c_2 = \frac{F_Z}{F_Y} \quad (5.19)$$

The ratio of the axial and radial force is primarily dependent on the lead angle with smaller influence by the rake of the cutting edge. In general, the axial force is the smallest of the three forces in milling, while in boring the axial (feed) force could be either higher or lower than the radial force, depending on the feed, depth-of-cut, lead angle, and corner radius of the insert. Equation 5.18 can be reduced to

$$M = \frac{9.55 \cdot 10^6 \cdot P}{R \cdot N} L \quad (5.20)$$

for small diameter (<100 mm) and small lead angle (<15°) milling cutters. Rake angles greater than 10° will contribute significantly to the cutting force ratios. All three forces and especially the tangential and radial forces (estimated through models) should be considered for all other cases in Equation 5.18 (Figure 5.107).

Example 5.10 Consider three different shrink-fit holders for 6, 12, and 20 mm diameter tool shanks. Their nose characteristics are given in Table 5.22. The clamping length of the tool shank is considered the maximum in the toolholder. Estimate the clamping pressure and torque, the chucking axial force, and the temperature required to expand the toolholder bore to insert the tool shank for an interference fit of 0.005, 0.010, and 0.020 mm.

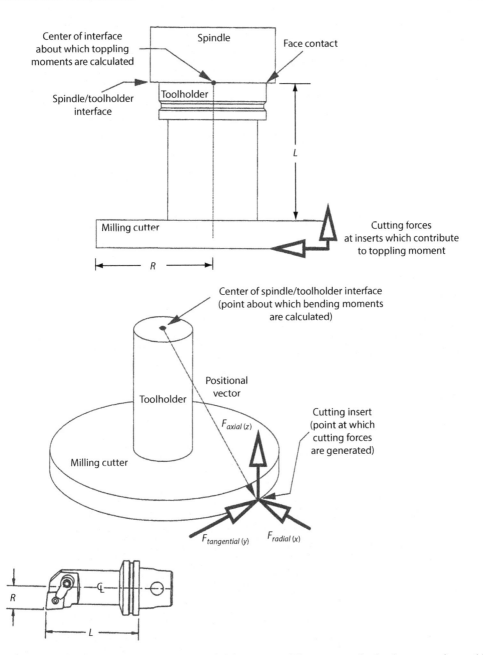

FIGURE 5.107 The maximum allowable tangential force on a milling cutter or boring bar or turning tool is a function of dimensions L and R and the type and shrink-fit size of toolholder interface.

Solution: The analysis for the shrink-fit interface to evaluate the clamping pressure, torque and the interfacial stresses can be carried out using Equations 5.3 through 5.5. The interference fit is the initial differences in diameters between the inner radius of the toolholder nose and the tool shank diameter. The clamping pressure is calculated from Equation 5.3 using different interferences and the toolholder characteristics given in Table 5.22. The pressure and hence the stresses increase as the size of the tool shank decreases and as the wall thickness of the toolholder nose increases. The chucking torque is lower for carbide tools than for HSS tools. The chucking axial force is high for all sizes and interferences. Comparing the results in Table 5.22 with those in Figure 5.79 for shrink-fit indicate that the manufacturers are expecting lower interference fit for the smaller sizes of tools.

TABLE 5.22
Nominal Runout Values in Microns (μm) for Different Assemblies

Tool Size	Interference (mm)	Tool Size			
		6 mm	12 mm	20 mm	20 mm
ID (mm)		6	12	20	20
OD (mm)		23	27	38	51
L (mm)		26	37	42	42
P-HSS (MPa)	0.005	150	57.7	29.3	38
T-HSS (N m)		33	72	116	150
F-HSS (N)		11	12	11.6	15
P-Carb (MPa)		134	52.8	27	34.4
T-Carb (N m)		30	66	107	136
F-Carb (N)		9.9	11	10.7	13.6
Δθ (°C)		107	44	19	19
P-HSS (MPa)	0.010	301	116	59	76
T-HSS (N m)		66	145	232	300
F-HSS (N)		22	10624	23	30
P-Carb (MPa)		268	106	54	69
T-Carb (N m)		59	132	215	272
F-Carb (N)		20	22	22	27
Δθ (°C)		168	76	38	38
P-HSS (MPa)	0.020	602	231	117	152
T-HSS (N m)		133	290	463	600
F-HSS (N)		44	48	46	60
P-Carb (MPa)		536	211	109	138
T-Carb (N m)		118	265	429	544
F-Carb (N)		39	44	43	55
Δθ (°C)		293	137	75	75

Notes: P-HSS, Clamping pressure with HSS tool shank; P-Carb, Clamping pressure with Carbide tool shank; T-HSS, Transmitted Torque with HSS tool shank; F-HSS, Transmitted Axial Force with HSS tool shank with; L length of contact in the holder bore; α, Thermal expansion coefficient for steel 0.0000075 mm/mm/°F. E_{steel} = 207,000 MPa and $E_{carbide}$ = 400,000 MPa.
Poisson's ratios for steel and carbide material are considered 0.29.
The coefficient of friction is considered 0.15.

This is the main reason for controlling the tool shank diameter with h6 tolerance or better (h5 for small size tooling). For example, the h6 tolerance specifies a variation of 0 to −0.008 mm for a 6 mm tool shank diameter; this means that the chucking pressure and torque would vary significantly by comparing the results between the 0.005 and 0.010 mm interference-fits since the toolholder nose is designed to accept tool shanks in the range of 5.992–6 mm. Therefore, there is a concern with small size tooling using a shrink-fit chuck. The variation of the toolholder bore size among different manufacturers results in different chucking torques. Note that the coefficient of friction of 0.15 used in Table 5.22 has a good safety factor as described in Equation 5.5. The toolholder nose is heated by a temperature ΔT. Assuming a uniform temperature field, the initial shape of the nose remains the same after heating. The result of the heating is an increase of the inner bore by an amount $\delta r = \alpha \Delta T r_i$ and the outer diameter of the nose by an amount $\delta r = \alpha \Delta T r_o$. The temperature difference ΔT required to expand the bore by 0.005 mm larger than the interference (between the tool shank and the toolholder nose) specified in the Table 5.22 is calculated and given in Table 5.22. The temperature is higher for the smaller diameter tool shank interfaces as expected.

5.8 PROBLEMS

Problem 5.1 Consider the boring bar (B) in example 3.4. The boring bar is integral to a CAT-50 holder. The toolholder–spindle CAT-50 interface stiffness is modeled by a linear spring and a torsional spring. The linear spring constant is 3×10^9 N/m, while the torsional spring constant is 4×10^7 N m/rad with the joint location (pivot point) 40 mm in the spindle nose. The length of the toolholder taper in the spindle is 100 mm. The spindle stiffness at the front (by the toolholder interface) is 2×10^8 N/m. Estimate the deflection at the tool point.

Problem 5.2 A 40 mm diameter hole is being bored to 44 mm diameter with a single point boring bar. The workpiece material is medium-carbon steel. A carbide insert is used in the indexable steel boring bar. The boring bar is integral to a CAT-50 toolholder and extends 200 mm. The cross section of the solid boring bar is cylindrical with 30 mm diameter. The cutting speed for the bar is 80 m/min while the feed is 0.2 mm/rev. The toolholder–spindle interface stiffness is characterized by a linear spring and a torsional spring. The linear spring constant is 3×10^9 N/m, while the torsional spring constant is 4×10^7 N m/rad with the joint location 40 mm in the spindle nose. The length of the toolholder taper in the spindle is 100 mm. The spindle stiffness at the front (by the toolholder interface) is 2×10^8 N/m. Estimate the deflection at the tool point in order to determine the bore size quality. The material properties are

$$E_{steel} = 206,700 \text{ MPa}$$

REFERENCES

1. E. I. Rivin, Tooling structure: Interface between cutting edge and machine tool, *CIRP Ann.* **49**:2 (2000) 591–634.
2. E. I. Rivin, Trends in tooling for CNC machine tools: Tool-spindle interfaces, *Manuf. Rev.* **4** (1991) 264–274.
3. ISO 286-2:2010, Geometrical product specifications (GPS)—ISO code system for tolerances on linear sizes—Part 2: Tables of standard tolerance classes and limit deviations for holes and shafts.
4. S. Baier, Part I: Spindles and their relationship to high-speed toolholders, *Moldmaking Technol.* (August 2003). http://www.moldmakingtechnology.com.
5. M. Tsutsumi, A. Miyakawa, and Y. Ito, Topographical representation of interface pressure distribution in a multiple bolt-flange assembly—Measurement by means of ultrasonic waves, ASME paper 81-DE-7, 1981.
6. S. Hazem, J. Mori, M. Tsutsumi, and Y. Ito, A new modular tooling system of curvic coupling type, *Proceedings of the 26th International Machine Tool Design and Research Conference*, MacMillan, New York, 1987, pp. 261–267.
7. Anon., *Designing With Kennametal*, Kennametal, Inc., Latrobe, PA, 1978.
8. R. H. Thornley and I. Elewa, The static and dynamic stiffness of interference shrink-fitted joints, *Int. J. Mach. Tools Manuf.* **28** (1988) 141–155.
9. ISO 1947:1973, System of cone tolerances for conical workpieces from C = 1:3 to 1:500 and lengths from 6 to 630 mm.
10. ANSI/ASME B5.50-2009, 7/24 taper tool to spindle connection for automatic tool change.
11. ISO 7388-1:2007, Tool shanks with 7/24 taper for automatic tool changers—Part 1: Dimensions and designation of shanks of forms A, AD, AF, U, UD and UF.
12. DIN 69871-1:1995-10, Steilkegelschäfte für automatischen Werkzeugwechsel—Teil 1: Form A, Form AD, Form B und Ausführung mit Datenträger.
13. DIN 2080-1:1978-12, Steilkegelschäfte für Werkzeuge und Spannzeuge, Form A.
14. JSA JIS B 6339:1992-01, Machining Centers—Tool Shanks and Retention Knobs.
15. C. Xu, J. Zhang, P. Feng, D. Yu, and Z. Wu, Characteristics of stiffness and contact stress distribution of a spindle–holder taper joint under clamping and centrifugal forces, *Int. J. Mach. Tools Manuf.* **82–83** (2014) 21–28.
16. J. S. Zenker, Design and application of high speed spindles for high speed machining, Presented at the *SME Hi-Speed Machining Seminar*, Minneapolis, MN, February 15, 1994.

17. E. I. Rivin, Tooling structures—A weak link in machining centers and FMS: Some ways for improve-ment, *International Conference on Manufacturing Systems and Environment*, vol. 21, JSME, Tokyo, Japan, May 28, 1990.

18. J. S. Agapiou, Selection of tool holding system for a machine tool spindle, *The First International Machine Tool Conference: The Dominance of Spindle Performance*, SME, Dearbon, MI, May 2003.

19. Jacobs P. Thomas, *Characterization of the Machine Tool Spindle to Toolholder Connection*, M.S. Thesis, University of North Carolina at Charlotte, Charlotte, NC, 1999.

20. M. Tsutsumi, Y. Anno, and N. Ebata, Static characteristics of 7/24 tapered joint for machining center, *Bull. JSME* **26** (1983) 461–467.

21. M. Weck and I. Schubert, New interface machine/tool: Hollow shank, *CIRP Ann.* **43** (1994) 345–348.

22. M. Tsutsumi, M. Ohya, T. Aoyama, S. Shimizu, and S. Hachiga, Deformation and interface pressure distribution of 1/10 tapered joints at high rotational speed, *Int. J. Jpn Soc. Prec. Eng.* **30** (1996) 23–28.

23. Anon., *Big-Plus Spindle System*, Big Daishowa Seiki Co. LTD, Osaka, Japan, Catalog EX-48, 1998.

24. E. Abele, Y. Altintas, and C. Brecher, Machine tool spindle units, *CIRP Ann.* **59** (210) 781–802.

25. Anon., *3-lock System*, Nikken Kosakusho Works, LTD., Osaka, Japan, Catalog 3341, nd.

26. J. S. Agapiou, E. Rivin, and C. Xie, Toolholder/spindle interfaces for CNC machine tools, *CIRP Ann.* **44** (1995).

27. Anon., *Showa D-F-C Shank*, Showa Tool Co., LTD, Tottori, Japan, Catalog DF-9808.

28. W. Kelch, "HSK tooling system for high speed machining, *Proceedings of the 6th International Machine Tool Conference* (*IMTC*), Osaka, Japan, 1994, pp. 126–148.

29. M. Tsutsumi, T. Kuwada, T. Aoyama, S. Shimizu, and S. Hachiga, Static and dynamic stiffness of 1/10 tapered joints for automatic changing, *Int. J. Jpn. Soc. Prec. Eng.* **29** (1995) 301–306.

30. I. Schubert et al., *Interface Machine/Tool: Testing & Optimization, Final Report on the Research Project*, WZL Laboratory for Machine Tools and Applied Economics, Aachen University, Aachen, Germany, March 31, 1994.

31. M. Weck and T. Reinartz, Determination of the load capacity of HSK interfaces, *Prod. Eng.* **11** (2004) 99–102.

32. DIN 69893-1 2011-5, Kegel-Hohlschäfte mit Plananlage—Teil 1: Kegel-Hohlschäfte Form A und Form C; Maße und Ausführung.

33. DIN 69063-1:2008-08, Werkzeugmaschinen—Aufnahmen für Kegel-Hohlschäfte—Teil 1: Für Form A und Form C nach DIN 69893, Anschlussmaße.

34. H. Hanna, J. S. Agapiou, and D. A. Stephenson, Modeling the HSK toolholder-spindle interface, *ASME J. Manuf. Sci. Eng.* **124** (2002) 734–744.

35. E. Kocherovsky, *HSK Handbook*, Intelligent Concept, West Bloomfield, MI, 1999.

36. J. S. Agapiou, P. Bandyopadhyay, C. H. Shen, and D. A. Stephenson, Operational practices for using HSK tool-holder-spindle systems in machining applications, *Trans. NAMRI/SME* **30** (2001) 231–236.

37. Anon., *KM Modular Quick-Change Tooling Systems for Lathes and Machining Centers*, Kennametal Inc., Latrobe, PA, Catalog A93-49 (105) E3, nd.

38. ISO 26622-1:2008, Modular taper interface with ball track system—Part 1: Dimensions and designation of shanks.

39. ISO 26622-2:2008, Modular taper interface with ball track system—Part 2: Dimensions and designation of receivers.

40. Anon., *New KM4X™ Spindle Connection from Kennametal Delivers Maximum Spindle Power*, Kennametal Inc., Latrobe, PA, 2012.

41. S. Chen, T. J. Long, C. Gey, and R. F. de Souza, A systems approach for successful titanium machining, *Mod. Mach. Shop*, April 13, 2012. http://www.mmsonline.com.

42. *NC5 Tooling System*, Nikken Kosakusho Works, LTD., Osaka, Japan, Catalog 8205.

43. DIN ISO 10889-1:2006-01, Werkzeughalter mit Zylinderschaft—Teil 1: Zylinderschaft, Aufnahme-bohrung; Technische Lieferbedingungen.

44. ISO 10889-1:2004, Tool holders with cylindrical shank—Part 1: Cylindrical shank, location bore—Technical delivery conditions.

45. T. R. Kim, S. M. Wu, and K. F. Eman, Identification of joint parameters for a taper joint, *ASME J. Eng. Ind.* **111** (1989) 282–287.

46. D. M. Shamime and Y. C. Shin, Ánalysis of No.50 taper joint stiffness under axial and radial loading, *Trans. NAMRI/SME* **28** (1999) 111–116.

47. D. M. Shamime, S. W. Hong, and Y. C. Shin, An in situ modal-based method for structural dynamic joint parameter identification, *J. Mech. Eng.* **214**:C5 (2000) 641–653.

48. J. H. Wang and S. B. Horng, Investigation of the tool holder system with a taper angle 7:24, *Int. J. Mach. Tools Manuf.* **34** (1994) 1163–1176.

49. A. Erturk, H. N. Ozguven, and E. Budak, Analytical modeling of spindle–tool dynamics on machine toolsusing Timoshenko beam model and receptance coupling for the prediction of tool point FRF, *Int. J. Mach. Tools Manuf.* **46** (2006) 1901–1912.

50. T. L. Schmitz, K. Powell, D. Won, G. S. Duncan, W. G. Sawyer, and J. C. Ziegert, Shrink fit tool holder connection stiffness/damping modeling for frequency response prediction in milling, *Int. J. Mach. Tools Manuf.* **47** (2007) 1368–1380.

51. C. Xu, J. Zhang, P. Feng, D. Yu, and Z. Wu, Characteristics of stiffness and contact stress distribution of a spindle–holder taper joint under clamping and centrifugal forces, *Int. J. Mach. Tools Manuf.* **82–83** (2014) 21–28.

52. J. Agapiou, A methodology to measure joint stiffness parameters for toolholder/spindle interfaces, *Trans. NAMRI/SME* **33** (2004) 503–510.

53. T. L. Schmitz et al., Improved milling capabilities through dynamics prediction: Three component spindle-holder-tool model, *Proceedings of 2005 NSF DMII Grantees Conference*, Scottsdale, AZ, 2005.

54. T. L. Schmitz, M. Davies, and M. Kennedy, Tool point frequency response prediction for high-speed machining by RCSA, *ASME J. Manuf. Sci. Eng.* **123** (2001) 700–707.

55. I. Mancisidor, A. Urkiola, R. Barcena, J. Munoa, Z. Dombovari, and M. Zatarain, Receptance coupling for tool point dynamic prediction by fixed boundaries approach, *Int. J. Mach. Tools Manuf.* **78** (2014) 18–29.

56. U. V. Kumar and T. L. Schmitz, Spindle dynamics identification for receptance coupling substructure analysis, *Prec. Eng.* **36** (2012) 435–443.

57. T. L. Schmitz and K. S. Smith, *Machining Dynamics: Frequency response to Improved Productivity*, Springer, New York, 2009.

58. V. Ganguly and T. L. Schmitz, Spindle dynamics identification using particle swarm optimization, *J. Manuf. Proc.* **15** (2013) 444–451.

59. A. S. Delgado, E. Ozturk, and N. Sims, Analysis of non-linear machine tool dynamic behavior, *Proc. Eng.* **63** (2013) 761–770.

60. D. L. Lewis, Factors for successful rotating tool operation at hi speeds, Presented at the *SME Hi-Speed Machining Clinic and Tabletop Exhibit*, Schaumburg, IL, May 10–11, 1994.

61. DIN 6499:2002-04, Spannzangen mit Einstellwinkel 8° für Werkzeugspannung—Spannzangen, Spannzangenaufnahmen, Spannmuttern.

62. ISO 15488:2003, Collets with 8 degree setting angle for tool shanks—Collets, nuts and fitting dimensions.

63. A. Mannan and B. Lindstrom, Investigations in the static and dynamic performance of different end mill locking systems, *CIRP Ann.* **30** (1981) 265–268.

64. M. Tsutsumi, Chucking force distribution of collet chuck holders for machining centers, *J. Mech. Working Technol.* **20** (1989) 491–501.

65. F. Mason, Tools run true in hydraulic toolholders, *Am. Mach.* (July 1994) 59–61.

66. M. Koch, Toolholders: An important connection between spindle and cutting tool, *Moldmaking Technol.* (September 2003).

67. S. Woods, Toolholder turnaround, *Cutting Tool Eng.* **63**:7 (July 2011).

68. J. Lorincz, Mechanical tool holder provides a gripping solution, *Tool. Prod.* (November 2003). http://www.ctemag.com.

69. Anon., *Die & Mold Making—Application Guide*, Sandvik Coromant, Fair Lawn, NJ, 2000.

70. D. Korn, Toolholder change helps HMC realize full machining potential, *Mod. Mach. Shop* (October 12, 2006). http://www.mmsonline.com.

71. S. A. Webster, Beyond runout: Toolholders that grip, lock and raise the bar, *Manuf. Eng.* (January 2013) 75–82.

72. K. Hanson, Worry free—Antipullout toolholders reduce scrap, increase tool life and eliminate anxiety while milling, *Cutting Tool Eng.* (August 2013) 50–58.

73. D. Moore, Driving the tap, *Cutting Tool Eng.* (August 1995) 52–60.

74. E. Lenz, J. Rotberg, R. C. Petrof, D. J. Stauffer, and K. D. Metzen, Hole location accuracy in high speed drilling influence of chucks and collets, *CIRP Manufacturing Systems Symposium*, Ann Arbor, MI, May 1995.

75. J. Tlusty, High speed milling, *Proceedings of the 6th International Machine Tool Conference (IMTC)*, Osaka, Japan, 1994, pp. 35–60.

76. S. Baier, Part II: Spindles and their relationship to high-speed toolholders, *Moldmaking Technol.* (September 2003). http://www.moldmakingtechnology.com.

77. ANSI S2.19-1999 (R2004), Mechanical vibration - balance quality requirements of rigid rotors

78. ISO 1940-1:2003, Mechanical vibration—Balance quality requirements for rotors in a constant (rigid) state—Part 1: Specification and verification of balance tolerances.

79. G. K. Grim, J. W. Haidler, and B. J. Mitchell, Jr., *The Basics of Balancing*, Balance Technology Inc., Whitmore Lake, MI, nd.

80. M. H. Layne, *Detecting and Correcting the SME High Speed Machining Clinic and Tabletop Exhibit*, Chicago, IL, April 25–26, 1995.

81. M. H. Layne, On balance, *Cutting Tool Eng.* (August 1991) 36–41.

82. D. W. McHenry, Keeping your balance, *Moldmaking Technol.* (August 2004).

83. S. Zhou and J. Shi, Active balancing and vibration control of rotating machinery: A survey, *The Shock and Vibrat. Digest* 33(4) (July 2001) 361–371.

84. S. W. Dyer, B. K. Hackett, and J. Kerlin, Electromagnetically actuated rotating unbalance compensator, U.S. Patent No. 5,757,662, May 26, 1998.

85. S. Zhou and J. Shi, Optimal one-plane active balancing of a rigid rotor during acceleration, *J. Sound Vibrat.* **249** (2002) 196–205.

86 H. Fan, M. Jing, R. Wang, H. Liu, and J. Zhi, New electromagnetic ring balancer for active imbalance compensation of rotating machinery, *J Sound Vibrat.* **333** (2014) 3837–3858.

87. D. Chartier and G. VanWaes, Toolholders: The critical interface, Presented at the *SME High Speed Machining Clinic and Tabletop Exhibit*, Chicago, IL, April 25–26, 1995.

88. J. Y. H. Fuh, C. H. Chang, and M. A. Melkanoff, An integrated fixture planning and analysis system for machining processes, *Robotics Computer-Integr. Manuf.* **10**:5 (1993) 339–353.

89. Y. Rong and Y. Zhu, *Computer-Aided Fixture Design*, Marcel Dekker, New York, 1999.

90. E. G. Hoffman, *Jig and Fixture Design*, 2nd edn., Delmar Publishers Inc., Albany, NY, 1985.

91. W. E. Boyes and R. Subrin, *Low Cost Jigs and Fixtures & Gages for Limited Production*, SME, Dearborn, MI, 1986.

92. W. E. Boyes and R. Bakerjian, *Handbook of Jig and Fixture Design*, SME, Dearborn, MI, 1989.

93. B. Shirinzadeh, Issues in the design of the reconfigurable fixture modules for robotic assembly, *J. Manuf. Syst.* **12**:1 (1993) 1–13.

94. K. Nyamekye and S. S. Mudiam, A model for predicting the initial static gripping force in lathe chucks, *Int. J. Adv. Manuf. Technol.* **7** (1992) 285–291.

95. M. V. Gandhi and B. S. Thompson, Automated design of modular fixtures for flexible manufacturing systems, *J. Manuf. Syst.* **5**:4 (1986) 1–13.

96. R. J. Menassa and W. R. DeVries, Locating point synthesis in fixture design, *CIRP Ann.* **38** (1989) 165–169.

97. T. J. Drozda, R. E. King, and J. B. Creutz, *Jigs and Fixtures*, 3rd edn., SME, Dearborn, MI, 1989.

98. P. C. Miller, Workholding for CNC efficiency—Part I and II, *Tool. Prod.* (December 1994, 29–32 and January 1995) 23–28.

99. R. Okolischan and J. Camp, Modular fixturing slashes leadtime, *Cutting Tool Eng.* (June 1992) 99–102.

100. J. L. Colbert, R. Menassa, and W. R. DeVries, A modular fixture for prismatic parts in an FMS, *Proc. NAMRC* **14** (1986) 597–602.

101. B. Benhabib, K. C. Chan, and M. Q. Dai, A modular programmable fixturing system, *ASME J. Eng. Ind.* **113** (1991) 93–100.

102. J. H. Buitrago and K. Youcef-Toumi, Design of active modular and adaptable fixtures operated by robot manipulators, *Proceedings of USA-Japan Symposium on Flexible Automation*, Minneapolis, MN, July 1988, pp. 467–474.

103. D. Chakraborty, E. C. De Meter, and P. S. Szuba, Part location algorithms for an intelligent fixturing system—Part 1 and 2, *J Manuf. Syst.* **20** (2001) 124–148.

104. Lamb Technicon Corporation, *Intelligent Fixturing System* undated technical brochure, Warren, MI.

105. R. Wilson, Magnetic manufacturing technology," *Automotive Ind.* **184**:8 (August 2004) 39.

106. C. H. Shen, Y. T. Lin, J. S. Agapiou, G. L. Jones, M. A. Kramarczyk, and P. Bandyopadhyay, An innovative reconfigurable and totally automated fixture system for agile machining applications, *Trans. NAMRI/SME* **31** (2003) 395–402.

107. C. H. Shen, Y. T. Lin, J. S. Agapiou, P. A. Bojda, G. L. Jones, and J. P. Spicer, Reconfigurable workholding fixture, US Patent 6,644,637, November 11, 2003.

108. C. H. Shen, Y. T. Lin, J. S. Agapiou, and P. Bandyopadhyay, Reconfigurable fixtures for automotive engine machining and assembly applications, in: O. Dashchenko, Ed., *Transformable Factories and Reconfigurable Manufacturing in Machine Building*, Springer-Verlag, Berlin, Germany, 2005, Ch. 42.

109. M. V. Gandhi and B. S. Thompson, Phase change fixturing for flexible manufacturing systems," *J. Manuf. Syst.* **4**:1 (1985) 29–38.

110. M. V. Gandhi, B. S. Thompson, and D. J. Mass, Adaptable fixture design: An analytical and experimental study of fluidized-bed fixturing, *J. Mech.* **108**:15 (March 1986).

111. J. Abou-Hanna and K. Okamura, Mechanical properties of steel pellets in particulate fluidized bed fixtures, *J. Manuf. Syst.* **10**:4 (1991) 307–313.

112. G. A. Phillipson, Electropermanent magnetic fixtures: An attractive alternative, *Cutting Tool Eng.* (June 1993) 85–89.

113. K. Rong and S. Zhu, *Computer-Aided Fixture Design*, Marcel Dekker, New York, 1999.

114. Y. C. Chou and M. M. Barash, A mathematical approach to automatic design of fixtures: Analysis and synthesis, *Proceedings of ASME Winter Annual Meeting*, Boston, MA, 1986, pp. 11–27.

115. J. C. Trappey and C. R. Liu, An automatic workholding verification system, *Proceedings of International Conference on Manufacturing Technology of the Future (MSTF'89)*, Stockholm, Sweden 1989, pp. 23–24.

116. A. Markus, Z. Markusz, J. Farkas, and J. Filemon, Fixture design using prolog: An expert system, *Robot. Comput. Integr. Manuf.* **1**:2 (1984) 167–172.

117. S. Nnaji et al., A framework for a rule-based expert fixturing system for face planer surfaces on a CAD system using flexible fixtures, *J. Manuf. Syst.* **7**:3 (1988) 193–207.

118. M. R. Cutkosky and S. H. Lee, Fixture planning with friction for concurrent product/process design, *Proceedings of NSF Engineering Design Research Conference*, University of Massachusetts, Amherst, MA, 1989, pp. 613–628.

119. S. L. Jeng, L. G. Chen, and W. H. Chieng, Analysis of minimum clamping force, *Int. J. Mach. Tools Manuf.* **35** (1995) 1213–1224.

120. R. O. Mittal et al., Dynamic modeling of the fixture-workpiece system, *Robot. Computer-Integr. Manuf.* **8** (1991) 201–217.

121. X. J. Wan and Y. Zhang, A novel approach to fixture layout optimization on maximizing dynamic machinability, *Int. J. Mach. Tools Manuf.* **70** (2013) 32–44.

122. H. Asada and A. By, Kinematic analysis of workpart fixturing for flexible assembly with automatically reconfigurable fixtures, *IEEE J. Robot. Autom.* **RA-1** (1985).

123. Y. C. Chou, V. Chandu, and M. M. Barash, A mathematical approach to automatic configuration, of machining fixtures, analysis and synthesis, *ASME J. Eng. Ind.* **111** (1989) 299–306.

124. M. Mani and W. R. D. Wilson, Automated design of workholding fixtures using kinematic constraint synthesis, *Proc. NAMRC* **15** (1988) 427–432.

125. R. J. Menassa and W. R. DeVries, A design synthesis and optimization method for fixtures with compliant elements, in: P. H. Cohen and S. B. Joshi, Eds., *Advances in Integrated Product Design and Manufacturing*, ASME PED Vol. 47, ASME, New York, 1990, pp. 203–218.

126. H. Wanga, Y. Ronga, H. Li, and P. Shaun, Computer aided fixture design: Recent research and trends, *Comput.-Aided Des.* **42** (2010) 1085–1094.

127. I. Boyle, Y. Rong, and D. C. Brown, A review and analysis of current computer-aided fixture design approaches, *Robot. Comput. Integr. Manuf.* **27** (2011) 1–12.

128. J. Q. Xie, J. S. Agapiou, D. A. Stephenson, and P. Hilber, Machining quality analysis of an engine cylinder head using finite element methods, *J. Manuf. Proc.* **5** (2003) 170–184.

129. Y. Wang, J. F. Xie, Z. Wang, and N. Gindy, A parametric FEA system for fixturing of thin-walled cylindrical components, *J. Mater. Proc. Technol.* **205** (2008) 338–346.

130. S. Ratchev, K. Phuah, and S. Liu, FEA-based methodology for the prediction of part–fixture behaviour and its applications, *J. Mater. Proc. Technol.* **191** (2007) 260–264.

131. P. M. Ferreira and C. R. Liu, Generation of workpiece orientations for machining using a rule based system, *Robot. Comput. Integr. Manuf.* **4**:3/4 (1988) 543–555.

132. W. Jiang, Z. Wang, and Y. Cai, Computer-aided group fixture design, *CIRP Ann.* **37** (1988) 145–148.

133. B. Bidanda and P. H. Cohen, Development of a computer aided fixture selection system for concentric, rotational parts, *Advances in Integrated Product Design and Manufacturing*, ASME PED Vol. 47, ASME, New York, 1990, pp. 151–162.

134. J. S. Agapiou, E. Steinhilper, F. Gu, and P. Bandyopadhyay, Modeling machining errors on a transfer line to predict quality, *J. Manuf. Proc.* **5** (2003) 1–12.

6 Mechanics of Cutting

6.1 INTRODUCTION

The forces generated in metal cutting operations have long interested engineers. These forces determine machine power requirements and bearing loads; cause deflections of the part, tool, or machine structure; and supply energy to the machining system, which may result in excessive cutting temperatures or unstable vibrations. Measured cutting forces are also sometimes used to compare the machinability of materials, especially in cases in which tool life tests cannot be performed due to time constraints or limited material supplies. Cutting force measurements are also used for sensor-based control and monitoring of the cutting process for tool wear and failure.

This chapter gives a broad overview of cutting mechanics. Cutting force and chip thickness measurement methods are discussed in Section 6.2. Force components, empirical force models, and specific cutting power are discussed in Sections 6.3 through 6.5. A theoretical investigation of cutting mechanics requires consideration of deformation and frictional conditions, which are reviewed in Sections 6.6 and 6.7. Shear plane, shear zone, and finite element models for cutting forces are reviewed in Sections 6.8 through 6.12. The mechanics of discontinuous chip formation and built-up edge formation are discussed briefly in Sections 6.13 and 6.14.

6.2 MEASUREMENT OF CUTTING FORCES AND CHIP THICKNESS

Cutting forces are often measured in machinability testing and research. Measurements are usually made using specially designed dynamometers. Early researchers used a variety of hydraulic, pneumatic, and strain gage instruments [1,2]. More recently, however, piezoelectric dynamometers employing quartz load measuring elements have become standard for routine measurements [3–8]. Compared to other designs, quartz dynamometers have a high stiffness and broad frequency response; depending on mounting arrangements, bandwidths over 1 kHz are common. They are also very stable thermally and exhibit little static cross talk between measurements in different directions.

Most commonly, dynamometers are mounted between the tool or workpiece and a nonrotating part of the machine tool structure (Figure 6.1). This arrangement simplifies the wiring required to route signals to amplifiers, as well as the interpretation of measured data. Dynamometers of this type are called platform dynamometers. Traditionally, they have only been used to measure cutting forces in three directions, but recently four-component milling dynamometers that can also measure the torque on the cutter have become available. In some applications, however, it is desirable to mount the dynamometer in the spindle so that it measures forces in a rotating coordinate system [6,9–11] (Figure 6.2). This is most often true in boring and milling tests, since forces on individual inserts can be resolved without geometric transformations of the signals, and in tests on irregular parts, which present fixturing difficulties. Spindle-mounted dynamometers are also used when milling large workpieces; in this case, the weight of the workpiece on a table-mounted dynamometer can reduce the natural frequency of the measuring system to a value comparable to the tooth pass frequency [12]. The main difficulty in using spindle-mounted dynamometers in earlier work was that signals had been routed through slip rings to stationary amplifiers or recorders. Current dynamometers use radio frequency transmission or other noncontact methods to eliminate this issue. An additional difficulty is that adapters are required to mount the dynamometer into the spindle nose and the tool into the dynamometer; if it is desired to use the dynamometer on a variety of machines with different spindle mounting methods and different toolholders, a large inventory of adapters must be maintained. The adapters may also increase tool assembly overhang and reduce stiffness.

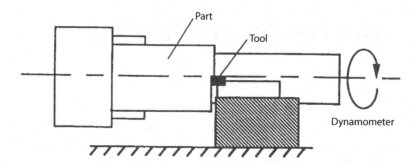

FIGURE 6.1   Platform dynamometer.

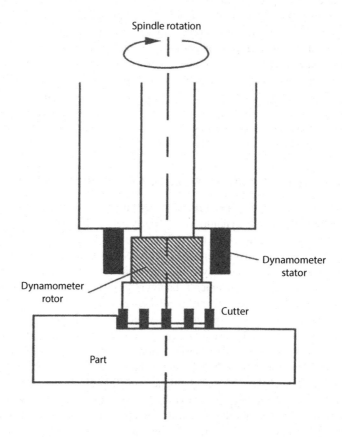

FIGURE 6.2   Rotor–stator type spindle dynamometer.

The frequency response of the measurement system is an important issue in testing at very high cutting speeds. Piezoelectric platform dynamometers typically have bandwidths between 800 and 2000 Hz. Filtering has been reported to increase effective dynamometer bandwidth [13–15], up to roughly 3500 Hz in a typical drilling case [16]. Special inserts with piezoelectric coatings have also been investigated to provide very high bandwidths (up to 40 MHz) for high speed milling and acoustic emission studies [17,18]. For accuracy, these inserts require careful dynamic calibration, since the signals they provide vary nonlinearly with the applied force. In tests at the highest cutting speeds, such as those produced by ballistic or gas gun test rigs, no dynamometer has a sufficiently high bandwidth, and forces must be inferred indirectly from the change in momentum estimated from workpiece velocity measurements [19–22].

The measurement of forces on production machinery presents special difficulties. Often, it is not possible to shut machines down for extended periods to install instrumentation, or to use fixtures that significantly change the location of the tool point or workpiece surface. In these cases, specially modified toolholders incorporating strain gages or piezoelectric force cells under the tool may be used to minimize installation time and changes in tool point position [23,24]. In other cases, single- or three-axis force transducers (e.g., piezoelectric cells) are installed in the base of fixtures or the spindle carrier (at the interface of the carrier and spindle housing, behind the coupling of turret housing, behind the outer race of the front spindle bearings, etc.) The spindle bearings can also be instrumented with strain gages to measure forces. Cutting power can be inferred from feed or drive motor current measurements [25,26]. Current measurements alone usually have limited bandwidth, and the calibration required to relate current levels to power is often difficult to perform on older equipment. Spindle and feed drive power measurements, coupled with additional vibration sensors, are often used in machine tool monitoring and advanced adaptive control systems [27,28].

Routine forces measurements are generally quite accurate. For commercial dynamometers, significant errors are likely to occur only if the cutting force is applied at a distance from the measuring elements; in this case, a large torque acts on the dynamometer surface, and some load-measuring elements may become overloaded. Dynamometer manufacturers provide technical documentation indicating acceptable force ranges and application distances for a particular instrument. In processes such as boring in which a large tool overhang is expected, specially instrumented tools may produce more accurate results than commercial platform dynamometers [29,30]. The accuracy of specially built dynamometers is most often limited by calibration errors, especially for dynamic measurements at higher frequencies, which require careful calibration [31].

Force measurements from repeated tests under nominally identical conditions typically show variations of roughly 10% around mean values, with the main cutting force (the tangential force in turning) being more repeatable than the other components [32]. These variations, however, are generally not due to instrument errors but rather due to variations in workpiece properties and tool cutting-edge geometry or wear.

The chip thickness, which determines the shear angle (Sections 6.6 and 6.8), is also sometimes measured in laboratory work. This parameter is more difficult to measure accurately than cutting forces. The average chip thickness is most commonly measured directly using a point micrometer or optical microscope. Depending on the material being machined, such direct measurements show a variability of between 10% and 25% about mean values [32]. The chip thickness can also be inferred from a comparison of the length of the chip and the cutting distance. This method is effective in interrupted cutting tests, in which discrete sections of known length are cut, and in tests in which a long, continuous chip is produced and the cutting time is accurately known. In the latter case, if the chip is tightly curled or tangled, the chip thickness can be estimated by weighing a known length of chip [2]. Measurements using these methods usually show less scatter than direct measurements and are regarded as more reliable than micrometer measurements [2,33]. Scatter in chip thickness measurements results from several sources, including the roughness of the free surface of the chip and variations in chip thickness resulting from shear localized (type 4) chip formation or (for soft metals) melting of the chip near the cutting edge. In the last two cases, it is not very critical to measure the average chip thickness accurately, since the parameter has less physical significance.

6.3 FORCE COMPONENTS

A number of coordinate systems can be used to resolve the cutting force into directional components. Most analyses use coordinate systems in which one axis is parallel to either the cutting edge or cutting velocity. For a simple oblique machining process, the resulting force components for mutually orthogonal coordinate directions for these two cases are shown in Figure 6.3. F_n and F_c are

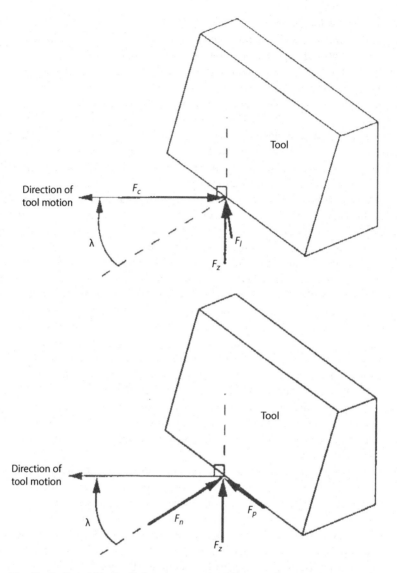

FIGURE 6.3 The F_c–F_l–F_z and F_n–F_p–F_z force systems in oblique cutting.

the components of force normal to the cutting edge and parallel to the cutting velocity, respectively; F_p is the force component parallel to the cutting edge; F_z is the force component normal to the plane defined by F_n and F_p; and F_l is the force component normal to the plane defined by F_c and F_z. F_c, F_l, F_n, and F_p are related by the inclination angle λ through the equations

$$F_n = F_c \cos\lambda + F_l \sin\lambda \qquad (6.1)$$

$$F_p = -F_c \sin\lambda + F_l \cos\lambda \qquad (6.2)$$

$$F_c = F_n \cos\lambda - F_p \sin\lambda \qquad (6.3)$$

$$F_l = F_n \sin\lambda + F_p \cos\lambda \qquad (6.4)$$

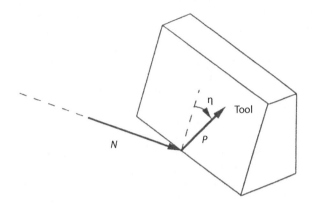

FIGURE 6.4 The forces N and P on the rake face of an oblique cutting tool.

F_n and F_p are always positive; the positive senses of λ and F_l should be defined so that both always have the same sign.

Some analyses use coordinate systems referred to the rake face of the tool. In this system (Figure 6.4), the most commonly defined force components are N and P, the force components normal and parallel to the rake face. N and P can be computed from the equations

$$N = F_n \cos\alpha - F_z \sin\alpha \tag{6.5}$$

$$P = \frac{F_n \sin\alpha + F_z \cos\alpha}{\cos\eta} \tag{6.6}$$

where α is the normal rake angle of the tool, measured in a plane normal to the cutting edge, and η is the chip flow angle, defined as the angle between the direction of chip flow and the normal to the cutting edge, measured in the plane of the rake face of the tool.

The chip flow angle can be measured directly using a microscope or high-speed camera mounted near the tool [34], but this is rarely done. More commonly, the apparent chip flow angles due to cutting forces or the chip width, η_f or η_w, are measured. η_f is calculated from cutting forces using the relation

$$\tan\eta_f = \frac{F_p}{F_n \sin\alpha + F_z \cos\alpha} \tag{6.7}$$

η_w is determined from chip width measurements (Figure 6.5) using the formula

$$\cos\eta_w = \frac{b_c \cos\lambda}{b} \tag{6.8}$$

In most cases, η, η_f, and η_w are equal to within the accuracy of the measurements required to compute them. In the remainder of this chapter, we will assume the equality of these parameters,

$$\eta = \eta_f = \eta_w \tag{6.9}$$

and use the notation η in equations. The assumption $\eta = \eta_f$ implies that the friction force P acts in the direction of chip flow. Equation 6.9 is inaccurate only when turning small diameter parts (i.e., those

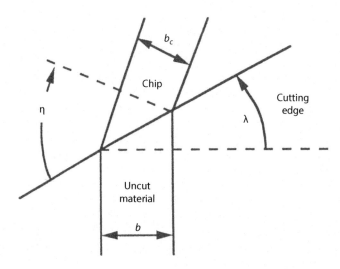

FIGURE 6.5   Relationship between width of cut b, width of chip b_c, inclination angle λ, and chip flow angle η in oblique cutting.

with diameter on the order of 2 cm), since in this case the curvature of the part influences the chip flow direction. In this case, the chip flows along an arc rather than a straight line so that the concept of a chip flow direction loses meaning. Since chip width measurements are difficult to perform accurately, values of η determined from Equation 6.7 are usually more repeatable than those determined from Equation 6.8.

For a straight-edged oblique cutter, η is determined by the inclination angle λ:

$$\eta = \arctan\left[C\tan\lambda\right] \tag{6.10}$$

Experiments usually yield values of C_η between 0.7 and 1.2. C_η is often assumed to be 1.0; in this case, Equation 6.10 is referred to as Stabler's rule [35]. This assumption implies that there is no width change in the chip, and thus, the chip is formed by plain strain deformation. Deviations from Stabler's rule are usually smaller than the typical scatter in measurements. It is most accurate when λ or the rake angle α is small; when this condition is not met, the relation $C_\eta = \tan\alpha$ is more accurate when cutting steel (Figure 6.6) [36]. For a turning tool with a nose radius, η varies with the depth of cut and nose radius as well as λ and can be calculated using more complicated relations [37–40] as discussed in Section 8.6. Under these conditions, η is usually referred to as the effective lead angle.

F_c, F_z, and F_l can be calculated from N and P using the relations

$$F_c = F_{nz}\cos\lambda + F_{pz}\sin\lambda \tag{6.11}$$

$$F_z = P\cos\eta\cos\alpha - N\sin\alpha \tag{6.12}$$

$$F_l = F_{nz}\sin\lambda - F_{pz}\cos\lambda \tag{6.13}$$

where

$$F_{nz} = P\cos\eta\sin\alpha + N\cos\alpha \tag{6.14}$$

$$F_{pz} = P\sin\eta \tag{6.15}$$

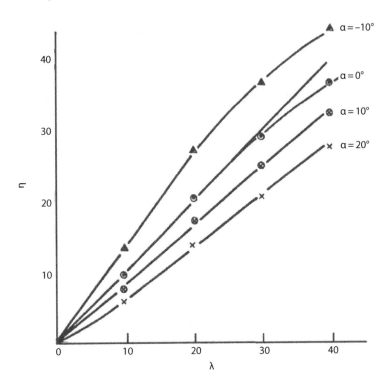

FIGURE 6.6 Relation between the chip flow angle η and inclination angle λ when cutting 1008 steel with tools with various rake angles. (After Brown, R.H., and Armarego, E.J.A., *Int. J. Mach. Tool Des. Res.*, 4, 9, 1964.)

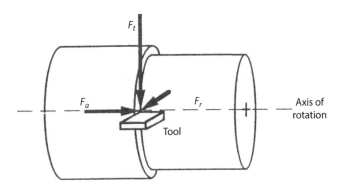

FIGURE 6.7 Forces on a turning tool.

In specific processes, force components are often related to the axes of motion of the machine tool. In turning and boring, the tangential (cutting), axial (feed), and radial force components F_t, F_a, and F_r shown in Figure 6.7 are used. F_t and F_a act in the cutting speed and feed directions, respectively. In milling the tangential (cutting), radial, and axial force components, F_t, F_r, and F_a shown in Figure 6.8 are used. If a spindle-mounted or advanced platform dynamometer is used, the torque M on the cutter can also be measured. The axial force in turning is equivalent to the radial force in milling, and the radial force in turning is equivalent to the axial force in milling. In drilling, the thrust or axial force Th, torque M, and radial force F_r shown in Figure 6.9 are used. F_r rotates with time and is measured by measuring orthogonal components in directions fixed with respect to the machine tool.

The feed and radial forces are proportional to the tangential cutting force, with the proportionality factor depending on the depth of cut (doc) and lead angle. In rough turning with zero or small

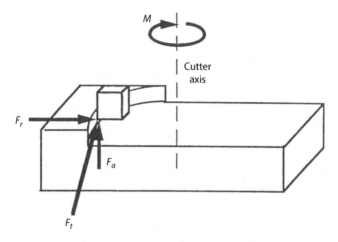

FIGURE 6.8   Forces on a cutting insert in face milling.

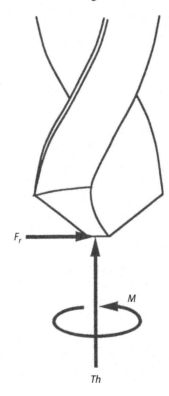

FIGURE 6.9   Forces in drilling.

lead angle the ratio $F_t{:}F_a{:}F_r$ is typically 4:2:1. The ratio of the axial and radial forces is reduced as the lead angle increases. In addition, the axial and radial forces approach or even exceed the tangential force in finishing operations where the doc is small. In peripheral (end) milling operations, the radial force is approximately 30%–50% of the tangential cutting force and the axial force is affected by the lead of the cutter and the edge corner geometry. In drilling, the total thrust Th is the sum of the chisel edge thrust and main cutting-edge thrust. The chisel edge thrust depends on the length of the chisel edge as explained in Chapters 4 and 8. For small-diameter drills, the thrust force is typically 1–1.5 times the tangential cutting force, and for larger drills (especially spade or indexable drills), it is typically about one half.

6.4   EMPIRICAL FORCE MODELS

Cutting force estimates are often used to determine machine power requirements and bearing loads, and to design fixtures. In practice, forces are usually estimated using a variety of similar empirical equations [41–48].

For a given material, cutting forces depend most strongly on the feed rate and the width of cut. For a straight oblique cutting edge, the forces depend linearly on the width of cut, provided this parameter is much larger than the feed per revolution. For a turning tool with a nose radius, which has curved cutting edges, the dependence on the corresponding parameter, the effective length of the cutting edge (or cutting-edge engagement), is also almost linear. (Equations for the effective cutting-edge engagement for various processes are given in Chapter 2.) Cutting forces are therefore often divided by the width of cut or effective edge length to determine the force per unit length of cutting edge. This parameter, which is important in determining if the tool will chip or break, increases most strongly with the feed rate. Increasing the rake angle generally reduces cutting forces; similarly, forces are usually lower for a tool with a formed-in chip breaker than for a flat-faced tool, since the effective rake angle is larger. The cutting forces increase significantly with increasing flank wear. Varying the cutting speed has a minor effect on cutting forces for most materials and cutting conditions. Speed effects are important, however, when cutting materials with temperature-sensitive material properties, such as titanium alloys and some steels, and when cutting over a wide speed range. Speed effects are also important in cutting soft metals at very low speeds, since a built-up edge may form and increase the effective rake angle, or when cutting at very high speeds when the momentum of the chip may become significant.

In orthogonal and oblique cutting with a straight cutting edge, the force per unit width of cut is usually assumed to vary exponentially with the feed rate and cutting speed, and to vary linearly or sinusoidally with the rake angle. For example, the forces per unit width of cut normal and parallel to the rake face of the tool can be calculated from [49]

$$\frac{N}{b} = C_1 a^{a_1} V^{b_1} (1 - \sin \alpha)^{c_1} \tag{6.16}$$

$$\frac{P}{b} = C_2 a^{a_2} V^{b_2} (1 - \sin\alpha)^{c_2} \tag{6.17}$$

where
$\quad$ N and P are the forces normal and parallel to the tool rake face
$\quad$ a is the uncut chip thickness
$\quad$ b is the width of cut
$\quad$ V is the cutting speed
$\quad$ α is the normal rake angle
$\quad$ C_1, a_1, b_1, c_1, C_2, a_2, b_2, and c_2 are parameters that depend on the tool/workpiece material combination

Typical values of these parameters for some common work materials are given in Table 6.1. Other force components such as F_c, F_n, F_z, F_l, and F_p can be calculated from the estimates of N and P using the equations summarized in Section 6.3 or directly from equations similar to Equations 6.16 and 6.17, which are sometimes given in tooling catalogs or handbooks.

For tools with a nose radius, forces are often calculated by multiplying the area of material being cut by an empirical cutting pressure, which are, in turn, modeled using exponential equations. This approach is used in many simulation programs and is discussed in Section 8.6.

Exponential equations generally provide accurate force estimates. They tend to overestimate the force at low feed rates, leading to an overestimate of power requirements for machines used for

TABLE 6.1

Coefficients in Equations 6.16 and 6.17 When Cutting Common Materials with Tungsten Carbide Tools

Material	C_1	a_1	b_1	c_1	C_2	a_2	b_2	c_2
Gray cast iron (180 BHN)	1106.6	0.011	0.760	1.277	436.2	−0.042	0.595	0.065
Nodular cast iron (280 BHN)	1820.7	−0.092	0.649	0.849	1526.9	−0.091	0.400	−0.082
1018 Steel (cold drawn)	2032.4	−0.080	0.730	1.425	889.8	0.000	0.641	0.958
10L45 Steel (cold drawn)	1571.8	−0.101	0.682	0.853	567.4	−0.096	0.515	−0.491
319 Aluminum (cast)	673.2	0.083	0.936	1.109	288.9	0.200	0.912	0.000
2024-T6 Aluminum (cold drawn)	863.1	−0.018	0.800	1.007	566.2	0.000	0.688	0.000

Note: These values give N and P in N/mm^2 when V is given in m/min, a in mm, and α in degrees. They are valid for cutting speeds below 300 m/min, uncut chip thicknesses between 0.1 and 0.5 mm, and rake angles between −10° and 0° (iron and steel) or 0° and 20° (aluminum alloys).

finishing cuts, but the practical significance of this error is usually small. A useful feature of these equations is that they yield reasonable force estimates even when extrapolating beyond the range of cutting conditions used to generate baseline data. Such extrapolation should be ideally avoided but may be necessary, in some cases, to reduce testing requirements.

One limitation of the exponential models is that all variables appear independently, making it impossible to model interactions between variables. This limitation can be overcome by fitting polynomial equations that include cross-product terms. Such equations are easily fit to data from factorial experimental designs. The results often show statistically significant interactions between the feed rate and rake angle and, for some materials, between the cutting speed and all other inputs [16,47,48]. The gain in accuracy over exponential equations, however, is generally small except over narrow ranges of inputs (e.g., for low feed rates comparable to the chamfer on the tool edge). Equations including interaction terms also often yield unreasonable (or even negative) force estimates when extrapolated to conditions outside those used to gather the data.

Force components related to the cutting power, such as the tangential cutting force F_t in turning or the torque M in drilling, can also be estimated from tabulated values of the specific cutting power using the equations given in the next section. Values of the specific cutting power for common work materials can be obtained from machinability databases and tooling catalogs; values for some materials are given in Table 2.1. Estimates obtained in this way are not as accurate as those obtained from more complicated empirical relations but are often adequate for cutting power and bearing load estimates and can often be made without performing cutting tests.

6.5 SPECIFIC CUTTING POWER

The specific cutting power u_s, also called the unit cutting power, specific cutting energy, or unit cutting energy, is the power required to machine a unit volume of the work material. It is easily computed from measured cutting forces and is often used to compare the machinability of different materials, especially when comparative tool life data is unavailable.

The rotational speed is much larger than the feed speed in most cutting operations so that only the torque or tangential force on the rotating element of the system must be considered in computing u_s. In *turning* and *boring*, the power consumption is given by the product of the cutting speed and tangential cutting force, VF_t. The volume of material removed per unit time is given by Equation 2.6 so that u_s is given by

$$u_s = \frac{F_t}{fd} \tag{6.18}$$

In *drilling*, the power is determined by the product of the torque M and the rotational speed ω, and the volume of material removed per unit time is $D^2 f \omega /8$, where D is the drill diameter and f is the feed. u_s is thus given by

$$u_s = \frac{8M}{D^2 f} \tag{6.19}$$

M can be measured directly using a drilling dynamometer. In *milling*, the cutting forces and the cutting power vary continuously. Instantaneous values of u_s can be computed from measured cutting forces or torque. For a three-component platform dynamometer, the tangential force F_t can be computed from measured forces if a single tooth cutter (or fly cutter) is used (Figure 6.10). The instantaneous specific cutting power in this case is given by

$$u_s = \frac{F_t}{fd} \tag{6.20}$$

For a four-component platform and spindle-mounted dynamometers, the torque on the cutter can be measured directly, and the instantaneous value of u_s is given by

$$u_s = \frac{2M}{Dfd} \tag{6.21}$$

Values for multi-tooth cutters can be calculated by summing results on individual inserts.

Typical values of u_s for various materials are listed in Table 2.1. u_s varies somewhat with cutting conditions. It usually increases as the feed decreases because the force component due to flank friction becomes increasingly significant. The specific energy is affected significantly by the cutting-edge geometry (sharp, hone, or K-land as discussed in Chapter 4). u_s increases if the undeformed chip thickness, a, becomes smaller than the hone radius or K-land width due to increased plowing rather than shearing deformation. It also increases as the rake angle is decreased, since this increases the cutting force. Finally, it normally decreases slightly as the cutting speed is increased. In using u_s to rate the machinability of various materials, similar cutting conditions should be used in tests on all materials.

In turning tests, u_s is sometimes divided into deformation and frictional energy components, u_d and u_f, which determine the proportions of the total energy expended in deforming the chip

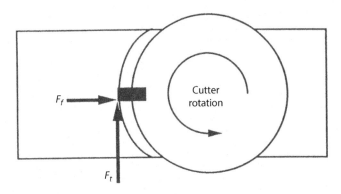

FIGURE 6.10 Forces on the tool in fly milling. The tangential force F_t, feed force F_f, and axial force F_a (directed out of the page) can be measured with a three-component platform dynamometer when the tool is at full engagement.

and in overcoming friction between the tool and the chip, respectively. If Stabler's rule, $\eta = \lambda$, is assumed, u_f and u_d are given by

$$u_f = \frac{P}{a_c \cdot b} \tag{6.22}$$

$$u_d = u_s - u_f \tag{6.23}$$

where
  P is the force parallel to the rake face of the tool (Section 6.3)
  a_c, a, and b are the chip thickness, uncut chip thickness, and chip width (Section 2.2)

u_f has the greater influence on tool life; an increase in u_f will usually increase temperatures along the tool–chip interface and lead to more rapid tool wear.

6.6 CHIP FORMATION AND PRIMARY PLASTIC DEFORMATION

The preceding sections have discussed the empirical study of cutting forces. Developing theoretical models based on solid or continuum mechanics requires consideration of deformation and frictional conditions in cutting.

  Much of the plastic deformation in machining occurs in the formation of chips, which can be produced in a great variety of shapes and sizes. The first task in analyzing cutting processes is to identify the basic types of chips produced by distinct physical mechanisms. Four such basic types are shown in Figure 6.11, obtained by taking a cross section perpendicular to the cutting edge as illustrated in Figure 6.12. The first three types were defined by Ernst [50] and Merchant [51] in their early development of the theory of orthogonal cutting; the fourth was widely observed somewhat

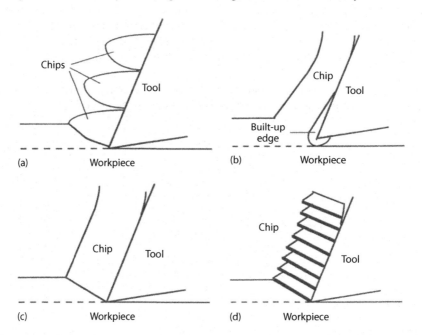

FIGURE 6.11 Four basic types of chips: (a) discontinuous, (b) continuous, (c) continuous with built-up edge (BUE), and (d) shear localized.

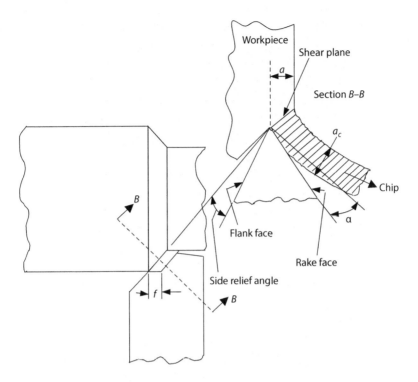

FIGURE 6.12 Cross section of chip formation perpendicular to cutting edge.

later as cutting speeds increased and new work materials were introduced. Type 1 chips are discontinuous chips of appreciable size. These are formed by a fracture mechanism when brittle materials such as beta brass are cut at low cutting speeds. Type 2 chips are continuous chips formed without a built-up edge on the tool. This is the most desirable type of chip. Type 3 chips are continuous chips formed with a built-up edge; these are usually encountered when machining soft, ductile metals, although many materials form a built-up edge at low cutting speeds. Type 4 chips are macroscopically continuous chips consisting of narrow bands of heavily deformed material alternating with larger regions of relatively undeformed material. These shear-localized (or catastrophic shear) chips can be formed when the yield strength of the work material decreases with temperature; under the proper conditions, rapidly heated material in a narrow band in front of the tool can become much weaker than the surrounding material, leading to localized deformation [52–54]. This type of chip is obtained when cutting hardened and stainless steels and titanium alloys at high cutting speeds [22,55–60].

This system of classification is adequate for idealized analyses of cutting but does not reflect some readily observed details of chip formation. Micrographs of etched and polished chips (Figure 6.13), especially specimens obtained from quick-stop devices, show that even macroscopically continuous chips are not uniform, as might be suggested from Figure 6.11. Most chips are smooth on the side formed by contact with the tool but exhibit a rough texture on the free surface. Grains in the chips are deformed in characteristic directions, which may change through the thickness of the chip, particularly at low cutting speeds [61]. Deformation is also often concentrated in narrow bands. This is true of plastic deformation of metals in general but suggests that type 1 and type 4 chips are formed by a similar mechanism, and that the macroscopic differences between these chips result from differences in the scale and degree of concentration.

Deformation conditions have been most widely studied and are best understood for continuous chip formation. In this type of chip formation, plastic deformation occurs in two regions: the primary deformation zone, which stretches from the tool tip to the free surface of the workpiece,

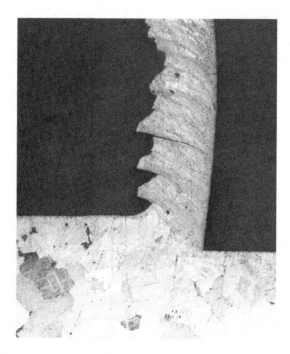

FIGURE 6.13 Typical metallographic section of a zinc chip, showing localized deformation and irregular free surface. (Courtesy of R. Stevenson, General Motors Research, Warren, MI.)

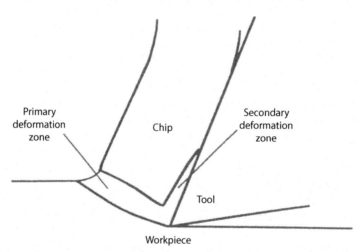

FIGURE 6.14 Deformation zones in cutting.

and the secondary deformation zone at the tool–chip interface (Figure 6.14). This section describes deformation conditions in the primary deformation zone; the secondary deformation zone and tool–chip friction are discussed in Section 6.7.

The primary deformation zone can be studied experimentally using a number of methods. The most useful early data comes from tests using "quick stop" devices [62–71]. In these devices, the tool is fixed to the machine tool through some mechanism that permits it to be withdrawn from the workpiece very rapidly, abruptly interrupting the cut and leaving a partially formed chip on the workpiece. The chip sample can then be sectioned and examined; the boundaries of the deformation zones and the strains imparted to the chip can be identified by studying the workpiece grain structure or the deformation of a previously inscribed grid. The tool can never be withdrawn instantaneously, and given the small dimensions of the chip and deformation zone, even a minimal deceleration period can introduce errors. Also, the method provides only a picture of the deformation at a particular instant and gives no information on time-varying phenomena. This method, however, can be used at normal cutting speeds, and permits data to be obtained at the middle of the chip, rather than the edges. The chip formation process can also be interrupted at very low cutting speeds by simply stopping the feed in a planer-like test rig [61,72–74].

The primary deformation zone can also be observed by taking high-speed films or videos of the side of the workpiece [62,75–82]. The earliest studies used very low cutting speeds (on the order of 1 cm/min). As better cameras have become available, results for normal cutting speeds have been reported. The major drawback of this method is that the side of the cut is observed. This is an anomalous region where side spread may occur so that the conditions there may differ significantly from those prevailing through the bulk of the deformation zone. Similar methods combining still photography to freeze cutting action at an instant have also been proposed [83] but are subject to the same limitations.

The stress distribution in the primary deformation zone can be studied using photoelastic methods [84]. These techniques can only be applied to transparent (e.g., plastic) work materials and yield largely qualitative information on the deformation characteristics of ductile metals. Chip formation can also be observed in detailed machining studies using special devices inside a scanning electron microscope [85–88]; this method is limited to relatively soft metals and very low cutting speeds.

Overall, the evidence from all methods is qualitatively consistent. The primary deformation zone is a narrow zone in which large strains are imparted to the chip. Most evidence indicates that the average thickness of the zone is on the order of one-tenth of the thickness of the chip. At low speeds, the zone is triangular in shape with the apex near the cutting edge, but as the cutting speed is increased this triangularity is reduced and the thickness of the zone becomes more constant. Most evidence also indicates that the curvature of the chip results directly from the primary deformation, rather than being produced by a subsequent process, which produces residual stresses. It is found that the curvature of chips does not change as their outer layers are dissolved in an acid bath [89]; if residual stresses were responsible for the curvature, some change would be expected. Finally, it is often observed that the deformation zone extends below the tool point, imparting a residual stress to the machined surface.

The strains imparted in the deformation zone are among the largest observed in any deformation process [2,57,90–93]. The magnitude of the strain depends on how the strain is defined. In metal cutting, the geometric shear strain γ calculated using the platelet model of chip formation (Figure 6.15) is often used. In this model, the strain is given by [2]

$$\gamma = \frac{\Delta S}{\Delta y} = \tan(\phi - \alpha) + \cot\phi \tag{6.24}$$

where
  α is the rake angle
  ϕ is the shear angle, which depends on the rake

The cutting ratio r_c of the uncut chip thickness a to chip thickness a_c (see Section 6.8 that follows):

$$\tan\phi = \frac{r_c \cdot \cos\alpha}{1 - r_c \cdot \sin\alpha} \tag{6.25}$$

$$r_c = \frac{a}{a_c} = \frac{\sin\phi}{\cos(\phi - \alpha)} \tag{6.26}$$

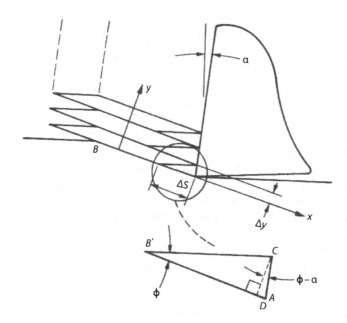

FIGURE 6.15 Platelet model for estimating shear strain in cutting. (After Shaw, M.C., *Metal Cutting Principles*, Oxford University Press, Oxford, U.K., 1984, Chs. 3, 7–9, and 12.)

Equation 6.24 can also be written as

$$\gamma = \frac{\cos \alpha}{\cos(\phi - \alpha) \cdot \sin \phi} \tag{6.27}$$

Strains calculated using these formulas for cutting steel, brass, and aluminum are plotted in Figure 6.16 [33]. As can be seen, strains are on the order of 100% and are highest at low rake angles.

The strains are imparted in a narrow zone that is rapidly traversed so that the machining strain rates are also larger than those observed in any process except ballistic impact. The shear strain rate for the platelet model of chip formation (Figure 6.15) is given by [2]

$$\dot{\gamma} = \frac{\cos \alpha}{\cos(\phi - \alpha)} \cdot \frac{V}{\Delta y} \tag{6.28}$$

where
  V is the cutting speed
  Δy is the thickness of the platelets

When cutting steel at moderate cutting speeds, Δy is typically less than 25 μm. Strain rates measured in quick-stop tests when cutting steel at various cutting speeds are plotted in Figure 6.17 [62]. Although the strain rate varies considerably with the rake angle, the rates on the order of 10^5 s^{-1} are observed when cutting speeds are up to 250 m/min. The strain rate increases roughly linearly with cutting speeds so that strain rates on the order of 10^6 s^{-1} occur when machining at speeds above 1000 m/min.

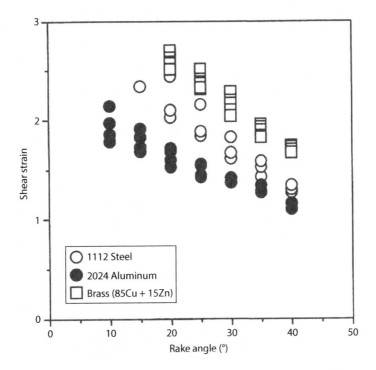

FIGURE 6.16 Shear strain when cutting steel, aluminum, and brass with tools with varying rake angles. (Data of Eggleston, D.M. et al., *ASME J. Eng. Ind.*, 81, 263, 1959.)

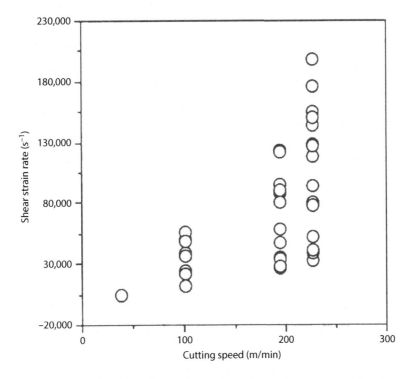

FIGURE 6.17 Shear strain rate in cutting when cutting steel at various cutting speeds. (Data from Kececioglu, D., *ASME Trans.*, 80, 158, 1958.)

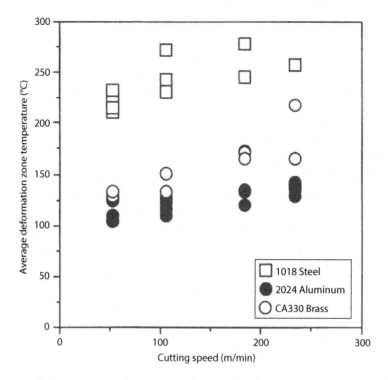

FIGURE 6.18 Average shear zone temperatures when cutting steel, brass, and aluminum at various cutting speeds. (Data from Stephenson, D.A., *ASME J. Eng. Ind.*, 113, 121, 1991.)

Average primary deformation zone temperatures, estimated from infrared measurements [94] for cutting steel, aluminum, and brass at various cutting speeds, are shown in Figure 6.18. Temperatures vary with the yield stress of the work material and are generally between 100°C and 300°C for soft metals cut at moderate speeds.

The shear flow stress of the work material in the primary deformation zone is usually larger than that measured in low-speed tension or torsion tests [93,95,96]. Figure 6.19 illustrates this effect for free-machining steel [97]. A number of unusual mechanisms, such as dependence of the flow stress on hydrostatic pressure [98,99], have been proposed to account for the increase, but it now seems to be plausibly explained by strain and strain rate hardening. The strains in the primary deformation zone are sufficient to produce saturation strain hardening, and most metals also exhibit strain rate hardening, especially at strain rates over $10^4\,s^{-1}$. Flow stresses measured in low-speed cutting tests appear to be consistent with those measured high-speed compression tests when strain, strain rate, and temperature effects are accounted for (Figure 6.20) [100].

The combination of strains, strain rates, and temperatures in the primary deformation zone, together with the frictional conditions to be discussed in the next section, cannot be reproduced in controlled material tests. As a result, the most difficult step in analyzing the mechanics of machining is describing the stress–strain behavior of the work material. Useful analyses must be based on realistic constitutive assumptions, and tests must be defined to estimate parameters in constitutive models.

Machining tests have sometimes been proposed for characterizing material flow properties at high strains and strain rates [101–103]. Since the strains, strain rates, and temperatures are not uniform throughout the deformation zone, some model of chip formation must be assumed to interpret the results. These tests are therefore useful mainly for estimating parameters for particular models.

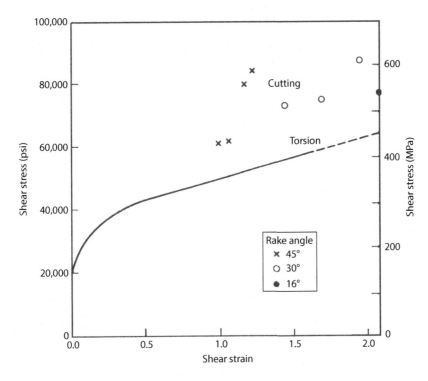

FIGURE 6.19 Shear stress of steel measured in cutting and torsion. (After Shaw, M.C. and Finnie, I., *ASME Trans.*, 77, 115, 1955.)

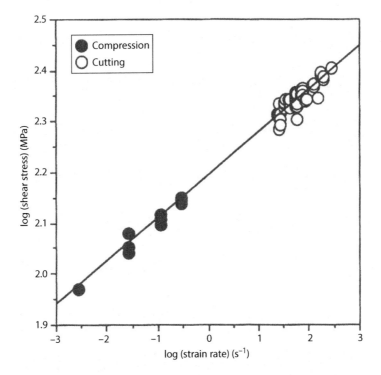

FIGURE 6.20 Shear stress of zinc in cutting and compression, corrected for strain and strain rate effects. (After Stevenson, R., and Stephenson, D.A., *ASME J. Eng. Mater. Technol.*, 117, 172, 1995.)

6.7   TOOL–CHIP FRICTION AND SECONDARY DEFORMATION

Friction between the tool and the chip in metal cutting influences primary deformation, built-up edge formation, cutting temperatures, and tool wear. An understanding of tool–chip friction is also necessary to develop accurate models for cutting forces and temperatures, since frictional stresses and heat fluxes are often used as boundary conditions [104,105].

The simplest way to characterize tool–chip friction is to define an effective friction coefficient, μ_e, as the ratio of the cutting force P parallel to the tool rake face to the force normal to the rake face, N:

$$\mu_e = \frac{P}{N} \tag{6.29}$$

μ_e is sometimes converted to a friction angle, β, given by

$$\beta = \arctan(\mu_e) \tag{6.30}$$

N and P can be estimated from cutting force measurements as discussed in Section 6.3. μ_e in cutting is usually larger than friction coefficients measured in conventional sliding friction tests; values above 1.0 are not uncommon. Increased friction in cutting results in part because the surface of the chip is newly formed and thus atomically clean. μ_e usually increases with the rake angle (Figure 6.21) [106] and also varies with the cutting speed. The friction coefficient often reaches a maximum over a narrow range of cutting speeds; in this range, the chip adheres strongly to the tool and may form a built-up edge (Figure 6.22) [60].

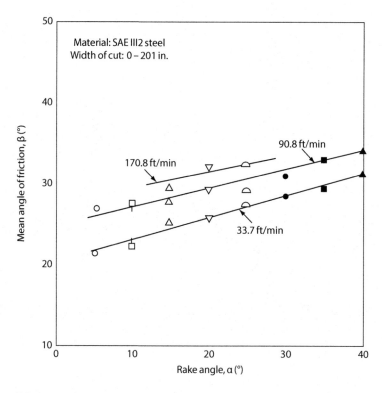

FIGURE 6.21   Friction angle when cutting steel with tools with varying rake angles. (After Bailey, J.A., *Wear*, 31, 243, 1975.)

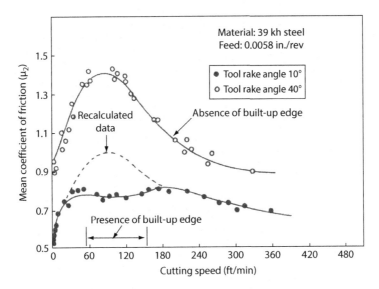

FIGURE 6.22 Friction angle when cutting steel at various cutting speeds, showing speed range over which built-up edge occurs. (After Bailey, J.A., *Wear*, 31, 243, 1975.)

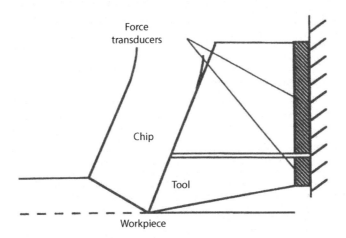

FIGURE 6.23 Composite tool for investigating the force distribution along the tool–chip interface.

The friction coefficient does not provide information on the distribution of stresses and strains across the tool–chip contact. Several experimental methods can be used to study the contact in more detail [107]. The most useful method employs special two-part tools (Figure 6.23) [108–112]. In this method, the tool holder is modified so that the forces on the two parts can be measured independently. By varying the relative sizes of the parts and noting how the forces acting on them change, the gradient of the force distribution (the stress distribution) can be measured, at least over parts of the interface. This method cannot be used to measure stresses at or very near the cutting edge because the parts of the tool cannot be made arbitrarily small.

Similar information can be obtained in tests with controlled contact length tools (Figure 6.24) [113–115]. In this method, contact stresses are inferred from differences in force components measured with tools with large recessed ground along the rake face to limit the tool–chip contact length to prescribed values. These results are not as useful as two-part tool results since limiting the contact length creates an artificial deformation geometry. The use of controlled contact tools greatly reduces the variability of force measurements, indicating that the significant scatter normally observed in repeated tests with conventional tools results from variations in tool–chip contact conditions.

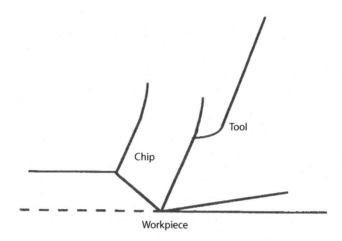

FIGURE 6.24 Tool with a restricted contact length.

Qualitative information on material behavior at the interface can be obtained from photoelastic studies [116–123] and from experiments in which soft metals such as lead are machined using transparent sapphire tools [124–127]. The normal stress distribution on the tool can also be inferred from the deformation of the cutting edge [128]. Average friction coefficients computed from measured forces, using standard cutting tests or special test rigs, can be used to infer the effects of process conditions, tool wear, and edge preparations on friction [129–132]. Finally, information on frictional strains can be obtained by examining chip samples from quick-stop tests and from the high-speed video studies discussed in the previous section.

The total contact length, which is an important parameter in calculating cutting temperatures, can be measured by all these methods. In addition, it can be estimated by examining the wear scar on the tool [33,129,130,133,134] and by ultrasonic methods [135]. The contact length in orthogonal cutting can also be calculated from slip-line solutions [132,136] or more complicated numerical models [105,137]; however, cutting data is normally needed to estimate empirical parameters in the analytical results.

Most methods yield consistent results [70,106,107]. The experimental observations are summarized in Figure 6.25. The total contact length L_c depends most strongly on the chip thickness a_c.

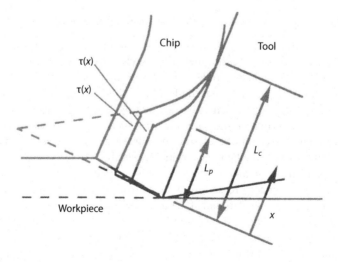

FIGURE 6.25 Tool–chip interface conditions.

Some research [115,120] indicates that the contact length is proportional to the product of the chip thickness and the effective friction coefficient:

$$L_c = C_f \mu_e a_c \tag{6.31}$$

C_f varies between 1.0 and 1.5 for common test conditions. This equation can be simplified to

$$L_c = C_l a_c \tag{6.32}$$

where C_l is a constant of proportionality. Reported values for C_l range from 1.75 [32] to 2.0 [110] for ductile metals and approximately 1.5 [94] for brittle materials such as cast iron. Measured values of the ratio of the contact length to the chip thickness show considerable scatter (Figure 6.26), and Equations 6.31 and 6.32 should be regarded as rough approximations.

The tool–chip contact consists of two parts, a region of sticking or seizure friction near the cutting edge, in which the work material is deformed in shear by high frictional stresses, and a region of sliding friction away from the cutting edge, in which such deformation does not occur (Figure 6.25). Most researchers indicated that the two regions are of roughly equal length, and a half sticking, half sliding contact has been assumed in research on cutting temperatures [94] and acoustic emissions [138]:

$$\frac{L_p}{L_c} = 0.5 \tag{6.33}$$

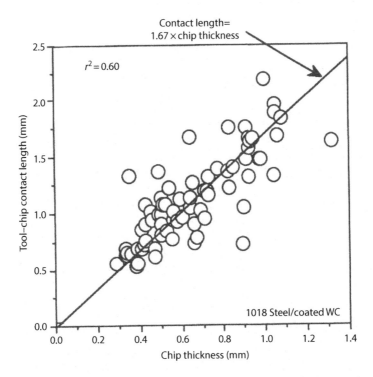

FIGURE 6.26 Relation between tool–chip contact length and chip thickness for 1018 steel workpieces cut at speeds up to 200 m/min. (Data from Stephenson, D.A., *ASME J. Eng. Mater. Technol.*, 111, 210, 1989.)

Experiments yield values of this ratio between 0.5 and 0.7, depending on the cutting conditions and tool angles.

The normal contract stress distribution $q(x)$ is found to be well approximated, at least over part of the contact, by a polynomial relationship. Based on photoelastic results, Zorev [120] proposed a single equation for the entire contact:

$$q(x) = Q \left[1 - \frac{x}{L_c} \right]^n \qquad (6.34)$$

where Q_c and n are empirical constants. Zorev's results, however, were not accurate near the cutting edge due to secondary contact with the machined surface. Composite tool tests [108–112] usually suggest a two-part normal stress distribution in which the normal stress is constant in the plastic region and follows a polynomial variation in the sliding region:

$$q(x) = Q \quad x \le L_p \qquad (6.35)$$

$$q(x) = Q \left[1 - \frac{x}{L_c} \right]^n \quad x > L_p \qquad (6.36)$$

As noted earlier, however, the two-part tool method is also inaccurate near the cutting edge, and stresses must be extrapolated over part of the sticking zone. From the data available, Equations 6.35 and 6.36 describe the normal stress distribution more accurately than Equation 6.34 under most conditions.

Most data indicates that a two-part relation similar to Equations 6.35 and 6.36 is needed to describe the frictional stress distribution $t(x)$:

$$t(x) = k \qquad x = L_p \qquad (6.37)$$

$$t(x) = \mu q(x) \qquad x = L_p \qquad (6.38)$$

In these relations, μ is the true friction coefficient for the sliding region, which is generally larger than the effective coefficient μ_e, and k is the yield strength of the chip in shear, which is typically comparable to the tensile strength of the undeformed work material.

The character of the contact also appears to be influenced by a number of factors that are difficult to control or to take into account theoretically, such as the presence of surface contaminants, the purity of the work material, and the cutting time [125]. Finally, it should be noted that cutting fluids often have little direct effect on tool–chip friction. Neat oils penetrate part of the contact at low cutting speeds, reducing the sliding friction coefficient. Streams of cryogenically cooled fluid directed at the base of the chip may have a similar effect at higher speeds [139,140]. The dilute water-based fluids normally used at higher cutting speeds, however, function primarily as coolants under most conditions.

6.8   SHEAR PLANE AND SLIP-LINE THEORIES FOR CONTINUOUS CHIP FORMATION

The simplest theoretical models for cutting forces are based on shear plane and slip-line field analyses. In both of these approaches, the process is assumed to be steady in time and the tool is assumed to be infinitely sharp. The best-known analyses are applicable only to the 2D case of orthogonal cutting (Figure 1.1) and assume that the work material is a rigid, perfectly plastic material.

Orthogonal cutting is a simple 2D case of machining, which can be approximated in planing, shaping, and end turning of a thin-walled tube, or plunge turning of a thin disk. The shear plane model of orthogonal cutting (Figure 6.27) was first proposed by Piispanen [141,142], Ernst [50,143], and Merchant [51,144,145]. They assumed that the chip is formed by shear along a single plane inclined at an angle ϕ (the shear angle) with respect to the machined surface. This assumption is consistent with deformation patterns observed in many chip samples from quick-stop tests. ϕ can be calculated from the cutting ratio, r_c, using Equations 6.25 and 6.26. The geometric strain in the chip is given by Equation 6.27. The shear stress along the shear plane is assumed to be equal to the flow stress of the material in shear, k, and the resultant cutting force R is assumed to be directed at an angle β with respect to a normal to the rake face of the tool. β is the friction angle related to the effective friction coefficient μ_e through Equation 6.30.

It is also useful to define average sliding velocities along the shear plane, V_s, and along the rake face of the tool, V_c. Neglecting the effects of secondary shear at the tool–chip interface, V_s and V_c can be calculated from (Figure 6.28)

$$\frac{V}{\cos(\phi - \alpha)} = \frac{V_s}{\cos\alpha} = \frac{V_c}{\sin\phi} \tag{6.39}$$

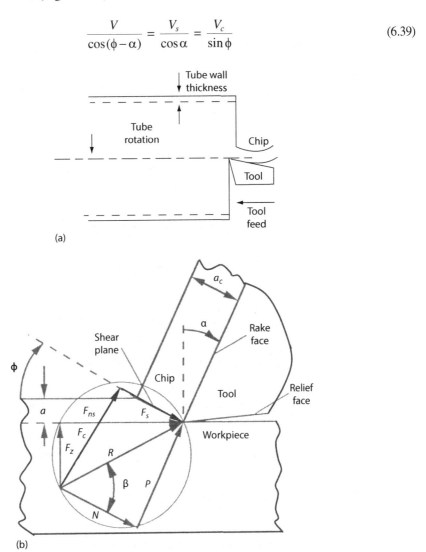

(a)

(b)

FIGURE 6.27 (a) End turning of a thin-walled tube to simulate orthogonal cutting; (b) Ernst and Merchant's shear plane theory of orthogonal cutting.

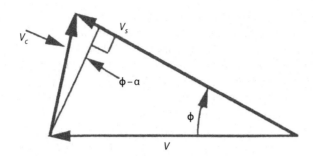

FIGURE 6.28   Velocity diagram for orthogonal cutting.

Conservation of mass also gives

$$Va = V_c a_c \tag{6.40}$$

Equations 6.39 and 6.40 can be combined to derive Equation 6.26.

Under these assumptions, an expression for the cutting and thrust forces, F_c and F_z, in orthogonal cutting can be derived geometrically:

$$F_c = \frac{kab \cos(\beta - \alpha)}{\sin \phi \cos(\phi + \beta - \alpha)} \tag{6.41}$$

$$F_z = \frac{kab \sin(\beta - \alpha)}{\sin \phi \cos(\phi + \beta - \alpha)} \tag{6.42}$$

where F_z, the force normal to the cutting speed, is equivalent to the axial (feed) force F_a in conventional turning. These relations can be combined with Equations 6.18, 6.21, and 6.22 to derive alternate equations for the unit deformation and frictional power consumptions, u_d and u_f:

$$u_d = \frac{(F_c \cdot \cos \phi - F_z \cdot \sin \phi) \cdot \cos \alpha}{a \cdot b \cdot \cos(\phi - \alpha)} \tag{6.43}$$

$$u_f = \frac{(F_c \sin \alpha + F_z \cos \alpha) \cos \alpha}{ab \cos(\phi - \alpha)} \tag{6.44}$$

The sum of u_d and u_f can be proven to be equal to u_s as expected and discussed in Section 6.5.

To apply these equations, the flow stress k and friction angle β must be specified. Both parameters are most accurately specified using cutting data. k and β can be calculated from measured cutting forces and chip thickness using the equations

$$k = \frac{(F_c \cos \phi - F_z \sin \phi) \sin \phi}{ab} \tag{6.45}$$

$$\beta = \arctan \left[\frac{F_z + F_c \tan \alpha}{F_c - F_z \tan \alpha} \right] \tag{6.46}$$

Typical values of k and β can be identified from a short series of tests. For ductile work materials such as low carbon steels and aluminum alloys, k is usually found to be roughly equal to the low-speed yield stress, Y, which in turn is equal to roughly one-third the penetration hardness, HB, in kg/mm^2. If tests cannot be performed, the assumptions $k = HB/3$ and $\beta = 45°$ can be used for rough calculations for these materials.

In addition to k and β, ϕ must be determined to calculate forces from Equations 6.41 and 6.42. Merchant noted that the cutting power is given by VF_c and assumed that ϕ would assume a value, which minimized this power. Under this assumption, assuming k and β do not vary with ϕ, differentiating Equation 6.41 with respect to ϕ yields

$$\phi = \frac{\pi}{4} - \frac{\beta}{2} + \frac{\alpha}{2} \tag{6.47}$$

The best-known slip-line field model was derived by Lee and Shaffer [146]. They proposed the slip-line field shown in Figure 6.29, in which all deformation occurs on a plane inclined at an angle ϕ with respect to the plane of cut and the material is stressed to the yield point in the triangular region ABC. From equilibrium considerations, the angle ACB can be shown to be equal to 45° so that the shear angle is given by

$$\phi = \frac{\pi}{4} - \beta + \alpha \tag{6.48}$$

This equation yields $\phi = 0$, which is not possible, when $\beta = 45°$ and $\alpha = 0$; they proposed a second solution assuming built-up edge formation for this condition.

These analyses provide physical insight into the cutting process and, as discussed in the next chapter, have been the basis for successful analyses of steady-state cutting temperatures. They are also simple and can be readily checked against independent experimental evidence. A number of researchers have compared the predicted relationships between the angles ϕ, β, and α with measured data [33,147–153]. In most cases, the agreement between measured and predicted values is poor for a broad data set, although both relations agree well with measurements for some materials.

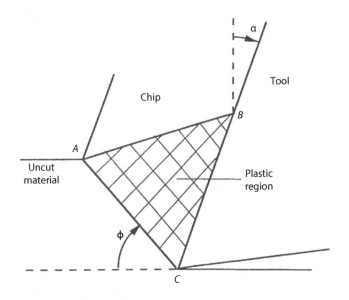

FIGURE 6.29 Lee and Shaffer's slip-line field solution for orthogonal cutting.

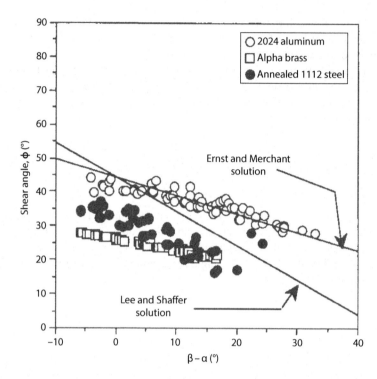

FIGURE 6.30 Comparison of typical cutting data with the Ernst and Merchant and Lee and Shaffer solutions for the relation between ϕ and $(\beta - \alpha)$. (After Eggleston, D.M. et al., *ASME J. Eng. Ind.*, 81, 263, 1959.)

A typical comparison is shown in Figure 6.30 [33]. These solutions are therefore of value mainly because they provide physical insight and a convenient framework for interpreting experimental data, and not as quantitatively accurate methods for predicting cutting forces.

A number of explanations for the poor agreement between these analyses and experimental data have been proposed. The minimum energy assumption used by Merchant is frequently questioned. As discussed in Section 6.11, however, this assumption is equivalent to the use of an upper bound theorem and is not inconsistent with the other assumptions. The most likely sources of the poor agreement are the material assumptions, namely, that the work material is rigid, perfectly plastic and that friction on the rake face of the tool can be characterized by a simple friction coefficient.

Many later analyses have taken more general material assumptions into account. Both Piispanen and Merchant proposed shear plane solutions in which the material flow stress depends on the normal pressure on the shear plane. A large number of slip-line solutions have also been published. The best-known Palmer and Oxley's model for a work-hardening material [154] and various models assuming sticking friction [155–159]; many of these are applicable to controlled contact tools (Figure 6.24). Shaw [2] and Johnson and Mellor [160] have reviewed the slip-line solutions published before 1970, and a recent CIRP review covers more recent work [161]. New shear plane and slip-line analyses are still reported [132,153,162,163]. It is difficult to judge the accuracy of the later analyses, since generally they have not been compared with independent experimental data.

6.9   SHEAR PLANE MODELS FOR OBLIQUE CUTTING

The analyses reviewed in Section 6.7 are applicable only to orthogonal cutting. The shear plane model has been extended to three dimensions by several researchers [144,164–166]. Slip-line analyses for oblique cutting, which yield similar results, have also been developed [167]. (Although most analyses yield similar numerical results, the equations differ considerably in form; also, in many cases different conventions are used for the positive direction of forces and the positive sense of the inclination angle λ, so that care should be taken in combining results from different analyses.) The angle definitions and shear angle solutions derived for orthogonal cutting can be used in the 3D analyses provided all angles are measured in the plane normal to the cutting edge. To emphasize this point, many researchers use the subscript n to denote normal angles, that is, α_n for the normal rake angle and ϕ_n and β_n for the shear and friction angles defined by Equations 6.25 and 6.30. In the equations in this section, the subscript will be omitted with the understanding that all angles are measured in the normal plane.

Assumptions regarding the direction of the shear and frictional forces are also required to derive equations. Most commonly, it is assumed that the friction force acts in the direction of chip flow and that the shear force acts in the direction of the shear strain. These additional assumptions probably contribute errors, which are small compared to those associated with the assumptions already made in the orthogonal analyses. Under these assumptions, the cutting forces in oblique cutting are given by [167]

$$F_c = \frac{kab\left[\cos(\beta-\alpha)\cos\lambda + \sin\beta\sin\lambda\tan\eta\right]}{\sin\phi\cos(\phi+\beta-\alpha)} \tag{6.49}$$

$$F_z = \frac{kab\sin(\beta-\alpha)}{\sin\phi\cos(\phi+\beta-\alpha)} \tag{6.50}$$

$$F_l = \frac{kab\left[\cos(\beta-\alpha)\sin\lambda - \sin\beta\cos\lambda\tan\eta\right]}{\sin\phi\cos(\phi+\beta-\alpha)} \tag{6.51}$$

As would be expected, in the orthogonal case in which $\lambda = c = 0$, Equations 6.49 and 6.50 reduce to Equations 6.41 and 6.42, and Equation 6.51 yields $F_l = 0$. Once F_c, F_z, and F_l have been determined, F_n, F_p, N, and P can be calculated from the equations given in Section 6.3.

Estimates of η, β and k are necessary to apply these equations. As in orthogonal cutting, reliable estimates are best made from a short series of cutting tests. η can be estimated from Equations 6.7 through 6.9. β can be determined from Equation 6.29 if P and N are determined from measured values of F_c, F_p, and F_z using Equations 6.5 and 6.6. k is most easily from the deformation power per unit volume of material cut, u_d. Since u_d is equal to the product of k and the shear strain, γ,

$$k = \frac{u_d}{\gamma} = \frac{\dfrac{F_c}{a\cdot b} - \dfrac{P}{a_c\cdot b}}{\gamma} \tag{6.52}$$

where P can be calculated from Equation 6.6 and γ for oblique cutting is given by [2]

$$\gamma = \frac{\cot\phi + \tan(\phi-\alpha)}{\cos\eta_s} \tag{6.53}$$

where η_s is the shear flow angle, given by

$$\eta_s = \arctan\left[\frac{\tan\lambda\cdot\cos(\phi-\alpha) - \tan\eta\cdot\sin\phi}{\cos\alpha}\right] \tag{6.54}$$

If tests cannot be performed, the penetration hardness of the work material can be used to obtain a rough estimate of k as described in the previous section.

6.10 SHEAR ZONE MODELS

The most serious limitation of the shear plane theory is that it is based on restrictive material assumptions. The basic assumption that the deformation occurs on a single plane produces a velocity discontinuity along which the strain rate is infinite. Although geometric arguments can be used to compute an average effective strain rate, this assumption makes it difficult to include strain rate hardening, which is known to be a significant factor in high rate deformation processes, in the analysis. Also, since force but not moment equilibrium considerations are used to derive force relations, most shear plane models also assume that the shear stress on the shear plane is uniform and that friction along the rake face of the tool can be characterized by a constant friction coefficient. As discussed in Section 6.7, the frictional assumption in particular is not consistent with experimental data.

More general material assumptions can be used if it is assumed that deformation occurs in a narrow band centered on the shear plane (Figure 6.31). Under this assumption, the strain rate is finite. Moreover, moment equilibrium can also be included so that variations in stress along the shear plane and tool rake face can also be considered.

The best-known shear zone model of the cutting process is Oxley's theory [166,168–172]. Oxley's analysis is complex and includes a numerical minimization scheme to determine the chip thickness and tool–chip contact length. A detailed description is given in his book [173]. Based on data from quick-stop tests such as those reported by Kececioglu [62], he assumes that the shear zone thickness is roughly one-tenth the shear zone length. Quick-stop data is also used to support assumptions about work material streamlines within the deformation zone. Based on these assumptions, the strain and strain rate at every point in the primary deformation zone can be calculated. Similar assumptions are used to compute the strain and strain rate distributions in the secondary deformation zone along the tool rake face. Stresses in both deformation zones are then calculated using an exponential constitutive equation,

$$\sigma = k\sqrt{3} = C\gamma^n \tag{6.55}$$

where both C and n depend on the velocity modified temperature, T_{mod}, defined by [174]

$$T_{mod} = T \cdot \left[1 - v\,\frac{\dot{\gamma}}{\dot{\gamma}_0} \right] \tag{6.56}$$

where
T is the deformation zone temperature
$\dot{\gamma}$ is the strain rate
v and $\dot{\gamma}_0$ are constants

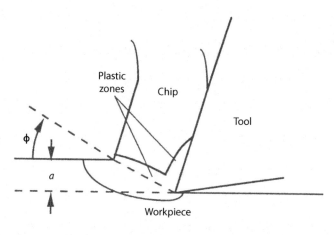

FIGURE 6.31 Shear zone model of Oxley.

T can be calculated using procedures based on analytical or finite element temperature models, which will be discussed in the next chapter. Both compression and cutting tests are used to determine the variation of C and n with T_{mod}. Early results were presented mainly for hot rolled steel work materials; a typical example is shown in Figure 6.32. The velocity modified temperature was initially used to analyze low-speed, isothermal deformations, and its use in analyzing machining is sometimes questioned [169,170]. Oxley has shown that the model generally yields accurate results for steel work materials, and subsequent researchers have found it accurate for hybrid and simulation force models [175–177]. The model has also been extended to more general material constitutive models [178,179].

A similar analysis was developed by Usui [180–184]. As shown in Figure 6.33, Usui's original model [180–182] is a 3D shear plane analysis, which includes secondary cutting edge and nose radius effects. The chip thickness is predicted using a minimum energy approach, and the work

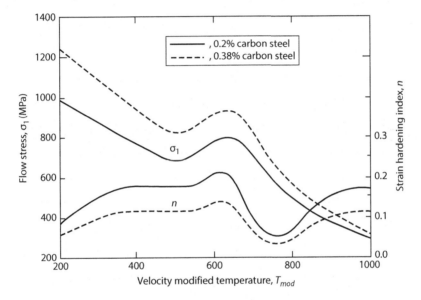

FIGURE 6.32 Variation of σ and n with T_{mod} for two plain carbon steels. (After Oxley, P.L.B., *The Mechanics of Machining*, Ellis Horwood, Chicester, U.K., 1989.)

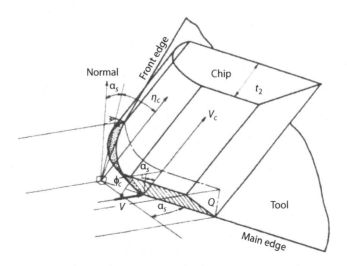

FIGURE 6.33 Usui's model of chip formation for a tool with a nose radius larger than the feed rate. (After Usui, E., and Hirota, A., *ASME J. Eng. Ind.*, 100, 229, 1978.)

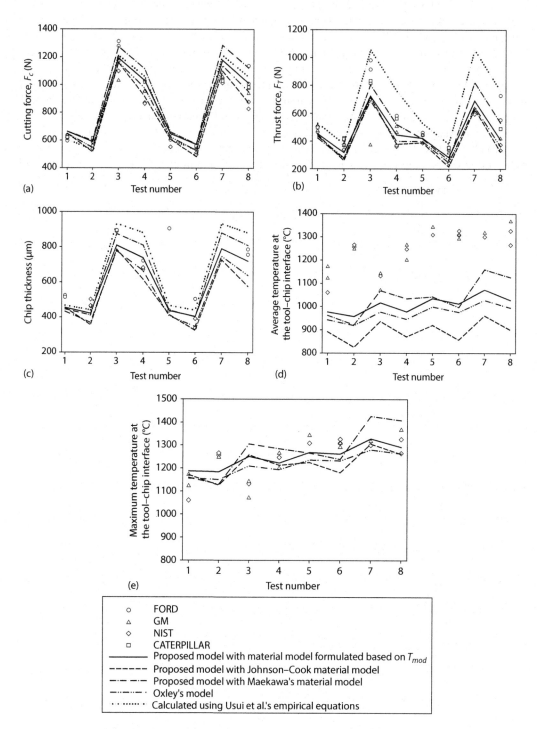

FIGURE 6.34 Comparison of the predictions of Oxley's with and without material modifications, Usui's model, and experimental data from four laboratories. (a) Cutting force, (b) thrust force, (c) chip thickness, (d) average temperature at the tool–chip interface, and (e) maximum temperature at the tool–chip interface. (After Abidi-Sedeh, A.H., and Madhavan, V., *Mach. Sci. Technol.*, 6, 379, 2002.)

material stress–strain response is modeled based on orthogonal cutting data taken under similar cutting conditions. The results are applicable to turning with a conventional single point tool, plain milling, and groove cutting. Based on the computed forces and assumptions about the stress distributions in the contact zones, the rates of deformation and frictional heat generation are calculated and used as inputs to a finite-difference scheme to calculate the temperature distribution in the work material and tool near the cutting edge. The temperature and contact conditions are then used to estimate tool wear rates for carbide cutting tools. In later work, Usui and Shakashiri [183] replaced the orthogonal cutting data with finite element calculations; for these calculations, the material stress–strain response is modeled using results from high-speed, high-temperature tension tests. A similar analysis assuming multiple thin shear planes has also been reported [185].

Both Oxley's and Usui's models are complex and require specialized material test data and substantial computer code development for application. Because of these requirements, they have not been widely applied outside of the research community. Oxley's model is found to agree well with experimental data [178,186,187], particularly when modifications are made to include a larger range of material models (Figure 6.34). Both the Oxley and Usui models are more complete than earlier solutions in that they include temperature effects, and thus, they can be used to investigate phenomena such as tool wear [182–184] and shear localized chip formation [188], which cannot be studied using simpler approaches. The development of increasingly powerful finite element models of machining, many based on commercially available solvers, has inhibited the broader development of thin shear zone models; however, finite element models have incorporated some of the underlying physical algorithms and constitutive assumptions, such as Usui's tool wear estimation algorithm [189] and Oxley's constitutive model [190].

6.11 MINIMUM WORK AND UNIQUENESS ASSUMPTIONS

Before describing finite element models of chip formation, two basic assumptions that have been widely made in analytical models should be considered: the assumption that the chip thickness will assume a value, which minimizes the work required to form the chip, and the assumption that a unique deformation geometry exists for given cutting conditions and tool angles.

The main difficulty in modeling the cutting process is that the size and shape of the chip are not determined by geometry and must be deduced from some physical condition. In analytical models, the most commonly used condition has been the minimum energy criterion proposed by Merchant. Although this criterion has intuitive appeal, it has been widely questioned in the literature, generally on the grounds that it does not yield an accurate answer [191,192]. Other conditions, such as conditions based on work hardening, have also been proposed [193], but they seem to yield accurate results only for a limited range of materials [194].

The validity of the minimum energy criterion depends on the assumed constitutive model of the work material. Applying the minimum energy criterion is equivalent to seeking an upper bound solution and is therefore valid only when an equivalent limit theorem can be derived for the material model being considered. Limit theorems are usually derived from variational formulations in which the equations of motion, stress–strain equations, and boundary conditions are not solved explicitly. Instead, an approximate solution is sought by finding a stationary value of an appropriate functional of the stress or strain field. The functional must be chosen so that the Euler equations and natural boundary conditions obtained by setting its first variation equal to zero are the field equations and boundary conditions of the problem. In practice, appropriate functionals of the strain (or, equivalently, displacement or velocity) fields often have units of energy, and the stationary value can be shown to be a minimum so that an approximate solution can be found by minimizing a scalar quantity with units of energy. This approach yields the minimum energy condition used by Merchant for a rigid, perfectly plastic material but not for more general viscoplastic materials [195]. Similar conclusions can be drawn from variational principles used to formulate finite element models [196,197]. This shows that the use of the minimum energy criterion in the simplest shear plane theories is consistent with the material model assumed and that the main source of inaccuracy in these models is most likely the material model itself.

A further question to be considered is whether the cutting forces and chip dimensions are uniquely determined by the current cutting conditions and tool angles, or whether they depend on initial conditions as well. Uniqueness is implicitly assumed in the shear plane theories but was questioned at an early date by Hill [191], who developed an analysis to determine permissible ranges of the shear angle rather than a unique value. Similar analyses have also been reported by Roth [198] and Dewhurst [158]. No experimental evidence for nonuniqueness, beyond the general variability of the process, has ever been presented. The only early experimental investigations of the question were those reported by Ota [199], whose results exhibited too much scatter to be conclusive, and Low [200,201], who found that the chip thickness did not vary significantly with initial conditions. More recent experiments [202] using a CNC machining center, which permitted more accurate control of initial conditions, showed no significant variation of chip thickness or cutting forces with initial conditions when machining zinc and aluminum. The uniqueness assumption therefore appears to be well supported by the available experimental evidence.

6.12   FINITE ELEMENT MODELS

Due to its potential to accommodate more general geometric and material assumptions, there has been a long interest in using finite element analysis (FEA) to analyze machining. These methods hold promise not only for computing cutting forces, but quantities such as tool stresses and cutting temperatures that are difficult to measure. This section focuses on general issues and force and stress modeling; finite element modeling of temperatures, residual stresses, and chip breaking are discussed in Chapters 7, 10, and 11, respectively.

The earliest finite element models were reported in the 1970s [203–206]. As computing power increased in the 1980s, improved models were developed [183,196,197,207–213]. Particularly, complete analyses were developed by Strenkowski and Caroll [208], Strenkowski and Moon [211], and Iwata et al. [196]; these were comprehensive steady-state models with strain rate and temperature effects, which predicted cutting forces, chip dimensions, and the strain, strain rate, and temperature distributions in the cutting zone. Progress continued in the 1990s, as numerical issues were addressed by researchers with backgrounds in structural analysis or metal forming analysis [186,187,189,190,214–221]. In recent years, there has been a great increase in finite element modeling of machining, driven by increased computing power and improved models, including commercially released and supported programs. Mackerle's bibliographies of papers on finite element analysis in machining [222,223] list over 1000 items, 370 of which were published between 1996 and 2002. A recent comprehensive CIRP review [161] summarizes work to 2013.

Currently available models can be used to investigate or predict cutting forces, stress and temperature distributions, tool wear rates, chip breakage, and residual stresses. Although most are restricted to the 2D case of orthogonal cutting, 3D simulations of processes such as drilling (Figure 6.35), milling, and gear cutting are becoming more common [224–226].

In discussing finite element modeling of machining, it is helpful to concentrate on three broad areas:

1. The formulation, which determines what problems the model will be best suited to, as well as what numerical issues will have to be addressed
2. Material assumptions, specifically the workpiece constitutive model, tool–work friction model, thermal property assumptions, and procedures for estimating required parameters from cutting or non-cutting tests
3. Accuracy based on comparison with measured data

There are two basic formulations, Lagrangian and Eulerian, as illustrated in Figures 6.36 and 6.37. In the Lagrangian formulation (Figure 6.36), the workpiece is fixed in space, and the tool is fed into it, imposing displacement boundary condition, which drives plastic deformation. This approach is well suited to incipient and transient analyses and to the study of chip breaking and burr formation. But it is not well suited to studying steady-state machining, since calculations must be carried out over

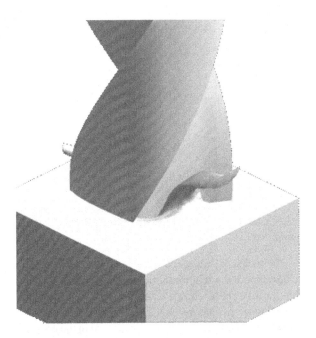

FIGURE 6.35 FEM of a drilling process to study the chip formation and torque. (Courtesy of WZL Aachen, Aachen, Germany.)

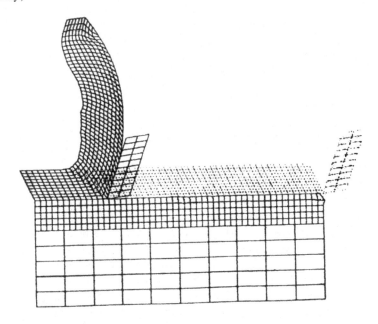

FIGURE 6.36 Finite element mesh for a Lagrangian analysis. (After Strenkowski, J.S., and Caroll III, J.T., *ASME J. Eng. Ind.*, 107, 349, 1985.)

significant intervals; since small time steps must normally be used to ensure accuracy, this can create a large computational load even for the relatively simple case of orthogonal cutting. Determining a criterion for the separation of the chip from the workpiece was an issue in early Lagrangian analyses, but in recent work remeshing strategies combined with physically meaningful ductile fracture criteria have reduced concerns in this area. In Eulerian formulations (Figure 6.37), the tool is viewed as fixed in space and the work material is treated as a fluid flowing through a control volume in front of it. This eliminates

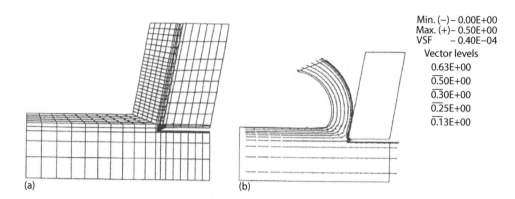

Min. (−) – 0.00E+00
Max. (+) – 0.50E+00
VSF   – 0.40E−04
Vector levels
0.63E+00
0.50E+00
0.30E+00
0.25E+00
0.13E+00

FIGURE 6.37   Finite element mesh (a) and velocity field (b) for an Eulerian analysis. (After Strenkowski, J.S. and Moon, K.J., *ASME J. Eng. Ind.*, 112, 1990.)

the need to integrate from the incipient stage to obtain steady-state results, reducing the computational load. Fewer elements are also reportedly required for equivalent accuracy [208]. A disadvantage is that the dimensions of the chip must be specified in advance; to produce a predictive model, an additional physical condition is required to determine the chip dimensions. Most commonly, a minimum energy principle or more general variational principle is used as discussed in the previous section. In this regard, the Eulerian models are similar to fluid-type variational models based on assumed chip flow lines [195]. Since computing power has steadily increased, almost all models developed since 1995 have employed Lagrangian formulations. A distinction in formulation can also be made between coupled and uncoupled analyses for models, which include temperature effects; this issue will be discussed in Chapter 7.

Most current models assume combined sliding and sticking friction at the tool–chip interface, although they differ in methods of determining where sticking beings. Computed temperature and other results are sensitive to fictional assumptions, so there has been considerable recent research on methods to accurately specify friction parameters [104,105]. The material constitutive response is usually modeled as elastic–plastic; although a great variety of functional forms are employed [161], almost all plasticity models account for strain hardening, strain rate hardening, and thermal softening. Power law models and the Johnson–Cook material model are especially widely used. Iteration is required to include thermal softening in models in which the thermal and mechanical solutions are uncoupled. Many models assume constant thermal properties, but some do account for variations with temperature as discussed in Chapter 7. Determination of material parameters for specific models often presents difficulty. Commercially released codes have defined test procedures, while research codes and those based on general purpose solvers often do not. In either case, significant effort for sample preparation and testing is normally required to generate a model for a specific alloy. Model parameters for widely studied alloys can often be found in the literature.

Assessing the accuracy of models presents difficulties due to the number of empirical constants generally employed and the difficulty of accurately measuring some predicted quantities. Large data sets of measured forces should be used to validate computed forces for models that have a large number of empirical material parameters. Validation of field quantities, such as strain and temperature distributions, is often difficult since these quantities cannot be measured accurately. Accuracy is most easily assessed if procedures for estimating material parameters are well defined and large data sets of quantities that can be accurately and repeatably measured are available. Such quantities include forces, chip thickness, and the tool–chip interface temperature; quantities that are more difficult to measure repeatably but that still provide insight into accuracy include residual stresses, the tool–chip contact length, and chip-breaking characteristics.

As an example, we consider the widely used finite element model of machining reported by Marusich and Ortiz [218] and marketed commercially, with subsequent development, as the AdvantEdge program by Third Wave Systems, Inc. [227]. It is a 2D explicit Lagrangian model employing adaptive remeshing and subcycling between uncoupled mechanical and thermal solutions. A two-stage rate hardening constitutive model is employed:

$$\left(1+\frac{\dot{\varepsilon}^p}{\dot{\varepsilon}_0^p}\right)=\left(\frac{\overline{\sigma}}{g(\varepsilon^p)}\right)^{m_1} \quad \text{if } \dot{\varepsilon}^p \leq \dot{\varepsilon}_t^p \tag{6.57}$$

$$\left(1+\frac{\dot{\varepsilon}^p}{\dot{\varepsilon}_0^p}\right)\left(1+\frac{\dot{\varepsilon}_t^p}{\dot{\varepsilon}_0^p}\right)^{m_2/m_1-1}=\left(\frac{\overline{\sigma}}{g(\varepsilon^p)}\right)^{m_2} \quad \text{if } \dot{\varepsilon}^p > \dot{\varepsilon}_t^p \tag{6.58}$$

where
  $\overline{\sigma}$ is the effective Mises stress
  g is the flow stress
  ε^p is the accumulated plastic strain
  $\dot{\varepsilon}^p$ is the plastic strain rate
  m_1 and m_2 are the low and high strain rate sensitivity exponents, respectively
  $\dot{\varepsilon}_0^p$ is the reference plastic strain rate
  $\dot{\varepsilon}_t^p$ is the threshold plastic strain rate separating the two regimes

A power hardening law with linear thermal softening is adopted for the flow stress:

$$g(\varepsilon^p)=\sigma_0\left[1-\alpha\left(T-T_0\right)\right]\left(1+\frac{\varepsilon^p}{\varepsilon_0^p}\right)^{1/n} \tag{6.59}$$

where
  n is the hardening exponent
  T is the current temperature
  T_0 is a reference temperature
  σ_0 is the yield stress at T_0

As noted earlier, the thermal and mechanical solutions are uncoupled. The temperature is treated as fixed in each stress calculation; temperatures are recomputed based on heat generation and used to update stress calculations iteratively until convergence is achieved.

A maximum plastic strain criterion is used for chip separation from the substrate; fracture occurs when the plastic strain ε^p reaches a critical value ε_f^p given by

$$\varepsilon_f^p \approx 2.48e^{-1.5p/\overline{\sigma}} \tag{6.60}$$

where p is the hydrostatic pressure. Combined sliding and sticking friction is assumed along the tool–chip interface. Sliding is assumed to occur until a maximum frictional stress is reached, after which sticking occurs and the stress is constant.

Material plasticity parameters are estimated from high-speed compression tests at different temperatures [227]. The sliding friction coefficient is determined from low-speed machining tests, or estimated based on experience. Marusich's initial paper did not describe experimental validation, but subsequent papers by AdvantEdge developers and users [228–232] have compared computations with measured values for a variety of steels and aluminum and titanium alloys.

Another widely used model is DEFORM, marketed by Scientific Forming Technologies Corp., Columbus, OH. It supports 2D and 3D analyses to predict cutting forces, chip properties, cutting temperatures, and tool stresses. It also uses a modified Lagrangian formulation, employs automatic remeshing, and supports a broad range of general material constitutive assumptions and damage models [225,233].

In general, the predicted forces and chip characteristics from the most comprehensive finite element models are reasonably accurate; typical results are shown in Figures 6.38 through 6.41.

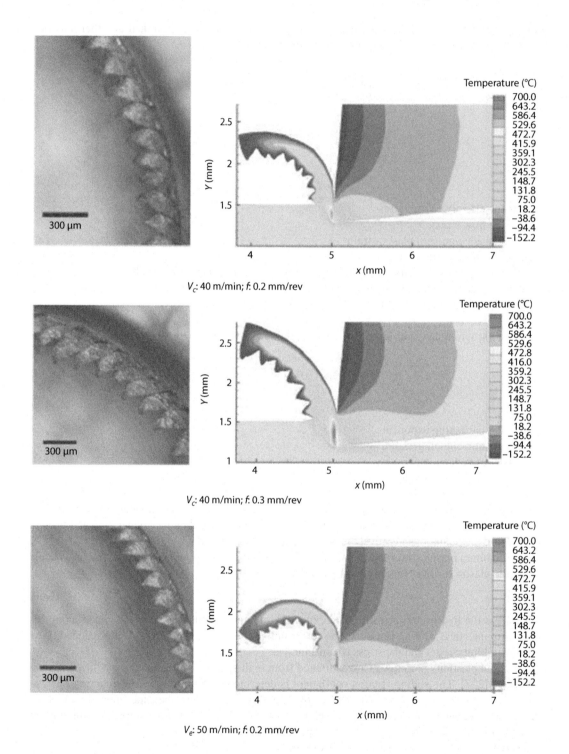

V_c: 40 m/min; f: 0.2 mm/rev

V_c: 40 m/min; f: 0.3 mm/rev

V_e: 50 m/min; f: 0.2 mm/rev

FIGURE 6.38   Comparison of measured and computed serrated chip forms and cutting forces for cryogenic machining of Ti–6Al–4V. (After Davoudinejad, A. et al., Finite Element Simulation and Validation of Chip Formation and Cutting Forces in Dry and Cryogenic Cutting of Ti–6Al–4V, *Proceedings of NAMRI/SME 43*, Charlotte, NC, June 2015.)   (*Continued*)

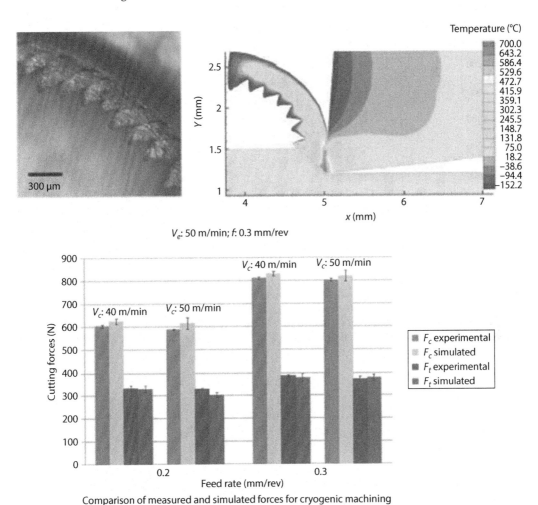

FIGURE 6.38 (*Continued*) Comparison of measured and computed serrated chip forms and cutting forces for cryogenic machining of Ti–6Al–4V. (After Davoudinejad, A. et al., Finite Element Simulation and Validation of Chip Formation and Cutting Forces in Dry and Cryogenic Cutting of Ti–6Al–4V, *Proceedings of NAMRI/SME 43*, Charlotte, NC, June 2015.)

The accuracy of predicted cutting temperatures and residual stresses is more difficult to assess due to limited experimental comparisons as discussed in Chapters 7 and 10.

6.13 DISCONTINUOUS CHIP FORMATION

The models described in Sections 6.7 through 6.11 are applicable to continuous chip formation. Although continuous chips are most commonly encountered in practice, many materials form discontinuous chips under some or all cutting conditions. The transition from continuous to discontinuous chip formation depends on the thermophysical properties and metallurgical state of the work material as well as on the dynamics of the machine structure and cutting process. Brittle materials such as cast iron and beta brass often form discontinuous chips. Many metals that have high hardness, hexagonal close-packed structure, and a low thermal conductivity (e.g., titanium alloys) may also form discontinuous chips. Other metals that form continuous chips under most conditions may

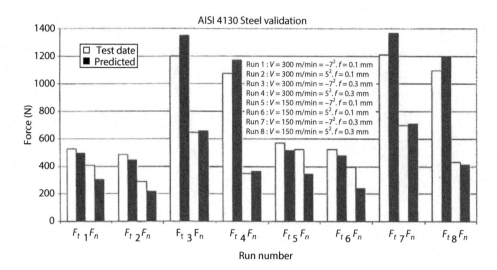

FIGURE 6.39 Comparison of measured cutting forces with forces predicted using a finite element model when cutting 4130 steel cut with coated carbide tools. (After Kalidas, S., Cost Effective Tool Selection and Process Development through Process Simulation, *Proceedings of 2002 Third Wave AdvantEdge User's Conference*, Atlanta, GA, April 2002, Paper 2.)

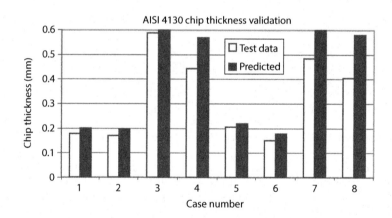

FIGURE 6.40 Comparison of measured and FEA-calculated chip thicknesses when cutting 4130 steel cut with coated carbide tools. (After Marusich, T.D. et al., A Methodology for Simulation of Chip Breakage in Turning Processes Using an Orthogonal Finite Element Model, *Proceedings of Fifth CIRP International Workshop on Modeling Machining Operations*, West Lafayette, IN, 2002, pp. 139–148.)

form discontinuous chips at low cutting speed; for example, many low carbon steels form discontinuous chips at cutting speeds less than 1 cm/s.

The formation of discontinuous chips was discussed qualitatively by several early researchers [141,234,235]. Although the mechanism of discontinuous chip formation varies somewhat for different work materials [236–238], it can be explained in broad terms with reference to Figure 6.42, which shows the stages of formation of discontinuous chips when cutting beta brass. The chip initially begins to form with a relatively high shear angle. As deformation continues, the chip slides against or adheres to the rake face. This increases the frictional force and causes the shear angle to decrease and the material to bulge. As the shear angle decreases the strain along the shear plane

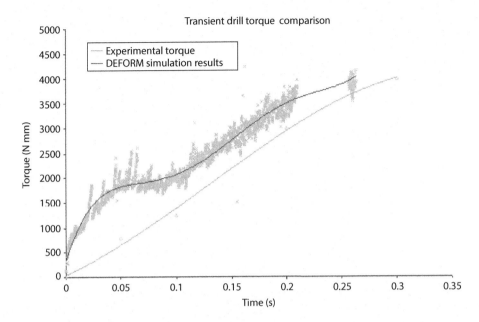

FIGURE 6.41 Comparison of drilling torque computed using the using finite element model shown in Figure 6.35 with experimental values. (Courtesy of WZL Aachen, Aachen, Germany.)

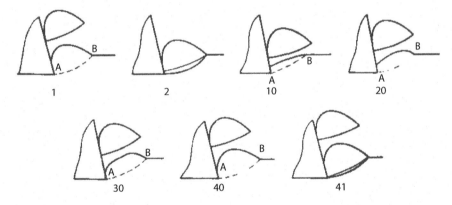

FIGURE 6.42 Formation of discontinuous chips when cutting beta brass at a speed of 1.25 cm/min. (After Cook, N.H. et al., *ASME Trans.*, 76, 153, 1954.) The numbers beneath successive stages are frame numbers from high-speed films; the tool feed rate is 0.0005 in/frame.

increases until a critical value is reached, producing a ductile shear fracture. The process then starts over. Although this description is inexact, it emphasizes the importance of tool–chip friction and the ductility of the work material in discontinuous chip formation. When cutting steels, discontinuous chips are more likely to occur at low cutting speeds because tool–chip temperatures are low, increasing the strength of the work material and thus frictional stresses near the cutting edge. Similarly, the use of low or negative rake angles, which increases the effective friction coefficient, also promotes discontinuous chip formation. The importance of two additional factors is not clearly indicated by this description. Discontinuous chips are also more likely to be formed when the rigidity of the tool or part is low and when the work material contains inhomogeneities. Low system stiffness increases the elastic strain energy stored in the system, especially at high feed rates, and

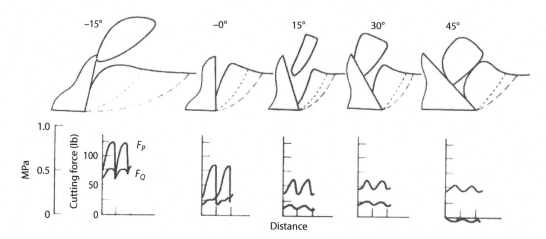

FIGURE 6.43 Variation of cutting forces with time for discontinuous chip formation when cutting beta brass with tools with varying rake angles. (After Cook, N.H. et al., *ASME Trans.*, 76, 153, 1954.)

promotes crack propagation and fracture. Similarly, inhomogeneities in the work material produce stress concentrations, which promote crack nucleation and propagation. Both of these factors are discussed in more detail by Shaw [2].

Discontinuous chip formation is inherently unsteady and produces periodic cutting forces. At low cutting speeds, the amplitude of the force variations is large, especially for low or negative rake angles (Figure 6.43). At higher speeds, however, the amplitude of variation is smaller. When cutting brittle materials such as cast iron at conventional speeds, cutting force variations manifest themselves as apparent noise in measured signals, but individual peaks are not discernible unless the sampling rate is very high.

The analysis of the mechanics of discontinuous chip formation has long presented difficulties. Some work on predicting the initial and fracture shear angles from energy considerations has been reported [234,239,240], but the results have not agreed well with experimental data. Finite element models for discontinuous chip formation have been developed [241], and similar models shear localized chip formation [242] may also be adaptable to discontinuous chip formation. Although this work provides additional physical insight, results depend on a number of material parameters that are difficult to estimate without cutting tests so that cutting forces for materials that form discontinuous chips must generally be modeled empirically.

6.14 BUILT-UP EDGE FORMATION

The built-up edge is an accumulation of heavily strained work material, which collects on the cutting edge under proper conditions. It is an undesirable feature for several reasons. It reduces machining accuracy by changing the effective feed rate. It also reduces the quality of the machined surface because it periodically breaks off and reforms, introducing irregularities into the surface. The periodic breakage can also lead to chipping of the cutting edge. Finally, built-up edge may also promote the thermal cracking of the tool.

Built-up edge formation usually occurs at low cutting speeds. These speeds were much more typical of industrial practice in the middle of the last century, and much of the literature on the formation and avoidance of the built-up edge dates from this period. (Built-up edge formation is still a problem in low-speed processes such as drilling and end milling, especially for soft, ductile work material such as aluminum alloys, and in minimum quantity lubrication machining.) The early work, much of which concentrated on steel work materials, has been reviewed in detail by Ernst and Martellotti [243] and Heginbotham and Gogia [244]. Nakayama [245] has more recently clarified the relation of the built-up edge to cutting forces, temperatures, and surface finish. A number of researchers have described built-up edge formation for work materials other than steel [246–248], and Trent [249] has published a detailed discussion of the metallurgical aspects of the phenomenon.

Built-up edge formation is similar to discontinuous chip formation in that it is a time-varying process, which depends heavily on tool–chip friction and the ductility of the work material. In fact, the built-up edge can be viewed as a partially formed discontinuous chip around which the undeformed work material flows. The time-varying aspect of the phenomenon and the importance of tool–chip friction were demonstrated by Heginbotham and Gogia's quick-stop experimental studies using steel workpieces (Figure 6.44). They noted that the built-up edge starts as an embryonic structure to which successive layers adhere as cutting progresses, until it eventually attains a size and shape characteristic of the cutting conditions. The size and shape may vary considerably; Heginbotham and Gogia identified four distinct types of built-up edge for steel workpieces that occur over specific speed ranges. Ultimately, the built-up edge breaks off and the process of formation repeats itself.

Physically, the built-up edge comprises heavily strained and hardened material. Trent [249] reports that the microhardness of the built-up edge can be more than twice that of the surrounding chip and that the strains involved in its formation are so high that they cannot be estimated because the grain structure of the material is no longer discernible. This observation further underscores the importance of the ductility of the work material in built-up edge formation, since less ductile work materials will fracture before such a structure can evolve.

As noted earlier, the built-up edge increases the effective rake angle of the tool and reduces cutting forces. A number of slip-line solutions have been published for cutting with a built-up edge [146,244]; these are similar to those for conventional continuous chip formation, except that they include a roughly triangular region of dead material at the tool point that does not deform. Built-up edge formation has also been simulated using finite element models [250,251].

Since the built-up edge is an undesirable feature, however, it is probably of greater interest to determine how it may be avoided. The standard method of reducing or eliminating built-up edge formation is to increase the cutting speed, which increases the tool–chip interface temperature and reduces the strength of the work material near the cutting edge. Other effective methods include

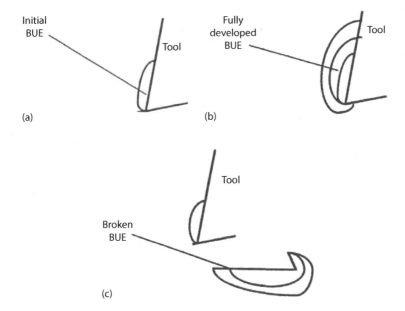

FIGURE 6.44 Formation of built-up edge by addition of layers of work material. (a) Initial BUE. (b) Fully developed BUE composed of layers of work material. (c) Broken BUE and reformation of initial BUE.

applying a lubricant or tool coating to reduce tool–chip friction, increasing the rake angle to reduce stresses at the tool point, and making the work material less ductile through cold work or the addition of inhomogeneities such as lead or sulfur particles in steel.

6.15 EXAMPLES

Example 6.1 In a turning operation on a mild steel tube with a 2.5 mm thickness, an end-cutting test was performed using a tool with a zero lead angle (which represents an orthogonal cutting test). The feed was 0.3 mm/rev, the rake angle of the tool was zero, and the cutting speed was 200 m/min. Several parameters were measured during this end-turning operation. The average chip thickness was measured with a point micrometer and found to be 0.7 mm. The tool–chip contact length (on the rake face) was estimated to be 0.5 mm. The cutting forces were measured with a dynamometer mounted under the toolholder. The tangential cutting and axial (feed) forces were 900 and 600 N, respectively.

Calculate (a) the mean angle of friction on the tool face, (b) the mean shear stress produced in cutting the workpiece, (c) the mean frictional stress at the tool face, (d) the estimated average strain and strain rate in the chip formation if the thickness of the shear zone is assumed to be 0.02 mm, and (e) the percentage of the total energy estimated to go into secondary deformation zone to overcome the friction at the tool–chip interface.

Solution:

(a) The mean friction coefficient on the rake face is given by Equation 6.30 using Figure 6.27 as

$$\left. \mu = \tan\left(\beta - \alpha\right) = \frac{F_Z}{F_c} \Rightarrow \beta - \alpha = \tan^{-1}\left(\frac{F_Z}{F_c}\right) \right\} \Rightarrow \beta = \tan^{-1}\left(\frac{600}{900}\right) = 33.7°$$
$$\text{But } \alpha = 0°$$

(b) The mean shear stress of the workpiece material during metal cutting (assuming uniform stress distribution on the shear plane) is

$$\tau_s = \frac{F_s}{A_s} = \frac{F_c \cos\phi - F_Z \sin\phi}{\dfrac{A_o}{\sin\phi}} = \frac{(F_c \cos\phi - F_Z \sin\phi)\sin\phi}{a \cdot b}$$

$$= \frac{(900\cos 23.3° - 600\sin 23.2°)\sin 23.2°}{(0.3)\cdot(2.5)} = 310\,\text{N/mm}^2$$

The cutting ratio from Equation 6.26 is $r_c = a/a_c = 0.3/0.7 = 0.429$. The mean shear angle between the direction of cutting speed and the shear plane is calculated from Equation 6.25

$$\tan\phi = \frac{r_c \cos\alpha}{1 - r_c \sin\alpha} = \frac{0.429\cos 0°}{1 - 0.429\sin 0°} = 0.429 \Rightarrow \phi = 23.2°$$

(c) The mean frictional stress at the tool–chip interface is

$$\tau_f = \frac{P}{A_f} = \frac{R\sin\beta}{A_f} = \frac{\sqrt{F_c^2 + F_Z^2}\,\sin\beta}{b \cdot l_f} = \frac{\sqrt{900^2 + 600^2}\,\sin 33.7°}{(2.5)(0.5)} = 480\,\text{N/mm}^2$$

(d) Shear strain, γ, at the shear zone is given by either Equation 6.24 or Equation 6.27

$$\gamma = \cot\phi + \tan(\phi - \alpha) = \cot(23.2°) + \tan(23.2° - 0) = 2.76$$

or

$$\gamma = \frac{\cos\alpha}{\sin\phi\cos(\phi - \alpha)} = \frac{\cos(0°)}{\sin(23.2)\cos(23.2 - 0)} = 2.76$$

also

$$u = \tau_s \cdot \gamma \Rightarrow \gamma = \frac{u_d}{\tau_s} = \frac{\text{Deformation power}}{\text{Shear stress along shear plane}}$$

or

$$\gamma = \frac{V_s}{V_n}$$

V_n is the velocity normal to shear plane as shown in Figure 6.28.
The shear strain is given by Equation 6.28 as

$$\dot{\gamma} = \frac{\gamma}{\Delta t} \quad \text{and} \quad \Delta t = \frac{\Delta y}{V_n} = \frac{\Delta y}{V \cdot \sin\phi} = \frac{0.02\,(\text{mm}) \cdot 60(\text{s/min})}{200,000\,(\text{mm/min}) \cdot \sin(23.2)} = 1.5 \cdot 10^{-5}\,\text{s}$$

where Δy is the thickness of the shear zone as illustrated in Figure 6.15

$$\dot{\gamma} = \frac{\gamma}{\Delta t} = \frac{2.76}{1.5 \cdot 10^{-5}\,\text{s}} = 1.84 \cdot 10^5\,\text{s}^{-1}$$

(e) The fraction of friction energy is given by

$$\frac{u_f}{u_s} = \frac{PV}{F_c V} = \frac{Pr_c}{F_c} = \frac{\sqrt{F_c^2 + F_z^2}\sin\beta r_c}{F_c} = \frac{\sqrt{900^2 + 600^2}\sin 33.7° 0.429}{900} = 0.29$$

The friction energy accounts for 29% of the total energy.

Example 6.2 Estimate the tangential (cutting) and feed forces for the turning operation in Example 6.1 if the rake angle is changed from 0° to −8°. The average chip thickness was measured to be 0.85 mm.

Solution: The mean shear angle should be changed because the rake angle of the tool was changed and can be calculated from Equation 6.24. The cutting ratio calculated from Equation 6.25 is $r_c = a/a_c = 0.3/0.85 = 0.353$. Therefore, the shear angle is

$$\tan\phi = \frac{r_c\cos\alpha}{1 - r_c\sin\alpha} = \frac{0.353\cos(-8°)}{1 - 0.353\sin(-8°)} = 0.368 \Rightarrow \phi = 20.2°$$

The shear angle is reduced from 23.2° to 20.2° as expected since the rake angle was reduced. The cutting forces can be estimated using the mean shear strength of the work material calculated for the zero rake angle tool in Example 6.1. The mean friction angle can be estimated from Equation 6.30, or even the same mean friction angle calculated when cutting with zero rake angle in previous example can be used. The cutting force can be estimated from the shear force. Therefore,

$$\tau_s = \frac{F_s}{A_s} = \frac{F_s}{\dfrac{A_o}{\sin\phi}} = \frac{F_s \sin\phi}{ab} \Rightarrow F_s = \frac{\tau_s\, ab}{\sin\phi}$$

$$= \frac{(310)(0.3)(2.5)}{\sin 20.2°} = 673\,\text{N/mm}^2$$

The cutting and feed forces can be calculated from the force vector diagram in Figure 6.11 through the resultant force R

$$F_s = R\cos(\phi + \beta - \alpha) \Rightarrow R = \frac{F_s}{\cos(\phi + \beta - \alpha)}$$

$$= \frac{673}{\cos(20.2° + 26.6° + 8°)} = 1168\,\text{N}$$

Hence, the cutting and feed forces are equal to

$$F_c = R\cos(\beta - \alpha) \Rightarrow F_c = 1168\cos(26.6° + 8°) = 961\,\text{N}$$

$$F_Z = R\sin(\beta - \alpha) \Rightarrow F_Z = 1168\sin(26.6° + 8°) = 663\,\text{N}$$

6.16   PROBLEMS

Problem 6.1 In a turning operation on a cast iron tube with a lead angle of zero (which represents an orthogonal cutting test), the following conditions were noted:
   Tube thickness = 3 mm
   Feed = 0.2 mm/rev
   Rake angle on the tool/insert body = −5°
   The chip thickness was measured with a point micrometer to be 1.0 mm
   The cutting forces were measured with a dynamometer mounted under the toolholder. The tangential (cutting) and feed forces were 800 and 400 N, respectively. Calculate the shear angle, the mean angle of friction on the tool face, and the mean shear strength of the workpiece material.

Problem 6.2 In an orthogonal cutting operation, the rake angle is 8° and the coefficient of friction using a coated insert is 0.4. Determine the percentage change in chip thickness if an uncoated inset is used, which results in double friction compared to coated insert.

Problem 6.3 An orthogonal cutting operation is being carried out using a 5 mm thick tube under the following conditions: feed = 0.1 mm/rev, chip thickness = 0.2 mm, cutting speed = 120 m/min, and rake angle = 10°. The forces were measured such as cutting force = 500 N and thrust force = 200 N. Calculate the percentage of the total energy dissipated in the shear plane during cutting.

Problem 6.4 In a turning operation of steel 1040 tube with a lead angle of zero (which represents an orthogonal cutting test), the following conditions were noted:

Tube thickness = 6 mm, feed = 0.2 mm/rev, and cutting speed 100 m/min. The selected tool has a rake angle of 10°. The chip thickness was measured with a point micrometer to be 0.4 mm. The specific cutting energy of the material is 2300 N/mm².

Estimate the mean shear angle, the mean angle of friction on the tool face, and the mean shear stress of the workpiece material.

Show that the sum of the deformation power and friction power is equivalent to the total power during cutting.

Estimate the average strain and strain rate in the chip formation if the thickness of the shear zone is estimated to be 0.025 mm.

REFERENCES

1. J. T. Nicolson, Experiments with a lathe-tool dynamometer, *ASME Trans.* **25** (1904) 627–684.
2. M. C. Shaw, *Metal Cutting Principles*, Oxford University Press, Oxford, U.K., 1984, Chs. 3, 7–9, and 12.
3. G. F. Micheletti, B. F. von Turkovich, and S. Rossetto, Three force component piezo-electric dynamometer (ITM mark 2), *Int. J. Mach. Tool Des. Res.* **10** (1970) 305–315.
4. G. H. Gautschi, Cutting forces in machining and their routine measurement with multi-component piezoelectric force transducers, *Proceedings of 12th International Machine and Tool Design Research Conference*, Birmingham, U.K., 1971, pp. 113–120.
5. K. J. Pedersen and J. Hoffmann, General analysis of dynamometers for metal cutting, *Proc. NAMRC* **14** (1986) 189–193.
6. B. Bill, *Measuring with Crystals*, Verlag Moderne Industrie, Landsberg/Lech, Germany, 2002, pp. 15–25.
7. S. Yaldız, F. Unsac-ar, H. Saglam, and H. Isık, Design, development and testing of a four-component milling dynamometer for the measurement of cutting force and torque, *Mech. Syst. Signal Process.* **21** (2007) 1499–1511.
8. G. Totis and M. Sortino, Development of a modular dynamometer for triaxial cutting force measurement in turning, *Int. J. Mach. Tools Manuf.* **51** (2011) 34–42.
9. K. Nakazawa, Improvement of adaptive control of milling machine by non-contact cutting force detector, *Proceedings of the 16th International Machine and Tool Design Research Conference*, Birmingham, U.K., 1975, pp. 109–116.
10. Y. Ikezaki, T. Takeuchi, and M. Sakamoto, Cutting force measurement of a rotating tool by means of optical data transmission, *CIRP Ann.* **33** (1984) 61–64.
11. G. Totis, G. Wirtz, M. Sortino, D. Veselovac, E. Kuljanic, and F. Klocke, Development of a dynamometer for measuring individual cutting edge forces in face milling, *Mech. Syst. Signal Process.* **24** (2010) 1844–1857.
12. J. Tlusty, D. Y. Jang, and Y. S. Tarng, Measurement of the milling force over a wide frequency range, *Proc. NAMRC* **14** (1986) 273–280.
13. W. A. Knight and M. M. Sadek, The correction for dynamic errors in machine tool dynamometers, *CIRP Ann.* **19** (1971) 237–245.
14. Y. L. Chung and S. A. Spiewak, A model of high performance dynamometer, *ASME J. Eng. Ind.* **116** (1994) 279–288.
15. N. Tounsi and A. Otho, Dynamic cutting force measuring, *Int J. Mach. Tools Manuf.* **40** (2000) 1157–1170.
16. D. N. Dilley, Accuracy, vibration, and stability in drilling and reaming, Sc.D. thesis, Washington University, St. Louis, MO, 2003.
17. B. Bischoff, M. Hallen, T. Moser, T. Shi, D. Frohrib, and S. Ramalingam, Real time tool condition sensing part II: Fracture detection using a new transducer/failure identification system, in: M. K. Tse and D. Dornfeld, Eds., *Sensors for Manufacturing*, ASME, New York, 1987, pp. 69–77.
18. T. Shi, S. Ramalingam, D. A. Frohrib, and T. Moser, Tool fracture detection in the sub-millisecond time frame using machining inserts with integral sensors, *Proceedings of USA-Japan Symposium on Flexible Automation*, ASME, New York, 1988, pp. 1025–1033.
19. G. Arndt, Ultra high speed machining: Notes on metal cutting at speeds up to 7300 ft/s, *Proceedings of the 11th International Machine and Tool Design Research Conference*, Birmingham, U.K., 1970, p. 533.

20. J. P. Kottenstette and R. F. Recht, An ultra-high-speed machining facility, *Proc. NAMRC* **9** (1981) 318–325.

21. J. P. Kottenstette and R. F. Recht, Ultra-high-speed machining experiments, *Proc. NAMRC* **10** (1982) 263–270.

22. R. Komanduri, D. G. Flom, and M. Lee, Highlights of the DARPA advanced machining research program, *ASME J. Eng. Ind.* **107** (1985) 325–335.

23. J. F. Pearson, W. D. Syniuts, and N. H. Cook, Development of an instrumented toolholder transducer, *ASME PED Vol. 2*, ASME, New York, 1980, pp. 71–80.

24. P. Y. Sun, Y. K. Chang, T. C. Wang, and P. T. Liu, A simple and practical piezo-electric shank type three-component dynamometer, *Int. J. Mach. Tool Des. Res.* **22** (1982) 111–124.

25. Y. Altintas, Prediction of cutting forces and tool breakage in milling from feed drive current measurements, *ASME J. Eng. Ind.* **114** (1992) 386–392.

26. Y.-H. Jeong and D.-W. Cho, Estimating cutting force from rotating and stationary feed motor currents on a milling machine, *Int. J. Mach. Tools Manuf.* **42** (2002) 1559–1566.

27. Anon., T-MAC Tool Monitoring System, Caron Engineering, Wells, ME, nd.

28. Anon., Acuvim II Series Intelligent Power Monitor, Mod-tronic Instruments, Ltd., Brampton, Ontario, Canada, 2010.

29. R. A. Hallam and R. S. Allsopp, The design, development and testing of a prototype boring dynamometer, *Int. J. Mach. Tool Des. Res.* **2** (1962) 241–266.

30. G. M. Zhang and S. G. Kapoor, Development of an instrumented boring bar transducer, *Proc. NAMRC* **14** (1986) 194–200.

31. E. Korkmaz, B. Bediz, B. A. Gozen, and O. B. Ozdoganlar, Dynamic characterization of multi-axis dynamometers, *Precis. Eng.* **38** (2014) 148–161.

32. D. A. Stephenson, Material characterization for metal cutting force modeling, *ASME J. Eng. Mater. Technol.* **111** (1989) 210–219.

33. D. M. Eggleston, R. Herzog, and E. G. Thomsen, Observations on the angle relationships in metal cutting, *ASME J. Eng. Ind.* **81** (1959) 263–279.

34. I. S. Jawahir and C. A. van Luttervelt, Recent developments in chip control research and applications, *CIRP Ann.* **42** (1993) 659–693.

35. G. Stabler, The fundamental geometry of cutting tools, *Proc. Inst. Mech. Eng.* **165** (1951) 14–21.

36. R. H. Brown and E. J. A. Armarego, Oblique machining with a single cutting edge, *Int. J. Mach. Tool Des. Res.* **4** (1964) 9–25.

37. L. V. Colwell, Predicting the angle of chip flow for single-point cutting tools, *ASME Trans.* **76** (1954) 199–204.

38. K. Okushima and K. Minato, On the behavior of chip in steel cutting, *Bull. Jpn. Soc. Precis. Eng.* **2** (1959) 58–64.

39. H. T. Young, P. Mathew, and P. L. B. Oxley, Allowing for nose radius effects in predicting the chip flow direction and cutting forces in bar turning, *Proc. Inst. Mech. Eng.* **201C** (1987) 213–226.

40. H. J. Fu, R. E. DeVor, and S. G. Kapoor, A mechanistic model for the prediction of the force system in face milling operations, *ASME J. Eng. Ind.* **106** (1988) 81–88.

41. B. K. Srinivas, The forces in turning, SME Technical Paper MR82-947, 1982.

42. K. Nakayama, M. Arai, and K. Takei, Semi-empirical equations for three components of resultant cutting force, *CIRP Ann.* **32** (1983) 33–35.

43. C. A. Brown, A practical method for estimating machining forces from tool-chip contact area, *CIRP Ann.* **32** (1983) 91–93.

44. G. Subramani, R. Suvada, S. G. Kapoor, and R. E. DeVor, A model for the force system for cylinder boring process, *Proc. NAMRC* **15** (1987) 439–446.

45. J. W. Sutherland, G. Subramani, M. J. Kuhl, R. E. DeVor, and S. G. Kapoor, An investigation into the effect of tool and cut geometry on cutting force system prediction models, *Proc. NAMRC* **16** (1988) 264–272.

46. Shareef, N. Wiemken, and J. Lis, Development of mathematical model for prediction of forces in machining, SME Technical Paper MS90-262, 1990.

47. D. A. Stephenson and J. S. Agapiou, Calculation of main cutting edge forces and torque for drills with arbitrary point geometries, *Int J. Mach. Tools Manuf.* **32** (1992) 521–538.

48. D. A. Stephenson and J. W. Matthews, Cutting forces when turning and milling cast iron with silicon nitride tools, *Trans. NAMRI/SME* **21** (1993) 223–230.

49. D. A. Stephenson and P. Bandyopadhyay, Process independent force characterization for machining simulation, *Trans. ASME, J. Eng. Mater. Technol.* **119** (1997) 86–94.

50. H. Ernst, Physics of metal cutting, *Machining of Metals*, American Society for Metals, Metals Park, OH, 1938, p. 24.
51. M. E. Merchant, Mechanics of the metal cutting process. I. Orthogonal cutting and a type 2 chip, *J. Appl. Phys.* **16** (1945) 267–275.
52. R. F. Recht, Catastrophic thermoplastic shear, *ASME J. Appl. Mech.* **31** (1964) 186–193.
53. J. C. Lemaire and W. A. Backofen, Adiabatic instability in orthogonal cutting of steel, *Met. Trans.* **3** (1972) 477.
54. J. Q. Xie, A. E. Bayoumi, and H. M. Zbib, A study on shear banding in chip formation of orthogonal machinng, *Int. J. Mach. Tools Manuf.* **36** (1996) 835–847.
55. R. Komanduri, T. A. Schroeder, J. Hazra, B. F. von Turkovich, and D. G. Flom, On the catastrophic shear instability in high-speed machining of ANSI 4340 steel, *ASME J. Eng. Ind.* **104** (1982) 121–131.
56. R. Komanduri, Some clarification on the mechanics of chip formation when machining titanium alloys, *Wear* **76** (1982) 15.
57. D. Lee, The effect of cutting speed on chip formation under orthogonal machining, *ASME J. Eng. Ind.* **107** (1985) 55–63.
58. B. M. Manyindo and P. L. B. Oxley, Modelling the catastrophic shear type of chip when machining stainless steel, *Proc. Inst. Mech. Eng.* **200C** (1986) 349–358.
59. T.J. Burns and M.A. Davies, A nonlinear dynamics model for chip segmentation in machining, *Phys. Rev. Lett.* **79** (1997) 447–450.
60. J. Barry and G. Byrne, The mechanism of chip formation in machining hardened steels, *ASME J. Manuf. Sci. Eng.* **124** (2002) 528–535.
61. R. Stevenson, The morphology of machining chips formed during low speed quasi-orthogonal machining of CA360 brass and a model for their formation, *ASME J. Eng. Ind.* **114** (1992) 404–411.
62. D. Kececioglu, Shear strain rate in metal cutting and its effects on shear flow stress, *ASME Trans.* **80** (1958) 158–168.
63. W. F. Hastings, A new quick-stop device and grid technique for metal cutting research, *Annals CIRP* **15** (1967) 109.
64. J. Ellis, R. Kirk, and G. Barrow, The development of a quick-stop device for metal cutting research, *Int. J. Mach. Tool Des. Res.* **9** (1969) 321.
65. P. K. Phillip, Study of the performance characteristics of an explosive quick-stop device for freezing cutting action, *Int. J. Mach. Tool Des. Res.* **11** (1971) 133.
66. C. Spaans, A treatise on the streamlines and the stress, strain, and strain rate distributions, and on stability in the primary shear zone in metal cutting, *ASME J. Eng. Ind.* **94** (1972) 690–696.
67. R. H. Brown and R. Komanduri, An investigation of the performance of a quick-stop device for metal cutting studies, *Proceedings of 13th MTDR Conference*, 1973, pp. 225–231.
68. R. H. Brown, A double shear-pin quick stop device for rapid disengagement of a cutting tool, *Int. J. Mach. Tool Des. Res.* **16** (1976) 115–121.
69. J. T. Black and C. R. James, The hammer QSD-quick stop device for high speed machining and rubbing, *ASME J. Eng. Ind.* **103** (1981) 13–21.
70. E. M. Trent, Metal cutting and the tribology of seizure: I. Seizure in metal cutting, *Wear* **126** (1988) 29–45.
71. G. Boothroyd and W. A. Knight, *Fundamentals of Machining and Machine Tools*, 2nd edn., Marcel Dekker, New York, 1989, Chs. 2, 3.
72. M. R. Shankar, S. Chandrasekar, W. D. Compton, and A. H. King, Characteristics of aluminum 6061-T6 deformed to large plastic strains by machining, *Mater. Sci. Eng. A* **410–411** (2005) 364–368.
73. S. Swaminathan, M. R. Shankar, S. Lee, J. Hwang, A. H. King b, R. F. Kezar, B. C. Rao, T. L. Brown, S. Chandrasekar, W. D. Compton, and K. P. Trumble, Large strain deformation and ultra-fine grained materials by machining, *Mater. Sci. Eng. A* **410–411** (2005) 358–363.
74. S. Swaminathan, T. L. Brown, S. Chandrasekar, T. R. McNelleya, and W. D. Compton, Severe plastic deformation of copper by machining: Microstructure refinement and nanostructure evolution with strain, *Scripta Materialia* **56** (2007) 1047–1050.
75. N. H. Cook and M. C. Shaw, A visual metal cutting study, *Mech. Eng.* **73**(11) (1951).
76. W. B. Palmer and P. L. B. Oxley, Mechanics of orthogonal machining, *Proc. Inst. Mech. Eng.* **173** (1959) 623–654.
77. A. I. Isayev and V. N. Gorbunova, A new filming method for investigating the process of plastic deformations in the zone of chip formation, *Res. Film* **3**:6 (1960) 349–356.
78. T. H. C. Childs, A new visio-plasticity technique and a study of curly chip formation, *Int. J. Mech. Sci.* **18** (1971) 373–387.

79. J. H. L. The, High speed films of the incipient cutting process in machining at conventional speeds, *ASME J. Eng. Ind.* **99** (1977) 262–268.

80. V. Kalhori, M. Lundblad, and L. E. Lindgren, Numerical and experimental analysis of orthogonal metal cutting, *Manuf. Sci. Eng.*, ASME MED Volume 6-2, ASME, New York, 1997.

81. R. Rhorer, E. Whitenton, T. Burns, S. Mates, J. Heigel, A. Cooke, J. Soons, and R. Ivester, Machining process measurements: A titanium machining example, *Proceedings of the Annual Meeting of the Society for Experimental Mechanics,* Albuquerque, NM, June 1–4, 2009.

82. E. Whitenton, J. Heigel, and R. Ivester, Measurement and characterization of dynamics in machining chip segmentation, *Proceedings of 11th CIRP Conference on Modeling of Machining Operations,* Gaithersburg, MD, 2008, pp. 237–246.

83. J. R. Crookall and D. B. Richardson, Use of photographed orthogonal grids and mechanical quick-stopping techniques in machining research, *Photography in Engineering*, The Institution of Mechanical Engineers, London, U.K., 1969, paper 4.

84. S. Ramalingam, A photoelastic study of stress distribution during orthogonal cutting part II: Photoplasticity observations, *ASME J. Eng. Ind.* **93** (1971) 538–544.

85. A. C. Bell, S. Ramalingam, and J. T. Black, Dynamic metal cutting studies as performed on the SEM, *Proc. NAMRC* **1** (1973) 99–110.

86. K. Iwata and K. Ueda, Crack nucleation and its propagation in discontinuous chip formation performed within a scanning electron microscope, *Proceedings of the NAMRC 3*, SME, Dearborn, MI, 1975, pp. 603–617.

87. B. F. von Turkovich and M. Field, Survey on material behavior in machining, *CIRP Ann.* **30** (1981) 533–540.

88. K. Iwata, K. Ueda, and K. Okuda, A study of mechanisms of burrs formation in cutting based on direct SEM observation, *J. Jpn. Soc. Precision Eng.* **48** (1982) 510–515.

89. R. S. Hahn, Some observations on chip curl in metal cutting process under orthogonal cutting conditions, *ASME Trans.* **75** (1953) 538–544.

90. J. K. Russel and R. H. Brown, Deformation during chip formation, *ASME J. Eng. Ind.* **87** (1965) 53–56.

91. T. C. Hsu, An analysis of the plastic deformation due to orthogonal and oblique cutting, *J. Strain Anal.* **15** (1966) 375–379.

92. S. Ramalingam, Deformation in orthogonal cutting, *ASME J. Eng. Ind.* **113** (1991) 121–128.

93. B. Wang, Z. Liu, and Q. Yang, Investigations of yield stress, fracture toughness, and energy distribution in high speed orthogonal cutting, *Int. J. Mach. Tools Manuf.* **73** (2013) 1–8.

94. D. A. Stephenson, Assessment of steady-state metal cutting temperature models based on simultaneous infrared and thermocouple data, *ASME J. Eng. Ind.* **113** (1991) 121–128.

95. B. F. von Turkovich, Shear stress in metal cutting, *ASME J. Eng. Ind.* **92** (1970) 151–157.

96. J. T. Black, Flow stress model in metal cutting, *ASME J. Eng. Ind.* **101** (1979) 403–415.

97. M. C. Shaw and I. Finnie, The shear stress in metal cutting, *ASME Trans.* **77** (1955) 115–125.

98. M. E. Merchant, Mechanics of the metal cutting process. II. Plasticity conditions in orthogonal cutting, *J. Appl. Phys.* **16** (1945) 318–324.

99. S. Santhanam and M. C. Shaw, Flow characteristics for the complex stress state in metal cutting, *CIRP Ann.* **34** (1985) 109–111.

100. R. Stevenson and D. A. Stephenson, The mechanical behavior of zinc during machining, *ASME J. Eng. Mater. Technol.* **117** (1995) 172–178.

101. P. L. B. Oxley and M. G. Stevenson, Measuring stress/strain properties at very high strain rates using a machining test, *J. Inst. Metals* **95** (1967) 308.

102. M. G. Stevenson and P. L. B. Oxley, An experimental investigation of the influence of strain-rate and temperature on the flow stress properties of a low carbon steel using a machining test, *Proc. Inst. Mech. Eng.* **185** (1970) 741.

103. A. E. Bayoumi and M. N. Hamdan, Characterization of dynamic flow stress-strain properties through machining tests, *ASME Man. Rev.* **1** (1988) 130–135.

104. L. Filice, F. Micari, S. Rizzuti, and D. Umbrello, A critical analysis on the friction modelling in orthogonal machining, *Int. J. Mach. Tools Manuf.* **47** (2007) 709–714.

105. D. Ulutan and T. Ozel, Determination of tool friction in presence of flank wear and stress distribution based validation using finite element simulations in machining of titanium and nickel based alloys, *J. Mater. Proc. Technol.* **213** (2013) 2217–2237.

106. J. A. Bailey, Friction in metal machining: mechanical aspects, *Wear* **31** (1975) 243–275.

107. J. A. Arsecularatne, On tool-chip interface stress distributions, ploughing force and size effect in machining, *Int. J. Mach. Tools Manuf.* **37** (1997) 885–899.

108. M. B. Gordon, Instrument for measuring the friction forces in metal cutting, *Mach. Tooling* **36**:7 (1965) 30–32.

109. M. B. Gordon, The applicability of the binomial law to the process of friction in the cutting of metals, *Wear* **10** (1967) 274–290.

110. S. Kato, K. Yamaguchi, and M. Yamada, Stress distribution at the interface between tool and chip in machining, *ASME J. Eng. Ind.* **94** (1972) 683–689.

111. G. Barrow, W. Graham, T. Kurimoto, and Y. F. Leong, Determination of the rake face stress distribution in orthogonal machining, *Int. J. Mach. Tool Des. Res.* **22** (1982) 75–85.

112. T. H. C. Childs and M. I. Mahdi, On the stress distribution between the chip and tool during metal turning, *CIRP Ann.* **38** (1989) 55–58.

113. T. C. Hsu, A study of the normal and shear stresses on a cutting tool, *ASME J. Eng. Ind.* **88** (1966) 51–64.

114. P. W. Wallace and G. Boothroyd, Tool forces and tool-chip friction in orthogonal machining, *J. Mech. Eng. Sci.* **6** (1964) 74–87.

115. A. Bhattacharyya, On the friction process in metal cutting, *Proceedings of Sixth International Machine Tool Design and Research Conference*, 1963, pp. 491–505.

116. E. G. Coker and K. C. Chakko, An account of some experiments on the action of cutting tools, *Proc. Inst. Mech. Eng.* **117** (1922) 567–621.

117. M. Okoshi and S. Fukui, Studies of cutting action by means of photoelasticity, *J. Soc. Precis. Mech. Jpn.* **1** (1934) 598.

118. L. C. Andreev, Photoelastic study of stresses in a cutting tool by means of cinematography, *Vestn. Mashinostr.* **38**:5 (1958) 54–57 (in Russian).

119. W. B. Rice, R. Salmon, and W. D. Syniuta, Photoelastic determination of cutting tool stresses, *Trans. Eng. Inst. Canada* **4** (1960) 20–23.

120. N. N. Zorev, Interrelationship between shear process occurring along the tool face and on shear plane in metal cutting, *International Research in Production Engineering*, ASME, New York, 1963, pp. 42–49.

121. E. Usui and H. Takeyama, A photoelastic analysis of machining stresses, *ASME J. Eng. Ind.* **82** (1966) 303–308.

122. W. Kattwinkel, Untersuchung an Schneiden Spannender Werkzeuge mit Hilfe der Spannungsoptik, *Industrie Anzeiger* **60** (1957) 29–36.

123. H. Chaandrasekaran and D. V. Kapoor, Photoelastic analysis of tool-chip interface stresses, *ASME J. Eng. Ind.* **87** (1965) 495–502.

124. E. D. Doyle, J. G. Horne, and D. Tabor, Frictional interactions between chip and rake face in continuous chip formation, *Proc. R. Soc. (Lond.)* **A366** (1979) 173–183.

125. P. K. Wright, J. G. Horne, and D. Tabor, Boundary conditions at the chip-tool interface in machining: comparisons between seizure and sliding friction, *Wear* **54** (1979) 371–390.

126. P. K. Wright, Frictional interactions in machining: comparisons between transparent sapphire and steel cutting tools, *J. Metals Technol.* **8** (1981) 150–160.

127. A. Bagchi and P. K. Wright, Stress analysis in machining with the use of sapphire tools, *Proc. R. Soc. (London)* **A409** (1987) 99–113.

128. G. W. Rowe and A. B. Wilcox, A new method of determining the pressure distribution on a steel cutting tool, *J. Iron Steel Inst.* **209** (1971) 231–232.

129. W. Grzesik, D. Kowalczyk, and K. Zak, A new mechanistic friction model for the oblique cutting with tool wear effect, *Tribol. Int.* **66** (2013) 49–53.

130. W. Grzesik, J. Rech, and K. Zak, Determination of friction in metal cutting with tool wear and flank face effects, *Wear* **317** (2014) 8–16.

131. H. Puls, F. Klocke, and D. Lung, Experimental investigation on friction under metal cutting conditions, *Wear* **310** (2014) 63–71.

132. Y. Karpat and T. Ozel, Identification of friction factors for chamfered and honed tools through slip-line field analysis, *Proceedings of ASME MSEC Conference*, Ypsilanti, MI, 2006, paper MSEC2006-21058.

133. M. Y. Friedman and E. Lenz, Investigation of the tool-chip contact length in metal cutting, *Int. J. Mach. Tool Des. Res.* **10** (1970) 401–416.

134. R. L. Woodward, Determination of plastic contact length between chip and tool in machining, *ASME J. Eng. Ind.* **99** (1977) 802–804.

135. C. Spaans, A comparison of an ultrasonic method to determine the chip/tool contact length with some other methods, *CIRP Ann.* **19** (1971) 485–490.

136. S. Ramalingam and P. V. Desai, Tool-chip contact length in orthogonal machining, ASME Paper 80-WA/Prod-23, 1980.

137. W. F. Hastings, P. Mathew, and P. L. B. Oxley, A machining theory for predicting chip geometry, cutting forces, etc. from work material properties and cutting conditions, *Proc. R. Soc. (Lond.)* **A371** (1980) 569–587.

138. E. Kannatey-Asibu and D. A. Dornfeld, Quantitative relations for acoustic emissions in orthogonal cutting, *ASME J. Eng. Ind.* **103** (1981) 330–340.

139. M. Mazurkiewicz, Z. Kubala, and J. Chow, Metal machining with high-pressure water-jet cooling assistance—A new possibility, *ASME J. Eng. Ind.* **111** (1989) 7–12.

140. Increasing machine tool productivity with high pressure cryogenic coolant flow, Manufacturing Technology Directorate, Wright Laboratory, Air Force Systems Command Report WL-TR-92-8014, May 1992.

141. V. Piispanen, Lastunmuodostumisen Teoriaa, *Teknillinen Aikakauslehti* **27** (1937) 315–322 (in Finnish).

142. V. Piispanen, Theory of formation of metal chips, *J. Appl. Phys.* **19** (1948) 876–881.

143. H. Ernst and M. E. Merchant, Chip formation, friction, and high quality machined surfaces, *Surface Treatment of Metals*, ASM, Cleveland, OH, 1941, pp. 299–378.

144. M. E. Merchant, Basic mechanics of the metal-cutting process, *ASME J. Appl. Mech.* **11** (1944) A168–A175.

145. M. E. Merchant, Mechanics of the metal cutting process. II. Plasticity conditions in orthogonal cutting, *J. Appl. Phys.* **16** (1945) 318–324.

146. E. H. Lee and B. W. Shaffer, The theory of plasticity applied to a problem of machining, *ASME J. Appl. Mech.* **18** (1951) 405–412.

147. M. C. Shaw, N. H. Cook, and I. Finnie, Shear angle relationships in metal cutting, *ASME Trans.* **75** (1953) 273–288.

148. J. H. Creveling, T. F. Jordan, and E. G. Thomsen, Some studies of angle relationships in metal cutting, *ASME Trans.* **79** (1957) 127–138.

149. H. D. Pugh, Mechanics of the cutting process, *Proceedings of IME Conference on Technology of Engineering Manufacture*, London, U.K., 1958, p. 237.

150. S. Kobayashi and E. G. Thomsen, Metal cutting analysis—I. Re-evaluation and new method of presentation of theories, *ASME J. Eng. Ind.* **84** (1962) 63–70.

151. E. J. A. Armarego, A note on the shear angle relation in orthogonal cutting, *Int. J. Mach. Tool Des. Res.* **6** (1966) 139–141.

152. N. Ueda and T. Matsuo, An investigation of some shear angle theories, *CIRP Ann.* **35** (1986) 27–30.

153. A. G. Atkins, Modelling metal cutting using modern ductile fracture mechanics: Quantitative explanations for some longstanding problems, *Int. J. Mech. Sci.* **45** (2003) 373–396.

154. W. B. Palmer and P. L. B. Oxley, Mechanics of orthogonal machining, *Proc. Inst. Mech. Eng.* **173** (1959) 623–654.

155. W. Johnson, Some slip-line fields for swaging or expanding, indenting, extruding and machining for tools with curved dies, *Int. J. Mech. Sci.* **4** (1962) 323–347.

156. E. Usui and K. Hoshi, Slip-line fields in metal machining which involve centered fans, *Int. Res. Prod. Eng.*, ASME, New York, 1963, p. 61.

157. H. Kudo, Some new slip line solutions for two-dimensional steady-state machining, *Int. J. Mech. Sci.* **7** (1965) 43–55.

158. P. Dewhurst, On the non-uniqueness of the machining process, *Proc. R. Soc. Lond.* **A360** (1978) 587–610.

159. L. De Chiffre and T. Wanheim, Chip compression relationships in metal cutting, *Proc. NAMRC* **9** (1981) 231–234.

160. W. Johnson and P. B. Mellor, *Engineering Plasticity*, Van Nostrand Reinhold, London, U.K., 1973, pp. 467–493.

161. P. J. Arrazola, T. Ozel, D. Umbrello, M. Davies, and I. S. Jawahir, Recent advances in modelling of metal machining processes, *CIRP Ann.* **62** (2013) 695–718.

162. F. C. Appl and S. Saleem, Prediction of shear angle using minimum energy principle and strain softening, *Trans. NAMRI/SME* **19** (1991) 113–120.

163. T. Shi and S. Ramalingam, Slip line solution for orthogonal cutting with a chip breaker and flank wear, *Int. J. Mech. Sci.* **33** (1991) 689–704.

164. M. C. Shaw, N. H. Cook, and P. A. Smith, The mechanics of three-dimensional cutting operations, *ASME Trans.* **74** (1952) 1055–1064.

165. N. N. Zorev, *Metal Cutting Mechanics*, Pergamon Press, Oxford, U.K., 1966, Chapter 6 (Translated by H. S. H. Massey).

166. G. C. I. Lin and P. L. B. Oxley, Mechanics of oblique machining: Predicting chip geometry and cutting forces from work-material properties and cutting conditions, *Proc. Inst. Mech. Eng.* **186** (1972) 813–820.

167. W. A. Morcos, A slip line field solution of the free continuous cutting problem in conditions of light friction at chip-tool interface, *ASME J. Eng. Ind.* **102** (1980) 310–314.

168. R. G. Fenton and P. L. B. Oxley, Mechanics of orthogonal machining: Allowing for the effects of strain rate and temperature on tool-chip friction, *Proc. Inst. Mech. Eng.* **178** (1969) 417–438.

169. R. G. Fenton and P. L. B. Oxley, Mechanics of orthogonal machining: Predicting chip geometry and cutting forces from work-material properties and cutting conditions, *Proc. Inst. Mech. Eng.* **184** (1970) 927–942.

170. P. L. B. Oxley and W. F. Hastings, Minimum work as a possible criterion for determining the frictional conditions at the tool/chip interface in machining, *Phil. Trans.* **282** (1976) 565–584.

171. W. F. Hastings, P. Mathew, and P. L. B. Oxley, A machining theory for predicting chip geometry, cutting forces etc. from work material properties and cutting conditions, *Proc. R. Soc. Lond.* **A371** (1980) 569–587.

172. P. L. B. Oxley, Machinability: A mechanics of machining approach, *On the Art of Cutting Metals—75 Years Later*, ASME PED Vol. 7, ASME, New York, 1982, pp. 37–83.

173. P. L. B. Oxley, *The Mechanics of Machining*, Ellis Horwood, Chicester, U.K., 1989.

174. C. W. MacGregor and J. C. Fisher, Tension tests at constant true strain-rates, *ASME J. Appl. Mech.* **13** (1946) A11.

175. X. P. Li, K. Iynkaran, and A. Y. C. Nee, A hybrid machining simulator based on predictive machining theory and neural network modeling, *J. Mater. Proc. Technol.* **89–90** (1999) 224–230.

176. H. Z. Li, W. B. Zhang, and X. P. Li, Modelling of cutting forces in helical end milling using a predictive machining theory, *Int. J. Mech. Sci.* **43** (2001) 1711–1730.

177. N. Fang and I. S. Jawahir, An analytical predictive model and experimental validation for machining with grooved tools incorporating the effects of strains, strain-rates, and temperatures, *CIRP Ann.* **51** (2002) 83–86.

178. A. H. Abidi-Sedeh and V. Madhavan, Effect of some modifications to Oxley's machining theory and the applicability of different material models, *Mach. Sci. Technol.* **6** (2002) 379–395.

179. D. I. Lalwani, N. K. Mehta, and P.K. Jain, Extension of Oxley's predictive machining theory for Johnson and Cook flow stress model, *J. Mater. Proc. Technol.* **209** (2009) 5305–5312.

180. E. Usui, A. Hirota, and M. Masuko, Analytical prediction of three dimensional cutting process. Part 1. Basic cutting model and energy approach, *ASME J. Eng. Ind.* **100** (1978) 222–228.

181. E. Usui and A. Hirota, Analytical prediction of three dimensional cutting process. Part 2. Chip formation and cutting force with conventional single-point tool, *ASME J. Eng. Ind.* **100** (1978) 229–235.

182. E. Usui, T. Shirakashi, and T. Kitagawa, Analytical prediction of three dimensional cutting process. Part 3. Cutting temperature and crater wear of carbide tool, *ASME J. Eng. Ind.* **100** (1978) 236–243.

183. E. Usui and T. Shirakashi, Mechanics of machining—From "descriptive" to "predictive" theory, *On The Art of Cutting Metals—75 Years Later*, ASME PED Vol. 7, ASME, New York, 1982, pp. 13–35.

184. E. Usui and T. Shirakashi, Analytical prediction of cutting tool wear, *Wear* **100** (1984) 129–151.

185. E. Shamoto, M. Kato, N Suzuki, and R. Hino, Analysis of three-dimensional cutting process with thin shear plane model, *Trans. ASME, J. Manuf. Sci. Eng.* **135** (2013) 041001–041008.

186. M. Shatla, C. Kerk, and T. Altan, Process modeling in machining. Part I: Determination of flow stress data, *Int. J. Mach. Tools Manuf.* **41** (2001) 1511–1534.

187. M. Shatla, C. Kerk, and T. Altan, Process modeling in machining—Part II—Applications of flow stress data to predict process variables, *Int. J. Mach. Tools Manuf.* **41** (2001) 1659–1680.

188. B. M. Maniyindo and P. L. B. Oxley, Modeling the catastrophic shear type of chip when machining stainless steel, *Proc. Inst. Mech. Eng.* **C200** (1986) 349–358.

189. Y.-C. Yen, J. Soehner, H. Weude, J. Schmidt, and T. Altan, Estimation of tool wear of carbide tool in orthogonal cutting using FEM simulation, *Proceedings of Fifth CIRP International Workshop of Modeling of Machining Operations*, West Lafayette, IN, 2002, pp. 149–160.

190. K. W. Kim and H.-C. Sin, Development of a thermo-viscoplastic cutting model using finite element method, *Int. J. Mach. Tools Manuf.* **36** (1996) 379–397.

191. R. Hill, The mechanics of machining: A new approach, *J. Mech. Phys. Solids* **3** (1954) 47–53.

192. C. Rubenstein, A note concerning the inadmissibility of applying the minimum work criterion to metal cutting, *ASME J. Eng. Ind.* **105** (1983) 294–296.

193. P. K. Wright, Predicting the shear plane angle in machining from work material strain-hardening characteristics, *ASME J. Eng. Ind.* **104** (1982) 285–292.

194. A. Bagchi, Discussion on a previously published paper by P. K. Wright, *ASME J. Eng. Ind.* **105** (1983) 129–131.

195. D. A. Stephenson and S. M. Wu, Computer models for the mechanics of three-dimensional cutting processes I. Theory and numerical method, *ASME J. Eng. Ind.* **110** (1988) 32–37.

196. K. Iwata, K. Osakada, and Y. Terasaka, Process modeling of orthogonal cutting by the rigid-plastic finite element method, *ASME J. Eng. Mater. Technol.* **106** (1984) 132–138.

197. J. T. Carroll III and J. S. Strenkowski, Finite element models of orthogonal cutting with application to single point diamond turning, *Int. J. Mech. Sci.* **30** (1988) 899–920.

198. R. N. Roth, The range of permissible shear angles in orthogonal machining allowing for variable hydrostatic stress on the shear plane and variable friction angle along the rake face, *Int. J. Mach. Tool Des. Res.* **15** (1975) 161–177.

199. T. Ota, A. Shindo, and H. Fukuola, An investigation on the theories of orthogonal cutting, *Trans. Jpn. Soc. Mech. Eng.* **24** (1958) 484–493 (in Japanese).

200. A. H. Low and P. T. Wilkinson, An investigation of non-steady-state cutting, Report 45, National Engineering Laboratory, Glasgow, Scotland, 1962.

201. A. H. Low, Effects of initial conditions in metal cutting, Report 65, National Engineering Laboratory, Glasgow, Scotland, 1962.

202. R. Stevenson and D. A. Stephenson, The effect of prior cutting conditions on the shear mechanics of orthogonal machining, *ASME J. Manuf. Sci. Eng.* **120** (1998) 13–20.

203. K. Okushima and Y. Kakino, The residual stress produced by metal cutting, *CIRP Ann.* **10** (1971) 13–14.

204. B. E. Klamecki, Incipient chip formation in metal cutting—A three-dimensional finite-element analysis, Ph. D. Thesis, University of Illinois, Urbana, IL, 1973.

205. A. O. Tay, M. G. Stevenson, and G. de Vahl Davis, Using the finite element method to determine temperature distributions in orthogonal machining, *Proc. Inst. Mech. Eng.* **188** (1974) 627–638.

206. P. D. Muraka, G. Barrow, and S. Hinduja, Influence of the process variables on the temperature distribution in orthogonal machining using the finite element method, *Int. J. Mech. Sci.* **21** (1979) 445–456.

207. R. T. Sedgwick, Numerical modeling of high-speed machining processes, *High Speed Machining*, ASME PED Vol. 12, ASME, New York, 1984, pp. 141–155.

208. J. S. Strenkowski and J. T. Caroll III, A finite element model of orthogonal metal cutting, *ASME J. Eng. Ind.* **107** (1985) 349–354.

209. J. S. Strenkowski and G. L. Mitchum, An improved finite element model of orthogonal metal cutting, *Proc. NAMRC* **15** (1987) 506–509.

210. T. H. C. Childs and K. Maekawa, A computer simulation approach towards the determination of optimum cutting conditions, *Strategies for Automation of Machining,* ASM, Materials Park, OH, 1987, pp. 157–166.

211. J. S. Strenkowski and K. J. Moon, Finite element prediction of chip geometry and tool/workpiece temperature distributions in orthogonal metal cutting, *ASME J. Eng. Ind.* **112** (1990).

212. K. Komvopoulos and S. A. Erpenbeck, Finite element modeling of orthogonal metal cutting, *ASME J. Eng. Ind.* **113** (1991) 253–267.

213. T. Tyan and W. H. Yang, Analysis of orthogonal metal cutting processes, *Int. J. Num. Methods Eng.* **34** (1992) 365–389.

214. A. J. Shih and H. T. Y. Yang, Experimental and finite element predictions of residual stresses due to orthogonal metal cutting, *Int. J. Num. Methods Eng.* **36** (1993) 1487–1507.

215. Z. C. Lin and W. C. Pan, A thermoelastic-plastic large deformation model for orthogonal cutting with tool flank wear—Part I: Computational procedures, *Int. J. Mech. Sci.* **35** (1993) 829–840.

216. Z. C. Lin and W. C. Pan, A thermoelastic-plastic large deformation model for orthogonal cutting with tool flank wear—Part II: Machining application, *Int. J. Mech. Sci.* **35** (1993) 841–850.

217. G. S. Sekhon and J. L. Chenot, Numerical simulation of continuous chip formation during non-steady orthogonal cutting, *Eng. Comput.* **10** (1993) 31–48.

218. T. D. Marusich and M. Ortiz, Modelling and simulation of high-speed machining, *Int. J. Num. Methods Eng.* **38** (1995) 3675–3694.

219. T. Ozel and T. Altan, Process simulation using finite element method—Prediction of cutting forces, tool stresses, and temperatures in high-speed flat eng milling process, *Int. J. Mach. Tools Manuf.* **40** (2000) 713–738.

220. T. Obikawa and E. Usui, Computational machining of titanium alloy—Finite element modeling and a few results, *ASME J. Manuf. Sci. Eng.* **118** (1996) 208–215.

221. Y. B. Guo and C. R. Liu, 3D FEA modeling of hard turning, *ASME J. Manuf. Sci. Eng.* **124** (2002) 189–199.

222. J. Mackerle, Finite-element analysis and simulation of machining: A bibliography (1976–1996), *J. Mater. Proc. Technol.* **86** (1999) 17–44.

223. J. Mackerle, Finite element analysis and simulation of machining: An adendum, a bibliography (1966–2002), *Int. J. Mach. Tools Manuf.* **43** (2003) 103–114.

224. T. D. Marusich, S. Usui, R. Aphale, N. Saini, R. Li and A.J. Shih, Three-dimensional finite element modeling of drilling processes, *Proceedings of ASME MSEC Conference*, Ypsilanti, MI, October 2006, paper MSEC2006-21059.

225. Z. L. Wang, Y. J. Hu, and D. Zhu, DEFORM-3D base on machining simulation during metal machining, *Key Eng. Mater.* **579–580** (2013) 197–201.

226. W. Liu, D. Rena, S. Usui, J. Wadell, and T. D. Marusich, A gear cutting predictive model using the finite element method, *Proc. CIRP* **8** (2013) 51–56.

227. T. D. Marusich, *Third Wave AdvantEdge Theoretical Manual*, Version 4.3, Third Wave Systems, Inc., Minneapolis, MN, 2002.

228. T. D. Marusich and E. Askari, Modeling residual stress and workpiece quality in machined surfaces, Third Wave Systems, Inc., Minneapolis, MN, 2002.

229. T. D. Marusich, C. J. Brand, and J. D. Thiele, A methodology for simulation of chip breakage in turning processes using an orthogonal finite element model, *Proceedings of Fifth CIRP International Workshop on Modeling Machining Operations*, West Lafayette, IN, 2002, pp. 139–148.

230. S. Kalidas, Cost effective tool selection and process development through process simulation, *Proceedings of 2002 Third Wave AdvantEdge User's Conference*, Atlanta, GA, April 2002, Paper 2.

231. M. Lundblad, Influence of cutting tool geometry on residual stress in the workpiece, *Proceedings of 2002 Third Wave AdvantEdge User's Conference*, Atlanta, GA, April 2002, Paper 7.

232. A. Davoudinejad, E. Chiappini, S. Tirelli, M. Annoni, and M. Strano, Finite element simulation and validation of chip formation and cutting forces in dry and cryogenic cutting of Ti–6Al–4V, *Proceedings of NAMRI/SME 43*, Charlotte, NC, June 2015.

233. R. Seshadri, I. Naveen, S. Sharan, M. Viswasubrahmanyam, K. S. VijaySekar, and M. P. Kumar, Finite element simulation of the orthogonal machining process with Al 2024 T351 aerospace alloy, *Procedia Eng.* **64** (2013) 1454–1463.

234. M. Field and M. E. Merchant, Mechanics of formation of the discontinuous chip in metal cutting, *ASME Trans.* **71** (1949) 421–430.

235. N. H. Cook, I. Finnie, and M. C. Shaw, Discontinuous chip formation, *ASME Trans.* **76** (1954) 153–162.

236. W. B. Palmer and M. S. M. Riad, Modes of cutting with discontinuous chips, *Proceedings of the 8th International MTDR Conference*, Manchester, U.K., 1967, pp. 259–279.

237. R. Komanduri and R. H. Brown, On the mechanics of chip segmentation in machining, *ASME J. Eng. Ind.* **108** (1968) 33–51.

238. N. Ueda and T. Matsuo, An analysis of saw-toothed chip formation, *CIRP Ann.* **31** (1982) 81–84.

239. T. H. C. Childs and G. W. Rowe, Physics in metal cutting, *Rep. Prog. Phys.* **36** (1973) 223–288.

240. W. K. Luk and R. C. Brewer, An energy approach to the mechanics of discontinuous chip formation, *ASME J. Eng. Ind.* **86** (1964) 157–162.

241. T. Obikawa, H. Sasahara, T. Shirakashi, and E. Usui, Application of computational machining method to discontinuous chip formation, *ASME J. Manuf. Sci. Eng.* **119** (1997) 667–674.

242. J. Q. Xie, A. E. Bayoumi, and H. M. Zbib, FEA modeling and simulation of shear localized chip formation in metal cutting, *Int. J. Mach. Tools Manuf.* **38** (1998) 1067–1087.

243. H. Ernst and M. Martelloti, The formation and function of the built-up edge, *Mech. Eng.* **57** (1935) 487.

244. W. B. Heginbotham and S. L. Gogia, Metal cutting and the built-up nose, *Proc. Inst. Mech. Eng.* **175** (1961) 892–917.

245. K. Nakayama, M. C. Shaw, and R. C. Brewer, Relationship between cutting forces, temperatures, built-up edge and surface finish, *CIRP Ann.* **14** (1966) 211–223.

246. H. Takeyama and T. Ono, Basic investigation of built-up edge, *ASME J. Eng. Ind.* **90** (1968) 335–342.

247. K. Iwata, J. Aihara, and K. Okushima, On the mechanism of built-up edge formation in cutting, *CIRP Ann.* **19** (1971) 323.

248. H. Bao and M. G. Stevenson, An investigation of built-up edge formation in the machining of aluminum, *Int. J. Mach. Tool Des. Res.* **16** (1976) 165–178.

249. E. M. Trent, Metal cutting and the tribology of seizure: II. Movement of work material over the tool in metal cutting, *Wear* **128** (1988) 47–64.

250. E. Usui, K. Maekawa, and T. Shirakashi, Simulation analysis of built-up edge formation in machining low carbon steel, *Bull. Jpn. Soc. Precis. Eng.* **15** (1981) 237–242.

251. D. H. Howerton, J. S. Strenkowski, and J. A. Bailey, Prediction of built-up edge formation in orthogonal cutting of aluminum, *Trans. NAMRI/SME* **17** (1989) 95–102.

7 Cutting Temperatures

7.1 INTRODUCTION

When metal is cut, energy is expended in deforming the chip and in overcoming friction between the tool and the workpiece. Almost all of this energy is converted to heat [1–4], producing high temperatures in the deformation zones and surrounding regions of the chip, tool, and workpiece (Figure 7.1).

Cutting temperatures are of interest because they affect machining performance. Temperatures in the primary deformation zone, where the bulk of the deformation involved in chip formation occurs, influence the mechanical properties of the work material and thus the cutting forces. For this reason most of the more complete analyses of the mechanics of cutting use temperature-dependent constitutive models. Temperatures on the rake face of the tool have a strong influence on tool life. As temperatures in this area increase, the tool softens and either wears more rapidly through abrasion or deforms plastically itself. In some cases constituents of the tool material diffuse into the chip or react chemically with the cutting fluid or chip, leading ultimately to tool failure. Since cutting temperatures increase with the cutting speed, temperature-activated tool wear mechanisms limit maximum cutting speeds for many tool–work material combinations. An understanding of temperatures in this region therefore provides insight into the requirements for tool materials and coatings. Finally, temperatures on the relief face of the tool affect the finish and metallurgical state of the machined surface. Moderate levels of these temperatures induce residual stresses in the machined surface due to differential thermal contraction, while high levels may leave a hardened layer on the machined part.

This chapter describes the measurement of cutting temperatures, empirical observations on temperatures, analytical and numerical temperature models, temperatures in interrupted cutting and drilling, and thermal expansion.

7.2 MEASUREMENT OF CUTTING TEMPERATURES

Cutting temperatures are more difficult to measure accurately than cutting forces. The cutting force is a vector completely characterized by three components, while the temperature is a scalar field, which varies throughout the system and which cannot be uniquely described by values at a few points. For this reason no simple analog to the cutting force dynamometer exists for measuring cutting temperatures; rather, a number of measurement techniques based on various physical principles have been developed. Particular methods generally yield only limited information on the complete temperature distribution.

7.2.1 TOOL–WORK THERMOCOUPLE METHOD AND RELATED TECHNIQUES

The tool–work thermocouple method (Figure 7.2), first developed in the 1920s [5–7], uses the tool and workpiece as the elements of a thermocouple. The hot junction is the interface between the tool and the workpiece, and the cold junction is formed by the remote sections of the tool and workpiece, which must be connected electrically and held at a constant reference temperature. At least one leg of the circuit must be insulated from the machine tool, although in practice both legs are often insulated to eliminate noise from rotating elements of the system [8].

This method can only be used when both the tool and workpiece are electrical conductors, and thus cannot be used with many ceramic cutting tools. The thermoelectric power of the circuit is

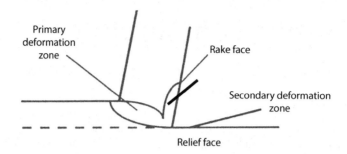

FIGURE 7.1 Areas of the cutting zone in which cutting temperatures are of practical interest.

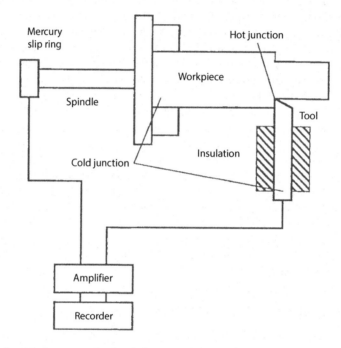

FIGURE 7.2 Tool–work thermocouple circuit for measuring cutting temperatures.

usually low and must be estimated by calibrating the circuit against a reference thermocouple. The calibration is critical to obtaining accurate results [1,9–13]. Most calibration methods involve comparing the emf produced by the tool and chip with that produced by a standard thermocouple when both are heated by a metal bath, welding torch, or induction coil (Figure 7.3). It is also often difficult to maintain the cold junction at a constant temperature; this is particularly true when small indexable tool inserts are used, since in this case a secondary hot junction may arise at the interface between the insert and the toolholder. This error can be minimized through a compensation circuit or by making electrical connections using wires made of materials that have a low thermoelectric power when coupled with the insert material [8,9]. The requirement that one leg of the circuit be insulated from the machine tool can also create difficulties; the presence of the insulating material often reduces the stiffness of the machining system and leads to chatter in high speed tests.

It is usually assumed that the tool–work thermocouple method measures an average interfacial temperature, although interpretation has sometimes been questioned [14]. Based on an analysis of the potential field [8], the physical interpretation of the measured voltage depends on the thermoelectric characteristics of the materials involved. For sharp tools, the quantity measured in the

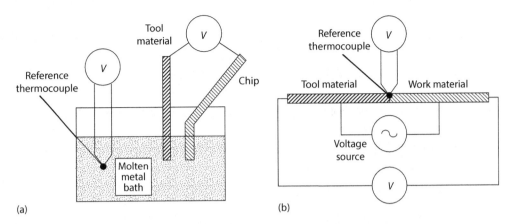

FIGURE 7.3 Methods of calibrating tool–work thermocouples. (a) Metal bath method described by Gottwein. (From Gottwein, K., *Maschienenbau*, 4, 1129, 1925.) (b) Electric heating method described by Braiden. (From Braiden, P.M., The Calibration of Tool/Work Thermocouples, *Proceedings of Eight International MTDR Conference*, Birmingham, U.K., 1967, pp. 653–666.)

tool–work thermocouple method is the average thermal emf at the tool–chip interface. If the thermoelectric emf varies linearly with the temperature difference between the hot and cold junctions, this emf corresponds to the average interfacial temperature. For nonlinear temperature-emf relations, however, the temperature obtained by substituting the measured emf into the temperature-emf relation differs from the average interfacial temperature. For carbide tools and most common work materials, the difference between these temperatures is on the order of five percent, and it can be assumed that the tool–work thermocouple method measures the average interfacial temperatures. For material combinations with more nonlinear thermoelectric characteristics, however, this interpretation entails a more significant error.

Despite these difficulties, the tool–work thermocouple method has a number of advantages. The results are repeatable and correlate well with tool wear for carbide and high-speed steel tools. The measurements also show good time response, making the method suitable for measuring temperatures in thermally transient processes such as for milling and for monitoring temperatures as an indication of tool wear [15]. The instrumentation required can also be built into the machine tool and operated reliably without constant readjustment. For these reasons the tool–work thermocouple method is well suited for monitoring temperatures in routine machinability testing of common tool/work material combinations.

Methods similar to the tool–work thermocouple technique include those in which insulated wires are embedded in the tool (Figure 7.4) or workpiece (Figure 7.5). In the first case, the hot junction is formed at the point of contact between the workpiece and the wire embedded in the tool; in the second case, the hot junction is formed when the tool cuts through the wire. In either case, the cold junction is formed in the same manner as in the tool–work thermocouple method. If the wire material is properly selected (e.g., a copper wire for a tungsten carbide tool [16]), the thermoelectric power of the circuit can be increased, improving the signal-to-noise ratio of the measurement. Also, the interpretation of the measurement is not an issue, since the temperature measured is clearly the temperature at the point where the hot junction is formed. If a number of tests are performed with wires at varying locations, the distribution of temperatures along the tool rake or relief face can be measured. These methods have been used most widely in drilling tests [16–21] but have also been applied in orthogonal cutting studies [22–24]. For drilling, the embedded wire can be replaced by an insulated, embedded foil, which allows mapping the temperature distribution across the drill lip in a single test [25]. The chief disadvantages of these methods are that they require careful calibration and tedious specimen preparation and data reduction. In the case of wires embedded in the workpiece, the time required for the tool and wire to come to thermal equilibrium also limits the maximum cutting speed that can be used.

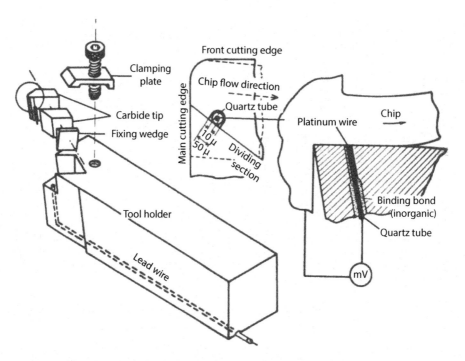

FIGURE 7.4 Insulated platinum wire embedded in a cutting insert to form a thermocouple with the chip. (After Usui, E. et al., *ASME J. Eng. Ind.*, 100, 236, 1978.)

FIGURE 7.5 Insulated copper wire embedded in a cast iron workpiece to form a thermocouple with a drill. (After Agapiou, J.S., and Stephenson, D.A., *ASME J. Eng. Ind.*, 116, 54, 1994.)

7.2.2 Conventional Thermocouple Methods

Conventional thermocouples can be embedded in the tool [11,16,26–32] or workpiece [11,33–38] to map temperature distributions. This approach has not been widely applied because of the extensive specimen preparation required. Since temperature gradients near the cutting zone are steep, its accuracy is limited by the placement accuracy of the thermocouples. The resolution and accuracy of the measurements are also limited by the bead size of the thermocouple, by the difficulty of obtaining good thermal contact between the thermocouple bead and specimen, and by the fact that the temperature field is disturbed by the presence of the holes required to insert the thermocouples. Nonetheless these methods can yield valuable information on the effects of process parameters or cooling strategies in many applications.

Advances in thin film fabrication have led to the development of cutting inserts with integral thin film thermocouple (TFT) arrays for temperature measurements [39,40]. These sensors have very fast time response and can be used for PCBN and other tool materials that are not easily fitted with conventional thermocouples. This method has great promise but require specialized tool fabrication capabilities.

Conventional thermocouples can also be used to measure temperatures at points in the tool remote from the cutting zone (Figure 7.6). These remote measurements can then be used to back-calculate assumed temperature distributions along the cutting edge based on theoretical temperature fields [41–48]. (Full field infrared cameras can also be used for this purpose [49–51].) The factors that limit the accuracy of such inverse methods are well known [52] and include the placement accuracy and temperature gradient effects of the embedded thermocouple methods. An added difficulty is the extrapolation effect, which tends to magnify small errors in the remote measurements themselves. For these reasons, it is often difficult to obtain repeatable results with this approach. An advantage, however, is that the required thermocouples can be built into the toolholder, making the method attractive for routine measurements and process monitoring.

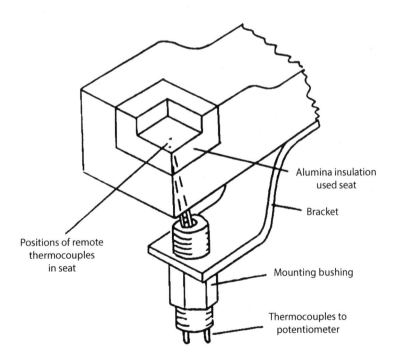

FIGURE 7.6 Toolholder with thermocouple welded to the seat to measure the temperature on the underside of a cutting insert. The alumina insulation is included to provide a controlled boundary condition and would not be used in routine testing. (After Groover, M.P., and Kane, G.E., *ASME J. Eng. Ind.*, 93, 603, 1971.)

7.2.3 METALLURGICAL METHODS

Metallic tool materials often undergo metallurgical transformations or hardness changes, which can be correlated to temperature. This fact makes it possible to map temperature distributions in tools by sectioning the tools after cutting and performing metallographic or microscopic examinations. Appropriate procedures have been reported for both high-speed steel [53–58] and cemented carbide [59,60] tools. These methods require postmortem measurements and are thus difficult to apply to routine testing. The limited results that have been published, however, have shed valuable light on the nature of tool temperature distributions and the location of regions of maximum temperature (Figure 7.7).

In a related technique, split tools can be coated with powders that melt at specific temperatures to map isotherms within the tool [61]. This method can be used with cermet and ceramic tools but does not appear to have been widely applied.

7.2.4 INFRARED METHODS

Cutting temperatures can also be estimated by measuring the infrared radiation emitted from the cutting zone. The best known early studies were carried out by Schwerd [62], Reichenbach [29], and Boothroyd [63,64]. Reichenbach used a point sensor aligned with a small drilled hole to measure both shear plane and relief face temperatures. Since reliable point sensors have been available for some time, they have been applied by a number of researchers to measure rake and relief face temperatures in both cutting and grinding [65–75] (Figure 7.8). (In a more recent study, a point sensor mounted at the end of a fiber optic cable routed through the oil hole of a drill was used to measure hole surface temperatures in drilling [76].) Boothroyd took full-field infrared photographs of the cutting zone in low speed experiments. Due to limited film sensitivity, he preheated his samples to a high temperature to generate a strong infrared signal. Later full-field infrared systems used electronic detectors rather than film, but still had limited spatial resolution and time response. They were initially used to map steady-state grinding temperatures [77,78] but were not used until recently in cutting tests since the hot chip tended to dominate the signal, obscuring other features.

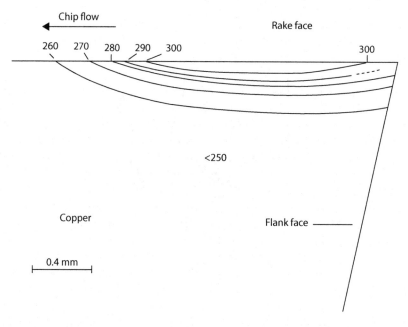

FIGURE 7.7 Temperature distribution in a high speed steel tool used to cut copper, measured by a metallographic method. (After Wright, P.K., *ASME J. Eng. Ind.*, 100, 131, 1978.)

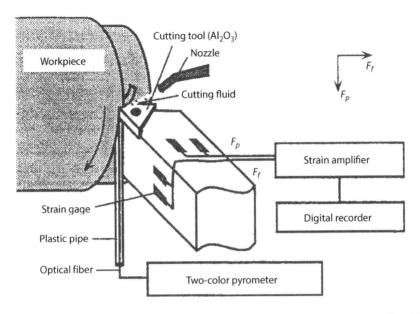

FIGURE 7.8 Fiber optic two-color pyrometer used to measure rake face temperatures. (After Al Huda, M. et al., *ASME J. Manuf. Sci. Eng.*, 124, 200, 2002.)

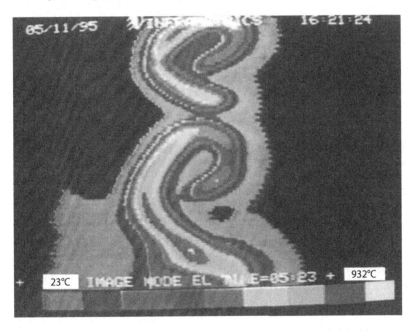

FIGURE 7.9 Full-field infrared thermal image of chip formation when machining 1045 steel at 170 m/min. (Courtesy of I. S. Jawahir, University of Kentucky, Lexington, KY.)

The technology has continued to advance, however, and new systems with much finer resolution and sensitivity are now widely used to map temperature distributions in orthogonal and ballistic cutting tests [79–85] (Figure 7.9).

Infrared measurements are limited to exposed surfaces and cannot be used to directly measure temperatures in the interior of the chip. Because a finite amount of infrared energy is necessary for an accurate measurement, the spatial resolution of point measurements in older systems was often limited to an area on the order of 0.10–0.25 mm². This limitation has become less serious as more sensitive detectors have become available, some of which have spatial resolutions below 5 μm in some setups [85]. As noted before, the signal-to-noise ratio of measurements from many areas of the cutting zone may also be limited by the fact that the chip is normally much hotter than other areas and can dominate the infrared signal. This becomes less of an issue as spatial resolution improves. Most importantly, it is necessary to estimate the emissivity of the target to convert measured infrared intensities to temperatures. Estimating the emissivity is often difficult because it varies with both the temperature and the surface finish, so that assigning a value to specific points in the cutting zone is problematic without extensive calibration [80,82]. If strict conditions on the variation of emissivities at different wavelengths are satisfied, the emissivity problem can be eliminated by using instruments that measure intensities at two infrared wavelengths simultaneously [71,75].

Because of these limitations infrared measurements have traditionally been difficult to perform accurately and repeatably. The chief source of difficulty is the variation of the target emissivity over broad ranges of temperature and surface finish. The most accurate infrared methods appear to be those that measure intensities through pre-drilled holes, since in this case the emissivity over a narrow area can be accurately calibrated and the problem of the signal being overwhelmed by the infrared image of the chip can be eliminated.

7.2.5 Other Methods

Temperature distributions can also be estimated by coating specimens with thermosensitive paints [86–89]. The limitations of this method are well documented [11]; it appears it is at best suitable for making qualitative comparisons. Similarly, the temperature of steel chips can be estimated in a crude fashion by noting their color, which gives an indication of the degree of temperature-dependent oxidation, which they have undergone. This is also a qualitative method suitable mainly for comparisons over limited ranges of cutting conditions.

In a recently described method [90], the heat flux into an internally cooled tool was estimated by measuring the temperature rise in the cooling fluid and used to estimate tool interface temperatures based on an inverse solution.

7.3 FACTORS AFFECTING CUTTING TEMPERATURES

The process parameter with the greatest influence on cutting temperatures is the cutting speed. Since cutting forces generally don't vary strongly with cutting speed, increasing the cutting speed increases the rate at which energy is dissipated through plastic deformation and friction, and thus the rate of heat generation in the cutting zone. Increasing the feed rate also increases heat generation and cutting temperatures. For moderate ranges of these variables, the cutting speed has a greater influence, and the tool–chip interface temperature increases with the square root of the cutting speed but the third root of the feed [1]:

$$T_{int} \propto V^{0.5} a^{0.3} \tag{7.1}$$

Other parameters that affect the cutting force, such as the depth of cut and the rake angle, also influence cutting temperatures; changes in these parameters that increase the cutting force normally slightly increase cutting temperatures.

At very high cutting speeds, the tool–chip interface temperature does not vary as indicated in Equation 7.1, but rather approaches the melting temperature of the work material asymptotically. Kramer [91] reported that tool–chip interface temperatures approached the melting temperature of the chip when machining Inconel 718 at speeds up to 1200 m/min. Kottenstette [71] observed a leveling off of interfacial temperatures and evidence of chip melting when machining hardened 4340 steel at speeds greater than 3000 m/min. Chip melting also appears to occur when machining 1100 aluminum at higher speeds [40], and when cutting 6061 Al at speeds over 700 m/min.

Material properties also strongly influence cutting temperatures. Cutting temperatures are higher for harder work materials because cutting forces and thus energy dissipation are increased. For materials of similar hardness, cutting temperatures increase with ductility, since more ductile materials can absorb more energy through plastic deformation. At a machining speed of roughly 300 m/min, the tool–chip interface temperature is usually roughly 400°C for an aluminum alloy with a Brinnel hardness of 100, 750°C for brittle gray cast iron with a Brinnel hardness of 200, and over 1000°C for ductile mild steel with a Brinnel hardness of 200.

Thermal properties of the work material that influence cutting temperatures include the thermal conductivity k and heat capacity ρc. Temperatures generally decrease as these parameters increase, since an increase implies that heat is more readily conducted away through the workpiece (higher k) or that the temperature increases more slowly for a given heat input (higher ρc). Increasing the thermal conductivity and heat capacity of the tool material also reduces temperatures, although the effect does not appear to be as marked as for the work material.

Another parameter that affects the thermal aspects of cutting is the ratio of the rate of material removal per unit depth of cut to the thermal diffusivity of the work material, $Va\rho c/k$, which is referred to as the thermal number in the metal cutting literature. In the general literature on heat transfer, it is defined as the Peclet number, Pe, which reflects the relative importance of mass transport and conduction. For low values of Pe, the speed of the tool with respect to the workpiece is relatively small compared to the thermal diffusivity of the material, and a comparatively large portion of the heat generated in the deformation zone conducts into the workpiece. For large values of Pe, on the other hand, almost all of the cutting remains in the chip and is transported out of the cutting zone. In conventional machining with a sharp tool, Pe is generally on the order of 10^4 and the workpiece often heats up appreciably during cutting. In high-speed machining, Pe is more often on the order of 10^5, and the workpiece normally remains much cooler [49].

Finally, peak tool–chip interface temperatures are influenced by the tool nose radius and included angle [92]. Increasing the nose radius reduces the peak temperature by reducing the maximum uncut chip thickness and distributing frictional energy more evenly over the cutting edge. Reducing the included or wedge angle (by increasing the rake or relief angles) increases the peak temperature by reducing the area through which heat can diffuse from the cutting edge through the tool.

7.4 ANALYTICAL MODELS FOR STEADY-STATE TEMPERATURES

The difficulty of measuring cutting temperatures led to early research interest in analytical models for predicting temperatures. The best known early analyses were simple solutions for plane heat sources based on the shear plane model of cutting mechanics (Section 6.8) [64,93–97]. Although differing in details, these models were all based on the physical assumptions illustrated in Figure 7.10. The assumed work material enters at initial temperature θ_i and is heated by two plane heat sources of strength P_s and P_f, representing heating due to plastic deformation along the shear zone and frictional heating along the tool rake face. The material is assumed to emerge from the first source as a

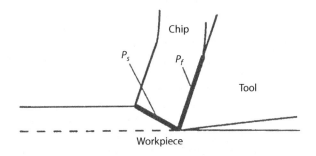

FIGURE 7.10 Two-source model of the cutting zone used in simple analyses of cutting temperatures.

chip with a uniform temperature θ_s and to undergo a further temperature rise θ_f in interacting with the second, frictional heat source. These models assume a steady state and neglect flank friction, which would introduce a third heat source.

In Loewen and Shaw's model [96], θ_s is given by

$$\theta_s = \frac{\Gamma_1 P_s}{\rho cabV} + \theta_i = \frac{\Gamma_1 u_d}{\rho c} + \theta_i \tag{7.2}$$

where P_s is the deformation power, which for orthogonal cutting is given by

$$P_s = V\left(F_c - P\frac{a_c}{a}\right) = F_s V_s \tag{7.3}$$

and u_d is the specific deformation power (Equation 6.23). Γ_1 is the proportion of the deformation energy entering the chip. F_c, P, L_c, L_p, a, and a_c are defined in Figures 6.3, 6.4, 6.24, and 6.26. a is the undeformed chip thickness and b is the width of cut. Based on Jaeger's solution for a plane heat source sliding on a half space [98], Γ_1 can be estimated from the equation

$$\Gamma_1 = \frac{1}{1 + 1.328\sqrt{\dfrac{k\gamma}{\rho cVa}}} \tag{7.4}$$

where

k, ρ, and c are the thermal conductivity, density, and specific heat of the work material
γ is the strain in the chip defined by Equation 6.24

Similarly, the frictional temperature rise θ_f is given by

$$\theta_f = \frac{0.377\,\Gamma_2 P_f}{bk\sqrt{L_2}} \tag{7.5}$$

where

$$L_2 = \frac{V_c L_c \rho c}{4k} \tag{7.6}$$

L_c is the tool–chip contact length
P_f is the frictional power, which for orthogonal cutting is given by

$$P_f = PV\frac{a}{a_c} \tag{7.7}$$

and Γ_2 is the proportion of frictional energy flowing into the chip. Again using Jaeger's friction slider solution, Γ_2 can be estimated by

$$\Gamma_2 = \frac{P_f\dfrac{B}{bk_t} + \theta_i - \theta_s}{P_f\dfrac{B}{bk_t} + 0.377\dfrac{P_f}{bk\sqrt{L_2}}} \tag{7.8}$$

where

$$B = \frac{2}{\pi}\left[\sinh^{-1}\left[A_r\right] + A_r \sinh^{-1}\left(\frac{1}{A_r}\right) + \frac{1}{3}A_r^2\right] + \frac{2}{\pi}\left[\frac{1}{3A_r} - \frac{1}{3}\left(A_r + \frac{1}{A_r}\right)\sqrt{1+A_r^2}\right] \tag{7.9}$$

where
k_t is the thermal conductivity of the tool material
A_r is the aspect ratio of the contact area, given by

$$A_r = \frac{b}{2L_c} \quad \text{(Orthogonal tool)}$$

$$= \frac{b}{L_c} \quad \text{(lathe tool with nose radius)} \tag{7.10}$$

The workpiece material properties k and ρc and the tool thermal conductivity k_t are assumed to vary with temperature in this model. To account for variations in material properties with temperatures, calculations are carried out iteratively from an initial guess temperature, with successive estimates of temperatures used to calculate new thermal properties until convergence was achieved.

For many metals, k varies approximately linearly with temperature, while ρc varies quadratically with temperature:

$$k(\theta) = k_0 + k_1\theta \tag{7.11}$$

$$\rho c(\theta) = \rho c_0 + \rho c_1\theta + \rho c_2\theta^2 \tag{7.12}$$

Values of the constants k_0, k_1, ρc_0, ρc_1, and ρc_2 for some common workpiece materials are listed in Table 7.1 [99]. Values of the tool thermal conductivity k_t for common tool materials are given in Table 7.2 [100–103].

The mean temperature rise of the chip surface along the tool face, u_T, is equal to the sum of the mean shear plane temperature rise and the mean temperature rise due to friction in the tool–chip contact area:

$$\theta_T = \theta_s + \theta_f \tag{7.13}$$

Other simple analytical models for cutting temperatures use similar equations for calculating temperatures. The basic model of Trigger and Chao [94,95] is equivalent to the Loewen–Shaw model.

TABLE 7.1
Values of Parameters in the Thermal Property Equations, Equations 7.11 and 7.12, for Common Workpiece Materials

Material	k_0	k_1	ρc_0	ρc_1	ρc_2
1018 steel	53.6	−0.027	3,124,025	6480	−4.32
1070 steel	49.3	−0.025	3,000,000	9370	−7.0
2024 Al	112.5	0.256	2,437,866	1515	−0.255
CA330 brass	105.3	0.108	3,236,407	2993	−3.74
Gray cast iron	49.0	−0.013	3,997,673	812.5	−4.27

Source: Stephenson, D.A., *ASME J. Eng. Ind.*, 113, 121, 1991.
Note: These values yield k in J/(s m °C) and ρc in J/(m³ °C) when θ is given in °C.

TABLE 7.2

Thermal Conductivities of Common Tool Materials at Approximately 100°C in J/(s m °C)

Material	k_t
M10 HSS	35.5
T1 HSS (tempered)	36.4
C2 tungsten carbide	75
TiC cermet	22.5
Cold pressed Al_2O_3	15
Al_2O_3–TiC	18.3
Hot pressed silicon nitride	18.3

Sources: Touloukian, Y.S. et al., *Thermocphysical Properties of Matter*, IFI/Plenum, New York, 1970, Vol. 1, pp. 1194–1196, Vol. 4, pp. 1282–1284; Roberts, G.A. and Cary, R.A., *Tool Steels*, 2nd edn., ASM, Metals Park, OH, 1980, p. 709; Tay, A.O. et al., *Int. J. Mach. Tool Des. Res.*, 16, 335, 1976; Bordui, D., Third Generation Silicon Nitride, *Ceramic Cutting Tools and Applications*, SME, Dearborn, MI, 1989.

Boothroyd's model [64] uses Weiner's solution [104] for an inclined plane heat source on an infinite body to obtain a different equation for Γ_1:

$$\Gamma_1 = \frac{\mathrm{erf}\sqrt{Y_L}}{4Y_L} + \left[1 + Y_L\right]\mathrm{erfc}\sqrt{Y_L} - \frac{\exp\left[-Y_L\right]}{\sqrt{\pi}}\left(\sqrt{Y_L} + \frac{1}{2\sqrt{Y_L}}\right)$$

$$- \frac{\exp\left[-Y_L\right]}{\sqrt{\pi}}\left(\sqrt{Y_L} + \frac{1}{2\sqrt{Y_L}}\right) \tag{7.14}$$

where

$$Y_L = \frac{Va\rho c \tan\phi}{4k} \tag{7.15}$$

In Boothroyd's model, the plane heat source on the rake is replaced by a distributed source, and the frictional temperature rise is calculated using a finite difference procedure. This is difficult to implement in practice, and other researchers have substituted alternative analytical schemes [102]. Other equations for the proportions of the heat flowing into the workpiece, chip, and tool have also been published [105–109] but do not appear to have been widely applied.

These models agree better with measured temperatures than the shear plane models agree with measured forces. One of the earliest experimental comparisons was carried out by Nakayama [110], who used measured temperatures on the face of a tube being cut to estimate the proportion of cutting heat entering the workpiece. He found that Hahn's partition function [93], which is an extension of Jaeger's solution accounting for oblique motion, agreed best with his data. Similar results have been reported by Boothroyd [64], who found that Weiner's solution was most accurate. More recently [99], temperatures calculated using four simple models were compared with tool–chip interface temperatures measured using the tool–work thermocouple method and shear zone temperatures estimated from infrared measurements. These results showed that the models based on Loewen and Shaw's analysis agreed well with measured values for materials that form a continuous chip, and that models based on Boothroyd's analysis overestimated measured temperatures. Typical results for low carbon steel and aluminum workpieces are shown in Figures 7.11 and 7.12.

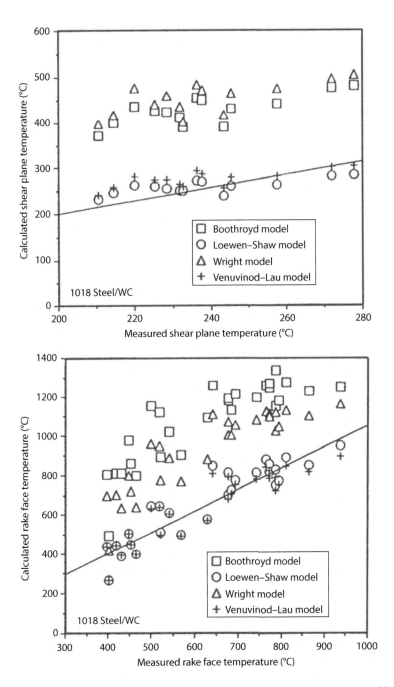

FIGURE 7.11 Comparison of measured average shear plane and rake face temperatures with values calculated using the models of Loewen and Shaw [96], Boothroyd [64] (as modified by Tay et al. [102]), Wright et al. [113], and Venuvinod and Lau [112]. The work material is 1018 steel cut at speeds up to 240 m/min. (After Stephenson, D.A., *ASME J. Eng. Ind.*, 113, 121, 1991.)

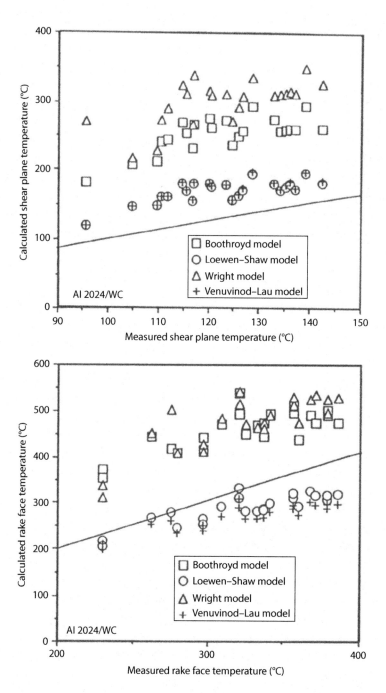

FIGURE 7.12 Plots of measured versus calculated temperatures corresponding to those shown in Figure 7.11 for 2024 aluminum workpieces. (After Stephenson, D.A., *ASME J. Eng. Ind.*, 113, 121, 1991.)

Boothroyd's model would yield better agreement if temperature-dependent thermal properties were used. All models overestimated temperatures when cutting cast iron, which form discontinuous chips. Leshock and Shin [13] found that temperatures predicted using the Loewen–Shaw model generally agreed with measured values for 4140 steel, although the model underestimated temperatures at low feeds, probably due to the neglect of flank friction.

A number of researchers have extended the basic models to apply to more general cutting conditions or to yield more complete information on temperature distributions. Chao and Trigger [111] extended their model to calculated relief face temperatures by adding a third heat source due to flank friction. Venuvinod and Lau [112] derived a model for temperatures in oblique cutting, which reduces to the equivalent solutions of Chao and Trigger [94,95] and Loewen and Shaw [96] for orthogonal cutting conditions. Average tool–chip interface temperatures calculated using this model agreed reasonably well with tool–work thermocouple measurements from tests on mild steel workpieces. Chao and Trigger [95], Wright [113], and Venuvinod and Lau [112] have also extended the analysis to predict the temperature distribution along the rake face of the tool, rather than an average interfacial temperature. Wright's calculations compared well with temperature distributions measured by a metallurgical method in limited tests on mild steel workpieces. Grzesik and Nieslony [114] investigated more general heat partition models applicable to multilayered coated tools. The development of advanced finite element models as described in the next section has reduced interest in more general analytical models in recent years.

7.5 FINITE ELEMENT AND OTHER NUMERICAL MODELS

Finite element models for cutting temperatures were first investigated in the 1970s [115,116]. The best known early analysis was reported by Tay [116]; this model formed the basis for much subsequent work [102,117–120]. The chief advantage of these models was they predicted complete temperature distributions, rather than average temperatures over one or two surfaces. The input requirements, however, made them difficult to apply accurately to a broad set of cutting conditions. Tay's original model [116] and Muraka's model [117,118] required measured cutting forces, chip properties, tool–chip contact lengths, and work material velocity distributions to predict temperatures; considerable experimental effort is required to generate this information. Subsequent researchers eliminated the need for measured velocity fields, so that temperature distributions could be based on cutting force and chip property inputs equivalent to those used in the simpler analytical models [119]. Since this information is predicted by the more complete models of the cutting mechanics, these analyses are suitable for use in coupled force and temperature models such as those reported by Oxley [121], Usui and Shirakashi [122,123], and Strenkowski and Moon [124]. In these coupled models, initial predictions of forces and chip properties were used as inputs to temperature calculations, which in turn were used to modify work material flow characteristics; the two sets of calculations were carried out iteratively until convergence is achieved. This trend has been continued in the recent development of finite element models for machining, reviewed in Section 6.11. The most complete current models are all coupled thermal and mechanical analyses, some of which include wear and coolant effects [125–132].

Other numerical methods, such as the finite difference method [34,133] and boundary element method [134,135], have also been used to compute cutting temperatures. These methods often require less computing time than finite element calculations, but cannot be used with as broad a range of tool geometries and field equations. A wide variety of specialized numerical methods for calculating temperature fields within tools have also been developed by researchers investigating inverse methods for monitoring temperatures using remote measurements [44,45], the thermal contact resistance between the insert and tool holder [136], machining errors due to thermal expansion [137,138], and temperatures in interrupted cutting [139–142].

Marusich and Ortiz's analysis [126,143] is a representative example or current finite element models for machining temperatures. The governing thermal equation in this model is a weak form of the energy balance equation,

$$\int_{B_t} \rho c \dot{T} \eta dV + \int_{\partial B_{tq}} h\eta dS = \int_{B_t} \mathbf{q} \cdot \nabla \eta dV + \int_{B_t} s\eta dV \qquad (7.16)$$

where
  B_t is the boundary of a given deforming volume at time t (Neumann boundary)
  ρ is the density
  c is the specific heat
  T is the spatial temperature field
  η is an admissible virtual temperature field
  h is the outward heat flux through the surface
  $\mathbf{q}$ is the heat flux
  s is the distributed heat source density

The rate of heat supply due to plastic deformation in chip formation is estimated as

$$s = \beta \dot{W}^p \qquad (7.17)$$

where
  $\dot{W}^p$ is the plastic power per unit deformed volume (equivalent to u_d in Equation 6.22)
  β is a coefficient of the order of 0.9

The rate of frictional heat generation on the rake face is given by

$$h = -\mathbf{t} \cdot \|\mathbf{v}\| \qquad (7.18)$$

where
  $\mathbf{t}$ is the frictional traction
  $\|\mathbf{v}\|$ is the jump in velocity across the tool–chip interface (i.e., the relative sliding speed)

Based on infinite half-space solutions, the heat flux from Equation 7.18 is partitioned between the chip and tool using the formula

$$\frac{h_1}{h_2} = \frac{\sqrt{k_1 \rho_1 c_1}}{\sqrt{k_2 \rho_2 c_2}} \qquad (7.19)$$

where
  h_1 and h_2 are the heat fluxes going to the chip and tool, respectively
  k_1, ρ_1, and c_1 are the thermal properties of the chip
  k_2, ρ_2, and c_2 are the thermal properties of the tool

The tool and workpiece are assumed to obey Fourier's heat conduction law, which for material frame indifference is written in the form

$$\mathbf{q} = -\mathbf{D} \cdot \nabla \mathbf{T} \qquad (7.20)$$

where **D** is the spatial conductivity tensor. For an isotropic lattice, **D** is related to the Cauchy–Green deformation tensor, $\mathbf{B}^e$, through

$$\mathbf{D} = -k\mathbf{B}^e \qquad (7.21)$$

Inserting the finite element interpolation into Equation 7.16 results in the semi-discrete system of equations

$$\mathbf{C}\dot{\mathbf{T}} + \mathbf{K}\mathbf{T} = \mathbf{Q} \qquad (7.22)$$

where
  T is the array of nodal temperatures
  C is the heat capacity matrix
  K is the conductivity matrix
  Q is the heat source array

The components of **C**, **K**, and **Q** are given by

$$C_{ab} = \int_{B_t} \rho c N_a N_b dV \qquad (7.23)$$

$$K_{ab} = \int_{B_t} D_{ij} N_{a,i} N_{b,j} dV \qquad (7.24)$$

$$Q_a = \int_{B_t} s N_a dV - \int_{\partial B_{tq}} h_\alpha N_a dV \qquad (7.25)$$

where
  N_a and N_b are shape functions
  $N_{a,i}$ and $N_{b,j}$ are normal contact forces
  h_α, $\alpha = 1,2$ has the appropriate value for the chip or tool as in Equation 7.18

Integrating Equation 7.21 using a forward Euler algorithm yields

$$\mathbf{T}_{n+1} = \mathbf{T}_n + \Delta t \dot{\mathbf{T}}_n \qquad (7.26)$$

$$\mathbf{C}\mathbf{T}_n + \mathbf{K}_n \mathbf{T}_n = \mathbf{Q}_n \qquad (7.27)$$

The use of the lumped heat capacity matrix, **C**, eliminates much equation solving and reduces the computational load.

It is difficult to assess the accuracy of finite element temperature models because the temperature distributions they predict cannot be easily measured. For some earlier models, calculations were compared with limited point temperature measurements from conventional thermocouples [115], tool–work thermocouple measurements [124], or partial tool temperature distributions measured using metallurgical techniques [119]. More recently, full field infrared methods have become more common for validation [81,131]. Typical comparisons of temperatures calculated using finite element models with measurements are shown in Figures 7.13 and 7.14. In both cases, there is general qualitative agreement, but significant quantitative differences. This is to be expected given that

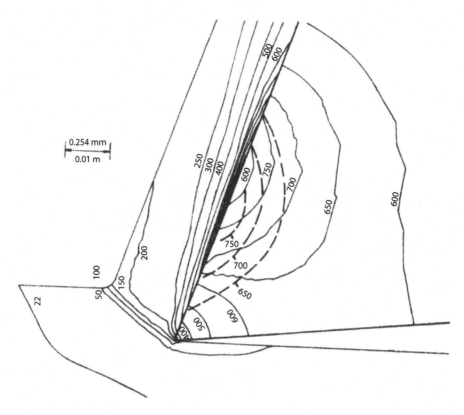

FIGURE 7.13 Temperature distributions in the workpiece, chip, and tool calculated using the finite element model developed by Stevenson, Wright, and Chow [119]. The broken lines in the tool show isotherms measured using a metallographic method. The work material is 12L14 steel cut at a speed of 106 m/min.

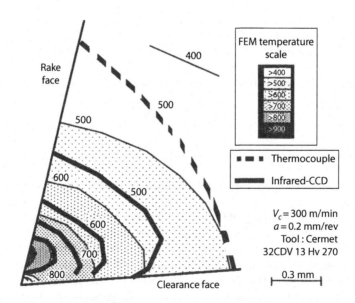

FIGURE 7.14 Comparison of tool temperature field calculated using a finite element model developed by M'Saoubi et al. [81] with full-field infrared temperature measurements.

all models, however general the geometries and boundary conditions they can accommodate, still employ significant simplifying assumptions for physical quantities not easily measured or specified. (One example is the use of Equation 7.19 to partition heat in Marusich and Ortiz's model.) Based on experience with earlier analytical models, steady-state finite element models, which include temperature-varying thermal properties, would be expected to be reasonably accurate for dry cutting. More validation and modeling experience is necessary to judge the quantitative accuracy of more general models.

7.6 TEMPERATURES IN INTERRUPTED CUTTING

The temperature models discussed in Sections 7.4 and 7.5 are in most cases applicable to steady-state processes such as orthogonal cutting and simple turning and boring operations. In most industrial processes the cutting speed or area of cut vary continuously, so that a thermal steady state is never reached. Examples include turning operations on complex parts using CNC machine tools and interrupted cutting processes such as milling. The thermal aspects of interrupted cutting have been more widely studied than other nonsteady-state processes [37,38,42,79,144–146] and are reviewed in this section. Models for transient temperatures in contour turning operations are reviewed in Section 8.2.

In the simplest case of interrupted cutting, (Figure 7.15), the tool cuts through short lengths of metal interrupted by open regions of air. In this case the temperature response in the tool will clearly be periodic. In more general interrupted processes the lengths of metal and air cut may vary with time, as may the feed rate or depth of cut, and the temperature response is oscillatory but not necessarily periodic.

Cyclic variations in temperatures can lead to tool failure due to thermal fatigue. Thermal fatigue cracks develop in the tool as the surface layer expands and contracts in response to temperature variations and eventually grow large enough to cause chipping of the cutting edge. This problem is sometimes observed when milling steel with tungsten carbide cutters. The development of thermal cracks has been studied by Zorev and several other researchers [147–151]. As would be expected, these cracks grow more rapidly as peak temperatures during the heating cycle increase and as the difference between peak and low temperatures increases. Thus, cracks are more likely to occur at high cutting speeds when cutting with a coolant. If these conditions cannot be avoided, thermal

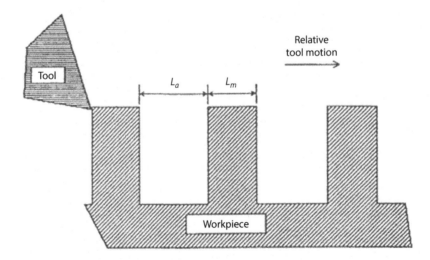

FIGURE 7.15 A simple interrupted cutting process in which the tool alternatively cuts through constant lengths of metal and air. (After Stephenson, D.A., and Ali, A., *ASME J. Eng. Ind.*, 114, 127, 1992.)

cracking can be reduced by changing the tool material grade to increase its thermal shock and fatigue resistance [152].

Peak temperatures are lower in interrupted cutting than in continuous cutting under the same conditions, leading to a reduction in chemical or diffusion wear. This type of wear is further reduced by the fact that peak temperatures are reached for short rather than extended periods. As illustrated in Figure 7.16 [79], peak temperatures depend more on the length of metal cut in continuous segments,

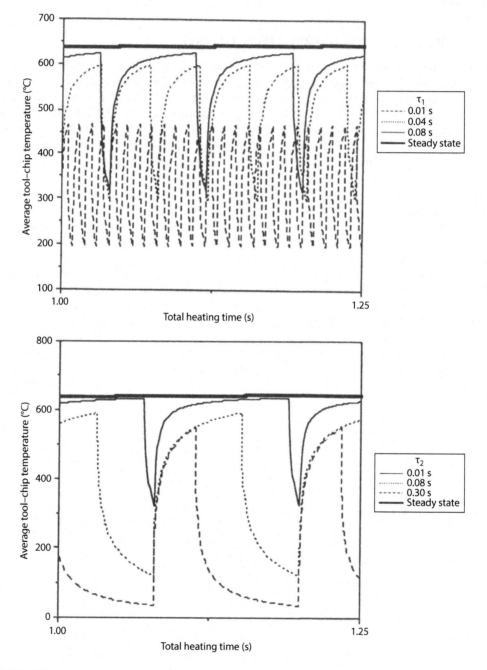

FIGURE 7.16　Variation in average interface temperatures in interrupted cutting when the length of the heating cycle (τ_1) and cooling cycle (τ_2) are varied. (After Stephenson, D.A., and Ali, A., *ASME J. Eng. Ind.*, 114, 127, 1992.)

rather than on the length of noncutting intervals. This indicates that the reduction in temperature is due primarily to interruption of the heating cycle during a transient phase, rather than from cooling between heating cycles. Physically, the temperature distribution at the tool–chip interface will not reach a steady state before the subsurface region of the tool near the cutting edge does; this requires a finite amount of energy, which cannot be supplied instantaneously, resulting in a period of transient (and lower) interfacial temperatures. If the ratio of the length of the transient period to the cutting speed increases with increasing cutting speed, it is possible that peak temperatures will decrease with cutting speed above critical speeds. Palmai [153] discussed this possibility and derived equations for calculating critical speeds from material properties. He also suggested that this phenomenon would explain the well-known results of Salomon [154,155], who reported an increase in tool life with increasing cutting speed in very high-speed milling tests. This has not been observed in experiments at cutting speeds up to 300 m/min [79,156]. Ming et al. [146] reported a decrease in cutting temperatures with increasing speed for end milling tests at speeds up to 700 m/min but used an indirect temperature measurement method, which would be sensitive to increased convective cooling at higher spindle speeds.

7.7   TEMPERATURES IN DRILLING

Thermal conditions in drilling also differ significantly from those in orthogonal cutting, turning, and boring. The chip is formed at the bottom of the hole and remains in contact with the drill over a comparatively long distance, which increases tool temperatures. Temperatures in drilling are further increased by the fact that the drill point moves slowly into the portion of the work material being heated by chip formation; in turning, the work material approaching the cutting edge is generally cooler. Temperatures in drilling often do not reach a steady state, but rather increase with hole depth. For difficult-to-machine materials such as powder metals [157], this phenomenon limits the maximum hole depth, which can be drilled without excessive tool wear. Increasing temperatures and the potential accumulation of hot chips at the bottom of the hole are serious problems in deep hole drilling, driving the use of horizontal setups and high pressure, through tool coolant systems. Finally, the cutting speed varies with the radius of the drill, so that temperatures vary across the cutting edges. Speeds and temperatures are highest near the outer corner or margin of the drill, and temperature-activated margin wear often limits maximum spindle speeds.

Drilling temperatures have been studied both experimentally and theoretically for many years. The experimental work carried out before the mid-1960s has been reviewed by DeVries [20]. The most common method of measuring drill temperatures has been to embed thermocouples in the drill, often routing them through the oil holes of pressure-fed drills (Figure 7.17) [16,32,157]; fiber optic cables connected to photosensors have also been routed through oil holes [76]. Tool–work and thin wire or foil thermocouple methods (see Figure 7.5) [17–21,25], inverse methods based on thermocouples mounted in the workpiece [46–48], temperature sensitive paints [89], and metallurgical methods [57,58] have also been used. Typical signals from an oil hole thermocouple are shown in Figure 7.18 [16]; they exhibit the characteristic increase in temperature with drilling depth. Differences in cutting temperatures between flutes on two-flute drills have also been observed experimentally [19], indicating that the cutting edges of the drill do not wear evenly and cut different amounts of material as drilling progresses; differences in edge temperatures have also been observed with asymmetric drills such as indexable drills, which do not have identical cutting edges.

The first analytical models for drill temperatures were reported in the 1960s. Early researchers [17,158,159] approximated the drill as a semi-infinite body and applied temperature analyses for orthogonal cutting to discrete segments of the cutting edge to compute the variation of temperature across the cutting edge. Later researchers treated the drill as finite and computed temperatures using finite difference [21,160] or finite element [161,162] methods. In these analyses, the proportion of shear zone and frictional heat entering the drill was still computed using steady-state solutions such as Jaeger's solution for a friction slider on a half space. More recently, transient partition

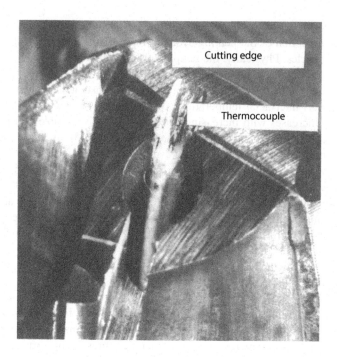

FIGURE 7.17 Thermocouple routed through the oil hole of a drill and welded near the cutting edge. (After Agapiou, J.S., and Stephenson, D.A., *ASME J. Eng. Ind.*, 116, 54, 1994.)

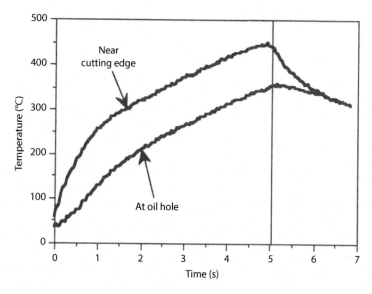

FIGURE 7.18 Temperatures measured when drilling cast iron with a carbide drill using thermocouples welded at the oil hole and near the cutting edge. The spindle speed is 1000 rpm and the feed rate is 0.457 mm/rev. (After Agapiou, J.S., and Stephenson, D.A., *ASME J. Eng. Ind.*, 116, 54, 1994.)

functions that reflect the fact that a larger percentage of this heat enters the tool in drilling than in other processes have been used. This approach has been used to develop a method of computing temperatures for drills with arbitrary point geometries based only on the drilling torque and material properties [16,163,164]. Despite the number of simplifying assumptions required in these analyses, their results agreed well with limited thermocouple measurements (Figure 7.19).

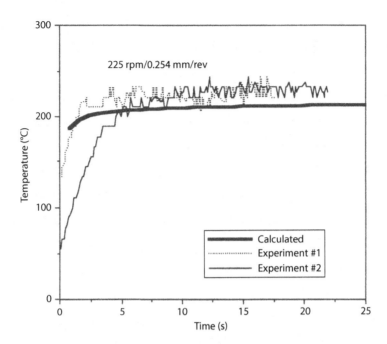

FIGURE 7.19 Comparison of drilling temperatures calculated using the model of Agapiou [15,123,124] with values measured by embedding a copper wire in the workpiece (Figure 7.5). (After Agapiou, J.S., and Stephenson, D.A., *ASME J. Eng. Ind.*, 116, 54, 1994.)

As in other cutting processes, temperatures in drilling are most strongly affected by the spindle speed and the feed rate. Among geometric factors, the point angle has the largest influence. As the point angle is increased, the length of the drill cutting edge decreases, and temperatures increase as a roughly constant amount of heat diffuses into a smaller body. Increasing the helix angle, which reduces the drilling torque, does not seem to affect the drill temperature as strongly as would be expected since the bulk of the heat conducted into the drill comes from frictional contact rather than from shearing of the work material.

7.8 THERMAL EXPANSION

Thermal expansion can produce significant dimensional and form errors in precision machining processes. In many cases, errors are caused by the accumulation of hot chips on flat surfaces of the machine tool. A variety of methods for controlling these errors, including modifications of the machine design to eliminate flat surfaces, the use of coolants to ensure chip removal, and the use of constant-temperature fluid baths to control temperatures throughout the system, are used in practice [165]. Errors resulting from the conduction of cutting heat into the tool or workpiece are less common and have been less widely studied. The available work [137,138,166,167] indicates that the expansion of the tool generally produces more significant errors than the expansion of the workpiece, largely because tool temperatures are usually higher than workpiece temperatures. The thermal expansion of the tool can be reduced by using brazed or clamped, rather than bonded, cutting tools [136], and by using toolholders cooled by cold fluids pumped through internal cooling passages [168]. The expansion of the workpiece produces more significant errors in precise hole-making processes such as cylinder boring. In these processes, thermal errors may be comparable to the mechanical errors produced by cutting forces [35,169]. If adequate chip control can be ensured, thermal errors in boring can be reduced by increasing the cutting speed, since as the cutting speed increases the proportion of heat entering the workpiece decreases. Partly for this reason, carbide

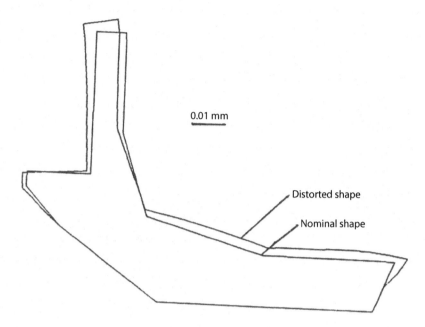

FIGURE 7.20 Computed dimensional and form errors due to thermal expansion when turning a hardened steel wheel spindle. (After Stephenson, D.A. et al., *ASME J. Eng. Ind.*, 117, 542, 1995.)

boring cutters have been replaced by ceramic or PCBN cutters designed for use at high cutting speeds in most automotive engine boring operations.

The thermal expansion of the workpiece has traditionally been a more serious problem in grinding than in conventional cutting operations. It has become a more significant issue recently in dry cutting operations. Hard turning, for example, is often carried out with ceramic tools, which must be used dry. Since the cutting speed is relatively low and the specific cutting energy is high, this can result in a significant heat flux into the workpiece, and significant thermal distortions as shown in Figure 7.20 [170]. Dimensional and form errors due to this source can be minimized by machining part features with critical tolerances first, to avoid heating the part when cutting noncritical features, and by letting parts cool before gaging is performed in operations in which a compensation system is used to correct for tool wear. Thermal distortion of the workpiece is also a serious issue in dry drilling applications [47,171–175]. In these operations the workpiece often heats up and expands ahead of the drill; when the drill is retracted, the workpiece contracts, reducing the drilled hole diameter. (In some cases witness marks are visible in the hole, indicating that significant contraction occurs prior to drill retraction.) In high precision operations, errors due to this source are commonly reduced by cooling the workpiece with cold air. Internal cooling of the drill using a heat pipe has also been explored [176].

Thermal expansion of the workpiece may also cause significant dimensional errors in MQL machining as discussed in Section 15.6.

7.9 EXAMPLES

Example 7.1 Using the Example 6.1 (in previous chapter), estimate the maximum temperature at the chip–tool interface.

Solution: The maximum temperature at the chip–tool interface along the tool rake face is estimated using Equation 7.13. This requires the estimation of the temperature rise along the primary shear plane (Equation 7.2) and the frictional temperature rise (Equation 7.5). The deformation power along the shear plane in orthogonal cutting is calculated as

$$P_s = F_s V_s = \left(F_c \cos\phi - F_z \sin\phi \right) \frac{\cos\alpha}{\cos(\phi - \alpha)} V$$

The shear angle ϕ was calculated in Example 6.1 to be $23.2°$.

$$P_s = \left(F_c \cos\phi - F_z \sin\phi \right) \frac{\cos\alpha}{\cos(\phi - \alpha)} V = \frac{\left(900\cos 23.2° - 600\sin 23.2° \right)\left(200 \text{ m/min} \right)}{\cos(23.2°)}$$

$$P_s = 128,571 \text{ N m/min} = 2,143 \text{ W} = 2,143 \text{ J/s}$$

The friction power at the chip–tool interface is equal to the total power minus the sheer power (as discussed in the orthogonal case of a simplified problem where the friction power at the interface of the tool flank with the workpiece is neglected).

$$P_f = F_c V - P_s = (900 \text{ N})(200 \text{ m/min}) - 128,571 = 51,429 \text{ N m/min} = 857 \text{ W}$$

The proportion of the primary shear energy entering the chip is Γ_1 (given by Equation 7.4). The workpiece material properties k and ρc and the tool thermal conductivity k_t vary with temperature as described in Equations 7.11 and 7.12. To account for variations in material properties with temperatures, calculations are carried out iteratively from an initial guess temperature, with successive estimates of temperatures used to calculate new thermal properties until convergence is achieved. This will be illustrated here. Using the properties of 1070 steel in Tables 7.1 and 7.2 for the tool, Γ_1 at room temperature is

$$\Gamma_1 = \frac{1}{1 + 1.328\sqrt{\dfrac{[48.55 \text{ J/(s m °C)}](2.762)}{[3,274,800 \text{ J/(m}^3 \text{ °C)}](3.333 \text{ m/s})(0.0003 \text{ m})}}} = 0.867$$

The temperature rise along the primary shear plane is

$$\theta_s = \frac{\Gamma_1 P_s}{\rho c a b V} + \theta_i$$

$$= \frac{(0.866)(2,143 \text{ J/s})}{[3,274,800 \text{ J/(m}^3\text{°C)}](0.0003 \text{ m})(0.00025 \text{ mm})(3.333 \text{ m/s})} + 30°C$$

$$\theta_s = 257°C$$

The workpiece properties should be adjusted for an average temperature of the part at the primary shear zone since part of the heat generated along the shear plane is conducted into the workpiece adjacent to the shear zone where the next shear plane occurs. Therefore, by estimating the workpiece material properties k and ρc using Equations 7.11 and 7.12 at $(\theta_s + \theta_i)/2 = 144°C$, the temperature at the primary shear zone is estimated to be $\theta_s = 208°C$. One iteration is sufficient for adjusting the workpiece material properties.

The temperature rise of the chip at the interface with the tool is given by Equation 7.5. The constant L_2 is calculated using Equation 7.6.

$$L_2 = \frac{V_c \, L_c \rho c}{4k} = \frac{Va \, L_c \rho c}{4a_c k} = \frac{(3.333 \text{ m/s})(0.3 \text{ mm})(0.5 \text{ mm})[3,274,800 \text{ J/(m}^3\,°C)]}{4(0.7 \text{ mm})[48.55 \text{ J/(s m } °C)]}$$

$$L_2 = 12.05$$

The B and Γ_2 factors are calculated using Equations 7.9 and 7.8, respectively, so

$$B = 1.424 \text{ and } \Gamma_2 = 0.841.$$

The frictional temperature rise is given by Equation 7.5

$$\theta_f = \frac{0.377 \, \Gamma_2 P_f}{bk\sqrt{L_2}} = \frac{0.377(0.9)(857 \text{ J/s})}{(2.5 \text{ mm})[48.55 \text{ J/(s m } °C)] \, (\sqrt{12.05})} = 645°C$$

Therefore, the maximum temperature of the chip at the chip–tool interface is estimated from Equation 7.13

$$\theta_T = \theta_s + \theta_f = 238°C + 645°C = 893°C$$

The workpiece material properties should be adjusted for an average chip temperature. Considering the average chip temperature to be 450°C, the L_2, B, and Γ_2 factors are recalculated to be

$$L_2 = 27.22, \, B = 1.424, \text{ and } \Gamma_2 = 0.860$$

the maximum temperature of the chip at the chip–tool interface is estimated to be

$$\theta_T = \theta_s + \theta_f = 238°C + 560°C = 798°C$$

Example 7.2 Estimate the maximum temperature at the chip–tool interface if the cutting speed is reduced from 200 to 40 m/min in Example 7.1. The cutting and axial forces were increased about 4%. The chip thickness was measured and increased to 0.75 mm. The tool–chip contact length was 0.57 mm.

Solution: The deformation power along the shear plane is calculated in orthogonal cutting as in Example 7.1. The shear angle ϕ is calculated (as in Example 6.1) to be 21.8°.

$$P_s = (F_c \cos\phi - F_z \sin\phi)\frac{\cos\alpha}{\cos(\phi - \alpha)} V = \frac{(936\cos 21.8° - 624\sin 21.8°)(40 \text{ m/min})}{\cos(21.8°)}$$

$$P_s = 27,456 \text{ N m/min} = 458 \text{ W} = 458 \text{ J/s}$$

The friction power at the chip–tool interface is

$$P_f = F_c V - P_s = (936 \text{ N})(40 \text{ m/min}) - 27,456 = 166 \text{ W}$$

The proportion of the primary shear energy entering the chip is

$$\Gamma_1 = \cfrac{1}{1 + 1.328\sqrt{\cfrac{[45.7 \text{ J/(s m }°\text{C)}](2.9)}{[4,200,449 \text{ J/(m}^3 °\text{C)}](0.667 \text{ m/s})(0.0003 \text{ m})}}} = 0.744$$

The workpiece properties are considered at the average temperature of 144°C, (obtained in Example 7.1). The temperature rise along the primary shear plane is

$$\theta_s = \frac{(0.744)(458 \text{ J/s})}{[4,200,449 \text{ J/(m}^3 °\text{C)}](0.0003 \text{ m})(0.00025 \text{ mm})(0.667 \text{ m/s})} + 30°\text{C}$$

$$\theta_s = 193°\text{C}$$

The temperature rise of the chip at the interface with the tool is estimated by Equation 7.5. The factors L_2, B, and Γ_2 are calculated using Equations 7.4, 7.9, and 7.8, respectively. The workpiece properties are considered at the average temperature of 300°C (based on the results in Example 7.1).

$$L_2 = 4.71, B = 1.351 \text{ and } \Gamma_2 = 0.690.$$

The frictional temperature rise is given by Equation 7.5

$$\theta_f = \frac{0.377 \, \Gamma_2 \, P_f}{bk\sqrt{L_2}} = \frac{0.377(0.878)(166 \text{ J/s})}{(2.5 \text{ mm})[41.8 \text{ J/(s m }°\text{C)}](\sqrt{4.71})} = 191°\text{C}$$

Therefore, the maximum temperature of the chip at the chip–tool interface is estimated from Equation 7.13

$$\theta_T = \theta_s + \theta_f = 193°\text{C} + 191°\text{C} = 384°\text{C}$$

The comparison of the results in this example with those in Example 7.1 illustrates the importance of the frictional temperature rise along the rake face of the tool. The temperature at the tool–chip interface increased drastically with increased cutting speed as compared to the temperature along the primary shear plane.

Example 7.3 Assume a 200 mm diameter steel tube is turned down to 195 mm. The inside diameter and the length of the tube are 180 and 350 mm, respectively. The cutting conditions are the same as in Example 7.2 (40 m/min cutting speed and 0.3 mm/rev feed). A sharp tool is used with very small nose radius. The lead angle of the tool is zero. Estimate the temperature effect on the physical dimensions of the workpiece for dry machining.

Solution: The greatest proportion of the heat generated in the shear zone is carried away by the chip, whereas a smaller portion is conducted back into the workpiece raising its temperature. Part of the heat flowing into the workpiece is removed with the chip generated in the following revolution. It becomes difficult to calculate the exact amount of heat carried away with the chips leaving the shear zone, as well as the temperature of the workpiece, which is dependent on the cutting parameters (speed and feed) and the physical properties of the work material. Let us assume that this turning case is represented by the orthogonal case in Example 7.1 (since the radial force is small). The portion of the heat flowing in the workpiece is $(1 - \Gamma_1)$ that was estimated in Example 7.1 to be 0.242. The average temperature rise in the workpiece is given by the equation,

$$\Delta\theta = \frac{(1-\Gamma_1)P_s t_m}{mc} = \frac{(1-0.744)(458 \text{ J/s})(300 \text{ mm})/[(0.3 \text{ mm/rev})(64 \text{ rpm})]}{(\text{Workpiece volume})(7800 \text{ kg/m}^3)(502 \text{ J/kg}\,^\circ\text{C})} = 19^\circ\text{C}$$

where
   t_m is the machining (cutting) time that the tool is in contact with the workpiece
   m is the mass of the workpiece (or the mass of the machining section); it is an application specific assumption
   c is the specific heat of the workpiece material

The workpiece size will expand with increasing temperature. The change in diameter (D) and length (L) is related to its change in temperature ($\Delta\theta$) by the following equation,

$$\Delta D = \Delta\theta D\alpha = (19^\circ\text{C})(180 \text{ mm})(0.0000117 \text{ mm/}^\circ\text{C mm}) = 0.040 \text{ mm}$$

$$\Delta L = \Delta\theta L\alpha = (19^\circ\text{C})(300 \text{ mm})(0.0000117 \text{ mm/}^\circ\text{C mm}) = 0.068 \text{ mm}$$

where α is the coefficient of thermal expansion for the workpiece material. The aforementioned equation assumes a uniform temperature distribution throughout the workpiece that is not the case in metal cutting. Actual measurements have indicated the part to change temperature in a nonuniform manner. As the part temperature distribution is changing, not only size change occurs, but its geometry is distorted as well. Although the temperature uniformity assumption is generally incorrect, the earlier analysis is useful for quick and rough estimates of thermal growth. FEA can be also used to obtain better results.

7.10   PROBLEM

Problem 7.1 The pump bore of an aluminum transmission case is rough bored at 400 m/min cutting speed and 0.2 mm/rev feed. The depth of cut is 1.5 mm. A diamond insert tool with zero nose radius, 8° rake angle, and zero lead angle (to minimize the radial force) is used. The diameter of the bore is 300 mm and the thickness of the wall at that section is 5 mm. The chip thickness was measured to be 0.4 mm. The tool–chip contact length was measured in the microscope to be 0.7 mm (which can be also estimated from Equation 6.31). Estimate the maximum tool–chip interface temperature and the possible bore expansion during boring.

REFERENCES

1. M. C. Shaw, *Metal Cutting Principles,* Oxford University Press, Oxford, U.K., 1984, Ch. 12.
2. A. O. Schmidt, W. W. Gilbert, and O. W. Boston, A thermal balance method and mechanical investigation for evaluating machinability, *ASME Trans.* **67** (1945) 225–232.
3. G. I. Epifanov and P. A. Rebinder, Energy balance of the metal cutting process, *Dok. Akad. Nauka USSR* **66** (1949) 653–656 (in Russian; Henry Brutcher Translation No. 2394).
4. M. B. Bever, E. R. Marshall, and L. B. Ticknor, The energy stored in metal chips during orthogonal cutting, *J. Appl. Phys.* **24** (1953) 1176–1179.

5. H. Shore, Thermoelectric measurement of cutting tool temperature, *J. Washington Acad. Sci.* **15** (1925) 85–88.

6. K. Gottwein, Die Messung der Schneidentemperatur beim Abdrehen von Flusseisen, *Maschienenbau* **4** (1925) 1129–1135.

7. E. G. Herbert, The measurement of cutting temperatures, *Proc. Inst. Mech. Eng.* (1926) 289–329.

8. D. A. Stephenson, Tool–work thermocouple temperature measurements—Theory and implementation issues, *ASME J. Eng. Ind.* **115** (1993) 432–437.

9. K. J. Trigger, R. K. Campbell, and B. T. Chao, A tool-work thermocouple compensation circuit, *ASME Trans.* **80** (1958) 302–306.

10. P. M. Braiden, The calibration of tool/work thermocouples, *Proceedings of Eight International MTDR Conference*, Birmingham, U.K., 1967, pp. 653–666.

11. B. Alvelid, Cutting temperature thermo-electrical measurements, *CIRP Ann.* **18** (1970) 547–354.

12. G. Barrow, A review of experimental and theoretical techniques for assessing cutting temperatures, *CIRP Ann.* **22** (1973) 203–211.

13. C. E. Leshock and Y. C. Shin, Investigation of cutting temperature in turning by a tool-work thermocouple technique, *ASME J. Manuf. Sci. Eng.* **119** (1997) 502–508.

14. E. M. Trent, *Metal Cutting*, Butterworths, London, U.K., 1977, p. 58.

15. N. F. Shillam, Machine tool control systems, US Patent 3,646,839, March 7, 1972.

16. J. S. Agapiou and D. A. Stephenson, Analytical and experimental studies of drill temperatures, *ASME J. Eng. Ind.* **116** (1994) 54–60.

17. M. Tsueda, Y. Hasegawa, and Y. Ishida, The study of cutting temperature in drilling (1) on the measuring method of cutting temperature, *Trans. JSME* **27** (1961) 1423–1430 (in Japanese).

18. S. Nishida, S. Ozaki, S. Nakayama, T. Shiraishi, and K. Nagura, Study on drilling II—Lip temperature, *J. Mech. Lab. Jpn.* **8** (1962) 59–60 (in Japanese).

19. M. Tsueda, Y. Hasegawa, N. Nisina, and T. Hirai, Research on the cutting temperature of a drill point (2) the unequal temperature distribution along two cutting edges, *Trans. JSME* **28** (1962) 1076–1083 (in Japanese).

20. M. F. DeVries, Drill temperature as a drill performance criterion, ASTME Technical Paper MR68-193, 1968.

21. K. Watanabe, K. Yokoyama, and R. Ichimiya, Thermal analyses of the drilling process, *Bull. Jpn. Soc. Precis. Eng.* **11** (1977) 71–77.

22. K. F. Meyer, Untersuchung an Keramischen Schneidstoffen, *Industrie-Anzeiger* **11** (1962).

23. E. Usui, T. Shirakashi, and T. Kitagawa, Analytical prediction of three-dimensional cutting process Part 3. Cutting temperature and crater wear of carbide tool, *ASME J. Eng. Ind.* **100** (1978) 236–243.

24. M. Hirao, Determining temperature distribution on flank face of cutting tool, *J. Mater. Shaping Technol.* **6** (1989) 143–148.

25. M. Bono and J. Ni, A method for measuring the temperature distribution along the cutting edges of a drill, *ASME J. Manuf. Sci. Eng.* **124** (2002) 921–923.

26. D. L. Rall and W. H. Giedt, Heat transfer to, and temperature distribution in, a metal-cutting tool, *ASME Trans.* **78** (1956) 1507–1512.

27. M. B. Hollander and J. E. Eglund, Thermocouple technique investigation of temperature distribution in the workpiece during metal cutting, ASTME Research Report No. 7, 1957.

28. M. C. Shaw, N. H. Cook, and P. A. Smith, Report on the cooling characteristics of cutting fluid, ASTME Research Report No. 19, 1958.

29. G. S. Reichenbach, Experimental measurement of metal-cutting temperature distributions, *ASME Trans.* **80** (1958) 525–540.

30. T. I. El-Wardany, E. Mohammed, and M. A Elbestawi, Cutting temperature of ceramic tools in high speed machining of difficult-to-cut materials, *Int. J. Mach. Tools Manuf.* **36** (1996) 611–634.

31. D. Umbrello, L. Filice, S. Rizzuti, and F. Micari, On the evaluation of the global heat transfer coefficient in cutting, *Int. J. Mach. Tools Manuf.* **47** (2007) 1738–1743.

32. R. Li and A. J. Shih, Spiral point drill temperature and stress in high-throughputdrilling of titanium, *Int. J. Mach. Tools Manuf.* **47** (2007) 2005–2017.

33. A. H. Quereshi and F. Koenigsberger, An investigation into the problem of measuring the temperature distribution on the rake face of a cutting tool, *CIRP Ann.* **14** (1966) 189–199.

34. V. A. Ostafiev, A. A. Cherniavskaya, and V. A Sinopalnikov, Numerical calculation of non-steady-state temperature fields in oblique cutting, *CIRP Ann.* **32** (1983) 43–46.

35. G. Subramani, M. C. Whitmore, S. G. Kapoor, and R. E. DeVor, Temperature distribution in a hollow cylindrical workpiece during machining: Theoretical model and experimental results, *ASME J. Eng. Ind.* **113** (1991) 373–380.

36. W. Grzesik, Experimental investigation of the cutting temperature when turning with coated indexable inserets, *Int. J. Mach. Tools Manuf.* **39** (1999) 355–369.
37. S. Lin, F. Peng, J. Wen, Y. Liu, and R. Yan, An investigation of workpiece temperature variation in end milling considering flank rubbing effect, *Int. J. Mach. Tools Manuf.* **73** (2013) 71–86.
38. D. K. Aspinwall, A. L. Mantle, W. K. Chan, R. Hood, and S. L. Soo, Cutting temperatures when ball nose end milling γ-TiAl intermetallic alloys, *CIRP Ann.* **62** (2013) 75–78.
39. L. Li, B. Li, X. Li, and K. F. Ehmann, Experimental investigation of hard turning mechanisms by PCBN tooling embedded micro thin film thermocouples, *ASME J. Manuf. Sci. Eng.* **135** (2013) 041012–041013.
40. A. Bastia, T. Obikawa, and J. Shinozuka, Tools with built-in thin film thermocouple sensors for monitoring cutting temperature, *Int. J. Mach. Tools Manuf.* **47** (2007) 793–798.
41. M. P. Lipman, B. E. Nevis, and G. E. Kane, A remote sensor method for determining average tool-chip interface temperatures in metal cutting, *ASME J. Eng. Ind.* **89** (1967) 333–338.
42. K. K. Wang, S. M. Wu, and K. Iwata, Temperature responses and experimental errors for multitooth milling cutters, *ASME J. Eng. Ind.* **90** (1968) 353–359.
43. M. P. Groover and G. E. Kane, Continuous study in the determination of temperature in metal cutting using remote thermocouples, *ASME J. Eng. Ind.* **93** (1971) 603–608.
44. D. W. Yen and P. K. Wright, A remote temperature sensing technique for estimating the cutting interface temperature distribution, *ASME J. Eng. Ind.* **108** (1986) 252–263.
45. J. G. Chow and P. K. Wright, On-line estimating of tool/chip interface temperatures for a turning operation, *ASME J. Eng. Ind.* **110** (1988) 56–64.
46. B. L. Tai, D. A. Stephenson, and A. J. Shih, An inverse heat transfer method for determining workpiece temperature in minimum quantity lubrication deep hole drilling, *ASME J. Manuf. Sci. Eng.* **134** (2012) 021006-1.
47. B. L. Tai, A. J. Jessup, D. A. Stephenson, and A. J. Shih, Workpiece thermal distortion in minimum quantity lubrication deep hole drilling—Finite element modeling and experimental validation, *ASME J. Manuf. Sci. Eng.* **134** (2012) 011008-1.
48. B. L. Tai, D. A. Stephenson, and A. J. Shih, Workpiece temperature during deep-hole drilling of cast iron using high air pressure minimum quantity lubrication, *ASME J. Manuf. Sci. Eng.* **135** (2013) 031019-1.
49. D. A. Stephenson, An inverse method for investigating deformation zone temperatures in metal cutting, *ASME J. Eng. Ind.* **113** (1991) 129–136.
50. D. O'Sullian and M. Cotterell, Workpiece temperature measurement in machining, *Proc. Inst. Mech. Eng., J. Eng. Manuf.* **216B** (2002) 135–139.
51. P. Kwon, T. Schiemann, and R. Kountanya, An inverse estimation scheme to measure steady-state tool-chip interface temperatures using an infrared camera, *Int. J. Mach. Tools Manuf.* **41** (2001) 1015–1030.
52. J. V. Beck, B. Blackwell, and C. R. St. Clair, *Inverse Heat Conduction: Ill-Posed Problems*, Wiley Interscience, New York, 1985, pp. 13–36, 154–156.
53. P. K. Wright and E. M. Trent, Metallurgical methods of determining temperature gradients in cutting tools, *J. Iron Steel Inst.* **211** (1973) 364–368.
54. E. F. Smart and E. M. Trent, Temperature distribution in tools used for cutting iron, titanium and nickel, *Int. J. Prod. Res.* **13** (1975) 265–290.
55. P. K. Wright, Correlation of tempering effects with temperature distribution in steel cutting tools, *ASME J. Eng. Ind.* **100** (1978) 131–136.
56. B. Mills, D. W. Wakeman, and A. Aboukhashaba, A new technique for determining the temperature distribution in high speed steel cutting tools using scanning electron microscopy, *CIRP Ann.* **29** (1980) 73–77.
57. B. Mills, T. D. Mottishaw, and A. J. W. Chisolm, The application of scanning electron microscopy to the study of temperatures and temperature distributions in M2 high speed steel twist drills, *CIRP Ann.* **30** (1981) 15–20.
58. A. Thangaraj, P. K. Wright, and M. Nissle, New experiments on the temperature distribution in drilling, *ASME J. Eng. Ind.* **106** (1984) 242–247.
59. P. A. Dearny and E. M Trent, Wear mechanisms of coated carbide tools, *Metals Tech.* **9** (1982) 60–75.
60. P. A. Dearny, New technique for determining temperature distribution in cemented carbide cutting tool, *Metals Technol.* **10** (1983) 205–210.
61. S. Kato, Y. Yamaguchi, Y. Watanabe, and Y Hiraiwa, Measurement of temperature distribution within tool using powders of constant melting point, *ASME J. Eng. Ind.* **98** (1976) 607–613.
62. F. Schwerd, Ueber die Bestimmung des *Temperaturfeldes* beim Spanablauf, *Z. VDI* **77** (1933) 211–216.
63. G. Boothroyd, Photographic technique for the determination of metal cutting temperatures, *Br. J. Appl. Phys.* **12** (1961) 238–242.

64. G. Boothroyd, Temperatures in orthogonal metal cutting, *Proc. Inst. Mech. Eng.* **177** (1963) 789–802.

65. E. Lenz, Die Temperaturmessung in der Kontaktzone Span-Werkzeug beim Drehvorgang, *CIRP Ann.* **13** (1966) 201–210.

66. E. Lenz, Die Temperaturverteilung in der Kontaktzone Span-Werkzeug beim Drehen von Stahl mit Hartmetallwerkzeugen, *CIRP Ann.* **14** (1966) 137–144.

67. O. D. Prins, The influence of wear on the temperature distribution at the rake face, *CIRP Ann.* **19** (1971) 579–584.

68. B. T. Chao, H. L. Li, and K. J. Trigger, An experimental investigation of temperature distribution at tool-flank surface, *ASME J. Eng. Ind.* **83** (1961) 496–504.

69. A. E. Focks, F. E. Westerman, P. E. Rentschler, J. Kemphaus, T. W. Shi, and M. Hoch, Heat flow patterns in superhard tools when cutting superalloys, *Proc. NAMRC* **13** (1985) 394–401.

70. T. Ueda, A. Hosokawa, and A. Yamamoto, Measurement of grinding temperature using infrared radiation pyrometer with optical fiber, *ASME J. Eng. Ind.* **108** (1986) 247–251.

71. J. P. Kottenstette, Measuring tool-chip interface temperatures, *ASME J. Eng. Ind.* **108** (1986) 101–104.

72. J. Lin, S.-L. Lee, and C.-I. Wang, Estimation of cutting temperature in high speed machining, *ASME J. Eng. Mater. Technol.* **114** (1992) 289–296.

73. E. Belotserkovsky, O. Bar-Or, and A. Katzir, Infrared fiberoptic temperature monitoring during machining procedures, *Meas. Sci. Technol.* **5** (1994) 451–453.

74. P. Mueller-Hummel and M. Lahres, Infrared temperature measurement on diamond-coated tools during machining, *Diam. Relat. Mater.* **3** (1994) 765–769.

75. M. Al Huda, K. Yamada, A. Hosokawa, and T. Ueda, Investigation of temperature at tool-chip interface in turning using two-color pyrometer, *ASME J. Manuf. Sci. Eng.* **124** (2002) 200–207.

76. M. Sato, T. Aoki, H. Tanaka, and S. Takeda, Variation of temperature at the bottom surface of a hole during drilling and its effect on tool wear, *Int. J. Mach. Tools Manuf.* **68** (2013) 40–47.

77. L. Kops and M. C. Shaw, Thermal radiation in surface grinding, *CIRP Ann.* **31** (1982) 211–214.

78. L. Kops and M. C. Shaw, Application of infrared radiation measurements in grinding studies, *Proc. NAMRC* **14** (1983) 390–396.

79. D. A. Stephenson and A. Ali, Tool temperatures in interrupted metal cutting, *ASME J. Eng. Ind.* **114** (1992) 127–136.

80. L. Wang, K. Saito, and I. S. Jawahir, Infrared temperature measurement of curled chip formation in metal cutting, *Trans. NAMRI/SME* **24** (1996) 33–38.

81. R. M'Saoubi, C. Le Calvez, B. Changeux, and J. L. Lebrun, Thermal and microstructural analysis of orthogonal cutting of a low alloyed carbon steel using an infrared-charge-coupled device camera technique, *Proc. Inst. Mech. Eng., J. Eng. Manuf.* **216B** (2002) 153–165.

82. G. Sutter, L. Faure, A. Molinari, N. Ranc, and V. Pina, An experimental technique for the measurement of temperatures fields for the orthogonal cutting in high speed machining, *Int. J. Mach. Tools Manuf.* **43** (2003) 671–678.

83. R. M'Saoubi and H. Chandrasekaran, Experimental tool temperature distributions in oblique and orthogonal cutting using chip breaker geometry inserts, *ASME J. Manuf. Sci. Eng.* **128** (2006) 606–610.

84. R. W. Ivester, Tool temperatures in orthogonal cutting of alloyed titanium, *Proc. NAMRI/SME* **39** (2011) 253–258.

85. G. Suttera and N. Ranc, Temperature fields in a chip during high-speed orthogonal cutting—An experimental investigation, *Int. J. Mach. Tools Manuf.* **47** (2007) 1507–1517.

86. F. Penzig, Sichtbarmachen von Temperaturfeldern Durch Temperaturabhaengigen Farbenstriche, *Z. VDI* **83** (1939) 69.

87. H. Schallbroch and M. Lang, Messung der Schnittemperatur Mittles Temperaturanzeigender Farbenstriche, *Z. VDI* **87** (1943) 15–19.

88. U. Koch, Experimental and theoretical analysis of lathe tool temperature distribution in oblique cutting, *Proceedings of 11th MTDR Conference*, Manchester, U.K., 1970, Vol. 1, pp. 533–540.

89. U. Koch and R. Levi, Some mechanical and thermal aspects of twist drill performance, *CIRP Ann.* **19** (1971) 247–254.

90. S. Shu, K. Cheng, H. Ding, and S. Chen, An innovative method to measure the cutting temperature in process by using an internally cooled smart cutting tool, *ASME J. Manuf. Sci. Eng.* **134** (2012) 061018-1.

91. B. M. Kramer, On tool materials for high speed machining, *ASME J. Eng. Ind.* **109** (1987) 87–91.

92. A. Anagonye and D. A. Stephenson, Modeling cutting temperatures for turning inserts with various tool geometries and materials, *ASME J. Manuf. Sci. Eng.* **124** (2002) 544–552.

93. R. S. Hahn, On the temperature developed at the shear plane in the metal cutting process, *Proceedings of the First U.S. National Conference on Applied Mechanics*, ASME, New York, 1951, pp. 661–666.

94. K. J. Trigger and B. T. Chao, An analytical evaluation of metal cutting temperatures, *ASME Trans.* **73** (1951) 57–68.

95. B. T. Chao and K. J. Trigger, Temperature distribution at the tool-chip interface in metal cutting, *ASME Trans.* **77** (1955) 1107–1121.

96. E. G. Loewen and M. C. Shaw, On the analysis of cutting-tool temperatures, *ASME Trans.* **76** (1954) 217–231.

97. A. C. Rapier, A theoretical investigation of the temperature distribution in the metal cutting process, *Br. J. Appl. Phys.* **5** (1954) 400–405.

98. J. C. Jaeger, Moving sources of heat and the temperature at sliding contacts, *Proc. Roy. Soc. NSW* **76** (1942) 203–224.

99. D. A. Stephenson, Assessment of steady-state metal cutting temperature models based on simultaneous infrared and thermocouple data, *ASME J. Eng. Ind.* **113** (1991) 121–128.

100. Y. S. Touloukian et al., *Thermocphysical Properties of Matter*, IFI/Plenum, New York, 1970, Vol. 1, pp. 1194–1196, Vol. 4, pp. 1282–1284.

101. G. A. Roberts and R. A. Cary, *Tool Steels*, 2nd edn., ASM, Metals Park, OH, 1980, p. 709.

102. A. O. Tay, M. G. Stevenson, G. de Vahl Davis, and P. L. B. Oxley, A numerical method for calculating temperature distributions in machining, from force and shear angle measurements, *Int. J. Mach. Tool Des. Res.* **16** (1976) 335–349.

103. D. Bordui, Third generation silicon nitride, *Ceramic Cutting Tools and Applications*, SME, Dearborn, MI, 1989.

104. J. H. Weiner, Shear-plane temperature distribution in orthogonal cutting, *ASME Trans.* **77** (1955) 1331–1341.

105. G. Vieregge, Energieverteilung und Temperatur bei der Zerspannung, *Werkstatt und Betrieb* **86** (1953) 691–703.

106. W. C. Leone, Distribution of shear-zone heat in metal cutting, *ASME Trans.* **76** (1954) 121–125.

107. P. R. Dawson and S. Malkin, Inclined moving heat source model for calculating metal cutting temperatures, *ASME J. Eng. Ind.* **106** (1984) 179–186.

108. N. A. Abukhshim, P. T. Mativenga, and M. A. Sheikh, Investigation of heat partition in high speed turning of high strength alloy steel, *Int. J. Mach. Tools Manuf.* **45** (2005) 1687–1695.

109. N. A. Abukhshim, P. T. Mativenga, and M. A. Sheikh, Heat generation and temperature prediction in metal cutting: A review and implications for high speed machining, *Int. J. Mach. Tools Manuf.* **46** (2006) 782–800.

110. K. Nakayama, Temperature rise in the workpiece during metal cutting, *Bull. Fact. Eng. Yokohama Nat. Univ.* **5** (1956) 1–10.

111. B. T. Chao and K. J. Trigger, Temperature distribution at the tool-chip and tool-work interface in metal cutting, *ASME Trans.* **80** (1958) 311–320.

112. P. K. Venuvinod and W. S. Lau, Estimation of rake temperatures in free oblique cutting, *Int. J. Mach. Tool Des. Res.* **26** (1986) 1–14.

113. P. K. Wright, S. P. McCormick, and T. R. Miller, Effect of rake face design on cutting tool temperature distributions, *ASME J. Eng. Ind.* **102** (1980) 123–128.

114. W. Grzesik and P. Nieslony, A computational approach to evaluate temperatue and head partition in machining with multilayer coated tools, *Int. J. Mach. Tools Manuf.* **43** (2003) 1311–1317.

115. W. M. Mansour, M. O. M. Osman, T. S. Sankar, and A. Mazzawi, Temperature field and crater wear in metal cutting using a quasi-finite element approach, *Int. J. Prod. Res.* **11** (1973) 59–68.

116. A. O. Tay, M. G. Stevenson, and G. de Vahl Davis, Using the finite element method to determine temperature distributions in orthogonal machining, *Proc. Inst. Mech. Eng.* **188** (1974) 627–638.

117. P. D. Muraka, Prediction of temperatures in orthogonal machining using the finite element method, PhD thesis, University of Manchester, Manchester, U.K., 1977.

118. P. D. Muraka, G. Barrow, and S. Hinduja, Influence of the process variables on the temperature distribution in orthogonal machining using the finite element method, *Int. J. Mech. Sci.* **21** (1979) 445–456.

119. M. G. Stevenson, P. K. Wright, and J. G. Chow, Further developments in applying the finite element method to the calculation of temperature distributions in machining and comparisons with experiment, *ASME J. Eng. Ind.* **105** (1983) 149–154.

120. T. Kagiwada and T. Kanauchi, Numerical analysis of cutting temperatures and flowing ratios of generated heat, *JSME Int. J., Series III* **31** (1988) 624–633.

121. P. L. B. Oxley, *The Mechanics of Machining*, Ellis Horwood, Chicester, U.K., 1989.

122. E. Usui and T. Shirakashi, Mechanics of machining—From "Descriptive" to "Predictive Theory," *On the Art of Cutting Metals—75 Years Later*, ASME PED ASME, New York, 1982, Vol. 7, pp. 13–35.

123. E. Usui and T. Shirakashi, Analytical prediction of cutting tool wear, *Wear* **100** (1984) 129–151.

124. J. S. Strenkowski and K. J. Moon, Finite element prediction of chip geometry and tool/workpiece temperature distributions in orthogonal metal cutting, *ASME J. Eng. Ind.* **112** (1990).

125. K. W. Kim and H.-C. Sin, Development of a thermo-viscoplastic cutting model using finite element method, *Int. J. Mach. Tools Manuf.* **36** (1996) 379–397.

126. T. D. Marusich and M. Ortiz, Modelling and simulation of high-speed machining, *Int. J. Num. Methods Eng.* **38** (1995) 3675–3694.

127. T. Ozel and T. Altan, Process simulation using finite element method—Prediction of cutting forces, tool stresses, and temperatures in high-speed flat end milling process, *Int J. Mach. Tools Manuf.* **40** (2000) 713–738.

128. T. Obikawa and E. Usui, Computational machining of titanium alloy—Finite element modeling and a few results, *ASME J. Manuf. Sci. Eng.* **118** (1996) 208–215.

129. G. List, G.Sutter, and A. Bouthiche, Cutting temperature prediction in high speed machining by numerical modelling of chip formation and its dependence with crater wear, *Int J. Mach. Tools Manuf.* **54–55** (2012) 1–9.

130. A B. M. Hadzley, R. Izamshah, A. S. Sarah, and M. N. Fatin, Finite element model of machining with high pressure coolant for Ti-6Al-4V alloy, *Proc. Eng.* **53** (2013) 624–631.

131. P. J. Arrazola, T. Ozel, D. Umbrello, M. Davies, and I. S. Jawahir, Recent advances in modelling of metal machining processes, *CIRP Ann.* **62** (2013) 695–718.

132. S. Pervaiza, I. Deiab, E. M. Wahbac, A. Rashid, and M. Nicolescu, A coupled FE and CFD approach to predict the cutting tool temperature profile in machining, *Proc. CIRP* **17** (2014) 750–754.

133. A. J. R. Smith and E. J. A. Armarego, Temperature prediction in orthogonal cutting with a finite difference approach, *CIRP Ann.* **30** (1981) 9–13.

134. C. L. Chan and A. Chandra, A boundary element method analysis of the thermal aspects of metal cutting processes, *ASME J. Eng. Ind.* **113** (1991) 311–319.

135. A. Chandra and C. L. Chan, Thermal aspects of machining: A BEM approach, *Int. J. Solids Struct.* **31** (1994) 1657–1693.

136. S. Darwish and R. Davies, Investigation of the heat flow through bonded and brazed metal cutting tools, *Int. J. Mach. Tool Des. Res.* **29** (1989) 229–237.

137. R. Ichimiya and K. Kawahara, Investigation of thermal expansion in machining operations (elongations of tool and workpiece in turning), *Bull. JSME* **14** (1971) 1363–1371.

138. R. Ichimiya and Y. Usuzaka, Analysis of thermal expansion in face-cutting operations, *ASME J. Eng. Ind.* **96** (1974) 1222–1229.

139. D. E. McFeron and B. T. Chao, Transient interface temperatures in plain peripheral milling, *ASME Trans.* **80** (1958) 321–329.

140. K. K. Wang, K. C. Tsao, and S. M. Wu, Investigation of face-milling tool temperatures by simulation techniques, *ASME J. Eng. Ind.* **91** (1969) 772–780.

141. H. Wu and J. E. Mayer, Jr., An analysis of thermal cracking of carbide tools in intermittent cutting, *ASME J. Eng. Ind.* **101** (1979) 159–164.

142. R. Radulescu and S. G. Kapoor, An analytical model for prediction of tool temperature fields during continuous and interrupted cutting, *Materials Issues in Machining and the Physics of Machining Processes*, TMS, Materials Park, OH, 1992, pp. 147–165.

143. T. D. Marusich, *Third Wave AdvantEdge Theoretical Manual*, Version 4.3, Third Wave Systems, Inc., Minneapolis, MN, 2002.

144. T.-C. Jen, J. G. Gutierrez, and S. Eapen, Non-linear numerical analysis in interrupted cutting tool temperatures, *Numer. Heat Transfer A: Appl.* **39** (2001) 1–20.

145. I. Lazoglu and Y. Altintas, Prediction of tool and chip temperature in continuous and interrupted machining, *Int. J. Mach. Tools Manuf.* **42** (2002) 1011–1022.

146. C. Ming, S. Fanghong, W. Haili, Y. Renwei, Q. Zhenghong, and Z. Shquiao, Experimental research on the dynamic characteristics of cutting temperature in the process of high speed milling, *J. Mater. Process. Technol.* **138** (2003) 468–471.

147. N. N. Zorev, Machining steel with a carbide tool in interrupted heavy-cutting conditions, *Russ. Eng. J.* **43** (1963) 43–47.

148. N. N. Zorev, Standzeit und Leistung der Hartmetall-Werkzeuge beim Unterbrochenen Zerspanen des Stahls mit Grossen Zerspanungsquerschnitten, *CIRP Ann.* **11** (1963) 201–210.

149. T. Hoshi and K. Okushima, Optimum diameter and position of a fly cutter for milling 0.45 C steel, 195 BHN and 0.4 C steel, 167 BHN at light cuts, *ASME J. Eng. Ind.* **87** (1965) 442–446.

150. P. M Braiden and D. S. Dugdale, Failure of carbide tools in intermittent cutting, *Materials for Metal Cutting*, ISI Special Publication 126, ISI, London, U.K., 1970, pp. 30–34.

151. A. J. Pekelharing, Cutting tool damage in interrupted cutting, *Wear* **62** (1978) 37–48.
152. S. F. Wayne and S. T. Buljan, The role of thermal shock on tool life of selected ceramic cutting tool materials, *J. Am. Ceram. Soc.* **75** (1989) 754–760.
153. Z. Palmai, Cutting temperature in intermittent cutting, *Int. J. Mach. Tools Manuf.* **27** (1987) 261–274.
154. C. Salomon, Verfahren zur Bearbeitung von Metallen oder bei einter Bearbeitung durch Schneidende Werkzeuge sich aehnlich Verhaltenden Werkstoffen, German Patent 523594, 1931.
155. R. I. King, *Handbook of High Speed Machining Technology*, Chapman and Hall, New York, 1985, pp. 3–4.
156. P. Lezanski and M. C. Shaw, Tool face temperatures in high speed milling, *ASME J. Eng. Ind.* **112** (1990) 132–135.
157. J. S. Agapiou, G. W. Halldin, and M. F. DeVries, On the machinability of powder metallurgy austenitic stainless steels, *ASME J. Eng. Ind.* **110** (1988) 339–343.
158. R. P. Hervey and N. H. Cook, Thermal parameters in drill tool life, ASME Paper 65-Prod-15, 1965.
159. M. F. DeVries, U. K. Saxena, and S. M. Wu, Temperature distributions in drilling, *ASME J. Eng. Ind.* **90** (1968) 231–238.
160. U. K. Saxena, M. F. DeVries, and S. M. Wu, Drill temperature distributions by numerical solutions, *ASME J. Eng. Ind.* **93** (1971) 1057–1065.
161. K. J. Fuh, Computer aided design and manufacture of multifacet drills, PhD thesis, University of Wisconsin, Madison, WI, 1987, Ch. 8.
162. R. Li and A. J. Shih, Tool temperature in titanium drilling, *ASME J. Manuf. Sci. Eng.* **129** (2007) 740–749.
163. J. S. Agapiou and M. F. DeVries, On the determination of thermal phenomena during a drilling process Part I—Analytical models of twist drill temperature distributions, *Int. J. Mach. Tools Manuf.* **30** (1990) 203–215.
164. J. S. Agapiou and M. F. DeVries, On the determination of thermal phenomena during a drilling process Part II—Comparison of experimental and analytical twist drill temperature distributions, *Int. J. Mach. Tools Manuf.* **30** (1990) 217–226.
165. N. Ikawa, R. R. Donaldson, R. Komanduri, W. Koenig, P. A. McKeown, E. T. Moriwaki, and I. F. Stowers, Ultraprecision metal cutting—The past, the present and the future, *CIRP Ann.* **40** (1991) 587–594.
166. R. G. Watts and E. R. McClure, Thermal expansion of workpiece during turning, ASME paper 68-WA/Prod-24, 1968.
167. Y. Takeuti, S. Zaima, and N. Noda, Thermal-stress problems in industry I: On thermoelastic distortion in machining metals, *J. Therm. Stresses* **1** (1978) 199–210.
168. B. Krauskopf, Diamond turning: Reflecting demands for precision, *Manuf. Eng.* **24**:5 (May 1984) 90–100.
169. N. N. Kakade and J. G. Chow, Computer simulation of bore distortions for engine boring operation, *Collected Papers in Heat Transfer*, ASME HTD 1989, Vol. 123, pp. 259–265.
170. D. A. Stephenson, M. R. Barone, and G. F. Dargush, Thermal expansion of the workpiece in turning, *ASME J. Eng. Ind.* **117** (1995) 542–550.
171. D. M. Haan, S. A. Batzer, W. W. Olson, and J. W. Sutherland, An experimental study of cutting fluid effects in drilling, *J. Mater. Proc. Technol.* **71** (1997) 305–313.
172. M. J. Bono and J. Ni, The effects of thermal distortions on the diameter and cylindricity of dry drilled holes, *Int. J. Mach. Tools Manuf.* **41** (2001) 2261–2270.
173. M. J. Bono, Experimental and analytical issues in drilling, PhD thesis, Mechanical Engineering, University of Michigan, Ann Arbor, MI, 2002.
174. S. Kalidas, S. G. Kapoor, and R. E. DeVor, Influence of thermal effects on hole quality in dry drilling, Part 1: A thermal model of workpiece temperatures, *ASME J. Manuf. Sci. Eng.* **124** (2002) 258–266.
175. S. Kalidas, S. G. Kapoor, and R. E. DeVor, Influence of thermal effects on hole quality in dry drilling, Part 2: Thermo-elastic effects on hole quality, *ASME J. Manuf. Sci. Eng.* **124** (2002) 267–274.
176. T. C. Jen, G. Gutierrez, S. Eapen, G. Barber, H. Zhao, P. S. Szuba, J. Labataille, and J. Manjunathaiah, Investigation of heat pipe cooling in drilling applications. Part I: Preliminary numerical analysis and verification, *Int. J. Mach. Tools Manuf.* **42** (2002) 643–652.

8 Machining Process Analysis

8.1 INTRODUCTION

Analysis methods have seen increasing use for machining process design and improvement in the automotive, aerospace, construction equipment, and cutting tool industries. Broadly, three types of analyses are performed: force, power, and cycle time analyses using kinematic simulations (or mechanistic models), structural analysis for clamping and fixturing using finite element methods, and detailed chip formation analyses using finite element models. This chapter describes kinematic simulations and structural finite element analyses. Finite element chip formation analyses, which are used especially in the cutting tool and aerospace industries, are described in Sections 6.12, 7.5, and 10.7.

Kinematic simulations of machining processes are used to calculate cycle times and time histories of cutting forces and power. The inputs required include the part and tool geometries, tool paths, and cutting pressures for the tool–workpiece material of interest, which may be measured in tests or estimated from finite element calculations. The tool geometries and tool paths are preferably read directly from CAD and CAM systems. Based on this information, the kinematic motions of the tool with respect to the workpiece as a function of time can be simulated, and the instantaneous area of material being cut (the interference between the tool and workpiece) at any time can be computed from the tool path and part geometry. Forces are computed by multiplying the instantaneous area by the measured cutting pressures, and power is computed as the product of the cutting speed and the appropriate cutting force component. Accurate cycle time calculations require consideration of the transient response of the spindle and slides, times required for tool changes, pallet rotations, coolant purges, etc. These characteristics can be obtained from machine tool manufacturer's specifications, or measured if hardware is available. Computed force and power histories can be used to level forces and reduce cycle time with minimal testing, and as inputs to finite element analyses. These analyses may be integrated with CAD and CAPP programs to simplify geometric and process condition input [1]. Commercially supported programs of this type include Third Wave Systems' Production Module programs [2] and MillSim from Manufacturing Laboratories, Inc. [3].

In the research literature, kinematic simulation programs are often called mechanistic models [4–6]. This was the term favored by pioneering researchers of the subject, notably S. Kapoor and R. E. Devor and their coworkers at the University of Illinois, to distinguish them from purely empirical models. Engineers who apply these programs in industry often call them CAE analyses, process simulations, or process models [1,4,7,8]. The term "kinematic simulation" is more descriptive but is not widely used.

Kinematic simulations for turning, boring, face and end milling, and drilling are described in Sections 8.2 through 8.5. Models for other processes, such as tapping [9,10], thread milling [11], and crankshaft turnbroaching [12] are similar to those presented and are not reviewed in detail. The use of baseline cutting force data to develop cutting pressure equations is described in Section 8.6. Typical applications are reviewed in Section 8.7.

Structural finite element analysis is used to estimate workpiece distortions due to clamping and machining. The objective of the analysis is to minimize such distortions for critical features, which may be accomplished by stiffening the part or fixture in directions of heavy loading, modifying the tool path or cutter geometry to direct forces in stiff or noncritical directions, choosing clamping and locating schemes, which minimize clamping distortion and support compliant portions of the part, and minimizing clamping forces. If the required input information is readily available, finite element analysis permits a wider variety of options to be investigated more quickly and cheaply than through prototype part and fixture tests.

Technical issues in structural finite element analysis are discussed in Section 8.8, and typical application examples are given in Section 8.9.

8.2 TURNING

Turning (Figure 8.1) is one of the simplest cutting processes to simulate because the geometry and kinematic motions of the tool and workpiece are easily described. Moreover, the benefits of simulation are relatively easy to quantify when turning large volumes of parts on CNC lathes, since simulation can be used to reduce cycles times and thus the number of machines and capital investment required.

The earliest work on turning simulation was based on the shear plane cutting theory [13,14]. As noted in the introduction, in more recent work [15–20], cutting forces are calculated by multiplying measured cutting pressures by the calculated uncut chip area. A common approach [15] is to relate cutting forces to the normal cutting pressure, K_n, the effective friction coefficient, K_f, and the effective lead angle, γ_{Le} (Figure 8.2). K_n and K_f are defined by

$$K_n = \frac{N}{A_c} \tag{8.1}$$

$$K_f = \frac{P}{N} \tag{8.2}$$

where
  N and P are the force components normal and parallel to the rake face of the tool (Figure 6.4)
  A_c is the uncut chip area, which in most cases is adequately approximated by

$$A_c = fd \tag{8.3}$$

where
  f is the feed per rotation
  d is the depth of cut

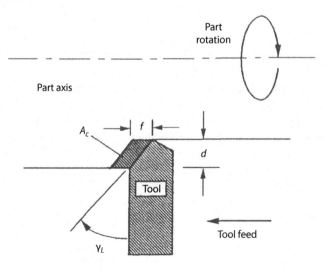

FIGURE 8.1 Uncut chip area A_c in turning.

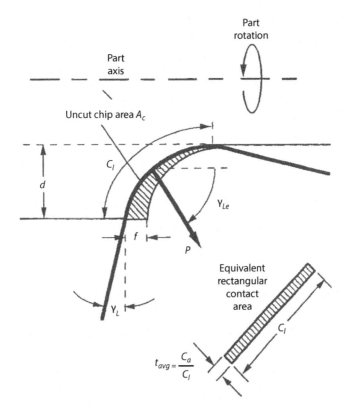

FIGURE 8.2 Definition of the effective lead angle γ_{Le}. (After Stephenson, D.A., and Bandyopadhyay, P., *ASME J. Eng. Mater. Technol.*, 119, 86, 1997.)

This equation overestimates A_c for tools with a nose radius; for shallow depths of cut with such tools, more exact equations for A_c are available [18]. γ_{Le} determines the direction of the friction force and depends on the lead angle of the toolholder, the feed rate, depth of cut, and tool nose radius. The calculation of γ_{Le} and the estimation of K_n and K_f from baseline test data are discussed in Section 8.6. In terms of these parameters, the tangential, axial, and radial cutting forces (Section 6.3) are given by [21]

$$F_t = K_n A_c \left[\cos\alpha_b \cos\alpha + K_f \left(\cos\gamma_{Le} \sin\alpha + \sin\gamma_{Le} \sin\alpha_b \right) \right] \tag{8.4}$$

$$F_a = K_n A_c \left[-\cos\alpha_b \sin\alpha + K_f \left(\cos\gamma_{Le} \cos\alpha \right) \right] \tag{8.5}$$

$$F_r = K_n A_c \left[-\sin\alpha_b + K_f \left(\sin\gamma_{Le} \cos\alpha_b \right) \right] \tag{8.6}$$

In these equations,

α_b is the back rake angle

α is the normal rake angle, which is related to the inclination angle, λ, and to the side rake, back rake, and lead angles, α_s, α_b, and γ_L (discussed in Chapters 4 and 6) through [22]

$$\lambda = \tan^{-1}[\tan\alpha_b \cos\gamma_L - \tan\alpha_s \sin\gamma_L] \tag{8.7}$$

$$\alpha = \tan^{-1}[\cos\lambda(\tan\alpha_s \cos\gamma_L + \tan\alpha_b \sin\gamma_L)] \tag{8.8}$$

These equations can be used to compute time histories of cutting forces in contour turning operations with varying speeds, feeds, or depths of cut. This is usually done by computing forces at fixed time steps based on the instantaneous cutting conditions at that time step, often determined from CAD/CAM data. Typical time-varying force plots are shown in Section 8.7. This information can be used to identify part of the tool path in which forces are particularly high or low. The tangential force can also be multiplied by the cutting speed to determine the instantaneous cutting power, which can be compared to the machine tool's available power to identify portions of the cut over which stalling may occur. The calculations can then be used to modify tool paths, for example, by adjusting the feed rate, to level the cutting forces and match the instantaneous power consumption to the machine's capability. In mass production applications for complex parts, this process can be used to significantly reduce cycle times.

The analysis in this section is applicable to quasi-steady processes and does not explicitly address tool wear, chip breaker, and vibration effects. The effect of tool wear on forces can be accounted for by multiplying K_n and K_f by correction factors [17]; for carbide tools, factors of 1.3 for K_n and 1.5 for K_f are typically used. The effect of chip breaking grooves in tools can be handled by modified or separately measured K_n and K_f values, or by defining an equivalent flat tool geometry based on groove parameters [20]. When deflections and vibrations cannot be neglected, a dynamic simulation, which takes into account the compliances of the tool and part, is required [18,19]. In this case the deflections are used to modify the calculated uncut chip area and thus the instantaneous force. The geometric component of the machined surface finish can also be calculated in dynamic simulations [19], although the wear component of the roughness is often more significant (Chapter 10). Finally, cutting temperatures can be calculated from simulated cutting forces using transient temperature models [23–25]. These more advanced simulations can in principle be used to investigate the effects of changes in process conditions on dynamic stability, accuracy, and tool life. They are not widely used at present, however, since methods of efficiently generating required inputs and unambiguously interpreting results have not been established.

8.3 BORING

There are two classes of boring operations as discussed in Section 4.6 and shown in Figure 8.3: single point boring, which is used as a finishing operation in cases in which bore roundness and the location of the bore axis are critical, and multipoint boring, which is used as a roughing or semi-finishing operation to remove metal more rapidly. Multi-toothed boring is used especially in automotive engine manufacture.

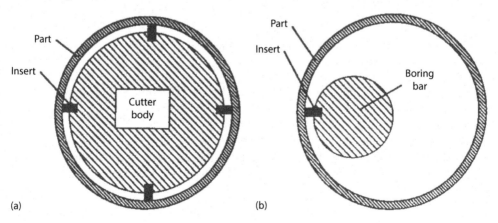

(a)   (b)

FIGURE 8.3   (a) Multipoint boring; (b) single-point boring.

Single-point boring is kinematically equivalent to turning, and the variation of static cutting forces for this case can be calculated using the equations similar to those given for turning in the previous section [16,26]. These equations are adequate for estimating power requirements and bearing loads. The deflections of the boring bar and part, which determine bore accuracy in this case, are of particular importance in single-point boring. Depending on the rigidities of the various structural elements, either the boring bar, part, or both must be modeled as compliant in this case. In Zhang and Kapoor's model [27], the part was treated as rigid while the bar was compliant. Stability limits for chatter predicted using this model were found to agree well with the results of turning experiments. Zhang and Kapoor have also published a random excitation model for calculating the surface finish, which is applicable to finish boring [28,29]. This model was found to agree well with the data from boring tests on steel specimens.

In multipoint boring, the feed per tooth f_t can be substituted for the feed per revolution f in turning or single-point boring equations to calculate the uncut chip area and effective lead angle. With this substitution, these equations can be used to estimate power requirements and bearing loads in rough boring, although radial runout between inserts has some effect on peak forces [30]. In practice, drilling simulations (as described in Section 8.5) are also used to calculate rough boring forces and power. When calculating deflections, the bar is usually treated as rigid [31–34] unless the bore being machined is interrupted, leading to asymmetric loading. The part deflection can be calculated either by using simulated cutting forces as inputs to a series of displacement solutions using a finite element model of the part [32,34], or directly by multiplying the simulated forces and the compliance matrix of the finite element model [33]. Errors due to thermal distortion have also been simulated [33–35]. The heat entering the part is calculated from the cutting forces or power (as explained in Chapter 7) or estimated to be some fraction of the total power (e.g., 10%), and used as an input for calculating the part temperature and thermal deflection distributions. Cylindricity errors calculated using Subramani's analysis [33,35] have been compared to measurements from experiments on a cast iron engine block; the agreement between simulated and measured errors was reasonable (Figure 8.4). Simulation results indicate that the thermal and mechanical error components may be comparable. In many production applications, however, thermal errors are eliminated using flood coolants. In practice, it is found that distortions due to clamping and inaccuracies in the cast bore (especially those due to core shift in sandcast bores) are as significant as cutting deflections. The clamping distortions can often be modeled, but casting errors generally cannot. It may be possible to model the effect of casting inaccuracies by running a series of analyses at the limits of the casting tolerance bands, but this does not appear to have been widely done. The machined surface finish is generally not modeled in multipoint boring since the bores are single-point finish bored, microsized, and/or honed in subsequent operations.

Simulations of turning and boring operations are useful to optimize the process parameters such as speed, feed, doc, and number of passes with respect to dimensional and surface quality specifications. Improved boring process designs developed through simulation could in principle reduce the number of boring passes, eliminate the need for some of these subsequent processes (i.e., microsizing or honing operations), and increase tool life in those which remain, but this does not appear to have been demonstrated in practice.

8.4 MILLING

Milling is more difficult to simulate than turning and boring because the kinematics are more complicated and because the cutting forces are periodic and excite harmonic vibrations. As discussed in Section 2.6, milling processes can be classified as face milling and peripheral (end) milling operations. Because practical concerns differ, simulations are often applicable only to a particular class of processes.

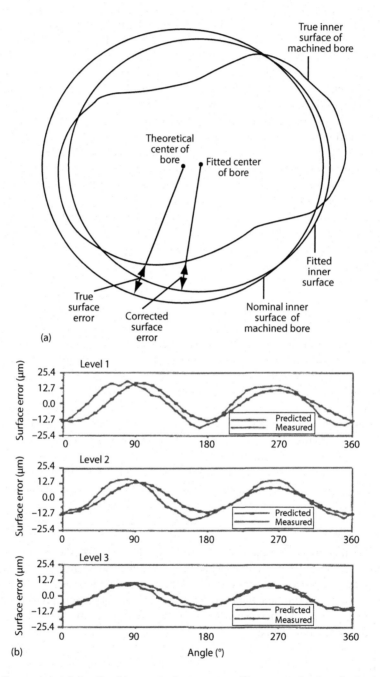

(a)

(b)

FIGURE 8.4 Comparison of simulated bore roundness errors with measured values for boring a cast iron cylinder. The feed rate and depth of cut are 0.254 mm/rev and 0.764 mm, respectively. Tools with a positive side rake angle, 5° lead angle, and 1.192 mm nose radius were used. (a) Bore center location error and (b) surface errors. (After Subramani, G. et al., *ASME J. Eng. Ind.*, 115, 15, 1993.)

8.4.1 Face Milling

In quasi-static analyses of face milling [21,36–40], the variation of the uncut chip area with the cutter rotation is approximated by a sinusoidal function as discussed in Section 2.6, and forces are estimated by multiplying the uncut chip area by measured cutting pressures. Forces must be calculated for each insert in the cutter as a function of the cutter rotation angle; total forces are obtained by summing the forces on the inserts that are cutting. In Fu's well-known model [38], the tangential, radial, and axial forces acting on insert i when the cutter rotation angle is ϕ (Figure 8.5) are given by

$$
\begin{Bmatrix} F_t(i,\phi) \\ F_r(i,\phi) \\ F_a(i,\phi) \end{Bmatrix} = K_t C_l(i,\phi) d_e(i,\phi) \begin{bmatrix} 1 + K_r \dfrac{\cos\eta_{Le}\ \tan\alpha_r}{\cos\alpha_a} \\[2mm] -\tan\alpha_r + K_r \dfrac{\cos\eta_{Le}}{\cos\alpha_a} \\[2mm] -\dfrac{\tan\alpha_a}{\cos\alpha_r} + K_r \dfrac{\sin\eta_{Le}}{\cos\alpha_a \cos\alpha_r} \end{bmatrix} \tag{8.9}
$$

where

K_t and K_r are empirical cutting pressures in the tangential and radial directions
γ_{Le} is the effective lead angle
α_a and α_r are the axial and radial rake angles of the cutter
η_{Le} is the effective chip flow angle

Methods for calculating K_t, K_r, and γ_{Le} are discussed in Section 8.6. $d_e(i, \phi)$ is the effective axial depth of cut, given by

$$
d_e(i,\phi) = d - R\sin\eta_t + R\sin\eta_t \sin\theta_i(\phi) \tag{8.10}
$$

where

η_t is the spindle tilt angle
$\theta_i(\phi)$ is the angular position of insert i at cutter rotation angle ϕ, given by

$$
\theta_i(\phi) = \phi + \theta_i(0) \tag{8.11}
$$

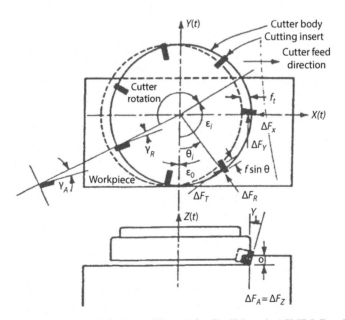

FIGURE 8.5 Cutting forces on an insert in face milling. (After Fu, H.J. et al., *ASME J. Eng. Ind.*, 106, 81, 1984.)

where $\theta_i(0)$ is the initial spacing angle of insert i. For evenly spaced inserts,

$$\theta_i(0) = \frac{i-1}{n_t} 360° \quad \text{(evenly spaced inserts)} \tag{8.12}$$

For unevenly spaced inserts, $\theta_i(0)$ values must be specified as inputs. $C_l(i, \phi)$ is the uncut chip thickness for insert i at a rotational angle of ϕ, given approximately by

$$C_l(i,\phi) \approx f_t(i)\sin\theta_i(\phi) + \varepsilon(i) - \varepsilon(i-1) \tag{8.13}$$

where f_t is the feed per tooth for insert i,

$$f_t(i) = f_r \frac{\theta(i,\phi) - \theta(i-1,\phi)}{360°} \tag{8.14}$$

and $\varepsilon(i)$ and $\varepsilon(i-1)$ are the radial runouts (or setup errors) at inserts i and $i-1$, which must be input. For large setup errors, Equation 8.13 becomes inaccurate near insert entry and exit, and a more complicated equation must be used to accurately estimate forces [38]. When the forces on individual inserts have been calculated, the total milling forces can be calculated by summing forces on all inserts cutting at a particular time:

$$\begin{Bmatrix} F_x(\phi) \\ F_y(\phi) \\ F_z(\phi) \end{Bmatrix} = \sum_1^{n_t} \delta(i,\phi) \begin{bmatrix} \cos\theta_i(\phi) & -\sin\theta_i(\phi) & 0 \\ \sin\theta_i(\phi) & -\cos\theta_i(\phi) & 0 \\ 0 & 0 & 1 \end{bmatrix} \begin{Bmatrix} F_t(i,\phi) \\ F_r(i,\phi) \\ F_a(i,\phi) \end{Bmatrix} \tag{8.15}$$

In this equation, $\delta(i, \phi)$ is a parameter equal to 1 for inserts that are cutting, and 0 to inserts that are not. $\delta(i, \phi)$ must be determined from geometric constraints, which depend on the part geometry; for complex parts a cross section of the CAD model is analyzed using a bookkeeping algorithm to assign $\delta(i, \phi)$ values.

As in turning simulation, quasi-static models of face milling are used to compute time histories of milling forces and power. Typical examples are shown in Section 8.7. Simulated time histories are useful in estimating power requirements and in designing milling cutters to reduce forces in critical areas, especially for complex parts if the simulation is integrated into a CAD/CAM system so that the part geometry file can be read directly [41]. Forces predicted in quasi-static simulations can be used as inputs to analyses in which the workpiece and/or spindle are modeled as compliant to calculate deflections [42]. Deflection calculations can be used to predict dimensional and flatness errors in rough milling. Quasi-static forces can also be used as inputs to dynamic analyses to calculate dynamic forces and stability limits [43–45]. For stable cutting conditions, static and dynamic forces do not differ significantly. Dynamic simulations are therefore most useful for predicting stability limits.

8.4.2 END MILLING

Simulation can be used in end milling to optimize the cutting conditions, tool path, and the number of passes for deep cuts or thin-walled structures. A number of quasi-static simulations of end milling forces have been reported [46–57]. Kline's well-known model [51,52] is reviewed in this section. In this model, the cutting edge of each flute of the cutter is divided into N axial elements

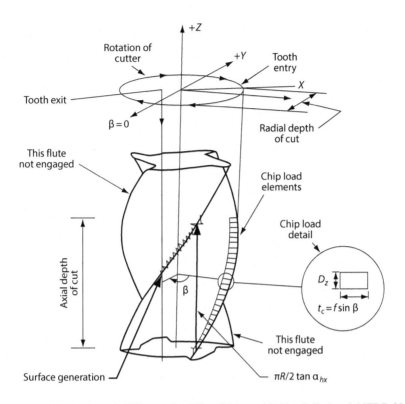

FIGURE 8.6 Uncut chip area in end milling. (After Kline, W.A., and DeVor, R.E., *Int. J. MTDR*, 23, 123, 1983.)

of height Δz (Figure 8.6). When spindle runout effects are included, the tangential and radial forces on the cutter at time t, $F_t(t)$, and $F_r(t)$ are given by

$$F_t(T) = \sum_{i=1}^{N} \left[\sum_{j=1}^{n_t} K_t t_{cij}(t) \Delta z \right]$$ (8.16)

$$F_r(t) = K_r F_t(t)$$ (8.17)

where
n_t is the number of flutes on the cutter
K_t and K_r are empirical cutting pressures equivalent to those used in face-milling models (Section 8.6)
$t_{cij}(t)$ is the uncut chip thickness for flute j at element i at time t, given by

$$t_{cij}(t) = f_t \sin \beta_{ij}(t) + dr_{ij}(t) + n_j$$ (8.18)

$\beta_{ij}(t)$ is the angle of engagement of the ith element of flute j at time t, given by

$$\beta_{ij}(t) = -\theta(t) + (k-1)\gamma + z_i \frac{\tan \alpha_{hx}}{r_{ij}(t)}$$ (8.19)

where
γ is the angular spacing between flutes
α_{hx} is the helix angle of the cutter

$d_{rij}(t)$ is the change in radius from the current rotation to the previous rotation for flute j at element i, given by

$$d_{rij}(t) = r_{ij}(t) - r_{i-1}, j(t) \tag{8.20}$$

where $r_{ij}(t)$ is the instantaneous radius for flute i at element j, given by

$$r_{ij}(t) = R + \rho \cos \left[\delta \xi - \xi + (j-1) \frac{2\pi}{n_t} \right] \tag{8.21}$$

$$\delta \xi = z_i \frac{tab\alpha_{hx}}{r_{ij}(t)} \tag{8.22}$$

where
 ξ is an angle of rotation of the cutter from a prescribed zero point (Figure 8.7)
 z_i is the axial height of the ith element above the free end of the cutter

$$z_i = \left(i - \frac{1}{2} \right) \Delta z \tag{8.23}$$

where

$$\Delta z = \frac{D_z}{N} \tag{8.24}$$

and D_z is the axial depth of cut. A somewhat similar formulation of the analytical modeling of end-milling forces is discussed by Altintas [58].

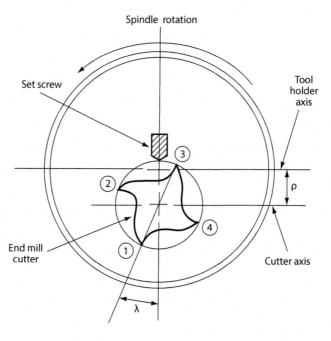

FIGURE 8.7 Top view of end-milling cutter showing the angle ξ. (After Kline, W.A., and DeVor, R.E., *Int. J. MTDR*, 23, 123, 1983.)

Provided the runout can be accurately specified, quasi-steady forces calculated using this analysis agree well with measured values (Figure 8.8). End milling simulations can be adapted to calculate quasi-static forces in slab milling by restricting the region of contact between the cutter and workpiece [41].

As discussed by Smith and Tlusty [46], quasi-static force results can be used as initial inputs to dynamic simulations to predict dynamic forces [59], cutter and part deflections [59–62], and

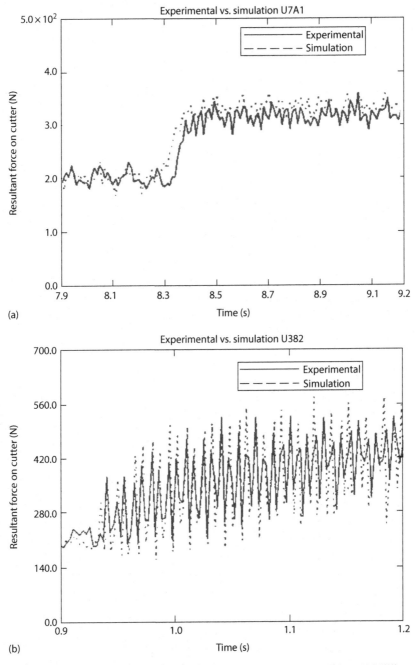

FIGURE 8.8 Comparison of simulated and measured cutting forces in end milling: (a) Milling a 1018 steel workpiece with a High Speed Steel cutter with variable feed rate; (b) Milling a 2024 aluminum workpiece with a High Speed Steel cutter with variable depth of cut. (After Kolartis, F.M., and DeVries, W.R., *ASME J. Eng. Ind.*, 113, 176, 1991.)

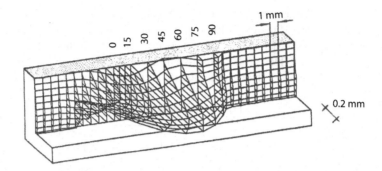

FIGURE 8.9 Simulated surface error in a pocketing cut in end milling. (After Tlusty, J., Effect of end milling deflections on accuracy, *Handbook of High Speed Machining Technology*, Chapman & Hall, New York, 1985, pp. 140–153.)

machined surface characteristics [63,64]. Simulations with tool wear modeling have also been reported [57,63]. The stiffness of the cutter is lower in end milling than in face milling and most other processes, and in fact is often comparable to the part stiffness. Therefore, both the part and cutter are treated as flexible in the most detailed analyses [62]. Also, since deflections are often comparable to the stock removal and thus can significantly influence the instantaneous uncut chip area and cutting forces, the coupling between dynamic forces and deflections must also be taken into account to accurately simulate the process under general conditions.

End milling simulations have been applied primarily in airframe and aircraft engine manufacture [60]. In these applications thin, fin-like features are often produced in multiple passes on CNC milling machines. Simulations can be used to predict part deflections (Figure 8.9) and to modify tool paths to compensate for these deflections, leading to significant reductions in the number of passes required, especially in pocketing and thin-wall applications.

8.4.3 BALL END MILLING

The ball-end milling process is similar to end milling, but cutting occurs at the curved end, rather than the straight periphery, of the cutter (Figure 8.10). A number of simulations of ball-end milling have been reported [65–69]. These analyses are similar to straight end milling simulations, although a different geometric analysis is required to calculate the uncut chip area as explained in Example 2.6. Ball-end milling is used primarily to machine three-dimensional contours on molds, dies, turbine blades, and similar parts. As in straight end milling, simulation can be used in principle to adjust tool paths to compensate for anticipated cutter deflections and to significantly reduce machining times. Simulated cutting forces in ball-end milling compare well with measured values from tests on mild steel and titanium alloy parts [65,67–68]. Simulated surface errors have also been compared with measured values [66]; the agreement in this case is not as good because the reported models do not take the runout of the cutter into account.

8.5 DRILLING

Drilling differs from operations such as turning and face milling in that the cutting speed is low, the tool geometry is more complex, and high-speed steel tools are still common. The approach used in simulating other processes, in which forces are calculated by multiplying the uncut chip area by measured-cutting pressures, can be applied to drilling [70,71] but is not well suited to studying some aspects of the process. This is due in part to the influence of the drill point geometry on process performance; for a given spindle speed and feed rate, drills with different point configurations yield the same total uncut chip area but often produce significantly different

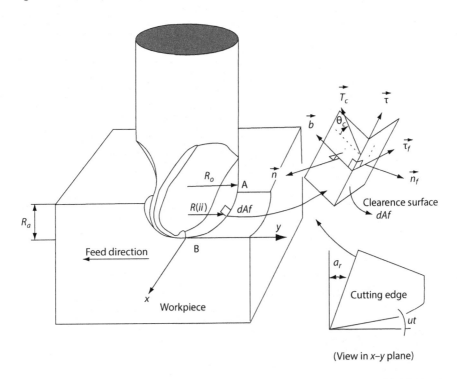

(View in x–y plane)

FIGURE 8.10 Uncut chip area in ball end milling. (After Yucesan, G., and Altintas, Y., Mechanics of Ball End Milling Process, *Manufacturing Science and Engineering*, ASME PED Vol. 64, ASME, New York, 1993, pp. 543–551.)

forces and torques. In view of this fact, drilling forces can be more effectively simulated by dividing the drill's cutting edges into small segments, which can be treated as oblique cutting edges, determining the forces on each element by oblique cutting measurements or calculations, and summing the results to calculate total loads. This approach does not require drilling tests whose results are applicable to specific point geometries, and thus permits investigation of the effect of varying the point geometry based on relatively few experiments. It also permits investigation of phenomena, which involve only part of the cutting edge, such as drilling through cross holes or inclined exit surfaces.

As shown in Figure 8.11, a typical drill has three types of cutting edges. The main cutting edges or cutting lips account for most of the torque and power consumption, a significant portion of the axial thrust force, and the radial forces caused by cross holes, inclined exits, and point asymmetry. The central chisel edge generally produces much of the axial thrust force and affects the drill's centering accuracy and buckling behavior. The chisel edge is absent on a few point geometries such as inverted points and those found on indexable drills. The marginal cutting edges form the machined surface of the hole and may contribute to the torque. For simulation, the main cutting edges are of greatest interest, although the chisel edge must be considered to accurately predict the thrust force.

A number of methods of calculating main cutting edge forces based on oblique machining models have been reported [72–81]. In these analyses the main cutting edges are divided into N segments. The radial distance from the drill axis to the center of the ith segment, r_i, is

$$r_i = r_0 + \left(i - \frac{1}{2} \right) \frac{R - r_0}{N} \tag{8.25}$$

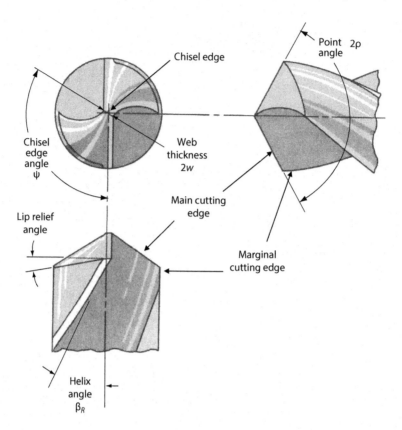

FIGURE 8.11 The point angle 2ρ, web thickness $2w$, chisel edge angle ψ, and outer helix angle β_R and cutting edges on a conventional drill.

where r_0 is the radius from the drill axis to the chisel edge. For conventional drills with web thickness $2w$ and chisel edge angle ψ (Figure 8.11),

$$r_0 = \frac{w}{\cos(\Psi - 90°)} \tag{8.26}$$

The cutting speed V_i at segment i is

$$V_i = \Omega r_i \tag{8.27}$$

The uncut chip thickness at a given radius, $t_{1j}(r_i)$, depends on the drill point angle and the number of flutes. It may vary from flute to flute at a given radius on drills with asymmetric point geometries, particularly if some flutes cut over only part of the total radius (e.g., indexable drills). For drills with symmetric point geometries,

$$t_{1j}(r_i) = \frac{f}{n_f} \sin\left[\rho(r_i)\right] \tag{8.28}$$

where $\rho(r_i)$ is the point angle at radius r_i. The rake and inclination angles of the cutting edges also vary with radius. For a symmetric drill with straight cutting edges, the rake and inclination angles at radius r_i, $\alpha(r_i)$ and $\lambda(r_i)$, are given by [82]

$$\alpha(r_i) = \tan^{-1}\left[\frac{\tan\left[\beta(r_i)\right]\cos\left[\theta(r_i)\right]}{\sin\left[\rho(r_i)\right] - \cos\left[\rho(r_i)\right]\sin\left[\theta(r_i)\right]\tan\beta_R}\right] \tag{8.29}$$

$$\lambda(r_i) = \sin^{-1}\left\{\sin\left[\rho(r_i)\right]\sin\left[\theta(r_i)\right]\right\} \tag{8.30}$$

where $\theta(r_i)$ is the web angle at radius r_i, which for drills with straight cutting edges is given by

$$\theta(r_i) = \sin^{-1}\left(\frac{w}{r_i}\right) \tag{8.31}$$

$\beta(r_i)$ is the helix angle at radius r_i, given by

$$\beta(r_i) = \tan^{-1}\left[\frac{r}{R}\tan\beta_R\right] \tag{8.32}$$

β_R is the outer helix angle, which is specified in the drill design (Figure 8.11). For drills with asymmetric points or curved cutting edges, more complicated relations must be substituted for Equations 8.28 and 8.31 [78].

Once the geometric characteristics of the various elements have been calculated, the force components F_c, F_z, and F_l (Section 6.3) acting on each element can be calculated from

$$F_{cji} = F_c(V(r_i), t_{1j}(r_i), \alpha(r_i), \lambda(r_i))\, \Delta L(r_i) \tag{8.33}$$

$$F_{zji} = F_z(V(r_i), t_{1j}(r_i), \alpha(r_i), \lambda(r_i))\, \Delta L(r_i) \tag{8.34}$$

$$F_{lji} = F_l(V(r_i), t_{1j}(r_i), \alpha(r_i), \lambda(r_i))\, \Delta L(r_i) \tag{8.35}$$

where

$$\Delta L(r_i) = \frac{R - r_0}{N\sin\left[\rho(r_i)\right]} \tag{8.36}$$

The functional forms of the relations can be determined using an oblique cutting theory [73–75], empirically from end turning test data [77–79], or through FEA calculations as discussed in the next section. Alternatively, empirical equations for the forces normal and parallel to the rake face, N and P, can be used and converted to the components F_c, F_z, and F_l using Equations 6.11 through 6.15 [80].

When the force components acting on each element have been determined, the drilling torque M and main cutting edge contribution to the thrust force F_{tmce} can be calculated from

$$M = \sum_{j=1}^{n_f}\sum_{i=1}^{N} dM_{ij} \tag{8.37}$$

$$F_{tmce} = \sum_{j=1}^{n_f}\sum_{i=1}^{N} dF_{tij} \tag{8.38}$$

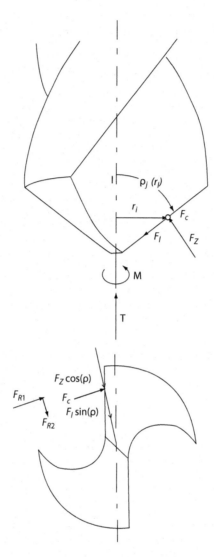

FIGURE 8.12 Relation between cutting forces at a point on the main cutting edge and drilling thrust, torque, and radial forces. (After Stephenson, D.A., and Agapiou, J.S., *Int. J. Mach. Tools Manuf.*, 32, 521, 1992.)

where (Figure 8.12)

$$dM_{ij} = r_i F_{cij} \tag{8.39}$$

$$dF_{tij} = F_{zij} \sin[\rho_j(r_i)] - F_{lij} \cos[\rho_j(r_i)] \tag{8.40}$$

If the oblique cutting relations, Equations 8.33 through 8.35, are accurate, torques calculated using Equation 8.37 generally agree well with measured values (Figure 8.13). Thrust values calculated using Equation 8.38 agree reasonably well with measured values for drills without a chisel edge and for drilling with a pilot hole with a diameter larger than the drill's web thickness; in the more general case, however, the main cutting edges may account for less than half the total thrust, so that Equation 8.38 significantly underestimates measured values (Figure 8.14). Accurate thrust

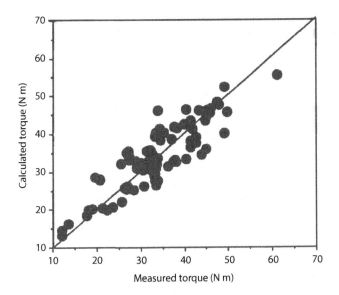

FIGURE 8.13 Comparison of simulated and measured steady-state torques for drilling cast iron with tungsten carbide drills with varying point geometries. (After Stephenson, D.A., and Agapiou, J.S., *Int. J. Mach. Tools Manuf.*, 32, 521, 1992.)

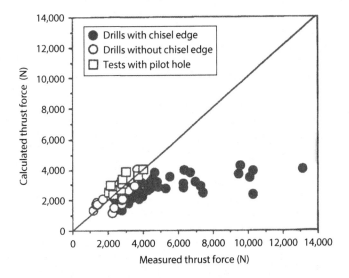

FIGURE 8.14 Comparison of simulated main cutting edge thrust contributions with the total thrust measured in cast iron drilling tests. The simulation underestimates the thrust in cases in which the chisel edge thrust is significant. (After Stephenson, D.A., and Agapiou, J.S., *Int. J. Mach. Tools Manuf.*, 32, 521, 1992.)

contributions in this case can be obtained by adding an estimate of the chisel edge contribution to the thrust, F_{tce}, to the main cutting edge contribution:

$$F_t = F_{tmce} + F_{tce} \tag{8.41}$$

A number of analyses for calculating F_{tce} have been reported [74,83–87]. In these analyses the chisel edge cutting action is usually modeled as some combination of an orthogonal cut and a wedge indentation. The results are applicable to restricted ranges of conditions and require tests to estimate empirical constants. More broadly applicable results can be obtained by fitting empirical equations to chisel edge thrust contributions measured in tests with and without pilot holes. F_{tce} is well approximated by a relation of the form

$$F_{tce} = C_{ce} f^a L_{ce}^b H^c \tag{8.42}$$

where

C_{ce}, a, b, and c are empirical constants
L_{ce} is the chisel edge length, which is usually equal to two times r_0
H is the Brinell hardness of the work material

Separate equations are required for each type of chisel edge geometry (straight, bevel, crown, etc.). Typical values for drilling gray cast iron are given in Table 8.1.

For drills with asymmetric points, the drill experiences a net radial force normal to the drill axis. The components of this force due to a particular flute, F_{R1j} and F_{R2j} (Figure 8.12), can be calculated from the relations

$$F_{R1j} = \sum_{i=1}^{N} dF_{R1ji} \tag{8.43}$$

$$F_{R2j} = \sum_{i=1}^{N} dF_{R2ji} \tag{8.44}$$

where

$$dF_{R1ji} = F_{cji} \tag{8.45}$$

$$dF_{R2ji} = F_{zji} \cos[\rho(r_i)] + F_{lji} \sin[\rho(r_i)] \tag{8.46}$$

The net radial force rotates with the drill, so that the component in a direction X fixed with respect to the workpiece, $F_X(t)$, varies sinusoidally with time:

$$F_X(t) = A \cos(\Omega t) \tag{8.47}$$

TABLE 8.1
Estimates of Parameters in Equation 8.42 for Drilling Pearlitic Gray Cast Iron with Tungsten Carbide Drills

Parameter	Drill Point Geometry	
	Conventional/Split	Helical
C_{ce}	121.5	32.1
a	0.97	1.00
b	1.10	0.42
c	0.54	0.82

Note: These values predict Th_{ce} in N when C_e is given in mm, f in mm/rev, and H in kg/mm^2. Computed values are valid for two-fluted drills; for three-fluted drills, the estimate should be multiplied by 1.5.

The relation between the amplitude A and F_{R1j} and F_{R2j} depends on the number of flutes on the drill. For two-fluted drills,

$$A = \left[\left(F_{R11} - F_{R12} \right)^2 + \left(F_{R21} - F_{R22} \right)^2 \right]^{1/2} \tag{8.48}$$

More complicated relations are required for three- and four-fluted drills [78]. Radial forces calculated using these relations agree reasonably well with measured values [78].

Drilling torque and force simulations can be used to estimate drill loads and power requirements as functions of the point geometry, feed rate, and spindle speed. Torque and force simulations are especially useful for complicated tools, which produce holes with multiple diameters, such as combined drilling/counterboring tools or tapered reamers (Figure 8.15), or when the workpiece exhibits special features, which produce unbalanced loads, such as cross holes, irregular cast holes, or inclined exit surfaces (Figure 8.16). To investigate these applications, calculations must be

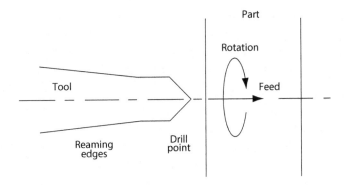

FIGURE 8.15 Drilling a tapered hole in a thin part with a combined drilling and reaming tool. (After Stephenson, D.A. and Bandyopadhyay, P., *ASME J. Eng. Mater. Technol.*, 119, 86, 1997.)

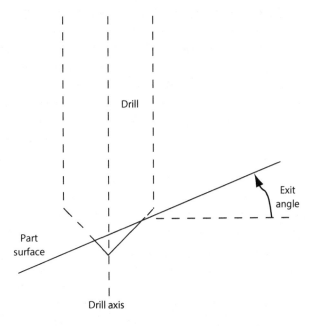

FIGURE 8.16 Drilling a through hole with an inclined exit surface.

carried out over a series of time steps as the drill advances, with the limits on the summations in Equations 8.33, 8.34, 8.39, and 8.40 being adjusted based on geometric considerations to reflect which potions of the cutting edges are actually cutting. In this case, simulated results can be used to determine bearing load and power requirements for various tool designs and feed/speed cycles, with the ultimate goal of reducing the cycle time or number of passes required. Force and torque simulations can also be used as inputs to drill temperature models (Section 7.7) to estimate the drill tip temperature as a function of hole depth.

An important issue in precision and deep-hole drilling is the bending and wandering of the drill tip, which can lead to hole straightness errors and reduced drill life. Predicting these errors accurately is difficult because the drill interacts with and is supported by the hole wall. Simulations that can provide more than a relative comparison of expected errors for different process designs can probably be developed only by combining models of the rigidity of freely vibrating drills [79,88–93] with experimental data on the effective stiffness of the supporting wall.

8.6 FORCE EQUATIONS AND BASELINE DATA

All the process simulations described in this chapter use baseline cutting force data to predict steady-state cutting forces. The collection of baseline data is often the critical step in applying simulation in practice. The accuracy of the baseline data determines the accuracy of all simulated outputs. Moreover, the effort required to generate this data determines whether the use of simulation in a particular application will reduce development time and testing requirements.

The baseline data required for each tool/workpiece material combination consists of measured cutting forces for a range of cutting conditions, which can be used to fit cutting force or pressure equations as functions of controllable inputs over the test range. The input conditions varied usually include the cutting speed, feed rate, and depth of cut, and in some cases may also include tool angles, the tool nose radius, the edge preparation, and the chip breaker geometry.

In turning, boring, and face milling, in which indexable inserts are commonly used, forces are usually characterized by two cutting pressures and an effective lead angle (Figure 8.2). The advantage of this approach is that variations in the direction of the resultant cutting force for tools with different nose radii are accounted for by the effective lead angle. This reduces baseline testing requirements since the nose radius does not have to be included as an independent variable in test plans. Some analyses, such as those described for milling in Section 8.4, use the tangential and radial cutting pressures, K_t and K_r, defined by

$$K_t = \frac{F_t}{A_c} \tag{8.49}$$

$$K_r = \frac{F_r}{A_c} \tag{8.50}$$

where

F_t and F_r are the tangential and radial cutting forces (Section 6.3)
A_c is the uncut chip area (Section 8.2)
K_t is proportional to the unit cutting power u_s used to estimate the cutting power (Sections 2.2 and 6.5)

Other analyses, such as those described for turning in Section 8.2, use the normal and frictional cutting pressures, K_n, and K_f, defined in Equations 8.1 and 8.2. K_f is the effective friction coefficient (Section 6.7).

Several equations for calculating the tangential, axial, and radial cutting forces from various combinations of cutting pressures have been reported. Equation 8.9 (Section 8.4) relates K_t and K_r to these cutting forces for face milling; these equations can be used for turning or boring by substituting $A_c = d_e(i, \phi)C_1(i, \phi)$ in the leading term. K_n and K_f are related to the tangential, axial, and radial force components by Equations 8.4 through 8.8 in Section 8.2.

γ_{Le}, the effective lead angle, determines the direction of the friction force and is similar to the chip flow angle used in more basic analyses [94–96]. For relatively deep cuts in which the ratio of the depth of cut to the tool nose radius is large (i.e., greater than 5), the effective lead angle is approximately equal to the lead angle. For other cases a numerical procedure is required to calculate γ_{Le}. Most procedures differ in assumptions about the distribution of frictional pressure along the cutting edge. Fu's method [38] assumes the frictional pressure is uniformly distributed; Subramani's method [31], which is more accurate, assumes the pressure distribution is proportional to the chip thickness distribution. A detailed description of this method is given by Kuhl [15]. A curve fit to results computed using Subramani's method yields the following relation for the effective lead angle [7]:

$$\tan \gamma_{Le} = 0.5053 \ \tan \gamma_L + 1.0473 \frac{f}{r_n} + 0.4654 \frac{r_n}{d} \tag{8.51}$$

where
  r_n is the tool nose radius
  f is the feed per revolution
  d is the depth of cut, all given in mm

This relation is valid for $-5° < \gamma_L < 45°$, $0.5 < d < 5.0$ mm, $0.1 < f < 1.0$ mm, $0.4 < r_n < 4.4$ mm, and $r_n > f$. For conditions outside this range, the exact procedure should be used.

The cutting pressures, K_n, K_f, K_t, and K_r, vary with the uncut chip thickness, tool angles, and cutting speed. The uncut chip thickness has the strongest effect; generally pressures increase at low chip thicknesses due to the increased importance of flank friction. K_n and K_f are usually calculated from empirical equations of the general form [7]

$$K_n = C_n t_{avg}^{a_n} V^{b_n} \left(1 - \sin \alpha\right)^{c_n} \tag{8.52}$$

$$K_f = C_f t_{avg}^{a_f} V^{b_f} \left(1 - \sin \alpha\right)^{c_f} \tag{8.53}$$

where
  C_n, a_n, b_n, c_n, C_f, a_f, b_f, and c_f are empirical constants
  t_{avg} is the average uncut chip thickness computed assuming an equivalent rectangular contact
    (Figure 8.2)

Values of these parameters for five tool/work material combinations are given in Table 8.2. In the literature, the rake angle dependence is often neglected, and the cutting speed dependence is sometimes neglected, so that K_n and K_f are given by Equations 8.52 and 8.53 with $c_n = c_f = 0$ or with $b_n = b_f = c_n = c_f = 0$. In these cases separate equations are required for each cutter geometry. A number of similar cutting pressure equations have been reported [4,16,21,46,97–100].

TABLE 8.2

Parameter Estimates for Five Common Tool/Work Material Combinations

Work Material	Gray Cast Iron	Nodular Iron	390 Al	356 Al	1018
Steel Hardness (BHN)	170	270	110	73	163
Tool Material	Silicon Nitride	Coated WC	PCD	PCD	Coated WC
C_n	1227	1730	470	356	2119
a_n	−0.338	−0.336	−0.243	−0.475	−0.231
b_n	−0.121	−0.089	0.060	−0.040	−0.080
c_n	1.190	1.183	1.065		1.149
C_f	0.405	0.304	0.303	0.453	0.453
a_f	−0.363	−0.258	−0.306	−0.368	−0.095
b_f	0	0	0.025	−0.085	−0.090
c_f	−2.126	−1.392	−1.810		−0.233

Source: Stephenson, D.A., and Bandyopadhyay, P., *ASME J. Eng. Mater. Technol.*, 119, 86, 1997.

When the depth of cut is large compared to the tool nose radius, t_{avg} is approximately equal to $f \cos \gamma_L$, where γ_L is the lead angle (see Equation 2.3). In other cases, t_{avg} must be calculated using numerical methods used for the effective lead angle. A curve fit to numerical results yields the equation [7]

$$t_{avg} = 0.5334 \left(\frac{r_n}{f} \right)^{0.0921} \left(\frac{r_n}{d} \right)^{-0.3827} \left(f \cos \gamma_L \right)^{1.0317} \tag{8.54}$$

As for the effective lead angle, this relation is valid for $5° < \gamma_L < 45°$, $0.5 < d < 5.0$ mm, $0.1 < f < 1.0$ mm, $0.4 < r_n < 4.4$ mm, and $r_n > f$.

For processes such as drilling and end milling, which are often carried out using solid tools, the cutting edges of the tool can be divided into small segments, which can be approximated as straight-edged oblique cutting tools (Figure 1.1). In these cases cutting pressures are sometimes not used; rather, empirical equations for specific cutting force components such as Equations 8.33 through 8.35 are used [65,78,79]. Often these equations are fit to data from end turning tests on thin-walled tubes (Figure 8.17). For a given tool/workpiece material combination, forces in end turning depend on the cutting speed, feed rate, cutting edge width, and tool rake and inclination angles. Polynomial equations or exponential equations of the type described in Section 6.4 can be used; the exponential equations generally yield more accurate results, especially when extrapolating to cutting conditions outside the test range. If equations are fit to the normal and frictional cutting forces, N and P, they can be related to cutting pressures. In this case cutting pressure data from turning or face milling tests can be used for materials combinations for which no end turning data is available [7] as discussed as follows.

Cutting forces calculated from baseline cutting pressure or force equations and the effective lead angle agree reasonably well with experimental data [7]. Typical errors for the tangential, axial, and radial forces in turning are 5%, 10%, and 10%, respectively. The accuracy is best for roughing cuts and decreases as the feed and depth of cut are reduced; this is especially true for the radial component of the force. The accuracy of baseline cutting force or pressure equations carries over into process simulations in which they are used to compute forces, especially those components that

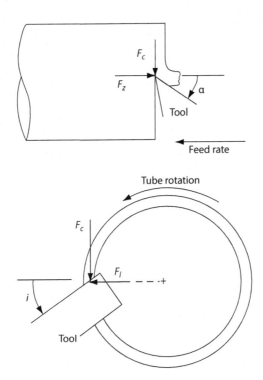

FIGURE 8.17   End turning of a tube to simulate oblique cutting.

determine cutting power and depend primarily on the tangential forces. Typical examples for drilling and milling processes are shown in Figures 8.18 and 8.19 [7,101].

For new materials, collecting baseline data involves measuring forces in a series of tests in which the cutting speed, feed rate, depth of cut, and (in many cases) the tool geometry are varied. Generally full or half fraction factorial experiments using three or more levels of feed and at least two levels of the cutting speed and depth of cut are required; this results in test plans of between 25 and 50 tests per insert type or tool geometry. On CNC machines, testing effort can be reduced by writing programs to perform a series of short cuts at various combinations of the cutting speed and feed rate; the entire test plan can then be completed quickly by running the program for each depth of cut and tool geometry. All test conditions should be repeated at least once; if measured forces vary by more than roughly ten percent between repeated runs, it is likely that there is an error in the program or that the tool is wearing rapidly or chipping due to excessive cutting speeds or feeds.

Gathering baseline data for several material combinations can involve a significant testing effort and, as noted at the beginning of this section, can reduce or even eliminate the benefit of applying simulations. At present, baseline testing requirements make simulation most suitable for mass production applications involving relatively few material combinations, and less attractive for batch production operations involving many materials. Some research on reducing testing requirements has been reported [7,77,97,98]; this has focused on process-independent testing methods which produce data useful for simulating more than one process. This work suggests that turning, boring, and face milling processes can be simulated from a common set of turning or milling data [7,97], or even from orthogonal cutting data similar to that used for drilling simulation [98].

Although this section has focused on experimental measurement of cutting pressures, recently finite element calculations have replaced the baseline experimental in some cases. In this approach

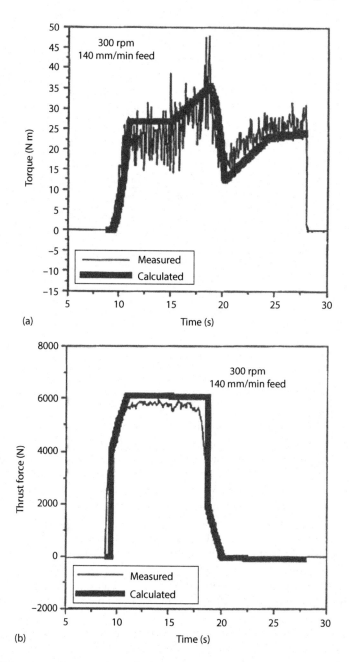

FIGURE 8.18 Comparison of measured and simulated forces and torque for the combined drilling and reaming process shown in Figure 8.15. The spindle speed is 300 rpm and the feed rate is 140 mm/min. (a) Torque and (b) thrust force. (After Stephenson, D.A., and Bandyopadhyay, P., *ASME J. Eng. Mater. Technol.*, 119, 86, 1997.)

finite element calculations are carried out for a range of input parameters (as in a series of experiments) and force equations are curve-fit to the results. This is a particularly attractive approach for engineers who have an accurate constitutive model for FEA for the material in question and limited time or access to equipment to perform experiments. There is a limited research literature on the use of finite element calculations in process simulation [53], but no study of the comparative accuracy of FEA-generated vs. experimentally measured cutting pressures seems to have been reported.

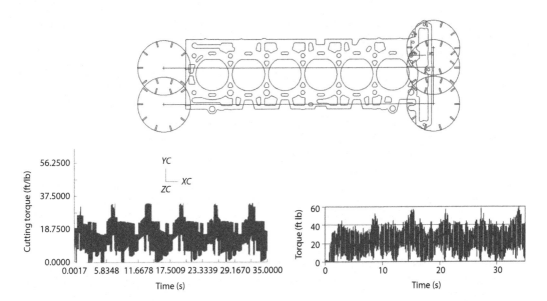

FIGURE 8.19 Comparison of torques calculated using a face milling simulation and measured for cutting the joint face of an aluminum engine head. (From Stephenson, D.A., Machining Process Analysis at General Motors Powertrain, *Proceedings of 2004 Third Wave AdvantEdge User's Conference*, Gaithersburg, MD, March 2004, Paper 2.) The calculated torque history plot is shown on the left, below the part geometry and cutter path; the measured torque history is shown on the right. The average calculated and measured torques are 31.6 and 33.2 N m, respectively.

8.7 PROCESS SIMULATION APPLICATION EXAMPLES

In mass production applications, machining process simulation can be used both when purchasing new equipment and when processing or reprocessing parts for existing equipment. For new equipment, simulation can be used to define machine horsepower and bearing requirements, and to create tool paths which make full use of available power during roughing cuts and which do not contain excessive noncutting time. For complex parts, processing steps can be moved between operations to balance work content. Accurate process simulations can be used to evaluate a wider range of potential machine types, process sequences, and tool paths than would be possible through physical testing, and to select the option which minimizes investment. In processing a new part on existing equipment, simulation is used to tailor the process to the machinery; in this case, the machine's horsepower and transient response characteristics are fixed constraints rather than options to be selected. It is also frequently necessary to analyze the existing process to accommodate a part change (which often involves adding machined content) or volume increase without additional investment. Reasonable objectives of this type can often be met by increasing feeds during roughing cuts, increasing cutting speeds if new tool grades are available, using rapid traverse through noncutting portions of tool paths, etc. Simulation in current processes helps to increase productivity, part quality, and reduce tooling and production costs.

Machining process simulation is also used to compute time histories for input to finite element analyses as discussed below. Force and torque calculations are also useful in designing special drills and milling cutters. Finally, simulations can be of benefit for visualization and training. Since production machining equipment has heavy guarding and often high volume coolant delivery, waste time during tool motion is often not directly observable, but can be detected through graphic simulation. Engineers with relatively little experience can also use simulations to learn CNC code, the effects of various tool angles, and similar material without cutting test parts.

Machining simulation is not currently useful for directly solving tool life and chip control problems. There are numerous models which purport to predict tool life as a function of cutting speed, feed, etc. for various materials. In most cases, these models assume that tool life ends when a defined level of flank wear occurs. In practice, however, tool life depends on part tolerances. A tool will be removed from service when a part's dimension, surface finish, or edge condition is out of specification; this will occur at various levels of flank wear depending on the part geometry and part tolerances. It might be possible to tailor a tool life model to a narrow group of operations in which the end of tool life is consistently known; this would require extensive experimentation, however, and it is not clear what additional information the simulation would provide. Simulation can, however, provide qualitative insight into tool life issues, since measures which reduce cutting forces generally increase tool life. Similarly, chip control problems are best investigated by increasing or interrupting feeds, changing tool angles, and testing inserts with various chip breaking patterns; the physical complexity of the chipbreaking process, and the fact that such problems often arise from variability in the incoming material, make it difficult to add value through simulation.

Figures 8.20 through 8.24 show typical applications of process simulations [8,101–103]. In Figure 8.20, two potential methods of milling and engine block front face are investigated. The results show that the larger cutter gives better flatness error at a constant cycle time since the cutter path

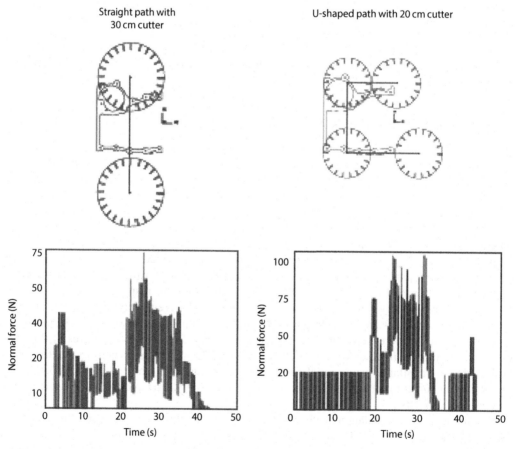

FIGURE 8.20 Analysis of two options for milling an engine block front face. (From Stephenson, D.A., Casting and Machining Process Analysis at GM Powertrain, SAE Technical Paper 2002-01-0622, 2002.) (a) Large cutter with straight path and (b) small cutter with U-shaped path. For a fixed cycled time, case (a) yields 30% better flatness accuracy than case (b), since the shorter tool path permits use of a lower feed rate, resulting in lower forces normal to the cut surface.

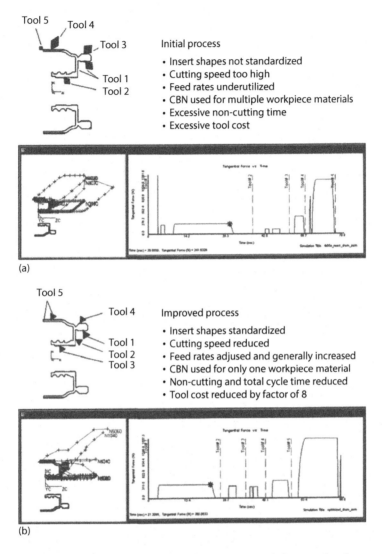

Tool 5 Tool 4

Tool 3

Initial process

Tool 1
Tool 2

- Insert shapes not standardized
- Cutting speed too high
- Feed rates underutilized
- CBN used for multiple workpiece materials
- Excessive non-cutting time
- Excessive tool cost

(a)

Tool 5

Tool 4

Improved process

Tool 1
Tool 2
Tool 3

- Insert shapes standardized
- Cutting speed reduced
- Feed rates adjused and generally increased
- CBN used for only one workpiece material
- Non-cutting and total cycle time reduced
- Tool cost reduced by factor of 8

(b)

FIGURE 8.21 Analysis to optimize a turning operation on a transmission reaction drum assembly. (From Stephenson, D.A., Machining Process Analysis at General Motors Powertrain, *Proceedings of 2004 Third Wave AdvantEdge User's Conference*, Gaithersburg, MD, March 2004, Paper 2.) (a) Initial process. (b) Improved process.

is shorter and a lower feed rate can be used. This result is obvious once the analysis is performed but was not obvious beforehand. Routine analytical evaluations of this kind can be done more quickly than cutting tests, and provide more physical insight into the process.

Figure 8.21 shows an analysis of a multi-tool turning operation on a transmission reverse drum assembly, which consists of a mild steel housing containing a welded hardened bearing race. The operation was exhibiting excessive tool cost and downtime. Investigation of the initial operation (Figure 8.21a) showed that CBN inserts were being used on all materials, not just the hardened race, and that speeds and feeds were incorrect. In addition, insert shapes were not standardized, increasing tool room inventory. Analysis was used to design a new operation (Figure 8.21b) with standardized inserts in which CBN tooling was used only for the hardened race. Speed was reduced, and feed rates were adjusted (generally increased) to level forces and reduce cycle time. The new operation reduced tool costs by a factory of 8 with a slightly lower cycle time. Downtime was also reduced since less frequent tool changes were required.

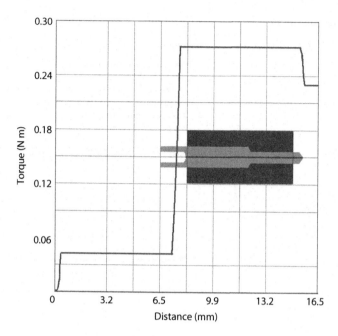

FIGURE 8.22 Torque history plot for a step drill designed to replace two single-diameter drills. Since a tool change is eliminated, the step drill can run at a lower feed rate without increasing the cycle time. (From Stephenson, D.A., Casting and Machining Process Analysis at GM Powertrain, SAE Technical Paper 2002-01-0622, 2002.)

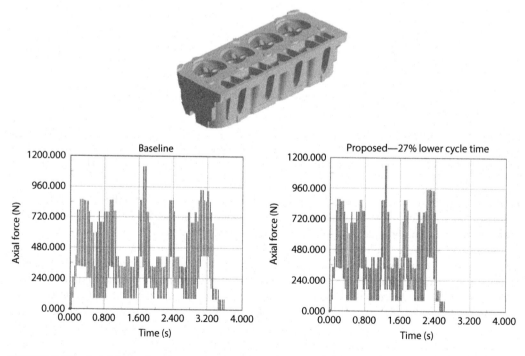

FIGURE 8.23 Axial force plots for baseline and proposed tool paths and milling cutters for an engine head cubing operation. (From White, D.R., Machining Applications at GM Powertrain, *Proceedings of 2002 Third Wave AdvantEdge User's Conference*, Atlanta, GA, April 2002, Paper 6.)

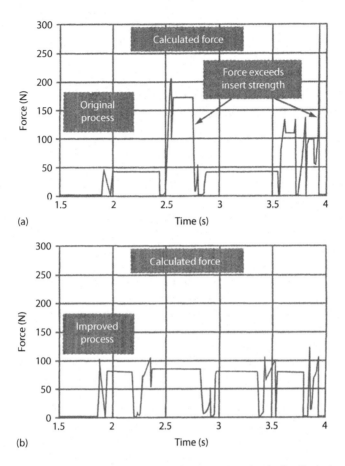

FIGURE 8.24 Analysis to optimize a turning operation on an engine hydraulic lash adjuster. (a) Initial process. (b) Improved process. (From Simulated Action, *Cutting Tool Eng.*, 56:3, 69, March 2004.)

Figure 8.22 shows a torque history plot for a step drill designed to replace two single-diameter drills. The single diameter drills broke frequently due to material buildup and chip clogging. Since a tool change was eliminated, the step drill could be run at a lower feed rate to eliminate these problems. Analysis was used to determine the maximum feed at which target thrust and torque values would not be exceeded. When implemented the step drill eliminated three hours of downtime due to tool breakage per week.

Figure 8.23 shows an engine head casting cubing application. Simulation was used to reprocess this operation to accommodate a volume increase without adding an additional machining center. Cutters were commonly removed because parts were out of dimension, due in part to axial deflection of the milling spindle as tool wear increased. A number of new cutters with different insert and pocket geometries were investigated analytically, and a new cutter design which yielded the same axial force at a higher feed rate (and lower cycle time) was identified. Axial forces also increased more slowly with insert wear for this design, leading to a reduction in tooling costs as well as cycle time in production.

Figure 8.24 shows force simulations from a boring operation on an engine head hydraulic lash adjuster. This component was made of 1013 steel and measured 27 mm long by 12 mm in diameter. The bore depth was 24 mm, and the diameter varied from 8 to 9.5 mm. An 8 mm boring bar with a triangular coated carbide insert was initially used, but exhibited an unacceptable failure rate. Simulation showed that the initial tool path resulted in load spikes on the insert.

Feed rates were adjusted to level the forces and eliminate the load spikes while maintaining cycle time. Peak loads on the insert were reduced by a factor of two, greatly reducing tool breakage.

8.8 FINITE ELEMENT ANALYSIS FOR CLAMPING, FIXTURING, AND WORKPIECE DISTORTION APPLICATIONS

Structural finite element analysis is used to estimate workpiece distortions due to both clamping and machining forces.

Clamping distortions are more easily calculated. The inputs required are the clamping and locating points and the clamping forces. Finite element models of both the part and fixture structure are generally required; attempts to replace the fixture with equivalent boundary conditions, such as springs or displacement constraints, save computing time but generally yield less accurate results. The part finite element model used for structural design is usually adequate, although some mesh refinement near the clamping points may be needed. If a fixture model is not available, one must be created for the support and clamping elements in contact with the part. The interfaces between the part and fixture should be modeled using contact elements with friction for optimum accuracy. A static analysis in which the clamping loads are applied at the clamping points yields a distortion prediction. The major unknowns are usually the friction coefficients at the contact points; these can be determined experimentally if the solution is sensitive to these variables. In overconstrained clamping schemes (i.e., for four-point locating schemes on planes), locator dimensional variations are also significant. This type of analysis is most often used for thin-walled, compliant parts; in this case it is rarely necessary to model additional elements of the machine tool structure.

When modeling distortions induced by machining forces, more elements of the system must generally be considered. As shown in Figure 8.25, machining forces act between the tool and part, and may cause deflections of two broad structural assemblies: the tool, toolholder, and machine tool structure on one side, and the part and fixture on the other. To compute deflections, cutting force histories must be estimated, often using the kinematic simulations described in the preceding sections, and applied to structural finite element models of both assemblies [104]. In some operations, however, the compliance of one element of the system (tool/toolholder/machine structure or part/fixture) may be much larger than the other, so that the other element can be treated as rigid. As noted below, sequential or iterative analyses may be required in applications in which machining significantly changes the structural compliance of the part or in which cutting force and deflections are coupled.

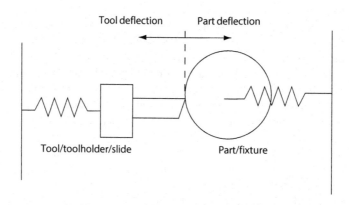

FIGURE 8.25 Components of dimensional errors in machining due to deflections.

In most turning and multipoint boring simulations, the tool can be assumed to be rigid and the part compliant [18,32–34]. In this case the compliance of the part is usually determined from a finite element model. Deflections are then calculated from

$$U = K^{-1}F \tag{8.55}$$

where
    U is the nodal displacement vector
    K^{-1} is the compliance matrix of the model
    F is the nodal force vector

Simulated cutting forces can be used to construct the nodal force vector at any given time. Two approaches, illustrated in Figure 8.26, can then be used to calculate deflections. In the first, the simulated force vector is used as an input to a finite element analysis [32,34]. As the tool moves with respect to the workpiece, new force vectors are calculated at regular intervals and used as inputs to further analyses. In the second approach, the compliance matrix of the part is extracted from an initial analysis and stored [33]; the simulation program can then calculate deflections from Equation 8.55 at each time step without the need for further finite element analyses. The second approach requires much less computing time, since only one finite element analysis is needed, and should be used whenever the compliance of the part does not change significantly as a result of cutting. The first approach is required, however, in cases in which machining removes enough material to reduce the part's stiffness. In this case, the finite element model must be updated in addition to the force vector at each time step.

In face milling, the workpiece and fixture are often more rigid than the tooling and machine structure. In this case the deflections are calculated by multiplying simulated forces by the compliance of the tooling. The compliance of the tooling is usually determined by the axial or radial compliance of the spindle, which can be measured or calculated from a finite element model. The variation of the spindle compliance with spindle speed must be taken into account in variable speed applications. In face milling thin-walled aluminum parts, the tooling is more rigid than the part and fixture, and analyses similar to those described for turning and boring are required.

In end milling, either the tool or workpiece may be treated as compliant. In the first case the tool is usually modeled as a simple spring-mass-dashpot system characterized by effective mass, stiffness, and damping coefficients estimated from experimental data. These models are similar to

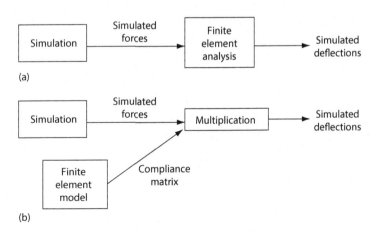

FIGURE 8.26 (a) Simulated forces used as inputs to a series of finite element analyses to calculate deflection; (b) multiplication of simulated forces with a stored compliance matrix to calculate deflections.

models for face milling with a radially compliant spindle. In the case in which the tool is treated as rigid, the compliance of the part is usually estimated from a finite element model. Deflections are calculated from finite element analyses using simulated forces as inputs. Some work has been reported on modeling both the tooling and workpiece as compliant and accounting for the dependence of cutting forces on deflections [61]. In this case, the compliances are modeled as in simpler analyses and iterative methods are used to compute forces.

Verifying the accuracy of simulation models for deflections is difficult because deflections are difficult to measure accurately on the machine in the clamped condition, and because the stiffness and damping characteristics of the system often must be determined experimentally and thus may effectively "calibrate" the model. Most models have only been compared with post-process data on part quality, often for very simple parts. This method assumes that all deviation of the machined surface from specified dimensions is due to deflection, although in reality machine motion errors, residual stresses, and other factors may also contribute significantly to these deviations. Bearing these considerations in mind, simulated deflections in boring and end milling generally agree well with post-process part variations (Figures 8.4 and 8.27).

Simulation models for deflections are most useful for designing tooling and selecting process conditions to maximize part accuracy. In the case of compliant parts, they can also be used to ensure that fixtures are supporting the part adequately over areas where high cutting forces are expected. Deflection models are not as easy to implement as static force simulations, since additional inputs to characterize the stiffness of the machining system are required. The results are also not as general, and in fact are valid only for a specific application.

Simulations which predict deflections can also be used to analyze the dynamic response of the machining system under given conditions, and specifically to determine if the system will be stable dynamically. This subject is discussed in Chapter 12 and in review papers [46,60].

8.9   FINITE ELEMENT APPLICATION EXAMPLES

Few production applications of structural finite element analyses have been reported in the literature [101,102,105,106]. In most of these, only clamping distortions have been considered, and analysis results have been used primarily for qualitative comparisons, that is, for ranking a group of design options or fixturing concepts from best to worst. Such qualitative rankings are quite reliable provided no gross modeling errors were made. It is more difficult to use the results for quantitative assessments, for example, to determine if a given operation will produce a part within tolerance [105]. Finite element studies generally underestimate distortions when compared to measurements, partly because finite element models are generally stiffer than real parts, and partly because elements of the system which are compliant in reality are assumed to be rigid in the analysis. As a general guideline, predicted distortions or stresses should be multiplied by a safety factor when making quantitative assessments; a safety factor between 1.5 and 2.0 is typically adequate for routine analyses for which proper element types and boundary conditions are known.

Figures 8.28 through 8.33 show examples of finite element analyses applied to machining fixturing problems. Figure 8.28 shows a clamping study for a torque converter housing. The original fixture design called for clamping on a large bore in the center of the part. This bore has little supporting structure, however, and clamping distortions in this area can lead to large flatness deviations due to the asymmetric geometry of the part. FEA calculations prior to prototype build showed that much lower flatness deviations would result from clamping on a smaller side bore with more supporting structure.

Figure 8.29 shows a sensitivity study for a channel plate milling fixture, which had exhibited periodic flatness variations. The results showed that it is critical to ensure that the center clamp was properly seated and engaged during each cycle. Analysis can be of great benefit in eliminating potential error sources when troubleshooting complex problems; in this example, the error in

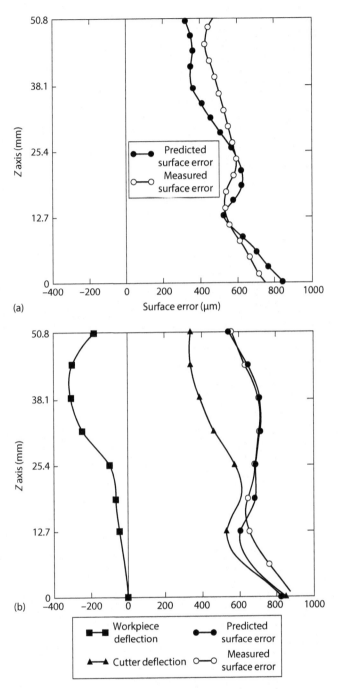

FIGURE 8.27 Surface errors in end milling of aluminum with a high-speed steel cutter: (a) rigid workpiece; (b) compliant workpiece. (After Kline, W.A. et al., *Int. J. MTDR*, 22, 7, 1982.)

question occurred once or twice per day, so that considerable engineering attention and continuous gaging would have been required to solve the problem without analysis.

Figure 8.30 shows an analytical study of the effect of varying clamping pressure on the milled deck face flatness for an engine block. The station in question had both front and rear clamps actuated by independent hydraulic cylinders. Since the front of the block is structurally weaker than

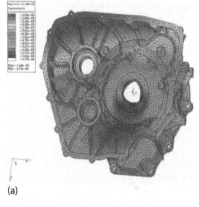

- 3-jaw chuck on large pilot bore
- Baseline analysis performed to estimate Z-datum movement during
- Baseline-datum deflection

(a)

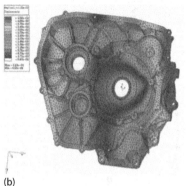

- New chuck location in the small diameter
- Z-datum deflection reduced by factor of 3

(b)

FIGURE 8.28 Finite element clamping study for a transmission converter housing. (a) Clamping scenario 1 and (b) Clamping scenario 2. (From Stephenson, D.A., Machining Process Analysis at General Motors Powertrain, *Proceedings of 2004 Third Wave AdvantEdge User's Conference*, Gaithersburg, MD, March 2004, Paper 2.)

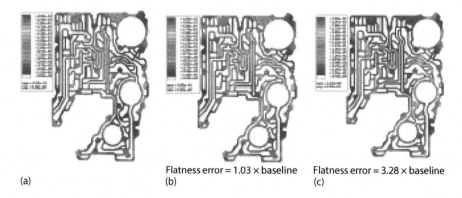

(a)

Flatness error = 1.03 × baseline
(b)

Flatness error = 3.28 × baseline
(c)

FIGURE 8.29 Analysis to determine the sensitivity of part flatness to locating and fixturing errors when finish milling a channel plate. (a) Baseline, (b) Center locators missing, and (c) Top clamp and center locators missing. (From Stephenson, D.A., Casting and Machining Process Analysis at GM Powertrain, SAE Technical Paper 2002-01-0622, 2002.)

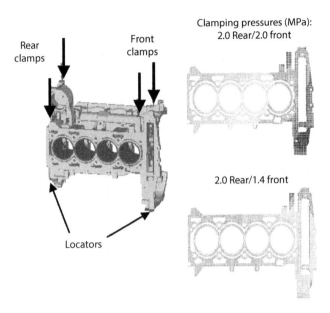

FIGURE 8.30 Analytical study of varying clamping pressures to control flatness in engine block deck milling. (From Stephenson, D.A., Casting and Machining Process Analysis at GM Powertrain, SAE Technical Paper 2002-01-0622, 2002.)

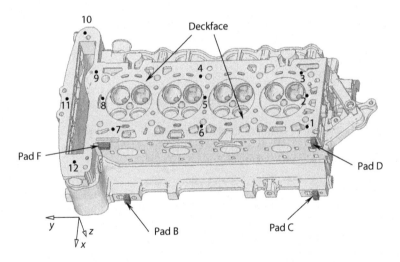

FIGURE 8.31 Twelve points on an engine head deckface to estimate distortion (supporting pads also shown). (From Xie, J.Q. et al., *SME J. Manuf. Proc.*, 5, 170, 2003.)

the rear, reduced distortion (and improved flatness) can be achieved by using unbalanced clamping forces, with the highest force applied by the rear clamps. For the best combination of pressures, the flatness was reduced to less than 10% of the baseline level for equal loading.

Figures 8.31 and 8.32 show an analytical study of the deckface flatness for a cylinder head due to (1) pressing in powder metal valve seats and guides and (2) part clamping at the milling station. The objective of the analysis was to determine whether a single-pass milling operation could be used to finish the deckface, or if both a semi-finish and finish pass were required. The overall deckface flatness prediction of 38 μm due to seat and guide interference fit was within

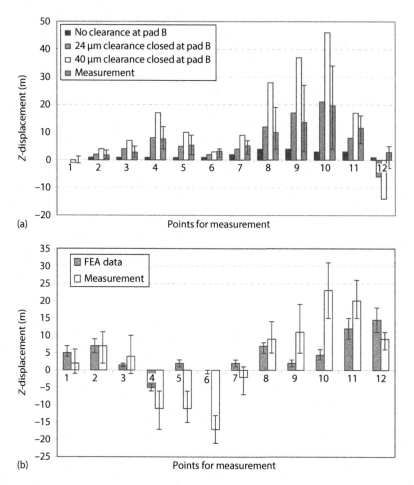

FIGURE 8.32 (a) Comparison of deckface displacements due to seat and guide pressing computed using FEA with measurements. (b) Comparison of deckface displacements due to clamping with different initial clearances. (From Xie, J.Q. et al., *SME J. Manuf. Proc.*, 5, 170, 2003.)

5%–10% of the measured values [103]. The deckface distortion due to clamping forces using several cases of locating clearance in the 4–2–1 locating scheme was evaluated to be in the range of 10–50 μm depending on the relative height error among the four support pads. The static deckface distortion due to the cutting forces during milling was analyzed and found to be about 10–15 μm. These results were used to decide if a single-pass milling process was feasible for this operation, and if the milling operation should be carried out before or after seat and guide pressing by comparing the required deckface flatness part quality to the computed 38 μm pressing distortion.

Figure 8.33 shows a study of locating schemes for an engine block. For the operation in question the part is to be located on its rear face and clamped from the top; the cylinder head banks are to be milled, and the shape and location of the locating pads, which minimized the bank face flatness deviation, must be determined. Finite element calculations show that using a round central pad and two square pads greatly reduces the flatness deviation when compared to various combinations of square pads, and that the best diameter for the round pad is 150 mm. Qualitative analytical comparisons of this type done prior to prototype build can improve accuracy while reducing testing costs and lead time.

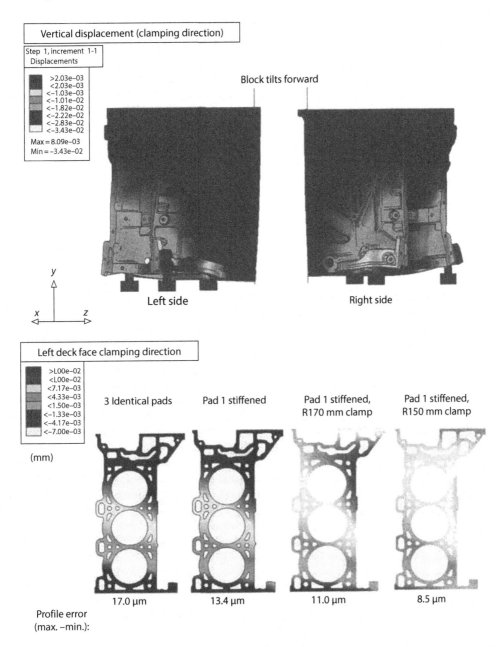

FIGURE 8.33 Analytical study to determine the best geometry for locating pads in an engine block deck milling operation. (From Stephenson, D.A., Machining Process Analysis at General Motors Powertrain, *Proceedings of 2004 Third Wave AdvantEdge User's Conference*, Gaithersburg, MD, March 2004, Paper 2.)

8.10   EXAMPLES

Example 8.1 Calculate the forces acting on an end-milling cutter with diameter of 25 mm and two straight flutes performing an up-milling process. The axial and radial depths of cuts are 10 and 4 mm, respectively. The feed per tooth (chip load) is 0.1 mm, and the spindle speed is 8000 rpm.

Solution: The end-milling operation is illustrated in Figure 2.11 and a cross section of up-milling with a two-fluted cutter is shown in Figure 8.34a. The uncut chip thickness changes from zero to

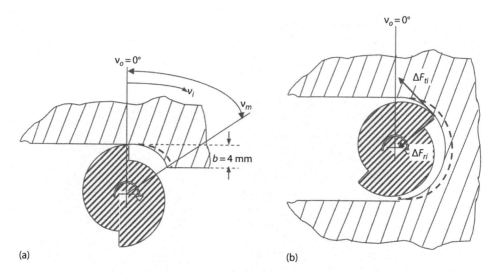

FIGURE 8.34 (a) Cross section of up-milling with a two-fluted cutter. (b) End-mill fully immersed in the workpiece.

maximum as discussed in Section 2.6 and illustrated in Figures 2.12 and 2.14. The resultant force on the cutter can be decomposed into three forces, feed (horizontal) force F_f, normal force F_n (in the direction of the width of cut), and axial force F_a as illustrated in Figures 6.8 and 8.5. The resultant force on each cutting edge can be decomposed into three forces, the tangential force F_t (tangent to the tool perimeter in the direction of the peripheral cutting velocity), the radial force F_r (along the radius of cutter), and the axial force F_a (normal to the other two forces along the axis of the tool). The tangential force is the main cutting force. The other two forces are estimated from the tangential force as $F_r = 0.3\ F_t$ and $F_a = F_r \tan \lambda$, where λ is the helix of the cutter. If the helix angle is 45°, the axial force is equal to the radial force. The undeformed chip thickness is given by Equation 2.21 or $a_i = f \sin \nu_i$. The tangential force as a function of the angle of rotation relative to the workpiece is

$$F_{ti} = u_s d_a f \sin \nu_i \tag{8.56}$$

where u_s is the specific cutting energy of the workpiece material (see Chapters 2 and 6). If the cutter is performing up-milling (Figure 8.34a), as in the case of peripheral milling in Figure 2.14, the tooth engages over the arc of the width of cut. The engagement angle is determined by Equations 2.21 through 2.24. So $\nu_o < \nu_I < \nu_m$, where in the above case the $\nu_o = 0°$ and $\nu_m = 47°$.

The tangential and radial forces acting on the cutting edges can be decomposed into components in the feed direction, F_x, and normal to feed direction, F_y, in the workpiece coordinate system through the engagement angle:

$$F_x = \sum_{i=1}^{n_t} \left(F_{ti} \cos \nu_i + F_{ri} \sin \nu_i \right)$$

$$= u_s d_a f \sum_{i=1}^{n_t} \left(\sin \nu_i \cos \nu_i + 0.3 \sin^2 \nu_i \right) \tag{8.57}$$

$$F_y = \sum_{i=1}^{n_t} \left(F_{ti} \sin v_i - F_{ri} \cos v_i \right)$$

$$= u_s d_a f \sum_{i=1}^{n_t} \left(\sin^2 v_i - 0.3 \sin v_i \cos v_i \right) \tag{8.58}$$

where

n_t is the number of teeth engaged in the workpiece at an instant
d_a is the axial depth of cut (equivalent to d in Chapter 2)
b is the radial depth of cut in Figure 2.11 for end-milling

The maximum forces are obtained when the maximum number of teeth are engaged in the workpiece. One rotation of the cutter is divided into K intervals that define the angular integration. The forces are periodic as shown in Figure 8.35. The resultant force (F_{tot}) is also given in the graph. Every 360° the teeth pass through the same location with respect to the engagement angle v. An abrupt drop of these forces will occur if all cutting edge are disengaged from the cut because the radial width of the cut is small or the number of edges around the cutter is small (i.e., 2-flute end-mill). A small computer program can be generated to calculate the F_x and F_y force components using Equations 8.57 and 8.58. These forces can be calculated at small increments of the angle of cutter rotation v_i. The angle of engagement for 4 mm width of cut is $0° < v_I < 47°$, and the forces are plotted in Figure 8.35. The forces are maximum at the exit of cut because the chip thickness starts with zero and gradually increases to maximum in up-milling (Figure 8.34).

Example 8.2 Calculate the forces acting on a slot milling operation for the same tool and conditions used as in Example 8.1. Estimate the cutting torque and power required by the spindle.

Solution: The analysis of a slot milling operation (with the end mill fully immersed in the workpiece as shown in Figure 8.34b) is similar to that in partial immersion in previous example.

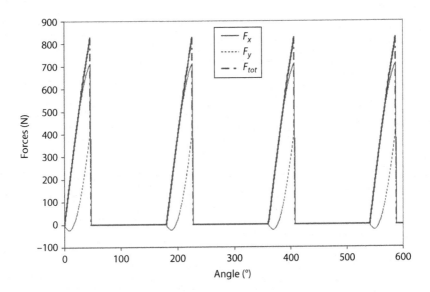

FIGURE 8.35 Forces on the cutter in Example 8.1.

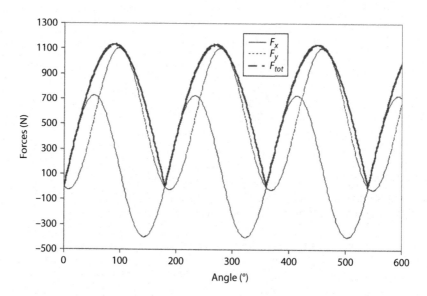

FIGURE 8.36 Forces on the teeth in Example 8.2.

Since the cutter diameter is fully engaged in the workpiece, $0° < \nu_I < 180°$ and the forces are given by Equations 8.57 and 8.58. The forces acting on the teeth are estimated and shown in Figure 8.36. The forces are sinusoidal because the chip thickness starts with zero and gradually increases to a maximum at 90° rotation and gradually reduces to zero at 180°. The total force is zero only at the zero chip load location occurring at $\nu_i = 0°$.

The instantaneous cutting torque and power on the spindle are

$$M = \frac{D}{2} \sum_{i=1}^{n_t} F_{ti} \quad \text{and} \quad P = V \sum_{i=1}^{n_t} F_{ti} \tag{8.59}$$

where
F_t is the tangential force
D the cutter diameter

The torque and power have the same trend as the tangential force given by Equation 8.57 as illustrated in Figure 8.37. The average spindle power obtained from the instantaneous graph or Equation 2.29 is 7.2 kW.

Example 8.3 Calculate the forces acting on an end-milling cutter with diameter of 25 mm and two helical smooth edges performing an up-milling process. The axial and radial depths of cuts are 10 and 9 mm, respectively. The feed per tooth (chip load) is 0.1 mm, and the spindle speed is 8000 rpm.

Solution: A typical end-milling cutter with helical flutes is shown in Figure 2.11. The helix provides a gradual engagement and disengagement of the cutting edge with the radial depth of cut contact. An unwound cutting edge has a straight line contact along the axial depth of cut inclined

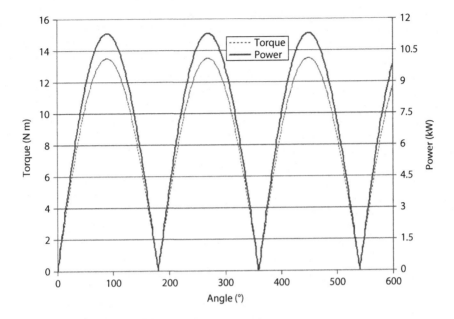

FIGURE 8.37 Variation of torque and power in Example 8.2.

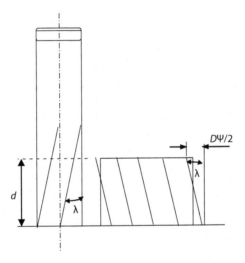

FIGURE 8.38 Definition of the angle ψ in Example 8.3.

at the helix angle λ. Therefore, any point (element j) on the cutting edge above the free end of the cutter will lag by an angle ψ (as illustrated in Figure 8.38):

$$\psi_j = \frac{2z_j \tan \lambda}{D} \tag{8.60}$$

where
    D is the diameter of the cutter
    z_j is the distance of element j along the cutting edge from its free end

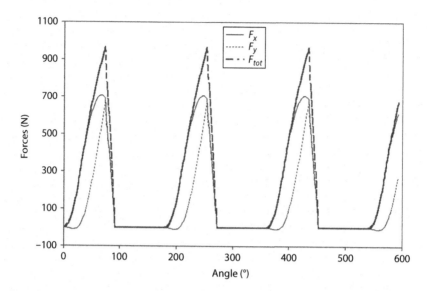

FIGURE 8.39 Variation of forces in Example 8.3.

The cutting edge is divided into N elements (N is the number of axial integration steps). Therefore, the immersion angle of any point (element) along the cutting edge at distance z above the free end of the cutter will have an immersion angle of $(\nu - \psi)$. The chip thickness will be different for each point along the cutting edge according to the immersion angle (as illustrated in Figure 8.6), compared to a constant chip thickness with straight flute end mills discussed in Example 8.1. The analysis of each point along the cutting edge is similar to that discussed in Example 8.1 with an immersion angle of $(\nu - \psi)$.

The cutting edge of each flute of the cutter is divided into N axial elements of height Δz (Figure 8.6). The tangential and radial forces on the cutter at time t, $F_t(t)$, and $F_r(t)$, are given by Equations 8.16 and 8.17, respectively (see Section 8.4). The F_x and F_y forces are calculated by Equations 8.57 and 5.58 by summing the elemental forces along each of the flutes:

$$F_x = \sum_{j=1}^{N}\left[\sum_{i=1}^{n_t} F_{ti}\cos\left(\nu_i - \psi_i\right) - F_{ri}\sin\left(\nu_i - \psi_i\right)\right] \tag{8.61}$$

where $(\nu_i - \psi_i)$ is the angle of the cutter rotation for each element in relation to the zero reference point ($\nu_i = 0°$ in Figure 8.34). This means that while the cutter rotates at an angle ν_i, each element j along the cutting edge will rotate at an angle $(\nu_i - \psi_i)$. The variation of the forces for a helical end mill is shown in Figure 8.39. The ratio F_x/F_y is smaller for a helical tool compared to a straight fluted one. The forces increase at a slower rate with a helical tool even though the resultant force is higher for a helical tool because the length of contact is longer as illustrated by the contact angle in the horizontal axis.

8.11 PROBLEMS

Problem 8.1 Calculate the instantaneous forces acting on the end-milling cutter in Example 8.1 when performing a down-milling process. Explain the significant difference in the cutting force response between down-milling and the up-milling case discussed in Example 8.1.

Problem 8.2 Calculate the instantaneous forces acting on the end-milling cutter in Example 8.1 assuming the end mill has four flutes. Compare the cutting force response of the four-flute cutter with that of two-flute cutter in Example 8.1.

REFERENCES

1. P. Khurana, D. Sormaz, and R. Khetan, Integration of CAD, CAPP and process modeling using XML technologies, *Proceedings of ASME MSEC Conference*, October 8–11, 2006, Ypsilanti, MI, Paper MSEC2006-21066.
2. Anon., *Third Wave Production Module*, Third Wave Systems, Inc., Minneapolis, MN, nd.
3. Anon., *MilSim*, Manufacturing Laboratories, Inc., Las Vegas, NV, nd.
4. K. F. Ehman, S. G. Kapoor, R. E. DeVor, and I. Lazoglu, Machining process modeling: A review, *ASME J. Manuf. Sci. Eng.* **119** (1997) 655–663.
5. E. J. A. Armarego, J. A. Arsecularatne, P. Mathew, and S. Verezub, A CIRP survey on available predictive performance models of machining operations, *Proceedings of 52nd Annual CIRP Assembly*, San Sebastian, Spain, August 2002, pp. 61–76.
6. P. J. Arrazola, T. Ozel, D. Umbrello, M. Davies, and I. S. Jawahir, Recent advances in modelling of metal machining processes, *CIRP Ann.* **62** (2013) 695–718.
7. D. A. Stephenson and P. Bandyopadhyay, Process independent force characterization for machining simulation, *ASME J. Eng. Mater. Technol.* **119** (1997) 86–94.
8. Delphi Corporation, Simulated action, *Cutting Tool Eng.* **56**:3 (March 2004) 69–73.
9. A. P. S. Dogra, S. G. Kapoor, and R. E. DeVor, Mechanistic model for tapping process with emphasis on process faults and hole geometry, *ASME J. Manuf. Sci. Eng.* **124** (2001) 18–25.
10. T. Cao and J. W. Sutherland, Investigation of thread tapping load characteristics through mechanistics modeling and experimentation, *Int. J. Mach. Tools Manuf.* **42** (2002) 1527–1538.
11. A. C. Araujo, J. L. Silveira, M. B. G. Jun, S. G. Kapoor, and R. E. DeVor, A model for thread milling cutting forces, *Int. J. Mach. Tools Manuf.* **46** (2006) 2057–2065.
12. R. B. Schroeter, C. M. Bastos, and J. M. C. Filho, Simulation of the main cutting force in Crankshaft turn broaching, *Int. J. Mach. Tools Manuf.* **47** (2007) 1884–1892.
13. W. R. DeVries and M. S. Evans, Computer graphics simulation of metal cutting, *CIRP Ann.* **33** (1984) 81–88.
14. C. J. Li and W. R. DeVries, The effect of shear plane length models on stability analysis in machining simulation, *Proc. NAMRC* **16** (1988) 195–201.
15. M. J. Kuhl, The prediction of cutting forces and surface accuracy for the turning process, MS thesis, Mechanical Engineering, University of Illinois, Urbana, IL, 1987.
16. J. W. Sutherland, G. Subramani, M. J. Kuhl, R. E. DeVor, and S. G. Kapoor, An investigation into the effect of tool and cut geometry on cutting force system prediction models, *Proc. NAMRC* **16**, 1988, 264–272.
17. D. A. Gustafson, The effect of tool geometry and wear on the cutting force system in turning, MS thesis, Mechanical Engineering, University of Illinois, Urbana, IL, 1990.
18. W. J. Endres, J. W. Sutherland, R. E. DeVor, and S. G. Kapoor, A dynamic model of the cutting force system in the turning process, *Monitoring and Control for Manufacturing Processes,* ASME PED Vol. 44, ASME, New York, 1990, pp. 193–212.
19. G. M. Zhang, S. Yerramareddy, S. M. Lee, and S. C.-Y. Lu, Simulation of intermittent turning processes, *ASME J. Eng. Ind.* **113** (1991) 458–466.
20. G. Parakkal, R. Zhu, S. G. Kapoor, and R. E. Devor, Modeling of turning process cutting forces for grooved tools, *Int. J. Mach. Tools Manuf.* **42** (2002) 179–191.
21. F. M. Gu, S. G. Kapoor, R. E. DeVor, and P. Bandyopadhyay, An approach to on-line cutter runout estimation in face milling, *Trans. NAMRI/SME* **19** (1991) 240–247.
22. N. H. Cook, *Manufacturing Analysis*, Addison-Wesley, Reading, MA, 1966, p. 73.
23. D. A. Stephenson, T. C. Jen, and A. S. Lavine, Cutting tool temperatures in contour turning: Transient analysis and experimental verification, *ASME J. Manuf. Sci. Eng.* **119** (1997) 494–501.
24. T.-C. Jen and A. Anagonye, An improved transient model of tool temperatures in metal cutting, *ASME J. Manuf. Sci. Eng.* **123** (2001) 30–37.
25. T.-C. Jen, S. Eapen, and G. Gutierrez, Nonlinear numerical analysis in transient cutting tool temperatures, *ASME J. Manuf. Sci. Eng.* **125** (2003) 48–56.
26. F. Atabey, I. Lazoglu, and Y. Altintas, Mechanics of boring processes—Part I, *Int. J. Mach. Tools Manuf.* **43** (2003) 463–476.

27. G. M. Zhang and S. G. Kapoor, Dynamic modeling and analysis of the boring machining system, *ASME J. Eng. Ind.* **109** (1987) 219–226.

28. G. M. Zhang and S. G. Kapoor, Dynamic generation of machined surfaces, Part 1: Description of a random excitation system, *ASME J. Eng. Ind.* **113** (1991) 137–144.

29. G. M. Zhang and S. G. Kapoor, Dynamic generation of machined surfaces, Part 2: Construction of surface topography, *ASME J. Eng. Ind.* **113** (1991) 145–153.

30. F. Atabey, I. Lazoglu, and Y. Altintas, Mechanics of boring processes—Part II—Multi-insert boring heads, *Int. J. Mach. Tools Manuf.* **43** (2003) 477–484.

31. G. Subramani, R. Suvada, S. G. Kapoor, and R. E. DeVor, A model for the prediction of force system for cylinder boring process, *Proc. NAMRC* **15** (1987) 439–446.

32. W. E. Sneed, Chip removal simulation to predict part error and vibration, *Proceedings of ASME Computers in Engineering Conference*, New York, NY, 1987, Vol. 2, pp. 447–455.

33. G. Subramani, S. G. Kapoor, and R. E. DeVor, A model for the prediction of bore cylindricity during machining, *ASME J. Eng. Ind.* **115** (1993) 15–22.

34. N. N. Kakade and J. G. Chow, Finite element analysis of engine bore distortions during boring operation, *ASME J. Eng. Ind.* **115** (1993) 379–384.

35. G. Subramani, M. C. Whitmore, S. G. Kapoor, and R. E. DeVor, Temperature distribution in a hollow cylindrical workpiece during machining: Theoretical model and experimental results, *ASME J. Eng. Ind.* **113** (1991) 373–380.

36. P. C. Subbarao, R. E. DeVor, S. G. Kapoor, and H. J. Fu, Analysis of cutting forces in face milling of high silicon casting aluminum alloys, *Proc. NAMRC* **10** (1982) 289–296.

37. Z. Ruzhong and K. K. Wang, Modeling of cutting force pulsations in face milling, *CIRP Ann.* **32** (1983) 21–26.

38. H. J. Fu, R. E. DeVor, and S. G. Kapoor, A mechanistic model for the prediction of the force system in face milling operation, *ASME J. Eng. Ind.* **106** (1984) 81–88.

39. F. M. Gu, S. G. Kapoor, R. E. DeVor, and P. Bandyopadhyay, A cutting force model for face milling with a step cutter, *Trans. NAMRI/SME* **20** (1992) 361–367.

40. A. D. Spence and Y. Altintas, A solid modeller based milling process simulation and planning system, *ASME J. Eng. Ind.* **116** (1994) 61–69.

41. S. Jain and D. C. H. Yang, A systematic analysis of the milling operation, *Computer-Aided Design and Manufacture of Cutting and Forming Tools*, ASME PED Vol. 40, ASME, New York, 1989, pp. 55–63.

42. H. Schulz, Optimization of precision machining by simulation of the cutting process, *CIRP Ann.* **42** (1993) 55–58.

43. S. J. Lee and S. G. Kapoor, Cutting process dynamics simulation for machine tool structure design, *ASME J. Eng. Ind.* **108** (1986) 68–74.

44. H. S. Kim and K. F. Ehmann, A cutting force model for face milling operations, *Int. J. Mach. Tools Manuf.* **33** (1993) 651–673.

45. M. A. Elbestawi, F. Ismail, and K. M. Yuen, Surface topography characterization in finish milling, *Int. J. Mach. Tools Manuf.* **34** (1994) 245–255.

46. S. Smith and J. Tlusty, An overview of modeling and simulation of the milling process, *ASME J. Eng. Ind.* **113** (1991) 169–175.

47. F. Koenigsberger and A. J. P. Sabberwal, An investigation into the cutting force pulsations during the milling process, *Int. J. MTDR* **1** (1961) 15.

48. J. Tlusty and P. MacNeil, Dynamics of the cutting forces in end milling, *CIRP Ann.* **24** (1975) 248.

49. R. E. DeVor, W. A. Kline, and W. J. Zdeblick, A mechanistic model for the force system in end milling with application to machining airframe structures, *Proc. NAMRC* **8** (1980) 297–303.

50. W. A. Kline, R. E. DeVor, and J. R. Lindberg, The prediction of cutting forces in end milling with application to cornering cuts, *Int. J. MTDR* **22** (1982) 7–22.

51. W. A. Kline and R. E. DeVor, The effect of runout on cutting geometry and forces in end milling, *Int. J. MTDR* **23** (1983) 123–140.

52. W. A. Kline, R. E. DeVor, and I. Shareef, The prediction of surface accuracy in end milling, *ASME J. Eng. Ind.* **104** (1982) 272–278.

53. T. Ozel and T. Altan, Process simulation using finite element method—Prediction of cutting forces, tool stresses and temperatures in high-speed flat end milling, *Int. J. Mach. Tools Manuf.* **40** (2000) 713–738.

54. J. H. Ko, W.-S. Yun, D.-W. Cho, and K. F. Ehmann, Development of a virtual machining system, Part 1: Approximation of the size effect for cutting force prediction, *Int. J. Mach. Tools Manuf.* **42** (2002) 1595–1605.

55. W.-S. Yun, J. H. Ko, D.-W. Cho, and K. F. Ehmann, Development of a virtual machining system, Part 2: Prediction and analysis of machined surface error, *Int. J. Mach. Tools Manuf.* **42** (2002) 1607–1615.

56. W.-S. Yun, J. H. Ko, H. U. Lee, D.-W. Cho, and K. F. Ehmann, Development of a virtual machining system, Part 3: Cutting process simulation in transient cuts, *Int. J. Mach. Tools Manuf.* **42** (2002) 1617–1626.

57. K. D. Bouzakis, R. Paraskevopoulou, G. Katirtzoglou, S. Makrimallakis, E. Bouzakis, and K. Efstathiou, Predictive model of tool wear in milling with coated tools integrated into a CAM system, *CIRP Ann.* **62** (2013) 71–74.

58. Y. Altintas, *Manufacturing Automation—Metal Cutting Mechanisms, Machine Tool Vibrations, and CNC Design*, Cambridge University Press, Cambridge, U.K., 2000.

59. F. M. Kolartis and W. R. DeVries, A mechanistic dynamic model of end milling for process controller simulation, *ASME J. Eng. Ind.* **113** (1991) 176–183.

60. J. Tlusty, Effect of end milling deflections on accuracy, *Handbook of High Speed Machining Technology*, Chapman and Hall, New York, 1985, pp. 140–153.

61. D. Montgomery and Y. Altintas, Dynamic peripheral milling of flexible structures, *ASME J. Eng. Ind.* **114** (1992) 137–145.

62. D. Montgomery and Y. Altintas, Mechanism of cutting force and surface generation in dynamic milling, *ASME J. Eng. Ind.* **113** (1991) 160–168.

63. F. Ismail, M. A. Elbestawi, R. Du, and K. Urbasik, Generation of milled surface including tool dynamics and wear, *ASME J. Eng. Ind.* **115** (1993) 245–252.

64. M. A. Elbestawi, F. Ismail, and K. M. Yuen, Surface topography characterization in finish milling, *Int. J. Mach. Tools Manuf.* **34** (1994) 245–255.

65. M. Y. Yang and H. D. Park, The prediction of cutting force in ball-end milling, *Int. J. Mach. Tools Manuf.* **31** (1991) 45–54.

66. E. M. Lim, H. Y. Feng, and C. H. Menq, The prediction of dimensional errors for machining sculptured surfaces using ball-end milling, *Manufacturing Science and Engineering*, ASME PED Vol. 64, ASME, New York, 1993, pp. 149–156.

67. G. Yucesan and Y. Altintas, Mechanics of ball end milling process, *Manufacturing Science and Engineering*, ASME PED Vol. 64, ASME, New York, 1993, pp. 543–551.

68. C. G. Sim and M. Y. Yang, The prediction of the cutting force in ball end milling with a flexible cutter, *Int. J. Mach. Tools Manuf.* **33** (1993) 267–284.

69. S. J. Wou, Y. C. Shin, and H. El-Mounayri, Ball end milling mechanistic model based on a voxel-based geometric representation and a ray casting technique, *J. Manuf. Proc.* **15** (2013) 338–347.

70. V. Chandrasekharan, S. G. Kapoor, and R. E. DeVor, A mechanistic approach to predicting the cutting forces in drilling: With application to fiber-reinforced composite materials, *ASME J. Eng. Ind.* **117** (1995) 559–570.

71. V. Chandrasekharan, S. G. Kapoor, and R. E. DeVor, A mechanistic model to predict the cutting force system for arbitrary drill point geometry, *ASME J. Manuf. Sci. Eng.* **120** (1998) 563–570.

72. E. J. A. Armarego and C. Y. Chen, Drilling with flat rake face and conventional twist drills—1. Theoretical investigation, *Int. J. MTDR* **12** (1972) 17–35.

73. E. J. A. Armarego and S. Wiriyacosol, Thrust and torque prediction in drilling from a cutting mechanics approach, *CIRP Ann.* **28** (1979) 87–91.

74. A. R. Watson, Drilling model for cutting lip and chisel edge and comparison of experimental and predicted results, Parts I–IV, *Int. J. MTDR* **25** (1985) 347–404.

75. D. A. Stephenson and S. M. Wu, Computer models for the mechanics of three-dimensional cutting processes Part II: Results for oblique end turning and drilling, *ASME J. Eng. Ind.* **110** (1988) 38–43.

76. K. J. Fuh, Computer-aided design and manufacture of multifacet drills, PhD thesis, Mechanical Engineering, University of Wisconsin, Madison, WI, 1987, Ch. 6.

77. D. A. Stephenson, Material characterization for metal cutting force modeling, *ASME J. Eng. Mater. Technol.* **111** (1989) 210–219.

78. D. A. Stephenson and J. S. Agapiou, Calculation of main cutting edge forces and torque for drills with arbitrary point geometries, *Int. J. Mach. Tools Manuf.* **32** (1992) 521–538.

79. D. M. Rincon, Coupled force and vibration modeling of drills with complex cross sectional geometries, PhD thesis, Mechanical Engineering, University of Michigan, Ann Arbor, MI, 1994, Ch. 3.

80. H. T. Huang, C. I. Weng, and C. K. Chen, Prediction of thrust and torque for multifacet drills (MFD), *ASME J. Eng. Ind.* **116** (1994) 1–7.

81. K. Sambhav, P. Tandon, and S. G. Dhande, Force modeling for generic profile of drills, *ASME J. Manuf. Sci. Eng.* **136** (2014) 041019-1.

82. D. F. Galloway, Some experiments on the influence of various factors on drill performance, *ASME Trans.* **79** (1957) 191–231.

83. R. A. Williams, A study of the basic mechanics of the chisel edge of twist drills, *Int. J. Prod. Res.* **8** (1970) 325–343.

84. R. A. Williams, A study of the drilling process, *ASME J. Eng. Ind.* **96** (1974) 1207–1215.

85. S. Bera and A. Bhattacharyya, On the determination of the torque and thrust during drilling of ductile materials, *Proceedings of Eight International MTDR Conference*, Birmingham, U.K., 1967, pp. 879–892.

86. C. A. Mauch and L. K. Lauderbaugh, Modeling the drilling process—An analytical model to predict thrust force and torque, *Sensors and Controls in Manufacturing*, ASME PED Vol. 20, Dallas, TX, 1990.

87. A. Paul, S. G. Kapoor, and R. E. DeVor, Chisel edge and cutting lip shape optimization for improved twist drill point design, *Int. J. Mach. Tools Manuf.* **45** (2005) 421–431.

88. E. Magrab and D. E. Glisin, Buckling loads and natural frequencies of twist drills, *ASME J. Eng. Ind.* **196** (1984) 196–204.

89. O. Tekinalp and A. G. Ulsoy, Modeling and finite element analysis of drill bit vibrations, *ASME J. Vib. Acoust Stress Reliab. Des.* **111** (1989) 148–155.

90. O. Tekinalp and A. G. Ulsoy, Effects of geometric and process parameters on drill transverse vibrations, *ASME J. Eng. Ind.* **112** (1990) 189–194.

91. M. McColl and R. Ledbetter, CAD applied to twist drills, *Proc. Inst. Mech. Eng.* **207B** (1993) 251–256.

92. J. Z. Yan, V. Jaganathan, and R. Du, A new dynamic model for drilling and reaming processes, *Int. J. Mach. Tools Manuf.* **42** (2002) 299–311.

93. P. V. Bayly, M. T. Lamar, and S. G. Calvert, Low-frequency regenerative vibration and the formation of lobed holes in drilling, *ASME J. Manuf. Sci. Eng.* **124** (2002) 275–285.

94. L. V. Colwell, Predicting the angle of chip flow for single point cutting tools, *ASME Trans.* **76** (1954) 199–204.

95. H. T. Young, P. Mathew, and P. L. B. Oxley, Allowing for nose radius effects in predicting the chip flow direction and cutting forces in bar turning, *Proc. Inst. Mech. Eng.* **C201** (1987) 213–226.

96. J. Wang, Development of a chip flow model for turning operations, *Int. J. Mach. Tools Manuf.* **41** (2001) 1265–1274.

97. D. A. Stephenson and J. W. Matthews, Cutting forces when turning and milling cast iron with silicon nitride tools, *Trans. NAMRI/SME* **21** (1993) 223–230.

98. E. Budak and Y. Altintas, Prediction of milling force coefficients from orthogonal cutting data, *Manufacturing Science and Engineering*, ASME PED Vol. 64, ASME, New York, 1993, pp. 453–459.

99. S. Jayaram, S. G. Kapoor, and R. E. DeVor, Estimation of the specific cutting pressures for mechanistic cutting force models, *Int. J. Mach. Tools Manuf.* **41** (2001) 265–281.

100. D. Germaina, G. Fromentinb, G. Poulachonb, and S. Bissey-Bretona, From large-scale to micromachining: A review of force prediction models, *J. Manuf. Proc.* **15** (2013) 389–401.

101. D. A. Stephenson, Machining process analysis at general motors powertrain, *Proceedings 2004 Third Wave AdvantEdge User's Conference*, Gaithersburg, MD, March 2004, Paper 2.

102. D. A. Stephenson, Casting and machining process analysis at GM powertrain, SAE Technical Paper 2002-01-0622, 2002.

103. D. R. White, Machining applications at GM powertrain, *Proceedings of 2002 Third Wave AdvantEdge User's Conference*, Atlanta, GA, April 2002, Paper 6.

104. D. T.-Y. Huang and J.-J. Lee, On obtaining machine tool stiffness by CAE techniques, *Int. J. Mach. Tools Manuf.* **41** (2001) 1141–1163.

105. J. S. Agapiou, E. Steinhilper, F. Gu, and P. Bandyopadhyay, Modeling machining errors on a transfer line to predict quality, *SME J. Manuf. Proc.* **5** (2003) 1–12.

106. J. Q. Xie, J. S. Agapiou, D. A. Stephenson, and P. M. Hilber, Machining quality analysis of an engine cylinder head using finite element methods, *SME J. Manuf. Proc.* **5** (2003) 170–184.

9 Tool Wear and Tool Life

9.1 INTRODUCTION

Tool wear and failure mechanisms are of great practical interest because they affect machining costs and quality. Tools that wear slowly have comparatively long and predictable service lives, resulting in reduced production costs and more consistent dimensions and surface finish. Tools that fail rapidly and unpredictably increase costs and scrap rates. For these reasons, tool life is the most common criterion used to rate cutting tool performance and the machinability of materials.

An understanding of the tool life requires an understanding of the ways in which tools fail. Broadly, tool failure may result from wear, plastic deformation, or fracture. Tool wear may be classified by the region of the tool affected or by the physical mechanisms that produce it. The dominant type of wear in either case depends largely on the tool material. The common types of wear and wear measurement methods are discussed in Sections 9.2 and 9.3; tool wear mechanisms and material aspects of wear are discussed in Sections 9.4 and 9.5. Tools deform plastically or fracture when they are unable to support the loads generated during chip formation. The physical mechanisms leading to plastic deformation are similar to those resulting in wear and are also discussed in Sections 9.4 and 9.5. Common fracture mechanisms are discussed in Section 9.9. The general discussions in these sections are applicable primarily to continuous turning operations. Special wear and fracture concerns in drilling and in milling are discussed in Sections 9.10 and 9.11.

A primary goal of metal cutting research has been to develop methods of predicting tool life from a consideration of tool failure mechanisms. One of the objectives in this chapter is to provide a qualitative understanding of the basic physics of tool wear so that the gradual wear of a tool can be described quantitatively. Tool life testing and methods for predicting tool life are discussed in Sections 9.6 through 9.8. In considering these topics, it is important to bear in mind the distinction between tool wear and tool life. It is often possible to predict tool wear rates based on test data or physical considerations, but this does not translate into a prediction of tool life in a general sense, since tool life depends strongly on part requirements. In practice, tools are removed from service when they no longer produce an acceptable part. This may occur when the part's dimensional accuracy, form accuracy, or surface finish are out of tolerance, when an unacceptable burr or other edge condition is produced, or when there is an unacceptable probability of gross failure due to an increase in cutting forces or power. Tools used under the same conditions in different operations may have quite different usable lives depending on critical tolerances or requirements. Because of this fact, methods of predicting tool life are useful primarily for comparative purposes, for example, in ranking expected levels of tool life for different work materials, tool materials, or cutting conditions; realistically, they cannot be expected to yield an accurate estimate of tool life in a given application unless prior application data for similar parts is available.

Because of the practical importance of tool life, the subject may be discussed from a variety of viewpoints in a general survey of machining theory and practice. The purpose of this chapter is to clarify the physical mechanisms, which lead to tool failure and to identify general strategies for reducing failure rates and increasing tool life. The discussion is meant to complement related discussions on tool life concerns when selecting tool materials (Chapter 4) and when rating machinability (Chapter 11).

9.2 TYPES OF TOOL WEAR

The principle types of tool wear, classified according to the regions of the tool they affect, are shown in Figure 9.1 [1–12]. Wear occurs on both the rake and relief faces of the tool due to several mechanisms as discussed in Sections 9.3. Wear on the relief face is called *flank wear* (Figure 9.1a) and results in the formation of a wear land (Figures 9.2 and 9.3). Rubbing of the wear land against the machined surface damages the surface and produces frictional heating and flank forces, which increase deflections and reduce dimensional accuracy. As discussed in the next section, flank wear most commonly results from abrasion of the cutting edge. The extent of flank wear is characterized by the average or maximum land width (Figure 9.2). The flank wear rate changes with time as shown in Figure 9.4. After an initial wearing in period corresponding to the initial rounding of the cutting edge, flank wear increases slowly at a steady rate until a critical land width is reached, after which wear accelerates and becomes severe. Flank wear progress can be monitored in production by examining the tool or (more commonly) by tracking the change in size of the tool or machined part. The flank wear land is generally of uniform width, with thicker sections occurring near the ends. Flank wear can be minimized by increasing the abrasion resistance of the tool material and by the use of hard coatings on the tool.

Rake face or *crater wear* (Figure 9.1b) produces a wear crater on the tool face (Figures 9.2 and 9.5). Moderate crater wear usually does not limit tool life; in fact, crater formation increases the effective rake angle of the tool and thus may reduce cutting forces. However, excessive crater wear weakens the cutting edge and can lead to deformation or fracture of the tool, and should be avoided because it shortens tool life and makes resharpening the tool difficult. The extent of crater wear is characterized by the crater depth KT. Crater wear also varies with time in a manner similar to flank wear. As discussed in Section 9.4, severe crater wear usually results from temperature-activated diffusion or chemical wear mechanisms. Crater wear can be minimized by increasing the chemical stability of the tool material or by decreasing the tool's chemical solubility in the chip; this can be done by applying coatings as discussed in Chapter 3. Reducing the cutting speed is also effective in controlling crater wear.

Tools used in rough turning often develop *notch wear* (Figures 9.1c and 9.6) on the tool face, especially at the point of contact between the tool and the unmachined part surface or free edge of the chip. Depth of cut notching usually results from abrasion [13] and is especially common when

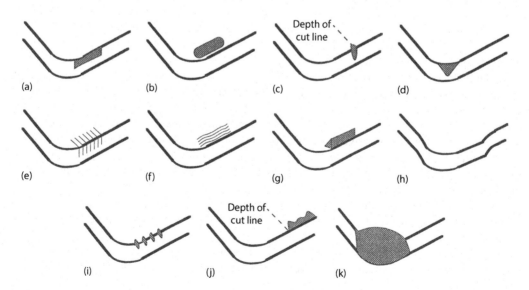

FIGURE 9.1 Types of wear on cutting tools: (a) flank wear; (b) crater wear; (c) notch wear; (d) nose radius wear; (e) comb (thermal) cracks; (f) parallel (mechanical) cracks; (g) built-up edge; (h) gross plastic deformation; (i) edge chipping or frittering; (j) chip hammering; (k) gross fracture.

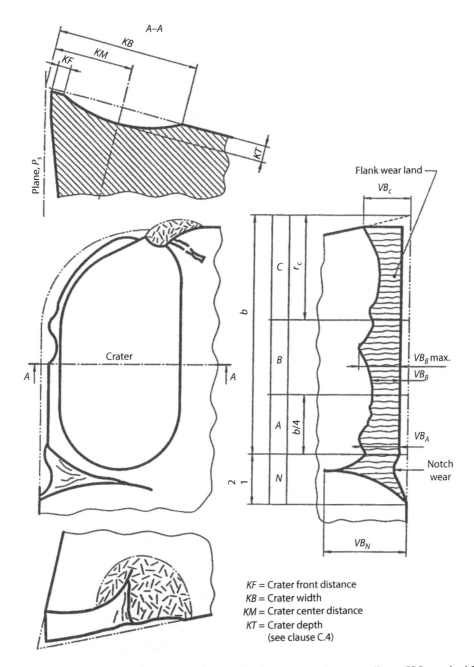

FIGURE 9.2 Characterization of flank wear land and rake face wear crater according to ISO standard 3685. (From Tool Life Testing with Single-Point Turning Tools, ISO Standard 3685:1993(E), 1993.)

cutting parts with a hard surface layer or scale, or work hardening materials, which produce an abrasive chip (e.g., stainless steels and nickel-based alloys). Notch wear may also result from oxidation if a coolant is used, or by chemical reactions or corrosion at the interface between the tool and the atmosphere [7]. Severe notch wear makes resharpening the tool difficult and can lead to tool fracture, especially with ceramic tools. Notch wear can be reduced by increasing the lead angle, which increases the area of contact between the tool and part surface, by varying the depth of cut in multipass operations, and by increasing the hot hardness and deformation resistance of the tool material.

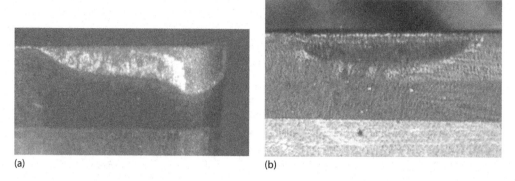

(a)   (b)

FIGURE 9.3   (a) Severe flank and nose radius wear on a carbide insert used to machine 390 Al. (b) Flank wear on a polycrystalline cubic boron nitride tool used to cut powder metal steel.

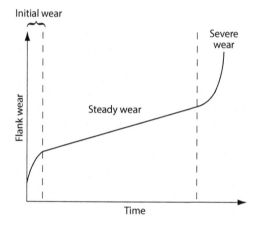

FIGURE 9.4   Variation of the flank wear rate with cutting time, showing the initial wear, steady wear, and severe wear periods.

FIGURE 9.5   Crater wear on a coated carbide insert used to machine a nickel alloy.

FIGURE 9.6   Notch wear on a coated carbide insert used to machine a nickel alloy.

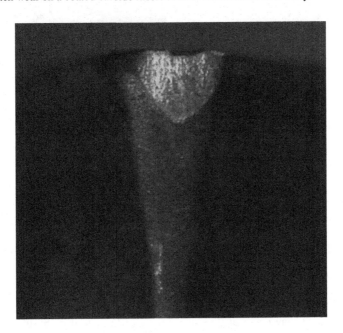

FIGURE 9.7   Nose radius wear on a coated carbide insert used to machine a nickel alloy.

Nose radius wear (Figures 9.1d, 9.3, and 9.7) occurs on the nose radius of the tool, on the trailing edge near the end of the relief face. It resembles a combined form of flank and notch wear, and results primarily from abrasion and corrosion or oxidation [14]. Severe nose radius wear degrades the machined surface finish.

Thermal and mechanical cracking (Figures 9.1e, f and 9.8) usually results from cyclic loading of the tool in interrupted cutting or when machining materials that generate high tool–chip temperatures. Two types of cracks may occur: cracks perpendicular to the cutting edge, which usually result from cyclic thermal loads, especially when a coolant is used, and cracks parallel to the cutting edge, which usually result from cyclic mechanical loads. Crack formation leads to rapid tool fracture or chipping. Tool failure due to cracking in interrupted cutting is discussed in more detail Section 9.11.

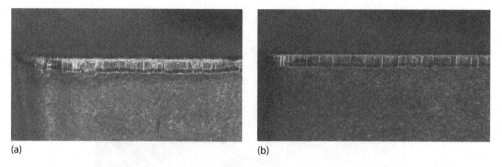

FIGURE 9.8 Thermal cracks in coated carbide inserts used to mill compacted graphite iron. (a) Wet machining. (b) Dry machining.

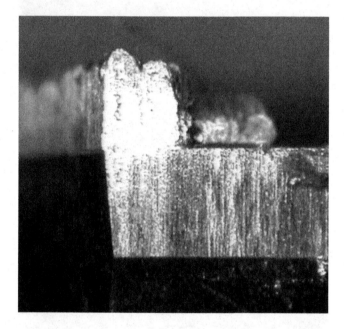

FIGURE 9.9 Edge buildup on a polycrystalline diamond tool used to machine aluminum.

Edge buildup (Figures 9.1g and 9.9) most often occurs when cutting soft metals, such as aluminum alloys, at low cutting speeds. It results when metal adheres strongly to the cutting edge, building up, and projecting forward from it. As discussed in Section 9.10, buildup is also a serious problem in drilling operations; it may occur at the outer corner of the spiral point drills because the chip becomes thin at this point. Built-up edge (BUE) formation is undesirable because it changes the effective depth of cut (or hole diameter) and because it is often unstable, leading to poor surface finish and tool chipping. The mechanics of BUE formation is described in Section 6.14. BUE formation can be minimized by using a more positive rake angle, tools with smooth surface finishes (<0.5 μm R_a), coolant with increased lubricity, higher pressure coolants directed on the rake face, and higher cutting speeds.

Plastic deformation (Figures 9.1h and 9.10) of the cutting edge occurs when the tool is unable to support the cutting pressure over the area of contact between the tool and chip. Cutting edge deformation usually occurs at high feed rates, which produce high cutting edge loads, or at higher cutting speeds, since the hardness of the tool decreases with increasing cutting temperature. Excessive cutting edge deformation results in a loss dimensional accuracy, poor surface finish, and severe flank wear or tool fracture.

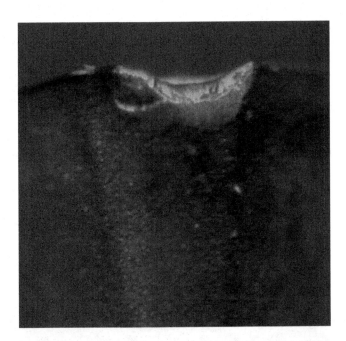

FIGURE 9.10   Plastic deformation of a carbide tool edge used to machine a nickel alloy.

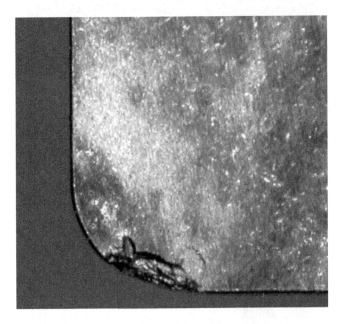

FIGURE 9.11   Edge chipping on a polycrystalline cubic boron nitride insert used to machine hard steel.

Edge chipping or *frittering* (Figures 9.1i and 9.11) occurs when cutting with brittle tool materials, especially ceramics and polycrystalline compacts, or when cutting work materials, which include hard or abrasive particles, such as metal matrix composites or aluminum-silicon alloys. Vibration due to excessive cutting forces or low system stiffness can also lead to chipping. Chipping results in poor surface finish and increased flank wear and may lead to tool breakage. As discussed in Section 9.7, chipping can often be controlled by changing the tool edge preparation or by increasing the fracture strength of the tool material.

Chip hammering (Figures 9.1j and 9.12) occurs when cutting materials form a tough or abrasive chip (e.g., stainless steels, Ni-based alloys). It occurs when the chip curls back and strikes the tool face away from the cutting edge. It leads to chipping or pitting of the tool surfaces and in extreme cases to tool fracture. Chip hammering results from improper chip control and can often be eliminated by changing the lead angle, depth of cut, feed rate, or tool nose radius to alter the chip flow direction.

Tool fracture or breakage (Figures 9.1k and 9.13) results in the catastrophic loss of the cutting edge and a substantial portion of the tool. The causes of fracture are discussed in Section 9.9. General strategies for eliminating fracture include reducing cutting forces, using stronger or more rigid tooling setups, and using tools with increased fracture toughness.

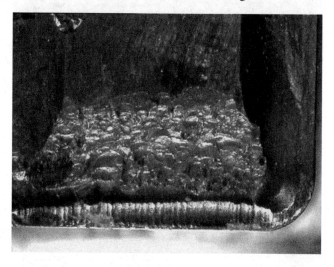

FIGURE 9.12   Chip hammering and cratering on a coated carbide insert used to machine a nickel alloy.

FIGURE 9.13   Fractured edge on a polycrystalline cubic boron nitride insert used to machine hard steel.

9.3 MEASUREMENT OF TOOL WEAR

Flank and crater wear are the most important and thus the most widely measured forms of tool wear. Flank wear is most commonly used for tool wear monitoring since it occurs in virtually all machining operations.

Tool wear is most commonly measured by examining the wear scar on the tool using a microscope or (less commonly) a stylus tracing instrument. In older practice, toolmaker's microscopes with calibrated eyepieces or micrometer stages were widely used. More recently, general purpose digital microscopes have become standard for routine measurements [15,16] (Figures 9.3 and 9.5 through 9.13). A disadvantage of these instruments, however, is that they normally provide only two-dimensional information. More advanced microscopes using confocal or fringe projection methods, as well as laser interferometers, have been developed for general purpose 3-D contouring or for tool surface finish or edge preparation measurements [17–21]. Stylus instruments similar to profilometers (Chapter 10) were used in the older literature for 3D wear scar mapping [10].

Since the initially sharp cutting edge becomes rounded and the original edge is destroyed, it is sometimes necessary to establish a datum line from which flank wear progress can be measured. Video imaging procedures provide a means of comparing sharp and worn edges, which establish and preserve a datum line. Similarly, photographs of the cutting edge at intervals during the tool's life can be used to record flank wear progress. Other techniques, such as gradually lapping down the tool from the top edge and optical contour mapping, can also be used [22,23].

On the tool flank, the average and maximum values of the flank wear land width, VB_{avg} and VB_{max}, are normally measured (Figure 9.2) [9]. (If significant notch wear is present, the notch depth is measured separately, and VB_{max} is defined as the maximum width of the central section of the wear land.) The volume of material worn off the flank, v_w, can be calculated approximately from the average land width, VB_{avg}, using the equation [24]

$$v_w = \frac{VB_{avg}^2 b \tan \theta}{2} \tag{9.1}$$

where
  θ is the relief angle
  b is the width of cut

As discussed in Section 9.6, most standard tool life tests use flank wear criteria to define the end of tool life.

On the rake face, the crater width KB, crater depth KT, and crater land width KF are most commonly measured (Figure 9.2). The total volume of the wear crater, v_{cr}, can be calculated approximately from the equation [24]

$$v_{cr} = \frac{2b(KB - KF)KT}{3} \tag{9.2}$$

The reduction of the volume of the cutting tool is sometimes used to indicate the degree of tool wear. Methods used for measuring the wear volume include (1) geometric determination of the flank and crater wear volumes using Equations 9.1 and 9.2; (2) weighing the tool [25,26]; (3) 3D optical microscopy as described earlier, and (4) radiotracer methods [27] as described as follows.

The specific wear rate η can be calculated by assuming that the volume of tool material worn away is proportional to the area of the contact surface, A, and the length of sliding, L_s:

$$\eta = \frac{1}{A} \frac{dv}{dL_s} \tag{9.3}$$

The specific wear rate for flank wear, η_f, is [28,29]

$$\eta_f = \sin\theta \frac{dVB}{dL_s} \tag{9.4}$$

where θ is the relief angle. The specific wear rate for crater wear, η_{cr}, is

$$\eta_{cr} = \frac{2}{3} \frac{dKT}{dL_c} \tag{9.5}$$

where L_c is the tool–chip contact length.

The wear crater grows both in depth (KT) and width ($KB - KF$). As crater wear progresses, the width of the land between the crater and cutting edge, KF, and the location of the maximum depth, KM, changes. As KF decreases and the crater boundary merges into the cutting edge, the effective rake angle α_{eff} approaches

$$\alpha_{eff} = \alpha_0 + \arctan\left(\frac{KT}{KM}\right) \tag{9.6}$$

where α_0 is the nominal rake angle. As KT decreases, the edge weakens and catastrophic failure becomes more likely.

A great variety of on-line tool wear sensing methods based on optical, pneumatic, electric, displacement, and force measurements have been proposed. A detailed discussion of these methods is beyond the scope of this book and can be found in review articles [30–33]. Of the available methods, those based on force and power measurements seem to be the most practical. The axial and radial forces in turning are much more sensitive to flank wear than the tangential force, so that the ratio of the axial or radial force to the tangential force is often strongly correlated to tool flank wear. Optical methods, which would seem to be well suited for this task, are affected in wet cutting operations by coolant mists and films on the tool.

Beginning in the 1950s, radiotracer methods were developed for measuring instantaneous wear rates [27,34–37]. In these methods, the tool was irradiated so that some of its constituents became radioactive (e.g., the carbon in WC tools). A Geiger counter was then placed next to the chip stream to monitor how much material was being removed from the tool as a function of time. These methods had excellent resolution and provided early convincing evidence of diffusion wear of tools. They are difficult and expensive, however, and are no longer used.

9.4 TOOL WEAR MECHANISMS

The physical mechanisms that produce the various types of wear described in Section 9.2 depend on the materials involved, tool geometry, and the cutting conditions, especially the cutting speed.

Adhesive and abrasive wear are the most significant types of wear at lower cutting speeds. *Adhesive* or *attritional wear* [3,7,8] occurs when small particles of the tool adhere or weld to the chip due to friction and are removed from the tool surface. It occurs primarily on the rake face of the tool and contributes to the formation of a wear crater. Adhesive wear rates are usually low, so that this form of wear is not normally practically significant. However, significant adhesive wear may accompany built-up edge (BUE) formation, since the BUE is also caused by adhesion and can result in chipping of the tool. As noted in the previous section, this occurs primarily when cutting soft work materials at low speeds, and in drilling. Methods for reducing BUE formation are discussed in Section 6.14.

Abrasive wear occurs when hard particles abrade and remove material from the tool. The abrasive particles may be contained in the chip, as with adhering sand in sand-cast parts, iron carbides from foundry chill in cast iron, martensite, austenite, and other hard phases in steels, free silicon particles in aluminum-silicon alloys, and fibers in metal matrix composites [38,39]. They may also result from the chip form or from a chemical reaction between the chips and cutting fluid, as with powder metal steels (which form a powdery chip) or cast irons alloyed with chromium. Abrasion occurs primarily on the flank surface of the tool. Abrasive wear by hard particles entrained in the cutting fluid is sometimes called *erosive wear*. Abrasive wear is usually the primary cause of flank wear, notch wear, and nose radius wear, and as such is often the form of wear that controls tool life, especially at low to medium cutting speeds. Abrasive wear also affects crater wear [40].

Both adhesive and abrasive wear can be described quantitatively by an equation of the form [41]

$$v = \frac{k_w N L_s}{H} \tag{9.7}$$

where
   v is the volume of material worn away
   k_w is the wear coefficient
   N is the force normal to the sliding interface
   L_s is the distance slid
   H is the penetration hardness of the tool

This equation shows that an effective method of controlling wear is to increase the hardness H of the tool. This can be done directly, by using tools made of harder materials, or indirectly, by coating the tool with a hard surface layer. Equation 9.7 also indicates that reducing cutting forces, which reduces N, should also reduce wear due to these mechanisms. This is most easily done by increasing the tool rake angle, although this may also reduce the cutting edge strength and lead to tool deformation or chipping. Finally, Equation 9.7 indicates that reducing the sliding distance, generally by increasing the feed, can also be effective in controlling abrasive wear.

As the cutting speed is increased, adhesive and abrasive wear rates increase for two reasons. First, the distance slid in a given time, L_s, also increases with the cutting speed. Second, increasing the cutting speed increases cutting temperatures. The hardness of the tool material typically drops with increasing temperature. This phenomenon, known as *thermal softening*, not only leads to increased abrasive wear, but may also result in plastic deformation of the cutting edge. Typical hardness-temperature relations for cemented carbides and ceramics are shown in Figure 9.14. Tools made of materials other than high speed steel are composed of hard particles held together by a metallic or ceramic binder; the binder, rather than the hard particles, controls thermal softening behavior. This form of wear can be reduced by reducing the binder content, or altering the binder composition increase the tool's hot hardness, or by reducing the cutting speed to reduce cutting temperatures.

As the cutting speed is increased further, temperature-activated wear mechanisms become dominant. These include diffusion, oxidation, and chemical wear. These forms of wear depend on the chemical compatibility of the tool and workpiece materials; the cutting speeds at which these forms of wear become significant depend in addition on the tool–chip interface temperature and the melting temperature of the chip. Generally these forms of wear can be reduced by machining at lower speeds to reduce cutting temperatures, or by coating the tool with a hard layer of chemically inert material to act as a buffer between the tool substrate and the chip, cutting fluid, and atmosphere [14].

In *diffusion* or *solution wear*, a constituent of the tool material diffuses into or forms a solid solution with the chip material. This weakens the tool surface and results in a wear crater on the rake face of the tool. Severe cratering ultimately leads to tool failure due to breakage. The diffusion wear

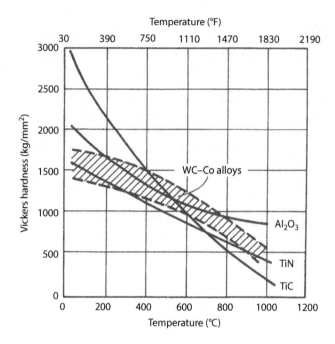

FIGURE 9.14 Variation of penetration hardness with temperature for cemented carbide and ceramic work materials. (After Santhanam, A.T. et al., Cemented Carbides, *Metals Handbook*, 10th edn., Vol. 2, ASM, Materials Park, OH, 1990, pp. 950–977.)

rate depends primarily on the solubility of the tool material in the work material and the contact time between the tool and chip at elevated temperatures, and increases exponentially as the cutting temperature increases. Diffusion wear can be reduced by changing tool materials to a less soluble grade.

Oxidation occurs when constituents of the tool (especially the binder) react with atmospheric oxygen. It most often occurs near the free surface of the part, where the hot portion of the tool in and around the tool–chip contact region is exposed to the atmosphere. Oxidation often results in severe depth-of-cut notch formation and can be recognized by the fact that the tool material is typically discolored in the region near the notch. Oxidation of wear debris or particles of the work material may also result in the production of hard oxide particles, which increase abrasive wear. Oxidation does not occur with aluminum oxide-based ceramic tools.

Chemical wear or *corrosion*, caused by chemical reactions between constituents of the tool and the workpiece or cutting fluid, produces both flank and crater wear, with flank wear dominating as the cutting speed is increased. Chemical wear scars are smooth compared to wear scars produced by other mechanisms and may appear to be deliberately ground into the tool. This type of wear is commonly observed when machining highly reactive materials such as titanium alloys [42]. Chemical wear may also result from reactions with additives (e.g., free sulfur or chlorinated EP additives) in the cutting fluid. (EP additives, in fact, are used to reduce adhesive wear by producing controlled chemical wear [14].) The surface layer of the tool is changed to the reaction product, which is typically soft and wears rapidly by abrasion. Changing the tool material (or coating) or the additives in the cutting fluid will often reduce this type of wear.

As noted before and in Chapter 4, coating the tool with a thin layer of hard, chemically inert material can reduce abrasive, diffusion, oxidation, and chemical wear and permit the tool to be used at higher cutting speeds. The coating itself generally wears by abrasion. The wear rate is relatively low until the coating is worn through (Figure 9.15), after which other forms of wear attack the substrate of the tool; since coated tools are used at higher cutting speeds than uncoated tools, tool

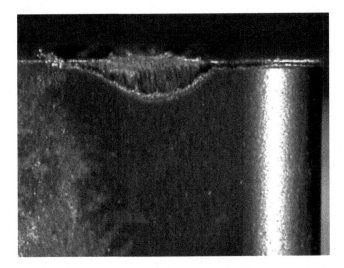

FIGURE 9.15 Coating failure on a carbide insert used to machine a nickel alloy.

FIGURE 9.16 Coating spalling on the flank face of a coated carbide insert used to mill compacted graphite iron.

failure usually follows rapidly. Coatings may also fail by *spalling* due to mechanical or thermal fatigue (Figure 9.16). Advances in coating application methods have reduced the severity of spalling in recent years, but it is still encountered in interrupted cutting. Coating failure can be prevented by changing the coating materials or thickness, by machining dry, and by changing the exit angle in interrupted cuts.

Methods for diagnosing and reducing various forms of tool wear are summarized in Table 9.1. Reducing the cutting speed is effective in reducing many types of wear, but since this reduces the metal removal rate, it should be considered only as a last resort in most operations. It should be noted, however, that reducing the cutting speed has no impact on productivity in many transfer machining operations, specifically those in which the cutting time is substantially lower than the cycle time of the line. In transfer machine operations, the cycle time should be determined by optimizing the speed and feed in critical operations; in the other operations, the cutting speed should be reduced to use all of the available cutting time and increase tool life.

9.5 TOOL WEAR: MATERIAL CONSIDERATIONS

The types and mechanisms of tool wear were discussed from a general viewpoint in Sections 9.2 and 9.4. The dominant form of wear in a given application depends on a number of factors, the most important of which are the tool material, work material, and cutting speed. Not all types or mechanisms of wear are observed with all tool materials, and chemical, oxidation, and diffusion

TABLE 9.1

Mechanisms, Characteristics, and Countermeasures for Common Types of Tool Wear and Failure

Type of Wear	Mechanism	Characteristics	Countermeasures
Flank wear	Abrasion	Even wear scar	Use harder tool material
			Use coated tool
			Filter cutting fluid
			Clean parts
			Refine part microstructure
			Reduce feed rate (*)
	Thermal softening		Reduce speed (*)
	Cutting above center height	Poor finish	Check insert height
	Edge deformation	Edge deformed	See below under "Edge deformation"
	Feed too low	Poor finish	Increase feed
Crater wear	Diffusion	Rapid wear rate	Reduce cutting speed (*)
			Improve cooling ability of coolant
			Increase coolant volume and pressure
			Direct coolant toward chip/tool interface
	Chemical wear	Smooth wear scar	Change tool or coating material or coolant
	Feed too low	Poor finish	Increase feed rate
Notch wear	Abrasion	Occurs at part free surface	Vary depth of cut
			Use harder tool
			Increase lead angle
	Oxidation	Discoloration	Change coolant
			Reduce speed (*)
			Change tool material or coating
Nose radius wear	Abrasion	Rough, uneven scar	Reduce feed (*)
			Use harder tool
			Increase nose radius
Edge cracking	Thermal fatigue	Cracks normal to edge	Reduce cutting speed (*)
			Machine dry
			Use tougher tool
	Mechanical fatigue	Cracks parallel to edge	Reduce feed (*)
			Use tougher tool
Built up edge	Adhesion	Poor surface finish	Increase cutting speed
			Increase rake angle
			Increase lubricity of coolant
Edge deformation	Overload	Occurs rapidly	Reduce feed (*)
	Thermal softening		Use harder tool
			Reduce speed (*)
			Increase coolant supply
			Use low-friction coating
Edge chipping	Abrasion (hard spots)		Inspect work material
			Use tougher tool
			Use stronger edge preparation
			Reduce feed (*)
	Improper chip breaker		Increase chipbreaker land width

(Continued)

TABLE 9.1 (*Continued*)
Mechanisms, Characteristics, and Countermeasures for Common Types of Tool Wear and Failure

Type of Wear	Mechanism	Characteristics	Countermeasures
	Vibration	Noise, chatter marks	Increase system stiffness
			Use tougher tool
			Use stronger edge preparation
			Reduce depth of cut (*)
	Intermittent coolant		Increase coolant supply or cut dry
	Improper seating		Check seat condition
	Adhesion	Built up edge	See above under "built-up edge"
Chip hammering	Improper chip flow	Damage away from edge	Change chip flow direction (change lead angle or tool nose radius)
Gross fracture	Overload	Occurs rapidly	Use tougher tool
			Increase nose radius
			Use stronger edge preparation
			Reduce feed (*)
			Reduce depth of cut (*)
	Vibration	Noise, chatter marks	Increase system stiffness
			Use tougher tool
			Decrease nose radius
			Use stronger edge preparation
			Reduce depth of cut (*)

Note: This table presents general information which is most directly applicable to continuous turning and boring. Countermeasures marked with an asterisk (*) reduce the metal removal rate and should be used only as a last resort in general purpose machining.

wear are particularly sensitive to the materials involved. In this section the common types of wear encountered when machining with tools of specific materials are briefly discussed. The discussion focuses in particular on how various failure modes limit the range of application of tools made of specific materials. A detailed description of all possible failure mechanisms, especially those due to chemical wear, is beyond the scope of this book but can be found in many of the references cited.

High Speed Steel (HSS) tools most commonly fail by abrasion, often aggravated by thermal softening, by plastic deformation, and by adhesion and edge build-up [43]. The hot hardness of HSS decreases rapidly at temperatures above roughly 540°C [44]; at cutting speeds that produce temperatures in excess of this level, rapid abrasive wear and plastic deformation occur. This limits the usable speed for HSS tools to roughly 35 m/min when machining soft steels. Adding cobalt to HSS increases the hot hardness and permits their use at somewhat higher cutting speeds (up to 50 m/min). When machining aluminum alloys and other nonferrous metals, which generally melt at temperatures below 600°C, thermal softening does not usually limit tool life. For these materials, abrasion due to hard particles in the work material (e.g., Si particles in hypereutectic Al–Si alloys and SiC fibers in metal-matrix composites), built-up edge formation, or burring limit tool life. Built-up edge formation is a particularly serious problem for HSS tools because the major constituent of the tool material, iron, is a metal with a comparatively high chemical affinity with, and thus tendency to adhere to, common work materials. Coating HSS tools with TiN or other thin ceramic layer can increase their resistance to abrasion and BUE formation, permitting their use at higher speeds. The tempering temperature of a HSS tool also has a major influence on its wear resistance and performance [45]. HSS tools generally fail by thermal softening before temperatures are reached at which diffusion or chemical wear become significant. Also, tools made of HSS generally do not chip because of the material's high fracture toughness.

Sintered Tungsten Carbide (WC) tools may fail by abrasion, edge chipping, plastic deformation, and diffusion, oxidation, and chemical wear. Abrasion generally results from hard inclusions in the work material, although carbide grain pullout by attritional wear can produce abrasion when machining soft metals [46]. Abrasion and plastic deformation accelerate at cutting temperatures above roughly 700°C, the temperature range in which the hot hardness of most grades begins to decrease rapidly (Figure 9.14) [2]. (The hot hardness of specific carbide grades depends on the carbide grain size and binder content, and micrograin grades in particular are effective at higher temperatures.) For low carbon steels and cast irons, temperatures of this magnitude occur at cutting speeds above roughly 100 m/min [22,47]. Chipping most commonly results from hard inclusions in the work material or from tool vibrations caused by inadequate system stiffness.

Diffusion wear occurs when cutting ferrous work materials with uncoated WC tools when the cutting temperature exceeds 750°C [8], which for low carbon steels usually occurs at cutting speeds above 150 m/min. Steel has a high affinity for carbon, which diffuses out of the hard WC particles and into the chip stream. Diffusion can be inhibited by adding a small amount of TiC or TaC to the WC, since this lowers the solubility of the carbide phase in iron [7,48]. Steel cutting carbide grades (C5 to C8 in the ANSI system) therefore commonly include such additives. Diffusion may also be observed when cutting cast irons with WC tools, but at higher cutting speeds (roughly 200 m/min) because iron produces shorter chips and thus lower tool–chip interface temperatures.

Chemical wear may result from reactions with the work material or the cutting fluid. Fluids containing high concentrations of free sulfur additives are particularly likely to produce chemical wear; in this case, the cobalt binder in the tool reacts with the sulfur to form cobalt sulfide, a soft salt that is rapidly removed by abrasion [1]. Cutting fluids containing chemically combined rather than free sulfur should be used to eliminate this mechanism of wear. When machining titanium alloys, which are highly reactive, the carbon in both WC and PCD tools may react with the chip to form a titanium carbide interlayer, which promotes rapid diffusion wear [42].

Oxidation wear also usually attacks the cobalt binder of WC tools [1]. Atmospheric oxygen penetrates the tool–chip contact primarily at the free surface of the workpiece, producing a wear notch. In the interior of the cut, oxidation may result from air entrained in the cutting fluid, since most coolant application methods result in aeration of the fluid. Once oxygen combines with the binder, the WC particles are removed rapidly from the weakened tool matrix. Oxidation wear can be distinguished from other forms of chemical wear by discoloration of the tool material near the wear scar. Oxidation is generally insignificant at temperatures below 700°C [14].

Coated WC tools exhibit a two-stage wear process [2,49–51]. While the coating is intact, abrasive wear due to hard particles in the workpiece predominates. The wear rate is lower than for uncoated tools because thermal softening is less pronounced. Eventually the coating fails either through excessive abrasive wear or through spalling, which occurs mainly in interrupted processes such as milling. Since coated tools are used at higher cutting speeds than uncoated tools, rapid cratering of the substrate due to diffusion or chemical wear usually follows coating failure.

Coating life depends on the coating material, thickness, and method of deposition. The two most common classes of coating materials are TiN-based (gold) coatings and Al_2O_3-based (black) coatings. (As discussed in detail in Chapter 4, other coating materials are also used, and multiphase coatings consisting of layers of various materials are common.) Al_2O_3-based coatings are harder (Figure 9.14) and more chemically inert at high temperatures and provide better abrasion and cratering resistance; they are preferred when cutting cast and nodular irons. Gold coatings, however, reduce friction and thus cutting temperatures and flank wear at higher cutting speeds. They provide longer tool life when cutting steels at high cutting speeds (Figure 9.17) [2]. Gold coatings are also more commonly used when cutting aluminum alloys because they are less likely to produce a built-up edge.

The coating thickness must be controlled within a narrow range to maximize tool life. If the coating is too thin, it will wear rapidly and fail due to abrasion. If the coating is too thick, however, differences in the thermal expansion coefficient between the coating and substrate will lead to large

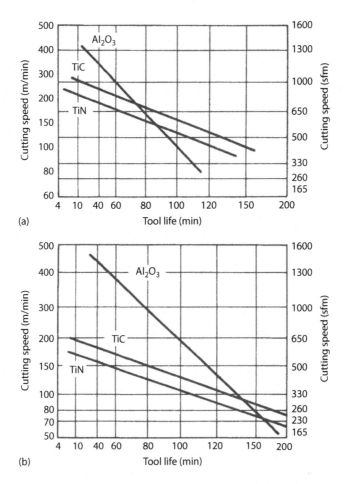

FIGURE 9.17   Variation of tool life with cutting speed for cemented carbide tools coated with various materials when cutting 1045 steel (a) and gray cast iron (b). Al_2O_3 coatings provide superior abrasion resistance and yield the longest tool lives at low cutting speeds, but TiN and TiC coatings reduce tool–chip friction, cutting temperatures, and crater wear and provide increased tool life at higher cutting speeds. (After Santhanam, A.T. et al., Cemented Carbides, *Metals Handbook*, 10th edn., Vol. 2, ASM, Materials Park, OH, 1990, pp. 950–977.)

thermoelastic stresses in the coating and cause coating failure due to spalling. The optimum coating thickness is usually on the order of 0.010 mm [2]. More details on coating materials and methods are given in Chapter 4.

Cermet tools [52–56] typically fail by mechanisms similar to those for WC tools, that is, by abrasion, plastic deformation, edge chipping (often accompanied by BUE), diffusion and chemical wear, and depth-of-cut notching due to oxidation. The discussions of abrasion, plastic deformation, and diffusion, chemical, and oxidation wear given before for WC tools are applicable for cermet tools as well. However, cermet tools are often manufactured with nickel rather than cobalt binders, so that the reactions involved in chemical and oxidation wear differ from those encountered with WC tools.

As discussed in Chapter 4, TiC-, TiN-, or TiCN-based cermets in both hard and tough grades have been developed. Generally, these materials have hot hardness characteristics comparable to WC, are chemically stable at higher temperatures, and have a lower fracture toughness and thermal shock resistance. Because of their greater chemical stability, they are used at higher cutting speeds than WC tools, and under appropriate conditions provide better tool life at comparable cutting speeds. The use of higher cutting speeds also reduces built-up edge formation, which can lead to edge chipping and limit tool life for hard cermets at lower cutting speeds. Because of their lower fracture toughness, edge chipping is a common failure mechanism for cermet tools; it is controlled by avoiding the use of positive rake angles and by using honed or chamfered edge preparations. Fracture concerns also limit the allowable depth of cut and prevent the use of many cermet grades in rough machining, interrupted cutting, and in the machining of hard steels. Cracking due to thermal shock also occurs when cermets are used with a cutting fluid; to avoid failures due to this mechanism, cermet tools should be used dry or with copious flood coolant.

Alumina-based (or oxide) ceramic tools fail by abrasive notch and flank wear, mechanical cracking, plastic deformation, edge chipping, gross fracture, thermal shock, and diffusion or chemical wear [57–62].

Alumina tools have a high hardness at temperatures up to 1000°C and thus exhibit little abrasive wear when used to cut cast iron, nickel-based superalloys, and aluminum-silicon alloys at speeds up to 300 m/min [58]. Alumina tools have found limited success in the high-speed machining of steels due to their poor thermal shock resistance and fracture strength. Polycrystalline cubic boron nitride (PCBN) tool are often a better choice for these applications.

Chipping and fracture failures are common with alumina tools due to the brittleness of the material. These may result from hard inclusions in the work material, high cutting forces encountered when machining at low speeds or with heavy depths of cut, vibration, improper exit conditions in interrupted cutting, or thermal shock, especially in interrupted cutting applications. Because of their low fracture resistance, alumina tools are often used only for finish or semi-finish cuts at high cutting speeds. Fracture due to vibration can be reduced by limiting the depth of cut and by using extremely rigid tooling and fixturing setups. Chipping and fracture can be controlled by using honed or chamfered edge preparations (see Chapter 4). The fracture toughness and thermal shock resistance of alumina tools can be increased by adding ZrO_2, TiC, TiN, or SiC whiskers to the tool [55,56], and these grades are commonly recommended for interrupted cutting. As discussed in the following, however, these additives reduce the tool's chemical stability. Finally, since even toughened grades of alumina have limited thermal shock resistance, alumina ceramics are best suited for dry, continuous cutting. The use of coolants invariably leads to fracture due to thermal cracking, and thermal cracking is also observed in dry interrupted cutting [58,59]. As in the case of excessive notch or flank wear, the best course of action under these conditions is often to switch to PCBN tooling.

Alumina is chemically stable at temperatures up to 1200°C [58]. Diffusion and chemical wear are most often observed when machining steels, which produce temperatures in excess of this level. Chemical stability is a particular problem for toughened grades containing TiC, TiN, and SiC whiskers [56]. The titanium additives are soluble in iron at high temperatures and produce a glassy phase, which wears rapidly. Similarly, SiC whiskers also react chemically with iron at high temperatures, leaving holes in the tool matrix, which contribute to rapid wear or fracture of the tools. As a result, plain rather than toughened grades are best suited for continuous cuts on steel.

TiC- and TiN-based ceramics fail by mechanisms similar to those observed for alumina-based ceramics. The major difference is that these materials have a high chemical affinity for iron, so that diffusion and chemical wear are more serious problems when cutting ferrous work materials [58].

Silicon Nitride tools [58,59,63–65] have a lower hot hardness, but higher fracture toughness than alumina-based ceramic tools. They are well suited for machining cast iron in both continuous and interrupted operations. Sintered grades (Section 4.2) are also used to machine Ni-based alloys.

In continuous cutting applications, Si_3N_4 tools fail primarily by abrasive wear. Abrasion may result from hard inclusions in the work material, especially adhering sand or carbide inclusions in sand castings, or by thermal softening. Thermal softening occurs at temperatures above 900°C with hot pressed Si_3N_4 tooling because hot pressed tools contain a glassy binder phase, which begins to melt at this temperature [59]. This limitation makes hot pressed Si_3N_4 tooling unsuitable for machining steels and nickel alloys under most conditions, although sintered grades are effective in some operations. The use of thin CVD coatings on Si_3N_4 substrates can increase chemical and abrasive wear resistance and shows some promise for overcoming this limitation.

In interrupted cutting, Si_3N_4 tooling often fails by edge chipping or fracture. Due to its higher fracture toughness, chipping is usually less of a problem than for alumina tools [66] and can often be controlled by using honed and (especially) chamfered cutting edge preparations.

Si_3N_4 tooling is not used to machine aluminum alloys due to the high solubility of silicon in aluminum. Similarly, it is not used to machine titanium alloys because titanium reacts with nitrogen. As with alumina tooling, many Si_3N_4 grades have poor thermal shock resistance and should be used dry.

Polycrystalline Cubic Boron Nitride (PCBN) tools are hard and chemically stable at high temperatures and have excellent thermal shock resistance [67,68]. They most commonly fail by abrasion, edge chipping, chemical wear, and thermal shock.

Because of the high hardness of PCBN, abrasive wear rates are low. For conventional cast irons, the wear rate may be imperceptible, particularly for tools with large nose radii. This makes PCBN especially attractive for precision operations such as engine cylinder boring. Although these operations can also be performed using silicon nitride tooling, the use of PCBN can lead to significantly higher tool life.

Abrasive wear rates are higher when machining hardened iron or steel components, chilled irons, cobalt and nickel-based superalloys, and powder metals [67,69,70]. Abrasion generally produces flank rather than notch wear in these applications. The wear rate is often controlled by the binder phase, which softens thermally. Abrasive wear rates may be reduced by using harder grades with ceramic rather than metallic binders, and by applying copious flood coolants in continuous cutting operations, especially finishing operations. Chemical wear also attacks the binder phase in PCBN tools; it can also be reduced by using ceramic rather than metallic binders in most cases, or by switching to alumina tools in continuous cuts.

Edge chipping is common in interrupted cutting. Chipping may also result in both continuous and interrupted cutting from vibrations caused by inadequate system stiffness, especially when cutting hardened irons or steels. Sharp cutting edges and positive insert geometries should be avoided in these applications; negative rake tools with chamfered edge preparations will provide better tool life. Thermal shock also occurs in interrupted cutting and results in thermal cracking of the tool, especially if a coolant is applied. Thermal shock can be reduced by cutting dry or by increasing the coolant supply.

As with ceramic tools, a variety of additional problems may be encountered when machining soft materials or when cutting at low speeds with PCBN tools. These include edge build-up, excessive flank wear, cratering or spalling, and poor surface finish. The measures for counteracting these problems summarized in Table 9.1 also generally work for PCBN tools, but the most economic solution may be to use higher cutting speeds and lower feed rates, or (for soft materials) to switch to carbide or cermet tooling.

Polycrystalline diamond (PCD) tools are the hardest tools available and provide excellent abrasive wear resistance. They are particularly well suited for machining nonferrous materials such as aluminum-silicon and magnesium alloys. Since these materials melt at low temperatures and produce low cutting pressures, tool lives on the order of months can be achieved in high volume production, and maximum cutting speeds are limited by spindle capabilities rather than tool material limitations.

PCD tooling is not as well suited for machining ferrous alloys due to the high chemical affinity of iron for carbon. At higher cutting speeds, carbon in diamond tools diffuses into the chip, resulting in edge weakening and fracture. This limits the maximum speed at which PCD tools can be used for these materials to the range typical of WC tooling. As noted before, PCD tools are also not well suited for cutting titanium alloys because of the high chemical affinity of titanium to carbon [42], which can lead to crater wear at higher speeds [71].

PCD tools may also fail by edge chipping when cutting harder materials due to their limited fracture toughness. Chipping problems are aggravated by the fact that positive rake angles and sharp edge preparations must normally be used to prevent edge build-up. Toughened PCD grades and diamond-coated WC grades suitable for machining ferrous alloys at low cutting speeds have been developed, but to date it has often been more economical to use WC or cermet tooling in these applications.

9.6 TOOL LIFE TESTING

A number of standardized tool life tests have been developed to help rank the performance of cutting tool materials or the machinability of workpiece materials. These include the ISO standard tests for single-point turning [9], face milling [72], and end milling [73], equivalent tests defined by national standard organizations, the ASTM bar turning test [74], and the Volvo end milling test [75,76]. Various nonstandard tests such as drilling tests are also used for these purposes.

The standard tests strictly define the tool and workpiece geometries, cutting conditions, machine tool characteristics, and tool life criteria needed to construct repeatable tool life curves. They typically use a maximum average flank wear criterion to define tool life; in the ISO turning test, for example, tool life is assumed to be over when the average flank wear reaches 0.5 mm under roughing conditions ($VB = 0.5$ mm in Figure 9.2). The flank wear criterion is used because flank wear is normally the most desirable form of wear (given that some form of wear is unavoidable); the flank wear rate is relatively low and repeatable, so that tools that fail due to flank wear have comparatively long and consistent lives. (Other failure criteria based on crater wear, notch wear, surface finish, chip control, edge condition, or dimensional variations may also be used for tool–work material combinations for which flank wear is not the critical failure mode.) The use of a flank wear criterion, however, requires periodic measurement of the wear land width during testing. Since these measurements must be made off-line with a microscope, such measurements contribute significantly to the time required to perform the tests. Constructing tool life curves at two cutting speeds using the ISO turning test, for example, often requires roughly 40 h of machine time.

The standard tests are applicable mainly to steel and iron work materials cut with HSS or WC tools. The standard material used as a baseline in many tests is 1045 steel. These tests were developed by steel and insert manufacturers and are useful primarily for ranking the machining performance of various WC or HSS tool grades, or the machinability of various steel and iron alloys. The standard tests are generally not used for softer work materials such as aluminum alloys, or for coated carbide, ceramic, or polycrystalline tool materials. In the case of softer work materials, the flank wear rate may be quite low, and tool failures may more often result from edge buildup and microchipping, surface finish degradation, or burr formation; also, relative machinability rankings may be of less interest since these materials are relatively easy to machine. The tests are not used for advanced tool materials cutting nonferrous work materials because a prohibitive amount of material would have to be machined to produce the required level of flank wear.

Specialized tests such as drilling tests are often used on materials such as powder metals or special steel alloys, which are difficult to produce in large quantities in bar form [77,78]. Drilling tests are attractive for these applications because they require relatively small amounts of material. An added advantage is that direct wear measurements of the tool are often not required. In many cases drills are tested to failure, and the number of holes required to produce failure is the primary test output; the drill thrust force, torque, hole surface finish, or hole diameter may be monitored, however, as an indication of the progress of drill wear through the tests. When direct tool wear measurements are not required, these tests require substantially less machine time (e.g., 8–12 h) than the standard ISO tests.

Many accelerated wear tests have also been proposed. In these tests the cutting speed is increased significantly beyond the general application range for the tool/work material combination to produce rapid tool failures. One well known test of this type is the face turning test [24] in which a turning tool is used to cut the face of a cylinder rotating at a constant spindle speed. The tool starts cutting at the center and is fed outward. The cutting speed increases with the radial distance from the center, until failure occurs; the radius being cut at failure is used as a relatively measure of machinability. The disadvantage of this and similar tests is that tool failure often results from thermal softening or diffusion or chemical wear, rather than from flank wear as is usually desired in practice.

In conducting tool life tests, it is essential to use consistent tool and work materials. When possible, a single batch of cutting tools should be used for all tests to minimize variations between tools due to differences in microstructure, heat treatment, and grinding geometry. This requirement becomes increasingly important as the complexity of the tool increases; it is most critical when testing drills and end mills. It is also desirable to conduct tool life tests under dry cutting conditions, as this eliminates variation due to coolant concentration and condition. The scatter inherent in tool life testing can be illustrated by considering a round robin test in which the tool life was measured by 10 laboratories using steel from a single heat, tools from a single batch, and identical tool holders, but different machines [76]. The scatter in this data was statistically significant, and the longest tool life measured was more than twice as long as the shortest tool life. The scatter was much greater at the lowest cutting speed used in the test, since at the low speed several tool wear mechanisms were active. Most tool life test results converge at higher cutting speeds since the tool life becomes short and temperature-activated wear mechanisms dominate.

It should be emphasized that both the standard and specialized tests are useful largely for comparison purposes, that is, for ranking the machinability of a group of work materials or the cutting performance of a group of tool grades. (Machinability ratings derived from standard tool life tests may also be useful for cost estimating in job shop production.) They do not generally provide an accurate indication of the tool life to be expected in most production operations. As noted in the introduction, tool life in a production operation depends on the part tolerances, and is effected by the part geometry (and rigidity), equipment condition, toolholding and fixturing method in addition to cutting conditions. Also, the dominant wear mechanisms in production may vary significantly from the flank wear observed when cutting dry with uncoated tools. Realistic estimates of tool life in production operations can only be obtained from pilot test results or experience with similar applications.

9.7   TOOL LIFE EQUATIONS

Since tool life has a strong economic impact in production operations, the development of quantitative methods for predicting tool life has long been a goal of metal cutting research. As noted in Section 9.1, tool life depends as much on part requirements as on the tool material and cutting conditions, making it difficult to develop general methods of predicting tool life. A number of simple tool life equations at least qualitatively applicable to many machining operations, however, have been developed.

The most widely used tool life equation is the Taylor tool life equation [79], which relates the tool life T in minutes to the cutting speed V through an empirical tool life constant, C_t:

$$VT^n = C_t \qquad (9.8)$$

Where
   T is specified in minutes
   C_t is the cutting speed which yields 1 min tool life

The exponent n determines the slope of the tool life curve (Figure 9.18; see also Figure 1.19) and depends primarily on the tool material; typical values are 0.1–0.17 for HSS tools, 0.2–0.25 for uncoated WC tools, 0.3 for TiC or TiN coated WC tools, 0.4 for Al_2O_3-coated WC tools, and 0.4–0.6 for solid ceramic tools [5,80–82]. The constant C_t varies widely with the tool material, work material, and tool geometry and is typically on the order of 100 m/min for rough machining of low carbon steels.

The basic Taylor equation reflects the dominant influence of the cutting speed on tool life, but does not account for the smaller but significant effects of the feed rate and the depth of cut. For this reason, a modified version of Taylor's equation, called the extended Taylor equation [81], is often used:

$$VT^n f^a d^b = K_t \qquad (9.9)$$

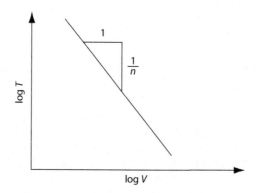

FIGURE 9.18 A Taylor tool life curve, showing linear variation between tool life and cutting speed in a log–log plot. The slope of the curve is negative and equal in magnitude to the inverse of the exponent n in the Taylor tool life equation.

For HSS tools, typical values of n, a, and b are 0.17, 0.77, and 0.37, respectively, when T is given in minutes, V in ft/min, and d and f in inches [83]. K_t varies considerably with the rake angle of the tool but is typically approximately 500 for mild steels and 200 for cast iron [81,82]. Values of K_t, n, a, and b for specific tool grades and common work materials are sometimes tabulated in tooling catalogs. In general, $n < a < b$ because cutting speed has a larger influence on tool life than feed and depth of cut, assuming that tool life is limited by thermal damage and abrasion as expected for a properly designed cutting process. However, in cases in which chipping and fracture failures are common, the exponents a and b are smaller than the exponent n.

The extended Taylor equation treats the influences of the feed rate and depth of cut independently. An alternative equation proposed by Colding [84] combined the influence of these parameters by assuming tool life depends on the effective chip thickness, ECT:

$$T = A V^m \text{ECT}^p \tag{9.10}$$

ECT is equivalent to the average chip thickness computed in simulation programs (Chapter 8) and depends on the feed, depth of cut, lead angle, and tool nose radius. Equation 9.10 involves fewer experimental constants than the extended Taylor equation and yields more broadly applicable results since tool nose radius and lead angle effects are accounted for in the ECT calculation.

Many similar relations have been proposed [6]. Simple tool life equations are useful mainly for comparative purposes. They are useful for ranking the general machining performance of insert grades or similar alloys as discussed in the previous section. Taylor tool life tests performed according to defined standards [9] are also useful in ensuring results reported by different laboratories are consistent; this is an important concern in large machinability testing programs involving several independent laboratories [76]. Tool life predictions based on these relations, however, are generally not quantitatively reliable for the reasons discussed in connection with standard tool life tests in the previous section. The results are most accurate quantitatively for general purpose, low speed machining using conventional machine tools, in which part dimensional tolerances are typically on the order of 0.1 mm. The results are least accurate for high speed machining, in which tool failure often results from crater wear; in precision machining, in which dimensional tolerances are much smaller; and in mass production, in which tool lives on the order of hours rather than minutes are desirable.

First- and second-order mathematical models for tool life can also be fit to experimental data [82,85–88]. Unlike the empirical equations discussed earlier, which are intended to be general, the first- and second-order mathematical models are narrowly applicable to the range of conditions used to generate the test data. Using the cutting speed, feed rate, and depth of cut as independent variables, a first-order mathematical model has the form

$$\ln T = b_0 + b_1 \ln V + b_2 \ln f + b_3 \ln d \tag{9.11}$$

while a second-order model has the form [86–88]

$$\begin{aligned}\ln T = b_0 &+ b_1 \ln V + b_2 \ln f + b_3 \ln d \\ &+ b_{11} (\ln V)^2 + b_{22} (\ln f)^2 + b_{33} (\ln d)^2 \\ &+ b_{12} \ln V \ln f + b_{13} \ln V \ln d + b_{23} \ln f \ln d\end{aligned} \tag{9.12}$$

Equations of this type are used to select optimum economic cutting conditions for a narrow range of applications involving a single tool–work material combination; in contrast to the empirical equations described before, they can be used for complex cutting operations in addition to simple operations such as single-point turning.

The tool life equations described earlier treat tool life as a deterministic property. In reality, tool life exhibits scatter and is probabilistic in nature. Sources of tool life variability in practice include (1) variation in the workpiece material hardness, microstructure, composition, and surface characteristics; (2) variation in the cutting tool material, geometry, and grinding methodology; (3) variations in coolant concentration and condition; and (4) vibrations due to the compliance of the machine tool structure, workpiece, fixture, toolholder, and tool body. Various probabilistic tool life models based on response surface methods and similar techniques have been proposed [89–98], but most are difficult or inefficient to apply in practice. For example, one common approach is to study tool reliability, or the probability of failure-free operation for a specified time in a specified manufacturing environment. In cases in which tools fail randomly, the reliability is described by either an exponential or Weibull distribution [92,93,99]. This requires large number of samples, between 50 and 100, which are generally not available because tools are often removed at regular safe intervals.

9.8 PREDICTION OF TOOL WEAR RATES

A number of researchers have developed models for predicting the rate of tool wear due to specific mechanisms [11,29,100–110].

The best known analysis was developed by Usui and coworkers [100] based on the shear zone model of cutting mechanics discussed in Section 6.10. This model predicts the wear rates due to adhesive, abrasive, and (indirectly) diffusion wear for a carbide tool when cutting steel. Based on earlier work [101,102] and assumptions about the thermal softening behavior of carbide tools, they derive a differential equation relating the incremental volume of material worn away at a given point of the tool, dv, and the incremental distance slid, dL_s:

$$dv = C_1 q \exp\left[-\frac{C_2}{\theta} \right] dL_s \tag{9.13}$$

where
   q is the normal stress at the point in question
   θ is the interfacial temperature
   C_1 and C_2 are empirical constants
   C_1 is proportional to the tool hardness

Based on composite tool measurements like those discussed in Section 6.7, Usui assumed the stress on the rake face of the tool was given by an additional exponential equation:

$$q(x)=\exp\left\{D_1\left[1-\frac{x}{L_c}\right]\right\} \tag{9.14}$$

where
  x is the distance from the cutting edge
  L_c is the tool–chip contact length
  D_1 is an empirical constant

He assumed the stress on the flank face of the tool was constant, and measured temperatures on both the rake and flank faces using the embedded wire method described in Section 7.2.

Under all these assumptions, the constants C_1, C_2, and D_1 can be estimated from a series of orthogonal cutting tests, and the total volume of material worn away as a function of time can be calculated by substituting Equation 9.13 into Equation 9.14 and integrating with respect to x and time. Usui presents data showing that the functional form of Equation 9.14 is reasonable, and that the equation generally yields accurate results when cutting low carbon steels with an uncoated carbide tool (Figures 9.19 and 9.20).

This model is useful because it illustrates the importance of the tool hardness (through C_1) and interfacial temperature on the wear rate. It was initially difficult to apply in practice because

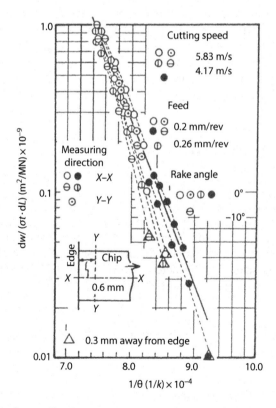

FIGURE 9.19 Variation of normalized crater wear rate with cutting temperature when cutting 1020 steel with a P20 carbide tool. The dashed lines represent theoretical predictions. (After Usui, E. et al., *ASME J. Eng. Ind.*, 100, 236, 1978.)

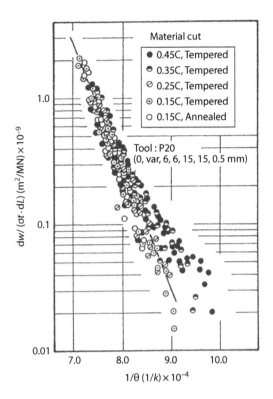

FIGURE 9.20 Variation of normalized crater wear rate with cutting temperature when cutting various low carbon steels with a P20 carbide tool. (After Usui, E. et al., *ASME J. Eng. Ind.*, 100, 236, 1978.)

the extensive tests required estimating temperature distributions and empirical constants. More recently, testing requirements have been reduced by estimating the stress and temperature distributions using finite element analyses [110,111].

Models have also been derived to calculate wear rates due to diffusion [103–105,109]. The best known is Kramer and Suh model [103], which is applicable to WC tools used to cut steel. For this material combination, they assume the crater wear rate is governed by the equation

$$\frac{dW}{dt} = K\left[-D\frac{dc}{dy} + CV_y\right]_{y=0} \tag{9.15}$$

where
   K is the ratio of the molar volumes of the tool material and the workpiece material, respectively
   D is the diffusivity of the slowest diffusing tool constituent into the work material
   c is the concentration of the tool material in the chip
   C is the equilibrium concentration or solubility of the tool material in the work material
   V_y is the bulk velocity of the work material normal to the rake face (Figure 9.21)

This equation identifies the thermodynamic properties of the tool–work material combination, which control the diffusion wear rate. It is difficult to apply because the equilibrium concentration C is difficult to specify [106]. It has been used qualitatively to rank the cutting performance of various tool coatings [42].

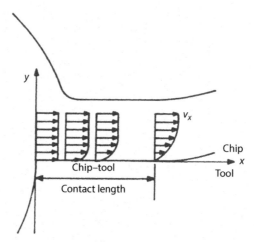

FIGURE 9.21 Definition of variables used in Equation 9.30. (After Kramer, B.M., and Suh, N.P., *ASME J. Eng. Ind.*, 102, 303, 1980.)

9.9 TOOL FRACTURE AND EDGE CHIPPING

Tool fracture occurs when the tool is unable to support the cutting force over the tool–chip contact area. When fracture occurs near the cutting edge and results in the loss of only a small part of the tool, it is referred to as edge chipping or frittering. Chipped tools produce poor surface finishes but are often still usable. Fracture that occurs away from the cutting edge and results in the loss of a substantial portion of the tool is referred to as gross fracture or breakage; when this occurs, the tool is unusable and must be changed.

Since fracture results when the tool is loaded beyond its structural limit, it can be prevented either by reducing the cutting load or by increasing the structural strength of the tool. Since reducing the cutting load usually requires reducing the metal removal rate and thus the productivity of the process, increasing the structural strength of the tool should be attempted first. The structural strength of the tool can be increased by increasing the tool nose radius, using negative rather than positive rake angles, using a thicker insert, and increasing the lead angle; these measures spread the load over a larger volume of the tool material and reduce maximum tool stresses. Switching to a tougher or more fracture resistant tool grade may also be effective. The property of the tool that controls fracture toughness is the transverse rupture strength, which varies with the chemical composition and binder content. Generally, increasing the binder content of the tool increases the fracture toughness (Figure 9.22). Adding strengthening agents such as whiskers or titanium phases to ceramic tools also increases fracture strength. As discussed in Section 9.4, however, tools with increased binder contents or strengthening agents may have reduced resistance to abrasive and chemical wear. Edge chipping can often be controlled by using honed or chamfered, rather than sharp, edge preparations. The optimum chamfer length varies with the tool material but is usually comparable to the feed rate; recommendations for chamfer lengths and angles are often given in tooling catalogs.

If strengthening methods are ineffective, load reduction measures must be used. In determining which measures to use, it is useful to distinguish between initial fractures, which occur at the onset of cutting, and regular or intermittent fractures, which occur either predictably or unpredictably after the tool has cut for some time.

Initial fracture usually results from overloading or from the use of cracked or damaged tools. When tools consistently break at the onset of cutting, the cutting force is too large for the edge configuration. This type of fracture can be prevented by reducing the feed rate or depth of cut, or in some cases by increasing the cutting speed. When initial fractures occur unpredictably,

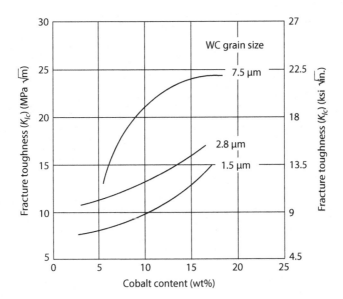

FIGURE 9.22 Variation of fracture toughness with binder content for cemented carbide tools with various grain sizes. (After Santhanam, A.T. et al., Cemented Carbides, *Metals Handbook*, 10th edn., Vol. 2, ASM, Materials Park, OH, 1990, pp. 950–977).

they may result from tool damage and particularly from cracks or nicks in the tool resulting from improper tool handling. This is particularly common for brittle ceramic or cermet tools and can be prevented by handling tools carefully before use and by avoiding excessive tool clamping pressures.

Regular fractures occur at predictable periods and usually result from tool wear in continuous cutting or from fatigue in interrupted cutting. In continuous cutting, progressive flank wear increases stresses in the tools, while crater and notch wear reduce cutting edge strength. In interrupted cutting, thermal or mechanical fatigue can result in cracks in the cutting edge, which ultimately lead to fracture. The methods for reducing wear rates and crack formation discussed in Section 9.3 should reduce the occurrence of regular fractures.

Intermittent fractures occur at irregular intervals and may result from hard inclusions in the work material or from intermittent vibration or chatter problems. If hard inclusions such as carbide inclusions in steel or adhering sand in sand castings are responsible, the best course of action is to inspect incoming workpieces and discard or repair those containing inclusions. If this is not practical or effective and such fractures cannot be tolerated, the feed rate or depth of cut should be reduced or a tougher tool grade should be used. Vibration problems are usually easily recognized because they create excessive noise or leave characteristic marks on the machined surface; they are most commonly eliminated by increasing the stiffness of the tool, tool holder, or part fixture, by changing the cutting speed, or by reducing the depth of cut. Machine tool vibrations are discussed in detail in Chapter 12.

Tool fracture is generally of the brittle type and result from excessive tensile stresses in the tool. Finite element can be used to calculate the stress distribution within the tool for the given edge and boundary conditions [59,110,112]. The inputs to these analyses are the normal and frictional stresses at the tool–chip interface, the tool geometry, and the transverse rupture strength of the tool; the outputs are the tensile stress distribution in the tool and, ultimately, the maximum cutting loads, which can be sustained without exceeding the maximum allowable tensile stress for a given tool geometry (Figure 9.23). These analyses may permit the prediction of the maximum allowable feed rate for a given tool geometry, and may be useful in determining optimum edge conditions (particularly chamfer lengths and angles) for the given edge loadings. Since the transverse rupture strength of

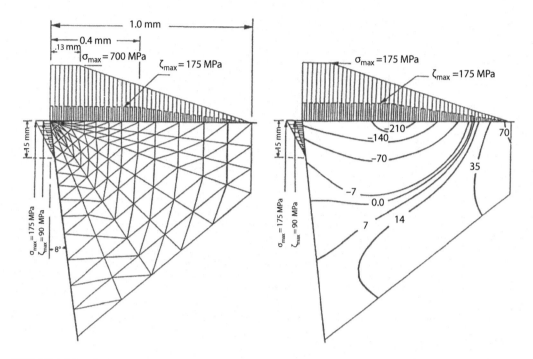

FIGURE 9.23 Finite element model of the tip region of a cutting tool, and contours of maximum principal stresses for a given edge loading. (After Tlusty, J., and Masood, Z., *ASME J. Eng. Ind.*, 100, 403, 1978.)

the tool depends on temperature, coupled analyses which determine both the stress and temperature distributions in the tool yield the most accurate predictions.

9.10   DRILL WEAR AND BREAKAGE

Characteristic wear patterns for conventional twist drills are shown in Figures 9.24 and 9.25. Four types of wear are commonly observed: chisel edge wear, lip wear, margin wear, and crater wear [113–116].

Chisel edge wear occurs at the central chisel edge of the drill and results from abrasion or plastic deformation. This form of wear may double or triple the thrust force on the drill and affects the drill's centering accuracy.

Lip or flank wear occurs along the relief face of the drill's cutting lips and results from abrasion, plastic deformation, insufficient flank relief caused by improper point grinding, or excessive

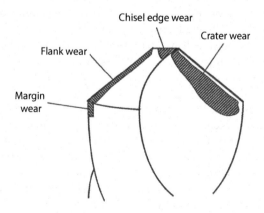

FIGURE 9.24 View of drill point showing regions where flank, nose radius, margin, and crater wear occur. The drill is rotating clockwise when viewed from the top.

FIGURE 9.25 Photomicrograph of a drill point showing chisel edge and flank wear.

dwelling at the bottom of blind holes. Lip wear increases the thrust force, power consumption, and maximum lip temperature, which in turn leads to increased thermal softening and further wear. Lip wear also increases the size of the burr produced when drilling through holes. When lip wear becomes excessive, the drill may cease to cut and fail by chatter or breakage.

Margin or land wear occurs at the outer corner of the cutting lip or on the land, which contacts the drilled surface. Margin wear results from abrasion, thermal softening, or diffusion wear. Excessive margin wear results in poor hole size control and surface finish. Generally, margin wear produces an undersized hole unless it is accompanied by built-up edge formation or centering errors.

Crater wear occurs on the flute surface and results from thermal softening or diffusion wear. Moderate crater wear is usually not of concern, but excessive cratering weakens the drill's cutting edge and can lead to edge deformation, chipping, or breakage.

At low to medium cutting speeds, abrasive wear is usually dominant and most wear occurs at the outer corner near the margins. At higher cutting speeds, in addition to abrasive wear, workpiece deposits may form on the drill lips and the margins, especially when drilling aluminum. Generally, the tool life decreases as the speed is increased when drilling abrasive materials. The feed rate should be maintained above a minimum level to avoid rubbing along the drill lips.

When wear is caused primarily by abrasives in the work material, the number of holes to failure can be estimated from an abrasive wear model. In general, the number of holes to failure N depends directly on the feed f and is inversely proportional to the cutting speed V:

$$N = C_1 \frac{VB \cdot f}{V^a} \qquad (9.16)$$

The time to failure T is independent of the feed and inversely proportional to the cutting speed:

$$T = C_2 \frac{VB}{V^{a+1}} \qquad (9.17)$$

In these equations, C_1, C_2, and a are empirical constants [117]. VB is the width of flank wear.

Methods for reducing various types of drill wear are summarized in Table 9.2. Generally, chisel edge wear results from excessive force and can be reduced by reducing the feed, while margin and

TABLE 9.2

Mechanisms, Characteristics, and Countermeasures for Common Types of Drill Wear and Failure

Type of Wear	Mechanism	Characteristics	Countermeasures
Flank wear	Abrasion	Even wear scar	Use harder drill
			Filter fluid
			Increase feed
	Thermal softening		Reduce speed (*)
Margin wear	Thermal softening	Hole undersized	Use harder drill
			Reduce speed (*)
			Use corner radius or spiral point
Crater wear	Diffusion	Wear crater present	Reduce speed (*)
			Use harder drill
			Use coated drill
			Increase helix angle
Chisel edge wear	Abrasion		Reduce feed
			Use harder drill
			Thin web
Corner chipping	Interrupted hole	Occurs at breakthrough	Use corner radius
			Reduce feed at exit (*)
			Use three- or four-fluted drill
	Vibration	Margin damaged	Increase system rigidity
			Use bushing
			Reduce feed (*)
Drill breakage	Overload	Occurs immediately	Reduce feed (*)
			Reduce helix angle
	Misalignment	Margin damaged	Check alignment
			Use bushing
	Centering error	Occurs near entry	Use self-centering point
			Spotface or centerpunch
			Reduce feed at entry (*)
			Pilot drill (*)
	Chip packing	Flutes clogged	Peck drill
			Increase flute space
			Use parabolic flute
			Increase helix angle
			Increase coolant supply filter coolant
	Vibration	Noise, margin damaged	Increase system rigidity
			Use tougher drill
	Feed surging	Occurs at exit	Reduce feed at exit (*)
	Flank or Heel rubbing	Noise, heel damaged	Regrind point to increase clearance
Hole oversize	Edge buildup		Increase spindle speed
			Use corner radius or spiral point
			Use coated drill
	Poor alignment	Margin worn	Check alignment
			Use bushing

Note: Countermeasures marked with an asterisk (*) reduce the metal removal rate and should be used only as a last resort in general purpose machining.

crater wear result from excessive heat and can be reduced by reducing the spindle speed. Cutting temperatures at a given cutting speed can be reduced by using flood coolants, or by using coolant fed drills with compressed (and sometimes chilled) air or pressurized coolants. Flank wear due to abrasion can be reduced by increasing the feed, provided this does not reduce accuracy or cause chip clogging. Increasing the feed reduces the distance slid per unit time or hole depth, increasing the number of holes that can be drilled to failure; this initially surprising effect is reflected in Equation 9.16 and is observed in tests [118]. When drilling through holes, wear is increased due to torsional vibrations and feed surging at breakthrough [115]; reducing the feed rate prior to breakthrough can increase drill life on CNC machines. For blind holes, wear increases with the dwell time at the bottom of the hole as the feed is reversed. The dwell time cannot be adjusted on many machines, but on CNC machines equipped with high acc/dec feed slides, excessive dwelling should be avoided.

In many cases it is more economical to change the drill material rather than reduce the spindle speed or feed rate. Increasing the drill's hot hardness will reduce most forms of wear and is especially effective in reducing margin wear. A significant increase in drill life can thus usually be obtained by using HSS-Co or solid WC drills instead of HSS drills. Coating drills with TiN reduces both margin and crater wear. Because margin wear is reduced, coated drills produce holes of constant diameter throughout their lives.

Drill breakage and chipping may result from overloading, hard spots in the work material, chip packing, misalignment, spindle runout, vibration, hole interruption, or improper point grinding [119]. Measures for eliminating breakage due to these causes are also summarized in Table 9.2.

Overloading usually produces an immediate failure and can be eliminated by reducing the feed rate. Similarly, as discussed in Section 9.8, breakage due to hard spots in the work material can be reduced by reducing the feed rate if hard spots cannot be detected by improved inspection of incoming work.

Chip packing occurs when the chips become lodged in the flute, so that further chip evacuation is impossible. It occurs especially when drilling deep holes. Chip packing can be prevented by changing the flute profile (e.g., to a parabolic flute) to increase flute space, by changing (generally increasing) the helix angle, by decreasing the feed rate, by changing the orientation of the spindle from vertical to horizontal, and on special machines by drilling from the underside rather than the top of the part. On properly equipped machines, using high pressure through spindle coolant is also effective. Finally, on CNC machines, the use of a pecking cycle in which the drill is periodically withdrawn from the bottom of the hole can also eliminate chip packing, although this increases the cycle time and may increase flank wear.

If the spindle is misaligned or runs out, the drill will often exhibit markings due to impact or rubbing on the margin. In this case the runout and alignment of the spindle should be checked and adjusted as necessary. Spindle runout is especially critical when using solid WC drills; for these drills, the runout typically should not exceed 0.025 mm.

Excessive vibration can usually be detected by the noise it produces or by marks on the machined surface. As discussed in Chapter 12, vibration can be eliminated by increasing the stiffness of the toolholder, spindle, or part, or by reducing the spindle speed. On CNC machines, reducing the feed rate at breakthrough can reduce vibration when drilling through holes.

Hole interruption due to inclined exits or intersecting holes induces vibration and can lead to chipping, breakage, straightness errors, and burr formation. These effects can be minimized by using three- or four-fluted drills rather than two-fluted drills [120], by grinding a corner break or corner radius on the drill point, by using a guide bushing, and by reducing the feed rate at breakthrough on CNC machines.

If a drill point is ground improperly and has inadequate clearance, the heel of the flank will drag, producing noise and often drill breakage. Used tools will have worn spots on the heel in this case. This problem, which is less common for CNC ground drills, can be eliminated by correcting the grinding procedure to provide adequate clearance.

Drills are sometimes broken in practice by running the spindle backwards or by indexing the fixture before retracting the drill from the part. These problems, which result from programming errors, are easily corrected.

9.11   THERMAL CRACKING AND TOOL FRACTURE IN MILLING

In milling, tools are subjected to cyclic thermal and mechanical loads and may fail by mechanisms not observed in continuous cutting. Two common failure mechanisms unique to milling are thermal cracking and exit failure.

The cyclic variations in temperature in milling induce cyclic thermal stresses as the surface layer of the tool expands and contracts. This can lead to the formation of thermal fatigue cracks near the cutting edge. In most cases such cracks are perpendicular to the cutting edge and begin forming at the outer corner of the tool, spreading inward as cutting progresses (Figure 9.8). The growth of these cracks eventually leads to edge chipping or tool breakage. This problem is especially common when milling ferrous parts with WC tools [121–127]. Thermal cracks form and grow more rapidly as peak temperatures during the cutting cycle increase and as the difference between peak and low temperatures increases. As discussed in Section 7.6, peak temperatures during the heating cycle increase as the cutting speed and the length of the heating cycle (i.e., the cutter immersion) increase. The difference between the peak and low temperatures increases when a cutting fluid is applied.

Thermal cracking can be reduced by reducing the cutting speed (or spindle rpm) or by using a tool material grade with a higher thermal shock resistance [127]. In applications in which a coolant is applied, adjusting the coolant volume can also reduce crack formation. An intermittent coolant supply or insufficient coolant volume promote crack formation; if a steady, copious volume of coolant cannot be supplied, tool life can often be increased by switching to dry cutting.

When using coated WC tooling, the coating may fail by spalling due to a similar thermal fatigue mechanism. This type of coating failure can be prevented by changing to a more shock resistant coating grade or by eliminating the coolant in wet operations. Coolants should be avoided with many coating grades; for example, coolant should not be used when milling cast iron with aluminum oxide coated tools. Tooling catalogs normally provide coolant application guidelines for specific grades.

Edge chipping is also common in milling. Chipping may occur when the tool first contacts the part (entry failure) or, more commonly, when it exits the part (exit failure). Entry and exit failures have been most widely studied for face milling steel with WC tooling, but fracture mechanisms and appropriate countermeasures are applicable to other material combinations.

Entry failures most commonly occur when the outer corner of the insert strikes the part first (Figure 9.26) [125]. This is more likely to occur when the cutter rake angles are positive rather than negative. Entry failure is therefore most easily prevented by switching from positive to negative rake cutters. When entry failure occurs with negative rake cutters, it can be prevented by using a cutter with a lower lead angle [125], or by reducing the cutter diameter or moving the cutter center toward the part to increase the entry angle.

Exit failures usually occur when an exit burr or foot is formed as discussed in Chapter 11. As shown in Figure 9.27, the interaction between the burr and the tool as the tool exits can induce large stresses near the cutting edge. The measures for reducing burr formation discussed in Chapter 11 are effective in preventing exit failures. Exit failures can also be prevented by using a larger diameter cutter to increase the exit angle, and by switching from negative to positive rake cutters to increase edge strength.

Failure of aluminum oxide ceramic tools in intermittent cutting of hard steels occurs due to damage accumulation in the tool and can be modeled as a fatigue process [128].

Finally, the methods discussed in Section 9.9 for preventing fractures in general, such as the use of chamfered or honed edge preparations, tougher tool grades, and reduced feeds or depths of cut, are also effective in preventing both entry and exit failures.

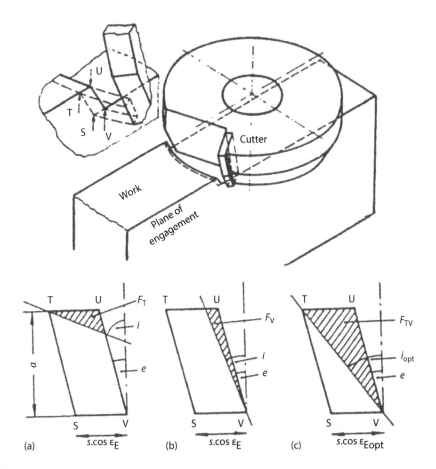

FIGURE 9.26 Initial contact between the workpiece and insert in face milling, described by defining the parallelogram STUV, where S is the outer corner of the insert. The insert should strike the workpiece at point U first to minimize the chance of breakage. When possible, the lead and entry angles should also be adjusted to produce the maximum possible engagement (c) before points T and V enter the workpiece, rather than restricted axial (a) or radial (b) engagement. (After Opitz, H., and Beckhaus, H., *CIRP Ann.*, 18, 257, 1970.)

9.12 TOOL WEAR MONITORING

Tool wear is one important factor contributing to the variation of cutting forces and surface finish. In transfer line applications, worn tools are often changed on a statistical basis at a rate dictated by the shortest life expectancy for multi-tool operations. In such cases, significant useful life of the tools may be wasted and system productivity may be reduced. Alternatively, an in-process tool wear monitoring and control approach may be used to predict in-process tool wear throughout the life of specific tools. This is a common approach in aerospace machining and in CNC-based high volume production systems.

Tool wear can be monitored by either direct or indirect measurements. Direct methods involve the measurement of the wear and evaluating the volumetric loss from the tool due to tool wear using radio-active, optical, laser scan micrometer, or electrical resistance sensors [129]. Indirect measurements of tool wear are made by relating tool wear to other cutting parameters such as part quality, surface roughness, cutting force or torque, motor power (or current), acoustic emissions, or vibration [130–133].

A substantial amount of research has been conducted on machining process monitoring and control [134–145]. Research studies and several production applications have shown that there is a good

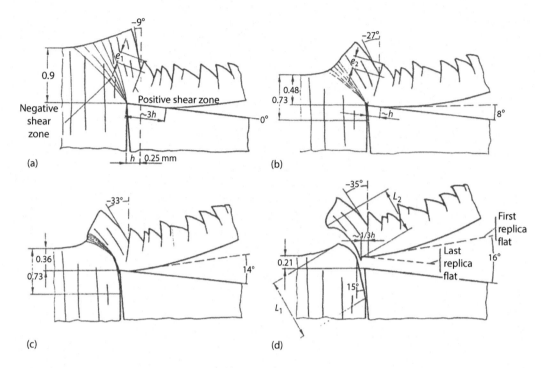

FIGURE 9.27 Progression of foot formation at exit in milling (from (a) through (d)). The formation of the foot induces a large tensile stress on the tool face which can lead to fracture. (After Pekelharing, A.J., *Wear*, 62, 37, 1978.)

correlation between many measurable quantities and flank wear [142,143]. All monitoring systems use sophisticated signal processing techniques or wear estimation. Monitoring system performance is highly dependent on the quality of the transducer data available and the signal processing technique. In many cases multiple sensors are used to improve the reliability of the monitoring system [144].

Several tool monitoring have been developed to protect the machine tool from potential damage due to collision, missing tools, tool breakage, or tool wear as illustrated in Figure 9.28. Spindle power monitors are now common on contemporary machine tools. Sensitivity can be a problem, especially with small diameter tools, because of the inertia and friction present in the system. Strain gages mounted in the spindle bearings and piezoelectric load cells mounted under the tool are more sensitive to cutting forces. Newer wireless sensor technologies provide much higher flexibility for measuring forces, vibration, etc., because they can be installed in a rotary components of the spindle. Additional information on force-based monitoring systems is reviewed in Section 6.2.

9.13 EXAMPLES

Example 9.1 Estimate the parameters for Taylor's too life equation (Equation 9.8) for a new grade carbide indexable insert. An alloy steel bar is being machined using a turning operation with the carbide insert to evaluate the tool life in accordance with the ISO 3685 standard. The cutting conditions used in this test are: 2.54 mm depth of cut and 0.18 mm/rev feed.

Solution: Several turning tests were performed under extensive controllable cutting conditions to minimize variations and scatter in tool life. Several inserts were evaluated at four different cutting speeds: 90, 120, 150, and 180 m/min. Two inserts from the same production batch were tested at 90 m/min, four inserts were tested at 120 m/min, six inserts were tested at 150 m/min, and eight inserts were tested at 180 m/min. More inserts were tested at higher speeds because the tool life

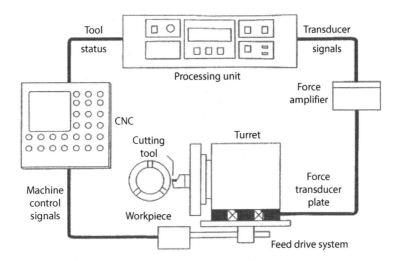

FIGURE 9.28 Tool wear monitoring system.

was shorter and a larger variation was expected due to more rapid tool wear. The flank wear was measured at different time intervals during turning. The average tool wear data obtained from the repeated tests is summarized in Table 9.3 and plotted in Figure 9.29.

The flank wear limit based on the response in Figure 9.29 and previous experience was assumed to be 0.5 mm. The tool life in minutes corresponding to this wear limit was obtained by taking the intercepts of the wear curves with this limit as shown in Figure 9.29. The tool life results are given in Table 9.4 and plotted in Figure 9.30. The tool life data is plotted on log–log paper in Figure 9.31. The data yields a straight line on the log–log scale since

$$V \cdot T^n = C_t \Rightarrow \log V + n \log T = \log C_t \Rightarrow \log V = \log C_t - n \log T$$

The slope of the straight-line n is equal to

$$n = \frac{\log V_1 - \log V_4}{\log T_4 - \log T_1} = \frac{\log 90 - \log 150}{\log 3 - \log 50} = 0.25$$

TABLE 9.3
Experimental Tool Wear Data from Turning Operation

Tool Test Number	Cutting Speed (m/min)	Cutting Time (min)	Flank Wear (mm)	Tool Test Number	Cutting Speed (m/min)	Cutting Time (min)	Flank Wear (mm)
1	180	1	0.14	3	120	12	0.37
1	180	2	0.35	3	120	15	0.47
1	180	3	0.49	3	120	20	0.71
1	180	4	0.64	4	90	6	0.11
2	150	2	0.18	4	90	12	0.21
2	150	4	0.3	4	90	25	0.27
2	150	6	0.46	4	90	35	0.31
2	150	8	0.66	4	90	40	0.37
3	120	2	0.11	4	90	45	0.42
3	120	4	0.16	4	90	50	0.5
3	120	6	0.23	4	90	55	0.6
3	120	8	0.29	4	90	60	0.78

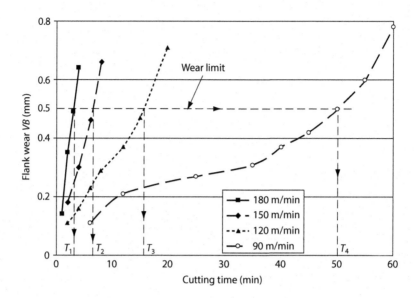

FIGURE 9.29 Graph of tool wear data for Example 9.1, showing method of intercepts for determining tool life for a given cutting speed and wear limit.

TABLE 9.4
Tool Life at Different Speeds

Test	Cutting Speed (m/min)	Tool Life (min)
4	90	50
3	120	16
2	150	6
1	180	3

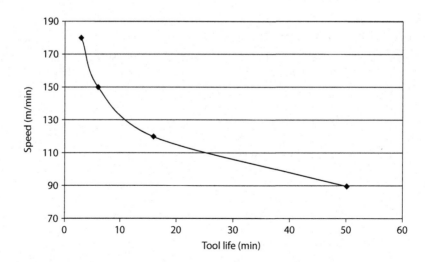

FIGURE 9.30 Linear plot of tool life versus cutting speed for Example 9.1.

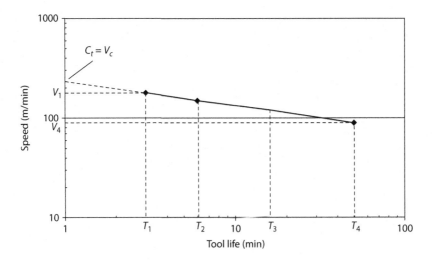

FIGURE 9.31 Log–log plot of tool life versus cutting speed for Example 9.1.

While the intercept of the line with the vertical axis at Time 1 gives the constant C_t since

$$\log V = \log C_t - n \log T \Rightarrow \log V = \log C_t - n \log(1)$$
$$\Rightarrow \log V = \log C_t - n(0) \Rightarrow C_t = V \text{ for } T = 1$$

The value of constant C_t can be also obtained from the data by

$$VT^n = C_t \Rightarrow C_t = VT_1^n = VT_2^n = VT_3^n = VT_4^n$$
$$\Rightarrow C_t = (180)(3)^{0.25} = 237$$

Hence, the Taylor's tool life equation for this case is

$$V \cdot T^{0.25} = 237$$

Example 9.2 The following table of tool life values was obtained from several tool life tests carried out by drilling steel with carbide drills. The drill diameter is 8 mm and the hole depth 25 mm. The mean of the tool life corresponding to different speeds and feeds are given in the table.

	V (m/min)	F (mm/rev)	T (min)
1.	70	0.15	120
2.	70	0.25	105
3.	140	0.15	24

Compute the parameters of the extended Taylor tool life equation, Equation 9.9, for this data.

Solution: The extended Taylor tool life Equation is

$$VT^n f^a = K_t$$

Let us tabulate: $V_1 = V_2 = 70$ and $V_3 = 140$, $f_1 = f_3 = 0.15$ and, $f_2 = 0.25$. The following equality is obtained using case 2 and 3 in the above data:

$$\frac{V_3}{V_1} = \frac{T_1^n}{T_3^n} \quad \text{or} \quad \frac{140}{70} = \left(\frac{120}{24}\right)^n \Rightarrow n = \frac{\ln 2}{\ln 5} = 0.43$$

Similarly,

$$\frac{f_1^a}{f_2^a} = \frac{T_2^n}{T_1^n} \quad \text{or} \quad \left(\frac{0.15}{0.25}\right)^a = \left(\frac{100}{120}\right)^{0.43} \Rightarrow a = 0.43\frac{\ln 0.83}{\ln 0.5} = 0.11$$

$$K_t = VT^n f^a = 70 \times 120^{0.43} \times 0.15^{0.11} = 445$$

Hence, the tool life equation is $VT^{0.43}f^{0.11} = 445$.

This equation may be expressed in two different forms in a rectangular coordinate system as shown in Figure 9.32. These graphs illustrate the considerations for the choice of cutting speeds and

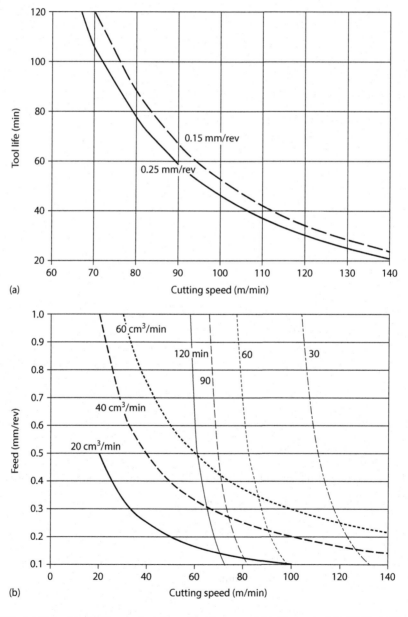

FIGURE 9.32 (a) Plot of tool life as a function of cutting speed for Example 9.2. (b) Plot of tool life and MRR as a function of the cutting speed and feed rate for Example 9.2.

feeds as explained in more detail in Chapter 13. Both graphs illustrate that the selection of the cutting speed is more important than the feed. An increase in cutting speed of 100% (from 70 to 140 m/min) can decrease the tool life by as much as 80% as shown in Figure 9.32a. However, an increase in feed of 67% (from 0.15 to 0.25 mm/rev) decreases the tool life by only 13%. The effect of the combination of speed and feed on tool life is shown in Figure 9.32b. For example, if the cutting speed of 110 m/min is selected with a requirement of 30 min tool life, the feed should selected to be below 0.6 mm/rev. If the speed is increased to 130 m/min, the feed must be reduced to a maximum of 0.13 mm/rev for the same tool life of 30 min. This figure illustrates that the effect of speed on tool life is more significant than the effect of feed on tool life, although both affect the material removal rate.

Example 9.3 Estimate the improvement in tool life if the clearance (relief) angle of a tool is increased from 5° to 11°. Assume that the tool is considered worn at a specified amount of flank wear VB_e.

Solution: As explained in Chapter 4, the cutting edge geometry (as shown in Figure 9.33) and strength are affected by both the rake angle (α) and primary clearance angle (θ). The tool performance improves by increasing the rake angle (assuming the workpiece material and process allow) or the clearance angle assuming the cutting edge does not fracture as discussed in Chapter 4. The width of flank wear is related to the depth of flank wear (defined as *DB* in the feed direction as illustrated in Figure 5.27). The relationship of the width of flank wear to the tool geometry is:

$$\frac{VB}{DB} = \cot\theta - \tan\alpha$$

Figure 9.34 shows the relationship between the width of flank wear land *VB* and the clearance angle for zero and 8° rake tools. Assuming that the flank wear characteristics stay the same for different clearance angles, the width of flank wear is proportional to the cot θ (see above equation). Therefore, the width of the flank wear is reduced significantly by increasing the primary clearance angle. The rake angle has no influence on the width of flank wear as shown in Figure 9.34. However, large clearance and rake angles will weaken the cutting edge significantly and the reliability of the tool will be reduced as explained in Chapter 4. *VB/DB* decreases with increasing clearance angle, which results in longer tool life even though the cutting tool edge length *DB* is reduced significantly. *DB* is important in turning or boring operations and can be compensated for if a relation between *DB* to tool life is known.

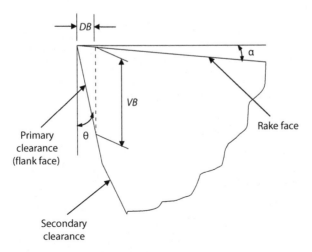

FIGURE 9.33 Definition of the rake angle (α) and primary clearance angle (θ).

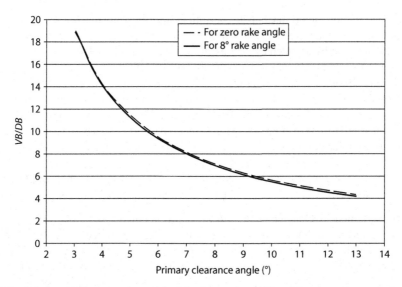

FIGURE 9.34 Relationship between the width of flank wear land VB and the clearance angle for zero and 8° rake tools for the data in Example 9.3.

9.14 PROBLEMS

Problem 9.1 The parameters of the Taylor tool life equation for a carbide tool used in steel machining were estimated to be $n = 0.4$ and $C_t = 450$. Calculate the percentage increase or reduction in tool life if the cutting speed is (a) reduced by 20% and (b) increased by 30%.

Problem 9.2 Determine the constants of Taylor tool life equation, n and C_t, for the three tool materials shown in Figure 9.35.

Problem 9.3 Tool life tests were conducted with a carbide tool at 3 mm depth of cut and 0.5 mm feed using three different workpiece materials. The tool life was evaluated at several speeds as summarized in Table 9.5.

1. Plot the Taylor tool life curves in logarithmic scale.
2. Determine the Taylor tool life equation ($VT^n = C_t$) for each workpiece material.

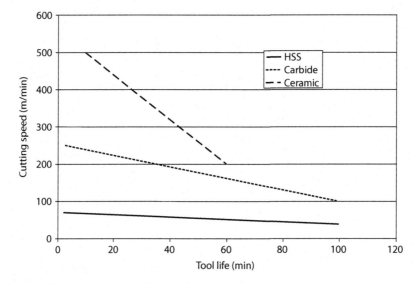

FIGURE 9.35 Tool life as a function of cutting speed for three tool materials, for Problem 9.2.

TABLE 9.5
Tool Life Data for a Carbide Cutting Tool with Three Different Workpiece Materials

	Cutting Speed (m/min)	Tool Life (min)
A. Gray 30 cast iron	30	350
	45	115
	60	53
	75	29
	90	17.5
	105	11.5
B. 1020 CRS	105	900
	120	520
	135	330
	150	200
	165	145
	180	100
C. Pearlitic malleable (BHN 180)	60	580
	75	270
	90	150
	105	80
	120	50
	135	35
	150	24
	180	13

3. Describe some of the shortcomings of Taylor's equation: $VT^n = C_t$?
4. Referring to the 1020 CRS data in Table 9.5,
   (a) Find the cutting speed (V) for 60 min of tool life,
   (b) Estimate the tool life which would result if the cutting speed (V) is increased 20%.

REFERENCES

1. C. F. Barth, Turning, *Handbook of High Speed Machining Technology*, Chapman and Hall, New York, 1985, pp. 173–196.
2. A. T. Santhanam, P. Tierney, and J. L. Hunt, Cemented carbides, *Metals Handbook*, 10th edn., Vol. 2, ASM, Materials Park, OH, 1990, pp. 950–977.
3. P. K. Wright and A. Bagchi, Wear mechanisms which dominate tool life in machining, *J. Appl. Metalworking* **1** (1981) 15–23.
4. G. T. Smith, *Advanced Machining: The Handbook of Cutting Technology*, IFS/Springer Verlag, London, U.K., 1989, pp. 186–190, 227–258.
5. M. C. Shaw, *Metal Cutting Principles*, Oxford University Press, Oxford, U.K., 1984, Ch. 11.
6. E. J. A. Armarego and R. H. Brown, *The Machining of Metals*, Prentice-Hall, New York, 1969, Ch. 7.
7. E. M. Trent, *Metal Cutting*, Butterworths, London, U.K., 1977, Ch. 6.
8. N. H. Cook, Tool wear and tool life, *ASME J. Eng. Ind.* **95** (1973) 931–938.
9. Tool life testing with single-point turning tools, ISO Standard 3685:1993(E), 1993.
10. M. R. Abou-Zeid and S. E. D. M. K. Oweis, Cutting tool wear, SME Technical Paper TE81-956, 1981.
11. I. Ham, Fundamentals of tool wear, ASTME Technical Paper MR68-617, 1968.
12. V. C. Venkatech and M. Satchithanandam, A discussion on tool life criteria and total failure causes, *CIRP Ann.* **29** (1980) 19–22.
13. M. Lee, J. Horne, and D. Tabor, The mechanism of notch formation at the depth of cut line of ceramic tools machining nickel base superalloys, *Proceedings of the International Conference on Wear of Materials*, ASME, New York, 1979, pp. 460–469.

14. T. Kurimoto and G. Barrow, The influence of aqueous fluids on the wear characteristics and life of carbide cutting tools, *CIRP Ann.* **31** (1982) 19–23.
15. A. Attanasio, E. Ceretti, C. Giardini, and C. Cappellini, Tool wear in cutting operations: Experimental analysis and analytical models, *ASME J. Manuf. Sci. Eng.* **135** (2013) 051012-1.
16. Anon., Eclipse LV100D Industrial Microscope, Nikon Corporation, Tokyo, Japan, nd.
17. Anon., VHX-2000 Series, Keyence Corporation of America, Itasca, IL, nd.
18. Anon., *Typical Interferometer Setups*, Zygo Corporation, Middlefield, CT, 2003.
19. Anon., *μsurf Expert*, Nanofocus AG, Oberhausen, Germany, nd.
20. Anon., *MikroCAD 3D Surface Measurement*, GFM Messtechnik GmbH, Berlin, Germany, nd.
21. Anon., *3D Measurement of the Real Edge Contour*, Alicona GmbH, Graz, Austria, nd.
22. H. Takeyama and R. Murata, Basic investigation of tool wear, *ASME J. Eng. Ind.* **85** (1963) 33–38.
23. S. M. Wu and R. N. Meyer, Optical contour mapping of cutting tool crater wear, *Int. J. MTDR* **6** (1966) 153–170.
24. N. H. Cook, *Manufacturing Analysis*, Addison-Wesley, Englewood Cliffs, NJ, 1966, pp. 60–61, 69.
25. S. Popov, *Nucleonics* **19**:76 (1961).
26. P. F. McGoldrick and M. A. M. Hijazi, The use of a weighing method to determine a tool wear algorithm for end-milling, *Proceedings of 20th International MTDR Conference*, 1979, pp. 345–349.
27. S. N. Mukherjee and S. K. Basu, Multiple regression analysis in evaluation of tool wear, *Int. J. MTDR* **7** (1967) 15–21.
28. R. A. Etheridge and T. C. Hsu, The specific wear rate in cutting tools and its application to the assessment of machinability, *CIRP Ann.* **17** (1970) 107–117.
29. A. Ber and M. Y. Friedman, On the mechanism of flank wear in carbide tools, *CIRP Ann.* **15** (1967) 211–216.
30. G. F. Michelletti, W. Koenig, and H. R. Victor, In process tool wear sensors for cutting operations, *CIRP Ann.* **25** (1976) 483–496.
31. N. H. Cook, Tool wear sensors, *Wear* **62** (1980) 49–57.
32. J. I. El Gomayel and K. D. Bregger, On-line tool wear sensing for turning operations, *ASME J. Eng. Ind.* **108** (1986) 44–47.
33. N. Constantindes and S. Bennet, An investigation of methods for the on-line estimation of tool wear, *Int. J. Mach. Tools Manuf.* **27** (1987) 225–237.
34. M. E. Mercahnt, H. Ernst, and E. J. Krabacher, Radioactive cutting tools for rapid tool life testing, *ASME Trans.* **75** (1953).
35. B. Colding and L. G. Erwall, Wear studies of irradiated carbide cutting tools, *Nucleonics* **11** (1953) 46–49.
36. N. H. Cook and A. B. Lang, Criticism of radioactive tool life testing, *ASME J. Eng. Ind.* **85** (1963).
37. G. Lume and B. Anderson, A study of the wear processes of cemented carbide cutting tools by a radioactive tracer technique, *Int. J. MTDR* **10** (1970) 79–93.
38. H. Chen and A. T. Alpas, Wear of aluminum matrix composites reinforced with nickel-coated carbon fibres, *Wear* **192** (1996) 186–198.
39. F. Bergman and S. Jacobson, Tool wear mechanisms in intermittent cutting of metal matrix composites, *Wear* **179** (1994) 89–93.
40. W. Kim and P. Kwon, Understanding the mechanisms of crater wear, *Trans. NAMRI/SME* **29** (2001) 383–390.
41. E. Rabinowicz, *Friction and Wear of Materials*, Wiley, New York, 1965, Chs. 6 and 7.
42. B. M. Kramer, On tool materials for high speed machining, *ASME J. Eng. Ind.* **109** (1987) 87–91.
43. P. K. Wright and E. M. Trent, Metallurgical appraisal of wear mechanisms and processes on high-speed-steel cutting tools, *Metals Technol.* (January 1974) 1323.
44. T. J. Drozda and C. Wick, *Tool and Manufacturing Engineers Handbook*, Vol. 1: *Machining*, SME, Dearborn, MI, 1988, Ch. 1.
45. B. F. Von Turkovich and W. E. Henderer, On the tool life of high speed steel tools, *CIRP Ann.* **27** (1978) 35–38.
46. X. Wang and P. Y. Kwon, WC/Co tool wear in dry turning of commercially pure aluminium, *ASME J. Manuf. Sci. Eng.* **136** (2014) 031006-1.
47. D. A. Stephenson, Assessment of steady-state metal cutting temperature models based on simultaneous infrared and thermocouple measurements, *ASME J. Eng. Ind.* **113** (1991) 121–128.
48. E. M. Trent, Some factors affecting wear on cemented carbide tools, *Proc. Inst. Mech. Eng.* **166** (1952) 64–73.
49. P. A. Dearnley and E. M. Trent, Wear mechanisms of coated carbide tools, *Metals Technol.* **9** (1982) 60–75.
50. P. A. Dearnley, Rake and flank wear mechanisms of coated cemented carbides, *Surf. Eng.* **1** (1985) 43–58.

51. A. T. Santhanam and G. P. Grab, Innovations in coated carbide cutting tools, *Tool Materials for High-Speed Machining*, ASM, Materials Park, OH, 1987, pp. 67–76.

52. W. W. Gruss, Cermet cutting tools: Turning with small tools made of cermet grade, *Ceramic Cutting Tools and Applications*, SME, Dearborn, MI, 1989, Paper 4.

53. R. D. Gantt, Finish boring steel bearing liners with cermets, *Ceramic Cutting Tools and Applications*, SME, Dearborn, MI, 1989, Paper 6.

54. K. Kitajima, I. Matsune, Y. Tanaka, and K. Kishimoto, Cutting performance of solid cermet end mill, SME Technical Paper TE89-164, 1989.

55. C. Wick, Cermet cutting tools, *Manuf. Eng.* **99**:6 (December 1987) 35–40.

56. H. K. Toenshoff, H. G. Wobker, and C. Cassel, Wear characteristics of cermet cutting tools, *CIRP Ann.* **43** (1994) 89–92.

57. W. W. Gruss, Turning steel with ceramic cutting tools, *Tool Materials for High Speed Machining*, ASM, Material Park, OH, 1987, pp. 105–115.

58. S. Lo Casto, E. Lo Valvo, and V. F. Ruisi, Wear mechanisms of ceramic tools, *Wear* **160** (1993) 227–235.

59. H. K. Toenshoff and S. Batsch, Wear mechanisms of ceramic cutting tools, *Ceram. Bull.* **67** (1988) 1020–1025.

60. S. T. Buljan and S. F. Wayne, Wear and design of ceramic cutting tool materials, *Wear* **133** (1989) 309–321.

61. M. Masuda et al., Cutting performance and wear mechanism of alumina-based ceramic tools when machining austempered ductile iron, *Wear* **174** (1994) 147–153.

62. G. K. L. Goh et al, Transitions in wear mechanisms of alumina cutting tools, *Wear* **201** (1996) 199–208.

63. J. Vleugels et al., Machining of steel with sialon ceramics: Influence of ceramic and workpiece composition on tool wear, *Wear* **189** (1995) 32–44.

64. E. Besenyi, Performance of silicon nitride cutting tool inserts prepared by gas pressure sintering, *Int. J. Refract. Met. Hard Mater.* **11** (1992) 121–126.

65. B. Bitterlich, S. Bitsch, and K. Friederich, SiAlON based ceramic cutting tools, *J. Eur. Ceram. Soc.* **28** (2008) 989–994.

66. H. K. Tonschoff and B. Denkena, Tribological aspects of interrupted cutting with ceramic tool materials, *Trans. NAMRI/SME* **19** (1991) 191–198.

67. C. Wick, Machining with PCBN tools, *Manuf. Eng.* **101**:1 (July 1988) 73–78.

68. K. Shintani and Y. Fujimura, Effective use of CBN tool in fine cutting (Continuous and Intermittent Turning), *Strategies for Automation of Machining*, ASM, Materials Park, OH, 1987, pp. 117–126.

69. G. Dawson and T. R. Kurfess, Wear trends of PCBN cutting tools in hard turning, White Paper, Hardinge Inc., Horseheads, NY, 2001.

70. K. Liu et al., CBN tool wear in ductile cutting of tungsten carbide, *Wear* **255** (2003) 1344–1351.

71. D. J. Schrock, D. Kang, T. R. Bieler, and P. Y. Kwon, Phase dependent tool wear in turning Ti–6Al–4V using polycrystalline diamond and carbide inserts, *ASME J. Manuf. Sci. Eng.* **136** (2014) 041018-1.

72. Tool Life Testing in Milling—Part 1: Face Milling, ISO Standard 8688-1, 1989.

73. Tool Life Testing in Milling—Part 2: End Milling, ISO Standard 8688-2, 1989.

74. Anon., Standard Method for Evaluating Machining Performance of Ferrous Metals Using and Automatic Screw/Bar Machine, ASTM Standard E618-07, 2013.

75. Anon., The Volvo Standard Machinability Test, Standard 1018.712, Volvo Laboratory for Manufacturing Research, Trollhattan, Sweden, 1989.

76. A. J. DeArdo, C. I. Garcia, R. M. Laible, and U. Eriksson, A better way to assess machinability, *Am. Mach.* **137**:5 (May 1993) 33–35.

77. J. J. Fulmer and J. M. Blanton, Enhanced machinability of P/M parts through microstructure control, SAE Technical Paper 940357, 1994.

78. Y. Matsushima, M. Nakamura, H. Takeshita, S. Akiba, and M. Katsuta, Improvement of drilling machinability of microalloyed steel, *Kobelco Technol. Rev.* **17** (April 1994) 38–43.

79. F. W. Taylor, On the art of cutting metals, *ASME Trans.* **28** (1907) 31–350.

80. T. Floyd, High efficiency turning, *Cutting Tool Eng.* (March 1993) 55–58.

81. N. E. Woldman and R. C. Gibbons, *Machinability and Machining of Metals*, McGraw-Hill, New York, 1951, pp. 47–53.

82. M. Kronenberg, *Machining Science and Application*, Pergamon Press, Oxford, U.K., 1966, pp. 170–204.

83. O. W. Boston, *Metal Processing*, Wiley, New York, 1941.

84. B. N. Colding, A three-dimensional, tool-life equation—Machining economics, *ASME J. Eng. Ind.* **81** (1959) 239–250.

85. H. J. J. Kals and J. A. W. Hijink, A computer aid in the optimization of turning conditions in multi-cut operations, *CIRP Ann.* **27** (1978) 465–469.

86. S. Rossetto and A. Zompi, Stochastic tool life model, *ASME J. Eng. Ind.* **103** (1981) 126–130.

87. S. Ramalingam, Tool life distribution, ASME Paper 77-WA/Prod.-40, 1977.

88. W. J. Zdeblick and R. E. DeVor, Open loop adaptive control methodology, *Proceedings of Seventh NAMRC*, SME, Dearborn, MI, 1979, pp. 300–306.

89. S. M. Wu, Tool life testing by response surface methodology, Parts I and II, *ASME J. Eng. Ind.* **86** (1964) 105–116.

90. R. Vilenchich, K. Strobele and R. Venter, Tool life testing by response surface methodology coupled with a random strategy approach, *Proceedings of the 13th International MTDR Conference*, 1972, pp. 261–266.

91. E. Kuljanic and T. F. R. Rijeka, Random strategy method for determining tool life equations, *CIRP Ann.* **29** (1980) 351–356.

92. S. Ramalingam and J. D. Watson, Tool-life distributions, Part 1: Single-injury tool-life model, and Part 2: Multiple-injury tool-life model, *ASME J. Eng. Ind.* **99** (1977) 523–531.

93. B. F. Von Turkovich and W. E. Henderer, On the tool life of high speed steel tools, *CIRP Ann.* **27** (1978) 35–38.

94. S. M. Pandit and C. H. Kahng, Reliability and life distribution of ceramic tools by data dependent systems, *CIRP Ann.* **27** (1978) 23–27.

95. S. Rossetto and R. Levi, Fracture and wear as factors affecting stochastic tool-life models and machining economics, *ASME J. Eng. Ind.* **99** (1977) 281–287.

96. A I. Daschenko and V. N. Redin, Control of cutting tool replacement by durability distributions, *Int. J. Adv. Manuf. Technol.* **3** (1988) 36–60.

97. J. M. Pan, W. J. Kolarik and B. K. Lambert, Mathematical model to predict system reliability to tooling for automated machining systems, *Int. J. Prod. Res.* **24** (1986) 493–505.

98. S. Rossetto and A. Zompi, A stochastic tool-life model, *ASME J. Eng. Ind.* **103** (1981) 126–130.

99. L. M. Leemis, Lifetime distribution identities, *IEEE Trans. Reliab.* **R35** (1986) 170–174.

100. E. Usui, T. Shirakashi, and T. Kitegawa, Analytical prediction of three dimensional cutting process Part 3: Cutting temperature and crater wear of carbide tool, *ASME J. Eng. Ind.* **100** (1978) 236–243.

101. M. C. Shaw and S. O. Dirke, On the wear of cutting tools, *Microtechnic* **10** (1956) 187.

102. K. J. Trigger and B. T. Chao, The mechanism of crater wear of cemented carbide tools, *ASME Trans.* **78** (1956) 1119.

103. B. M. Kramer and N. P. Suh, Tool wear by solution: A quantitative understanding, *ASME J. Eng. Ind.* **102** (1980) 303–309.

104. E. Kannatey-Asibu, Jr., A transport-diffusion equation in metal cutting and its application to analysis of the rate of flank wear, *ASME J. Eng. Ind.* **107** (1985) 81–89.

105. V. C. Venkatesh, On a diffusion wear model for high speed steel tools, *ASME J. Eng. Ind.* **100** (1978) 436–441.

106. B. M. Kramer, An analytical approach to tool wear prediction, PhD thesis, Mech. Eng., Massachusetts Institute of Technology, Cambridge, MA, 1979.

107. S. Park, S. G. Kapoor, and R. E. DeVor, Microstructure-level model for the prediction of tool failure in WC-Co cutting tool materials, *ASME J. Manuf. Sci. Eng.* **128** (2006) 739–748.

108. K.-M. Li and S. Y. Liang, Predictive modeling of flank wear in turning under flood cooling, *ASME J. Manuf. Sci. Eng.* **129** (2007) 513–519.

109. M. Braglia and D. Castllano, Diffusion theory applied to tool-life stochastic modeling under a progressive wear process, *ASME J. Manuf. Sci. Eng.* **136** (2014) 031010-3.

110. P. J. Arrazola, T. Ozel, D. Umbrello, M. Davies, and I. S. Jawahir, Recent advances in modelling of metal machining processes, *CIRP Ann.* **62** (2013) 695–718.

111. Y.-C. Yen, J. Soehner, H. Weude, J. Schmidt, and T. Altan, Estimation of tool wear of carbide tool in orthogonal cutting using FEM simulation, *Proceedings of Fifth CIRP International Workshop of Modeling of Machining Operations*, West Lafayette, IN, 2002, pp. 149–160.

112. J. Tlusty and Z. Masood, Chipping and breakage of carbide tools, *ASME J. Eng. Ind.* **100** (1978) 403–412.

113. D. F. Galloway, Some experiments on the influence of various factors on drill performance, *ASME Trans.* **79** (1957) 191–231.

114. S. Sonderberg, O. Vingsbo, and M. Nissle, Performance and failure of high speed steel drills related to wear, *Wear of Materials-1981*, ASME, New York, 1981, pp. 456–467.

115. K. Subramanian and N. H. Cook, Sensing of drill wear and prediction of drill life, *ASME J. Eng. Ind.* **99** (1977) 295–301.

116. W. E. Henderer, Relationship between alloy composition and tool-life of high speed steel twist drills, *ASME J. Eng. Mater. Technol.* **114** (1992) 459–464.

117. J. Alverio, J. S. Agapiou, and C. H. Shen, High speed drilling of 390 aluminum, *Trans. NAMRI/SME* **18** (1990) 209–215.

118. R. T. Coelho, S. Yamada, D. K. Aspinwall, and M. L. H. Wise, The application of polycrystalline diamond (PCD) tool materials when drilling and reaming aluminum based alloys including MMC, *Int. J. Mach. Tools Manuf.* **35** (1995) 761–774.

119. Anon., G. T. E. Valenite Corporation, Pocket drilling guide, Troy, MI, 1987.

120. J. S. Agapiou, An evaluation of advanced drill body and point geometries in drilling cast iron, *Trans. NAMRI/SME* **19** (1991) 79–89.

121. N. N. Zorev, Machining steel with a carbide tool in interrupted heavy-cutting conditions, *Russ. Eng. J.* **43** (1963) 43–47.

122. N. N. Zorev, Standzeit und Leistung der Hartmetall-Werkzeuge beim Unterbrochenen Zerspanen des Stahls mit Grossen Zer-spanungsquerschnitten, *CIRP Ann.* **11** (1963) 201–210.

123. T. Hoshi and K. Okushima, Optimum diameter and position of a fly cutter for milling 0.45 C steel, 195 BHN and 0.4 C steel, 167 BHN at light cuts, *ASME J. Eng. Ind.* **87** (1965) 442–446.

124. P. M. Braiden and D. S. Dugdale, Failure of carbide tools in intermittent cutting, *Materials for Metal Cutting*, ISI Special Publication 126, ISI, London, U.K., 1970, pp. 30–34.

125. H. Opitz and H. Beckhaus, Influence of initial contact on tool life when face milling high strength materials, *CIRP Ann.* **18** (1970) 257–264.

126. A. J. Pekelharing, Cutting tool damage in interrupted cutting, *Wear* **62** (1978) 37–48.

127. S. F. Wayne and S. T. Buljan, The role of thermal shock on tool life of selected ceramic cutting tool materials, *J. Am. Ceram. Soc.* **75** (1989) 754–760.

128. X. Cui, J. Zhao, Y. Zhou, and G. Zheng, Damage mechanics analysis of failure mechanisms for ceramic cutting tools in intermittent turning, *Eur. J. Mech. A/Solids* **37** (2013) 139–149.

129. T. Matsumura and E. Usui, Self-adaptive tool wear monitoring system in milling process, *Trans. NAMRI/SME* **29** (2001) 375–382.

130. L. Dan and J. Mathew, Tool wear and failure monitoring techniques for turning—A review, *Int. J. Mach. Tools Manuf.* **30** (1990) 579–598.

131. G. Byrne, D. Dornfeld, I. Inasaki, G. Ketteler, W. Konig, and R. Teti, Tool condition monitoring (TCM)—The status o research and industrial application, *CIRP Ann.* **44**:2 (1995) 541–567.

132. D. E. Dimla Sr., Sensor signals for tool-wear monitoring in metal cutting operations—A review of methods, *Int. J. Mach. Tools Manuf.* **40** (2000) 1073–1098.

133. X. Li, A brief review: Acoustic emission method for tool wear monitoring during turning, *Int. J. Mach. Tools Manuf.* **42** (2002) 157–165.

134. S. M. Pandit and S. Kashou, A data dependent systems strategy of on-line tool wear sensing, *ASME J. Eng. Ind.* **104** (1982) 217–223.

135. P. K. Ramakrishna Rao et al., On-line wear monitoring of single point cutting tool using vibration techniques, *NDE Science & Technology, Proceedings of 14th World Conference on Non-Destructive Testing*, New Delhi, India, December 8–13, 1996, pp. 1151–1156.

136. J. J. Park and A. G. Ulsoy, On-line flank wear estimation using an adaptive observer and computer vision, Part 1: Theory, *ASME J. Eng. Ind.* **115** (1993) 31–36.

137. J. J. Park and A. G. Ulsoy, On-line flank wear estimation using an adaptive observer and computer vision, Part 2: Experiment, *ASME J. Eng. Ind.* **115** (1993) 37–43.

138. R. G. Landers, A. G. Ulsoy, and R. Furness, Monitoring and control of machining operations, in: O. Nwokah, Ed., *Mechanical Systems Design Handbook*, CRC Press, Boca Raton, FL, 2001.

139. L. Wang, M. G. Mehrabi, and E. K. Asibu, Hidden Markov model-based tool wear monitoring in turning, *ASME J. Manuf. Sci. Eng.* **124** (2002) 651–658.

140. L. Wang, M. G. Mehrabi, and E. K. Asibu, Tool wear monitoring in reconfigurable machining systems through wavelet analysis, *Trans. NAMRI/SME* **29** (2001) 399–406.

141. M. C. Lu and E. Kannatey-Asibu, Analysis of sound signal generation due to flank wear in turning, *International ME Congress & Exposition*, Orlando, FL., 2000.

142. P. K. Ramakrishna Rao et al., Acoustic emission technique as a means for monitoring single point cutting tool wear, *NDE Science & Technology, Proceedings 14th World Conference on Non-Destructive Testing*, New Delhi, India, December 8–13, 1996, pp. 2513–2518.

143. G. F. Michelleti, Tool wear monitoring through acoustic emission, *CIRP Ann.* **38** (1989) 99–102.

144. C. Scheffer and P. S. Heyns, Wear monitoring in turning operations using vibration and strain measurements, *Mech. Syst. Signal Process.* **15** (2001) 1185–1202.

145. D. E. Dimla Sr. and P. M. Lister, On-line metal cutting tool condition monitoring. I: Force and vibration analyses, *Int. J. Mach. Tools Manuf.* **40** (2000) 739–768.

10 Surface Finish, Integrity, and Flatness

10.1 INTRODUCTION

Many parts are machined to produce surfaces with consistent dimensions, forms, and finishes for locating, sealing, or similar applications. In many cases, especially finishing operations, surface flatness, and finish requirements restrict the range of tool sizes, geometries, and feed rates that can be used. Moreover, since the machined surface finish becomes rougher and less consistent as the tool wears, stringent finish requirements may also limit tool life and thus strongly influence machining productivity and tooling costs. Surface flatness and waviness also change with tool wear because the distortion of the part may change due to increased cutting forces.

There are two components to the machined surface finish. The first is the ideal or geometric finish, which is defined as the finish that would result from the geometry and kinematic motions of the tool. The ideal roughness can be calculated from the feed rate per tooth, the tool nose radius, and the tool lead angle. The ideal finish is the predominant component of the finish in operations in which tool wear and cutting forces are low, for example, when machining aluminum alloys with diamond tools. The second component is the natural (or inherent) finish, which results from tool wear, vibration and dynamics of the cutting process, work material effects such as residual stresses, inhomogeneity, built-up edge formation (BUE), and rupture at low cutting speeds. Unlike the ideal finish, the natural finish is difficult to predict in general. It is often the predominant component of the finish when machining steels and other hard materials with carbide tooling, or when machining inhomogeneous materials such as cast iron or powder metals.

The flatness and profile (or waviness) of a machined surface can also be thought of as having ideal (geometric) and natural components. In many applications the geometric flatness and profile deviation is comparatively small due to fixture design and machine calibration (offsets), and the overall machined profile or flatness error depends largely on the natural component, specifically on deflections of the part due to clamping or cutting loads and to distortion due to residual stress. This is true in many automotive applications involving thin walled aluminum parts, and in airframe applications in which large parts are machined from rolled stock. The natural flatness is difficult to predict in general but can be estimated based on previous experience for similar parts in the same machine. Often some of the distortion is repeatable, but a fraction remains unpredictable [1,2]. The repeatable portion of the distortion is often not easily eliminated without major changes in the fixture or cutter, but may still be reduced by compensation without major changes. Engineering surfaces manufactured by different processes have different topographies. For example, a milled surface is spatially inhomogeneous, a ground surface may have pits and troughs, and a honed surface has cross-hatched grooves. Such surface features often cannot be modeled analytically; therefore, statistical parameters have traditionally been used to describe their influence on part function. It is important to use the most suitable surface characterization method that links the process to functionality.

The first part of this chapter describes methods of measuring and predicting the machined surface finish. The measurement of surface finish is discussed in Section 10.2; the discussion concentrates on stylus-type measurements, which are most common in practice. The surface finish in turning, boring, milling, drilling, reaming, and grinding operations is discussed in Sections 10.3 through 10.6. These sections consider only the texture of the machined surface. Of equal importance in many applications is the surface integrity, which can be defined broadly as the metallurgical and mechanical state of the machined surface. Surface integrity can be assessed using microhardness measurements or microstructural analyses, which reveal microcracks, phase transformations, melted and redeposited layers, and similar features. Residual stresses in machined surfaces are discussed in Section 10.7. White layer formation in steels and surface burning in grinding are considered in Sections 10.8 and 10.9. Surface integrity is influenced by thermal effects in cutting (Chapter 7).

Finally, surface flatness measurement is discussed in Section 10.10, and flatness compensation methods for face-milling operations are described in Section 10.11.

10.2 MEASUREMENT OF SURFACE FINISH

10.2.1 STYLUS MEASUREMENTS

The finish of machined surfaces is most commonly measured with a stylus-type profile meter or profilometer, an instrument similar to a phonograph that amplifies the vertical motion of a stylus as it is drawn across the surface [3–5]. The output of the profilometer is a two-dimensional profile of the traced surface segment, amplified in the directions both normal and along the surface to accentuate surface contours and irregularities (Figure 10.1). On a gross scale, the surface profile of a nominally smooth surface gives an indication of the surface's shape, waviness, and roughness (Figure 10.2). The *shape* of the surface is the macroscopic surface contour. Shape errors may result

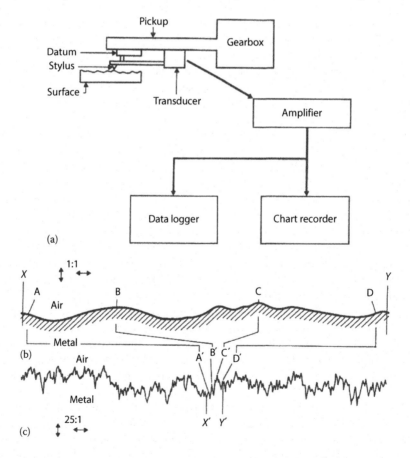

FIGURE 10.1   Schematic illustration of a profilometer (a), and typical unamplified (b), and amplified (c) surface profile traces. (After Thomas, T.R., Stylus Instruments, Chapter 2, in: Thomas, T.R., ed., *Rough Surfaces*, Longman Group, Harlow, Essex, U.K., 1982.)

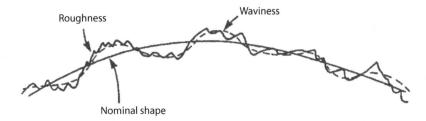

FIGURE 10.2   The shape, waviness, and roughness of a surface. (After Whitestone, D.J., *Surfaces and Their Measurement*, Kegan Page Science, London, U.K., 2002, Chapters 2, 7, and 8.)

from errors in the machine tool guideways or machined part, distortions due to clamping forces or subsequent heat treatment, and from tool wear. For nominally flat surfaces, the shape is referred to as the slope or lay of the surface. *Waviness* refers to variations in the surface with relatively long wavelengths or, equivalently, lower frequencies. Waviness may result from clamping errors, errors in the tool or cutter geometry, or vibrations of the system. As discussed in Section 10.4, spindle tilt in face-milling operations also produces a waviness or shape error. *Roughness* is the term for surface profile variations with wavelengths shorter than those characteristic of waviness. As discussed in the previous section, roughness has a geometric component dependent on the feed rate, tool nose radius, tool lead angle, and cutting speed, as well as a natural component resulting from tool wear, inhomogeneities in the work material, higher frequency vibrations of the machining system, and damage to the surface caused by chip contact. In measurement practice, roughness, waviness, and lay are generally distinguished by cutoff wavelengths.

Many parameters have been proposed to characterize surface roughness [3,4,6–11]. The commonly used parameters, together with standard measurement methods, are defined in national and international standards, which are broadly equivalent. The American national standard is ASME B46.1-2009 [12]; a series of ISO and DIN standards cover the same subjects, the most relevant being ISO 4288:1996, ISO 11562:1996, ISO 4287:1997, DIN 4768:1990, ISO 4287, and ISO 13565-2 [13–18]. These standards define the parameters most often used for inspection and tolerancing of machined surfaces: the average roughness R_a, maximum peak height R_p, maximum valley depth R_v, peak to valley height R_t, average maximum profile height R_z, maximum roughness depth R_{max}, and bearing ratio t_p.

The parameters R_a, R_v, R_p, and R_t are all defined with respect to a centerline of a filtered stylus trace (Figure 10.3). Filtering is performed to remove the slope and waviness components of the trace. Once filtering is performed, the centerline is determined as the mean line of the surface profile.

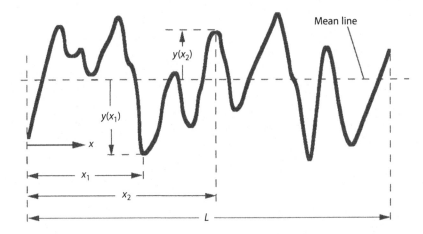

FIGURE 10.3   Definition of parameters used to compute the roughness measures R_a, R_p, R_v, and R_t.

Profile deviations from the centerline are designated y. The average roughness R_a is defined as the average absolute deviation of the workpiece from the centerline:

$$R_a = \frac{1}{L} \int_0^L |y(x)| dx \qquad (10.1)$$

R_p is the maximum deviation of a peak above the centerline encountered within the sampling length (Figure 10.3):

$$R_p = \text{Max } y(x), \quad 0 < x < L \qquad (10.2)$$

where
    x is the distance along the trace
    L is the sampling length

Similarly, R_v, the maximum depth of valley below the centerline, is defined as

$$R_v = |\text{Min } y(x)|, \quad 0 < x < L \qquad (10.3)$$

R_t, the maximum peak to valley deviation or total profile height, is equal to

$$R_t = R_p + R_v \qquad (10.4)$$

The average maximum profile height R_z and maximum roughness depth R_{max} (Figure 10.4) are defined in ASME B46.1 [12] as the average and maximum values of the profile heights over five successive sampling intervals:

$$R_z = \frac{1}{5} \sum_{i=1}^{5} R_{zi} \qquad (10.5)$$

and $R_{max} = \max(R_{zi})$.

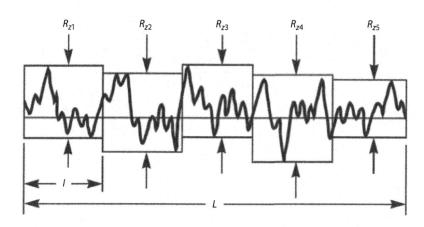

FIGURE 10.4 Definition of the average maximum profile height R_z and maximum roughness depth R_{max}.

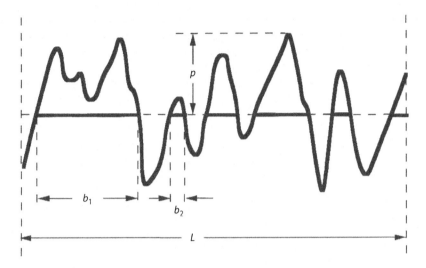

FIGURE 10.5 Definition of parameters used to compute the bearing ratio t_p.

The bearing ratio t_p (Figure 10.5) is a function of the depth p below the highest peak and is defined as the ratio of the total length of the profile below the depth p to the total trace length L:

$$t_p = \frac{\sum_{i=1}^{N} b_i}{L} \tag{10.6}$$

where N is the number of segments over which the profile height exceeds p over the entire trace length.

R_a is the most commonly specified roughness parameter and is well suited for monitoring the consistency of a machining process. R_a increases with severe tool wear and is also sensitive to changes in the condition of the coolant or work material. R_z is similar to R_a but is more sensitive to the presence of high peaks or deep valleys and gives better sense of deviations from the mean. R_t, R_p, R_v, and similar parameters are sensitive to the presence of high peaks and deep scratches, which would affect oil loss on sliding or sealing surfaces [10]. R_t is a good indicator of potential crack propagation, which is more likely when there are many high peaks and low valleys. The bearing area curve provides information on how the surface profiles, which affects the tribology of bearing surfaces. It indicates how well two surfaces might mate for sealing or wear (e.g., meshing gears and cylinder bores). The related parameter P_c (representing the number of peaks above a given height in a given length or area) has been used as an indicator for the performance of surfaces designed for plating, painting, forming, and lubricant retention.

Other surface parameters have been used to characterize certain functional surfaces. The motif parameters [14,19,20] have been used in the European automotive industry to describe significant components of the surface roughness and waviness not adequately characterized by more traditional methods; these parameters correlate well with the functions of specific mechanical parts. In addition, the DIN R_k parameters [21,22], originally developed for the assessment of honed surfaces of cylinder bores, have been adapted to characterize general bearing area curves. Five parameters are used: R_k, R_{pk}, R_{vk}, M_{r1}, and M_{r2}. R_k is the core roughness depth, R_{pk} is the reduced peak height and describes the top portion of the surface that wears away first, and R_{vk} is the valley depth that indicates the oil retaining capability of deep troughs.

Surface profiles can also be analyzed using normal distribution statistics such as standard deviation, skew, and kurtosis. In this analysis, the standard deviation provides a measure of the variability of the surface, and the skew and kurtosis are useful in characterizing surfaces for bearing and locking applications [3,5]. Skew can distinguish between asymmetrical profiles of the same R_a and R_t. Negative skew indicates a predominance of valleys, while positive skew indicates a predominance of peaks. Plateau honed surfaces generally have negative skew, since peaks are removed; ground surfaces typically have zero skew, while turned, EDMed, or bead blasted surfaces typically have positive skew.

Parameter variation in surface roughness characterization may be caused by the surface topography or by the measurement and data processing methods [11]. Variations due to the latter cause may result from improper selection of instrument factors such as the stylus geometry and sampling interval and length. ISO Standard 11562 [14] describes the metrological characteristics of phase correct filters for the measurement of surface profiles, and specifies how to separate the long and short wave content of a surface profile. The stylus radius also has a significant influence on the measurement of surface geometry, especially for smooth surfaces. For ground surfaces, the stylus diameter is typically 0.01 mm. A stylus with a diameter of 0.005 mm may be used for finer finishes, while a 0.02 mm diameter stylus is used for rougher surfaces generated by turning, milling, and other operations. There are two types of stylus systems: skidded and skidless. The former has a skid or foot with a much larger diameter that rides on the surface at the same time as the stylus and defines the tracing datum for the stylus measurement. The stylus of a skidless instrument is free to move up and down with respect to the instrumentation datum of the tracing unit, which provides better accuracy. The stylus force (preload) should be between 50 and 200 mg to avoid surface damage and maintain measurement consistency.

The sampling or cutoff length has a strong influence on measurement results [12]. It determines the longest wavelength included in the profile analysis and distinguishes roughness irregularities from waviness and form error irregularities. The sampling length is controlled by the profile filter, a low-pass filter that removes the longer wavelength components of the surface profile. Standard cutoff values from ISOs 11562, 12085, and 13565-2 [14,18,20] are 0.08, 0.25, 0.8, 2.5, and 8 mm. 2RC phase corrected Gaussian filters are normally used. When a filtering limit for short wavelength (high frequency) profile components is not used in the instrument, a limit is determined by the stylus geometry, the electrical amplifier characteristics, or the sampling interval. The cutoff length affects the R_a value because the R_a generally increases with increasing cutoff length. In most cases, the 0.8 mm length cutoff is preferred.

The evaluation length is the length over which the trace and the roughness parameters are calculated; it is usually smaller than the traversing length so that the stylus starting and stopping intervals are not included in the analysis. The evaluation length should be at least five times the sampling length for statistical purposes.

Practical engineering surfaces normally contain non-stationary or special features such as spikes, pits, and troughs. As a result, the statistical parameters used to characterize surface roughness vary with the number of samples, correlation between samples, and the number, type, and location of surface features measured in the samples [11]. Measurement errors may result from both instabilities in the tracing speed and the kinematic characteristics of the measurement system in relation to the tracing speed [23,24]. The measurement datum and data processing datum could contribute to parameter variations; the significance of these errors is surface dependent [25,26]. Varying the sampling interval within a proper range does not have much effect on amplitude-related parameters but does affect spacing, slope, and curvature parameters significantly [27,28].

It is often problematic to correlate measurements from different profilometers, especially those from different manufacturers. In addition, profilometers typically show more measurement variation than dimensional gages. This in part reflects more real physical variation in surface properties versus gross dimensions. As a result, it is often difficult to achieve statistical capability for surface texture parameters in mass production applications.

10.2.2 OTHER METHODS

The main limitations of profilometer methods for characterizing surfaces are that profilometer measurements are two-dimensional and do not provide information on the three-dimensional characteristics of the surface, and that the measurements are time-consuming.

A number of alternative methods for characterizing surface topography have developed in response to these limitations. In order to characterize surfaces in three dimensions, special scanning or tracking profilometers, which measure a series of parallel traces, have been developed (Figure 10.6) [29]. Numerous optical techniques have also been developed as reviewed by Whitehouse [3]. In machining and abrasive applications, optical white light interferometers (Figure 10.7) and similar submicron

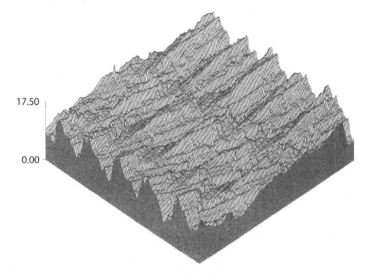

FIGURE 10.6   Three-dimensional profile of a typical end milled surface, measured using a traversing profilometer. (After Stout, K.J. et al., *Atlas of Machined Surfaces*, Chapman & Hall, London, U.K., 1990.)

FIGURE 10.7   White light interferometer image of a honed surface. (Courtesy of Zygo Corporation, Middlefield, CT.)

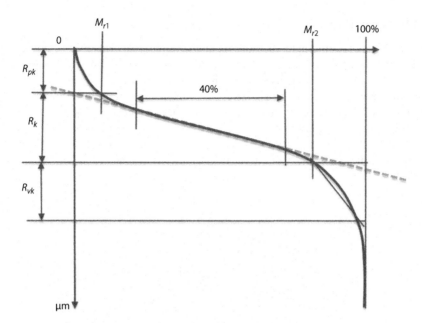

FIGURE 10.8 Bearing curve parameters as defined in ISO 13565-2 STD.

resolution instruments are sometimes used to analyze critical surfaces for roughness features and surface damage [30]. The ISO 25178-6 3D [31] roughness parameters (S_a, S_k, S_{pk}, etc., as shown in Figure 10.8) are extracted using a standard filtering for analysis of the surface. In certain cases, the 3D STD parameters are more effective than the 2D parameters in describing the surface characteristics in applications such as plateau honing of cylinder bores [32,33].

10.3 SURFACE FINISH IN TURNING AND BORING

A turned or bored surface usually shows a uniform roughness distribution without significant waviness (Figure 10.9). In this case the surface finish is adequately characterized for most purposes by the average roughness parameters R_t and R_a.

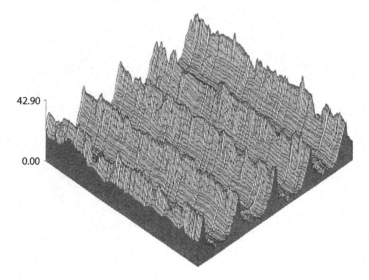

FIGURE 10.9 Three-dimensional profile of a typical bored surface, showing regular grooving in the feed direction. (After Stout, K.J. et al., *Atlas of Machined Surfaces*, Chapman & Hall, London, U.K., 1990.)

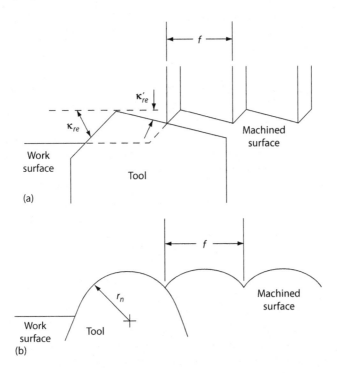

FIGURE 10.10 Schematic illustrations for determining the geometric component of roughness when turning with a sharp-nosed tool (a) and a tool with a nose radius (b).

As discussed in Section 10.1, there are two components to the machined surface finish: the ideal or geometric finish, and the natural finish. For turning and single-point boring, the geometric roughness is easily calculated from the tool angles and feed. For a sharp-nosed tool (Figure 10.10), the geometric peak-to-valley roughness R_{tg} is given by [34]

$$R_{tg} = \frac{f}{\cot \kappa_{re} + \cot \kappa'_{re}} \tag{10.7}$$

where
    f is the feed
    κ_{re} and κ'_{re} are defined in Figure 10.10

Since the contact region between the tool and part is triangular, the average geometric roughness R_{ag} is equal to one-fourth the peak-to-valley roughness R_{tg}:

$$R_{ag} = \frac{R_{tg}}{4} = \frac{f}{4(\cot \kappa_{re} + \cot \kappa'_{re})} \tag{10.8}$$

For a tool with a nose (or corner) radius (Figure 10.10), the depth of cut is often smaller than the nose radius, especially in finish turning and boring. In this case the geometric roughness is independent of the tool angles κ_{re} and κ'_{re} and is determined by the feed per revolution f and nose radius r_n [5,34]:

$$R_{tg} = \frac{f^2}{8r_n} \tag{10.9}$$

$$R_{ag} = \frac{0.0321f^2}{r_n} \qquad (10.10)$$

These equations provide lower bound for the roughness obtained in practice and indicate that smoother surfaces can be generated by using a smaller feed, larger tool nose radius, and larger tool lead angle. In practice, all three methods are used.

The roughness in multipoint boring is generally higher than in single-point boring due to radial runout between inserts, which causes some inserts to cut deeper than others and produce deeper troughs relative to the highest peaks. Axial runout between inserts also affects finish parameters by shifting the positions and heights of successive peaks. The increase in roughness compared to single point boring is generally proportional to the radial runout. Multipoint boring also produces a higher roundness error than single-point boring as discussed in Section 4.6.

Wiper radius or flat inserts (Figure 10.11) are effective in improving surface finish in turning and facing operations. The wiper radius geometry consists of a crown with a large radius adjacent to the standard corner radius. It reduces the effective cutting edge angle and increases the effective lead angle, and cuts the peak of the feed mark. It can be used for both positive and negative rake inserts in both roughing and finishing operations as explained in Chapter 4. As shown in Figure 10.12, wiper radius inserts can reduce average roughness by a factor of two or more; the degree of improvement increases with the feed rate. Wiper flat inserts have a flat edge parallel to the machined surface (zero effective cutting edge angle) and produce a smooth surface through similar to that produced by corner flat milling cutters (discussed in the next section). The feed rate per revolution should be less than the flat length. *Trigon and 80° rhombic inserts* produce a perfect corner radius with no need for adjustment of the tool offset. *Square, triangular and 55° rhombic inserts* do not produce a perfect corner radius, which may affect workpiece dimensions.

The natural component of the roughness results from tool wear, machine motion errors, inconsistencies in the work material, discontinuous chip formation, and machining system vibrations.

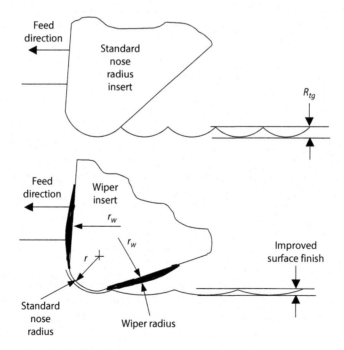

FIGURE 10.11 Illustration of how the surface finish is improved with a wiper insert versus the standard nose radius insert.

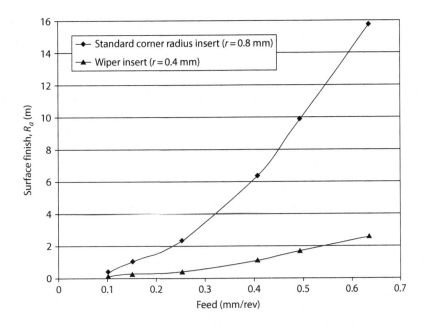

FIGURE 10.12 Comparison of the surface finish generated by a wiper insert versus the same type and size insert without wiper. The standard insert has 0.8 mm nose radius compared to 0.4 mm for the wiper insert.

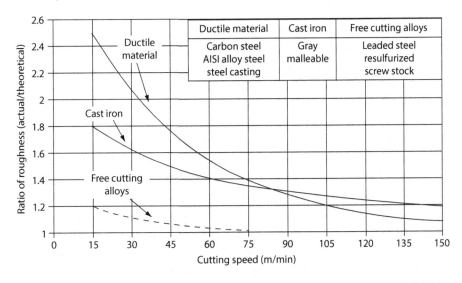

FIGURE 10.13 Variation of measured roughness with cutting speed. (Data courtesy of Carboloy Inc., Warren, MI.)

This component of roughness is difficult to predict. When the work material forms continuous chips and tool wear and system vibrations are not excessive, the natural and geometric components of roughness are generally of equal magnitude, and the actual roughness R_a is usually less than twice the geometric roughness. The natural component of the roughness becomes smaller as the cutting speed is increased, so that the actual roughness approaches the geometric roughness (Figure 10.13). This is particularly true for soft materials prone to BUE formation; lower speeds promote BUE, which tears and galls the machined surface. The ratio of the actual to the geometric roughness is also reduced by increasing the rake angle and by proper coolant application, which can affect surface texture by reducing both BUE and tool wear. The machined surface finish can also be affected by the surface finish of the cutting edge. Polishing cutting edges generally improves surface finish; polishing an uncoated insert can result in a 10%–20% improvement in surface finish. Ceramic inserts can perform as well as coated carbides when they are new, but the surface finish deteriorates much more drastically with worn ceramics than worn carbide inserts due to differences in the wear patterns of these materials.

The surface roughness of hard-turned steel has been found to be comparable to that of ground surfaces [35–38]. The tool, toolholder, and machine requirements for machining hardened materials are somewhat different than those for softer materials as discussed in Chapters 2 and 11. Generally, the equations used for predicting roughness in finish turning cannot be extrapolated to predict the surface roughness in hard turning. The relationship between surface roughness and feed is not monotonic in hard turning, and the overall variation of the roughness is much smaller than for ordinary steels [38].

In summary, the surface roughness produced in turning and boring depends on the feed rate, tool geometry, tool wear, and work and tool material characteristics. The surface roughness produced by these operations can be reduced by

1. Increasing the tool nose radius, provided chatter does not result
2. Decreasing the end cutting edge angle
3. Decreasing the feed rate, although this reduces the material removal rate
4. Decreasing the depth of cut
5. Increasing the tool lead angle (for tools with no nose radius), provided chatter does not result
6. Increasing the cutting speed, provided chatter and excessive tool wear do not result
7. Increasing the rake angle
8. Using the proper cutting fluid and application method
9. Improving workpiece material by adding free-machining additives or increasing the hardness.

10.4 SURFACE FINISH IN MILLING

The discussion of surface finish in turning and boring in the previous section is qualitatively applicable to milling. The finish in milling is also affected by a number of factors not present in turning; these factors result mainly from differences in tooling and process kinematics.

Face milling is carried out using multi-toothed cutters. Due to setup or grinding errors, the cutting edges all cut at slightly different feed rates and depths of cut, especially with indexable cutters. Vibration due to the interrupted nature of the process and changes in cutter position caused by spindle or cutter runout produces further variations in the effective feed rate and depth of cut of each cutting edge. The effective feed rate also varies with the angle of the cutting edge from the feed direction. As a result of all these effects, the surface finish in milling is less uniform than that in turning and single-point boring, in which the machined surface is formed by a single cutting edge cutting at a comparatively constant feed rate and depth of cut (Figure 10.14). The finish has a spatial variation, which is not reflected in simple roughness parameters such as R_t and R_a. Measurements of R_t or R_a at specified points on the workpiece surface can be used to monitor the process for consistency; for more general applications, three-dimensional methods of assessing the surface finish should be considered.

When face milling with radiused inserts, the surface finish depends on the insert radius and on the effective feed rate. The geometric component of the average peak to valley cusp height, R_{tg}, and average roughness, R_{ag}, measured in the feed direction can be calculated approximately using the formulas

$$R_{tg} = \frac{f_{eff}^2}{8r_n} \tag{10.11}$$

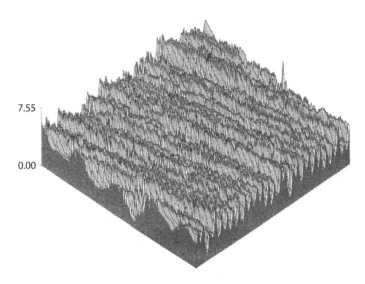

FIGURE 10.14 Three-dimensional profile of a typical fly milled surface, showing curvature of grooves in the feed direction. For face milling with multi-toothed cutters, the groove height would also vary periodically. (After Stout, K.J. et al., *Atlas of Machined Surfaces*, Chapman & Hall, London, U.K., 1990.)

$$R_{ag} = \frac{0.0321 f_{eff}^2}{r_n} \tag{10.12}$$

where
r_n is the nose radius of the inserts
f_{eff} is the effective feed per tooth, given by

$$f_{eff} = f_t \cos\left(\frac{z}{R}\right) \tag{10.13}$$

where
f_t is the feed per tooth
R is the cutter radius
z is the distance between the measurement trace and the centerline of the cutter (Figure 10.15)

These equations are most accurate for a single-toothed cutter (fly mill) with no spindle runout; for multi-toothed cutters, the degree of approximation involved depends on cutter setup errors and runout. The direction as well as the location of the surface trace affects roughness results for milled surface. The surface finish measured along the feed direction is rougher than that measured perpendicular to the feed direction. In finish face milling, fine finishes are often produced using cutters with corner chamfered inserts or wiper inserts (Figure 10.16) [5]. When using corner chamfer inserts, the chamfer should be parallel to the machined surface (perpendicular to the cutter axis), and the feed per tooth should be less than the chamfer length (see Chapter 4). Under these conditions the finish is generated by the cutting edge that cuts deepest. As the cutter grinding or setup accuracy improves, all cutting edges tend to cut at the same level, producing an overlapping effect that improves the machined surface finish. Wiper inserts are crowned inserts mounted behind the regular cutting inserts on a cutter. When properly applied, they improve surface roughness and waviness by deforming (or smearing) the roughness cusps produced by the regular inserts.

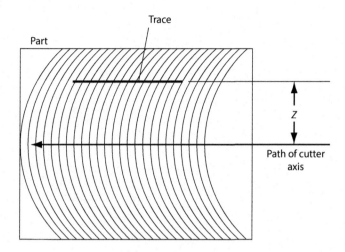

FIGURE 10.15 Schematic illustration for determining the effective feed rate for traces parallel to the feed direction.

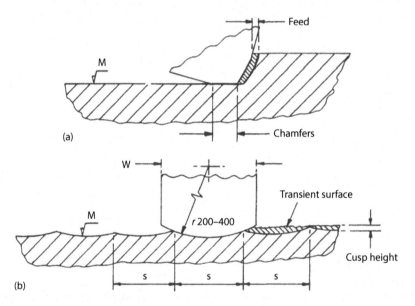

FIGURE 10.16 Milling with corner chamfer (a) and wiper (b) inserts. (After Smith, G.T., *Advanced Machining*, Springer-Verlag, New York, 1989, pp. 198–223.)

Wipers usually have a crown radius between 200 and 400 mm. As rules of thumb, wipers should cut at a depth between 0.05 and 0.1 mm; the feed rate should be roughly half the wiper length. When large cutters with many wipers are used, the resulting finish depends significantly on cutter setup face runout errors. Wiper inserts can produce finer finishes than corner chamfer inserts. They are used primarily on materials such as cast iron or brass, which produce short chips; when used with steel, they can impede chip flow, generating large tangential and axial forces, which can lead to chatter.

In face-milling applications on transfer machines, the spindle is tilted slightly in the direction of feed to provide relief or "dish" behind the cut (Figure 10.17). Spindle tilt is also sometimes used in CNC production systems on machines, which perform dedicated straight line milling operations.

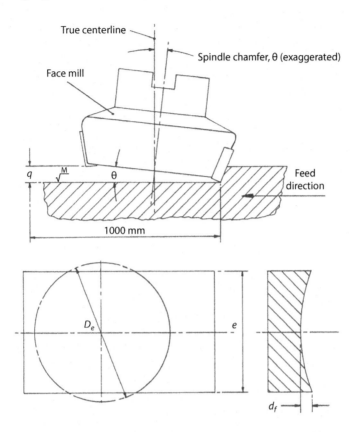

FIGURE 10.17 Waviness error produced by spindle tilt or camber in face milling. (After Smith, G.T., *Advanced Machining*, Springer-Verlag, New York, 1989, pp. 198–223.)

Spindle tilt produces a concave machined surface and results in a flatness error. The depth of the concavity d_f can by calculated from the Kirchner-Schulz formula [5]

$$d_f = \tan\theta \left[\frac{D_e}{2} - \left(\frac{D_e^2}{4} - \frac{e^2}{4} \right)^{1/2} \right] \tag{10.14}$$

where
  D_e is the effective diameter of the cutter
  e is the width of the workpiece
  θ is the spindle tilt angle

In peripheral milling (slab or end milling), the final finish is also produced by multiple cutting edges and is generally less uniform than in turning. The geometric component of the average peak-to-valley roughness height, R_{tg}, can be calculated approximately from Equation 10.11 with the feed per tooth f_t substituted for the effective feed f_{eff} [39]. More exact equations that take into account the trochoidal nature of the tool motion with respect to the workpiece are available [40–41], but the increase in accuracy is usually small compared to the natural roughness component. Reducing spindle runout and cutter grinding or setup errors reduces the average roughness, since reducing these errors causes all cutting edges to cut at a more uniform depth. The radial cutting force will contribute to the shape of the surface (a static form error perpendicular to the surface) because end mills are generally the most flexible part in the machine tool system. The form error on the surface is more complex for mills with helical rather than straight flutes [42–46]. The tool's static deflection is affected by the number of teeth cutting simultaneously. If there is only one tooth in contact with the workpiece at any time, the surface form error is small in end milling for a straight flute cutter because the uncut chip thickness becomes zero when the cutting edge passes a line perpendicular to the machine surface.

10.5 SURFACE FINISH IN DRILLING AND REAMING

The machined surface in drilling is produced by a combined cutting and rubbing action; the marginal cutting edges near the drill point initially cut the surface, and the margins and lands rub and burnish the surface as the drill enters the hole [5]. The component of the finish due to cutting depends primarily on the feed rate per revolution, while the rubbing component depends on the margin design, land width, and the plastic response (hardness and ductility) of the work material. Relatively little can be said quantitatively about the finish in drilling; the process is generally regarded as a roughing operation, so that the drilled surface roughness is not normally monitored as an indication of process consistency and has not been the subject of much research. Holes requiring fine finishes are normally finished by boring, reaming, burnishing, or honing. The finish produced by boring was discussed in Section 10.3. Gun drilling can also produce fine surface finishes since the guide pads burnish the hole wall. Similarly, drills with double or triple margins (or "G" drills) produce a significantly better surface finish (30%–60% smoother) than the standard two-flute drills. Finally, cases in which indexable drills can be used, wiper inserts can be incorporated to improve hole surface finish.

BUE formation, which is common in drilling, degrades the hole surface finish and increases the hole size error. Methods of controlling BUE formation are discussed in Section 6.14; increasing the cutting speed or using through-the-tool coolant (when practical) are often the most effective control measures.

Like drilling, *reaming* produces a surface by a combined cutting and rubbing action. The reamed surface finish depends most strongly on the reamer's chamfer geometry, land width, and relief geometry. A double-chamfer geometry will produce a smoother finish than a standard 45° entry chamfer. The finish also improves with an increase in the land width and a decrease in the land relief angle. However, the use of double chamfers, wide lands, and low land relief angles increases the reamer's susceptibility to chatter and BUE formation. As with gun drills, gun reamers or single-flute reamers with carbide or diamond pads produce very smooth surface finishes. Reamer manufacturers provide technical data for selecting optimum values of these design parameters for finish reamers as a function of the reamer diameter and work material.

In many reaming operations, especially those involving tapered reamers, the reamed surface roughness can be further reduced by allowing the reamer to dwell from one to three seconds at the bottom of the stroke before retracting the tool. This approach, however, can reduce tool life and increase the likelihood of BUE and white layer formation.

10.6 SURFACE FINISH IN GRINDING

Grinding is often used to produce parts with fine surface finishes and tight tolerances. The surface finish and tolerance are related in grinding, since tolerances comparable to the average roughness are often specified. When the wheel is dressed frequently so that the wheel wear is not a significant variable, the ground surface finish depends primarily on the grinding conditions, wheel type, and wheel dressing method.

A number of equations for the ideal or geometric roughness in grinding have been reported; typical results are summarized by Malkin [47]. The actual roughness obtained, however, is normally much larger than the ideal roughness. In practice, the roughness can be predicted more reliably from empirical equations.

Measured roughness values in *cylindrical plunge grinding* are usually well characterized by an equation of the form (Figure 10.17) [47]

$$R_a = R_1 \left[\frac{v_w a}{v_s} \right]^x \tag{10.15}$$

where

v_w is the workpiece velocity
v_s is the wheel velocity
a is the depth of cut
R_1 and x are empirical coefficients for a given wheel type

The exponent x is usually between 0.15 and 0.6. The depth of cut, a, in Equation 10.15 is the depth of cut during the spark-out phase of the grinding process, which is smaller than the nominal stock removal. Increasing the spark-out time generally improves the ground surface finish; if complete spark-out occurs, so that the depth of cut diminished to zero before the wheel is retracted, the final roughness is reduced by about one-half. The quantity in brackets in Equation 10.15 is the average uncut chip thickness.

The finish in straight *surface grinding* is found empirically to depend only on the speed ratio v_w/v_s; the roughness increases with this ratio, and to a first approximation is independent of depth of cut. In *creep feed grinding*, roughness is generally proportional to $(v_w/v_s)\sqrt{a}$.

Dressing parameters strongly influence the ground surface finish since they determine the topography of the wheel surface. For *single-point dressing* (Figure 10.18), the wheel topography depends on the dressing lead s_d and dressing depth a_d; if these parameters are known, Equation 10.15 for the roughness in cylindrical plunge grinding can be refined by replacing the coefficient R_1 with a different empirical coefficient R_2 multiplied by a term including these factors [48]:

$$R_a = R_2 s_c^{1/2} a_d^{1/4} \left[\frac{v_w a}{v_s} \right]^x \tag{10.16}$$

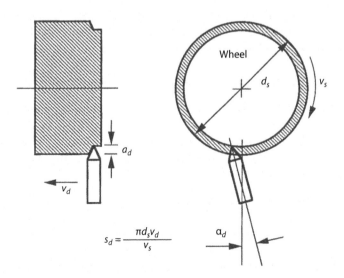

FIGURE 10.18 Single-point diamond dressing.

For *rotary diamond dressing*, the wheel roughness is found to depend on the angle δ at which the diamonds initially cut the wheel surface; in this case Equation 10.15 can be refined to include this parameter [49]:

$$R_a = R_3 \delta^{1/3} \left[\frac{v_w a}{v_s} \right]^x$$ (10.17)

where R_3 is an empirical coefficient.

The earlier equations are generally valid for a newly dressed wheel, although their accuracy is sensitive to the dressing conditions and geometry. However, as the wheel wears, they do not predict the surface finish accurately. Wheel breakdown is very difficult to predict because it is affected by several parameters, which are difficult to quantify. Roughness values from these equations provide lower bounds for the surface roughness; they are most useful in predicting the effect of varying grinding conditions on the roughness when the wheel type and work material are held constant.

The condition of the dressing tool also influences the ground surface finish; an optimum combination of the dressing lead, dressing depth, and dressing tool sharpness generally gives the best surface finish. A finer finish is obtained when the dressing tool is sharp rather than dull. The nature of the dressing cycle also has an influence. In single-point dressing, for example, dressing should be carried out in a single pass, with the tool being withdrawn from the wheel prior to retraction; if the tool is allowed to cut on both the feed and retraction strokes, interfering threads will be cut on the wheel, which will result in a nonuniform finish on the ground workpiece [50].

The properties of the wheel that have the strongest influence on the ground surface finish are the wheel grit size, grit spacing, effective diameter, and hardness. Smoother surface finishes are usually obtained with fine grained wheels; as the wheel grit size increases, the effective spacing of cutting edges decreases, so that roughness peaks are more closely spaced and thus shorter. For similar reasons, the ground roughness also decreases with increasing grain density. With other factors held constant, the average chip thickness and ground roughness both decrease with increasing effective wheel diameter. Finally, smoother finish is usually obtained as the wheel hardness is increased. For example, wheels with hard shellac bonds are used to produce smooth finish on hardened workpieces [34]. As discussed in Chapter 4, the effective hardness of the wheel depends to some extent on the grinding conditions as well as the wheel structure.

Finally, it should be noted that the ground surface finish deteriorates markedly if chatter occurs. The methods of controlling chatter in general cutting processes discussed in Chapter 12 are broadly applicable to grinding. Since the specific cutting energy and contact forces are high in grinding, system stiffness is a particular concern, and the condition of the main spindle bearings and related elements should be closely monitored [50].

10.7 RESIDUAL STRESSES IN MACHINED SURFACES

Machined surfaces often exhibit residual stresses, which are induced both by differential plastic deformation and by surface thermal gradients [34,51–53]. Stresses due to plastic deformation are obviously mechanically induced, but those due to thermal gradients may reflect phase transformations or chemical reactions. These stresses increase with tool wear, since both deformation forces and tool–workpiece frictional heating increase. For a sharp tool, significant residual stresses typically do not occur at depths much greater than 50 μm below the surface; for worn tools, however, significant stresses may occur at 5 or 10 times this depth.

Residual stresses are undesirable for two reasons. First, residual stresses give rise to residual strains, which can cause significant distortions in thin-walled workpieces. This has long been an issue in airframe structure and turbine blade machining. (A related phenomenon occurs when machining thin-walled die castings; in this case machining often results in deformation due to removal of a cast residual stress layer.) Second, tensile surface residual stresses reduce the fatigue resistance of the surface [54]. This is a particular issue for parts subject to cyclic loading, such as bearing races, cam lobes, and rotating turbine components. A standard method for countering both these effects is to remove most material in initial machining passes, leaving a small amount of stock to remove in a final pass to clean up any distortion and significant surface stresses. This approach adds cost but is effective when properly executed.

In most cases, plastic deformation results in compressive residual stress, while thermal gradients in the absence of phase transformations result in tensile residual stresses. This is illustrated in a simplified fashion in Figure 10.19 [55]. Plastic deformation of the surface generally produces an initial tensile stress behind the tool, which results in a compressive stress after elastic recovery due to overstraining. The workpiece surface layer is heated by plastic deformation and tool–workpiece friction and is therefore at a higher temperature than the substrate of the workpiece. When the part cools, the surface layer, which had expanded at a high temperature, contracts more than the substrate. The differential contraction results in a residual tensile stress in the surface. When surface temperatures are particularly high, such as when cutting hard materials with high specific energies or when cutting using a worn tool, the slope of the temperature gradient into the workpiece and peak residual stresses increases. The situation is more complex if phase transformations or chemical reactions occur in the surface layer. It should be noted that applying a flood coolant can eliminate high workpiece surface temperatures and thus tensile residual stresses in the machined surface; this is a standard method for combating residual stress, especially in grinding.

Residual stresses and micro-hardness distributions in machined surfaces have been documented since the early 1950s, leading to the development of models for determining the depth of

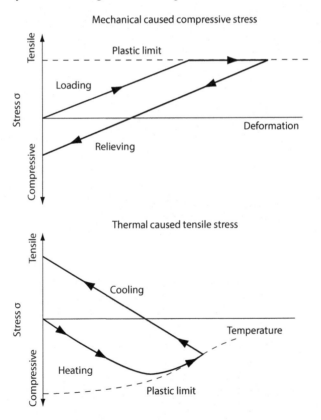

FIGURE 10.19 Simplified schematic illustration of the formation of residual stresses in machined surface layers. Plastic deformation tends to give rise to compressive stress, while thermal gradients tend to give rise to tensile stresses. (After Reinhardt, M. et al., Modeling of Difficult to Cut Materials, *Proceedings of 2002 Third Wave AdvantEdge User's Conference*, Atlanta, GA, April 2002, Paper 5.)

the hardened layer [56]. More recently, finite element methods of machining have been used to predict residual stresses from computed stress and temperature distributions [57–67]. The accuracy of these predictions depends on the accuracy of constitutive models and the stress and temperature. In early studies in which residual stresses computed using commercially released finite element programs have been compared to x-ray diffraction measurements, the agreement was inconsistent [63,64]. More recent work has shown better general correlation between calculations and measurements [67]. For aerospace alloys, accuracy currently appears to be better for nickel than for titanium alloys [68,69].

Residual stress formation can be reduced by reducing plastic deformation, frictional heating, and temperature gradients at the tool–workpiece interface; this can be done by applying flood coolants and avoiding the use of dull tools as discussed before and by reducing the cutting speed or increasing the relief angle on the tool. For materials that are susceptible to surface damage, such as nickel alloys, down-milling rather than up-milling is normally used to avoid dragging the tool margin over the machined surface (Section 2.6).

Residual stress formation is a particular problem in grinding due to the high cutting speed, high specific energy, and blunt relief geometry of the cutting edges characteristic of the process. Residual stresses in grinding are most effectively controlled by reducing the grinding wheel speed, increasing the work speed, or improving oil or coolant application.

10.8   WHITE LAYER FORMATION

White layer formation occurs when cutting ferrous materials, especially steels [5,47,70–72]. The white layer is a surface layer, which has undergone microstructural alterations caused by excessive surface temperatures and air hardening. It is resistant to standard etchings, so that it appears white under an optical microscope (or featureless in a scanning electron microscope.) The white layer has the same chemical composition as the substrate, but due to its different microstructure it has different mechanical properties, and most significantly increased hardness. In grinding, the white layer is reported to be composed of a hard layer of untempered martensite with a softer layer of tempered martensite often forming beneath it [73], although a variety of austenitic and other microstructures appear to occur in other cases [74–76]. Regardless of its microstructure, the white layer acts as a stress riser, which significantly reduces the fatigue life of the part. White layer formation should therefore be avoided if possible, and any white layer detected on the part should be removed by additional processing.

White layers occur in processes in which high surface temperatures are generated in the absence of sufficient cooling by cutting fluids, specifically in grinding [47,50,75,77–79], hard turning [70,75,76], drilling [5,80], and electro-discharge machining [70]. In drilling, white layer formation results from the heating produced by both the cutting action of the point and the rubbing of the margin [5]. Although much of the affected layer may be removed by the drill as it advances into the workpiece (Figure 10.20), some sections near the hole entry and at significant depths may remain following completion of the operation. The white layer in drilling not only reduces the fatigue resistance of the workpiece but interferes with subsequent operations such as reaming, threading, and honing. White layer formation in drilling can be controlled by reducing the feed rate or cutting speed, avoiding the use of worn drills, and improving cooling through the use of through-the-tool coolant. In grinding, white layer formation results from frictional heating of the workpiece surface by the abrasive grains. It can be controlled through the same measures used to reduce residual stress formation; the measures discussed in the next section to reduce surface burning are also effective. In hard turning, white layer formation increases with tool wear and is combated by reducing tool wear rates or changing tools more frequently.

Anomalous layers similar to the white layers in steels have also been observed when machining Ti- and Ni-based alloys, brass, and copper [81,82]. Ni alloys are subject mainly to mechanical damage owing to their high melting points and transformation temperatures; in these alloys, the white layer is characterized by a severely distorted microstructure. In Ti alloys, white layer is also

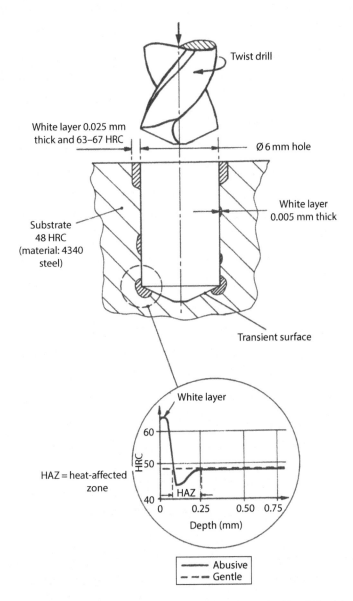

FIGURE 10.20 White layer formation when drilling steel. (After Smith, G.T., *Advanced Machining*, Springer-Verlag, New York, 1989, pp. 198–223.)

characterized by severe plastic deformation but is usually accompanied by thermal damage. Ti alloys machined with worn tools may also exhibit alpha case, a brittle layer caused by oxidation and more commonly associated with titanium casting [83]. As with steels, white layer formation increases with tool flank wear for these materials.

10.9 SURFACE BURNING IN GRINDING

Surface burning concerns often limit the maximum wheel speed or stock removal in grinding operations. Surface burning has been studied in detail mainly for carbon and alloy steel workpieces [47,50,78,79]. Burning is accompanied by metallurgical and chemical phenomena such as oxidation (which produced the characteristic burn marks on the surface), tempering, residual stresses, and phase transformations; apart from aesthetic concerns, burning should be avoided because these

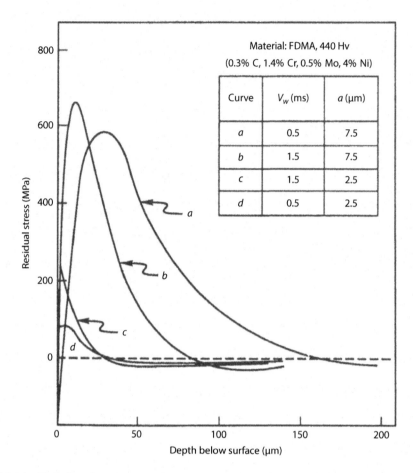

FIGURE 10.21 Variation of hardness with workpiece surface depth when grinding steel with and without surface burning. (After Malkin, S., *Grinding Technology: Theory and Applications of Machining with Abrasives*, Ellis Horwood, Chichester, U.K., 1989, Chapters 4, 6, and 7.)

phenomena lead to a reduction in fatigue life. When no oxide layer is visible (e.g., when it is removed during spark-out), the occurrence of burning can be detected through microhardness measurements (Figure 10.21) or ferrographic analysis of the grinding swarf [78]. Burning is obviously a thermal problem, which can be controlled by reducing the workpiece surface temperature [47]. Thermal analyses indicate that burning occurs when a critical temperature, which depends on the specific grinding energy, is exceeded. The critical specific energy for burning, u^* is found to depend almost linearly on a factor computed from the grinding conditions (Figure 10.22):

$$u^* \propto d_e^{1/4} a^{-3/4} v_w^{-1/2} \tag{10.18}$$

where
  d_e is the effective wheel diameter
  a is the depth of cut
  v_w is the workpiece speed

Altering grinding conditions to increase the value of this factor increases cooling and the critical specific energy for burning. The critical specific energy is larger in creep-feed grinding than

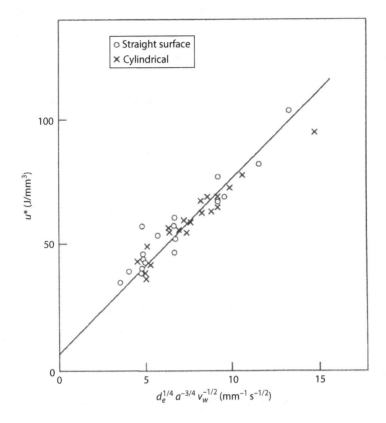

FIGURE 10.22 Variation of critical specific energy for surface burning u^* with $d_e^{1/4}a^{-3/4}v_w^{-1/2}$. (After Malkin, S., *Grinding Technology: Theory and Applications of Machining with Abrasives*, Ellis Horwood, Chichester, U.K., 1989, Chapters 4, 6, and 7.)

in other types of grinding because the stock removal is comparatively large. Burning can also be reduced by monitoring the wheel condition to ensure that the wheel remains sharp (which reduces frictional heating); force or power measurement systems may be used to monitor wheel sharpness. Burning can also be reduced by switching from water-based to oil-based grinding fluids. Oil-based fluids reduce heat build-up because they have better lubricity [50] and are also typically used at lower wheel speeds. Burn limits and residual stress predictions can also be computed using grinding simulation programs and used to suitably adjust infeed cycles, the number of passes, and wheel characteristics to maximize productivity [84].

10.10 MEASUREMENT OF SURFACE FLATNESS

Machined surfaces used for sealing or motion control often have a specified flatness tolerance. As shown in Figure 10.23, flatness is defined as the minimum separation between two parallel planes containing the entire surface profile. Since the entire profile is not normally measured with high resolution, the sampling method and resolution of individual measurements both influence measured flatness values. Therefore, the flatness and profile/waviness requirements on the part should be considered in the selection of the process even though the proper surface finish can be achieved.

There are several contact and non-contact methods for measuring the flatness or profile of a surface. Automated flexible methods use coordinate measuring machines (CMM). All CMMs have a mechanical or optical probe attached to the third moving axis of the machine. An electronic touch

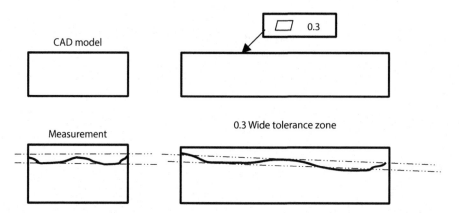

FIGURE 10.23 Illustration of surface flatness.

trigger probe, the most common probe, uses a stylus for contact with the surface; it can either measure individual points or scan/drag the stylus along the surface of the part taking points at specified intervals. The scanning method is often more accurate, since it samples more points, and is also faster. Both the specific points measured in the point mode and the path scanned in the scanning mode affect the reported measurement. These points or paths should be carefully chosen to reflect real areas of concern for part function.

Non-contact methods, such as optical, laser, or white light interferometry, take thousands of measurements per second to generate a 3D model of the part surface. Such systems can cover large areas, achieve high throughput, provide operator-independent results, and capture information which often better controls a process. White light interferometers [85] offer high resolution but often have a limited field of view and require image stitching to cover larger areas; stitching generally requires precise alignment to prevent some images from going out of focus. A recently developed 3D holographic laser method [86] offers submicron level resolution in the normal direction while covering a much larger area than white-light interferometers. A 3D measured surface model is usually compared to the corresponding CAD model. A 3D comparison can provide a color map of the CAD model displaying deviations between the 3D model and the physical part.

Correlation among the different methods is not always straightforward and may be influenced by the fixturing, residual stresses, and details of the machining process. As an example, a flatness comparison was carried out between contact probe CMM methods and a non-contact laser holographic system method on a channel face of an aluminum valve body [87]. The flatness of the surface was measured in a CMM using 27 individual points, with a CMM scanning along specified channels, and with the laser holographic system. The laser system gave a higher flatness reading, partly due to the increase density of data, but also due to increased error due to reflection of light by the milled surface. The results from the aforementioned three methods based on several tests are given in Table 10.1. A trace from the 3D holographic laser method is shown in Figures 10.24.

The flatness repeatability for identical parts varies depending on the factors influencing the surface distortion and the magnitude of the natural flatness component. The flatness variation among

TABLE 10.1

Surface Flatness Measurements Using Three Methods

Method	Flatness (μm)
Individual points	30
Scanning stylus	36
3D holographic laser	43

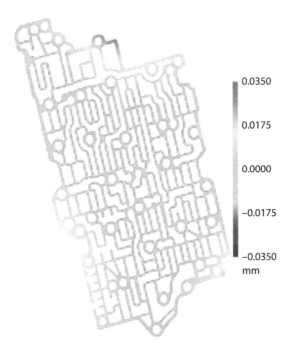

0.0350

0.0175

0.0000

−0.0175

−0.0350
mm

FIGURE 10.24   Illustration of the of the channel surface flatness using the 3D holographic laser method.

identical parts can be as high as 50 μm for a 100 μm flatness specification on a 70 cm × 40 cm surface area. However, if the machine temperature and tool wear are controlled, the flatness variation can generally be maintained within 10–20 μm in similar cases.

10.11   SURFACE FLATNESS COMPENSATION IN FACE MILLING

Flat surfaces are often machined by face milling, especially in high volume applications. The standard methods of controlling flatness in these applications include fixture and cutter optimization and the use of multiple passes, including a low DOC finishing pass. Limited feed rate optimization, in which the feed rate is slowed over sections of the tool path with low part stiffness, is also used. Other tool path compensation methods have been described in the research literature to further improve flatness, with the goal of reducing the impact of fixture variation and eliminating passes in multipass operations. These methods are described in this section.

Several cutting parameters including the depth of cut, feed rate, and tool path can be used to compensate the surface flatness. Most compensation methods rely on process models for individual machining processes, particularly the static process simulation models described in Chapter 8. These models could incorporate the following process variations: (1) geometric errors due to casting variation and machine/spindle/tool/fixture errors, (2) static errors resulting from clamping and cutting distortions, and (3) dynamic errors resulting from the thermal distortion of the part and machine tool, tool wear, and chatter.

A measured casting thickness variation (instead of the casting process tolerance) should be used in these models to estimate the depth of cut variation affecting force variation along the tool path, which can be used with the part/fixture compliance matrix to estimate the profile and flatness error. Another source of surface flatness/profile error is the variation of the machine geometry over time, which is more complex to model. In face milling, the cutter axis is sometimes tilted so that the trailing edge of the cutter does not mark the cut surface as discussed in Section 10.4. This results in a concave surface profile variation in the direction perpendicular to the cutter path.

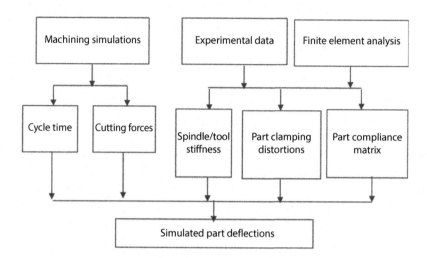

FIGURE 10.25 Procedure to estimate part deflections and optimize the milling process.

The geometric flatness and profile errors can be estimated from the motion errors, cutting forces and fixture part compliance, and static stiffness of toolholder/spindle system (see Chapters 4 and 8). The approach is illustrated in Figure 10.25 for calculating part deflections (Chapter 8) and optimizing the cycle time (Chapter 13). The compliance matrix of the part and the spindle/tool stiffness can be extracted either experimentally or through FEA of the part and fixture as discussed in Chapter 8.

10.11.1 Tool Path Direction Compensation

The tool path direction has an important influence on both cycle time and surface flatness when face milling large surfaces. Cutting forces interact with the part and fixture to cause deflections, and due to spatial and directional variations in part and fixture stiffness, applying forces from different direction (by changing the tool path) results in different flatness and profile errors. The milling process, up or down milling, also affects the forces on the part for a similar reason. The removal of residual stress is a progressive process, which is affected by the tool path direction and local DOC. Tool path direction compensation (TPDC) is performed on a macroscale because the face mill cutter diameter is usually large in order to cover as much surface area as possible.

The cutter diameter in relation to the part surface size determines all possible tool paths and the corresponding number of passes at a given level required to cover the surface. Visually the surface appears to have seam line(s) between passes when more than one pass is required. The seam line sometimes results in a surface step depending on the tool-part stiffness and process vibration. Therefore, the tool path direction and the number of passes should be optimized based on the cycle time (as discussed in Chapter 13) and surface flatness/profile error incorporating part/fixture stiffness and cutter diameter effects.

Because tool path changes affect the entire area covered by the face-milling cutter, only the average profile and flatness of the surface can be improved by such changes. It is generally not possible to identify a linear or simple curve representation for the surface profile error along the tool path and across the width of the cutter that would permit compensation using an exact mirror image of the profile [86].

As an example application of the TPDC method, consider face milling of a transmission valve body (Figure 10.24). Traditionally, a two-pass (rough and a finish) face-milling process is used for this application. Cutting forces for different tool path directions were obtained through process simulation (as explained in Section 8.4). The face-milling operation was simulated using a 125 mm face-milling cutter at various tool paths to reduce the cycle time and improve flatness. In this case,

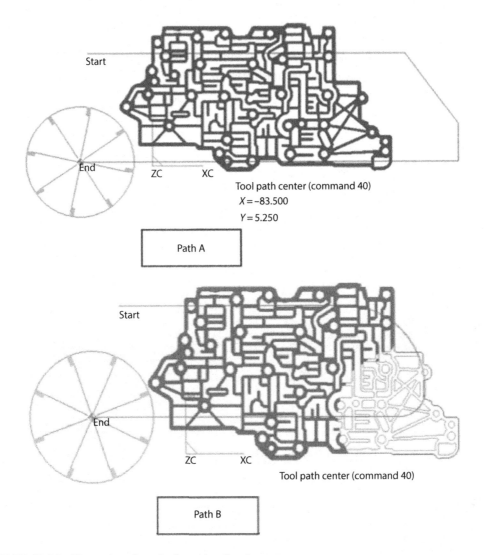

Tool path center (command 40)
$X = -83.500$
$Y = 5.250$

Path A

Tool path center (command 40)

Path B

FIGURE 10.26 Illustration of two horizontal tool paths.

the simulation provided the cutting forces and the surface profile was estimated from the part compliance matrix as explained in Section 8.8. The tool path A for the rough and finish passes is shown in Figure 10.26. In this case, the cutter traveled out of the surface at the end of the first pass and re-entered the surface for the second pass. The power and normal force requirements for the finish pass with 0.3 mm DOC are summarized in Table 10.2. A similar optimized tool path B is shown in the figure, with results also summarized in Table 10.2. The mill makes a curvic interpolation into the horizontal path to connect the two straight paths to mill the full width of the part. The cycle time for the milling path A is 25% longer than the optimized path B while the flatness is similar for both paths. The optimized path B is preferable because it requires lower spindle power and cutting force in addition to a cycle time reduction. Single pass face milling was also evaluated to reduce the number of passes for tool path B. The single pass removed 1.5 mm nominal stock available in the part as compared to the 1.2 and 0.3 mm, respectively, for the rough and finish passes in the previous cases. The measured flatness of the surface was 48 μm as shown in Figure 10.27, which is 50% higher than for the two-pass process. The largest profile error was at the right section of the part due to lower part-fixture support stiffness.

TABLE 10.2

Surface Flatness Measurements for Two Different Tool Paths for the Finish Pass (0.3 mm DOC)

	Tool Paths	
	A	B
Power, average (kW)	12	8
Power, max. (kW)	15	12.5
Force, average (N)	45	38
Force, max. (N)	60	53
Flatness (μm)	32	31
Cycle time (s)	4.7	3.51

FIGURE 10.27 Surface flatness for tool path B with single-pass (1.5 mm DOC).

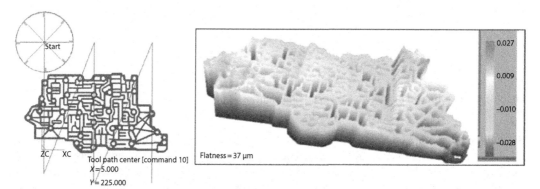

FIGURE 10.28 Illustration of vertical tool path and its surface flatness.

The tool path was changed from horizontal to vertical to further investigate the tool path influence on the flatness. This tool path covers the full surface in three vertical passes as illustrated in Figure 10.28; the mill travels out of the part at the end of each pass to avoid large contact of the milling cutter with the surface, which would result in larger forces and deflections. The entry of each pass is from the same side of the part. The horizontal path required two passes with a single seam line instead of two seam lines in the vertical path. The measured surface flatness for the vertical path was 37 μm as shown in Figure 10.28. The 20% average improvement due to the change in the tool path direction is significant to single pass process capability. One explanation for this flatness improvement is the higher part/fixture rigidity along the shorter path.

10.11.2 Depth of Cut Compensation

The milling path along a plane can be compensated normal to the plane by cutting shallower on convex surface sections (high spots) and deeper on concave surface sections (low spots) [86]. Since the depth of cut compensation (DOCC) method can reduce the error components only along to the feed direction, it is necessary to find a surface profile that can best represent the surface shape in the region the cutter passes over, in order to determine a proper compensation cutting profile. The method is further explained in Figure 10.29. Depth of cut compensation is best suited to applications in which the surface profile error is convex.

The procedure for estimating the surface profile deviation from the zero (part locating) plane consists of the following steps:

1. The part surface flatness and profile are measured based on a large number of points.
2. The surface profile is analyzed along the feed direction.
3. A number of critical points N are identified on the surface where there is a significant profile deviation/error along the feed direction.
4. A zero plane is defined passing through the three ($C1$, $C2$, $C3$) support/clamping points in the fixture supporting the part.
5. The deviations of the N surface points from the zero plane are estimated, as shown in Figure 10.30, and their average value is computed using

$$[\Delta Z_{xy}] = [Z_{xy}] - [Z_{pxy}] \tag{10.19}$$

6. The milling cutter is programmed using linear ramping in the X/Y and Z axis between points of slope change in the surface profile.

The matrix Z incorporates all the Y-points covered by the width of each pass. All the Z-points at each X-location covered by a single pass are grouped.

$$Z_{x_i m} = \{Z_{x_i y_i}, Z_{x_i y_{i+1}}, \ldots, Z_{x_i y_{i+j}}\} \tag{10.20}$$

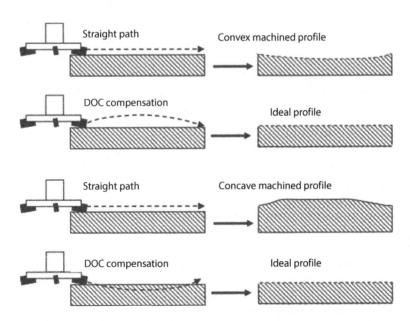

FIGURE 10.29 Schematic diagram of DOCC method for convex or concave profiles after machining.

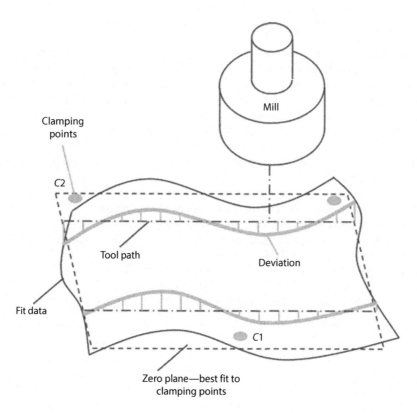

FIGURE 10.30 Schematic of DOCC method.

The user evaluates either the average of the Z-points at each X_i-location or the best fit function through all the Z-points along the tool path. If the peaks or valleys along the tool path can be compensated with the DOCC approach, the milling cutter is programmed using linear ramping to reduce the peaks and valleys. In addition, surface decomposition using a discrete cosine transformation (DCT) can be used to extract the surface profile from the measured data [86]. A DCT expresses a sequence of finite data points in terms of a sum of cosine functions at different frequencies.

In simple terms, the DOC reduction is smoothly programmed as a mirror image of the profile error. Smooth transitions between DOCs are made by either programming curve or multiple straight segments of different DOC to approximate the curvature. The DOCC averages out the deviations by cutting more on the high spots and less on the low spots along the tool path surface profile. The compensation is performed on a macroscale because of the cutter diameter effect in relation to local profile errors. However, the reverse profile deviations don't produce a flat tool path relative to the part finished surface in clamped position. When a large diameter face mill is driven into the valleys, or climbs over the hills, the finished surface can be different than the theoretically calculated one. Therefore, the compensation analysis serves as a guide to reduce the average profile error. The ramping step 6 of the method can start from in to out or from out to in, depending on the surface error along the feed axis.

This DOCC method minimizes the average profile deviations from the zero part locating plane to improve the flatness. However, this method requires a slight spindle tilt in the direction of the tool path to provide relief behind the cut to avoid back cutting and gouging, especially on concave surfaces. Concave profile errors cannot generally be corrected using the depth compensation due to the face-milling cutter size limitation. However, shallow concave surfaces can be compensated if the spindle-part tilt process is used to avoid trailing tool marks.

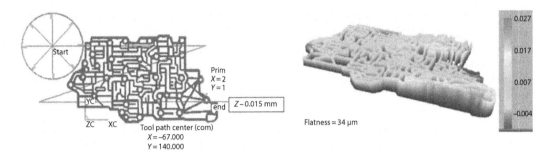

FIGURE 10.31 Single pass milling (1.5 mm DOC) process using DOCC and SPTC and its surface flatness.

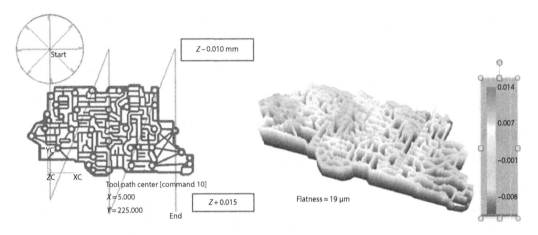

FIGURE 10.32 Milling with DOCC in the third pass.

The DOCC method was applied at the end of the second pass in the previous single-pass horizontal path (Figure 10.31) after the zero plane passing through the three support/clamping points of the fixture was estimated in order to evaluate the surface error from that plane. In this case, the tool path was also modified so that both passes are in the same direction to avoid any significant step on the seam line between passes since spindle-part tilt compensation (SPTC, to be discussed in the following) was also used. The surface was compensated by cutting 0.015 mm deeper as illustrated in Figure 10.31. The flatness of the above process using both DOCC and SPTC methods was reduced to 34 μm, which represented a 30% improvement compared to the flatness in Figure 10.27.

The vertical single pass tool path in Figure 10.28 was also evaluated with DOCC. Since a significant profile error is at the right section of the surface, the DOC for the third vertical pass was compensated −0.010 mm at the start and 0.015 mm at the end, as shown in Figure 10.32. The flatness was reduced to 19 μm with DOCC. The above vertical tool path with DOCC for the third pass process resulted in better surface flatness than the horizontal tool path with DOCC and SPTC.

10.11.3 Tool Feed Compensation

The tool feed compensation (TFC) method focuses on cutting force–induced errors, which can be directly controlled by the optimization of feed and speed in relation to the cycle time. The feed per tooth can be optimized to improve the surface flatness because cutting forces exerted on the part surface are proportional to feed. The TFC method has been effective in cases where certain sections of the part have low stiffness such as cantilever sections (farther away from the fixture supports) or sections with large span between supports. In such cases, the feed is reduced over low stiffness sections of the part, and increased in regions with higher stiffness. This approach can yield reduction in flatness error without increasing machining time. However, it produces spatial variations in surface finish, which may not be acceptable in finishing operations.

A process simulation is often used to estimate the cutting forces at different feeds, which are used together with the part compliance matrix to identify the feed for acceptable surface deformations. The structural stiffness or compliance of the part in the clamped state is used in simulations to evaluate feed rate changes to minimize surface deflection over critical sections to improve flatness. The surface flatness can be estimated using Equation 8.55 based on the estimated distorted surface points. Sections 8.8 and 8.9 review some of the details of the method. If the feed for acceptable flatness is too low for the tool effectiveness, the DOC is reduced instead by considering a multipass process to reduce the cutting force per pass. In face milling, the part and fixture are often more rigid than the tooling and machine structure. In this case, the deflections are estimated by multiplying simulated forces (at specified feeds) by the compliance of the tooling. The compliance of the tooling can be estimated from the axial or radial compliance of the spindle, which can be measured or estimated from a finite element model. When milling thin-walled parts, the tooling is more rigid than the part and fixture, and the compliance matrix of the part/fixture is multiplied with the simulated forces to minimize deflections by optimizing feed changes along the tool path.

10.11.4 Spindle-Part Tilt Compensation

The spindle-part tilt compensation (SPTC) method is used to prevent contact of the face-milling cutter trailing edges with the machined surface. If the face-milling cutter is perfectly aligned with infinite stiffness, a hash-marked surface pattern is produced as shown by the solid line marks in Figure 10.33. Undesirable crosshatches (shown by the dotted line marks in Figure 10.33) are produced because the heel of the trailing edges generally wipes at a very light DOC the area that is just machined. This is due to the following reasons: (1) spindle misalignment relative to the normal to the machine axis; (2) milled surface springback due to the relaxation from clamping and machining forces and the removal of residual stresses; (3) milling cutter/spindle deflection, and (4) radial and axial cutting tool runout. The crosshatches are more obvious in a single pass milled surface because the cutting forces are larger compared to a two pass process with a small DOC for the finish pass.

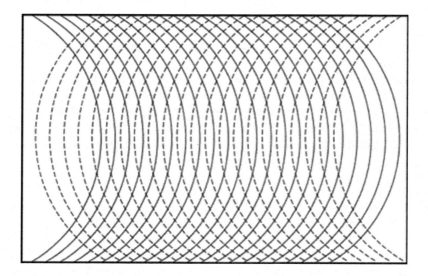

FIGURE 10.33 Horizontal tool path marks on a milling surface.

In transfer line applications, the spindle is tilted by a small angle in the direction of the feed to provide relief as explained in Section 10.4 and Equation 10.14. The SPTC method was developed for four- or five-axis machining centers [87]. The steps of this method are as follows:

1. The part is tilted along the direction of the feed (or tool path) at a certain very small angle θ, by rotating the table (table rotary axis perpendicular to the tool path).
2. The coordinate system is rotated by the same amount around the rotary axis.
3. The milling process path is programmed in the direction of the tilt (as planned without any tilt).

This method creates a controlled angle θ between the trailing edge of the face mill and the finished part surface as shown in Figure 10.34. The SPTC method provides a concave machined surface and results in a waviness error as illustrated in Figure 10.34. The depth of concavity d_f can be calculated from Equation 10.14. e is either the width of the part (for single pass when $D_e > e$) or the width of the passes ($D_e < e$). Therefore, the objective for the SPTC method is to estimate the minimum angle to avoid contact of the trailing edges while minimizing the waviness error d_f. As a starting point for a new application, the allowable d_f can be estimated as 20%–40% the flatness tolerance. This provides a tilt angle θ, and the clearance at the trailing edges can be calculated from the equation

$$s = D \tan(\theta) \tag{10.21}$$

This estimate of s is based on a rigid spindle-cutter system. However, the stiffness of the spindle-cutter assembly is limited and some deflection due to cutting forces is expected. Therefore, the

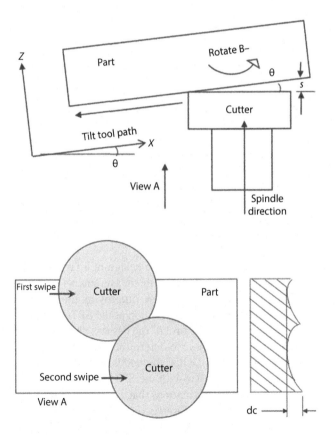

FIGURE 10.34 Schematic illustration of the SPTC method.

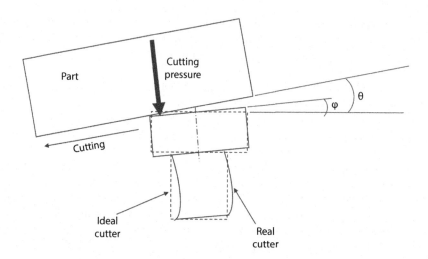

FIGURE 10.35   Schematic illustration of the SPTC method's tilt including tool deflection.

deflection of the trailing edges toward the part's finished surface is represented by the angle φ in Figure 10.35. As long as $\theta > \varphi$, and the resultant clearance at the trailing edges is sufficient to avoid contact, the estimated θ is selected. If $\theta \leq \varphi$, either the tilt angle θ is increased, assuming the waviness error is not too large compare to the flatness requirement, or a two pass milling process is used so that the low cutting forces of the finish pass reduced the flatness error. Therefore, the allowable tilt angle θ is obtained either from Equation 10.14 or Equation 10.21 and the actual relief angle is estimated as $\theta - \varphi$.

10.11.5   Surface Flatness Compensation Methods Characteristics

Seam lines in multipass paths and especially with the TPDC method are significant because a seam line step can pass the flatness specification, even if it is greater than 0.01 mm, but may still impact part functionality. Therefore, the surface finish should be measured in the areas of seam lines to properly define the errors across seam lines. In general, a 0.01 mm seam line step is acceptable for most applications. In addition, the compensation values among adjacent passes should be controlled to be smaller than the seam line tolerance. Therefore, flatness compensation via the tool path compensation method is often significantly constrained.

The DOCC is very effective only when applied with SPTC, otherwise the two axis ramping allows back cutting. Generally, the SPTC method is required to control the surface flatness characteristics. In addition, a 10° inclination of the cutter inserts is often helpful in preventing wiping marks on the surface.

FEA analysis of surface distortion is useful in the design of a face-milling process and fixture; a trial-and-error approach based on the surface profile measurements and the application of appropriate compensation methods can beneficial in further improving the flatness assuming the profile errors are analyzed and well understood. The application of the SPTC method to avoid back cutting in machining centers is very critical to the control of the flatness and successful application of other compensation methods. The main objective is to determine the minimum tilt angle for the milling cutter to optimize the surface flatness. The DOCC method is possible on convex profile surfaces and light concave surfaces when applied together with the SPTC method. This approach is successful when the profile error indicates surface deformations that can be covered by the cutter diameter. The diameter of the cutter in relation to the surface area is very important for the optimum tool path with respect to cycle time and surface flatness. The optimum tool path for flatness improvement is different than the optimum tool path for cycle time. Therefore, the TPDC method should be integrated with feed and speed optimization analysis, the surface distortion from FEA. The influence of TFC method is very effective at the design stage of the machining process together with the optimization of the other cutting parameters.

The TPDC and TFC methods are primarily implemented during the design of the face-milling process through off-line modeling. The SPTC and DOCC methods require measurements of the surface profile and a limited trial-and-error approach. All of the above compensation methods can be considered during the trial-and-error approach depending on the repeatable profile error on the surface and all the other constraints on the cycle time, surface finish, the ratio of the cutter diameter to part size, and the characteristics of the machine tool.

As explained in the surface finish sections and several other chapters in this book, the fixture, clamping, and cutting geometry contribute significantly to part surface flatness. In some cases, the proper selection of the cutter geometry (as discussed in Chapter 4 for face-milling cutters) will help reduce forces significantly, which results in better flatness. The fixture and clamps should be designed to hold the part stably during milling including entering and exiting the part. Surface flatness can usually be improved by reducing spindle deflection and using a larger diameter face cutter or a more positive rake cutter. As a result there are few rules typically followed to fine tune the setup including cutter selection and avoid major concerns:

1. The fixture should be designed to hold the part correctly and the correct milling cutter size should be selected.
2. The cutter diameter should be 20%–50% larger than the workpiece width of cut. When several tool path passes are required (because the part surface width is larger than 80% the cutter diameter), the diameter/width of cut relationship should be approximately 1.3–1.6 to maintain the proper chip formation.
3. A low or zero lead angle is preferable for weak structures or thin walled parts to keep the axial force low. A 45° lead angle is good for general purpose face milling because it balance the axial and radial forces and provides smooth entry and exit in the part surface. Lead angle relates well with secondary edge orientation when using square inserts. A 10° or larger lead angle is preferable to avoid contact of the secondary edge with the machined surface unless a smaller angle is necessary for surface finish or flatness requirements.
4. The insert geometry should be properly selected for the operation. Wiper inserts can be used to improve surface finish, but care should be taken to ensure that measures for improving finish (such as wipers and secondary edges) do not negatively affect flatness.
5. The cutter insert density (cutter pitch) should be optimized based on the workpiece material, chip formation, and the cutting forces.
6. The tool path should be selected to keep the milling cutter continuously in contact with the part and avoid dwell on the part to reduce chatter tendency.

A single pass process will finish a cast up the uneven cast surface. Following, the finish pass will remove the distortion due to removal of the residual stresses during the first pass and the residual stresses due to the cutting process of the first pass. The tool path could be changed between the two passes. Different cutter and inserts geometries can be used for each pass in the two pass process if necessary. Therefore, the flatness of a double pass milling is generally better than that of a single pass if all parameters are optimized. Similarly, the surface finish from a double pass process is better than that from a single pass process.

10.12   EXAMPLES

Example 10.1 A turning insert with a sharp corner is used in a finishing operation. The lead angle of the tool (or side-cutting-edge angle, SCEA) is 20°. The minor cutting edge angle (ECEA) is 10°. What maximum feed is allowed if the surface finish requirement is 3 μm?

Solution: Equation 10.8 gives the average geometric roughness of a surface in turning with a sharp-nosed tool.

$$f = 4R_a[\cot \kappa_{re} + \cot \kappa'_{re}]$$

$$= 4(0.003)[\cot(90° - 30°) + \cot(10°)]$$

$$= 0.075 \text{ mm}$$

The feed was calculated by assuming that the geometric roughness generated by the insert is equal to the surface finish requirement. This neglects the natural part of the surface finish as discussed in Section 10.3. The maximum feed should be less than the ideal roughness calculated above. Therefore, if the natural part of the finish is assumed to be 40% of the geometric (based on previous experience for the particular cutting conditions, workpiece, coolant, and machine tool), the maximum feed should be

$$f_{max} = \frac{0.075}{1.4} = 0.054 \text{ mm}$$

Example 10.2 What corner radius should be used on an insert to give an arithmetical mean surface roughness of 1 μm? The feed selected for this turning operation is 0.1 mm/rev. What speed range should be used if the workpiece material is aluminum?

Solution:

(a) The average geometric roughness of a surface in turning with a nose radius tool is given by Equation 10.10.

$$r_n = \frac{0.0321 \cdot f^2}{R_a} = \frac{0.0321 \cdot (0.1)^2}{0.001} = 0.321 \text{ mm}$$

Therefore, the standard corner radius size 1 (corresponding to 0.4 mm) should be used. It is understood that the surface finish with the 0.4 mm radius should be higher than 1 μm since the natural part of the roughness is not included in the above calculation. If it is assumed that the natural part of the roughness is, in the worse case, equal to the geometric, the size 2 nose radius (corresponding to 0.8 mm) should be used so that the average geometric roughness is significantly lower than the upper limit of 1 μm. The surface roughness can be also maintained to 1 μm by either reducing the feed or using the same insert geometry with a wiper. However, reducing the feed will result in lower productivity, while the wiper design will result in lower tool life.

(b) The highest speed range of the machine should be used to turn this aluminum workpiece in order to minimize BUE formation and the natural part of the surface roughness.

Example 10.3 Compare the approaches for improving surface flatness for face milling the transmission valve body discussed in Section 10.11.

Solution: All the four surface flatness compensation methods (TPDC, DOCC, TFC, and SPTC) presented in Section 10.11 are applied in the valve body and validated by milling the part in a machining center. The fixture and the part are illustrated in Figure 10.36. 1, 2, and 3 locators and clamps in the Z-dir, 4 and 5 supports in the X-dir, and 6 push locator in the Y-dir are illustrated in Figure 10.36b. The part/fixture stiffness at sections RT and RB (Figure 10.35a is lower than the rest

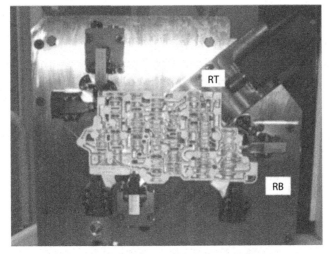

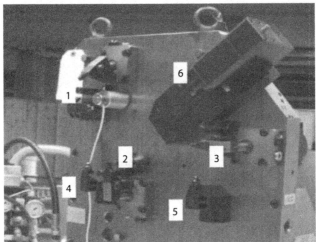

FIGURE 10.36 Illustration of the locating and clamping the part using the 3–2–1 locating scheme on the fixture.

of the part since only three supports are used in the Z-direction and the part is square with an ear at the RB section. The fourth axis in the CNC machine is the rotary table that allows the application of the SPTC method. The summary of the results is provided in Table 10.3. Multiple compensation methods are combined to improve the flatness and most of them are illustrated in Section 10.11. In all the cases, the SPTC method is utilized as long as the tool part is perpendicular to the axis of the rotary table. The methods are combined in some cases to compensate for problematic part regions due to insufficient support or stiffness. The Z-dir cutting force is shown for the TPDC cases in Figure 10.37 for comparison purposes. There is a significant deference in the force profile among these cases. However, the static part surface deflection from the simulation didn't show a significant difference among these cases since the variation was with 0.01–0.02 mm. The TFC method is applied in combination with the TPDC method (the last two cases in Table 10.3) to further improve the flatness by reducing the profile error at the right section of the surface. Therefore, the feed is lowered by 30%–40% to reduce proportionally the cutting forces. This method is applied to both the single pass horizontal and vertical paths along the right section of the surface. The flatness improvement is 6 μm on the average; it is not as significant as the improvement obtained with DOCC method.

TABLE 10.3

Surface Flatness Comparison for All Cases

Tool Paths	Passes	Compensation Methods Applied	Flatness (μm)	Case Figure
A Hor	2	TPDC & SPTC	32	10.25
B Hor	2	TPDC & SPTC	31	10.25
B Hor	1	TPDC & SPTC	48	10.26
Ver	1	TPDC	37	10.27
Hor	1	DOCC & SPTC	34	10.30
Ver	1	DOCC	19	10.31
B Hor	1	TPDC, TFC, SPTC	27	
Ver	1	TPDC & TFC	32	

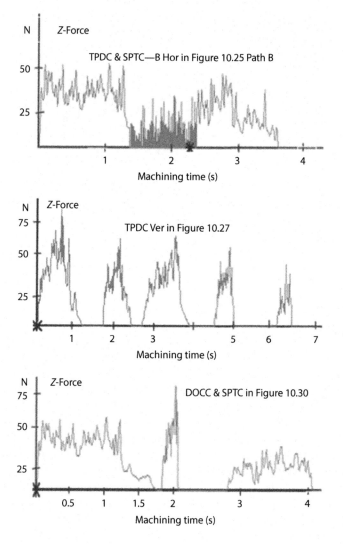

FIGURE 10.37 Cutting force for three cases in Table 10.3.

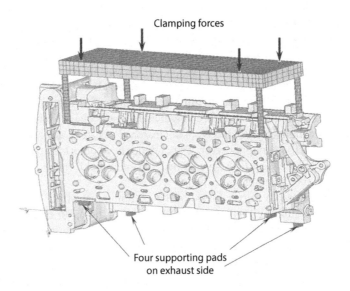

Clamping forces

Four supporting pads
on exhaust side

FIGURE 10.38 Simplified clamping fixture with four clamping pins and their connector.

Example 10.4 Investigate the surface flatness characteristics of an engine head deckface (in Figure 8.31) to improve the flatness or reduce the number of passes.

Solution: This example is used in Section 8.9 to discuss how FEA is used to simulate part distortions [88]. As illustrated in Figure 8.31, 12 points on the deckface are used to determine the surface flatness. These points were also measured in a CMM for comparison with predictions. Using Figure 10.25, the machining simulations, FEA, and cutting tests are performed to simulate flatness improvements. As mentioned in Section 8.9, one key to the success of the Deckface flatness prediction is the accuracy of the station level error models such as geometric errors (casting variation, spindle/tool errors, part pad height errors, etc.), static errors (clamping distortion, cutting distortion, spindle/cutter deflection), and dynamic errors (cutting distortions).

Figure 10.38 illustrates the fixture model including the four pads and the corresponding clamps. The 4–2–1 locating scheme is used in the fixture to machine the deckface (a 4–2–1 scheme is often used for large prismatic parts subjected to large cutting forces, since a standard 3–2–1 scheme would leave one or two corners of the part unsupported). The locating model is defined by the four locating pads in the primary plane and two pins (round and diamond) in the secondary and tertiary planes. An equalizer was used in the fixture to evenly distribute the clamping load at each of the four clamping rods (provided by a single hydraulic cylinder). Only simplified clamping rods and their connector were modeled since this study is mainly to focus on cylinder head deformation rather than fixture behavior. But the compliance of the clamping rods is considered by modeling the fixture as deformable material. The clamping force is 2300 N/clamp.

Some variation of the actual pad heights in the part is expected through the machining process. The four primary locators in the fixture will also probably have some relative error from the nominal value of zero (about 5–15 μm) due to the manufacturing and assembly errors of fixture components. Since three points define a plane, the fourth point will not in general lie in that plane. As a consequence, there will be some gap in one of the four pads (between the locating pad and its corresponding locating point on exhaust face) when the head is rested on the fixture before the clamping load is applied. This gap could be closed under clamping loads if there is sufficient deformation of the head body in that section. Also, both head body and fixture pads are considered as deformable bodies. To accurately simulate this, contact pairs between head body and clamping pins and between head body and supporting pads with actual shapes and sizes are created in this model.

The actual geometry size for the contacts are modeled using 0.3 value for the coefficient of friction at the contact pads and clamping locations. The distortion of the part during clamping relative to a fixed coordinate system estimated by FEA is the combined effect of rigid body motion of the part due to localized elastic deformation at the contacts and the part deformation itself due to clamping forces. Three cases are modeled: case 1 is the nominal where all pads were in contact, cases 2 and 3 have three pads in contact with 24 and 40 μm clearance at the fourth pad B. The 12 points on the deckface are also used to define the deckface height and flatness due to clamping distortion.

The results from the models, with and without clearance at pad B, are compared to understand the significance of this static clamping error. The cylinder head deformed elastically under clamping forces as well as when the deckface is machined to its nominal form and location. And then the part is unclamped and the clamping deformation will spring back to cause the deckface to have a form error. This form error is a mirror image of the clamping distortion. The deckface deflection at the 12 points due to clamping predicted from the FEA is shown in Figure 10.39 for all cases including the measurement from the actual cutting. It can be seen that deck face flatness is significantly affected by the initial clearance. The deflection of the 12 points on deckface for the nominal case (with zero relative pad height error) is very small. The deflection of the case with 40 μm gap at pad B is large and significant to flatness. The 40 μm gap is reduced to 16 μm after clamping at 2300 N/clamp. The gap at pad B of 24 μm is closed after clamping deformation. It can be seen that flatness error increased significantly (by a factor of up to 15) as the gap at pad B increased from zero to 40 μm. The FEA simulation shows that after the initial clearance is completely closed, further increasing clamping force will have much less significant impact on deck face flatness. The above results illustrate the significance of minimizing the relative height error between the four support locations. In addition, it illustrates the concern of using 4–2–1 locating scheme versus the standard 3–2–1 scheme, although as noted above it is often necessary for large prismatic parts. The cutting distortion is estimated through simulation as in the previous example and the variation of the 12 points is obtained for different conditions.

The flatness error is defined for each part as the range (maximum minus the minimum value) of the height of the surface at 12 points. The Monte Carlo method is adapted to stack-up all anticipated and predicted deviations/errors during manufacturing of the selected part features. One advantage of the Monte Carlo method is that it lets the user enter a probability distribution for each feature error. The 3DCS model accepts multiple statistical distributions to define the variation for each input and output variable. The feature errors are then stacked-up in the computer using 3DCS software.

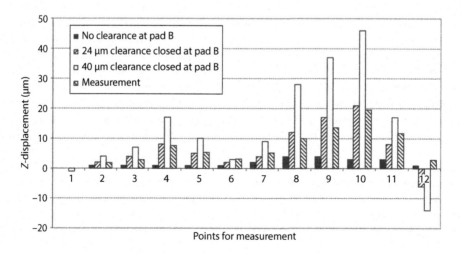

FIGURE 10.39 Comparison of deckface displacements due to clamping with different initial clearances.

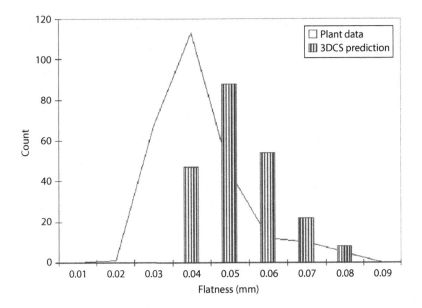

FIGURE 10.40   Comparison of flatness using 219 heads after finish milling operation.

The Monte Carlo simulation will run a specified number of parts. All the expected process errors such as pad height error, clamping and cutting distortion, and spindle tilt were modeled in 3DCS software [89].

The comparison of the quality model predicted flatness versus the plant measurements for the 219 parts is shown in Figure 10.40. The two distributions have about the same shape, but the quality model prediction is approximately 0.010 mm higher than the plant measurements. We believe that one of the geometric errors estimated for the machine is larger than the actual machine error for the machining station and should be reduced. The agreement for the flatness prediction is much better than for the prediction of the surface heights at individual points because flatness zeros out local variations caused by geometric and dynamic errors. Considering the limitations of the input data (values which had to be estimated), we believe that the model provides fairly good agreement with the measured SPC data.

A sensitivity analysis is performed to evaluate the contributions of the pad height error, clamping, cutting, and spindle tilt on the deckface height for the 12 points. The results are shown in Figure 10.41. The different patterns in the bar chart identify the individual contributions to changes in the surface profile from the various tolerances or factors. The figure shows that the sensitivities are different at each different point. Nevertheless, some general trends can be identified: (1) the clamping contribution is very significant for most of the points, (2) the spindle tilt is significant at the points far from the centerline of the cutter as expected, and (3) the contribution from cutting deflection tends to be the smallest of the four factors. The relative contributions to variance are shown in Figure 10.42. This figure shows that most of the contributions to the variation come from two sources, both of which are related to the variations in machining the locating pads. This figure also shows that clamping distortion also can make a significant contribution to the part-to-part variations in the height of the deckface. At nearly all the points on the deckface, the clamping distortion is much larger than the cutting distortion. Finally, since a simplified fixturing model has been used in this study, it is expected that full modeling of the fixture can further reduce the simulation error since the full model would account the whole compliance of the fixturing system. In addition, accuracy of the static simulation could also be improved with additional dynamic analysis to single out the most important characteristics of the workpiece-fixture system.

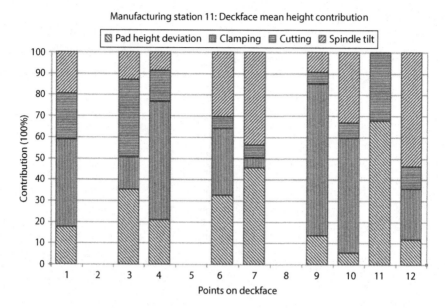

FIGURE 10.41 Comparison of mean shift contributions predicted through modeling.

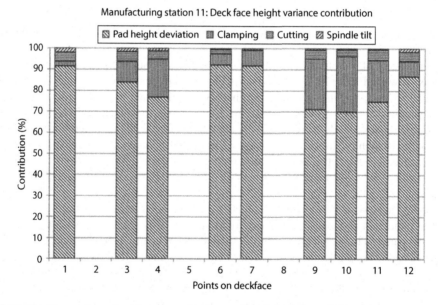

FIGURE 10.42 Comparison of variance contribution predicted through modeling.

Example 10.5 Investigate the CNC process for the spindle-part tilt compensation using a tilt angle for the cutter of 0.012°.

Solution: This example requires to define the NC program to rotate the coordinate system around the rotary axis by a shift and a rotation. The steps are

    a. Table rotation to adjust the part tilt
    b. Translate the center of the part to adjust the part reference
    c. Rotate coordinate system
    d. Perform the single pass milling using straight path movements

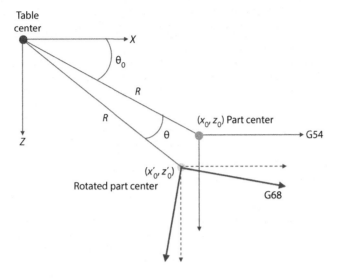

FIGURE 10.43 Illustration of part coordinate system rotation.

Suppose the original G54 origin is set in (x_0, z_0) measured from the rotary table center (see Figure 10.43). We have

$$\begin{cases} \theta_0 = \arctan \dfrac{z_0}{x_0} \\ R = \sqrt{x_0^2 + z_0^2} \end{cases} \tag{10.22}$$

When the coordinate rotated around the table center, the origin shifts to (x_0', z_0'). We have

$$\begin{cases} x_0' = R\cos(\theta_0 + \theta) \\ z_0' = R\sin(\theta_0 + \theta) \end{cases} \tag{10.23}$$

The shift is

$$\begin{cases} \Delta x_0 = x_0' - x_0 \\ \Delta z_0 = z_0' - z_0 \end{cases} \tag{10.24}$$

Using Equations 10.22 through 10.24 into a spread sheet, the shift $(\Delta x_0, \Delta z_0)$ is calculated.

Input	θ	0.012
	x_0	−118.600
	z_0	169.500
⇓	θ_0	124.981
	R	206.872
	x_0'	−118.635
	z_0'	169.475
Output	Δx_0	−0.035
	Δz_0	−0.025

In this case, the CNC command G68X-0.035Z-0.25J-0.012 shifts the original G54 ($X = -0.035$, $Z = -0.025$) to the dot-line coordinate; and rotates $-0.012°$ to the solid line slanted coordinate. Looking down from Y axis, the table rotates clockwise positive. The coordinate transformation takes counter clockwise positive. The sign convention is just the opposite for table rotation and coordinate rotation. Therefore, G68 takes in a negative θ for input.

10.13 PROBLEMS

Problem 10.1 A 250 mm diameter milling cutter using 16 square inserts with a nose radius of 0.8 mm is used to finish a 180 mm wide surface. The centerline of the cutter is offset by 20 mm from the centerline of the part. The feed per tooth is 0.1 mm. Determine the average geometric roughness of the surface along the centerline of the part and 20 mm away from the sides of the part.

Problem 10.2 A 30 mm diameter indexable drill uses a special trigon insert with 0.4 mm nose radius at the outer corner. The feed is 0.2 mm/rev. Determine the average geometric roughness of the surface in the hole. How can the surface finish of the hole be improved using an indexable drill of this type?

REFERENCES

1. J. Agapiou et al., A predictive modeling methodology for part quality from machining lines, *NAMRC-XXXI Conference*, SME, Hamilton, Ontario, Canada, 2003, pp. 629–636.
2. J. Agapiou et al., Modeling machining errors on a transfer line to predict quality, *J. Manuf. Process.* **5**:1 (2003) 1–12.
3. D. J. Whitehouse, *Surfaces and Their Measurement*, Kegan Page Science, London, U.K., 2002, Chapters 2, 7, and 8.
4. T. R. Thomas, Stylus instruments, Chapter 2, in: T. R. Thomas, Ed., *Rough Surfaces*, Longman Group, Harlow, Essex, U.K., 1982.
5. G. T. Smith, *Advanced Machining*, Springer-Verlag, New York, 1989, pp. 198–223.
6. J. Peters, P. Vanherck, and M. Sastrodinoto, Assessment of surface topography analysis techniques, *CIRP Ann.* **28** (1979) 539–554.
7. E. G. Thwaite, Measurement and control of surface finish in manufacture, *Precision Eng.* **3** (1981) 97–104.
8. T. R. Thomas, Characterization of surface roughness, *Precision Eng.* **6**:4 (October 1984) 207–217.
9. G. H. Schaffer, The many faces of surface texture, *American Machinist and Automated Manufacturing*, June 1988, pp. 61–68.
10. P. M. Noaker, The well-honed competitive edge, *Manuf. Eng.* **115**:3 (September 1995) 53–59.
11. W. P. Dong, P. J. Sullivan and K. J. Stout, Comprehensive study of parameters for characterizing three-dimensional surface topography I: Some inherent properties of parameter variation, *Wear* **159** (1992) 161–171.
12. ASME B46.1-2009, Surface Texture (Surface Roughness, Waviness, and Lay).
13. ISO 4288:1996, Geometrical product specifications (GPS)—Surface texture: Profile method—Rules and procedures for the assessment of surface texture.
14. ISO 11562:1996, Geometrical product specifications (GPS)—Surface texture: Profile method—Metrological characteristics of phase correct filters.
15. ISO 4287:1997, Geometrical product specifications (GPS)—Surface texture: Profile method—Terms, definitions and surface texture parameters.
16. DIN 4768:1990, Determination of roughness parameters R_a, R_z, R_{max} by means of stylus instruments; terms, measuring conditions.
17. ISO 4287:2009, Geometrical product specifications (GPS)—Surface texture: Profile method—Terms, definitions and surface texture parameters.
18. ISO 13565-2:1996, Geometrical product specifications (GPS)—Surface texture: Profile method; Surfaces having stratified functional properties, Part 2: Height characterization using the linear material ratio curve.
19. J. Boulanger, The MOTIFS method: An interesting complement to ISO parameters for some functional problems, *Int. J. Mach. Tools Manuf.* **32** (1992) 203–209.

20. ISO 12085:1996, Geometrical product specifications (GPS)—Surface texture: Profile method—Motif parameters.
21. U. Schneider, A. Steckroth, N. Rau, and G. Hubner, An approach to the evaluation of surface profiles by separating them into functionally different parts, *Surf. Topogr.* **1** (1988) 71–83.
22. R. A. Aronson, The secrets of surface analysis, *Manufacturing Engineering*, October 1995, pp. 57–64.
23. J. I. McCool, Assessing the effect of stylus tip radius and flight on surface topography measurements, *ASME J. Tribol.* **106** (1986) 203–211.
24. A. Konczakowski and M. Shiraishi, Sampling error in a/d conversion of the surface profile signal, *Precision Eng.* **4** (1982) 49–60.
25. E. J. Davis and K. J. Stout, Stylus measurement techniques: A contribution to the problem of parameter variation, *Wear* **83** (1982) 49–60.
26. H. Dagnall, *Exploring Surface Texture*, Rank Taylor Hobson, Leicester, U.K., 1986.
27. T. Y. Lin, P. J. Sullivan, K. J. Stout, and K. L. Lo, Measurement variation of surface profile parameters: A viewpoint from standardisation, *Proceedings of the Sixth Annual Congress of the American Society for Precision Engineering*, Santa Fe, NM, October 13–18, 1991, pp. 145–148.
28. T. G. King, RMS skew and kurtosis of surface profile height distributions: Some aspects of sample variation, *Precision Eng.* **2** (1980) 207–215.
29. K. J. Stout, E. J. Davis, and P. J. Sullivan, *Atlas of Machined Surfaces*, Chapman & Hall, London, U.K., 1990.
30. W. Grabon, W. Koszela, P. Pawlus, and S. Ochwat, Improving tribological behavior of piston ring–cylinder liner frictional pair by liner surface texturing, *Tribol. Int.* **61** (2013) 102–108.
31. ISO 25178-6:2010, Geometrical product specifications (GPS)—Surface texture: Areal—Part 6: Classification of methods for measuring surface texture.
32. M. Yousfi, S. Mezghani, I. Demirci, and M. El Mansori, Comparative study between 2D and 3D characterization methods for cylinder linear plateau honed surfaces, *Proceedings of the NAMRI/SME*, Vol. 42, 2014.
33. Z. Dimkovski, C. Anderberg, R. Ohlsson, and B.-G. Rosén, Characterisation of worn cylinder liner surfaces by segmentation of honing and wear scratches, *Wear* **271**:3–4 (2011) 548–552.
34. G. Boothroyd and W. A. Knight, *Fundamentals of Machining and Machine Tools*, 2nd edn., Marcel Dekker, New York, 1989, pp. 166–173.
35. T. Ekstedt, The challenge of hard turning, *Carbide Tool J.* **19**:5 (1987) 21–24.
36. W. Konig, R. Komanduri, and H. K. Toenshoff, Machining of hard materials, *CIRP Ann.* **33** (1984) 417–427.
37. T. I. El-Wardany and M. A. Elbestawi, Performance of ceramic tools in hard turning, *Proceedings of the Sixth International Conference of PEDAC*, Alexandria, Egypt, 1992.
38. T. I. El-Wardany, M. A. Elbestawi, M. H. Attia, and E. Mohamed, Surface finish in turning of hardened steel, *Engineered Surfaces*, ASME PED-Vol. 62, ASME, New York, 1992, pp. 141–159.
39. M. C. Shaw, *Metal Cutting Principles*, Oxford University Press, Oxford, U.K., 1984, pp. 487–519.
40. E. J. A. Armarego and R. H. Brown, *The Machining of Metals*, Prentice-Hall, Englewood Cliffs, NJ, 1969, pp. 187–188.
41. N. H. Cook, *Manufacturing Analysis*, Addison-Wesley, Reading, MA, 1966, pp. 79–80.
42. S. Smith and J. Tlusty, An overview of modeling and simulation of the milling process, *ASME J. Eng. Ind.* **113** (1991) 169–175.
43. W. A. Kline et al. The prediction of surface accuracy in end milling, *ASME J. Eng. Ind.* **104** (1982) 272–278.
44. E. Budak and Y. Altintas, Flexible milling force model for improved surface error predictions, *Proceedings of 1992 Engineering System Design and Analysis Conference*, Istanbul, Turkey, ASME PD-Vol. 47-1, ASME, New York, 1992, pp. 89–94.
45. J. W. Sutherland and R. E. DeVor, An improved method for cutting force and surface error prediction in flexible end milling systems, *ASME J. Eng. Ind.* **108** (1986) 269–279.
46. Y. Altintas, *Manufacturing Automation, Metal Cutting mechanics, Machine Tool Vibrations, and CNC Design*, Cambridge University Press, Cambridge, U.K., 2000, pp. 70–72.
47. S. Malkin, *Grinding Technology: Theory and Applications of Machining with Abrasives*, Ellis Horwood, Chichester, U.K., 1989, Chapters 4, 6, and 7.
48. S. Malkin and T. Murray, Comparison of single point and rotary dressing of grinding wheels, *Proc. NAMRC* **5** (1977) 278.
49. T. Murray and S. Malkin, Effects of rotary dressing on grinding wheel performance, *ASME J. Eng. Ind.* **100** (1978) 297.

50. R. S. Hahn, Trouble-shooting surface finish/integrity grinding problem, *Proceedings of Milton C. Shaw Grinding Symposium*, ASME PED-Vol. 16, ASME, New York, 1985, pp. 409–414.
51. E. K. Henriksen, Residual stresses in machined surfaces, *ASME Trans.* **73** (1951) 69–76.
52. C. R. Liu and M. M. Barash, Variables governing patterns of mechanical residual stress in a machined surface, *ASME J. Eng. Ind.* **104** (1982) 257–264.
53. E. Brinksmeier, J. J. Cammett, W. Konig, P. Leskovar, J. Peters, and H. K. Tönshoff, Residual stresses—measurements and causes in machining processes, *CIRP Ann.* **31** (1982) 491–510.
54. R. A. Roggie and J. J. Wert, Influence of surface finish and strain hardening on near surface residual stress and the friction and wear behavior of A2.D2 and CPM-10V tool steels, *International Conference on Wear of Materials*, Vol. 8, Orlando, FL, 1991, pp. 497–501.
55. M. Reinhardt, L. Markworth, and S. Knodt, Modeling of difficult to cut materials, *Proceedings of 2002 Third Wave AdvantEdge User's Conference*, Atlanta, GA, April 2002, Paper 5.
56. Y. Matsumoto, M. M. Barash, and C. R. Liu, Residual stress in the machined surface of hardened steel, *High Cutting Speed*, SME, Dearborn, MI, 1986, pp. 193–204.
57. J. Mackerle, Finite-element analysis and simulation of machining: A bibliography (1976–1996), *J. Mater. Process. Tech.* **86** (1999) 17–44.
58. K. Okushima and Y. Kakino, The residual stress produced by metal cutting, *CIRP Ann.* **10** (1971) 13–14.
59. C. Wiesner, Residual stresses after orthogonal machining of AISI 304: Numerical calcuation of the thermal component and comparison with experiments, *Metall. Trans. A* **23** (1992) 989–996.
60. T. Shirakashi, T. Obikawa, H. Sasahara, and T. Wada, Analytical prediction of the characteristics within machined surface layer (First Report, The analysis of the residual stress distribution), *J. Jpn. Soc.Precision Eng.* **59** (1993) 1695–1700.
61. A. J. Shih and H. T. Y. Yang, Experimental and finite element predictions of residual stresses due to orthogonal metal cutting, *Int. J. Numer. Methods Eng.* **36** (1993) 1487–1507.
62. R. Liu and Y. B. Guo, Finite element analysis of the effect of sequential cuts and tool-chip friction on residual stresses in a machined layer, *Int. J. Mech. Sci.* **42** (2002) 1069–1086.
63. M. Lundblad, Influence of cutting tool geometry on residual stress in the workpiece, *Proceedings of 2002 Third Wave AdvantEdge User's Conference*, Atlanta, GA, April 2002, Paper 7.
64. M. Russel, Modeling for machining, *Proceedings of 2002 Third Wave AdvantEdge User's Conference*, Atlanta, GA, April 2002, Paper 10.
65. C. Shet and X. Deng, Residual stresses and strains in orthogonal metal cutting, *Int. J. Mach. Tools Manuf.* **43** (2003) 573–587.
66. J. D. Thiele and S. N. Melkote, Effect of cutting edge geometry and workpiece hardness on surface residual stresses in finish hard turning of AISI 52100 steel, *ASME J. Manuf. Sci. Eng.* **122** (2000) 642–649.
67. M. N. A. Nasra, E.-G. Nga, and M. A. Elbestawia, Modelling the effects of tool-edge radius on residual stresses when orthogonal cutting AISI 316L, *Int. J. Mach. Tools Manuf.* **47** (2007) 401–411.
68. T. Ozel and D. Ulutan, Prediction of machining induced residual stresses in turning of titanium and nickel based alloys with experiments and finite element simulations, *CIRP Ann. Manuf. Technol.* **61** (2012) 547–550.
69. P. J. Arrazola, A. Kortabarria, A. Madariaga, J. A. Esnaola, E. Fernandez, C. Cappellini, D. Ulutan, and T. Özel, On the machining induced residual stresses in IN718 nickel-based alloy: Experiments and predictions with finite element simulation, *Simul. Modell. Pract. Theory* **41** (2014) 87–103.
70. Y. K. Chou and C. J. Evans, White layers and the thermal modeling of hard turned surfaces, *Int. J. Mach. Tools Manuf.* **39** (1999) 1863–1881.
71. S. S. Bosheh and P. T. Mativenga, White layer formation in hard turning of H13 tool steel at high cutting speeds using CBN tooling, *Int. J. Mach. Tools Manuf.* **46** (2006) 225–233.
72. X. Sauvage, J. M. Le Breton, A. Guillet, A. Meyer, and J. Teillet, Phase transformations in surface layers of machined steels investigated by x-ray diffraction and Mössbauer spectrometry, *Mater. Sci. Eng.* **A362** (2003) 181–186.
73. M. C. Shaw and A. Vyas, Heat affected zones in grinding steel, *CIRP Ann.* **43** (1994) 279–282.
74. S. Kompella S. P. Moylan, and S. Chandrasekar, Mechanical properties of thin surface layers affected by material removal processes, *Surf. Coat. Technol.* **146–147** (2001) 384–390.
75. H. Tönshoff, H.-G. Wobker, and D. Brandt, Hard turning: Influences on the Workpiece's properties, *Trans. NAMRI/SME* **23** (1995) 215–220.
76. A. M. Abrao and D. K. Aspinwall, The surface integrity of turned and ground hardened bearing steel, *Wear* **196** (1996) 279–284.

77. R. S. Hahn, Grinding, *Handbook of High Speed Machining Technology*, Chapman & Hall, New York, 1985, pp. 348–354.
78. D. Cantillo, S. Calabrese, W. R. DeVries, and J. A. Tichey, Thermal considerations and ferrographic analysis in grinding, *Grinding Fundamentals and Applications*, ASME PED-Vol. 39, ASME, New York, 1989, pp. 323–333.
79. R. Snoeys, M. Maris, and J. R. Peters, Thermally induced damage in grinding, *CIRP Ann.* **27** (1978) 571–581.
80. B. J. Griffiths, White layer formations at machined surfaces and their relationship to white layer formation at worn surfaces, *ASME J. Tribol.* **107** (1985) 165–171.
81. L. E. Samuels, *Metallographic Polishing of Metals*, ASM, Metals Park, OH, 1981.
82. M. Ravi Shankar, B. C. Rao, S. Lee, S. Chandrasekar, A. H. King and W. D. Compton, Severe plastic deformation (SPD) of titanium at near-ambient temperature, *Acta Mater.* **54** (2006) 3691–3700.
83. C. Ohkubo, I. Watanabe, J. P. Ford, H. Nakajima, T. Hosoi, and T. Okabe, The machinability of cast titanium and Ti-6Al-4V, *Biomaterials* **21** (2000) 421–428.
84. G. Xiao, R. Stevenson, I. M Hanna, and S. A. Hucker, Modeling of residual stress in grinding of nodular cast iron, *ASME J. Manuf. Sci. Eng.* **124** (2002) 833–839.
85. R. Leach, L. Brown, X. Jiang, R. Blunt, M. Conroy, and D. Mauger, Guide for the measurement of smooth surface topography using coherence scanning interferometry—Measurement Good Practice Guide No. 108, National Physical Laboratory, Queen's Printer for Scotland, Teddington, Middlesex, Crown, 2008.
86. T. L. Bruce, D. A. Stephenson, and A. J. Shih, Improvement of surface flatness in face milling based on 3-D holographic laser metrology, *Int. J. Mach. Tools Manuf.* **51** (2011) 483–490.
87. J. Gu and J. Agapiou, Approaches for improving surface flatness for face milling, *Trans. NAMRI/SME* **42** (2014) 542–553.
88. J. S. Agapiou, E. Steinhilper, F. Gu, and P. Bandyopadhyay, Modeling machining errors on a transfer line to predict quality, *SME J. Manuf. Proc.* **5** (2003) 1–12.
89. J. Q. Xie, J. S. Agapiou, D. A. Stephenson, and P. M. Hilber, Machining quality analysis of an engine cylinder head using finite element methods, *SME J. Manuf. Proc.* **5** (2003) 170–184.

11 Machinability of Materials

11.1 INTRODUCTION

"Machinability" can refer to either the ease or difficulty of machining a material, as when discussing machinability ratings [1–10], or to the body of knowledge and practice accumulated on the machining of a particular material, as in machinability handbooks or databases [2,6,11].

When used in the first sense, machinability is often regarded as a material property and assigned a numerical rating. As shown schematically in Figure 11.1, however, the ease or difficulty of machining a particular part is affected by a variety of factors. These include the chemical composition and thermomechanical properties of the work material, the rigidities of the machine tool, part, and fixture, the tool material, and the cutting speeds and feeds. Machinability is thus in reality a property of a machining system operating under a given set of conditions.

Although quantitative machinability ratings should not be misconstrued as material properties, they are often useful for comparison or ranking purposes. For example, they may be used to rank a list of material choices for a given part from most to least machinable in a general sense. In these cases the primary consideration is usually the overall machining cost, which in many cases depends most strongly on the tool life, which can be achieved under the assumed production conditions. Machinability ratings are also useful for general cost-estimating calculations. In job shop production, for example, the cost of machining a part made of an unfamiliar material can be estimated by comparing its machinability rating with that of a material with which the shop has more experience.

As noted before, the term "machinability" also refers to the accepted machining practices for a given material [2,6,11]. Machinability data collected in handbooks or computer databases consists of recommended cutting speeds, feed rates, and depths of cut for specific work materials. Separate values are normally given for different operations and tool material grades. Machinability data is generally gathered from production experience and summarizes machining conditions, which yield acceptable tool life and part quality under common operating conditions. The recommended speeds and feeds represent only initial starting conditions, which should be modified as needed to optimize machining performance in a given application. Since machining practice changes over time due to new developments in tool materials and machine tool capabilities, machinability data must be updated frequently to remain current. Extensive current machinability data is available from laboratories specializing in machinability testing and from cutting tool manufacturers in the form of insert grade advisers [12–14].

Common machinability criteria, tests, and indices are discussed in Section 11.2. Chip control and burr formation, which have a significant impact on the machinability of soft, ductile alloys, are discussed in Sections 11.3 and 11.4. General machining practices and machinability concerns for selected common engineering materials are described in Section 11.5. This material is intended to supplement the discussion on tool material applications in Chapters 4 and 9.

11.2 MACHINABILITY CRITERIA, TESTS, AND INDICES

The most commonly used criteria for assessing machinability are [2,6–8] as follows:

1. *Tool life or tool wear rates.* This is the most meaningful and common machinability criterion for ferrous materials [15,16]. Tool wear affects both the quality and cost of the machined part. Machinability is said to increase when tool wear rates decrease (or tool life increases) under the machining conditions of interest. Ratings based on wear rates are generally applicable to a restricted range of cutting conditions; when conditions change,

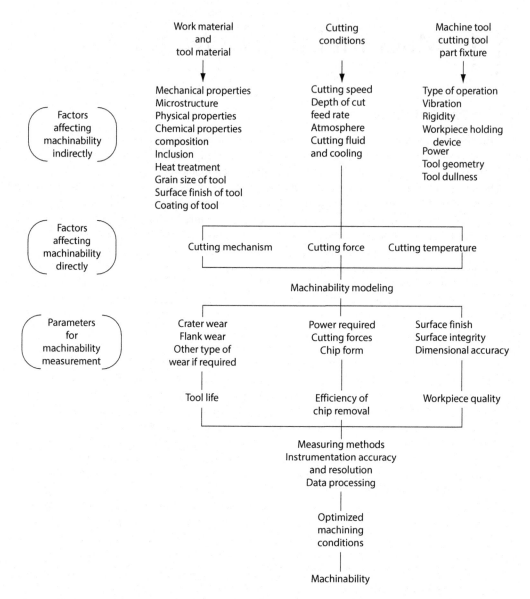

FIGURE 11.1 Factors affecting machinability. (After ElGomayel, J., Fundamentals of the Chip Removal Process, SME Technical Paper MR77-256, 1977.)

for example, when the cutting speed is substantially increased or decreased, the dominant tool wear mechanism and tool wear rate may change. This fact is particularly relevant when ranking the machinability of a group of materials under different cutting conditions.

2. *Chip form and tendency to burr.* Materials that produce short chips, which are easily managed and disposed of, are more machinable than those which produce long, unbroken chips or small, powder-like chips. The chip form is particularly important for applications such as drilling for which chip breaking and disposal concerns may limit production rates. Ductile materials that form long, unbroken chips also have a greater tendency to form burrs, especially as the tool wears. Chip form and burr behavior are often used to assess the machinability of soft, ductile alloys, especially aluminum alloys, since these materials are often machined with diamond tools and do not generate a great deal of wear. Chip control issues and burr formation are discussed in the next two sections.

Less commonly used machinability criteria include the following [8,17]:

3. *Achievable surface finish.* Generally, machinability increases as the surface finish achievable under a given set of cutting conditions improves. As discussed in Chapter 10, the average roughness is the most common parameter used to assess surface quality in machining tests. This criterion is most useful in ranking different classes of materials, for example, conventional versus powder metal steels. This criterion is not applicable to roughing cuts.

4. *Achievable tolerance.* Machinability increases as the tolerance that can be achieved under a given set of cutting conditions decreases. As with the surface finish criterion, this criterion is most useful when comparing different classes of materials. It is more commonly used to assess the machinability of woods or plastics than metals; for metals, it is most commonly used in medium- and high-production applications. Tolerance criteria are also often used to define the end of tool life in tests involving form tools (e.g., the ASTM bar turning test [18]).

5. *Functional or surface integrity.* Materials with high susceptibility to metallurgical damage, such as residual stress formation, surface galling, or hardened or white layer formation, are less machinable than materials with low susceptibility to such damage.

6. *Cutting forces or power consumption.* Machinability increases as cutting forces and power consumption decrease. Lower cutting forces generally imply lower tool wear rates, better dimensional accuracy (due to decreased deflections), and increased machine tool life (due to reduced loads on bearings and ways). Cutting force or power consumption measurements require much less material and machine time than tool life tests, and are particularly suitable for screening the machinability properties of a large group of materials to identify a few promising candidates for tool life tests. Cutting force measurements from short series of tests, however, have limited resolution and detect only gross differences in machinability.

7. *Cutting temperature.* Cutting temperature measurements such as tool–work thermocouple measurements (Chapter 7) can be used to compare the machinability of materials under a given set of cutting conditions [19]. Machinability increases as cutting temperatures decrease because many tool wear mechanisms are temperature-activated; the average temperature is therefore generally correlated with tool life, and thus with machinability. Some tool materials such as ceramics are comparatively insensitive to tool temperatures, so that small increases in temperature do not affect or may even improve their performance. In a related method, temperature and strain maps that correlate with machinability rankings are used [20]. Like cutting force measurements, these measurements require less time and material than tool life tests and are particularly well suited for screening purposes but have limited resolution.

8. *Mechanical properties.* Mechanical properties of work material, such as the hardness, yield strength, ductility, and strain-hardening exponent, have occasionally been correlated to machinability. The correlation exists within a group or class of metals. For example, steels in the same class generally show negative correlation between machinability and both hardness and yield stress. Generally, low values of hardness and yield stress are favorable. The same is true with respect to hardness within the general group of cast irons. A hardness effect is not as obvious for aluminum alloys, however; for cast aluminum alloys, machinability correlates more strongly with ductility. Hardness tests are used to evaluate work material uniformity between lots to ensure consistent machinability. Machinability within some classes of materials, such as heat-resistant alloys and austenitic stainless steels, decreases with increasing work hardening rate. A significant increase in hardness will occur in a thin layer of the machined surface for these materials. The depth of the work-hardened layer and the level of hardness should be kept to minimum especially if the feed per edge is close to the depth of the work-hardened layer. In such cases, the most positive and sharp cutting edge should be selected to reduce the cutting energy.

Tool life is the most widely used machinability criterion, and as a result tool life tests are often used to assess machinability. Standardized tool life tests are described in Section 9.6. Traditionally, variants of the ISO turning test [21] or ASTM bar turning test [18] have most commonly been used for machinability assessment. These are very general tests, which do not uniquely define all cutting parameters and test procedures. They do, however, provide an excellent framework for developing simpler test procedures optimized for specific applications, such as the simplified test strategy shown in Figure 11.2. Since these tests are time-consuming and require a large supply of uniform material, there has long been interest in developing test methods, which require less machine time and material. Commonly used tests of this type are the Volvo end milling test [22,23] and various drilling tests [24–27], which are also described in Section 9.6. Less common tests, which are described in detail by Smith [5], include face and taper turning tests, in which the cutting speed increases linearly with time due to the workpiece geometry (with the maximum speed achieved before tool failure used as the measure if machinability) [28,29]; tests in which the cutting speed is increased in steps, rather than continuously (permitting the use of standard bar specimens); and various nonmachining tests in which combinations of physical or microstructural properties thought to be related to tool wear rates are measured.

Test purpose	Machinability evaluation for a family of work materials.
	Evaluation of other parameters (cutting fluids, tool materials, coatings, etc.).

Machine tools	Must be capable of maintaining constant cutting speed by variable spindle rpm.
	Must have adequate horsepower to maintain the cutting speed under the cutting load both new and worn tools.
	CNC lathes of suitable size and power are considered as the ideal test bed.

Work material	Must be obtained from one supplier and from the same batch or heat.
	All test pieces must have the same diameter and length.

Toolholder	Use one repeatable holder if possible.
	Otherwise, the same style from the same supplier must be used.

Tool material	Use an ANSI insert style, all from the same supplier, the same batch of material, and the same batch of manufacturing.

Cutting fluid	"Dry"—do not use coolant if possible.
	"Wet"—specify the coolant type, concentration, flow rate, pressure, temperature, and the way it is supplied at the cutting zone (i.e., nozzle location, nozzle size, filtration system).

Test procedure	Specify the cutting conditions.
	Specify the machining and data collection procedure.

Measuring equipment	Specify the type of instruments.
	Specify the procedure used for the measurements.

Cutting conditions	Select the speeds and feeds based on the objectives of the tests.
	Slightly higher values than those recommended (i.e., *Machinability Data Handbook*) would accelerate the testing time.

Measuring parameters	Tool wear/tool life/cutting forces/cutting power/cutting temperature/surface finish/chip formation/dimensional accuracy.

FIGURE 11.2 Suggested strategy for machinability testing.

The limited evidence available indicates most machining tests, whether based on turning, drilling, or milling, yield equivalent ranking results for HSS and uncoated carbide tools [30].

In mass production applications in which transfer machines and other serial production systems are used, tools are often changed according to a regular schedule (e.g., after one or two shifts of production). In this case small increases in tool life have less impact on production costs than the maximum production rate, which can be achieved while maintaining some minimum tool life. For these operations accelerated tool wear tests at various metal removal rates are sometimes used to assess machinability. Tests of this type have not been standardized because the required minimum tool life varies considerably from application to application.

One factor that should be considered in machinability testing is the consistency of the work material. As shown in Figure 11.1, machinability is affected by composition, hardness, microstructure, and inclusion morphology, all of which are influenced by the fabrication history of the material. If a uniform lot of work material of sufficient size for testing is not available, different grades with different fabrication histories may be used in the test, introducing confounding effects into the results. The size of the test workpiece should be standardized when possible to eliminate variation due to specimen temperature rise caused by heat entering the workpiece; this is especially important when drilling tests are used to assess machinability. Test conditions should be selected so that chatter does not occur. The characteristics of the material should be uniform through the core. If the fabrication procedure, composition, or heat treatment produce a material with a serious change in microstructure from outer to inner diameters, a more suitable sample size and machining process should be selected.

To this point we have discussed machinability from a comparative (ranking) perspective. For cost estimating purposes, there has long been interest in developing absolute or quantitative measures for machinability [1]. The most commonly used measure is the machinability rating or index, I_m, defined by [13]

$$I_m = 100 \times \frac{(V_{60})_{\text{mat}}}{(V_{60})_{\text{ref}}} \tag{11.1}$$

where

    $(V_{60})_{\text{mat}}$ is the cutting speed at which the material being rated yields a 60 min tool life for a specified feed rate, depth of cut, tool material, and tool geometry

    $(V_{60})_{\text{ref}}$ is the cutting speed, which yields a 60 min tool life for a reference material with a defined machinability rating of 100 under the same conditions

This is the machinability index most commonly used in tables of material properties included in engineering handbooks. In the United States, SAE B1113 sulfurized free-machining steel has traditionally been used as the reference material, although in recent years SAE 1045, 1018 [31], or 1212 [7] have been used. The index is also less commonly calculated using 20 or 90 min tool life speeds; all three definitions yield equivalent results for common work materials cut with HSS or uncoated carbide tools.

Numerous other machinability indices, often computed from more complicated equations, which include metallurgical or chemical content parameters, have been proposed [1,32,33]. These parameters were more meaningful in an absolute sense when the range of cutting materials available was restricted to HSS and uncoated WC. The development of coated carbide, cermet, ceramic, PCBN, and PCD tools has greatly reduced the absolute accuracy of such indices, so that they are now useful primarily for comparative or ranking purposes.

11.3 CHIP CONTROL

Chip control is an important issue in many machining operations, particularly turning, boring, and drilling operations involving ductile work materials such as low-carbon steels and aluminum alloys. Chip control usually involves two tasks: breaking of chips to avoid the formation of long, continuous chips, which can become entangled in the machinery, and removal of chips from the cutting zone to prevent damage to the machined surface and errors due to thermal distortion. Because a very large spectrum of chip forms are encountered in practice, no general theory of chip control is available. Rather, a number of strategies applicable to particular operations or materials, usually based on test results, have been developed. This section summarizes the more common approaches and is intended to provide physical insight into chip breaking and control issues. More detailed descriptions of specific strategies are given in review articles on the subject [34–40].

In turning, chip breaking is usually accomplished by directing the chip toward an obstacle to produce a bending stress sufficient to fracture it as shown in Figure 11.3. If the chip is modeled as a beam, which can sustain a maximum bending strain of ε_b, the chip will break when [34]

$$h_c \cdot \left[\frac{1}{R_L} - \frac{1}{R_0} \right] \geq \varepsilon_b \tag{11.2}$$

where
h_c is the distance between the neutral bending plane in the chip and the chip surface (usually roughly half the chip thickness [41])
R_0 and R_L are the natural and obstructed radii of chip curl

The quantitative accuracy of this relation is difficult to assess because an appropriate fracture strain ε_b is difficult to estimate and the inherent radius of curl R_0 varies considerably in experiments; however, it does reflect qualitative dependencies, which can be exploited to promote chip breaking.

Equation 11.2 indicates that chip breaking is promoted when ε_b is reduced and when the distance to the neutral plane, h_c, is increased. ε_b can be reduced by imparting prior cold work to the part or by adding alloying elements that reduce ductility. For steel, ε_b can sometimes be reduced by increasing the cutting speed, since this increases the chip temperature and may produce a brittle blue oxide layer on the chip surface. h_c can be increased by increasing the feed rate and

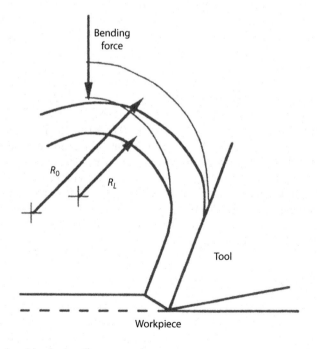

FIGURE 11.3   Chip breaking by bending.

thus the chip thickness. These options are not always feasible in practice, but the underlying physical concepts are in accord with practical experience, since chip breaking is normally more of a problem in finishing operations on relatively ductile (e.g. stamped) parts than in roughing cuts on cold worked (e.g. forged) parts.

Equation 11.2 also indicates that a continuous chip will not break if it is permitted to curl naturally, that is, if $R_L = R_0$. The chip must therefore be guided toward an obstruction to produce a bending strain. This can be accomplished by properly directing chip flow. In turning, the chip flow direction depends on the radius of the workpiece, the lead and back rake angles of the tool, and the ratio of the depth of cut to the tool nose radius [42–47]. In turning, as shown in Figure 11.4, common strategies include directing the chip flow parallel to the workpiece axis, so that the chip breaks against the workpiece surface, or imparting side flow to the chip so that it breaks against the tool holder. The first type of breakage can be promoted by using a small lead angle and tool nose radius; the second type is accomplished by using larger lead and back rake angles and a large tool nose radius.

More effective chip breaking in turning can be ensured by clamping an obstruction-type chip breaker to the insert (Figure 11.5) or by using inserts with formed-in chip breakers. Formed-in chip breakers come in a great variety of forms but may be classed as groove-type or pattern-type (Figure 11.6).

Obstruction-type breakers work by imparting curvature to the chip to produce bending [48–53]. The placement of the obstruction is important to effective operation; it must be moved closer to the cutting edge as the feed is reduced. The advantage of obstruction chip breakers is that they can be adjusted to operate effectively over a wide range of feed rates. The main disadvantage is that they require careful setup and adjustment. (Research on automatic adjustment of obstruction chip breakers has been reported [54], but commercial systems of this type are not available.)

Groove-type chip breakers have various functions. The groove may guide the chip toward an obstruction and produce a thick section in the chip, which promotes breaking (Figure 11.7). More commonly, the groove imparts a bending stress by decreasing the chip's radius of curvature. The dimensions and position of the groove must be selected properly for effective operation. As shown in Figure 11.8, the geometry of the groove is characterized by the land width, groove width, groove depth, back wall height, and groove geometry. The land width is the most critical parameter and must be adjusted based on the feed rate. Smaller land widths are used in finish cutting to ensure that the chip flows into the groove; larger land widths are used for roughing cuts to increase the strength of the cutting edge. The groove depth and back wall height are not as critical as the land width but should also be increased for roughing cuts to increase the bending stress.

The main advantages of groove-type chip breakers are that they require no setup and that they increase the effective rake angle and reduce the tool–chip contact length. They can therefore reduce

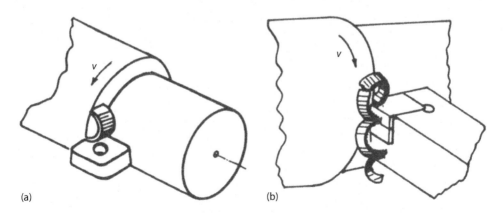

(a)   (b)

FIGURE 11.4 Breaking of the chip against the workpiece surface (a) or tool holder (b). (After Kaldor, S. et al., *ASME J. Eng. Ind.*, 101, 241, 1979.)

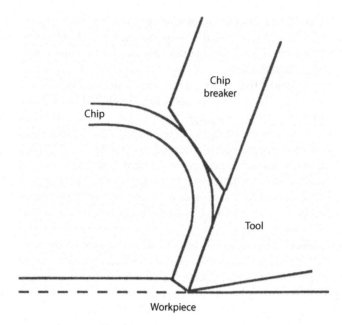

FIGURE 11.5   Obstruction chip breaker.

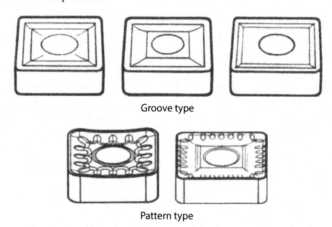

FIGURE 11.6   Typical chip breaking patterns formed into inserts.

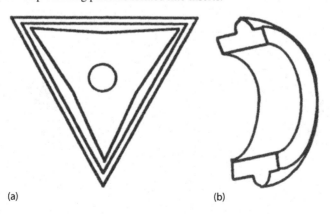

(a)   (b)

FIGURE 11.7   (a) Insert with formed-in chip breaking groove. (b) Chip with thickened section produced by grooved insert. (After Kaldor, S. et al., *ASME J. Eng. Ind.*, 101, 241, 1979.)

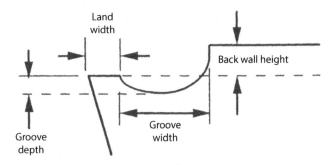

FIGURE 11.8 Parameters characterizing chip breaking groove.

cutting forces and increase tool life when compared to flat-faced, negative rake inserts. The main disadvantage is that they are effective over relatively narrow ranges of feeds and cannot be adjusted. Double-groove designs are sometimes used to increase the effective range of operation in semi-finish and finish machining; this approach is not used in rough machining because it reduces edge strength.

Integral pattern-type chip breakers come in a variety of forms, which consist primarily of irregularly shaped depressions and raised dots. They function by a number of mechanisms and may serve to guide chip flow, impart bending stresses, and induce stress concentrations. The number of possible patterns is infinite, and reliable inserts are available for many common operations. As with groove-type chip breakers, the main advantage of pattern-type chip breakers is that they require no setup, and the main disadvantage is that they are effective over a limited range of feed rates.

The range of applicability of particular chip breaker designs cannot be accurately predicted by analysis. In practice, therefore, recommendations of insert manufacturers must be relied upon in selecting a design for a particular application. Insert manufacturers provide two types of charts to guide insert selection: charts of recommended inserts for particular combinations of the feed and depth of cut (Figure 11.9), and charts of the effective chip-breaking range for a particular insert design (Figure 11.10). These charts are based on limited tests, usually involving only one work material, tool nose radius, or cutting speed. They provide useful initial insert designs. In high volume operations, additional tests should be conducted to accurately establish chip breaking limits for the work material and cutting conditions involved.

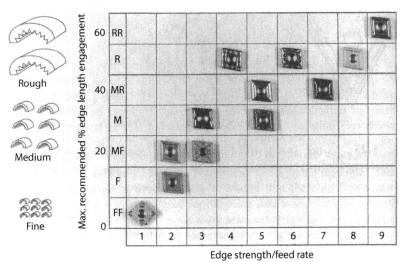

FIGURE 11.9 Chart of chip breaking patterns recommended for cutting steel at various feed rates and depths of cut.

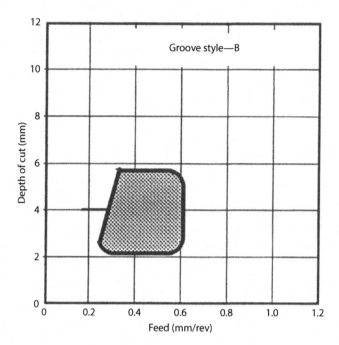

FIGURE 11.10 Typical chip breaking chart for a particular insert. Separate charts are required for each work material and cutting speed.

Chip breaking and removal are also important in drilling, since the allowable penetration rate is often limited by chip-breaking characteristics. Chip breaking can be promoted by modifying the flute profile or drill point [55]. Common modifications, shown in Figure 11.11, include special flute profiles which compress the chip to increase bending stresses, grooves in the flute or flank profile to split and produce thick sections in the chip, and stepped cutting edges to split the chip. Chip breaking in drilling can also be promoted by using multifaceted point geometries with cusps, which split the chip, and by using indexable drills with chip-breaking inserts.

Active chip-breaking strategies can also be used for chip control. The most common strategy is to vary the feed rate to periodically interrupt the cut or to produce thick sections in the chip, which will break more readily. On numerically controlled machines the feed made be increased or halted at prescribed points to control the chip length [56,57]. A cam-driven, hydraulic, or piezoelectric system may also be added to make the tool vibrate in the feed direction at an amplitude comparable to the feeds [58,59]. This latter strategy is particularly common in the drilling of small holes, where axial vibration of the drill is sometimes used to break chips [60]. The chief limitations of these methods are that they reduce the stability of the machining process and produce an irregular surface finish. High pressure (6,000–40,000 psi) jets of water-based fluids or cryogenic cutting fluids, directed at the base of the tool–chip interface, have been used to break chips when cutting difficult-to-machine materials such as titanium alloys or stainless steels [61–66]. Conventional high pressure coolant systems (1000–1500 psi) promote chip breaking in many applications. Research on chip pulling strategies, in which the chip is caught and collected as it forms, has also been reported [67].

Even if chips are broken, they may cause difficulties if they are not cleared from the machine tool or machined part. Chips are hot when they emerge from the cutting zone. If they collect on the machine tool or within the machined part, they can produce thermal distortions, which will lead to dimensional or form errors in the machined part. This mechanism can be a source of significant errors in precision machining processes such as the boring of engine cylinders, and in MQL aluminum machining. Chips that remain in the cutting zone may also be recut in subsequent

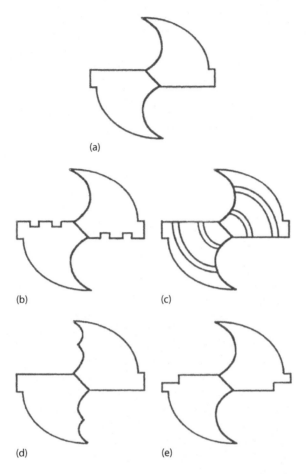

FIGURE 11.11 Drill point and flute modifications to improve chip breaking: (a) standard drill, (b) chip-breaking grooves in flute, (c) chip-breaking grooves in flank, (d) cusped flute, and (e) stepped cutting edge.

passes, which reduces tool life since the chip material is heavily work-hardened and often abrasive. Chip recutting may also lead to unacceptable surface damage in nickel alloys and similar damage-sensitive materials. The common method of evacuating chips is to use a stream of coolant to carry chips into a chip conveyor system. Some care must be taken to avoid damaging tools if they have poor thermal shock resistance and are normally used dry; this is the case for some ceramic tools and for oxide-coated carbide tools. In these cases, compressed air can be used instead of a water-based coolant. Chip removal can also be improved by feeding pressurized coolant through the spindle for a rotary tool or through the tool holder for a turning insert; this is only an option on machines equipped with high pressure, high volume coolant systems.

11.4   BURR FORMATION AND CONTROL

A phenomenon similar to the formation of chips is the formation of burrs at the end of a cut, especially in milling and drilling. In most cases, burrs are partially formed chips left at points along the workpiece too weak to support the forces involved in complete chip formation. Burrs are undesirable because they present a hazard in handling machined parts and can interfere with subsequent assembly operations. Burrs are also associated with certain types of tool wear such as notch formation in turning and exit fracture in milling [68].

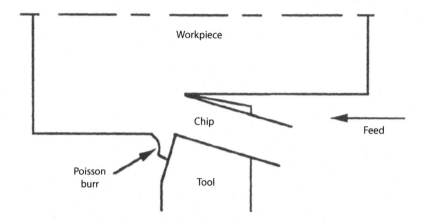

FIGURE 11.12   Poisson burr.

The best known early studies of burr formation were carried out by Gillespie et al. [69–71], who identified three principle types of burrs. The first is the Poisson or compression burr (Figure 11.12), which results from the workpiece material's tendency to bulge in the direction parallel to the cutting edge when compressed by flank forces on the tool. For large flank forces, such as those typical of worn tools, burrs of considerable size can be produced by this mechanism. The second and most common type of burr is the rollover burr (Figures 11.13 and 11.14), produced when a partially formed chip bends in the direction of the cutting velocity at the end of the cut. The third is the tear or breakout burr (Figure 11.15), formed when work material ruptures rather than deforms at the end of the cut. Gillespie and Blotter [69] analyzed the formation of Poisson and rollover burrs and derived equations for their approximate size under given cutting conditions. The quantitative accuracy of these equations is difficult to assess due to the difficulty of specifying some inputs, but they are useful in identifying measures that can minimize burr size.

Alternative but similar classification schemes have also been proposed [72–74]. Beginning in the 1990s, Dornfeld and coworkers conducted extensive studies on burr formation and control through the Consortium on Deburring and Edge Finishing (CODEF) [75–79]. The current research understanding of burr analysis, control, and removal methods is summarized in recent review articles [80,81].

The mechanical property that determines a given work material's tendency to burr is ductility, which is quantified by the percent elongation or strain to fracture (equivalent to the fracture strain ε_b introduced in the previous section). Considering basic burr formation mechanisms, it is clear that

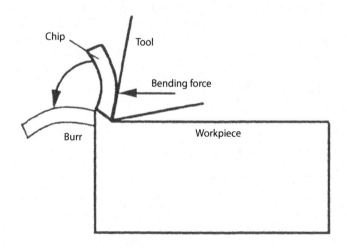

FIGURE 11.13   Rollover burr formed by bending.

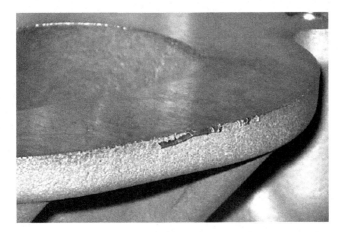

FIGURE 11.14   Rollover burr on a milled aluminum part.

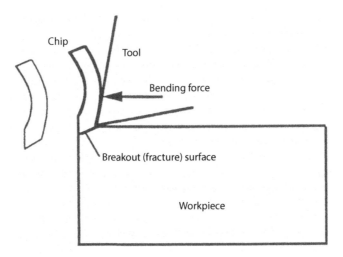

FIGURE 11.15   Workpiece tearing or breakout.

materials that can deform to large strains before fracturing are much more likely to form compression or rollover burrs, while materials that fracture at low strains will often break out before they burr. Therefore, burr formation is a particular problem for highly ductile materials, including pure metals (e.g., pure Cu, SHG Mg, and Al-00 aluminum), some soft low carbon steels (e.g., 1008 and 1010), and highly ductile aluminum alloys (such as A356). For some of these materials, burring and chip-breaking characteristics are the major factors determining tool life and machinability.

More specific studies of burr formation in milling and drilling have also been reported. In milling, the most widely studied burr is the foot or negative shear burr, which is a special type of tearout burr associated with chipping of the cutting edges of milling tools [82–86]. As with more common rollover burrs, they are formed by deformation in the direction of the cutting velocity, but the deformation and rupture are produced by shearing rather than bending. Tool failures due to foot formation can be eliminated by changing the disengage angle of the milling cutter so that the force is not directed perpendicular to the exit surface. Changes in tool geometry, such as the use of a larger axial rake angle, positive rather than negative radial rake angles, and smaller insert nose radii, can also be effective. Finally, burr size can be reduced by inclining the exit surface of the workpiece with respect to the machined surface to increase its bending strength (Figure 11.16), although this is seldom an option in existing operations.

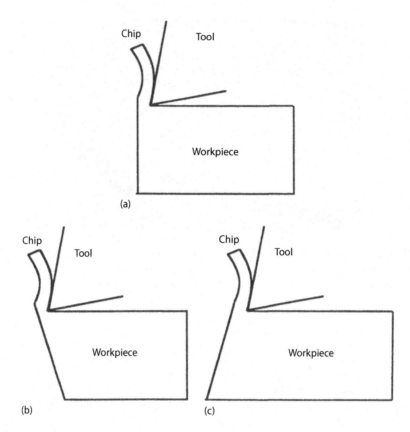

FIGURE 11.16 (a) Normal, (b) acute, and (c) oblique exit surfaces. The bending and shearing strength of the workpiece is greatest for the oblique surface and smallest for the acute surface.

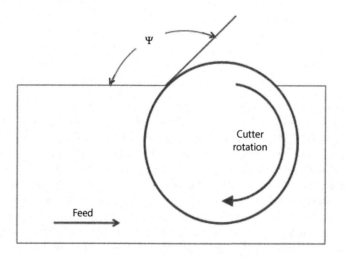

FIGURE 11.17 Definition of the in-plane exit angle Ψ.

As with negative shear burrs, the formation of rollover burrs in face milling is influenced by the exit geometry, and specifically by the in-plane exit angle Ψ defined in Figure 11.17 [76,79,87–89]. When possible, it is advisable to keep this angle below a maximum value, generally between 100° and 120°, to avoid thinning of the uncut cross section of the workpiece at exit. In-plane exit calculations can be added to kinematic simulations of face milling (Section 8.4) and used to evaluate alternate cutter paths to reduce burrs [79,87]. Controlling the exit angle is simplest when milling rectangular surfaces without interruptions; unfavorable exit conditions are often unavoidable when machining complex cast surfaces.

Exit burrs in drilling are generally of the rollover type. They typically have an irregular appearance, but for highly ductile materials may deform into a cap or crown structure (Figure 11.18) [77]. As shown in Figure 11.19, the most common type forms when the annulus of work material cut away as the drill exits become too weak to support the thrust force. Changes in drill point geometry can reduce the size of these burrs [90–93]. One effective method is to use a spiral-pointed drill (Figure 11.20) rather than a conventional conical point drill. This drill not only increases the bending strength of the annulus throughout the exit phase, but also reduces the thrust force near the end of exit and directs a larger fraction of it in the direction parallel to the exit surface. The size of these burrs can sometimes also be reduced by using inverted or fishtail point drills, which exit at the outer margin rather than the center, or by imparting axial vibrations to the drill [60].

Burrs are a more serious problem for worn cutting tools, since cutting forces increase with tool wear. Measures that increase tool life will therefore also often reduce burr size. Since the strength of the workpiece will always decrease to zero near the end of the cut, however, burrs of some size will inevitably be formed in many machining operations, and deburring operations

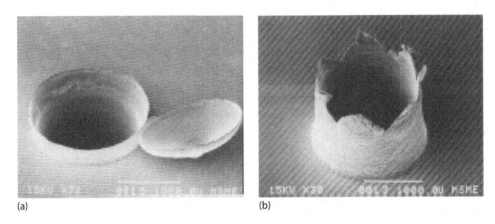

(a)　　　　(b)

FIGURE 11.18 (a) Cap and (b) crown burrs formed in through drilling of 304L stainless steel. (After Kim, J. et al., *Int. J. Mach. Tools Manuf.*, 41, 923, 2001.)

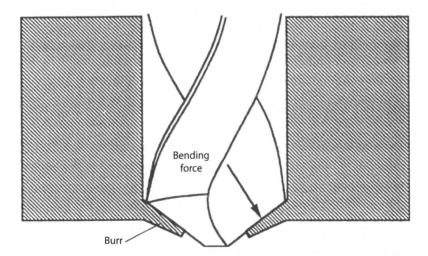

FIGURE 11.19 Formation of rollover burr at exit in drilling with a conventional drill.

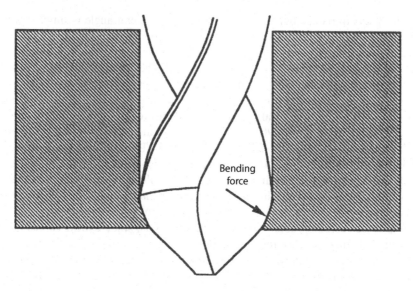

FIGURE 11.20 Formation of exit burr in drilling with a spiral-pointed drill. The burr size is smaller than for a conventional drill since the bending force is smaller and the bending.

will often be required before machined parts can be assembled. Manual deburring is common in small lot production. In mass production, high pressure water jet, wire brush, vibratory, and centrifugal methods are often used. Many robotic and abrasive methods have been developed, but their application has been limited due to cost and environmental concerns. More detailed descriptions of specific deburring methods are given in Section 2.12, deburring handbooks, and review articles [71,72,80].

11.5 MACHINABILITY OF ENGINEERING MATERIALS

The machinability of any alloy family is primarily influenced by its base metal; aluminum alloys are generally highly machinable, ferrous alloys exhibit moderate machinability, and high temperature alloys have poor machinability. The microstructure of the metal also plays an important role in machinability. Some important microstructural parameters include the grain size, the number and size of inclusions, and the type(s) of metallic structures present. In general, hard structures and fine grains result in lower tool life. Machinability improves for softer structures and coarse grains. Small hard inclusions in the matrix promote abrasive tool wear. Soft inclusions are often beneficial to tool life. Materials that work-harden rapidly have lower machinability than those that do not. The chemical composition of a metal is very important and has a complex effect on machinability within the metal family. The primary manufacturing process of a material or component will also affect its machinability.

The remainder of this section discusses specific factors affecting the machinability of common classes of ferrous and nonferrous alloys.

11.5.1 MAGNESIUM ALLOYS

Magnesium alloys are the most machinable of the common structural metals in terms of tool life, cutting forces, power consumption, and surface finish. Because magnesium alloys have a hexagonal structure and low ductility, cutting forces and power consumption are typically over 50% lower than for aluminum alloys under comparable conditions [2,93–96]. Magnesium alloys also have a low melting point (typically less than 650°C), so that cutting temperatures are low [97]. Due of their brittle nature, they form short, segmented chips, which restricts the length of the contact between the tool and chip. Finally, magnesium has a high thermal conductivity and does not alloy with steel. For all these reasons, tool wear rates due to thermal softening and diffusion are low for WC and HSS tooling. Most information indicates tool life when machining magnesium alloys dry is roughly five times longer than when machining aluminum wet under otherwise comparable conditions [93,95]. Tool lives on the order of hundreds of hours, or dozens of production shifts in mass production, were achievable in the early 1960s [98]. More recently, PCD tools have been used to machine cast magnesium parts [99], resulting in tool lives approaching one million parts in some milling and drilling operations [73]. Magnesium alloys acquire fine surface finishes under normal cutting conditions; average roughnesses down to 0.1 μm can be obtained by turning or milling at high or low speeds, with or without cutting fluids [94], so that grinding and polishing operations are often unnecessary.

There are two chief concerns in machining magnesium: the risk of fire and built-up edge (BUE) formation.

Magnesium burns when heated to its melting temperature. In machining magnesium, fires are most likely to occur when thin chips or fines with a high surface area-to-volume ratio are produced and allowed to accumulate. The source of ignition may be frictional heating caused by a tool, which is dull, broken, improperly ground, or allowed to dwell at the end of a cut, or sparks generated externally to the machining process. To minimize the risk of fire, the following practices should be observed [93–95]:

1. Sharp tools with the largest possible relief angles should be used.
2. Heavy feed rates, which produce thick chips, should be used when possible.
3. Allowing tools to dwell against the part at the end of the cut should be avoided.
4. Chips should be collected and disposed of frequently.
5. Appropriate coolants should be used when fine feeds are necessary.

Since magnesium chips react with water to form magnesium hydroxide and free hydrogen gas, water-based coolants have traditionally been avoided. The accepted practice was to machine dry when possible and to use a mineral oil coolant when necessary [100,101]. Dry machining magnesium parts in high volumes, however, presented long-term housekeeping problems, especially when processes such as drilling or tapping, which produced fine chips, were used. Airborne fines could collect over time in areas such as space heaters and electric motor vents, leading to unexpected fires [100]. Beginning in the 1980s, water-based coolants that can be used with magnesium without excessive hydrogen generation have been developed. These coolants are now used in production in some installations and reportedly increase tool life and reduce the risk of fire when compared to dry machining. Grinding or polishing of magnesium should be avoided when possible but, when necessary, should be carried out using dedicated machinery equipped with a wet-type dust collection system [93].

Built-up edge formation is observed when machining cast magnesium–aluminum alloys dry with HSS or carbide tools. The metallurgical factors controlling BUE formation are discussed in Ref. [96]. BUE formation can be reduced or eliminated by applying a mineral oil coolant or by switching to PCD tooling. Flank buildup due to adhesion may also occur at cutting speeds over 600 m/min, especially when cutting dry; this may result in increased cutting forces, chatter, and poor surface finish [102].

Extensive tables of recommended speeds, feeds, depths of cut, and tool geometries for various operations are available from magnesium suppliers [93,103]. Generally, turning and boring are carried out at speeds between 700 and 1700 m/min, feed rates greater than 0.25 mm/rev, and depths of cut up to 12 mm using tools with positive (up to 20°) rake angles, large relief angles, and honed edges. Face milling may be carried out at speeds up to 3000 m/min, feed rates between 0.05 and 0.5 mm/tooth, and depths of cut up to 12 mm. Finer feeds and higher speeds are used in finishing. Milling cutters should generally have positive cutting geometries and roughly one-third fewer teeth than cutters of comparable diameter used to machining other metals.

11.5.2 Aluminum Alloys

Aluminum alloys are also among the most machinable of the common metals [2,104]. Cutting forces are generally low, and because aluminum is a good heat conductor and most alloys melt at temperatures between 500°C and 600°C, cutting temperatures and tool wear rates are also low. Tool life approaching one million parts can be obtained in some mass production operations using PCD tooling. When cut under proper conditions with sharp tools, aluminum alloys acquire fine finishes through turning, boring, and milling, minimizing the necessity for grinding and polishing operations.

Aluminum is commonly machined with HSS, carbide, and PCD tooling. Silicon nitride-based ceramic tools are not used with aluminum because of the high solubility of silicon in aluminum. The major machinability concerns with aluminum alloys include tool life, chip and burr characteristics, edge build-up, and surface finish. Tool life is a concern especially with alloys containing hard inclusions such as aluminum oxide, silicon carbide, or free silicon (e.g., the hypereutectic aluminum-silicon alloys discussed in the following).

Two major classes of commonly machined aluminum alloys are cast alloys, used especially in automotive powertrain and component manufacture, and wrought or cold worked alloys, used in airframe manufacture and similar structural applications.

The most commonly machined cast aluminum alloys by volume are cast aluminum-silicon alloys, which are used extensively in automotive applications. From a machining viewpoint it is common to distinguish between eutectic alloys, containing 6%–12% silicon, and hypereutectic alloys containing generally 17%–23% silicon [2,105,106]. In the eutectic alloys (e.g., 319, 356, 380, 383, and most piston alloys), silicon is encountered only in a eutectic phase. In the hypereutectic alloys (e.g., 390), the silicon content exceeds the eutectic limit, so that particles of free silicon are contained in the matrix. From a tool life viewpoint, most eutectic alloys, when properly tempered, present few difficulties, so that long tool life can be achieved at relatively high cutting speeds. Speeds up to 450 m/min can be used when turning with carbide tools [2], and speeds as high as 4000 m/min can be achieved in some milling applications with PCD tooling. Of the common eutectic alloys, the most easily machined are 319 and 380. The presence of hard inclusions limits achievable cutting speeds and tool life when machining hypereutectic alloys. The free silicon particles are much harder than the surrounding matrix and produce accelerated abrasive wear with conventional tools and wear and chipping with PCD tools. Cutting speeds are typically limited to roughly 100 m/min when turning with uncoated carbide tools [2], and to 1000 m/min when milling with PCD tools. The carbide grades used for cast iron are suitable for machining aluminum alloys. In drilling, wear becomes excessive at speeds above 200–400 m/min with carbide drills, although diamond coated drills may be used at speeds up to 600 m/min [106,107]. Sand casting produces a coarser microstructure with larger silicon particles than permanent mold casting, and thus yields poorer tool life and lower allowable cutting speeds; die casting yields the finest microstructure [106–108]. Ductility is generally lower for sand rather than semi-permanent mold or die cast alloys.

Iron, which is present in most cast alloys, also forms hard inclusions, which have a negative impact on tool life and machinability. Iron is deliberately added to die casting alloys to prevent die soldering; the addition of up to 0.2% manganese to these alloys refines the iron structure and improves tool life [104]. Many machinability concerns for cast alloys containing other alloying elements, such as magnesium, zinc, or copper, are identical to those for the corresponding wrought alloys discussed in the following.

356 aluminum, a corrosion resistant alloy, is sometimes used in place of 319 or 380 aluminum in marine, wheel, and high performance engine applications. It contains less copper and approximately 1% more silicon than these alloys, and consequently wears tools more rapidly by abrasion. Most data indicates that the wear rate increases by 50%–100% compared to 319 Al. Wear is a particular issue in tapping and milling operations. Roll taps rather than cut taps may be used to improve tap life, as is also the case with die cast alloys. Since the material is highly ductile,

burr formation often limits tool life in milling, even at modest insert wear levels [109]. Chip control problems are also experienced in drilling operations. Proper tempering (to T6) can improve machinability; adding strontium to the alloy is also reported to be effective. Generally, this material must be machined at lower speeds and higher feeds that other hypoeutectic alloys to achieve adequate tool life.

Wrought or cold-worked alloys are classified according to primary alloying elements and temper. A four digit classification system is used in which the first digit indicates the primary alloying element. An excellent summary of the grade contents and machinability concerns was given by Chamberlain [104]; the discussion in the following paragraphs is taken largely from this source.

Grades designated 1xxx are commercially pure grades without significant alloy content. Grades designated 3xxx have manganese as the principle alloying element. 1xxx and 3xxx grades are soft grades, which cannot be heat treaded and which have similar machining characteristics. They generally present few tool life problems but are subject to chip control, built-up edge, and surface finish and integrity problems as discussed below.

Grades designated 2xxx have copper as the primary alloying element, although other elements (e.g., magnesium) may be specified. Copper forms hard particles in the matrix, which increase tool wear; proper tempering should be used to ensure that the copper is taken into solution. 2xxx alloys generally have the best machinability of the wrought aluminum alloys. 2011, which contains bismuth and lead, and 2024 are common free-machining grades, which yield excellent tool life and chip characteristics in turning and drilling applications. Grades designated 5xxx have magnesium as the principle alloying element. These are soft alloys that generally are almost as easily machined as the 2xxx grades. Grades designated 7xxx have zinc as the principle alloying element but may also contain specified levels of copper, magnesium, chromium, and zirconium. These are also soft grades, which are generally easily machined, although they produce continuous chips, which may present disposal problems at high cutting speeds.

Silicon and magnesium are the principle alloying element in grades designated 6xxx. These grades are more difficult to machine because the magnesium and silicon are tied up in hard Mg_2Si particles. 6262 is the easiest material in this series to machine because it contains bismuth added specifically to improve machinability. A commonly encountered grade in this series is 6061, used for welding and corrosion resistance applications, as well as for forged wheels and military vehicle structures; it is a relatively poor heat conductor and must be cut at speeds lower than those used for 2024 to avoid melting of the chip and consequent surface finish and dimensional accuracy problems.

Grades designated 4xxx have silicon as the principle alloying element. These grades, which are used primarily in forgings and for welding and brazing, are generally the most difficult to machine because they contain free silicon. Grades designated 8xxx include lithium alloys and some miscellaneous grades with typically yield intermediate tool life.

As noted earlier, aside from tool life, the major concerns in machining aluminum are chip and burr control and surface finish and integrity. Chip control is a particular concern in drilling and in turning at high cutting speeds. Drills with high helix angles and open (parabolic) flute designs are best suited for aluminum alloys because they minimize chip obstruction and packing. Polishing flutes in the direction of chip flow is also effective in preventing these problems, as is the use of through-the-tool coolant. In turning, chip control problems and BUE occur for alloys that form continuous chips, such as the 7xxx series of wrought alloys. Other alloys, notably most cast alloys (except 356) and wrought alloys of the 2xxx series, generally form short, segmented chips, which are more easily disposed of; even for these alloys, however, flood coolant is usually required for chip control at higher cutting speeds.

Surface finish problems occur especially when machining soft alloys such as 319 and the 1xxx and 3xxx series wrought alloys. For these grades, BUE may form on the cutting edge, and the material itself may tear rather than form a chip, producing and uneven or gouged machined surface. These problems are more serious for carbide than for PCD tools. The most common approach for minimizing these problems is to increase the cutting speed, provided this does not result in chip control difficulties; other effective strategies include using flood coolant and sharp, uncoated tools with high positive rake angles and polished rake faces. TiN coatings, especially for drills, are also often found to be effective in reducing friction and edge build-up when machining cast alloys.

Special tooling and processing concerns for Minimum Quantity Lubrication (MQL) machining of aluminum alloys are discussed in Chapter 15.

11.5.3 METAL MATRIX COMPOSITES

Metal matrix composites (MMCs) are metals reinforced with particles, fibers, or whiskers of harder materials. In most cases, the metal matrix is a cast or sintered aluminum alloy, although steel, magnesium, and titanium MMCs are also available [110,111]. Common reinforcing materials include boron, silicon carbide (SiC), aluminum oxide, graphite, tungsten, and various proprietary fibers such as KAOWOOL or SAFFIL [112]. In general, MMCs are more difficult to machine than the unreinforced base material due to their increased abrasiveness. In the case of whisker- and fiber-reinforced composites, delamination of the fibers from the matrix during machining may present additional concerns [113]. Composites reinforced with chopped or short fibers are less abrasive and more easily machined than those reinforced with longer fibers.

Aluminum MMCs reinforced with boron and SiC are machined using practices similar to those used for hypereutectic cast aluminum-silicon alloys (e.g., 390 aluminum), which were discussed in the previous section [110]. Due to the abrasive nature of these materials, PCD [114–116], diamond tipped [117–119], or CVD diamond-coated tools [120] normally yield longer tool life than carbide tooling [121]. Typically, these tools provide close tolerances and excellent surface finish and integrity when conventional aluminum machining practices are followed with slight modifications. The tool life, however, is reduced due to increased abrasive wear and tool chipping. Since abrasive wear due to the fibers is the primary failure mechanism, cutting fluids have limited effect on tool life or achievable finish [121], although they may aid in swarf disposal and inhibit BUE formation. BUE formation is a particular concern with certain base matrix materials (e.g., 6061 aluminum). Aluminum MMCs reinforced with proprietary fibers other than SiC can also generally be turned and milled using standard aluminum machining practices with reasonable results. In most cases longer tool life is achieved with these materials than with SiC reinforced MMCs because the proprietary fibers are less abrasive than SiC [112].

Aluminum MMCs reinforced with alumina are more difficult to machine than SiC-reinforced materials. Relatively little data on machining of these materials has been published [110]. In some cases carbide rather than PCD tooling may provide better tool life due to its increased toughness and reduced tendency to chip.

Holemaking operations such as drilling, reaming, and tapping are typically the most critical cutting operations with composite materials. Feed is the most significant parameter affecting drill life and tool forces. An increase in the feed will result in an increase in tool forces and significant improvement in tool life. Feeds must generally be higher than 0.10 mm/rev to achieve acceptable tool life. High pressure/high volume coolant should be used to eject SiC particles generated during hole production.

Delamination of fiber-reinforced materials is a particular concern in drilling, especially for through holes. Although the tendency to delaminate is to a large extent a material property depending on the strength of the bond between the matrix and reinforcing fiber, delamination can be inhibited by using sharp drills with high helix angles. For more abrasive materials this greatly limits the tool life, which can be achieved with conventional HSS drills, and carbide drills are normally used.

Magnesium MMCs are machined using practices similar to those used for aluminum MMCs. Titanium MMCs are more difficult to machine than aluminum MMCs and are often machined using lasers or other nontraditional methods, rather than the conventional chip-forming metal cutting processes.

Steel MMCs are typically composed of roughly 25%–45% ultra-hard titanium carbide grains by volume dispersed in a high-carbon-high-chromium tool steel matrix. They are more difficult to machine than the base steel materials. Steel MMC parts are most readily processed by machining to the desired shape in the annealed state (at 44–50 R_c hardness) using cutting tools and conditions similar to those used for the base material with similar hardness. Following this, the parts are heat treated and tempered to 65–70 R_c; the critical dimensions are then ground.

11.5.4 Copper Alloys

Copper and its alloys, the brasses and bronzes, are in most cases easily machinable, although notable exceptions exist. They typically have melting points lower than 1050°C; temperatures of this magnitude are sufficient to cause softening of many types of tools but are rarely achieved at the tool point due to the high thermal conductivity of these materials and the fact that they are frequently cut at lower cutting speeds. Brasses, for example, are often machined in wire or bar form on automatic screw machines (e.g., in the manufacture of electrical components, plumbing components, and hydraulic line fittings); the small stock diameter limits the maximum attainable cutting speed. They are typically machined using HSS or carbide tooling. PCD tooling could in principle be used to machine many copper alloys but does not appear to have been widely used in practice for common alloys.

Pure copper and brasses are normally machined in wrought or drawn bar form. Pure copper is a difficult material to machine because it is highly ductile. It produces high cutting forces, especially at low cutting speeds, which may lead to excessive deflection, vibration, and breakage of tools, especially drills [2]. Its high ductility leads to difficulties in breaking chips and to excessive burr formation. It also yields a poor machined finish at low cutting speeds. Pure copper should be cut at relatively high cutting speeds (greater than 200 m/min) using sharp tools; deburring is often required following machining. The machinability of copper improves with increasing cold work, since this reduces ductility. Alloying electrical grades with sulfur or tellurium to form free-cutting grades also improves chip breaking characteristics and machinability.

Alloying copper with zinc to form brass greatly improves its machinability. Tool forces are reduced for 70/30 (single phase or α) brass and especially 60/40 (two-phase or α–β) brass [2]. α-brass yields discontinuous chips and a poor surface finish at low cutting speeds; other types of brass (e.g., red and cartridge brass) produce continuous chips, which are easier to break than those obtained with pure copper but which still may present disposal problems. Chip breaking is improved by the addition of lead, which precipitates into soft inclusions, which provide lubrication and weak points in the chip. Leaded free machining brasses (e.g., CA360 brass), generally containing 2%–3% lead by weight, are regarded as easily machinable materials. They are used especially in turning at cutting speeds up to 300 m/min, although much higher speeds can be achieved with larger diameter stock using carbide or PCD tooling.

Due to regulatory changes that prohibit the use of lead in plumbing applications, there has been significant recent interest in machining lead-free brasses [122–124]. These commonly contain bismuth or similar free machining additives. They are best machined at higher cutting speeds using coated carbide or PCD tooling; common carbide coatings include titanium diboride and diamond-like carbon. They have reduced machinability compared to leaded brasses and may present particular difficulty in drilling and threading operations. These materials are more easily machined at higher cutting speeds than traditional brass alloys, so increasing cutting speeds is a common approach to solving machining problems when this is an option. When converting older equipment or machining smaller diameter parts, changing tool coatings or coolants is often the best available approach.

Cast copper alloys are commonly used for fittings, bearings, and bushings. From a machinability viewpoint, these alloys can be divided into three groups [125]. The first group includes single-phase alloys rich in copper but containing lead. As with leaded free-machining brass, these are easily machined because they contain lead particles that facilitate chip breaking. The second group includes grades with two or more phases, in which the secondary phases are harder or more brittle than the matrix. Materials in this group include many aluminum bronzes, silicon bronzes, and high tin bronzes. The brittle phases improve chip breakability, so that these materials generally yield short, broken chips. Nonetheless, the hard phases also increase abrasive tool wear, so that the machinability of these grades is classified as moderate. The third group includes high strength manganese bronzes and aluminum bronzes with high iron or nickel contents. These materials are difficult to machine because they produce continuous chips and contain hard phases, which increase tool wear.

11.5.5 CAST IRON

Iron is usually machined in cast form and is used extensively in the manufacture of automotive components and machine tool structures. Three basic types of cast iron are in wide use for these applications: gray iron, malleable iron, and nodular or ductile iron. Two other forms, white iron and compacted graphite iron, are also used in specialized applications.

Gray cast iron [126–130] often consists primarily of pearlite (typically 85% of the matrix) and ferrite, although primarily ferritic grades also exist and most grades contain some steadite or austenite. It contains between 3% and 5% graphite in flake or lamellar form. Flake graphite reduces ductility, facilitates chip breaking, and acts as a natural internal lubricant. Gray iron therefore produces comparatively small cutting forces for its hardness and forms short, easily broken chips. Its limited ductility also reduces the tool–chip contact length and limits maximum cutting temperatures. For these reasons it is usually classified as an easily machinable material. It is used extensively for engine blocks and heads, brake rotors, and other automotive components because of its good machinability, low cost, and high internal damping.

Gray iron castings are often used in mass production applications. For a fixed iron chemistry, tool life in these applications correlates strongly with the casting hardness, which depends on the cooling rate and metallurgical state. Casting hardness increases as the cooling rate is increased. Generally, machining processes in such applications are designed to provide acceptable tool life for a range of casting hardness values. If shakeout times are reduced, resulting in more rapid cooling, castings with hardness values exceeding the specified limit may be produced. When this is the case, the foundry practice must be altered to reduce the incoming hardness or the machining processes must be redesigned to provide acceptable tool life with harder iron. Inspection of incoming castings to ensure that their hardness is within an acceptable range can help reduce intermittent tool life problems caused by occasional hard castings. Annealing castings prior to machining can also significantly increase tool life and machinability.

Other machinability problems encountered with gray iron can often be traced to one of two causes: hard inclusions formed during solidification, and sand adhering to or entrained in the cast layer. These phenomena typically result from poorly designed or controlled casting process and are best eliminated through changes in foundry practices. Hard inclusions include iron carbide phases and martensite inclusions. Both are often caused by excessively rapid cooling or foundry chill, which is a particular problem at the corners or in thin sections of castings, and by segregation during solidification. Hard inclusions increase abrasive wear rates and may also contribute to tool chipping or breakage. They are best eliminated at the foundry by redesigning the casting process so that all sections of the casting cool at the same rate; this can be accomplished by altering gating, changing shakeout practices, redesigning castings to eliminate thin sections, and (less commonly) by adding alloying elements to the iron. Adhering sand or dross in the cast layer can be eliminated by properly cleaning castings, generally by shot blasting. Another common source of machining problems is dimensional variation in incoming castings, caused by core shifts, foundry swell, or casting distortion; the amount of stock to be removed in roughing cuts may be increased beyond process capabilities in the case of excessively distorted castings. As with other defects, casting swell or distortion should be solved at the foundry by altering core assembly methods and by careful inspection of finished castings.

Alloying elements have an impact on the machinability of gray iron, although their influence is usually less pronounced than that of the casting hardness. Adding copper or tin to the iron usually refines the pearlite structure and reduces carbide formation, resulting in an increase in machinability. Adding chromium or nickel to castings to increase their abrasive wear resistance (e.g., in the cylinders of engine blocks) increases casting hardness and abrasive tool wear. Adding phosphorus in concentrations of about 0.15% results in steadite formation, which increases abrasive wear rates and reduces machinability.

Machining iron at high speeds in mass production applications can result in a buildup of dust, composed primarily of graphite, in the machining system. The graphite dust has a negative impact on machine life, and in some cases the dust may contain particles of alloying elements such as chromium, which react with the cutting fluid to form abrasive oxides, which accelerate spindle bearing and way wear. Frequent cleaning of machinery and the use of flood coolants (when the tool material permits) to reduce dust dispersal are effective in reducing machine failures due to this cause.

The machinability of pearlitic gray irons generally improves as the pearlite structure becomes finer, although the alloying elements added to refine pearlite may counteract this effect. Ferritic gray irons are softer and more easily machined than pearlitic irons, although they may be subject to built-up edge formation. Free ferrite may also react with the binder in PCBN tooling, greatly reducing tool life. Steadite and austenite phases increase hardness and result in increased abrasive wear and a reduction in machinability.

Gray iron can be machined using coated or uncoated carbides, alumina and silicon nitride ceramics, or PCBN tooling [131–133]. It can also be finished turned using cermets. Coated carbides can be used for turning and milling at speeds up to 150 m/min. Flood coolants are normally used with carbide tooling to reduce dust dispersal. Tool life on the order of 1000 or 2000 parts can be achieved with carbide tooling. Silicon nitride tooling is used in turning, milling, and boring at speeds between 800 and 1300 m/min. Hot pressed silicon nitride tools are used without coolants to prevent tool failures due to thermal shock. Chamfered edge preparations are normally used to prevent tool chipping, especially in interrupted operations. Tool life between 2,000 and 10,000 parts can be achieved with silicon nitride tooling. Depths of cut should be greater than the thickness of the cast layer; when this is not possible, alumina-based ceramic tooling should be used for enhanced wear resistance. PCBN tooling is used especially in milling and finish boring at speeds similar to those for silicon nitride. Coolants are often used to prevent dust dispersal and thermal expansion of the workpiece but is not required for acceptable tool life. Tool life of over 10,000 parts (often up to 50,000 parts) can be achieved with PCBN tooling. Negative rake cutting geometries are used for turning, boring, and milling operations with all these tool materials. Gray iron can be drilled using HSS, HSS-Co, or carbide drills. For HSS drills, the cutting speed is usually kept below 25 m/min to provide adequate tool life (>1000 parts). Micrograin carbide drills with through the tool coolant can be used to improve drill life (often to greater than 30,000 parts at a cutting speed of 80 m/min) and drilled hole quality [134–136].

Malleable iron, sometimes called ARMA steel, is more ductile than gray iron and generally harder to machine. It produces a longer chip–tool contact length and higher cutting temperatures. A frequently quoted general rule is that malleable iron yields 25% higher tool life than free cutting steel under comparable conditions [130]. Traditionally, malleable iron has been machined using coated carbide tools; a black oxide coating is normally preferred because it increases abrasive wear resistance. Hot pressed silicon nitride tooling has traditionally not been used for malleable iron because cutting temperatures exceed the melting temperature of the glassy-phase binders in these tools. Coated silicon nitride grades, however, are reportedly suitable for machining malleable iron [137], and sintered grades also suitable for this material. Relatively little information on the machinability of malleable iron is available in the literature, although tooling catalogs provide some information on suitable speeds, feeds, and depths of cuts for specific tool grades. Much of the information discussed in the following for ductile iron can be applied in a general sense to malleable irons. Pearlitic or decarburized structures at or near the surface of ferritic malleable iron castings reduce machinability.

In *ductile* or *nodular iron*, the graphite has a spherical or nodular, rather than flake, structure. As the name implies, nodular irons have significant ductility (with elongations between 5% and 15%). They are used in applications requiring fatigue resistance, for example, for crankshafts, camshafts, bearing caps, steering knuckles, clutch housings, axle carriers, and wind turbine hubs.

Due to their increased ductility, nodular irons produce longer tool–chip contact lengths and higher cutting temperatures than gray iron. The nodular structure of the graphite also inhibits its dispersal at the tool–chip interface and thus its lubricating effect. As a result they are more difficult to machine than gray irons; in fact, nodular iron is comparable to mild steel in machinability. The machinability of nodular irons depends on their microstructure, alloy content, hardness, and ductility.

Nodular irons may have a primarily pearlitic, ferritic, or austenitic structure. Pearlitic structures are common and generally result in intermediate hardness and ductility. Irons with a primarily ferritic structure are softer and more ductile than pearlitic grades and provide roughly equivalent machinability. In pearlitic grades, the graphite nodules are usually encased in ferrite. Increasing the thickness of the ferrite layer around the nodule appears to reduce machinability in general, perhaps by impeding graphite dispersal and lubrication. Grades containing steadite and (particularly) austenite are harder and more abrasive than other grades and are the most difficult to machine.

Austempered ductile iron (ADI), a heat-treated grade sometimes used in place of forged steel, is a high strength material with a ferrite and austenite matrix [138,139]. It is more difficult to machine than nodular iron due to its increased strength, which increases both adhesive and abrasive wear. In addition, it also forms segmented chips more easily, which can increase dynamic force variations. The machinability of ADI is reportedly similar to that of hardened steel (HRC 30) [138]. When possible, it should be rough machined prior to heat treatment, leaving a small stock allowance for finishing after austempering. Information on wear rates, chip forms, forces, and surface finish capabilities for ADI is summarized in Ref. [138].

The effect of alloying elements on the machinability of nodular iron is similar to that for gray irons [126]. Adding copper or tin reduces hardness and refines the pearlite structure and results in an increase in machinability. Adding phosphorous promotes steadite formation and reduces machinability. Adding chromium, nickel, and manganese increases abrasive tool wear and reduces machinability. The amount of inoculant added has a significant effect on tool life [140].

Increasing either hardness or ductility reduces machinability. The influence of ductility appears to be stronger than that of hardness within the normally encountered ranges of these variables. For example, a grade with a Brinnel hardness of 280 and elongation of 6% typically yields a longer tool life than any grade with a Brinnel hardness of 270 and elongation of 10%.

Nodular irons are normally turned or milled with coated carbide or PCBN tooling. Hot pressed silicon nitride tooling is not used due to excessive cutting temperatures. In contrast to gray iron, positive rake tooling may be used with some nodular irons to reduce cutting forces and chatter. Black oxide coatings are preferred because they increase abrasive wear resistance. Tool life generally increases with the application of a flood coolant provided this does not result in coating failure due to thermal shock. The allowable cutting speeds in turning, boring, milling, and drilling with carbide tools are typically roughly half those quoted earlier for gray iron. The allowable speeds for PCBN tools, which are used especially with harder grades, are higher, but much lower than those used to cut gray iron with silicon nitride tools [132,133,141]; recommendations for specific tool grades are available from tooling manufacturers.

Unlike gray irons, nodular irons are subject to flank buildup and burr formation [142–144]. The strategies discussed in Section 11.4 are effective in reducing burr formation. Flank buildup can be reduced or eliminated by decreasing the feed rate, using positive rake angles, increasing clearance angles, and by using a flood coolant.

Foundry defects such as chill, adhering sand, and swell, affect the machinability of nodular irons in the same manner as gray iron as discussed before.

Compacted graphite iron (CGI) has a structure intermediate between gray and ductile iron and is used especially for diesel engine blocks and heads and for bearing caps, where it provides increased strength. Most data indicates its machinability of CGI is closer to that of nodular rather than gray iron [130,145–150]. This is in part because complex castings nominally made of compacted graphite iron may exhibit a range of microstuctures, including nodular structures in thin sections. In high volume applications, it is normally assumed that CGI components require one third more capital investment than for gray iron due to reduced cutting speeds. As with austempered ductile iron, CGI exhibits increased adhesive and abrasive wear compared to gray iron [148]. Temperatures are also higher in CGI machining, with thermal cracking and other heat-related tool failures being common. In drilling, the use of curved, contoured chisel edges rather than standard straight chisels reportedly improves tool life [151]. In face milling, PCBN tooling is often avoided due to oxidation and diffusion reactions between the work material and the binder in the tool [152], although ceramic binder grades may be effective in these applications. The sulfur content of CGI is reported to have a significant effect on machinability through the formation of soft MnS layers at the tool surface (as in gray iron) [149]. However, CGI is produced by adding magnesium to liquid iron to consume both sulfur and oxygen; the presence of Mg in CGI greatly inhibits MnS formation [150]. Increased titanium content also reduces CGI machinability, so that Ti content in CGI should be specified to be less than 0.01% [150]. Other chemical and metallurgical effects are summarized by Dawson et al. [150].

White irons and corrosion-resistant *high silicon gray irons* are the most difficult cast irons to machine because they are brittle and abrasive [130]. Alloyed white irons containing nickel are generally ground to finished dimensions, but can be turned and bored if rigid tooling setups and ceramic tools are used. Electro-discharge and electro-chemical machining methods should also be considered for these materials. High silicon irons have high hardness and are also normally ground to finish dimensions. Adding carbon or phosphorous reportedly improves their machinability but degrades other mechanical properties, which may be critical in applications for which they are considered.

11.5.6 Carbon and Low Alloy Steels

Steels vary greatly in chemical content and microstructure. The next two sections discuss the machinability of carbon, low alloy, and stainless steels, concentrating primarily on wrought or bar products. The factors affecting the machinability of these materials are well understood in a general sense. Cast, hardened, and high alloy steels are not discussed in detail. These materials are harder than carbon and low alloy steel and in general are much more difficult to machine. An indication of the tool life that can be expected for these materials as compared to low alloy steels is provided by tables of machinability indices [31,153]. Recommendations for machining specific alloys can be obtained from specialized studies [154–157], review articles [158], or from steel or tooling manufacturers who have experience with the material in question.

The machinability of steel depends on hardness, chemistry, microstructure, mechanical state, and work-hardening characteristics. As with most materials, machinability decreases with increasing hardness. Increasing the hardness increases cutting forces and temperatures and stresses at the tool point; for HSS and sintered carbide tools, excessive temperatures and stresses may lead to plastic collapse of the cutting edge. Graphs of limiting speeds and feed rates for preventing plastic collapse at specified hardness levels are available [2] (Figure 11.21). The mechanical state of the material, particularly the degree of prior cold work, influences cutting forces and chip breaking characteristics. Annealed steels produce lower cutting forces and temperatures [159], but due to increased ductility they may produce chips, which are more difficult to break; cold worked materials produce higher forces but more easily broken chips. Alloys that work harden rapidly produce more rapid tool wear, especially notch wear, than those that harden more slowly. As discussed in the following, chemistry and microstructure influence the distribution of hard particles within the matrix and thus abrasive tool wear rates. Macro inclusions (>150 μm) negatively affect machinability and can result in sudden tool failure Al_2O_3 is especially hard and abrasive; FeO and MnO have a similar but less severe effect.

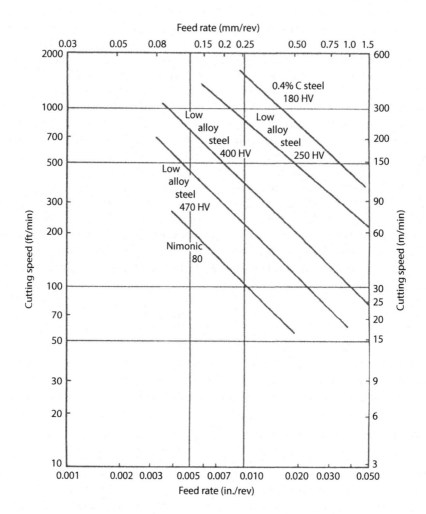

FIGURE 11.21 Conditions of deformation of cemented carbide tools when cutting steels of different hardness. (After Trent, E.M., *Metal Cutting*, Butterworths, London, U.K., 1977, pp. 139–180.)

Carbon steels may be classified as low carbon (containing less than 0.3%C; AISI grades 1005–1029), medium carbon (containing between 0.3% and 0.6% carbon; AISI grades 1030–1059), and high carbon (containing over 0.6%C; AISI grades 1060–1095) [153]. As-rolled low carbon steels consist primarily of ferrite, although other phases, particularly pearlite, are present. The pearlite content (and, generally, the hardness) increases with increasing carbon content. As-rolled medium carbon steels have a pearlite–ferrite structure and are predominately pearlitic when the carbon content exceeds 0.4%, although ferritic grades with dispersed graphite are also encountered [160]. As-rolled high carbon steels have a pearlitic matrix, with cementite, the hardest component of steel, predominating for steels with carbon content greater than 0.8%. In general, as-rolled carbon steels become harder and less machinable as the carbon content increases. Heat treating carbon steels may result in the formation of bainite or tempered martensite phases. Although these phases are harder than pearlite or ferrite, they serve to help break chips, so that tempering generally increases machinability.

Carbon steels are most commonly turned, bored, and milled with coated carbide tooling. As discussed in Chapter 4, steel-cutting carbide grades contain TiC to reduce crater wear; TiN-based (gold) coatings are also often used to reduce cratering due to diffusion. Negative rake tools with honed edge preparations are most commonly used, and turning inserts usually have molded-in chip breakers. Carbon steels can also be finish turned with cermets, turned and bored with alumina-based ceramic tooling, and turned, bored, and milled with PCBN tooling. Drilling, end milling, and tapping of carbon steels may be performed using HSS, HSS-Co, or solid carbide tooling; rotary tools are often coated with TiN-based coatings to reduce cratering. Most carbon steel machining operations are carried out using a flood coolant to control tool and workpiece temperatures, although crankshaft alloys such as 15V35 may be drilled using MQL.

Low carbon steels generally have the highest machinability of the carbon steels. They are typically turned at speeds of roughly 200 m/min and feed rates of roughly 0.15 mm/rev. Milling is carried out at somewhat higher speeds and lower feed rates. Drilling is carried out at a variety of speeds that are generally below 20 m/min for a HSS drill. The feed rate often has a stronger influence on maximum penetration rates than the speed because it influences drill breakage. In the as-rolled or annealed conditions, machinability is generally best for steels with carbon contents between 0.15% and 0.25% [153]. At lower carbon levels, the material is more ductile and adheres to the tool, increasing the difficulty of breaking chips and resulting in built-up edge formation, which leads to poor surface finish. Very low carbon grades such as AISI 1010 or 1008 are notoriously susceptible to these difficulties in the annealed condition. Moreover, since these materials often contain controlled scrap when obtained in bar form, they may show significant variation in trace chemistry and ductility in mass production operations unless the consistency of incoming material is ensured through careful inspection. Cold working decreases ductility and usually increases machinability; it is particularly effective in eliminating chip breaking problems and BUE formation.

Medium carbon steels are harder than low carbon steels and yield higher cutting forces under equivalent conditions. They also contain higher proportions of harder phases, particularly cementite, which accelerates abrasive tool wear. As a result, they are machined at lower cutting speeds than low carbon steels; speeds should be progressively decreased as the carbon content is increased [153]. Annealing and normalizing these steels coarsens the pearlite structure and improves machinability. Medium carbon steels are less subject to chip breaking difficulties and BUE formation than low carbon steels.

High carbon steels contain a greater proportion of hard phases (e.g., cementite) than medium carbon steels and therefore produce higher cutting forces, temperatures, and abrasive tool wear rates. As a result, they are machined at lower cutting speeds and feed rates; as with medium carbon steels, the allowable cutting speed for acceptable tool life decreases with increasing carbon content. As noted earlier, grades containing more than 0.8% carbon have excess cementite in the matrix in the as-rolled or air-cooled condition and are particularly difficult to machine. Annealing tends to consolidate the cementite into larger particles and typically improves machinability [153]. A spheroidized microstructure also improves the machining characteristics of higher carbon steels. Assembling carbides into spheroids improves the cutting action and reduces tool wear. Coarse spheroidite or pearlite structures provide better machinability than fine structures. High carbon steels tend to air harden, so that surface integrity problems such as residual stress and white layer formation are a particular concern if the coolant volume is insufficient or inconsistent. High carbon steels typically present few chip control problems and are not prone to BUE formation. Hardened high carbon steels contain abrasive martensite and bainite phases and are the most difficult of the plain carbon steels to machine [161]. In addition, they are prone to white layer formation [162–164] as discussed in Chapter 10. When hardening is required, steels should be rough machined prior to hardening, and hard turned or ground to their final dimension following hardening.

Adding alloying elements to steel generally increases hardness in the as-rolled and annealed conditions, resulting in decreased machinability [153]. Many alloying elements are added to increase strength or wear resistance, and combine with carbon to form very hard, abrasive carbides, which reduce tool life; examples include chromium, nickel, and manganese [2]. Not surprisingly, the impact of alloying elements on machinability depends on the alloy content; low or lean alloy steels machine much more like carbon steels than corresponding high alloy grades [165]. A spheroidized structure in alloy steels results in improved overall machining performance because it reduces hardness and arranges the hard carbide phase into spheroids, which reduces the abrasive action of the carbides. Tempered alloy steels produce better surface finishes than annealed steels because the tempered structure reduces or eliminates BUE formation. As noted before, machinability tables can be used to assess the tool life to be expected from specific high alloy grades as compared to plain carbon grades. Formulas for computing an equivalent carbon content from the alloy content can also be used for this purpose [166]. A detailed discussion of the machinability of specific alloy families, related to their metallurgy, is given by Finn [153].

There are alloying elements, however, which are added to steel specifically to increase machinability at conventional and high speeds [167]. These include lead, sulfur, manganese sulfide, phosphorous, calcium, bismuth, selenium, and tellurium [2,154,168]. Typically, these additives, which are often used in combination, result in insoluble inclusions in the matrix. In addition to the content, the size, shape, and distribution of the inclusions affect machinability. These inclusions cause the metal matrix to deform more easily and facilitate crack propagation, resulting in reduced cutting forces, enhanced chip breakability, and improved surface finish. The resulting grades are designated variously as free machining, free cutting, or enhanced machining steels.

Sulfurized free-machining steels have a high sulfur content. The sulfur is often added in the form of manganese sulfide, MnS, a solid lubricant that forms inclusions in the matrix [2,153]. During cutting, the MnS coats and lubricates the rake face of the tool, reducing friction, tool–chip temperatures, and tool wear rates. The inclusions also enhance chip breaking. Depending on the amount of MnS, other mechanical properties of the steel such as corrosion resistance, ductility, toughness, formability, and weldability may be negatively impacted. A wide variety of sulfurized grades, which trade off one or more of these properties for enhanced machinability, are available.

Leaded free-machining steels contain lead, which also forms inclusions in the steel matrix, which serve to lubricate the rake face of the tool and to break chips. Most leaded grades also include other free machining additives such as MnS and phosphorous; as a result, they generally provide better machinability than non-leaded sulfurized grades. Leaded steels have become less common as concerns about occupational exposure to lead have grown. This has led to interest in the development of non-leaded free-machining steels (e.g., grades containing bismuth). In mass production applications in which a number of machines are connected to a single recirculating cooling system, strict care should be taken to change the coolant at regular intervals if leaded steels are machined. If this is not done, lead may build up to unacceptable levels in the coolant sump.

Calcium is sometimes added to deoxidizing agents in the final stages of steel production. Calcium deoxidation reduces abrasive tool wear and improves machinability, especially for high carbon grades [153]. As with other enhanced machining grades, the increase in machinability results from the formation of soft inclusions in the matrix. The chemical composition of the inclusions depends on the other deoxidizing elements used.

Microalloyed steels [26,169–172] contain small amounts of hard elements such as vanadium and tungsten. They are most often intended for use in an as-forged condition in applications that normally require forged and heat treated (quenched and tempered) steels, for example, for high-strength crankshafts. The aim of microalloying is to increase strength and wear resistance in the as-forged condition significantly, so that subsequent quenching and tempering is unnecessary. Most test results indicate that as-forged microalloyed steels are significantly more machinable than forged, quenched, and tempered carbon grades. In these tests, the hardness of the microalloyed steel is generally significantly lower than that of the quenched and tempered steels.

11.5.7 STAINLESS STEELS

Stainless steels contain a high proportion of chromium, generally in excess of 11%. They are normally difficult to machine due to their high tensile strength, high ductility, high work hardening rate, low thermal conductivity, and abrasive character. This combination of properties often results in high cutting forces, temperatures, and tool wear rates, as well as a susceptibility to notch wear, chip breaking difficulties, BUE formation, and poor machined surface finish [2,173].

Alloying elements can be added to reduce some of these difficulties, however, resulting in free-cutting grades with comparatively good machinability.

Stainless steels are usually classified into four categories depending on the primary constituent of the matrix: ferritic, martensitic, austenitic, and duplex (combined ferritic/austenitic). They may also be classified based on their heat treatment (precipitation-hardenable versus non-precipitation hardenable alloys) or machining characteristics (free machining versus non-free machining grades).

Ferritic stainless steels are alloyed primarily with chromium, although molybdenum, titanium, or niobium may be added to some grades to improve corrosion resistance or as-welded properties. Ferritic alloys are generally more machinable than other alloys [173,174]. They are used in an annealed or cold worked condition but are not heat treated, so that their hardness is comparatively low. Their machinability generally decreases with increasing chromium content.

In addition to chromium, martensitic alloys may contain molybdenum or nickel to increase strength. The machinability of martensitic stainless steels is influenced by hardness, carbon content, nickel content, and metallurgical structure [174]. As with most materials, increasing hardness typically reduces tool life and machinability. Increasing the carbon content increases the proportion of abrasive chromium carbides in the matrix and reduces tool life and machinability. Increasing the nickel content increases the annealed hardness and also reduces machinability. The metallurgical factor that has the strongest influence on machinability is the proportion of free ferrite in the matrix; generally machinability increases with free ferrite content.

Austenitic stainless steels contain nitrogen and nickel or manganese in addition to chromium. They exhibit high strength, ductility, and toughness and are typically more difficult to machine than ferritic or martensitic stainless steels. Specific difficulties encountered when machining austenitic stainless steels include high wear rates due to high cutting forces and temperatures, BUE formation, chip control problems, poor surface integrity (hardened machined surfaces), and a tendency to chatter [175–177]. Poor tool life is related to the annealed hardness, which increases with increasing nitrogen content [177]. Increasing the carbon content increases the work-hardening rate and also decreases machinability. Abrasive carbon/nitrogen compounds may form in the matrix and reduce tool life; these can be controlled by adding titanium or niobium. As with other stainless steels, hardness increases and machinability decreases with increasing nickel content. Imparting moderate cold work to the material typically increases machinability by reducing the tendency for BUE formation and improving the machined surface finish and integrity.

Duplex alloys have a chemistry similar to austenitic stainless steels but are generally more difficult to machine due to their high annealed strength [178,179]. They have high abrasive wear rates due to hard oxide phases and are also prone to BUE formation. Machining duplex alloys can be particularly challenging because no standard enhanced-machining grades are available.

Enhanced or free machining ferritic, martensitic, and austenitic stainless steels contain alloying elements intended to improve machinability [180–182]. Common machinability additives include sulfur, selenium, tellurium, lead, bismuth, and phosphorous. As with carbon and low alloy steels, these additives are effective because they form compounds that have low solubility in the matrix and precipitate as inclusions, which serve to lubricate the tool–chip interface and to break chips. Sulfur is the most commonly used additive, followed by selenium, although bismuth and calcium are also effective [182]. Lead is among the most effective additives but is subject to exposure concerns as discussed for brasses and low carbon steels [180]. Machinability can also be improved by varying the deoxidizing agents used in steelmaking to control oxide inclusions in the matrix.

General guidelines for machining stainless steels include [173]

- Using lower cutting speeds and metal removal rates than for carbon steels
- Using rigid tooling and fixturing to avoid chatter
- Maintaining feed above a minimum level to avoid poor surface integrity
- Using sharp tools with a fine finish to avoid BUE formation
- Using proper cutting fluids with sufficient flow rates for heat removal

Detailed guidelines for machining various alloy families with HSS and carbide tools are given in the literature [173,176–179]. Stainless steels may also be machined with ceramic and PCBN tooling; guidelines for the application of specific tool grades can be obtained from tooling manufacturers. For PCBN tooling, proper edge preparations and system rigidity are particularly critical.

11.5.8 POWDER METAL (P/M) MATERIALS

Powder metal (P/M) parts are near-net shaped parts produced by sintering metal powders under pressure. They generally require little machining, although grinding, drilling, and threading operations are often necessary to achieve tight tolerances and produce features such as transverse or threaded holes. P/M parts may be made of many materials such as aluminum, iron, nickel, and copper. Carbide, PCBN [183], and oxide ceramic tooling is used depending on the application. Machining practices for these materials are similar to those of cast or wrought materials with similar chemical compositions, but are affected by unique features of P/M materials including porosity, variations in density and thermal conductivity, and increased abrasiveness. This section concentrates primarily on the machinability of P/M structural steels. The influence of properties such as porosity on machinability is similar for sintered metals made of other materials.

Generally, increasing the density of P/M materials increases machinability. Optimum machinability is often achieved for materials with densities between 90% and 95% of the theoretical maximum [184]. In this case the machinability of P/M parts is comparable to that of corresponding cast or wrought parts. Machinability increases with density because porosity decreases. Increased porosity has a negative effect on machinability for a number of reasons [184,185]. Porosity causes discontinuous contact between the tool and workpiece, resulting in an interrupted cutting action, which increases dynamic and localized stresses on the tool. As a result, the tool wears more rapidly and may chip because it is subject to shock loading and increased vibration. Porosity also affects thermal properties. Increasing porosity reduces thermal conductivity, resulting in higher cutting temperatures and increased chemical (thermally-activated) wear. Materials with densities greater than 95% of the theoretical maximum are often produced by forging; in this case, increased hardness often offsets the effect of reduced porosity and results in decreased machinability.

The increased cutting temperatures caused by high porosity may also accelerate abrasive wear. Abrasive wear rates are generally higher even for low porosity P/M materials than for corresponding cast or wrought materials because P/M materials form a powdery chip, which may include free particles of hard materials. As temperatures increase, constituents of the machining debris may oxidize, resulting in a higher concentration of abrasive oxide particles.

Optimum machinability can also be obtained by machining materials in the presintered condition [184]. Presintering is performed to volatilize and burn off lubricants used in pressing, prior to heating parts under pressure to amalgamate the metal powder. Presintered parts are easier to machine than sintered parts because they are typically softer and less abrasive. However, machining in the presintered condition reduces achievable accuracy (tolerance) and finish characteristics.

Finally, the machinability of P/M parts can be increased through the use of free machining additives [184,186,187]. Additives are effective for P/M materials for essentially the same reasons they are effective in cast irons and wrought steels; they result in the formation of inclusions in the material matrix, which serve to lubricate the tool–chip contact and reduce cutting friction and temperature. Additives also generally reduce cutting forces, reduce BUE formation, improve surface finish, and increase chip breakability. For P/M steels, manganese sulfide is the most common and effective additive, followed by sulfur and lead. Bismuth, graphite, tellurium, and selenium are also usually effective. Adding copper is also effective when it results in a significant increase in thermal conductivity. Iron–copper P/M steels are used extensively in automotive applications for bearing caps, connecting rods, and valve seats and guides; care should be taken in these applications in reducing copper content to reduce costs, as this may result in greatly reduced tool life. Many contemporary valve seat alloys are infused with copper to reduce porosity and increase heat conduction; these alloys have high hardness, however, and are normally difficult to machine.

11.5.9 TITANIUM ALLOYS

Titanium alloys have a high strength-to-weight ratio and are used in many aerospace, armor plate, and biomedical applications [188–192]. Pure titanium undergoes a metallurgical transformation at about 830°C, changing from a hexagonal close-packed structure (alpha phase) to a body-centered cubic structure (beta phase) [193]. Adding alloying elements can significantly change the transformation temperature, so that the beta phase is stabilized and can be retained at room temperature. There are four main groups of titanium alloys: unalloyed titanium, which sometimes contains small amounts of oxygen or iron for increased strength; alpha alloys, which contain alpha phase stabilizers such as aluminum, oxygen, nitrogen, and carbon; alpha–beta alloys, which contain both alpha and beta phases and which are alloyed with both alpha and beta stabilizers; and beta alloys, which are alloyed with beta stabilizers such as molybdenum, vanadium, niobium, copper, and silicon. Most alloys used in aerospace applications, for example, Ti–6Al–4V, are alpha–beta alloys. Generally increasing the beta phase content increases strength and decreases machinability, although in some cases increasing beta phase content improves chip formation and results in an increase in machinability. Chip form and machinability are also affected by heat treatment [194].

Titanium alloys are regarded as difficult to machine for the following reasons [192,193,195]:

- They maintain high strength at high temperatures, increasing cutting forces and tool stresses.
- They produce thin chips, which increases cutting temperatures and stresses at the tool cutting edge.
- They have comparatively low thermal conductivities, which further increases cutting temperatures.
- They have a high chemical reactivity with almost all tool materials at elevated temperatures.
- Due to their thermomechanical properties, they often produce discontinuous or shear localized chips.
- They have a low modulus of elasticity, which can lead to excessive deflection of the workpiece and to chatter.
- They are susceptible to surface damage during machining, and yield poor machined surface finishes under many conditions.
- Their chips may ignite during machining due to the high cutting temperatures and sparks are often generated, especially in dry cutting applications.

Because of all these factors, tool life and allowable machining rates are lower for titanium alloys than for most other metals. In particular, the choice of tool materials is limited due to high chemical reactivities. In most cases, titanium alloys are machined using HSS or uncoated carbide tools. High cobalt high speed steels (HSS-Co grades) typically yield better tool life than straight HSS grades [193,195]. Among straight grades, highly alloyed materials such as T5, T15, M33, and the M40 typically perform better than general purpose grades such as M1, M2, M7, and M10 [195]. For carbide tooling, straight grades such as C2 (ISO K20) perform better than the steel cutting grades such as C8 (ISO grade PO1). Micrograin C2 carbide grades are commonly used for end mills. The steel cutting grades, which are alloyed with TiC and TaC, wear more rapidly by diffusion; possible metallurgical reasons for this are discussed by Trent [2]. Older coated carbide tools wore more rapidly than uncoated grades in many applications [196], although recently developed, multilayer coatings are more effective [188], and coatings are also generally effective in drilling. Among advanced materials, ceramics such as sialon and alumina-based materials are not suitable for machining titanium due to rapid chemical and abrasive wear [196]. PCBN and PCD tooling reportedly provide acceptable performance [193,195] but are subject to chemical and diffusion wear, which generally attacks the binder in the tool [189].

As noted before, titanium alloys produce serrated or shear-localized chips under most cutting conditions [193]. As discussed in Section 6.6, shear-localized chip formation results from the high strength, low thermal diffusivity, and temperature-softening behavior of the material. This type of chip formation can lead to large variations in cutting forces and induce chatter. The poor thermal properties of titanium may also lead to poor surface integrity in the machined surface due to the occurrence of steep temperature gradients and differential cooling of the surface layer. Titanium also adheres strongly to common tool materials, resulting in high friction and a tendency for surface damage due to deformation and BUE formation.

As a result of these material constraints, the following general guidelines should be followed in machining titanium alloys [189,192,195]:

- Low cutting speeds should be used to limit cutting temperatures.
- High feed rates should be maintained to avoid surface damage.
- High coolant volumes should be maintained to reduce temperatures and clear chips; high pressure coolants are generally required when machining with PCD and PCBN tooling.
- Sharp tools with positive rake angles and proper clearance should be used to avoid BUE formation.
- Dwelling of the tool against the workpiece should be avoided to reduce surface damage.
- Rigid tooling and fixturing setups should be used to avoid excessive workpiece deflections and chatter.

Turning should be carried out using carbide tooling when possible [195]. Negative rake tools are used for roughing with carbide tools, while positive rakes are used for finish turning and with HSS tooling.

Milling titanium alloys is more difficult than turning or boring because chips may adhere to the tool during non-cutting periods in interrupted cutting, resulting in tool chipping or breakage [197]. Milling operations are often carried out at lower cutting speeds than turning or boring. Climb milling is preferred to conventional milling when possible to reduce tool chipping [195]. When face milling large surfaces, the spindle should be tilted so that the trailing cutting edges do not rub against the machined surface.

Drilling should be carried out using sharp drills with high point angles [198] or spiral point geometries [195]. Flood or (preferably) through-the-tool coolant should be used to dissipate heat. Dwelling of the drill in the bottom of the hole should be avoided; positive feed mechanisms are useful in avoiding dwell. Better tool life is obtained with solid carbide or carbide-tipped than with HSS drills; when HSS drills are used, chromium or oxide coatings are often effective in reducing galling at the margins.

More detailed recommendations on tool geometries for turning, milling, and drilling, as well as recommended practices for reaming, tapping, grinding, and nontraditional machining processes, are discussed in review articles [188–190,192,195].

11.5.10 Nickel Alloys

Nickel alloys are generally difficult to machine. They have high strength and ductility, work harden rapidly, and contain hard, abrasive phases [2,188,199–202]. Because of this combination of properties, they are subject to many of the machining difficulties encountered with austenitic stainless steels.

The work-hardening characteristics of nickel are the source of many machining difficulties [199,202]. Since they work harden rapidly, they produce high tool stresses and temperatures, which can accelerate tool wear. Work-hardened nickel chips are also abrasive and produce notch wear. The work-hardening rate is highest for annealed or hot-worked materials; one common method of improving machinability is to impart cold work to the material prior to machining when possible. The effect of work hardening can also be reduced by using sharp tools with positive rake angles. High feed rates and depths of cut are also recommended to reduce damage to the machined surface caused by excessive frictional heating. The high forces generated during machining can also lead to distortion of the part and to chatter. To prevent distortion, it is often advisable to rough parts to near the finished dimension, stress relieve them, and then finish machine. This is particularly true for the Group D alloys discussed in the following. Microstructural features also have some influence on machinability. As with cast iron and free-machining steels, the presence of graphite or sulfide phases in the matrix improves machinability by reducing friction and improving chip breaking. Conversely, the presence of hard phases such as carbides, nitrides, oxides, and silicates increases abrasive tool wear and reduces tool life.

From a machining viewpoint, nickel alloys may be classified into five groups [199]: Group A, consisting of alloys containing more than 95% nickel (e.g., commercially pure nickel grades); Group B, consisting of most nickel–copper alloys (e.g., Monel 400 and Nilo alloy 48); Group C, consisting of solid solution nickel–chromium and nickel–chromium–iron alloys (e.g., Inconel 600 and Incoloy 800); Group D, consisting of age-hardenable alloys (e.g., Incoloy 925 and Inconel 718); and Group E, consisting of the free machining Monel R-45 alloy. Group A alloys have moderate strength and are subject to BUE formation and chip control problems in the annealed or hot-worked state. When possible, they should be cold worked prior to machining. Group B alloys have higher strength and lower ductility than Group A alloys. They should be machined in the cold worked condition for optimum results. Group C alloys are subject to hard phase formation and should be cold drawn and stress relieved prior to machining. Group D alloys have high strength and hardness and are the most difficult to machine. They produce high cutting forces and are subject to distortion as discussed earlier. They should be rough machined in the unaged condition when possible; material that has been solution annealed and quenched or rapidly air cooled has the lowest hardness and best machinability. The Group E alloy is the most machinable of the nickel alloys and is suitable for high production applications in automatic bar machines.

Nickel alloys are most commonly machined using HSS and carbide tooling [203]. Because they have low chemical reactivity with nickel, alumina-based ceramics and PCBN tools are also suitable [204–206]. Straight alumina ceramics may fracture in interrupted cutting operations; research suggests that whisker-reinforced ceramics may be suitable for these operations [202,207]. As with steel, the high cutting temperatures generated makes hot pressed silicon-nitride tools unsuitable for machining nickel alloys; sintered grades, however, can be used for turning nickel alloys. Widespread use of ceramic tooling has been inhibited by the fact that many nickel alloys form shear-localized chips at high cutting speeds [208,209], which effectively limits the maximum cutting speeds that can be employed (often to below 30 m/min).

Carbide and sintered silicon nitride tools are most commonly used in turning operations. Positive effective rake angles, large nose radii, and molded-in chip breakers are used to prevent BUE formation, tool fracture, and chip control problems. Surface finish is influenced by the feed and depth of cut; an increase in depth of cut improves the surface finish generated with coated carbide tools and leads to deterioration in the finish when uncoated tools are used [210,211]. Milling is often carried out using HSS cutters because carbide cutting edges tend to chip or fracture in interrupted operations. Micrograin carbides are widely used, however, for end milling. Rigid tool and fixturing setups are required to maintain accuracy and prevent chatter. In ball end milling, the tilt angle of the cutter with respect to the workpiece surface has a significant influence on tool life [212]. Climb rather than conventional milling is used to minimize machined surface damage. General purpose drilling can be carried out using HSS [203] or solid carbide drills, which are often coated [213]. Drilling tends to produce hardened layers and other surface integrity issues at the bottom of holes. As with titanium alloys, positive feed equipment should be used to prevent dwelling of the drill at the bottom of the hole. Crankshaft drills and carbide-bladed spade drills provide better performance in deep hole applications. Specific recommendations on tool geometries and cutting conditions for these and other processes are summarized in review articles [199,201,202].

Neat oils or water-based flood coolants are generally used for nickel machining. High pressure coolant is less common due to the relatively low cutting speeds employed.

11.5.11 Depleted Uranium Alloys

Depleted uranium alloys are byproducts of enrichment processes in which the ^{235}U is extracted from natural uranium. They are machined on a large scale at high production rates to produce armor, calorimeter plates, ballistic penetrators, radiation shielding, gyroscope rotors, flywheels, sinker bars, and aircraft counterweights [214,215]. Three classes of alloys can be distinguished: dilute alloys, containing less than 0.4%wt of alloying elements; lean alloys, containing between 0.4 and 4%wt of alloying elements; and stainless alloys, containing more than 4%wt of alloying elements. Common alloying elements include titanium, which increases hardness, and niobium, which increases corrosion resistance. Metallurgically, these alloys usually consist of an orthorhombic α uranium matrix with carbides and oxides as principal inclusions [214]. Complex tetragonal (β) and body-centered cubic (γ) crystal structures also occur [215].

Although depleted uranium alloys are in many ways free cutting, they exhibit a number of material properties that create machining difficulties. These include

- High ductility, adhesiveness, and tendency to gall
- Abrasiveness
- High work hardening
- Low modulus
- Reactivity with tools and coolants
- Pyrophoricity
- Anisotropic thermal expansion
- Shape memory
- Toxicity and radioactivity

The negative consequences of some of these properties have been discussed with reference to other materials. High ductility, adhesiveness, and tendency to gall limit cutting speeds and can lead to BUE, chip breaking, and burring problems as discussed for pure copper, 356 Al, and 1008 and 1010 steel. High abrasiveness and work hardening can increase abrasive and notch tool wear as discussed for stainless steels and nickel alloys. Low modulus and high reactivity can limit tool material choices and achievable tolerances and promote chatter as discussed for titanium alloys. The countermeasures adopted to address these difficulties for other materials are broadly applicable to machining depleted uranium. The problems more specific to depleted uranium alloys are pyrophoricity, anisotropic thermal expansion, shape memory, and toxicity and radioactivity.

Pyrophoricity is the tendency to catch fire. Uranium alloys oxidize rapidly when exposed to air, and may generate self-sustaining fires when machined into thin chips at low feeds. (A similar problem occurs with magnesium alloys, but a spark or external heat source is required to produce a fire in the magnesium case.) To combat this problem, chips are normally collected in a pan and submerged beneath coolant to reduce temperatures. Water-based coolants are preferred. Mineral oil coolants can also be used, but in this case fires must be extinguished before temperatures exceeding the flash point of the oil are generated. In either case the chip pan should be emptied frequently so that large masses of chips do not accumulate. Finish machined uranium components are also often coated after fabrication to limit oxidation.

Due to its crystal structure, unalloyed uranium expands in two directions and contracts in the third when heated. Uranium alloys also exhibit shape memory, a tendency to return to their original shape after a temperature change. Because of this property, it is also advisable to avoid working the material in one direction, which can impart a preferred orientation to the crystal structure. Both the anisotropic thermal expansion and shape memory properties dictate the use of sharp tools and controlled temperature coolant to minimize distortion due to temperature effects. In precision applications, stock may be left for a light finish pass after roughing; the finish pass is often performed several days after roughing to allow the material to stabilize, and often also after heating to the β transition temperature followed by rapid cooling.

Depleted uranium is a toxic and radioactive material that can present serious exposure hazards. Detailed safety protocols are provided by regulatory agencies.

Depleted uranium can be machined on standard CNC machine tools. Turning, facing, single-point threading, and milling should be performed with sharp tools to limit heat generation and cold working. Carbide tooling is generally used. Drilling may be performed with solid carbide drills coated with a sulfur paste lubricant. Spiral taps with variable rakes and oxide coatings are used for internal threading. Grinding should be avoided but can be performed with silicon carbide wheels and water-based coolants when necessary. Recommended cutting speeds and feeds for these operations are summarized in Ref. [214]. Tolerances of ± 0.05 mm are reportedly achievable in turning and milling [215].

REFERENCES

1. E. J. A. Armarego and R. H. Brown, *The Machining of Metals*, Prentice-Hall, New York, 1969, pp. 246–253.
2. E. M. Trent, *Metal Cutting*, Butterworths, London, U.K., 1977, pp. 139–180.
3. M. C. Shaw, *Metal Cutting Principles*, 3rd edn., Oxford University Press, Oxford, U.K., 1984, pp. 6–7.
4. G. Boothroyd and W. A. Knight, *Fundamentals of Machining and Machine Tools*, 2nd edn., Marcel Dekker, New York, 1989, pp. 148–151.
5. G. T. Smith, *Advanced Machining*, Springer-Verlag, New York, 1989, pp. 167–198.
6. C. F. Barth, Turning, in: R. I. King, Ed., *Handbook of High Speed Machining Technology*, Chapman & Hall, New York, 1985, pp. 174–179.
7. T. J. Drozda and C. Wick, Eds., *Tool and Manufacturing Engineers Handbook*, 4th edn., Vol. 1, SME, Dearborn, MI, 1983, pp. 40–59.
8. N. H. Cook, What is machinability? *Influence of Metallurgy on Machinability*, ASM, Metals Park, OH, 1975, pp. 1–10.
9. C. Zimmerman, S. P. Boppama, and K. Katbi, Machinability test methods, *Metal Handbook*, Vol. 16: *Machining*, 9th edn., ASM, Materials Park, OH, 1989, pp. 639–647.
10. J. ElGomayel, Fundamentals of the chip removal process, SME Technical Paper MR77-256, 1977.
11. Metcut Research Associates, *Machinability Data Handbook*, Cincinnati, OH, 1982.
12. P. Balakrishnan and M. F. DeVries, A review of computerized machinability data base systems, *Proceedings of NAMRC10*, SME, Dearborn, MI, 1982, pp. 348–356.
13. L.-Z. Lin and R. Sandstrom, Evaluation of machinability data, *J. Testing Eval.* **22** (1994) 204–211.
14. S. V. Wong and A. M. S. Hamouda, The development of an online knowledge-based expert system for machinability data selection, *Knowledge Based Syst.* **16** (2003) 215–229.
15. M. P. Groover, A survey on the machinability of metals, SME Technical Paper MR76-269, 1976.
16. G. F. Micheletti, Work on machinability in the co-operative group C of CIRP and outside this group, *CIRP Ann.* **18** (1970) 13–30.
17. D. A. Stephenson, Tool-work thermocouple temperature measurements—Theory and implementation issues, *ASME J. Eng. Ind.* **115** (1993) 432–437.
18. Standard method for evaluating machining performance of ferrous metals using and automatic screw/bar machine, ASTM Standard E618-07, 2013.
19. M. Kikuhi, The use of cutting temperature to evaluate the machinability of titanium alloys, *Acta Biomater.* **5** (2009) 770–775.
20. I. Arriola, E. Whitenton, J. Heigel, and P. J. Arrazola, Relationship between machinability index and in-process parameters during orthogonal cutting of steels, *CIRP Ann.* **60** (2011) 93–96.

21. Tool life testing with single-point turning tools, ISO Standard 3685:1993(E), 1993.
22. The volvo standard machinability test, Standard 1018.712, Volvo Laboratory for Manufacturing Research, Trollhattan, Sweden, 1989.
23. A. J. DeArdo, C. I. Garcia, R. M. Laible, and U. Eriksson, A better way to assess machinability, *Am. Mach.* **137**:5 (May 1993) 33–35.
24. G. Lorenz, Comparative drill performance tests, *Machinability Testing and Utilization of Machining Data*, ASM, Metals Park, OH, 1979, pp. 147–163.
25. J. J. Fulmer and J. M. Blanton, Enhanced machinability of P/M parts through microstructure control, SAE Technical Paper 940357, 1994.
26. Y. Matsushima, M. Nakamura, H. Takeshita, S. Akiba, and M. Katsuta, Improvement of drilling machinability of microalloyed steel, *Kobelco Technol. Rev.* **17** (April 1994) 38–43.
27. J. S. Agapiou, G. W. Halldin, and M. F. DeVries, On the machinability of powder metallurgy austenitic stainless steels, *ASME J. Eng. Ind.* **110** (1988) 339–343.
28. A. Salaka, K. Vasilko, M. Selecka, and H. Danninger, New short time face turning method for testing the machinability of PM steels, *J. Mater. Proc. Tech.* **176** (2006) 62–69.
29. R. Lalbondre, P. Krishna, and G. C. Mohankumar, Machinability studies of low alloy steels by face turning method: An experimental investigation, *Procedia Eng.* **64** (2013) 632–641.
30. V. C. Venkatesh and V. Narayanan, Machinability correlation among turning milling and drilling processes, *CIRP Ann.* **35** (1986) 59–62.
31. Bar Products Group, American Iron and Steel Institute, *Steel Bar Product Guidelines*, Iron and Steel Society, Warrendale, PA, 1994, pp. 164–166.
32. A. Henkin and J. Datsko, Influence of physical properties on machinability, *ASME J. Eng. Ind.* **85** (1963) 321.
33. N. Boubekri, J. Rodriguez, and S. Asfour, Development of an aggregate indicator to assess the machinability of steels, *J. Mater. Proc. Tech.* **134** (2003) 159–165.
34. K. Nakayama, Chip control in metal cutting, *Bull. Jpn. Soc. Precision Eng.* **18**:2 (June 1984) 97–103.
35. C. Spaans, A systematic approach to three-dimensional chip curl, chip breaking and chip control, SME Technical Paper MR70-241, 1970.
36. W. Kluft, W. Konig, C. A. Van Luttervelt, K. Nakayama, and A. J. Pekelharing, Present knowledge of chip control, *CIRP Ann.* **28**:2 (1979) 441–455.
37. I. S. Jawahir, A survey and future predictions for the use of chip breaking in unmanned systems, *Int. J. Adv. Manuf. Tech.* **3**:4 (1988) 87–104.
38. I. S. Jawahir and C. A. van Luttervelt, Recent developments in chip control research and applications, *CIRP Ann.* **42** (1993) 659–693.
39. X. D. Fang, J. Fei, and I. S. Jawahir, A hybrid algorithm for predicting chip form/chip breakability in machining, *Int. J. Mach. Tools Manuf.* **36** (1996) 1093–1107.
40. A. K. Balaji, R. Ghosh, X. D. Fang, R. Stevenson, and I. S. Jawahir, Performance-based predictive models and optimization methods for turning operations and applications: Part 2—Assessment of chip forms/chip breakability, *J. Manuf. Proc.* **8** (2006) 144–158.
41. P. D. Liu, R. S. Hu, H. T. Zhang, and X. S. Wu, A study on chip curling and breaking, *Proceedings of 29th Matador Conference*, Manchester, England, 1992, pp. 507–512.
42. G. Stabler, The fundamental geometry of cutting tools, *Proc. Inst. Mech. Eng.* **165** (1951) 14–21.
43. L. V. Colwell, Predicting the angle of chip flow for single-point cutting tools, *ASME Trans.* **76** (1954) 199–204.
44. K. Okushima and K. Minato, On the behavior of chip in steel cutting, *Bull. Jpn. Soc. Precision Eng.* **2** (1959) 58–64.
45. R. H. Brown and E. J. A. Armarego, Oblique machining with a single cutting edge, *Int. J. Mach. Tool Des. Res.* **4** (1964) 9–25.
46. H. T. Young, P. Mathew, and P. L. B. Oxley, Allowing for nose radius effects in predicting the chip flow direction and cutting forces in bar turning, *Proc. Inst. Mech. Eng.* **201C** (1987) 213–226.
47. H. J. Fu, R. E. DeVor, and S. G. Kapoor, A mechanistic model for the prediction of the force system in face milling operations, *ASME J. Eng. Ind.* **106** (1988) 81–88.
48. K. Nakayama, A study on chip breaker, *Bull. JSME* **5** (1962) 142–150.
49. T. L. Subramanian and A. Bhattacharyya, Mechanics of chip breakers, *Int. J. Prod. Res.* **4** (1965) 37–49.
50. A. R. Trim and G. Boothroyd, Action of obstruction type chip former, *Int. J. Prod. Res.* **6** (1968) 227.
51. C. Spaans and P. F. H. J. Van Geel, Break mechanisms in cutting with a chip breaker, *CIRP Ann.* **19** (1970) 87–92.
52. B. Worthington, The operation and performance of a groove type chip breaker, *Int. J. Prod. Res.* **14** (1976) 529–558.

53. S. Kaldor, A. Ber, and E. Lenz, On the mechanism of chip breaking, *ASME J. Eng. Ind.* **101** (1979) 241–249.

54. P. K. Venuvinod and A. Djordjevich, Towards active chip control, *CIRP Ann.* **45** (1996) 83–86.

55. M. Ogawa and K. Nakayama, Effects of chip splitting nicks in drilling, *CIRP Ann.* **34** (1985) 101.

56. I. I. Shilin and K. G. Sadolevskaya, Chip breaking by interrupted feed, *Mach. Tooling* **36**:3 (1965) 30–32.

57. H. Takayama, H. Sekiguchi, and K. Takada, One solution for chip hazard in turning—Study on automatic programming for numerically controlled machines (1st Report), *J. Jpn. Soc. Precision Eng.* **36** (1970) 150–156.

58. P. F. Ostwald, Dynamic chip breaking: Can it overcome the surface finish problem? ASTME Paper MR67-228, 1967.

59. F. Rasch, Hydraulic chip breaking, *CIRP Ann.* **30** (1981) 333–335.

60. H. Takayama and S. Kato, Burrless drilling by means of ultrasonic vibration, *CIRP Ann.* **40** (1991) 83–86.

61. M. Mazurkiewicz, Z. Kubala, and J. Chow, Metal machining with high-pressure water-jet cooling assistance—A new possibility, *ASME J. Eng. Ind.* **111** (1989) 7–12.

62. Increasing machine tool productivity with high pressure cryogenic coolant flow, Manufacturing Technology Directorate, Wright Laboratory, Air Force Systems Command Report WL-TR-92-8014, May 1992.

63. A. G. Ringler, High velocity coolant distribution system for improved chip control and disposal, *Strategies for Automation of Machining: Materials and Processes*, ASM, Metals Park, OH, 1987, pp. 147–155.

64. J. D. Christopher, The influence of high pressure cryogenic coolant on tool life and productivity in turning, SME Technical Paper MR90-249, 1990.

65. R. R. Lindeke, F. C. Schoenig, A. K. Khan, and J. Haddad, Machining of $\tilde{\alpha}$ β titanium with ultra-high pressure through the insert lubrication/cooling, *Trans. NAMRI/SME* **19** (1991) 154–161.

66. S. Y. Honga, Y. Ding, and R. G. Ekkens, Improving low carbon steel chip breakability by cryogenic chip cooling, *Int. J. Mach. Tools Manuf.* **39** (1999) 1065–1085.

67. E. Shamoto, K. Yasuda, T. Aoki, N. Suzuki, and T. Koide, Studies on continuous chip disposal and chip-pulling cutting—Challenge to chip control by changing shape of rake face, *J. Jpn. Soc. Precision Eng.* **77** (2011) 520–524.

68. M. Arai and K. Nakayama, Boundary notch on cutting tool caused by burr and its suppression, *J. Jpn. Soc. Precision Eng.* **52** (1986) 864–866.

69. L. K. Gillespie and P. T. Blotter, The formation and properties of machining burrs, *ASME J. Eng. Ind.* **98** (1976) 66–74.

70. L. K. Gillespie, *Deburring Capabilities and Limitations,* SME, Dearborn, MI, 1976, Chapters 3–5.

71. L. K. Gillespie and R. E. King, Eds., *Robotic Deburring Handbook*, SME, Dearborn, MI, 1987.

72. K. Nakayama and M. Arai, Burr formation in metal cutting, *CIRP Ann.* **36** (1987) 33–36.

73. M. Hashimura, Y. P. Chang, and D. A. Dornfeld, Analysis of burr formation mechanism in orthogonal cutting, *ASME J. Manuf. Sci. Eng.* **121** (1999) 1–7.

74. G.-L. Chern, Study on mechanisms of burr formation and edge breakout near the exit of orthogonal cutting, *J. Mater. Proc. Tech.* **176** (2006) 152–157.

75. D. A. Dornfeld, J. S. Kim, H. Dechow, J. Hewson, and L. J. Chen, Drilling burr formation in titanium alloy, li-6AI-4V, *CIRP Ann.* **48** (1999) 73–76.

76. C.-H. Chu and D. A. Dornfeld, Geometric approaches for reducing burr formation in planar milling by avoiding tool exits, *J. Manuf. Proc.* **7** (2005) 182–195.

77. J. Kim, S. Min, and D. A. Dornfeld, Optimization and control of drilling burr formation of ANSI 304L and AISI 4118 based on drilling burr control charts, *Int. J. Mach. Tools Manuf.* **41** (2001) 923–936.

78. D. A. Dornfeld, Ed., *Publications of CODEF*, 2nd edn., Abrasives Mall, 2001.

79. D. A. Dornfeld and M. C. Avila, Eds., *Research Reports 2001–2002*, Laboratory for Manufacturing Automation, University of California, Berkeley, CA, 2002.

80. J. C. Aurich, D. Dornfeld, P. J. Arrazola, V. Franke, L. Leitz, and S. Min, Burrs—Analysis, control and removal, *CIRP Ann.* **58** (2009) 519–542.

81. P. J. Arrazola, T. Ozel, D. Umbrello, M. Davies, and I. S. Jawahir, Recent advances in modelling of metal machining processes, *CIRP Ann.* **62** (2013) 695–718.

82. A. J. Pekelharing, Exit failure in interrupted cutting, *CIRP Ann.* **27** (1978) 5–10.

83. S. Nakamura and A. Yamamoto, Influence of disengage angle upon initial fracture of tool edge, *Bull. Jpn. Soc. Precision Eng.* **19** (1985) 169–174.

84. T. C. Ramaraj, S. Santhanam, and M. C. Shaw, Tool fracture at the end of a cut I. Foot formation, *ASME J. Eng. Ind.* **110** (1989) 333–338.

85. S. L. Ko and D. A. Dornfeld, A study on burr formation mechanism, *Symposium on Robotics*, ASME DSC Vol. 11, ASME, New York, 1988, pp. 271–282.

86. S. L. Ko and D. A. Dornfeld, Analysis and modelling of burr formation and breakout in metal, *Mechanics of Deburring and Surface Finishing Processes*, ASME PED Vol. 38, ASME, New York, 1989, pp. 79–91.

87. O. Olvera and G. Barrow, Influence of exit angle and tool nose geometry on burr formation in face milling operations, *Proc. Inst. Mech. Eng., J. Eng. Manuf.* **212B** (1998) 59–72.

88. M. Hashimura, J. Hassamontr, and D. Dornfeld, Effect of in-plane exit angle and rake angle on burr, *Trans. ASME, J. Manuf. Sci. Eng.* **121** (1999) 13–19.

89. G.-L. Chern, Experimental observation and analysis of burr formation mechanisms in face milling of aluminum alloys, *Int. J. Mach. Tools Manuf.* **46** (2006) 1517–1525.

90. H. Ernst and W. A. Haggarty, The spiral point drill—A new concept in drill point geometry, *ASME Trans.* **80** (1958) 1059–1072.

91. S. Zaima, A. Yuki, and S. Kamo, Drilling of aluminum plates with special type point drill, *J. Jpn. Inst. Light Metals* **18** (1986) 269–276, 307–313.

92. S.-L. Ko and J.-K. Lee, Analysis of burr formation in drilling with a new-concept drill, *J. Mater. Proc. Tech.* **113** (2001) 392–398.

93. Dow Chemical, Machining magnesium, Dow Chemical Company, Midland, MI, 1982, Form No. 141-480-82.

94. D. Benjamin, Ed., *Metals Handbook*, Vol. 2: *Properties and Selection: Nonferrous Alloys and Pure Metals*, 9th edn., ASM, Metals Park, OH, 1979, pp. 549–551.

95. H. J. Morales, Magnesium, machinability and safety, SAE Technical Paper 800418, 1980.

96. M. Videm, R. S. Hansen, N. Tomac, and K. Tonnesen, Metallurgical considerations for machining magnesium alloys, SAE Technical Paper 940409, 1994.

97. K. T. Kurihara, H. Kato, and T. Tozava, Cutting temperature of magnesium alloys, *ASME J. Eng. Ind.* **103** (1981) 254–260.

98. O. Hoehne and D. Korff, The design and production of light metal castings at volkswagen, SAE Technical Paper 800B, 1964.

99. N. Tomac, K. Tonnessen, and F. O. Rasch, PCD tools in machining magnesium alloys, *Eur. Mach.*, May/June 1991, 12–16.

100. A. Spicer, J. Kasi, C. Billups, and J. Pajec, Machining magnesium with water base coolants, SAE Technical Paper 910415, 1991.

101. K. Tonnessen, N. Tomac, and F. O. Rasch, Machining magnesium alloys with use of oil-water emulsions, *Eighth International Colloquium, Tribology 2000*, Esslingen, Germany, January 1992, pp. 18.7-1–18.7-9

102. N. Tomac and K. Tonnessen, Formation of flank build-up in cutting magnesium alloys, *CIRP Ann.* **41** (1991) 79–82.

103. *Machining Magnesium*, Hydro Magnesium (now Norsk Hydro), Oslo, Norway, 1988, Booklet No. BA01186.

104. B. Chamberlain, Machinability of aluminum alloys, *Metals Handbook*, Vol. 2: *Properties and Selection: Nonferrous Alloys and Pure Metals*, 9th edn. ASM, Metals Park, OH, 1979, pp. 187–190.

105. R. E. DeVor and J. C. Miller, Machinability of high silicon cast aluminum alloy with carbide and diamond cutting tools, *Proceedings of NAMRC* **9**, SME, Dearborn, MI, 1981, pp. 296–304.

106. J. Alverio, J. S. Agapiou, and C.-H. Shen, High speed drilling of 390 aluminum, *Trans. NAMRI/SME* **18** (1990) 209–215.

107. W. Koenig and D. Erinski, Machining and machinability of aluminum cast alloys, *CIRP Ann.* **32** (1983).

108. J. L. Jorstad, Influence of aluminum casting alloy metallurgical factors on machinability, *Modern Casting* (December 1980) 47–51.

109. S. D. Jones and R. J. Furness, An experimental study of burr formation for face milling 356 aluminum, *Trans. NAMRI/SME* **25** (1997) 183–188.

110. H. E. Chandler, Machining of metal-matrix composites and honeycomb structures, *Metals Handbook*, Vol. 16: *Machining*, 9th edn., ASM, Materials Park, OH, 1989, pp. 893–901.

111. N. P. Hung, F. Y. C. Boey, K. A. Khor, C. A. Oh, and H. F. Lee, Machinability of cast and powder-formed aluminum alloys reinforced with SiC particles, *J. Mater. Proc. Tech.* **48** (1995) 291–297.

112. A. R. Chambers, The machinability of light alloy MMCs, *Composites Part A* **27A** (1996) 143–147.

113. N. P. Hung, N. L. Loh, and Z. M. Xu, Cumulative tool wear in machining metal matrix composites Part II: Machinability, *J. Mater. Proc. Tech.* **58** (1996) 114–120.
114. C. Lane, Machining characteristics of particulate-reinforced aluminum, *Fabrication of Particulate-Reinforced Metal Composites*, ASM, Materials Park, OH, 1990, pp. 195–201.
115. G. A. Chadwick and P. J. Heath, Machining metal matrix composites, *Metals Mater.* (February 1990) 73–76.
116. J. T. Lin, D. Bhattacharyya, and C. Lane, Machinability of a silicon carbide reinforced aluminum metal matrix composite, *Wear* **181–183** (1995) 883–888.
117. W. S. Ricci, S. E. Swider, and T. J. Moores, Mechanisms of tool wear when diamond machining composites, *Proceedings of 1987 Eastern Manufacturing Technology Conference*, Springfield, MA, 1987, pp. 3-190–3-214.
118. C. Lane, Drilling and tapping of SiC particle-reinforced aluminum, *A Systems Approach to Machining*, ASM, Materials Park, OH, 1993.
119. J. Bunting, Drilling advanced composites with diamond veined drills, *A Systems Approach to Machining*, ASM, Materials Park, OH, 1993, pp. 99–104.
120. C. Lane and M. Finn, Observations on using CVD diamond in milling MMC's, *Materials Issues in Machining and the Physics of Machining Processes*, TMS, Warrendale, PA, 1992, pp. 39–51.
121. N. P. Hung, S. H. Yeo, and B. E. Oon, Effect of cutting fluid on the machinability of metal matrix composites, *J. Mater. Proc. Tech.* **67** (1997) 157–161.
122. S. Li, K. Kondoh, H. Imai, and H. Atsumi, Fabrication and properties of lead-free machinable brass with Ti additive by powder metallurgy, *Powder Technol.* **205** (2011) 242–249.
123. H. Atsumia, H. Imai, S. Li, K. I. Kondoh, Y. Kousaka, and A. Kojima, High-strength, lead-free machinable α–β duplex phase brass Cu–40Zn–Cr–Fe–Sn–Bi alloys, *Mater. Sci. Eng. A* **529** (2011) 275–281.
124. C. Nobel, F. Klocke, D. Lung, and S. Wolf, Machinability enhancement of lead-free brass alloys, *Procedia CIRP* **14** (2014) 95–100.
125. D. Benjamin, Ed., Copper, *Metals Handbook*, Vol. 2: *Properties and Selection: Nonferrous Alloys and Pure Metals*, 9th edn., ASM, Metals Park, OH, 1979, pp. 388–390.
126. J. R. Davis, Ed., Machining of cast irons, *Metals Handbook*, Vol. 16: *Machining*, 9th edn., ASM, Materials Park, OH, 1989, pp. 648–665.
127. Gray iron, *Metal Handbook*, Vol. 1: *Properties and Selection: Irons and Steels*, 9th edn., ASM, Materials Park, OH, 1978, pp. 23–53.
128. R. T. Wimber, Machinability of pearlitic gray cast iron, *Proc. NAMRC* **9** (1981) 290–295.
129. T. H. Wickenden, Data on machinability and wear of cast iron, *SAE Trans.* **23** (1925) 181–189.
130. J. R. Davis, Ed., *ASM Specialty Handbook: Cast Irons*, ASM, Materials Park, OH, 1996, pp. 244–254.
131. D. Bordui, Third generation silicon nitride, *Ceramic Cutting Tools and Applications*, SME, Dearborn, MI, 1989.
132. C. Wick, Machining with PCBN tools, *Manuf. Eng.* **101**:1 (July 1988) 73–78.
133. M. Goto, T. Nakai, and S. Nakatani, Cutting performance of PCBN for cast iron, SAE Technical Paper MR91-176, 1991.
134. J. S. Agapiou, Design characteristics of new types of drill and evaluation of their performance drilling cast iron—I. Drills with four major cutting edges, *Int. J. Mach. Tools Manuf.* **33** (1993) 321–341.
135. J. S. Agapiou, Design characteristics of new types of drill and evaluation of their performance drilling cast iron—II. Drills with three major cutting edges, *Int. J. Mach. Tools Manuf.* **33** (1993) 343–365.
136. J. S. Agapiou, An evaluation of advanced drill body and point geometries in drilling cast iron, *Trans. NAMRI/SME* **19** (1991) 79–89.
137. Cerasiv GmbH, Anwendungstechnik: SPK-Schneidkeramik, SPK-Cermet, WURBON, SPK Tools, Florence, KY, 1995.
138. F. Klocke, C. Klöpper, D. Lung, and C. Essig, Fundamental wear mechanisms when machining austempered ductile iron (ADI), *CIRP Ann.* **56** (2007) 73–76.
139. M. V. de Carvalho, D. M. Montenegro, and J. de Oliveira Gomes, An analysis of the machinability of ASTM grades 2 and 3 austempered ductile iron, *J. Mater. Proc. Tech.* **213** (2013) 560–573.
140. DOE/ID/13319-T1, Clean iron production and machining technology project, report, University of Alabama, Birmingham, AL, July 1996.
141. W. Grzesik, P. Kiszka, D. Kowalczyk, J. Rech, and C. Claudin, Machining of nodular cast iron (PF-NCI) using CBN tools, *Procedia CIRP* **1** (2012) 483–487.
142. K. J. Trigger, L. B. Zylstra, and B. T. Chao, Tool forces and tool-chip adhesion in the machining of nodular cast iron, *ASME Trans.* **74** (1952) 1017–1027.

143. I. Ham, K. Hitomi, and G. L. Thuering, Machinability of nodular cast irons part I: Tool forces and flank adhesion, *ASME J. Eng. Ind.* **83** (1961) 142–154.
144. K. Hitomi and G. L. Theuring, Machinability of nodular cast irons part II: Effect of cutting conditions on flank adhesion, *ASME J. Eng. Ind.* **84** (1962) 282–288.
145. J. S. Agapiou, Development of gun-drilling MQL process and tooling for machining of compacted graphite iron (CGI), *Trans. NAMRI/SME* **38** (2010) 73–80.
146. M. Heck, H. M. Ortner, S. Flege, U. Reuter, and W. Ensinger, Analytical investigations concerning the wear behaviour of cutting tools used for the machining of compacted graphite iron and grey cast iron, *Int. J. Refractory Metals Hard Mater.* **26** (2008) 197–206.
147. V. Nayyar, J. Kaminski, A. Kinnander, and L. Nyborg, An experimental investigation of machinability of graphitic cast iron grades; flake, compacted and spheroidal graphite iron in continuous machining operations, *Procedia CIRP* **1** (2012) 488–493.
148. M. B. Da Silva, V. T. G. Naves, J. D. B. De Melo, C. L. F. De Andrade, and W. L. Guesser, Analysis of wear of cemented carbide cutting tools during milling operation of gray iron and compacted graphite iron, *Wear* **271** (2011) 2426–2432.
149. E. Abele, A. Sahm, and H. Schulz, Wear mechanism when machining compacted graphite iron, *CIRP Ann.* **51** (2002) 53–56.
150. S. Dawson, I. Hollinger, M. Robbins, J. Daeth, U. Reuter, and H. Schulz, The effect of metallurgical variables on the machinability of compacted graphite iron, SAE Technical Paper 2001-01-0409, 2001.
151. V. V. de Oliveira, P. A. de C. Beltrão, and G. Pintaude, Effect of tool geometry on the wear of cemented carbide coated with TiAlN during drilling of compacted graphite iron, *Wear* **271** (2011) 2561–2569.
152. M. Gastel, C. Konetschny, U. Reuter, C. Fasel, H. Schulz, R. Riedel, and H. M. Ortner, Investigation of the wear mechanism of cubic boron nitride tools used for the machining of compacted graphite iron and grey cast iron, *Int. J. Refractory Metals Hard Mater.* **18** (2000) 287–269.
153. M. E. Finn, Machining of carbon and alloy steels, *Metals Handbook*, Vol. 16: *Machining*, 9th edn., ASM, Materials Park, OH, 1989, pp. 666–680.
154. A. M. Abrao, D. K. Aspinwal, and M. L. H. Wise, Tool life and workpiece surface integrity evaluations when machining hardened AISI H13 and AISI E52100 steels with conventional ceramics and PCBN tool materials, SME Technical Paper MR950159, 1995.
155. A. Medvedeva, J. Bergström, S. Gunnarsson, P. Krakhmalev, and L. G. Nordh, Influence of nickel content on machinability of a hot-work tool steel in prehardened condition, *Mater. Des.* **32** (2011) 706–715.
156. S. Chinchanikar and S. K. Choudhury, Investigations on machinability aspects of hardened AISI 4340 steel at different levels of hardness using coated carbide tools, *Int. J. Refractory Metals Hard Mater.* **38** (2013) 124–133.
157. A. S. Kumar, A. R. Durai, and T. Sornakumar, Machinability of hardened steel using alumina based ceramic cutting tools, *Int. J. Refractory Metals Hard Mater.* **21** (2003) 109–117.
158. J. R. Davis, Ed., Machining of Tool Steels, *Metal Handbook*, Vol. 16: *Machining*, 9th edn., ASM, Materials Park, OH, 1989, pp. 708–732.
159. Y. Ozcatalbas and F. Ercan, The effects of heat treatment on the machinability of mild steels, *J Mater. Proc. Tech.* **136**:(1–3) (2003) 227–238.
160. S. Katayama and M. Toda, Machinability of medium carbon graphitic steel, *J. Mater. Proc. Tech.* **62** (1996) 358–362.
161. S. V. Subramanian, H. O. Gekonde, and J. Gao, Microstructural engineering for hard turning, *40th MWSP Conference*, ISS, Pittsburgh, PA, 1998.
162. S. Akcan et al., Characteristics of white layers formed in steel by machining, ASME MED Vol. 10, 1999, pp. 789–795.
163. Y. Chou and C. J. Evans, Process effects on white layer formation in hard turning, *Trans. NAMRI/SME* **26** (1998) 117–122.
164. Y. B. Guo and J. Sahni, A comparative study of the white layer by hard turning vs. grinding, *Int. J. Mach. Tools Manuf.*, Tuscaloosa, Alabama, **44**:2–3 (2004) 135–145.
165. D. W. Murray, Machining performance of steels, *Iron Steel Eng.* (April 1967) 123–128.
166. M. E. Finn, G. A Beaudoin, R. M. Nishizaki, S. V. Subramanian, and D. A. R. Kay, Influence of steel matrix on machinability of high carbon steels in automatic machining, *Strategies for Automation of Machining: Materials and Processes*, ASM, Materials Park, OH, 1987, p. 43.
167. S. V. Subramanian et al., Inclusion engineering of steels for high speed machining, *CIM Bull.* **91** (1998) 107–115.

168. S. V. Subramanian, D. A. R. Kay, and J. Junpu, Inclusion engineering for the improved machinability of medium carbon steels, *Proceedings of Symposium on Inclusions and Their Influence on Material Behavior*, Chicago, IL, 1988 (ASM Paper No. 8821-003).

169. J. H. Hoffman and R. J. Turonek, High performance forged steel crankshafts—Cost reduction opportunities, SAE Technical Paper 910139, 1991.

170. A. R. Chambers and D. Whittaker, Machining characteristics of mircoalloyed forging steels, *Metals Technol.* **11** (1984) 323–333.

171. V. Ollilainen, T. Lahti, H. Pontinen, and E. Heiskala, Machinability comparison with substituting microalloyed forging steel for quenched and tempered steels, *Fundamentals of Microalloying Forging Steels*, TMS, Warrendale, PA, 1987, pp. 461–474.

172. V. Sivaraman, S. Sankaran, and L. Vijayaraghavan, Machinability of multiphase microalloyed steel, *Procedia CIRP* **2** (2012) 52–59.

173. T. Kosa and R. P. Ney, Sr., Machining of stainless steels, *Metals Handbook*, Vol. 16: *Machining*, 9th edn., ASM, Materials Park, OH, 1989, pp. 681–707.

174. D. M. Blott, Machining wrought and cast stainless steels, *Handbook of Stainless Steels*, McGraw-Hill, New York, 1977, Section 24, pp. 2–30.

175. D. O'Sullivan and M. Cotterell, Machinability of austenitic stainless steel SS303, *J. Mater. Proc. Tech.* **124** (2002) 153–159.

176. H. Shao, L. Liu, and H. L. Qu, Machinability study on 3%Co–12%Cr stainless steel in milling, *Wear* **263** (2007) 736–744.

177. J. Paro, H. Hänninen, and V. Kauppinen, Tool wear and machinability of X5 CrMnN 18 18 stainless steels, *J. Mater. Proc. Tech.* **119** (2001) 14–20.

178. J. Nomani, A. Pramanik, T. Hilditch, and G. Littlefair, Machinability study of first generation duplex (2205), second generation duplex (2507) and austenite stainless steel during drilling process, *Wear* **304** (2013) 20–28.

179. J. Paro, H. Hänninen, and V. Kauppinen, Tool wear and machinability of HIPed P/M and conventional cast duplex stainless steels, *Wear* **249** (2001) 279–284.

180. D. Wu and Z. Li, A new ph-free machinable austenitic stainless steel, *J. Iron. Steel Res. Int.* **17** (2010) 59–63.

181. Z. Li and D. Wu, Effect of free-cutting additives on machining characteristics of austenitic stainless steels, *J. Mater. Sci. Tech.* **26** (2010) 839–844.

182. T. Akasawa, H. Sakurai, M. Nakamura, T. Tanaka, and K. Takano, Effects of free-cutting additives on the machinability of austenitic stainless steels, *J. Mater. Proc. Tech.* **143–144** (2003) 66–71.

183. R. M'Saoubi, T. Czotscher, O.Andersson, and D. Meyer, Machinability of powder metallurgy steels using PCBN inserts, *Procedia CIRP* **14** (2014) 83–88.

184. H. E. Chandler, Machining of powder metallurgy materials, *Metals Handbook*, Vol. 16: *Machining*, 9th edn., ASM, Materials Park, OH, 1989, pp. 879–892.

185. J. S. Agapiou, G. W. Halldin, and M. F. DeVries, Effect of porosity on the machinability of P/M 304L stainless steel, *Int. J. Powder Metall.* **25** (1989) 127–139.

186. J. S. Agapiou and M. F. DeVries, Machinability of powder metallurgy materials, *Int. J. Powder Metall.* **24** (1988) 47–57.

187. M. Hamiuddin and Q. Murtaza, Machinability of phosphorous containing sintered steels, *Mater. Chem. Phys.* **67** (2001) 78–84.

188. E. O. Ezugwu, J. Bonney, and Y. Yamane, An overview of the machinability of aeroengine alloys, *J. Mater. Proc. Tech.* **134** (2003) 233–253.

189. M. Rahman, Y. S. Wong, and A. R. Zareena, Machinability of titanium alloys, *JSME Int. J. Ser. C* **46** (2003) 107–114.

190. P.-J. Arrazola, A. Garay, L.-M. Iriarte, M. Armendia, S. Marya, and F. Le Maître, Machinability of titanium alloys (Ti6Al4V and Ti555.3), *J. Mater. Proc. Tech.* **209** (2009) 2223–2230.

191. C. Ohkubo, I. Watanabe, J. P. Ford, H. Nakajima, T. Hosoi, and T. Okabe, The machinability of cast titanium and Ti-6A-4V, *Biomaterials* **21** (2000) 421–428.

192. E. O. Ezugwu and Z. M. Wang, Titanium alloys and their machinability—A review, *J. Mater. Proc. Tech.* **68** (1997) 262–274.

193. A. R. Machado and J. Wallbank, Machining of titanium and its alloys—A review, *Proc. Inst. Mech. Eng.* **B204** (1990) 53–59.

194. N. Khanna, A. Garay, L. M. Iriarte, D. Soler, K. S. Sangwan, and P. J. Arrazola, Effect of heat treatment conditions on the machinability of Ti64 and Ti54M alloys, *Procedia CIRP* **1** (2012) 477–482.

195. H. E. Chandler, Machining of reactive metals, *Metals Handbook*, Vol. 16: *Machining*, 9th edn., ASM, Materials Park, OH, 1989, pp. 844–857.

196. P. A. Dearnley and A. N. Grearson, Evaluation of principle wear mechanisms of cemented carbides and ceramics used for machining titanium alloy IMI 318, *Mater. Sci. Tech.* **2** (1986) 47–58.

197. Basic Design Guide–RMI Titanium, RMI Co., Niles, OH.

198. H. Barish, Quality drills contribute to successful titanium tooling, *Cutting Tool Eng.* **40** (February 1988) 38–41.

199. R. W. Breitzig, Machining nickel alloys, *Metals Handbook*, Vol. 16: *Machining*, 9th edn., ASM, Materials Park, OH, 1989, pp. 835–843.

200. K. Uehare, High-speed machining Inconel 718 with ceramic tools, *CIRP Ann.* **42** (1993) 103–106.

201. I. A. Choudhury and M. A. El-Baradie, Machinability of nickel-base super alloys: A general review, *J. Mater. Proc. Tech.* **77** (1998) 278–284.

202. E. O. Ezugwu, Z. M. Wang, and A. R. Machado, The machinability of nickel-based alloys: A review, *J. Mater. Proc. Tech.* **86** (1999) 1–16.

203. E. O. Ezugwu and C. J. Lai, Failure modes and wear mechanisms of M35 high-speed steel drills when machining inconel 901, *J. Mater. Process. Tech.* **49** (1995) 295–312.

204. B. M. Kramer and P. D. Hartung, Theoretical considerations in the machining of nickel-based alloys, *Cutting Tool Materials*, ASM, Metals Park, OH, 1981, pp. 57–74.

205. V. Bushlyaa, J. Zhou, and J. E. Stahl, Effect of cutting conditions on machinability of superalloy Inconel 718 during high speed turning with coated and uncoated PCBN tools, *Procedia CIRP* **3** (2012) 370–375.

206. J. P. Costes, Y. Guillet, G. Poulachon, and M. Dessoly, Tool-life and wear mechanisms of CBN tools in machining of Inconel 718, *Int. J. Mach. Tools Manuf.* **47** (2007) 1081–1087.

207. A. D. Johnson, A. R. Thangaraj, and K. J. Weinmann, The influence of mechanical properties on the performance of Al_2O_3-SiCw ceramics in the machining of Inconel 718, *Materials Issues in Machining and the Physics of Machining Processes*, TMS, Warrendale, PA, 1992, pp. 187–202.

208. P. K. Wright and J. G. Chow, Deformation characteristics of nickel alloys during machining, *ASME J. Eng. Mater. Tech.* **108** (1986) 85–93.

209. R. Komanduri and T. A. Schroeder, On shear instability in machining nickel-based superalloy, *ASME J. Eng. Ind.* **108** (1986) 93–100.

210. I. A. Choudhury and M. A. EI-Baradie, Machinability assessment of Inconel 718 by factorial design of experiment coupled with response surface methodology, *J. Mater. Process. Tech.* **55** (1999) 30–39.

211. M. Alauddin, M. A. EI-Baradie, and M. S. Hashmi, Optimization of surface finish in end milling Inconel 718, *J. Mater. Process. Tech.* **56** (1996) 54–65.

212. D. K. Aspinwall, R. C. Dewes, E.-G. Ng, C. Sage, and S. L. Soo, The influence of cutter orientation and workpiece angle on machinability when high-speed milling Inconel 718 under finishing conditions, *Int. J. Mach. Tools Manuf.* **47** (2007) 1839–1846.

213. S. L. Soo, R. Hood, D. K. Aspinwall, W. E. Voice, and C. Sage, Machinability and surface integrity of RR1000 nickel based superalloy, *CIRP Ann.* **60** (2011) 89–92.

214. J. A. Aris, Machining of uranium and uranium alloys, *Metals Handbook*, Vol. 16: *Machining*, 9th edn., ASM, Materials Park, OH, 1989, pp. 874–878.

215. Users guide to MSC materials: Depleted uranium, Manufacturing Sciences Corporation, Oak Ridge, TN, nd.

12 Machining Dynamics

12.1 INTRODUCTION

As discussed in previous chapters, the machine tool, cutting tool, part, and fixture form a complex system consisting of several coupled structural elements. During cutting, high forces are generated and the energy is dissipated through plastic deformation and friction, leading to deflections of the structural components of the system and to vibrations (relative motion between the tool and workpiece). These vibrations should be minimized because they degrade machining accuracy and the machined surface texture; moreover, under unfavorable conditions they may become unstable, leading to chatter, which can cause accelerated tool wear and breakage, accelerated machine tool wear, and damage to the machine tool and part. Vibration is particularly a serious problem in fine-finishing operations such as grinding, and in processes such as boring and end milling that employ compliant tooling [1]. Unstable vibrations are a major factor limiting production rates in many operations, especially at high cutting speeds.

A number of analytical and theoretical methods have been developed to study the dynamic behavior of the machining system. The study of machining dynamics has two basic objectives: (1) to identify rules for designing stable machine tools and (2) to identify rules for choosing dynamically stable cutting conditions. The dynamic analysis of the machining system is complicated by the physical complexity of machine tool systems and the cutting process, the difficulty of estimating dynamic properties of joints between structural components, and the fact that the system is time varying since components move relative to each other during the process.

The design of machine tools, cutting tools, and tool holders was discussed in Chapters 3 through 5. Engineers have traditionally relied on the inherent rigidity of the machining system to control deflections, which cause dimensional errors in the machined part [2–4]. This rigidity and the internal damping of the system are also the primary design features, which control vibration amplitudes and stability. Although machine structural elements are designed for rigidity, they all exhibit structural vibration modes, which may be excited during cutting operations [5–7]. Moreover, chatter may also result from the vibration of the part or fixture, which often are more compliant than the machine structure.

This chapter describes analytical and experimental methods of investigating machining dynamics to address the two objectives stated earlier. Topics covered include analysis methods, dynamics of lumped parameter systems, types of machine tool vibrations, forced vibrations, self-excited vibrations, experimental methods for stability analysis, chatter detection, and passive and active methods for chatter suppression and control.

12.2 VIBRATION ANALYSIS METHODS

Several methods can be used to analyze the structural dynamics of machine tool systems [8–16]. These include kinematic simulations, lumped mass and massless beam methods, the receptance method with beams having distributed mass, and finite element analysis (FEA).

There are several simulation programs for the large-scale transient analysis of controlled mechanical systems such as machine tool systems. These programs can be used to perform kinematic and dynamic analyses of a machine structure. The kinematic simulation tools not only provide position, velocity, and acceleration information at each time step for massless mechanisms, but also all internal reaction forces at joints and constraints. They can also be used to perform dynamic analyses involving the mass properties of the moving bodies and forces acting on the bodies.

They also handle flexible bodies with nonlinear components, and they can efficiently perform static structural optimization based on the modal potential energy. They can be very useful for the static analysis of machine tools during design optimization because they can simulate the static structural deflections of several machine configurations and different load cases in a short time frame.

The structural elements of a machine tool system are continuous with distributed mass. For many analysis purposes, however, the various components can be treated as a collection of discrete or lumped masses connected by springs and dampers. This idealization is the basis for many experimental methods for analyzing the dynamics and stability of machine tools. The dynamics of lumped mass systems is described in detail in Section 12.3.

Among continuous or distributed mass methods, finite element methods are the most practically useful for analyzing machining systems because they permit use of the most realistic assumptions and because they can be used not only for dynamic analysis, but also for static and thermal analysis using the same model.

Two approaches can be used to apply the finite element method to predict the dynamic behavior of the machining system. In the first approach, a single large model of the whole system is employed. In the second approach, the overall system is subdivided into smaller subsystems (machine components), which are analyzed separately; the results of all the subsystems are linked together in an overall system structural response program using a generalized building block approach to predict the structural characteristics of the whole system. The latter method provides more flexibility because individual components of the machining system can be changed and analyzed individually, but requires more model building time.

Several element types have been used to model machine tools in the past, including plate, shell, beam, and solid elements. In more recent practice, automeshers using either tetragonal or cubic (hex) elements have been increasingly used. These greatly reduce model building time, although some manual refinement of the mesh is often still necessary. First-order elements are most commonly used, although in some cases (e.g., bending loads on cantilever-like structures) coarser meshes of second-order elements yield better results with less computing time.

Ideally, the machine tool structure's static and dynamic behavior should be evaluated through FEA before hardware is built. At this stage, several design variations can be compared and structural parameters can be optimized. Unfortunately, machine tool stiffness and damping ratios generally cannot be predicted, but must be measured, and a number of other nonlinear structural effects are difficult to account for in the design stage. In practice, therefore, FEA provides reasonable static stiffness assessments for major components, but the dynamic behavior of the entire system can only be modeled after hardware is built and system properties such as the dynamic stiffness, damping percentages, natural frequencies, and mode shapes are determined experimentally.

12.3 VIBRATION OF DISCRETE (LUMPED MASS) SYSTEMS

Machine tools are composed of several components and therefore can be considered multi-mass vibrators. Although the structural elements of a machine tool are geometrically complex and have continuously distributed masses, it is possible in some instances to simulate the behavior of a machine tool structure under the influence of dynamic loading by considering several discrete masses connected together. Although practical systems such as machine tools are multiple degree of freedom (MDOF) and have some degree of nonlinearity, they can be represented as superposition of single degree of freedom (SDOF) spring-mass-dashpot vibrators for many purposes. For example, the mechanical model in Figure 12.1a can represent a cutting process in which m is the mass of the tool and the damping coefficient c and spring stiffness k can be determined through modal analysis of the machine tool. x is the displacement of the center of the tool relative to the workpiece. The structure is assumed to be flexible in the x direction only. This reduces the model to a SDOF system. This SDOF model is appropriate, for example, when a thin-walled workpiece is the most flexible part of the structure and is compliant primarily in one direction.

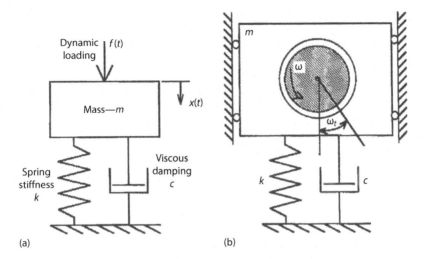

FIGURE 12.1 Free-body diagram for forced vibrations with harmonic excitation of a SDOF mass–spring–viscous damper system. (a) Mass subjected to forced vibration. (b) Unbalanced motor running at constant velocity.

The dynamic performance (or stiffness) of a linear mechanical system of this type can be described by a set of transfer functions (TFs) or by the resonance frequencies and their associated displacements at the different points, which are called associated modes of vibration.

Several methods have been proposed to examine the dynamic behavior and system stability, including: (1) the s-plane approach, (2) the frequency (j) plane approach, and (3) the time-domain approach [17–19].

The s-plane approach describes the response and performance of a system in terms of the complex frequency variable s. It uses the Laplace transformation to transform the differential equations representing the system motions into algebraic equations expressed in terms of the complex variable s. The TF of a linear system, representing the input–output relationship, is then obtained from the algebraic equations. The root locus method can then be used to evaluate the stability of the system by examining the poles and zeros of the TF on the s-plane.

The frequency response of a system is defined as the steady-state response of the system to a sinusoidal input signal. The output of a linear system to such an input is also sinusoidal with the same frequency in the steady state; it differs from the input wave form only in the amplitude and phase angle. This method is easily applied to experimentally determine the frequency response and the TF of a system. The TF in the frequency response method can be also obtained from the TF in the s-plane by replacing s with j, where $j = \sqrt{-1}$. The disadvantage of the frequency response method for analysis and design is the indirect link between the frequency domain and the time domain. The frequency domain approach is also limited in applicability to linear time-invariant systems; it is particularly limited when considering multivariable control systems due to the emphasis on the input–output relationship of TFs. The frequency response method analyzes a system based on the amplitude and phase relationships between sinusoidal curves, which is a natural approach since many machining vibrations result from rotating forces or imbalance.

In the time-domain method, the system response is described by a set of state variables. The state variables are those variables that determine the future behavior of a system when the present state of the system and the excitation signals are known [18–22]. The time-domain approach can be used not only for linear systems but for nonlinear, time-varying, and multivariate systems.

In this section, we review the analysis of lumped mass oscillators based on the frequency domain approach, the approach most commonly applied in experimental modal analysis. The dynamic behavior of SDOF discrete systems will be discussed first. This will provide the basis for a discussion of machine component interactions or MDOF systems.

12.3.1 Single Degree-of-Freedom (SDOF) Systems

The analysis of a SDOF system provides basic physical insight into the dynamic behavior of vibrating systems in general. The commonly analyzed system of this type, shown in Figure 12.1a, consists of a mass connected to ground by a spring and viscous damper in parallel. The system is excited by a periodic force and vibrates at the forcing frequency. The frequency response method is often used to analyze such a system, generally under the assumption that the driving forcing function is sinusoidal. Physically, a sinusoidal forcing function may result from imbalance in a motor running at constant angular velocity as shown in Figure 12.1b; the four rubber isolators on which the motor legs are mounted represent the spring of the system across which the vibration occurs. The equation of motion for the SDOF system shown in Figure 12.1 with mass m, spring stiffness k, and viscous damping is

$$m\ddot{x}(t) + c\dot{x}(t) + kx(t) = f(t) \tag{12.1}$$

where c is the coefficient of viscous damping, or damping constant, which has dimensions of force per unit velocity (N s/m or kg/s). $x(t)$, $\dot{x}(t)$, and $\ddot{x}(t)$ are the displacement, velocity, and acceleration of the externally applied load(s). The natural angular frequency of free undamped oscillations, ω_n (rad/s), and the damping ratio, ζ, the basic dynamic characteristics of an SDOF system, are

$$\omega_n = \sqrt{\frac{k}{m}} \quad \text{and} \quad \zeta = \frac{c}{c_c} = \frac{c}{2\sqrt{km}} \tag{12.2}$$

The damping ratio ζ is the ratio of the actual damping c to the critical damping c_c, which is the smallest value of c for which the free damped motion is non-oscillatory, that is, the level of damping that would just prevent vibration. The damping ratio is very small for machine tools; typically for machine tools $\zeta \leq 0.05$, which means that the actual damping is 5% of the amount that would just prevent vibration. The damped natural angular frequency and the resonance angular frequency of the system, ω_{dn} and ω_r, are related to undamped natural frequency by

$$\omega_{dn} = \omega_n \sqrt{1 - \zeta^2} \tag{12.3}$$

$$\omega_r = \omega_n \sqrt{1 - 2\zeta^2} \tag{12.4}$$

The resonance frequency is the frequency at which resonance occurs and the compliance increases. For machine tool systems, the differences between the frequencies ω_n, ω_{dn}, and ω_r are usually not significant because the damping ratio is small.

All mechanical systems exhibit damping, as evidenced by the fact that the amplitude of free vibrations diminishes with time. The effect of damping on the system response can be ignored when the damping ratio is small and the system is not excited at a frequency near resonance. Accurately modeling damping becomes most critical when the system is excited near resonance. Since the effect of damping is not generally known before the machine tool is analyzed, an attempt should be made to model damping until experiments prove it to be insignificant.

Damping forces arise from several sources, including friction between lubricated sliding surfaces, air or fluid resistance, electric damping, internal friction due to the imperfect elasticity of real materials, internal strain, looseness of joints, and other complex causes [6]. There are several forms of damping including viscous, hysteretic, and dry frictional damping. Each assumed model of damping represents different idealization of the actual damping and results in a different predicted response. For example, hysteretic and viscous damping limit resonant amplitudes, while dry frictional damping does not; viscous damping affects the frequency of the resonant peak, whereas hysteretic and dry frictional damping do not. The differences in the predicted response diminish as the difference between the forcing frequency and the natural frequency increases. Viscous damping is the simplest case to analyze mathematically because it is proportional to velocity. Therefore, the resisting forces of complex systems, such as machine tools, are often replaced by equivalent viscous dampers to simplify mathematical analysis. The equivalent damping coefficient is chosen to dissipate the same amount of energy per cycle as the actual resisting force. It is important to note that a large part of the energy dissipation occurs at the toolholder–spindle interface.

It is often convenient to characterize a linear system by its response to a specific sinusoidal input force given by

$$f(t) = Fe^{j \cdot \omega \cdot t} \tag{12.5}$$

where
  F is the forcing amplitude
  $j = \sqrt{-1}$
  t is the time
  ω is the exciting frequency in rad/s

The circular frequency ω is related to the frequency f in Hz through

$$f = \frac{\omega}{2\pi} \tag{12.6}$$

Modal analysis is performed using the Fourier transform $X(\omega)$ of the displacement $x(t)$:

$$X(\omega) = \int_{-\infty}^{\infty} x(t)e^{-j \cdot \omega \cdot t} dt \tag{12.7}$$

The Fourier transforms of the time derivative of a function can be determined by multiplying the Fourier transform of the function by $j\omega$:

$$\int_{-\infty}^{\infty} \dot{x}(t)e^{-j \cdot \omega \cdot t} dt = j\omega X(\omega)$$

$$\tag{12.8}$$

$$\int_{-\infty}^{\infty} \ddot{x}(t)e^{-j \cdot \omega \cdot t} dt = -\omega^2 X(\omega)$$

Taking the Fourier transform of both sides of Equation 12.1 therefore yields

$$\left(-\omega^2 m + j\omega c + k\right) X(\omega) = F(\omega) \tag{12.9}$$

where the forcing function $f(t)$ is given by Equation 12.5 and $F(\omega)$ is the Fourier transform of $f(t)$.

The steady-state response of this system, which is present as long as the forcing function is active, is given by

$$X(\omega) = G(\omega)F(\omega) \tag{12.10}$$

where

$$G(\omega) = \frac{X(\omega)}{F(\omega)} = \frac{1}{-\omega^2 m + j\omega c + k}$$

$$= \frac{1}{k} \cdot \frac{1}{1 - r^2 + 2j\zeta r} \tag{12.11}$$

$$r = \frac{\omega}{\omega_n} \tag{12.12}$$

$G(\omega)$, the "frequency response function" (FRF) of the system is the ratio of the complex amplitude of the displacement (which is a harmonic motion with frequency ω) to the magnitude F of the forcing function; In other words, it is the amplitude of vibration produced by a unit force at the frequency ω. The factor $1/k$ is the static flexibility or compliance of the system (the deflection for a unit force). The FRF is also called the magnification factor or TF in metal cutting, although in the general literature on vibration the TF is usually defined in terms of Laplace transforms (s-plane method).

The FRF is a complex quantity and contains both real and imaginary parts, given by

$$\mathrm{Re}\left[G(\omega)\right] = \frac{k - m\omega^2}{\left(k - m\omega^2\right)^2 + \left(c\omega\right)^2} = \frac{1}{k} \frac{1 - r^2}{\left(1 - r^2\right)^2 + \left(2\zeta r\right)^2}$$

$$\mathrm{Im}\left[G(\omega)\right] = \frac{-c\omega}{\left(k - m\omega^2\right)^2 + \left(c\omega\right)^2} = \frac{1}{k} \frac{-2\zeta r}{\left(1 - r^2\right)^2 + \left(2\zeta r\right)^2} \tag{12.13}$$

The real part represents the "mobility" of the system, while the imaginary part represents the "inertance." The magnitude of the FRF is given by

$$|G(\omega)| = \sqrt{\frac{|X(\omega)|}{|F(\omega)|}}$$

$$= \frac{1}{k\sqrt{\left(1 - r^2\right)^2 + \left(2\zeta r\right)^2}} \tag{12.14}$$

which represents the "dynamic compliance" of the system.

The fact that the imaginary part is negative indicates that the cosine component of the forced amplitude must lag the sine component. Therefore, the displacement X lags the disturbing force F by the phase angle $\phi(\omega)$ given by

$$\phi(\omega) = \tan^{-1}\left\{\frac{\mathrm{Im}\left[G(\omega)\right]}{\mathrm{Re}\left[G(\omega)\right]}\right\} = \tan^{-1}\left[\frac{2\zeta r}{1 - r^2}\right] \tag{12.15}$$

Plots of the magnitude $|G(\omega)|$ and phase angle $\phi(\omega)$ as functions of ω together define the system frequency response and are known as "Bode plots" (Figure 12.2) presented in polar coordinates. The phase ranges from 0° to 180° and the response lags the input by 90° at resonance.

The compliance and phase of the FRF are plotted as functions of the frequency ratio r in Figure 12.2. Resonance occurs when the response frequency equals the natural frequency of the system ($\omega = \omega_r \approx \omega_n$ for $\zeta \ll 1$). The amplitude of the response at resonance is $1/2k\zeta$ and is limited only by the amount of damping in the system. The static stiffness k can be determined from the compliance vs.

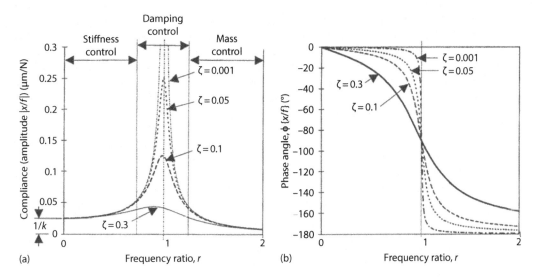

FIGURE 12.2 Response curves (the Bode plot, consisting of two graphs) for the system in Figure 12.1 using several damping ratios. (a) Compliance/amplitude (of FRF) vs. frequency. (b) Phase vs. frequency.

frequency curve and is equal to the reciprocal of the compliance at zero frequency (Figure 12.3). However, the dynamic compliance and stiffness differ from the static values. Low-frequency exciting forces cause displacements determined by the static compliance. The drive point mobility becomes the mobility of the spring at very low frequencies. The amplitude of the displacement (or the dynamic compliance) increases with increasing excitation frequency up to the resonance frequency. Any further increase in the excitation frequency will then result in a decrease in dynamic compliance. The mobility of the mass dominates at frequencies much larger than the resonance frequency. The compliance in the mass-controlled region at high frequencies approaches $1/\omega^2$. The stiffness and mass mobilities balance each other at frequencies near the resonance, so that only the mobility of the damper is left. Therefore, the system is said to be damping controlled at frequencies near the resonance frequency.

There are several different force-response relationships, summarized in Table 12.1, that are often used in vibration analysis [19]. The "impedance" and mobility are often of interest. The impedance of the SDOF system in Figure 12.1 is

$$\frac{F}{v} = c + j\left(m\omega - \frac{k}{\omega}\right) \tag{12.16}$$

where v is the velocity. The real part of the impedance is called the mechanical resistance. The imaginary part is called the mechanical reactance; $m\omega$ is the mechanical inertance term, while ω/k is the mechanical compliance term. The system's response is damping controlled if the mechanical resistance term is dominant (Figure 12.2); the system's response is stiffness controlled if the mechanical compliance term is dominant. If the mechanical inertance term is dominant, the system's response is mass controlled. The FRF can be presented as displacement (shown in Figure 12.2), velocity, or acceleration. Acceleration is currently the accepted method of measuring modal response as will be discussed later. When $\omega \ll \omega_n$ the frequency response is approximately equal to the asymptote shown in Figure 12.2, it is called the static stiffness line with slope of 0, 1, or 2 for displacement, velocity, and acceleration responses, respectively.

The real and imaginary parts of the FRF can be plotted as functions of frequency to identify the system as shown in Figure 12.3c and d. At resonance, the real part of the TF is zero, and the

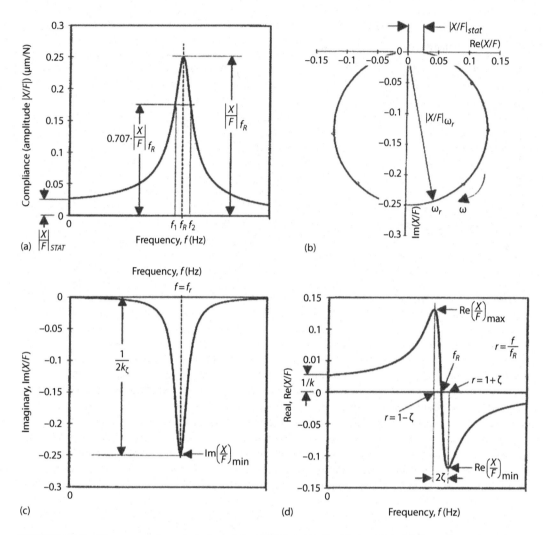

FIGURE 12.3 Frequency characteristics of the SDOF system in Figure 12.1. (a) Compliance/amplitude (of FRF) vs. frequency. (b) Real part vs. imaginary part (Nyquist plot). (c) Imaginary part vs. frequency. (d) Real part (of FRF) vs. frequency.

TABLE 12.1

Different Types of Frequency Response

FRF	Represents
Displacement/force	Receptance
Force/displacement	Dynamic stiffness
Velocity/force	Mobility
Force/velocity	Impedance
Acceleration/force	Inertance
Force/acceleration	Apparent mass

Source: Manufacturing Laboratories, Inc., MetalMAX.

imaginary part of the TF reaches a maximum. The imaginary component approaches zero at frequencies away from the resonance frequency. The damping ratio can be determined from the real curve, and the amplitude can be determined from the imaginary curve.

A third method of representing the FRF is to plot the real part versus the imaginary part to obtain a "Nyquist plot" (or a vector response plot). In this representation, also called the operative receptance locus, harmonic response locus, or harmonic receptance, the magnitude is plotted against the phase angle in polar form. Values of magnitude and phase angle at various frequencies represent points along the locus of the polar plot. This permits the characteristics of vibration to be combined in a single polar curve as shown in Figure 12.3b. The compliance or degree of flexibility is represented by the polar radius, while the phase is given by the angle between the polar vector and the positive real axis. This representation can be used for phase plane analysis, a graphical method for determining system stability [23–26]. For a SDOF system, the polar plot is almost circular when hysteretic damping is used; the diameter of the circle increases with decreasing damping ratio.

The effect of the damping ratio on the system response is shown in Figure 12.2, in which the magnitude of the frequency response (dynamic compliance) and the phase angle are plotted for various values of the damping ratio. Two characteristics are clear: (1) the compliance decreases almost proportionally with increasing damping for all frequencies; and (2) the maximum compliance at higher damping values (>0.2) occurs at a frequency lower than the resonant frequency. The phase angle is small for small values of the frequency ratio, and approaches 180° asymptotically for very large frequency ratios. This means that the amplitude of vibration is in phase with the exciting force for $r \ll 1$ and out of phase for $r \gg 1$. At resonance the phase angle is $-90°$ for all values of viscous damping.

There are several methods for determining the damping ratio [19,27,28] from measured data. It is easily determined using the half-power method with the compliance-frequency curve (Figure 12.3) using the formula

$$\zeta \approx \frac{\Delta f}{2 f_R}; \quad \Delta f = f_2 - f_1 \tag{12.17}$$

f_1 and f_2 represent the half power points, which can be measured from the plot. This method can be used for every resonance increase in an MDOF system. The damping ratio for a SDOF system can also calculated from the formula

$$\zeta = \frac{|G(0)|}{2|G(\omega_n)|} \tag{12.18}$$

Finally, the damping ratio can be estimated from the decay curve of the system as discussed in Ref. [28].

12.3.2 Multiple Degree-of-Freedom (MDOF) Systems

In general, the dynamic response of a cutting tool, machine tool, or workpiece-fixture structural system cannot be described adequately by a SDOF model as the response usually includes time variations of the displacement shape as well as amplitude. For some analysis purposes, machine tool structural elements can be idealized as discrete lumped masses connected by springs and dampers or a distributed parameter system. In this idealization the system is treated as an MDOF system, which is a natural extension of the SDOF system discussed earlier.

MDOF systems have one natural frequency for each degree of freedom (or mass). Each natural frequency has a corresponding characteristic deformation pattern or mode shape. The free vibration of the system can be described as a linear combination of vibrations in the individual modes.

The real and imaginary parts of the FRF of a MDOF system are shown in Figure 12.4. The polar plot for a system with multiple natural frequencies has multiple loops, one for each natural frequency in the frequency response curve. Polar plots for three milling machines in three directions are shown in Figure 12.5; they clearly show the structural characteristics for each machine in the selected directions.

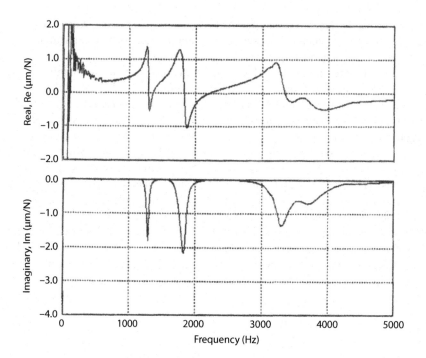

FIGURE 12.4   Real and imaginary parts of a milling machine (a MDOF system).

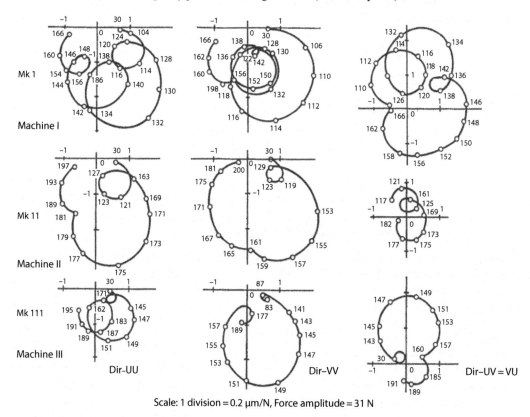

Scale: 1 division = 0.2 μm/N, Force amplitude = 31 N

FIGURE 12.5   Polar curve of the flexibility of three different milling machines in three directions. (From Schmitz, T.L., and Smith, K.S., *Machining Dynamics Frequency Response to Improved Productivity*, Springer, 2009.)

Polar curves are often used to identify the parameters of MDOF systems (i.e., the appropriate values for masses and spring and damper constants) as will be discussed in the following.

The equations of motion for a MDOF system in matrix form are

$$[M]\{\ddot{x}\}+[C]\{\dot{x}\}+x[K]\{x\}=\{f(t)\} \tag{12.19}$$

where [M], [C], and [K] are, respectively, the mass, damping, and stiffness matrices for the system. The elements m_{ij}, c_{ij}, and k_{ij} in the [M], [C], and [K] matrices represent the inertial, damping, and spring forces, respectively, acting at position i due to a unit acceleration of mass, velocity, and displacement, respectively, at position j, holding all other masses fixed. The equations of motion in the before system are generally coupled but can be uncoupled through coordinate transformations so that each equation in the matrix system can be solved separately. This is accomplished using a modal matrix [ψ] as the transformation matrix:

$$\{x\}=[\psi]\{q\} \tag{12.20}$$

where
$\{x\}$ is the displacement vector in the local coordinates
$\{q\}$ is the displacement vector in the principal (orthogonal) coordinates

A unique displacement vector called the "mode shape" exists for each distinct natural frequency and damping ratio. For a system with N degrees of freedom, [ψ] is formed by placing the eigenvectors $\{\psi_i\}$ for each mode i in the columns of an $N \times N$ matrix:

$$[\psi]=\left[\{\psi_1\}\{\psi_2\}\{\psi_3\}...\{\psi_N\}\right] \tag{12.21}$$

As discussed in the following, the eigenvector for a mode i represents the relative amplitude of vibration of the various masses, or mode shape, for vibrations at the natural frequency of the mode ω_{ni}.

The eigenvalues and eigenvectors for the system described by Equation 12.19 are determined by solving the corresponding undamped free vibration equations,

$$[M]\{\ddot{x}\}+[K]\{x\}=\{0\} \tag{12.22}$$

Values of a scalar λ that solve the equation

$$\left([K]-\lambda[M]\right)\{x\}=\{0\} \tag{12.23}$$

are sought; for an N-DOF system, there are N such values, determined by solving the associated characteristic equation,

$$\left|[K]-\lambda[M]\right|=0 \tag{12.24}$$

The resulting N eigenvalues λ_i are the squares of the natural frequencies of the system for each mode:

$$\lambda_i = \omega_{ni}^2 \tag{12.25}$$

For this homogeneous, undamped case, the eigenvalues are real. An eigenvector or mode shape $\{\psi_i\}$, called the normal mode vector, is associated with every undamped natural frequency ω_{ni}. The eigenvectors $\{\psi_i\}$ are determined by substituting the corresponding eigenvalue λ_i into Equation 12.22.

By substituting Equation 12.20 into Equation 12.19 and multiplying by the transpose of the modal matrix, Equation 12.19 can be transformed to

$$\left[M_q\right]\{\ddot{q}\}+\left[C_q\right]\{\dot{q}\}+\left[K_q\right]\{x\}=\left[\psi\right]^T\{f\} \tag{12.26}$$

where $[M_q]$, $[C_q]$, and $[K_q]$ are, respectively, the mass, damping, and stiffness matrices in principal coordinates (discussed in the following). The principle modes of vibration are orthogonal in the sense that, for two modes i and j with eigenvectors $\{\psi_i\}$ and $\{\psi_j\}$ and eigenvalues $\lambda_i \neq \lambda_j$, the following relationships hold:

$$\{\psi_i\}^T\left[M\right]\{\psi_j\}=\{\psi_j\}^T\left[M\right]\{\psi_i\}=0 \tag{12.27}$$

$$\{\psi_i\}^T\left[C\right]\{\psi_j\}=\{\psi_j\}^T\left[C\right]\{\psi_i\}=0 \tag{12.28}$$

and

$$\{\psi_i\}^T\left[K\right]\{\psi_j\}=\{\psi_j\}^T\left[K\right]\{\psi_i\}=0 \tag{12.29}$$

If these conditions are true, the matrices $[M_q]$, $[C_q]$, and $[K_q]$ are diagonal matrices and are referred to as the principal mass, damping, and stiffness matrices, respectively. They provide the mass, damping, and stiffness coefficients in normal coordinates. The mass and stiffness matrices can be diagonalized easily through the following transformations:

$$\left[\psi\right]^T\left[M\right]\left[\psi\right]=\left[M_q\right] \quad \text{and} \quad \left[\psi\right]^T\left[K\right]\left[\psi\right]=\left[K_q\right] \tag{12.30}$$

However, the damping matrix cannot generally be diagonalized simultaneously with the mass and stiffness matrices. When damping is present, the modes have phase relationships that complicate the analysis. The eigenvalues are real and negative or complex with negative real parts [29]. The complex eigenvalues occur as conjugate pairs and therefore their eigenvectors also consist of complex conjugate pairs. The treatment of damping is discussed in more detail in the following.

The eigenvalues and eigenvectors for free vibration of an undamped MDOF system are determined by solving Equation 12.22. The forced vibration of an undamped MDOF system can be analyzed based on Equation 12.26. The steady-state solution for the elements q_i in the displacement vector $\{q\}$ for harmonic excitation is

$$\{q_i\}=\frac{\{\psi_i\}^T\{f\}}{k_{qi}-m_{qi}\lambda_i} \tag{12.31}$$

The displacements in the local coordinates are given by Equation 12.20.

Damping of the machine tool must be considered when its frequency response is needed at or near a natural frequency. The solution of Equation 12.26 can be simplified by assuming that the damping matrix is linearly related to the mass and stiffness matrices:

$$\left[C\right]=a\left[M\right]+b\left[K\right] \tag{12.32}$$

where a and b are constants. This idealization is often used and is referred to as the "proportional damping" assumption because the damping matrix is proportional to the mass and stiffness matrices.

This assumption allows the equation of motion, Equation 12.19 or 12.26, to be uncoupled by the same transformation used for the undamped system [30]; therefore, the modal analysis method used before for an undamped system can be used to determine the eigenvalues and eigenvectors of a MDOF system with proportional damping. In this case the principal damping matrix is diagonalized by the transformation

$$[C_q] = [\psi]^T (a[M] + b[K])[\psi]$$

$$= a[M_q] + b[K_q] \tag{12.33}$$

The equations of motion are given by Equation 12.26; the solutions are

$$\{q_i\} = \frac{\{\psi_i\}^T \{f\}}{k_{qi} - m_{qi}\omega_i^2 + jc_{qi}\omega} \quad \forall i = 1,2,\ldots,N \tag{12.34}$$

The mode shapes of a system with proportional damping are the same as for an undamped system since they have the same modal matrix. Physically, this means that, when the system vibrates at a particular mode shape, every point reaches the maximum position at the same time. However, different modes have different phase relations relative to the exciting force for forced vibrations.

Caughey [31] showed that the proportionality condition, Equation 12.32, is sufficient but not necessary for a damped system to have principal modes. The required condition for obtaining principal modes in a damped system is that the transformation adapted before to diagonalize the damping matrix must also uncouple the equations of motion, Equation 12.26. This requirement is less restrictive than Equation 12.32.

A lightly damped system (specifically with damping ratios ζ_i less than 0.2 for all modes) can be analyzed assuming that the equations of motion are uncoupled by the modal matrix determined for the system without damping [29]. Under this assumption, the matrix $[\psi]$ is assumed to be orthogonal with respect to not only $[M]$ and $[K]$, but also $[C]$, that is, that Equation 12.28 holds true. It implies that any off-diagonal terms resulting from the operation

$$[C_q] = [\psi]^T [C][\psi] \tag{12.35}$$

are small and can be neglected. This assumption, called "modal damping," has great practical application since the damping ratios for machine tools generally satisfy the light damping condition. It should be emphasized that this concept is based on the normal coordinates for the undamped system, and that damping ratios must be specified in those coordinates.

Under this assumption, the resultant response of a MDOF system when an exciting force $\{f\}$ is applied in different points is given in terms of the N normal modes by equation

$$\{x\} = \sum_{i=1}^{N} \{\psi_i\}\{q_i\} \tag{12.36}$$

Also under this assumption, the cross FRF G_{ij} for the displacement $\{x_1\}$ of point i excited by a force $\{F_j\}$ at a point j is given by

$$G_{ij}(\omega) = \frac{X_i(\omega)}{F_j(\omega)} = \sum_{p=1}^{N} \frac{S_{ijp}}{1 - r_p^2 + 2j\zeta r_p} \tag{12.37}$$

where X_i and F_j are the amplitude of the corresponding displacement and force. If $i = j$ the direct frequency response is obtained. S_{ijp} is the cross ($i \neq j$) or direct ($i = j$) modal flexibility.

The case of arbitrary viscous damping, in which the earlier proportionality condition is not satisfied and the off-diagonal terms in matrix $[C_q]$ in Equation 12.35 cannot be neglected, can be analyzed by the transformation method developed by Duncan et al. [32] and later used by Foss [33]. Such systems are generally highly damped and have eigenvalues with significant imaginary parts. In this case, the real modal matrix $[\psi]$ cannot uncouple the equations of motion since the damping matrix does not satisfy the orthogonality condition. The Duncan/Foss approach transforms the N second-order equations of motion, Equation 12.19, into $2N$ uncoupled first-order equations. This leads to $2N$ eigenvalues, which are real and negative in the case of critically damped systems and complex with negative real parts in the case of underdamped systems (such as machine tools). This transformation is described in Refs. [32–34].

In metal cutting, the number of degrees of freedom and natural frequencies of interest from a practical viewpoint is generally limited. The machining system or any part of the system can often be simplified by reducing it to a few major modes of vibration; however, even then the determination of the compliance and damping characteristics for the system and its components is not simple because the system varies as the machine axes (i.e., column, table, spindle, workpiece, etc.) move. In addition, the metal cutting process itself influences the system vibration when the tool is in contact with the workpiece. In general, it is important to determine the natural frequencies of the machine tool or its major structural elements so that forced vibrations at resonant frequencies can be avoided. The frequencies of the disturbing forces generated during conventional cutting are typically limited and therefore only a limited number of modes with relatively low natural frequencies must be considered.

12.4   TYPES OF MACHINE TOOL VIBRATION

Machine tools are subject to three basic types of vibration: free, forced, and self-excited vibrations [35,36].

Free or *natural* vibrations (Figure 12.6a) occur when the stable system is displaced from its equilibrium position by shock; in this case, the system will vibrate and return to its original position in a manner dictated by its structural characteristics. Since machine tools are designed for high stiffness, this type of vibration seldom causes practical problems, and will not be considered further.

A *forced* vibration (Figure 12.6b) occurs when a dynamic exciting force is applied to the structure. Such forces are commonly induced by one of the following three sources: (1) alternating cutting forces such as those induced by (1a) inhomogeneities in the workpiece material (i.e., hard spots, cast surfaces, etc.), (1b) built-up edge (which forms and breaks off periodically), (1c) cutting forces periodically varying due to changes in the chip cross section, and (1d) force variations in interrupted cutting (i.e., in milling or turning a nonround or slotted part); (2) internal source of vibrations, such as (2a) disturbances in the workpiece and cutting tool drives (caused by worn components, i.e., bearing faults, defects in gears, and instability of the spindle or slides), (2b) out of balance forces (rotating unbalanced members, i.e., masses in the spindle or transmission), (2c) dynamic loads generated by the acceleration/deceleration or reversal of motion of massive moving components; and (3) external disturbances transmitted by the machine foundation [37]. Force vibration is discussed in more detail in Section 12.5.

"Self-excited vibration" or "chatter" (Figure 12.6c) is most often induced by variations in the cutting forces (caused by changes in the cutting velocity or chip cross section), which increase in amplitude over time due to closed loop regenerative effects. This is indicated by the control loop schematics in Figures 12.6c and 12.7. This type of vibration is referred to as chatter and is the least desirable type of vibration because the structure enters an unstable vibration condition.

Chatter is a complex phenomenon, which depends on the design and configuration of both the machine and tooling structures, on workpiece and cutting tool materials, and on machining regimes. The stiffness of the tool, spindle, workpiece, part, and fixture are important factors. The cutting

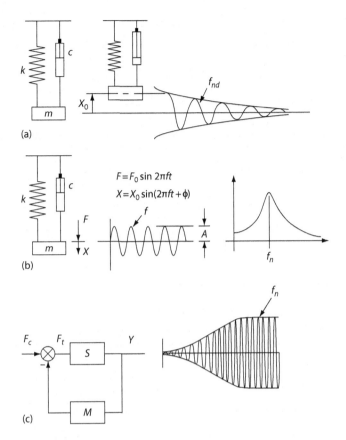

FIGURE 12.6 Schematics of various vibrations. (a) Free; (b) forced; (c) unstable self-excited. (Courtesy of D3 Vibrations Inc., Royal Oak, MI.)

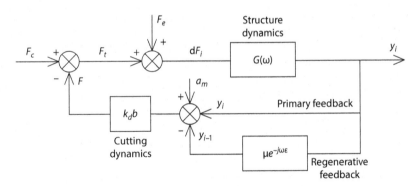

FIGURE 12.7 Block diagram representation of the machining process with noise input.

stiffness of the workpiece material is also an important factor; for example, steels have a greater tendency than aluminum to cause chatter. Cutting conditions, such as doc, width of cut, and cutting speed, greatly affect the onset of chatter. The chatter resistance of a machine tool is often expressed in terms of the maximum allowable width of cut b_{lim}. Forced vibrations are often easily detectable during the development stage or final inspection of a machine tool and can be reduced or eliminated. However, chatter occurrence may not be easily detected during the runoff stage unless the machine tool is thoroughly tested [38,39]. In addition, because it is a complex and typically nonlinear phenomenon, chatter may occur only under certain seemingly random conditions and may come and go sporadically. The elimination of chatter in a particular machining process can be a laborious exercise and frequently can be accomplished only by reducing the production rate in conventional spindles.

Chatter occurs because the damping of the machine tool system is not sufficient to absorb the portion of the cutting energy transmitted to the system. The practical significance of the chatter depends on the type of operation, whether it is a finishing or roughing operation, the surface finish requirements, tool wear characteristics, the allowable acoustic noise, and on its propagation to surrounding equipment. Chatter is never desirable because it can lead to accelerated machine tool wear, reduced part quality, and catastrophic tool failure. Chatter becomes more significant as cutting speeds increase since the exciting forces approach the natural frequencies of the system. It is often difficult to overcome chatter, but progress can be made through the proper selection of cutting conditions, improved design of the machine tool structure and spindle, and improved vibration isolation.

Chatter suppression typically refers to increasing the machining process stability. A related term, noise attenuation, is usually understood to mean improving the workpiece surface finish. In practice, the stability problem (chatter suppression) and the surface finish problem (noise attenuation) are coupled because the occurrence of vibration and especially chatter reduces machined surface quality.

Two approaches may be taken to solving chatter problems. The first is to choose or change cutting conditions such as the cutting speed, feed, doc, tool geometry, coolant, etc., to optimize the metal removal rate while operating in a stable regime. This is the test cuts approach (that detects and corrects). The second is to analyze the dynamic characteristics of the machining system to determine the stable operating range, and to suggest improvements to the system design, which can extend this range. The second approach is often called the stability chart method (prediction and avoidance) and will be explained in Sections 12.6 and 12.7. Active chatter suppression is discussed in Section 12.9.

12.5   FORCED VIBRATION

The cutting process generates forced vibrations through transient cutting forces, especially in interrupted cutting. The forcing frequency is usually the spindle rotational frequency and harmonics, or the tooth impact frequency and harmonics. In side or face milling the frequency of the forced vibration equals the product of the tool/spindle rotational frequency and the number of teeth on the tool (or tooth passing frequency). The forcing frequency is easily changed by adjusting the spindle rpm or number of teeth on the tool. Forced vibrations decay when the excitation is removed.

Figure 12.8 shows a flymilling operation with a single insert at 360 rpm. The top plot is a simplified time-domain forcing function showing the force turning on and off as the tooth enters and exits the cut. The bottom plot shows the frequency-domain forcing function for that time-domain force. The tooth passing frequency or fundamental forcing frequency (FFF) and harmonics are present because of this step-function. The FFF equals 6 Hz (1 tooth × 360 rpm ÷ 60 s/min) and the harmonics are integer multiples of the FFF (12, 18, 24, etc.).

Forced vibration was analyzed mathematically in Section 12.3 for single and multiple DOF systems. The amplitude of forced vibrations depends on the amplitude of the exciting force and on dynamic stiffness of the machine tool, cutting tool, and the workpiece, which are often an order of magnitude lower than the corresponding static values. The FRF of the part used for Figure 12.8 is given in Figure 12.9 and shows the compliance or flexibility of the part at every frequency. For a linear system the displacement of the part at each frequency can be determined from the FFF and FRF:

$$\text{Displacement (mm)} = \text{FFF (N)} \times \text{FRF (mm/N)} \tag{12.38}$$

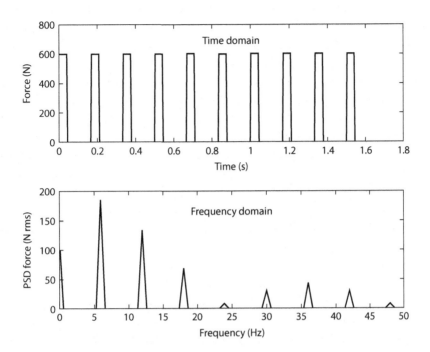

FIGURE 12.8 Step Forcing Function and PSD simulating a SDOF system of a flymill single insert rotating at 360 rpm during in a ½ slot. (Courtesy of D3 Vibrations Inc., Royal Oak, MI.)

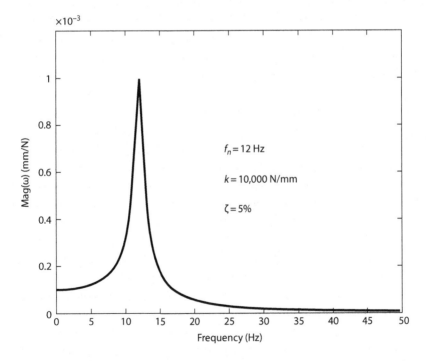

FIGURE 12.9 Frequency Response Function (FRF) of a SDOF part used for Figure 12.8. (Courtesy of D3 Vibrations Inc., Royal Oak, MI.)

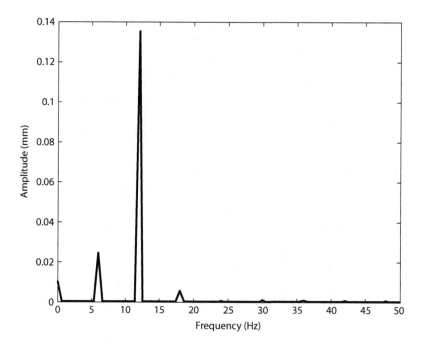

FIGURE 12.10   Estimated SDOF part displacement amplitude as a function of FFF (Figure 12.8) and FRF (Figure 12.9). (Courtesy of D3 Vibrations Inc., Royal Oak, MI.)

The displacement amplitude is shown in Figure 12.10, where the displacement is greatest at 12 Hz, which is the 1st harmonic of the forcing frequency. Even though the force is less at 12 Hz, the displacement is larger because the dynamic flexibility is near the resonant frequency. To characterize the ability to resist this kind of vibration, an index of machine tool dynamic stiffness is used:

$$C_d = \frac{F}{A_{max}} \tag{12.39}$$

where
   F is the disturbing force
   A_{max} is the resonant amplitude of vibration

Force vibrations can be produced by all machining operations but are especially important in finish machining operations in which surface waviness is unacceptable. Forced vibrations have their greatest practical impact when the exciting frequency is near one of the system's natural frequencies. For instance, if the rotational speed of a single-point boring bar (rev/s) is very close to or equal to a natural frequency of the bar, the bored surface will exhibit excessive waviness due to the vibration of the bar. The effect of the number of teeth on a milling cutter is shown in Figure 12.11; the frequency of the torque variations increases while the peak torque decreases as the number of teeth on the cutter increases, assuming that all cutters run at the same speed and feed rate.

   Forced vibrations due to periodic forces of defective machine components (such as unbalanced motors) typically become significant only when they excite a system resonance. In this case, the problem can generally be eliminated by changing the machine structure to change the resonant frequency.

   Force vibrations can lead to instability when machining at high speeds if the spindle speed is large enough for the tooth passing frequency to approach a dominant natural frequency of the system. For example, if the measured dominant natural frequency for a motorized spindle shaft is 500 Hz,

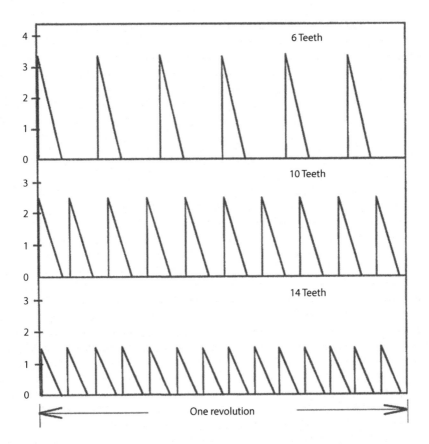

FIGURE 12.11 Variations in spindle torque or forces during climb (down) milling with a zero axial and radial rake cutter.

the spindle rpm corresponding to this natural frequency is 30,000 rpm. Similarly, if the measured dominant natural frequency for a long end mill with 6 teeth is 400 Hz, the critical spindle rpm N_{cr} producing the same tooth passing frequency is 4000 rpm based on the following equation:

$$N_{cr} = \frac{60 f_n}{n_t} = \frac{30 \omega_n}{\pi n_t} \qquad (12.40)$$

where n_t is the number of teeth on the cutter.

12.6 SELF-EXCITED VIBRATIONS (CHATTER)

Self-excited or self-induced vibrations occur because the dynamic cutting process forms a closed-loop system [35,40]. Disturbances in the system (i.e., vibrations that affect cutting forces) are fed back into the system and over time, under appropriate conditions, may result in instability. Self-excited vibrations do not result directly from external forces but draw energy from the cutting process itself; they arise due to the regeneration of surface waviness that occurs as the tool removes the chip from the surface that was created during the previous pass. The regenerative effect has become the most commonly accepted explanation for machine tool chatter [7,41–45].

The characteristic features of self-excited vibrations are (1) the amplitude increases with time, until a stable limiting value is attained; (2) the frequency of the vibration is equal to a natural frequency or critical frequency of the system, and (3) the energy supporting the vibration is obtained from a steady internal source.

Self-excited vibrations can be distinguished from forced vibrations by the fact that they disappear when cutting stops; forced vibrations exist and often persist whether or not the tool is engaged. The forced vibration caused by an interrupted cut can be distinguished from self-excited vibration by performing tests at two different cutting speeds. If the frequency of vibration changes and is equal to some multiple of the spindle speed in both tests, the vibration is forced; if the frequency of vibration does not change significantly, the vibration is self-excited.

Self-excited vibrations arise spontaneously and may grow in amplitude until some nonlinear effect provides a limit. Their occurrence is determined by the stability of the machine–tool–workpiece system. The cutting process is considered unstable if growing vibrations are generated; then the cutting tool may either oscillate with increasing amplitude or monotonically recede from the equilibrium position until nonlinear or limiting restraints appear. On the other hand, the cutting process is considered stable if the excited cutting tool approaches the equilibrium position in either a damped oscillatory fashion or asymptotically.

Self-excited vibrations in cutting result from variations in the chip thickness, doc, and cutting speed, which result in cutting force variations. The system is considered statically unstable if displacement-dependent forces are causing chatter. A system excited by velocity-dependent forces is said to be dynamically unstable.

Chip formation under unstable conditions (due to part or fixture deflection or tool/spindle deflection) results in variation of the cutting forces. Force variations lead to excessive vibration if the rigidity of the system is low. Force variations can lead to machine vibration, which in turn can cause additional force fluctuations by inducing variations in the uncut chip thickness. The variation of the uncut chip thickness due to vibrations during the previous pass (in turning and single-point boring) or previous tooth (in milling, reaming, and multi-tooth boring) cause additional force fluctuations. When the dynamic cutting force is out of phase with the instantaneous relative movement between the tool and workpiece, this leads to the development of self-excited vibration. This type of instability is called "regenerative chatter" because the vibration reproduces itself in subsequent revolutions through the generation of the waviness. It is the most practically significant form of self-excited vibration and is the main type discussed in the remainder of this section.

A second type of self-excited vibration, "nonregenerative chatter," occurs without undulation. There are other mechanisms that could lead to dynamic instability, such as the dependence of the cutting force on the cutting velocity. These types of chatter are relatively uncommon and have not been widely studied; one type of nonregenerative chatter, mode coupling, is discussed briefly at the end of this section.

12.6.1 Regenerative Chatter, Prediction of Stability Charts (Lobes)

The dynamic cutting process and the machine tool structure represent a three dimensional MDOF system as shown in Figure 12.12a. The dynamic machining process can be represented as a closed looped system by the block diagram shown in Figure 12.12b. Dynamic fluctuations of the cutting force and tool position relative to the workpiece occur in all machining processes because workpiece and tool are not infinitely stiff. This relative motion leaves an undulation on the machined surface with amplitude y_i. Oscillations on the surface left by the tool are removed by the succeeding tooth (in milling, reaming, and multi-tooth boring) or by the tool during the following revolution of the workpiece (in turning and single-point boring), resulting in a subsequent oscillation of amplitude y_{i-1}. The tooth cutting, a wavy surface, experiences a variable force that causes additional to the tool vibration. If the phase relation between the cutting force and surface oscillations is unfavorable, this can lead to vibrations of increasing amplitude, or regenerative chatter.

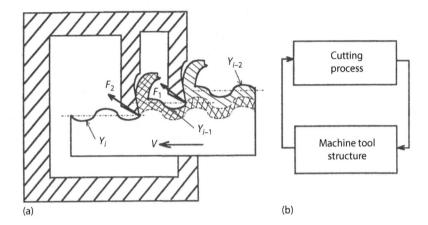

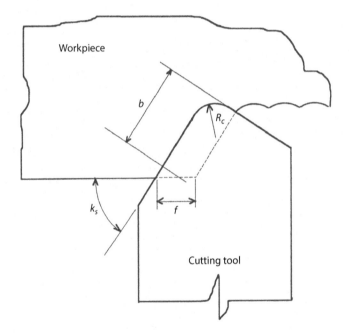

FIGURE 12.12 The dynamic machining process. (a) Closed-loop diagram of the structure, tool, and cutting process. (b) Diagram for deriving stability limits. (After Tlusty [40].)

FIGURE 12.13 Model of a dynamic cutting system (chip geometry variation for a single point cutting tool).

Variations in the uncut chip cross-section (shown in Figure 12.13) occur as a result of deflections of the tool or workpiece. The contact length between the tool and workpiece is a function of the doc, lead angle, and tool corner radius (Figure 12.13). The length of the contact for the cutting edge is

$$b = \frac{d - r_n + r_n \cos(k_s)}{\sin(k_s)} + \frac{\pi r_n k_s}{180} \tag{12.41}$$

The uncut chip thickness is given by Equation 2.2. Therefore, any movement of the tool or the workpiece will result in a corresponding change in the width and thickness of the chip (db and da, respectively). The changes in the chip cross-sectional area leads to proportional variation on the cutting force [d$F = f$(da)].

The conditions for the limit of stability of the machine tool structure and cutting process system can be explained with reference to Figure 12.12. Several assumptions and simplifications are made: (1) the cutting process takes place in one plane; (2) the machine tool structure is represented by a SDOF mechanical system; (3) the system is linear; (4) the direction of the variable component of the cutting force is constant and lies in the same plane as the cutting velocity; and (5) the variable component of the cutting force depends only on vibration in the direction normal to the cut surface Y.

The direction of principal vibration X, shown in Figure 12.14, forms an angle α with the normal to the machined surface Y. The direction of the cutting force F is inclined by an angle β from the normal Y. The average cutting speed is V and the width of cut is b. The chip thickness variation due to the surface wave Y_{i-1} produced by subsequent cuts depends on the phase lag ε with the surface wave Y_i left by the previous revolution. The number of waves between cuts is

$$n_p + \frac{\varepsilon}{2\pi} = \frac{f}{N} \tag{12.42}$$

at spindle speed N (in rev/min) and the chatter frequency f (Hz). n_p is the largest possible integer (number of whole waves) such that $\varepsilon/2\pi < 1$, and ε is the phase of inner modulation y_i to the outer modulation y_{i-1}. In other words, the number of cycles of vibration (waves) between subsequent revolutions is an integer plus a fraction. When the vibration frequency is a whole number multiple of the rotational speed ($\varepsilon = 0°$), the vibration allows the tool to follow the previous wavy surface and the self-excited vibration is not present. The maximum chip thickness variation occurs at $\varepsilon = 180°$.

Many researchers have studied machining stability, and several theories [3,4,40,46] have been proposed to explain self-excited chatter. The stability of chatter vibrations was initially analyzed by Tobias [47], Tlusty [48], and Merrit [49] using linear theory. Several nonlinearities such as multiple regeneration, interrupted cut, process damping, and large vibrations resulting in zero chip thickness are neglected in linear stability analyses. In this book, the well-known and typical theories of Tlusty and Tobias are briefly summarized.

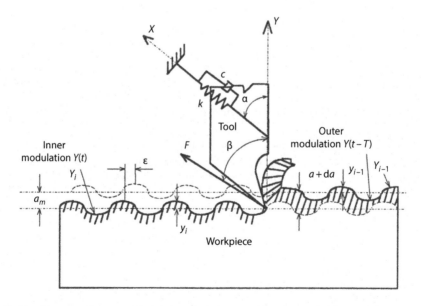

FIGURE 12.14   Diagram of regenerative chatter due to the cutting of an undulated surface.

12.6.2 Tlusty's Theory

Tlusty proposed a simple analysis that assumes that the dynamic cutting force is proportional to the undeformed chip thickness [4,46,49]. The vibration of the tool in the direction normal to the cut surface during the ith cut is

$$y_i = Y_i \sin \omega t = X_i \cos \alpha \sin \omega t \tag{12.43}$$

and the mean chip thickness variation is

$$a = a_m + \mathrm{d}a = a_m + y_{i-1} - y_i$$

$$= a_m + (x_{i-1} - x_i) \cos \alpha \tag{12.44}$$

where
 $\mathrm{d}a$ is the variable component of the chip thickness
 y_{i-1} is amplitude of the surface undulation

The magnitude of the dynamic force variation depends on the relative motion between the cutting tool edge and the workpiece surface and on the angle between the cutting force and the direction of principal vibration. The force on any tooth in a cutter is proportional to the chip thickness. Therefore, the variable force component or regenerative force is

$$\mathrm{d}F_i = k_d b \mathrm{d}a = k_d b (y_{i-1} - y_i)$$

$$= k_d b (x_{i-1} - x_i) \cos \alpha \tag{12.45}$$

where k_d is the specific dynamic cutting stiffness or specific force, which is assumed to be a material constant. k_d can be determined from the specific energy of the workpiece material as discussed in Chapter 6. The force depends not only on the feed per tooth and on the deflection of the cutter, but also on the surface that was left by previous teeth. The assumption that the cutting force is changing in phase with y is not really true and Equation 12.45 represents only an approximation of the real physical system. The generated vibration either grows or diminishes depending on several parameters (i.e., k_d, b, N). The variable force excites the vibration of the machine tool in the ith cut, represented by a SDOF system (Figure 12.15). The amplitude of vibration during the ith cut is given by Equations 12.9 and 12.10:

$$x_i = \mathrm{d}F_i \cos(\alpha - \beta) \frac{1}{k} \frac{1}{1 - r^2 + 2j\zeta r} \tag{12.46}$$

The relation between the amplitudes for the ith and $(i-1)$th cuts can be determined by substituting $\mathrm{d}F_i$ from Equation 12.45 into Equation 12.46:

$$\frac{y_i}{y_{i-1}} = \frac{x_i}{x_{i-1}} = \frac{G(\omega)}{G(\omega) + \dfrac{1}{k_d b}} \tag{12.47}$$

where

$$G(\omega) = \frac{Y(\omega)}{F(\omega)} = \frac{Y(\omega)}{F_x(\omega)} u = u G_d(\omega) \tag{12.48}$$

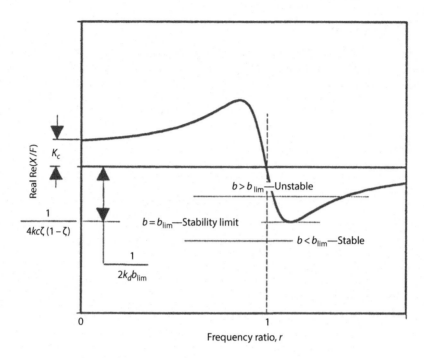

FIGURE 12.15   Schematic representation of the stability limit for a SDOF system.

$$u = \cos(\alpha)\cos(\alpha - \beta) \tag{12.49}$$

and

$$G_d(\omega) = \frac{1}{k}\frac{1}{1 - r^2 + 2j\zeta r} \tag{12.50}$$

where
   TF $G(\omega)$ represents the response in the direction Y to a force acting in the direction of the cutting
      force
   $G_d(\omega)$ is the direct TF defined in the direction X
   u is the directional orientation factor

Based on regenerative chatter theory [50], the state of the dynamic cutting process is described by

$$\left|\frac{y_i}{y_{i-1}}\right| = \begin{cases} >1 & \text{unstable} \\ =1 & \text{at stability limit} \\ <1 & \text{stable} \end{cases} \tag{12.51}$$

Regenerative instability will occur when the vibratory motion increases with time, in which case the
magnitude of Equation 12.47 is greater than 1 for some frequencies. The borderline of stability is
obtained when the magnitude of the expression is equal to unity. Therefore, equating the magnitude
of Equation 12.47 to 1,

$$\text{Im}\{G(\omega)\} = 0 \quad \text{and} \quad \text{Re}\{G(\omega)\} = \frac{-1}{2k_d b} \tag{12.52}$$

This equation is the simplest form of the condition for the limit of stability. The real part of the FRF is obtained from Equation 12.13:

$$\text{Re}\{G(\omega)\} = \frac{u}{k} \frac{1-r^2}{\left(1-r^2\right)^2 + \left(2\zeta r\right)^2} \tag{12.53}$$

with extreme values of

$$\text{Re}\{G(\omega)\}_{\min} = -\frac{u}{k} \cdot \frac{1}{4\zeta(1+\zeta)} = \frac{-1}{4 \cdot k_c \zeta(1+\zeta)} \tag{12.54}$$

for $r = \sqrt{1+2\zeta}$ and $\zeta \ll 1$ as shown in Figure 12.15. The maximum chip width for stable cutting (or in other words the limit of stability) can be calculated from Equations 12.47 and 12.52 assuming that the dynamic characteristics of the system (the cutting stiffness) are known. The cutting stiffness k_d is defined as the increment in the magnitude of the cutting force per unit increment of depth of cut at unit chip width.

The earlier analysis considers a phase shift between two subsequent cuts. The dynamic portion of the instantaneous chip thickness is

$$y_{i-1} - y_i = \mu y_i(t-\varepsilon) - y_i(t) + r(t) \tag{12.55}$$

where ε represents the time between the production of an undulated surface and the renewed cutting of that wavy surface (i.e., one revolution in turning as shown in Figure 12.14). $r(t)$ represents external disturbances, which affect the chip thickness. μ is the overlap factor that lies in the interval [0,1] (i.e., for plunge cutting $\mu = 1$ and for thread cutting $\mu = 0$). The amplitude of vibration of the $(i - 1)$th cut is

$$y_{i-1} = \mu y_i e^{-j \cdot \omega \cdot \varepsilon} \tag{12.56}$$

The stability condition can be redefined by substituting Equation 12.56 into Equation 12.47 to yield

$$G(\omega) \cdot \left(1 - \mu \cdot e^{-j \cdot \omega \cdot \varepsilon}\right) = \frac{-1}{k_d b} \tag{12.57}$$

There is full overlap between two consequent cuts in many operations ($\mu = 1$), which means that the full undulation produced during one revolution is removed in the next.

The process of self-excitation and regenerative chatter can be represented by the block diagram of the closed loop system shown in Figure 12.7. In this model, the external disturbances that affect the chip thickness $r(t)$ are not directly accounted for but two new force inputs are considered, $F_c(t)$ and $F_e(t)$, both of which represent noise affecting the process. $F_c(t)$ represents the cutting noise, while $F_e(t)$ represents noise from external sources. The cutting noise results from material inhomogeneity, workpiece out-of-roundness, and the force variations that occur during the chip formation process. The external noise results from rough bearings, spindle unbalance, floor vibrations, and similar sources. The force input representing the external noise $F_e(t)$ is not transmitted by the cutting tool. This force acts directly on the workpiece-machine structure and must induce motion of the structure before a tool force results. Analysis of this loop using feedback control theory also yields Equation 12.57. Therefore,

$$\left[\left(\text{Re}\{G(\omega)\}+\text{Im}\{G(\omega)\}\right)(\cos\varepsilon-j\sin\varepsilon-1)\right]k_db=1$$

$$\text{Re}\{G(\omega)\}(\cos\varepsilon-1)+\text{Im}\{G(\omega)\}\sin\varepsilon \qquad (12.58)$$

$$+j\cdot\left(\left[\text{Im}\{G(\omega)\}(\cos\varepsilon-1)\right]-\text{Re}\{G(\omega)\}\sin\varepsilon\right)=\frac{1}{k_db}$$

The imaginary part is zero, hence

$$\text{Im}\{G(\omega)\}(\cos\varepsilon-1)-\text{Re}\{G(\omega)\}\sin\varepsilon=0$$

$$\sin\varphi(\cos\varepsilon-1)-\cos\varphi\sin\varepsilon=0$$

$$\sin(\varphi-\varepsilon)=\sin\varphi \qquad (12.59)$$

$$\varphi=\frac{\pi+\varepsilon}{2}$$

The real part is equal to 1, hence

$$\left[\text{Re}\{G(\omega)\}(\cos\varepsilon-1)+\text{Im}\{G(\omega)\}\sin\varepsilon\right]k_db=1$$

$$b_{cr}=\frac{1}{k_d\left[\text{Re}\{G(\omega)\}(\cos\varepsilon-1)+\text{Im}\{G(\omega)\}\sin\varepsilon\right]} \qquad (12.60)$$

and since

$$\tan\phi=\frac{\text{Im}\{G(\omega)\}}{\text{Re}\{G(\omega)\}}=\frac{\sin\varepsilon}{\cos\varepsilon-1} \qquad (12.61)$$

this yields

$$b_{cr}=\frac{-1}{2k_d\,\text{Re}\{G(\omega)\}} \qquad (12.62)$$

This stability criterion was proposed by Merritt [49], who gave Tlusty [48] credit for prior publication. The limit of stability occurs at the critical width of cut; the minimum borderline of stability is given by

$$\left(b_{cr}\right)_{\text{min}}=\frac{-1}{2k_d\left(\text{Re}\{G(\omega)\}\right)_{\text{min}}}$$

$$\left(\text{Re}\{G(\omega)\}\right)_{\text{min}}=\left|\left(\text{Re}\{G(\omega)\}\right)_{\text{neg max}}\right| \qquad (12.63)$$

b_{cr} is a limiting case, below which no chatter will occur and above which chatter may occur. The earlier solution is valid only for negative values of Re$\{G(\omega)\}$ since b_{cr} is a physical quantity. This solution represents an orthogonal cutting case of a SDOF system that represents a turning process with zero lead angle well. For a tool with a lead angle κ and a cutting force at an angle σ (from a line normal to the machined surface), the stability lobes move up resulting in higher stability. The limit of stability is given by the oriented TF

$$\text{Re}\{G_1(\omega)\} = \sin^2\kappa\cos\sigma\,\text{Re}\{G(\omega)\} \tag{12.64}$$

The FRF is generally either that of the workpiece or the tool (the weakest of the two), typically the boring tool or spindle. The milling process is better treated in time domain simulation, but can be analyzed as a turning process by fixing the tool at a given position. Equations 12.63 and 12.64 can be used to simplify SDOF milling process with n_{tc} teeth in the cut in which case

$$(b_{cr})_{\text{min,milling}} = \frac{(b_{cr})_{\text{total}}}{n_{tc}} = \frac{(b_{cr})_{\text{min}}}{n_{tc}} \tag{12.65}$$

and $(b_{cr})_{\text{total}}$ is given by Equation 12.63. In the case of end milling, $\kappa = 90°$ and in the case of face milling $\kappa = 0°$. The direction of the cutting force σ is zero in milling.

In many cases, it is assumed that the structural dynamics of the machine tool can be represented by a SDOF system based on a second-order dynamic model as discussed earlier. Under this assumption, Re$\{G(\omega)\}_{\text{min}}$ is given by Equation 12.52 and b_{lim} is given by

$$b_{\text{lim}} = \frac{2k_c\zeta(1+\zeta)}{k_d} \cong \frac{2k_c\zeta}{k_d} \tag{12.66}$$

b_{lim} can also be calculated from Equation 12.62 because the Re$\{G(\omega)\}$ can be measured in Figure 12.3. If the value $\zeta = 0.05$ is adopted as a nominal value for machine tool structures [51], the critical value of the gain ratio is

$$b\frac{k_d}{k_c} = 0.105 \tag{12.67}$$

As a result, for the second-order structural dynamic model with $\zeta = 0.05$, the stability of the overall system will be improved if the critical gain ratio is greater than 0.105. This theory should be applied only when the modes of the machine tool structure are well separated as is usually the case in drilling and face milling and sometimes the case in turning.

A MDOF system can be analyzed using a similar approach because the real part of the TF of a complex system is simply a sum of the TFs of the individual modes involved [46]. Therefore,

$$G(\omega) = u_1G_{d1}(\omega) + u_2G_{d2}(\omega) + \cdots = \sum_{i=1}^{N}u_iG_{di}(\omega) \tag{12.68}$$

An example of this method is given in Figure 12.16 for three and four DOF systems.

For example, milling cutter can be assumed to have two orthogonal degrees of freedom as shown in Figure 12.17. Two modes of vibration are considered, one in the feed direction X and the other in the normal direction Y. The cutter is assumed to have n_{tc} teeth in the cut (in contact with the workpiece at any instant) and a zero helix angle. More than one tooth cuts at a time, and the orientation

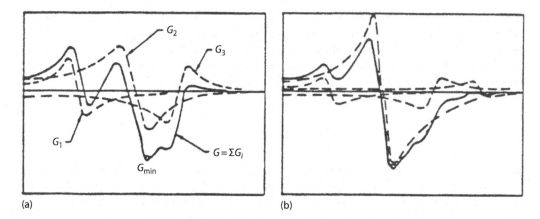

(a)   (b)

FIGURE 12.16 The real part of oriented transfer functions generated from various combinations of modes for a three (a) and a four (b) DOF system. (From Bayly, P.V. et al., Effects of Radial Immersion and Cutting Direction on Chatter Instability in End-Milling, *Proceedings of the ASME IMECE*, New Orleans, LA, No. IMECE2002-34116, ASME, 2002.)

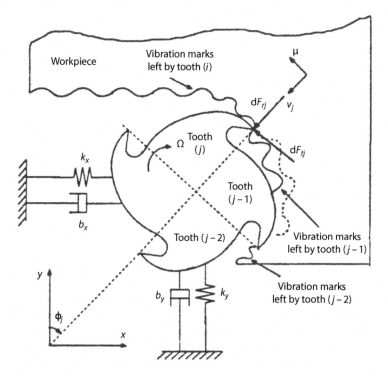

FIGURE 12.17 Dynamic model of milling with two degrees of freedom. (From Tlusty, J., and Rao, S.B., *Proc. NAMRC*, 6, 420, 1978.)

of the force F changes. The cutting forces excite the structure, causing deflections in the two perpendicular directions in the plane of the cut. These deflections are carried to rotating tooth number (j) in the radial or chip thickness direction by its projection onto v_j, $v_j = X \sin(v_j) + Y \cos(v_j)$; where the angle v_j is the instantaneous angular immersion of tooth j. The critical axial depth of cut is

$$b_{cr} = \frac{-1}{2k_d n_{tc}\left(u_x \operatorname{Re}\left\{G_x(\omega)\right\} + u_y \operatorname{Re}\left\{G_y(\omega)\right\}\right)} \tag{12.69}$$

where the directional orientation factors for X and Y, respectively, are

$$u_x = \cos \alpha \sin v = \sin(\beta + v)\sin v$$
$$u_y = \cos(\beta + v)\cos v \tag{12.70}$$

More details on the analysis of milling processes can be found in Refs. [52–55].

The procedure for determining the value of $b_{\lim}$ for a given oriented TF for every possible chatter frequency series corresponding to different spindle speeds N is as follows:

1. Select a frequency f.
2. Determine the values of $\text{Re}\{G(\omega)\}$, $\text{Im}\{G(\omega)\}$, and v.
3. Calculate the value of ε.
4. Determine the speed N from Equation 12.42 for several values of n_p (n_p = 1, 2, 3,...); for milling the frequency f is divided by the number of teeth in the milling cutter n_t in Equation 12.42.
5. Calculate $b_{\lim}$, from Equation 12.62 or Equation 12.69.
6. Rearrange and plot the pairs (b_{cr}, N) in ascending order of N.

A simple way to graph the lobes is to use the solution of the SDOF cutting process and plot $b_{cr}(\omega)$ versus the spindle RPM $N(\omega)$ using the equation

$$N(\omega) = \frac{60}{n_t} \frac{\omega}{-a\tan\left(b(\omega)k_d\text{Mag}\{G(\omega)\} + \text{Re}\{G(\omega)\}\right) \pm 2\pi n_p} \tag{12.71}$$

This is the solution of the SDOF system based on the stability condition given by Equation 12.57. A typical stability chart for a machine tool is shown in Figure 12.18. Three borderlines of stability are shown; these are called lobed, tangent, and asymptotic borderlines [39]. The lobed borderline of stability is the exact borderline obtained through the earlier procedure. The tangent borderline is a hyperbola touching all the unstable ranges of the lobed borderline. The asymptotic borderline is determined using $b = (b_{cr})_{\min} = b_{\lim}$ calculated from Equation 12.63. The asymptotic borderline of stability defines the maximum width of cut, which results in stable cutting at all spindle speeds. However, a larger width of cut can be used when the proper spindle speed is selected (between lobes). The lobed borderline indicates that the system exhibits conditional stability above one specific speed and below another specific speed. In addition it indicates that the danger of instability at

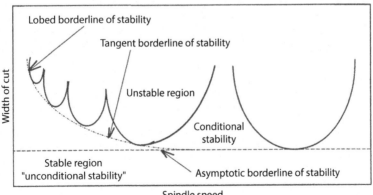

FIGURE 12.18   Typical stability chart for a machine tool.

fairly low speeds is reduced due to process damping [56–59]. The process damping effect can serve to increase the allowable width of cut for low spindle speeds. The process damping depends on the contact pressure, and the deformation of material under the flank face, which is affected by the clearance/relief angle and edge prep, cutting speed, and chatter frequency. The stable speed ranges (lying between the unstable ranges of speed) become wider at higher speeds. The lobed borderline of stability is therefore of practical use primarily at high cutting speeds. It is complex to apply in practice because it requires a different stability chart for each possible position of the movable elements of the machine tool. Commercially available software allows measurements of the frequency response of the cutter-toolholder system without cutting and without running the spindle. Such software converts the frequency response data to usable dynamically optimized cutting parameters using stability lobes [35,60]. In addition, a safety zone along the lobes can be plotted to compensate for the variation and uncertainty of some of the factors in the stability limit including the directional orientation factors. Consequently, the machining and especially milling of aluminum alloys in the aerospace community has shifted to higher speeds to improve dynamics and efficiency, while for hard to machine materials (such as steel alloys, titanium, Inconel, etc.), process designers are taking advantage of the process damping regime.

The stability lobe diagram has been one of the best utilities to determine the threshold values for dynamic stiffness because it is affected by the dynamic stiffness of the tool geometry, the radial and axial immersion of tool into workpiece, spindle speed, machine tool characteristics, as well as the workpiece material. The diagram is simpler for long and slender end-mills or boring tools since the dynamic response is often dominated by the spindle–toolholder–tool system. Generally, in these applications, low frequency chatter is associated with the structural modes of the machine, while the higher frequencies are associated with the spindle and tool modes.

The effect of the individual cutting parameters on stability lobes are as follows: (1) the limit of stability is controlled by the upper limit of chip width b in turning and by the axial depth of cut in milling. (2) The axial depth of cut in milling is affected by the number of teeth n_{tc} in the cut (or the radial immersion for the cutter discussed in Chapter 2). (3) The ADOC is proportional to the dynamic stiffness (or product of stiffness and damping) of the machine or workpiece structure and inversely proportional to the specific force. This means that the ADOC decreases with increasing workpiece material hardness since k_d increases. (4) The limit of stability in milling is determined by the limit of the MRR affected by the product of the ADOC and RDOC. (5) The spindle RPM, N, and the number of teeth in the cutter affect the stability lobes and the selection of b_{lim}. (6) The feed per tooth has a small effect on the limit of stability but does affect the amplitude of vibration in unstable regions.

12.6.3 Shear Plane Method

An earlier, less convenient method of stability analysis was the Shear Plane Method developed by Tobias and Fishwick [47,61,62]. This quasi-empirical method was intended to develop stability charts based on a mathematical representation of the machining process with the necessary constants obtained experimentally. The dynamic cutting force expression in this theory is

$$dF = K_1 \cdot da + K_2 \cdot df_r + K_3 \cdot dN \tag{12.72}$$

where da, df_r, and dN are changes of the chip thickness, feed rate, and rotational (spindle) speed from one state to another, which cause variations in the cutting force components. The coefficients K_1, K_2, and K_3 are defined by

$$K_1 = \left(\frac{\partial F}{\partial a} \right)_{df_r = dN = 0} \quad ; \quad K_2 = \left(\frac{\partial F}{\partial f_r} \right)_{da = dN = 0} \quad ; \quad K_3 = \left(\frac{\partial F}{\partial N} \right)_{da = df_r = 0} \tag{12.73}$$

Assuming that the influence of the vibration $x(t)$ on the cutting speed is negligibly small and at the critical state of stability the total damping is zero, the solution of the equation of motion for the weakest mode of the machine structure is:

$$\left(\frac{1}{Q\omega_n} + \frac{K_1}{k} F_2 + \frac{K^*}{kN}\right) = 0 \tag{12.74}$$

and

$$\omega^2 = \omega_n^2 \cdot \left(1 + \frac{K_1}{k} \cdot F_1\right) \tag{12.75}$$

with

$$\omega_n^2 = \frac{k}{m} \quad Q = \frac{k}{c \cdot \omega_n} = \frac{1}{2 \cdot \zeta} \tag{12.76}$$

and the coefficients

$$F_1 = 1 - \mu \cdot \cos\frac{\omega}{N} \quad F_2 = \frac{\mu}{\omega} \cdot \sin\frac{\omega}{N} \tag{12.77}$$

Equations 12.74 and 12.75 define the stability conditions for the particular mode of vibration; using these equations, the relations between Q and N and ω and N can be solved. The stability chart is generated using Q as the ordinate and the parameter n_tN/f_n as the abscissa. This chart defines stable and unstable regions similar to those defined in Figure 12.18 depending on the magnitude and sign of the penetration rate coefficient K^* [61]. The asymptotic borderline of stability is obtained when $K^* = 0$; the tangent and lobed borderline of stability are generated when $K^* > 0$; finally, when $K^* < 0$, the low unstable speed ranges in the stability chart are displaced downward and the risk of instability at low speeds increases. The parameters K_1 and K_2 are functions of the material, tool geometry, feed, chip width, and similar factors. These coefficients can be measured through dynamic experiments [3]. In general, one or more modes of vibration can become unstable, so that stability charts (Figure 12.18) should be generated for each mode; in practice, however, only a single mode of vibration is unstable in many operations [3], so that stability analysis can be carried out using a single stability chart. One of the fundamental problems in this approach is the prediction of the cutting forces while cutting. It should be also noted that a number of other different models have been proposed with varying degrees of success [62,63].

12.6.4 OTHER METHODS

The before analytical methods for predicting stability lobes are based on simplified linear models, which assume an average tooth position, force direction, and number of teeth in the cut, and which do not account for process damping. They cannot be used to predict limiting levels of chatter or to model nonlinear springs, backlash, and similar real phenomena. These methods are applicable mainly to operations such as turning, which resemble a basic orthogonal cutting operation in which the direction of the cutting forces and chip thickness do not vary with time. For rotating tooling applications such as milling, in which the chip thickness, cutting forces, and direction of excitation vary and are intermittent, a more realistic and more detailed method for generating stability charts is "time domain simulation" (TDS) [32,36,48,51,53–55,62–71]. Even though TDS is a computationally intensive method, it is attractive in many cases because it can include many of the complex characteristics of real machining systems, including the rotation of the spindle and the nonlinear support conditions for the spindle.

As an example, a milling cutter can be assumed to have two orthogonal degrees of freedom with one or more masses, springs, and dashpots along each direction as shown in Figure 12.17. The lumped parameters representing the modal stiffness, mass, and damping can be extracted from the TFs in two orthogonal directions. Velocity-dependent process damping and basic process nonlinearities arising when the tool exits the cut due to excessive vibration can be considered using TDS with this idealization. For example, Montgomery and Altintas [55] considered process damping and stiffness in a time domain simulation model, and predicted the surface finish produced by chatter in milling.

TDS can be used for SDOF or MDOF systems with closed-loop behavior described by a second-order system equivalent to Equation 12.1. In this method the cutter is advanced in small steps; 360 steps per natural period is required to match TDS results to frequency-domain solutions [66]. For each increment, the forces on each tooth are calculated and summed vectorially. The acceleration produced by the force is used to calculate displacements in both directions; the resulting differential difference equation is numerically integrated. This approach accounts for change in chip thickness throughout the cut including the entry and exit transients. The limiting width of cut is determined by running simulations at various widths of cut for each spindle speed of interest. Convergence on b_{lim} is achieved by considering increments and decrements of the axial depth of cut, based on the stability conclusion at each axial depth of cut considered, while gradually reducing the magnitude of the step changes. The peak-to-peak (PTP) amplitude of the displacement and force are calculated for each simulation and plotted versus the spindle speed for various axial depths of cut as shown in Figure 12.19 [53,56]. For each simulation, only the PTP amplitude is recorded for force and displacement.

Figure 12.19 was constructed for high speed milling with a 9.5 mm diameter two-flute carbide end-mill, which is the same system used to compute the results shown in Figure 12.4. The individual lines correspond to axial depths of cut from 0.5 to 7 mm in steps of 0.5 mm (14 axial depths of cut). The spindle speeds varied from 25,000 to 45,000 in 500 rpm steps (40 speeds), so that the

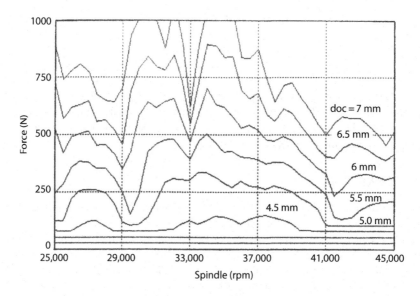

FIGURE 12.19 PTP diagram for a two flute, 9.5 mm diameter carbide end mill, illustrating an algorithm for optimizing the MRR.

information in the figure is a composite of 560 simulation runs. The graph plots the force in the X-direction vs. spindle speed. The stable zones are above 40,000 rpm and between 25,000 and 29,000 rpm. For axial depths of cut below 4 mm, all cuts were stable. The doc lines that have unstable speed ranges are marked. In the unstable regions, the PTP forces may become quite large. TDS can provide limited information about the surface roughness generated by considering the vibration and the trajectory of the teeth along the cutting path that the tool center follows.

In general, the identification of critical chatter frequencies in milling is not a trivial task either experimentally or theoretically. Milling force or acceleration power spectra typically exhibit several peaks [72]. Some result from tooth pass excitation, while others are due to regenerative effects or the natural frequency of the tool. In the case of milling, therefore, there are multiple chatter frequencies compared to the single well-defined chatter frequency typical of an unstable turning process. The analysis of the milling process is more complex because the direction and level of the cutting force changes with tool rotation and because the cut is interrupted. Tooth pass excitation results in a parametric excitation of the system, and the governing equation of motion is a delay-differential equation (DDE). Stability properties have been studied through analysis of these DDEs [73–91] and through frequency analysis of chatter signals [74,92,93]. Some chatter frequencies are related to unstable periodic motions about stable stationary cutting states. In this case, subcritical (Hopf) bifurcation occurs, as demonstrated experimentally by Shi and Tobias [94] and analytically by Stépán and Kalmár-Nagy [95]. Similarly, periodic doubling bifurcation is also a typical mode of stability loss in milling processes [81–83,91]. The nonlinear analysis of Stépán and Szalai [96] showed that this period doubling bifurcation is also subcritical.

"Low radial immersion milling" has become an increasingly important process due to advances in machining centers, which have enabled rapid adoption of high speed milling, and due to increased near-net shape manufacturing of aerospace components. This has driven the need to better understand stability boundary and the corresponding surface location error in interrupted cutting processes as the radial depth of cut (RDOC) or radial immersion is varied (Figure 12.20). The relationship between the direction of tool rotation and feed defines two types of partial immersion milling operations: up-milling and down-milling (Figure 12.21). Both operations can produce the same result, but their dynamic and stability properties are not the same. Partial immersion milling operations are characterized by the number N of teeth and the radial immersion ratio RDOC/D, where RDOC is the radial depth of cut and D the diameter of the tool (see Chapter 2). Several researchers have shown that the stability properties for low radial engagements differ from

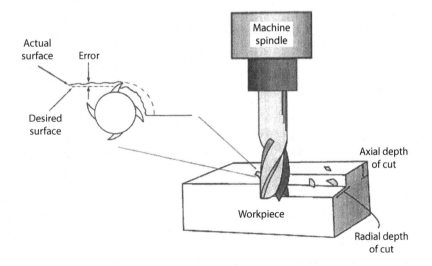

FIGURE 12.20 Schematic diagram showing the milling process is an interrupted cutting process.

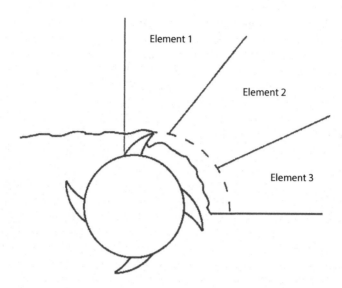

FIGURE 12.21 The temporal finite element approach divides the cutting motion into elements. (Courtesy of B. Mann, University of Florida, Fort Myers, FL.)

those observed when cutting a full slot [81–83,87,91,97]. For instance, Davies et al. examined interrupted cutting processes for an infinitesimal cutting period to formulate a dynamic map [81,98]. Stability predictions were made from the map transition matrix eigenvalues. Stepan, Insperger, and Balachandran have also developed an approach to formulate dynamic maps for stability prediction using a method called semi-discretization [72,82,99]. Bayly, Halley, and Mann have presented an approach called temporal FEA to examine interrupted cutting in turning and milling [87,100–104]. This method forms an approximate analytical solution by dividing the time in the cut into a finite number of elements (Figure 12.21). The approximate solution for the cutting motion is then matched with the exact solution for non-cutting time periods to obtain a discrete linear map. The formulated dynamic map is then used to determine stability and the steady-state surface location error (Figure 12.20). Figure 12.22 shows the effect of changing the radial immersion on the stability boundaries for three different radial depths of cut. One interesting result from the three plots is that a larger axial depth can be obtained for low radial immersion cutting processes.

12.6.5 Nonregenerative Chatter, Mode Coupling

One type of nonregenerative chatter occurs when the tool vibrates relative to the workpiece in at least two directions in the plane of cut as shown in Figure 12.23. The machining system is modeled by a two degree-of-freedom mass-spring system, with orthogonal axes of major flexibilities and a common mass. The characteristics of the vibration are such that the tool follows a closed elliptical path relative to the workpiece as indicated by the arrows in the contour in front of the tool. During the periodic motion of the tool from portion 1 to 2 along the contour, the resultant cutting force is in the opposite direction of the tool motion and energy is dissipated from the system. However, during the other half of the contour, from portion 2 to 1, where the motion and cutting force act in the same direction, energy is supplied in the system, which increases the vibratory energy of the tool. The force F tends to be larger over the lower portion of the elliptical contour than over the upper portion because it is located deeper in the cut; therefore, the input energy is larger than the total energy losses per cycle, which leads to increasing vibration amplitudes, thus producing negative effective damping. This type of instability is usually referred to as *mode coupling* [46].

The mode coupling principle may be explained by a simple form of a two degree-of-freedom system as shown in Figure 12.23. The system is assumed to be linear as long as the tool does not

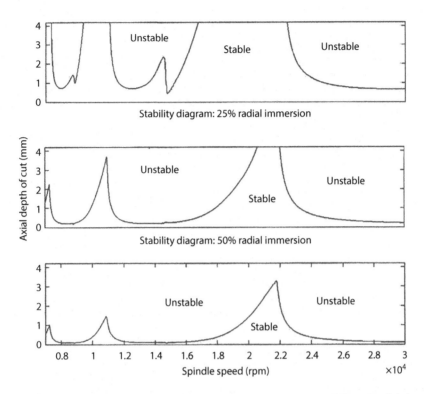

FIGURE 12.22 Example of low radial immersion stability lobes for three different radial depths of cut. These plots show it is possible to obtain a larger axial depth of cut in finishing operations. (Courtesy of B. Mann, University of Florida, Fort Myers, FL.)

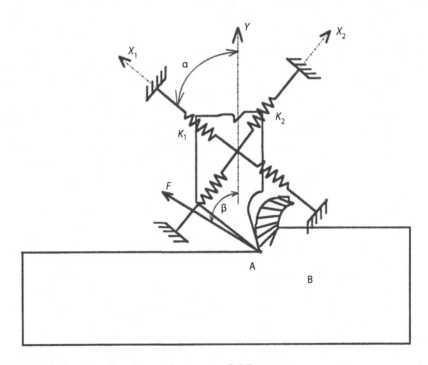

FIGURE 12.23 Mechanism of mode coupling in a two DOF system.

leave the cut. The width of cut for the threshold of stability depends directly on the difference between the two principal stiffness values, and chatter tends to occur when the two principal stiffnesses are close in magnitude. This type of nonregenerative chatter may occur in thread turning and shaping, operations during which the tool does not cut into the surface generated by the previous revolution of the workpiece. In addition, the regenerative mode may be hindered if the phase shift between two subsequent passes or teeth is constrained to zero ($\varepsilon = 0$), or even when milling with special tools such as those with alternating helix [105]. Nonregenerative chatter of this type cannot exist for a single-degree-of-freedom system [46]. The equations of motion of the system in directions X_1 and X_2 (as shown in Figure 12.23) are given by Equation 12.45. This type of vibration can be characterized by Equation 12.57 as explained in Refs. [106,107].

Mode coupling can be studied using the analysis for regenerative chatter with $\varepsilon = 0°$. The limiting width of the chip for nonregenerative instability is approximately twice that for regenerative chatter [46], so that

$$\left(b_{\lim}\right)_n \approx 2 \cdot \left(b_{\lim}\right)_r \tag{12.78}$$

Therefore, this type of instability will not occur for conditions where successive passes of the cutting tool overlap each other. The theory of mode coupling has been applied in boring and turning, operations in which two modes of the system tend to be close to each other in frequency.

12.7 CHATTER PREDICTION

As noted earlier, chatter is undesirable because it limits production rates and leads to poor machined surface finish and accelerated machine and cutting tool wear. Chatter detection and prediction are therefore of great practical importance in machine tool monitoring and dynamic performance testing [108]. Several methods have been proposed for chatter detection or prediction. Some of the parameters used for this purpose included (1) the cross-correlation coefficient between the dynamic thrust force and cutting forces [109], (2) the static deflection of a workpiece [110], (3) the RMS value of the vibration signal [111], (4) the peak of the power spectrum of the vibration signal [112], and (5) other parameters determined from spectral analysis of the force or vibration signal [36,56,113–118].

Chatter prediction actually means the prediction of the stability limits of the machining process. Obviously this information is important when determining optimal cutting conditions for programming automated machines. There are two prevailing techniques for this purpose: "limit chip analysis" and the "stability chart method."

Limit chip tests are conducted under conditions that have been standardized to some extent. The tests yield the limiting axial depth of cut at which chatter occurs. This method is the basis of ASME standard B5.54 for chatter limit evaluation of machining centers [38]. Care should be used in interpreting results of these tests because the critical depth of cut is influenced not only by the dynamic characteristics of the machine tool (as assumed in the standard), but also by the workpiece, fixture, toolholder, cutting tool characteristics, and cutting conditions. This method is therefore primarily useful for ranking the dynamic performance of various machines, rather than for predicting stability limits in specific applications.

Stability chart methods are based on a number of analysis techniques such as those discussed earlier in the regenerative chatter section. The most common approaches make use of experimentally measured TFs of the cutting process and the machine tool [50]. The limit of stability is computed using classical control theory. In a similar approach, Weck et al. [119] proposed using the measured stability lobes (shown in Figure 12.18) to select chatter-free spindle speeds. Since stability lobes are narrow at low cutting speeds, this approach is effective primarily in the higher cutting speed ranges. In another similar approach, the tooth passing frequency is equated to the chatter frequency, which is close to the natural frequency of the structure [120]. This method minimizes the phase between the inner and outer modulations and increases the chatter-free depth of cut. Below the limit of stability shown in Figure 12.18, the force and sound spectra are dominated by the tooth frequency, runout, harmonics, and noise. Above the limit of stability, a prominent chatter frequency appears; the spindle speed is selected so that the tooth passing frequency does not excite the structure at this frequency. This approach requires different stability charts for different cutting positions because the conditions under which chatter occurs vary with the cutting position. Commercial analysis systems are available for measuring the FRF of a system, generating a stability chart, and determining optimum process conditions [35]. A comprehensive review of the methods for turning, boring, drilling, milling, and grinding has been given by Altintas and Weck [121].

Another approach is to extract the vibration frequencies using non-contact tests during cutting such as sound recordings, displacements, forces, or accelerations during the cutting tests, and using an inverse machining dynamics model to extract the dynamic parameters of the structure. These parameters are then used in a machining dynamics model to determine the stability lobes [118,122,123]. This approach considers the changes in the dynamic parameters that may arise due to the spindle's rotation or preload and nonlinearities.

12.7.1 Experimental Machine Tool Vibration Analysis

The first task in analyzing machine tool vibration problems is to determine what type of vibration is present. Methods of distinguishing between forced and self-excited vibrations were discussed in Section 12.5. The machine structure has an infinite number of natural modes of vibration. The direction of relative motion between the cutting edge and the workpiece is determined by the principal vibration directions for each natural mode of vibration. The machine frame can be idealized as a sum of several elementary systems representing each natural frequency of the frame assuming that the natural frequencies are well separated (i.e., that no nearly equal pairs of natural frequencies are present). Each elementary system consists of an equivalent mass m, an equivalent linear spring with constant k, and an equivalent viscous damper with constant c. The stability of the cutting process is determined by the stability behavior of all active modes. It is generally necessary to consider only the least stable modes (the modes with the lowest natural frequencies) in most applications.

Due to the complexity of the phenomena involved, experimental methods are generally used in practice once machines have been built. Experimental analysis is used for two major purposes: (1) to measure the modes of vibration and mode shapes; and (2) to obtain a mathematical model of the structure [124]. Two types of tests are used for vibration measurement. In the first type (i), one parameter such as a force component, acceleration, or similar signal is measured during machine operation either while the spindle is rotating or while the cutting tool is engaged in the workpiece. In the second type (ii), the response output and an input are measured simultaneously. Type (i) tests are generally used to define a mathematical model of a process or for on-line machine control. These tests address only the second measurement objective (2) discussed earlier; a response measurement alone cannot be used to determine whether a particularly large response magnitude is due to strong excitation or to structural resonance. Type (ii) tests are commonly performed because the measurement of the input and output signals simultaneously can be used to completely define the system vibration characteristics. These tests can be used to address either of the measurement objectives (1) and (2) discussed before. In these tests, the structure or cutting tool is excited using an input signal, which is often stronger than the disturbances expected in normal service.

In order to determine the characteristics of the system from the measured response, it is necessary for the response to have the same frequency response characteristics (poles) as the system, or to correct the measured response to remove the poles and zeroes of the excitation [125]. In general, the measured response contains the poles of the system under study and of the input. Therefore, if the input force is not measured, it is not possible, without some prior knowledge about the input, to determine if the poles of the response are truly system characteristics. The poles of the response in the frequency range of interest are system characteristics if no poles or zeros exist in the force spectrum in this range. The input to the system should be measured to avoid this potential problem.

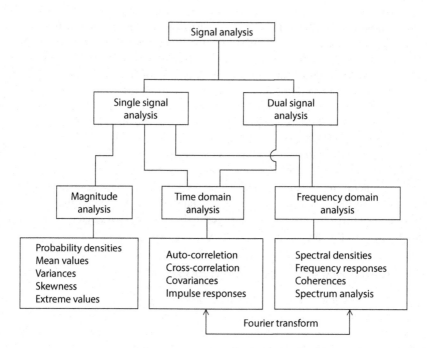

FIGURE 12.24 Signal analysis techniques.

The signal analysis techniques commonly used to quantify an experimentally measured signal are summarized in Figure 12.24. "Signal magnitude analysis" is used primarily for monitoring and provides information about a single measured signal. The other two techniques are used for both single and dual signal analysis.

The relationship between the input and output in frequency domain is described by the TF. The TF is an important dynamic characteristic of structures as discussed in previous sections. The measurement of the TF is also required for machine tool structure identification. The natural frequencies, damping ratios, and mode shapes can be determined from the TF.

12.7.2 MEASUREMENT OF TRANSFER FUNCTIONS

The problem of determining the *TF* of the cutting process or its inverse (generally called the dynamic cutting coefficient) is a central problem in machining dynamics theory and practice. The experimental methods developed to measure the TF of the cutting process can be broadly classified into two main groups: static methods and dynamic methods.

Static methods are based on measurements of static force components. Their analytical formulations were discussed in the previous section as the shear plane method and the incremental stiffness method. In static methods, the TF of the cutting process is determined indirectly because some system model is assumed in the analysis.

Dynamic methods measure the dynamic parameters directly and are subdivided into four classes: stiffness methods, dynamometer methods, frequency-response functions, and time series methods.

"Stiffness methods" measure the cutting process stiffness and damping based on the pulse response of the test rig measured both while cutting and while idling. The cutting process characteristics are calculated from differences in the measurements taken under both conditions [67,68]. This approach is accurate because the cutting process adds stiffness and damping to the machining system in addition to that available from the machine tool and toolholder structures.

"Dynamometer methods" allow measurement of the tool vibration and cutting forces during chatter conditions. In these methods, the cutting tool–toolholder system is excited using a vibration exciter while the toolholder is mounted on a dynamometer. The two main concerns with this method are: (1) the dynamic response of the dynamometer must be accounted for by using either compensating circuits [65] or a light weight dynamometer; (2) it is difficult to account for outer modulation and to combine it precisely in phase with the inner modulation signal measured nearly simultaneously on the tool [126–130].

In recent years, computerized "modal analysis" systems have become increasingly common [131–136]. These systems use the frequency-response function method to identify the natural frequencies, damping ratios, and mode shapes of the structure associated with specific modes of vibration. The methods used are collectively referred to as experimental modal analysis. The vibration measurements can be made while the machine tool, the cutting tool in the spindle, or the workpiece in the fixture are excited by known disturbances, which as noted earlier may be stronger than those expected in normal service. This method is known as the *Excitation Test Method* because the frequency-response functions are measured using excitation at single or multiple points [120,136]. It is important to include sufficient points in the test to completely describe all modes of interest. If the excitation points have not been chosen carefully or if enough response points are not measured, a particular mode may not be adequately represented. Frequency responses can be measured independently using single point excitation or simultaneously with multiple point excitations.

The measured vibration data of a machine tool (the TF data) are subjected to a range of curve-fitting procedures in an attempt to find the mathematical model, which provides the closest description of the observed response. After the modal frequencies are found, the machine may be excited at each such frequency to identify each mode shape. Based on this procedure, the weak mode of the machine can often be identified; this provides direction for improving the system design.

Single-point excitation is commonly used in modal analysis. The main purpose of the excitation test method is to exert an exciting force, which simulates the cutting force between the workpiece and tool at the same time the response is measured. This test should be performed under conditions as close to the practical working conditions as possible.

The measurement process involves three equally important components: (1) the correct support and excitation of the structure; (2) the correct measurement of the input force and displacement response; and (3) proper signal processing, which depends on the type of test used.

The structure is generally excited either by connecting it to a vibration generator or shaker, or by using some form of transient input [136]. The source signal can take several forms: harmonic sinusoidal (from an oscillator), random (from a noise generator), periodic (from a signal generator such as sinewave generator), or transient (from a pulse generating device or by applying an impact with a hammer). The magnitudes of the applied exciting force and the responses of the structure can be analyzed using either a spectrum (Fourier) analyzer or a frequency response analyzer. Responses measured may be accelerations, velocities, or displacements.

Types of devices used to excite the structure include mechanical exciters (out-of-balance rotating masses), electromagnetic exciters (often based on a moving coil in magnetic field) and electrohydraulic exciters, electrodynamic exciters, and impulse hammers [136,137]. The manner in which the exciter is attached to the structure should be chosen to ensure that the excitation and response act in the same direction.

The impulse response or hammer or impact test is a relatively simple means of exciting the structure. In this test, a small sensor is mounted at a location of interest and the structure is struck with an instrumented hammer. Both the input and response signals are measured. The hammer can be fitted with various tips and heads, which serve to extend the input frequency and force level ranges. The magnitude of the impact is determined primarily by the mass of the hammer head.

Microphones have been used successfully to measure the sound of the cutting process and detect chatter if the energy of the measured sound signal exceeds a certain threshold [138,139]. A Fourier analyzer is used to extract the component frequencies from the time-based sound signal. Software is available to identify the critical frequencies, such as tooth passing and runout frequencies, to recommend a new spindle speed for stable, chatter-free machining.

The time domain input and response signals are usually transformed into FRFs using a FFT analyzer [140]. All modal parameters are estimated from these FRFs, which in principle should provide an accurate representation of the actual structural response [133,141–143]. FFT analysis requires an analog to digital data conversion. The signals must be preprocessed (filtered) and digitized in a manner that prevents frequency aliasing; for this purpose, a sampling rate at least twice as large as the highest significant frequency component in the signal should be used. Spectrum analysis can be also performed to determine the frequency characteristics of the signal.

Modal parameters are extracted using a two-step process. In the first step, the TF is estimated using a FFT spectrum analyzer and the global parameters (natural frequencies and damping ratios) are identified. There are three common methods for parameters estimation: the simple quadrature peak pick, the resonant frequency detector fit, and the multi-function, multi-degree of freedom curve fit [125,136]. The magnitude of the TF can be determined by curve fitting based on the natural frequencies and damping ratios at each resonance. If the modes are not heavily coupled, a SDOF curve fit is acceptable for each mode. In many circumstances, however, modes are heavily coupled; under this condition, a MDOF curve fit must be used. The damping ratio is determined based on the methods discussed in Section 12.3.

In the second step the mode shapes are determined. The measurement of modal coefficients may be accomplished using one of two methods: quadrature peak pick or least-squares circle fit. Three main categories of system model can be identified: the spatial model (of mass, stiffness, and damping properties), the modal model (consisting of the natural frequencies and mode shapes), and the response model (consisting of a set of FRFs). The completion of the second step yields a minimum parameter linear mathematical model of the particular machine tool structure. The block diagram of a typical experimental setup is shown in Figure 12.25.

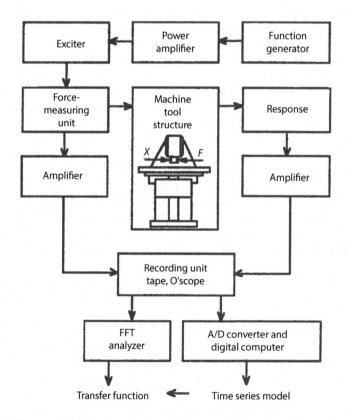

FIGURE 12.25 Basic setup for evaluating the dynamic characteristics of machine tools.

There are several methods for the identification of the stability lobes in the time domain [144,145]. Frequency domain methods have been used for stability analysis because more complex models and measurements can be adapted [146–149]. A comparison of the stability analysis in frequency and time domain was evaluated for milling process in Refs. [124,150].

Chatter and the cutting process TF can also be modeled using a "time series approach." Various approaches [151–156] can be used to model machine tool chatter as a self-excited random vibration based on time series methods. The time series, obtained by sampling the continuous vibration signal at uniform intervals, is a realistic modeling technique, which takes into account the unknown factors in chatter and its random nature. In the time series approach, the random vibration systems are excited by white noise when sampled at uniform intervals. The underlying structure is extracted first in the time domain by a parametric modeling procedure and sub-sequently transformed into frequency domain. The system is represented by linear stochastic systems and is represented by a discrete autoregressive moving average (ARMA) model of order (n,m), defined as

$$x_t - \phi_1 \cdot x_{t-1} - \phi_2 \cdot x_{t-2} - \cdots \phi_n \cdot x_{t-n} = a_t - \theta_1 \cdot a_{t-1} - \theta_2 \cdot a_{t-2} - \cdots \theta_m \cdot a_{t-m} \qquad (12.79)$$

where
x_t is the system's response series
a_t is a discrete white noise series
ϕ_i are the autoregressive parameters
θ_i are the moving average parameters

The estimated model will best approximate the structure in terms of the mean and covariance because it satisfies the conditions

$$E\left[a_t\right] = 0 \quad \text{and} \quad E\left[a_t \cdot a_{t-k}\right] = \delta_k \cdot \sigma^2 \qquad (12.80)$$

where
δ_k is the Kronecker delta function
E is the expected value operator

The strategy used to determine the adequate orders of the model n and m is discussed in Refs. [157,158].

Once the model is known, the modal natural frequencies and damping ratios can be determined by solving the characteristic equation [152]

$$\lambda^n - \sum_{j=1}^{n} \phi_j \cdot \lambda^{n-j} = 0 \qquad (12.81)$$

for the n roots λ. The ARMA model allows the definition of the TF in the general form

$$T(B) = \frac{1 - \sum_{k=1}^{n-1} \theta_k \cdot B^k}{1 - \sum_{k=1}^{n} \phi_k \cdot B^k} \qquad (12.82)$$

where B is the backshift operator, that is, $x_{t-k} = B^k \cdot x_t$.

The FRF is often measured with a hammer under static conditions. However, it has been shown that machine tool structure dynamics could change under rotating conditions [159]. The consideration of the spindle speed dependence on the FRF requires measurements of the FRF while the spindle is rotating, which is a challenging task as explained in Ref. [160]. Therefore, FEM analysis can be used to estimate the FRF in rotating conditions so that several aspects of high speeds are considered for the dynamic response [161]. The FEM models are calibrated using experimental dynamic parameters from another spindle and machine tool. FEA can be conveniently used for evaluating the stability diagrams of simple systems but becomes more involved for complex machine tool models having varying dynamic behavior. However, FEA can be helpful in analyzing the influences of the dynamic components on the machine side (spindle, tool/holder, tool geometry, etc.) on the process dynamics.

12.8 VIBRATION CONTROL

The dynamic behavior of a machining system can be improved by reducing the intensity of the sources of vibration for the machine tool, toolholder, and cutting tool. Several sources, primarily stiffness and damping, have a significant impact on forced and self-excited vibrations. As explained in Chapter 3, stiffness affects the accuracy of machine tools by reducing structural deformations due to cutting forces, while damping accelerates the decay of transient vibrations. The stability of the cutting process against vibration and chatter can be improved by several approaches, which can be categorized as methods for selecting cutting parameters in the stable zone within the stability lobe diagram and methods of avoiding chatter by changing the system behavior and modifying the stability of the system [36,121,124,162]:

1. Optimizing the design of the machine tool using both analytical and experimental methods to provide maximum static and dynamic stiffness
2. Selecting the best toolholder device for the particular tool or application, and reducing the tool overhang length
3. Selecting the proper bearing types, configurations, and installation geometry to provide maximum stiffness and damping
4. Isolating the system from vibration forces and using active or passive dynamic absorbers
5. Increasing the effective structural damping and using tuned mass vibration dampers
6. Selecting optimum cutting conditions, especially the spindle speed; using high-speed milling to machine between stability lobes. Some commercial software packages simplify testing and offer automatic predictions of the stability lobe diagram [138,139]. Reducing the depth of cut to perform machining under the stability limit
7. If the wavelength of chatter marks is small, increasing process damping by reducing the surface cutting speed
8. Selecting special cutting tool geometries; minimize the length of the cutting edge(s) in contact with the part; Reduce the nosing radius of the insert; Increasing the rake angle at the cutting edge; For milling, reducing the number of teeth on the cutter
9. Increasing the part stiffness
10. For forced vibration, decreasing cutting force, increasing part stiffness, changing the tooth passing frequency to be far away from resonance frequency of structure (see Section 12.5)
11. For self-excited vibration, decreasing doc and number of teeth in cut or changing the tooth passing frequency to match resonance frequency (see Section 12.6)

12.8.1 STIFFNESS IMPROVEMENT

The static stiffness, the ratio of the deflection to the applied static force at the point of application, can be measured for all three coordinate axes of the machine tool. The main contributors to deformation between tool and workpiece are the contact deformations in movable and stationary joints between components of the machine structure and fixture, the toolholder–spindle interface, and tool–toolholder interface. Individual components of the structure are generally designed for high stiffness and many make a comparatively insignificant contribution to the deflection.

The stiffness of the structure is determined primarily by the stiffness of the most flexible component in the loading path. This component should be reinforced to enhance stiffness. Therefore, the contribution of each machine component to the overall deflection should be estimated analytically or experimentally [50].

The effective static stiffness of a machine tool may vary within wide limits as discussed in Chapter 3. The overall stiffness can be improved by placing the tool and workpiece near the main column, by using rigid tools, toolholders, and clamps, by using rigid supports and clamps in the fixture, and by securely clamping all machine parts that do not move with respect to each other. Stiff foundations or well-damped mountings are required.

The bearing design (size, type, distance between bearings, and shaft type as explained in Chapter 3) has a strong influence on the static and dynamic behavior of machine tools. For example, a solid shaft increases the spindle stiffness, but its substantially greater mass reduces the spindle natural frequency, which is undesirable for high speed spindles.

Although a decrease in rigidity in a machine tool is generally undesirable, it may be tolerated when it leads to a desirable shift in natural frequencies (especially in high speed machining) or is accompanied by a large increase in damping or by a beneficial change in the ratio of stiffnesses along two orthogonal axes, which can result in improved nonregenerative chatter stability [3,50]. However, reducing the tool–workpiece compliance is not always possible in practice, and other approaches, discussed next, should be examined before productivity is sacrificed by reducing the depth of cut to ensure stable operation.

The static and dynamic analysis of the cutting tool and toolholder is important. For example, the analysis of an end mill or boring tool with a simple geometry can be determined using Equation 4.1 and the material in Example 4.6. An FEM of an end mill or boring tool can be used to perform the dynamic (modal) analysis to determine the natural frequencies of more complex tool bodies. Cutting tools should be optimized with the proper geometry, material, length-to-diameter ratio, and toolholder interface as explained in Chapters 4 and 5. The static and dynamic stiffness for several toolholder interfaces is discussed in Chapter 5. Therefore, several approaches could be used to evaluate and improve the vibration of end mill or boring tools by increasing rigidity.

12.8.2 ISOLATION

Vibration isolation is the reduction of vibration transmission from one structure to another via some elastic device; it is an important and common component of vibration control. Vibration isolation materials, such as rubber compression pads, metal springs, and inertia blocks, may be used [6,163]. Rubber is useful in both shear and compression; it is generally used to prevent transmission of vibrations in the 5–50 Hz frequency range. Metal springs are used for low frequencies (>1.5 Hz). Inertia blocks add substantial mass to a system, reducing the mounted natural frequency of the system and unwanted rocking motions, and minimizing alignment errors through an increase in inherent stiffness.

12.8.3 DAMPING AND DYNAMIC ABSORPTION

The overall damping capacity of a structure depends on the damping capacity of its individual components and more significantly on the damping associated with joints between components (i.e., slides and bolted joints [37]). The typical contributions of various components to the overall damping capacity of a machine tool are as follows: mechanical joints (log decrement 0.15), steel welded frames (structural damping 0.001), cast iron frames (0.004), material damping in polymer-concrete (0.02), and granite (0.015) [6,37,164].

The overall damping of machine tools varies, but the log decrement is usually between 0.15 and 0.3. While structural damping is significantly higher for frame components made of polymer-concrete composites or granite (as discussed in Chapter 3), the overall damping does not change significantly since the damping of even these materials is small compared to damping due to joints. A significant damping increase can be achieved by filling internal cavities of the frame components with special materials (i.e., replicated internal viscous dampers [164]) as discussed in Chapter 3. Resonant structural vibrations can be reduced by applying a dynamic absorber or layers of damping material on the surfaces of the structure [163].

The effect of increasing the damping ratio from 0.02 to 0.2 on the stability of a lathe having a resonant frequency of 60 Hz is shown in Figure 12.26 [165]; the asymptotic borderline of stability is increased from 0.04 to 0.48 mm, which corresponds to an increase of a factor of 12 in the effective cutting stiffness.

A dynamic absorber or tuned mass damper is an alternative form of vibration control. It consists of a secondary mass attached to the primary vibrating component via a spring, which can be either damped or undamped. This secondary mass oscillates out of phase with the main mass and applies an inertial force (via the spring), which opposes the main mass. For maximum effectiveness, the natural frequency of the vibration absorber is tuned to match the frequency of the exciting force. Auxiliary mass dampers can be used on machine columns, spindles, and rams.

Tuned dynamic vibration absorbers have been used with considerable success in milling and boring applications [37,166–168]. A tunable tool provides a controlled means of adjusting the dynamic characteristics of the tool in a particular frequency range. A tuning system minimizes the broad-band dynamic response of the tool without requiring cutting tests or trial-and-error tuning. A very common vibration damper used in boring bars consists of an inertial weight or a spring-mounted lead slug fitted into a hole bored into the end of boring bar; bars so equipped are often called antivibration boring bars. The weight helps damp bar motion and prevents chatter. The chatter resistance of boring bars can also

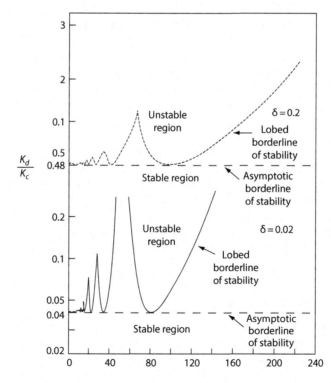

FIGURE 12.26 Effect of the damping ratio on the stability chart for a typical SDOF structure with a natural frequency of 60. (From Ling, C.C. et al., *ASME Manuf. Sci. Eng.*, MED-Vol. 2-1/MH-Vol. 3-1, 149, 1995.)

be increased by using different materials in the bar structure as discussed in Chapter 3 [169]. Impact dampers can be also installed in the toolholder, spindle, or ram to absorb vibration energy.

Active control of structures can be also used to suppress vibration and chatter in machining. Actively controlled dynamic absorbers use sensors and force actuators together in a closed-loop control system to alter the dynamic characteristics of a structure so that it possesses greater damping and stiffness characteristics [170,171]. Accurate system identification (both in terms of sufficient model order and parameter accuracy), hybrid high-speed control, and proper power force actuators are essential elements of any successful structural or tool vibration control system. These methods are discussed in more detail in the next section.

12.8.4 TOOL DESIGN

Reductions of both forced and self-excited vibrations for multiple cutting edges tool structures (such as milling cutters, end-mills, and reamers) have been achieved by employing non-standard cutting tools such as unequally spaced cutting edges (nonuniform tooth pitch), variable axial rake, variable helix cutting edges (alternating helix on adjacent flutes), lip height error, or serrated and undulated edges as discussed in Chapter 4 [46,171–175]. Nonuniform tooth pitch cutters are often called white nose cutters.

All these tool design variations increase stability by reducing the effective tooth passing frequency and disturbing the regeneration of surface waviness (the phase between the inner and outer modulations). Their effects are most pronounced over particular ranges of cutting speeds which depend on the cutter geometry (L/D ratio, lead angle, helix angle, etc.), the configuration of the particular part feature to be machined, and the cutter path (up milling vs. down milling, end-mill diameter vs. diameter of circular interpolation, etc.). Alternating the helix increases pitch variations along the axial doc, improving both the dynamic performance and operative speed range of the cutter. An optimal distribution of spacings or helix angle variations between the teeth can be found, but the non-uniform pitch or helix cutters can be used within a limited speed range constrained by the dominant structural mode. Theoretical analysis of stability for milling with such special cutters is described in Refs. [46,49].

Sharp tools are more likely to chatter than slightly blunted tools. Therefore, a lightly honed cutting edge can be used to avoid chatter. Negative rakes and small clearance angles minimize chatter occurrence due to process damping. It has been demonstrated that reduction in the relief angle and an increase in flank wear of the cutting edge increases process damping [177–181]. In general, it is important to use the smallest possible tool nose radius that gives acceptable tool life, because a small nose radius can alleviate the regenerative effect [182]. Finally, combination tools which generate complex surfaces should incorporate design features which inhibit chatter. For example, combined reaming, countering, and chamfering tools, require a cylindrical land not only along the reaming and counterboring margin sections, but also a narrow land along the chamfer section to prevent chatter on the chamfered surface, which could occur with a conventional sharp chamfering design.

12.8.5 VARIATION OF PROCESS PARAMETERS

Cutting conditions, especially the cutting speed, directly affect chatter generation. At a given spindle speed, the widths and depths of cut are often limited by the chatter threshold. As indicated in the stability lobe diagram (Figure 12.18) a small increase or decrease in speed may stabilize the cutting process. Small changes in the cutting speed are particularly effective in increasing stability in milling operations. In addition, speed and feed affect process damping [183–185].

Automatic regulation of the spindle speed for stable cutting can be used in CNC machine tools. If it is known that there is a gap in the stability lobe diagram, then that frequency is used while increasing the doc until the maximum machine/spindle power is reached [36,119,120,185–187]. This approach is currently only applicable to uniform pitch cutters. Chatter suppression systems employing this principle are discussed in the next section. Case studies indicate that the use of spindle speed variations provides more flexibility than the use of variable pitch cutters [188].

12.9   ACTIVE VIBRATION CONTROL

Active machine tool control systems use feedback from sensors to regulate the tool position and cutting conditions (speed and feed) to prevent or suppress chatter. Several quantities can be measured for this purpose, including the relative displacement between tool and workpiece, the cutting force, the tool acceleration, and the sound spectrum of the process.

The first active chatter control system was built by Comstock [189], who used the relative displacement between tool and workpiece, measured by an inductive transducer, as the control variable. A block diagram of the control scheme is shown in Figure 12.27. The noise signals F_e and F_c act on the machine structure to produce motion y_m. y_i is the displacement of workpiece relative to the cutting tool. The term "chase control" is used for this control scheme because a variation in the relative tool and workpiece position gives rise to an error, which causes the controller to supply a corrective motion y_t to the cutting tool; in effect, the tool attempts to catch or "chase" the workpiece and cancel the error. It is assumed that the displacement transducer measures the tool to workpiece centerline distance rather than the tool to workpiece surface distance. All except gross surface irregularities can be eliminated through use of a proximity gage with a sensing head whose dimensions are large compared to the spacing of the irregularities. Signals corresponding to gross irregularities and workpiece out of roundness can be eliminated through use of a high-pass filter. By removing both surface ripple and out of roundness signals from the control signal, the controller will respond only to workpiece motion. The stability criterion for the controlled system is given by Equation 12.57. Comparison of this stability criterion to that of the uncontrolled process indicates that the increase in stability achieved is almost proportional to increase of the controller gain, as would be expected for an ideal controller. The case of a non-ideal controller is discussed in Ref. [51].

The basic advantage of chase mode control is that it is capable of attenuating noise and improving stability at the same time. However, the instrumentation required may present problems. A proximity gage, used to measure tool to workpiece relative motion, may interfere with the cutting system (chips and tooling) or become contaminated by cutting fluids and chips. Additionally, a proximity gage actually measures the relative distance with respect to the workpiece surface rather than the actual tool to workpiece centerline distance; signal components due to surface irregularities, unless eliminated by filtering, will cause the controlled tool to follow the irregular surface rather than maintain a constant position with respect to the workpiece centerline. In general, the lower the order of the controller, the more effective filtering it will provide. In the case of a first-order controller, the corner frequency need be no higher than the frequency of the machine structural mode responding to the noise.

A second active control system was built by Nachtigal and Cook [190]. The block diagram of this feed-forward control scheme, which is called "predictive control," is shown in Figure 12.28. The cutting noise F_c is the monitored parameter. The principal advantage of a predictive mode control system is the relative simplicity of the instrumentation, since cutting force gaging is not physically dependent on the geometry of the workpiece. However, there are two basic disadvantages to this

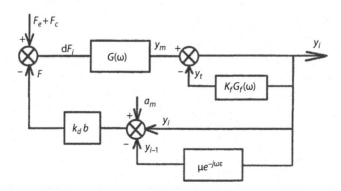

FIGURE 12.27   Block diagram of the chase mode control system.

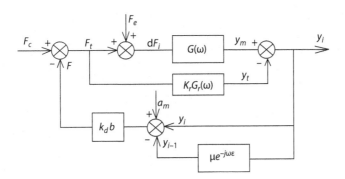

FIGURE 12.28   Block diagram of the predictive control system.

scheme: (1) it is not effective for controlling the external noise force F_e, and (2) in order to reduce the effects of the cutting noise F_c, the controller must accurately simulate the dynamics of the machine-workpiece structure. This is difficult because the structure will generally have several prominent vibration modes and, moreover, its dynamic characteristics change as a function of the position of the moving components of the machine tool. A predictive control scheme is particularly sensitive to changes in the dynamic characteristics of the machine tool because it is not based on a feedback loop.

The chase control scheme was reported to increase stability by a factor of 35, while the predictive control scheme was reported to increase it by a factor of 15. In the chase control scheme, stability basically depends on the dynamic order of the controller and, at least theoretically, a factor of 1000 increase is easily attainable. In contrast, the predictive control scheme cannot achieve an increase in stability of more than a factor 20 even if an adaptive controller is used. Another advantage of the chase control scheme over the predictive control scheme is its effectiveness in attenuating both F_c and F_e noise effects.

An advantage of feed-forward (predictive) control schemes is that the tool motion begins essentially at the same time as the workpiece motion. Thus, theoretically, one should be able to reduce relative motion to zero. However, since it is an open loop method, the tool moves without prior knowledge of workpiece motion, introducing an additional source of uncertainty. A disadvantage of chase control schemes are that they require high controller gain in order to perform well, otherwise the control signal always lags the noise signal; the use of a high gain factor may cause stability problems in the control loop itself. The stability condition requires that $K_r G_r(\omega) < G(\omega)$.

Second-generation active machine tool control systems were developed by Nachtigal and coworkers [191–193]. The primary problem they confronted in all cases was the sensitivity of the controller to machine dynamic variations. In one attempt to overcome this limitation, the wideband active control scheme [192] was designed; it was actually the same as Nachtigal's original system but incorporated an additional lead compensator. The maximum increases in stability for this control scheme were less than those reported for the original system due to the effects of the lead function. Another attempt was based on the fact that changing dynamics can be efficiently accounted for using an adaptive control scheme [193]. This system adaptively varied the controller natural frequency to maintain a predetermined phase relationship between the cutting force and tool servo responses. This feature made the adaptive control scheme capable of maintaining a greater machining rate even with relatively large variations in the machine tool natural frequency. However, the increase in stability was not as large as in Nachtigal's original system, although it was greater than that of the wideband controller scheme.

Third-generation active control schemes were multivariable control schemes designed by Mitchell and Harrison [194]. They used the best features of each type of control scheme, and, simultaneously, eliminated some of their drawbacks. The predictive controller was an open loop, feed forward system, whereas the chase controller was a closed loop, feedback system; the multivariable controller was basically a closed loop, feed forward system. A block diagram of the control system, which employs a hardware observer, is shown in Figure 12.29. Observer theory was applied so

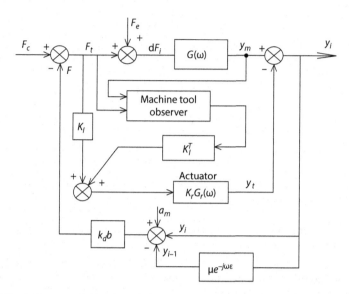

FIGURE 12.29 Block diagram of the controlled system employing a hardware observer.

that the active control scheme could effectively reduce chatter as well as attenuate noise effects. The state estimator or observer can be effective because it is not easy to determine the machine tool dynamic characteristics, which depend on the positions of movable elements, and because it is difficult to measure the necessary states of the rotating workpiece, such as workpiece center line position, velocity, or acceleration. An observer is actually a closed loop system itself which uses the same input signal as the observed system and compares its output with the observed system output. The observer tends to force the tool to follow the workpiece motion.

The actuator-observer control scheme takes advantage of the predictive properties of feed forward systems. Additionally, since it is a closed loop system, it tends to be insensitive to parameter variations. This scheme requires an actuator faster than the mode it has to control. The predictive type control requires an actuator of about the same speed as the controllable mode, whereas the chase controller can be slower than the mode it is controlling.

A number of other techniques have been also proposed with varying degrees of success. One common approach employs active suppression of chatter by programmed variation of the spindle speed [195–207]. Jemielniak and Widota [208] concluded, based on a simplified analysis of dynamic orthogonal cutting in which the time variation of the dynamic system within each spindle revolution was neglected, that the speed must be continuously varied in order to prevent the dynamic cutting process from locking itself into the most favorable phase for chatter. The continuously varying spindle speed damps the chatter vibration amplitudes, which in turn varies the phase shift between the inner and outer modulations on the cut surface [198]. Speed variations during machining disturb the regeneration of waviness responsible for chatter. Sinusoidal speed variation was found to be the most favorable form of oscillation for the spindle drive system for improving chatter resistance in milling [199]. Methods employing speed oscillation require a high performance, high torque delivery spindle drive system, which can deliver speed oscillations with a wide range of amplitudes and frequencies over a short time interval.

Altintas and Chan [188] developed an in-process chatter detection system, which used the measured sound spectrum as the control variable and suppressed chatter through spindle speed variation. A remotely positioned microphone was used to measure the sound pressure during end milling. The in-process chatter detection and avoidance algorithm is outlined in Figure 12.30. The machine operator inputs the chatter threshold factor for the sound spectrum, the nominal speed, and the amplitude and frequency of speed variation. The sound signals are sampled at a frequency at least five times higher than the possible chatter frequency. The search is carried out within a possible chatter

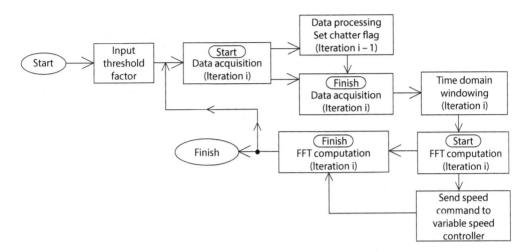

FIGURE 12.30 Chatter avoidance algorithm. (From Altintas, Y., and Chan, P.K., *Int. J. Mach. Tools Manuf.*, 32, 329, 1992.)

frequency range, and precautions are taken to avoid the tooth passing frequency harmonics. In addition, the average magnitude in a specified low frequency (chatter free) range is computed. Chatter is assumed to be present when the maximum magnitude exceeds the low frequency spectrum average by a factor greater than the threshold factor. If chatter is detected, speed variation signals are sent to the spindle speed controller; otherwise a constant speed signal is sent. The spindle speed is varied in a sinusoidal manner to disrupt the regenerative effect of chatter [209]. The difficulty of effectively varying the spindle frequency at high frequencies limits this technique's effectiveness.

In contrast to systems that vary the spindle speed during machining, Weck et al. [119] proposed the use of stability lobes (as shown in Figure 12.18) for selecting chatter free spindle speeds. Chatter was detected by comparing the maximum spindle torque with an allowable threshold. The spindle speed was varied slowly until one of the stable lobes was penetrated. Since stability lobes are narrow at low cutting speeds, this approach is most effective at higher cutting speeds.

A similar approach equates the tooth passing frequency to the chatter frequency, which is close to the natural frequency of the structure [120]. This approach is used in the CRAC or Harmonizer system developed by Tlusty et al. [210]. This system selects spindle speeds for which favorable phasing prevents chatter [56,186,187,211–213] and allows the axial depth of cut to be increased to maximize the metal removal rate. The Harmonizer system operates in such a way that if a frequency is detected in a audio spectrum, which is not the tooth passing frequency or a harmonic of the tooth passing frequency, then a new spindle speed is selected that would make the tooth passing frequency equal to the detected chatter frequency. Furthermore, the system is interfaced with the machine controller, which commands a spindle speed change. A few trial speeds may be required before an acceptable speed is found.

The principle of the Harmonizer approach is illustrated in Figures 12.31 and 12.32. The cut using the speed and doc indicated by *a* is unstable because it falls inside the stability lobes shown in Figure 12.31. The corresponding chatter frequency that would be detected is also marked a in the chatter frequency diagram. The Harmonizer system regulates the spindle speed and proportionally the feed (to keep the chip load constant) so that the second harmonic of the tooth passing frequency is equal to the detected chatter frequency; this corresponds to a horizontal move from *a* to *c* in the figure. Similarly, if the cut *b* was initially selected, the Harmonizer system will move the process to the cut marked *c*. This is the first function of the Harmonizer system. The second function is to maximize the MRR by increasing the axial doc in the stable zone as indicated in Figure 12.32. The two plots completely describe the spindle speed-chatter frequency relationship typical for most tools. Therefore, the Harmonizer system commands a speed change to the cut marked *b* and

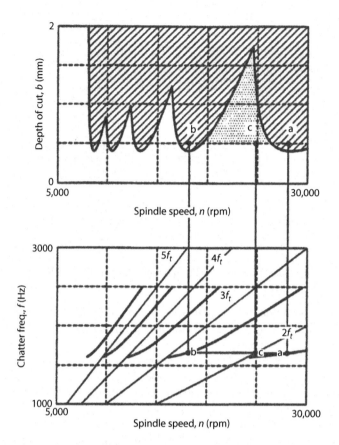

FIGURE 12.31 Stability lobe diagram and chatter frequency diagram showing the basic operation of the CRAC system. (From Winfough, W.R., and Smith, S., *Trans. NAMRI/SME*, 23, 163, 1995.)

then commands an increase in the axial doc until the cut is unstable as indicated by the cut marked *c*. At this point another spindle speed change is used to move to the cut marked *d*, which is stable. The doc is further increased until chatter occurs for the cut is indicated by *f*. Stable spindle speeds are searched at either sides if *f* (to *g* and *h*) but are not found as expected. Therefore, the conditions of the last stable cut (marked *e*) are recorded. The MRR at the conditions of cut e (obtained by reducing the spindle speed by 20% while increasing the doc by 20%) is greater than that using conditions of cut *a* or *b*. There are several obstacles to overcome in order to ensure effective and reliable audio detection. Audio detection has been found very effective when a frequency-based approach is used instead of a typical RMS or other time-based approaches [36,41].

A final group of chatter suppression schemes employ the "forecasting control" (FC) strategy. In this strategy, the machining process is considered to be unknown, and the onset of chatter is predicted by fitting autoregressive models to suitable forms of the acceleration signal. The models are then used as a basis for forecasting corrective signals [189]; when the process approaches instability, appropriate corrective actions can be taken before the problems actually occur. In this control strategy the dynamic characteristics of the machining process are determined on-line. Prior knowledge of the system is helpful but not required. The most significant advantage of using on-line identification of the process dynamics is that the estimated model can vary with time, so that the time varying characteristics of the machining process are not ignored under this scheme. An autoregressive moving average (ARMA) or AR stochastic model is generally used to represent the system [152]. This model is adequate, in a statistical sense, and is justified physically by the fact that, as long as the process is stable, it can be efficiently described by a linear differential equation with constant coefficients.

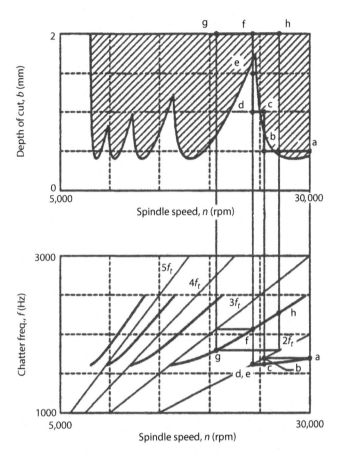

FIGURE 12.32 Stability lobe diagram and chatter frequency diagram showing an algorithm for optimizing the MRR. (From Winfough, W.R., and Smith, S., *Trans. NAMRI/SME*, 23, 163, 1995.)

In this case, the corresponding discrete-time system is a difference equation in the form of a proper ARMA model [157]. After the process dynamics have been determined some quantity that is important for the stability of the process is computed. The values of this quantity create a new time history; a new ARMA or AR model is used to describe the updated time series. Using the forecasting idea, as explained in [157], the values of this series are forecast and tendencies toward instability are detected. If such tendencies are present, corrective actions are taken to ensure stability.

A block diagram representation of the FC strategy is shown in Figure 12.33. It consists of five main functional blocks: the machining process, process identification, forecasting and parameter updating, decision and modification, and drive blocks. The decision and modification block decides if there is any need to issue new speed or feed commands and, if corrective actions are necessary, what their magnitudes should be. The FC strategy has been applied successfully in turning operations [214,215]. A nonlinear mathematical model for chatter has been also proposed in Ref. [216].

Recently, a commercial solution for chatter detection and spindle speed adjustment called Machining Navi has been developed for milling [217]. The chatter vibration is measured by either a piezoelectric accelerometer attached in the spindle or a microphone within the working zone. The system provides a stable spindle speed that the operator can use for chatter avoidance. The main drawback in this case, as in most cases aforementioned, has been the speed of system reaction as soon as chatter starts. This is not probably a significant concern for roughing operations, but it might be a problem in finish operations. Therefore, the user should apply: (1) predictive analytical methods to identify an optimum regime in the stability diagram so that active vibration control will correct

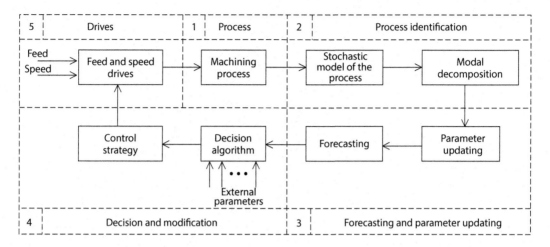

FIGURE 12.33 Block diagram of the forecasting control strategy.

light chatter; and (2) a chatter identification scheme in the active control system fast enough to identify chatter as it arises, before the vibration is fully developed in order to maintain part quality during the correction stage. In this case, the machine tool will stop to change the cutting conditions including the tool path on-line, if necessary, to avoid chatter.

12.10 EXAMPLES

Example 12.1 A facemilling operation is to be designed with maximum spindle speed of 5000 rpm. 50–100 mm diameter cutters can be used and the number of inserts varies between 4 and 8 (i.e., the 50 mm has 4 inserts, 75 mm has 5 or 6 inserts and the 100 mm has 7–8 inserts). It is assumed that self-excited chatter is not present. As mention in Section 12.5, the forcing frequency of a linear forced vibration in a machining process can be changed by the operator through the machining parameters, which can be used to improve the design of a process. The characteristics of the part were measured (using the modal or "tap" test as explained in Section 12.7) and the FRF is given in Figure 12.34. Assume the cutting tool is much stiffer than the part.

    a. Select the RPM that will place the fundamental forcing frequency (FFF) at 245 Hz.
    b. Determine the harmonics of the FFF
    c. For quality reasons, would you recommend using the maximum spindle rpm or the RPM found in part A for the various cutters with different numbers of teeth?

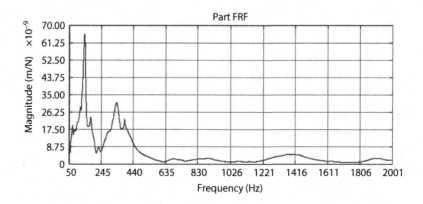

FIGURE 12.34 FRF of part in *X*-direction. (Courtesy of D3 Vibrations Inc., Royal Oak, MI.)

Solution:

a. The spindle RPM can be selected using Equation 12.40 so that the FFF is 245 Hz. The FFF of 245 Hz was selected to lie between the two peaks (modes) at $f_1 = 140$ Hz and $f_2 = 340$ Hz in the FRF to avoid excessive vibration. The corresponding RPM for the different number of teeth in the cutter is given in Table 12.2.

b. The first five harmonics of the FFF, obtained by multiplying the FFF by 2, 3, 4, 5, and 6, are given in Table 12.3.

c. The tooth passing frequencies at 5000 rpm for the various numbers of teeth, computed from Equation 12.40, are given in Table 12.4. Running the milling cutters with 4 and 5 teeth at this speed is not recommended because the corresponding tooth passing frequencies are near the second natural frequency of 340 Hz as shown in Figure 12.35. Running the cutters with 6, 7, or 8 teeth at this speed is recommended because the tooth passing frequencies are far from the second natural frequency.

TABLE 12.2
Tooling RPM for Cutters with Different Number of Teeth at an FFF of 245 Hz

Number of Teeth	RPM for 245 Hz
4	3675
5	2940
6	2450
7	2100
8	1838

TABLE 12.3
First Five Harmonics of the FFF in Example 12.1

FFF	245
First harmonic	490
Second harmonic	735
Third harmonic	980
Fourth harmonic	1225
Fifth harmonic	1470
Sixth harmonic	1715

TABLE 12.4
FFF for Tools with Different Numbers of Teeth in Example 12.1

Number of Teeth	FFF at 5000 rpm	Recommend
4	333	No
5	417	No
6	500	Yes
7	583	Yes
8	667	Yes

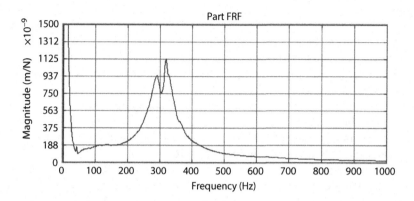

FIGURE 12.35 FRF of part in *X*-direction. (Courtesy of D3 Vibrations Inc., Royal Oak, MI.)

Example 12.2 The characteristics of a part were measured using a modal test; and the FRF is given in Figure 12.35. A combination tool consisting of a two-flute twist drill and a spotfacer was used in a machining center. A spindle speed that is a multiple of 500 rpm between 4,000 and 10,000 rpm must be selected. Assume self-excited vibration is not present.

a. Select the RPM range where the FFF, first and second harmonic fall at frequencies at which the compliance is below 375×10^{-9} m/N on the FRF.
b. If the force magnitude at the FFF is 100 N, at 1st harmonic is 40 N, and at the second harmonic is 15 N, estimate the displacement at each frequency and sum them for the following spindle speeds: 4000, 6000, and 8000 rpm.

Solution:

a. The tooth passing frequencies or fundamental forcing frequency (FFF), calculated from Equation 12.40 for a range of spindle RPMs at 500 rpm increments, are given in Table 12.5. The corresponding first and second harmonic frequencies for the FFF are also given in the table. The objective is to evaluate and reject the cutting speeds generating FFF and harmonics

TABLE 12.5

Calculation of the FFF and the First and Second Harmonics of the Tooling System in Example 12.2

Spindle RPM	FFF (Hz)	First Harmonic (Hz)	Second Harmonic (Hz)
4,000	133	**267**	400
4,500	150	**300**	450
5,000	167	**333**	500
5,500	183	**367**	550
6,000	200	400	600
6,500	217	433	650
7,000	233	467	700
7,500	**250**	500	750
8,000	**267**	533	800
8,500	**283**	567	850
9,000	**300**	600	900
9,500	**317**	633	950
10,000	**333**	667	1000

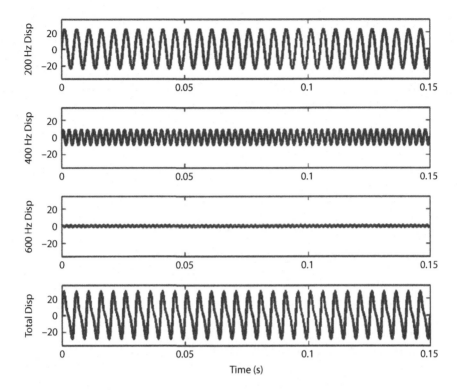

FIGURE 12.36 Time domain for the 6000 rpm displacement (μm) for the FRF of part in Figure 12.35. (Courtesy of D3 Vibrations Inc.)

frequencies within the bell curve of the natural frequency with a magnitude greater than 375 × 10^{-9} m/N in the FRF shown in Figure 12.35. The frequencies to avoid are between 240 and 375 Hz. Therefore, the speeds resulting in FFFs and their harmonics in the range of 240–375 Hz should be avoided. Any RPM between 6000 and 7000 rpm will reduce the amount of forced vibration. Therefore, a spindle speed between 6000 and 7000 rpm should be used.

b. The displacement at the three cutting speeds is estimated for each frequency (FFF, first harmonic, second harmonic) as illustrated in Table 12.6 for the three spindle speeds of 4000, 6000, and 8000 rpm. For example, the three tooth passing frequencies at 4000 rpm are FFF = 133 Hz, first harmonic = 267 Hz, and second harmonic = 400 Hz, and the corresponding compliances are estimated from Figure 12.36 and given in the table. The displacements are calculated based on the applied force and the compliance at a particular frequency: 0.00019 mm/N × 100 N = 0.019 mm, 0.00055 mm/N × 40 N = 0.022 mm, and 0.0002 mm/N × 15 N = 0.003 mm, respectively at FFF, first harmonic, and second

TABLE 12.6

Calculation of the Displacement FFF at Various RPMs for the Tooling System in Figure 12.35

Frequency (Hz)	133	200	267	400	533	600	800	Estimated Total Displacement (μm)
Compliance (m/N × 10^{-9})	190	230	560	230	90	40	20	
Displacement (μm) at 4000 rpm	19		22	3				44
Displacement (μm) at 6000 rpm		23		9		1		33
Displacement (μm) at 8000 rpm			56		4		0	60

harmonic. The total displacement of the cutter at 4000 rpm is the sum of the three above displacements, or 0.044 mm. The displacement at 6000 rpm is the lowest, 0.033 mm, as expected from the analysis in part A. Since the phasing of the frequencies is not considered, the sum value may be different than estimated; however this can still be a useful qualitative analysis. A time domain plot for the 6000 rpm displacement is shown in Figure 12.36 for the FFF, 1st harmonic, 2nd harmonic, and total.

Example 12.3 The FRF for 17 points in the Z-direction on the pump face of a transmission valve body structure was measured using a modal "tap" test. Compliance in the Z-direction controls the flatness of a critical surface on this part. The FRF is given in Figure 12.37. Select a spindle speed (RPM) for a 5-tooth milling cutter so the FFF and the first two harmonics don't fall on a resonant peak. The wall thickness of the pump face is narrow, so that only 1 tooth is generally engaged (i.e., this is a low immersion cut).

Solution: The high compliance frequencies are 585, 707, 1022, 1420, 1595, 2325, and 2968 Hz as shown in Figure 12.37. Harmonics are caused by the tooth impacts, and are found at integer multiples of fundamental forcing frequency (FFF). For example, the 585 Hz FFF for the surface has harmonics at 1170, 1755, etc., as shown in Figure 12.38. When trying to reduce forced vibrations, it is most important to have FFF and lower harmonics away from high response areas. The spindle speeds for the 5-tooth cutter having tooth passing frequencies at or near the high compliant frequencies of the surface can be identified using Equation 12.40. For example, 7000 rpm will result in a FFF of 583 Hz for the tool (7000 rpm × 5/60 = 583 Hz). If the FFF is 850 Hz, the first and second harmonics are 1700 and 2550 Hz, respectively. These three frequencies are plotted on the FRF plot in Figure 12.37 (illustrated by the marks A, B, and C). This figure shows that the force vibration will be avoided at these frequencies because they do not fall near FRF peaks. The spindle speed corresponding to these frequencies is 10,200 rpm.

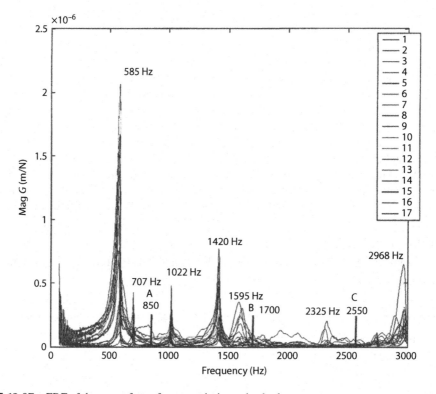

FIGURE 12.37 FRF of the pump face of a transmission valve body.

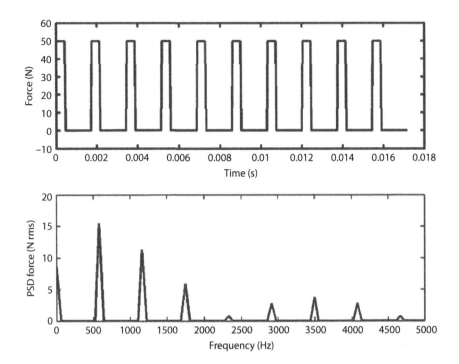

FIGURE 12.38 Simplified FFF of the face milling cutter force at 7000 rpm with five teeth used for the pump face of a transmission valve body.

Example 12.4 A face milling cutter with 30 inserts running at 1997 rpm introduces vibration into a machine tool. A microphone was used to measure time and frequency domain signals as shown in Figure 12.39. How can the problem be corrected?

Solution: The audio signal in the figure indicates that the third harmonic near 4000 Hz must be near a resonant frequency, leading to excessive forced vibration. The displacement at these frequencies is much higher than at the FFF, first harmonic, and second harmonic, although there is a higher force at 1000, 2000, and 3000 Hz. Note that the harmonic at 4000 Hz is four times the tooth impact (or FFF) of 998.5 Hz. Under normal vibration conditions the higher harmonic would be a small fraction in amplitude compared to the tooth impact frequency. Hence, it most likely this milling cut exhibits resonance at 3994 Hz. This indicates that the cutting speed (RPM) should be changed so that no harmonic falls near 4000 Hz. It is suggested to increase or reduce the spindle speed by between 5% and 20% [34] as long as instability does not develop at the new spindle speed. Therefore, if the spindle speed is decreased by 10% to 1800 rpm, the resonance vibration at this mode will be eliminated. However, as discussed in Section 12.5, it is sometimes difficult to identify the best speed if the system has several modes close to each other, since a spindle change may move another nearby harmonic into resonance.

Example 12.5 A full slot is milled in aluminum 7075 using a two-flute end-mill. The diameter of the end-mill is 20 mm and its length extending from the toolholder is 140 mm ($L/D = 7$). The maximum spindle speed of the CNC machine is 12,000 rpm. A stability chart for the tool mounted in the spindle is shown in Figure 12.40 for a full slot.

a. Select the RPM that provides the highest stable axial depth of cut (ADOC). What is this maximum doc?

b. What RPM would you select if the engineering process required you to have a 500 rpm cushion on either side of your rpm, and a DOC a minimum of 0.5 mm below the stability line?

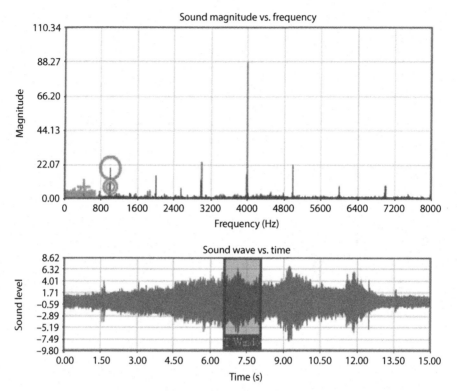

FIGURE 12.39 Audio signal of a vibration case using a 30 inserts PCD face mill in Example 12.4. (Harmonizer® Courtesy of Manufacturing Laboratories Inc., Las Vegas, Nevada.)

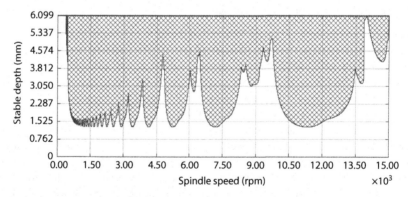

FIGURE 12.40 Stability chart for a two-flute ($L/D = 7$) end-milling a full slot in aluminum part in Example 12.5. (Courtesy of D3 Vibrations Inc., Royal Oak, MI.)

    c. If the spindle speed could be increased to 15,000 rpm maximum, could improvements be made in the process? What improvements?

    d. What would happen if the tool had three flutes?

Solution: This example shows how stability charts can be used to design a milling process. The Stability chart of the cutter for full-immersion slotting is shown in Figure 12.40. It was measured using a modal test on the end-mill clamped in the spindle through a toolholder. The modal parameters of the end-mill in the spindle converted to the displacement domain were identified using modal analysis software and are given in Table 12.7. The parameters are identical in both x and y directions

TABLE 12.7

Identified Modal Parameters of the End-Mill in the Spindle in Either *X* or *Y* Directions

Mode	*f* (Hz)	*K* (N/m)	ζ
1	424	1.26 E7	0.0205
2	492	2.36 E7	0.0268
3	641	2.38 E7	0.0233
4	882	8.33 E7	0.0188

because the FRF's in these directions are similar. This stability chart provides the minimum stable depth of cut as a function of spindle RPM. The shaded section inside the lobes is the unstable region. Note that there are hardly any stable pockets at speeds below 4000 rpm.

a. From the chart, the minimum stable depth of cut is approximately 1.2 mm. The most stable spindle speed below 12,000 rpm is approximately 9,750 rpm, where the stable depth of cut reaches a maximum of approximately 5 mm. This maximum doc will result in the maximum material removal rate without chatter.

b. The same stability pocket should be used, but the stable ADOC must be reduced so that the required conditions of a 500 rpm cushion on either side of the selected rpm, and an ADOC a minimum of 0.5 mm below the stability line are met. These conditions are met using a spindle speed of 9250 rpm and an ADOC of 2.5 mm.

c. The ideal stability pocket for spindle speed lies at about 14,000 rpm, with a stable ADOC of 6.1 mm. A 15,000 rpm maximum spindle speed would permit use of this more robust region, and provide a throughput improvement because both the speed and ADOC are higher than those discussed in part B. Selecting 2.5 mm ADOC at 15,000 rpm provides a 62% productivity gain and should yield similar quality; moreover, the process would be more robust because the width of the pocket between lobes is wider.

d. Increasing the number of flutes moves the stability lobes down and left in Figure 12.41. This is seen by comparing the stability chart for the 3-flute cutter, shown in Figure 12.41, to that of the 2-flute cutter in Figure 12.40. The achievable ADOC is not as high for the three-flute end-milling tool in this instance. There is a large robust region between 9,000 and 14,000 rpm; however, the ADOC is not as great as the two-flute. In this case, even if feed/tooth is held constant, the two-flute cutter would yield a higher MRR.

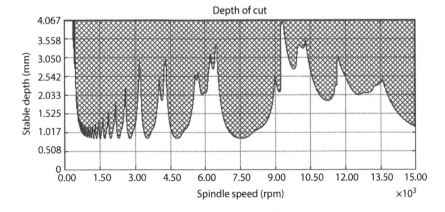

FIGURE 12.41 Stability chart for a three-flute (*L/D* = 7) end-milling a full slot in aluminum part in Example 12.5. (Courtesy of D3 Vibrations Inc., Royal Oak, MI.)

Example 12.6 A full slot is milled in aluminum 7075 using a two-flute end-mill. The diameter of the end-mill is 9.52 mm and its length extending from the toolholder is 105 mm ($L/D = 11$). The stability chart for the tool mounted in the spindle is shown in the top graph in Figure 12.42. The bottom graph in the figure shows the estimated chatter frequency at the chosen RPM. This helps distinguish between self-excited vibration (chatter) and the forced vibrations discussed in Examples 12.1 and 12.2.

    a. At what frequency will the tool chatter if it is rotating at 7500 rpm with an ADOC of 0.12 mm?

    b. What is the fundamental forcing (tooth passing) frequency (FFF)?

    c. What is the ratio of chatter frequency to the FFF?

Solution: The chatter frequency diagram is generated from the solution of a SDOF system using Equation 12.57, which has three unknowns: (b, ω, ε); ω is the tooth pass frequency, which is related to the spindle RPM [$\omega = 60/(n_t \text{ RPM})$]. In Figure 12.42, b is plotted versus RPM in the top graph, and the chatter frequency is plotted versus RPM in the bottom graph.

    a. The spindle speed is 7500 rpm and the ADOC is 0.12 mm. Since these cutting conditions lie in the shaded area in the lobe diagram, it is clear that the tool will chatter. The chatter frequency is estimated from the chatter frequency graph to be approximately 670 Hz. In addition, vibration analysis was performed in this case using an audio analysis (as explained in Section 12.9) and yielded a chatter frequency of 662 Hz while cutting as shown in Figure 12.43. This measurement agrees with the stability chart estimate.

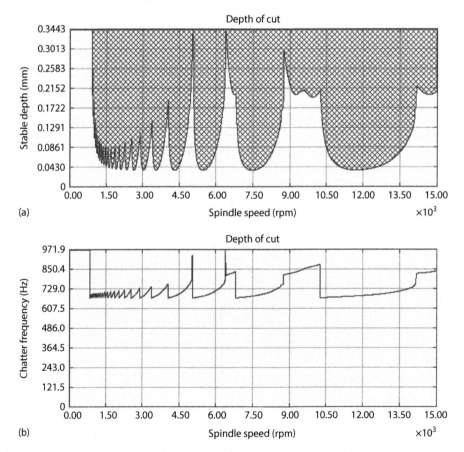

(a)

(b)

FIGURE 12.42 Stability chart and the corresponding chatter frequency charts for a two-flute ($L/D = 11$) end-milling a full slot in aluminum workpiece in Example 12.6. (a) Stability chart. (b) Chatter frequency chart. (Courtesy of D3 Vibrations Inc., Royal Oak, MI.)

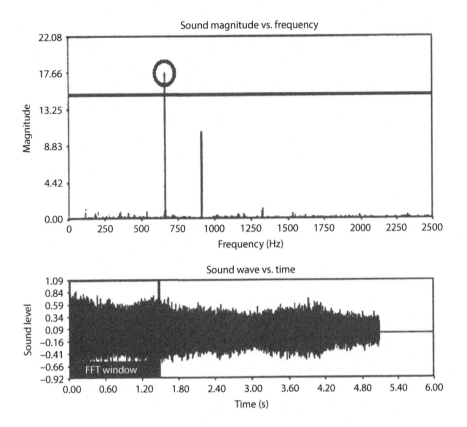

FIGURE 12.43 Audio signal of two-flute (*L/D* = 11) end-milling of aluminum at 7500 rpm and 0.125 mm ADOC showing a chatter frequency of 662 Hz in Example 12.6. (Courtesy of Manufacturing Laboratories Inc., Las Vegas, Nevada.)

b. The FFF of the two-flute cutter at 7500 rpm, calculated from Equation 12.40 (f_n = 2 × 7500/60), is 250 Hz.

c. The ratio of chatter frequency to the FFF is 670/250 = 2.68. This ratio is a non-integer multiple of the FFF, which indicates that the vibration is self-excited. A forced vibration would be an integer multiple (i.e., the FFF or one of its harmonics).

Example 12.7 The part and cutting process described in Example 12.3 is used in this example to illustrate how a knowledge of the material removal rate (MRR) based on the stability lobe diagram can be used to select cutting conditions. The MRR chart given in Figure 12.44 was generated based on the stability chart in Figure 12.40 using Equation 2.29 and assumes constant feed per tooth of 0.125 mm/tooth for the 20 mm diameter cutter. The stable ADOC at each spindle speed in the MRR calculation was obtained from Figure 12.40.

a. Determine the maximum MRR if the following maximum speeds are allowed: (a) 10,000 rpm, (b) 12,000 rpm, and (c) 15,000 rpm.

b. Determine the time required to remove 1000 cm³ of material for the three conditions above.

c. Compute the time improvement for a speed decrease from 12,000 to 10,000 rpm based on the stability chart. Note that the ADOC is about 1.5 mm from Figure 12.40.

d. If the process only required a 1.0 mm ADOC, what stable rpm is the most productive given a 12,000 rpm maximum spindle speed?

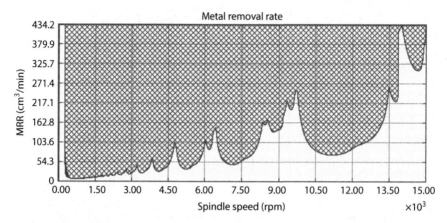

FIGURE 12.44 MRR stability chart for the 20 mm diameter two-flute ($L/D = 7$) end-milling of aluminum at 0.125 mm/tooth based on the stability chart in Figure 12.40 in Example 12.7. (Courtesy of D3 Vibrations Inc., Royal Oak, MI.)

Solution:

a. The maximum MRR for a given speed is estimated from the MRR stability chart by finding the RPM intercept with the stability boundary. The maximum MRR's for 10,000, 12,000, and 15,000 rpm are 110, 90, and 380 cm³/min, respectively. Note that a higher spindle speed is not always better, as the maximum MRR is not obtained at the maximum spindle rpm in this case.

b. The time required to remove 1000 cm³ of material is calculated as t_m = Vol/MRR. The machining times for the 10,000, 12,000, and 15,000 rpm are 9, 11.1, and 2.6 min, respectively.

c. The MRR at 12,000 rpm is approximately 90 cm³/min at the stability boundary; therefore, it takes 11.1 min to manufacture the part. By choosing a lower speed of 10,000 rpm where ADOC is larger and MRR is 110 cm³/min (greater than that at the higher speed of 12,000 rpm), the cycle time reduces by 2.1 min or a 19% gain.

d. 12,000 rpm is the most productive since it is the machine maximum and a 1.0 mm ADOC is below the stability limit at 12,000 rpm.

Example 12.8 A face milling operation is performed on a cast aluminum workpiece with $K_c = 2.8 \times 10^8$ N/m². A 125 mm diameter milling cutter with PCD inserts is used. The cutting parameters are: 4 mm ADOC, 7000 rpm maximum spindle speed, 40 kW maximum spindle power, 0.1–0.3 mm/tooth suggested feed range for the end-mill, and 8 s maximum allowable cycle time. The surface to be milled is 600 mm long and 100 mm wide. The feed distance for the cutter is 800 mm. The dynamic characteristics of both the part and the tooling was evaluated and the face mill was found to be stiffer than part in this case. The part dynamic characteristics are $f_n = 763$ Hz, $k_d = 2.2 \times 10^7$ N/m, and $\zeta = 0.01$ based on a SDOF system.

a. Analyze the dynamics of the system by plotting the TF and the stability chart using the dynamic characteristics of the weakest component of the machine–tool–workpiece system.

b. Determine the following cutting parameters: RPM, number of teeth for the face milling cutter, feed, and actual cycle time. (*Note*: The 125 mm face milling cutter is available with 6, 8, 10, and 12 inserts.)

Solution: When determining stability for high-speed applications, only the RPM, number of teeth in the cutter, number of teeth engaged in the workpiece, and ADOC and RDOC are significant factors in regenerative chatter. The product of the ADOC and RDOC affects stability.

a. The workpiece is represented by a SDOF system since a single natural frequency was measured in the modal test. The system dynamic characteristics are $f_n = 763$ Hz, $k_d = 2.2 \times 10^7$ N/m, and $\zeta = 0.01$ or 1.0% as give above. The real and imaginary parts of the TF and corresponding stability diagram are plotted on top in Figure 12.45. The real and the imaginary parts of the TF were generated using Equation 12.13. The stability lobes we generated for different numbers of teeth in the 125 mm face milling cutter using Equations 12.62 and 12.69 and the procedure discussed in Section 12.6. Since $b_{\lim}$ is not high enough on any of the tooth designs, the chosen RPM is important. The procedure requires the calculation of the spindle RPM from Equation 12.71 for each stability lobe $n_p = 1, 2, 3,\ldots$ for a range of frequencies around the dominant mode in the transfer function ($f_n = 763$ Hz).

b. The maximum robust RPM's for each tooth design were chosen to lie within the middle of the wider stability pocket between lobes and are summarized in Table 12.8. Since the workpiece is aluminum and diamond inserts are used in the face milling cutter, the RPM is not a limiting factor in this case. The torque (= force × cutter radius), or power (= force × velocity) could be limiting factors. Assuming middle range for the feed of 0.2 mm/tooth, the cutting force is estimated either from simulation (as explained in Chapter 8) or from the cutting power (as explained in Chapter 2). Cutting force results based on simulation are provided in Table 12.9 for the four teeth designs. The force per cutting edge is the same for all cutters since the geometry of the cutters (insert type and geometry, axial and radial rake angles, lead angle, etc.) is the same and only the number of teeth is different. A worn tool factor is included in the force estimates. The torque and power required by the spindle were calculated as explained in Chapter 2 for the different number of teeth and are given in Table 12.9. The power for the 10 and 12 tooth cutters was greater than 40 kW, so the cutting speed was dropped from 7000 to 6000 in these cases. The feed rate, cycle time, and MRR were calculated using the equations from Chapter 2; the results are given in Table 12.10. It is assumed that part quality and tool life are adequate for any number of teeth. Since the tools with 6, 8, 10 and 12 meet the cycle time requirement of 8 s, and the part quality is sufficient the problem is solved. If the production system is a transfer line where the machines are in serial, the feed rate is often decreased (as long as tool life is not affected) to reduce cutting forces while maximizing the cycle time, thus improving part quality and tool life. The feed and rpm can also be reduced together to improve tool life; however, it is important to check stability before altering the rpm.

Example 12.9 This example considers the solution of an intermittent chatter problem observed during a semi-finish boring operation. A 90 mm diameter bore was machined with a boring bar with three inserts. The spindle speed was 3600 rpm and the feed rate was 1060 mm/min. The spindle speed of 3600 rpm is equivalent to a 60 Hz exciting disturbance; a three-inserted boring bar operating at this speed excites the structure at a frequency of 180 Hz. Intermittent chatter was observed, which in some instances produced chatter marks (deflections) up to 1.27 mm in amplitude, even though the average depth of cut for the operation was roughly 0.380 mm. In many cases, the chatter diminished and died out as the tool advanced, but in some cases it grew in amplitude, resulting in spindle failure.

Several process changes were implemented in an attempt to eliminate the chatter. These included (1) using only one insert, rather than three, in the boring bar, and (2) changing the design of the front spindle bearings. The original spindle had a set of two 70 mm angular contact bearings behind a double row of roller bearings. In the modified design, a set of two 25° angular contact bearings with larger balls was substituted for the original double row of angular bearings with smaller balls. After these modifications were made, the system never chattered or failed. However, the modifications resulted in an unacceptably low production rate. A study was undertaken to identify the underlying cause of chatter so that further modifications could be made to increase the production rate without inducing chatter.

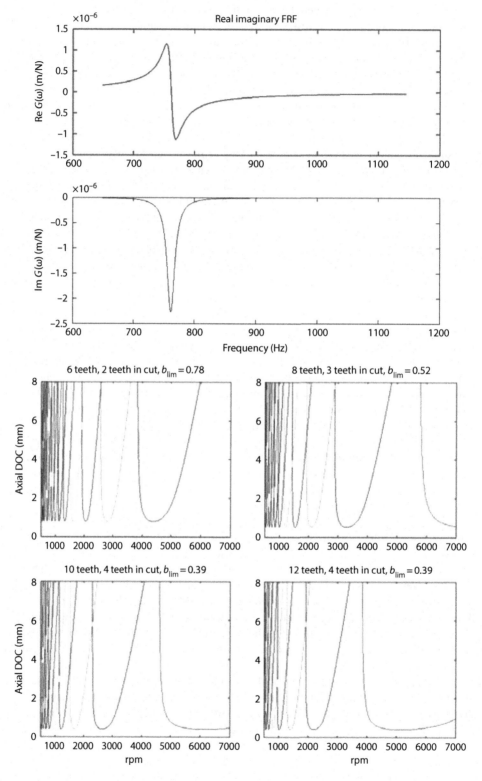

FIGURE 12.45 Transfer function of the workpiece (represented by its real and imaginary parts) and the stability charts a cutter with different number of teeth.

TABLE 12.8
Selected Cutting Speeds for the Corresponding Number of Teeth
for the Face Milling Cutter in Example 12.8

Number of Teeth	6	8	10	12
RPM	7000	5200	6000	6000
Angular Vel (rad/s)	733	544.5	628.3	628.3
Radial Vel (m/min)	2749	2042	2356	2356

TABLE 12.9
Selected Feed and Estimated Torque and Power for the Cutter with
Four Different Teeth Designs Defined in Table 12.8 for Example 12.8

Number of Teeth	6	8	10	12
Force/tooth (N)	224	224	224	224
Force/cutter (N)	448	672	896	896
Torque (N m)	28	42	56	56
Power (kW)	20.5	22.9	35.2	35.2

TABLE 12.10
Calculated Feed Rate, Cycle Time, and MRR for the Cutter with Four
Different Teeth Designs Defined in Table 12.8 for Example 12.8

Number of Teeth	6	8	10	12
Feed (mm/min)	8,400	8,320	12,000	14,400
Cycle time (s)	5.7	5.8	4.0	3.3
MRR (mm³/min)	663,600	440,960	408,000	489,600

The stiffness of each spindle was measured off-line by mounting on a bench and applying static and dynamic loads at the front of the spindle shaft and at the end of the boring bar. In addition, the stiffness of the boring bar was measured when mounted on the production machine. All the measurements were made while the spindle was not rotating. Both the static and dynamic measurements were made using a noncontact gage and a hammer. An impulse force was applied either at the front of the spindle shaft or at the end of the boring bar; a capacitance probe located opposite the excitation location was used to measure the resulting deflection. The frequency response of the system was analyzed to characterize the system dynamics and identify the source of the chatter.

Typical dynamic responses of the modified spindle on the bench and on the machine are shown in Figures 12.46 and 12.47. The results of the static spindle stiffness test are shown in Figure 12.48. The stiffness of the spindle with the tool is reduced by roughly 60%–75% compared to the stiffness of the spindle itself. The natural frequencies for both spindles are shown in Figure 12.49. The natural frequency of the spindle with the tool on the bench was slightly higher than the frequency measured in the machine. As expected, the chatter frequency (about 480 Hz) coincides with the natural frequency of the spindle with the tool in the machine station, which was between 470 and 500 Hz.

These results show that the original spindle is significantly stiffer than the modified spindle. The dynamic stiffness results shown in Figure 12.50 indicate that the original spindle had minimum

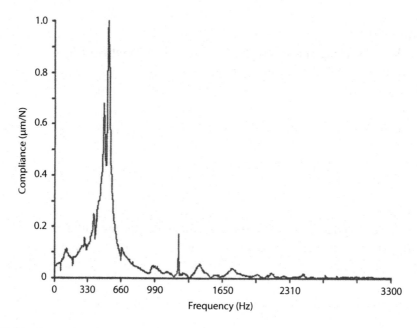

FIGURE 12.46   Dynamic stiffness response of the modified spindle.

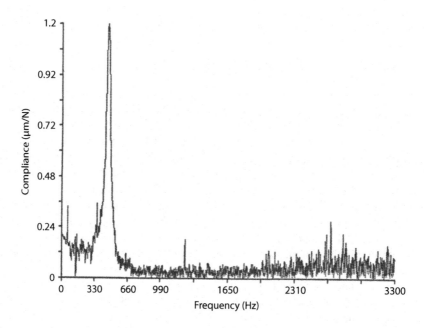

FIGURE 12.47   Dynamic stiffness response of the modified spindle installed on the machine.

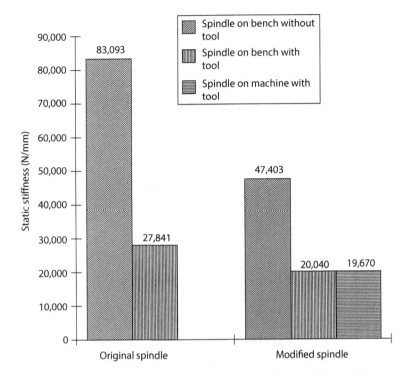

FIGURE 12.48 Comparison of the static stiffness for the original and modified spindles with boring tools on the bench and the modified spindle with tooling installed on the machine.

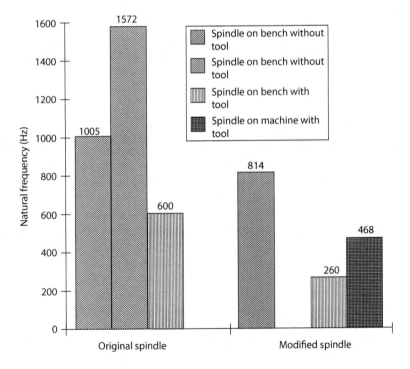

FIGURE 12.49 Comparison of the natural frequencies of the original and modified spindles with boring tools on the bench and the modified spindle with tooling installed on the machine.

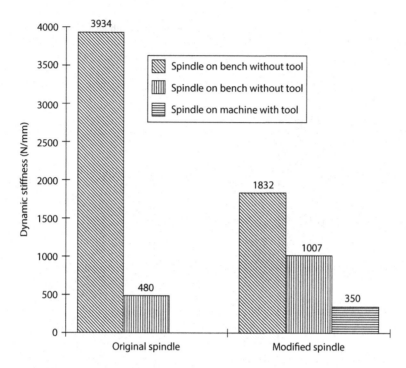

FIGURE 12.50 Comparison of the natural frequencies of the original and modified spindles with boring tools on the bench and the modified spindle with tooling installed on the machine.

natural frequencies of 1000 and 1570 Hz, compared to 800 Hz for the modified spindle. This result might have been anticipated from the bearing designs of the two spindles. The bearings in the original spindle had smaller balls than the bearings in the modified spindle; the smaller balls should have provided higher stiffness and increased heat generation. The static stiffness results indicated that changing the spindle design did not address the root cause of the chatter.

Dynamic stiffness was characterized by constructing a stability lobe diagram as shown in Figure 12.51. The stability lobes for the original and modified spindles with tooling are shown in Figures 12.52 and 12.53, respectively, for the bench tests and in Figure 12.54 for the modified spindle on the machine. Chatter vibration can be avoided by selecting the width of cut to be below the lobes or well between lobes farther to the right on the graph. The axial cumulative width of cut for one insert is located below the lobes in both figures, while that for three inserts is much closer to the lobes. The process may cross the limiting lobe curves (i.e., become unstable) due to variations in the spindle speed during operation when the three-insert boring bar is used. This explains why the chatter occurred when three inserts were used but did not occur when a single insert was used.

The following conclusions can be drawn from the earlier results:

- The static and especially the dynamic stiffness of the tool on the machine station is insufficient for the operation to be performed with a three-insert boring bar.
- The chatter frequency is the same as the lowest natural frequency of the boring bar.
- The system stability was improved using one insert instead of three because the cumulative width of cut was reduced significantly.
- It would be worthwhile to investigate using two inserts instead of one, since this might permit an increase in the production rate without resulting in chatter.

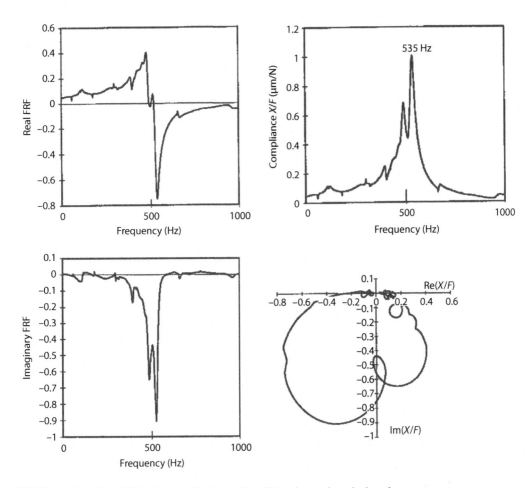

FIGURE 12.51 The FRF of the modified spindle with boring tool on the bench.

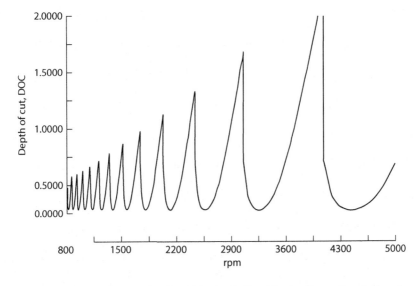

FIGURE 12.52 Stability lobe diagram for the original spindle with a boring tool on the bench.

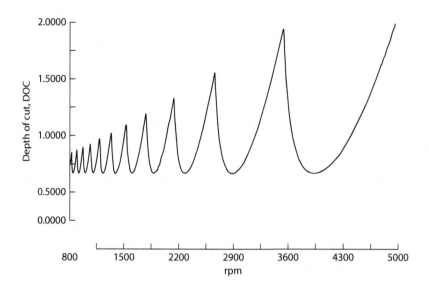

FIGURE 12.53 Stability lobe diagram for the modified spindle with a boring tool on the bench.

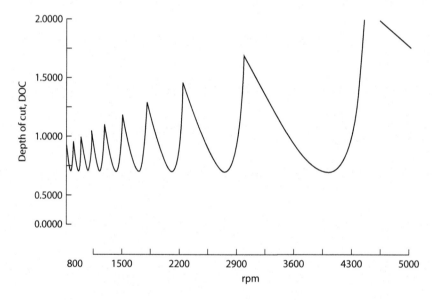

FIGURE 12.54 Stability lobe diagram for the modified spindle with a boring tool installed on the machine.

Example 12.10 As an example of the identification of modal parameters, we consider the characterization of the dynamic properties of a single spindle drill press.

An electrodynamic exciter was mounted between the spindle and the table. The drilling machine structure was excited in the Z-direction at point 1 as shown in Figure 12.55, using a pink noise random signal with bandwidth of 200 Hz generated by a function generator. The random vibration signal was measured with accelerometers at points 1, 2, 3, and 4 in Figure 12.55. The direct and cross TFs between the input force signal and the output displacement signals were estimated using a dual channel Fast Fourier Transform (FFT) Real Time Analyzer (RTA). Each set of data was analyzed using a transform size of 1024 points, a frequency range of 100 Hz (with a sampling rate of 2000 Hz), 32 sample averaging, and a Hanning window.

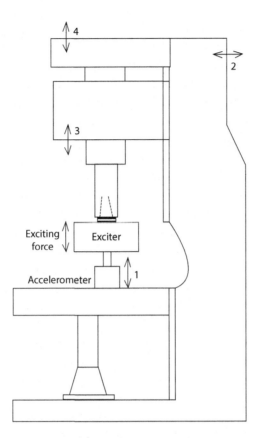

FIGURE 12.55 Schematic diagram of the experimental setup for vibration test on the drilling machine structure.

A 32 signal average was used to reduce the variance of the spectrum since the 3 dB bandwidth of the Hanning window is 0.36 Hz. The direct TF at point 1 is shown in Figure 12.56. The auto-spectrum of the input and output signals are shown in Figure 12.57. Although the spectrum of the input signal to the exciter from the function generator is almost flat over the frequency range of interest as shown in Figure 12.58, the spectrum of the actual input force signal to the structure is not quite flat. The coherency spectrum at point 1 (Figure 12.59) indicates that the TF is accurate over a relatively wide range of frequencies, that is, that noise effects are acceptably small except at very low frequencies.

To identify the dynamic characteristics of the structure, the real part of the TF was fit using a nonlinear least squares curve fitting routine based on the model in Equations 12.37 or 12.68. Initial values were estimated from visual inspection of the shape of the TF. Assuming a three DOF system, 30 points of the TF from 21 to 50 Hz were fit. The curve fit TF and the original TF are shown in Figure 12.60. The decomposition of the three modes is shown in Figure 12.61, from which the effect of each mode on the vibration at point 1 can be identified. The dynamic characteristics of the structure in the normal coordinate system were determined using the analysis described in Section 12.3 and are summarized in Table 12.11. The mode shapes of the structure are found by curve-fitting the real parts of all the cross TF's using the natural frequencies, stiffness coefficients, and damping factors in Table 12.11. The eigenvectors or mode shapes were calculated using Equation 12.34 and are summarized in Table 12.12 and illustrated in Figure 12.62.

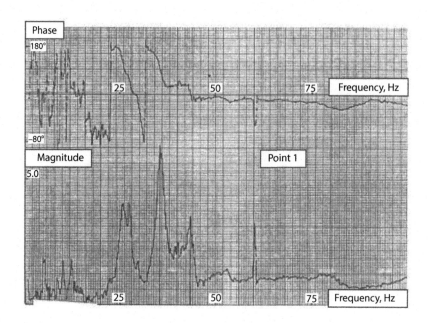

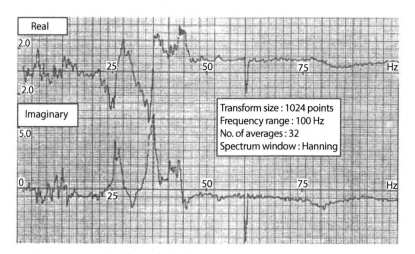

FIGURE 12.56   Characteristics of the FRF at point 1 on the drilling machine.

Example 12.11 Consider a single point boring bar used for both roughing and finishing a deep hole. The body of the bar is cylindrical with a diameter of 50 mm, a length of 350 mm ($L/D = 7$), and a lead angle of 30°. The bar is made of steel. The bar is clamped in a CAT-50 hydraulic toolholder as explained in Chapter 5. Estimate the limit of stability and the corresponding material removal rate (MRR) assuming that the bar acts as a SDOF system.

Solution: The stiffness of the boring bar including the tool–toolholder–spindle interface can be calculated as explained in Chapter 5. However, let us assume the bar is a cantilevered beam. In this case, the stiffness of the bar is derived using the material in Chapters 4 and 5, assuming a damping ratio of 0.04:

$$k_c = \frac{3E\pi D^4}{64L^3} = 4293 \text{ N/mm}$$

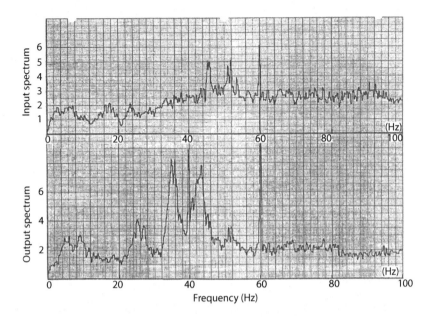

FIGURE 12.57 Autospectrum of input and output signals at point 1 on the drilling machine.

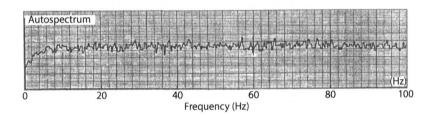

FIGURE 12.58 Autospectrum of the input signal to the exciter generated by the function generator for the identification of the drilling machine structure.

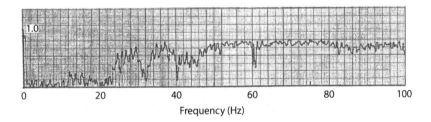

FIGURE 12.59 Coherency of TF at point 1 on the drilling machine.

and using Equation 12.66

$$b_{\lim} = \frac{2k_c \zeta (1+\zeta)}{k_d \sin^2 \kappa} = 0.67 \text{ mm}$$

The DOC of 0.65 mm is small for a roughing operation and could be increased by increasing the diameter of the boring bar or using a stiffer bar material as explained in Chapter 4. Changing the material of the boring bar to heavy metal with $E = 330,000$ MPa instead of $206,700$ MPa for steel,

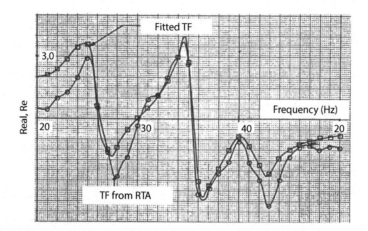

FIGURE 12.60   Curve fitting of TF at point 1 on the drilling machine.

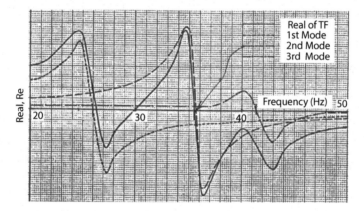

FIGURE 12.61   Decomposition of real part of TF at point 1 on the drilling machine.

TABLE 12.11

Modal Parameters of the Drilling Machine Structure Estimated by Curve Fitting and Decomposition

Mode	Natural Frequency	Damping Ratio	Stiffness
First	26	0.051	1.85
Second	35.6	0.026	2.62
Third	42	0.007	11.93

TABLE 12.12

Mode Shapes of the Drilling Machine Structure Estimated by Curve Fitting

Position	First Mode	Second Mode	Third Mode
1	1.0	1.0	1.0
2	0.013	0.016	−0.304
3	0.407	−0.5	0.493
4	0.409	−0.45	0.574

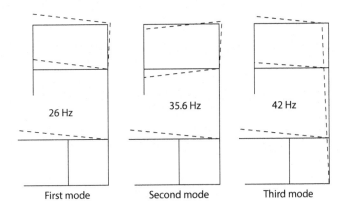

26 Hz

35.6 Hz

42 Hz

First mode   Second mode   Third mode

FIGURE 12.62   Mode shapes at point 1 on the drilling machine.

the b_{lim} = 1.07 mm. The above minimum DOC is sufficient for a finish cut even using the steel bar. Selecting a roughing feed of 0.2 mm/rev and a cutting speed of 100 m/min, the MRR is estimated using Equation 2.7 to be

$$MRR_{lim} = Q = Vfb_{lim} = 21,430 \, mm^3/min$$

12.11   PROBLEMS

Problem 12.1 Select a spindle speed (rpm) for a 4-flute end mill cutter so that the fundamental forcing frequency (FFF) and first two harmonics don't fall within 100 Hz of the main resonance peak in Figure 12.63. The spindle speed must be between 6,000 and 18,000 rpm. Plot the frequencies of the FFF and first two harmonics on Figure 12.63.

Problem 12.2 Dynamic test results for a carbide end mill in a vertical machining center are shown in Figure 12.64. The machine has a 14,000 rpm maximum spindle speed and the spindle power is not a constraint. Estimate the time required to machine a pocket 800 mm long, 400 mm wide, and 20 mm deep from a solid block workpiece using the maximum estimated material removal rate.

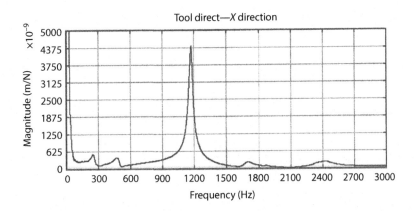

FIGURE 12.63   FRF for a four-flute end-mill used in Problem 12.1. (Courtesy of D3 Vibrations Inc., Royal Oak, MI.)

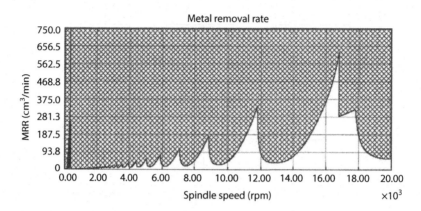

FIGURE 12.64   FRF for an end-mill used in Problem 12.2. (Courtesy of D3 Vibrations Inc., Royal Oak, MI.)

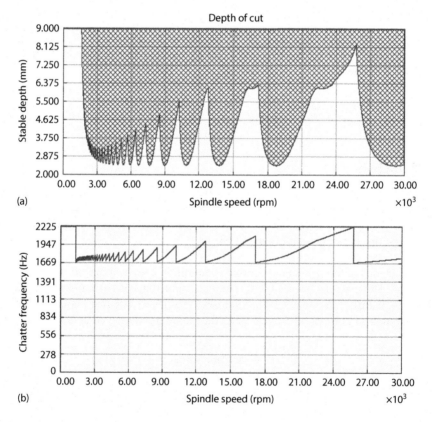

FIGURE 12.65   Stability chart for a 15 mm diameter two-flute carbide end-mill for full slot milling at 0.15 mm/tooth in aluminum workpiece. (a) Stability chart. (b) Chatter frequency chart. (Courtesy of D3 Vibrations Inc., Royal Oak, MI.)

Problem 12.3 A two-flute carbide end-milling cutter with an overhang of 87 mm out of the toolholder is used in an operation. The feed is 0.15 mm/tooth. The stability lobes and the MRR lobes were estimated and are shown in Figures 12.65 and 12.66, respectively. Determine: (a) the maximum axial depth of cut and MRR in Figures 12.65 and 12.66 if maximum spindle speeds of 18,000, 24,000, and 30,000 rpm are selected; (b) Name two physical changes that can be made to the cutter geometry or setup to alter Figure 12.65 to look more like Figure 12.67, to make the system more productive at 30,000 rpm.

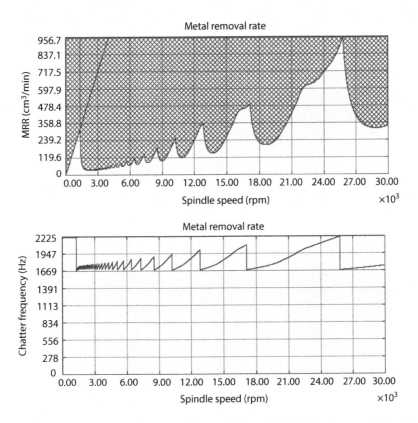

FIGURE 12.66 MRR stability chart for the two-flute carbide end-mill in aluminum workpiece based on the stability chart in Figure 12.65. (Courtesy of D3 Vibrations Inc., Royal Oak, MI.)

Problem 12.4 A pocket is to be machined in an aluminum structure ($K_c = 2.8 \times 10^8$ N/m²). An end-milling process will be used on a horizontal machining center capable of a maximum spindle speed of 15,000 rpm. The workpiece is determined to be stiff (2.3×10^9 N/mm) compared to the long end-mill (1.1×10^7 N/m) required to machine the deep pocket. Use the approximate values for the first bending mode of a 2-tooth end-mill obtained from experimental modal analysis ($f_n = 315$ Hz, $k = 1.1 \times 10^7$ N/m, $\zeta = 0.02$) to

a. Create an analytical FRF
b. Create stability lobes for a 1, 2, 3, and 4 tooth endmill assuming all four tools have the same dynamic values. (*Note*: for a full slot, 1 and 2 tooth end mills have 1 tooth in cut, and a 3 and 4 tooth end mills have 2 teeth in cut.)

Determine

1. What is the maximum negative real value of the FRF? Why is this important?
2. What is $b_{\lim}$ for each of the four tools?
3. What is the maximum axial depth of cut (ADOC) for each tool at 4000 and 5000 rpm?
4. The tool you are using is 25 mm in diameter and the feed rate is 0.05 mm/tooth. If it is suggested that you must use a two-flute tool with 0.05 mm/tooth feed rate, remove between 5 and 10 mm ADOC, and rotate between 4,000 and 10,000 rpm, then what RPM and ADOC will you choose to maximize the material removal rate (MRR) and stay at least 1000 rpm and 2 mm ADOC away from a stability line? What is the volume of material removed per minute?

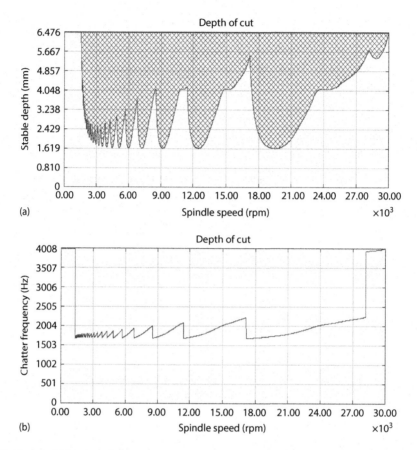

FIGURE 12.67 Stability chart for same carbide end-mill as in 12.65 with long overhang milling a full slot in aluminum workpiece. (a) Stability chart. (b) Chatter frequency chart. (Courtesy of D3 Vibrations Inc., Royal Oak, MI.)

Problem 12.5 A 25.4 mm slot is to be machined in a steel structure (cutting stiffness $K_c = 7 \times 10^8$ N/m^2). An end-milling operation will be designed for the full slot on a horizontal machining center capable of a maximum spindle speed of 15,000 rpm. A 4-flute end-mill is suggested with a feed of 0.1 mm/tooth. The workpiece is determined to be much stiffer than the long 25.4 diameter end-mill required to machine the slot. Use the approximate values for the first bending mode of a 2-tooth endmill obtained from experimental modal analysis ($f_n = 5000$ Hz, $k = 2 \times 10^7$ N/m, $\zeta = 0.05$) to:

a. Create an analytical FRF
b. Create stability lobes for the end-mill. (Note: for a full slot, a 4-flute endmill has 2 teeth in cut.)

Questions:

1. What is maximum negative real value of FRF? Why is this important?
2. What is b_{lim} for each of the two tools?
3. What is the maximum axial depth of cut (ADOC) for each tool at 15,000 rpm?

Problem 12.6 Derive the solution of Equation 12.71 for regenerative chatter used to plot the stability chart of b versus RPM.

Problem 12.7 Consider a 2-DOF milling system with the following characteristics: $k_1 = 4.4\text{E}6$ N/m, $f_{n1} = 60$ Hz, $\zeta_1 = 0.04$ and $k_2 = 5.6\text{E}6$ N/m, $f_{n2} = 66$ Hz, $\zeta_1 = 0.04$. Assume the directions of individual modes of vibration are measured to be 30° and 120° as illustrated in Figure 12.17. Calculate the critical limiting width of chip assuming the specific force for the workpiece material is $k_d = 2400$ N/mm^2.

REFERENCES

1. L. Andren, L. Hakansson, A. Brandt, and I. Claesson, Identification of motion of cutting tool vibration in a continuous operation—Correlation to structural properties, *Mech. Syst. Signal Process.* **18** (2004) 903–927.
2. J. Tlusty, *Technology of Machine Tools*, Vol. 3: *Machine Tool Mechanics*, Lawrence Livermore National Laboratory Rep. UCRL-52960-3; distribution by SME.
3. J. Tlusty and T. Moriwaki, Experimental and computational identification of dynamic structural models, *CIRP Ann.* **25** (1976).
4. J. Tlusty and F. Ismail, Dynamic structural identification tasks and methods, *CIRP Ann.* **29** (1980) 251–255.
5. Y. Altintas, *Manufacturing Automation—Metal Cutting Mechanics, Machine Tool Vibrations, and CNC Design*, Cambridge University Press, Cambridge, U.K., 2000.
6. E. I. Rivin, *Stiffness and Damping in Mechanical Design*, Marcel Dekker, New York, 1999.
7. D. J. Seagalman and E. A. Butcher, Suppression of regenerative chatter via impedance modulation, *J. Vibrat. Control* **6** (2000) 243–256.
8. H. Shinno and Y. Ito, Computer aided concept design for structural configuration of machine tools: Variant design using directed graph, *J. Mech. Transmis. Automat. Des.* **109**:3 (1987) 372–376.
9. H. Shinno and Y. Ito, Structural description of machine tools, 1st report: Description method and some applications, *Bull. Jpn. Soc. Mech. Eng.* **24**:187 (1981) 251–258.
10. H. Shinno and Y. Ito, Structural description of machine tools, 2nd report: Evaluation of structural similarity, *Bull. Jpn. Soc. Mech. Eng.* **24**:187 (1981) 259–265.
11. J. N. Dube, Computer aided design of machine tool structure with model techniques, *Comput. Struct.* **28**:3 (1988) 345–352.
12. J. G. Bollinger, Computer aided analysis of machine dynamics, *International Conference on Manufacturing Technology by The American Society of Tool and Manufacturing Engineers*, Dearborn, MI, September 1967, pp. 77–88.
13. R. C. Bahl and P. C. Pandey, Computer analysis of machine tool structures using flexibility method, *Mech. Eng. Bull.* **7** (1976) 42–51.
14. R. C. Bahl and P. C. Pandey, Comparative study of computer aided methods for the design of machine tool structures, *Mech. Eng. Bull.* **8** (1977) 33–38.
15. O. I. Averyanov and A. P. Bobrik, Methodology of forming standardized subassemblies for machine tools, *Soviet Eng. Res.* **3** (1983) 55–57.
16. A. Archenti, Model-based investigation of machining systems characteristics—Static and dynamic stability analysis, Thesis, KTH Production Engineering, Stockholm, Sweden, 2008.
17. R. C. Dorf, *Modern Control Systems*, 4th edn., Addison-Wesley, Reading, MA, 1986.
18. D. C. Karnopp, D. L. Margolis, and R. C. Rosenberg, *System Dynamics*, 2nd edn., Wiley, New York, 1990.
19. M. P. Norton, *Fundamentals of Noise and Vibration Analysis for Engineers*, Cambridge University Press, New York, 1994.
20. R. C. Dorf, *Time-Domain Analysis and Design of Control Systems*, Addison-Wesley, Reading, MA, 1965.
21. T. H. Glisson, *Introduction to System Analysis*, McGraw-Hill, New York, 1984.
22. H. P. Neff, *Continuous and Discrete Linear Systems*, Harper and Row, New York, 1984.
23. S. A. Tobias, *Machine Tool Vibrations*, Blackie and Son, London, U.K., 1965.
24. J. Tlusty and K. M. Polacek, Experiences with analysing stability of machine-tool against chatter, *Proceedings of the Ninth International MTDR Conference*, Pergamon Press, New York, 1968.
25. J. Tlusty, A method of analysis of machine tool stability, *Proceedings of the Sixth MTDR Conference*, Manchester, England, 1965, p. 5.
26. D. B. Welbourne and J. K. Smith, *Machine-Tool Dynamics: An Introduction*, Cambridge University Press, New York, 1970.
27. C. W. Bert, Material damping: An introduction review of mathematical models, measures and experimental techniques, *J. Sound Vibrat.* **29**:2 (1973) 129–153.
28. R. H. Lyon, *Machinery Noise and Diagnostics*, Butterworths, London, U.K., 1987.

29. S. Timoshenko, D. H. Young, and W. Weaver, Jr., *Vibration Problems in Engineering*, 4th edn., Wiley, New York, 1974.
30. J. W. S. Rayleigh, *Theory of Sound*, 2nd edn., Vol. 1, Dover Publications, New York, 1945.
31. T. K. Caughey, Classical normal modes in damped linear dynamic systems, *ASME J. Appl. Mech.* **27** (1960) 269–271; **32** (1965) 583–588.
32. R. A. Fraser, W. J. Duncan, and A. R. Collar, *Elementary Matrices*, Cambridge University Press, London, U.K., 1946.
33. K. A. Foss, Coordinates which uncouple the equations of motion of damped linear dynamic systems, *ASME J. Appl. Mech.* **25** (1958) 361–364.
34. J. Peters and M. Mergeay, Dynamic analysis of machine tools using complex modal method, *CIRP Ann.* **25** (1976) 257–261.
35. D. Dilley and T. Delio, Machine tool vibration monitoring using audio signal analysis, *First Annual Manufacturing Technology Summit* by SME, Dearborn, MI, August 10–11, 2004.
36. G. Quintana and J. Ciurana, Chatter in machining processes: A review, *Int. J. Mach. Tools Manuf.*, McGraw-Hill, New York, NY, **51** (2011) 363–376.
37. E. I. Rivin, Machine-tool vibration, Chapter 40, in: C. M. Harris, Ed., *Shock and Vibration Handbook*, 4th edn., McGraw-Hill, 1996.
38. ASME, Methods for performance evaluation of CNC machining centers, ASME Standard B5.54, 1992.
39. M. Weck, *Handbook of Machine Tools*, Vols. 1–4, Wiley, New York, 1984.
40. J. Tlusty, Analysis of the state of research in cutting dynamics, *CIRP Ann.* **27** (1978) 583–589.
41. J. Tlusty, *Manufacturing Processes and Equipment*, Prentice Hall, Englewood Cliffs, NJ, 2000.
42. F. C. Moon, *Dynamics and Chaos in Manufacturing Processes*, Wiley, New York, 1998.
43. G. Stépán, Modelling nonlinear regenerative effects in mental cutting, *Philos. Trans. R. Soc.* **359** (2001) 739–757.
44. A. M. Gouskov, S. A. Voronov, H. Paris, and S. A. Batzer, Cylindrical workpiece turning using multiple-cutter tool heads, *Proceedings of the ASME 2001 Design Engineering Technical Conference*, Pittsburgh, PA, Paper no. DETC2001/VIB-21431.
45. J. E. Halley, A. M. Helvey, K. S. Smith, and W. R. Winfough, The impact of high-speed machining on the design and fabrication of aircraft components, *Proceedings of the ASME 1999 Design Engineering Technical Conference*, Las Vegas, NV, Paper no. DETC99/VIB-8057.
46. J. Tlusty, Machine dynamics, Chapter 3, in: R. I. King, Ed., *Handbook of High-Speed Machining Technology*, Chapman & Hall, New York, 1985, pp. 48–153.
47. S. A. Tobias and W. Fishwick, The chatter of lathe tools under orthogonal cutting conditions, *ASME Trans.* **80** (1958) 1079–1088.
48. J. Tlusty and M. Polacek, Beispiele der Behandlung der selbsterregten Schwingung der Werkzeugmaschinen, *Proceedings of the Third Fo Ko Co*, Vogel-Verlag, Würzburg, Germany, October 1957, p. 131.
49. H. E. Merritt, Theory of self-excited machine-tool chatter: Contribution to machine-tool chatter research—1, *ASME J. Eng. Ind.* **87** (1965) 447–454.
50. J. Tlusty and S. B. Rao, Verification and analysis of some dynamic cutting force coefficient data, *Proc. NAMRC* **6** (1978) 420–426.
51. E. E. Mitchell and E. Harrison, Active machine tool controller requirements for noise attenuation, *ASME J. Eng. Ind.* **96** (1974).
52. Y. Antilas and E. Budak, Analytical prediction of stability lobes in milling, *Ann. CIRP* **44** (1995) 357–362.
53. S. Smith and J. Tlusty, Efficient simulation programs for chatter in milling, *Ann. CIRP* **42** (1993) 463–466.
54. J. Tlusty, W. Zaton, and F. Ismail, Stability lobes in milling, *Ann. CIRP* **32** (1983).
55. D. Montgomery and Y. Altintas, Mechanism of cutting force and surface generation in dynamic milling, *ASME J. Eng. Ind.* **113** (1991) 160–168.
56. W. R. Winfough and S. Smith, Automatic selection of the optimum metal removal conditions for high-speed milling, *Trans. NAMRI/SME* **23** (1995) 163–168.
57. J. Tlusty, S. Smith, and T. Delio, Stiffness, stability, and loss of process damping in high-speed machining, *Fundamental Issues in Machining*, ASME PED-Vol. 43, 1990, pp. 171–191.
58. C. T. Tyler and T. L. Schmitz, Process damping analytical stability analysis and validation, *Proc. NAMRC/SME* **40** (2012).
59. E. Budak and L. T. Tunc, A new method for identification and modeling of process damping in machining, *J. Manuf. Sci. Eng.* **131** (2009) 5–10.

60. T. Delio, *The MetalMAX Machine Tool Analyzer and Dynamic Machining Process Optimizer*, Manufacturing Laboratories (MLI), Las Vegas, NV, 2006.

61. S. A. Tobias and W. Fishwick, The vibrations of radial drilling machines under test and working conditions, *Proc. Inst. Mech. Eng.* **170** (1956) 232–256.

62. M. K. Das and S. A. Tobias, The relation between the static and the dynamic cutting of materials, *Int. J. MTDR* **7** (1967) 63–89.

63. M. M. Nigm, M. M. Sadek, and S. A. Tobias, Determination of dynamic cutting coefficients from steady state cutting data, *Int. J. MTDR* **17** (1977) 19–37.

64. J. Peters and P. Vanherck, Machine tool stability tests and the incremental stiffness, *Ann. CIRP* **17** (1969) 225–232.

65. J. Tlusty and F. Ismail, Special aspects of chatter in milling, *ASME J. Vibrat. Stress Rel. Des.* **105** (1983) 24–32.

66. W. J. Endres, Prediction stability in machining using time-domain simulation—A comparison to frequency analysis, Technical Report, MEAM Department, University of Michigan, Ann Arbor, MI, 1995.

67. H. J. J. Kals, On the calculation of stability charts on the basis of the damping and the stiffness of the cutting process, *CIRP Ann.* **19** (1971) 197–303.

68. J. Peters, P. Vanherck, and H. Van Brussel, The measurement of the dynamic cutting force coefficient, *CIRP Ann.* **20** (1971) 129–136.

69. W. A. Knight and M. M. Sadek, The correction for dynamic errors in machine tool dynamometers, *CIRP Ann.* **19** (1971) 237–245.

70. J. Gradisek, E. Govekar, and I. Grabec, Time series analysis in metal cutting: Chatter versus chatter-free cutting, *Mech. Syst. Signal Process.* **12** (1998) 839–854.

71. M. K. Khraisheh, C. Pezeshki, and A. E. Bayoumi, Time series based analysis for primary chatter in metal cutting, *J. Sound Vibrat.* **180** (1995) 67–87.

72. T. Insperger, G. Stépán, P. V. Bayly, and B. P. Mann, Multiple chatter frequencies in milling processes, *J. Sound Vibrat.* **262** (2002) 333–345.

73. G. Stépán, Modelling nonlinear regenerative effects in mental cutting, *Philos. Trans. R. Soc.* **359** (2001) 739–757.

74. T. L. Schmitz, M. A. Davies, K. Medicus, and J. Snyder, Improving high-speed machining material removal rates by rapid dynamic analysis, *CIRP Ann.* **50**:1 (2001) 263–268.

75. A. Halanay, Stability theory of linear periodic systems with delay, *Revue de Mathéematiques Pures et Appliquées* **6**:4 (1961) 633–653 (in Russian).

76. J. K. Hale, *Theory of Functional Differential Equations*, Springer-Verlag, New York, 1977.

77. I. Minis and R. Yanushevsky, A new theoretical approach for the prediction of machine tool chatter in milling, *ASME J. Eng. Ind.* **115** (1993) 1–8.

78. Y. Altintas and E. Budak, Analytical prediction of stability lobes in milling, *CIRP Ann.* **44**:1 (1995) 357–362.

79. E. Budak and Y. Altintas, Analytical prediction of chatter stability in milling—Part I: General formulation, *ASME J. Dyn. Syst. Meas. Control* **120** (1998) 22–30.

80. E. Budak and Y. Altintas, Analytical prediction of chatter stability in milling—Part II: Application of the general formulation to common milling systems, *ASME J. Dyn. Syst. Meas. Control* **120** (1998) 31–36.

81. M. A. Davies, J. R. Pratt, B. Dutterer, and T. J. Burns, Stability prediction for low radial immersion milling, *ASME J. Manuf. Sci Eng.* **124** (2002) 217–225.

82. T. Insperger and G. Stépán, Stability of high-speed milling, *Proceedings of the ASME 2000 DETC, Symposium on Nonlinear Dynamics and Stochastic Mechanics*, Orlando, FL, ASME AMD-241, pp. 119–123.

83. P. V. Bayly, J. E. Halley, B. P. Mann, and M. A. Davies, Stability of interrupted cutting by temporal finite element analysis, *Proceedings of the ASME 2001 Design Engineering Technical Conference*, Pittsburgh, PA, Paper no. DETC2001/VIB-21581.

84. J. Tian and S. G. Hutton, Chatter instability in milling systems with flexible rotating spindles—A new theoretical approach, *ASME J. Manuf. Sci. Eng.* **123** (2001) 1–9.

85. S. Smith and J. Tlusty, An overview of modeling and simulation of the milling process, *ASME J. Eng. Ind.* **113** (1991) 169–175.

86. T. Insperger, B. P. Mann, G. Stépán, and P. V. Bayly, Stability of up-milling and down-milling, Part 1: Alternative analytical methods, *Int. J. Mach. Tool Manuf.* **43** (2003) 25–34.

87. B. P. Mann, T. Insperger, P. V. Bayly, and G. Stépán, Stability of up-milling and down-milling, Part 2: Experimental verification, *Int. J. Mach. Tool Manuf.* **43** (2003) 35–40.

88. B. Balachandran, Non-linear dynamics of milling process, *Philos. Trans. R. Soc.* **359** (2001) 793–820.

89. M. X. Zhao and B. Balachandran, Dynamics and stability of milling process, *Int. J. Solids Struct.* **38** (2001) 2233–2248.
90. D. M. Esterling, Y. Ren, and Y. S. Lee, Time-domain chatter prediction for high speed machining, *Trans. NAMRI/SME* **30** (2002).
91. W. T. Corpus and W. J. Endres, A high-order solution for the added stability lobes in intermittent machining, *Proceedings of the ASME 2001 DETC, Symposium on Machining Processes*, Orlando, FL, ASME MED-11, 2001, pp. 871–878.
92. J. Gradisek, E. Govekar, and I. Grabec, Time series analysis in metal cutting: Chatter versus chatter-free cutting, *Mech. Syst. Signal Process.* **12**:6 (1998) 839–854.
93. J. Gradisek, E. Govekar, and I. Grabec, Using coarse-grained entropy rate to detect chatter in cutting, *J. Sound Vibrat.* **214**:5 (1998) 941–952.
94. H. M. Shi and S. A. Tobias, Theory of finite amplitude machine tool instability, *Int. J. MTDR* **24** (1984) 45–69.
95. G. Stépán and T. Kalmár-Nagy, Nonlinear regenerative machine tool vibration, *Proceedings of the ASME 1997 Design Engineering Technical Conference*, Sacramento, CA, Paper no. DETC97/VIB-4021.
96. G. Stépán and R. Szalai, Nonlinear vibrations of highly interrupted machining, *Proceedings of the Second Workshop of COST P4 WG2 on Dynamics and Control of Mechanical Processing*, Budapest, Hungary, 2001, pp. 59–64.
97. J. Tlusty, Dynamics of high-speed milling, *ASME J. Eng. Ind.* **108** (1986) 59–67.
98. M. A. Davies, J. R. Pratt, B. Dutterer, and T. J. Burns, The stability of low radial immersion machining, *CIRP Ann.* **49**:1 (2000) 37–40.
99. B. Balachandran, Non-linear dynamics of milling process, *Philos. Trans. R. Soc.* **359** (2001) 793–820.
100. P. V. Bayly, J. E. Halley, M. A. Davies, and J. R. Pratt, Stability analysis of interrupted cutting with finite time in the cut, *Proceedings of the ASME 2000 DETC*, Orlando, FL, ASME MED-11, 2000, pp. 989–994.
101. P. V. Bayly, J. E. Halley, B. P. Mann, and M. A. Davies, Stability of interrupted cutting by temporal finite element analysis, *ASME J. Manuf. Sci. Eng.* **125** (2003) 220–225.
102. P. V. Bayly, B. P. Mann, T. L. Schmitz, D. A. Peters, G. Stépán, and T. Insperger, Effects of radial immersion and cutting direction on chatter instability in end-milling, *Proceedings of the ASME IMECE*, New Orleans, LA, No. IMECE2002-34116, ASME, 2002.
103. B. P. Mann, P. V. Bayly, M. A. Davies, and J. E. Halley, Limit cycles, bifurcations, and accuracy of the milling process, *J. Sound Vibrat.* **277** (2004) 31–48.
104. B. P. Mann, K. A. Young, T. L. Schmitz, M. J. Bartow, and P. V. Bayly, Machining accuracy due to tool or workpiece vibrations, *Proceedings of the ASME IMECE*, Washington, DC, No. IMECE2003-41991 ASME, 2003.
105. J. Tlusty, F. Ismail, and W. Zaton, Use of special milling cutters against chatter, *Proc. NAMRC* **11** (1983) 408–415.
106. F. Koenigsberger and J. Tlusty, *Machine Tool Structures*, Vol. 1, Pergamon Press, Oxford, U.K., 1970.
107. J. Tlusty, Basic non-linearity in machining chatter, *CIRP Ann.* **30** (1981) 299–304.
108. H. K. Tonshoff et al., Developments and trends in monitoring and control of machining processes, *CIRP Ann.* **37** (1988) 611.
109. H. Ota and T. Kawai, Monitoring of cutting conditions by means of vibration analysis, *Trans. Jpn. Soc. Mech. Eng., Series C* **57** (1991) 2752 (in Japanese).
110. M. Rahman, In-process detection of chatter threshold, *ASME J. Eng. Ind.* 110 (1988) 44.
111. M. Higuchi and M. Doi, A study on the estimation of the chatter commencing in turning process, *Trans. J. Soc. Mech. Eng., Series C* **52** (1986) 1697 (in Japanese).
112. T. Matsubare and H. Yamamoto, Study on regenerative chatter vibration with dynamic cutting force, *J. Jpn. Soc. Precision Eng.* **50** (1984) 1079 (in Japanese).
113. E. Kondo, H. Ota, and T. Kawai, A new method to detect regenerative chatter using spectral analysis Part I: Basic study on criteria for detection of chatter, *ASME Manufacturing Science and Engineering* MED-Vol. **2-1**/MH-Vol. **3-1** (1995) 617–627.
114. N. Grossi, L. Sallese, A. Scippa, and G. Campatelli, Chatter stability prediction in milling using speed-varying cutting force coefficients, *Procedia CIRP 14, Sixth CIRP International Conference on High Performance Cutting, HPC2014*, Berkeley, CA, 2014, pp. 170–175.
115. C. D. Erdbrink and V. V. Krzhizhanovskaya, Identifying self-excited vibrations with evolutionary computing, *Procedia Computer Science, ICCS 2014, 14th International Conference on Computational Science*, Cairns, Australia, Vol. 29, 2014, pp. 637–647.

116. M. Zapciu, O. Cahuc, C. F. Bisu, A. Gérard, and J.-Y. K'nevez, Experimental study of machining system: Dynamic characterization, *J. Mach. Form. Technol.* **1**:3–4 (2009) 1–18.

117. R. Daud, N. K. Hasfa, S. H. Tomadi, M. A. Hassan, K. Kadirgama, M. M. Noor, and M. R. M. Rejab, Prediction of chatter in CNC machining based on dynamic cutting force for ball end milling, *Proceedings of the International MultiConference of Engineers and Computer Scientists 2009 (IMECS 2009)*, Vol. II, Hong Kong, China, March 18–20, 2009.

118. M. Zapciu, J.-Y. K'nevez, A. Gérard, O. Cahuc, and C. F. Bisu, Dynamic characterization of machining systems, *Int. J. Adv. Manuf. Technol.* **57** (2011) 73–83.

119. M. Weck, E. Verhaag, and M. Gather, Adaptive control for face-milling with strategies for avoiding chatter-vibrations and for automatic cut distribution, *CIRP Ann.* **24** (1975) 405.

120. S. Smith and J. Tlusty, Update on high-speed milling dynamics, *ASME J. Eng. Ind.* **112** (1990) 142–149.

121. Y. Altintas and M. Weck, Chatter stability of metal cutting and grinding, *CIRP Ann. Manuf. Technol.* **53**:2 (2004) 619–642.

122. N. Suzuki, Y. Kurata, T. Kato, R. Hino, and E. Shamoto, Identification of transfer function by inverse analysis of self-excited chatter vibration in milling operations, *Precision Eng.* **36**:4 (2012) 568–575.

123. L. Pejryd and M. Eynian, Minimization of chatter in machining by the use of mobile platform technologies, *Proceedings of the Fifth International Swedish Production Symposium (SPS12)*, Linkoping, Sweden, 2012, pp. 179–189.

124. T. L. Schmitz and K. S. Smith, *Machining Dynamics Frequency Response to Improved Productivity*, Springer, New York, NY, 2009.

125. R. J. Allemang and D. L. Brown, Machine-tool vibration, Chapter 21, *Shock and Vibration Handbook*, 4th edn., McGraw-Hill, New York, NY, 1996.

126. R. L. Kegg, Cutting dynamics in machine tool chatter-contribution to machine tool chatter, Research-3, *Trans. ASME J. Eng. Ind.*, November 1965, pp. 464–470.

127. H. Van Brussel and P. Van Herck, A new method for the determination of the dynamic cutting coefficient, *Proceedings of the 11th MTDR Conference*, Birmingham, England, 1970, pp. 105–118.

128. H. Opitz, Application of a process control computer for measurement of dynamic coefficients, *CIRP Ann.* **21** (1972).

129. E. J. Goddarol, The development of equipment for the measurement of dynamic cutting force coefficients, MSc thesis, UMIST Manchester, Manchester, England, April 1972.

130. J. Tlusty and B. S. Goel, Measurement of the dynamic cutting force coefficient, *Proc. NAMRC* **2** (1974).

131. R. Snoey and D. Roesems, Survey of modal analysis applications, *CIRP Ann.* **20** (1979) 497–510.

132. L. D. Mitchell, Improved methods for the fast Fourier transform (FFT) calculation of the frequency response function, *ASME J. Mech. Des.* **104** (1982) 277–279.

133. K. B. Elliott and L. D. Mitchell, The improved frequency response function and its effect on modal circle fits, *ASME J. Appl. Mech.* **106** (1984) 657–663.

134. R. J. Allemang, Experimental modal analysis bibliography, *Proceedings of the IMAC2 Conference*, Orlando, Florida, February 1984.

135. L. D. Mitchell, Modal analysis bibliography—An update—1980–1983, *Proceedings of the IMAC2 Conference*, Orlando, Florida, February 1984.

136. D. J. Ewins, *Modal Testing: Theory and Practice*, John Wiley & Sons Inc., reprinted 1991.

137. M. M. Sadek and S. A. Tobias, Comparative dynamic acceptance tests for machine tools applied to horizontal milling machines, *Proc. Inst. Mech. Eng.* **185** (1970–1971) 319–337.

138. Manufacturing Automation Laboratories Inc., CUTPRO 10.2 software, Vancouver, Canada, October 2015.

139. Manufacturing Laboratories, Inc., MetalMAX 10.3 software, Las Vegas, NV, December 2014.

140. J. W. Cooley and J. W. Tukey, An algorithm for the machine computation of complex function series, *Math. Comput.* **19** (1965) 297–301.

141. D. L. Brown, R. J. Allemang, R. Zimmerman, and M. Mergeay, parameter estimation techniques for modal analysis, SAE Paper 790221, 1979.

142. L. D. Mitchell, Improved methods for the fast Fourier transform (FFT) calculation of the frequency response function, *ASME J. Mech. Des.* **104** (1982) 277–279.

143. G. T. Rocklin, J. Crowley, and H. Vold, A comparison of H1, H2, Hv frequency response functions, *Proceedings of the IMAC Conference*, Vol. 1, 1985, pp. 272–278.

144. T. Insperger and G. Stepan, Stability chart for the delayed Mathieu equation, *R. Soc. Lond. Proc. Ser. A Math. Phys. Eng. Sci.* **458**:2024 (2002) 1989–1998.

145. F. A. Khasawneh and B. P. Mann, A spectral element approach for the stability analysis of time-periodic delay equations with multiple delays, *Commun. Nonlinear Sci. Numer. Simulat.* **18**:8 (2013) 2129–2141.

146. Y. Altintas, *Manufacturing Automation: Metal Cutting Mechanics, Machine Tool Vibrations, and CNC Design*, Cambridge University Press, Cambridge, U.K, 2012.

147. A. Otto, S. Rauh, M. Kolouch, and G. Radons, Extension of Tlusty's law for the identification of chatter stability lobes in multi-dimensional cutting processes, *Int. J. Mach. Tools Manuf.* **82–83** (2014) 50–58.

148. A. Otto and G. Radons, Stability analysis of machine-tool vibrations in the frequency domain, 12th IFAC Workshop on Time Delay Systems, At Ann Arbor, MI, June 28–30, 2015.

149. M. Eynian, Frequency domain study of vibrations above and under stability lobes in machining systems, *Procedia CIRP 14, sixth CIRP International Conference on High Performance Cutting (HPC2014)*, Berkeley, CA, 2014, pp. 164–169.

150. Y. Altintas, G. Stepan, D. Merdol, and Z. Dombovari, Chatter stability of milling in frequency and discrete time domain, *CIRP J. Manuf. Sci. Technol.* **1**:1 (2008) 35–44.

151. S. M. Pandit, T. L. Subramanian, and S. M. Wu, Modeling machine tool chatter by time series, *ASME J. Eng. Ind.* **97** (1975).

152. K. F. Eman and S. M. Wu, A comparative study of classical techniques and the dynamic data system (DDS) approach for machine tool structure identification, *Proc. NAMRC* **13** (1985) 401–404.

153. F. A. Burney, S. M. Pandit, and S. M. Wu, A new approach to the analysis of machine tool system stability under working conditions, *ASME J. Eng. Ind.* **99** (1977).

154. T. L. Subramanian, M. F. DeVries, and S. M. Wu, An investigation of computer control of machining chatter, *ASME J. Eng. Ind.* **98** (1976).

155. T. Moriwaki, Measurement of cutting dynamics by time series analysis technique, *CIRP Ann.* 22 (1973).

156. T. Moriwaki and T. Hoshi, System identification on digital techniques, new tool for dynamic analysis, *CIRP Ann.* **23** (1974).

157. S. M. Pandit and S. M. Wu, *Time Series and System Analysis with Applications*, Wiley, New York, 1983.

158. S. M. Wu, Dynamic data system: A new modeling approach, *ASME J. Eng. Ind.* **99** (1977) 708–714.

159. T. Schmitz, J. Ziegert, and C. Stanislaus, A method for predicting chatter stability for systems with speed-dependent spindle dynamics, *Trans. NAMRI/SME* **32** (2004).

160. E. Ozturk, U. Kumar, S. Turner, and T. Schmitz, Investigation of spindle bearing preload on dynamics and stability limit in milling, *CIRP Ann. Manuf. Technol.* **64** (2012) 343–346.

161. A. S. Delgado, E. Ozturk, and N. Sims, Analysis of non-linear machine tool dynamic behavior, *The Manufacturing Engineering Society International Conference, MESIC 2013, Procedia Engineering* 63 (2013) 761–770.

162. Q. Fu, Joint interface effects on machining system vibration, Thesis, KTH School of Industrial Engineering and Management, Stockholm, Sweden, 2013.

163. M. P. Norton, *Fundamentals of Noise and Vibration Analysis for Engineers*, Cambridge University Press, New York, 1994.

164. A. H. Slocum, E. R. Marsh, and D. H. Smith, A new damper design for machine tool structures: The replicated internal viscous damper, *Precision Eng.* **16** (1994) 539.

165. C. C. Ling, J. S. Chen, and I. E. Morse Jr., An experimental investigation of damping effects on a single point cutting process, *ASME Manuf. Sci. Eng.* MED-Vol. **2–1**/MH-Vol. **3–1** (1995) 149–164.

166. E. I. Rivin and W. D'Ambrogio, Enhancement of dynamic quality of a machine tool using a frequency response optimization method, *Mech. Syst. Signal Process.* 4 (1990) 495.

167. M. Wang, T. Zan, Y. Yang, and R. Fei, Design and implementation of nonlinear TMD for chatter suppression: An application in turning processes, *Int. J. Mach. Tools Manuf.* **50**:5 (2010) 474–479.

168. D. Sathianarayanan, L. Karunamoorthy, J. Srinivasan, G. S. Kandasami, and K. Palanikumar, Chatter suppression in boring operation using magneto-rheological fluid damper, *Mater. Manuf. Process.* **23**:4 (2008) 329–335.

169. E. I. Rivin and H. Kang, Enhancement of dynamic stability of cantilever tooling structures, *Int. J. Mach. Tools Manuf.* **32** (1992) 539.

170. B. W. Wong, B. L. Walcott, and K. E. Routh, Active vibration control via electromagnetic dynamic absorbers, *Proceedings of the Fourth IEEE Conference on Control Applications*, Albany, NY, 1995.

171. M. A. Marra, B. L. Walcott, K. E. Rough, and S. G. Tewani, H vibration control for machining using dynamic absorber technology, *Proceedings of 1995 ACC*, Seattle, WA, 1995, pp. 739–743.

172. J. Slavicek, The effect of irregular tooth pitch on stability in milling, *Proceedings of the Sixth MTDR Conference*, Pergamon Press, London, U.K., 1965.

173. P. Vanherck, Increasing milling machine productivity by use of cutter with non-constant cutting-edge pitch, *Proceedings of the Advanced MTDR Conference*, Vol. 8, Pergamon Press, London, U.K., 1967, pp. 947–960.

174. B. J. Stone, Advances in machine tool design and research, *Proceedings of the 11th International MTDR Conference A*, Pergamon Press, London, U.K., 1970, pp. 169–180.

175. H. Fazelinia and N. Olgac, New perspective in process optimization of variable pitch milling, *Int. J. Mater. Prod. Technol.* **35**:1–2 (2009) 47–63.

176. A. R. Yusoff, S. Turner, C. M. Taylor, and N. D. Sims, The role of tool geometry in process damped milling, *Int. J. Adv. Manuf. Technol.* (2010) 1–13.

177. L. T. Tunç and E. Budak, Effect of cutting conditions and tool geometry on process damping in machining, *Int. J. Mach. Tools Manuf.* **57** (2012) 10–19.

178. C. Y. Huang and J. J. Wang, Mechanistic modeling of process damping in peripheral milling, *J. Manuf. Sci. Eng.* **129** (2007) 12–20.

179. C. M. Taylor, N. D. Sims, and S. Turner, Process damping and cutting tool geometry in machining, *Trends in Aerospace Manufacturing 2009 International Conference, Materials Science and Engineering*, IOP Publishing, U.K., Vol. 26, 2011.

180. L. T. Tunc and E. Budak, Effect of cutting conditions and tool geometry on process damping in machining, *Int. J. Mech. Sci.* **57** (2012) 10–19.

181. A. R. Yusoff and S. Turner, The role of tool geometry in process damped milling, *Int. J. Adv. Manuf. Technol.* **50** (2010) 883–895.

182. W. Chen, Cutting forces and surface finish when machining medium hardness steel using CBN tools, *Int. J. Mach. Tool Manuf.* **40** (2000) 455–466.

183. M. Eynian and Y. Altintas, Analytical chatter stability of milling with rotating cutter dynamics at process damping speeds, *ASME J. Manuf. Sci. Technol.* **132** (2009) 354–361.

184. Y. Altintas, M. Eynian, and H. Onozuka, Identification of dynamic cutting force coefficients and chatter stability with process damping, *CIRP Ann. Manuf. Technol.* **57** (2008) 371–374.

185. N. D. Sims and M. S. Turner, The influence of feed rate on process damping in milling modeling and experiments, *Proc. Inst. Mech. Eng. B: J. Eng. Manuf.* **225** (2011) 799–810.

186. S. Smith and T. Delio, Sensor-based chatter detection and avoidance by spindle speed selection, *ASME J. Dyn. Syst. Meas. Control* **114** (1992) 486–492.

187. T. Delio, S. Smith, and J. Tlusty, Use of audio signals chatter detection and control, *ASME J. Eng. Ind.* **114** (1992) 486–492.

188. Y. Altintas and P. K. Chan, In-process detection and suppression of chatter in milling, *Int. J. Mach. Tools Manuf.* **32** (1992) 329–347.

189. T. R. Comstock, F. S. Tse, and Z. R. Lemon, Application of controlled mechanical impedances for reducing machine tool vibration, *ASME J. Eng. Ind.* **91** (1969).

190. C. L. Nachtigal and N. H. Cook, Active control of machine tool chatter, *ASME J. Eng. Ind.* **92** (1970) 238.

191. R. G. Klein and C. L. Nachtigal, The active control of a boring bar operation, ASME Paper 72-WA/AUT-19, 1972.

192. K. C. Maddux and C. L. Nachtigal, Wideband active chatter control scheme, ASME Paper 71-WA/AUT-10, 1971.

193. C. L. Nachtigal and K. C. Maddux, An adaptive active chatter control scheme, ASME Paper 72 WA/AUT-10, 1972.

194. E. E. Mitchell and E. Harrison, Design of a hardware observer for active machine tool control, *ASME, J. Dyn. Syst. Meas. Control* **99** (1977) 227.

195. T. Takemura, T. Kitamura, and T. Hoshi, Active suppression of chatter by programmed variation of spindle speed, *CIRP Ann.* **23** (1974) 121–122.

196. T. Inamura and T. Sata, Stability analysis of cutting under varying spindle speed, *CIRP Ann.* **23** (1974) 119–120.

197. T. Takemura, T. Kitamura, T. Hoshi, and K. Okushima, Active suppression of chatter by programmed variation of spindle speed, *CIRP Ann.* **23** (1974) 121–122.

198. J.-y. Yu, X. J. Han, and B.-d. Wu, Study on turning with continuously varying spindle speed, *Proc. ASME Manuf. Int.* **1** (1988) 327–331.

199. S. C. Lin, R. E. DeVor, and S. G. Kapoor, The effects of variable speed cutting on vibration control in face milling, *ASME J. Eng. Ind.* **112** (1990) 1–11.

200. T. Hosi et al., Study of practical application of fluctuating speed cutting for regenerative chatter control, *CIRP Ann.* **26** (1977) 175–180.

201. H. Zhang, J. Ni, and H. Shi, Machining chatter suppression by means of spindle speed variation, *Proceedings of the First S.M. Wu Symposium on Manufacturing Science*, Evanston, Illinois, 1994, pp. 161–175.

202. M. Zatarain, I. Bediaga, J. Mun, and R. Lizarralde, Stability of milling processes with continuous spindle speed variation: Analysis in the frequency and time domains, and experimental correlation, *CIRP Ann.* **57** (2008) 79–84.

203. H. H. Zhang, M. J. Jackson, and J. Ni, Spindle speed variation method for regenerative machining chatter control, *Int. J. Nanomanuf.* **3**:1–2 (2009) 73–99.

204. S. Seguy, T. Insperger, L. Arnaud, G. Dessein, and G. Peigne, On the stability of high-speed milling with spindle speed variation, *Int. J. Adv. Manuf. Technol.* **48**:9–12 (2010) 883–895.

205. S. Seguy, G. Dessein, L. Arnaud, and T. Insperger, Control of chatter by spindle speed variation in high-speed milling, *Adv. Mater. Res.* **112**:1 (2010) 179–186.

206. I. Bediaga, J. Munoa, J. Hernandez, and L. N. Lopez, An automatic spindle speed selection strategy to obtain stability in high-speed milling, *Int. J. Mach. Tools Manuf.* **49**:5 (2009) 384–394.

207. D. Barrenetxea, J. I. Marquinez, I. Bediaga, and L. Uriarte, Continuous workpiece speed variation (CWSV): Model based practical application to avoid chatter in grinding, *CIRP Ann. Manuf. Technol.* **58**:1 (2009) 319–322.

208. K. Jemielniak and A. Widota, Suppression of self-excited vibration by the spindle speed variation method, *Int. J. Mach. Tool Des. Res.* **23** (1984) 207–214.

209. T. Hos et al., Study of practical application of fluctuating speed cutting for regenerative chatter control, *CIRP Ann.* **26**:1 (1977) 175–180.

210. T. Delio, Method of controlling chatter in a machine tool, US Patent No. 5,170,358, December 1992.

211. S. Smith and W. R. Winfough, The effect of runout filtering on the identification of chatter in the audio spectrum of milling, *Trans. NAMRI./SME* **22** (1994) 173–178.

212. S. Smith and J. Tlusty, Stabilizing chatter by automatic spindle speed regulation, *CIRP Ann.* **41** (1992) 433–436.

213. S. Smith et al. Automatic chatter avoidance in milling using coincident stable speeds for competing modes, ASME PED-68 (1994) 737–747.

214. S. Y. Tsai, On line identification and control of machining chatter in turning through the dynamic data system methodology, PhD dissertation, Mech. Eng. Dept., University of Wisconsin, Madison, WI, 1983.

215. K. F. Eman, Machine tool system identification and forecasting control of chatter, PhD dissertation, Mech. Eng., University of Wisconsin, Madison, WI, 1979.

216. Y. Miyoshi and H. Nakazawa, Time series forecasts of chatter stability limits (1st report)—Establishment of foundational method, *Bull. Jpn. Soc. Precision Eng.* **13** (1979).

217. W. Anderson, Cut the chatter, *Cutting Tool Eng.* **67**:4 (2015).

13 Machining Economics and Optimization

13.1 INTRODUCTION

Economic considerations are obviously important in designing a machining process. There is generally more than one approach for machining a particular part; each approach will have an associated cost and level of part quality. The machine tool, tooling, and process technologies used in a given application should ideally result in a process that meets part quality and production rate requirements with minimum cost. The initial method of producing a new part, including the machine tools, cutting tool materials and geometries, speeds and feeds, and coolant, is generally determined from previous experience with similar parts, handbook recommendations, catalog data, or rules of thumb. These sources provide plausible starting points but rarely yield the most efficient approach. Furthermore, cutting processes have to deal with a multitude of variations and disturbances, which can only be partially considered during the design phase. Therefore, during production the demanded process stability is not always maintained resulting in a quality below predefined tolerances. In high volume operations, changes are continuously made as experience with the specific part is accumulated. This can be a tedious process and often results in comparatively inefficient practices being used over much of the production run. A more efficient methodology to optimize machining practices would be desirable to reduce the time required to identify the best process more systematically.

In today's manufacturing environment, highly automated and computer-controlled machines are used, and there is an economic need to operate them as efficiently as possible to maximize productivity. The goal is to integrate strategic process optimization concepts and intelligent controls in high speed and high performance machining including on-machine quality control through the implementation of learning modules that will simulate new-machining processes during the design, analysis, and prototyping for new products. The selection of optimum process parameters directly affects the productivity, manufacturing cost, and part quality [1]. The manufacturing cost per part is made up of several components, including the machining cost, tool cost, tool change and its setup costs, and handling cost. These costs vary most significantly with the cutting speed as shown in Figure 13.1. At a certain cutting speed the costs reach a minimum, which corresponds to the optimum cutting speed. A large amount of machining information is constantly generated in a shop or manufacturing plant, which if properly analyzed could clarify cost relations and simplify the identification of and optimum process.

Modeling and optimization of process parameters of a machining process is a complex task because it requires the knowledge of an effective optimization criterion (i.e., cost model, productivity, part quality, etc.) empirical equations describing realistic constraints, machine tool capabilities, and the selection of the optimization technique [1]. Empirical and mechanistic models are used to describe the relationship of input–output and in-process parameters. Computer-aided selection of limited machining parameters has become possible based on look-up tables or mathematical formulas. Systems that perform such tasks can transform the selection of machinability information from an experience-based activity to a more scientific discipline so that an efficient/optimum machining process can be identified as shown in Figure 13.2. The most common class of such systems is computer-aided process planning (CAPP) systems [2,3], which work with computer-aided design (CAD) systems to extract the relevant geometric features of a part and produce a process sequence

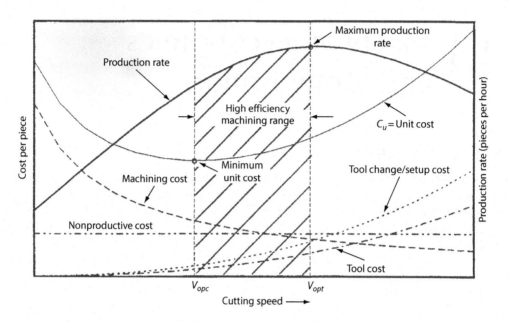

FIGURE 13.1 Machining cost and production rate versus cutting speed.

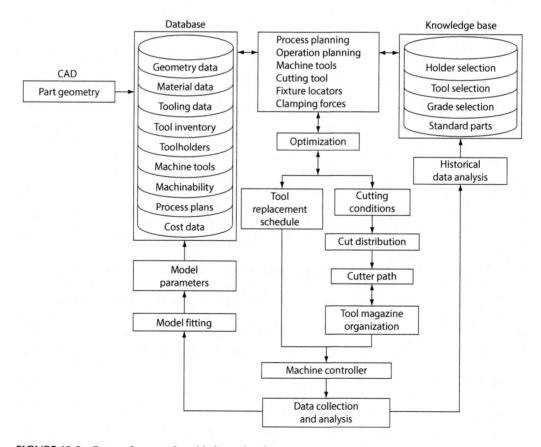

FIGURE 13.2 System framework and information flow.

or set of tool paths, which is optimal in some sense. The information reviewed in the previous chapters including analysis methods in Chapter 8 for process modeling and simulation are also useful for this purpose.

This chapter discusses the economics of machining operations with particular emphasis on analytical and computer methods. Computerized optimization systems (COSs) are discussed in Sections 13.2. Basic economic considerations are discussed in Section 13.3. Based on these considerations, analytical methods for optimizing various classes of machining systems and operations are discussed in Sections 13.4 through 13.7. In many ways, this discussion shows how information reviewed in previous chapters, especially Chapters 2, 3, 4, 9, 10, and 11, can be used to make the basic economic decisions, which are central to successful process design. Finally, four numerical examples are presented in Section 13.8.

13.2 ROLE OF A COMPUTERIZED OPTIMIZATION SYSTEM

This section discusses the general features and functions of a COS. There is a natural movement in the development of software to automate the process from product design to manufacturing. In addition, manufacturing complexity and loss of skilled manufacturing workers make the application of computers more attractive to define the machining process from CAD model. In describing the function of such systems, it is helpful to distinguish three levels of optimization corresponding to increasingly sophisticated analysis capabilities.

The common manufacturing data base of a CIM system typically consists of many kinds of interacting information sources. One of the important data bases for any CIM system is the one that stores machinability data (Figure 13.2). The basic role of a computerized machinability data base system is to select the cutting speed, feed rate, and depth of cut (if needed) for each operation in the process given the following characteristics of the machining operation: workpiece material, type of machining operation, machine tool, cutting tool, cut distribution, cutter path, tool replacement schedule, and operating parameters other than speed and feed. In addition to these basic functions, the *first level* of a COS can use computerized machinability systems to select the optimum cutting conditions based on economic criteria, select appropriate cutting tools from the available inventory, determine the type of cut (i.e., single- or multipass operations, number of passes, depth of cut (doc) for each pass), and select suitable machine tools from the available inventory. It is not uncommon to consider the surface finish and surface quality/integrity as manufacturing specifications, which eventually control the selection of cutting parameters. Often suitable models and the interconnection of virtual and actual results are necessary. This includes the joining of experimental results and simulations. FEM/FEA and analytical models are used (see Chapters 8 and 12) to simulate the cutting process together with experimental results to identify the basic process behavior and generate reliable models. In principle, the computer selected speeds and feeds can be inserted into the NC program automatically, but this has not been done on a widespread basis in practice. This level is sometimes defined as "operations-based system" because the user selects the operation including the tool path and cutting tools. Cutting speed, feed, and depth of cut are optimized based on the available data from the cutting tool manufacturer and the user previous knowledge. Software are used today to optimize cutting speeds for more efficient machining including for high speed machines including sequence of operations [4,5].

The *second level* of a COS optimizes the manufacturing process and sequence of operations (in the case of drilling, reaming, tapping, etc.) on a single or multistage machining system (e.g., dial machines, conventional transfer lines, flexible/convertible transfer lines, multi-axis machining centers with single or multiple spindles, and flexible manufacturing systems) using the specified part shape, size, and dimensional accuracy. Often the features to be machined are extracted from a CAD model using a feature extraction algorithm [6–8]. Such a planning system will reduce dependence on skilled human personnel for the development of an efficient part process sequence and tool magazine organization and can be designed to take advantage of the machine capabilities and system flexibility. The benefits of such an approach include gains in production time, machine utilization, and the number of parts produced between tool magazine reloadings, which results in lower manufacturing part costs. The second level of a COS sometimes is referred as CAPP system [2,3] (see Figure 13.3). Several CAPP systems are being developed as a link between design and manufacturing, filling the gap between CAD and computer aided manufacturing (CAM). While integrated CAD/CAM tools have been available, their costs have restricted their application to larger companies. Feature-based machining model has

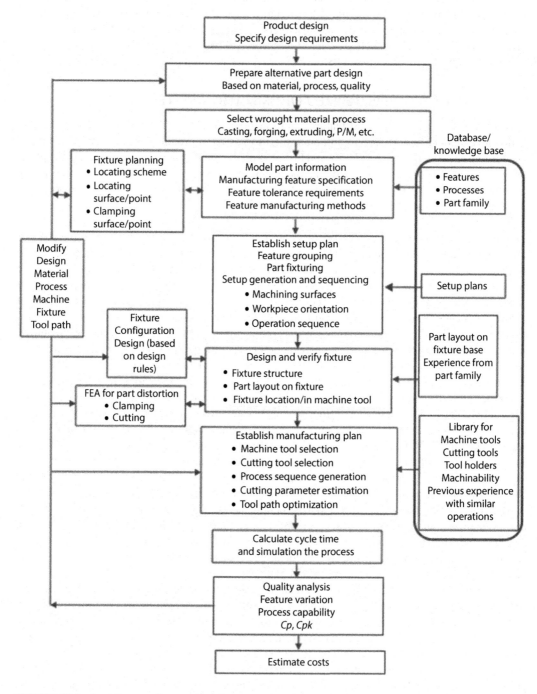

FIGURE 13.3   Input/output diagram for a CAPP system.

been used by several CAM software because they use knowledge-based machining (KBM) together with feature recognition rules and machining knowledge rules to assign cutting operations, cutting tools, cutting conditions, etc., to automatically create the machining process. The KBM captures user knowledge and shop expertise and allows frequently used tool path process parameters, machine tool on the floor, workpiece material, cutting tool geometries, and materials to be stored in a library that is available for later utilization for other parts [9–13].

The second level of a COS interacts with the first level. It uses detailed part geometry, machine motion, and tooling data files to calculate individual machine motions and cutting times based on the optimum cutting conditions identified by the first level of the COS. In this case, the CAD model is imported to be machined, the size of the starting part is specified, the user selects the machine to be used, and whether high speed or conventional conditions are used. The software will select the sequence of operations, cutting tools, tool path, cutting conditions, etc. This analysis can provide information on gross productivity, and estimate of the required investment, and detailed data on individual machine stations and station performance. Such computer programs analyze both the whole system and individual machining stations to balance work loads, increase productivity, and decrease investment. Compared to manual methods, such machine and part processing simulations can significantly reduce the total time required to develop and analyze manufacturing systems. Several variations of the machining system, the processing sequence, and tooling designs can be evaluated with minimal effort once the initial analysis of the system is completed.

The *third level* of a COS is used to determine optimum combination of all machining parameters including the machine tool characteristics (stiffness, dynamic parameters, accuracy, etc.), part fixture, cutting tool material and geometry, and cutting fluid. The optimization must include all system objectives as outputs and should use the entire range of working conditions as inputs. Approaches such as expert system concepts and neural networks for modeling manufacturing process parameters have been used to develop third-level COSs.

A knowledge-based expert system [14,15] includes a machinability data base (experimental results and theoretical predictions including information on machine tools, cutting tools, coolants, etc.), inference engine, knowledge base, and working memory. A wealth of symbolic (rules of thumb) knowledge gained through experience can be incorporated into the knowledge base. The knowledge is collected, analyzed, and systematically organized. The inference engine interacts sequentially with the databases to find the best combination of process parameters (selection of machine tools, cutting tools, coolant, etc.), from which the first or second level of the COS can be used to determine the optimum process.

Neural networks [16–21] process information through the interactions of a large number of simple processing elements or nodes. Knowledge is represented by the strengths of the connections between elements. Each piece of knowledge is a pattern of activity spread among many processing elements, and each processing element may be involved in the partial representation of many pieces of data. There is an input layer and an output layer to receive data and send information out to users. Neural networks have the ability to learn using an algorithm that maps input vectors to output vectors.

In addition, this level could analyze the part quality based on the machine tool characteristics as discussed in Chapter 16. Virtual quality analysis is performed based on the volumetric accuracy of the machine tool including the fixture and the details are provided in Chapter 16.

13.3 ECONOMIC CONSIDERATIONS

To reduce manufacturing costs, the direct concern of factory floor management has traditionally been the improvement of machine utilization and the reduction of part lead time. Machine time and part lead time are composed of

Machine time = Set-up time + Loading/Unloading time + Cutting time + Idle time

Part lead time = Set-up time + Loading/Unloading time + Cutting time + Waiting time + Moving time

Machine utilization and lead time efficiency are defined, respectively, as

$$\text{Machine utilization} = \frac{\text{Cutting time}}{\text{Machine time}}$$

$$\text{Lead time efficiency} = \frac{\text{Cutting time}}{\text{Lead time}}$$

The production cost is given by

The operating cost of the production

$$= \left(\text{Capital opportunity cost}\right) + \left(\text{Depreciation cost}\right) + \left(\text{Labor cost}\right)$$

$$= \left(\text{Capital investment}\right)\left(\text{Interest rate}\right)\left(\text{Processing time}\right)/\text{Utilization}$$

$$+ \left(\text{Capital investment}\right)\left(\text{Depreciation rate}\right)\left(\text{Processing time}\right)/\text{Utilization}$$

$$+ \left(\text{Labor cost rate}\right)\left(\text{Processing time}\right)$$

Reliability is important because, as the capability of machines increases, the failure of machines has a bigger effect on the material flow. This requires an analytical evaluation of routing changes for the workpiece during extended breakdowns. The part quality becomes important because a bad workpiece detected at an inspection station is either reprocessed or discarded. In the case of a discard, part processing expenses and raw material expenses are lost. Inspection cost may also increase as part quality decreases. Understanding the cause of poor quality and corrective measures (if they involve modifying machines) contribute additional costs.

13.4   OPTIMIZATION OF MACHINING SYSTEMS: BASIC FACTORS

The most important factors that influence a process are the workpiece material and finished part requirements, type and structure of machine tool, cutting tool material and geometry, toolholder, cutting fluid, and cutting conditions. Knowledge on these factors, generally based on experience, must be incorporated into the knowledge base.

The specific information for each group of factors necessary to implement a COS is discussed in the remainder of this section.

Workpiece and material: The material type, microstructure, hardness, ductility, machinability rating, heat treatment, and initial surface condition affect all the selection parameters discussed next. In addition, the surface finish and integrity, starting size and shape, final size and shape, tolerances, and lot size affect the selection parameters. Chapter 11 provides information on workpiece material.

Machine tool selection: For a given part size and geometry, a family of machine tools will be available from the database provided by the user. The operation selection and best tool orientation will screen the available machines further or, vise versa, the available machine tools will determine the best processing and tooling requirements.

The artificial intelligence approach requires that several machine tool performance parameters be available in the database. These parameters should describe the machine type (number of axes, spindle orientation, etc.), table type (rotary, indexing, etc.), table rotational speed, pallet change time, workpiece weight limit, capabilities of each axis (acceleration/deceleration, rapid traverse rate, maximum feed rate, maximum thrust, resolution increment, and full travel distance), spindle capabilities (speed vs. torque/power response, maximum speed, maximum thrust and radial forces, and acceleration/deceleration rate), tool change time, tool–spindle interface style, magazine capacity, coolant type(s), and accuracy and repeatability of the axes.

Cutting tool selection: Tool parameters to be selected include the tool type and style, material and grade, and geometry (tool angles). The tooling database should contain the tool description (type and style), diameter, geometry, chip breaker, material (grade), hardness, coating, number of teeth, tool performance with respect to chip ejection, hole location, and maximum allowable feed rate, surface speed or rpm, forces, and torque.

Toolholder selection: The proper selection of the toolholder is critical to tool performance as discussed in Chapter 5. The database should contain the available types of toolholders for the available cutting tools and their performance characteristics. The limitations of each toolholder type with respect to roundness and location as a function of the L/D ratio of the tool should be included. A comparative assessment of the toolholder stiffness and damping as a function of the L/D ratio of the tool should also be available.

Cutting fluid selection: The selection of the cutting fluid depends mainly on the operation, tool type, workpiece material, and cutting conditions. Some applications require coolant (i.e., gun drilling) while others prevent coolant application (i.e., ceramic tools). The coolant application method is also important as explained in Chapters 14 and 15.

Cutting conditions selection: The selection of cutting conditions depends on all of the factors listed before and can be optimized as will be discussed in the remainder of this chapter. The basic parameters to be selected are the cutting speed, feed rate, depth of cut, and length and width of cut, which determine cutting forces and tool life.

13.5 OPTIMIZATION OF MACHINING CONDITIONS

The proper selection of machining conditions—the cutting speed, feed rate, and depth of cut—has a strong impact on machining performance. The optimization of cutting conditions was initially investigated by Gilbert [22] in 1950 and is still an active area of research. Many new concepts and optimization procedures have been developed to determine optimum conditions, taking into account as many influencing variables as possible [23–26]. However, the use of these advanced techniques in machinability data base systems to date has been very limited, even though there is a strong economic need for such techniques.

Many investigations on the optimization of cutting conditions have been confined to single stage manufacturing, namely, on a single machine tool using a single cutting tool. Single stage optimizations may involve single pass or multipass operations. A single pass analysis typically addresses the problem of determining optimum values of the cutting speed and feed rate assuming that the depth of cut is given. A multipass analysis typically addresses the problem of determining optimum cutting speeds, feeds, docs, and number of passes simultaneously for a given total doc. Turning is the most commonly analyzed machining operation due to its simplicity. Since a product requiring only one machining operation is seldom found in practice, optimization of multistage manufacturing systems has been attempted by a few researchers.

The steps involved in optimizing cutting conditions include

- Formulating an objective function based on the desired economic criteria
- Defining all the constraints applicable to the machining system
- Minimizing/maximizing the objective function subject to the constraints

Formulation of the objective function based on the desired economic criterion has been investigated extensively in the literature. Economic criteria are normally functions of cost elements, metal removal rates, and tool life. Representative objective functions are related to only one criterion, such as the production cost, production time [27–31], profit rate [32,33], or to a combined criterion based on a weighted sum of these factors [34].

An objective function based on production cost does not consider the influence of production time. Maximum production objectives identify the cutting conditions that best balance the metal removal rate and tool life to produce the highest output. Minimum cost objectives identify the cutting conditions that best balance the metal removal rate and tool life for the lowest cost (see Figure 13.1). Maximum rate of profit criteria (or maximum efficiency criteria) has been proposed to include and eventually achieve a balance between the contributions of both minimum production cost and total time criteria into the objective function [32,33]. However, this criterion cannot adequately predict the optimum cutting conditions; it has been found that unless the profit margin is very high, the optimum conditions predicted by the maximum profit rate criterion lie close to the minimum production cost conditions [33]. Also, it cannot be readily applied since prior knowledge of the percentage of the product sale price (usually unknown) attributed to the particular operation being optimized is needed to determine the operation's profit rate. A more effective technique, especially for complex production systems, may be multiple criteria optimization [34], which considers interactions between the various criteria. A difficulty with this approach is that the order of importance of the different criteria must be specified, and this introduces a subjective element into the analysis.

Price estimation to manufacture a product is critical process since there is a large amount of variation that can occur based on the available production equipment or new equipment selection. Knowledge-based systems are used because the company's previous knowledge on machining other parts is available. The methods for estimating the cost and price including the knowledge-based systems are reviewed in Ref. [35].

13.6 FORMULATION OF THE OPTIMIZATION PROBLEM

The formulation of the optimization problem requires specification of equations to represent the economic and physical parameters of the machining process and the entire machine–tool–workpiece system. The physical parameters are obtained through tests, previous production runs, or existing data from machinability database systems.

13.6.1 FORMULATION OF OBJECTIVE FUNCTION

Both the production cost and total production time are considered in the formulation of the objective function. The equations for determining cutting parameters, metal removal rate, machining cost, and time are quite general and may be used for a variety of machining operations such as turning, milling, drilling, etc. The total production time for an operation is given by

$$T_{u_i} = t_{m_i} + \frac{t_{m_i}}{T_i} t_{\ell_i} + t_{cs} + t_{e_i} + t_{r_i} + t_{p_i} + t_{\alpha_i} + t_{d_i} + t_{x_i} \tag{13.1}$$

where
t_{m_i} is the cutting time
t_{ℓ_i}, t_{cs}, t_{e_i}, t_{r_i}, t_{p_i}, t_{α_i}, t_{d_i}, and t_{x_i} are the tool's loading/unloading, interchange, magazine traveling, approach, table index, acceleration, deceleration, and rapid travel location times, respectively, for the ith operation or pass

The tool's loading/unloading time, t_{ℓ_i}, represents the time to replace a worn or broken tool (not usable for further machining) in the machine. The interchange time, t_{cs}, represents the time required for a tool on the work spindle to be replaced from the tool magazine by the tool required for the next operation. All these times are required because the tool change time between consecutive operations in a NC machine is not constant but varies depending on the speed and acceleration/deceleration time of the spindle and machine slides, tool switch time, machine travel distance (which allows for tool change between operations), and the tool magazine indexing time. The assumption of constant tool-change time is often erroneous, especially when applied to high-speed machining centers for which the tool-change time is as significant or perhaps more important than the machining time. T_i is the life of the tool used in ith operation.

For a turning, boring, and drilling operations the cutting time is equal to

$$t_{m_i} = \frac{\pi D_i L_i}{1000 V_i f_i} \tag{13.2}$$

For a milling operation

$$t_{m_i} = \frac{\pi D_i (L+\varepsilon)_i}{1000 n_{t_i} V_i f_i} \tag{13.3}$$

For a tapping operation, Equation 13.2 is used with the thread pitch m_i substituted for the feed f_i. D is either the diameter of the machined surface or the rotary tool diameter. L, V, and f are the workpiece length to be machined, cutting surface speed, and feed of cutting tool, respectively. n_t and ε are, respectively, the number of inserts (edges) per cutter body and the overtravel of milling cutter on the workpiece (Chapter 2).

The machine slide's rapid travel time between two machine coordinate locations t_{x_i} is

$$t_{x_i} = \frac{V_s^2 + 7.2\alpha_s \Delta S}{3600 \alpha_s V_s} + t_{s_i} \tag{13.4}$$

where
$\Delta S = \sqrt{X_i^2 + Y_i^2 + Z_i^2}$
V_s and α_s are the speed and acceleration of the machine slides, respectively
ΔS is the machine rapid travel distance
X_i, Y_i, and Z_i are the machine traveling coordinates for the ith operation

The deceleration function of the slides consists of two parts, linear and nonlinear. A two part ramp-down is used to maximize speed during deceleration, yet minimize deceleration discontinuity, which causes system oscillation or shock. The time required to accomplish the nonlinear deceleration portion is called settling or stabilization time of the machine slides, t_{s_i}, which is a function of the servo control characteristics.

The time required for the spindle to accelerate or decelerate during the tool change is

$$t_{a_i} = t_{d_i} = \frac{V_i}{1.8\alpha D_i} \tag{13.5}$$

where α is the acceleration of the spindle. Linear accelerations and decelerations are assumed in the previous two equations.

The tool magazine indexing (traveling) time is almost proportional to the number of slots that the magazine has to travel:

$$t_{e_i} = \frac{|j_i - j_{i-1}| S_{cs}}{1000 V_{cs}} + \frac{V_{cs}}{1800 \alpha_{cs}}, \quad j = 1, 2, \ldots, M^T \tag{13.6}$$

where
S_{cs} is the distance between adjacent slots in the tool magazine
α_{cs} and M^T are the acceleration and the number of slots of the tool magazine, respectively

The tool switch time, t_{cs}, is the time required for a tool in a tool chain slot to be inserted into the spindle (Chapter 2).

A general tool life equation applied to all processes is

$$T_i(\underline{s}) = R_{ci} \prod_{k=1}^{p} s_{ik}^{a_{ki}} \quad \forall i \in A_c, \ \forall k \in A_b, \ s \in \{f, d_c, V, V_B, \ldots\} \tag{13.7}$$

where s_{ik} is the kth cutting parameter for the ith operation and R_c is a tool life constant.

The production cost for an operation is given by

$$C_{u_i} = C_o t_{m_i} + \frac{t_{m_i}}{T_i} (C_o t_{\ell_i} + C_{t_i})$$

$$+ C_o \cdot (t_{cs} + t_{e_i} + t_{r_i} + t_{p_i} + t_{\alpha_i} + t_{d_i} + t_{x_i}), \quad \forall i \in A_c \tag{13.8}$$

where C_o and C_t are the operating cost and tool cost. The relationships between the cutting speed and these costs are shown in Figure 13.1. The machining cost decreases with increasing cutting speed, while both the tool cost and the tool change/setup cost increase with increasing cutting speed. The cutting speed for minimum cost per part is determined by taking the partial derivative of Equation 13.8 with respect to speed and setting it equal to zero. The result is

$$V_{opc} = V_{\min \text{ cost}} = \frac{K_t}{f_h^a d_c^b \left[\left(\frac{1}{n} - 1\right)\left(t_\ell + \frac{C_t}{C_o}\right)\right]^n} = \frac{\left[(-a_3 - 1)\left(t_l + \frac{C_t}{C_o}\right)\right]^{1/a_3}}{\left(R_c f_h^{a_1} d_c^{a_2}\right)^{1/a_3}} \tag{13.9}$$

where f_h and d_c are the highest possible feed and depth of cut; the tool life constants a, b, n, a_1, a_2, and a_3 are from the tool life equations, Equations 9.9 and 13.7. When the maximum production rate is required, the time per part per operation, given by Equation 13.1, must be minimized. In this case, the tool change/replacement times are important. The highest permissible production rate is again found by selecting the highest possible feed and calculating the cutting speed from [36]

$$V_{opt} = V_{\max \text{ prod}} = \frac{K_t}{f_h^a d_c^b \left[\left(\frac{1}{n} - 1\right)t_\ell\right]^n} = \frac{\left[(-a_3 - 1)t_\ell\right]^{1/a_3}}{\left(R_c f_h^{a_1} d_c^{a_2}\right)^{1/a_3}} \tag{13.10}$$

The cutting speed for maximum production rate (V_{opt}) is always higher than that of minimum cost (V_{opc}) as shown in Figure 13.1. The speed range V_{opc} to V_{opt} is referred to as the high-efficiency machining range or "Hi-E" range, which means that any machining speed in this range is preferable from an economic standpoint. In cases in which the production rate is fixed (as in transfer lines and dial machines), the economic tool life or maximum production rate should not be used. Instead, the tool cost should be kept as low as possible, that is, by selecting the cutting conditions most suitable for the required production rate (cycle time). However, it is desirable to use V_{opt} in bottleneck operations. There are several cases that require the estimation of the optimum cutting parameters based on tool life or tool wear equations with constraints on the metal removal rate, thrust force, torque, etc. [23,37,38].

The total production time per part is

$$T_t = \sum_{i=1}^{n} T_{u_i} + t_h \qquad (13.11)$$

where
t_h is the loading and unloading time for the part
n is the number of operations or passes performed on the part in a given setup

The profit rate is given by

$$P = \frac{S - C_u}{T_u} \qquad (13.12)$$

where S is the sale price of the machined part. Until the maximum profit criterion was developed there had been no way to indicate how the optimum point should be chosen from the Hi-E range from V_{opc} to V_{opt}. The cutting speed for maximum profit lies between V_{opc} and V_{opt} and provides an economic balance between the cost and production rate [39,40].

The objective function is a weighted sum of the production cost and the production time:

$$U(f, V, d) = w_1 C_u + w_2 \lambda T_u \qquad (13.13)$$

where
w_1 and w_2 are the weighting coefficients representing the relative importance of the production cost and the total production time criteria, respectively
λ is a conversion factor (see the following)

It is usually assumed that the weighting factors should satisfy the condition

$$w_1 + w_2 = 1, \quad 0 \le w_1 \le 1, \quad 0 \le w_2 \le 1 \qquad (13.14)$$

Their specific values should be selected based on the relative importance of cost and production rate for the prevailing business conditions. Usually, the optimization problem is solved for different trial weight values to determine the optimum weight distribution for a particular case. These weight values can be used for optimizing processes for similar parts. Hence, the weighting coefficients are usually determined in part by analytical means and in part by experience.

The objective function is normalized using a constant multiplier, λ:

$$\lambda = \frac{C_{u,min}}{T_{u,min}} \qquad (13.15)$$

where $C_{u,min}$ and $T_{u,min}$ are the minimum production cost and total time, respectively, under the defined process constraints. The minimum cost and total time are determined by using the optimum values of the cutting speeds, V_{opc} and V_{opt}, in Equations 13.8 and 13.1 and the highest possible feed, f_h, in both equations. An analytical equation for λ cannot be developed because the highest feed is a function of depth of cut under the applicable process constraints.

13.6.2 Constraints

Physical limitations on cutting conditions due to the characteristics of the machine–tool–workpiece system should be identified from previous experience and taken into account in the optimization process. A number of deterministic constraints exist, such as allowable maximum cutting force, cutting temperature, depth of cut, speed, feed, machine power, vibration and chatter limits, and constraints on part quality. Some constraints, such as those on speed, feed, etc., are simple boundary constraints, while others must be computed from linear or nonlinear equations. The following general equation can be used:

$$G_{ij}(s) = B_{ij} \prod_{k=1}^{p} s_{ij}^{b_{kij}} \quad \forall i \in A_c, \quad \forall k \in A_b, \quad j \in 1, 2, \ldots, M_i^c \tag{13.16}$$

where b_{kij}, B_{ij}, and E_{ij} are empirical constants for the ith operation and jth constraint. Mathematical models for some of the commonly used constraints as a function of cutting parameters (cutting speed, feed, depth of cut, etc.) are given in Refs. [34,41,42]. Machining process modeling software are available for simulation to estimate force, torque, power, chatter stability, part and fixture deflection, and cutting tool deflection to evaluate constraints. Chapters 8 and 12 provide details on process modeling.

Depth of cut. In general, tool life is less affected by changes in the doc than by changes in either the feed or speed, particularly for roughing cuts. The most favorable compromise between the tool life and the metal removal rate in rough cutting is therefore often obtained when the highest permissible doc is used. However, a heavier doc could result in decreased tool life in semiroughing and finishing cuts. A 50% increase in the doc will typically produce only a 15% reduction in tool life when the doc exceeds 10 times the feed rate [38]. The effect of the doc is somewhat greater when the starting doc is less than 10 times the feed rate. The selection of the maximum doc is dependent on (1) tool material and geometry, (2) the cutting force, (3) the available machine horsepower, (4) the stability of the tool–work-machine system, (5) the required dimensional accuracy, and (6) surface finish requirements.

Limits due to material considerations (1) are usually provided by the tool manufacturer; the tool material and geometry determines the strength of the cutting edge and deflection characteristics of the tool under an applied force. Cutting force limits (2) become significant when deflection and chatter are of concern. Chatter is a major factor limiting the maximum depth of cut because it affects the surface finish, dimensional accuracy, tool life, and machine life. The prediction of the maximum chatter-free doc and the speed intervals in which resonant frequencies occur are explained in detail in Chapter 12.

Feed. The maximum allowable feed has a pronounced effect on both the optimum spindle speed and the production rate. Feed changes have a more significant impact on tool life than doc changes. In many cases, the largest possible feed consistent with the available machine power and surface finish requirements is desirable in order to increase machine utilization. It is often possible to obtain much higher metal removal rates without reducing tool life by increasing the feed and decreasing the cutting speed. This technique is particularly useful for roughing cuts, in which the maximum feed is dependent on the maximum force the cutting edge and the machine tool are able to withstand. The maximum permissible feed f_h, used to calculate the cutting speeds V_{opc} and V_{opt}, can be determined from

$$f_h = \min\left\{ f_{c,\max}, f_{s,\max}, f_{F,\max} \right\} \tag{13.17}$$

where $f_{c,\max}, f_{s,\max}$, and $f_{F,\max}$, are the maximum allowable feeds, respectively, due to the chip-breaking and surface finish requirements and to force limitations. In general, the maximum feed in a roughing operation is limited by the maximum force that the cutting tool, machine tool, workpiece, and fixture are able to withstand. The maximum feed in a finish operation is limited by the surface finish requirement (Chapter 10). Traditionally, multiple passes are used to mill a pocket at about 50% of the cutter width. On the other hand, single full-depth pass with 5%–10% cutter width and a two to three times faster feedrate can be obtained. This could reduce the cycle time by at least 50%.

Cutting speed. The cutting speed has a greater effect on tool life than either the doc or feed. When the cutting speed is increased by 50%, the tool life based on flank or crater wear typically decreased by 80%–90% [43]. The influence of cutting speed on tool life is discussed in detail in Chapter 9.

The cutting speed has only a secondary effect on chip-breaking when varied within conventional ranges. Depending on the type of tool and the workpiece materials, certain combinations of speed, feed, and doc are optimal from the point of view of chip removal. As discussed in Chapter 11, charts providing the feasible region for chip-breaking as a function of feed versus doc are often available from tool manufacturers for a specific insert or tool.

The constants in the deterministic constraint relation, Equation 13.16, must be identified from experimental data and are thus subject to inherent uncertainty. Experimental results contain only partial information because the number of experiments conducted is always limited, the measurements involve errors, and important (often unknown) factors may not be recorded or controlled. Also, physical constraints change during the life of the cutting tool and thus during the machining of each part, since the machining process changes with tool wear. Hence, most of the constraint equations, and especially the surface finish model, should be functions not only of the speed, feed, and doc but should include tool wear as an independent variable. The experimental evaluation of tool life, even in particularly simple conditions, may require a great deal of testing due to the inherent variability in tool wear processes. General constraints can be developed for a specific work material and tool geometry family for a specified range of cutting conditions [34,44–51].

13.6.3 Problem Statement

The machining optimization problem for a single pass, multipass, and single stage multifunctional system can be formulated as follows:

Minimize/Maximize the objective function $U_i(\underline{s})$

The objective function measures the effectiveness of the decisions about inputs as reflected in quantifiable outputs. A qualitative graph of minimum cost per piece and the maximum production rate (i.e., the minimum time per piece) is shown in Figure 13.1.

Subject to the constraints

$$G_{ij}\left(\underline{s}\right) \le 0 \ \forall i \in A_c \quad \text{and} \quad j \in 1,2,\ldots,M_i^c \tag{13.18}$$

The machining optimization problem for a multistage machining system (Figure 13.4) can be formulated as:

Minimize the objective function

$$U\left(\underline{s}\right) = \sum_{i=1}^{N} U_i(s), \quad \forall i \tag{13.19}$$

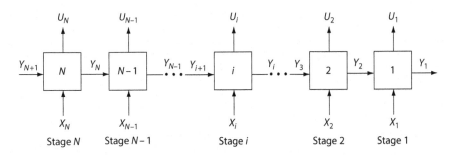

FIGURE 13.4 A multistage flow machining system.

Subject to the constraints

$$F_{ij}(s) = G_{ij}(s) - E_{ij} < 0, \quad \forall i, j \tag{13.20}$$

$$\Omega_i(s) = 0, \quad \forall i : T_{u_i} < t_x \tag{13.21}$$

The equality constraint, Equation 13.21, is applied only when the production time of the ith operation or station is lower than the system maximum cycle time t_x. E_{ij} are the upper bound values for the inequality constraints.

13.7 OPTIMIZATION TECHNIQUES

13.7.1 SINGLE-PASS OPERATION

The optimization problem for a single pass operation is straightforward because it can be reduced to the two-variable problem of determining the cutting speed and feed. Several methods of solution for this problem have been proposed by various investigators [51,52]. The simplest method is *linear programming* [53,54], which requires a linear objective function and linear constraints. In this method, the various constraints are expressed as power transforms, and the mathematical relations are linearized by logarithms. Standard linear programming procedures like the Simplex algorithm [55,56] can be used for optimization. *Nonlinear programming* methods may also be used. The sequential unconstrained minimization technique (SUMT) with the Davidon–Fletcher–Powell algorithm [27] can be used; this is a penalty function method. There are several problems with this approach, which can adversely affect results. Another approach is to use a logarithmic transformation [57] and solve the resulting equations using Rosen's gradient method with linear constraints. Some constraints are not readily transformed and must be approximated by piecewise linear functions, which significantly complicates the use of this search procedure [58]. The earlier methods are all deterministic approaches. A probabilistic approach using *change-constrained programming* has been suggested by Iwata et al. [59]; in this method, normal distributions for the constants in the constraints are used. The probabilistic nature of tool life affects the optimum cutting conditions and can be represented using a tool life probability distribution function [27,60,61]. Polynomial *geometric programming* can be used assuming an extended Taylor tool life equation [62–68]. Geometric programming can also be used based on quadratic polylog-nomials (QPL) [69]. "Probabilistic geometric programming" has been applied to multi-objective functions [70]. "Goal programming," a special type of linear programming developed to solve problems involving complex and usually conflicting multiple objectives, has been applied with both linear [71–73] and nonlinear [74] goals. "Genetic algorithms" (GA), based on the principles of natural biological evolution, have been used to optimize not only the cutting conditions but also system modeling and process optimization in automated process planning systems [38,75–77]. Finally, a "fuzzy optimization" approach can be used for selecting optimal machining conditions based on fuzzy set theory; this method manipulates the uncertainty present in the empirical equations or experimental data [78,79]. Genetic optimization of Fuzzy rules has been carried out with the help of an object-oriented genetic optimization library (GOL) [25]; the GOL can help the development of the fuzzy expert system for machining data selection. "Simulated annealing algorithms," based on the annealing process, have also been used to search for the optimal process parameters [23]. A computer-aided graphical approach for the optimization of the cutting conditions can be very helpful as illustrated in Example 13.4.

13.7.2 MULTIPASS OPERATION

The optimization problem for a multipass operation is generally more complex than for a single-pass operation [41,68,80–86]. A multipass process, particularly multipass turning application, is a four-variable problem in which the number of passes and doc for each pass must be determined through a dynamic programming procedure while the optimum cutting speed and feed for each pass is determined using single-pass optimization methods. Dynamic programming [87] is very useful in handling a multipass problem in which every cutting pass is independent of the previous passes, such as in turning operations. The details of the optimization are provided in Ref. [80].

13.7.3 SINGLE-STATION MULTIFUNCTIONAL SYSTEM (SSMS)

The productivity of a SSMS is dependent not only on the cutting time, but more importantly on the noncutting time; the part process sequence for the various operations, which significantly affects the part cycle time, should be considered in the optimization. There are three stages to planning part processes for such systems: (1) determining the cutter trajectory from geometric models of machining processes [88,89], (2) developing mechanistic models for the machining operations [90,91], and (3) applying optimization methodologies for the machining system. Several CAD/CAM software offer portion or most of the objectives mentioned before. CAM programs provide modeling capabilities and tool path verification. The optimization evaluates several factors, including speeds and feeds, depth of cut, chip thickness, and so on. This approach optimize the federate of the tool path.

A fundamental analysis of an SSMS was developed by Hitomi et al. [92,93], in which the optimum cutting speeds were analyzed based on a minimum production cost criterion with due-date constraints by neglecting all necessary constraints at each stage of the system. In a later publication, Hitomi and Ohashi [94] proposed a computational algorithm for determining the optimum sequence of operations for a random-select tool changer type (using the classic traveling salesman problem) by considering the tool-change time as a function of the tool location in the tool magazine and providing the time matrix for extracting tools, so that the noncutting time could be properly accounted for. Tang and Denardo [95] considered the minimization of the number of tool switches. The job sequence was considered fixed, and tooling decisions were the only decisions to be made.

A comprehensive optimization scheme for the part sequence of operations in SSMSs was proposed in Ref. [96]. The mathematical model and methodology for the determination of the optimum sequence of operations were analyzed with an equal emphasis on tool magazine organization for high-speed SSMSs. The analytical approach to SSMS optimization can be based on one of the following three schemes: (1) for mass production of the same part type with a cycle time constraint, (2) for mass production of the same part type without a cycle time constraint, and (3) for production of multiple part types of short or medium batch sizes. The details are available in Ref. [96].

13.7.4 MULTISTAGE MACHINING SYSTEM

The optimization problem for a multistage machining system is generally very complex compared to those for a single pass or multipass operation. The majority of high volume products have traditionally been produced using conventional transfer lines, although today flexible transfer lines (FTLs) or flexible manufacturing systems (FMSs) are very common. The productivity of a conventional transfer line, FTL, or FMS, is affected by the cutting conditions, the rapid machine travel rate, and the speed of the transfer mechanism.

FMSs used for multiproduct applications require the optimization of two functions, process planning and scheduling. However, the configuration of an FMS for a first-time installation or expansion is critical to process planning; an appropriate configuration must be chosen that satisfies both technical (design and dimension) and economic (operating cost, capital expenditure, and production quantity) requirements. The alternatives for the design and configuration of an FMS can be evaluated based on performance evaluation models or optimization models [97–103]. Performance models can be developed using static allocation models, analytical models based on queuing theory, or simulation models.

Balancing of a multistage manufacturing system, so that the total time required at each workstation is approximately the same, is an important aspect of optimization. Unfortunately, a perfect balance is not achievable in most practical situations. In this case, the slowest station determines the overall production rate of the line. By reducing the machining time by increasing the cutting speed and/or feed rates at stations with long process times, and increasing the machining time by reducing the cutting speed and feed combinations at stations with idle time, an improvement in cycle time and line balance can be achieved. Reducing idle time by reducing the cutting speed and feed combinations prolongs tool life, reduces tool failure, and reduces the probability of machine failure. Therefore, the use of all or a substantial portion of the idle time should result in higher system uptime and higher productivity.

A fundamental analysis of a multistage machining system was described by Hitomi [104–108] in which the optimum cutting speeds and the cycle time were analyzed based on one of the following three criteria: minimum production cost, maximum production rate, or optimum cycle time. He determined the optimum cutting speeds for both a single-product or multi-product production [107]. He also applied a simplifying approximation in order to optimize the cutting speed and feed [109]. The cycle time of the stations was controlled by the largest optimum production time of the individual stations.

The optimum cutting conditions for a job requiring multistage machining operations were determined by Rao and Hati [110] using a constrained mathematical programming approach based on one of the previous two optimization criteria, and by Iwata et al. [111,112], on the basis of the total minimum production cost while keeping the total production time as short as possible.

Discrete cutting speeds and feeds were used by Sekhon [113] in his optimization procedure, which is simpler since there is a finite number of speed and feed combinations. Iwata et al. [111,112] and Sekhon [113] did not consider the idle time for the various machining stations. Hitomi [103–108] and Rao and Hati [110] attempted to minimize the idle time through a trial-and-error method, which did not guarantee optimality. An effective optimization method for the cutting conditions should be based on the consideration of minimum idle time for all machining stations.

The analysis of a fundamental model and a methodology for determining optimum cutting conditions for each station of a flow-type multistage machining system are discussed in details in Refs. [114–116]. The optimization minimizes idle time at each station in order to reduce production costs and increase production output. This minimization is achieved using a zero idle time constraint in the optimization calculations.

13.7.5 CUTTING TOOL REPLACEMENT STRATEGIES

One major issue in machining systems analysis is tool balancing and tool regulation. The utilization of the machine tool can be improved by ensuring the availability of the required cutting tools when needed, eliminating machine interruptions due to manual tool changes and setup (during machine cycle), and monitoring tool wear and life in order to reduce scrap. This can be achieved by maintaining an adequate supply of usable tools, monitoring tool usage, wear and premature failure, and incorporating mathematical control algorithms for the automatic replacement of tools.

The most common method is forced replacement at specified intervals. These deterministic approaches minimize risk but do not optimize tool and machine. A tool replacement strategy based on a machining cost/production time criterion may be preferable [117–129]. The best tool replacement strategy will minimize the machine idle time due to tool change or tool magazine loading and tool failure while reducing tool redundancy and duplication in the tool magazine.

There are two approaches to the solution of the tool magazine system reliability problem: (1) the use of cutting tools of high reliability, and (2) the use of redundant tools. The use of redundant tools in the machine's magazine results in an increase in production tooling costs, but tends to increase productivity and reduce production costs. The mathematical solution for the optimum number of redundant tools in a magazine can be determined when the reliabilities of the individual tools are known [127,130–134].

13.7.6 Cutting Tool Strategies for Multifunctional Part Configurations

The majority of tools have a single function and are used for only one operation or are made to machine a single part feature. However, the use of multifunctional cutting tools has become necessary in single spindle manufacturing systems in order to increase productivity and/or quality and to reduce cost.

Multifunctional cutting tools include combination and multistep tools. The term "combination" is used for tools that are used in more than one distinct operation while the term "multistep" is used for tools employed for manufacturing a multistep configuration (e.g., a hole with multiple diameters at different depths) in one motion. For example, the valve guide and seat in an engine head may be manufactured using either a generated head tool or a plunge (form) tool. The generated head tool is considered a combination tool because it bores and/or reams the guide in one machine motion and then bores the seat in another. On the other hand, the plunge tool is a multistep tool since it machines the guide and seat in one motion.

The design of multistep tools was discussed in Chapter 4. Multistep tools are often used to drill or ream multi-diameter holes (see Figures 13.5, 4.64, 4.65, 4.79), and to drill and counterbore, drill and chamfer, ream and chamfer, or drill and thread simultaneously. These tools have more than one cutting edge along their axis and can machine a complex part feature in a single tool motion during which two or more tool steps generally cut at the same time. Stepped holes can be bored using either a multistep boring bar or a single point tool. Single point boring tools used to bore multistep holes can be treated as multistep tools for analysis purposes.

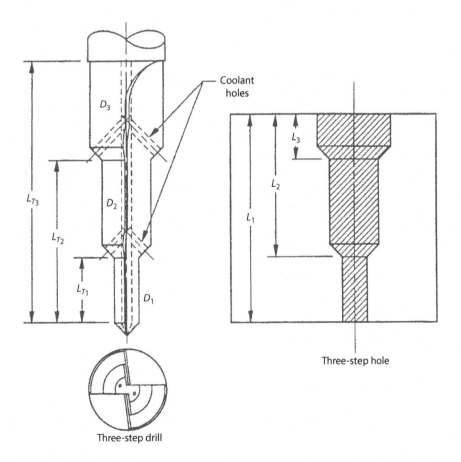

FIGURE 13.5 Configuration of a multistep drill and the corresponding generated hole.

   Combination tools are designed to be used for different distinct operations. Examples of such tools include generating head tools used for the machining of the valve seat and the guide in an engine head and drill-thread mill tools (Figure 4.81) used for drilling and thread milling a hole independently. Tools for various operations are being combined on one tool body to eliminate tool changes or improve the geometric relationship between features (e.g., perpendicularity or concentricity between features).

Multistep tools: Multistep tools fitted with indexable inserts or replaceable tool bits require a different mathematical analysis than solid or brazed tools, since the wear of one indexable step does not necessarily require the regrinding or disposal of the other unworn steps on the tool.

   The cutting conditions generally vary from step to step on a multistep tool. The tool steps can operate at a constant spindle speed (rpm), constant surface speed, varying surface speed or spindle rpm, and/or varying feed rates. The best strategy for a particular application depends on the tool (geometry, size, and material), operation economics, and the capability of the machine tool control system. It is especially common to vary the spindle rpm as different parts of the tool cut for the following reasons: (1) to reduce the surface speed for larger diameter steps to prevent excessive cutting temperatures and tool wear, (2) to optimize the speed when steps are made of different tool materials, and (3) to optimize part quality and surface finish for each step.

   The simplest method is to use constant spindle rpm and feed, which results in a different surface speeds at the various tool steps. On NC machines, it is also easy to operate at a constant surface speed by varying the spindle rpm according to the diameter of the step as each step cuts the workpiece. This approach should be used when there is a significant difference in tool step diameters. Current control systems allow speed or feed changes within 50–100 ms in a controllable manner. The least desirable strategy is to operate the various tool steps at different surface speeds and spindle rpm's. This approach is sometimes used as a compromise to provide acceptable cutting conditions for steps with significantly different diameters. In particular, this method is used for single step tools with a large length to diameter ratios, such as drills and reamers, in high-speed machining applications; it can eliminate the need for a bushing at the entrance by allowing the tool to enter the part at conventional (low) speeds and feeds, which are subsequently increased to the optimum values. The feed is usually kept constant for all the tool steps, except in special applications in which surface finish and chip ejection are significant concerns. The mathematical analysis of the variable feed approach, where a feed change is considered for at least one tool step, is similar to that for the variable speed approach. An analytical solution of the optimization (using production cost, production time, and/or composite objective functions) can be determined for constant rpm, constant surface speed, or variable speed and feed cases as discussed in Refs. [135,136].

Combination tools: The tool life equation for combination tools can be estimated from the tool life equations for similar conventional (single function) tools. The tool life for the different tool functions can be assumed to be independent. Tool life predictions for a combination tool can be obtained when the reliability of the different tool functions is either known or can be estimated from experience with single function tools as discussed earlier for multistep tools. The best tool design and cutting conditions are those which result in uniform or balanced tool life for all the tool functions. Mathematical relations between the different tool functions must be defined in order to analyze a combination tool [135].

13.8 EXAMPLES

Example 13.1 A gray iron casting with a diameter of 150 mm is rough turned to 144 mm for a length of 200 mm + 3 mm approach distance. An indexable carbide coated insert with eight corners is used in the turning operation with a cost of $20. The tool life equation is $VT^{0.2} = 230$ (metric units) for the before mentioned tool in the current workpiece material. The maximum allowable feed for the cutting tool is 0.35 mm at the maximum depth of cut of 4 mm. The time for loading/unloading an insert/tool is about 3 min. The part handling (load/unload) time is 26 s for the casting mounted on the fixture. The time for the tool to rapid travel across the part is 4 s. The operating cost is $80/h. Estimate the individual cost curves similar to Figure 13.1 and show the optimum cutting speed.

Solution: The production cost is given by Equation 13.8. There are four components in the cost equation: the machining cost, tool cost, tool change (loading/unloading the tool in the machine) cost, and nonproductive (including the part handling, rapid travel location times, etc.) cost given by the following equations, respectively:

$$C_o t_m, \quad C_o t_\ell \frac{t_m}{T}, \quad C_t \frac{t_{m_c}}{T}, \quad C_o\left(t_h + t_x + t_{cs}\right)$$

The difference between the machining time t_m and t_{mc} is that the t_m includes the approach time, while t_{mc} uses only the time that the part is in contact with the tool. The earlier four costs and the total cost (the sum of the four costs) are calculated for a range of cutting speeds using a spreadsheet. A plot is shown in Figure 13.6. The optimum cutting speed can be estimated either from the graph in the figure (corresponding to the lowest cost) or calculated from Equation 13.9 to be 127 m/min.

Example 13.2 Estimate the production cost for a rough turning operation on a free-machining steel bar 70 mm in diameter and 200 mm long, when 3 mm is to be removed from the O.D. of the bar using a carbide tool. Some of the characteristics of the turning lathe and cutting tool are given in the Table 13.1.

Solution: The production cost is given by Equation 13.8:

$$C_u = C_o t_m + \frac{t_m}{T}\left(C_o t_\ell + C_t\right) + C_o\left(t_{cs} + t_h + t_x\right)$$

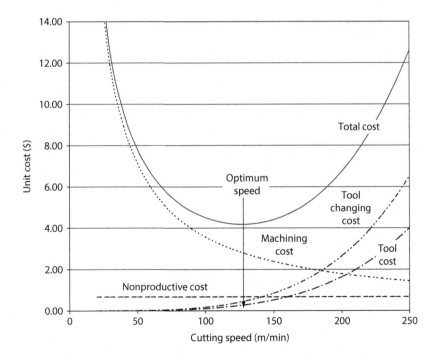

FIGURE 13.6 Various machining costs versus cutting speed.

TABLE 13.1

Tool and Economic Parameters for Example 13.2

Parameter	Value	Parameter	Value
Machine tool		Tool life equation, $VT^n = C_t$	
Spindle rpm	700 rpm	Tool life constant, C_t	500
Feed, f	0.3 mm/rev	Tool life exponent, n	0.25
Rapid feed rate, f_{rapid}	6000 mm/min	Cutting speed units	m/min
Operating cost, C_o	\$60/h	Tool interchange time between operations, t_{cs}	8 s
Part load/unload time, t_h	20 s	Number of cutting edges per insert, n_t	3
Tool load/unload time, t_ℓ	1 min		
Tool cost, C_{te}	\$9/insert		

The machining time is given by Equation 2.5 (or Equation 13.2) assuming 2 mm for the tool approach distance:

$$t_m = (L + L_e)/f_r = (L + L_e)/(f N) = (200 + 2)/(0.3 \times 700) = 0.962 \text{ min} = 58 \text{ s}$$

The cutting speed $V = \pi \cdot DN = \pi \cdot 70 \cdot 700/1000 = 154$ m/min.
The tool life equation $V \cdot T^{0.25} = 500$ or $T = 500^4 \cdot 154^{-4} = 111$ min

$$t_h = 20 \text{ s} = 0.333 \text{ min}$$

$$t_{cs} = 8 \text{ s} = 0.133 \text{ min}$$

$$t_\ell = 1 \text{ min}$$

The rapid travel time for the tool after the completion of the turning operation is

$$t_x = (L + L_t)/f_{rapid} = (200 + 80)/6000 = 0.047 \text{ min} = 2.8 \text{ s}$$

$$C_o = \$60/h = \$1/\text{min}$$

$$C_{te} = C_t/n_t = 9/3 = \$3 \text{ per cutting edge}$$

$$C_u = 1 \cdot 0.962 + \frac{0.962}{111}(1 \cdot 1 + 3) + 1 \cdot (0.133 + 0.333 + 0.047) = \$1.51$$

Example 13.3 Calculate the minimum production cost for Example 13.1 assuming that the maximum allowable feed for the insert is 0.5 mm/rev. The earlier mentioned tool life equation is modified to include the effect of feed with an exponent $a = 0.5$.

Solution: The extended Taylor equation is

$$VT^n f^a = K_t \quad \text{or} \quad VT^{0.25} f^{0.5} = 500$$

The feed for the turning operation has an upper bound of 0.5 mm/rev. The optimum cutting speed for this operation can be determined from Equation 13.9:

$$V_{opc} = \frac{K_t}{f_h^a \left[\left(\dfrac{1-n}{n} \right) \left(t_\ell + \dfrac{C_{te}}{C_o} \right) \right]^n} = \frac{500}{0.5^{0.5} \left[\left(\dfrac{1-0.25}{0.25} \right) \left(1 + \dfrac{3}{1} \right) \right]^{0.25}} = 380 \text{ m/min}$$

The minimum production cost that will occur is the optimum speed using the maximum allowable feed:

$$t_m = (L + L_e)/f_r = (L + L_e)/(f N) = (200 + 2)/(0.5 \times 1729) = 0.234 \text{ min} = 14 \text{ s}$$

The spindle rpm $N = V/(\pi \cdot D) = 380 \cdot 1000/(\pi \cdot 70) = 1729$ rpm.
    The tool life equation $VT^{0.25}f^{0.5} = 500$ or $T = 500^4 \cdot 380^{-4} \cdot 0.5^{-2} = 5.2$ min

$$C_u = 1 \cdot 0.234 + \frac{0.234}{5.2}(1 \cdot 1 + 3) + 1 \cdot (0.133 + 0.333 + 0.047) = \$0.928$$

Therefore, the production cost is reduced from \$1.51 to \$0.928 by using an optimum cutting speed.

Example 13.4 Optimize the cutting conditions for a multitool (multispindle) lathe operation for the part in Figure 13.7. The CNC lathe has two turrets, which allows two tools to cut the part simultaneously. The cutting parameters and cutting tool characteristics are given in Table 13.2. The cutting tools for turning and boring have the same tool life equation. The strategy of changing each tool individually will be considered. Calculate the minimum production cost and production time.

Solution: There are four operations, namely, OP1 (turning the 120 mm diameter), OP2 (turning the 70 mm diameter), OP3 (drilling the 25 mm hole), and OP4 (boring the 25.6 mm hole) as shown in Figure 13.7. OP1 and OP3 are performed simultaneously by the two independent turrets, as are OP2 and OP4. The spindle speed for the workpiece for each set of operations is determined based on the cycle times of the individual operations. The optimum cutting speeds for each individual operation can be estimated using Equations 13.9 and 13.10 based on the minimum production cost and time, respectively, using the maximum allowable feed for each tool. This analysis is conceptually similar to multistage machining system analysis (see Section 13.7) for the operations occurring

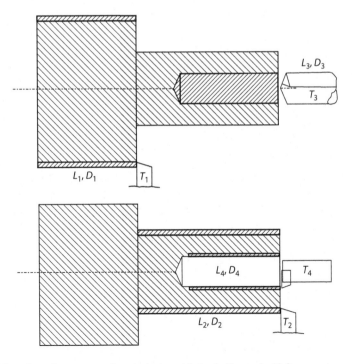

FIGURE 13.7 Manufacturing process for a twin-turret lathe in Example 13.3.

TABLE 13.2

Machining Parameters Used for Example 13.4

Parameter	Value	Parameter	Value	Parameter	Value
D_1	120 mm	f_{tmin}	0.12 mm/rev	Drill a_3	−3.7
D_2	70 mm	f_{tmax}	0.4 mm/rev	Drill R_c	3.0×10^8
D_3	25 mm	f_{bmin}	0.075 mm/rev	t_h	20 s
D_4	25.6 mm	f_{bmax}	0.15 mm/rev	$t_{\ell 1}, t_{\ell 2}, t_{\ell 4}$	3 min/edge
L_1	75 mm	f_{dmin}	0.15 mm/rev	$t_{\ell 3}$	4 min/edge
L_2	100 mm	f_{dmax}	0.3 mm/rev	t_{cs}	2 s
L_3	63 mm	Turn a_1	−1	t_x	0.05 min/operat
L_4	60 mm	Turn a_3	−3	C_o	$1/min
N_{min}	100 m/min	Turn R_c	3.4×10^6	$C_{te1}, C_{te2}, C_{te4}$	$4/edge
N_{max}	5000 m/min	Drill a_1	−1.4	C_{t3}	$30/drill/regrind

Note: a_1, a_3 = Exponents in tool life equation correspond to feed and speed parameters, respectively.

simultaneously by individual turrets. Therefore, operations occurring by different turrets should be balanced to minimize the idle time of either turret.

The optimum speed and the corresponding RPM and machining time for each operation based on minimum cost (using Equations 13.9, 2.1, and 13.2) are

$$V_{opc1} = 85 \text{ m/min}, \ V_{opc2} = 85 \text{ m/min}, \ V_{opc3} = 99 \text{ m/min}, \ V_{opc4} = 117 \text{ m/min}$$

$$N_{opc1} = 225 \text{ rpm}, \ N_{opc2} = 405 \text{ rpm}, \ N_{opc3} = 1265 \text{ rpm}, \ N_{opc4} = 1460 \text{ rpm}$$

$$t_{mc1} = 0.84 \text{ min}, \ t_{mc2} = 0.62 \text{ min}, \ t_{mc3} = 0.18 \text{ min}, \ t_{mc4} = 0.27 \text{ min}$$

The optimum speeds and corresponding machining times for each operation based on the minimum production time (using Equation 13.10) are

$$V_{opt1} = 112 \text{ m/min}, \ V_{opt2} = 112 \text{ m/min}, \ V_{opt3} = 162 \text{ m/min}, \ V_{opt4} = 156 \text{ m/min}$$

$$N_{opt1} = 298 \text{ rpm}, \ N_{opt2} = 511 \text{ rpm}, \ N_{opt3} = 2063 \text{ rpm}, \ N_{opt4} = 1937 \text{ rpm}$$

$$t_{mt1} = 0.63 \text{ min}, \ t_{mt2} = 0.49 \text{ min}, \ t_{mt3} = 0.11 \text{ min}, \ t_{mt4} = 0.21 \text{ min}$$

Since OP1 and OP3 occur together and $t_{mc1} > t_{mc3}$ but $N_{opc3} > N_{opt1}$, the spindle RPM $N_{opc1} = 225$ rpm is selected to control these two operations. However, the machining time for drilling the hole at 225 rpm is $t_{m3} = 0.756$ min and $t_{m3} > t_{mt1}$. The tool life for the tool in OP1 can be improved by reducing the feed below the maximum value used to estimate the V_{opt1}. Therefore, equating the machining time for OP1 to that of OP3 ($t_{m1} = t_{m3} = 0.756$ min), a feed of 0.34 mm/rev is selected ($f = 75$ mm/ [0.756 min × 298 rpm] = 0.34 mm/rev). The reduction of the feed from 0.4 to 0.34 mm/rev results in a tool life improvement from 6 to 7.3 min. The machining time and cost for these operations are

$$T_{u13} = t_{m3} + \frac{t_{m1}}{T_1} t_{\ell 1} + \frac{t_{m3}}{T_3} t_{\ell 3} + \left(t_{cs} + t_x \right)_{1+3}$$

$$T_{u13} = 0.756 + \frac{0.756}{7.3} 3 + \frac{0.756}{13,900} 4 + 2 \left(\frac{2}{60} + 0.05 \right) + \frac{20}{60} = 1.23 \text{ min}$$

$$C_{u13} = C_o t_{m3} + \frac{t_{m1}}{T_1}\left(C_o t_{\ell 1} + C_{t1}\right) + \frac{t_{m3}}{T_3}\left(C_o t_{\ell 3} + C_{t3}\right) + C_o\left(t_{cs} + t_x\right)_{1+3}$$

$$C_{u13} = 10.756 + \frac{0.756}{6.05}(13+4) + \frac{0.756}{13,900}(14+30) + 1\left(\frac{2}{60} + 0.05\right) = \$1.65$$

Similarly, OP2 and OP4 occur simultaneously and $t_{mc2} > t_{mc4}$ but $N_{opc4} > N_{opt2}$, so the spindle RPM $N_{opc2} = 511$ rpm is selected to control these two operations. However, the machining time for boring the hole at 511 rpm is $t_{m4} = 0.783$ min and $t_{m4} > t_{mt2}$. The tool life for the tool in OP2 can be improved by reducing the feed below the maximum value used to estimate the V_{opt2}. Therefore, equating the machining time for OP2 to that of OP4 ($t_{m2} = t_{m4} = 0.783$ min), a feed of 0.25 mm/rev is selected, which results in a tool life improvement from 14 to 22 min. The machining time and cost for these operations are

$$T_{u24} = t_{m4} + \frac{t_{m2}}{T_2}t_{\ell 2} + \frac{t_{m4}}{T_4}t_{\ell 4} + \left(t_{cs} + t_x\right)_{2+4}$$

$$T_{u24} = 0.783 + \frac{0.783}{22}3 + \frac{0.783}{327}3 + 2\left(\frac{2}{60} + 0.05\right) = 1.06\,\text{min}$$

$$C_{u24} = C_o t_{m4} + \frac{t_{m2}}{T_2}\left(C_o t_{\ell 2} + C_{t2}\right) + \frac{t_{m4}}{T_4}\left(C_o t_{\ell 4} + C_{t4}\right) + C_o\left(t_{cs} + t_x\right)_{2+4}$$

$$C_{u24} = 10.756 + \frac{0.756}{22}(13+4) + \frac{0.756}{327}(13+4) + 1\left(\frac{2}{60} + 0.05\right) = \$1.21$$

Finally, the total production time (Equation 13.11) and cost are

$$T_u = T_{u13} + T_{u24} + t_h = 1.23 + 1.06 + \frac{20}{60} = 2.63\,\text{min}$$

$$C_u = C_{u24} + C_{u24} + C_o t_h = 1.65 + 1.21 + 1\frac{20}{60} = \$3.20$$

Example 13.5 Optimize the cutting conditions for the single-pass turning operation defined in Table 13.3. The machining parameters and the coefficients and exponents for the tool life and constraint equations used in a single-pass turning operation are given in the table. The maximum and minimum values for feed, speed, power, and cutting force in the table are based on a knowledge of machine limitations and handbook information for the given workpiece and tool materials.

Solution: The solution is available in Ref. [137]. The contours for the production cost, production rate, and the constraints on the force F (600 and 900 N), surface roughness SR (5 and 8 µm), spindle power HP (4 kW), and temperature θ (400°C and 550°C), given by Equation 13.16, indicate that the optimum conditions are close to the conditions of low profit when emphasis is placed on the production cost by using a large value of the weight w_1. On the other hand, when the w_1 and w_2 values (of the objective function) are interchanged so that the larger emphasis is placed on the production time, the lower value contours of the objective function will move toward the higher profit rate contours.

TABLE 13.3

Machining Parameters Used for the Turning Example 13.5

Parameter	Value	Parameter	Value	Parameter	Value
L	203 mm	SR_{max}	8 m	R_c	1.396×10^9
D	152 mm	HP_{max}	5 kW	t_1	0.5 min/edge
V_{min}	30 m/min	F_{max}	1100 N	t_{cs}	0.2 min/tool
V_{max}	200 m/min	θ_{max}	500°C	t_x	0.13 min/pass
f_{min}	0.254 mm/rev	a_1	−1.16	C_o	\$0.1/min
f_{max}	0.762 mm/rev	a_2	−1.4	C_t	\$0.5/edge
SF_{max}	2 μm	a_3	−4	w_1	0.6
				w_2	0.4

Note: SF, surface finish; SR, surface roughness; HP, spindle power; F, cutting force; θ, Average cutting temperature at the tool–chip interface.

a_1, a_2, a_3 = Exponents in tool life equation correspond to feed, doc, and speed parameters, respectively, given by Equation 13.7.

When both weight coefficients are set equal to 0.5, the objective function contours are close to the higher profit rate contours. In general, a good practice would be to use weight coefficients $w_1 \geq w_2$ since in this case the optimum value lies in the area of maximum profit at the lower sale price. All the details are available in Ref. [137].

Example 13.6 Optimize conditions for a Single-Stage Multifunctional System (CNC machine). The optimum sequence of operations for a simple part made of gray cast iron, shown in Figure 13.8, is considered to demonstrate the SSMS analysis.

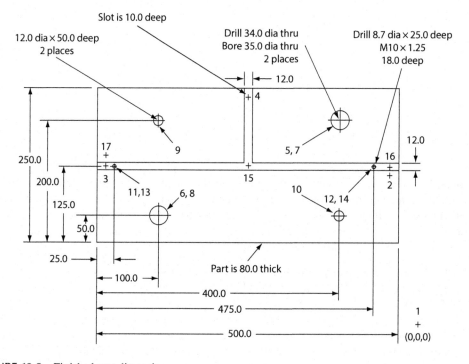

FIGURE 13.8 Finished part dimensions.

TABLE 13.4

The Optimum Sequence of Operations Used in Example 13.6

Tool Slot Location	Node Sequence	Tool Slot Location	Node Sequence
0	1	0	1
1	17	29	2
1	2	31	17
2	4	31	4
2	15	36	15
2	3	36	3
2	16	35	16
3	6	35	9
3	5	5	10
6	12	5	12
6	11	6	11
7	13	6	13
7	14	33	14
5	10	33	5
5	9	31	6
4	8	31	8
4	7	29	7
Total time:	5.029	Total time:	5.362

Solution: The analytical approach, the solution results, and the computational algorithms are discussed in Ref. [136]. All the operations and characteristic parameters for the cutting tools are provided in the tables of Ref. [136].

The problem was simulated for two different instants of the tools' location in the magazine, which will occur during production based on the previous tool arrangement. An evaluation of the optimum operations sequence, incorporating the tool organization in the magazine, is given in Table 13.4. The slot number in the magazine for the corresponding operation is also given in the table. When the tools for all the operations are located next to each other in the order of the optimum sequence found earlier without considering the terms t_{cs} and t_{ei}, the total time U^T to process a part is 5.029 min. However, when the tools are scattered in the magazine at a later production time, the total time U^T becomes 5.632 min. The part production time is 9.841 min and 10.444 min, respectively, for these two cases. The tools' location in the former case requires 6.1% shorter production (cycle) time than in the latter case.

Example 13.7 Optimization of a multistep tool. Consider two multistep solid carbide head drills with oil holes having three steps, $A_c = \{1, 2, 3\}$, with diameters D_1, D_2, and D_3, which are used to drill two different holes with lengths L_1, L_2, and L_3, respectively (see Figure 13.5). The work material is 390 aluminum. The same tool life equation is used for all steps on both tools. The tool life equation, the economic parameters for the process, and the operating constraints are provided in Refs. [135,136]. Both the constant rpm and surface speed approaches are evaluated. The first and second tool steps are assumed to operate at the same rpm. Optimize the cutting conditions.

Solution: Using the constant rpm approach for drill #1, the optimum cutting conditions are 5700 rpm and 0.263 mm/rev, which result in a tool life of 37.4 min (see Table 6 in Ref. [136]). Considering the constant rpm approach for drill #2, the optimum cutting conditions are 2800 rpm and 0.263 mm/rev, which result in a tool life of 20.3 min (see Table 13.6 in Ref. [136]). If a constant surface speed approach for drill #1 is considered, the tool life is increased to 52.4 min. The details of the solution are provided in Ref. [136].

The traditional manufacturing approach using three conventional (single step) tools for drilling the multistep hole produced with the multistep drill #1 is considered next. The number of holes drilled between regrinds by the single step tools is larger than the 1402 holes obtained with the multistep tool. However, the multistep drill reduced the production cycle time and cost by 65% and 60%, respectively. The breakeven point with respect to production cost is obtained at a production batch of 154 holes. Therefore, the multistep tool is superior at either high production rates or in medium or mass production applications. The details are provided in Ref. [136].

13.9 PROBLEMS

Problem 13.1 A 25 mm through hole is made in a medium carbon steel part using a two-flute drill with 140° point angle. The hole is made through a 50 mm thick plate. The tool life equation is $VT^{0.24} = 150$ (metric units) for this tool and workpiece material. The maximum allowable feed for the cutting tool is 0.15 mm per flute. The time for loading/unloading the drill in the machine tool is about 2 min. The part handling (load/unload) time is 20 s. The rapid feed rate for the machine is 25,000 mm/min. The drill cost is $100 with five maximum regrinds at $30 per regrind. The operating cost is $80/h. Estimate the individual cost curves similar to Figure 13.1 and show the optimum cutting speed.

Problem 13.2 Determine the optimum spindle rpm for minimum production cost in boring a 50 mm diameter hole 60 mm deep. The depth of cut is 3 mm and the maximum feed allowed is 0.2 mm. The tool life equation is $VT^{0.3}f^{0.6} = 450$. The load/unload time for the boring tool is 1 min. The tool cost is $4/edge. The operating cost is $80/h.

Problem 13.3 Determine the optimum spindle rpm for minimum production time in Problem 13.2.

Problem 13.4 The part in Figure 13.9 is being machined in a lathe using bar steel of $D_1 = 60$ mm. A triangular carbide insert is used for the rough turning operation with zero lead angle and a nose radius of 1.59 mm. The limits of this insert for feed and doc are 0.6 mm/rev and 5 mm, respectively. The feed limits for the drill and reamer per edge are 0.2 and 0.1 mm. A triangular carbide insert with a nose radius of 0.762 mm is also used for finish turning. The depth of cut for the finish operation is 0.4 mm. The surface finish requirement for turning the 44 mm diameter is 2 µm. The cost of the roughing insert is $7 with six useful edges, while the finishing insert costs $8 with three useful edges. The cost of the 12 mm-diameter two-flute carbide drill is $70, and on average it can be reground six times. The cost of the 12.5 mm-diameter four-flute carbide reamer is $100 and

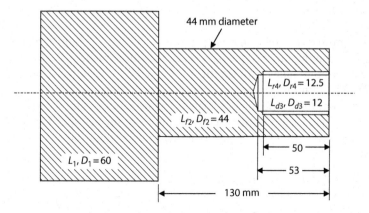

FIGURE 13.9 Configuration and finish dimensions for the part machined in Problem 13.4.

TABLE 13.5
Tool Life Constants for Problem 13.4

Tool Type	a	n	K_t
Rough Insert	0.5	0.38	150
Finish Insert	0.5	0.4	200
Drill	0.5	0.3	100
Reamer	0.5	0.5	180

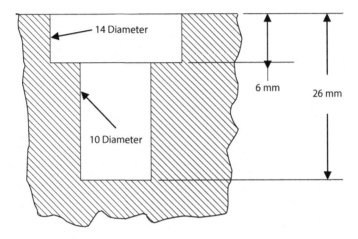

FIGURE 13.10 Configuration of the step hole machined in Problem 13.5.

on average it can be reground three times. The cost of regrinding is $15 for either the drill or the reamer. The operating cost is $60/h. The constants for the tool life equation $VT^n f^a = K_t$ are given in Table 13.5. The tool load/unload time is 2 min, and the tool index time is 5 s. Calculate the minimum production cost and the corresponding production time.

Problem 13.5 A step hole in a block of soft steel, shown in Figure 13.10, is drilled and counterbored with a step carbide drill (single tool). The hole diameters are 10 and 14 mm. The maximum allowable cutting speed for the tool is 100 m/min and the feed is 0.13 mm per tooth. A standard two-flute solid carbide step drill is used with a 120° point angle. The tool life is $VT^{0.35} f^{0.5} = 550$. The load/unload time for the boring tool is 2 min. The tool cost is $150 and the regrinding cost is $40 with four regrinds total. The operating cost is $60/h. Determine the optimum cutting conditions for machining the stepped hole with a single step tool.

REFERENCES

1. R. Venkata Rao, *Advanced Modeling and Optimization of Manufacturing Processes—International Research and Development*, Springer, London, U.K., 2011.
2. S. Yao, X. Han, Y. Rong, S. Huang, D. W. Yen, and G. Zhang, Feature-based computer aided manufacturing planning for mass customization of non-rotational parts, *23rd Computers and Information in Engineering (CIE) Conference*, Chicago, IL, September 2–6, 2003, DETC2003/CIE-48195.
3. Y. Rong, X. Han, S. Yao, and W. Hu, A computer aided production planning system for mass customization of non-rotational parts, SAE Technical Paper 2004-01-1248, 2004.
4. Vericut, VERICUT 7.4: CNC Machine Simulation Gets Simpler, Irvine, California (2015). http://cgtech.com /usa/Content-Downloads/VERICUT_Brochure_70.pdf.
5. Nanjing Swan Software Technology Company, Swansoft CNC Simulation Software (ssCNC) (2015). http://swansoftcncsimulator.com.

6. P. Holland, P. M. Standing, H. Long, and D. J. Mynors, Feature extraction from STEP (ISO 10303) CAD drawing files for metalforming process selection in an integrated design system, *J. Mater. Process. Technol.* **125–126** (2002) 446–455.

7. P. Pal, A. M. Tigga, and A. Kumar, Feature extraction from large CAD databases using genetic algorithm, *Comput. Aided Design—Geomet. Model Process.* **37**:5 (2004) 545–558.

8. L. Ladha and T. Deepa, Feature selection methods and algorithms, *Int. J. Comput. Sci. Eng. (IJCSE)* **3**:5 (2011) 1787–1797.

9. CIMdata, Geometric Brings Knowledge-Based CAM Software to Solid Edge, the Netherlands, 2014.

10. A. Mazakas, Knowledge-based CAD/CAM: Separating the hype from true machining intelligence, *MoldMaking Technol.*, March 2004.

11. I. Fidan and A. Elsawy, The development of a knowledge-based engineering tool for CNC machining, *Int. J. Eng. Ed.* **18**:6 (2002) 732–735.

12. S. Deb and K. Ghosh, An expert system based methodology for automating the set-up planning in computer-aided process planning for rotationally symmetrical parts, *Int. J. Adv. Manuf. Syst.* **10** (2007) 81–93.

13. S. K. Singh and S. Deb, A knowledge based approach for automatic process plan generation for machining, *IJAMS* **15**:1 (2014).

14. B. Gopalakrishnan, Computer integrated machining parameter selection in a job shop using expert systems, *J. Mech. Working Technol.* **20** (1989) 163–170.

15. L. Zeng and H. P. Wang, A patchboard-based expert-systems model for manufacturing applications, *Int. J. Adv. Manuf. Technol.* **7** (1992) 38–43.

16. G. Sathyanarayanan, I. J. Lin, and M. K. Chen, Neural network modeling and multiobjective optimization of creep feed grinding of superalloys, *Int. J. Prod. Res.* **30**:10 (1992) 2421–2438.

17. D. F. Cook and R. E. Shannon, A predictive neural network modeling system for manufacturing process parameters, *Int. J. Prod. Res.* **30**:7 (1992) 1537–1550.

18. Y. H. Pao, *Adaptive Pattern Recognition and Neural Networks*, Addison-Wesley, Reading, MA, 1989.

19. J. Wang, W. S. Tang, and C. Roze, Neural network applications in intelligent manufacturing: An updated survey, in: *Computational Intelligence in Manufacturing Handbook*, CRC Press LLC, Boca Raton, FL, 2001.

20. W. Sukthomya and J. Tannock, The training of neural networks to model manufacturing processes, *J. Intell. Manuf.* **16** (2005) 39–51.

21. M. Hazarika, S. Deb, U. S. Dixit, and J. P. Davim, Fuzzy set-based set-up planning system with the ability for online learning, *Proc. IMechE* **225 B** (2010) 247–263.

22. W. W. Gilbert, Economics of machining, *Machining-Theory and Practice*, ASM, Metals Park, OH, 1950, pp. 465–485.

23. B. Y. Lee, H. S. Liu, and Y. S. Tarng, Modeling and optimization of drilling process, *J. Mater. Process. Technol.* **74** (1998) 149–157.

24. M. Alauddin, M. A. El Baradie, and M. S. J. Hashmi, Optimization of surface finish in end milling Inconel 718, *J. Mater. Process. Technol.* **56** (1996) 54–65.

25. S. V. Wong and A. M. S. Hamouda, Development of genetic algorithm-based fuzzy rules design for metal cutting data selection, *Robot. Comput. Integr. Manuf.* **18** (2002) 1–12.

26. S. Deb and U. S. Dixit, Intelligent machining: Computational methods and optimization, in: J. P. Davim, Ed., *Machining-Fundamentals and Recent Advances*, Springer, London, U.K., 2008.

27. S. K. Hati and S. S. Rao, Determination of optimum machining conditions—Deterministic and probabilistic approaches, *ASME J. Eng. Ind.* **98** (1976) 354–359.

28. D. Ermer and S. Kromodihardjo, Optimization of multipass turning with constraints, ASME Paper 80-WA/Prod-22, 1980.

29. M. A. Shalaby and M. S. Riad, A linear optimization model for single pass turning operations, *Proceedings of the 27th International MATADOR Conference*, Manchester, England, 1988, pp. 231–235.

30. D. Ermer, Optimization of the constrained machining economics problem by geometric programming, *ASME J. Eng. Ind.* **93** (1971) 1067–1072.

31. K. Hitomi, Analysis of optimal machining speeds for automatic manufacturing, *Int. J. Prod. Res.* **27** (1989) 1685–1691.

32. S. M. Wu and D. Ermer, Maximum profit as the criterion in the determination of the optimum cutting conditions, *ASME J. Eng. Ind.* **88** (1966) 435–442.

33. G. Boothroyd and P. Rusek, Maximum rate of profit criteria in machining, *ASME J. Eng. Ind.* **98** (1976) 217–220.

34. J. S. Agapiou, The optimization of machining operations based on a combined criterion—Part I: The use of combined objectives in single-pass operations, *ASME J. Eng. Ind.* **114** (1992) 500–507.

35. Á. García-Crespo, B. Ruiz-Mezcua, J. López-Cuadrado, and I. González-Carrasco, A review of conventional and knowledge based systems for machining price quotation, *J. Intell. Manuf.* **22** (2011) 823–841.

36. E. J. A. Armarego and R. H. Brown, *The Machining of Metals*, Prentice-Hall, Englewood Cliffs, NJ, 1969, Chapter 9.

37. S. K. Choudhury and I. V. K. Appa Rao, Optimization of cutting parameters for maximizing tool life, *Int. J. Mach. Tools Manuf.* **39** (1999) 343–353.

38. W. T. Chien and C. S. Tsai, The investigation on the prediction of tool wear and the determination of optimum cutting conditions in machining 17-4PH stainless steel, *J. Mater. Process. Technol.* **140** (2003) 340–345.

39. M. F. DeVries, Basic machining economics, SME Technical Paper MR70-538, 1970.

40. E. J. A. Armerego and J. K. Russel, Maximum profit rate as a criterion for the selection of machining conditions, *Int. J. MTDR* **6** (1966).

41. D. Y. Jang, A unified optimization model of a machining process for specified conditions of machined surface and process performance, *Int. J. Prod. Res.* **30** (1992) 647–663.

42. Y. Altintas, P. Kersting, D. Biermann, E. Budak, B. Denkena, and I. Lazoglu, Virtual process systems for part machining operations, *CIRP Ann. Manuf. Technol.* **63** (2014) 585–605.

43. Carboloy, Inc., HI-E (HI Efficiency Machining) Technical Information Catalog, Detroit, MI, 1993.

44. J. W. Sutherland, G. Subramani, M. J. Kuhl, R. E. DeVor, and S. G. Kapoor, An investigation into the effect of tool and cut geometry on cutting force system prediction models, *Proc. NAMRC* **16** (1988) 264–272.

45. J. S. Agapiou and M. F. DeVries, On the determination of thermal phenomena during a drilling process. Part I—Analytical models of twist drill temperature distributions, *Int. J. Mach. Tools Manuf.* **30** (1990) 203–215.

46. S. Hinduja, D. J. Petty, M. Tester, and G. Barrow, Calculation of optimum cutting conditions for turning operations, *Proc. Inst. Mech. Eng.* **199B** (1985) 81–92.

47. D. A. Stephenson and S. M. Wu, Computer models for the mechanics of three-dimensional cutting processes, Part 1 and 2, *ASME J. Eng. Ind.* **110** (1988) 203–215.

48. D. A. Stephenson, Material characterization for metal cutting force modeling, *ASME J. Eng. Mater. Technol.* **111** (1989) 210–219.

49. D. A. Stephenson and J. S. Agapiou, Calculation of main cutting edge forces and torque for drills with arbitrary point geometries, *Int. J. Mach. Tools Manuf.* **32** (1992) 521–538.

50. J. S. Agapiou and D. A. Stephenson, Analytical and experimental studies of drill temperatures, *ASME J. Eng. Ind.* **116** (1994) 54–60.

51. A. M. Abuelnaga and M. A. El-Dardiry, Optimization methods for metal cutting, *Int. J. MTDR* **24** (1984) 11–18.

52. S. Bharathi Raja and N. Baskar, Investigation of optimal machining parameters for turning operation using intelligent techniques, *Int. J. Mach. Machinab. Mater.* **8**:1 (2010) 146–166.

53. D. A. Milner, Use of linear programming for machinability data optimization, ASME Paper No. 76-WA/Prod-5, 1976.

54. D. S. Ermer and D. C. Patel, Maximization of the production rate with constraints by linear programming and sensitivity analysis, *Proc. NAMRC* **2** (1974) 436–449.

55. J. A. Nelder and R. Mead, A simplex method for function minimization, *Comput. J.* **7** (1965) 308–313.

56. D. M. Olsson and L. S. Nelson, The Nelder-Mead simplex procedure for function minimization, *Technometrics* **17** (1975) 45–51.

57. K. Challa and P. B. Berra, Automated planning and optimization of machining processes: A systems approach, *Comput. Ind. Eng.* **1** (1976) 35–45.

58. D. L. Kimbler, R. A. Wysk, and R. P. Davis, Alternative approaches to the machining parameter optimization problem, *Comput. Ind. Eng.* **2** (1978) 195–202.

59. K. Iwata et al., A probabilistic approach to the determination of the optimum cutting conditions, *ASME J. Eng. Ind.* **94** (1972) 1099–1107.

60. C. Giardini, A. Bugini and R. Pagagnella, The optimal cutting conditions as a function of probability distribution function of tool life and experimental test numbers, *Int. J. Mach. Tools Manuf.* **28** (1988) 453–459.

61. A. K. Sheikh, L. A. Kendall, and S. M. Pandit, Probabilistic optimization of multitool machining operations, *ASME J. Eng. Ind.* **102** (1980) 239–246.

62. B. Gopalakrishnak and F. Al-Khayyal, Machine parameter selection for turning with constraints: An analytical approach based on geometric programming, *Int. J. Prod. Res.* **29** (1991) 1897–1908.

63. P. G. Petropoulos, Optimal selection of machining rate variables by geometric programming, *Int. J. Prod. Res.* **13** (1975) 390–395.

64. C. S. Beightler and D. T. Phillips, Optimization in tool engineering using geometric programming, *AIIE Trans.* **2** (1970) 355–360.

65. A. G. Walvekar and B. K. Lambert, An application of geometric programming to machining variable selection, *Int. J. Prod. Res.* **8** (1970) 241–245.

66. D. S. Ermer, Optimization of constrained machining economics problem by geometric programming, *ASME J. Eng. Ind.* **93** (1971) 1067–1072.

67. H. Eskicioglu, M. S. Nisli, and S. E. Kilic, An application of geometric programming to single-pass turning operations, *Proceedings of the International MTDR Conference*, Birmingham, U.K., 1985, pp. 149–157.

68. B. K. Lambert and A. G. Walvekar, Optimization of multipass machining operations, *Int. J. Prod. Res.* **9** (1978) 247–259.

69. C. L. Hough, Jr. and R. E. Goforth, Optimization of the second-order logarithmic machining economics problem by extended geometric programming, *AIIE Trans.* **13** (1981) 234–242.

70. V. I. Vitanov, D. K. Harrison, N. H. Mincoff, and T. V. Vladimirova, An expert system for the selection of metal-cutting parameters, *J. Mater. Process. Technol.* **55** (1995) 111–116.

71. R. H. Philipson and A. Ravindran, Application of goal programming to machinability data optimization, *ASME J. Mech. Des.* **100** (1978) 286–291.

72. R. M. Sundaram, An application of coal programming technique in metal cutting, *Int. J. Prod. Res* **16** (1978) 375–382.

73. G. W. Fischer, Y. Wei, and S. Dontamsetti, Process-controlled machining of gray cast iron, *J. Mech. Working Tech.* **20** (1989) 47–57.

74. P. C. Subbarao and C. H. Jacobs, Application of nonlinear goal programming to machining variable optimization, *Proc. NAMRC* **6** (1978) 298–303.

75. F. Cus and J. Balic, Optimization of cutting process by GA approach, *Robot. Comput. Integr. Manuf.* **19** (2003) 113–121.

76. E. E. Goldberg, *Genetic Algorithm in Searching, Optimization, and Machine Learning*, Addison-Wesley, Reading, MA, 1989.

77. W. J. Hui and Y. G. Xi, Operation mechanism analysis of genetic algorithm, *Control Theory Appl.* **13** (1996) 297–303.

78. D. Dubois, A fuzzy-set-based method for the optimization of machining operations, *Proceedings of the International Conference on Cybernetics and Society*, IEEE Systems, Man and Cybernetics Society, Atlanta, GA, October 1981, pp. 331–334.

79. X. D. Fang and I. S. Jawahir, Predicting total machining performance in finish turning using integrated fuzzy-set models of the machinability parameters, *Int. J. Prod. Res.* **32** (1994) 833–849.

80. J. S. Agapiou, The optimization of machining operations based on a combined criterion—Part II: Multipass operations, *ASME J. Eng. Ind.* **114** (1992) 508–513.

81. A. N. Shuaib, S. O. Duffuaa, and M. Alam, Optimal process plans for multipass turning operations, *Trans. NAMRI/SME* **20** (1992) 305–310.

82. I. Yellowley and E. A. Gunn, The optimal subdivision of cut in multi-pass machining operations, *Int. J. Prod. Res.* **27** (1989) 1573–1588.

83. J. R. Crookall and N. Venkataramani, Computer optimization of multipass turning, *Int. J. Prod. Res.* **9** (1971) 247–259.

84. H. J. J. Kals, J. A. W. Hijink, and A. C. H. Van der Wolf, Computer aid in the optimization of turning conditions in multi-cut operations, *CIRP Ann.* (1977) 465–471.

85. F. P. Tan and R. C. Creese, A generalized multi-pass machining model for machining parameter selection in turning, *Int. J. Prod. Res.* **33** (1995) 1467–1487.

86. K. P. Rajurkar, Optimization of multistage operations in EDM, *Proc. NAMRC* **16** (1986) 251–256.

87. G. L. Nemhauser, *Dynamic Programming*, Wiley, New York, 1966.

88. R. Srinivasan and C. R. Liu, On some important geometric issues in generative process planning, *Intelligent and Integrated Manufacturing Analysis and Synthesis*, ASME PED Vol. 25, ASME, New York, 1987, pp. 229–243.

89. N. Tounsi and M. A. Elbestawi, Optimized feed scheduling in three axes machining. Part I: Fundamentals of the optimized feed scheduling strategy, *Int. J. Mach. Tools Manuf.* **43** (2003) 253–267.

90. R. E. DeVor, W. J. Zdeblick, V. A. Tipnis, and S. Buescher, Development of mathematical models for process planning of machining operations, *Proc. NAMRC* **6** (1978) 395–401.

91. G. Chryssolouris and M. Guillot, *An A. I. Approach to the Selection of Process Parameters in Intelligent Machining*, ASME PED Vol. 33, ASME, New York, 1988, pp. 199–206.

92. K. Hitomi and I. Ham, Group scheduling technique for multiproduct, multistage manufacturing systems, *ASME J. Eng. Ind.* **99** (1977) 759–765.

93. K. Hitomi, N. Nakamura, and H. Tanaka, An optimization analysis of machining center, ASME Paper 76-WA/PROD-34, 1976.

94. K. Hitomi and K. Ohashi, Optimum process design for a single-stage multifuctional system, *ASME J. Eng. Ind.* **103** (1981) 218–223.

95. C. S. Tang and E. V. Denardo, Models arising from a flexible manufacturing machine, Part I: Minimization of the number of tool switches and Part II: Minimization of the number of switches instants, *Oper. Res.* **36** (1988) 767–784.

96. J. S. Agapiou, Sequence of operations optimization in single-stage multifunctional systems, *J. Manuf. Syst.* **10** (1992) 194–208.

97. H. Tempelmeier and H. Kuhn, *Flexible Manufacturing Systems—Decision Support for Design and Operation*, Wiley, New York, 1993.

98. H. J. Warnecke and R. Stenhilper, *Flexible Manufacturing Systems*, Springer Verlag, New York, 1985.

99. U. Nandkeolyar and D. P. Christy, Evaluating the design of flexible manufacturing systems, *Proceedings of the Third ORSA/TIMS Conference on Flexible Manufacturing Systems: Operations Research Models and Applications*, Amsterdam, the Netherlands, 1989, pp. 15–22.

100. C. H. Chung and I. J. Chen, A systematic assessment of the value of flexibility for an FMS, *Proceedings of the Third ORSA/TIMS Conference on Flexible Manufacturing Systems: Operations Research Models and Applications*, Cambridge, MA, 1989, pp. 27–37.

101. N. Alberti, U. La Mare, and S. N. La Diega, Cost efficiency: An index of operational performance of flexible automated production environments, *Proceedings of the Third ORSA/TIMS Conference on Flexible Manufacturing Systems: Operations Research Models and Applications*, Cambridge, MA, 1989, pp. 67–73.

102. I. Y. Alqattan, Systematic approach to cellular manufacturing systems design, *J. Mech. Working Technol.* **20** (1989) 415–424.

103. C. Doiteaux, B. Rochotte, E. Bajic, and J. Richard, FMS architecture for an optimum quality and process reactivity, *Proceedings of the 29th International MATADOR Conference*, Manchester, England, 1992, pp. 225–232.

104. K. Hitomi, Optimization of multistage machining system: Analysis of optimal machining conditions for the flow-type machining system, *ASME J. Eng. Ind.* **93** (1971) 498–506.

105. K. Hitomi, Analysis of production models Part II: Optimization of a multistage production system, *AIIE Trans.* **8** (1976) 106–112.

106. K. Hitomi, Optimization of multistage production systems with variable production times and costs, *Int. J. Prod. Res.* **15** (1977) 583–597.

107. K. Hitomi, Analysis of optimal machining conditions for flow-type automated manufacturing systems: Maximum efficiency for multi-product production, *Int. J. Prod. Res.* **28** (1990) 1153–1162.

108. K. Hitomi, Analysis of optimal machining conditions for flow-type automated manufacturing systems, *Int. J. Prod. Res.* **29** (1991) 2423–2432.

109. K. Hitomi, *Manufacturing Systems Engineering*, Taylor & Francis Ltd., London, U.K., 1979, pp. 184–193.

110. S. S. Rao and S. K. Hati, Computerized selection of optimum machining conditions for a job requiring multiple operations, *ASME J. Eng. Ind.* **100** (1978) 356–362.

111. K. Iwata, Y. Murotsu, and F. Oba, Optimum machining conditions for flow-type multistage machining systems, *CIRP Ann.* **23** (1974) 175–176.

112. K. Iwata, Y. Murotsu, F. Oba, and Y. Kawabe, Analysis of optimum loading sequence of parts and optimum machining conditions for flow-type multistage machining system, *CIRP Ann.* **24** (1975) 465–470.

113. G. S. Sekhon, Application of dynamic programming to multi-stage batch machining, *Comput. Aided Des.* **14** (1982) 157–159.

114. J. S. Agapiou, Optimization of multistage machining systems—Part I: Mathematical solution, *ASME J. Eng. Ind.* **114** (1992) 524–531.

115. J. S. Agapiou, Optimization of multistage machining systems—Part II: The algorithm and applications, *ASME J. Eng. Ind.* **114** (1992) 532–538.

116. K. S. Al-Sultan and M. A. Al-Fawzan, Determination of the optimal process means and production cycles for multistage production systems subject to process deterioration, *Prod. Plan. Control Manag. Oper.* **9**:1 (1998).

117. Y. S. Kim and W. J. Kolarik, Real-time conditional reliability prediction from on-line tool performance data, *Int. J. Prod. Res.* **30** (1992) 1831–1844.
118. A. Jeang and K. Yang, Optimal tool replacement with nondecreasing tool wear, *Int. J. Prod. Res.* **30** (1992) 299–314.
119. K. S. Park, Optimal wear-limit replacement wear-dependent failures, *IEEE Trans. Reliab.* **37** (1988) 293–294.
120. K. S. Park, Optimal wear-limit replacement under periodic inspections, *IEEE Trans. Reliab.* **37** (1988) 97–102.
121. J. Sharit and S. Elhence, Computerization of tool-replacement decision making in flexible manufacturing systems: A human-systems perspective, *Int. J. Prod. Res.* **27** (1989) 2027–2039.
122. P. Maropoulos and S. Hinduja, A tool-regulation and balancing system for turning centers, *Int. J. Adv. Manuf. Technol.* **4** (1989) 207–226.
123. S. Ramalingam, N. Balasubramanian, and Y. Peng, Tool life scatter, tooling cost and replacement schedules, *Proc. NAMRC* **5** (1977) 212–218.
124. U. La Commare, S. N. La Diega, and A. Passanati, Tool replacement strategies by computer simulation, *Proceedings of the Fourth International Conference on Production Control in the Metalworking Industry*, Belgrade, 1986.
125. U. La Commare, S. N. La Diega, and A. Passanati, Optimum tool replacement policies with penalty costs for unforeseen tool failures, *Int. J. MTDR* **23**:4 (1983), 237–243.
126. R. D. Yearout, P. Reddy, and D. L. Grosh, Standby redundancy in reliability—A review, *IEEE Trans. Reliab.* **35** (1986) 285–292.
127. P. D. T. O'Connor, *Reliability Engineering*, Hemisphere Publishing Corporation, New York, 1988.
128. S. O. Schall and M. J. Chandra, Optimal tool inspection intervals using a process control approach, *Int. J. Prod. Res.* **28** (1990) 841–851.
129. A. I. Daschenko and V. N. Redin, Control of cutting tool replacement by durability distributions, *Int. J. Adv. Manuf. Technol.* **3** (1988) 39–60.
130. C. P. Koulamas, Simultaneous determination of optimal machining conditions and tool replacement policies in constrained machining economics problems by geometric programming, *Int. J. Prod. Res.* **29** (1991) 2407–2421.
131. C. P. Koulamas, Total tool requirements in multi-level machining systems, *Int. J. Prod. Res.* **29** (1991) 417–437.
132. L. Zavanella, G. C. Maccarini, and A. Bugini, F.M.S. Tool supply in a stochastic environment: Strategies and related reliabilities, *Int. J. Mach. Tools Manuf.* **30** (1990) 389–402.
133. G. C. Maccarini, L. Zavanella, and A. Bugini, Production cost and tool reliabilities: The machining cycle influence in flexible plants, *Int. J. Mach. Tools Manuf.* **30** (1990) 415–424.
134. B. Mursec, F. Cus, and J. Balic, Organization of tool supply and determination of cutting conditions, *J. Mater. Process. Technol.* **100** (2000) 241–249.
135. J. S. Agapiou, Cutting tool strategies for multifunctional part configurations: Part I—Analytical economic models for cutting tools, *Int. J. Adv. Manuf. Technol.* **7** (1992) 59–69.
136. J. S. Agapiou, Cutting tool strategies for multifunctional part configurations: Part II—Discussion of experimental and analytical results, *Int. J. Adv. Manuf. Technol.* **7** (1992) 70–80.
137. J. S. Agapiou, The optimization of single or multipass machining operation based on a combined criterion, *ASME Conference on Advances in Manufacturing System Engineering*, PED-Vol. 37, San Francisco, CA, 1989, pp. 103–114.

14 Cutting Fluids

14.1 INTRODUCTION

The cutting fluid is an important component of the machining system in many applications. Cutting fluids are used in a large proportion of metal cutting operations to improve tool life, surface finish, and dimensional stability, and to help clear chips from the cutting zone. For many materials, cutting fluids are necessary to achieve acceptable part quality and tooling costs. However, cutting fluid acquisition, management, and treatment costs are a significant fraction of the overall operating expense in wet applications, and the fluids may also present an exposure risk to machine operators and require additional investment for enclosures, fire suppression, and air treatment.

Cutting fluids provide lubrication between the tool, chip, and workpiece at low cutting speeds, and cool the part and machine tool and clear chips at higher cutting speeds and in holemaking operations such as drilling and reaming. They also help prevent edge buildup and part rust in most circumstances. When properly applied, they permit the use of increased cutting speeds and feed rates, and improve chip formation, tool life, surface finish, and dimensional accuracy. A cutting fluid's cooling ability depends largely on the base fluid and the coolant volume. Chip flushing capabilities are determined by the operation geometry and the coolant application method. Lubrication is controlled by the chemical composition of the coolant and the application method. In recirculating systems, in which a large volume of coolant is continuously collected in a sump and reused, coolant performance over time is strongly influenced by coolant maintenance practices such as concentration monitoring, stabilization, tramp oil removal, rancidity control, and filtering.

In general, cutting fluids should pose no safety risk to the machine operator, have a long useful life, be waste-treatable, and be chemically inert with respect to the workpiece material to control rust, corrosion, or undesirable residues on machined surfaces. In addition to meeting these requirements, the coolant strategy in a given applications must significantly improve machining performance to justify the investment, infrastructure, and operating costs of the coolant system.

The types of coolants that are effective for broad classes of work materials and operations, for example, for turning aluminum alloys or milling steels, are generally understood. The selection and maintenance of a cutting fluid in a specific application, however, is often determined by experience and limited performance testing, and typically represents one of the more arbitrary decisions in process design. This is unfortunate, since the cutting fluid type, application method, pressure, and flow rate have as strong an influence on tool life and surface quality as parameters such as the tool grade, cutting speed, and feed rate, all of which are normally carefully optimized during process design. As an integral part of the system, the available coolant options should be considered in addition to the tooling and feed and speed variables in process development.

Studies have shown that occupational exposure to cutting fluids can lead to adverse health effects. This has led to increased interest in minimizing coolant usage and other measures to limit workforce exposure. The material safety data sheet (MSDS) describes the characteristics and safe usage practices for a given fluid and should be consulted prior to use. In the United States, the National Institute for Occupational Safety and Health (NIOSH) and Department of Labor provide documentation on the prevalence of hazards, the existence of safety and health risks, and the adequacy of control methods [1,2].

This chapter provides a broad overview of common cutting fluids, application methods, filtering, maintenance and waste treatment practices, and health and safety issues. It also provides an overview of recent research on the development of dry and near-dry machining technologies. The most widely used near-dry machining method, minimum quantity lubrication (MQL), is covered in detail in Chapter 15.

14.2   TYPES OF CUTTING FLUIDS

Cutting fluids are commonly classified as neat or cutting oils, water-based fluids, gaseous fluids, air-oil mists, and cryogenic fluids. Water-based fluids include emulsifiable oils, semi-synthetic fluids, and synthetic fluids.

14.2.1   NEAT OILS

Neat oils are mineral, animal, vegetable, or synthetic oils used without dilution with water [3–10]. Petroleum-based mineral oils, including light solvents, neutral oils, and heavy bright and refined oils, have traditionally been most common due to their low cost [3]. Fatty animal oils are used largely as compounding oils as discussed in the following. Vegetable oils used include palm oil, rapeseed (canola) oil, and coconut oil. They are more expensive than mineral oils but are sometimes required in environmentally sensitive applications, especially in Europe [7], for example, in machining titanium and stainless steel and in applications requiring the German WGK-0 or NWG classification [8]. Synthetic esters or fatty alcohols made from renewable sources are also used as substitutes for mineral oils, especially in MQL applications [11].

Mineral oils may be classified as straight or compounded, with compounded being the more common. Straight oils are base oils without active additives; compounded oils consist of the base oil mixed with polar or chemically active additives. Common polar additives include fatty animal oils such as lard oil or tallow, and vegetable oils such as palm oil or castor oil derivatives. These additives, which are used in concentrations between 10% and 40%, increase the cutting oil's wetting ability and penetrating properties and thus improve lubricity. Common active additives include chlorine, sulfur, and phosphorous compounds; they are surface reactive and form a metallic film at the tool surface, which acts as a solid lubricant. Additives must be chosen to be chemically compatible with the work material; improper additives can produce staining or corrosion of the part.

Cutting oils are more effective as lubricants than coolants [12]. They are used extensively in grinding and honing operations, where they permit higher metal removal rates with better finish and less surface damage than water-based fluids [13]. Due to their limited cooling capabilities, they are not used in high-speed machining and restricted to relatively low speed operations such as broaching, tapping, gear hobbing, and gun drilling, and in machining nickel alloys and other hard metals.

Cutting oils are stable and provide excellent rust protection. They are relatively costly, however, have limited applicability to high-speed applications, present a potential smoke and fire risk, and have been associated with contact dermatitis and other operator health issues [14]. Due to their high cost and fire risk, they are generally used only on individual machines, and not in large recirculating systems.

Vegetable oils have high flash point temperatures compared to high viscosity mineral oils (typically 230°C versus 165°C), so that smoke formation is less of a concern. They also provide better lubricity (because of the dipolar nature of their molecules) and are often used as a polar additive to enhance the lubricity of mineral oils. Even though vegetable oils are more expensive than mineral oil–based formulations, they may be cost effective in some applications due to reduced consumption (less dragout than mineral oils) and improved tool life and productivity [15]. Natural vegetable oils have limited shelf life due to hydrolization and bacterial degradation and are often treated to produce esters-based or other modified oils to improve stability.

14.2.2   WATER-BASED FLUIDS

Water-based fluids are dilute emulsions or solutions of oils in water, which provide less lubrication but better cooling and chip clearing abilities than neat oils. Water cools two to three times faster than mineral oils and can retain more than twice the amount of heat. They are used extensively in higher speed operations and large recirculating systems. There are three basic types: *soluble oils, semi-synthetics*, and *synthetics* [3–9,16].

Soluble oils are special types of mineral or other base oils emulsified in water at concentrations typically between 5% and 20%, with lower concentrations (less than 10%) being most common in general-purpose machining. (The term "soluble oil" is in fact incorrect, since oils are not soluble in water, but is in common use.) Soluble oil concentrates contain severely refined base oils (30%–85%), emulsifiers, and performance additives such as extreme pressure (EP) additives, stabilizers, rust inhibitors, defoamers, and bactericides. Base oils have traditionally been primarily naphthenic or paraffinic mineral oils, with naphthenic oils being more common due to their lower cost, although ester-based oils and vegetable oils [17,18] may also be emulsified. The oil viscosity is typically 100 SUS at 100 F (100/100 oils); higher viscosity oils provide better lubricity but are more difficult to emulsify. Emulsifiers are added to form stable dispersions of oil and water; emulsifier particles are located around the oil droplets to give them a negative charge that will bond them with the water molecules. The size of the emulsified oil droplets is very critical to fluid performance; it is easier for the smaller emulsion sizes to penetrate the interface of the cutting zone. Most emulsions have droplet sizes between 2 and 50 μm and give the fluid a translucent white or blue-white appearance. "Pearlescent" microemulsions have droplet sizes between 0.1 and 2 μm; they provide better lubricity but have a tendency to foam and entail greater waste-treatment difficulties. EP additives have the same function as chemically active additives in neat oils. Stabilizers inhibit the breakdown of the emulsion, which occurs over time as metal particles in the fluid combine with the emulsifiers. Bacteriocides are added to control bacterial growth and rancidity, although there is a trend toward reduced usage due to operator exposure concerns [19].

There has been increased interest in the development of new environmentally friendly cutting fluids, such as vegetable-oil based coolants, which can be derived from renewable resources [15,17,18]. More recently the term "green" metalcutting fluids has been used [20] to include not only vegetable-based soluble oils, but also other fluids that do not contain any chlorine, sulfur, or phosphorus, since these components can present waste disposal concerns. The greenest cutting fluids are vegetable-based neat oils used in MQL applications, since they leave behind very little residue and near-dry chips. Some synthetic and petroleum-based nontoxic and biodegradable oils are also considered green cutting fluids.

The emulsifier may be anionic (negative chemical charge; e.g., carboxylates, sulfonates, phosphates, and phosphonates), cationic (positive chemical charge amine salts, phosphonium compounds), or nonionic (neutral chemical charge; e.g., polymeric ethers, esters, and amides). Light-duty soluble oils use soap-sulfonate emulsifier systems. Moderate to heavy-duty soluble oils contain chlorinated paraffins or sulfurized fat emulsifiers. The percentage of the base oil, emulsifier, EP additives, and other additives by weight in general purpose light duty soluble oils is 83%, 15%, 0%, and 2%, respectively; for medium-duty soluble oils the corresponding percentages are 70%, 15%, 10%, and 5%, respectively, while for heavy-duty soluble oils the percentages are 30%, 20%, 40%, and 10%, respectively.

Soluble oils provide excellent cooling and reasonable lubricity performance at low concentrations, have inherent rust preventive properties, and can incorporate additional performance additives. Their disadvantages include their hard water sensitivity [21], susceptibility to microbial attack, tendency to foam, potential to cause contact dermatitis, and disposal difficulty.

Semi-synthetic cutting fluids are fine emulsions in which the base concentrate is a mixture of mineral oil and additional chemicals, which may include emulsifiers, couplers, corrosion inhibitors, EP additives, and bacteriocides and fungicides. They combine features of both soluble oils and synthetics. The base concentrate usually contains between 5% and 30% mineral oil and 30%–50% water. The fluid itself typically contains 50%–70% water. Semi-synthetics require higher emulsifier concentrations than soluble oils, especially when oil-soluble chemical additives are used. They produce smaller emulsion particle sizes than soluble oils (from 0.01 to 0.1 μm), which gives them improved lubricity and makes them transparent or translucent. They can be formulated either to emulsify or reject tramp oils. The advantages of semi-synthetics include rapid heat dissipation, excellent wettability, cleanliness, and resistance to rancidity and bacteria; bacterial resistance is enhanced by the small emulsion size and relatively low concentration of mineral oil. Compared to synthetic fluids, they provide better rust prevention, improved flexibility in incorporating additives (since both water- and oil-soluble additives can be used), and fewer waste treatment concerns. Their disadvantages include foaming and residue concerns and susceptibility to contamination by tramp oils.

Synthetic cutting fluids are water-based fluids consisting of chemical lubricating agents, wetting additives, disinfectants, and EP additives. They are true solutions in water and contain no oil. They are used in relatively low concentration (typically 5%) and produce a particle sizes below 0.005 µm. Dilution affects the extreme-pressure lubricity of the fluid, so heavier cuts require a higher concentration. They are generally clear, but are often dyed to indicate their presence in water. Additives are chosen to be low-foaming and stable in water.

Synthetic fluids typically contain rust or corrosion inhibitors, lubricants, and fungicides. Rust inhibitors are necessary since no oil is present. Common inhibitors include borate esters and amine carboxylate derivatives. Both boundary and EP lubricants are added. Typical boundary lubricants include soaps, amides, esters, and glycols. Common EP additives include chlorinated and sulfurized fatty acid soaps and esters. Fungicides are necessary since synthetic fluids are susceptible to yeast and mold growth; bacteria are not a concern since the fluids contain no mineral oils and have a relatively high pH.

Synthetic fluids provide excellent cooling and produce fine surface finishes. Compared to emulsifiable oils, they provide better resistance to bacterial degradation and improved tramp oil rejection, stability, and workpiece visibility. They do not usually cause dermatitis. Their disadvantages include reduced lubricity due to absence of petroleum oils, a tendency to leave hard crystalline residues, high alkalinity, a tendency to foam, and disposal problems.

Figure 14.1 shows historical trends in the usage of cutting oils and water-based fluids [3]. Only straight oils were used prior to 1910. Soluble oils were introduced between 1910 and 1920 to improve cooling properties and fire resistance. Semi-synthetic fluids were introduced in the 1950s to further improve cooling and rust inhibition. Synthetics were first widely used in the 1970s, when the price of petroleum increased dramatically. Over the long term, straight oils have been used less as cutting speeds have increased; of the water-based fluids, soluble oils are still most common due to their low cost.

Water-based fluids commonly fail by loss of emulsion stability or wetting ability (which are affected by the metal machined, the material removal rate in relation to the volume of coolant, the type of water used in the system, the water evaporation rate, tramp oils, the pH level, and the loss of stabilizers such as biocides and odor control agents). In general, the most common causes of failure include heat, hard water, tramp oil, oxidative and reactive agents, and microbial growth [21]. The ASTM Standard E1497-94 [22] and similar NIOSH and OSHA publications [1,2] should be consulted for safe use guidelines.

14.2.3 Gaseous Fluids

In some operations gases rather than liquids are used as cutting fluids. This is true in particular in applications in which no fluid residue on the workpiece can be tolerated, as is the case in some medical and aerospace applications. Gaseous fluids used include air, helium, CO_2, argon, and nitrogen [6], with air being the most common due to its low cost. (Freons were used historically but are no longer permitted in the United States.) Air can be compressed or chilled to provide better cooling by forced convection. High pressure air streams can be used to eject chips (though not as effectively as a liquid) assuming the proper shrouding is used. Noise generation is a significant concern with high velocity or compressed air systems. CO_2 provides evaporative cooling when it is compressed and sprayed into the cutting zone [23–25]. In Germany CO_2 systems are sometimes called "CO_2 snow" due to the frost frequently left on the tool after their use.

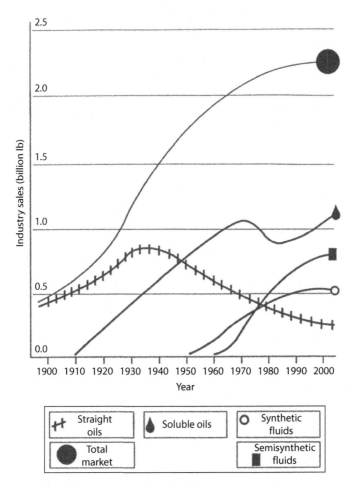

FIGURE 14.1 Historical usage of straight oil and water-based cutting fluids. (After Childers, J.C., The Chemistry of Metalworking Fluids, in: Beyers, J.P., ed., *Metalworking Fluids*, Marcel Dekker, New York, 1994, pp. 165–189.)

14.2.4 Air–Oil Mists (Aerosols)

Air–oil mists, consisting of small droplets of water-based oil mixed with air, have been successfully applied as cutting fluids in many applications. Historically, they were used in high-speed applications with small areas of cut, for example, end-milling applications [3]. With the development of through-spindle coolant systems, they have been increasingly restricted to low-speed cutting applications such as drilling, gear machining, and sawing operations. There are two methods of producing mists: aspirator methods and direct-pressure methods [4]. In aspirator systems, a stream of air is directed past an open tube containing oil, creating a partial vacuum, which draws oil droplets into the air stream. In direct-pressure systems, compressed air is directed through the oil to create a mist. Oil mists are best suited to applications in which flood coolant is impractical, for example, in applications in which the cutting zone is relatively inaccessible, and on large machines. Their major disadvantages are a tendency for the nozzles to clog and exposure of the operator to mist inhalation, which has adverse health consequences as discussed in Section 14.6. Lean air–oil mists are also used in minimum quantity lubrication (MQL) systems as described in Chapter 15.

14.2.5 CRYOGENIC FLUIDS

Cryogenic coolants (with boiling points below −150°C) are sometimes used to machine and grind difficult-to-machine materials [26–32]. Liquid nitrogen (LN_2) is the most common cryogenic coolant, although liquid argon and liquid CO_2 are also used [26,28,30]. Cryogenic coolant application is especially effective when machining materials for which cutting temperatures are a major concern, for example, titanium alloys, compacted graphite iron, and hard steels. They are also used for stainless steel and nickel alloys, but in these applications an additional oil line may be required to provide lubrication. When properly applied, cryogenic fluids can be used to increase both tool life and metal removal rates compared to conventional and high pressure water-based coolants. Cryogenic fluids are generally used with carbide tooling, and depending on the application method may require pre-cooling of the tool to prevent damage due to thermal shock. Another operational limitation of older systems was the tendency of the nozzles to ice over due to condensation of water vapor from the surrounding atmosphere, but countermeasures have reduced this issue on current systems. Cryogenic fluids are more expensive to apply than air–oil mists, and it may lead to part warping. However, they are vaporized during use and thus do not leave a residue or require treatment, which is a significant benefit in environmentally sensitive applications.

14.3 COOLANT APPLICATION

The effectiveness of cutting fluids depends to a large extent upon the method of their delivery into the cutting zone. There are four basic methods of applying coolant: low pressure flood application, high pressure flood application, through-tool application, and mist application (see Figure 3.61). In *low pressure flood application* systems, coolant is delivered through nozzles over the work zone at line water pressures. Coolant may be applied through either fixed or flexible piping, with fixed systems being common on dedicated systems and flexible tubing being more typical of general purpose machines. In either case coolant nozzles should be directed ahead of the cut, with sufficient additional coolant being supplied to cover the workpiece and back of the cutter, especially in milling operations. Insufficient or intermittent coolant application in milling operations can lead to thermal fatigue wear of the cutter. If the coolant volume is sufficient, low pressure flood application is effective in clearing chips and cooling the part to maintain dimensional tolerance but has limited lubricating effectiveness.

In *high pressure flood application*, coolant is directed through nozzles to impinge ahead of the cutter at higher pressures. The inlet nozzle pressure varies widely with the orifice diameter but is typically between 5 and 100 bar (75 and 1500 psi). Considerably higher pressures are used in some grinding operations and in impingement chip breaking jet systems. Coolant is applied though rigid piping, with nozzles typically being mounted on a ring around the spindle nose in boring, milling, and drilling (Figure 3.61), on the toolholder behind or through the insert in turning, and on a rigid pipe in front of the wheel in grinding [33]. High pressure application provides more effective lubricating and chip clearing capabilities but generates mist, which may present a hazard to the operator if not properly collected and filtered. There is also an increased tendency for the coolant to become aerated and foam.

In *through-tool coolant* systems, coolant is supplied through the spindle or a coolant inducer to coolant passages in the tool under high pressure. Typical application pressures are 35–100 bar (500–1500 psi). Through tool coolant is required to clear chips in many high throughput drilling and deep hole drilling operations and is also used in grinding (through porous wheels). It requires special seals and pumps, which generally must be built into the machine tool. It is very effective in cooling the cutting edge and clearing chips in holemaking operations, and in cooling the wheel in grinding. Effective seals must be installed and maintained to prevent coolant from leaking into the spindle bearings. In drilling, the coolant must be effectively filtered to prevent debris from clogging the tool coolant passages in recirculating systems. Disadvantages of high pressure coolant include increased maintenance to ensure that seals, pumps, and rotary unions do not fail, a tendency to generate mists, and increased foaming.

As discussed in the previous section, mist application is applied using either aspirator or direct pressure systems and is used especially in drilling, gear machining, and sawing operations.

Generally, the coolant pressure should be sufficient to penetrate the vapor barrier pocket generated around the cutting edge. The force with which the fluid penetrates the cutting zone (through the vapor barrier) is somewhat proportional to the coolant velocity and a critical factor for the design of the process. However, the velocity is proportional to the square root of the pressure. Therefore, doubling the coolant pressure results in a roughly 40% increase of the coolant force as explained in Chapter 4. Hence, it is preferable to increase the coolant volume rather than the pressure to optimize coolant performance in most applications. However, for some difficult-to-machine materials, coolant pressures of 20–30 bar (2000–3000 psi) have been found to be effective. Some machines are equipped with a standard coolant pump and a high-pressure pump for flexibility for various applications.

During machining, airborne mist is generated from cutting fluids. Many factors including the machining conditions, cutting tool design, and fluid application method influence mist generation. Mist generation is usually controlled through mechanical means (i.e., enclosures, ventilation), chemical means (i.e., anti-mist polymer additives, mist suppression at the source, foam reduction, formulations with low oil concentrations, avoiding contamination with tramp oil, etc.) [34,35], minimizing fluid delivery pressure and flow rate, the proper design and operation of the cutting fluid delivery system [1], or the use of dry or near-dry machining methods. Once generated, mists are normally controlled or abated through ventilation and filtering as discussed in the next section.

Regardless of the method used to apply coolant, sufficient volume must be supplied to provide adequate cooling and chip clearing capability. Coolant volume is measured by the flow rate in gallons or liters per minute. As a reasonable rule of thumb, the coolant volume should be 1–2 gal/min for each HP of cutting energy (or 5–10 L/min for each kW) for general machining [36], and 2–4 gal/min per HP (10–20 L/min/kW) for grinding [4]. As will be discussed in the next section, an adequate sump capacity is required to ensure proper filtering and minimize foaming; as a general rule of thumb, the sump capacity should be 3–10 times the flow rate per minute [36]. More detailed recommendations of the sump capacity are 5 times the flow rate for steel machining, 7 times the flow rate for cast iron and aluminum machining, 10 times the flow rate for grinding, and 10–20 times the flow rate for high stock removal machining and grinding [4].

14.4 FILTERING

Coolants in recirculating systems entrain and transport a number of impurities, including chips, airborne contaminants, hydraulic and machine way oils, residues left on the part from previous operations, etc. [37]. These impurities must be removed to maintain coolant performance. A variety of methods and systems are used to remove contaminants from cutting fluids, which can be broadly classified as either separation methods or filtration methods. Often two or more methods are used in sequence to increase effectiveness.

Common separation methods include settling tanks, centrifuges, cyclones, and magnetic separators [4,37]. A *settling tank* is a large tank with two or more baffles (Figure 14.2); as the fluid moves under and over successive baffles, tramp oils and lighter impurities rise to the surface, where they can be skimmed off, and chips and other heavy debris settle on the bottom, where they can be similarly removed. The effectiveness of the tank depends on the settling time or the volume of the tank divided by the inlet flow rate; if the settling time is too short, not all impurities will settle out, there will be an increased tendency to foam since coolant bubbles will not have time to burst, and coolant temperature will be more difficult to control [21,33,38]. Settling times should be 5 min for small systems [4], 7–10 min for general purpose systems [39], and 10–15 min for large systems [4,37]. Centrifuges and cyclones separate debris from the coolant through centrifugal action [4,37].

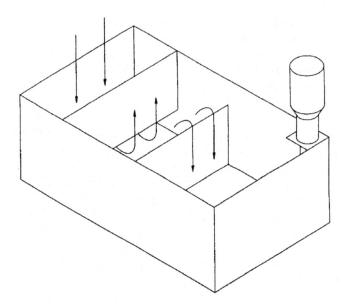

FIGURE 14.2   Settling tank. (After Brandt, R.H., Filtration Systems for Metalworking Fluids, in: Beyers, J.P., ed., *Metalworking Fluids*, Marcel Dekker, New York, 1994, pp. 273–303.)

In a *centrifuge,* dirty coolant is introduced between rotating bowls or cone-shaped disks. Chips and other heavy debris are forced toward the center of the bowls or disks, and tramp oils with low specific gravity are forced outward. The bowls or disks eventually fill with debris and require periodic cleaning. In a *cyclone* (Figure 14.3), dirty coolant is directed into a cone-shaped vessel; the resulting rotary motion of the fluid forces chips and debris outward, so that relatively clean fluid emerges from the center of the device [4,37,38]. In a *magnetic separator* the coolant is directed past a rotating magnetic drum. Chips and fines stick to the drum and are removed by a blade on the opposite side of the drum, generally out of the coolant bath [37]. Magnetic cyclones and belt systems are also used [40]. These systems are obviously effective mainly for ferrous chips, although they will also remove abrasive grains adhering to ferrous fines in grinding systems.

Since separators are not effective in removing impurities with specific gravities near that of the coolant, they are normally used in conjunction with filtering systems. These systems may use either disposable or permanent porous media; the degree of filtering achieved is determined by the pore size of the medium. Common *disposable medium filtering systems* include bag, cartridge, and roll systems (Figure 14.4) [4,37]. The filter medium in these systems may be made of paper, cotton, wool, synthetic fibers, or felted materials; they are often pre-coated with fine particles such as cellulose fibers or diatomaceous earth. In a bag system, coolant is passed through two or more fabric bags with successively smaller mesh sizes. The bags fill with chips and must be changed periodically. Cartridge systems are similar but use cylindrical cartridges, similar to an automotive oil filter, rather than bags [41]. Flat-bed roll filters are common in large recirculating systems. In this approach, coolant is passed through a sheet of filter media. As filtering progresses and the filter becomes clogged, the fluid level in the tank rises, eventually triggering a float, which indexes the roll to expose fresh media. Disposable media filters may be driven by gravity, a vacuum, or pressure. In a gravity system, fluid flow is maintained by gravity. In a vacuum system (Figure 14.5), a negative pressure is applied to the back of the medium to accelerate flow. In a pressurized system positive pressure is applied to the fluid to force it through the medium. Pressurized systems operate at higher pressures and can produce the largest flow rates. Both gravity and pressurized systems are used with flat-bed roll filters. Pressurized systems are also commonly used with cartridge filters.

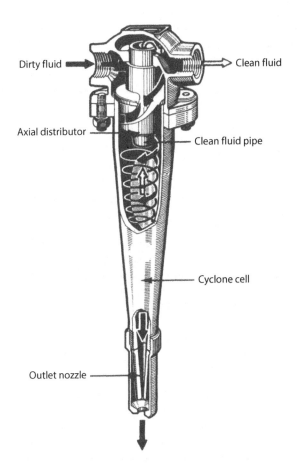

Dirty fluid → ⇒ Clean fluid

Axial distributor

Clean fluid pipe

Cyclone cell

Outlet nozzle

FIGURE 14.3 Chip cyclone. (After Drozda, T.J., and Wick, C., eds., *Tool and Manufacturing Engineers Handbook*, Vol. I: *Machining*, 4th edn., SME, Dearborn, MI, 1983, pp. 4.1–4.34.)

FIGURE 14.4 Disposable filtering media. (After Brandt, R.H., Filtration Systems for Metalworking Fluids, in: Beyers, J.P., ed., *Metalworking Fluids*, Marcel Dekker, New York, 1994, pp. 273–303.)

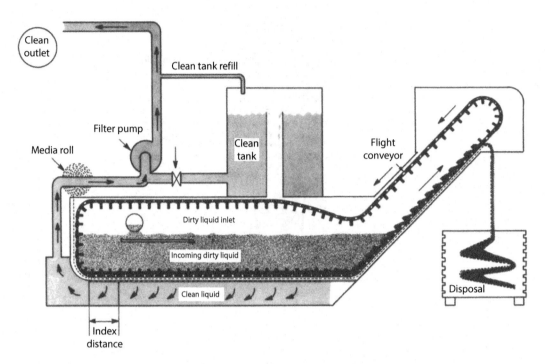

FIGURE 14.5 Vacuum media flat-bed roll filtering system. (After Drozda, T.J. and Wick, C., eds., *Tool and Manufacturing Engineers Handbook*, Vol. I: *Machining*, 4th edn., SME, Dearborn, MI, 1983, pp. 4.1–4.34.)

In *permanent media filtering systems*, a wire mesh or permanent fabric is used as the filtering medium [42]. As in disposable media systems, a porous pre-coating material is often used to increase effectiveness. Chips and other debris are periodically removed by a cleaning unit; cleaning may be controlled by a timer or activated by a float trigger. Common system configurations include those employing circular meshes, flat belts or screens, and stacked disks separated by spacers. Figure 14.6 shows one especially common approach, in which a wedge-wire mesh belt is used in a flat-bed configuration. Permanent media filters require higher initial investment but may be more economical over time, since disposable media costs are avoided; they also generate waste in cake form, which may be more economically disposed of than the contaminated fabrics produced by disposable media systems. Large sand or diatomaceous earth filter systems are sometimes used for neat oil applications, especially grinding. In these systems oil seeps through a bed of fine earth, which traps entrained particles. Periodic back-flushing is required to remove accumulated debris.

Aside from cost, factors to be considered in choosing a filtering strategy are the level of filtration necessary, required system capacity, and the effect of the filter medium on the coolant [4]. The level of filtration is generally expressed as the largest dimension of contaminant, which can be tolerated. Generally, standard chip removal and cutting fluid filtration systems remove debris over 50 μm in size fairly easily. Greater care and additional filtering stages are required to remove finer debris [43]. A common rule of thumb for general purpose machining is that coolant should be filtered to 1/10th the tolerance band [33]. This can lead to very stringent requirements for precision operations. In large systems employing through-tool coolant, it is common to filter to between 5 and 10 μm. The required system capacity depends on the coolant flow rate. Minimum capacities can also be computed from the sump volume; large systems typically pump at least three times the sump capacity in 24 h [38]. Bag filters generally have capacities between 25 and 50 gal/min/ft^2 of filter area (1000–2000 L/min/m^2) [37].

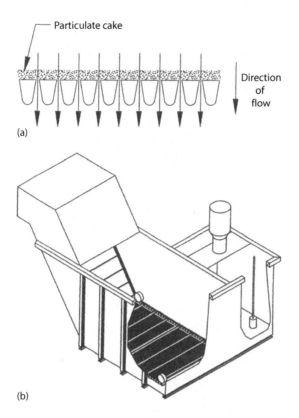

FIGURE 14.6 (a) Wedge-wire permanent media filtering principle. (b) Flat-bed wedge wire filtering system. (After Brandt, R.H., Filtration Systems for Metalworking Fluids, in: Beyers, J.P., ed., *Metalworking Fluids*, Marcel Dekker, New York, 1994, pp. 273–303.)

Cartridge filters typically have a capacity of approximated 1 gal/min/ft of cartridge length (1 L/min for 6 cm) [41]. Flat-bed systems have much higher capacities and are most common for large installations. Permanent media systems have the same flow capacity as disposable media systems of comparable pore size. Some types of filters can reduce fluid effectiveness by removing oxidization inhibitors, detergents, and other additives as well as contaminants [4]. Filter system manufacturers provide information on flow rates, filtering levels, and application ranges for specific systems.

Another frequently important component of the fluid filtration system is the *Ventilation System* to control mist and airborne contaminants [44–46]. Of major concern are atomized mists with particle diameters of tenths of a micron, which are most common in high-speed machining and grinding operations [42] with high pressure coolant. When mist generation cannot be reduced, effective building ventilation and source isolation through machine enclosures ventilated to appropriate mist collectors must be implemented. The ANSI B-11 Technical Report 2 and ACGIH provide guidelines for proper ventilation systems [47] consisting of physical barriers or ductwork. The key points discussed in the ANSI B-11 Report are summarized in Figure 14.7. Similar guidelines for isolation, ventilation, and filtration in recirculating air systems are provided by the U.S. Occupational Safety & Health Administration (OSHA) [1,2]. Systems that recirculate air (rather than venting it outside) normally require submicron mist filtering using high-efficiency particulate air (HEPA) filters. HEPA filters combined with spark arrest systems [48,49] are required for aluminum MQL machining systems to remove fine mists and mitigate fire and explosion risks [50].

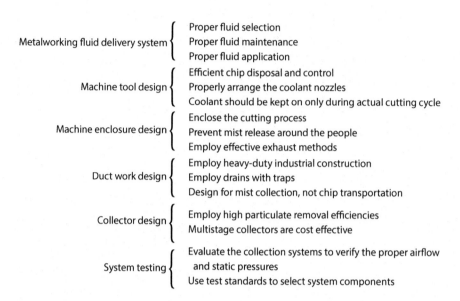

Metalworking fluid delivery system
- Proper fluid selection
- Proper fluid maintenance
- Proper fluid application

Machine tool design
- Efficient chip disposal and control
- Properly arrange the coolant nozzles
- Coolant should be kept on only during actual cutting cycle

Machine enclosure design
- Enclose the cutting process
- Prevent mist release around the people
- Employ effective exhaust methods

Duct work design
- Employ heavy-duty industrial construction
- Employ drains with traps
- Design for mist collection, not chip transportation

Collector design
- Employ high particulate removal efficiencies
- Multistage collectors are cost effective

System testing
- Evaluate the collection systems to verify the proper airflow and static pressures
- Use test standards to select system components

FIGURE 14.7 Key points discussed in the ANSI B-11 Report. (From ACGIH, *Industrial Ventilation: A Manual of Recommended Practice*, 22nd edn., *American Conference of Governmental Industrial Hygienists*, Committee on Industrial Ventilation, Cincinnati, OH, 1995.)

14.5 CONDITION MONITORING AND WASTE TREATMENT

The chemical composition and other properties of the cutting fluid should be monitored regularly to maintain acceptable performance. Common parameters monitored include concentration, percent solid and tramp oil, bacteria and fungi count, water quality, and temperature [4,51–55].

In water-based systems, the *concentration* of the base oil or synthetic concentrate should be monitored to ensure that it is within acceptable limits. If the concentration is too low, the fluid will not perform properly and corrosion of the workpiece may be a concern; if it is too high, the system is being run uneconomically, and the fluid will have an increased tendency to foam. In mineral oil–based fluids, excessive concentration may also promote bacterial growth. Concentration is most often measured using a refractometer, a device that correlates the diffraction of a beam of light by a drop of sample fluid to concentration [39,56]. Laboratory tests such as oil split methods and chemical titration are also used. Concentration can be difficult to measure accurately if tramp oil contamination is not controlled, since refractometers do not distinguish tramp oils from base oils well.

The *percent solid* and *tramp oil* are indications of the effectiveness of filtering and of maintenance issues. Excessive solid entrainment, which often produces poor surface finish, indicates that filtering is not effective and should be corrected. Similarly, excessive tramp oil may indicate the presence of a broken hydraulic line, bad seal, malfunctioning way lubrication unit, or other component failure. Excessive tramp oil contamination results in pumping problems, poor filter performance and life, and destabilization of the cutting fluid emulsion. Regular machine maintenance minimizes contamination of the cutting fluid with free debris and tramp oils leaking from spindles, slides, and gearboxes. Both percent solid and tramp oil are measured by allowing a fluid sample to stand in a graduated cylinder; over a period of hours, the tramp oil will rise to the top, and solids will settle in the bottom [39]. Centrifuges and filters can also be used for this purpose.

Bacteria and fungi in the coolant should be monitored and controlled. Standard tests for microbial content include plate count and dipslide tests; other methods, including enzyme catalase and dissolved oxygen monitoring, are less commonly used [57]. Excessive microbial content generates objectionable odors and may cause corrosion, changes in fluid chemistry, and filter blockage. In rare instances there may also be adverse health effects as discussed in the next section. In addition, bacteria consume cutting fluid components, and the by-products of this activity can lower the mix pH. Typically, well-maintained coolant systems have bacterial counts below one million per milliliter. Methods of controlling microbial growth include filtration, the use of chemical biocides, and a variety of less common pasteurization, radiation, and microwave treatments. The best method to control bacteria and fungi is to maintain a clean system and limit oil and water contamination. Excessive use of biocides should be avoided to limit occupational exposure to toxic concentrates.

Important factors in *water quality* are hardness, pH, and chemical content [4,58]. Fluid effectiveness can be compromised if its hardness exceeds 200 ppm dissolved minerals and organics, or if it contains excessive levels of chloride, sulfate, or phosphate ions [4,36]. Hard water increases concentrate usage and may lead to corrosion and residue problems. Hardness is a particular concern if wetting agent additives are used [4]. Conversely, when the water is too soft (total hardness less than 80 ppm) the fluid will have an increased tendency to foam [39,59]. The ideal water hardness range for metalworking fluid mixes is between 80 and 125 ppm. Treatment of incoming water or an alternative water source is required for excessively hard or soft water. The fluid's pH can be measured using a pH meter or pH paper. New coolants generally have a pH between 8 and 9 [4,39]. In this range the fluid is alkaline, which discourages microbial growth. The spoilage range of most coolants is between 7 and 8. Low pH may result in increased corrosion, objectionable odors, and destabilization of the coolant. If the pH is above 9.5, it may irritate the skin of operators [4].

Corrosion is a problem not only of water-diluted cutting fluids but also in dry machining [52,55]. Corrosion results from chemical reactions and increases with increasing temperature in the presence of moisture and oxygen in the atmosphere. Moisture condenses on a workpiece and acts as an electrolyte to form a galvanic cell. The concentration and type of additives used to provide protection against corrosion depend upon the type of metal(s) involved (including ferrous and nonferrous), the cutting fluid, and the anticipated chemical reactions. Some of the factors affecting corrosion are the pH of cutting fluid (>9 protects ferrous metals, although this level can adversely affect nonferrous metals such as aluminum and brass), impurities in the water (e.g., high concentration of ions, >100 ppm chloride, >100 ppm sulfate, or >50 ppm nitrate and conductivities > 4 mS/cm are considered aggressive waters), high bacteria counts, and unstable emulsions and fluid concentrations.

The cutting fluid *temperature* should be controlled in operations, which generate large amounts of heat, and in precision operations in which thermal expansion may produce significant dimensional errors. Temperature control to within 1°C–2°C is normally sufficient [39]. Temperature control can be accomplished using natural convection currents or forced air circulation, or circulating the coolant through a chiller integrated into the tank or sump. Fluids cooled to 10°C are effective in difficult operations.

Waste treatment is an important consideration in selecting a coolant [53,60–62]. The manner and degree to which used coolants must be treated are governed by local regulations. Generally, straight oils are reprocessed by distillation or vacuum distillation prior to reuse or burning. Soluble oils are treated by adding chemicals such as sulfuric acid, polyelectrolytes, or salts to deactivate the emulsifier to separate the concentrate [7,56]. Chemical separation is not effective for semisynthetics and synthetics [7], which must generally be treated using special filtration methods [56]. A more thorough discussion of this subject is given in Refs. [56,60].

14.6   HEALTH AND SAFETY CONCERNS

Occupational exposure to cutting fluids in liquid or mist form can have a number of adverse health effects [1,2,16,63–66]. The common exposure mechanisms are dermal (skin) contact and inhalation; less common mechanisms include ingestion either orally or through an open cut. The resulting health effects may include toxicity, dermatitis, respiratory disorders, microbial infections, and cancer.

14.6.1 Toxicity

Short- or long-term exposure to cutting fluids and fluid additives, especially biocides and fungicides, may induce toxic reactions [4]. Acute toxicity may occur through accidental ingestion through splashes or handling food with dirty hands, absorption through cuts, or inhalation of mists; chronic toxicity may result from long-term exposure to mists. Coolants and additives are regulated by local laws; in the United States, coolant suppliers must provide a material safety data sheet (MSDS), and biocides are regulated as pesticides by the EPA. The documentation required by regulation provides information on permissible exposure levels, necessary protective equipment, and countermeasures in case of ingestion. Coolants used in recirculating systems may leach heavy metals or lead from the work material [56]; if this is a possibility, coolants should be monitored for metal content and changed as needed to avoid unacceptable exposure levels.

14.6.2 Dermatitis

Dermal contact with cutting fluids can cause contact or allergic dermatitis [2,4–6,67]. Dermatitis is the most common health problem associated with cutting fluids; estimates are that it afflicts between 0.3% and 1% of machinists in the United States [68]. There are two common mechanisms for skin irritation: blocking of hair follicles by fines or other debris in used fluid, and removal of protective oils from the skin, particularly when the coolant pH is high (over 9) [4]. The first mechanism leads to condition called oil acne, an old ailment first reported in the United States in 1861 [7]. The second mechanism has become common in recent years with the increased use of water-based fluids and is also associated with some types of additives, notable biocides [4]. Dermatitis is more easily prevented than treated; common prevention methods include ensuring that fluid concentration levels are not too high, providing gloves, hand creams, and low pH soaps to operating personnel, transferring operators who show particular sensitivity to dermatitis, substituting alternative additives for those suspected of causing a problem, improved filtering, and education on the workforce on proper hygiene and methods of minimizing exposure [4,59].

14.6.3 Respiratory Disorders

Exposure to cutting fluid mists has often been reported to cause acute respiratory difficulties including coughs, increased airway secretions, asthma, bronchitis, and airway constriction, which can result in shortness of breath [2,59,69–74]. When these symptoms persist or get worse over time, the condition is sometimes called occupational asthma [69]. Since these difficulties result from mist exposure, they occur primarily in water-based fluid applications [72]. As discussed in detail in Ref. [69], a number of cutting fluid components or additives may act as sensitizing agents. Machine enclosures and proper mist filtering can reduce workforce exposure to this risk [2].

Another respiratory disorder associated with cutting fluid mists is hypersensitive pneumonitis (HP) [69,75,76], which may ultimately result in pulmonary fibrosis or similar build-up in the lungs of sensitive individuals [69,76]. Bacteria have been proposed as sensitizing agent, although other causes have also been suggested. Switching biocides to alter the bacterial makeup of the fluid may help, but the safest course appears to be to transfer affected operators to a job that does not entail mist exposure.

14.6.4 Microbial Infections

Coolant sumps often sustain populations of bacteria, yeasts, and fungus [57]. Most microbes feed off mineral oils, so microbial growth is a particular problem for coolants with high mineral oil content, either through high concentration or free oil contamination, and less of a problem for synthetic fluids, which contain no mineral oil. Vegetable oils are also subject to microbial attack and spoilage. Other factors contributing to bacterial growth are a bacteria-rich water supply, poor housekeeping, and low fluid pH.

Excessive microbial growth in the sump is regarded as a problem mainly because it produces objectionable odors [59,77] and leads to fluid breakdown. It has traditionally been felt that microbial infection is unlikely [4]. Most coolant systems are comparatively nutrient-poor and unlikely to support true human pathogens, which generally require a nutrient-rich environment. Isolated exceptions are reported, however [57]. Microbial infections may be a problem for operators whose immune systems are compromised. There have also been reports of flu-like outbreaks associated with the bacteria that causes Legionnaire's disease. Control measures recommended in these cases included diligent microbial monitoring and avoiding allowing sumps to stagnate for long periods of time.

14.6.5 CANCER

Long-term exposure to cutting fluids has been associated with increased incidence of several types of cancer [2,66,78–80], including skin, scrotal, laryngeal, rectal, pancreatic, and bladder, and digestive cancers. Exposure routes include both dermal contact and inhalation. Details of specific correlations are given in Ref. [66]. Broadly, risks are increased when cutting with neat mineral oils rather than water-based fluids, and in grinding operations, which generate more mists.

There is some indication that greater risk is associated with exposure to older oils [2]. Prior to the 1950s, the mineral oils used were primarily raw or mildly refined petroleum and shale oils, which contained significant concentrations of substances now known to be carcinogenic; in addition, work practices at that time resulted in more severe dermal exposure [81]. Since the mid-1970s, and especially after 1985, substantial changes have been made in fluid composition, which reduce concentrations of harmful substances [2,66], and work practices have been modified to reduce exposure. As noted in the OSHA best practices manual [2], the substantial changes in industry practice over the last decades have likely reduced cancer risks from metalworking fluid exposure, but more data is required for a more definite conclusion.

14.7 DRY AND NEAR-DRY MACHINING METHODS

There has been increasing interest in reducing or eliminating the use of cutting fluids in machining. This would be of benefit for three reasons. First, it would reduce or eliminate exposure of operators to health risks as discussed in the previous section. Second, it would reduce machining costs. One frequently cited study by an automotive company indicates that 16% of the cost of machined parts is directly attributable to cutting fluids (including fluid management, disposal, and equipment) [82]. While this percentage varies for different applications, there is no question that the costs associated with the purchase, maintenance, and disposal of cutting fluids are invariably significant. Third, large central coolant systems require in-ground trenches and sumps, which greatly limit equipment flexibility.

Some processes are readily carried out without a cutting fluid for some workpiece materials; for example, cast iron parts can be machined dry under conventional cutting conditions if the tool is oriented properly for chip ejection; aluminum parts can be milled and turned without coolants at high cutting speeds using PCD tooling [83], and low carbon and hardened steels are also often machined dry. As discussed in Chapter 17, high-speed gear milling and hobbing are performed dry. Titanium medical implants are also often machined dry to avoid contamination [84]. As some of these applications illustrate, dry machining is often advantageous in interrupted cutting operations such as milling or hobbing at high speeds, due to reduced thermal fatigue of the tool (especially for insufficient coolant volume). For most materials, however, dry machining is more difficult to perform effectively, especially at higher cutting speeds. This is especially true for harder materials such as stainless steels and nickel alloys [29,85,86] Without coolant, it becomes difficult to effectively clear chips (especially in holemaking operations), control dimensional distortion due to part heating, and prevent build-up on the tool. Special machine architectures that shed chips passively and low-friction tool coatings help address some of these problems [87]. It is unlikely, however, that coolants will be completely eliminated in the near future, especially in holemaking and grinding processes.

Significant reduction in coolant use, however, is achievable in many operations [88]. Strategies include reducing cycle times to reduce the amount of coolant used per part, extending coolant life to reduce waste disposal volumes per unit time, and reducing coolant flow rates [89]. When true dry machining is not practical, significant benefits can still be achieved using these methods or by using minimum quantity lubrication (MQL), a near-dry machining method discussed in detail in the Chapter 15. MQL has many of the advantages of the dry machining but also some of the disadvantages of wet machining. MQL implementation has led to tooling innovations and chip removal practices that are generally applicable to dry machining.

14.8 TEST PROCEDURE FOR CUTTING FLUID EVALUATION

The availability of new coolant types, and increased interest in reducing coolant use per part, has increased the need for a common procedure for evaluating or comparing cutting fluids. The current practice has been to evaluate two or more cutting fluids in the same machine, under the same operating and cutting conditions, and to measure cutting forces, spindle power, part quality (surface finish, dimensional control, etc.), or tool wear as the main criteria for performance comparison. Recommended procedures for cutting fluid performance testing or Machinability Test Guidelines are available for turning (plunging), drilling, milling, and grinding processes [16]. Such tests require the specification of the cutting tool, cutting conditions, cutting fluid, criteria for ending the test (e.g., tool wear considering uniform flank wear along the cutting edge or grinding ratio G).

Some of the critical areas to be evaluated for longer-term tests are tool life, odor, foam, rust, residue, stability, filterability, dermatitis, biological control, pH, machine cleanliness, and overall usage rate as a function of the condition of the coolant. Information about solving cutting fluid problems and extending the life of the cutting fluids is summarized in the Refs. [16,55,59].

As noted before, cutting fluid costs often account for 10%–15% of the total machining cost, while cutting tool costs typically account for a smaller component (on the order of 5% in high volume machining). In many operations, reducing or eliminating cutting fluid use if possible and accepting somewhat lower tool life may lead to lower overall machining costs.

REFERENCES

1. United States National Institute for Occupational Safety and Health (NIOSH), *What You Need to Know About Occupational Exposure to Metalworking Fluids*, DHHS (NIOSH) Publication No. 98-116, Cincinnati, OH, March 1998.
2. Occupational Safety & Health Administration (OSHA)/U.S. Department of Labor, *Metalworking Fluids: Safety and Health Best Practices Manual*, Washington, DC, January 1999.
3. J. C. Childers, The chemistry of metalworking fluids, in: J. P. Beyers, Ed., *Metalworking Fluids*, Marcel Dekker, New York, 1994, pp. 165–189.
4. T. J. Drozda and C. Wick, Eds., *Tool and Manufacturing Engineers Handbook*, Vol. I: *Machining*, 4th edn., SME, Dearborn, MI, 1983, pp. 4.1–4.34.
5. J. K. Howel, W. E. Lucke, and J. C. Steigerwald, Metalworking fluids: Composition and use, *Proceedings of the AAMA Metalworking Fluids Symposium*, Dearborn, MI, March 1996, pp. 13–22.
6. Sandvik Coromat, Inc., *Modern Metal Cutting*, Sandvik Coromat, Inc., Sandviken, Sweden, 1996, Chapter XII, pp. 22–41.
7. J. V. Owen, Picking a coolant, *Manuf. Eng.* **120**:5 (May 1998).
8. G. Littlefair, Trends in cutting fluid, *Metalwork. Prod.*, **15** (December 11, 2000), 26–27.
9. H. Limper and K. Cavanaugh, What do you want your coolant to do? *Am. Mach.* February 20, 2008. http://americanmachinist.com.
10. J. Benes, Engineering metal working fluids, *Am. Mach.*, March 22, 2007. http://americanmachinist.com.

11. S. Suda, H. Yokota, I. Inasaki, and T. Wakabayashi, A synthetic ester as an optimal cutting fluid for minimal quantity lubrication machining, *CIRP Ann.* **51** (2002) 95–98.

12. C. Claudin, A. Mondelin, J. Rech, and G. Fromentin, Effects of a straight oil on friction at the tool–work material interface in machining, *Int. J. Mach. Tools Manuf.* **50** (2010) 681–688.

13. G. S. Cholakov, T. L. Guest, and G. W. Rowe, Lubricating properties of grinding fluids: 1. Comparison of fluids in surface grinding experiments, *Lubr. Eng.*, February 1992, 155–163.

14. ASTM, *ASTM Standard E1687–95: Determining Carcinogenic Potential of Virgin Base Oils in Metalworking Fluids*, American Society for Testing and Materials, Philadelphia, PA, 1997.

15. S. R. Steigerwald, Vegetable oil improves machining of medical devices, *Prod. Mach.* (November/December 2003) 44–46.

16. T. F. McClure, *Pollution Prevention Guide to using Metal removal Fluids in Machining Operations*, Developed for the United States Environmental Protection Agency at Cincinnati by IAMS and its Int. National Industrial Working Group, Cincinnati, OH, 1996.

17. E. Kuram, B. Ozcelik and E. Demirbas, Environmentally friendly machining: Vegetable based cutting fluids, in: J. P. Davim (Ed.), *Green Manufacturing Processes and Systems*, Materials Forming, Machining and Tribology, Springer-Verlag, Berlin, Germany, 2013.

18. S. A. Lawal, I. A. Choudhury, and Y. Nukman, Application of vegetable oil-based metalworking fluids in machining ferrous metals—A review, *Int. J. Mach. Tools Manuf.* **52** (2012) 1–12.

19. A. Richter, So long bactericide, *Cutting Tool Eng.* **66**:10 (October 2014).

20. B. J. Hogan, What makes a coolant green?, *Manuf. Eng.* (July 2010) 63–72.

21. R. B. Aronson, Fluid management basics, *Manuf. Eng.* **126**:6 (June 2001).

22. ASTM Standard E 1497-94: Safe use of water-miscible metalworking fluids, 1997, 43.

23. ChilAire—Setting New Standards, Coolclean Technologies, Eagan, MN, nd.

24. Anon., CO_2 Coolant System, *Prod. Mach.*, June 8, 2009. http://www.productionmachining.com.

25. C. Machai and D. Biermann, Machining of β-titanium alloy Ti–10V–2Fe–3Al under cryogenic conditions: Cooling with carbon dioxide snow, *J. Mater. Proc. Tech.* **211** (2011) 1175–1183.

26. Y. Yildiz and M. Nalbant, A review of cryogenic cooling in machining processes, *Int. J. Mach. Tools Manuf.* **48** (2008) 947–964.

27. S. Hong and Y. Ding, Cooling approaches and cutting temperatures in cryogenic machining of Ti-6Al-4V, *Int. J. Mach. Tools Manuf.* **41** (2001) 1417–1437.

28. B. D. Jerold and M. P. Kumar, The influence of cryogenic coolants in machining of Ti–6Al–4V, *ASME J. Manuf. Sci. Eng.* **135** (2013) 031005-1.

29. A. Shokrani, V. Dhokia, and S. T. Newman, Environmentally conscious machining of difficult-to-machine materials with regard to cutting fluids, *Int. J. Mach. Tools Manuf.* **57** (2012) 83–101.

30. B. D. Jerold and M. P. Kumar, Experimental comparison of carbon-dioxide and liquid nitrogen cryogenic coolants in turning of AISI 1045 steel, *Cryogenics* **52** (2012) 569–574.

31. C. Courbon, F. Pusavec, F. Dumont, J. Rech, and J. Kopac, Tribological behaviour of Ti6Al4V and Inconel 718 under dry and cryogenic conditions—Application to the context of machining with carbide tools, *Tribol. Int.* **66** (2013) 72–82.

32. A. Richter, Cryogenic machining systems can extend tool life and reduce cycle times, *Cutting Tool Eng.* **67**:2 (February 2015).

33. S. Malkin, *Grinding Technology*, Society of Manufacturing Engineers, Dearborn, MI, 1989, pp. 215–216.

34. S. Kalhan, Mist control by using shear stable antimist polymers, *Machining & Metalworking Conference*, Detroit, MI, September 4, 1999.

35. E. Gulari, C. W. Manke, S. Yurgelevic, and J. M. Smolinski, Suppression and management of mist with polymeric additives, *In the Industrial Metalworking Environment: Assessment and Control of Metal Removal Fluids*, American Automotive Manufacturing Association, Washington, DC, 1998, pp. 291–307.

36. M. Hoff, Critical coolant questions, *Manuf. Eng.* **124**:5 (May 2000).

37. R. H. Brandt, Filtration systems for metalworking fluids, in: J. P. Beyers, Ed., *Metalworking Fluids*, Marcel Dekker, New York, 1994, pp. 273–303.

38. H. Urdanoff and R. J. McKinley, Why do coolants fail? *Manuf. Eng.* **130**:5 (May 2003) 122–130.

39. G. Foltz, Cooling fluids: Forgotten key to quality, *Manuf. Eng.* **130**:1 (January 2003) 66–69.

40. J. Mackowski, Magnetic filters keep coolant clean, *Manuf. Eng.* **124**:4 (April 2000).

41. C. Likens, Keep good coolant from going bad, *Am. Mach.*, April 2000. http://americanmachinist.com.

42. A. Shanley, Mist collection: The pressure is on, *Cutting Tool Eng.* **55**:7 (July 2003).

43. A. Richter, Recovery processes—Options and considerations for processing and removing metal chips, *Cutting Tool Eng.* **55**:7 (July 2003).

44. W. J. Johnston, Metal removal fluid mist control: System considerations, *In the Industrial Metalworking Environment: Assessment and Control of Metal Removal Fluids*, American Automotive Manufacturing Association, Washington, DC, 1998, pp. 213–221.

45. J. B. D'Arcy, D. Hands, and J. J. Hartwig, Comparison of machining fluid aerosol concentrations from three different particulate sampling and analysis methods, *The Industrial Metalworking Environment: Assessment and Control*, American Automotive Manufacturing Association, Washington, DC, 1998, pp. 196–199.

46. S. J. Cooper, D. Leith, and D. Hands, Vapor generation in laboratory and industrial mist collectors, *The Industrial Metalworking Environment: Assessment and Control of Metal Removal Fluids*, American Automotive Manufacturing Association, Washington, DC, 1998, pp. 233–238.

47. ACGIH, *Industrial Ventilation: A Manual of Recommended Practice*, 22nd edn., *American Conference of Governmental Industrial Hygienists*, Committee on Industrial Ventilation, Cincinnati, OH, 1995.

48. Handte MF-SORP—Dry Filter, Camfil Handte APC GmbH, Tuttlingen, Germany, nd.

49. Dry filter systems for applications using minimum quantity lubrication (MQL), Keller USA Inc., Fort Mill, SC, 2009.

50. Anon., Minimum quantity lubrication for machining operations, Report BGI/GUV-I 718 E, DGUV (German Federation of Institutions for Accident Insurance and Prevention), Berlin, Germany, November 2010.

51. J. J. Werner, A practical analysis of cutting fluids, *Lubr. Eng.* **40**:4 (April 1984) 234–238.

52. G. Foltz, Corrosion and metalworking fluids, *Prod. Mach.*, May 15, 2003. http://www.productionmachining.com.

53. D. Elenteny and J. Manfreda, Maximizing coolant endurance and economy, *Am. Mach.*, February 7, 2006.

54. B. Reynolds and D. Fecher, Metalworking fluid management and best practices, *Prod. Mach.*, October 18, 2012.

55. Metalworking fluid troubleshooting guide, *Mod. Mach. Shop*, **75**:10 (March 3, 2009) 68.

56. Optimising the use of metalworking fluids, Guide GG199, Environmental Technology Best Practice Programme, U.K., September 1999.

57. L. A. Rossmoore and H. W. Rossmoore, Metalworking fluid microbiology, in: J. P. Beyers, Ed., *Metalworking Fluids*, Marcel Dekker, New York, 1994, pp. 247–271.

58. E. E. Heidenreich, Metalworking fluids—A practical approach to recycling, SME Technical Paper MR84-918, 1984.

59. G. J. Foltz, Metalworking fluid management and troubleshooting, in: J. P. Beyers, Ed., *Metalworking Fluids*, Marcel Dekker, New York, 1994, pp. 305–337.

60. P. M. Sutton and P. N. Mishra, Waste treatment, in: J. P. Beyers, Ed., *Metalworking Fluids*, Marcel Dekker, New York, 1994, pp. 367–394.

61. N. Hilal, G. Busca, F. Talens-Alesson, and B. P. Atkin, Treatment of waste coolants by coagulation and membrane filtration, *Chem. Eng. Process.* **43** (2004) 811–821.

62. C. Cheng, D. Phipps, and R. M. Alkhaddar, Treatment of spent metalworking fluids, *Water Res.* **39** (2005) 4051–4063.

63. C. R. Mackerer, Health effects of oil mists: A brief review, *Toxicol. Ind. Health* **5** (1989) 429–440.

64. P. J. Beattie and B. H. Strohm, Health and safety aspects in the use of metalworking fluids, in: J. P. Beyers, Ed., *Metalworking Fluids*, Marcel Dekker, New York, 1994, pp. 411–422.

65. D. Park, The occupational exposure limit for fluid aerosol generated in metalworking, *Safe Health Work* **3** (2012) 1–10.

66. G. M. Calvert, E. Ward, T. M. Schnorr, and L. J. Fine, Cancer risks among workers exposed to metalworking fluids: A systematic review, *Am. J. Ind. Med.* **33** (1998) 282–292.

67. C. G. T. Mathias, Contact dermatitis and metalworking fluids, in: J. P. Beyers, Ed., *Metalworking Fluids*, Marcel Dekker, New York, 1994, pp. 395–410.

68. E. O. Bennet, *Dermatitis in the Metalworking Industry*, Pamphlet SP-ll, STLE, Park Ridge, IL, 1983.

69. D. J. P. Basett, Review of acute respiratory health effects, *Proceedings of the AAMA Metalworking Fluids Symposium*, Dearborn, MI, March 1996, pp. 147–152.

70. S. M Kennedy, I. A. Graves, D. Kriebel, E. A. Eisen, T. J. Smith, and S. R. Woskie, Acute pulmonary responses among automobile workers exposed to aerosols of machining fluids, *Am. J. Ind. Med.* **15** (1989) 627–641.

71. K. D. Rosenman, M. J. Reilly, D. Kalinowski, and F. Watt, Occupational asthma and respiratory symptoms among workers exposed to machining fluids, *Proceedings of the AAMA Metalworking Fluids Symposium*, Dearborn, MI, March 1996, pp. 143–146.

72. M. S. Hendy, B. E. Beattie, and P. S. Burges, Occupational asthma due to an emulsified oil mist, *Brit. J. Ind. Med.* **42** (1985) 51–54.

73. H. J. Cohen, A study of formaldehyde exposures from metalworking fluid operations using hexahydro-1,3,4-tris (2-hydroxyethyl)-S-triazine, *Proceedings of the AAMA Metalworking Fluids Symposium*, Dearborn, MI, March 1996, pp. 178–183.

74. E. Brandy and D. Bennet, Occupational airway diseases in the metal-working industry, *Tribol. Int.* **18** (1985) 169–176.

75. R. B. Aronson, What's happening with coolants, *Manuf. Eng.* **130**:6 (June 2003) 81–88.

76. B. G. Shelton, W. D. Flanders, and G. K. Morris, *Mycobacterium* sp. as a possible cause of hypersensitivity Pneumonitis in machine workers, *Emerg. Infect. Dis.* **5**:2 (March–April 1999).

77. E. O. Bennett and D. L. Bennett, Cutting fluids and odors, *Clinic on Metalworking Coolants*, Dearborn, MI, 1987.

78. E. A. Eisen, P. E. Tolbert, M. F. Hallock, R. R. Monson, T. J. Smith, and S. R. Woskie, Mortality studies of machining fluid exposure in the automotive industry III: A case-control study of larynx cancer, *Am. J. Ind. Med.* **25** (1994) 185–202.

79. E. A. Eisen, P. E. Tolbert, R. R. Monson, and T. J. Smith, Mortality studies of machining fluid exposure in the automotive industry I: A standardized mortality ratio analysis, *Am. J. Ind. Med.* **10** (1992) 803–815.

80. E. A. Eisen, Case-control studies of five digestive cancers and exposure to metalworking fluids, *Proceedings of the AAMA Metalworking Fluids Symposium*, Dearborn, MI, March 1996, pp. 25–28.

81. C. M. Skisak, Metal working fluids, base oil safety, *Proceedings of the AAMA Metalworking Fluids Symposium*, Dearborn, MI, March 1996, pp. 44–49.

82. D. Graham, Going dry, *Manuf. Eng.* **124**:1 (January 2000).

83. D. Graham, D. Huddle, and D. McNamara, Machining dry is worth a try, *Mod. Mach. Shop*, **76**:5 (October 1, 2003) 79.

84. C. Koepfer, Coolant considerations for medical machining, *Mod. Mach. Shop*, **75**:11 (April 15, 2003) 84.

85. D. Dudzinski, A. Devillez, A. Moufki, D. Larrouquere, V. Zerrouki, and J. Vigneau, A review of developments towards dry and high speed machining of Inconel 718 alloy, *Int. J. Mach. Tools Manuf.* **44**:4 (2004) 439–456.

86. M. EL Mansoria and M. Nouari, Dry machinability of nickel-based weld-hardfacing layers for hot tooling, *Int. J. Mach. Tools Manuf.* **47** (2007) 1715–1727.

87. F. M. Kustas, L. L. Fehrehnbacher, and R. Komanduri, Nanocoatings on cutting tool for dry machining, *CIRP Ann.* **46** (1997) 39–42.

88. H. K. Toenschoff, F. Kroos, W. Sprintig, and D. Brandt, Reducing use of coolants in cutting processes, *Prod. Eng.* **1**:2 (1994) 5–8.

89. T. Kondo, Environmentally clean machining in Toyota, *Int. J. Jpn. Soc. Precision Eng.* **31** (1997) 249–252.

15 Minimum Quantity Lubrication

15.1 INTRODUCTION

Minimum quantity lubrication (MQL), in German *Minimalmengenschmierung* (MMS), is a near-dry machining method in which a fine aerosol mist of neat oil is substituted for water-based coolant [1–9]. In MQL, the oil is consumed in the process, rather than collected, filtered, and reused. MQL systems use oil flow rates measured in mL/h, rather than in L/min as in water-based coolant systems. Oil flow rates between 5 and 200 mL/h, depending on the work material and tool size, are used in most applications. MQL was developed in the 1990s, primarily in Germany and Japan [10,11]. During this time, the costs and potential exposure risks of water-based coolants led to interest in reducing coolant use, and early dry machining research had demonstrated the difficulty of effectively machining many materials without any lubricant whatsoever. MQL has since been widely applied, especially in the machining of airframe and other large structural components [12–15] and in automotive powertrain machining [16–25]. Typical work materials machined using MQL include aluminum alloys, carbon and low alloy steels, and gray and nodular cast irons.

Since it is a neat oil system, MQL provides better lubrication at the cutting edge than water-based fluids, reducing frictional heating of the tool and part and material buildup on the tool. Compared to flood coolant, however, MQL does not provide significant cooling and thermal stability and is not as effective in clearing and transporting chips. Achieving acceptable dimensional stability and chip handling requires a dimensional and measurement strategy, modifications to cutting tools, and changes to machining process parameters. The benefits of MQL machining must exceed the associated engineering and process monitoring costs to justify MQL implementation.

In appropriate applications, MQL implementation may be justified by increased productivity, capital cost avoidance, or variable cost avoidance. In many drilling operations, such as crankshaft oil hole or large structural applications [13,17], penetration rates are significantly higher for MQL than for alternative wet or dry processes, leading to an increase in throughput. Capital cost avoidance is a significant driver for applying MQL in high volume machining systems such as those used in automotive powertrain manufacture, since in these systems the capital associated with a coolant system (including piping, pumps, filters, and chip driers) may be a significant fraction of the total system cost [16,19,26]. Increased throughput in MQL operations may also contribute to capital cost avoidance. MQL also permits the avoidance of variable costs for filter media and chip drying, and often a reduction in costs for maintenance and coolant management resources, but these savings may be offset to some degree by increased tool costs.

Effectively implementing MQL requires engineering of the entire machining system, including the machine tools, fixtures, process parameters and sequence, and cutting tools. This is especially true when manufacturing complex components with high machined content using integrated machining systems. This chapter presents an overview of the engineering aspects of MQL machining, including MQL system types, lubricants, machine tool and cutting tool requirements, thermal management and dimensional control, air and chip handling, and current research areas.

15.2 MQL SYSTEM TYPES

MQL systems can be classified by mist delivery method and by the location of mist generation relative to the tool cutting edges.

15.2.1 EXTERNAL AND INTERNAL MIST DELIVERY

The mist may be delivered externally, through a nozzle or nozzles mounted on the spindle housing, fixture, or toolholder and aimed at the active cutting edge (Figure 15.1), or internally through the spindle and coolant channels inside the tool body (Figure 15.2) as with through-tool water-based coolants.

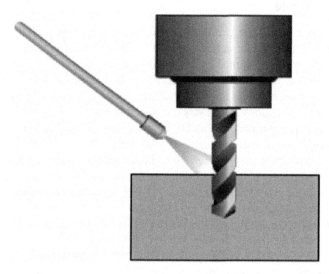

FIGURE 15.1   External MQL application. (Courtesy of UNIST, Grand Rapids, MI.)

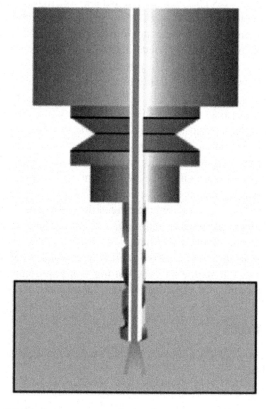

FIGURE 15.2   Internal MQL application. (Courtesy of UNIST, Grand Rapids, MI.)

External application is simpler to implement, since it does not require modification of the machine tool, and can also be used with relatively inexpensive solid tooling. It is used in many aerospace and structural applications, especially those involving retrofits of large machines [8,13]. It is also well suited for end and face milling operations, especially those using large diameter or high flute density tooling. In these applications, through tool delivery may require a significant increase in tooling cost for cutters with multiple internal coolant channels, and may lead to excessive oil consumption since mist is supplied continuously to all edges whether they are cutting or not. External application is also common in turning applications on CNC lathes. External delivery may not be robust if nozzles are easily displaced and is generally not effective for deeper cuts ($L/D > 3$ [1,2]) due to the "shadowing effect" or obstruction of the oil mist stream by the part or fixture when the tool is in cut [2,6,8]. In turning applications, nozzles are sometimes aimed at the flank rather than rake face of the tool to avoid shadowing [9,13]. External nozzles have limited adjustment options for tools with different lengths and diameters [8] and are thus less suitable for multi-tool processes. External nozzles also disperse oil over a wider area, and particularly on parts of the tool or workpiece remote from the cutting edges [6,8]. The percentage of oil delivered to a targeted spot decreases roughly linearly with distance to the target for external nozzles [27], so nozzles must generally be placed within 200 mm of the cutting edge [2].

Internal mist delivery is more robust, since it eliminates nozzle dislodgement or misaiming failures, but requires modifications to the machine tool and specialized tooling as discussed in Sections 15.4 and 15.5. Internal mist delivery is particularly common for drilling, especially deep holes, and for parts with high machined content requiring frequent tool changes, since it is difficult to design robust external nozzles systems for these applications. In Internal delivery, the mist is delivered directly to the cutting edges, eliminating the oil dispersal losses inherent with external nozzles. Internal delivery is difficult to apply to some tools with diameters less than 3 mm due to the difficulty of fabricating small tools with sufficiently large internal coolant channels [2].

15.2.2 One- and Two-Channel MQL Systems

In machining center applications, the mist may be generated externally and piped through the machine (one-channel MQL), or the oil and air may be piped through the spindle separately and mixed to form a mist in the toolholder (two-channel MQL) [1–6,8,9,20]. Both one- and two-channel systems can also be used with external nozzle delivery; in two-channel external application, oil and air are piped separately to the nozzle and mixed near the tip [6,8].

One-channel MQL systems (Figure 15.3) are simpler to implement since major machine tool modifications and special toolholders are not required. They are commonly used with external mist delivery and can also be used in through-tool applications. The mist is generated inside the MQL unit and in some cases may be supplied to the cutting tool through existing coolant ports or nozzles. There are two common methods of generating the mist in one-channel systems: the metering pump method and the pressurized tank method [1,6,8]. In the metering pump method, lubricant is supplied to the air stream by a pneumatic positive displacement micropump connected to an in-line air blast nozzle. The oil volume supplied for mist generation is regulated by controlling the pump speed. With current pump and nozzle arrangements, these systems produce mists with relatively large droplet sizes; the average size is typically over 10 μm, with a significant fraction of droplets having diameters between 20 and 40 μm [27,28]. Also, in current systems the mist is generated in discreet pulses and is not uniform over time intervals comparable to the pump frequency, although variations tend to even out over the length of the delivery path. Discrete pulsing may be ameliorated by pump systems designed to give continuous output. In the pressurized tank method (Figure 15.4), the lubricant tank is pressurized, and the mist is generated by mixing the oil and compressed air in a venturi nozzle or similar device. Oil delivery is adjusted by adjusting the supply pressure settings and by throttling elements in the piping. The best control can be achieved if the air pressure, lubricant pressure, and lubricant quantity can be adjusted separately. This method produces finer mists, with droplet sizes typically below 2 μm [6]. The mist is also more uniform over time. In some

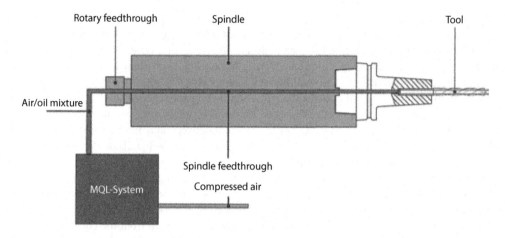

FIGURE 15.3   One-channel MQL delivery system. (From GROB, Grob Werke, Mindelheim, Germany.)

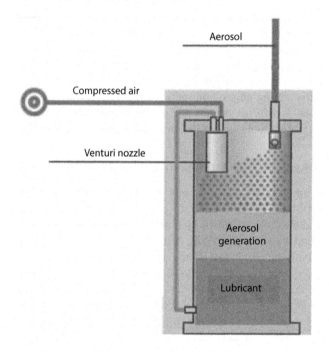

FIGURE 15.4   Mist generation using the pressurized tank method. (From Vogel SKF, Berlin, Germany.)

advanced systems employing a venturi with a particle profiling plate (Figure 15.5), baffling, or similar secondary apparatus [29–31], even finer mists with droplet sizes below 1 μm can be generated. In fine mist systems, oil droplets coalesce during passage through narrow nozzles or tool coolant channels to form larger droplets, which are more effective for lubrication [9].

The oil volume delivered at the tool point is difficult to control precisely in one-channel MQL because it depends on multiple factors as summarized in Figure 15.6 [32]. Since the oil is supplied to the tool in an air–oil mist, the rate of oil delivery depends on the air flow rate [3,6,8]. The mist supply passages should be free of obstructions, diameter changes, and sudden changes in direction to avoid restricting flow. Increased air pressure (5–7 bar vs. 4–6 bar for two-channel systems) is used in some applications to maintain adequate air flow. In internal delivery applications on machining centers, the oil volume also depends on the spindle rpm [3,4,6,8], since droplets are subject to inertial (centrifugal) forces when passing through the spindle and may adhere to the piping walls rather than being transported to the tool.

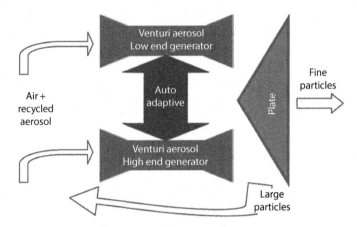

FIGURE 15.5 Schematic of a one-channel MQL system employing adaptive venturis and a particle profiling plate to generate mists with droplet sizes below 1 μm. (From Dropsa USA, Sterling Heights, MI.)

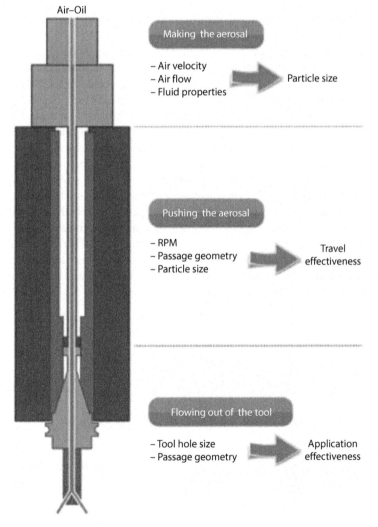

FIGURE 15.6 Factors affecting oil delivery in a one-channel MQL system. (Courtesy of UNIST, Grand Rapids, MI.)

The sensitivity of the mist to centrifugal forces depends on the droplet size; mists with smaller (lower mass) droplets are less affected since the inertial force on a droplet is smaller in relation to the air transport force. The droplet size depends on the mist generation method, air flow rate and velocity, and viscosity of the oil. Current systems employing a metering pump produce mists with large droplets and are best suited for external delivery applications, but can be used in through-tool applications at spindle speeds under 5000 rpm without oil starvation at the tool point. Standard pressurized tank systems produce smaller droplet sizes and can be used at spindle speeds up to 16,000 rpm [3,20]. Pressurized tank systems employing profiling plates or other secondary methods to further reduce mist size can reportedly be used at spindle speeds up to 20,000 rpm [30]. Air flow and oil delivery through the tool are affected by the tool's internal coolant passage geometry as discussed in Section 15.5.

The length of the path the mist must travel through the spindle or to external nozzles may be of concern in one-channel internal delivery applications with frequent tool changes, especially if significantly different oil dosage levels are used for successive tools; in these cases, the entire path volume may have to be purged before the tool cuts [9,20], leading to increased tool purge times and oil contamination of the workspace. Traditional one-channel systems typically have response times over 1 s, compared to 0.1–0.3 s for two-channel systems [20], although advanced systems have bypass modes to reduce response times [6,30]. One-channel systems also require oils with relatively low viscosities (up to 50 mm²/s [3,6]) to limit droplet sizes. They have fewer component wear and maintenance concerns than two-channel systems [6].

Two-channel MQL systems (Figure 15.7) are less common and are used primarily for internal mist delivery applications involving frequent tool changes or high spindle speeds, although they are also used in turning [6,33]. Two-channel systems permit more precise control of the lubricant delivery rate since a prescribed amount of oil is introduced into the air stream near the tool point. The oil delivery rate can be metered using either a positive displacement pump with controlled speed [5,8] or an internal pressurized oil circuit with a metering valve [4]. Since both the pumps and metering valves cycle at high frequencies, reliability of these components is a concern in two-channel systems. Typical mist droplet sizes are 5 μm for through-tool delivery [3] and 15–40 μm for external nozzle delivery [2]. Since the mist does not pass through the machine spindle, it is not subject to centrifugal separation in internal delivery applications. The oil delivery rate at the tool point is thus substantially independent of the spindle rpm and the air flow rate [3,6,8]. Since the mist is generated near the tool point, two-channel systems require shorter purge times for tool changes [9,20] and thus generally have shorter response times (typically 0.1–0.3 s depending on the tool size). Two-channel systems can use oils with viscosities up to 100 mm²/s [3,4,6]. In machining center applications, they require an HSK spindle interface, significant machine tool modifications, and special toolholders as discussed in Sections 15.4 and 15.5.

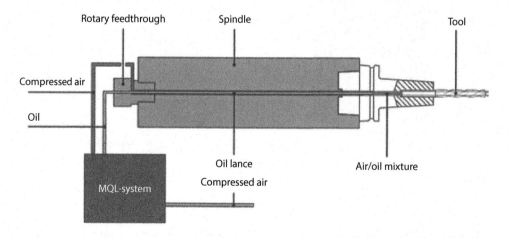

FIGURE 15.7   Two-channel MQL delivery system. (From GROB, Grob Werke, Mindelheim, Germany.)

TABLE 15.1

Comparison of One- and Two-Channel MQL Systems

One-Channel	Two-Channel
Oil quantity dependent on air flow	Oil quantity independent of air flow
Oil quantity dependent on spindle rpm	Oil quantity independent on spindle rpm
Longer reaction time	Short reaction time (0.1–0.3 s)
Required air pressure 4–5 bar	Required air pressure 4 bar
Max spindle rpm 16,000–20,000	Max spindle rpm 40,000
Max lubricant viscosity 50 mm²/s	Max lubricant viscosity 100 mm²/s
No wear components	Metering pump/value are wear items
One MQL unit feeding multiple spindles possible	Each spindle requires MQL unit
Standard toolholders for internal delivery possible	Requires special toolholders for internal delivery

Sources: Handbuch Minimalmengenschmierung (MMS), KometGroup GmbH, Besigheim, Germany, March 2008; *Minimum Quantity Lubrication (MQL) Systems in Metal Cutting*, Bielomatik Leuze GmbH, Neuffen, Germany, nd; Minimum quantity lubrication for machining operations, report BGI/GUV-I 718 E, Deutsche Gesetzlich Unfallversicherung (DGUV) Spitzenverband, Berlin, Germany, November 2010; Walker, T., *The MQL Handbook*, UNIST, Inc., Grand Rapids, MI, 2013.

Table 15.1 summarizes differences between one- and two-channel MQL Systems. Two-channel systems offer advantages in control of oil dosage, response time, and range of permissible oil viscosities over one-channel systems, and are reported to outperform one-channel systems in many applications [8]. However, few head-to-head comparisons of one- and two-channel systems have been reported in the technical literature. In one recent study, two-channel systems yielded better surface finish and dimensional control in aluminum powertrain component machining [34], but both one- and two-channel systems have been successfully used in production for these applications.

Most MQL systems operate at conventional shop air pressures between 4 and 6 bar without requiring special air filtering or drying. The air pressure may be increased to roughly 10 bar in deep hole drilling and similar applications [35] to improve chip evacuation. Higher pressures are also used in some one-channel applications to maintain adequate air flow and oil delivery.

15.3 MQL OILS

MQL requires oils with better lubrication properties than base oils in water-based coolants, since they must provide adequate lubrication at much lower application volumes. Since the oil is consumed in an MQL process, rather than collected for filtering and disposal as in water-based coolant systems, MQL oils should also be nontoxic and biodegradable to reduce operator exposure risks and environmental contamination [2]. Biodegradability is also encouraged by environmental regulations in some locations [36]. Both these considerations make derivatives of vegetable oils most suitable for MQL applications. Vegetable oils are inherently biodegradable [37] and generally have higher molecular weights than common mineral oils, giving them superior lubrication properties. Unmodified vegetable oils are sometimes used in MQL research [37–40]. In production applications, however, oil shelf life and stability over time are important considerations due to the low oil usage rate. Since straight vegetable oils have limited shelf life due to hydrolyzation and oxidation, oils produced by modifying vegetable base oils to improve stability are used in MQL systems. Most MQL oils in use today are synthetic esters or fatty alcohols [1,2,6,8] generally derived from vegetable oil sources, although other vegetable and mineral base oils are also used [9]. A detailed discussion of MQL oil compositions is given by Debnath et al. [37].

Synthetic esters were identified as promising MQL oils in early oil development work in Japan [5,36,41,42]. This work assessed the biodegradability, oxidation and storage stability, and cutting performance of candidate oils. Polyol esters with viscosities suitable for MQL application were found to outperform mineral oils and straight vegetable oils. Fatty alcohols, high molecular weight lubricants also derived from vegetable oil bases, were first used in early German MQL research [2]. Both types of oils are formulated to have high lubricating capability at low viscosities and acceptable vaporization and flash point properties [6,8].

Synthetic esters have low viscosities and relatively high flash points, are relatively thermally stable, and have low evaporation rates. They can be formulated with a range of viscosities. They have traditionally been used in the majority of MQL applications, in which the primary consideration is good lubrication to reduce friction and tool wear, and in which cooling by the oil is not required [1,2,6,8]. They evaporate relatively slowly and leave a thin residue on the part, which may enhance corrosion protection but which may also interfere with subsequent washing or assembly operations. Fatty alcohols are less effective as lubricants but have a lower flashpoint at a given viscosity than synthetic esters, and thus remove more heat by evaporation. They have been traditionally used in applications in which more oil cooling is desired [2,8]. This is especially the case for work materials such as cast iron, which have some self-lubricating capability [1,2,6]. They are also used in nonferrous applications in which prevention of built-up edge is important [6]. Fatty alcohols tend to be more fully consumed in the process and leave less residue on the part. Oils formulated from fatty acids derived from vegetable sources have also recently become available.

Regardless of base chemistry, recommended ranges of physical properties for MQL oils are [6] viscosity > 10 mm²/s at 40°C, flashpoint > 150°C, and evaporative loss < 65% at 250°C. The mist droplet size depends on the surface tension of the oil, which is usually correlated to viscosity. Too small a viscosity can result in very fine mist generation and excessive MQL emissions [8]. The oil temperature is sometimes controlled to maintain consistent viscosity and mist characteristics [20]. MQL oils should also have high wettability with respect to the tool and work materials for optimum lubrication [8]. Wetting properties usually improve with decreasing MQL oil viscosity.

15.4 MACHINE TOOLS FOR MQL

MQL implementation does not require significant machine tool modifications in retrofit applications using external nozzles [8]. Regular cleaning should be instituted to prevent the buildup of chip and dust, and proper ventilation or air filtering must be maintained as discussed in Section 15.7.

For high volume production systems, MQL implementation requires machine tools designed for dry cutting. Generally, these machine tools have a horizontal spindle architecture, steep internal walls, and an open bed so that chips may fall directly onto a chip conveyor (Figure 15.8). The work space is typically enclosed and sealed for dust [1]. Additional common design features of MQL machine tools include [1,2,5,6,8,16,18,19]

- Steep internal walls without protruding boltheads or other features on which chips can accumulate. The chips should fall unhindered to the chip conveyor. Typically, walls should be inclined greater than 50° to the horizontal. The walls may be made of stainless steel or painted with special paints to limit adhesion of oily chips. Hydraulic piping on fixtures and similar structures should be covered with smooth sheet metal guarding.
- An automated chip handling system to continuously remove chips from the work space for disposal.
- Automated air and dust extraction. Generally, a dry filter system as described in Section 15.7 is used.
- Automated cleaning features such as brushes or air knifes to periodically clear any accumulated chips.
- Provision for suspending the workpiece or rotating it to dump chips. This requirement favors the use of A-axis over B-axis table architectures, although B-axis machines can be used in many MQL applications.

FIGURE 15.8 Interior of a twin spindle MQL machining center, showing horizontal spindle *A*-axis architecture, steep, unobstructed stainless steel walls, and an open bed over a chip conveory. (From GROB, Grob Werke, Mindelheim, Germany.)

Additional design features suggested in the literature but not currently widely applied include [1,2,6]

- Epoxy concrete beds to limit thermal expansion of the machine tool structure due to heat buildup over time
- Thermal isolation of the interior walls of the work space from the machine structure to minimize structural thermal distortion over time.

Retrofitting wet machines for MQL machining generally requires significant machine tool modifications, especially for through-tool and two-channel implementations. For machining centers, these modifications include installation of the MQL unit, replacement of the rotary coupling on the back of the spindle, modifications to the spindle interface for automatic tool changes, and controls modifications. The rotary coupling on a wet machine must generally be replaced to seal an aerosol rather than a liquid, and to run dry (without lubrication from the coolant stream) in most cases [8]. For two-channel installations, a rotary union with separate air and oil channels and a central lance down the spindle must also be installed (Figure 15.9).

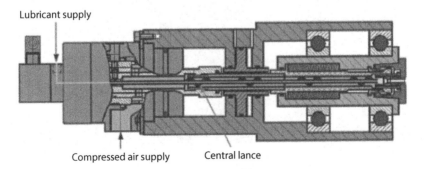

FIGURE 15.9 Rotary union and spindle modifications required for two-channel MQL in through-tool delivery applications. (From Handbuch Minimalmengenschmierung [MMS], KometGroup GmbH, Besigheim, Germany, March 2008.)

15.5 MQL CUTTING TOOLS

Standard solid cutting tools are used in external MQL applications. For internal (through tool) applications, tools designed for MQL are generally required. MQL tools are designed for consistent mist delivery and efficient removal of chips from the cutting edges. MQL tools must be engineered to operate at higher temperatures than wet tooling, and often at higher stresses as well since more aggressive cutting conditions are often used in MQL machining to minimize cutting time and part heating [2,5,9]. MQL tools differ from standard (wet) tooling in their shank interfaces, internal coolant passages, cutting geometry, surface finish, and coatings.

MQL tools used in through-tool machining center applications require a standard shank interface to ensure consistent mist delivery with automatic tool changes. In drilling applications for two-fluted drills, a 90° cone with a wide circular slot aligned with the coolant holes in the tool performs well [43–45] (Figure 15.10), although other designs for different coolant hole configurations have also been investigated [16,18,43,46]. The slot should be parallel to the line through the coolant holes, not skewed at an angle [18]. The cone is mounted against a mating feature in the toolholder (generally an adjustment screw). Toolholders require a method to connect to the MQL supply during tool changes, and in two-channel applications they require a mixing chamber as well [6] (Figure 15.11). In two-channel applications, a two-tube system with an adjustment screw is often used (Figure 15.12) [2–5]. In one-channel systems, which do not require a mixing chamber, a spring bushing with a connection tube may be used [46]. For CAT-V toolholders, similar components can be integrated into the retention knob for one-channel applications [46]. Tool change systems for ABS connections have also been developed [3].

The requirements for internal coolant passage geometry differ between MQL and wet tools as shown in Figure 15.13 [3,46]. Coolant passages should be straight and of constant diameter when possible, and required changes in diameter and direction should be gradual to avoid creating

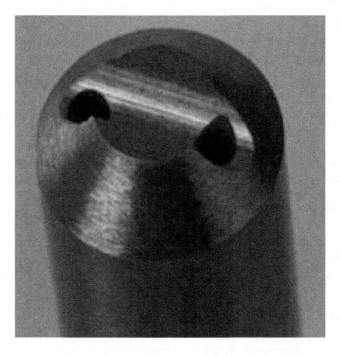

FIGURE 15.10 90° cone MQL shank interface with a wide lubricant slow aligned with the coolant holes. (After Hänle, P., and Schwenck, M., Optimisation of Cutting Tools Using CA-technologies, *Proceedings of the Eight CIRP International Workshop on Modeling of Machining Operations*, Chemnitz, Germany, 2005, pp. 463–468.)

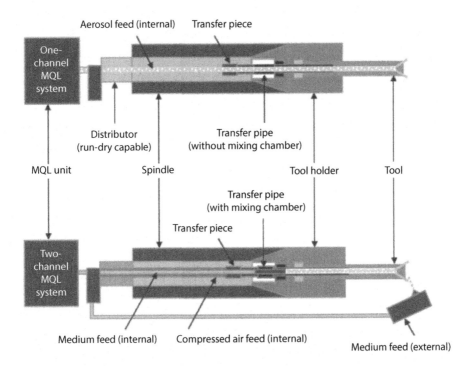

FIGURE 15.11 Special spindle and toolholder features for one- and two-channel MQL. (From Minimum quantity lubrication for machining operations, report BGI/GUV-I 718 E, Deutsche Gesetzlich Unfallversicherung [DGUV] Spitzenverband, Berlin, Germany, November 2010.)

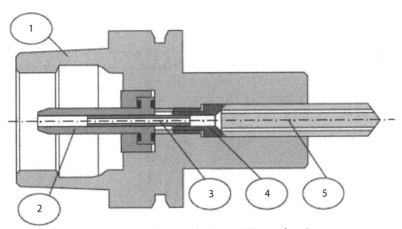

1. Toolholder 2. Coolant supply tube 3. MQL transfer tube
4. Adjustment screw 5. Tool

FIGURE 15.12 Transfer tube system for two-channel MQL tooling with automatic tool changes. (From Handbuch Minimalmengenschmierung [MMS], KometGroup GmbH, Besigheim, Germany, March 2008.)

stagnation points where oil separation and accumulation can occur. The passages should have as large a diameter as possible, generally larger than used in wet tooling, to facilitate air flow; this is especially important in one-channel applications since the oil delivery rate depends on the air flow rate, and the tool coolant channels are often the flow-limiting element in the system. MQL handbooks recommend a minimum coolant passage diameter of between 0.5 and 0.7 mm

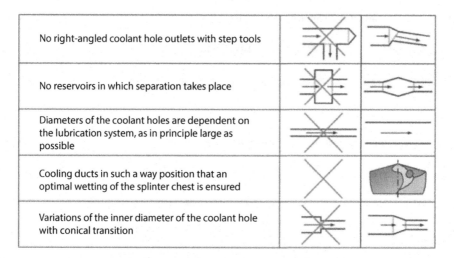

No right-angled coolant hole outlets with step tools		
No reservoirs in which separation takes place		
Diameters of the coolant holes are dependent on the lubrication system, as in principle large as possible		
Cooling ducts in such a way position that an optimal wetting of the splinter chest is ensured		
Variations of the inner diameter of the coolant hole with conical transition		

FIGURE 15.13   Rules for internal coolant passage design in MQL tooling. (After Blosser, M.A., Design and application of MQL tooling for internal machining, *SME Collaborate Conference*, Fort Worth, TX, October 16–18, 2007.)

for one-channel tooling [3,6]. For a given input pressure, air flow in MQL drills has been found to depend primarily on the number and diameter of coolant holes, with the tool length having a minor effect and the helix angle little effect [45]. Some advanced MQL drills have trigonal rather than round coolant holes to increase cross sectional while maintaining radial stiffness [7,8,43,47,48] (Figure 15.14).

For tools with multiple cutting edges to be supplied with MQL, it is common to use a central main supply channel with side supply channels to each edge drilled by EDM [3,46] (Figure 15.15). Side passages should be symmetric for symmetric tools and should exit as close to the cutting edge as possible. Side passages to edges on a given axial level should be drilled to the same branch point to ensure even delivery [46]. Side passages should connect to the main passage at an angle less than 30° relative to the main passage axis to ensure consistent oil delivery [46]. The ratio of the sum of the areas of all side passages to the area of the main

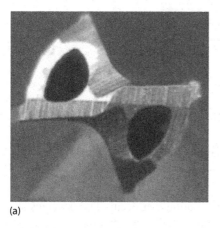

(a)

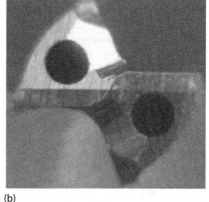

(b)

FIGURE 15.14   MQL drill with trigonal coolant passages (a), which permit increased air flow compared to conventional round passages (b). (After Hänle, P., and Gsänger, D., Prozesssichere Werkzeuge fürs Bohren mit MMS, *Werkstatt und Betrieb*, October 2003.)

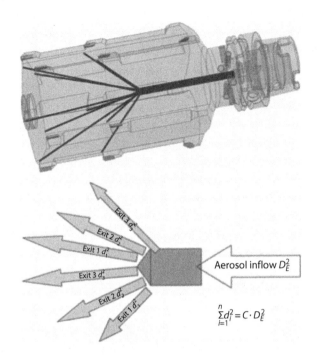

FIGURE 15.15 Multi-edge MQL tooling with multiple exit nozzles fed from a central internal passage. (After Blosser, M.A., Design and Application of MQL Tooling for Internal Machining, *SME Collaborate Conference*, Fort Worth, TX, October 16–18, 2007.)

supply passage, C in Figure 15.15, should be controlled for optimum oil delivery. In one-channel applications, C should be approximately 1 to ensure maximum air flow and oil delivery [3]; in two-channel applications, C should be approximately 0.7 to increase air velocity through the tool and to improve aerosol mixing [3,46]. In manufacturing these tools, it is important to control the EDM drilling process to avoid diameter variations in the side passages and overshoots in the connection zone to the main passages where oil may accumulate [19] (Figure 15.16). Currently, internal passage geometries are designed primarily based on experience, although limited work on applying CFD methods to the task has been described [3,43,45,48,49]. Coolant passages in tool assemblies can be inspected using air pressure measurements or test sprays on paper [16,18,19,50–52]. It is important to spray test complex tooling assemblies prior to use. Testing is often performed on the machine after a tool change but may also be performed off line using a test bench in high volume applications.

Drills used for MQL have different body and point geometries than standard wet tooling as summarized in Figures 15.17 and 15.18 [3,7,20,46]. MQL drills commonly have higher backtaper than conventional drills to account for increased thermal expansion during cutting [2,7,53]. This limits the number of regrinds, which can be performed before the drill loses size. MQL drills also commonly have thinner webs and wider flutes, and in some cases, flutes which increase in size along the axis, to facilitate chip ejection. Enlarging the flutes also reduces bending stiffness and drill stability [45], which is sometimes compensated for using double or triple margin designs [48,53]. Flute surfaces are commonly ground or polished to a high finish to inhibit buildup and chip accumulation [6,7,9]. The margin land width is typically reduced to reduce frictional heating between the margins and hole wall [2,3,7,46]. Point geometry modifications include optimized chisel edge configurations and special grinding of coolant hole gashes and chip exit features [3,7,46]. The web is thinned past the middle to produce a small, radiused chisel edge. Coolant holes are fully gashed for open air and oil flow to prevent material buildup in the center [46]. Smooth grinding of the transitions from the flute to the point eliminates constrictions, which may also cause material buildup during cutting.

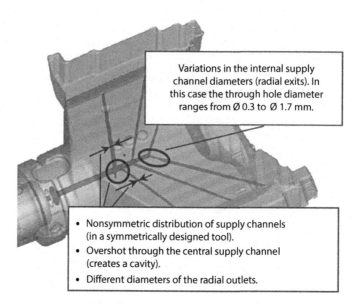

Variations in the internal supply channel diameters (radial exits). In this case the through hole diameter ranges from Ø 0.3 to Ø 1.7 mm.

- Nonsymmetric distribution of supply channels (in a symmetrically designed tool).
- Overshot through the central supply channel (creates a cavity).
- Different diameters of the radial outlets.

FIGURE 15.16 Design and manufacturing errors in a multi-port MQL tool. (After Stoll, A. et al., Lean and Environmentally Friendly Manufacturing—Minimum Quantity Lubrication [MQL] is a Key Technology for Driving the Paradigm Shift in Machining Operations, SAE Technical Paper 2008-01-1128, 2008.)

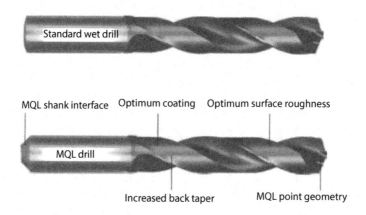

Standard wet drill

MQL shank interface   Optimum coating   Optimum surface roughness

MQL drill

Increased back taper   MQL point geometry

FIGURE 15.17 Differences in drill geometries between wet and MQL drills. MQL drills also commonly have larger, polished flutes for chip evacuation. (From Minimalmengenschmierung-Grundlagen, Gühring KG, Albstadt, Germany, 2013.)

The positioning of the coolant holes exits in drilling also influences MQL performance; coolant holes may be placed closer to the center of the drill than in wet machining to more effectively lubricate the inner cutting edges, and may also exit in the flute to lubricate chip ejection [1,2]. Generally, positive cutting geometries with light edge preparations are used for aluminum MQL. Edge preparations should be uniform across the entire edge in these applications to minimize chip constrictions and edge buildup [46]. Chamfered edges are sometimes used for ferrous MQL drills [53].

Coatings are commonly used on MQL cutting tools to reduce friction and material adhesion, improve tool life, and reduce tool temperatures and thermal expansion [1,2,3,7,46]. In aluminum MQL machining, heating of the drill is not a primary concern, and low-friction coatings are sometimes used to reduce frictional heating of the part and to improve chip evacuation. Diamond, DLC, TiB_2, MoS_2, and gold TiN-based coatings are commonly used in these applications [3,9,54]. Many aluminum MQL applications are performed with uncoated PCD and carbide tooling.

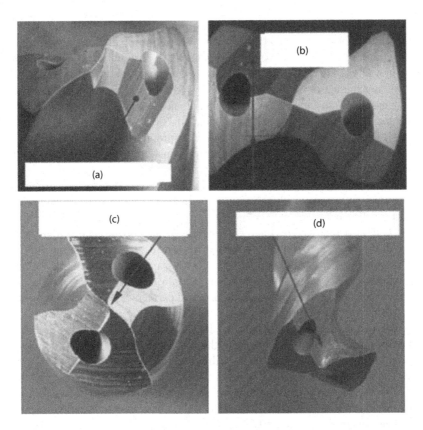

FIGURE 15.18 Point geometry modifications for MQL drills: (a) Open the coolant passage, (b) reduced web thickness, (c) point thinning beyond the middle, and (d) radius in the point thinning. (From Handbuch Minimalmengenschmierung [MMS], KometGroup GmbH, Besigheim, Germany, March 2008.)

Tool coatings are used in ferrous MQL machining for wear resistance and to provide a thermal barrier between the chip and tool. The thermal barrier function is especially important in drilling ferrous materials since significant heating of the drill can lead to thermal expansion and dimensional errors. TiAlN and similar hard coatings are typically used for ferrous MQL drills [3,53].

Obikawa [9] discussed MQL mist application in turning, and identified several methods for improving performance, including the use of flank as well as rake face nozzles, methods for placing nozzles close to the cutting edge, and the use of high air pressures with narrow nozzles to increase air velocity.

15.6 THERMAL MANAGEMENT AND DIMENSIONAL CONTROL

Since MQL does not provide significant cooling, it does not thermally stabilize the machining system; temperatures of machine tool components, cutting tools, and parts vary with time depending on ambient temperature variations, heating during the machining cycle, and machine transients during interruptions in production. Significant dimensional variations can result in the machine tool structure, spindle, part, and cutting tool as a result of varying temperatures, and a thermal management strategy is needed to manufacture parts with tight dimensional tolerances using an MQL process.

One basic concern is the transient thermal behavior of the machine tool itself. Machines are cold at the start of production and heat up over time as parts are cut, leading to thermal distortion of the machine structure and generally to growth of the spindle. There are two basic approaches to controlling these errors: warm-up cycles and probing-based machine compensation. In the warm-up cycle approach, cold machines execute a noncutting program, which cycles the machine through motions designed to establish a known thermal state. Shorter programs are also often executed after short interruptions in production. This method is generally robust, but the warm-up cycles typically last several minutes and reduce the productive time available on the machine. Warm-up cycles are also used in wet machining. In the probing-based compensation approach, a touch probe is periodically used to measure the distance between the spindle nose and a feature on the fixture and work table, with the measurements being used to adjust machine offsets to maintain consistent dimensions [16,18,19,55,56]. This approach can also be effective if the probing intervals are suitably chosen; often probing measurements are made frequently in the first hours after start of production, and less frequently during sustained production to save cycle time. Probing requires cycle time and may also contribute to downtime due to probing faults and reliability limitations. Material substitution approaches, such as the use of epoxy concrete rather than cast iron beds [1,2], can also be used to reduce machine tool thermal distortions, but have not been widely implemented in current applications.

In addition to the machine, the part and cutting tools heat up during machining, leading to additional thermal distortion. The part may heat up one to a few degrees C over a machining cycle [19], especially if the cycle contains drilling or boring operations, and uncoated tools may similarly expand due to increased temperatures. Part and cutting tool thermal variations can be controlled using process changes, waiting and cooling strategies, and sensor-based part temperature compensation.

Process changes are made to more effectively eject chips and to reduce cycle times [1,5,8,22,55]. Much of the heat retained in the part is transferred from hot chips when drilling or machining internal features of the part, so it is desirable to eject chips as rapidly as possible and to reduce frictional heat generation using the tooling modifications discussed for drills in the previous section (wider, polished flutes, narrower lands, etc.). As noted in Section 15.4, the machine tool and fixture should be designed to discourage chip accumulation on their surfaces. It is also desirable to reduce cycle times in internal machining operations such as drilling to limit contact time between the part and hot chips. Higher penetration rates are typically used in MQL than wet drilling, and milling and boring operations are also often performed more aggressively [2,5,8]. It is also common to machine high precision features at the beginning of operations when possible, before significant part heating has occurred [1,8,55]. Higher pressure or chilled air may also be used to reduce temperatures and promote chip ejection in MQL machining [9,35].

Part temperatures can also be controlled by installing cooling tunnels or areas between operations, and by monitoring the part temperature and not starting a cycle until the part cools to a prescribed level [1,55]. This is generally a robust strategy, but may reduce throughput by requiring waiting periods. Part temperature measurements can also be used to compensate tool paths, so that features and holes will be in the required position when the part cools [16,18,19,55,56]. The algorithms required to do this are complex and require appropriate controller capabilities; in addition, temperature sensor reliability can be an issue in this approach. Tool temperatures are reduced by the cooling and waiting strategy, and are compensated for to some extent in the sensor-based compensation approach. Dimensional errors cannot be compensated on fixed diameter tooling such as drills and reamers; these tools are often sized at the small end of the tolerance band in the cold state, so that they operate near the center or top of the band during steady production. When there are wide seasonal variations in plant ambient temperatures, seasonal tools (e.g., summer and winter drills with different diameters) may be required. It is often more efficient in these cases to install temperature control in the plant to limit ambient temperature changes [56].

15.7 AIR AND CHIP HANDLING

MQL operations generate an aerosol oil mist, metal dust, and metal chips [57]. These emissions should be properly managed to eliminate operator mist exposure and fire and explosion risks. In intermittent operations or retrofits on large machines, adequate ventilation and filtering of recirculated air should be provided, and a regular cleaning schedule should be implemented to avoid chip accumulation. In high volume installations, machines should be enclosed and provided with an automated extraction system.

The machine enclosure should be sealed to contain mists and dust and should be kept at a negative pressure with respect to the ambient air to prevent emissions from escaping during part loading and tool changes [6,8]. Its interior walls should be cleaned at regular intervals to remove adhering oil and dust not collected by the extraction system. Suitable cleaning methods include manual cleaning, low pressure air or water cleaning, and dry ice blasting [6].

Automated extraction systems remove mist and dust as they are generated from within the machine enclosure [6]. The air extraction rate should be sufficient to remove all emissions and to prevent mist escape during enclosure openings. An air speed over 20 m/s should be maintained in the extraction ductwork to prevent oil and dust accumulation. In many cases air is extracted downward through the chip conveyor. Extracted air is routed to a cleaning unit to remove dust and treat return air. Air returned to the work area normally requires high-efficiency particulate absorption (HEPA) filtering. Common dust extraction methods include precoated filter media methods [6,18,19,58] and spark arrest systems [59,60].

Chip conveyors, often connected to individual machine hoppers, are used for chip handling in high volume operations. As discussed in Section 15.4, the machine tools are designed with an open bed design to facilitate chip collection. MQL chips are normally dry and do not require special treatment prior to recycling or disposal.

15.8 MQL RESEARCH AREAS

15.8.1 Hard Alloy Machining and Grinding

Conventional MQL has limited cooling capability and as a result has not been widely used to machine harder alloys such as alloy and other hard steels, titanium alloys, and nickel alloys. The high cutting temperatures these materials produce would be expected to limit MQL tool life and metal removal rates. MQL is also not widely used in grinding due to similar thermal concerns. Due to the potential benefits of reducing or eliminating coolant usage, however, there has been considerable research interest in developing practical MQL processes for some hard alloy machining and grinding operations [37,40].

MQL has been investigated for turning hardened steels and for turning, drilling, and milling alloy and tool steels [61–65]. Most studies have used 1-channel external nozzle delivery and have compared their results primarily to dry cutting, since in practice these materials are often machined dry. MQL generally improved tool life and surface finish compared to dry cutting. In the few studies that included flood coolant results, MQL generally produced worse results.

One-channel MQL delivered through external nozzles has also been studied for turning, milling, and drilling titanium alloys [66–68]. In most cases results were compared only to dry cutting and showed an improvement in tool life and surface finish due to a reduction in adhesive wear and buildup. MQL performed well compared to flood cooling in drilling tests [66].

MQL has been used in finish turning and end milling tests on Inconel 718 [49,69–72]. Finish turning Inconnel 718 has been studied by Obikawa and coworkers [49,69–71] using nozzle geometries designed using CFD and special tool coatings (Figure 15.19). They optimized nozzles to minimize tool wear and lubricant consumption. MQL yielded an improvement in tool life over dry cutting in end milling tests [72].

MQL grinding research has concentrated on the development of nozzle geometries and oil additives to improve wheel life and reduce wheel loading in peripheral surface grinding [73–77], primarily for steels and ceramics. Most studies have concentrated on designing the best MQL grinding process (by selecting oils, wheel grades, and nozzle angles) without comparison to wet grinding. These MQL processes produce acceptable wheel life and surface quality but lower material removal rates than conventional grinding.

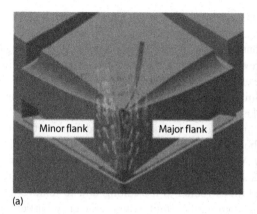 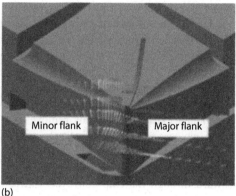

(a)   (b)

FIGURE 15.19 Mist flow visualizations for (a) normal and (b) oblique nozzles in turning tools. (After Obikawa, T., *Int. J. Mach. Tools Manuf.*, 49, 971, 2009.)

15.8.2 ALTERNATIVE CARRYING GASES AND COOLING STRATEGIES

Conventional MQL uses compressed air as the carrying gas for the oil mist. A number of researchers have investigated using alternative carrying gases (or gas–fluid mixtures) to provide additional cooling or to prevent chemical reactions at the tool–chip interface.

Wakabayashi et al. investigated aluminum MQL machining using air, oxygen, and nitrogen as carrying gases using drilling tests in a controlled atmosphere chamber [78]. The different gases were studied to determine their effect on the adsorption behavior of ester oil at the tool–chip interface. The best results were obtained with nitrogen since oxygen was found to have an unfavorable effect on process tribology.

Several methods of enhancing cooling in MQL by additions to the carrier gas have also been investigated. Liquid nitrogen was added to MQL sprays to improve cooling in early German research [10]. Chilled air is sometimes used for cooling in conventional MQL [9], and a system that uses chilled CO_2 as a carrying gas has been tested with positive results in grinding tool steel [79]. Specialized MQL systems also use chilled CO_2 to provide cooling and atmospheric isolation [80,81].

Systems that deliver water with the MQL mist to provide evaporative cooling have been widely investigated [9,38,75,78,82–86]. Distilled water is typically supplied at a rate comparable to the MQL oil rate for evaporative cooling. External nozzle delivery was used in most reported research. This method generally yields better performance than dry machining or conventional MQL, and results comparable or superior to external flood cooling in some drilling and grinding tests.

Cooling may also be provided using supercritical carbon dioxide (scCO_2) as a carrying gas [87–91]. In the supercritical phase, CO_2 has excellent solubility for aliphatic and most aromatic hydrocarbons and thus can carry metalworking lubricants in solution. Figure 15.20 shows a supercritical CO_2-based MQL system [89]. In this system liquid CO_2 is compressed to high pressure, heated in a chamber containing oil to supercritical temperature, and pumped to the tool point. Upon exiting an orifice on the tool, the CO_2 goes back to atmospheric pressure, providing significant cooling due to expansion, and the dissolved oil comes out of solution, providing lubrication. This approach has shown promise for improving tool life and MRR in titanium and hard ferrous alloy machining [88] and in rough machining nickel alloys [90].

The cryogenic machining systems described in Section 14.2 can also be used to increase cooling in dry and MQL machining applications.

15.8.3 MQL PROCESS MODELING

Research has also been conducted on modeling various aspects of the MQL processes. Studies have been carried out to specify appropriate frictional and thermal boundary conditions for finite element modeling of MQL processes [92–94]. Li and Liang [95,96] developed an analytical temperature

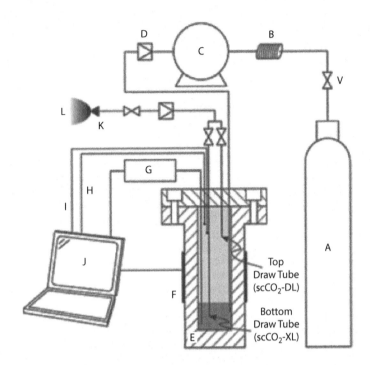

FIGURE 15.20 Schematic of supercritical CO_2 MQL delivery system. (A) Cylinder of food-grade CO_2, (B) cooling unit, (C) pump, (D) check valve, (E) high pressure vessel, (F) heating element, (G) soybean oil sump, (H) pressure transducer, (I) thermocouple, (J) computer, (K) nozzle, (L) supercritical CO_2 MQL spray, (V) on/off valve. (After Supekar, S.D., *J. Mater. Process. Technol.*, 212, 2652, 2012.)

model and a cutting force model including flank wear effects, which were verified in turning tests on steel with external nozzle mist application. MQL cutting temperatures have also been studied experimentally [28]. Finite element modeling has also been used to model thermal distortions in a vertical machining center in response to ambient temperature variations [97].

Limited computation fluid dynamics work on air flow in MQL tools [3,43,45,48,49,98] and on droplet formation in two-channel nozzles [99] has also been reported. Obikawa analyzed the travel distance of individual oil droplets in external nozzle applications [9], and analyses of droplet spreading in microlubrication systems have also been reported [100,101]. Further physical insight in these areas would be useful given the importance of air flow and mist properties in internal MQL applications.

15.8.4 Oil Additives and Ionic Fluids

Much research has been reported on enhancing MQL performance through additives to MQL oils. Additives such as nanoparticles of MoS_2 in soybean oil [102,103], Al_2O_3 in deionized water for grinding [104–106], SiO_2 nano-particles [107], carbon nanotubes in soybean oil [108], and ionic fluids [109–111] have been evaluated and shown promise in specific applications.

Ionic liquids, a promising and relatively new family of environmentally green chemicals, can be used in MQL to reduce friction at the tool–chip and tool–workpiece interfaces. Most functional ionic fluids are based on the cation of alkylated methylimidazolium salts. Several candidate ionic liquids were tested as additives (at 1% concentration by weight) in vegetable oil during MQL machining and yielded lower cutting forces and workpiece surface roughness compared to dry, flood, and conventional MQL coolant application [112]. Further studies have been in progress to explore the working principal of ionic liquids as metalworking fluids so that the proper combination of cations and anions can be suitably designed for specific applications, machining processes, cutting parameters and tool–workpiece material pairs [113,114].

REFERENCES

1. K. Weinert, I. Inasaki, J. W. Sutherland, and T. Wakabayashi, Dry machining and minimum quantity lubrication, *CIRP Ann.* **53**:2 (2004) 511–537.
2. Anon., Handbuch Minimalmengenschmierung-Eine Einführung, Willy Vogel AG/SKF, Berlin, Germany, 2004.
3. Anon., Handbuch Minimalmengenschmierung (MMS), KometGroup GmbH, Besigheim, Germany, March 2008.
4. Anon., Minimum Quantity Lubrication (MQL) Systems in Metal Cutting, Bielomatik Leuze GmbH, Neuffen, Germany, nd.
5. Anon., Machining machine tool innovation by iMQL System, Horkos Corp., Hiroshima, Japan, 2009.
6. Anon., Minimum quantity lubrication for machining operations, report BGI/GUV-I 718 E, Deutsche Gesetzlich Unfallversicherung (DGUV) Spitzenverband, Berlin, Germany, November 2010.
7. Anon., Minimalmengenschmierung-Grundlagen, Gühring KG, Albstadt, Germany, 2013.
8. T. Walker, *The MQL Handbook*, UNIST, Inc., Grand Rapids, MI, 2013.
9. T. Obikawa, Machining with least quantity lubrication, *Comprehensive Materials Processing.* Vol. 11: *Advanced Machining Technologies*, Elsevier, Amsterdam, the Netherlands, 2014, pp. 255–281.
10. U. Heisel, M. Lutz, D. Spath, R. Wassmer, and U. Walter, Application of minimum quantity cooling lubrication technology in cutting processes, *Prod. Eng.* **11**:1 (1994) 49–54.
11. T. Wakabayashi, H. Sato, and I. Inasaki, Turning using extremely small amount of cutting fluids, *JSME Int. J. (Ser. C)* **41** (1998) 143–148.
12. M. Tolinski, Near-dry machining cuts the heat, *Manuf. Eng.*, **139**:4 (October 2007), AAC19-25.
13. C. Boelkins, MQL: Lean and green, *Cutting Tool Eng.* **61**:3 (March 2009).
14. Anon., Less lubricant, more productivity, *Aerosp. Manuf.*, July 2010, 27.
15. BDL-1250/9D Caribide Drill Line, Peddinghaus Corp. USA, Bradley, IL, 2012.
16. R. Furness, A. Stoll, G. Nordstrom, G. Martini, J. Johnson, T. Loch, and R. Klosinski, Minimum quantity lubrication (MQL) machining for complex powertrain components, *Proceedings of 2006 ASME Manufacturing Science and Engineering Conference (MSEC)*, Ypsilanti, MI, Paper MSEC2006-21112.
17. A. Filipovic and D. A. Stephenson, Minimum quantity lubrication (MQL) applications in automotive powertrain machining, *Mach. Sci. Technol.* **10** (2006) 3–22.
18. A. Stoll, S. Silverson, and R. J. Furness, Environmentally friendly and low cost manufacturing—Implementation of MQL machining (minimum quantity lubrication), SAE Technical Paper 2007-01-1338, 2007.
19. A. Stoll, A. J. Sebastian, R. Klosinski, and R. J. Furness, Lean and environmentally friendly manufacturing—Minimum quantity lubrication (MQL) is a key technology for driving the paradigm shift in machining operations, SAE Technical Paper 2008-01-1128, 2008.
20. A. Richter, Tool spray, *Cutting Tool Eng.* **60**:3 (March 2008).
21. R. Kuchenmeister, Minimalmengenschmierung in der Motorblockfertigung, *Technik+Einkauf*, January 2009, pp. 48–49.
22. Anon., Making the leap to MQL, *Cutting Tool Eng.* **65**:9 (September 2013).
23. Anon., Umwelterklärung 2013—Salzgitter, Volkswagen AG, Wolfsburg, Germany, 2013.
24. Anon., Neues Motoren-Duo aus dem Werkteil Bad Cannstatt: Starke Antriebe aus der Wiege des Motorenbaus, Daimler AG, Stuttgart, Germany, 2014.
25. L. J. Tai, D. A. Stephenson, R. J. Furness, and A. J. Shih, Minimum quantity lubrication (MQL) in automotive powertrain machining, *Procedia CIRP* **14** (2014) 523–528.
26. D. Watts, Why aren't more manufacturers using minimum quantity lubrication? *Am. Mach.*, February 2, 2012. http://americanmachinist.com.
27. K.-H. Park, J. Olortegui-Yume, M.-C. Yoon, and P. Kwon, A study on droplets and their distribution for minimum quantity lubrication (MQL), *Int. J. Mach. Tools Manuf.* **50** (2010) 824–833.
28. T. Ueda, A. Hosokawa, and K. Yamada, Effect of oil mist on tool temperature in cutting, *ASME J. Manuf. Sci. Eng.* **128** (2006) 130–135.
29. Anon., High performance Near-dry machining solutions with MKD-DUAL, Dropsa USA Inc., Sterling Heights, MI, nd.
30. Anon., LubriLean—Minimal quantity lubrication, Brochure 1-5102-US, Willy Vogel AG, Berlin, Germany, 2005.
31. Anon., MQL system for internal tools lubrication, *Metalworking emagzaine*, January 2, 2005.
32. Anon., White Paper: Single-Channel MQL, UNIST, Inc., Grand Rapids, MI, 2014.

33. D. Deutges, Drehen mit 2-Knala-Minimalmengenschmierung, *Werkstatt und Betrieb*, September 2010, 138–139.
34. S. J. Kurgin, J. M. Dasch, D. L. Simon, G. C. Barber, and Q. Zou, A comparison of two minimum quantity lubrication delivery systems, *Ind. Lubr. Tribol.* **66** (2014) 151–159.
35. B. L. Tai, D. A. Stephenson, and A. J. Shih, Workpiece temperature during deep-hole drilling of cast iron using high air pressure minimum quantity lubrication, *ASME J. Manuf. Sci. Eng.* **135** (2013) 031019-1.
36. S. Suda, H. Yokota, I. Inasaki, and T. Wakabayashi, A synthetic ester as an optimal cutting fluid for minimal quantity lubrication machining, *CIRP Ann.* **51** (2002) 95–98.
37. S. Debnath, M. M. Reddy, and Q. S. Yi, Environmental friendly cutting fluids and cooling techniques in machining: A review, *J. Cleaner Prod.* **83** (2014) 33–47.
38. S. A. Lawal, I. A. Choudhury, and Y. Nukman, A critical assessment of lubrication techniques in machining processes: A case for minimum quantity lubrication using vegetable oil-based lubricant, *J. Cleaner Prod.* **41** (2013) 210–221.
39. S. A. Lawal, I. A. Choudhury, and Y. Nukman, Application of vegetable oil-based metalworking fluids in machining ferrous metals—A review, *Int. J. Mach. Tools Manuf.* **52** (2012) 1–12.
40. A. Shokrani, V. Dhokia, and S. T. Newman, Environmentally conscious machining of difficult-to-machine materials with regard to cutting fluids, *Int. J. Mach. Tools Manuf.* **57** (2012) 83–101.
41. T. Wakabayashi, I. Inasaki, S. Suda, and H. Yokota, Tribiological characteristics and cutting performance of lubricant esters for semi-dry machining, *CIRP Ann.* **52** (2003) 61–64.
42. S. Suda, T. Wakabayashi, I. Inasaki, and H. Yokota, Multifunctional application of a synthetic ester to machine tool lubrication based on MQL machining lubricants, *CIRP Ann.* **53** (2004) 61–64.
43. P. Hänle and M. Schwenck, Optimisation of cutting tools using CA-technologies, *Proceedings of Eight CIRP International Workshop on Modeling of Machining Operations*, Chemnitz, Germany, 2005, pp. 463–468.
44. P. Hänle and D. Gsänger, MMS-gerechte Gestaltung des Bohrerschaftes, *Werkstatt und Betrieb*, September 2003.
45. P. Hänle and D. Gsänger, MMS-gerechte Werkzeugauslegung, in: K. Weinert, Ed., Spanende Fertigung, 4. Ausgabe, Vukan Verlag, Essen, Germany, 2005, pp. 437–448.
46. M. A. Blosser, Design and application of MQL tooling for internal machining, *SME Collaborate Conference*, Fort Worth, TX, October 16–18, 2007.
47. P. Hänle and D. Gsänger, Prozesssichere Werkzeuge fürs Bohren mit MMS, *Werkstatt und Betrieb*, October 2003.
48. Anon., MQS Drill Series, Mitsubishi Materials Corp., Tokyo, Japan, 2011.
49. T. Obikawa, Y. Asano, and Y. Kamata, Computer fluid dynamics analysis for efficient spraying of oil mist in finish-turning of Inconel 718, *Int. J. Mach. Tools Manuf.* **49** (2009) 971–978.
50. E. Culp and J. Schreier, MQL cutting tools are automatically tested and inspected away from the machine, *ETMM*, February 1, 2012.
51. Anon., Mapal UNITEST-MQL, MAPAL Dr. Kress KG, Aalen, Germany, September 2011.
52. Anon., PCI test bench for MQL cutting-tool and tool-holder, PCI, Saint Etienne, France, May 2013.
53. Anon., MQL Power Long Drill, Nachi America, Inc., Macomb, MI, nd.
54. G. Fox-Rabinovich, J. M. Dasch, T. Wagg, K. Yamamoto, S. Veldhuis, G. K. Dosbaeva, and M. Tauhiduzzaman, Cutting performance of different coatings during minimum quantity lubrication drilling of aluminum silicon B319 cast alloy, *Surf. Coat. Technol.* **205** (2011) 4107–4116.
55. R. Quaile, Understanding MQL, *Mod. Mach. Shop*, December 13, 2004.
56. D. Korn, The many ways ford benefits from MQL, *Mod. Mach. Shop*, September 24, 2010.
57. J. E. Anderson, J. J. Szente, S. A. Mueller, B. R. Kim, M. M. Maricq, T. A. Loch, A. Stoll, G. P. Nordstrom, J. S. Johnson, and R. J. Furness, Particle and vapor emissions from dry and minimum quantity lubrication (MQL) machining of aluminum, *A&WMA Conference & Exhibition*, Indianapolis, IN, June 22–25, 2004, Paper 101.
58. Anon., Air extraction—Exhaust air purification—Safety technology for metal-removing machine tools and transfer lines, HANDTE Umwelttechnik GmbH, Tuttlingen, Germany, nd.
59. Anon., VARIO—Separation of fine dust particles, Keller Lufttechnik GmbH, Kirchheim unter Teck, Germany, nd.
60. Anon., Dry filter systems for applications using minimum quantity lubrication (MQL) and explosive dusts from dry processes or MQL extracted directly at the machine and separated safely, Keller Lufttechnik GmbH, Kirchheim unter Teck, Germany, September 2009.

61. J. Sharma and B. S. Sidhu, Investigation of effects of dry and near dry machining on AISI D2 steel using vegetable oil, *J. Cleaner Prod.* **66** (2014) 619–623.
62. S. Chinchanikar and S. K. Choudhury, Hard turning using HiPIMS-coated carbide tools: Wear behavior under dry and minimum quantity lubrication (MQL), *Measurement* **55** (2014) 536–548.
63. S. Chinchanikar and S. K. Choudhury, Machining of hardened steel—Experimental investigations, performance modeling and cooling techniques: A review, *Int. J. Mach. Tools Manuf.* **89** (2015) 95–109.
64. Y. S. Liao and H. M. Lin, Mechanism of minimum quantity lubrication in high-speed milling of hardened steel, *Int. J. Mach. Tools Manuf.* **47** (2007) 1660–1666.
65. M. Hadad and B. Sadeghi, Minimum quantity lubrication-MQL turning of AISI 4140 steel alloy, *J. Cleaner Prod.* **54** (2013) 332–343.
66. E. A. Rahim and H. Sasahara, A study of the effect of palm oil as MQL lubricant on high speed drilling of titanium alloys, *Tribol. Int.* **44** (2011) 309–317.
67. P. C. Priarone, M. Robiglio, L. Settineri, and V. Tebaldo, Milling and turning of titanium aluminides by using minimum quantity lubrication, *Procedia CIRP* **24** (2014) 62–67.
68. Z. Liu, Q. An, J. Xu, M. Chen, and S. Han, Wear performance of (nc-AlTiN)/(a-Si3N4) coating and (nc-AlCrN)/(a-Si3N4) coating in high-speed machining of titanium alloys under dry and minimum quantity lubrication (MQL) conditions, *Wear* **305** (2013) 249–259.
69. Y. Kamata and T. Obikawa, High speed MQL finish turning of Inconel 718 with different coated tools, *J. Mater. Process. Tech.* **192–193** (2007) 281–286.
70. T. Obikawa and Y. Kamata, MQL cutting of inconel 718 with a super lattice coating tool, *Key Eng. Mater.* **291–292** (2005) 433–438.
71. T. Obikawa, Y. Kamata, Y. Asano, K. Nakayama, and A. W. Otieno, Micro-liter lubrication machining of Inconel 718, *Int. J. Mach. Tools. Manuf.* **48** (2008) 1605–1612.
72. S. Zhang, J. F. Li, and Y. W. Wang, Tool life and cutting forces in end milling Inconel 718 under dry and minimum quantity cooling lubrication cutting conditions, *J. Clean. Prod.* **32** (2012), 81–87.
73. Y. Zhang, C. Li, D. Jia, D. Zhang, and X. Zhang, Experimental evaluation of MoS$_2$ nanoparticles in jet MQL grinding with different types of vegetable oil as base oil, *J. Cleaner Prod.* **87** (2015) 930–940.
74. M. Emamia, M. H. Sadeghia, and A. A. D. Sarhan, Investigating the effects of liquid atomization and delivery parameters of minimum quantity lubrication on the grinding process of Al$_2$O$_3$ engineering ceramics, *J. Manuf. Processes* **15** (2013) 374–388.
75. T. Tawakoli, M. J. Hadad, and M. H. Sadeghi, Investigation on minimum quantity lubricant-MQL grinding of 100Cr6 hardened steel using different abrasive and coolant–lubricant types, *Int. J. Mach. Tools. Manuf.* **50** (2010) 698–708.
76. D. Oliveira, L. G. Guermandi, E. C. Bianchi, A. E. Diniz, P. R. de Aguiar, and R. C. Canarim, Improving minimum quantity lubrication in CBN grinding using compressed air wheel cleaning, *J. Mater. Process. Tech.* **212** (2012) 2559–2568.
77. M. Hadad and B. Sadeghi, Thermal analysis of minimum quantity lubrication-MQL grinding process, *Int. J. Mach. Tools. Manuf.* **63** (2012) 1–15.
78. T. Wakabayashi, S. Suda, I. Inasaki, K. Terasaka, Y. Musha, and Y. Toda, Tribological action and cutting performance of MQL media in machining of aluminum, *CIRP Ann.* **56** (2007) 97–100.
79. J. A. Sanchez, I. Pombo, R. Alberdi, B. Izquierdo, N. Ortega, S. Plaza, and J. Martinez-Toledano, Machining evaluation of a hybrid MQL-CO$_2$ grinding technology, *J. Cleaner Prod.* **18** (2010) 1840–1849.
80. C. Felix, A cool way to cool, *Prod. Mach.*, May 17, 2007.
81. FCS-C500 System, Fusion Coolant Systems, Detroit, MI, 2013.
82. S. Bhowmick and A. T. Alpas, The role of diamond-like carbon coated drills on minimum quantity lubrication drilling of magnesium alloys, *Surf. Coat. Technol.* **205** (2011) 5302–5311.
83. Anon., MQL 2+ Channel System, Bielomatik Leuze Gmbh & Co, Neuffen, Germany, nd.
84. S. Zhang, J. F. Li, and Y. W. Wang, Tool life and cutting forces in end milling Inconel 718 under dry and minimum quantity cooling lubrication cutting conditions, *J. Cleaner Prod.* **32** (2012) 81e87.
85. R. Belentani, H. F. Júnior, R. C. Canarim, A. E. Diniz, A. Hassui, P. R. Aguiar, and E. C. Bianchi, Utilization of minimum quantity lubrication (MQL) with water in CBN grinding of steel, *Mater. Res.* **17** (2014) 88–96.
86. F. Itoigawa, T. H. C. Childs, T. Nakamura, and W. Belluco, Effects and mechanisms in minimal quantity lubrication machining of an aluminum alloy, *Wear* **260** (2006) 339–344.
87. A. F. Clarens, K. F. Hayes, and S. J. Skerlos, Feasibility of metalworking fluids delivered in supercritical carbon dioxide, *J. Manuf. Process.* **8** (2006) 47–53.
88. A. F. Clarens, D. J. Maclean, K. F. Hayes, Y.-E. Park, and S. J. Skerlos, Solubility of a metalworking lubricant in high-pressure CO2 and effects in three machining processes, *Trans. NAMRI/SME* **37** (2009) 645–652.

89. S. D. Supekar, A. F. Clarens, D. A. Stephenson, and S. J. Skerlos, Performance of supercritical carbon dioxide sprays as coolants and lubricants in representative metalworking operations, *J. Mater. Process. Tech.* **212** (2012) 2652–2658.

90. D. A. Stephenson, S. J. Skerlos, A. S. King, and S. D. Supekar, Rough turning Inconel 750 with supercritical CO_2-based minimum quantity lubrication, *J. Mater. Process. Tech.* **214** (2014) 673–680.

91. Anon., FCS-L500 System, Fusion Coolant Systems, Detroit, MI, 2013.

92. P. Faverjon, J. Rech, and R. Leroy, Coefficient and work-material adhesion during machining of cast aluminum with various cutting tool substrates made of polycrystalline diamond, high speed steel, and carbides, *ASME J. Tribol.* **135** (2013) 041602-1.

93. S. Kurgin, J. M. Dasch, D. L. Simon, G. Barber, and Q. Zou, Evaluation of the convective heat transfer coefficient for minimum quantity lubrication (MQL), *Ind. Lubr. Tribol.* **64** (2012) 376–386.

94. M. J. Hadad, T. Tawakoli, M. H. Sadeghi, and B. Sadeghi, Temperature and energy partition in minimum quantity lubrication-MQL grinding process, *Int. J. Mach. Tools Manuf.* **54–55** (2012) 10–17.

95. K.-M. Li and S. Y. Liang, Modeling of cutting temperature in near dry machining, *ASME J. Manuf. Sci. Eng.* **128** (2006) 416–424.

96. K.-M. Li and S. Y. Liang, Modeling of cutting forces in near dry machining under tool wear effect, *Int. J. Mach. Tools. Manuf.* **47** (2007) 1292–1301.

97. N. S. Mian, S. Fletcher, A. P. Longstaff, and A. Myers, Efficient estimation by FEA of machine tool distortion due to environmental temperature perturbations, *Precision Eng.* **37** (2013) 372–379.

98. Y. Iskandar, A. Tendolkar, M. H. Attia, P. Hendrick, A. Damir, and C. Diakodimitris, Flow visualization and characterization for optimized MQL machining of composites, *CIRP Ann.* **63** (2014) 77–80.

99. C. Diakodimitris, Y. R. Iskandar, P. Hendrick, and P. Slangen, New approach of gas–liquid computational fluid dynamics simulations for the study of minimum quantity cooling with airblast plain-jet injectors, *ASME J. Manuf. Sci. Eng.* **135** (2013) 041009-1.

100. I. Ghai, J. Samuel, R. E. DeVor, and S. J. Kapoor, Analysis of droplet spreading on a rotating surface and the prediction of cooling and lubrication performance of an atomization-based cutting fluid system, *ASME J. Manuf. Sci. Eng.* **135** (2013) 031003-1.

101. C. Nath, S. G. Kapoor, A. K. Srivastava, and J. Iverson, Study of droplet spray behavior of an atomization-based cutting fluid spray system for machining titanium alloys, *ASME J. Manuf. Sci. Eng.* **136** (2014) 021004-1.

102. J. Yan, Z. Zhang, and T. Kuriyagawa, Effect of nanoparticle lubrication in diamond turning of reaction bonded SiC, *Int. J. Autom. Technol.* **5** (2011) 307–312.

103. B. Shen, A. J. Malshe, P. Kalita, and A. Shih, Performance of Novel MoS_2 Nanoparticle based grinding fluids in minimum quantity lubrication grinding, *Trans. NAMRI/SME* **36** (2008) 357–364.

104. S. Khandekar, M. R. Sankar, V. Agnihotri, and J. Ramkumar, Nano-cutting fluid for enhancement of metal cutting performance, *Mater. Manuf. Processes* **27** (2012) 693–697.

105. B. Shen, A. Shih, and S. Tung, Applications of nanofluids in minimum quantity lubrication grinding, *Tribol. Trans.* **51** (2008) 730–737.

106. D. Setti, S. Ghosh, and P. V. Rao, Application of nano cutting fluid under minimum quantity lubrication (MQL) technique to improve grinding of Ti-6Al-4V alloy, *World Acad. Sci. Eng. Technol.* **70** (2012) 512–516.

107. A. Sarhan, Reduction of power and lubricant oil consumption in milling process using a new SiO2 nanolubrication system, *Int. J. Adv. Manuf. Technol.* **63** (2012) 505–512.

108. B. Shen and A. Shih, Minimum quantity lubrication (MQL) grinding using vitrified CBN wheels. *Trans. NAMRI/SME* **37** (2009) 129–136.

109. A. E. Jimenez and M. D. Bermudez, Ionic liquids as lubricants of titanium-steel contact, *Tribol. Let.* **33** (2009) 111–126.

110. B. Davis, J. K. Schueller, and Y. Huang, Study of ionic liquid as effective additive for minimum quantity lubrication during titanium machining, *Manuf. Lett.* **5** (2015) 1–6.

111. A. Libardi, S. R. Schmid, M. Sen, and W. Schneider, Evaluation of ionic fluids as lubricants in manufacturing, *J. Manuf. Proc.* **15** (2013) 414–418.

112. G. S. Goindi, S. N. Chavan, D. Mandal, P. Sarkar, and A. D. Jayal, Investigation of AISI 1045 steel, *All India Manufacturing Technology, Design and Research Conference (AIMTDR 2014)*, December 12–14, 2014, IIT Guwahati, Assam, India, pp. 277-1.

113. N. I. Doerr, I. C. Gebeshuber, D. Holzer, H. D. Wanzenboeck, A. Ecker, A. Pauschitz, and F. Franek, Evaluation of ionic liquids as lubricants, *J. Microeng. Nanoelectron.* **1** (2010) 29–34.

114. F. Zhou, Y. Liang, and W. Liu, Ionic liquid lubricants: designed chemistry for engineering applications, *Chem. Soc. Rev.* **38** (2009) 2590.

16 Accuracy and Error Compensation of CNC Machining Systems

16.1 INTRODUCTION

Machined part accuracy is affected by the machine tool, part and fixture stiffness, cutting tool characteristics, cutting conditions, ambient environment, and external vibrations (Figure 16.1). Many key factors such as resolution of the machine tool, the type of workpiece, the machining sequence and program, the cutting tools and cutting conditions, the fixture characteristics, etc., are generally determined based on previous experience, modeling, or analysis as discussed in previous chapters. However, once these parameters are defined, the accuracy and repeatability of the machine tool will determine its ability to precisely position the cutting tool tip at the specified part location. Therefore, the machine tool characteristics have a major contribution to part accuracy and thus are analyzed in the chapter.

Computer numerical controlled (CNC) machining systems are typically used to accurate machine parts according to part drawing specifications. A CNC machining system operates according to a part process using a sequence of commands (e.g., G-code) that moves a controllable cutting tool to machine the part clamped within a fixture, which is in turn positioned on a machine table. The control system often monitors the position of the tool relative to the part during the operation through the servomotor control.

The part program is generated by computer-aided part programming (CAD/CAM) systems. The post tool path is converted into CNC commands (as DNC operations) through processing of the CAM programs. This transfers the tool path that is originally described in the workpiece coordinate system (WCS) to the machine coordinate system (MCS) through inverse kinematic transformations. The CNC machine control unit (MCU) issues commands in the form of numeric data to the motors of the machine linear or rotary axis that positions the slides and tool according to the workpiece location in relation to the MCS. The computer-aided machining process may result in approximations because the vast majority of machines are set to use four decimal places for motion control. In addition, each machine axis is subject to following errors, leading to orientation and position errors of the tool [1], and the machine structure joints and motion errors in its quasi-static or dynamic state result in position and orientation errors of the tool [2,3]. Moreover, the thermal variation of the machine is another source of error [4]. It is well established that the thermal deformation of the workpiece and the machine tool structure has a significant effect on machining accuracy and may contribute to more than 50%–70% of the machining error, especially when parts are made of materials of low thermal conductivity and thermal capacity or high thermal expansion coefficient. Finally, the forces from the cutting process result in tool or workpiece deflection or distortion, causing another source of machined surface location error [5].

The two terms often used with quality are *error* and *accuracy*. *Error* represents deviations of the cutting tool point from the theoretical position on the part in order to machine part dimensions to specified tolerance. However, *accuracy* is defined as the part's conformance to the specifications in the manufacturing drawing or CAD model.

Section 16.2 describes machine tool errors. Machine tool accuracy assessment and performance evaluation are discussed in Sections 16.3 and 16.4. Section 16.5 examines various methods for

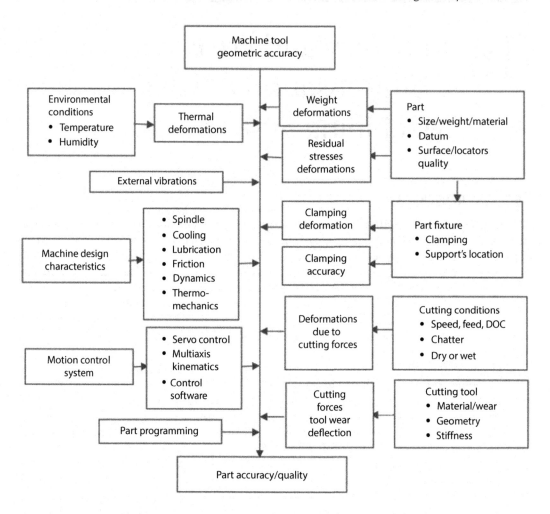

FIGURE 16.1 Factors affecting machine tool accuracy.

compensation or calibration of machine tools; first, the compensation strategies are reviewed followed by a discussion of the error modeling methods. All the various compensation methods are discussed so that the reader gains enough background to select a machine tool with compensation capabilities suitable for the target application(s).

16.2 MACHINE TOOL ERRORS

Error sources that affect the dimensional accuracy of finished parts are broadly classified as (1) those occurring during the design, manufacturing, and assembly of the structural machine components and (2) error sources present during machining. They can also be classified on factors attributed to (1) the machine itself and (2) the cutting process. Machine tool accuracy is determined by the positional errors of the cutting tool relative to the workpiece, which are defined as geometric, kinematic, dynamic, thermal, and processing errors. The understanding of the individual sources of error by assessing their magnitude and variation and how they are interacting in a multiaxis machine tool could be very helpful for improvements in the design, manufacturing, and assembly of the machine components, including the best approach to compensate such errors.

CNC machine tool frame is composed of a minimum of three orthogonal linear axes, generating the geometry of a Cartesian coordinate system. A typical three-axis machine consists of X-, Y-, and Z-axes (standard horizontal or vertical configuration) as explained in Section 3.3, and several of the figures in Chapter 3 illustrate various machine tool types and the design of each axis (see Figures 3.46 through 3.49). The machine's carriage motion over the linear guideways is described by rigid kinematics. Each translation of an axis is defined by six degrees of freedom, three translations, and three rotations. Therefore, each of the three linear axes has six possible errors as shown in Table 16.1 and defined in ISO 230-1 [6] and ASME B5.54 [7]. These six error components contribute to the positioning error of the slide within the machine tool's workspace. Figure 16.2 illustrates the component errors of the displacement accuracy for a single axis, which also occurs in the other linear axes; the five degrees of freedom are the three angular errors, pitch ($\varepsilon_z(x)$—tilt error around Z), yaw ($\varepsilon_y(x)$—tilt error around Y), roll error ($\varepsilon_x(x)$—tilt error around the X), and two straightness errors in the direction perpendicular to the direction of motion: these are along the vertical (straightness error of X in Y-direction, $\delta_y(x)$) and horizontal (straightness error of X in Z-direction $\delta_z(x)$) directions. The sixth degree of freedom is the translation (position) error $\delta_x(x)$ along the X-axis desired motion. The effect of angular errors on the cutting tool positioning error is the angular error times the Abbe offset as will be discussed in the following. A contribution of the roll error around the axis of motion to the other two axes is illustrated in Figure 16.3; it represents the Z-axis roll error in a horizontal machine tool where

$$\delta_{ez}(x) = x\varepsilon_z(z) \quad \text{and} \quad \delta_{ez}(y) = y\varepsilon_z(z) \tag{16.1}$$

TABLE 16.1

Description of the Six Degrees of Freedom Errors for X-Axis of Motion

Error Parameters	Direction of Error
Displacement—Linear position	X
Angular—Pitch	X
Angular—Yaw	X
Angular—Roll	Y and Z
Vertical straightness—XY	Y
Horizontal straightness—XZ	Z

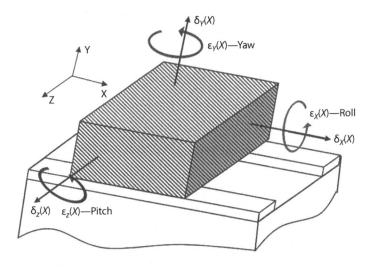

FIGURE 16.2 Illustration of the six degrees of freedom for a traveling machine carriage along the X-axis.

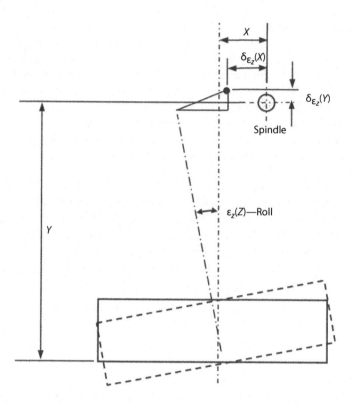

FIGURE 16.3 Illustration of the Z-axis roll error of the machine table.

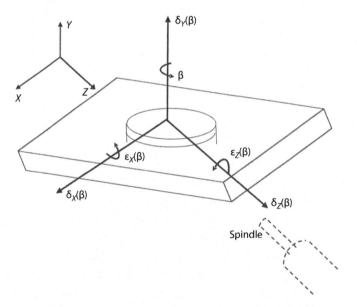

FIGURE 16.4 Illustration of the six degrees of freedom for a rotary axis (of a *B* table machine).

Similar error parameters are present in the other two linear axes of motion. Three more error parameters determine the orthogonality (squareness) errors of machine axes for the total of 21 error parameters. Likewise, a rotary axis will have six component errors similar to linear axis as illustrated in Figure 16.4. The three translational errors include axial motion along the *Y*-axis and two straightness errors affecting the radial error motion; the three angular errors include the rotational error about the Y-axis and rotations about the X- and Z-axes affecting the tilt (or wobble) motion for a B table. Since the rotary B table is sitting on the XZ plane, two additional squareness errors are present.

The total number of degrees of freedom on any given machine tool is generally calculated using the following simple equation [8]:

DOF = 6 (number of axes) + 4 (number of rotary axes) + 2 (number of translation axes)
      + DOF of the tool − 6

The −6 at the end of the equation represents the workpiece alignment error (location features and distortion due to clamping and residual stress), which is considered as setup error instead of machine tool accuracy error. For example, the traditional three-axis machine tool has 21 error parameters, while four-axis or five-axis machine tools have 32 and 43 minimum error parameters, respectively.

The deterministic error is affected by Abbe offset errors in pitch and yaw and by alignment errors of the machine axis. Due to pitch error in the machine's rail, the tool tip leads or lags behind the intended location. Ernst Abbe proposed the principle of alignment that explains that the scale of a linear measuring system should be collinear with the location of the desired position (displacement to be measured) on the workpiece (see Figures 3.46 and 3.47). If this is not the case, the measurement must be corrected for the associated Abbe error using the angular error of the corresponding axis. Therefore, the Abbe offset error is defined as

$$\delta = P\sin(\theta) \approx P\theta \tag{16.2}$$

where
δ is the Abbe offset error
P is the Abbe offset [9] (the distance between the axis displacement transducer and the point/line where the part is machined and measured)
θ is the angular misorientation (the tilt or curvature of the guideway or displacement transducer in rad) as illustrated in Figure 16.5

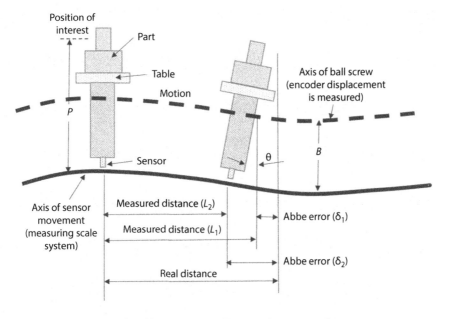

FIGURE 16.5 Sketch showing the effects of Abbe offset angular error on linear error.

The pitch tilt is a major contributor to Abbe error. This error results from the angular or tilt motions of the part relative to cutting tool. This error is determined by the distance of the ball screw or guideways from the workpiece or the cutting tool (see Figure 3.46). The Abbe offset errors are considered the greatest errors in medium to large machines. These errors are eliminated only if P is zero. Otherwise, it is minimized by making P or θ as small as possible. Finally, Abbe offset error can be measured (mapped out) and compensated so that θ becomes zero as it is further explained in Example 16.1.

The guideway tilt not only affects the Abbe offset error but in a lesser degree the straightness error, which is defined normal to the axis of motion as illustrated in Figure 16.6. The straightness error in the direction of the Y-axis due to the tilt error of the X-axis is

$$\delta_y(x) = \frac{X\theta_x}{8} \tag{16.3}$$

where
  x is the traveling length
  θ_x is the tilt angle of the carriage

This equation is valid for small values of θ. It is important to review the Bryan principle that is corollary to the Abbe principle that addresses the influence of the angular error on straightness measurements. The Bryan Principle states that "the straightness measuring system should be in line with the functional point whose straightness is to be measured. If this is not possible, either the guideways that transfer the straightness must be free of angular motion, or angular-motion data must be used to calculate the consequences of the offset" [9]. Therefore, the slide's angular errors should be incorporated in the straightness errors. The straightness error is defined as (1) the maximum or average that is based on the span of displacement selected to estimate the tilt angle or (2) a function of X if the actual straightness profile is measured. The maximum angle of the straightness profile is used in Figure 16.6. For example, the straightness profile is illustrated in Figure 16.7 for each of the X- and Y-axis; the δ_{yx} and δ_{xy} least squares straightness lines can be used to estimate either the actual $\delta_y(x)$ and $\delta_x(y)$ straightness errors or the corresponding tilt angles θ_x and θ_y at a specified span. The X- or Y-axis defined by the least squares method could have a slope to the reference coordinate system.

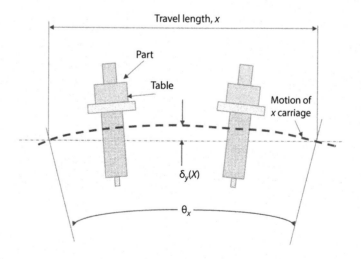

FIGURE 16.6   Sketch illustrating the straightness errors in X-axis.

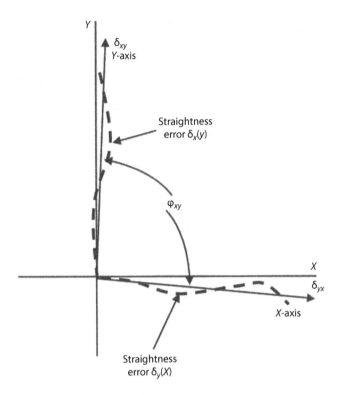

FIGURE 16.7 Sketch illustrating the measurement of orthogonality error between X- and Y-axis.

Orthogonality (equivalent to squareness) errors among the three axes, considering X–Y plane as the datum, are denoted by α_{xy}, α_{xz}, and α_{yz}, respectively, along the X–Y, X–Z, and Y–Z directions of error. The orthogonality error between the X- and Y-axis is defined along the X-axis as illustrated in Figure 16.7 and is given by the equation

$$\delta_{xy} = y\alpha_{xy} \quad \text{and} \quad \alpha_{xy} = \varphi_{xy} - \frac{\pi}{2} \tag{16.4}$$

The squareness angle is defined as the slopes of the straightness error profiles for the two axes. Therefore, the squareness angle is the deviation of the included angle between the two measured axes from a right angle as illustrated in Figure 16.8 for φ_{xy} between X- and Y-axis. The φ_{xy} is the included angle between the least squares fitting straightness lines that define the X- and Y-axis for the corresponding guideways. Therefore, the coordinate reference system is aligned with the X–Y datum and the X-axis of the least squares straightness line.

The positional error of a machine axis is dependent upon the geometrical, kinematic, moving loads, structural stiffness and dynamic response, dynamic loads, thermal distortions, and motion and software control errors of the system (see Figure 16.1) [10]. The positional error is determined by the linear and rotary axes position errors that are determined by the position accuracy, the repeatability, and reverse deviation of each axis. Therefore, the displacement accuracy of all axes (linear and rotary) can be described with mathematical models relating the kinematics and individual component quality characteristics as explained in Ref. [11]. The errors due to the process such as tool wear and deflection, workpiece deflection, workpiece location, and clamping errors are not considered in the machine tool errors even though they affect part quality. Geometrical accuracy of the machine and thus the *geometric errors* result from the basic design of the machine structure including component and assembly tolerances. The errors are treated as a function of

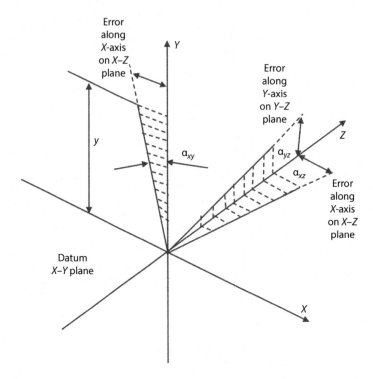

FIGURE 16.8 Sketch illustrating the orthogonality/squareness errors in a three-axis machine.

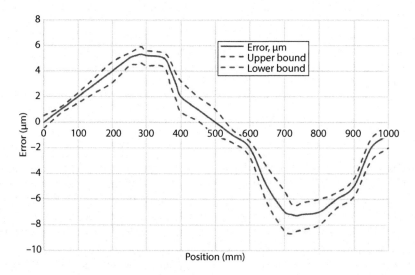

FIGURE 16.9 Linear positioning error in the direction of X-axis motion.

displacement because they vary along the length of each axis (see Figure 16.9). The manufacturing errors of the machine include dimensional errors of the individual machine structural components including ball screws, axis alignment errors, linear scales, as well as the gradual changes due to the wear of the moving components. The effect of the geometric inaccuracies results in squareness, parallelism, and straightness errors (which includes pitch, yaw, and roll errors; perpendicularity of the spindle; parallelism of the spindle and working surface; and for rotary axes—the radial, axial, and angular errors) among the machine's moving elements and axes. The limitation in meeting demand for increased precision in machine tools has been the quality of the machine tools used to machine the required components. A detailed description of the geometric errors is available in Ref. [12]. *Kinematic errors* are those rigid body errors that are affected by the relative motion errors of several moving machine components, especially during simultaneous motion of multiple linear or rotary axes that are used to form 2D or 3D profiles. Joints, couplings, and gears between machine components may cause deformations because of limited stiffness and generate kinematic errors. They take place during linear, circular, spline, Nurbs, or five-axis surface interpolation [13,14]. The interpolation method breaks the tool path into smaller segments to avoid any significant spikes or defects on the surface.

The *errors due to loads* on the machine are those affecting the nonrigid body behavior; for example, structural deformations of the guideways, table, and carriage due to the weight of the part, bending/tilt of the machine column or fixture due to high thrust force, deflection of spindle due to cutting loads, etc. The nature of the deformation by the part mass depends on the machine structure and location of load. The dead weight of the machine's components could affect the bend of the guideways at certain weak structural positions. Structural deformations could be a result of over-constrained mechanical systems. Some of the important nonrigid body motion errors are caused by weight shifting, improper alignment and assembly of machine's components, and leveling problems.

Dynamic errors are caused by cutting loads/forces (self-induced and force vibrations on the machine structure), deflections due to inertial forces during acceleration/deceleration of the axes around corners or complex profiles, hysteresis or backlash in the machine's drive system, control errors, and spindle error motion (axial and radial runout). The load-induced errors are affected by the forces resulting from cutting process and the static and dynamic stiffness of all machine's components that are within the structural loop. These errors are greater during heavy cuts, which are the reason a rough pass followed by a light finishing pass is used for quality surfaces or features. These errors are affected by the cutting force dynamics and can be minimized by optimizing the cutting conditions, fixture, and the damping characteristics of the machine tool structure. The systematic error of an axis is very different during forward and backward travel, which defines the hysteresis or backlash in the system, which is affected by the friction forces, the play in the sliding components due to wear or looseness of components, excessive strain on the ball screw, and encoder hysteresis. The control system errors are affected by the parameters controlling the tightness of the control such as proportional loop gain, integrated loop gain, damping characteristics in the loop, look-ahead function, and feed-forward feedback loop. Control of the acceleration/deceleration of the spindle and linear or rotary axes is very important for high-speed machining to minimize spikes on machining surfaces. Machine tool damping characteristics, as explained in Chapter 12, are also important to machining performance. Higher resonance frequencies for the machine and its components supporting the workpiece and cutting tool result in shorter loop response times.

Thermal errors are caused by internal or external heat or cold sources. They are the result of temperature gradients in the structure of a machine or part, which generate dimensional changes in the machine tool structure or the part that affects machine tool accuracy. They are very important because they account for 30%–70% of the total dimensional and shape errors in a part (because the systematic geometric and kinematic errors are often compensated). Sources of heat-induced errors include (1) the heat generated by the machine's moving and operational elements, (2) the heat generated through the cutting process at the tool–workpiece interface, (3) cooling or heating induced by the cooling system, and (4) temperature control (or changes) in the ambient environment with time. The machine sources such as the spindle, drive motors, pumps, gears and hydraulic fluid, and friction in bearings, ball screws, guideways, and machine axes affect the heat distribution through the machine structure. The heat from the cutting process is transferred, especially by the hot chips, within the workspace, workpiece, fixture, and spindle. The temperature fluctuation in the machine structure by the aforementioned components results in deformation or expansion of the machine elements. Thermal gradients within the structure are usually transient and therefore cause time-dependent errors. The heat generated by the various sources travel through the machine structure by conduction, convection, and radiation and therefore is impossible to estimate the average heat distribution and temperature in the machine tool elements and hence the distortion of these elements. For example, a temperature rise results in spindle growth, thermal expansion of the ball screws, thermal distortion of the column and guideways, workpiece and fixture distortion due to thermal expansion, and thermal expansion of the axes scales. The thermal deformation of the machine elements are a major contribution to machine uncertainty.

Thermal distributions and temperature sensitivity in the machine components are controlled by [15] (1) avoiding large thermal drift through the machining process through either (a) material selection and structural design or (b) the control of the source to steady-state temperature level and (2) compensation. Some of the heat sources are controllable to a steady-state temperature level by using various means to remove heat, but it is difficult to completely eliminate the contribution of the earlier sources especially the heat generated at the cutting zone. Temperature change of the machine tool structure and workpiece will affect the dimensional characteristics. Temperature compensation is possible since the error can be estimated using empirical and analytical modeling or FEA methods. The dimensional change is estimated by the equation

$$\Delta L = L \cdot \alpha \cdot \Delta T \tag{16.5}$$

where
  L is the nominal length
  α is the thermal coefficient of expansion
  ΔT is the deviation from standard temperature

The thermal expansion coefficient is 12 μm/m °C for steel and 23 μm/m °C for aluminum. For example, the linear error of an axis is initially significantly high (0.03–0.1 mm during cold machine start), while it drops to less than 0.01 mm after it reaches the steady state. The temperature usually rises exponentially from room temperature to the steady state after a period of time that varies from 60 to 200 min. Figure 16.10 shows the temperature rise in the spindle during the warm-up cycle together with the spindle shaft displacement along the spindle Z-axis; the longer the thermal stabilization for a machine tool component, such as the spindle, the higher the warm-up temperature (or temperature rise) and the larger the thermal expansions or deformation of the component,

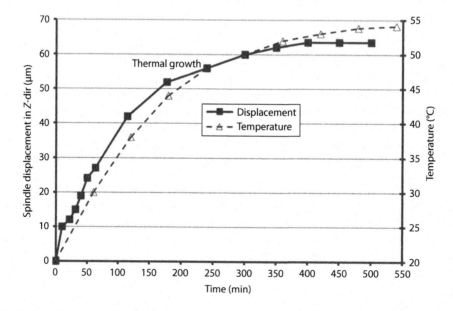

FIGURE 16.10 Spindle temperature response and thermal growth of the shaft front relative to warm-up time.

as clearly shown in Figure 16.10. The error is smaller if machine tool reaches steady-state condition with small temperature changes across the structure. Likewise, the part temperature during machining should be controlled, especially for larger parts, to avoid large thermal drift through its machining process.

Processing errors are caused by dynamic and thermal loading conditions, workpiece residual stresses, and sometimes deformation of the workpiece and fixture due to cutting loads and clamping loads. They are difficult to estimate them. Cutting tool deflections due to tool and spindle stiffness and cutting force affect the surface profile and roughness and the depth of cut. Therefore, the workpiece itself is a major consideration for processing errors since it is subject to errors due to loads.

These errors consist of systematic and random components. No matter how well a machine tool is designed or manufactured (as explained in Chapter 3), there is a limit to the accuracy that could be achieved for a new machine. Thermal deformations, cutting force deformations, and fixture and workpiece deformations cannot be fully accounted for at the design stage of the machine or the process.

In the earlier days of CNC machine tools, quasi-static errors accounted for about 70% of the total error, but current compensation approaches have reduced most of the geometric errors. However, dynamic errors cannot be compensated completely and therefore are considered the most significant errors. Our recent experience with large- and medium-sized machine tools indicates that the error budget is distributed as follows among the sources of error: 40%–50% on static/geometric, 30%–40% on dynamic (thermal, part, and tool deflection, etc.), 10%–20% on environment, and 5%–10% on measurement. The static error (65%–70%) is due to axes positioning and squareness errors, especially for larger machine tools. Machine tools, however, are sensitive to a variety of random influences, such as a crush, varying thermal state, possible contamination, wear of moving elements, etc., which may result either in progressive errors (such as drift, which progressively degrade their accuracy over time) or in dramatic, permanent changes in performance.

16.3 MACHINE TOOL ACCURACY CHARACTERIZATION

The accuracy or process capability of a machine tool is very important because it directly affects the accuracy of the part. However, the part accuracy is affected by several other parameters including environmental effects; part characteristics; and cutting tool, fixture, and cutting conditions, as discussed in Section 10.2 and shown in Figure 16.1. The overall accuracy of a machine tool is complex because it is made of several components and it depends on several factors such as (1) the displacement accuracy for each axis, (2) the repeatability of positioning in each axis, (3) the resolution of each built-in measuring system, (4) the accuracy of the pallet/table, and (5) the volumetric accuracy. According to ASME standards [7], the structural loop of the machine, defined as the assembly of the mechanical components moving to position the tool at the specified position at the workpiece, is affected significantly by the aforementioned factors. The structural loop is very complex to analyze because it consists of all the machine components such as frame, guideways, bearings, all spindle components, drives for the axes, table/pallet, holder and cutting tool, and part fixture. Therefore, the influence of the earlier components and their interaction with the thermal sources defines the the accuracy of a machine tool. The accuracy of an axis of rotation, such as spindle or machine rotary axes, is relatively complex because it relates to the errors of motion. The accuracy of the work zone is affected by the displacement (1D uncertainty), planar (2D uncertainty), or volumetric (3D uncertainty) accuracies. In general, the volumetric accuracy includes (1) the displacement accuracy of all axes (both linear and rotary), (2) the geometric accuracy of each axis, (3) squareness of pallet/table, and (4) the dynamic performance of the structure, servo-drives, continuous path control, interpolating errors, etc. The geometric accuracy is determined by the geometric errors consisted by a deterministic portion and a probabilistic distribution of random variation that defines the repeatability (or bandwidth of uncertainty). The repeatability depends on (1) the stability of the measuring system affected by thermal, electrical, or mechanical drift, (2) friction in the machine moving components (i.e., guideways), (3) backlash and hysteresis in the drive components, (4) the stiffness characteristics of the structure, (5) the stability of the machine's structure to thermal effects (ability to resist thermal deformations), (6) digitization error, and (7) software limitations.

Machine tools are ultimately deterministic in performance because individual geometric errors can be interpreted as a deterministic value. The accuracy specification quantifies how well the cutting tool locates at a given location within the workspace. However, there is a probabilistic distribution of random variation due to environmental effects or thermal drifts and the friction within the machine axes. The maximum uncertainty or repeatability includes the systematic plus random errors. If the stochastic/random variation of the error is within the desired tolerance volume, the deterministic error can be eliminated if the error can be predicted and compensated [16]. On the other hand, if the random variation of errors is too large compared to the part tolerances, a significant number of parts machined on this machine will exceed the specified tolerance requirement. Therefore, the random characteristics of the machine tools must be analyzed in order to improve machining accuracy.

Over the past three decades, various national and international standards have been developed to assess the hardware performance of machine tools in order to provide a one-to-one correspondence between machine errors and machined part errors. These standards are used as performance evaluation methods that allow comparison between different machine tools. Because the positional accuracy and repeatability are the characteristic parameters describing the performance of a machine tool, a number of standards and guidelines exist outlining how to evaluate the accuracy of a machine tool. Multiple standards exist that manufacturers use to test CNC machine position accuracy such as JIS B6336 and JIS B6192 (Japanese), VDI/DGQ 3441 (German), BSI BS 4656 (British), NMTBA (United States), ASME B5.54-92 (United States), and ISO 230-2 (European), and within these standards changes are always considered. Most standards are related to ISO 230-2:1997. The accuracy standard typically reflects the standard of the country where the manufacturer is located. The same machine will have different positional accuracies when tested against any of the various standards because these standards differ in their analysis procedures and in some of the key parameter definitions. Even though all standards are acceptable, it is important to understand the difference among each other and how the different calculated values compare to each other especially when buying machines made at different countries. The standards are compared in Refs. [17–19], and a summary is provided in Table 16.2. A numerical analysis is described in Ref. [17] to evaluate how similar parameters compare to each other under different conditions and determine if conversion factors exist allowing parameters from different standards to be directly compared. The results indicated that even for the standards with identical parameters but different mathematical algorithms, there is a significant difference (greater than 14%); therefore, the conversion among standards should be done very carefully if it is necessary. The JIS B6192 is equivalent to ISO 230-2:2006 [20] without the measurement uncertainty. The JIS B6336

TABLE 16.2

Differences between Positional Standards

STD Organization	NMTBA	ISO	VDI/DGQ	BSI	ASME	JIS
STD Number		230-2	3441	BS 4656	B5.54	B6336
Year Adopted	1972	2006	1977		2005	1986
Statistics Based	Yes	Yes	Yes	Yes	Yes	No
Unidirectional or Bidirectional	Uni/Bi	Bi	Bi	Bi	Bi	Uni
Terminology	Bidirectional accuracy	Accuracy	Positional uncertainty	Accuracy	Accuracy	Maximum positional deviation
Value Reporting	±	Absolute	±	Absolute	±	±
Relative Values	±125	100	±45	100	±45	±25

uses worse position along axis with no statistics, which produces best apparent accuracy and then reports result as ±. The NMTBA standard emphasizes unidirectional measurement of accuracy and derives bidirectional position accuracy by statistically combining the forward accuracy, the reverse accuracy, and worst amount of lost motion to arrive at a total value. VDI/DGQ has no accuracy term but positional uncertainty is close to ISO 230-2 position accuracy. ISO 230-2 position accuracy is larger than JIS, smaller than NMTBA and about the same as the positional uncertainty term in VDI/DGQ; it is reported as an absolute number, not as a ± value. Other international standards equivalent to ISO 230-2:1997 are the GB/T 17421.2 (2000) (China), JIS B6192 (1999) (Japan), BS ISO 230-2 (1999) (England), and DIN ISO 230-2 (2000) (Germany). The difference between the ISO 230-2:1997 and the ISO 230-2:2006 is the measurement uncertainty statement that is added to the 2006 version.

Generally, every machine tool builder lists, as part of a machine's specification, accuracy, and repeatability figures. However, the method or standard utilized to measure these figures is not provided, which makes very difficult for users to compare machine tools or decide the part quality expectations. The most important steps to take for obtaining CNC machine position accuracy are as follows:

1. Request positional accuracy from the manufacturer for each CNC axis. This involves a request for values according to ISO 230-2:2006. If VDI/DGQ Standard is used, request for positional uncertainty values. The values should be measured bidirectionally and should be reported as the total absolute amount (not as ± value).
2. Review product brochures or technical specification literature for reported position accuracy and note the standard used.
3. Review runoff documentation and note values for position accuracy that were demonstrated and the procedure followed for the evaluation.
4. Review current or recent past part machining capability information to estimate the positional accuracy based on six-sigma calculation of feature dimensional capability.

16.4 MACHINE TOOL PERFORMANCE EVALUATION

There are several reasons for performing machine tool characterization tests before purchasing a machine tool for specific application(s). Evaluating the machine on the quality of a specific part is not always adequate if the machine must handle a variety of parts. A flexible machine must be capable of interchanging parts and machining processes while maintaining part quality uniformly across the part variety. Volumetric accuracy is critical to minimize changeover time, especially if the machine has unique programs with offsets to compensate for accuracy errors, since that information will be lost when new part program is used. If the machine is accurate over the entire working volume, and the volumetric accuracy is maintained, the machining center will retain its performance for a variety of applications. It is important to verify that the machine's dynamic volumetric accuracy within the full envelope is 70%–80% tighter than the part tolerance. Brochure data and builder specifications are often unreliable, and information such as volumetric performance is unavailable. Machine tool characterization efforts have improved the last decade and allow for the fair comparison and rating of machine tools with common tests and terminology. They identify specific areas of concern through testing to avoid potential manufacturing difficulties from the outset. Machine tool error data can be merged with other variation data to allow error budgeting or process capability prediction. They improve the control and optimization of machining systems including the process, components, tools, and fixtures.

The machine tool and its controls should be evaluated at the manufacturer's site to demonstrate conformance to specifications (accuracy and repeatability) provided for the order. Testing the equipment that utilizes a specified procedure is very important especially if the machine is expected to produce certain high-quality parts. Attention should be provided to the test environment (floor and thermal) because it could be very critical to machine performance. The specified test using the same measuring instruments should also be repeated in the customer's premises after machine installation because shipping and site installation can have a detrimental effect on accuracy and repeatability. On-site acceptance tests are very important. This performance check is the best approach to evaluate the machine (1) to manufactured specifications and (2) to check it periodically for maintenance. There are several elements of machine tool performance as shown in Figure 16.11.

A tremendous effort has been dedicated by many researchers internationally in the machine tool and manufacturing industries to develop a robust and cost-effective method to evaluate the accuracy of a machine tool. This has been even a larger challenge for evaluating the accuracy of large machine tools with several meters volumetric space. Suggested acceptance test procedures are provided in the following even though the accuracy assessment of a machine tool often requires complex sets of measurements. Unfortunately, there is no reliable short-cut method (or single parameter) that can be used to assess the volumetric accuracy of machine tools. Even though most of the tests are performed without the spindle running, under unloaded conditions, and after the

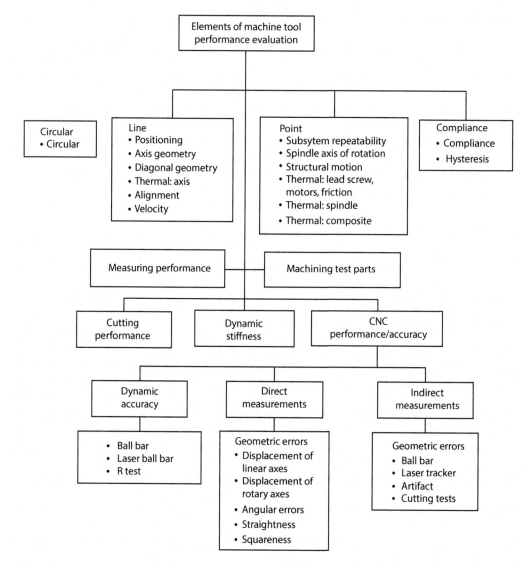

FIGURE 16.11 Consideration for the various element for machine tool performance.

machine has reached thermal equilibrium at a control room temperature, a few tests evaluating the spindle performance and under cutting loads could be very important. The specific details of testing are found in the various standards [7,20,21]. In addition, some of the quick or abbreviated tests can be used as interim checks (between calibration intervals) to track performance decline over time information. The tests include measuring

1. *Static Positional Accuracy* of the 21 rigid body errors (the systematic geometry errors) in any three-axis machine tool (six errors per axis plus three squareness errors). ISO 230-2 [20] provides standard procedure for the testing of machine tool axes. A linear displacement error is the positioning error in the same direction as the axis direction. Figure 16.9 illustrates the position error for the X-axis of a machine tool as measured at certain intervals. Laser interferometer systems with nanometer-level resolution are used for comprehensive accuracy assessment of machine tool axes because they have high and consistent accuracy. They are very accurate, stable, and traceable (based on standard wavelength of light) compared to other systems using electronic targets to measure pitch, yaw, and straightness errors. Comprehensive software packages are available to report the errors according to the international standards. Laser interferometers are available in 1, 3, 5, and 6 degree-of-freedom configurations. Some laser systems include modules for linear, angular, rotary axis, straightness, squareness, and flatness measurements, as well as dynamic measurement (velocity, vibration) capability. The 1D is the simplest laser interferometer for measurements of linear and angular displacement and flatness in only one machine tool axis. A 6D system provides simultaneous measurement of all axis errors (one linear axis error, horizontal and vertical straightness, angular errors (pitch and yaw), and roll error from a single setup.) The 5D is similar to 6D without the measurement capability of roll error. This approach requires precise mounting of the laser and reflector, so reflected light hits the detector over the entire length of stage travel in that axis as illustrated in Figure 16.12. The setup for measuring straightness is shown in Figure 16.12. Generally, different measurement setups are required for most errors to be identified. Glass scales are also used for positional errors. Crossed-grid encoders are precision devices for 2D position measurements. The measuring range is limited by the size of the grid encoder (commonly less than 250 mm diameter). Angular errors are measured using electronic levels or angular interferometers. The roll of horizontal axes is measured by levels. The

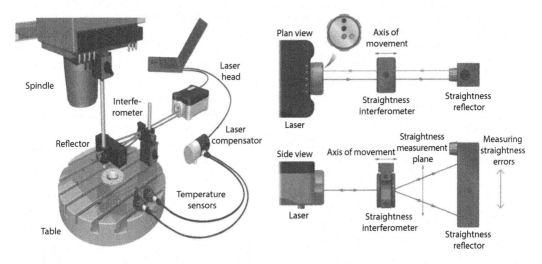

FIGURE 16.12 Illustration of a typical setup for straightness measurement on a moving bed using a laser interferometer. (Courtesy of Renishaw, Hoffman Estates, IL.)

roll of vertical axes is not measured directly. Straightness errors are measured by straight edges or interferometers. Autocollimator or rotary encoders are used to measure the position and squareness of rotary axes [6,22] but detecting all the errors of a rotary axis is a difficult problem. Rotational errors of rotary axes are measured with either laser or talyvel system. Swivel check devices are available for evaluating the performance of swivel axes and tilt tables. Squareness is measured using various methods such as reference squares and dial indicator, telescoping ball bar test, laser alignment with an optical square, and laser interferometer straightness measurement with an optical square (pentaprism). Talyvel electronic level systems are used for a wide variety of measurements such as straightness, flatness, or absolute level (checking columns for squareness to guideways, guideways for straightness and twist, surface flatness of table(s) because they are very accurate, stable, and repeatable). The static positional accuracy can be evaluated by measuring either the individual errors or all the points together based on the 3D volumetric positioning errors as will be discussed in the third subsection. The telescoping ball bar system characterizes circular motion of two or more axes as will be discussed in the next subsection.

One of the major concerns when measuring the machine tool accuracy is the time frame required for the measurement. Thermal expansion/contraction complicates measurements because the machine heats and cools during stops for making the measurements. Therefore, the shorter the time frame for the measurements, less thermal changes occur and the lower the temperature drift in the machine tool structure resulting in lower variation in the measurements. In addition, laser wavelength compensation for environmental conditions is necessary; air temperature, pressure, and humidity sensors are used for wavelength compensation.

A 6D laser measurement system provides simultaneous measurement of all axes errors from a single setup; it measures all six errors for each linear axis in a single measurement scan. A three-axis machine requires only six measurement runs to complete the data set instead of 36 runs (forward and reverse data for 18 error parameters). However, additional setup time and equipment are required to measure the roll in the vertical axis. This laser could reduce the measurement to about 8 h instead of several days for smaller machines.

2. *Dynamic Accuracy* depends on quantities such as backlash (by measuring the position accuracy from opposite directions), spindle compliance (radial compliance in X- and Y-directions, and axial compliance in the Z-direction), spindle runout (radial and axial motion), rotary table angular and tilt compliance (measure compliance and hysteresis of the table), pallet changer load, load-carrying capacity, drift (effect of thermal growth), and control system functionality (continuous path contouring using various interpolation methods such as linear, circular, etc., cutter compensation, and finally control resolution). For precision and super-finishing operations, static positioning accuracy is not sufficient because contouring features require the volumetric positioning accuracy and the dynamic accuracy at higher acceleration and deceleration rates using advanced servo algorithms. Laser interferometer systems have the capability to provide some dynamic analysis with respect to displacement, velocity, and acceleration data to quantify certain machine error characteristics such as (1) positional stability and encoder performance, (2) preload and hysteresis of the ball screw and nut mechanisms, (3) resonance characterization of spindles and drive motors, (4) control loop optimization, and (5) feedrate and interpolation accuracies and stability [23]. The circular test is widely accepted as an indirect evaluation of the geometric accuracy or contouring performance of two orthogonal linear axes as described in ASME B5.54, ISO 230-4, and ASME B89.3.4 [7,24–26].

The *telescoping magnetic ball bar (TBB)* device (see Figure 16.13) has been popular because it accurately measures the actual circular path (affected not only by the scale, straightness, and squareness errors but by the backlash, reversal spikes, servo mismatch, and cyclic errors due to the coordination needed to produce smooth interpolation motion)

FIGURE 16.13   Illustration of ball bar test in a machine tool. (Courtesy of Renishaw, Hoffman Estates, IL.)

and compares it with the programmed path. The TBB device (also called "double ball bar") was developed by Jim Bryan, Lawrence Livermore Laboratories, and was adopted in the ASME B5.54, ASME B5.57, and ISO 230-2 standards. It is essentially a displacement transducer, held between two precision spheres as illustrated in Figure 16.13 [23,24,26]; one sphere is mounted on the magnetic cup fixed to the spindle, while the other is mounted on the magnetic cup fixed to the table; the relative displacement between the two spheres is continuously measured by a transducer built into the TBB to determine the actual bar length around the travel path. The transducer is able to follow radially around the center of the circular path while measuring the radial error and any irregularities in this motion along angular positions on the circular path. The TBB method can only perform circular tests due to its nature and the stroke of the transducer is small (about 3 mm). A circular path part program is used to drive two axes of the machine tool in a continuous circle around the precision balls located on the machine table. It is often used to measure simultaneous movements of at least two linear and one rotary axis for evaluating circular interpolation accuracy. A software package is used to compute geometric errors and detect inaccuracies induced by the machine's controller and servo drive systems. The presence of any errors will distort the circular path from a perfect circle. While the TBB can generate a full circle on a plane normal to its post (i.e., X–Y plane), it can also generate a 220° arcs on perpendicular planes (X–Z and Y–Z in this case); this allows the evaluation of the circular test in three planes in a single setup. TBBs are a practical, convenient, and comprehensive diagnostic tool for assessing the contouring characteristics (on top of accuracy) for machine tools including five-axis machines. It provides an overall assessment of machine performance (e.g., circular deviation), and it can calculate control loop errors (servo mismatch, reversal spikes, backlash) and guideway errors (squareness, straightness, stick-slip, planar bidirectional positional accuracy) within the area enclosed by the ball bar test. The user has a choice of several report formats according to ISO and ASME standards and the diagnostics features of at least 19 different errors (although these are all not identical to 21 errors of a three-axis machine tool) according to the literature [25,26]. Figure 16.14 provides diagnostics report of a machine tool positional performance, and the circular deviations indicate the contribution factors. This is an indirect method of measuring machine tool error parameters. It is recognized around the world as the chosen method for measuring a CNC machine tool's circularity performance, and it can be used as a fast diagnostic check for machines' linear axis performance. It has been also used to identify kinematic errors on five-axis machines [27]; it reveals the maximum error within the measured volume but not where it occurs. Several TBB measurements may be required to estimate the squareness for the full travel range. The squareness error is identified by an elliptical rather than circular shaped trace as shown in Figure 16.14; the direction of the circular interpolation should not affect the orientation of the trace. However, straightness errors of an axis will result in elliptical trace, and it is difficult to separate different errors without multiple measurements with the TBB. In addition,

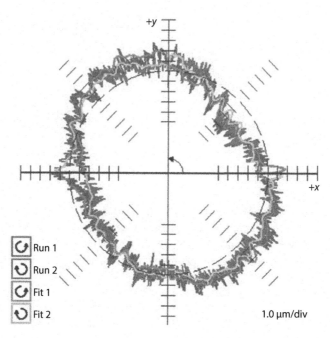

FIGURE 16.14 Schematic of ball bar diagnostics report of a machine tool positional performance. (Courtesy of Renishaw, Hoffman Estates, IL.)

the squareness error calculated by the commercial TBB software can be more accurately estimated by incorporating the straightness error data for the full travel range [28].

The TBB circular tests have been used for evaluating all possible combinations of linear and rotary axes driven simultaneously of various types of five-axis machine tools, and total TBB dynamic tests have been proposed to identify dynamic differences between linear and rotary axes. More importantly, the velocity gains of the position control loops of all servo-controlled linear and rotary axes can be tuned synchronously to eliminate gain mismatch errors [29,30]. The ISO/DIS 10791-6 [31] describes the measurement of the geometric errors with TBB for various types of five-axis machine tools including a tilting rotary table. The TBB can simply generate a number of measurement directions in the workspace. For example, measurement in three directions can be performed independently, and the radii of circular interpolations for each measurement are different [32,33]. The TBB has been used to assess the motion errors of rotary axis in a five-axis machine or a machine with a trunnion type [34,35]. TBB tests were conducted under simultaneous three-axis motion in four patterns in order to estimate eight position independent geometric errors associated with rotary axes [35–37]. It is better to have the two rotary axes move simultaneously during the measuring test, while the other three linear axes are kept stationary in order to avoid other error sources influencing the result. The motion equations of the two rotary axes moving simultaneously while the translational axes are stationary are important to realize a circular trajectory and estimate the eight geometric errors. Even though various length TBBs are available, as yet no error map for the whole workspace can be measured.

A TBB designed with adjustable preload (called "loaded ball bar") was used as an ordinary TBB system to evaluate the machine tool accuracy characteristics with a load applied to the structure [38]. The preload in the TBB affects the static behavior of the system in the same way as the load of the cutting tool. This test device helps to evaluate the machine's tool elastic deflection in different directions on top of errors identified by an ordinary TBB. It is also possible to track different error patterns to the applied load.

TBBs with resolution of 0.1 μm and measurement accuracy of ±0.5–2 μm are available. However, the telescoping measurement range is limited to few millimeters while the lengths of the commercial bars range from 50 to 600 mm using extensions with increments of 50 mm [23,26]. The inability to perform circular tests with smaller radii could be a concern for certain applications. The evaluation of the ball bar circularity test in nine machines indicated a 10–30 μm error in the X–Y plane as illustrated in Figure 3.42.

The *laser ball bar* (LBB) method has been developed for measuring the distance between the spheres in the spindle and table by a laser interferometer [39–43]. Therefore, respective distances to the tool center point are determined by an interferometer. The span of the system is between 0.3 and 1 m [41,44]; it is designed on the same principle as the TBB with a laser at one end of the telescoping tube and a retro-reflector at the other end with precision spheres at both ends. It measures (1) X-, Y-coordinates to generate the circular part, which is considered a 2D measurement compared to the TBB measuring the radius (1D measurement) or (2) X-, Y-, Z-coordinates of arbitrary points in the workspace and the reach of the LBB. A laser tracker with a passive tracing mechanism that has been developed on the telescoping principle is acting as an LBB since it utilizes telescoping tubes with two spheres supported by magnet nests, one on the device on the machine table and the other on the spindle side (see Figure 16.15) [45–47].

The *telescoping laser tracker* (TLT) is a self-tracking laser interferometer with a resolution of 1 nm. The telescoping tube keeps the beam pointing at the retro-reflector and determines the distances, and its working range is 275 mm to 1 m. Its angular working range is −35° to 80° in the vertical direction and 360° around the horizontal plane depending on the location of the base on the table. The TLT measures the relative distance between the two spheres for each arbitrary location along the reach of the ball bar. The sequential multilateration method is used with this type of LT. The results can be processed as those of laser tracker (which will be discussed later) since the X-, Y-, Z-coordinates of the points are recorded. In small machines, the TLT is repositioned twice to measure the machine within its entire working space. A TLT is capable of measuring rotary table characteristics

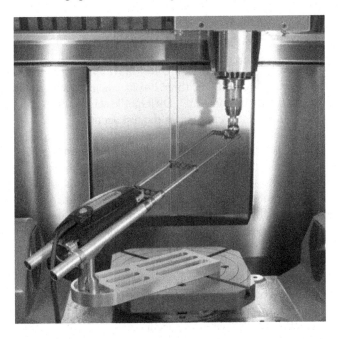

FIGURE 16.15 Illustration of the laser ball bar test on a rotary table of five-axis machine tool. (Courtesy of Etalon AG, Braunschweig, Germany.)

FIGURE 16.16 Illustration of a radial probe (trinity probe head) and radial test method of a rotary table of five-axis machine tool. (Courtesy of IBS Precision Engineering, Eindhoven, the Netherlands.)

using the same approach as the traditional TBB. Finally, ball bars are limited to small and middle range workspaces by their minimum length.

The *radial test* (R test) was developed by researches using a precision sphere with multiple individual probes simultaneously to measure values in all three directions especially for five-axis machine tools [6,48–50]. The system is designed in two different configurations: (1) the location of the master ball attached on the machine's table is determined accurately in 3D (*X*, *Y*, and *Z*) simultaneously by a probe (with three displacement sensors) on the spindle as illustrated in Figure 16.16 [51] and (2) the location of the sphere on the spindle tool is measured by the three displacement sensors spaced at evenly 120° apart in a fixture on the rotary table [52]. Different kinematic tests can be developed for specific machines, but the most common is the test according to ISO 10791-6 [53]. This type of test has been included in the most recent draft version of ISO 230-1 [6]. With this system, only one measurement is necessary in order to measure radial, tangential, and axial relative displacements with respect to the movement of the rotary axis [10,51,53]. In three- or four-axis machines, the circular trajectories are implemented using two linear axes synchronized with the rotation of a rotary axis at a specified speed. With respect to five-axis machines, the circular path verifies the interpolation motion of the three linear axes and the two rotary axes while controlled simultaneously at constant speed [54]. The measuring uncertainty of the R Test according to the literature is 1 μm. The R test (like the TBB) measures the actual location of the programmed rotary axis range of locations. Software determines the location and squareness errors and more importantly the dynamic characteristics of the rotary table(s) [24,51].

The *cross-grid scale method* can also be used to evaluate the static and dynamic characteristics of a machine tool's contouring accuracy (as illustrated in Figure 16.17). It can perform circular interpolation or any free-form tests in two axes with small radii less than 120 mm. This method is a contact-free measurement. It consists of one or two grid plates

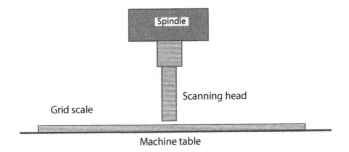

FIGURE 16.17 Schematic of contouring error measurement using a cross-grid scale.

(2D scale) with graduation and a scanning head [55]. The scanning head (located on the spindle) moves over the grid plate (aligned on the table) without making mechanical contact. The grid encoder can not only perform the circular contour test in the same way as a TBB but can also detect additional errors that are too subtle for the TBB test; for example, excessive lead or lag errors gain adjustments and interpolation and other control problems and are more easily separated from the geometric errors at small traced circles. In addition, it allows tracing a straight diagonal line across the grid plate to evaluate the linear interpolation or to test 90° turns at high feed rates [56,57]. Motion errors due to servo control system are often more easily identified on noncircular paths [58]. Grid plates have also been used to measure geometric errors [59].

Errors caused by thermal expansion and distortion affect the dynamic accuracy because they may affect the squareness and straightness. Therefore, the earlier mentioned tests can be used under transient thermal conditions (cold vs. hot) to evaluate the machine's accuracy and more importantly determine the warm-up period necessary before cutting parts.

3. *Volumetric Positional Accuracy* is defined by the RMS (root mean square) of the maximum errors of each axis. In the past, measurements of only the linear displacement errors, $\delta_x(x)$, $\delta_y(y)$, and $\delta_z(z)$, were acceptable. Improvements in ball screws and linear encoders during the last few decades have minimized linear errors. Therefore, the angular, squareness, and straightness errors have become major contributors to positioning errors. Volumetric accuracy can be defined as the maximum error between any two points within the swept volume of the machine tool. Volumetric accuracy essentially is a very significant term because it considers the "Abbe offset error" (as discussed in Section 16.2) since this method effectively measures where the tool/spindle is located away from the slide's scale or encoder. It provides a comprehensive understanding of the condition of the machine tool's accuracy and repeatability with respect to volumetric positioning. Traditional techniques for generating an error map handle one axis at a time, using devices such as laser interferometers or artifacts. The selection of the target positions for measurement is discussed in ISO 230-2.

The 3D volumetric positioning accuracy errors represent vector errors with three components determined by the linear displacement errors and the geometrical errors of the machine (including straightness, squareness, pitch, yaw, and roll). In this case, the volumetric error is defined as the maximum of the RMS using the total errors in each axis direction. As explained in Ref. [60], the error in each axis direction is defined either by a polynomial function or by individual points to describe the error model as

$$\delta_x(x,y,z) = \delta_x(x) + \delta_x(y) + \delta_x(z)$$

$$\delta_y(x,y,z) = \delta_y(x) + \delta_y(y) + \delta_y(z) + y\alpha_{xy} \tag{16.6}$$

$$\delta_z(x,y,z) = \delta_z(x) + \delta_z(y) + \delta_z(z) + z\alpha_{yz} + z\alpha_{xz}$$

The maximum error in each axis is

$$\delta_{xm}(x,y,z) = \text{Max}\{\delta_x(x,y,z)\} - \text{Min}\{\delta_x(x,y,z)\}$$

$$\delta_{ym}(x,y,z) = \text{Max}\{\delta_y(x,y,z)\} - \text{Min}\{\delta_y(x,y,z)\} \qquad (16.7)$$

$$\delta_{zm}(x,y,z) = \text{Max}\{\delta_z(x,y,z)\} - \text{Min}\{\delta_z(x,y,z)\}$$

The volumetric error is the RMS of the earlier maximum errors along each axis:

$$E_v(x,y,z) = \sqrt{(\delta_{xm}(x,y,z))^2 + (\delta_{ym}(x,y,z))^2 + (\delta_{zm}(x,y,z))^2} \qquad (16.8)$$

Angular errors are not included in this equation but can be incorpora   ted   knowing the machine configuration because they affect the Abbe offsets [60,61]. The errors in the equation are measured either directly using a laser interferometer or indirectly. The traditional method of machine tool accuracy is the direct method of measuring all the geometric error parameters in Equation 16.6. Each axis of travel is measured with a laser interferometer, one at a time. This technique may require multiple setups of the laser to complete all the different measurements and is very time-consuming; hence, it is also subject to thermal drift.

ISO 230-1:2012 [6] defines volumetric accuracy as the "maximum range of relative deviations between actual and ideal position in X-, Y-, and Z-axis directions and maximum range of orientation deviations for A-, B-, and C-axis directions for X-, Y-, and Z-axis motions in the volume concerned." One of the indirect measurements has been a 3D grid within the work zone to generate error mapping. It is based on the multiaxis linear and rotational movements on the machine points distributed in the workspace. These measurements include the Abbe offsets. Measuring the individual errors for squareness and straightness is complex and time-consuming as opposed to the body diagonal displacement errors to define the volumetric positioning error. In this case, the volumetric accuracy test could be carried out according to either the ISO 230-6 [21] or ASME [7] standards. Using *four body diagonals* (as illustrated in Figure 16.18) and possibly the face diagonals of a rectangular prism defined by the workspace of the machine tool, the volumetric accuracy can be estimated faster than by traditional methods. This test measures deviations from a perfect geometrical structure of the diagonal lengths and is sensitive to most of the geometric error components. Therefore, this test assesses the geometry of the machine tool structure to supplement the linear displacement accuracy test of the individual axes. An error map of the tool point along the four diagonals at the selected interval positions is generated throughout the volume of the machine, including simultaneous motions of the axes, linear, and rotational. Hence, it is an effective method to represent the volumetric error especially for machines with more than three axes. The body diagonal measurement is similar to the laser linear displacement measurement of an axis; the laser beam points along each of the body diagonal directions (PPP, NPP, NNP, and PNP) by moving x-, y-, and z-axis together as described in ISO 230-6 and ASME B5.54. There are several laser interferometer systems for this measurement, each with its own software [62–65]. The laser is positioned at the O edge of the diagonal on the machine's table and the mirror at the other end of the diagonal on the spindle and are aligned perfectly (see Figure 16.18). The X-, Y-, and Z-axes are moved simultaneously from the zero position to each increment along the three axes in order to move from one target point to the next along the diagonals (i.e., diagonal O-PPP in Figure 16.18). When the axes move, the laser measures

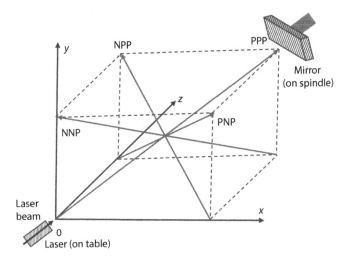

FIGURE 16.18 Schematic of the four body diagonals method to evaluate the volumetric accuracy of the workspace.

the combined effects of the three axes. Each diagonal is measured forward and reverse (bidirectional) so that four positive and four negative body diagonal measurements are available for analysis. It is a fast method to evaluate the accuracy of a machine tool. However, the data collected are not sufficient to identify the individual geometric error parameters of the axes and the source of the errors. The body diagonal method identifies the accuracy of a machine tool, but it does not help the user to pinpoint the cause of the problem(s) in order to correct it. For example, the performance of a three-axis machine tool was evaluated before and after calibration using the body diagonal method; the measurement of four diagonals with a laser interferometer before and after compensation is shown in Figure 16.19. The diagonal error was as high as 0.05 mm and was reduced to about 0.005 mm after calibration.

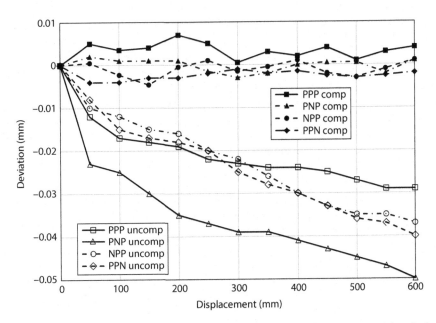

FIGURE 16.19 Measurement of four diagonals with laser interferometer before (uncomp) and after (comp) calibration of a machine tool.

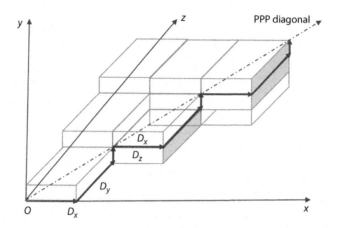

FIGURE 16.20 Schematic of the step diagonal displacement measurement method.

The *sequential step diagonal vector* method was developed to relate the measurements to the positioning errors [61,66–70]. This method moves one axis at the time and stops to collect data at each individual X, Y, and Z moves to reach each increment (a zig-zag path) along the diagonals as shown in Figure 16.20. The spindle moves X only, stops and collects data, then moves Y, stops to collect data, and finally moves Z only and stops to collect data; this process is repeated until all the incremental points along each diagonal are measured. The measured displacement errors are sensitive to the errors that are both parallel and perpendicular to the direction of the linear axis. This method generates three times more data than the body diagonal method. In this case, the laser Doppler displacement meter (LDDM) laser interferometer is used for the measurements [62]. The sequential step diagonal method acquires 12 sets of data instead of four sets with the body diagonal method. The 12 sets of data are sufficient to estimate the errors in Equation 16.6, the three displacement (linear), six straightness, and three squareness errors.

The most critical issues with the body diagonal and step diagonal tests are as follows: (1) the alignment error of the laser to mirror directions, (2) the machine's angular errors must be very small since they are neglected in the diagonal measurements, and (3) if there is a large linear error in one axis, the diagonal methods will not correctly identify the error since it is distributed into the remaining axes of the machine tool [71,72]. In addition, the recent development indicated that even if the step diagonal or body diagonal test results show improvements, the machine's accuracy could be poor because the length change of the diagonals by one source of error in the machine can be cancelled out by another error parameter since the diagonals are determined by the square root of the squares of the three-axis motions [73].

A new formulation of the step diagonal method has been proposed by first measuring independently the linear positioning errors and then applying the step diagonal method to identify the contribution of the straightness and squareness error, to eliminate the major concern of the alignment error of the laser to mirror directions [74]. The new formulation of laser step diagonal measurement was further developed to accurately identify 3D volumetric errors even under the existence of setup errors [67]. Therefore, since the traditional body diagonal or step diagonal methods cannot reliably determine the machine tool accuracy (especially the linear positioning errors), the available software in the market for the step diagonal method allows the user to acquire additional conventional linear laser measurements along two of the machine's X-, Y-, and Z-axis. Therefore, the current step diagonal test could take longer than the originally developed one-day quick check for evaluating the accuracy of a machine tool.

The body diagonal errors for each diagonal and direction are defined from Figure 16.19 as $\delta(ppp)$, $\delta(nnn)$, $\delta(npp)$, $\delta(pnn)$, $\delta(pnp)$, $\delta(npn)$, $\delta(ppn)$, and $\delta(nnp)$. The three letters ppp and nnn indicate the body diagonal direction with respect to X-, Y-, and Z-axes all positive or negative, respectively. The pnp indicates positive, negative, and positive with respect to X-, Y-, and Z-axes. The error for each diagonal is defined as

$$E\left(\frac{ppp}{nnn}\right) = \text{Max}\left[\delta(ppp), \delta(nnn)\right] - \text{Min}\left[\delta(ppp), \delta(nnn)\right]$$

$$E\left(\frac{npp}{pnn}\right) = \text{Max}\left[\delta(npp), \delta(pnn)\right] - \text{Min}\left[\delta(npp), \delta(pnn)\right]$$

$$E\left(\frac{pnp}{npn}\right) = \text{Max}\left[\delta(pnp), \delta(npn)\right] - \text{Min}\left[\delta(pnp), \delta(npn)\right] \quad (16.9)$$

$$E\left(\frac{ppn}{nnp}\right) = \text{Max}\left[\delta(ppn), \delta(nnp)\right] - \text{Min}\left[\delta(ppn), \delta(nnp)\right]$$

and the volumetric error as defined in ISO 320-6 and ASME B5.52 is

$$E_v = \text{Max}\left[E\left(\frac{ppp}{nnn}\right), E\left(\frac{npp}{pnn}\right), E\left(\frac{pnp}{npn}\right), E\left(\frac{ppn}{nnp}\right)\right] \quad (16.10)$$

or as

$$E_{sv} = \text{Max}\left[\delta(ppp), \delta(nnn), \delta(npp), \delta(pnn), \delta(pnp), \delta(npn), \delta(ppn), \delta(nnp)\right]$$
$$- \text{Min}\left[\delta(ppp), \delta(nnn), \delta(npp), \delta(pnn), \delta(pnp), \delta(npn), \delta(ppn), \delta(nnp)\right] \quad (16.11)$$

which includes the squareness error, which Equation 16.9 does not. The volumetric error for 10 different CNC machines was evaluated in Ref. [61] using their corresponding positioning errors and the three definitions of volumetric error in Equations 16.8, 16.10, and 16.11. It was concluded that the laser body diagonal measurement according to Equation 16.10 is a quick check for the E_v. In addition, the E_{sv} is a more accurate representation of volumetric accuracy than using E_v in the laser body diagonal method. Table 16.3 indicates the

TABLE 16.3

Comparison of Volumetric Error among 10 CNC Machine Tools

	Machine Tool Number—Error (μm)									
	1	2	3	4	5	6	7	8	9	10
$E_v(x, y, z)$—Equation 16.8	43	59	49	63	77	62	44	63	78	64
E_v—Equation 16.10	16	33	34	38	45	32	16	42	33	27
E_{sv}—Equation 16.11	30	33	33	33	46	34	27	45	54	44
$E_v(x,y,z)/E_v$	2.7	1.8	1.4	1.64	1.7	2	2.8	1.5	2.4	2.4
$E_v(x,y,z)/E_{sv}$	1.4	1.8	1.5	1.9	1.7	1.8	1.6	1.5	1.5	1.5

Source: Svoboda, O. et al., Machine Tool 3D Volumetric Positioning Error Measurement under Various Thermal Conditions, *ISPMM 2006 the Third International Conference on Precision Mechanical Measurements*, 2006.

volumetric errors using the individual positional errors of the axes by Equation 16.8 and the body diagonal measurements by Equations 16.10 and 16.11. The average conventional volumetric error for the 10 machines is 100% and 60% higher than the corresponding body diagonal volumetric error without and with the squareness errors, respectively. This is an indication that the body diagonal volumetric error measurement (requiring lower measurement time) will be less than 50% of the conventional volumetric error as has been reported in the literature.

The *telescoping ball bar test* could be used at the three orthogonal planes to generate sphericity results from the three individual circularity results in a single setup even though it is not supported in the standards [23,75,76]. In addition, using special ball mounts (for center pivot and tool cup extension), the ball bar is able to rotate in planes through the center pivot axis at maximum of about 220° arc; this allows TBB measurements in three or more orthogonal planes in a single setup to analyze the volumetric error. In addition, if the TBB is driven by the spindle around the pivot at three different levels, that is, the angle between the ball bar and the pole is 0°, 45°, and −20°, the bar length is measured at eight positions at each level as shown in Figure 16.21. The range of the ball bar length at different positions will be the volumetric error for the machine tool in the corresponding space:

$$E_{Bv} = \text{Max}\left[\delta_0(i), \delta_{45}(i), \delta_{-20}(i)\right] - \text{Min}\left[\delta_0(i), \delta_{45}(i), \delta_{-20}(i)\right], \quad i = 1, 2, \ldots, 8 \quad (16.12)$$

Researches have used the TBB and cross-grid encoder to evaluate the geometric error components of multiaxis machine tools. TBB can provide volumetric information as illustrated in Figure 16.22.

The *telescoping laser tracker (TLT)* has a large stroke and is designed to measure the volumetric error through mapping using the multilateration principle (which is discussed further in the following). For example, it requires two setups in opposite quadrants on the table for a four-axis machine tool with a rotary axis since two more can be obtained by rotating the table 180°. An LBB located on a gimbal to provide the vertical and horizontal rotation with the corresponding encoders represents a 3D LBB because it measures the position of the bar in the spherical coordinate system. The minimum length of the telescoping tube is the main limitation of the LBB, TLT, and 3D LBB devices because it determines the dead space of the workspace.

More recently, *a laser tracker (LT)* has been proposed as a measuring system for volumetric accuracy by tracking a retro-reflector to detect its location in a spherical coordinate

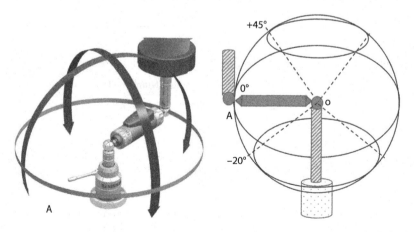

FIGURE 16.21 Schematic of the ball bar measurement in three orthogonal planes in a single setup. (Courtesy of Renishaw, Hoffman Estates, IL.)

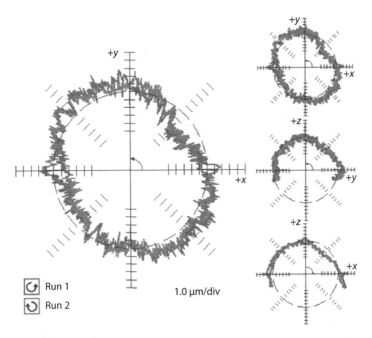

FIGURE 16.22 Schematic of ball bar volumetric diagnostics report of a machine tool performance. (Courtesy of Renishaw, Hoffman Estates, IL.)

FIGURE 16.23 Illustration of the laser tracker: (a) on a rotary table of five-axis machine tool, (b) on a large machine tool. (Courtesy of Etalon AG, Braunschweig, Germany.)

system (see Figure 16.23). It is mainly intended for large machine tools due to the difficulty in using the other discussed methods. The tracker includes a laser, either interferometer (IFM) or absolute distance meter (ADM) system, a tracking mirror mechanism, a position detector with a motor control unit, and a spherical mounted retro-reflector (SMR). An IFM measures a relative distance (since it cannot determine an absolute position in 3D space without having a known starting point first) with extreme precision, while an ADM measures absolute distance but cannot achieve the speed of an IFM for dynamic measurements. The ADM was incorporated into LTs to avoid resetting the system every time the beam is interrupted.

The tracker is mounted either inside the machine on the table or outside the table on a tripod and follows the retro-reflector target positioned on the spindle as a cutting tool. The tracking laser interferometers follow a retro-reflector in a spatial grid to generate error map. The laser IFM automatically tracks the retro-reflector and records the spatial displacement and the direction of the laser beam at each grid position throughout the workspace [6,76–80].

Conventional laser trackers use a mirror mounted on a two-axis gimbal mechanism to steer the beam to the target. The displacement measuring beam remains at the center of the target reflector by the tracking mirror mechanism to provide the radial coordinate. Two angular encoders provide the horizontal and vertical angles of a spherical coordinate system. These data are then combined with a distance from the laser to calculate X-, Y-, Z-coordinates.

This indirect method was originally developed to measure the entire working envelope for large machine tools (workspace greater than 1–2 m shown in Figure 16.23b) with five or six axis as more rapidly than other methods. LTs reduce the measurement time significantly (as reported to about 1 day) depending on the workspace and the density of the grid (as illustrated by the target points in Figure 16.24). The manual process is still time-consuming, error prone, and tedious. However, triggering techniques are available with active target to reduce the measurement time and improve accuracy. Many measurements are repeated (30–100 times) for each of the measuring points in the map grid and averaged. An active target is a motorized SMR steering the mirror the internal reflector in multiple directions to avoid breaking the beam. Usually, the tracker monitors the beam at the center of SMR by splitting off part of the returning beam to a position sensing detector (PSD). The active target is positioned to a grid of either 200–300 statistically randomized multiaxis points, or 500–1000 points in a 3D grid (of a chosen dot pitch of 50 mm in Figure 16.24) within the workspace; Figure 16.24 shows the systematic measurement of vector error components at a random number of points in the volume. Positioning error maps are generated from the 3D volumetric positioning errors.

Even though ADMs have been advancing rapidly, their accuracy cannot match interferometers. Combining the absolute measurement from ADM with the interferometer, the absolute interferometer (AIFM) was developed [81]; in this case, the concern of the uncertainty due to the movement during the integration time of the ADM system is eliminated. Hence, using the AIFM, the relative motion of the retro-reflector is monitored by the IFM, and the longer integration time by the ADM will not affect the uncertainty of the AIFM system that will be within the uncertainty of the ADM (typically 5–10 μm).

The major tracker limitations are the characteristics of the gimbal rotation axes. Axis misalignment and manufacturing errors in mirror mechanism introduce significant measurement uncertainties. The angular measurement uncertainty directly contributes to

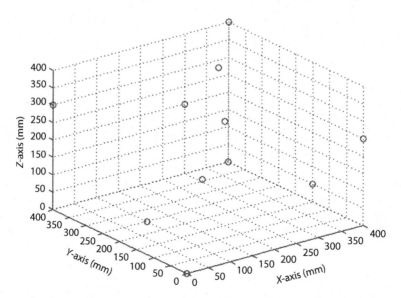

FIGURE 16.24 Illustration of measurement error at a random number of points in the 3D grid volume.

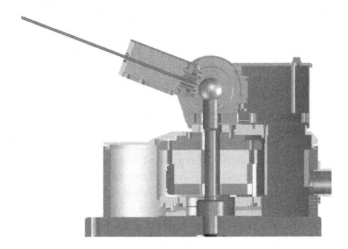

FIGURE 16.25 Illustration of the laser tracker in Figure 16.22 using an interferometer moving on a gimbal. (Courtesy of Etalon AG, Braunschweig, Germany.)

the measuring uncertainty of the target coordinates. Also, the measurement uncertainty increases linearly with the beam length as in any laser interferometer. Therefore, the uncertainty of the aforementioned LT types is significantly above 5 μm with a maximum permissible error (MPE) of 10 μm. LT with the interferometer moving on a gimbal mounted around a high-quality stationary sphere (LTS), as illustrated in Figure 16.25 [82,83], eliminates the inaccuracy of rotation axes in two-axis gimbal; this type of LT measures directly in reference to the stationary sphere. Therefore, the radial measurement uncertainty decreases over the former discussed common LTs since the LTS uses only the interferometer uncertainty [69–70]. Generally, LTs could collect 1000–2000 points/s while the movement of the retro-reflector is tracked.

Some LTs measure the distance directly in spherical coordinates, while other LTs (such as LTS, TLT, or even conventional systems) use the interferometric measurement (only the distance) together with multilateration method [79,84–87]. This is based on the GPS principle where the space coordinates for each measured point is determined by different base stations (see Figure 16.26). The multilateration and sequential multilateration methods are

FIGURE 16.26 Illustration of the multilateration method using four laser trackers to define each point. (Courtesy of Etalon AG, Braunschweig, Germany.)

used to enhance the accuracy of 3D coordinate measurements to minimize the uncertainty error. The multilateration approach acquires the same data from at least four LTs. Hence, the sequential multilateration principle is more cost-effective because a single LT is used in multiple locations to measure the distances of the spatial points (common targets in the grid in the workspace) to estimate the 3D positions and the corresponding errors. The LTS system uses multilateration, and it is reported that the measuring time for a small- to middle-sized machine tools is reduced to 6 h for the first time to generate the setup and then to less than 4 h using commercial software [82]. The sequential multilateration method for machine tool error mapping has advantages (as explained in the literature) because it reduces the measuring uncertainty compared to LTs measuring the spherical coordinates of the target points. Several researchers showed that the uncertainty also depends on the LT positions relative to target points [79–84]; a wider distribution of LT positions around the table will improve the estimation of the accuracy of a machine tool.

The manufacturers of the laser trackers provide software to estimate the 21 error parameters for a three-axis machine for the specific control system in the machine tool. Figure 16.27 provides the estimation of 6 out of the 21 error parameters at several locations along the X- and Y-axis using an LT. The laser tracker can be used to generate a grid pattern error table for three-, four-, five-, or six-axis machine tools with a span of up to 50 m [82–87]. The angular error of rotary axis can be measured using long and short extensions for the target so that the created vector is compared with the designed vector. Volumetric accuracy usually results in 3–5 times greater values than the positioning accuracy of a single axis as illustrated in Figure 3.42 for several machine tools. Hence, volumetric accuracy is often more significant than the repeatability of individual axes because it incorporates most of the errors including the Abbe offset. The error map provides the machine error at the cutting tool point based on the available measurements of nearby points in the workspace grid.

The angular encoders in the LT and the beam steering mechanism could lead to errors in the measured coordinates due to misalignment and wear. Therefore, the LT manufacturers provide correction factors through calibration. The calibration of LTs [76,88–93] is very critical for estimating the error model parameters to compensate the laser tracker and evaluate coordinate uncertainty. LTs are subject to alignment errors due to manufacturing tolerances and mechanical design. Their performance is affected

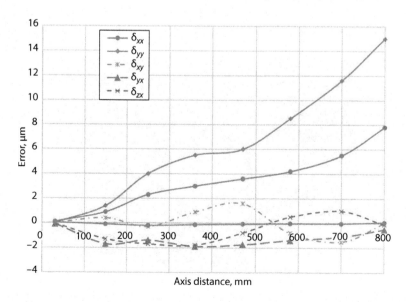

FIGURE 16.27 Linear positioning errors along the X- and Y-axis measured in a five-axis machine tool.

by several sources including (1) static/quasi-static sources that include geometric (e.g., the wavelength of the laser radiation, the internal alignment of the gimbal axes and the linearity and alignment of the internal angular measuring encoders) and nongeometric (e.g., compensation for atmospheric effects and thermally induced distortions such as thermal expansion of the instrument and its mount), (2) dynamic sources (e.g., system vibrations), (3) measuring strategy and sequence, and (4) the mathematical algorithm used to define the parameters. The most important error sources are the internal mechanical and optical alignments and the quality and alignment of the angular encoders. Specified tests are designed to evaluate the capabilities of these instruments by (1) measuring a known length (reference length) in various positions and orientations that are sensitive to error sources of the laser tracker and (2) ranging tests to evaluate the IFM or ADM capability [76,88–91]. The standard B89.4.19 does not test the algorithms, temperature compensation, and the retro-reflector. NIST and NPL assessed the magnitude of 200 mm radial error motions of several LTs including LTS [94]; their results indicated that the maximum permissible error (MPE) for two trackers was ±0.46 μm and ±4.16 μm, while the two-sigma specification for a third tracker was ±10 μm. The MPE for the LTS was ±0.25 μm. The observed errors of the SMRs in a 200 mm radius from all the systems (the three trackers and the LTS) were substantial portion of the total uncertainty specification provided by the corresponding manufacturers, allowing small room for other possible sources of error not considered by the previous test [94].

Finally, the speed and accuracy of LTs distinguish them from other methods especially as applied to the performance evaluation of large machine tools. LTs are very versatile measuring systems with software that automatically calculates and proposes the positions for the LT and the retro-reflector. In addition, the individual machine tool errors can be estimated based on the volumetric grid errors. The manufacturer of the LT supports the necessary software to generate the point cloud with collision-free measurement plan based on the CAD drawing and the kinematics of the machine tool. In addition, manufacturers of LTs with multilateration method provide software (using Monte Carlo simulation) to identify the best tracker locations and the uncertainty expected for a specific workspace.

4. *Measurement of a Master Part or Spatial Reference System* [95–97] with known dimensions can account the geometric errors. On-machine measurements (OMMs) are used with a contact trigger probe on the spindle to detect contact between the stylus and an artifact to read the machine axes positions. This method can be considered a hybrid method because the part/artifact is measured off-machine for calibration and on-machine to estimate the volumetric accuracy or geometric error parameters. The ease of repositioning of the artifact on a fixture/pallet is very important and is often ensured using kinematic coupling. OMMs can be integrated with mathematical kinematic error models to relate artifact errors to machine tool errors. There are several specific test artifacts that can be measured using the machine tool touch probe and the coordinate points are used in a software to compare them to the calibrated values. The errors are calculated to determine the accuracy of a machine tool. The performance of the machine in measuring these test artifacts can be used as an interim measure of capability as recommended for coordinate measuring machines (CMMs) [11,97–102]. This direct measurement method can be fast depending on the complexity of the artifact that affects the number of points required for the measurements. There are a number of simplified master parts including simple shapes, tetrahedrons, ball bars [96,97], ball arrays, and 1D or 2D or 3D ball plates (see Figure 16.28) [103–108], 2D plates and 3D shapes with holes or spheres [104,106,109] (see Figures 16.29 and 16.30), 3D step gages [110,111], identical well-calibrated production parts, and machine checking gages (step gage, etc.). Calibrated 2D ball plates with 36 ceramic spheres creating a quadratic grid with a mesh size of 100 mm have been used to measure the relative error at the machine tool positions of the grid from the individual

FIGURE 16.28 Illustration of a 2D artifact with precision spheres on a rotary table of five-axis machine tool. (Courtesy of IBS Precision Engineering, Eindhoven, the Netherlands.)

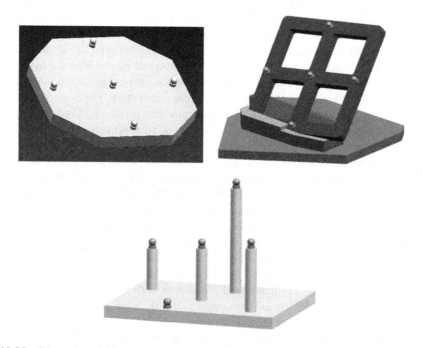

FIGURE 16.29 Illustration of 2D and 3D artifacts with precision spheres.

FIGURE 16.30 Illustration of 2D and 3D artifacts with precision holes. (From Watts, W.A., Artifact and Method for Verifying Accuracy of a Positioning Apparatus, U.S. Patent No. 5,313,410, May 17, 1994; Asanuma, S., Inspection Master Block and Method of Producing the Same, U.S. Patent No. 6,782,730, August 31, 2004.)

sphere locations [63,104]. Two-dimensional cross-grid encoders and displacement sensors have been used as a reference artifact by several projects [112–116]. Measurements of grid plates artifact integrated with vision system have been found attractive for 2D measurements [117]. A cube array artifact composed of eight cubes was proposed [111] for estimating the positioning errors of a three-axis machine tool by approximating error components as polynomial functions and considering the effects of backlash errors. Recently, another method was developed, using an uncalibrated artifact with a set of balls and a 3D probe, for evaluating the geometric errors of machine tools [118,119]. An uncalibrated artifact can be measured in a CMM to verify the machine performance and extract the machine tool errors. For example, a production part can be made the master part after it is measured on a certified CMM. The evaluation of the axes in five-axis machine tool is possible by OMM of any part including using several of the earlier mentioned artifacts [108,120,121]. Some of the artifacts have been used to measure the machine tool error parameters either directly (i.e., straight edges, squareness standards) or indirectly by mapping the errors [104,122]. For example, a fixed-length ball bar is often used at several positions and orientations on the CMM table to evaluate volumetric accuracy.

The artifact methods require (1) setup of artifact on the pallet, (2) probing the artifact at specified position(s) in the direction normal to the surface to collect data for the controlled features, and (3) data analysis to determine machine performance. Depending on the artifact design characteristics, this method can be used to evaluate the machine's positioning and squareness errors (i.e., using a calibrated tetrahedron, ball bar, and kinematic models to identify the geometric error parameters). For example, the tetrahedron artifact (shown in Figure 16.31) is made from 400 mm carbon fiber composite links with metallic magnetic end caps; sphere centroid measurements are used to calculate the linear axis X, Y, Z, squareness X–Y, X–Z, Y–Z, and volumetric accuracy of the machine tool. Often, it is preferable to evaluate the machine performance using calibrated workpieces of similar dimension and geometry as an artifact instead of one of the earlier simple artifacts; the errors can be used to possibly identify the machine tool errors affecting the specific part quality. The specific test artifacts are regularly measured using the CMM, and the performance of the CMM in measuring these test artifacts is used as an interim measure of capability as recommended by ISO Standard 10360-2 and 10360-3 [123,124]; these acceptance and verification tests given in the standard are applicable to three-, four- and five-axis (with a rotary table as the fourth axis) machine tools with a touch probe. When artifacts are used to evaluate machine

FIGURE 16.31 Illustration of the tetrahedron artifact in a machine tool measured with a touch probe. (From Omari, M. et al., Machine Tool Accuracy Quick Check in Automotive Tool and Die Manufacturing, *International Conference on Smart Machining Systems*, Gaithersburg, MD, March 2007.)

performance, the procedure requires the use of the artifact immediately after a machine calibration to obtain the baseline performance of the machine using ISO 15530-3 [125,126]. The artifact method will not identify all dynamic machining errors because the spindle load and part distortion due to cutting forces and clamping pressures, error due to part locating surfaces, and thermal errors induced by the machining process are not present during gaging. The thermal effects can be evaluated if the artifact is measured rapidly after the machining of a part is completed or during the machine's warm-up cycle. In addition, the probing errors must be very small (at least an order of magnitude smaller than the machine measured errors) for the measurements to be accurate. Generally, 3D touch probes are used with measuring repeatability 1–5 μm and a compensation map for the stylus sphere. A special stylus may be required to probe some of the more complex artifacts. Inspection errors caused by the cusp shape of part can be reduced using various approaches [127]. Other sensors may be used for OMM in order to improve the speed of measurements, eliminate the dynamic limitations of the touch probe, and improve the machine tool efficiency [128]. The artifact must also be very stable and highly accurate. The uncertainty of the artifact should also be small as expressed in the following form [129]:

$$U = k + c10^{-6}L \qquad (16.13)$$

where
    L is the distance of the feature to be measured in the artifact (in mm)
    k is the systematic or inherent artifact error that is not length dependent (in μm)
    c is a constant that defines the artifact's length-dependent error

In one case [87], error mapping using a ceramic hole plate 960 mm × 960 mm with a grid spacing of 60 mm was compared with a commercial multilateration laser tracker. The results indicated that the difference among the two methods of measuring the volumetric accuracy of a machine is less than 1 μm for the translational errors and 1 arc-s for the rotational axis errors.

The method represents the measure of a very stable and highly accurate artifact on a machine tool to collect accuracy and repeatability data. The accuracy of the artifact should be at least two to three times (if not an order of magnitude) better than the expected machine accuracy. Therefore, periodic off-machine measurements (in a certified CMM) to ensure that the artifact changes overtime do not affect machine tool calibration accuracy are required. Self-calibration approaches have been used for this purpose [130–132]. Artifacts have been used as a means to quickly check a few parameters, which are good indicators of machine geometric changes. Measuring the artifact on each machine periodically provides a trend to show the continuing capability of the machine. It ensures the machine tool is always within the allowable tolerances (i.e., 20% uncertainty of a manufacturing tolerance). The results are used to predict when equipment calibrations are required or to compare results pre- and post-collision to provide a quick post-collision analysis. An artifact can also be used in a production environment to determine the part drift temperature changes. The machine tool can be compensated based on the master part measurements to eliminate any thermal effects.

Finally, an artifact verifies machine performance prior to machining expensive products, supports the "buy-off" of new machine tools, provides a simple, quick, and cost-effective method to independently validate the quality of a machine, and provides a post error/collision test methodology. It also provides information on thermal stability and can be used for fault determination and root cause analysis. Artifacts are well suited to testing multiple machines prior to purchase.

5. *Machining Test Parts* require planning because the variety of workpieces and features to be machined, as well as the workpiece size, geometry, weight, and material, will affect the machine tool performance. Machining test parts may incorporate (1) a standard part for the validation of machine accuracy or specific operation accuracy, (2) a part to evaluate specific operations, and (3) specific production parts to evaluate the machine's overall performance/capability. It is a composite machine tool evaluation test because it addresses multiple error sources at the same time. The objective is to evaluate machine tool cutting performance using a CMM as a master gage measuring the workpiece accuracy. Since CMMs can operate in manufacturing facilities with temperature-varying environments, they can be easily used as a master gage to measure machined parts and identify machine tool errors. Any other gage can be used to measure parts as long as it is calibrated or mastered using a CMM [133]. Therefore, it involves the interaction of the particular part process (including cutting conditions, tool design, fixture, etc.) with the machine's structural design characteristics including stiffness, damping, and other dynamic features (i.e., heat distribution). The advantages of this method are that it can be applied to any machine tool, that it is a quick test, and that it gives a good demonstration of the machine's overall machining performance. The drawbacks have been (1) the level of required measurement precision for the part and (2) the difficulty of using the measured part results to fully characterize each contributing error source in the machine so that error compensation methods can be used. On the other hand, the deviations from nominal part can be used for correcting the machine tool program for subsequent production parts. This method requires on-line transfer of feature errors to the CNC program through the control's registers.

The standard NAS "circle-diamond-square" (CDS) composite cutting test part with inverted cone as specified in details in the ASME B5.54, National Aerospace Standard (NAS), and ISO [7,116,134–136] is widely accepted as a final performance test for multiaxis machines. The ISO 10791-7 standard [136] specifies the machining of standard dimension workpieces for three-, four-, and five-axis machine tools with reference to the finish cutting conditions in ISO 230-1 [6]; it specifies methods for assessing the accuracy and repeatability of the machine tool using test parts. The CAD/CAM model of the CDS workpiece can be used to program one or more of the several cutting patterns for the specific machine tool configuration using simple straight edge cutting. The CDS part is designed to be cut from aluminum or other more stable material; the procedure is to machine a square

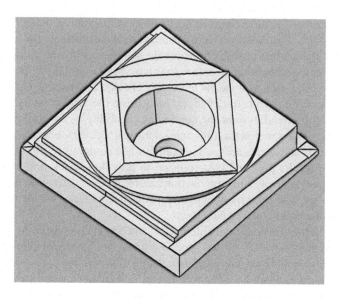

FIGURE 16.32   Illustration of the NAS "circle-diamond-square" with inverted cone cutting test part.

as a base, then a circle on top of the square followed by a diamond shape on top of the circle, and finally a cone using a straight end mill (see Figure 16.32). The inverted cone is used for five-axis machines and the location of the workpiece on the table is very important [122,137]. The measurements provide (1) dimensional accuracy, flatness, squareness, parallelism, and surface finish (using the outside 160 mm square surface) and location accuracy and spindle characteristics using the four counterbore holes and the center bore; (2) angular deviation (using the 5° ramp); (3) dimensional accuracy, squareness, parallelism, and surface finish (using the center diamond); (4) dimensional accuracy, roundness, and finish (using the 108 mm diameter circle), (5) characteristics of servo response and slide way friction by visual inspection of the surface finish (using the 5° ramp and .005″ taper cuts), and (6) contouring performance of five-axis machining (using the inverted cone). More details on workpieces for five-axis machines are described in Refs. [135,136]. Therefore, the kinematic errors of a three-axis machine and some of five-axis machines can be clearly measured on the CDS workpiece.

Several other shapes related or not to CDS have been machined in five-axis machining centers to evaluate the machine's accuracy and the rotary axes' performance [27,138–143]. The strategy behind this method is to utilize, because of its accuracy, a CMM to obtain precise measurement of the machined part. Sometimes, nonrotational machining is used to copy and trace the machine trajectory on the workpiece, minimizing the machining error [138,144]. The profiles of the machined workpiece are measured, and the geometric errors of the machining center are extracted. The errors except the geometric accuracy of a machining center can be minimized by using light cutting conditions, which do not significantly affect the desired machined workpiece profile or features.

Geometric errors can be evaluated using machined workpieces of specific geometry as has been used by machine tool builders. For example, boring tests are used to evaluate roundness and cylindricity accuracy of the spindle moving along the Z-axis. Milling tests are used with large-diameter cutters to mill the perimeter of a square part at about 60–80 mm width of cut to estimate flatness and profile accuracy. A face milling test of a four-sided square part (about 300 × 300 mm) on a four-axis machine is used to evaluate the squareness and parallelism of the opposite sides. Boring four holes in a square pattern (at about 200 mm apart) can be used to evaluate the positioning accuracy including hole diameter consistency (if performing automatic tool change of the same tool each time the boring is conducted) by measuring the linear distance between any two holes and the hole diameter in X- and Y-axis. Side cutting a square part on the four sides moving one axis at the time or in a separate test moving two axes together by cutting a square at 30° inclination can be used to evaluate the straightness, parallelism, and squareness by measuring the four sides of the square part. Circular side cutting with end mill can be used to evaluate the roundness. Five-axis machines are evaluated using either simple or complex workpieces. ISO 10791-7 describes simple boring tests for five-axis machines. The geometric errors measured of such special parts can be used to estimate some of the kinematic errors of the machine tool under cutting loads [122].

6. *Other Static Performance Tests* are described in detail in Ref. [12]; a summary is provided in Table 16.4. These errors are also summarized in ISO 230-1 [6]. A comparison of some of these errors measured according to Table 16.4 for various small- and medium-size machine tools is provided in Figures 16.33 and 3.41. These figures indicate that machine performance may vary significantly from the expected. The spindle analyzer shown in Figure 16.34 can be used to evaluate the spindle performance with respect to out of roundness of spindle rotation, axial thermal drift, shift as function of speed, etc. The measurement of this or similar devices using spheres is in accordance with ASME B5.54, ISO 230-3, and 230-7 standards [7,22].

Table 16.5 provides a comparison of machine tool errors as tested on the manufacturer's floor, user floor, and the tolerance specs. The errors are smaller when the measurements are done on the manufacturer's floor based on the initial setup and calibration. However, the same errors increased significantly after the same machine is relocated. Therefore, a reevaluation of the machine setup and possibly recalibration are required after a relocation.

TABLE 16.4
Brief Description of the Tests Performed for Machine Tool Performance Evaluation

Test	Purpose/Procedure
Linear position	Uses laser interferometer to measure positioning accuracy and reversal error
Volumetric performance	Uses laser interferometer to measure the four body diagonals of the working volume. This test gives an overall picture of the positioning accuracy and often impossible to mask errors. Results are usually 3–5 times greater than the positioning accuracy of a single axis.
Repeatability	Uses laser interferometer to measure bidirectional positioning repeatability
Ball bar circularity test	Uses LVDT in a telescoping bar to measure contouring performance and squareness of axes
Periodic error	Measures positioning errors for small-scale (less than the pitch of lead screw) moves
Drift test	Measures the effect of thermal growth over time on repeatability by cycling the machine in rapid traverse mode and periodically checking positioning accuracy
Spindle runout	Uses two noncontacting capacitance probes at 90° to measure radial error in the spindle at various speeds
Spindle analyzer	Uses three noncontacting capacitance probes in X-, Y-, Z-directions to measure radial runout and axial growth
Static spindle compliance	Measures stiffness of spindle, monitoring force with a load cell, and displacement with proximity sensors
Rotary table angular compliance	Uses load cell, displacement indicator, and angle bracket to measure compliance and hysteresis of the rotary table
Rotary table tilt compliance	Uses load cell, displacement indicator, and angle bracket to measure tilting compliance and hysteresis of the rotary table
Pallet changer load	Measures angular error in the work zone due to loading the external pallet changer with 800–1000 lb

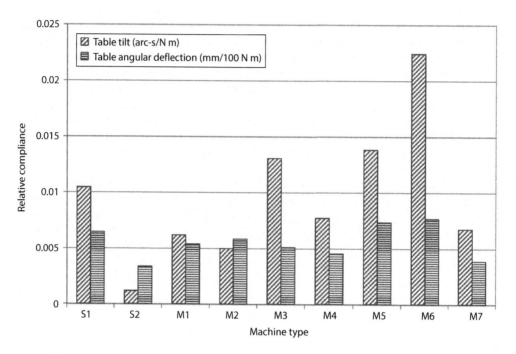

FIGURE 16.33 Relative static compliance of the machine rotary table. S, small-size machines; M, medium-size machines.

FIGURE 16.34 Illustration of a spindle analyzer device. (Courtesy of IBS Precision Engineering, Eindhoven, the Netherlands.)

TABLE 16.5

Comparison of Machine Tool Performance at the Manufacturer Floor versus after Its Installation on User's Floor

Machine Tool Errors		Mfg. Floor	User Floor	Tolerance/Spec
X-Axis	Max. pitch error, arc-s		5.00	
	Max. yaw error, arc-s		105.90	
	Max. roll error, arc-s			
Y-Axis	Max. pitch error, arc-s		10.60	
	Max. yaw error, arc-s		4.10	
	Max. roll error, arc-s			
Z-Axis	Max. pitch error, arc-s		9.20	
	Max. yaw error, arc-s		8.10	
	Max. roll error, arc-s			
X-Axis	Straightness in Y, mm	0.0030	0.0160	0.03/m
	Straightness in Z, mm	0.0030	0.0120	0.03/m
Y-Axis	Straightness in X, mm	0.0020	0.0160	0.0050
	Straightness in Z, mm	0.0020	0.0190	0.0050
Z-Axis	Straightness in X, mm	0.0020	0.0230	0.0050
	Straightness in Y, mm	0.0020	0.0150	0.0050
X-Axis	Max. Pos. error, mm	0.0028	0.0850	0.002–0.003
	Max. Bidirect. Rep., mm	0.0009	0.0760	0.001–0.0015
X-Axis	Max. Pos. error, mm	0.0017	0.0290	0.002–0.003
	Max. Bidirect. Rep., mm	0.0004	0.0020	0.001–0.0015
Z-Axis	Max. Pos. error, mm	0.0024	0.0030	0.002–0.003
	Max. Bidirect. Rep., mm	0.0004	0.0050	0.001–0.0015

16.5 METHOD FOR COMPENSATING THE DIMENSIONAL ACCURACY OF CNC MACHINING SYSTEM

16.5.1 Error Reduction and Compensation Strategies

The improvement of CNC machine tool accuracy has been the subject of great attention for several decades. The strategies to reduce the errors that adversely affect the accuracy of a machine tool can broadly be divided into three categories, namely, error avoidance, root cause error correction, and error compensation [10] as illustrated in Figure 16.35. The error avoidance approach tries to minimize the sources contributing to machine tool inaccuracy, which requires a statically and dynamically stable machine tool. Historically, this has been achieved by reducing rigid body errors by carefully manufacturing and assembling the machine tool structure. Most of the geometric errors of the machine tool are primarily misalignment of the machine's axis, position, and straightness errors of each axis, ball screw or rack-and-pinion errors due to wear, metal fatigue, or foundation problems. These errors are considered a function of displacement because they vary along the length of each axis. Permanent and systematic design-related geometric errors are relatively common in general purpose CNC machine tools. On the other hand, the geometric errors of precision machine tools are minimized by design because they have considerable effect on part dimensional accuracies and reduce their contribution to machine's inaccuracy [12]. Since the 1980s, an alternative method known as error budgeting has been used. An error budget analysis can be used at the design stage to predict the total error of the machine system. This approach can identify the individual sources of error including their magnitude and direction. It is systematic method for predicting the degree of difficulty of achieving the target error. The design of such

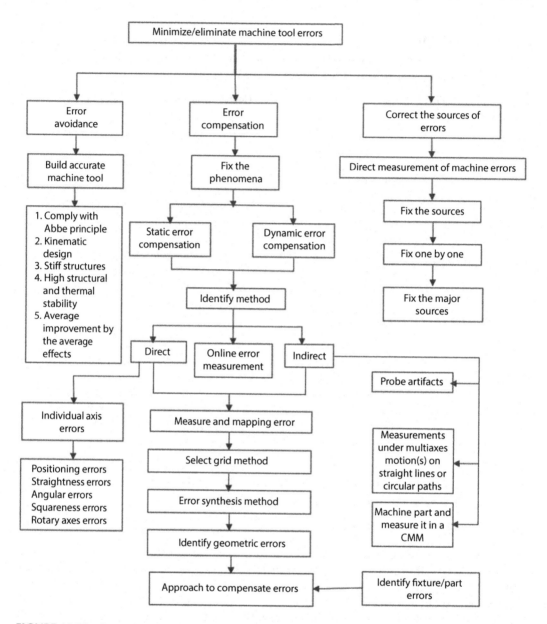

FIGURE 16.35 Strategies to reduce machine tool errors.

machine is very often costly in addition to the complexity of the control system for the traditional manufacturing floor. In addition, the refinement of the machine tool structure beyond a certain level of accuracy is prohibitively expensive.

The second strategy emphasizes the correction of the sources of the errors requiring direct measurement of the individual errors for each axis, then the error will be fixed one by one, at least for prioritized errors. This method is effective only when the source of the problem is identifiable. The problem-solving process may take a significant amount of time when the problem is complex, as when dealing with multiaxes machine tools with two or three dozen possible individual errors.

The third approach is to correct the error or the phenomena without identifying the source(s) of the error. This method is very effective when the sources are numerous and not well understood and when it would be time-consuming to evaluate each axis' errors. In CNC programming, this method is called compensation [145]. This method requires the measurement of the errors, modeling or generation of an error map, and compensation. However, this method is not applicable after a machine tool crash, resulting in large errors. This strategy uses either direct or indirect measuring methods as illustrated in Figure 16.35. The direct methods evaluate the individual errors for each machine axis. Over the last three decades, error models have been developed based on kinematic analyses of the machine tool axes to quantify the quasi-static errors in the workspace as a function of the individual machine tool geometric errors using rigid body kinematics as explained in Refs. [11,12] and many others found in Refs. [10,146]. The indirect methods use measurements that superpose the single errors, such as error mapping compensation techniques, to increase the machine accuracy at lower cost. This strategy designs a machine with as much as possible minimum error at competitive cost; it assumes that the errors are either measurable/systematic or predictable and they can be compensated. Error correction is performed through a calibration procedure and can be repeated at specified intervals (yearly or every few months) because systematic errors still exist, due to machine wear, looseness of assembled components, accidents, and uncertainties in earlier calibrations. Therefore, the error mapping compensation technique is considered to be the key approach to eliminate the machine's quasi-static errors because it provides a correction for the rigid body errors. Dynamic errors can be compensated if they are present during the measurement of the errors and are repeatable over similar or different operating conditions. In the direct method, the errors are first measured and then are mathematically combined to determine the cutting tool error at each point in the work zone. The error is accounted by the control system for every move for the cutting tool to the correct position. Machine tool controllers allow for compensation values to be applied during the machine's movement to correct existing positioning errors in real time, which represents active compensation. The error compensation strategy will ensure accuracy if the machine tool's geometrical stability is good over all the operating conditions encountered during the machining of various parts.

Generally, the principles of machine design to achieve high accuracy are as follows: (1) comply with Abbe principle, (2) use kinematic or semi-kinematic design, (3) make the machine structure stiff and stable, (4) improve accuracy by averaging effects, and (5) use error reduction and compensation techniques. Even though the geometric machine errors may be compensated, the thermal deformations of the machine structure and the workpiece, which are due to both internal and external heat sources, for example, ball screw/nut and changes in the local air temperature, respectively, cannot be fully eliminated at the design stage and therefore require real-time compensation [4,147–149]. Due to the difficulty of direct measurement of the relative thermal displacement $\delta(x, y, z, t)$ between the tool and the workpiece during machining, control systems based on *inductive* and *deductive* approaches are used. In the inductive approach, the components of the vector $\delta(x, y, z, t)$ are empirically related to other variables, usually temperature, which is easily measured at some strategic points on the structure. During operation, the measured temperatures are used to estimate the position error $\delta(x, y, z, t)$ and activate the control system [150–153]. An artificial neural network has been used for thermal error analysis and modeling [154]. Fuzzy clustering and gray correlation theory, establishing compensation model separately with multiple linear regression model, distributed lag model, and support vector machines have also been evaluated [155–157]. Since the empirical base function bears no physical similarity to the actual phenomena, these solutions are unreliable outside of the range of tested inputs. In addition, the information contained in the discrete temperature measurements is incomplete, and therefore, the problem is not uniquely defined.

To overcome these problems, an alternative deductive approach was followed by many investigators. In this approach, numerical models are used to describe the thermal and deformation processes, which take place in the structure [158–161]. Thermal loop analysis was proposed to describe the thermal behavior of an entire machine tool using a path across the assembly of its mechanical components [163]; thermal error models (based on thermal modal analysis) are used to define the thermal links along the thermal loop. These approaches also suffer from a number of drawbacks because (1) the magnitudes of the heat sources are determined from off-line calibrations that result in poor predictions, (2) the numerical models are either inaccurate or too slow to be used in real-time control applications, and (3) the nonlinear thermoelastic behavior of the structural joint is neglected.

A *generalized modeling* was proposed [4] to eliminate the limitations of the inductive and deductive methods, by incorporating an inverse heat conduction solver to estimate the heat input to the structure in real time. It also provides the framework for developing a generic, multivariable control system applicable to nonlinear structures. This method requires (1) modeling and monitoring the temperature and thermal distortion of the machining system including the part; (2) characterization of the transient thermal field and the thermal deformation pattern in typical part; and (3) identification of the thermal boundary conditions and heat input to the machining system in order to compensate the thermal deformation of the machine tool structure. However, this method is not yet either fully implemented or validated.

The measurement of the geometric errors as a function of machine tool temperature at certain locations has been evaluated and used for error compensation. Assuming the method of measuring the specific machine tool geometric errors is fast enough to be repeated sequentially while the machine tool is warming up or axes are loaded through intermittent cutting tests, so time-dependent variation in geometric error parameters can be recorded, which is mostly caused by thermal deformation induced by spindle warm-up, slides movements, or environmental changes. In one case, a radial basis function neural network was used to predict the volumetric positioning errors estimated by the "vector" diagonal method as a function of temperature distribution of the machine [162]. Compensation of spindle growth and feed ball screws for thermal deformations is feasible if sensors are used to measure the temperature along the ball screw or spindle at strategic locations [152]. Several of the direct and indirect measurement methods discussed in the previous section can be used to measure the machine tool errors at different temperatures of the machine tool structure and components (using thermocouples in strategic locations within the machine tool). Such measurements enable compensation of thermal deformations. A master part is measured at different temperatures in a CMM and the deviations from nominal are recorded at each gage point. This part is then inspected in the machine tool at the corresponding temperature of production environment and immediately will compensate for any change in the thermal conditions during machining. Modeling the thermal behavior of machine tools using FEM and finite difference methods has been helpful to estimate some of the thermal deformations for compensation of machine tool errors especially when they are used together with measurements of geometric errors as a function of temperature [149,163–168]. Specific cutting tests can also be used to identify axes geometric errors that are affected by thermally induced variation [169].

Many further methods for enhancing the accuracy of CNC machine tools have been proposed [34,169–172]. An integrated geometric error modeling, identification, and compensation method for machine tools was developed based on the individual axes errors [173–179]. It identifies the 21 translational geometric error parameters associated with linear motion axes based on a laser interferometer and 6 angular geometric error parameters for each rotation axis based on a ball bar. An identification method is used based on the model to recognize these geometric errors, which are compensated by correcting corresponding NC codes. The results showed that the integrated method was effective and applicable in multiaxis machine tools. More recently, laser calibration data analysis software has been provided by a few vendors, which brings new levels of functionality and flexibility to compensation methods by using laser interferometer calibration systems combined with rotary axis calibrators to evaluate rotary axis positioning performance (with ±1 arc-s accuracy) in addition to linear axis measurement. The use of a laser interferometer has been the most common method to fully assess geometric errors as discussed in the previous section. Unfortunately, this technique has disadvantages in that it requires several days (resulting in significant downtime and cost) and uses a volumetric error synthesis model.

Different techniques of applying compensation have been developed and evaluated to identify a cost-effective method. However, it has been a great challenge to identify a cost-effective geometric accuracy error measurement and compensation methods applicable to a variety of machine tool types from different manufacturers. The compensation algorithm program is integrated by the machine tool builder using a specific method for error measurement. Some machines use an embedded software module that collects the machine tool accuracy error information. This module will update the positioning values based on encoder feedback so that a new corrected position is fed to the drive system. If the machine tool does not have an integrated compensation program, the NC program to machine a part can be corrected by a post processor off-line or online using a desktop PC [180,181] to incorporate the offsets. In this case, (1) the CAD/CAM part code is adjusted by a conversion software using the error lookup table before the post processor generates the G-code and (2) the desktop PC is interfaced to the post processor to accept the G-code tool path that is converted to generate the corrected NC program [182]. The complete measuring curves and the estimated parametric errors of the individual axes can be used instead of a lookup error table in the conversion software. The conversion software could be based (1) on the actual lookup error table for interpolating the nominal values based on the error locations available in the table or (2) on a kinematic model and the individual errors of each axis.

For the last two decades, the linear and volumetric error modeling has been the major approaches to generate accurate error models for machine tool error compensation to minimize the relative error between the cutting tool and the workpiece [10]. Linear and volumetric positioning error lookup tables are used for compensation. The error model predicts and controls the total error to a minimum for the machining system. The linear error is incorporated in the conventional calibration, which does not fully correct for angular errors. The traditional approach used a single axis software compensation using an array of points to generate a compensation table or function for each axis. Linear (or pitch error) compensation lookup tables can be used inside the controller; the compensation values are at discrete distances, and interpolation is used to find the compensation value between measured points. The error map may include unidirectional errors, bidirectional errors (includes backlash), and straightness depending on the compensation software. The software will use the error compensation lookup table to convert X-, Y-, and Z-axis coordinates in the part program. Many machine tool controllers support linear error compensation for each axis using error tables, but this approach only compensates for some of the error sources.

Volumetric compensation has the ability to correct all the machine tool geometric errors. Software for volumetric compensation enables more extensive measurements to be used to populate the compensation tables for the machine's axes, ensuring consistent performance throughout the working volume. Some volumetric compensation systems correct only for the 21 parameter errors of the three linear axes. Five-axis machines are calibrated volumetrically by either (1) calibrating all axes together or (2) calibrating the three linear axes volumetrically and the rotary axes separately; the calibration procedure provides either (a) 21 parameter errors plus the tool vector or (b) all the 43 parameter errors [183]. In three-axis machine tool, the software corrects for tool tip center point, while in five-axis application, it corrects for the tool tip center point and tool vector. Some five-axis machines may use a laser tracker to generate an error map for the three linear axes and the radial test for the rotary axes errors. Hence, the error modeling strategy is crucial to minimize the machine tool errors. The error model should include measurements of geometric errors, quasi-static thermal effects, and dynamic errors including contouring accuracy. Recently, some newer machines have offered an option for volumetric compensation. This type of compensation corrects the machine geometric performance, which is consistent throughout the whole volume of the machine tool.

Volumetric error compensation (VEC) has become an industry-accepted method because it is a complete error mapping of the machine tool through its entire working volume. The method has been successfully applied to CMMs and specialized large machine tools. Since the geometric error is a very important component of volumetric error, its characterization and mapping have been used either by defining generic kinematic models or by individual points along each axis of machine tools. If the complete kinematic model of the machine is available, the volumetric errors can be predicted based on the individual axes errors since the Abbe offset is considered. Many researchers developed various methods for VEC; several of these methods are discussed in Refs. [10,146,184]. The VEC has shown improvement in the range of 40%–90% depending on the method and instrument(s) used to measure the errors and then apply the necessary software/algorithm to compensate. LBB and LTS have shown up to 90% error improvement, while traditional LT has shown improvement of 40%–70%.

The implementation details of VEC will depend on the type of CNC machine and controller and on the software used to implement it. A 3D error grid can be used for the calibration of three-axis machine tool; the 3D grid is measured either with laser tracker, 6D laser interferometers, or an artifact as explained in the previous section. The manufacturers of measuring systems for determining machine tool errors provide the corresponding software tools for the particular machine tool controller compensation system to quickly and reliably convert the individual measurements into the file format accepted by the controller. In this case, dedicated compensation schemes for the specific machine tool structure design are developed by the manufacturer of the machine tool together with the manufacturer of the measuring system. Each machine controller volumetric compensation option is different and requires a custom software package specific to the controller manufacturer. The measurement of the machine tool errors and the compensation process define the calibration of the machine tool. The measurement time for generating the error map could vary from 4 h to 3 days depending on the calibration method, the measuring equipment, and the operator experience with the calibration procedure and equipment [10,184]. While the calibration methods have merit, they are not commonly used in industry today at regular intervals because the measurement of the individual errors or volumetric errors with laser interferometer is time-consuming and requires highly skilled personnel, which makes the method too expensive to be used periodically in a manufacturing system that includes several or dozen CNC machining centers. Volumetric compensation is more effective but not yet fully adapted in production.

Therefore, gaging with touch probes has been used for measuring master part(s) located on a fixture in machining centers. The difference among the actual and nominal dimension can be compensated either by correcting the individual feature location or by using the conventional offset method and adjusting the location of individual part faces. The conventional offset method requires a special spreadsheet to evaluate the average offset of the features at each face or part rotation. In addition, the inspection time for an artifact may be long.

A *fiducial calibration system* (FCS) is a method that allows for the accuracy of a CMM to be transferred to a large machine tool for large part. An FCS improves part accuracy without requiring thermal or geometrical models of the machine [185]. An FCS transfers responsibility for the metrology reference from the machine tool to the workpiece through the use of fiducials (targets) on/in the part. Fiducials are either installed on or machined into the part, and their positions are measured with CMM or other optical 3D scanner or laser tracker (for large parts) in a thermally controlled environment. The part is attached on a fixture on the machine tool, and the positions of the fiducials are measured using a sensor (such as a touch trigger probe or optical/vision system). The total error is measured after initial machining, and a mathematical transformation is used in the NC part program, altering the original programmed coordinates to correct for the errors. This method eliminates the need for machine thermal and geometrical error maps and part alignment with machine coordinate system. The accuracy of placing the fiducials on/in the part is not as critical as the accuracy of measuring them. The technique does not correct for rapidly changing dynamic and thermal errors. In addition, the release of residual stresses in the workpiece during machining will affect the location of the fiducials and therefore the corrections estimated based on their original location. Some of the fundamental questions about how much improvement in precision of the finished part can be expected and what machines and fiducials are required for given quality are discussed in Ref. [186].

All in all, complete error correction has been tedious. There are two aspects in the concept of cutting tool location control: (1) control the tool to a nominal location and (2) control the tool to an accurate location. Case 1 "Control the tool to a nominal location" is intuitive. If one wishes to drill a hole in a part at (x, y) location with L depth, one can command the tool to move to point (x, y, z_1) and then (x, y, z_2) in the coordinate system, where $L = z_2 - z_1$. A hole programmed to the nominal position may not be machined accurately enough to meet the part print specification (with respect to location and orientation) due to possible machine axes errors (e.g., if the calibration is outdated), table center location error, coordinate origin location error, and part/fixture alignment errors. It is a challenge to make an accurate hole in an error-prone environment, especially when the errors are numerous, and their nature may not be well understood. Therefore, compensation offset could be very effective. In this example, the hole is measured by a CMM machine and has a location error $(\Delta x, \Delta y)$. The compensation method simply offsets the machine tool coordinate by the opposite direction of the error values $(-\Delta x, -\Delta y)$. Then the hole machined in the next cycle is accurate. The compensation method is effective for any type of repeatable (consistent) error, including table errors, fixture errors, spindle errors, or any other unknown errors.

Finally, measuring a finished part in a CMM to determine the compensation parameters for the machine tool and overcome some of the limitations of either measuring the part or artifact on the machine can be very effective. The most common method of compensating CNC machines is to make small adjustments to the work offsets of the machine [187]. Work offset (also known as work coordinate system or fixture offset) is a method of positioning the cutting tool based on the machine zero position. Work offsets may be adjusted from its nominal values to compensate the position of its associated machined features, while allowing the tool path to be programmed to the nominal part print position of the feature. The work offset method relaxes the monitoring ranges for quasi-static and some dynamic errors and periodic calibration. Rather, it requires only simple CNC programming and CMM production measurement data of machined parts. The availability of the CMM on the manufacturing floor has enabled its use for multidimensional statistical process control and process characterization of production parts. A logical step is to integrate the feeding of part measurement information to the machine tool for direct process control. Therefore, the availability of the part quality information is used by the work offset technique that provides the ability to quickly and appreciably reduce the impact of machining errors on part quality.

Simulation of the compensation method is often important to evaluate part quality and performance of multiaxis machine tool. A virtual machine tool (VMT) model superimposes all the effects modeled for part accuracy as previously explained; the compensation algorithm is another important segment in modeling because they can be implemented in different ways. VMT can be used to evaluate the effect of individual error parameters on part quality and measurement results since the decomposition of the individual error contributors is too complex [188]. The model allows estimation of the location of the tool tip for a given programmed movement. The machine tool error compensation model incorporates the method of measuring the accuracy errors and possibly other analytical models such as static rigidity models or thermal models, together with the kinematic model. VMT can also perform error cause diagnosis and compensation verification of the part program. Software is being used to verify dimensional accuracy and optimize tool paths for better finishes on surfaces and edges [189]. Virtual simulation has become mature because machine tools have been successfully modeled from component level to predict major characteristics. This capability can be used to evaluate the advantages and disadvantages of different measurement methods or standards [17]. The Monte Carlo approach is used to consider hundreds or thousands of measurement points generated by assuming a normal or a skewed distribution to represent systematic errors and random variation. Simulation technology is used to improve the accuracy of machine tools at the design stage or for optimizing the tool path positions by incorporating the current machine tool accuracy errors and repeatability. Virtual evaluation of manufacturing process allows the user to analyze machine tool variables including fixture and part structural and material characteristics to improve part quality. Therefore, the correlation of key part features quality in the manufacturing process to machine tool accuracy is highly desirable. In one instance, a VMT was developed to analyze the combined or independent effect of the individual geometric and servo errors on standard five-axis machine tool to predict part dimensional errors [190]. The VMT model included ideal tool path trajectory, machine tool control and geometric errors, machine to part coordinate transformation, and part feature analysis. It is expected to observe differences between simulated and measurement accuracy data [180,191] due to thermal and dynamic effects. The kinematic model of a machine tool integrated with the error model, as will be discussed in the next section, can be used to make (1) virtual measurement of a part quality based on geometric and other available errors, (2) error cause diagnosis, and (3) compensation verification of the part program.

16.5.2 Error Modeling Methods

Error compensation or calibration is an effective method of improving the accuracy of machine tools and part quality. The axis drive systems in Figures 3.46 and 3.47 show two common methods of measuring and correcting the errors in axis motion, either (1) a digital rotary encoder attached to the motor or ball screw axis or (2) a glass linear scale. The concerns with a rotary encoder are mismatch of the angular encoder position and linear ball screw position (because the ball screw is not manufactured perfectly), friction, heat from the motor and ball nut, or reversal error. The glass linear scale eliminates several of the errors assuming (1) it is mounted so that it is largely independent from the thermal expansion of the machine tool and (2) the scale is attached closer to the table or guideway to reduce the Abbe offset error. However, machine tool manufacturers use rotary encoder feedback to compensate for ball screw manufacturing errors and thermal errors by measuring the displacement of ball screw errors at different system temperatures. The calibration values are greatly affected by the error modeling and error measurement techniques. Therefore, how to establish an accurate error model and accurately detect the geometric errors of machine tool are particularly important. Methods for measuring the machine tool error or accuracy were discussed in Section 16.4. The error modeling method is formulated based on the error measurement methods. As explained in the previous sections, the manufacturers of the specific devices for measuring the machine tool errors also provide the necessary software to estimate a portion or all of the individual axes errors.

For many years, researchers have been modeling multiaxis machine tools to develop mathematical models of the machines that use functional errors of machine components to derive the resultant error in tool and workpiece point deviation. Several researchers used the multibody system (MBS) theory to calculate the resultant machine tool error based on the individual axes errors. The objective of the kinematic model is to evaluate the position and orientation of the cutting tool in the workpiece coordinate system by superimposing the error motions of each axis. Several studies used a vertical machining center with or without rotary table with the MBS method [28,148,172,173–179]. However, since horizontal machining centers are very common in high-volume manufacturing, the error of four-axis horizontal machine tool (as illustrated in Figure 3.13) is analyzed using the normal kinematic modeling process. The total error motion of the machine is represented by homogeneous transformation matrix (HTM) that is a combination of rotational and translational errors [10,192] and represents the coordinate transformation from the coordinate system of the structural frame to that of the reference coordinate system. The kinematic chain of the horizontal machine tool is shown in Figure 16.36. Consider the

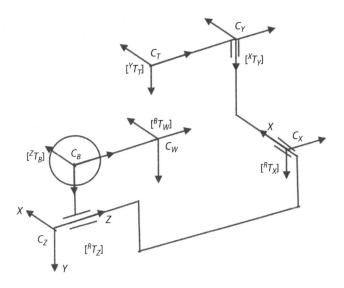

FIGURE 16.36 Kinematic chain for the workpiece and cutting tool of a four-axis horizontal machine tool with B table.

reference frame as the coordinate system located at the machine bed. ^RT_X represents the motion of the X-axis in the reference frame, ^XT_Y represents the motion of the Y-axis (spindle slide) with respect to the X-axis, and ^YT_S is the spindle coordinate frame with respect to the spindle slide frame, where the spindle nose is the point at which the error motion measurements are made. $^ST_{TL}$ is the tool coordinate frame with respect to the spindle. ^RT_Z is the Z-axis upper carriage coordinate frame with respect to the reference frame, $^ZT_{RT}$ is the rotary table coordinate frame with respect to the carriage, and $^{RT}T_W$ is the workpiece coordinate frame with respect to the rotary table frame. The HTM representing the X-axis position error of the slide with respect to its ideal position (see Figure 16.2) is

$$
\left[{}^RT_X\right] =
\begin{bmatrix}
1 & -\varepsilon_z(x) & \varepsilon_y(x) & 0 \\
\varepsilon_z(x) & 1 & -\varepsilon_x(x) & 0 \\
-\varepsilon_y(x) & \varepsilon_x(x) & 1 & 0 \\
0 & 0 & 0 & 1
\end{bmatrix}
\begin{bmatrix}
1 & 0 & 0 & x+\delta_x(x)+X_o \\
0 & 1 & 0 & \delta_y(x)+Y_o \\
0 & 0 & 1 & \delta_z(x)+Z_o \\
0 & 0 & 0 & 1
\end{bmatrix}
$$

$$
=
\begin{bmatrix}
1 & -\varepsilon_z(x) & \varepsilon_y(x) & x+\delta_x(x)+X_o \\
\varepsilon_z(x) & 1 & -\varepsilon_x(x) & \delta_y(x)+Y_o \\
-\varepsilon_y(x) & \varepsilon_x(x) & 1 & \delta_z(x)+Z_o \\
0 & 0 & 0 & 1
\end{bmatrix}
\tag{16.14}
$$

where
    x is the nominal X-axis position
    X_o, Y_o, Z_o are the offsets of the slide from home position

The orthogonality/squareness errors are often neglected but for high-precision machines they should be considered because they affect the angular errors. The HTM squareness error between X- and Y-axis is also considered for the HTM representing the Y-axis position of the column with respect to its ideal position.

$$
\left[{}^{X}T_{Y}\right] = \begin{bmatrix} 1 & -\alpha_{xy} & 0 & 0 \\ \alpha_{xy} & 1 & 0 & 0 \\ 0 & 0 & 1 & 0 \\ 0 & 0 & 0 & 1 \end{bmatrix} \begin{bmatrix} 1 & -\varepsilon_z(y) & \varepsilon_y(y) & 0 \\ \varepsilon_z(y) & 1 & -\varepsilon_x(y) & 0 \\ -\varepsilon_y(y) & \varepsilon_x(y) & 1 & 0 \\ 0 & 0 & 0 & 1 \end{bmatrix}.
$$

$$
\begin{bmatrix} 1 & 0 & 0 & \delta_x(y)+X_1 \\ 0 & 1 & 0 & y+\delta_y(y)+Y_1 \\ 0 & 0 & 1 & \delta_z(y)+Z_1 \\ 0 & 0 & 0 & 1 \end{bmatrix} = \begin{bmatrix} 1 & -\varepsilon_z(y)-\alpha_{xy} & \varepsilon_y(y) & \delta_x(y)-y\alpha_{xy}+X_1 \\ \varepsilon_z(y)+\alpha_{xy} & 1 & -\varepsilon_x(y) & y+\delta_y(y)+Y_1 \\ -\varepsilon_y(y) & \varepsilon_x(y) & 1 & \delta_z(y)+Z_1 \\ 0 & 0 & 0 & 1 \end{bmatrix}
$$

$$(16.15)$$

The HTMs for the stationary spindle, since it only rotates concentrically and the dynamic radial and tilt errors due to the heat and rpm cannot be corrected quickly enough for compensation, are

$$
\left[{}^{Y}T_{S}\right] = \begin{bmatrix} 1 & 0 & 0 & 0 \\ 0 & 1 & 0 & 0 \\ 0 & 0 & 1 & 0 \\ 0 & 0 & 0 & 1 \end{bmatrix} \tag{16.16}
$$

and the vector for the cutting tool, which is clamped to the spindle, has the tip coordinates

$$
\left[{}^{S}T_{TL}\right] = \begin{bmatrix} X_t \\ Y_t \\ Z_t \\ 1 \end{bmatrix} \tag{16.17}
$$

where
the X_t and Y_t could be considered zero (assuming the spindle is aligned to the axes)
the Z_t is the tool offset

The structural loop reference point to cutting tool is the product of the individual point location

$$
\left[{}^{R}T_{TL}\right] = \left[{}^{R}T_{X}\right] \cdot \left[{}^{X}T_{Y}\right] \cdot \left[{}^{Y}T_{S}\right] \cdot \left[{}^{S}T_{TL}\right] \tag{16.18}
$$

The workpiece structural loop reference point position to workpiece/fixture position relative to the ideal position is the Z-slide, B-rotary axis, and for workpiece/fixture components:

$$
\left[{}^{R}T_{W}\right] = \left[{}^{R}T_{Z}\right] \cdot \left[{}^{Z}T_{RT}\right] \cdot \left[{}^{RT}T_{W}\right] \tag{16.19}
$$

where

$$
\left[{}^{R}T_{Z}\right] =
\begin{bmatrix}
1 & 0 & \alpha_{xz} & 0 \\
0 & 1 & -\alpha_{yz} & 0 \\
-\alpha_{xz} & \alpha_{yz} & 1 & 0 \\
0 & 0 & 0 & 1
\end{bmatrix}
\cdot
\begin{bmatrix}
1 & -\varepsilon_{z}(z) & \varepsilon_{y}(z) & \delta_{x}(z)+X_{2} \\
\varepsilon_{z}(z) & 1 & -\varepsilon_{x}(z) & \delta_{y}(z)+Y_{2} \\
-\varepsilon_{y}(z) & \varepsilon_{x}(z) & 1 & z+\delta_{z}(z)+Z_{2} \\
0 & 0 & 0 & 1
\end{bmatrix}
$$

$$
=
\begin{bmatrix}
1 & -\varepsilon_{z}(z) & \varepsilon_{y}(y)+\alpha_{xz} & \delta_{x}(yz)+z\alpha_{xz}+X_{2} \\
\varepsilon_{z}(z) & 1 & -\varepsilon_{x}(y)-\alpha_{yz} & \delta_{y}(z)-z\alpha_{yz}+Y_{2} \\
-\varepsilon_{y}(z)-\alpha_{xz} & \varepsilon_{x}(z)+\alpha_{yz} & 1 & z+\delta_{z}(z)+Z_{2} \\
0 & 0 & 0 & 1
\end{bmatrix}
\tag{16.20}
$$

When the table rotates by an amount β about the Y-axis, the HTM transforming the coordinates of a point for the Z-coordinate frame into rotary table RT frame is

$$
\left[{}^{Z}T_{RT}\right] =
\begin{bmatrix}
\cos\beta & 0 & \sin\beta & 0 \\
0 & 1 & 0 & 0 \\
-\sin\beta & 0 & \cos\beta & 0 \\
0 & 0 & 0 & 1
\end{bmatrix}
\begin{bmatrix}
1 & -\varepsilon_{z}(\beta) & \varepsilon_{y}(\beta) & \delta_{x}(\beta) \\
\varepsilon_{z}(\beta) & 1 & -\varepsilon_{x}(\beta) & \delta_{y}(\beta) \\
-\varepsilon_{y}(\beta) & \varepsilon_{x}(\beta) & 1 & \delta_{z}(\beta) \\
0 & 0 & 0 & 1
\end{bmatrix}
$$

$$
=
\begin{bmatrix}
\cos\beta-\varepsilon_{y}\sin\beta & -\varepsilon_{z}\cos\beta+\varepsilon_{x}\sin\beta & \varepsilon_{y}\cos\beta+\sin\beta & \delta_{x}\cos\beta+\delta_{z}\sin\beta \\
\varepsilon_{z} & 1 & -\varepsilon_{x} & \delta_{y} \\
-\sin\beta-\varepsilon_{y}\cos\beta & \varepsilon_{z}\sin\beta+\varepsilon_{x}\cos\beta & -\varepsilon_{y}\sin\beta+\cos\beta & -\delta_{x}\sin\beta+\delta_{z}\cos\beta \\
0 & 0 & 0 & 1
\end{bmatrix}
$$

$$
\tag{16.21}
$$

The HTM vector for the fixture with the workpiece is

$$
\left[{}^{RT}T_{W}\right] =
\begin{bmatrix}
X_{w} \\
Y_{w} \\
Z_{w} \\
1
\end{bmatrix}
\tag{16.22}
$$

where X_{w}, Y_{w}, and Z_{w} are the fixture/workpiece linear positional errors from the nominal reference position. The offsets X_{o}, Y_{o}, Z_{o}, X_{1}, Y_{1}, Z_{1}, X_{2}, Y_{2}, Z_{2}, X_{t}, Y_{t} in the earlier equations may be considered zero since they are most likely incorporated for compensation in the controller during the initial setup of the machine tool. However, since the assembly of the machine components is not perfect, the machine tool errors will cause a relative error between the workpiece and the cutting tool. The spatial relationship between the workpiece and the cutting tool is defined as

$$
\left[{}^{R}T_{W}\right] = \left[{}^{R}T_{TL}\right]\cdot\left[C_{v}\right] \quad\text{or}\quad \left[C_{v}\right] = \left[{}^{R}T_{TL}\right]^{-1}\cdot\left[{}^{R}T_{W}\right]
\tag{16.23}
$$

where C_v is the error correction vector. In another approach, the error correction vector between the cutting tool and workpiece positions is

$$\begin{bmatrix} e_x \\ e_y \\ e_z \\ 0 \end{bmatrix} = \begin{bmatrix} ^R T_X \end{bmatrix} \cdot \begin{bmatrix} ^X T_Y \end{bmatrix} \cdot \begin{bmatrix} ^Y T_S \end{bmatrix} \cdot \begin{bmatrix} X_t \\ Y_t \\ Z_t \\ 1 \end{bmatrix} - \begin{bmatrix} ^R T_Z \end{bmatrix} \cdot \begin{bmatrix} ^Z T_{RT} \end{bmatrix} \cdot \begin{bmatrix} X_w \\ Y_w \\ Z_w \\ 1 \end{bmatrix} \tag{16.24}$$

These equations assume that the workpiece offsets from reference are known. However, the work-piece/fixture positional error is not known, and the relationship between the cutting tool location error vector and the workpiece is required. This relationship can be obtained using the ideal case of the machine tool, where the machine tool errors are zero and the point vector on the workpiece and the cutting edge vector on the cutting tool should be coincident with each other. Therefore

$$\begin{bmatrix} ^R T_W \end{bmatrix} = \begin{bmatrix} ^R T_{TL} \end{bmatrix} \quad \text{or} \quad \begin{bmatrix} ^{RT} T_W \end{bmatrix} = \begin{bmatrix} \begin{bmatrix} ^R T_Z \end{bmatrix} \cdot \begin{bmatrix} ^Z T_{RT} \end{bmatrix} \end{bmatrix}^{-1} \begin{bmatrix} ^R T_X \end{bmatrix} \cdot \begin{bmatrix} ^X T_Y \end{bmatrix} \cdot \begin{bmatrix} ^Y T_S \end{bmatrix} \cdot \begin{bmatrix} ^S T_{TL} \end{bmatrix}$$

$$\begin{bmatrix} X_w \\ Y_w \\ Z_w \\ 1 \end{bmatrix} = \begin{bmatrix} \begin{bmatrix} ^R T_Z \end{bmatrix} \cdot \begin{bmatrix} ^Z T_{RT} \end{bmatrix} \end{bmatrix}^{-1} \cdot \begin{bmatrix} ^R T_X \end{bmatrix} \cdot \begin{bmatrix} ^X T_Y \end{bmatrix} \cdot \begin{bmatrix} ^Y T_S \end{bmatrix} \cdot \begin{bmatrix} X_t \\ Y_t \\ Z_t \\ 1 \end{bmatrix} \tag{16.25}$$

This relationship should be adjusted for the specific angle β of table rotation as used in Equation 16.25. For $\beta = 0$ and $\beta = 90°$, the workpiece vector is, respectively,

$$\begin{bmatrix} X_w \\ Y_w \\ Z_w \\ 1 \end{bmatrix} = \begin{bmatrix} x + X_t \\ y + Y_t \\ Z_t - z \\ 1 \end{bmatrix} \quad \text{and} \quad \begin{bmatrix} X_w \\ Y_w \\ Z_w \\ 1 \end{bmatrix} = \begin{bmatrix} z - Z_t \\ y + Y_t \\ x + X_t \\ 1 \end{bmatrix} \tag{16.26}$$

The error between the cutting tool and the workpiece due to machine geometry errors is calculated by substituting Equation 16.26 into Equation 16.24. X_t and Y_t can be considered zero assuming; the spindle is aligned perfectly to the Z-axis. In addition, the contribution of the Y- and Z-axes to X-direction error is that of the straightness and the Abbe errors in the Y- and Z-directions, respectively. Therefore, the average Abbe error (as explained in Equation 16.2) should be estimated and considered with the straightness error in the previous equations by replacing δ with δ',

$$\delta'_y(x) = \delta_y(x) + x\theta_{xy} \quad \text{and} \quad \delta'_z(x) = \delta_z(x) + x\theta_{xz} \tag{16.27}$$

Validation of the aforementioned analysis and equations is carried out through MATLAB® software. The modeling analysis is quite simple, robust, and easily applied to analyzing and synthesizing the geometric errors of various machine tools to identify either the geometric volumetric workspace errors or workpiece errors. Therefore, if the individual machine tool errors are measured, this analysis can be easily performed. However, the geometric errors could vary significantly along each axis (as explained earlier in the chapter) and therefore either (1) the maximum total error value for each geometric error parameter between any two positions in a given displacement (usually at the full traverse of the slide) or (2) each geometric (linear or angular) error is expressed as a polynomial function of the position of the axis (X, Y, Z) [98,107] such that $\delta_x(X)$ is defined as $\delta_x(X, Y, Z)$, etc. The function will represent the systematic error and may include the uncertainty if several measurements of the same position are available.

The geometric errors in the analytical equations (for compensation purposes as derived for a specific machine tool structure as those in Equation 16.24) are measured by several of the methods discussed in Section 16.4, such as laser interferometer, laser tracker, artifacts with touch probe, and R test, methods.

16.5.3 ERROR COMPENSATION OFFSET METHODS

The machine coordinate system (MCS) defines the home position of the axes. The work coordinate system (WCS) defines the part location within the workspace. The WCS provides the origin and orientation of a point selected either on the part or fixture; it can also be the same as the location of the part coordinate system in CAD. A tool path is programmed in the work offset coordinate system. A part with multiple sides to be machined requires at least one WCS per side for machining or rotation of the machine table to access all sides.

The machine offsets are used to define the relationship of the fixture and cutting tool to the home position. An offset coordinate system consists of nominal values and offset values for the part features. The nominal values are the linear and rotary positions from the MCS. The offset is the adjustment to improve part dimensional accuracy by taking into account varying tool length, table errors, fixture errors, machine tool errors, etc. For example, it is difficult to manufacture a perfect fixture or place the fixture at an exact location on the table; therefore, the offset is used to define its exact location from the nominal position. There are several different types of work offset systems, as shown in Figure 16.37.

A part is located on the fixture using the manufacturing locators and then clamped. The fixture is installed on a rotary table. One of the manufacturing locators (or fixture center) is often a round pin. The offset system (offset adjustment) is associated with a coordinate system positioned at the fixture center. The majority of machine offsets are generally relative to the manufacturing datum in order to consider all the machine tool system errors relative to the part print datum. On many occasions, the part print datum for a number of features is not the same as the manufacturing datum.

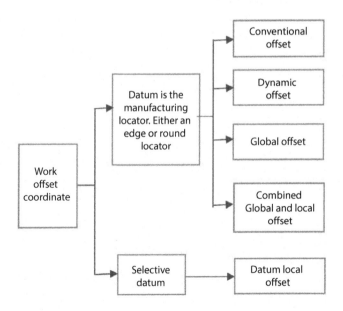

FIGURE 16.37 Several types of offset coordinate systems.

In that case, a tolerance allocation analysis should be carried out so that such features have in-process tolerance (are adjusted) relative to the manufacturing datum. This approach is sufficient for the majority of machined features.

In some cases, it is necessary to adjust a feature directly to a datum other than the manufacturing datum. A common example is a pair of dowel holes used to locate one component to another. In that case, the spatial relationship between the holes is critical, and it may be necessary to offset the secondary dowel hole relative to the primary dowel hole. The datum local offset strategy allows this type of direct adjustment.

Conventional Offset System. The conventional offset system is widely used in CNC machining since it can be used on any machine tool. Conventional offset coordinate systems are used for individual part faces, providing an individual offset for each part face. A conventional offset is defined for a group of features in a single table rotary position. If the center of table rotation is not the program zero location for each face, a separate program zero point will be assigned for each face. In addition, a face may have more than one set of offsets. This, of course, means multiple face offsets will be required since each face requires at least one set of three program zero assignment values (X_i, Y_i, Z_i). Mathematically, a conventional offset is defined as

$$\begin{cases} X_i = x_{ni} + \Delta x_i \\ Y_i = y_{ni} + \Delta y_i \\ Z_i = z_{ni} + \Delta z_i \end{cases} \tag{16.28}$$

where
(X_i, Y_i, Z_i) is the coordinate system origin for each ith rotation/face
(x_{ni}, y_{ni}, z_{ni}) are the nominal values
$(\Delta x_i, \Delta y_i, \Delta z_i)$ are the linear offsets

For instance, consider a part that has features to be machined in three faces as shown in Figure 16.38. This part is machined in a four-axis machine tool with B-rotary table using a single fixture. It has features to be machined at table positions $B_1 = 0$, $B_2 = 90$, and $B_3 = 270$ as shown in Table 16.6. The features of milled surfaces and drilled holes are identified as S and H, respectively.

There are at least three fixture offsets defined as G54, G55, and G57. Data collected from features S100, H101, and H102 are used to calculate the offset for WCS G54. Likewise, data collected from features S201, S202, and H201–H210 are used to calculate the offset for WCS G55. Finally, data collected from features S403, S405, and H430 are used to calculate the offset for WCS G57. The three WCS act separately and are compensated separately as shown in Table 16.7.

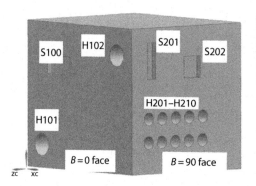
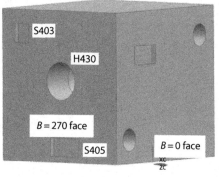

FIGURE 16.38 Example of a part's machined features to illustrate the offset coordinate systems.

TABLE 16.6

Features Machined in Various Table Rotary Positions

B Table Rotary Position	Feature Description	Coordinate System
$B_1 = 0°$	S100, H101, H102	G54
$B_2 = 90°$	S201, S202, H201–H210	G55
$B_3 = 270°$	S403, S405, H430	G57

TABLE 16.7

Parameters of Conventional Offset Compensation

B Table Rotary Position	Coordinate System	Conventional Offset
$B_1 = 0°$	G54	$\Delta x_0, \Delta y_0, \Delta z_0$
$B_2 = 90°$	G55	$\Delta x_{90}, \Delta y_{90}, \Delta z_{90}$
$B_3 = 270°$	G57	$\Delta x_{270}, \Delta y_{270}, \Delta z_{270}$

Dynamic Offset System. The location of the part on the fixture will result in at least some small error from the locating features. Fixtures are not manufactured accurately enough for perfect prediction of locating surface positions and alignments on the table. Measuring the position of locating surfaces on a new fixture may show that the location surfaces do not precisely match dimensions specified on the fixture drawing (although the work holding device will still function properly). In addition, the part center may never match the table center of rotation. Of course, the knowledge of the exact position of each location surface on the fixture from table center is most important. As the table rotates, the fixture orbits around the pallet's center and a new fixture offset is required to create a suitable workpiece coordinate system.

The dynamic offset method is a breakthrough in compensation theory (Fanuc, 2004) because it translates the fixture coordinate system origin from table zero rotary position relative to the other table positions. Regardless of which face of the workpiece is being machined, the programmed coordinates come from the same place. In contrast to the conventional offset method, which treats each machined face as a separate entity, the dynamic offset compensates all the machined features with a single set of offsets ($\Delta Px_0, \Delta Py_0, \Delta Pz_0, \Delta B$). These offsets are translated to each face for four program zero assignment parameters (Wx_i, Wy_i, Wz_i, WB).

The dynamic offset is defined by a rotational translation and can be described as a system of four equations:

$$\begin{cases} Wx = Tx + (Px_0 + \Delta Px_0)\cos B_i - (Pz_0 + \Delta Pz_0)\sin B_i \\ Wy = Ty + Py_0 + \Delta Py_0 \\ Wz = Tz + (Px_0 + \Delta Px_0)\sin B_i + (Pz_0 + \Delta Pz_0)\cos B_i \\ Wb = \Delta B \end{cases} \qquad (16.29)$$

where
(Wx_i, Wy_i, Wz_i) is the *i*th coordinate origin
Wb is the rotary offset for all orientations
B_i is the table nominal position for the *i*th workpiece orientation
(Tx, Ty, Tz) is the *B* table linear nominal position
(Px_0, Py_0, Pz_0) is the fixture linear nominal position relative to the zero table rotary position $B = 0$
$\Delta Px_0, \Delta Py_0, \Delta Pz_0, \Delta B$ are the fixture offsets, respectively in the *X*-, *Y*-, and *Z*-directions when $B = 0$

TABLE 16.8

Parameters of Dynamic Offset Compensation

B Table Rotary Position	Coordinate System	Dynamic Offset Parameters
$B_1 = 0°$	G54	$\Delta Px_0, \Delta Py_0, \Delta Pz_0, \Delta B$
$B_2 = 90°$	G55	
$B_3 = 270°$	G57	

Considering the same example in Figure 16.38, the data collected from all the features machined in the three faces are used to calculate the four offset parameters ΔPx_0, ΔPy_0, ΔPz_0, and ΔB (see Table 16.8). This method requires a perfect alignment of the table center, otherwise the programmed coordinates will not be correct. There is no easy way to deal with table position imperfections. The table center may drift from its original position and cause a degradation in quality.

Global Offset System. The global offset method is an innovation built upon the dynamic offset by incorporating the table center errors in compensation theory [193–195]. All of the machined features are adjusted as a pattern using six offset parameters; the global offset improves part accuracy by considering all the quasi-static errors in a machine including the table and fixture errors. Global offsets translate the table and fixture coordinate system origin from table zero rotary position relative to the other table positions. Optimal compensation is done through software, based on the part measurement data. Even though the pallet is attached to the bed of the machine tool in a fixed predetermined position using precision locating cones in the base of the pallet, there is still an error in manufacturing the locating features in the top plate of the pallet in relation to the locating cones. When the fixture is located on the pallet using the fixed predetermined locating features, the fixture errors are translated into workpiece machining errors. The positioning repetitive indexing accuracy (consistency) of a pallet could be as good as 0.002 mm but pallet errors including linear, angular, and flatness errors will be present. The location of the part on the fixture will result in some small error due to manufacturing errors of the locating features for fixture to pallet and part to fixture location. It is impossible to align the part center to the rotary axis. Therefore, the global offset is developed to compensate all the errors present after a setup is qualified within the machine tool. In addition, the global offset can be used to compensate errors that develop as the machine wears over time, or in the event of damage to the fixture or table.

The global offset is calculated from the equations:

$$\begin{cases} Wx = (Tx + \Delta Tx) + (Px_0 + \Delta Px_0)\cos B_i - (Pz_0 + \Delta Pz_0)\sin B_i \\ Wy = Ty + Py_0 + \Delta Py_0 \\ Wz = (Tz + \Delta Tz) + (Px_0 + \Delta Px_0)\sin B_i + (Pz_0 + \Delta Pz_0)\cos B_i \\ Wb = \Delta B \end{cases} \qquad (16.30)$$

where $\Delta Tx, \Delta Ty, \Delta Tz$ are the table center offsets in the X-, Y-, and Z-directions to compensate the linear table errors. Since the Y table error is collinear to the Y fixture error, the ΔTy is superimposed in the fixture error and eliminated from Equation 16.30. Therefore, the six offset parameters to be estimated in Equation 16.30 are ΔPx_0, ΔPy_0, ΔPz_0, ΔTx, ΔTz, and ΔB. These offset parameters are applied to the global offset.

It is important to understand that at the $B = 0$ rotary table position, the table and fixture offset errors are easily superimposed because their coordinates are aligned with each other. However, when the table rotates to positions other than $B = 0$, the fixture/table offset coordinate system rotates a specified angle B_i. In this case, the global offset changes relative to the angle B_i. The global offset incorporates both the fixture and table offsets through Equation 16.30.

TABLE 16.9
Global Offset Compensation

B Table Rotary Position	Coordinate System	Global Offset Parameters
$B_1 = 0°$	G54	$\Delta Tx_0, \Delta Tz_0, \Delta Px_0, \Delta Py_0, \Delta Pz_0, \Delta B$
$B_2 = 90°$	G55	
$B_3 = 270°$	G57	

Referencing the part in Figure 16.36 with features on three faces, the global offset is determined for the machine tool. The machining errors for each feature on the part are determined from the CMM measurements. These errors arc used to estimate the six parameters in Equation 16.30. The model parameters are used to define the global offset values that are used to compensate each feature in the corresponding coordinate system as shown in Table 16.9. Three offset coordinates are used for this part. The parameters for the G54, G55, and G57 are defined using Equation 16.30 and, respectively, replacing angle B by $B_1 = 0$, $B_2 = 90$, and $B_3 = 270$. The measurement data collected from all machined features are used in the global offset Equation 16.30 to estimate the six parameters ΔPx_0, ΔPy_0, ΔPz_0, ΔTx, ΔTz, and ΔB. Substituting these parameters in Equation 16.30, the global offset compensates all three coordinate systems for all the features in the earlier example. The details of the global offset method are discussed in Reference 196.

Global and Local Offset System. The local offset is an additional refinement (fine adjustment) to the global offset. It is a linear offset that is applied locally to an individual feature or group of features and does not consider additional rotary error. The local offset is defined by incorporating additional adjustment values in a specific table angular position. The local offset provides additional flexibility to improve the dimensional accuracy of certain part feature(s). Hence, it very effectively modifies the global offset for specific feature(s) with a tighter part print tolerance. It is given by

$$\begin{cases} Wx_{Tj} = Wx + \Delta Lx_j \\ Wy_{Tj} = Wy + \Delta Ly_j \\ Wz_{Tj} = Wz + \Delta Lz_j \\ Wb_j = Wb \end{cases} \tag{16.31}$$

where
$(\Delta Lx_j, \Delta Ly_j, \Delta Lz_j)$ is the local offset
$(Wx_{Tj}, Wy_{Tj}, Wz_{Tj})$ is the overall coordinate offset for each feature

The overall compensation value includes the global and local offsets for each coordinate rotational position of the table/fixture or feature. The global offset is obtained through a least squares fit of all the equations generated by organizing the CMM data for each of the part features in Equation 16.30. The local offset value is the residual deviation value after the global offset calculation. More specifically, the global offset will compensate the repetitive quasi-static errors including table and fixture errors. It will also average out some of the thermal (nonlinear) errors and part to part variation if CMM data from several identical parts are available. The local offset is intended to compensate as necessary for errors due to part or fixture distortion, machine tool nonlinear errors, etc.

For instance, a few local offsets can be assigned to a portion of the features for the part shown in Figure 16.36. First, assign a local offset to the three features machined on face $B_1 = 0$. Then, assign a second local offset for features H201 and H210 on face $B_2 = 90$. The remainder of the part features has global offset only. The procedure will use the CMM data for all part features to estimate the global offset for the part. Since all the features on face $B_1 = 0$ are incorporated in a single local offset, a single G54 coordinate system is used to compensate these features as shown in Table 16.10.

TABLE 16.10

Global and Local Offset

B Table Rotary Position	Coordinate System	Global Offset	Local Offset
$B_1 = 0°$	G54	$\Delta Tx_0, \Delta Tz_0$	$\Delta Lx_1, \Delta Ly_1, \Delta Lz_1$
$B_2 = 90°$	G55	$\Delta Px_0, \Delta Py_0$	
$B_2 = 90°$	G56		$\Delta Lx_2, \Delta Ly_2, \Delta Lz_2$
$B_3 = 270°$	G57	$\Delta Pz_0, \Delta B$	

However, for face $B_2 = 90$, the global offset is used to compensate all the features excluding H201 and H210 using the coordinate system G55. The two features H201 and H210 use the global offset together with the local offset in the coordinate system G56. Finally, the features on face $B_3 = 270$ will use the global offset using the coordinate system G57.

16.6   EXAMPLES

Example 16.1 Assume the x-axis of a machine tool has a straightness error because the guideway is curved as illustrated in Figure 16.6. The curvature of the x-axis in the y-direction $[\delta_y(x)]$ is measured to be 0.01 mm. The length of the slide is 1 m. The offset of the part feature to be machined from the slide is 600 mm. Estimate the magnitude of the Abbe error.

The maximum angular error of the slide is estimated from Equation 16.3 to be

$$\theta_x = \frac{8\,\delta_y(x)}{X} = \frac{8\,(0.01)}{1000} = 0.00008\,\text{rad}$$

The Abbe offset error given by Equation 16.2 is about 0.024 mm, and it should be considered as an offset in the part machining program.

$$\delta = P\,\sin\!\left(\frac{\theta_x}{2}\right) = 600\,\sin(0.00004) = 0.024\,\text{mm}$$

Example 16.2 Using the four-axis horizontal machining center analyzed in Section 16.5.2, estimate (1) the error at the cutting tool coordinate locations (0, 0, 0) at $\beta = 0$ and (10, 150, 8) at $\beta = 90°$ with a tool offset (0, 0, 50), and (2) generate a table for several locations in the workspace before and after calibration using the machine tool error values provided in Table 16.11.

Equations 16.14 through 16.26 are set up in MATLAB® software together with the input error parameters given in Table 16.11. Equation 16.24 is considered if the table is rotated so that the proper relationship of the ideal case from Equation 16.25 is used. The solution of the error vector (x, y, z) in Equation 16.24 with $\beta = 0$ is

$$e_x = \delta_x(x) + \delta_x(y) - \delta_x(z) - \delta_x(\beta) - y\alpha_{xy} - y_t\,\alpha_{xy} - z_t\,\alpha_{xz} + y\varepsilon_z(\beta) - y\varepsilon_z(x) + y\varepsilon_z(z) + y_t\,\varepsilon_z(\beta)$$

$$- y_t\,\varepsilon_z(x) - y_t\varepsilon_z(y) + y_t\,\varepsilon_z(z) + z\varepsilon_y(\beta) + z\varepsilon_y(z) - z_t\,\varepsilon_y(\beta) - z_t\,\varepsilon_y(x) + z_t\,\varepsilon_y(y) - z_t\varepsilon_y(z) \tag{16.32}$$

$$e_y = \delta_y(x) + \delta_y(y) - \delta_y(z) - \delta_y(\beta) + x_t\,\alpha_{xy} + z_t\,\alpha_{yz} - x\varepsilon_z(\beta) - x\varepsilon_z(z) - x_t\,\varepsilon_z(\beta) + x_t\varepsilon_z(x) + x_t\,\varepsilon_z(y)$$

$$- x_t\varepsilon_z(z) - z\varepsilon_x(\beta) - z\varepsilon_x(z) - z_t\,\varepsilon_x(x) - z_t\,\varepsilon_x(y) + z_t\,\varepsilon_x(z) + z_t\,\varepsilon_z(\beta) \tag{16.33}$$

TABLE 16.11

Comparison of Four-Axis Machine Tool Performance Before and After Calibration

Machine Tool Errors		Symbol	Before Compensation Error, μm or μrad	After Compensation Error, μm or μrad
X-Axis	Max. pitch, arc-s	$\varepsilon_z(x)$	43	6
	Max. yaw, arc-s	$\varepsilon_y(x)$	24	4
	Max. roll, arc-s	$\varepsilon_x(x)$	19	4
Y-Axis	Max. pitch, arc-s	$\varepsilon_z(y)$	18	2
	Max. yaw, arc-s	$\varepsilon_x(y)$	8	2
	Max. roll, arc-s	$\varepsilon_y(y)$	5	2
Z-Axis	Max. pitch, arc-s	$\varepsilon_x(z)$	37	2
	Max. yaw, arc-s	$\varepsilon_y(z)$	36	5
	Max. roll, arc-s	$\varepsilon_z(z)$	19	3
X-Axis	Straightness in Y, mm	$\delta_y(x)$	10	3
	Straightness in Z, mm	$\delta_z(x)$	18	5
Y-Axis	Straightness in X, mm	$\delta_x(y)$	5	2
	Straightness in Z, mm	$\delta_z(y)$	3	3
Z-Axis	Straightness in X, mm	$\delta_x(z)$	3	3
	Straightness in Y, mm	$\delta_y(z)$	7	2
X-Axis	Max. Pos. error, mm	$\delta_x(x)$	8	4
X-Axis	Max. Pos. error, mm	$\delta_y(y)$	8	6
Z-Axis	Max. Pos. error, mm	$\delta_z(z)$	13	6
	Squareness X–Y,	α_{xy}	64	9
	Squareness X–Z,	α_{xz}	40	4
	Squareness Y–Z,	α_{yz}	85	6
B-Axis	Tilt error motion of B around X-axis	$\varepsilon_{z\beta}$	60	4
	Angular positioning error motion of B	$\varepsilon_{y\beta}$	35	4
	Tilt error motion of B around Z-axis	$\varepsilon_{x\beta}$	35	5
	Radial error motion of B in X-dir	$\delta_{x\beta}$	8	4
	Radial error motion of B	$\delta_{y\beta}$	17	5
	Radial error motion of B in Z-dir	$\delta_{z\beta}$	23	7

$$e_z = \delta_z(x) + \delta_z(y) - \delta_z(z) + \delta_x(\beta) + x\,\alpha_{xz} + x_t\,\alpha_{xz} - y\,\alpha_{yz} - y_t\,\alpha_{yz} + x\varepsilon_y(\beta) + x\varepsilon_y(z) + x_t\,\varepsilon_y(\beta) - x_t\varepsilon_y(x)$$
$$- x_t\,\varepsilon_y(y) + x_t\varepsilon_y(z) + y\varepsilon_x(x) - y\varepsilon_x(y) - y\varepsilon_z(\beta) + y_t\,\varepsilon_x(x) + y_t\,\varepsilon_x(y) - y_t\,\varepsilon_x(z) - z_t\,\varepsilon_z(\beta)$$
$$(16.34)$$

The solution of the error vector (x, y, z) in Equation 16.24 with $\beta = 90°$ is

$$e_x = \delta_x(x) + \delta_x(y) - \delta_x(z) - \delta_z(\beta) - y\,\alpha_{xy} - y_t\,\alpha_{xy} - z_t\,\alpha_{xz} - y\varepsilon_x(\beta) - y\varepsilon_z(x) + y\varepsilon_z(z) - y_t\varepsilon_x(\beta)$$
$$- y_t\,\varepsilon_z(x) - y_t\varepsilon_z(y) + y_t\,\varepsilon_z(z) + z\varepsilon_y(\beta) + z\varepsilon_y(z) - z_t\,\varepsilon_y(\beta) - z_t\,\varepsilon_y(x) + z_t\,\varepsilon_y(y) - z_t\,\varepsilon_y(z)$$
$$(16.35)$$

$$e_y = \delta_y(x) + \delta_y(y) - \delta_y(z) - \delta_y(\beta) + x_t\,\alpha_{xy} + z_t\,\alpha_{yz} + x\varepsilon_x(\beta) - x\varepsilon_z(z) + x_t\,\varepsilon_x(\beta) + x_t\varepsilon_z(x)$$
$$+ x_t\,\varepsilon_z(y) - x_t\varepsilon_z(z) - z\varepsilon_x(z) - z\varepsilon_z(\beta) - z_t\,\varepsilon_x(x) - z_t\,\varepsilon_x(y) + z_t\,\varepsilon_x(z) + z_t\,\varepsilon_z(\beta)$$
$$(16.36)$$

$$e_z = \delta_z(x) + \delta_z(y) - \delta_z(z) + \delta_x(\beta) + x\alpha_{xz} + x_t\alpha_{xz} - y\alpha_{yz} - y_t\alpha_{yz} + x\varepsilon_y(\beta) + x\varepsilon_y(z) + x_t\varepsilon_y(\beta) - x_t\varepsilon_y(x)$$
$$- x_t\varepsilon_y(y) + x_t\varepsilon_y(z) + y\varepsilon_x(x) - y\varepsilon_x(y) - y\varepsilon_z(\beta) + y_t\varepsilon_x(x) + y_t\varepsilon_x(y) - y_t\varepsilon_x(z) - z_t\varepsilon_z(\beta)$$

$$(16.37)$$

1. Assuming a particular feature in the part is located at point (10, 150, 8) at β = 90° with a tool offset (0, 0, 50), the expected machine tool error is (−0.037, 0.001, −0.007) based on the machine tool geometric error parameters in Table 16.11. For a feature located at (0, 0, 0) at β = 0, the expected machine tool error is (−0.0045, 0.0005, −0.015).
2. Using Table 16.11 and Equations 16.32 through 16.37, the machine tool error components are estimated and summarized in Tables 16.12 and 16.13, respectively, for the zero degrees and 90° rotary table positions. While the vector error is above 0.1 mm before calibration, it is reduced below 0.015 mm after calibration. Therefore, the volumetric accuracy improvement after calibration is a roughly 80% reduction in this case.

Example 16.3 The effectiveness of the global compensation method versus dynamic compensation method is illustrated in this example using a four-axis machine tool to machine two bores and drill several holes in an aluminum cylinder block. This machine is selected because it has a large table center location error due to an accident. In this case, either the table must be repaired or the affected machined features must be compensated until the table is corrected. The global offset compensation method is reapplied to compensate the machine tool errors. A cylinder block is loaded on the fixture that sits on top of the table. Table 16.14 lists the machined features along with the nominal value of the feature location. The features in Table 16.14 are machined in either the B table zero or 180° positions. The percent deviation of the actual CMM values from the nominal values in relation to the feature tolerance [% Deviation = (Actual Value − Nominal Value)/tolerance range] is provided in Figure 16.39. The average deviation in relation to tolerances of the machined part is 47%, while the deviation for some of the features is as high as 100%. Global offset compensation can be used to reduce the errors. The CMM data from a single part are loaded into an excel-based global offset calculation software (GCOMP) to estimate

TABLE 16.12

Reduction of Geometric Errors through Compensation of Individual Error Parameters (Given in Table 16.11) for a Four-Axis Horizontal Machining Center at Zero Table Rotation

Workspace Coordinates			Before Calibration Error, μm			After Calibration Error, μm		
X	Y	Z	e_x	e_y	e_z	e_x	e_y	e_z
10	150	8	−8.1	−0.9	−34.6	−2.9	2.2	−6.2
0	0	0	−4.5	0.5	−15.0	−1.8	2.4	−5.0
100	100	100	−0.2	−14.6	−17.7	−1.7	1.0	−4.6
200	200	200	4.1	−29.7	−20.4	−1.6	−0.5	−4.2
300	300	300	8.4	−44.8	−23.1	−1.5	−1.9	−3.8
0	0	300	16.8	−21.1	−15.0	1.0	0.3	−5.0
0	300	0	−12.9	0.5	−56.4	−4.2	2.4	−7.7
300	0	0	−4.5	−23.2	18.3	−1.8	0.3	−1.1
0	0	600	−4.5	−46.9	51.6	3.7	−1.9	−5.0
0	600	0	−21.3	0.5	−97.8	−6.6	2.4	−10.4
600	0	0	−4.5	−46.9	51.6	−1.8	−1.9	2.8

TABLE 16.13
Reduction of Geometric Errors through Compensation of Individual Error Parameters (Given in Table 16.11) for a Four-Axis Horizontal Machining Center at 90° Table Rotation

Workspace Coordinates			Before Calibration Error, μm			After Calibration Error, μm		
X	Y	Z	e_x	e_y	e_z	e_x	e_y	e_z
10	150	8	−37.4	1.1	−7.3	−7.2	2.3	4.9
0	0	0	−19.5	1.8	16.0	−4.8	2.3	6.0
100	100	100	−24.7	−6.4	10.8	−5.6	1.9	6.5
200	200	200	−29.9	−14.5	5.6	−6.4	1.5	7.0
300	300	300	−35.1	−22.6	0.4	−7.2	1.1	7.5
0	0	300	1.8	−27.4	16.0	−2.1	0.5	6.0
0	300	0	−56.4	1.8	−32.9	−9.9	2.3	3.6
300	0	0	−19.5	6.6	49.3	−4.8	2.9	9.9
0	0	600	23.1	−56.5	16.0	0.7	−1.3	6.0
0	600	0	−93.3	1.8	−81.8	−15.0	2.3	1.2
600	0	0	−19.5	11.4	82.6	−4.8	3.5	13.8

TABLE 16.14
Feature Measurement Data Before and After Global Offset Compensation

Part Feature Name	Nominal Value mm	Actual Value Before Compensation mm	Actual Value After Compensation mm
Hole A—X	128.50	128.48	128.49
Hole A—Y	189.08	189.10	189.08
Hole B—X	−26.80	−26.78	−26.79
Hole B—Y	188.70	188.72	188.71
Hole C—X	41.00	41.05	41.01
Hole C—Y	295.50	295.52	295.50
Hole D—X	53.40	53.42	53.42
Hole D—Y	236.20	236.26	236.22
Chamfer E—Top	415.67	415.80	415.58
Chamfer E—Bot	437.77	438.06	437.83
Chamfer F—Top	300.71	300.86	300.66
Chamfer F—Bot	322.81	323.09	322.86
Chamfer G—Bot	77.19	76.91	77.13
Chamfer G—Top	99.29	99.16	99.34
Chamfer H—Bot	−44.79	−45.06	−44.87
Chamfer H—Top	−22.69	−22.85	−22.61
Chamfer I	−54.94	−55.18	−54.98

the six parameters in Equation 16.30 [196]. The estimated global offset parameters are provided in Table 16.15. Simulation software is also used based on Equation 16.30 to calculate the theoretical corrected dimensional values for the part features after the offset parameters are implemented to provide the user the estimated improvement in part quality. The optimization algorithm considers the objective function, convergence criteria, number of equations available based on the machined/measured

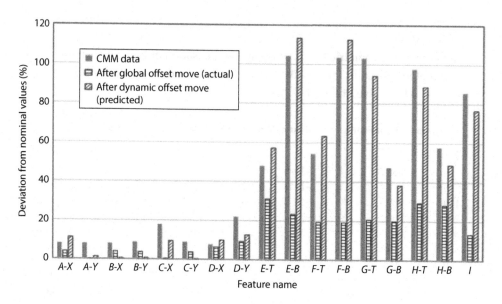

FIGURE 16.39 Comparison of global and dynamic offset compensation in Example 16.3.

TABLE 16.15

Calculated Global Offset and Dynamic Offset Model Parameters

Parameter	Global Offset Values mm	Dynamic Offset Values mm
ΔTx	0.03	0
ΔTz	−0.214	0
ΔPx_0	0.008	0.007
ΔPy_0	−0.023	−0.023
ΔPz_0	−0.001	−0.025
ΔB	−0.01	−0.003

features, regression functions to use, and method of optimization. The variation of model parameters continues until one of the convergence criteria is met.

The simulation also predicts part dimensional errors as a percentage of the feature tolerance after using the global offset. The global offset parameters in Table 16.15 are loaded into the CNC machine to reduce the errors. Another part is then machined and measured in the CMM. The actual feature locations before and after targeting using GCOMP are given in Table 16.14. Figure 16.40 shows the CMM data from the machined part with global offset as compared to the simulation values. The data shown in Figure 16.40 are based on two parts; the first part is used to calculate the offset parameters, the second part is used to validate the accuracy of the offset parameters. The average difference among the predicted and actual data is less than 1%. Figure 16.39 summarizes the "before" and "after" global offset measurement values. The average deviation of all the features machined in the operation is 14% with global offset as compared to 47% without compensation, a quality error reduction of 70%. The min deviation is reduced by 73%, while the max deviation is also reduced by approximately 70%. It is noted that the deviations of the actual and simulated values from the nominal may be positive or negative, depending on whether the measurement value is above or below the nominal part print value of the feature. For clarity, the absolute value of all deviations is shown in the graphs.

The simulated deviation can be larger or smaller than the measured deviation of an actual machined part because some dynamic machine tool errors including part to part variation are not

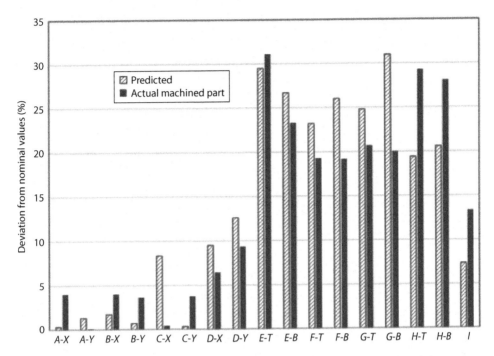

FIGURE 16.40 Comparison between simulations predicted improvement and actual improvement in Example 16.3.

taken into account in the analysis of the global offset especially if only one machined part is used for the estimation of the global offset parameters. This is observed in Figure 16.40. However, experience shows that if the measurements of several machined parts are considered in the analysis, the actual measured deviations after compensation will be smaller since the nominal part to part variation and some of the dynamic errors will be included in the analysis. It can be observed that after compensating the errors, the quality error is significantly reduced.

Using the previously mentioned predictive capabilities of the GCOMP software, it is possible to test a scenario in which the machine is adjusted using the fixture linear moves and table rotary move only, in other words, a dynamic offset. As shown in Table 16.15, there is a significant table center correction ΔTz prescribed by the GCOMP software. Table 16.15 provides the recommended offsets without the table center adjustment as an active variable. The predicted improvement based on the dynamic offset parameters is shown in Figure 16.39, along with the actual improvement provided by the global offset adjustment. It can be seen that, in this case, the global offset adjustment provided a significant improvement in quality over the dynamic offset adjustment. The improvement with the dynamic offset is very small since the major error source happened to be the machine table error. This is because the machine in question has a significant table center error, on the order of 0.200 mm, and the dynamic offset is ineffective in this case.

Example 10.4 The following example is used to illustrate the effectiveness of the global–local compensation method versus conventional compensation method. In this case, a four-axis CNC machine with C table is used to cut an array of 15 slots around a 150 mm bore in a transmission converter housing as shown in Figure 16.41. The part is located on a fixture on a rotary table, and the bore is machined concentric to the table rotation axis so that each slot is milled with a single axis of motion along the X-axis. The slots have equal spacing (24° apart).

One of the important control features is the location of the slot with respect to a specified location on the bolt circle. Therefore, the X location (normal to the radius of the bolt circle) error for each slot is critical. A significant X position deviation is found after machining the first two parts as shown

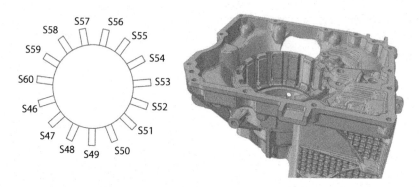

FIGURE 16.41 Machined slots in a transmission converter housing.

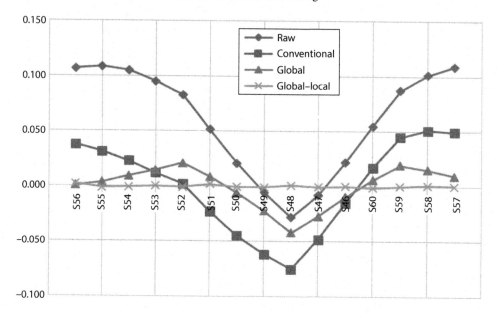

FIGURE 16.42 Comparison among compensation methods for slot machining.

in Figure 16.42. The slots are listed consecutively according to their position in the circle. In addition, the error is found to be repeatable between the two parts. The deviation ranges from −0.028 to 0.109 mm, a significant variation, and has a bell shape. The specific machine has a significant rotary table error that can be corrected either by physically repairing the problem or by applying error compensation. It is also observed that the error is not linear among the 15 slots. The nonlinearity in the error of the symmetric slots could be attributed to the table encoder, the distortion of the part under clamping, or tooling pressure. The mechanical repair procedure (possibly replacing the encoder) is time-consuming and typically limited to an accuracy of approximately 10 μm. The distortion of the part due to machining or clamping pressure is also complicated to address with simple fixture modifications and few trials; the trial-and-error process would be extremely time-consuming and may not fully resolve the quality problem. Therefore, the error compensation approach could be satisfactory and a better option, at least for the short term, as long as the compensation is sufficient to maintain the spacing of the slots within the tight tolerance.

The conventional offset is the first compensation method applied to reduce the location error. Additional parts are machined using the conventional offset. This method reduces the deviation range from −75 to +51 μm, which shifts the bell shape to the center as shown in Figure 16.40. This indicates that the conventional offset averages the error among the 15 slots, but there is no improvement in the deviation range. This method shifts the mean but does not improve the variation. Therefore, the global offset is evaluated using Equation 16.30. The global offset reduces the deviation range from -0.042 μm to $+0.021$ mm, which reduces the range as well as shifts the bell shape to the center as shown in Figure 16.40. The global offset offers obvious improvement over conventional offset, since the deviation range is reduced by 50%. The shape of the graph still indicates a reflection point, and it seems that the shape of the deviations can be eliminated using the global together with the local offset method. This is implemented using Equation 16.31. The global–local offset reduces the deviation range significantly (as illustrated in Figure 16.40), and the bell shape is eliminated, indicating that the nonlinearity has been removed. The previous example indicates that the global offset can effectively compensate the first source of error, while the local offset compensates the second source of error.

REFERENCES

1. S. Lavernhe, C. Tournier, and C. Lartigue, Kinematical performance prediction in multi-axis machining for process planning optimization, *Int. J. Adv. Manuf. Technol.* **37** (2008) 534.
2. T. R. Thomas, B. Bringmann, and W. Knapp, Machine tool calibration: Geometric test uncertainty depends on machine tool performance, *Prec. Eng.* **33** (2009) 524–529.
3. L. Andolfatto, S. Lavernheb, and J. R. R. Moroni, Evaluation of servo, geometric and dynamic error sources on five axis high-speed machine tool, *Int. J. Mach. Tools Manuf.* **51** (2011) 787–796.
4. S. Fraser, M. H. Attia, and M. O. M. Osman, Control-oriented modeling of thermal deformation of machine tools based on inverse solution of time-variant thermal loads with delayed response, *ASME J. Manuf. Sci. Eng.* **126** (2004) 286–296.
5. G. T. Smith, T. Schmitz, J. Ziegert, J. Canning, and R. Zapata, Case study: A comparison of error sources in high-speed milling, *Prec. Eng.* **32** (2008) 126–133.
6. ISO 230-1:2012, *Test Code for Machine Tools—Part 1: Geometric Accuracy of Machine Operating Under No-Load or Quasi-Static Conditions*, 3rd edn., An International Standard by International Standard Organization, Genève, Switzerland, 2012.
7. ASME B5.54-2005, *Methods for Performance Evaluation of Computer Numerically Controlled Machining Centers*, An American National Standard, ASME International, New York, NY, 2005.
8. D. Maxham, The practical approach to volumetric error compensation, *Quality Digest*, **30** (February 22, 2010) 35–40.
9. J. Bryan, The Abbe principle revisited: An updated interpretation, *Prec. Eng.* **1** (1979) 129–132.
10. H. Schwenke, W. Knapp, H. Haitjema, A. Weckenmann, R. Schmitt, and F. Delbressine, Geometric error measurement and compensation of machines—An update, *CIRP Ann.* **57** (2008) 660–675.
11. D. N. Reshetov and V. T. Portman, *Accuracy of Machine Tools*, ASME Press, New York, 1988.
12. M. Weck, *Handbook of Machine Tools, Vol. 4, Metrological Analysis and Performance Tests*, Translated by H. Bibring, Wiley, New York, 1984.
13. L. Huiying and H. Hongxia, Overview on common interpolation methods used in CNC systems, *2010 International Conference on Computer Application and System Modeling* (ICCASM 2010), Taiyuan, China, pp. V11-341–V11-344.
14. S. Suh, S. Kang, D. Chung, and I. Stroud, *Theory and Design of CNC Systems*, Springer-Verlag, London, U.K., 2008.
15. ANSI/ASME B89.6.2-1973, *Temperature and Humidity Environment for Dimensional Measurement*, An American National Standard, ASME International, New York, NY, R2003.
16. K. G. Ahn and D. W. Cho, Analysis of the volumetric error uncertainty of a three-axis machine tool by beta distribution, *Int. J. Mach. Tools Manuf.* **40** (2000) 2235–2248.
17. B. Mullany, *Evaluation and Comparison of the Different Standards Used to Define the Positional Accuracy and Repeatability of Numerically Controlled Machining Center Axes*, University of North Carolina, Charlotte, NC, October 2007.
18. R. G. Wilhelm, N. Srinivasan, and F. Farabaugh, A comparison of international standards for performance testing of machining centers, *First International Machining and Grinding Conference*, SME, September 1995.
19. S. Klabunde and R. Schmidt, How accurate is your machining center? *Mod. Mach. Shop*, **71** (March 1, 1998) 26–31.
20. ISO 230-2:2006, *Test Code for Machine Tools—Part 2: Determination of Accuracy and Repeatability of Positioning Numerically Controlled Axes*, 4th edn., An International Standard by International Standard Organization, Genève, Switzerland, 2006.

21. ISO 230-6:2002, *Test Code for Machine Tools—Part 6: Determination of Positioning Accuracy on Body and Face Diagonals (Diagonal Displacement Tests)*, 1st edn., An International Standard by International Standard Organization, Genève, Switzerland, 2002.

22. ISO/DIS 230-7:2013, *Machine Tools—Test Code for Machine Tools—Part 7: Geometric Accuracy of Axes of Rotation*, An International Standard by International Standard Organization, Genève, Switzerland, 2013.

23. Renishaw Company, *QC20-W Wireless Ballbar System Description and Specifications*, 1211 L-8014-1588-04-A, Renishaw, Inc., Hoffman Estates, IL, 2011.

24. Renishaw QC20-W Wireless Ballbar System Description and Specifications, 1211 L-8014-1588-04-A, Renishaw, Inc., Hoffman Estates, IL, 2011.

25. ISO 230-4:2005, *Test Code for Machine Tools—Part 4: Circular Test for Numerically Controlled Machine Tools*, An International Standard by International Standard Organization, Genève, Switzerland, 2005.

26. ASME B89.3.4M, Axes of Rotation: Methods for Specifying and Testing, ASME, New York, 2004.

27. S. Ibaraki and W. Knapp, Indirect measurement of volumetric accuracy for three-axis and five-axis machine tools: A review, *Int. J. Automat. Technol.* **6** (2012) 110–124.

28. H.-H. Lee, D.-M. Lee, and S.-H. Yang, A technique for accuracy improvement of squareness estimation using a double ball-bar, *Meas. Sci. Technol.* **25** (2014) 094009.

29. W. T. Lei, I. M. Paung, and C.-C. Yu, Total ballbar dynamic tests for five-axis CNC machine tools, *Int. J. Mach. Tools Manuf.* **49** (2009) 488–499.

30. M. S. Uddin, S. Ibaraki, A. Matsubara, and T. Matsushita, Prediction and compensation of machining geometric errors of five-axis machining centers with kinematic errors, *Prec. Eng.* **33** (2009) 194–201.

31. ISO 10791-6, *Test Conditions for Machining Centers—Part 6: Accuracy of feeds, speeds and interpolations*, An International Standard by International Standard Organization, Genève, Switzerland, 2014.

32. S. Ibaraki, Y. Kakino, T. Akai, N. Takayama, I. Yamaji, and K. Ogawa, Measurement of error motions on five-axis machine tools by ball bar measurements (1st Report)—classification of motion error components and development of the modified ball bar device (DBB5), *J. Japan Soc. Precis. Eng.*, **76** (2010) 333–337.

33. M. Tsutsumi, S. Tone, N. Kato, and R. Sato, Enhancement of geometric accuracy of five-axis machining centers based on identification and compensation of geometric deviations, *Int. J. Mach. Tools Manuf.* **68** (2013) 11–20.

34. W. Wang, Y. Zhang, and J. Yang, Double ballbar measurement for identifying kinematic errors of rotary axes on five-axis machine tools, *Adv. Mech. Eng.* (2013) Article ID 198487.

35. S. H. H. Zargarbashi and J. R. R. Mayer, Assessment of machine tool trunnion axis motion error, using magnetic double ball bar, *Int. J. Mach. Tools Manuf.* **46** (2006) 1823–1834.

36. D. M. Lee, Z. Zhu, K.-I. Lee, and S. H. Yang, Identification and measurement of geometric errors for a five-axis machine tool with a tilting head using a double ball-bar, *Int. J. Prec. Eng. Manuf.* **12** (2011) 337–343.

37. S. Xiang, J. Yang, and Y. Zhang, Using a double bar to identify position-independent geometric errors on the rotary axes of five-axis machine tools, *Int. J. Adv. Manuf. Technol.* **70** (2014) 2071–2082.

38. A. Archenti, M. Nicolescu, G. Casterman, and S. Hjelm, A new method for circular testing of machine tools under loaded condition, *Procedia CIRP* **1** (2012) 575–580.

39. J. Ziegert and C. Mize, The laser ball bar: A new instrument for machine tool metrology, *Prec. Eng.* **16**:4 (1994) 259–267.

40. T. Schmitz and J. Ziegert, Dynamic evaluation of spatial CNC contouring accuracy, *Prec. Eng.* **24** (2000) 99–118.

41. K.-C. Fan, H. Wang, F.-J. Shiou, and C.-W. Ke, Design analysis and applications of a 3D laser ball bar for accuracy calibration of multiaxis machines, *J. Manuf. Syst.* **23** (2004) 338–338.

42. D. Marin and N. Predincea, Circular path for CNC machine tools, *Proc. Manuf. Syst.* **8** (2013) 29–34.

43. C. Wang, Non-circular contouring measurement for servo tuning and dynamic performance of a CNC machine, *Proceedings of the 2002 Japan-USA Symposium on Flexible Automation*, Hiroshima, Japan, July 15–17, 2002.

44. T. Schmitz, Rapid prototyping using high speed machining, *Prototyping Technology International*, UK & International Press, Surrey, U.K., 2002, pp. 66–69.

45. J. Thiel, Squeezing out the last micron volumetric compensation improves production accuracy of 5-axis machining centers, *Optik&Photonik, Appl. Rep.* **2** (2014) 55–57.

46. H. Schwenke, *Hunting the Limits of Accuracy: State of the Art for Volumetric Compensation of Machine Tools*, CEO Etalon AG, Braunschweig, Germany, 2014.

47. C. T. Schneider, LaserTracer—A new type of self tracking laser interferometer, *IWAA2004,* CERN, Geneva, Switzerland, October 4–7, 2004.

48. S. Weikert and W. Knapp, R-Test, a new device for accuracy measurements on five axis machine tools, *CIRP Ann.* **53** (2004) 429–432.

49. B. Bringmann, *Improving Geometric Calibration Methods for Multi-Axis Machining Centers by Examining Error Interdependencies Effects,* Thesis, ETH Zurich, 2007.

50. B. Bringmann and W. Knapp, Model-based 'Chase-the-Ball' calibration of a 5-axis machining center, *CIRP Ann.* **55** (2006) 531–534.

51. *Machine Tool Inspection & Analyzer Solutions,* IBS Precision Engineering, Eindhoven, the Netherlands.

52. G. Morifino, System and Process for measuring, compensating and testing numerically controlled machine tool heads and/or tables, European Patent Application EP1549459 A2, July 6, 2005.

53. ISO 10791-6, *Test Conditions for Machining Centers—Part 6: Accuracy of Feeds, Speeds and Interpolations,* Genève, Switzerland, 1998.

54. S. Ibaraki, C. Oyamaa, and H. Otsubo, Construction of an error map of rotary axes on a five-axis machining center by static R-test, *Int. J. Mach. Tools Manuf.* **51** (2011) 190–200.

55. KGM Grid Encoder, Dr. Johannes Heidenhain GmbH, Traunreut, Germany, 2015.

56. Z. Pandilov, N. Durakbasa, and V. Dukovski, Improving the contouring accuracy of a HSC linear motor milling machine, *J. Mach. Eng.* **11** (2011) 130–137.

57. D. Marin and N. Predincea, Circular path for CNC machine tools, *Proc. Manuf. Syst.* **8** (2013) 29–34.

58. S. Ibaraki, Y. Kakino, K. Lee, Y. Ihara, J. Braasch, and A. Eberherr, Diagnosis and compensation of motion errors in NC machine tools by arbitrary shape contouring error measurement, *Laser Metrology & Machine Performance V,* Vol. 34, WIT Transactions on Engineering Sciences, Southampton, U.K., 2001.

59. H. Liu, B. Li, X. Wang, and G. Tan, Characteristics of and measurement methods for geometric errors in CNC machine tools, *Int. J. Adv. Manuf. Technol.* **54** (2011) 195–201.

60. C. Wang, Current issues on 3D volumetric positioning accuracy: Measurement, compensation and definition, *SPIE 2008, Seventh International Symposium on Instrumentation and Control Technology: Measurement Theory and Systems and Aeronautical Equipment,* October 2008.

61. O. Svoboda, P. Bach, G. Liotto, and C. Wang, Definitions and correlations of 3D volumetric positioning errors of CNC machining centers, *IMTS 2004 Manufacturing Conference,* Chicago, IL, September 8–10, 2004.

62. C. Wang, Laser doppler displacement measurement, *Laser Optronics* **9**:6 (1987) 69–71.

63. R. Flynn, Reducing engineering time for compensation of large CNC machines, *J. CMSC* (August 2011) 20–25.

64. C. Wang, *Machine Tool Errors, Measurement, Compensation and Servo Tuning Using MCV-5000 Aerospace Laser Calibration System,* Optodyne, Rancho Dominguez, CA.

65. J. H. Shen, J. G. Yang, and C. Wang, Analysis on volumetric positioning error development due to thermal effect based on the diagonal measurement, *ICADAM 2008 Conference on Advanced Design and Manufacture to Gain a Competitive Edge,* Sanya, China, pp. 293–302.

66. O. Svoboda, P. Bach, G. Liotto, and C. Wang, Machine tool 3D volumetric positioning error measurement under various thermal conditions, *Proceedings of the 3rd International Conference on Precision Mechanical Measurements (ISPMM 2006),* Urumqi, China, 2006.

67. S. Ibaraki and T. Hata, A new formulation of laser step diagonal measurement—Three-dimensional case, *Prec. Eng.* **34** (2010) 516–525.

68. Y. Jianguo, R. Yongqiang, G. Liotto, and G. C. Wang, Theoretical derivations of 4 body diagonal displacement errors in 4 machine configurations, *LAMDAMAP Seventh International Conference,* Cranfield, Bedfordshire, U.K., 27–30 June, 2005.

69. C. Wang and G. Liotto, A theoretical analysis of 4 body diagonal displacement measurement and sequential step diagonal measurement, *Laser Metrology and Machine Performance VI,* Vol. 44, WIT Transactions on Engineering Sciences, Southampton, U.K., 2003.

70. ISO 10360-2, *Geometrical Product Specifications (GPS)—Acceptance and Reverification Tests for Coordinate Measuring Machines (CMM)—Part 2: CMMs Used for Measuring Linear Dimensions,* Genève, Switzerland, 2009.

71. J. A. Soons, Analysis of the step-diagonal test, *Laser Metrology and Machine Performance VII, Seventh International Conference and Exhibition on Laser Metrology, Machine Tool, CMM & Robotic Performance, LAMDAMAP,* Bedford, U.K., 2005, pp. 126–137.

72. M. A. V. Chapman, Limitations of laser diagonal measurements, *Prec. Eng.* **27** (2003) 401–406.

73. O. Svoboda, Testing the diagonal measuring technique, *Prec. Eng.* **30** (2006) 132–144.

74. S. Ibaraki, W. Goto, and A. Matsubara, Issues in laser step diagonal measurement and their remedies, *Proceedings of the 2006 ISFA International Symposium on Flexible Automation,* Osaka, Japan, July 10–12, 2006.

75. Y. Abbaszadeh-Mir, J. R. R. Mayer, G. Cloutier, and C. Fortin, Theory and simulation for the identification of the link geometric error for a five-axis machine tool using a telescoping magnetic ball-bar, *Int. J. Prod. Res.* **40** (2002) 4781–4797.

76. ASME B89.4.19, *Performance Evaluation of Laser-Based Spherical Coordinate Measurement Systems,* An American National Standard, ASME International, New York, NY 2006.

77. H. Schwenke, M. Franke, J. Hannaford, and H. Kunzmann, Error mapping of CMMs and machine tools by a single tracking interferometer, *CIRP Ann.* **54** (2005) 475–478.

78. K. Umetsu, R. Furutnati, S. Osawa, T. Takatsuji, and T. Kurosawa, Geometric calibration of a coordinate measuring machine using a laser tracking system, *Meas. Sci. Technol.* **16** (2005) 2466–2472.

79. S. Ibaraki, K. Takeuchi, T. Yano, T. Takatsuji, S. Osawa, and O. Sato, Estimation of three-dimensional volumetric errors of numerically controlled machine tools by a tracking interferometer, *J. Mech. Eng. Automat.* **1** (2011) 313–319.

80. S. Aguado, D. Samper, J. Santolaria, and J. J. Aguilar, Volumetric verification of multiaxis machine tool using laser tracker, *Sci. World J.* (2014) Article ID 959510.

81. Leica Absolute Interferometer, *A New Approach to Laser Tracker Absolute Distance Meters,* White Paper, Leica Geosystems, Unterentfelden, Switzerland, 2012.

82. H. Schwenke, *Hunting the Limits of Accuracy: State of the Art for Volumetric Compensation of Machine Tools,* Etalon AG, Braunschweig, Germany, nd.

83. C. T. Schneider, LaserTracer—A new type of self tracking laser interferometer, *IWAA2004,* CERN, Geneva, Switzerland, October 4–7, 2004.

84. H. Schwenke, R. Schmitt, P. Jatzkowski, and C. Warmann, On-the-fly calibration of linear and rotary axes of machine tools and CMMs using a tracking interferometer, *CIRP Ann.* **58** (2009) 477–480.

85. J.-M. Linares, J. Chaves-Jacob, H. Schwenke, A. Longstaff, S. Fletcher, J. Flore, E. Uhlmann, and J. Wintering, Impact of measurement procedure when error mapping and compensating a small CNC machine using a multilateration laser interferometer, *Prec. Eng.* **38** (2014) 578–588.

86. K. Wendt, M. Franke, and F. Hartig, Measuring large 3D structures using four portable tracking interferometers, *Measurement* **45** (2012) 2339–2345.

87. S. Moustafa, N. Gerwien, F. Haertig, and K. Wendt, Comparison of error mapping techniques for coordinate measuring machines using the plate method and laser tracer technique, *XIX IMEKO World Congress, Fundamental and Applied Metrology,* September 6–11, 2009, Lisbon, Portugal.

88. K. M. Nasr, A. B. Forbes, B. Hughes, and A. Lewis, ASME B89.4.19 standard for laser tracker verification—Experiences and optimisations, *Int. J. Metrol. Quality Eng.* **3** (2012) 89–95.

89. B. Muralikrishnan, D. S. Sawyer, C. J. Blackburn, S. D. Phillips, B. R. Borchardt, and W. T. Estler, ASME B89.4.19 performance evaluation tests and geometric misalignments in laser trackers, *J. Res. Natl. Inst. Standards Technol.,* 2009.

90. S. Sandwith and R. Lott, Automating laser tracker calibration and technique comparison, *Large Volume Metrology Conference 2007* (LVMC), Liverpool, U.K.

91. R. Madhavan, E. W. Tunstel, and E. R. Messina, Performance evaluation and benchmarking of intelligent systems, in: R. Eastman et al., *Performance Evaluation of Laser Trackers,* Springer, New York, NY, 2009, Section 12.4.

92. J. Wang, J. Guo, G. Zhang, B. Guo, and H. Wang, The technical method of geometric error measurement for multi-axis NC machine tool by laser tracker, *Meas. Sci. Technol.* **23** (2012).

93. B. Hughes, A. Forbes, A. Lewis, W. Sun, D. Veal, and K. Nasr, Laser tracker error determination using a network measurement, *Meas. Sci. Technol.* **22** (2011).

94. B. Muralikrishnan, V. Lee, C. Blackburn, D. Sawyer, S. Phillips, W. Ren, and B. Hughes, Assessing ranging errors as a function of azimuth in laser trackers and tracers, *Meas. Sci. Technol.* **24** (2013).

95. J. Mou and R. C. Liu, A method for enhancing the accuracy of CNC machine tools for on-machine inspection, *J. Manuf. Syst.* **11** (1992) 229–237.

96. J. S. Agapiou and H. Du, Assuring the day-to-day accuracy of coordinate measuring machines—A comparison of tools and procedures, *J. Manuf. Process.* **9** (2007) 109–120.

97. M. Omari, D. Ajao, G. G. Kampmann, and I. Schmadel, Machine tool accuracy quick check in automotive tool and die manufacturing, *International Conference on Smart Machining Systems,* Gaithersburg, MD, March 2007.

98. L. Arriba et al., Methods and artifacts to calibrate large CMMs, *SMT Progress Report, Trimek Annex Report*, Project SMT4-CT97-2183, Bundesallee, Germany, July 1999.

99. S. Asanuma, Inspection master block and method of producing the same, U.S. Patent No. 6,782,730, August 31, 2004.

100. ASME B89.4.1, *Methods for Performance Evaluation of Coordinate Measuring Machines*, ASME, New York, 1997.

101. W. A. Watts, Artifact and method for verifying accuracy of a positioning apparatus, U.S. Patent No. 5,313,410, May 17, 1994.

102. E. P. Morse, Artifact selection and its role in CMM evaluation, IDW-2002, May 8, 2002.

103. G. X. Zhang and Y. F. Zang, A method for machine geometry calibration using 1-D Ball array, *CIRP Ann.* **40** (1991) 519–522.

104. T. Liebrich, B. Bringmann, and W. Knapp, Calibration of a 3D-ball plate, *Prec. Eng.* **33** (2009) 1–6.

105. E. Trapet, J. Aguilarmartin, J. Yague, H. Spaan, and V. Zeleny, Self-centering probes with parallel kinematics to verify machine tools, *Prec. Eng.* **30** (2006) 165–179.

106. B. Bringmann, A. Kung, and W. Knapp, A measuring artefact for true 3D machine testing and calibration, *CIRP Ann.* **54** (2005) 471–474.

107. H. Kunzmann, A uniform concept for calibration, acceptance test, and periodic inspection of coordinate measuring machines using reference objects, *CIRP Ann.* **39** (1990) 561–564.

108. S.-H. Suh, E.-S. Lee, and S.-Y. Jung, Error modeling and measurement for the rotary table of five-axis machine tools, *Int. J Adv. Manuf. Technol.* **14** (1998) 656–663.

109. E. Trapet and F. Wiudele, A reference object based method to determine the parametric error components of coordinate measuring machines and machine tools, *Measurement* **9** (1991) 17–22.

110. X. Mao, B. Li, H. Shi, H. Liu, X. Li, and P. Li, Error measurement and assemble error correction of a 3D-step-gauge, *Front. Mech. Eng. China* **2** (2007) 388–393.

111. J. P. Choi, B. K. Min, and S. J. Lee, Reduction of machining errors of a three-axis machine tool by on-machine measurement and error compensation system, *J. Mater. Process. Technol.* **155–156** (2004) 2056–2064.

112. W. Gao and A. Kimura, A three-axis displacement sensor with nanometric resolution, *CIRP Ann.* **56** (2007) 529–532.

113. Z. Du, S. Zhang, and M. Hong, Development of a multi-step measuring method for motion accuracy of NC machine tools based on cross grid encoder, *Int. J. Mach. Tools Manuf.* **50** (2010) 270–280.

114. K. Lee, S. Ibaraki, A. Matsubara, Y. Kakino, Y. Suzuki, S. Arai, and J. Braasch, A servo parameter tuning method for high-speed NC machine tools based on contouring error measurement, *Laser Metrology and Machine Performance VI*, WIT Press, Ashurst, Southampton, U.K., 2003.

115. K. Nagaoka, A. Matsubara, T. Fujita, and T. Sato, Analysis method of motion accuracy using NC system with synchronized measurement of tool-tip position, *Int. J. Automat. Technol.* **3** (2009) 394–400.

116. W. Zhu, Z. Wang, and K. Yamazaki, Machine tool component error extraction and error compensation by incorporating statistical analysis, *Int. J. Machine Tools Manuf.* **50** (2010) 798–806.

117. S. Ibaraki and Y. Tanizawa, Vision-based measurement of two-dimensional positioning errors of machine tools, *J. Adv. Mech. Des. Syst. Manuf.* **5** (2011) 315–328.

118. J. R. R. Mayer, Five-axis machine tool calibration by probing a scale enriched reconfigurable uncalibrated master balls artefact, *CIRP Ann.* **61** (2012) 515–518.

119. T. Erkan and J. R. R. Mayer, A cluster analysis applied to volumetric errors of five-axis machine tools obtained by probing an uncalibrated artefact, *CIRP Ann.* **59** (2010) 539–542.

120. S. Ibaraki, T. Iritani, and T. Matsushita, Calibration of location errors of rotary axes on five-axis machine tools by on-the-machine measurement using a touch-trigger probe, *Int. J. Mach. Tools Manuf.* **58** (2012) 44–53.

121. Q. Bi, N. Huang, C. Sun, Y. Wang, L. Zhu, and H. Ding, Identification and compensation of geometric errors of rotary axes of five-axis machine by on-machine measurement, *Int. J. Mach. Tools Manuf.* **89** (2015) 182–191.

122. N. A. Mchichi and J. R. R. Mayer, Axis location errors and error motions calibration for a five-axis machine tool using the SAMBA method, *Procedia CIRP* **14** (2014) 305–310.

123. ISO 10360-2, *Geometrical Product Specifications (GPS)—Acceptance and Reverification Tests for Coordinate Measuring Machines (CMM)—Part 2: CMMs Used for Measuring Linear Dimensions*, International Organization for Standardization (ISO), 2009.

124. ISO 10360-3, *Geometrical Product Specifications (GPS)—Acceptance and Reverification Tests for Coordinate Measuring Machines (CMM)—Part 3: CMMs with the Axis of a Rotary Table as the Fourth Axis*, International Organization for Standardization (ISO), 2000.

125. ISO/TS 15530-3, *Geometrical Product Specifications—Coordinate Measuring Machines: Technique for Determining the Uncertainty of Measurement—Use of Calibrated Workpieces or Standards, Part 3 Use of Calibrated Workpieces or Measurement Standards*, International Organization for Standardization (ISO), 2011.

126. P. D. Steven, Performance evaluations, in: J. A. Bosch, Ed., *Coordinate Measuring Machines and Systems*, Marcel Dekker, New York, 1995.

127. M.-W. Cho, G.-H. Kim, T.-I. Seo, Y.-C. Hong, and H. H. Cheng, Integrated machining error compensation method using OMM data and modified PNN algorithm, *Int. J. Mach. Tools Manuf.* **46** (2006) 1417–1427.

128. L. Mears, J. T. Roth, D. Djurdjanovic, X. Yang, and T. Kurfess, Quality and inspection of machining operations: CMM integration to the machine tool, *ASME J. Manuf. Sci. Eng.* **131** (2009).

129. H. Kunzmann, E. Trapet, and F. Waldele, Results of the international comparison of ball plate measurements in CIRP and WECC, *CIRP Ann.* **44** (1995) 479–482.

130. E. Loewen, Self-calibration: Reversal, redundancy, error separation, and 'absolute testing', *CIRP Ann.* **45** (1996) 617–634.

131. J. Ye, An exact algorithm for self-calibration of two-dimensional precision metrology stages, *Prec. Eng.* **20** (1997) 16–32.

132. O. Sato, S. Osawa, and T. Takatsuji, Calibration of two dimensional grid prates with the reversal measuring technique, *Proceedings of the Asian Symposium for Precision Engineering and Nanotechnology (ASPEN)*, Kitakyushu, Japan, 2009.

133. Equator, The Versatile Gauge, Renishaw, Inc., Hoffman Estates, IL, 2013.

134. AIA/NAS, NAS 979, Uniform Cutting Tests—NAS Series Metal Cutting Equipment Specifications, 1969.

135. AIA/NAS, NAS 979, Uniform Cutting Tests—NAS Series Metal Cutting Equipment Specifications, Revision/2nd Edition.

136. ISO 10791-7, Test Conditions for Machining Centers—Part 7: Accuracy of finished test pieces.

137. C. Hong, S. Ibaraki, and A. Matsubara, Influence of position-dependent geometric errors of rotary axes on a machining test of cone frustum by five-axis machine tools, *Prec. Eng.* **35** (2011) 1–11.

138. Y. Morimoto, K. Nakato, and M. Gontani, Accuracy evaluation of 5-axis machining center based on measurements of machined workpiece—Evaluation of accuracy of 5-axis controlled machining center, *Int. J. Automat. Technol.* **6** (2012) 675–681.

139. M. Gebhardt, W. Knapp, and K. Wegener, 5-Axis Test piece—Influence of machining position, *Proceedings of 2012 Machine Tool Technologies Research Foundation (MTTRF) Annual Meeting*, Iga City, Japan, pp. 299–304.

140. M. Givi and J. R. R. Mayer, Validation of volumetric error compensation for a five-axis machine using surface mismatch production tests and on-machine tough probing, *Int. J. Mach. Tools Manuf.* **87** (2014) 89–95.

141. Y. Morimoto, Study on accuracy compensation of machining center based on measurement results of machined workpiece, *Proceedings of the Fifth International Conference on Leading Edge Manufacturing in 21st Century*, 2009, No. 09-207, pp. 55–60.

142. S. Bossoni, Geometric and dynamic evaluation and optimization of machining centers, Dissertation, ETH Zurich, 2009.

143. S. Ibaraki, M. Sawada, A. Matsubara, and T. Matsushita, Machining tests to identify errors of five-axis machining tools, *Prec. Eng.* **34** (2010) 387–398.

144. S. L. Chen, T. H. Hsieh, W. Y. Jywe, C. H. She, and J. N. Lee, Kinematic errors evaluation of five-axis machine tool using direct cutting method, *Adv. Sci. Lett.* **8** (2012) 158–163.

145. H. Neiyuan, N. Jun, and Y. Jingxia, Generalized model formulation technique for error synthesis and error compensation on machine tools, SME Standard MS98-193, SME, Dearborn, MI, 1998.

146. R. Ramesh, M. A. Mannan, and A. N. Poo, Error compensation in machine tools—A review. Part I: Geometric, cutting-force induced and fixture-dependent errors, *Int. J. Mach. Tools Manuf.* **40** (2000) 1235–1256.

147. R. Ramesh, M. A. Mannan, and A. N. Poo, Error compensation in machine tools—A review. Part II: Thermal errors, *Int. J. Mach. Tools Manuf.* **40** (2000) 1257–1284.

148. Q. Cheng, C. Wu, P. Gu, W. Chang, and D. Xuan, An analysis methodology for stochastic characteristic of volumetric error in multiaxis CNC machine tool, *Math. Prob. Eng.* (2013) Article ID 863283.

149. J. Mayr et al., Thermal issues in machine tools, *CIRP Ann.* **61** (2012) 771–791.

150. M. H. Attia, Intelligent machine tool systems-computational aspect of the control of the thermal deformation problem, *Proceedings of the IEEE Seventh International Conference on Intelligent Engineering Systems,* Assiut, Egypt, 2003.

151. M. Weck, O. Schuze, F. Michels, and R. Bonse, Optimization of machine tools performance and accuracy, *ASME Symposium on Intelligent Machine Tool Systems*, International Mechanical Engineering Congress and Exposition, November 1994, pp. 895–908.

152. M. Pajor and J. Zapłata, Compensation of thermal deformations of the feed screw in a CNC machine tool, *Adv. Manuf. Sci. Technol.* **35** (2011).

153. J. S. Chen, J. X. Yuan, J. Ni, and S. M. Wu, Real-time compensation for time-variant volumetric errors on a machining center, *ASME J. Eng. Ind.* **115** (1993) 472–479.

154. H. L. Zeng, Y Sun, and H. Y. Zhang, Thermal error compensation on machine tools using rough set artificial neural networks, *Word Congress on Computer Science and Information Engineering CSIE2009*, Los Angeles, CA, 2009.

155. C. Chen, Y. Zhang, and H. Chen, Selection and modeling of temperature variables for the thermal error compensation in servo system, *Tenth International Conference on Electronic Measurement & Instruments ICEMI2011*, Cheng Du, China, August 16–18, 2011.

156. E. M. Miao, Y. Y. Gong, T. J. Cheng, and H. D. Chen, Application of support vector regression to thermal error modeling of machine tools, *Opt. Prec. Eng.* **21** (2013) 980–986.

157. E. M. Miao, Y. Y. Gong, P. C. Niu, C. Z. Ji, and H. D. Chen, Robustness of thermal error compensation modeling models of CNC machine tools, *Int. J. Adv. Manuf. Technol.* **69** (2013) 2593–2603.

158. P. Schellekens, J. Soons, H. Spann, V. Look, E. Trapet, J. Dooms, H. De Ruiter, and M. Maisch, Development of methods for the numerical correlation of machine tools, Report to the Commission of the European Communities, EUR 15377 EN 1-191, 1993.

159. T. Sata, Y. Takeuchi, and N. Okubo, Control of the thermal deformation of a machine tool, *Proceedings of the 16th International MTDR Conference,* 1975, pp. 203–208.

160. T. Sata, Y. Takeuchi, M. Sakamoto, and M. Weck, Improvement of working accuracy on NC lathe by compensation for the thermal expansion of tool, *CIRP Ann.* **30** (1981) 445–449.

161. T. Moriwaki, Thermal deformation and its on-line compensation of hydraulically supported precision spindle, *CIRP Ann.* **37** (1988) 393–396.

162. H. Zhang, J. Yang, Y. Zhang, J. Shen, and C. Wang, Measurement and compensation for volumetric positioning errors of CNC machine tools considering thermal effect, *Int. J. Adv. Manuf. Technol.* **55** (2011) 275–283.

163. J. Zhu, Robust thermal error modelling and compensation for CNC machine tools, Thesis, University of Michigan, Ann Arbor, MI, 2008.

164. M. Gebhardt, A. Schneeberger, W. Knapp, and K. Wegener, Measuring, modeling and compensating thermally caused location errors of rotary axes, *Proceedings of the MTTRF 2013 Annual Meeting*, San Francisco, CA.

165. C. Brecher, M. fey, and M. Wennemer, Volumetric thermo-elastic machine tool behavior, *Prod. Eng.* **9** (2005) 119–124.

166. J. Jedrzejewski, W. Modrzycki, Z. Kowal, W. Kwasny, and Z. Winiarski, Precise modeling of HSC machine tool thermal behaviour, *J. Achieve. Mater. Manuf. Eng.* **24** (2007) 245–252.

167. J. X. Yuan and J. Ni, The real-time error compensation technique for CNC machining systems, *Mechatronics* **8** (1998) 359–380.

168. S. Ibaraki and Y. Ota, A machining test to evaluate geometric errors of five-axis machine tools with its application to thermal deformation tests, *Procedia CIRP* **14** (2014) 323–328.

169. S. M. Wang, Y. Han-Jen, and L. Hung-Wei, A new high-efficiency error compensation system for CNC multi-axis machine tools, *Int. J. Manuf. Technol.* **28** (2006) 518–526.

170. M. Rahou, A. Cheikh, and F. Sebaa, Real time compensation of machining errors for machine tools NC based on systematic dispersion, *World Acad. Sci. Eng. Technol.* **56** (2009).

171. M. K. Yeung, High performance machining automation, SAE Technical Paper 2004-01-1245, 2004.

172. S. Zhu, G. Ding, S. Qin, J. Lei, L. Zhuang, and K. Yan, Integrated geometric error modeling, identification and compensation of CNC machine tools, *Int. J. Mach. Tools Manuf.* **52** (2012) 24–29.

173. A. C. Okafor and Y. M. Ertekin, Derivation of machine tool error models and error compensation procedure for three axes vertical machining center using rigid body kinematics, *Int. J. Mach. Tool Manuf.* **40** (2000) 1199–1213.

174. K. Kim and M. K. Kim, Volumetric accuracy analysis based on generalized geometric error model in multi-axis machine tools, *Mech. Mach. Theory* **26** (1991) 207–219.

175. A. W. Khan and C. Wuyi, Systematic geometric error modeling for workspace volumetric calibration of a 5-axis turbine blade grinding machine, *Chin. J. Aeronaut.* **23** (2010) 604–615.

176. J. H. Lee, Y. Kiu, and S. H. Yang, Accuracy improvement of miniaturized machine tool: Geometric error modeling and compensation, *Int. J. Mach. Tool Manuf.* **46** (2006) 1508–1516.

177. M. T. Lin, Y. T. Lee, W. Y. Jywe, and J. A. Chen, Analysis and compensation of geometric error for five-axis CNC machine tools with tilting rotary table, *IEEE/ASME International Conference on Advanced Intelligent Mechatronics (AIM2011)*, Budapest, Hungary, 2011.

178. V. S. B. Kiridera and P. M. Ferreira, Kinematic modeling of quasistatic errors of three-axis machining centers, *Int. J. Mach. Tool Manuf.* **34** (1994) 85–100.

179. W. Tian, W. Gao, W. Chang, and Y. Nie, Error modeling and sensitivity analysis of a five-axis machine tool, *Math. Prob. Eng.* (2014) Article ID 745250.

180. M. Rahman, Modeling and measurement of multi-axis machine tools to improve positioning accuracy in a software way, Dissertation, University of Oulu, Finland, 2004.

181. M. Vahebi Nojedeh, M. Habibi, and B. Arezoo, Tool path accuracy enhancement through geometrical error compensation, *Int. J. Mach. Tools Manuf.* **51** (2011) 471–482.

182. A. P. Longstaff, S. Fletcher, and A. Myers, Volumetric compensation for precision manufacture through a standard CNC controller, *Twentieth Annual Meeting of the American Society for Precision Engineering*, Norfolk, VA, October 9–14, 2005.

183. K. Lau and H. Qiao, Practical approaches to volumetric accuracy enhancement for large machine tools (VALMT), *CTMA Conference*, EMO, Hannover, Germany, September 19–24, 2011.

184. J. Dallam, Volumetric accuracy for large machine tools, *NCMS-CTMA Symposium*, Detroit, MI, 2009.

185. S. Smith, B. A. Woody, and J. A. Miller, Improving the accuracy of large scale monolithic parts using fiducials, *CIRP Ann.* **54** (2005) 483–486.

186. B. A. Woody, S. Smith, R. J. Hocken, and J. A. Miller, A technique for enhancing machine tool accuracy by transferring the metrology reference from machine tool to the workpiece, *ASME J. Manuf. Sci. Eng.* **129** (2007) 636–643.

187. J. Gu, J. S. Agapiou, and S. Kurgin, CNC machine tool work offset error compensation method, *NAMRC/MSEC 2015 Conference*, Charlotte, NC, June 2015.

188. A. Hansel, K. Yamazaki, and K. Konishi, Improving CNC machine tool geometric precision using manufacturing process analysis techniques, *Procedia CIRP* **14** (2014) 263–267.

189. Vericut Module: CNC Machine Simulation, CGTech, Irvine, CA, 2014.

190. R. R. Fespeman, S. P. Moylan, and M. A. Donmez, A virtual machine tool for the evaluation of standardized 5-axis performance tests, NIST Report, Gaithersburg, MD, 2012.

191. B. Bringmann, J. P. Besuchet, and L. Rohr, Systematic evaluation of calibration methods, *CIRP Ann.* **57** (2008) 529–532.

192. A. H. Slocum, *Precision Machine Design*, Prentice-Hall, Englewood Cliffs, NJ, 1992.

193. J. Gu and S. K. Kurgin, Global offset compensation for a CNC machine, U.S. patent 8,676,373, March 18, 2014.

194. J. Gu, S. K. Kurgin, and P. J. Deeds, Electronic system and method for compensating the dimensional accuracy of a 4-axis CNC machining system using global offsets, U.S. patent 8,509,940, August 13, 2013.

195. J. Gu, S. K. Kurgin, and P. J. Deeds, Electronic system and method for compensating the dimensional accuracy of a 4-axis CNC machining system using global and local offsets, U.S. patent 8,712,577, April 29, 2014.

196. J. Gu, J. S. Agapiou, and S. Kurgin, Global offset modeling and compensation for machine tool accuracy improvement, *NAMRC/MSEC 2016 Conference*, Blacksburg, Virginia, June 2016.

17 Gear Machining

17.1 INTRODUCTION

Gears are common mechanical components used for power transmission, speed regulation, and motion control. Although gears may be cast, extruded, or roll formed for noncritical applications, gear tooth forms and other dimensions must be precisely machined in applications requiring high load capacity, long fatigue life, and minimal vibration. Many common mechanisms, such as automotive transmissions and stationary gearboxes, contain multiple machined gear sets, and high-precision gears are critical components in jet engine, helicopter transmission, wind turbine, and oilfield applications [1–5]. Most precision gears are machined from ferrous alloys or powder metals.

Due to their unique geometry and required precision, gears are manufactured using specialized machining and finishing methods, some of which are also used for splines. The choice of methods for a particular application depends on the type of gear, workpiece material, available equipment, and required precision. Gear machining has traditionally required specialized gear cutting equipment, and while this is still true for high-volume and high-precision applications, improvements in five-axis machining centers and multitasking machines have increased options for flexible batch machining of general-purpose gears [6].

Because of their wide field of application and variety, designing and manufacturing gears is a specialized field with a large technical literature [1,7–10]. This chapter provides only a broad overview of gear machining and finishing methods for common gear types, concentrating on tooling and process details and ranges of applications. Gear design and inspection methods, heat treatment, and dedicated gearmaking machines are not discussed in detail, although for complex gears the design, machining, heat treatment, and inspection processes are interrelated and constrained by machine capabilities.

17.2 GEAR TYPES AND GEOMETRY

17.2.1 GEAR TYPES

Gears come in a large variety of types and can be classified in several ways. From a machining viewpoint, a kinematic classification based on the axes of rotations of gear sets is most useful. Gears are usually used in pairs or sets, and the relation between the axes of rotation of individual gears in a set influences the required orientation and form of the gear teeth and the potential machining options. Common gear types used in this classification system include parallel axis, intersecting axis, and crossed axis gears.

In *parallel axis gears*, the rotational axes of the drive and driven gears are parallel. Common types of parallel axis gears include spur gears, helical gears, double helical (herringbone) gears, and rack gears (Figure 17.1). (A rack gear is a spur gear with an infinite radius.) Parallel axes gears are also called *cylindrical gears* since they have a cylindrical pitch surface. They are ubiquitous in mechanical systems and are the most common type of gear.

In *intersecting axis gears*, the gear axes are not parallel but intersect. Common types include straight bevel gears, spiral bevel gears, and face gears (Figure 17.2). Bevel gears with equal numbers of teeth and axes at right angles are also called miter gears. Bevel and spiral bevel gears have a conical pitch surface.

In *crossed axis gears*, the gear axes are not parallel and do not intersect. Common types include hypoid bevel gears, worm gears, and crossed helical gears (Figure 17.3). Hypoid bevel gears have

(a)   (b)

(c)   (d)

FIGURE 17.1 Common types of parallel axes gears: (a) spur gears, (b) helical gears, (c) double helical (herringbone) gear, and (d) rack gear with pinion.

larger pinions and contact areas than corresponding spiral bevel gears, resulting in increased load-carrying capability, and are commonly used in automotive rear differentials.

17.2.2 GEAR GEOMETRY AND ACCURACY CLASSES

The basic geometry of spur gears is shown in Figure 17.4. Gear tooth profiles are designed to ensure smooth power transmission by minimizing speed variations and vibrations and to minimize internal stresses due to cyclic contact loads. The most commonly used tooth profile for spur gears is the involute profile, based on the involute of a circle, the curve drawn by a point on a taut cord unwinding from a base circle [9]. (Most gears do not use a true involute profile but incorporate modifications for clearance or load distribution.) For gear sets with involute tooth profiles, the common normal to the tooth profiles at contact points throughout the mesh cycle passes through a fixed pointed on the line of centers, the pitch point, and the velocity ratio of the gears is constant through the mesh cycle and equal to the ratio of the gear diameters. The pitch point also determines the pitch diameter of the gears. The line of action passes through the pitch point and is tangent to both base circles. The pressure angle is the angle between the line of action and the common tangent to the pitch circles. Standard gears are manufactured with fixed pressure angles, with 14.5° and 20° being common values.

From a machining viewpoint, gear sizes are characterized by the diametrical pitch (in the English system) or the module (in the metric system). The diametrical pitch is the number of teeth on a gear divided by the pitch diameter in inches. Conversely, the module is the pitch diameter in mm divided by the number of teeth. For a fixed pitch diameter, the gear tooth size increases with decreasing diametrical pitch or increasing module. (Due to this inverse definition, the product of the diametrical pitch and module on the same gear is 25.4.) In many cases, gear tooling, especially form tooling, is designed for a restricted range of diametrical pitches or modules.

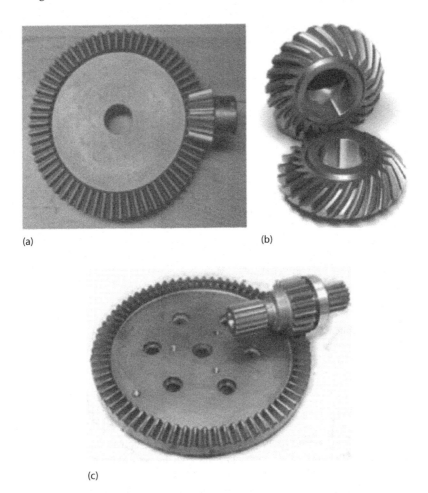

(a) (b)

(c)

FIGURE 17.2 Common types of intersecting axes gears: (a) spur or straight bevel gear, (b) spiral bevel gear, and (c) face gear.

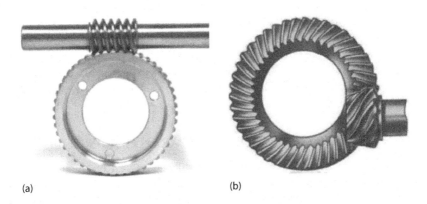

(a) (b)

FIGURE 17.3 Common types of nonintersecting axes gears: (a) worm gear and (b) hypoid bevel gear.

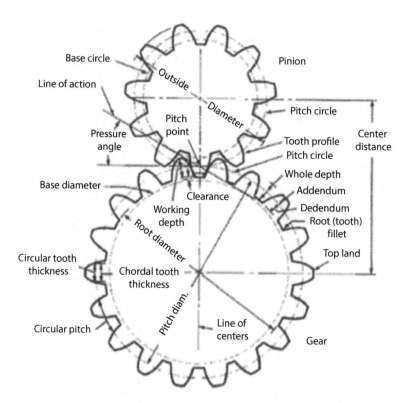

FIGURE 17.4   Gear geometry nomenclature.

Because small variations in the tooth profile can affect fatigue life and noise generation, gears are classified by accuracy class, with each class having defined tolerances on profile errors and tooth-to-tooth variation defined by standards. The current U.S. National Standard for spur and helical gears is ANSI/AGMA 2015-1-A01, published by the American Gear Manufacturers Association (AGMA) [11]. This standard is a replacement of earlier standards intended to more compatible with ISO standards [11,12] and uses the grading system of ISO 1328-1 [13]. It establishes 10 accuracy grades, labeled A2–A11, with tolerances increasing and accuracy decreasing as the grade number increases. Form error tolerances generally increase by a factor of $\sqrt{2}$ for each increase in grade number. Accuracy grades are also classified into accuracy groups, with grades A2–A5 being classified as high accuracy, A6–A9 as medium accuracy, and A10 and A11 as low accuracy. Corresponding AGMA and ISO standards establish accuracy grades for bevel gears [14,15].

The older Q accuracy grade numbers defined in earlier AGMA standards, such as AGMA 390.01 and ANSI/AGMA 2000-A88, are still widely used. In contrast to the current AGMA and ISO methods, in this system gears with a higher Q class number had increased precision; a Q7 or Q8 grade gear is a medium accuracy gear, whereas a Q13 gear would be a high precision or master gear. Limited graphs comparing Q and A accuracy class tolerances are included in the current standard [11], and Radzevich [1] provides a table relating equivalent ranges of AGMA Q-grades and DIN (ISO) accuracy classes.

17.3   TOOTH MACHINING METHODS FOR PARALLEL AXIS GEARS

Gear teeth are machined by a primary process, and depending on the required precision, they may require a secondary process to meet finish or dimensional tolerances. A number of primary processes can be used to machine gear teeth in parallel axis gears. These can be classified as form-cutting methods, in which the cutting tool has the form of a gear tooth, and generating methods, in

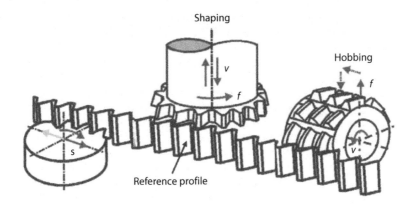

FIGURE 17.5 Generated reference rack gear in hobbing and shaping.

which the tooth flank is generated by relative rotation between the tool and blank during cutting. In generating methods, the rotation of the blank is often referred to as the generating rotation or motion. The most common form cutting methods are broaching and form milling; common generating methods include hobbing and shaping. As shown in Figure 17.5 [3,16], in hobbing and shaping the tool geometry and process kinematics create a straight rack gear-shaped generating surface. Parallel axes gears can also be ground from solid blanks using form wheels. Finally, spur and helical gears can be milled on five-axis machining centers as discussed in Section 17.5.

17.3.1 BROACHING

As described in Section 2.8, in broaching operations a tool with a series of cutting edges translates across a part to generate a flat or profiled surface with a constant cross section in the plane normal to the direction of broach travel. Successive cutting teeth are set at increasing heights so that each removed a controlled amount of stock. Broaches are widely used to cut internal splines or keyways and can also cut internal or external spur and helical gear teeth. Three types of gear broaching methods are common: internal broaching of splines or gear teeth, external or surface broaching of gear teeth or short gear tooth segments on disks, and pot broaching of spur gears [17–23]. As discussed in Section 2.8, broaching processes are not very flexible but can be very productive and are thus well suited to high-volume applications. Most broaching applications produce medium accuracy grade gears.

Internal gear teeth and splines are broached using a round or stick broach, which consists of a rod with a series of rings of peripheral tooth cutting edges arranged on its axis [17–23] (Figure 17.6). The broach is pushed or pulled through the gear blank, with each successive row of cutting edges removing more material to complete the gear. Full form broaches have an initial section of round edges to true the incoming hole, followed by roughing, semifinishing, and finishing edges with increasingly small stock removal, and in some cases by redundant finishing edges or burnishing buttons to improve finish. Generally, the material is removed by increasing the diametrical pitch between rows in a method called diametrical stepping or nibble broaching, although side shaving edges may be used in finishing sections to improve the final surface finish [17] (Figure 17.7). Helical gear teeth are produced by rotating the broach as it passes through the blank. Integral broaches for smaller diameters (less than roughly 5 cm) are typically made from solid HSS, with M2, M4, T15, and cobalt grades being common [24,25]. Carbide-tipped HSS broaches and small solid carbide broaches are also used. Larger broaching tools are often assembled by mounting shells of cutting elements on a central mandrel or hub. Larger assembled or built-up broach tooling is often held stationary while the gear blank indexes across it, although the tooling rotates for helical cuts; this arrangement is sometimes used to broach automotive ring gears. Carbide or P/M HSS cutting elements can be easily integrated

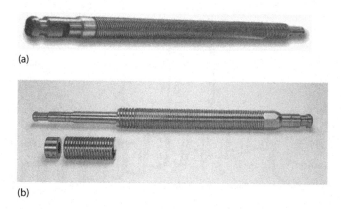

(a)

(b)

FIGURE 17.6   Internal gear broaching tools. (From (a) Mitsubishi, (b) Nachi.)

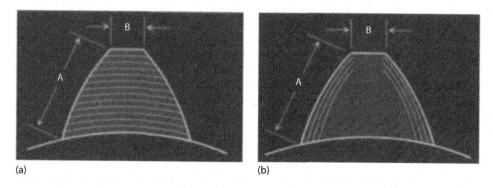

(a)   (b)

FIGURE 17.7   Broach material removal by successive cutting edges using the diametrical stepping (a) or side shaving (b) methods. (After Kosal, D., Broaching of Gears, *Gear Technol.*, 56, March/April 1977.)

into built-up broach tooling, often as planned replaceable wear items. In a common arrangement, illustrated in Figure 17.6b, an integral broach is equipped with a replaceable carbide shell finishing section to extend tool life. Built-up broaches require precision assembly to ensure accuracy. Broach tools are often nitrided or coated with TiN-, TiAlN-, TiCN-, or AlCrN-based coatings to improve life [22,26,27]. Internal broaching is normally carried out using neat oil or water-based flood coolants, but dry and MQL broaching are possible for some materials, and broach manufacturers offer coatings for dry and MQL applications [21,28,29]. There has also been recent interest in hard broaching, a finishing process in which a part is broached following heat treatment to remove distortions [22,30,31]; assembled tools with coated carbide cutting edges are used for these applications [32,33]. Internal broaching speeds are typically low and depend somewhat on the type of broaching machine (horizontal vs. vertical, pull vs. push, etc.). Top broaching speeds for standard hydraulically driven machines are roughly 15 m/min [21]. Hard broaching, in which stock removal levels are much lower, can be performed at speeds from 60 to 80 m/min [32,33].

Surface broaches are used to cut external gear teeth [1,19–21] (Figure 17.8). They are well suited to producing short gear segments on parts or gear segments with irregular tooth spacing. Smaller broaches may be of solid slab design, but larger broaches are usually built up with replaceable cutting sections. As with internal broaches, full form surface broaches usually have a stepped or nibbling rise profile in the roughing and semifinishing sections, followed by side shaving edges in the finishing section. Rigid fixturing is required to ensure part accuracy. In some applications, straddle broaching, in which two broaches cut simultaneously on opposite sides of the part, is used to balance forces and increase diametrical accuracy. As with internal broaches, the cutting edges may be made from HSS or tungsten carbide coated with TiN, AlCrN, and similar coatings.

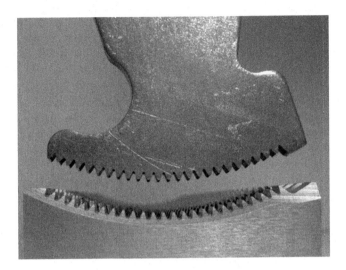

FIGURE 17.8   Surface broach (bottom) used to cut a segment of teeth on a part (top). (From Wolverine Broach Co., Inc., Harrison Charter Township, MI.)

Peak surface broaching speeds are roughly 35 m/min on older hydraulically driven machines and 100 m/min on electromechanically driven machines [21]. Improvements in milling methods have provided additional process options for parts that have traditionally been surface broached. Tooling and machine tool companies have also introduced indexable short stroke broaches and programming software for surface broaching simple tooth forms on CNC and mill-turn machines [34,35].

Pot broaching is a high-volume process for manufacturing external spur gears by pushing or pulling gear blanks through a tunnel-like tooling assembly (Figure 17.9) [17–20,23,36,37]. Push-type machines are used for shorter broaches, while pull-type machines are used for longer broaches due to buckling concerns with the push bar. There are three main types of pot broaching tooling: ring type, stick type, and a combination of ring and stick type [19,36,37]. Ring-type tooling consists of a split broach holder and annular rings of cutting edges in the form of relieved gear teeth. The rings are positioned using locating bars and radial slots that engage a key in the broach holder to control radial position. The rings are typically HSS and are precision ground; as the broach wears,

FIGURE 17.9   Stick-type pot broaching tool. (From Nachi of America, Inc., Greenwood, IN.)

rings from later sections of the tooling are ground to a larger diameter and reused in the earlier stages. The ring-type tooling provides relatively high accuracy. In stick-type tooling, sticks with rows of cutting elements like surface broaches are mounted in a split broach holder using precision bushings and locating bars. The sticks may be made of HSS or carbide. Stick-type broaches are not as accurate as the ring-type broaches but are used where accuracy permits, especially on parts with relatively shallow tooth profiles. The sticks can also be staggered to provide a smoother force profile. A combination of ring- and stick-type tooling is used in special applications such as parts requiring both gear teeth and slots [36]. Pot broaches are normally used with flood coolant. Pot broaching is a very productive process, with typical production rates of hundreds of parts per hour and broach lives of tens of thousands of parts between resharpenings. It can be used only for relatively simple gears without attached hubs, flanges, or other interfering features.

Broaching is a high force process and is often of interest to estimate broaching loads for fixture design or machine capacity studies. Psenka reported that broach loads could be estimated by multiplying the area of material being cut by an empirical constant [19]; the constant is equivalent to the specific cutting power described in Section 6.5. More recently Sutherland and coworkers reported a mechanistic force model for gear broach similar to the simulation models described in Chapter 8 [38]. This model is based on a geometric analysis of the instantaneous area of cut for the complex tooling and was validated by comparison with forces measured by instrumenting a production broaching machine. Finite element models with Lagrangian formulations are also well suited to simulating broaching operations.

17.3.2 FORM MILLING

In form milling, the gear tooth form is cut into a stationary or slowly rotating blank using a disk-shaped peripheral milling cutter with gear tooth–shaped cutting edges [1,3,39,40]. As noted in Chapter 1, form milling is one of the oldest methods of gear machining, with involute gear cutters being introduced in England in the 1840s.

In many applications, form milling is performed with solid HSS or carbide-tipped cutters (Figure 17.10). Typical surface speeds and feeds for HSS cutters for steel blanks are 15 SMM and 0.08 mm/tooth [1]. Carbide form cutters run at about 2.5 times the cutting speed of HSS cutters [39]. Standard involute cutters are available, which cover narrow ranges of modules or diametrical pitches. For precision work, special cutters ground to the exact tooth form are used as described by Radzevich [1]. Gear milling has traditionally been performed dry or with neat oil lubrication.

(a)   (b)

FIGURE 17.10   Disk-type form milling cutters: (a) solid HSS and (b) indexable carbide.

More recently, indexable gear milling cutters, consisting of a steel cutter body and replaceable coated carbide cutting edges, have been introduced [39–42]. One challenge in the development of these cutters has been repeatability in locating the inserts to ensure accuracy. Current advanced edge clamping methods permit positioning with a repeatability of 13 μm to datum [40] and are capable of producing gears to accuracy class A8 or A9 [39]. Duplex cutters that cut two rows of teeth at once have also been developed to reduce cycle time in roughing passes [41,43]. Standard inserts are manufactured to close tolerances, and custom inserts can be made for precision work [41]. Inserts normally have oxide- or AlCrN-based coatings applied using low-temperature PVD processes [42,43]. Indexable gear milling cutters are normally run dry, since flood coolant typically reduces tool life through thermal cyclic fatigue at the high cutting speeds normally used [39], although coatings suitable for wet milling at medium speeds are available [42]. Limited research on MQL gear milling has been reported [44]. Gear milling is commonly used in CNC machining center or mill-turn applications and for larger racks and gears [1,42,45]. For large precision gears, such as those used for wind turbines, milling may be used as a roughing process prior to finish grinding [43]; in some applications, such as those involving hardened blanks, milling may produce more favorable surface integrity than rough grinding [46].

Gears can be milled in one pass with a form cutter or in two passes with a stocking or gashing cutter for roughing followed by a finishing form cutter.

17.3.3 Hobbing

Hobbing is a generating cutting process in which a cutting tool shaped like a worm gear is fed axially across a rotating gear blank (Figure 17.11) [1,16,47,48]. The blank and hob speeds are held in constant relationship to generate the tooth form. Teeth are cut one or a few rows at a time, and the blank indexes between passes to complete the gear. Hobbing is a versatile and productive process, which can deliver high-dimensional and surface quality. Hobbing is normally performed on special hobbing machines but may also be performed on CNC machining and turning centers [1,48,49]. It is a very common method for machining spur and helical gears, especially in high volumes.

A hob, as noted earlier, is a worm gear, which is gashed along its axis to produce cutting edges. Gashing is generally either parallel to the hob axis or normal to the gear tooth profile. The cutting edges may be radial, that is, neutral rake, or they may have a small positive hook or rake angle.

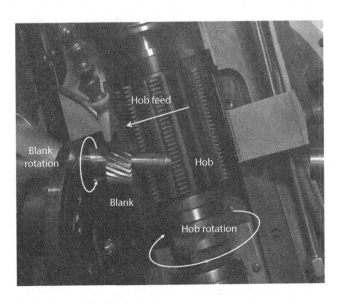

FIGURE 17.11 Hobbing an external helical gear.

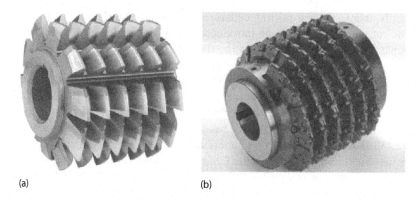

(a)   (b)

FIGURE 17.12   Gear hobs: (a) shell hob and (b) indexable hob.

The hob teeth are relieved to ensure efficient cutting. Hobs typically have straight-sided rather than involute-shaped cutting edges [1] and may also be ground with protuberances or ramps to cut clearances and other secondary features [50]. Hobs may be single thread or start, which cut only one row of gear teeth at a time, or multi-start (double or triple thread) to cut two or three rows. Although integral hobs are used for small gears, most hobs for larger gears are of either the shell or indexable type (Figure 17.12). Shell hobs are mounted on an arbor, which may be straight or tapered, and secured with a key. Straight arbors are most common; tapered arbors provide more rigid mounting but require a thicker shell. The shell material may be HSS, P/M HSS, or solid carbide [51–53], although cermet and ceramic hobs have also been tested [54,55]. As with gear milling cutters, shell hobs are often coated with TiN-, TiAlN-, or AlCrN-based coatings [42,52]. Indexable hobs have steel cutter bodies fitted with replaceable carbide inserts, often with special substrates and coatings similar to those used for shell hobs [41,45]. As with gear milling tools, the repeatability of insert position in indexable hobs influences the accuracy and quality of the machined gear, with radial positioning errors having a more negative effect on gear quality than axial positioning errors [56]. Current positioning accuracy for hobs is similar to that for indexable gear milling tools as discussed in the previous subsection. Due to the finite size of inserts and edge clamping requirements, indexable hobs cannot be used for smaller gear teeth, less than roughly module 3 [41]. Typical gear hobbing speeds are 50–75 m/min for soft steel blanks for general production and 100–200 m/min for high-volume applications [1,57]. Hobbing feeds vary depending on hob size and in roughing versus finishing cuts, but are typically 1.0–2.0 mm/rev for a 75 mm diameter blank [1].

Hobbing may be carried out as a single-pass process or as a two-pass process with roughing and finishing cuts, either with different hobs or with separate roughing and finishing shells set up on the same arbor [52]. Either conventional or climb cutting may be used for the hob; in general, climb hobbing yields better tool life, but conventional cutting provides a better surface finish [48]. For spur and helical gears, the hob is fed across the blank face. There are three common methods of infeed for the hob: axial or conventional feed, radial feed, and tangential feed [1,48]. In the axial feed method, the hob is set to full depth and fed across the face in the axial direction. In the radial feed method, the hob is fed radially into the blank to cut the full depth. Axial feed provides a more constant force profile and is normally used when there is sufficient clearance. Radial feed is used when the blank has an attached hub or other protruding feature that would interfere with the axial feed motion. In tangential feed, a conical hob is fed tangentially to the gear blank; this method is used in combination with axial feed on large, coarse-pitch gears [48].

In traditional practice, hobbing was performed using neat oil lubrication. Most current high-volume production applications are performed with water-based coolant or dry. Dry hobbing [42,51,54,57,58] was introduced in the 1990s for automotive-sized gears and has become increasingly common with improvements in tool materials and coatings. As with MQL machining, it requires changes in both tooling and machine architecture. Machines must be designed for thermal stability and chip management without coolant, using many of the design principles for MQL machining centers discussed in Section 15.4 [54,57]. Dry hobbing was initially applied using TiN-coated carbide hobs to enable higher cutting speeds (over 250 m/min) [58], since the tool grades available at the time were susceptible to thermal fatigue in interrupted wet cutting operations. With continued advances in tool, HSS and carbide hobs coated with TiAlN or AlCrN have been developed for dry hobbing. Generally, dry hobbing is better suited to smaller gears and high cutting speeds, while wet hobbing is used for moderate speeds and for larger gears with longer cutting cycles and more potential for heat build-up in the tooling or blank [42]. However, new tooling materials and coatings have enabled the used of dry hobbing speeds in some wet hobbing applications [51]. Water-cooled hobs have also been used to improve heat management in wet and dry hobbing of larger gears [59].

Although hobbing can produce high-dimensional and surface finish accuracy compared to form cutting methods, it is still often followed by a finishing process such as shaving or grinding in precision applications. Hobbed gears must also often be chamfered or deburred following hobbing; recently, combination tooling assemblies with separate hobbing and chamfering sections have been developed [51].

Kinematic simulations of hobbing based on 3D CAD models are used to predict chip cross sections, gear geometry, and profile generation from process and tooling parameters [60]. Process models based on finite element calculations [16,61–64] or mechanistic models like those described in Chapter 8 [65] have also been reported. All models predict hob forces and chip properties. The reported FEA models, some based on commercial machining codes such as DEFORM and AdvantEdge (Sections 6.12 and 7.5), also predict hob stresses and temperatures (Figure 17.13). Most models are validated using single-tooth cutting experiments on instrumented blanks.

17.3.4 SHAPING

In the shaping process, the gear tooth form is generated by a reciprocating gear-shaped cutter engaging a rotating blank (Figure 17.14) [1,16,47,66]. Either pinion- or rack-shaped cutters can be used. The rotation of the blank and motions of the cutter simulates two gears that mesh together.

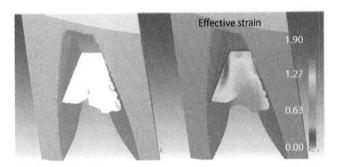

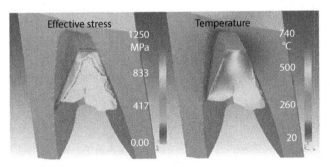

FIGURE 17.13 Finite element simulation of hob stresses and temperatures. (After Bouzakis, K.-D. et al., *ASME J. Manuf. Sci. Eng.*, 124, 42, 2002.)

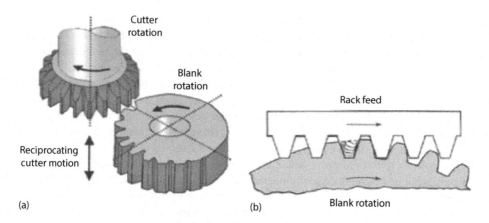

(a)   (b)

FIGURE 17.14   (a) Rotary shaping and (b) rack shaping.

The cutters cut on the downstroke and are retracted on the upstroke. The cutter is fed radially into the blank over a number of strokes until it reaches full depth, after which a generating feed motion is usually used to finish the gear as discussed in the following. Shaping requires less tool clearance than hobbing and is used especially for applications in which hobbing is not feasible due to geometry or restrictions on tool overtravel. Examples include internal gears, gears with shoulders or hubs that would interfere with hob travel, and cluster and double helical gears [1,67]. Shaping may also provide cycle time advantages over hobbing when machining narrow face width gears due to reduced overtravel requirements [1]. Since shaper cutters are simpler and less expensive than hobs, the process is also well suited for small-lot production of large gears [66]. Shaping is a lower speed and generally less productive process than hobbing but can produce equivalent dimensional and surface quality, with typical AGMA Q8 grade achievable accuracy [68,69].

As noted earlier, shaping may be carried out using pinion or rack style cutters (Figure 17.15). Pinion cutters may be further classified as disk type, which are mounted on arbors, and shank types, which have an integral shank and are used especially for smaller-diameter internal gears [1,70]. Wafer-type cutters, consisting of a replaceable thin cutting ring held between clamping and backup rings [71,72], are also used to eliminate resharpening costs and setup adjustments after a tool change. Rack cutters are used especially for larger gears. Rack cutters are more easily fabricated and less expensive than pinion cutters or hobs, and, in some cases, they can be resharpened by the end user using standard grinding equipment. Due to ease of fabrication, rack cutters are also well suited to applications with modified or special tooth profiles [67]. Shaper cutters are typically made of HSS with TiN, TiAlN, or TiCN coatings [66,72,73].

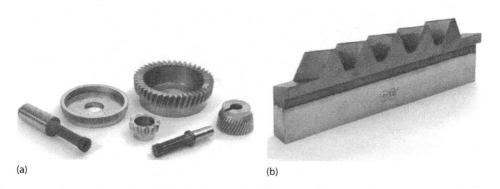

(a)   (b)

FIGURE 17.15   Shaper cutters: (a) pinion style and (b) rack style. (Courtesy of Star SU, Hoffman Estates, IL.)

A shaping process cycle consists of an infeed phase, in which the cutter is fed over a number of strokes to full depth, followed by one or more generative rotations of the gear without infeed to cut all teeth to the same depth. In a single-cut process, a single generative rotation is used to finish the gear; in the more common two-cut process, two generative rotations are used, with the cutting speed often being increased for the finishing (second) rotation [47,74]. Typical shaping stroke rates are 1000–2000/min for pinion cutters and 100–200/min for rack cutters [1,74]. Infeed per stroke cycle is typically 0.2–0.5 mm depending on gear size and material [1,47,74]. Continuous radial infeed with lower feed per stroke in the finishing cuts can be used with CNC gear shapers [74]. Typical shaper cutter speeds are 25–100 m/min, depending on the gear material and the face width or length of stroke [1,47,74]; narrower face width gears have lower average speeds due to cutter acceleration and deceleration.

Kinematic or geometric simulations can be used to design shaper cutters and specify grinding parameters and to study undercutting and clearance conditions [16,75–77]. As discussed earlier for broaching and in Section 6.5, the shaping force can also be estimated from the cross-section cut and the specific cutting power of the blank material. Like hobbing, the shaping process can also be modeled using commercial finite element codes to study chip formation, stresses and temperature, and wear behavior as reviewed by Bouzakis et al. [16].

17.3.5 Form Grinding from the Solid

Gear teeth can also be ground using form wheels or generative processes. Grinding is usually a finishing process and is discussed in more detail in Section 17.6.3. There are applications, however, in which gear teeth are ground from solid blanks, usually using form wheels. Form or profile wheels, like form milling cutters, are disk wheels with the involute tooth form dressed into the cutting surfaces, which cut the tooth form directly into the blank without generative rotation [1]. The wheel reciprocates axially with respect to the blank and is fed into the blank using a prescribed infeed cycle.

Form grinding from the solid is well suited to grinding fine pitch threads with little required stock removal [1] and can also be faster and more accurate than alternative processes for gears machined on superalloy parts [78]. Form grinding with electroplated CBN wheels, which are run at higher wheel speeds than conventional wheels and not dressed in cycle, can offer cycle time advantages over alternatives processes [79]. The accuracy and gaging capabilities of current CNC gear grinders also make profile grinding from the solid attractive for large gears, which require finish grinding, since setup requirements can be reduced in many cases [80,81]. Profile grinding using relatively inexpensive off-the-shelf wheels can also reduce lead times for special tooling and is well suited to producing gears with special tooth profiles [43,79].

17.4 BEVEL AND HYPOID GEAR MACHINING

Due to their geometric complexity, specifically the fact that their tooth dimensions and tooth spacing vary with radius, bevel gears usually cannot be cut with form tools and are machined using generating processes. A variety of processes were used on older mechanical bevel gear generators as summarized by Hünecke [82]. As an example, Figure 17.16 shows the operating principle of bevel generation using reciprocating cutters guided by a crowning gear with the same cone angle and diametrical pitch [83]. In this process, gear teeth were machined with two rapidly reciprocating cutters as the gear blank slowly rotated to generate the flank; when one set of cuts was completed, the blank was indexed to the next gap until the gear was completed. In more recent machines, the physical crown gear has been eliminated, but the generating reference surface for many methods is a virtual crown gear [3,84].

The introduction of six-axis CNC gear cutting machines has led to the development of more flexible and productive bevel gear machining processes. Commonly applied processes include

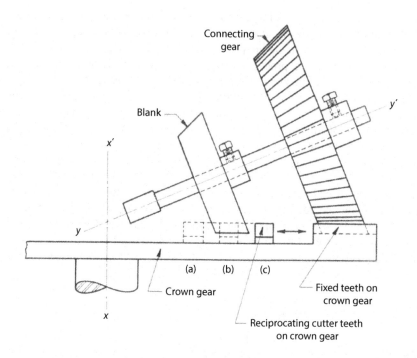

FIGURE 17.16 Bevel gear generation on a mechanical gear generating machine. (After DeGarmo, P.E., *Materials and Processes for Manufacturing*, 5th edn., Macmillan, New York, 1979, pp. 796, 814–819.)

peripheral milling, face milling, and face hobbing [1,3,84,85]. Peripheral milling can be used only for straight bevel gears, while face milling and face hobbing can be used for spiral bevel and hypoid gears. These processes are carried out using inserted form- or stick-blade circular cutter heads, often with proprietary tooling concepts and geometries; leading manufacturers of machines and tooling include the Gleason Works (Rochester, NY), Klingelnberg AG (Zurich), and Yutaka Seimitsu Kogyo, Ltd. (Seto-shi, Aichi Pref., Japan) [86]. Bevel, spur, and helical gears can also be machined by contour milling on five-axis machining centers and multitasking machines as discussed in Section 17.5. Depending on the tooth cutting method, bevel and hypoid gears can be finish machined by skiving, grinding, or lapping as discussed in Section 17.6.

17.4.1 Peripheral Milling

Straight bevel gears can be generated by peripheral milling using stick-blade cutter heads. Representative processes of this type include the Gleason Coniflex and Coniflex Plus processes [1,3,87] and the Klingelnberg Hypoflex process [82]. In the Coniflex process, with a completing process setup (Figure 17.17), the reciprocating blades shown in Figure 17.16 are replaced by a pair of ground HSS peripheral milling cutters, which reciprocate to machine the opposing flanks of the gear tooth; when one set of cuts is completed, the blank indexes to the next gap. On some types of machines only one cutter could be mounted at a time (semicompleting process), doubling the number of indexes and cycle time required to complete a gear. In the Coniflex Plus process, the ground HSS milling cutters are replaced with an inserted blade cutter head with offset blades to machine each side of the gear tooth (Figure 17.18). Coated tungsten carbide blades are employed. The blades are positively seated to increase accuracy and reduce setup time. The Hypoflex process also uses cutter heads with inserted coated carbide stick blades and can be set up to machine one or both gear flanks each index (completing vs. semi-completing processes). The Coniflex process was introduced as a wet milling process,

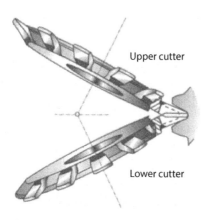

FIGURE 17.17   Coniflex process for peripheral milling straight bevel gears. (Courtesy of the Gleason Works, Rochester, NY.)

FIGURE 17.18   Coniflex Plus inserted blade cutter head. (Courtesy of the Gleason Works, Rochester, NY.)

while both the Coniflex Plus and Hypoflex processes, which operate at higher cutting speeds enabled by carbide tooling, are dry machining processes.

Compared to earlier processes, these processes offer higher metal removal rates, increased flexibility, and reduced setup times for small-lot production. Higher cutting speeds are enabled by the use of peripheral milling, which can generate higher cutting speeds than reciprocating slotter cutters, and especially by the use of tungsten carbide blades in the Coniflex Plus and Hypoflex processes. Stadtfeld reports that the use of carbide cutters increases production rates by at least a factor of three over earlier methods [87]. Increased accuracy, flexibility, and setup reduction are enabled by the use of CNC controls, although dedicated setup equipment may be required in some applications.

17.4.2 FACE MILLING

Face milling of spiral bevel and hypoid gears is performed with a circular cutter with blades inserted on the face rather than the periphery (Figure 17.19) [1,3,84,86,88–95]. The process kinematics and typical cutters are shown in Figure 17.20. Face milling is a single indexing process in which the gear blank is held stationary and the rotating cutter is fed into the blank face. For a forming rather than a generating cut, the cutter is fed axially to full depth and withdrawn; for a generating cut, the cutter is plunged to full depth at the heel roll position first, then rotated about the generating gear axis so that their blades work their way from the heel to the toe position to generate the flanks [84]. The cutter machines one gap at a time, and blank indexes between cuts to machine successive teeth. The blades on the cutter are positioned to cut either one or both flanks and have precisely defined

FIGURE 17.19   Face milling a spiral bevel gear.

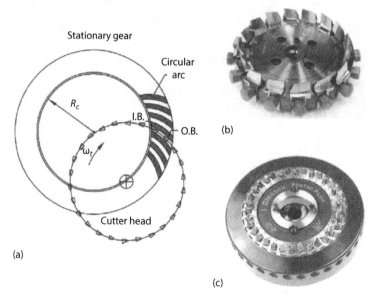

FIGURE 17.20   (a) Process kinematics for face milling. (After Fan, Q., *ASME J. Mech. Des.*, 129, 31, 2007.) (b) HSS blade cutter head. (c) Carbide blade cutter head. (b and c: Courtesy of the Gleason Works, Rochester, NY.)

geometries, setup procedures, and grinding methods [84,88,94]. Gear teeth produced by face milling generally have a tapered tooth depth, with the root being deeper at the toe than the heel, and a circular arc-shaped slot of constant width [84,85,89,90,94]. Kinematic simulations are used for tooth surface generation modeling and contact simulation and analysis [9,91,92,95] during the design of both the gear and the manufacturing method.

There are two common processes used for face milling: the five-cut process and the completing process [3,84,94]. The older five-cut process uses five separate operations, two on the gear and three on the pinion, to produce the gear set. The gear is roughed out with an alternating blade roughing cutter (which machines both sides of the gap with inside and outside blades), and finished in a second operation with an alternate blade finishing cutter. The pinion is roughed with an alternating blade cutter, then finished with two separate operations to cut each flank. In the completing process, both sides of the tooth slot are machined from a solid blank in one operation, with a cutter having alternate blades for the inner and outer flanks, and sometimes a third (bottom) blade. Separate cutters are used for the gear and pinion.

Cutter systems using the five-cut process, such as the Gleason HARDAC process [3], generally have inserted HSS blades with flood or neat oil coolants. Cutter systems for the completing process use inserted coated carbide stick blades. Common systems of this type include the Gleason RSR and PENTAC-FM systems [3] and the Klingelnberg ARCON system [94]. High-speed dry machining (Powercutting) was introduced by Gleason in the 1990s, and dry machining has become standard for completing cutter systems on contemporary bevel gear cutting machines due to its increased productivity over lower-speed wet processes. The carbide blades are typically made of extra-fine K-grade carbides coated with TiN-, TiAlN-, or AlCrN-based coatings [94,96]. A detailed discussion of blade geometries and rake angles is given by Stadtfeld [88]. Tooth flank deviations from target geometry can be reduced using error correction algorithms on CNC gear cutting machines [97]. Face-milled bevel gears can be finished by skiving, grinding, or lapping [84,94].

17.4.3 FACE HOBBING

Face hobbing [3,84,85,88–90,93,94] is a continuous indexing process, which, like face milling, uses circular inserted blade cutters with blades mounted on the face. The general process kinematics and details of the cutter tooth orientation are shown in Figure 17.21 [84,90,95,98,99]. In the continuous

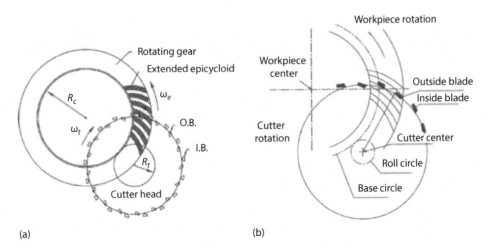

(a)   (b)

FIGURE 17.21   (a) Process kinematics for face hobbing. (After Fan, Q., *ASME J. Mech. Des.*, 129, 31, 2007.) (b) Details of cutter blade orientation. (After Simon, V., Advanced design and manufacture of face-hobbed bevel gears, *Proceedings of the ASME IMECE*, Lake Buena Vista, FL, November 13–19, 2009, Paper IMECE2009-10237.)

indexing method, the blank rotates as many time per cutter rotation as the cutter has starts as the cutter is fed into the blank to full depth. The teeth are machined by cutters with inside and outside blades; as an outside blade and the following inside blade machine one slot, the following blade group enters the next slot. The resulting teeth have constant tooth depth and slots with varying width and an extended epicycloid shape [3,90,98,99]. The process kinematics are in reality more complex than shown in Figure 17.21 since a secondary generating motion is required to produce localized contact of the gear tooth surfaces [90], as described in detailed kinematic and tooth-contact analyses of specific face hobbing processes [95,98–107].

Most current face hobbing systems are based on the Klingelnberg Cyclo-palloid system and earlier Oerlikon Spiroflex-Spirac systems [3,90,94,102] and used circular cutter heads. Systems differ in their cutter geometries and methods of controlling longitudinal tooth crowning [102,108]. There are several current inserted blade cutter systems, such as the Klingelnberg SPIRON system [93] and the Gleason PENTAC-FH, TRI-AC, and CYCLO-CUT systems [3,108]. In most systems, both the gear and pinion are generated, although in the earlier Spirac system, which is still widely used, only the pinion is generated [102].

Face hobbing cutters have more complex blade geometries and setup procedures than face milling cutters. The original Cyclo-Palloid system uses interlocking cutters with inside and outside blades rotating in different centers [102,108]. The outer and inner cutters machine the outside and inside flanks in a completing process, and the center distance between rotational axes is adjusted to control length crowning. In earlier practice, separate, one-sided (semicompleting) cutters could be used for the inner and outer flanks. This increased the cycle time, but reduced the setup time and increased cutter stiffness compared to interlocking cutters. In the Oerlikon system, length crowning is controlled by tilting the outer cutter (Spiroflex method) [102]. On contemporary free-form bevel gear generators, both inside and outside blades can be mounted on the same cutter heads provided their orientation is strictly controlled based on the gear design; in this case, cutter tilt and blade geometry are used to control length crowning in a completing process [3]. Some earlier systems also had a third bottom blade [94]. Although some systems, such as the first generation CYCLO-CUT system, used HSS blades and wet cutting processes [108], current face hobbing cutters typically use dry cutting and coated carbide blades similar to those used for face milling [88,94]. The same cutters are used for hard finishing by skiving in some cases; in these applications, CBN finishing blades are often used [108].

Face-hobbed bevel gears can be finished by skiving or lapping [84,93,94]. Grinding is sometimes used but is not generally recommended for finishing face-hobbed gears due to their epicycloid slot shape; grinding would produce circular arc-shaped flank surfaces with uneven case depth and potential noncleanup concerns [94]. Grinding methods for face-hobbed bevel gears also have a long cycle time compared to lapping.

Although the tooth geometries produced by face milling and face hobbing differ, they produce gears of equivalent quality and durability with proper finish machining. Face hobbing is a more productive process due to continuous indexing and multiple start tooling and typically offers a significant cycle time advantage over face milling for the rough machining operation. Face milling is better suited to high load-carrying applications, which require finish grinding [93].

17.5 FIVE-AXIS MACHINING OF GEARS

Gear machining has traditionally required dedicated machine tools, although some simpler processes could be performed on lathes or milling machines, and methods for hobbing and broaching with traditional tooling have been developed for machining and turning centers [34,49]. Continuing improvements in CNC machining centers and multitasking machines, together with path generation software and tooling systems, have led to increased capability for gear machining on multiaxis CNC equipment. New machining methods have been developed for both parallel axis and bevel gear applications. These are generally called five-axis gear machining methods since five-axis capability is required for complex gear geometries, although simpler spur gears require only four-axis capability, and multitasking machines may use more than five axes. The methods described for parallel axis gears can typically be used for straight bevel gears and in some cases extended to spiral bevel and other more complex gear types. All methods are intended to provide reduced tooling inventory and cycle and lead time savings over traditional methods for small- and medium-lot production.

17.5.1 PARALLEL AXIS GEARS

Coarse pitch spur and helical gears have long been milled on hobbing machines using involute-shaped end milling cutters [3]. On machining centers and multitasking machines, a wide range of spur and helical gears can be generated using standard milling tooling and special tool path programming software.

One common strategy is to use standard profile milling tooling and generate the tool path using free-form milling algorithms like those used for airfoil machining [6,109,110]. The advantages of these methods are that they use off-the-shelf solid or indexable coated carbide tooling with predictable tool life. They are especially well suited to small-lot production of gears of different sizes or types. Due to its flexibility, free-form milling also permits machining of special gear types and root forms, which are not manufacturable by standard gear machining methods [6,110]. The software design is critical since it influences both gear quality and cycle time. As discussed by Klocke and coworkers [109], the elements of the machining strategy, which must be incorporated into the path generation software, include the path trajectory (e.g., radial or axial profiling), the "lineness" or path density required to generate the flank (i.e., the number of passes on each tooth face), and the indexing strategy (single or continuous) (Figure 17.22). Trade-offs between quality and cycle time are required to generate an optimal strategy.

In integrated solutions using standard tooling, the path generation software is integrated with a CAD or solid modeling program and in some cases with on-machine inspection software. Examples include the Gear Engineer/DepoCAM system from DEPO GmbH (Marienfeld, Germany) [111,112] and the gearMILL software system developed by DMG Mori Pfronten [6]. The DEPO system, which is used on DEPO's high-speed multiaxis machining centers, was developed based on mold and die machining algorithms and achieves cycle time reductions through path optimization and high cutting speeds. Gear types supported include external and internal spur and helical gears

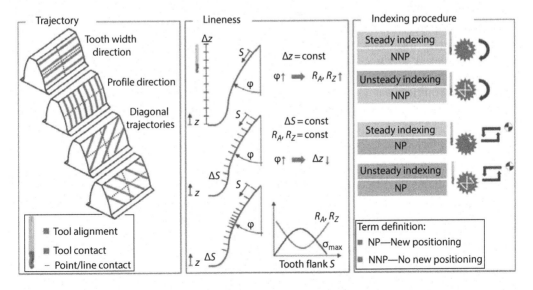

FIGURE 17.22 Elements of machining strategies for free-form milling of gears. (After Klocke et al. [109].)

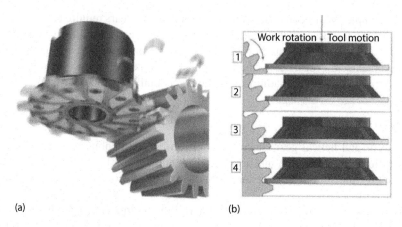

(a)   (b)

FIGURE 17.23   (a) Interpolation of a helical gear by a face milling cutter (source: Sandvik). (b) Interpolation strategy for InvoMilling. (After Hyatt, G., *Procedia CIRP*, 14, 72, 2014.)

as well as double helical and straight and spiral bevel gears. It can be used for rough and finish machining in the same setup and for hard machining of gears after heat treatment. The achievable gear quality is reportedly equivalent to AGMA grade Q12 (roughly ISO class A4/A5). The gear MILL system was developed using a surface-based solid modeling software and can be applied for the machining of common gear types and irregular gear forms on mill-turn machines. It offers an advantage in combined setup and machining times over hobbing for small-lot production, especially for gears with small face widths. The achievable gear quality is reportedly equivalent to AGMA grade Q13 (roughly ISO class A3).

Alternative methods use restricted tooling options to simplify programming and reduce cycle times over general purpose tooling. One well-known example is the InvoMilling system developed by Sandvik and DMG/Mori-Seiki [6,110,113–117]. This system uses indexable dish-shaped face milling cutters with standard carbide inserts to generate gears by interpolation on standard mill-turn machines (Figure 17.23). In this system, the contact between the tool face and the gear's involute profile is described by a chord bisecting the tool face, and radial tool paths, typically from the tip to the root, are used. It is a combination of slot milling and multiaxis turn milling. Simultaneous motion of the X and B axes or with the Y and B axes enables the milling cutter to follow an involute path. Unlike end mills, the face cutting tool does not scallop the surface with each pass, so larger stepovers (reduced "lineness") can be used without affecting surface quality. The system provides a cycle time advantage over free-form milling methods for this reason. A relatively small set of tools is required to cover a wide range of modules and gear types, including special tooth forms.

CNC plunge milling has also been investigated as a flexible alternative to hobbing for face gears [118]. Multitasking machines are also capable of performing traditional hobbing processes.

17.5.2 Bevel Gears

The free-form milling methods discussed for parallel axis gears in the previous subsection can also generally be applied to bevel gear machining. In addition, research into geometric modeling and path generation methods specifically for machining spiral bevel and hypoid gears on machining centers and multitasking machines has been reported [119,120]. An integrated system for spiral bevel gear machining on vertical machining canters was developed by Mitsui Seiki, with the software developed by CNC Software, Inc., Tolland, CT, the producers of the Mastercam CAM package [121]. This system uses end mills for roughing cuts and ball end mills for finishing and reportedly produces quality levels equivalent to AGMA class Q10–Q14 (roughly ISO classes A3–A6).

Spiral bevel gears can also be machined on five-axis machining centers using a mixture of standard and special tooling. The upGear system, developed by Voith Turbo AG and Heller Machine tools [122,123], permits pre-milling of blanks and gear tooth cutting operations to be carried out on a single machining center in two setups. Gear teeth are rough machined using standard milling cutters with round inserts, then finish machined using a special crown or dish milling cutter. Roughing can reportedly be performed at metal removal rates up to 400 cc/min, and the finish crown milling cutter interpolates the flank surfaces more rapidly than methods using standard end mills.

17.6 GEAR TOOTH FINISHING METHODS

The previous three sections described methods for machining gear teeth from solid blanks. Although gears machined using these methods may have acceptable quality for general purpose applications, in many cases further finish machining must be performed to meet dimensional or surface finish quality requirements. These methods may be applied to gears in the soft state to correct for machining deflections and to improve the surface finish or to hardened gears to remove heat treatment distortion.

17.6.1 SHAVING

Shaving (Figure 17.24) is a soft finishing process for external and internal spur and helical gears [1,16,47,124–126]. Shaving resembles shaping in that it uses a pinion or rack tool shaped like a gear nominally conjugate to the work gear. Shaving is a corrective process, which reduces errors in index, tooth profile, and runout. It also improves gear surface finish and can correct helix angle errors for narrow face width gears. The cutter can be ground to include form modifications to reduce gear noise or load concentrations in service. It can be used for gears with hardness less than about 450 BHN (47 R_c) [1]. Shaving is a wet machining process; coolant is required to minimize tool wear, improve surface finish, and clear chips. There are two basic methods of shaving: rotary shaving and rack shaving.

Rotary shaving uses a precision ground gear-shaped cutter [1,47,125]. The cutter and gear are run together, with the cutter generally being driven for small parts. During the shaving cycle, the gear reciprocates with respect to the cutter and is fed incrementally toward the cutter axis. At the end of

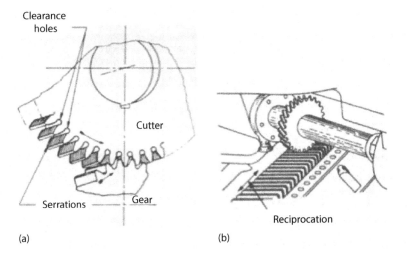

(a)   (b)

FIGURE 17.24   (a) Rotary shaving. (After Dugas, J.P., The Process of Gear Shaving, *Gear Technol.*, 40, January/February 1986.) (b) Rack shaving. (After Radzevich, S., *Dudley's Handbook of Practical Gear Design and Manufacture,* 2nd edn., CRC Press, Boca Raton, FL, 2012, pp. 4–20, 419–475, 491–525.)

each stroke the direction of feed is changed and the cutter rotation is reversed. The cutter teeth are serrated with several small rectangular ridges to produce cutting edges. The cutter and gear axes are set out of parallel by a few degrees so that the gears to run together like a pair crossed helical gears; this mismatch causes the cutter to shave thin layers of metal off the work gear to correct flank errors. The mismatch or shaft angle controls the cutting action of the cutter; a higher shaft angle usually increases metal removal rate, while low angles give better helix angle control. Shaft angles for external gears are typically between 8° and 15°, with lower shaft angles being used for internal gears [1].

Rotary shaving methods are classified based on the direction of work gear reciprocation or traverse [47,124,126]. The common methods are axial or conventional shaving, in which the gear traverses parallel to its axis; diagonal shaving, in which the traverse direction is at an acute angle to the gear axis; tangential or underpass shaving, in which the traverse is perpendicular to the gear axis; and plunge shaving, in which there is no traverse. Tangential and plunge shaving are short cycle time methods widely used in high-volume production.

In *rack shaving* of external gears, the work gear is rolled back and forth against a reciprocating rack cutter with serrated edges [1]. The rack must have a face width greater than the gear and a length and stroke longer than the gear circumference. The rack axis is not normal to the gear axis but inclined a few degrees to produce a mismatch and cutting action analogous to that produced by the shaft angle in rotary shaving.

A typical rotary shaving cutter is shown in Figure 17.25. Cutters are normally made of M2 or P/M HSS and are often not coated since they are designed for multiple resharpenings. (Limited research on PVD coatings indicates that they can not only improve tool life but also reduce gear quality due to uneven coating thickness [16].) Design considerations for shaving cutters are discussed by Radzevich [1]. A number of kinematic and tooth contact analyses of shaving have been reported; these analyses have been used to develop automatic design systems for shaving cutters [127–129], to investigate crowning and tooth form generation affected by both machine settings and cutter modifications [130–133], and to design the serrations on the cutter [134]. Limited finite element modeling of shaving to predict forces and temperatures has also been reported [16,135]. Validation of finite element models has presented difficulties since single edge (serration) test geometries are not available [16].

Shaving is a highly productive process with short cycle times and long edge life between resharpenings and has been widely applied in automotive transmission and other high-volume production applications for many years. In more recent practice, shaving has sometimes been replaced by grinding or other hard finishing operations due to requirements for increased gear accuracy and noise reduction.

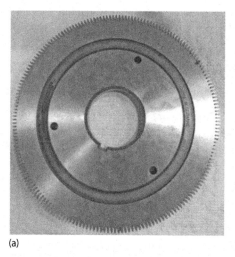

(a)

(b)

FIGURE 17.25   Rotary shaving cutter (a) with detail of tooth serrations (b).

17.6.2   Skiving (Hard Finishing)

17.6.2.1   Skiving: Hard Recutting Processes

Skiving is a hard finishing process in which the gear is recut after heat treatment to remove distortions [1,16,136–138]. For cylindrical gears, the skiving process is sometimes called skive hobbing [136] or hard carbide rehobbing [139,140]. Analogous hard recutting methods are used for bevel gears [84]. The gears are first machined in the soft state, heat treated, and then recut in the hardened state, generally using the same process employed for rough machining but with a modified tool with a stronger edge geometry. Skiving may be performed as a finishing process or for high accuracy cylindrical gears as a pre-finishing process prior to honing or a light final grinding operation [1]. Hard broaching, an analogous hard recutting process for broached gears, is discussed in Section 17.3.1.

For hobbed cylindrical gears, skiving is performed with a skiving hob, which has a large negative rake angle (Figure 17.26). The rake is typically −15° to −30° [1,136,141], depending on the tool geometry and gear module, with smaller negative rakes being used for coarse-pitch gears. The negative rake improved edge strength and creates a more favorable edge contact condition as the cutting edges enter the workpiece. The hobs and process are also designed so that the teeth cut only the flanks and not the root fillet area [136,138,141]. In older practice, HSS or carbide-tipped hobs were used with neat oil lubrication [136,137], but in most current applications high-precision finishing grade carbide hobs and dry cutting are used [137,140,142], Hobs are generally coated with TiN [141,142], TiCN [140], or similar coatings, especially for harder gears (>600H_v [142]). The process can be used for gears with hardness between 48–65 R_c [137,140,141]. Skive hobbing speeds are equal to or higher than finish hobbing feeds and are typically between 100 and 200 m/min [16]. A rigid, high-precision hobbing machine with sensors to position the hob between the gear teeth for even stock division is required [137–139]. Skiving can achieve quality levels of A2–A5 and surface quality comparable to grinding [137]. Simulation models have been reported for calculating forces and temperatures and the effects of tooling errors on gear quality [138,139,143]. Skiving has a more efficient cutting action than grinding and is especially attractive as an alternative to grinding for large gears which do not require the highest precision, and for small module gears, which may be difficult to grind [143].

In bevel gear machining using circular cutter heads, skiving is performed by replacing the carbide blades used for soft machining with negative rake skiving blades [84]. Straight bevel gears are usually ground when hard machined [82] but may be skived using a peripheral milling cutter. For Coniflex Plus cutters [87], skiving blades are three-face ground with a −20° rake angle and cut only the tooth flanks. Skiving is also attractive for face-hobbed spiral bevel and hypoid gears, since as discussed in Section 17.4.3 they are generally not ground due to their epicycloid slot shape. For CYCLO-CUT and PENTAC cutter systems, skiving blades are brazed CBN with a −20 rake [108]. Skiving is performed at 120 m/min with 0.1 mm stock removal in this system. Skiving can be

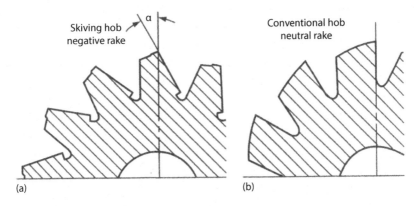

FIGURE 17.26   Negative rake skiving hob (a) compared to neutral rake radially ground conventional hob (b). The rake angle on a skiving hob is usually between −15° and −30°.

performed dry, but wet skiving is more common because it delivers better tool life with current blade materials [108]. In these systems, skiving is applied as an alternative to grinding or lapping, especially for small batch production of larger gears; the main advantage is that the same machines, and in many cases the same cutter heads, can be used for both rough and finish machining, reducing machine investment and tooling inventory [84].

17.6.2.2 Skiving: Other Processes

In gear tooth machining for cylindrical gears, the term "skiving" refers to a soft machining process for producing periodic axisymmetric shapes using a bevel gear generating machine, which is used especially as a more productive alternative to shaping for internal gears [16,144–146]. This process is also called power skiving. As noted in Section 2.8, the term "skiving" may also refer to a crankshaft turnbroaching process involving side cutting or "cheeking" of the counterweights [147].

17.6.3 Grinding

Grinding is the most common method of finishing high-precision gears, especially in the hardened state. There are three common methods of finish grinding: form or profile grinding, indexing generating grinding, and continuous generating grinding [1,78,148–153] (Figure 17.27). Finish grinding is used either to remove heat treatment distortion or to polish the gear tooth surface by removing 40–80 μm of stock on each flank. An optimum grinding process can achieve surface finish of 0.07 μm without burn marks.

These processes are applied mainly to ferrous work materials and use grinding wheels made of a restricted range of abrasives [1,81,150,151,154]. Conventional abrasive wheels are most commonly made of aluminum oxide or aluminum oxide mixed with sol–gel ceramic grains; mixed grain wheels are referred to as ceramic wheels. Ceramic grains are microcrystalline alumina produced by the sol–gel process [154–157]; the ceramic grains have high friability and fracture into multiple small cutting edges to maintain an efficient cutting geometry so that ceramic wheels can typically be run at higher speeds and removal rates than plain alumina wheels. Recently introduced wheels containing "Cubitron," a controlled shape alumina sol–gel, have a more efficient cutting action than randomly oriented abrasives and have enabled further gains in productivity [155,158,159]. Superabrasive gear grinding wheels are normally composed of vitrified or single-layer electroplated CBN [1,81,160].

As discussed in Section 4.11 on grinding wheels, the factors to be considered in selecting a wheel include grit size, wheel structure and hardness, bond friability, and dressing method. Wheel life between dressings is often controlled by surface finish requirements. Almost all gear grinding operations are wet, and proper coolant application is required to prevent wheel loading and surface burn.

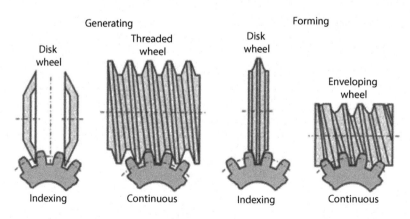

FIGURE 17.27 Cylindrical gear grinding processes.

17.6.3.1   Form Grinding

Form or profile grinding uses a wheel shaped like the desired tooth form to machine the tooth without a generating motion. In form grinding from the solid, which was discussed as a rough machining method in Section 17.3.5, a disk-shaped wheel similar to a form milling cutter is used in a radial infeed single indexing process. Finish form grinding can be performed using a variety of single indexing methods as shown in Figure 17.28 [151]. In restricted access applications, gear-shaped wheels rotating on axes radial to the work gear can be used to machine one or both flanks of the tooth gap (Figure 17.28a and b). More commonly, disk-shaped wheels inclined by the helix angle with respect to the gear axis are used. These wheels can be used for single- or double-flank grinding. In the single-flank approach (Figure 17.28c), the wheel axis is tilted with respect to the tooth flank; a larger tilt angle (I) should be used to reduce axial bending of the wheel surface. In the more common double-flank approach (Figure 17.28d), a full width wheel finishes both sides simultaneously. Wheels mounted on multiple spindles can be used for either single- or double-sided grinding to machine multiple gaps per index as shown in Figure 17.28e and f. In all methods, the wheel reciprocates with respect to the gear and is fed incrementally toward the gear axis. Form grinding can also be performed using continuous indexing with a globoid worm shaped enveloping grinding wheel as shown in Figure 17.27 [151]. This is a high-volume production method since the wheel geometry is part specific.

Form grinding can be carried out using vitrified conventional or CBN abrasive wheels or with electroplated CBN wheels [1,81,150,151,154,155,161]. Vitrified CBN wheels are used especially in

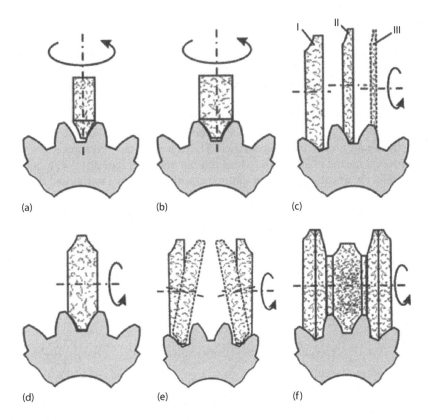

(a)   (b)   (c)

(d)   (e)   (f)

FIGURE 17.28   Single indexing form grinding methods. (a) Grinding pins, (b) grinding wheels, (c) three versions of single flank grinding, (d) two-flank profile grinding wheel, (e) application with two single grinding spindles to reach two teeth at different locations by tilting of the grinding wheels, and (f) multi-groove grinding wheel. (After Karpuschewski, B. et al., *CIRP Ann.*, 57, 621, 2008.)

hard grinding applications. Electroplated CBN wheels do not require dressing and thus offer a cycle time advantage in high-volume applications, although they require increased wheel inventory due to the reconditioning cycle for used wheels.

Form grinding is a relatively high metal removal rate process since the wheel contacts the tooth over the entire flank rather than over a small contact zone as in generating grinding. It can be used for either internal or external gears and is especially well suited to finishing small volume large gears, since it can often be performed using dressable off-the-shelf wheels [160,162]. Form grinding with vitrified wheels is also well suited to producing special tooth forms or flank modifications since the required features can be dressed into the wheel using a CNC dresser. Form grinding is also attractive for finishing hardnened internal gears, which may be difficult to process by other methods.

17.6.3.2 Indexing Generating Grinding

For cylindrical gears, indexing generating grinding can be performed with either disk or dish-shaped wheels as shown in Figure 17.29 [1,151,163]. In either case, the process generates the profile form by a finite number of profiling cuts. For the disk wheel method, the wheel is dressed with the root form of the gear, and the gear rolls about its axis as the wheel (or gear) reciprocates sideways to generate the flank form. This method is also called the conical wheel method. In the dish-shaped wheel method, two wheels that are concave toward the gear flank contact the flank over a narrow range as the gear rolls past the grinding area. The wheels can be set at a nonzero engagement angle, rather than being set up to be parallel; the nonzero angle typically improves accuracy by reducing the contact area but increases cycle time by increasing the running distance [163]. Generally, the conical wheel method is more productive due to its larger engagement and shorter travel distances, but the dish wheel method produces lower forces and a more accurate gear. A number of force analyses and finite element models of both processes have been developed to predict forces, temperatures, and surface finish [151,164]. Both processes can be performed on external or internal spur or helical gears, although internal applications require sufficient room for the wheel and clearance for wheel travel.

For face-milled spiral bevel gears, generating grinding is performed using the flared cup wheel method [1,152,153,165]. Figure 17.30 shows the general process and generating motions. A circular wheel with a flared edge ground on its face is used. The edge geometry is computed based on the gear design and process kinematics [165]. The wheel is set at an angle, typically 30°, to the gear. Like a face milling cutter, the wheel generates the entire gap with a combined rolling and cutter translation

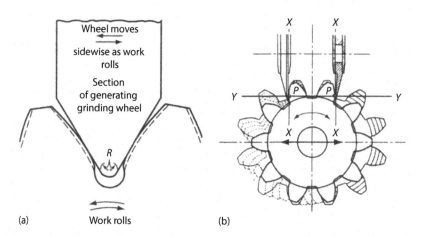

FIGURE 17.29 Disk or conical wheel (a) and saucer wheel (b) methods of generating grinding for cylindrical gears. (After Radzevich, S., *Dudley's Handbook of Practical Gear Design and Manufacture*, 2nd edn., CRC Press, Boca Raton, FL, 2012, pp. 4–20, 419–475, 491–525.)

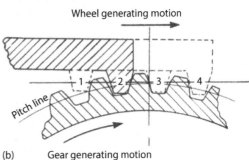

(a)   (b)   Gear generating motion

FIGURE 17.30   (a) Flared cup grinding of a spiral bevel pinion. (b) Cross section through wheel profile showing generating motions. (After Dodd, H., and Kumar, D.V., *Gear Technol.*, 30, 46, November/December 1985.)

generating motion. Figure 17.30b shows how the tooth form is swept by the generating motions from initial contact (position 1) through completion (position 4). The process is usually completed in one rotation of the gear in high-volume applications, but multiple rotations (2–6) are used for precision and aerospace parts [153]. Grinding can be carried out using conventional alumina or CBN wheels, with CBN generally used for hardened gears. For conventional wheels, Stadtfeld recommends an 80-grit, sintered, aluminum oxide abrasive with an open-pore, soft-ceramic bond [153]. Formed rather than generated, face-milled spiral bevel gears can be ground using an analogous form grinding method (without generating motions).

17.6.3.3   Continuous Generating Grinding

Continuous generating grinding of cylindrical gears is performed with a threaded wheel similar to a worm gear as shown in Figure 17.31 [1,78,149–151,166,167]. The process is also called threaded wheel grinding. The gear rotates continuously and traverses axially across the grinding wheel, which is fed in at the end of each stroke. The basic tool geometry and kinematic motions are similar to hobbing, but since the wheel moves at higher speeds than a hob (over 35 m/s peripheral speed, and often 80–100 m/s [151]), more precise and rapid machine motion and indexing controls are required [168]. Threaded wheel grinding is a rapid and high metal removal rate process due to continuous indexing and multiple cutting edges and is the most common grinding method for high-volume production of hardened gears.

   Threaded wheel grinding is performed with conventional alumina or ceramic wheels or with electroplated CBN wheels for hardened gears [151,168]. Development work on vitrified CBN threaded wheels has also been reported [169]. Electroplated CBN wheels may have separate roughing and finishing sections with different grits. Wheels may be single or multiple thread, with multiple thread wheels being common in high-volume operations. On CNC grinding machines, small modifications to the wheel motion can be employed for flank modification, crown control, or error compensation [151,170]. Axial or diagonal shifting strategies can also be employed to compensate for wheel wear or to reduce gear noise [151]. The process can be used for external spur and helical gears from about 0.2–10 module, at diameters up to 1 m, with helix angles up to 45° for smaller gears [1,151,166,167].

   Threaded wheel grinding is a very productive process, with metal removal rates much higher than other grinding processes and limited in many applications only by gear quality [167]. This is partly due to high wheel speeds and multiple thread tooling and also results from the fact that multiple cutting edges act on the blank throughout the cycle. Typical feeds per revolution of the workpiece are 1.8 mm (roughing) and 0.6 mm (finishing); a typical stock removal level is 0.1 mm/pass [1].

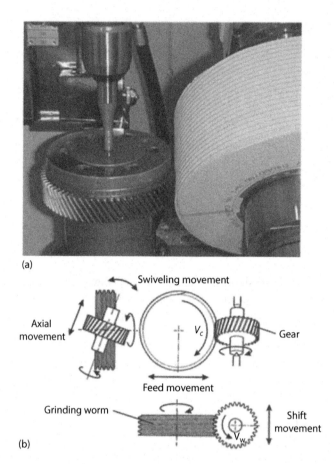

FIGURE 17.31 (a) Threaded wheel grinding. (b) Process kinematics. (After Karpuschewski, B. et al., *CIRP Ann.*, 57, 621, 2008.)

Kinematic simulations [171] can be used to simulate form deviations and corrective measures. Finite element models and process models based on modified hobbing simulations have been developed [151,167,172], primarily to study process temperatures and burn limits. The burn limit calculation is based on the critical specific energy input approach discussed in Section 10.9. Recent well-validated analytical results reported by Klocke et al. [167] showed that simulations can accurately predict force and load profiles and the effects of process parameters such as wheel speed and axial feed.

17.6.4 HONING

Gear honing [1,151,173–175] is kinematically similar to shaving in that it involves running the gear in a crossed-axis relationship with a hard, gear-shaped tool. In the honing process, an abrasive tool is used, which may be fabricated from bonded conventional grinding wheel abrasives or manufactured by electroplating a single layer of diamond to a precision-machined form [151,174]. Electroplated tools, which do not require dressing, are used primarily in high-volume applications. The gear and tool are run together under high contact pressure with a low viscosity honing oil. The crossed-axis relationship leads to mismatch, which promotes material removal by abrasion of the tool against the gear. The typical stock allowance is 10–15 μm on each flank [151]. Although external gears can be honed with a tool shaped like a second external gear, the high contact pressures required for honing make the use of internal ring gear tooling attractive [151,173] (Figure 17.32).

FIGURE 17.32 Honing an external gear using a honing ring. (From Sauter, Bachmann AG, Netstal, Switzerland.)

An internal ring gear or honing ring has higher stiffness and can sustain higher contact loads without loss of accuracy, enabling higher material removal rates. The internal setup also reduces honing machine size for a fixed gear size. Honing rings are available in conventional resin-bonded [176,177] or ceramic-bonded [178] abrasives; ceramic-bonded rings have higher stiffness and metal removal capability. Additional kinematic motions in the tangential or longitudinal feed directions can increase contact pressures and improve metal removal rates [151]; one well-known variant employing this method is spheric honing [173]. Tooth corrections, such as helix crowning and tip relief, can also be incorporated in the tooling [174].

Honing, which is usually used on hardened gears, can be used after grinding or in place of grinding. Honing is a low-speed and low-temperature process, which does not induce significant residual stresses [151,179] and thus can improve gear fatigue life. Honed surfaces also have a random surface pattern without feed lines that reduces gear noise.

Honing is also used to improve the finish and dimensional accuracy of gear bore holes and faces [180–182]. The related hybrid process of electrochemical honing has been investigated for finishing spiral bevel gears [183].

17.6.5 Lapping

In the lapping process, the gear is run against a mating gear or tool under a light load while an abrasive compound or slurry is pumped into the mesh [184–189] (Figure 17.33). For cylindrical gears, the gear is commonly run against a cast iron or nylon lap machined to mate with the gear. A nylon lap, which becomes embedded with the lapping abrasive, produces a smoother surface finish [184]. An axial traversing motion of the gear is used to distribute the abrasive. For bevel gears, a ring and pinion gear set are typically run together under limited speed and torque from different positions to distribute the abrasive [185]. The lapping compound is a light oil or water-based gel with suspended abrasive grains; common abrasives are silicon carbide, aluminum oxide, and boron carbide [190].

FIGURE 17.33 Lapping a spiral bevel gear set. (From Sauter, Bachmann AG, Netstal, Switzerland.)

Lapping can be performed after grinding or in place of grinding. It primarily improves surface finish and the surface contact pattern, reducing gear noise and improving fatigue life. Lapping is a rapid process and is well suited to mass production. It is the common method for finishing face-hobbed spiral bevel gears in mass production due to its geometric and cycle time advantages over alternative grinding methods.

Lapping can be modeled as an abrasive wear process based on the sliding wear equation (Equation 9.7, Section 9.4) [187]. Modeling of tooth contact stresses for hypoid gear lapping has also been reported [188].

REFERENCES

1. S. Radzevich, *Dudley's Handbook of Practical Gear Design and Manufacture*, 2nd edn., CRC Press, Boca Raton, FL, 2012, pp. 4–20, 419–475, 491–525.
2. G. Goch, W. Knapp, and F. Härtig, Precision engineering for wind energy systems, *CIRP Ann.* **61** (2012) 611–634.
3. H. J. Stadtfeld, *Gleason Kegelradtechnolgie*, Expert Verlag, Renningen, Germany, 2013, pp. 113–124, 267.
4. G. S. Vasilash, Inside ford sharonville transmission: Producing 19 million gears a year, *Mod. Mach. Shop*, July 29, 2014.
5. M. Albert, Gearing up to make big gears, *Mod. Mach. Shop*, **84** (August 18, 2011).
6. G. Hyatt, M. Piber, N. Chaphalkar, O. Kleinhenz, and M. Mori, A review of new strategies for gear production, *Procedia CIRP* **14** (2014) 72–76.
7. SAE Gear and Spline Technical Committee, *Gear Design, Manufacturing, and Inspection Manual (SAE AE-15)*, Society of Automotive Engineers, Warendale, PA, 1990.
8. R. H. Ewert, *Gears and Gear Manufacture: The Fundamentals*, Chapman & Hall, New York, 1997, pp. 155–162.
9. F. L. Litvin and A. Fuentes, *Gear Geometry and Applied Theory*, 2nd edn., Cambridge University Press, Cambridge, U.K., 2004, pp. 268–272, 627–696.
10. S. Radzevich, *Theory of Gearing: Kinematics, Geometry, and Synthesis*, CRC Press, Boca Raton, FL, 2012.

11. ANSI/AGMA 2015-1-A01 (R2008), *Accuracy Classification System—Tangential Measurements for Cylindrical Gears*, American Gear Manufacturers Association, Arlington, VA, 2002.
12. E. Lawson, New ANSI/AGMA accuracy standards for gears, *Gear Technol.*, March/April 2004, 22–26.
13. ISO 1328-1:2013, Cylindrical gears—ISO system of flank tolerance classification—Part 1: Definitions and allowable values of deviations relevant to flanks of gear teeth, 2013.
14. AGMA 2009-B01, *Bevel Gear Classification, Tolerances and Measuring Methods*, American Gear Manufacturers Association, Arlington, VA, 2009.
15. ISO 17485:2006, Bevel gears—ISO system of accuracy, 2006.
16. K.-D. Bouzakis, E. Lili, N. Michailidis, and O. Friderikos, Manufacturing of cylindrical gears by generating cutting processes: A critical synthesis of analysis methods, *CIRP Ann.* **57** (2008) 676–696.
17. D. Kosal, Broaching of Gears, *Gear Technol.*, March/April 1977, 48–56.
18. National Broach and Machine, Gear manufacturing methods—Forming the teeth, *Gear Technol.*, January/February 1987, 40–45, 48.
19. J. A. Psenka, Cutting tools/broaches, in *Gear Design, Manufacturing, and Inspection Manual (SAE AE-15)*, Society of Automotive Engineers, Warendale, PA, 1990, pp. 269–282.
20. C. Van De Motter, The art and science of broaching, *Gear Solutions*, March 2007, 30–35, 50.
21. G. Schneider, Cutting tool applications Chapter 14: Broaches and broaching, *Am. Mach.*, **154** (November 24, 2010).
22. W. R. Stott, State-of-the-art broaching, *Gear Technol.*, August 2011, 62–66.
23. M. Egrin, Conventional broaching, *Gear Solutions*, March 2014, 51–53.
24. E. Tarney and J. Beckman, Material properties and performance considerations for high-speed steel gear-cutting tools, *Gear Technol.*, July/August 2001, 17–21.
25. E. Tarney, High speed steel properties: Different grades for different requirements, *Gear Technol.*, September/October 2004, 54–57.
26. August-Berghaus, *Internal Broaching Tools*, August Berghaus GmbH & Co. KG, Remscheid, Germany, 2015.
27. Mitsubishi Materials, Helical broaches produce internal ring gears, *Mod. Mach. Shop*, **86** (August 20, 2013).
28. Nachi of America, *Gear Tools—Spur Broach*, Nachi of America, Greenwood, IN, 2014.
29. Nachi-Fujikoshi, *Broach for MQL*, Nachi-Fujikoshi, Toyama, Japan, 2015.
30. V. Schulze, H. Meier, T. Strauss, and J. Gibmeier, High speed broaching of case hardening steel SAE 5120, *Procedia CIRP* **1** (2012) 431–436.
31. H. Meier, K. Ninomiya, D. Dornfeld, and V. Schulze, Hard broaching of case hardened SAE 5120, *Procedia CIRP* **14** (2014) 60–65.
32. Nachi of America, *Gear Tools—Hard Broach*, Nachi of America, Greenwood, IN, 2014.
33. August-Berghaus, *Harträumwerkzeuge*, August Berghaus GmbH & Co. KG, Remscheid, Germany, nd.
34. C. Felix, Broaching on CNC machines, *Prod. Mach.*, December 23, 2011.
35. Broaching, *Gear Technol.*, January/February 2015.
36. J. A. Psenka, Making precision tooth forms with pot broaching, *Better Broaching Practice*, SME, Dearborn, MI, 1984, pp. 117–120.
37. R. S. Kusz, Pot broaching: High-production gear cutting, *Better Broaching Practice*, SME, Dearborn, MI, 1984, pp. 123–126.
38. J. W. Sutherland, E. J. Salisbury, and F. W. Hoge, A model for the cutting force system in the gear broaching process, *Int. J. Mach. Tools Manuf.* **37** (1997) 1409–1421.
39. A. Habeck, Progress in gear milling, *Gear Technol.*, January/February 2013, 13–16.
40. F. Berardi, Higher gear, *Cutting Tool Eng.* **66** (August 2014).
41. M. Jaster, Full speed ahead, *Gear Technol.*, May 2012, 25–30.
42. M. Jaster, Heavy-duty demands, *Gear Technol.*, May 2013, 22–27.
43. W. R. Stott, Delivering big gears fast, *Gear Technol.*, May 2013, 32–35.
44. D. Fratila, Evaluation of near-dry machining effects on gear milling process efficiency, *J. Cleaner Prod.* **17** (2009) 839–845.
45. W. R. Stott, Big gears better and faster, *Gear Technol.*, January/February 2011, 37–50.
46. B. Karpuschewski, M. Toefke, M. Beutner, and W. Spintig, Surface integrity aspects of milled large hardened gears, *Procedia CIRP* **13** (2014) 37–42.
47. R. Endoy, *Gear Hobbing, Shaping, and Shaving*, SME, Dearborn, MI, 1990, pp. 27–74, 127–138.
48. D. Gimpert, The gear hobbing process, *Gear Technol.*, January/February 1994, 38–44.
49. P. Zelinski, Hobbing on a turning center, *Mod. Mach. Shop*, **82** (June 2, 2009).

50. Starcut Sales Inc., Tooth forms for Hobs, *Gear Technol.*, **2** (March/April 1985) 32–33.
51. T. J. Maiuri, Hob tool life technology update, *Gear Technol.*, March/April 2009, 50–59.
52. O. Winkel, New developments in gear hobbing, *Gear Technol.*, March/April 2010, 45–52.
53. B. Karpuschewski, H.-J. Knoche, M. Hipke, and M. Beutner, High performance gear hobbing with powder-metallurgical high-speed-steel, *Procedia CIRP* **1** (2012) 196–201.
54. E. P. Kovar, Dry gear hobbing, *Gear Technol.*, July/August 1995, 39–41.
55. B. Vicenzi, L. Risso, and R. Calzavarini, High performance milling and gear hobbing by means of cermet tools with a tough (TI,W,Ta)(C,N)-C0,Ni,W composition, *Int. J. Ref. Met. Hard Mater.* **19** (2001) 11–16.
56. M. Svahn, L. Vedmar, and C. Andersson, The influence of tool tolerances on the gear quality of a gear manufactured by an indexable insert hob, *Gear Technol.*, July 2014, 48–53.
57. H. Kage, A new design for dry hobbing gears, *Gear Solutions*, June 2004, 18–24.
58. L. Ophey, Gear hobbing without coolant, *Gear Technol.*, November/December 1994, 20–24.
59. D. Witte, A revolution in hobbing technologies, *Gear Solutions*, April 2006, 38–43.
60. D. Vasilis, V. Nectarios, and A. Aristomenis, Advanced computer aided design simulation of gear hobbing by means of three-dimensional kinematics modeling, *ASME J. Manuf. Sci. Eng.* **129** (2007) 911–918.
61. K.-D. Bouzakis, S. Kombogiannis, A. Antoniadis, and N. Vidakis, Gear hobbing cutting process simulation and tool wear prediction models, *ASME J. Manuf. Sci. Eng.* **124** (2002) 42–51.
62. K.-D. Bouzakis, O. Friderikos, and I. Tsiafis, FEM-supported simulation of chip formation and flow in gear hobbing of spur and helical gears, *CIRP J. Manuf. Sci. Technol.* **1** (2008) 18–26.
63. W. Liu, D. Rena, S. Usui, J. Wadell, and T. D. Marusich, A gear cutting predictive model using the finite element method, *Procedia CIRP* **8** (2013) 51–56.
64. S. Stark, M. Beutner, F. Lorenz, S. Uhlmann, B. Karpuschewski, and T. Halle, Heat flux and temperature distribution in gear hobbing operations, *Procedia CIRP* **8** (2013) 456–461.
65. F. Klocke, C. Gorgels, R. Schalaster, and A. Stuckenberg, An innovative way of designing gear hobbing processes, *Gear Technol.*, May 2012, 48–53.
66. W. L. Janninck, Shaper cutters-design & application—Part 1, *Gear Technol.*, March/April 1990, 35–44.
67. P. G. Miller, Gear generating using rack cutters, *Gear Technol.*, October/November 1984, 17–19.
68. W. L. Janninck, Shaper cutters-design & application—Part 2, *Gear Technol.*, May/June 1990, 38–45.
69. D. Korn, Big machining for big machinery, *Mod. Mach. Shop*, March 19, 2013.
70. N. C. Ainsworth, Design implications for shaper cutters, *Gear Technol.*, July/August 1996, 30–33.
71. E. Haug, The wafer shaper cutter, *Gear Technol.*, March/April 1989, 26–28.
72. J. C. Crocket, New cutting tool developments in gear shaping technology, *Gear Technol.*, January/February 1993, 14–21.
73. Star SU, *Gear Shaper Cutters*, Star SU LLC, Hoffman Estates, IL, nd.
74. J. Lange, Innovative CNC gear shaping, *Gear Technol.*, January/February 1994, 16–28.
75. J.-D. Kim and D.-S. Kim, Development of software for the design of a pinion cutter, *J. Mater. Proc. Technol.* **68** (1997) 76–82.
76. C.-B. Tsay, W.-Y. Liu, and Y.-C. Chen, Spur gear generation by shaper cutters, *J. Mater. Proc. Technol.* **104** (2000) 271–279.
77. O. Alipiev, S. Antonov, and T. Grozeva, Generalized model of undercutting of involute spur gears generated by rack-cutters, *Mechanism and Machine Theory* **64** (2013) 39–52.
78. J. M. Lange, Gear grinding techniques parallel axes gears, *Gear Technol.*, March/April 1985, 34–48.
79. B. W. Cluff, Profile grinding gears from the solid…is it practical? *Gear Technol.*, May/June 1997, 20–25.
80. Y.-P. Shih and S.-D. Chen, Free-form flank correction in helical gear grinding using a five-axis computer numerical control gear profile grinding machine, *ASME J. Manuf. Sci. Eng.* **134** (2012) 041006-1.
81. D. Graham, M. Hitchiner, and P. Plainte, Advances in abrasive technology for grinding gears from solid, *Gear Solutions*, December 2013, 47–55.
82. C. Hünecke, The road leads straight to hypoflex, *Gear Technol.*, March/April 2010, 54–57.
83. P. E. DeGarmo, *Materials and Processes for Manufacturing*, 5th edn., Macmillan, New York, 1979, pp. 796, 814–819.
84. H. J. Stadtfeld, The basics of spiral bevel gears, *Gear Technol.*, January/February 2001, 31–38.
85. R. G. Hotchkiss, The theory of modern bevel gear manufacturing, *Gear Design, Manufacturing, and Inspection Manual (SAE AE-15)*, Society of Automotive Engineers, Warendale, PA, 1990, pp. 263–268.
86. D. B. Dooner, Hobbing of bevel and hypoid gears, *Proceedings of the ASME IDETC/CIE Conference*, Portland, OR, August 4–7, 2013, Paper DETC2013-12899.
87. H. J. Stadtfeld, CONIFLEX plus straight bevel gear manufacturing, *Gear Solutions*, August 2010, 40–55.

88. H. J. Stadtfeld, The new freedoms: Three- & four-face ground bevel gear cutting blades, *Gear Technol.*, September/October 2007, 50–55.

89. H. J. Stadtfeld, Tribology aspects in angular transmission systems—Part VII: Hypoid gears, *Gear Technol.*, June/July 2011, 66–72.

90. F. L. Litvin, Synthesis of spiral bevel gears, *Gear Technol.*, March/April 1991, 33–35.

91. P.-Y. Wang and Z.-H. Fong, Fourth-order kinematic synthesis for face-milling spiral bevel gears with modified radial motion (MRM) correction, *ASME J. Mech. Des.* **128** (2006) 457–467.

92. P.-Y. Wang and Z.-H. Fong, Mathematical model of face-milling spiral bevel gear with modified radial motion (MRM) correction, *Math. Comput. Model.* **41** (2005) 1307–1323.

93. H. Müller and J. Thomas, Face off—Face milling vs. face hobbing, *Gear Solutions*, September 2007, 49–60.

94. T. J. Maiuri, Bevel and hypoid gear cutting technology update, *Gear Technol.*, July 2007, 28–39.

95. Q. Fan, Enhanced algorithms of contact simulation for hypoid gear drives produced by face-milling and face-hobbing processes, *ASME J. Mech. Des.* **129** (2007) 31–37.

96. F. Klocke and A. Klein, Tool life and productivity improvement through cutting parameter setting and tool design in dry high-speed bevel gear tooth cutting, *Gear Technol.*, May/June 2006, 41–48.

97. Q. Fan, R. S. DaFoe, and J. W. Swanger, Higher-order tooth flank form error correction for face-milled spiral bevel and hypoid gears, *ASME J. Mech. Des.* **130** (2008) 072601-1.

98. V. Simon, Advanced design and manufacture of face-hobbed bevel gears, *Proceedings of the ASME IMECE*, Lake Buena Vista, FL, November 13–19, 2009, Paper IMECE2009-10237.

99. Q. Fan, Computerized modeling and simulation of spiral bevel and hypoid gears manufactured by gleason face hobbing process, *ASME J. Mech. Des.* **128** (2006) 1315–1327.

100. M. Lelkes, J. Marialigeti, and D. Play, Numerical determination of cutting parameters for the control of klingelnberg spiral bevel gear geometry, *ASME J. Mech. Des.* **124** (2002) 761–771.

101. Q. Fan, Kinematical simulation of face hobbing indexing and tooth surface generation of spiral bevel and hypoid gears, *Gear Technol.*, January/February 2006, 30–38.

102. Y.-P. Shih, Z.-H. Feng, and G. C. Y. Lin, Mathematical model for a universal face hobbing hypoid gear generator, *ASME J. Mech. Des.* **129** (2007) 38–47.

103. P.-Y. Shih and Z.-H. Fong, Flank modification methodology for face-hobbing hypoid gears based on ease-off topography, *ASME J. Mech. Des.* **129** (2007) 1294–1302.

104. Q. Fan, Tooth surface error correction for face-hobbed hypoid gears, *ASME J. Mech. Des.* **132** (2010) 011004-1.

105. I. Gonzalez-Perez, A. Fuentes, and K. Hayasaka, Analytical determination of basic machine-tool settings for generation of spiral bevel gears from blank data, *ASME J. Mech. Des.* **132** (2010) 101002-1.

106. K. Kawasaki and I. Tsuji, Analytical and experimental tooth contact pattern of large-sized spiral bevel gears in cyclo-palloid system, *ASME J. Mech. Des.* **132** (2010) 041004-1.

107. V. V. Simon, Manufacture of optimized face-hobbed spiral bevel gears on computer numerical control hypoid generator, *ASME J. Manuf. Sci. Eng.* **136** (2014) 031008-1.

108. H. J. Stadtfeld, Cyclocut bevel gear production, *Gear Solutions*, December 2011, 37–49.

109. F. Klocke, M. Brumm, and J. Staudt, Quality and surface of gears manufactured by free-form milling with standard tools, *Gear Technol.*, January/Februray 2015, 64–69.

110. N. Chaphalkar, G. Hyatt, and N. Bylund, Analysis of gear root forms: A review of designs, standards and manufacturing methods for root forms in cylindrical gears, *Gear Solutions*, February 2014, 49–56.

111. Anon., Cutting gear on a machining center, *Gear Technol.*, November/December 2009, 14–16.

112. DEPO, *Successful 5-Axis Milling of Gears*, DEPO GmbH & Co. KG, Marienfeld, Germany, 2013.

113. M. Albert, Changing the landscape of gear production, *Mod. Mach. Shop*, September 3, 2013.

114. S. Scherbarth, Tooth milling cutter and method for milling the teeth of toothed gear elements, U.S. Patent Application Publication US 2013/0322974A1, December 5, 2013.

115. Sandvik Coromant, *InvoMilling Agile Gear Manufacturing*, AB Sandvik Coromant, Sandviken, Sweden, 2014.

116. A. Habeck, Shifting gears, *Cutting Tool Eng.* **65** (February 2013).

117. A. Richter, Flexible gear milling, *Cutting Tool Eng.* **64** (June 2012).

118. X. Yang and J. Tang, Research on manufacturing method of CNC plungemilling for spur face-gear, *J. Mater. Proc. Technol.* **214** (2014) 3013–3019.

119. S. H. Suh, W. S. Jih, H. D. Hong, and D. H. Chung, Sculptured surface machining of spiral bevel gears with CNC milling, *Int. J. Mach. Tools Manuf.* **41** (2001) 833–850.

120. K. Kawasaki, I. Tsuji, Y. Abe, and H. Gunbara, Manufacturing method of large-sized spiral bevel gears in cylco-palloid system using multi-axis control and multi-tasking machine tool, *Gear Technol.*, August 2011, 56–61.

121. M. Albert, Cutting spiral bevel gears on a five-axis machining center, *Mod. Mach. Shop*, September 23, 2009.
122. Heller, Heavy-duty machine center for gear milling, *Am. Mach.*, July 21, 2010.
123. G. Wermeister, Milling tools for bevel gears, *Gear Solutions*, April 2011, 49–52.
124. J. P. Dugas, The process of gear shaving, *Gear Technol.*, January/February 1986, 40–48.
125. J. P. Dugas, Gear shaving basics—Part I, *Gear Technol.*, November/December 1997, 26–30.
126. D. Kosal, Gear shaving basics—Part II, *Gear Technol.*, January/February 1998, 45–48.
127. J.-D. Kim and D.-S. Kim, The development of software for shaving cutter design, *J. Mater. Proc. Technol.* **59** (1996) 359–366.
128. S. P. Radzevich, Design of shaving cutter for plunge shaving a topologically modified involute pinion, *ASME J. Mech. Des.* **125** (2003) 632–639.
129. S. P. Radzevich, On satisfaction of the fifth necessary condition of proper part surface generation in design of plunge shaving cutter for finishing of precision involute gears, *ASME J. Mech. Des.* **129** (2007) 969–980.
130. F. L Litvin, Q. Fan, D. Vecchiato, A. Demenego, R. F. Handschuh, and T. M. Sep, Computerized generation and simulation of meshing of modified spur and helical gears manufactured by shaving, *Comput. Methods Appl. Mech. Eng.* **190** (2001) 5037–5055.
131. R.-H. Hsu and Z.-H. Fong, Analysis of auxiliary crowning in parallel gear shaving, *Mech. Mach. Theory* **45** (2010) 1298–1313.
132. V.-T. Tran, R.-H. Hsu, and C.-B. Tsay, Tooth contact analysis of double-crowned involute helical pairs shaved by a crowning mechanism with parallel shaving cutters, *Mech. Mach. Theory* **79** (2014) 198–216.
133. A. Fuentes, H. Nagamoto, F. L. Litvin, I. Gonzalez-Perez, and K. Hayasaka, Computerized design of modified helical gears finished by plunge shaving, *Comput. Methods Appl. Mech. Eng.* **199** (2010) 1677–1690.
134. R.-H. Hsu and Z.-H. Fong, Serration design for a gear plunge shaving cutter, *ASME J. Manuf. Sci. Eng.* **133** (2011) 021004-1.
135. F. Klocke and T. Schröder, Gear shaving—Process simulation helps to comprehend an incomprehensible process, *Gear Technol.*, September/October 2006, 46–54.
136. W. E. Loy, Hard gear processing with skiving hobs, *Gear Technol.*, March/April 1985, 9–14.
137. B. W. Cowley, Micro skiving: Precision finishing of hardened small diameter fine module/pitch gears, splines, and serrations, *Gear Solutions*, September 2013, 65–69.
138. A. Antoniadis, N. Vidakis, and N. Bilalis, A simulation model of gear skiving, *J. Mater. Proc. Technol.* **146** (2004) 213–220.
139. P. Khurana, D. King, K. Marseilles, and S. Sengupta, Modeling of helical gear carbide re-hobbing process, *Proceedings of the ASME MSEC Conference*, Detroit, MI, June 9–13, 2014, Paper MSEC2014-3973.
140. F. Young, Carbide rehobbing—A new technology that works, *Gear Technol.*, May/June 1994, 16–22.
141. W. E. McElroy, Using hobs for skiving; A pre-finish and finishing solution, *Gear Technol.*, May/June 1993, 43–45.
142. T. Sugimoto, A. Ishibashi, and M. Yonekura, Performance of skiving hobs in finishing Induction hardened and carburized gears, *Gear Technol.*, May/June 2003, 34–41.
143. A. Antoniadis, Gear skiving—CAD simulation approach, *Comput. Aided Design* **44** (2012) 611–616.
144. Klingelnberg, Reliable and efficient skiving, *Gear Technol.*, September 2011, 11–13.
145. E. Weppelmann and J. Brogni, A breakthrough in power skiving, *Mod. Mach. Shop*, March 1, 2014.
146. H. J. Stadtfeld, Power skiving of cylindrical gears on different machine platforms, *Gear Technol.*, January/February 2014, 52–62.
147. Hegenscheidt-MFD, *Universal Turn/Skiving Machine Type CTS 650*, Hegenscheidt-MFD GmbH & Co. KG, Erkelenz, Germany, nd.
148. Y. Sharma, Gear grinding fundamentals, *Gear Technol.*, September/October 1989, 26–35, 38–39.
149. S. B. Rao, Grinding of spur and helical gears, *Gear Technol.*, July/August 1992, 20–31.
150. R. Burdick, Parallel axis gear grinding: Theory & application, *Gear Technol.*, November/December 2000, 77–81.
151. B. Karpuschewski, H.-J. Knoche, and M. Hipke, Gear finishing by abrasive processes, *CIRP Ann.* **57** (2008) 621–640.
152. H. Dodd and D. V. Kumar, Technological fundamentals of CBN bevel gear finish grinding, *Gear Technol.*, November/December 1985, 30–37, 46.
153. H. J. Stadtfeld, Guidelines for modern bevel gear grinding, *Gear Technol.*, August 2008, 42–53.
154. S. Kendjelic, Grinding guidelines for superior sharpness, *Gear Solutions*, January 2005, 25–31.
155. M. K. Krueger, S. C. Yoon, D. Gong, S. B. McSpadden Jr., L. J. O'Rourke, and R. J. Parten, New technology in metalworking fluids and grinding wheels achieves tenfold improvement in grinding performance, *Coolants/Lubricants for Metal Cutting and Grinding Conference*, Chicago, IL, June 7, 2000.

156. M. G. Schwabel and P. E. Kendall, Alumina abrasive grains produced by sol-gel technology, *Am. Ceram. Soc. Bull.* **66** (1991) 1596–1598.

157. J. Sung, Sol-gel alumina abrasive grain, U.S. Patent 5215552, June 1, 1993.

158. W. Graf, Cubitron II: Precision-shaped grain (psg) turns the concept of gear grinding upside down, *Gear Solutions*, May 2014, 37–44.

159. J. Erdmann, Brad foote and 3M collaborate on testing of ground parts, *Gear Technol.*, March/April 2014, 22–24.

160. M. Hitchiner, Complex form grinding technology for advanced abrasive technology, *Gear Solutions*, September 2013, 37–42.

161. H.-Y. You., P.-Q. Ye, J.-S. Wang, and X.-Y. Deng, Design and application of CBN shape grinding wheel for gears, *Int. J. Mach. Tools Manuf.* **43** (2003) 1269–1277.

162. A. Türich, C. Kobialka, and D. Vucetic, Innovative concepts for grinding wind power energy gears, *Gear Technol.*, June 2009, 39–44.

163. E. I. Podzharov, Gear grinding with dish wheels, *Gear Technol.*, September/October 1999, 51–55.

164. C. Haifeng, T. Jinyuan, and Z. Wei, Modeling and predicting of surface roughness for generating grinding gear, *J. Mater. Proc. Technol.* **213** (2013) 717–721.

165. T. Krenzer, CNC bevel gear generators and flared cup gear grinding, *Gear Technol.*, July/August 1993, 18–24.

166. D. Richmond, Continuous generation grinding, *Gear Solutions*, September 2010, 32–35.

167. F. Klocke, M. Brumm, and J. Reimann, Continuous generating gear grinding, *Gear Solutions*, May 2012, 31–44.

168. T. Emura, L. Wang, M. Yamanaka, H. Nakamura, Y. Kato, and Y. Teshigawara, A PC-based synchronous controller for NC gear grinding machines using multithread CBN wheel, *ASME J. Mech. Des.* **123** (2001) 590–597.

169. J. Reimann, F. Klocke, M. Brumm, A. Mehr, and K. Finkenwirth, Technological potential and performance of gears ground by dressable CBN tools, *Gear Technol.*, March/April 2014, 54–59.

170. A. Türich, Producing profile and lead modifications in threaded wheel and profile grinding, *Gear Technol.*, January/February 2010, 54–62.

171. G. Gravel, Simulation of deviations in hobbing and generation grinding, *Gear Technol.*, September/October 2014, 56–60.

172. F. Klocke, M. Brumm, and J. Reimann, Modeling of surface zone influences in generating gear grinding, *Procedia CIRP* **8** (2013) 21–26.

173. N. W. Wright and H. Schriefer, Basic honing & advanced free-form honing, *Gear Technol.*, July/August 1997, 26–33.

174. C. Malrzenell and H. K. Tönshoff, Properties of tooth surfaces due to gear honing with electroplated tools, *Gear Technol.*, November/December 2001, 43–49.

175. Sauter Bachmann, *Honing of Tooth Flanks*, Sauter, Bachmann AG, Netstal, Switzerland, nd.

176. Cleason, *Honing Rings*, The Gleason Works, Rochester, NY, nd.

177. Fassler, *Resin-Bonded Honing Rings*, MDC Max Daetwyler AG, Bleienbach, Switzerland, nd.

178. Fassler, *Ceramic-Bonded Honing Rings*, MDC Max Daetwyler AG, Bleienbach, Switzerland, nd.

179. Y. B. Guo and A. W. Warren, Microscale mechanical behavior of the subsurface by finishing processes, *ASME J. Manuf. Sci. Eng.* **127** (2005) 333–338.

180. G. Flores, A. Wiens, and O. Stammen, Honing of gears, *Gear Technol.*, August 2014, 60–66.

181. J. Goodman and J. Gaser, Softhoning for hard-turned steel gears, *Gear Solutions*, June 2005, 24–31.

182. S. Keshavan, Alternatives to gear grinding, *Gear Solutions*, June 2011, 41–46.

183. J. H. Shaikh and N. K. Jain, Modeling of material removal rate and surface roughness in finishing of bevel gears by electrochemical honing process, *J. Mater. Proc. Technol.* **214** (2014) 200–209.

184. M. Nakae, K. Hidaka, Y. Ariura, T. Matsunami, and M. Kohara, Gear finishing with a nylon lap, *Gear Technol.*, September/October 2005, 38–45.

185. J. Masseth, M. Kolivand, H. Blancke, and N. Wright, Studying the effects of lapping process on hypoid gears surface finish and transmission errors, SAE Technical Paper 2007-01-2229, 2007.

186. J. Masseth and M. Kolivand, Lapping and superfinishing effects on surface finish of hypoid gears and transmission errors, *Gear Technol.*, September/October 2008, 72–78.

187. Q. Jiang, C. Gosselin, and J. Masseth, Simulation of hypoid gear lapping, *ASME J. Mech. Des.* **130** (2008) 112601-1.

188. B. Y. Wei, X. Z. Deng, and Z. D. Fang, The lapping mechanism of hypoid gears, *J. Mater. Proc. Technol.* **209** (2009) 3001–3008.

189. Sauter Bachmann, *Lapping of Bevel Gears*, Sauter, Bachmann AG, Netstal, Switzerland, nd.

190. Lapmaster, *Lapping Grinding Compounds*, Lapmaster International LLC, Mount Prospect, IL, nd.

Index

A

Abrasive/erosive wear, 162, 166, 171, 538–539, 547
Absolute distance meter (ADM) system, 853–854
Absolute interferometer (AIFM), 854
Active chip-breaking, 632
Adhesive/attritional wear, 538–540
ADI, *see* Austempered ductile iron
AdvantEdge program, 428
AGMA, *see* American Gear Manufacturers Association
AIFM, *see* Absolute interferometer
Air–oil mists, 787–788, 806
Alloys
    aluminum alloys, 640–642
    carbon and low alloy steels, 647–650
    cast iron, 645–647
    copper alloys, 643–644
    depleted uranium alloys, 656–657
    magnesium alloys, 638–639
    MMCs, 642–643
    nickel alloys, 654–656
    P/M materials, 652–653
    stainless steels, 650–652
    titanium alloys, 653–654
Al_2O_3-based (black) coatings, 544
Alumina-based (or oxide) ceramic tools, 546, 655
Aluminum alloys, 171, 547, 638, 640–642
Aluminum oxide (Al_2O_3), 168, 174, 255–256, 560, 920
American Gear Manufacturers Association (AGMA), 900, 908, 916
American National Acme threaded (automotive) shank, 290
Arbor toolholder, 329
ARMA steel, *see* Malleable iron
Association for Manufacturing Technology (AMT), 108
Austempered ductile iron (ADI), 646–647
Austenitic stainless steels, 625, 651
Automated Manufacturing Research Facility (AMRF), 108
Automated tool change (ATC) system, 142–143, 145
Automatically Programmed Tools (APT), 21
Autoregressive moving average (ARMA) model, 705, 714–715

B

Ball-track clamping mechanism, 315
Bed-of-nails (BON) methods, 368–369
Bellville springs, 306, 315
Bessemer process, 16
Boothroyd's model, 460–463
Boring and Trepanning Association (BTA), 35–36, 228–229, 236
Boron carbide (BC), 174, 925
BTA (STS) drilling, 35
Built-up edge (BUE), 534, 538, 575
    cutting mechanics, 434–436
    drilling and reaming, 590

HSS, 543
    lubricating coating materials, 177
    magnesium alloys, 639
    tool–chip friction, 412–413
    tool coatings, 172
    type 2 and 3 chips, 405
Burrs
    cap/crown structure, 637
    CODEF, 634
    deburring, 637–638
    ductility, 634–635
    exit burrs, 637
    foot/negative shear burr, 635–636
    in-plane exit angle, 636
    Poisson/compression burr, 634
    rollover burr, 634, 637
    size, 635–636
    spiral-pointed drill, 637
    tear/breakout burr, 634

C

Cannon boring
    gun boring, 4
    military acceptance tests, 5
    rack-and-pinion feed mechanism, 6
    steam-engine boring mill, 6
    System Gribeauval, 5
Carbide tool materials, 72, 165
Carbon and low alloy steels
    calcium, 650
    high carbon steels, 649
    leaded free-machining steels, 650
    low carbon steels, 649
    machinability of steel, 647
    medium carbon steels, 649
    microalloyed steels, 650
    sulfurized free-machining steels, 650
Cardmatic, 21
Cast iron
    ADI, 646
    CGI, 646–647
    coated carbides, 645
    ductile/nodular iron, 646
    gray cast iron, 644–645
    high silicon gray irons, 647
    malleable iron, 645
    PCBN tooling, 645–646
    white irons, 647
CAT-V toolholder
    boring tooling assemblies, 287
    drilling tooling assemblies, 288
    tapping and slot milling tooling assemblies, 289
Cauchy–Green deformation tensor, 465
Cellular manufacturing system (CMS), 99–100, 107

Printed in the United States
by Baker & Taylor Publisher Services